HANDBOOK OF STRUCTURAL AND MECHANICAL MATRICES

OTHER McGRAW-HILL REFERENCE BOOKS OF INTEREST

Handbooks

American Institute of Physics • AMERICAN INSTITUTE OF PHYSICS HANDBOOK
Avallone and Baumeister • MARKS' STANDARD HANDBOOK FOR MECHANICAL ENGINEERS
Brady and Clauser • MATERIALS HANDBOOK
Callender • TIME-SAVER STANDARDS FOR ARCHITECTURAL DESIGN DATA
Chopey and Hicks • HANDBOOK OF CHEMICAL ENGINEERING CALCULATIONS
Condon and Odishaw • HANDBOOK OF PHYSICS
Croft, Watt, and Summers • AMERICAN ELECTRICIANS' HANDBOOK
Dean • LANGE'S HANDBOOK OF CHEMISTRY
Fink and Beaty • STANDARD HANDBOOK FOR ELECTRICAL ENGINEERS
Fink and Christiansen • ELECTRONICS ENGINEERS' HANDBOOK
Gaylord and Gaylord • STRUCTURAL ENGINEERING HANDBOOK
Harris and Crede • SHOCK AND VIBRATION HANDBOOK
Hopp and Hennig • HANDBOOK OF APPLIED CHEMISTRY
Juran • QUALITY CONTROL HANDBOOK
Maynard • INDUSTRIAL ENGINEERING HANDBOOK
Merritt • STANDARD HANDBOOK FOR CIVIL ENGINEERS
Perry • ENGINEERING MANUAL
Perry and Green • PERRY'S CHEMICAL ENGINEERS' HANDBOOK
Rohsenow, Hartnett, and Ganić • HANDBOOK OF HEAT TRANSFER FUNDAMENTALS
Rohsenow, Hartnett, and Ganić • HANDBOOK OF HEAT TRANSFER APPLICATIONS
Rosaler and Rice • STANDARD HANDBOOK OF PLANT ENGINEERING
Seidman and Mahrous • HANDBOOK OF ELECTRIC POWER CALCULATIONS
Shigley and Misckhe • STANDARD HANDBOOK OF MACHINE DESIGN
Tuma • HANDBOOK OF PHYSICAL CALCULATIONS
Tuma • ENGINEERING MATHEMATICS HANDBOOK

Encyclopedias

CONCISE ENCYCLOPEDIA OF SCIENCE AND TECHNOLOGY
ENCYCLOPEDIA OF ELECTRONICS AND COMPUTERS
ENCYCLOPEDIA OF ENERGY
ENCYCLOPEDIA OF ENGINEERING
ENCYCLOPEDIA OF PHYSICS

Dictionaries

DICTIONARY OF SCIENTIFIC AND TECHNICAL TERMS
DICTIONARY OF MECHANICAL AND DESIGN ENGINEERING
DICTIONARY OF COMPUTERS

HANDBOOK OF STRUCTURAL AND MECHANICAL MATRICES

DEFINITIONS ▪ TRANSPORT MATRICES
STIFFNESS MATRICES ▪ FINITE DIFFERENCES
FINITE ELEMENTS ▪ GRAPHS AND TABLES
OF MATRIX COEFFICIENTS

Jan J. Tuma, Ph.D.
Professor of Engineering
Arizona State University

McGraw-Hill Book Company

New York St. Louis San Francisco Auckland Bogotá
Hamburg London Madrid Mexico Milan Montreal
New Delhi Panama Paris São Paulo Singapore
Sydney Tokyo Toronto

Library of Congress Cataloging-in-Publication Data

Tuma, Jan J.
Handbook of structural and mechanical matrices.

Bibliography: p.
Includes index.
1. Structures, Theory of—Matrix methods—
Handbooks, manuals, etc. I. Title.
TA642.T86 1987 624.1′7 87-2585
ISBN 0-07-065433-6

1234567890 DOC/DOC 89210987

ISBN 0-07-065433-6

The editors for this book were Harold B. Crawford, Alice Goehring, and Peggy Lamb, the designer was Naomi Auerbach, and the production supervisor was Thomas G. Kowalczyk. It was set in Baskerville by University Graphics, Inc.

Printed and bound by R. R. Donnelley & Sons.

Information contained in this work has been obtained by McGraw-Hill, Inc., from sources believed to be reliable. However, neither McGraw-Hill nor its authors guarantees the accuracy or completeness of any information published herein and neither McGraw-Hill nor its authors shall be responsible for any errors, omissions, or damages arising out of use of this information. This work is published with the understanding that McGraw-Hill and its authors are supplying information but are not attempting to render engineering or other professional services. If such services are required, the assistance of an appropriate professional should be sought.

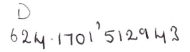

To the memory
of
Ing. Dr. Zdeněk Bažant
(1879–1954)
Professor of Structural Mechanics
Czech Institute of Technology
at Prague

Once in a lifetime one may have the rare privilege to attend lectures of a truly great teacher of engineering. I had this privilege.

J.J.T.

CONTENTS

Preface *ix*

PART I STATIC ANALYSIS OF BARS 1

1 Notation, Signs, Basic Relations 3
2 Transport Matrices, Free and Interactive Bars 25
3 Flexibility and Stiffness Matrices, Free and Interactive Bars 49
4 Free Straight Bar of Order One, Constant Section 77
5 Free Straight Bar of Order One, Variable Section 131
6 Free Circular Bar of Order One, Constant Section 155
7 Free Parabolic Bar of Order One, Constant Section 175
8 Straight Bar of Order One Encased in Elastic Foundation, Constant Section 185
9 Free Beam-Column of Order One, Constant Section 205
10 Beam-Column of Order One Encased in Elastic Foundation, Constant Section 227
11 Free Straight Bar of Order Two, Constant Section 247
12 Free Circular Bar of Order Two, Constant Section 263
13 Straight Bar of Order Two Encased in Elastic Foundation, Constant Section 287
14 Free Beam-Column of Order Two, Constant Section 299
15 Beam-Column of Order Two Encased in Elastic Foundation, Constant Section 309
16 Bar Analysis by Finite Differences 319

PART II DYNAMIC ANALYSIS OF BARS 325

17 Notation, Signs, Basic Relations 327
18 Lumped-Mass Models, Single Degree of Freedom 333
19 Lumped-Mass Models, Several Degrees of Freedom 347
20 Straight Bar of Order One, Constant Section 353
21 Straight Bar of Order Two, Constant Section 369

PART III STATIC ANALYSIS OF PLATES 379

22 Notation, Signs, Basic Relations 381
23 Circular Isotropic Plate of Constant Thickness 395
24 Rectangular Isotropic Plate of Constant Thickness 415
25 Plate Analysis by Finite Differences 431
26 Plate Analysis by Finite Elements 453

PART IV STATIC ANALYSIS OF SHELLS 475

27 Notation, Signs, Basic Relations 477
28 Circular Cylindrical Shell of Constant Thickness 487
29 Spherical Shell of Constant Thickness 501
30 Shell Analysis by Finite Differences and Finite Elements 515

APPENDIX A GLOSSARY OF SYMBOLS 529

APPENDIX B PROPERTIES OF NORMAL SECTION 537

APPENDIX C REFERENCES AND BIBLIOGRAPHY 543

Index 547

PREFACE

This handbook presents in one volume a concise collection of the major matrices encountered in the preparation of micro-, mini-, and main-frame computer programs for the analysis of structural and mechanical systems. It was prepared to serve as a *professional, users-oriented, desktop reference book* for engineers and architects.

The subject matter is divided into four major parts arranged in a logical sequence, each covering a distinct class of structural and mechanical elements.

The first part (Chapters 1–16) presents the matrix models of straight, circular, and parabolic bars and of straight interactive bars subjected to mechanical and thermal causes in a state of static equilibrium.

The second part (Chapters 17–21) shows the lumped- and distributed-mass matrix models of straight bars in a state of free and forced vibration.

The third part (Chapters 22–26) displays the matrix models of circular and rectangular plates subjected to mechanical and thermal causes. The finite-segment, finite-difference, and finite-element matrices are included.

The fourth part (Chapters 27–30) introduces the matrix models of cylindrical, spherical, and conical shells subjected to axisymmetrical causes. Again, the finite-segment, finite-difference, and finite-element matrices are included.

The form of presentation follows the telescopic pattern of this author's *Engineering Mathematics Handbook* and shows the same special features, facilitating an easy and rapid location of the desired information.

1. Each page presents the information in a graphical arrangement pertinent to the specific type of material, designated by a title and section number. Consequently, each page is a table.
2. Left and right pages of the book present related or similar material, with all matrices and their load functions arranged in logical sequences.
3. Matrix coefficients are expressed in analytical forms and, if applicable, their series expansions and graphs are included.
4. A consistent system of symbols is used throughout the book, allowing a rapid familiarization with the material and ensuring an easy use of this material.

Even a casual reader will observe that this book is not just a mechanical compilation of available formulas but represents an organized effort to present the matrix analysis in a new and unified form.

The preparation and organization of the material presented in this book spans a period of 12 years, during which the author has been assisted by many individuals and has relied on an extensive wealth of reference material forming a body of knowledge known as matrix structural and mechanical analysis. The space limitation prevents the inclusion of a complete list of references, yet a great effort was made to credit those sources which were directly used or were used for comparative purposes.

In closing, the author expresses his gratitude to his former assistants Dr. L. A. Hill, Dr. A. J. Celis, Dr. K. S. Havner, Mr. C. Martin, Dr. J. W. Gillespie, Dr. J. W. Harvey, Dr. Ch. O. Heller, Mr. H. C. Boecker, Mr. J. W. Exline, Dr. S. E. French, Mr. T. L. Lassley, Dr. H. S. Yu, Dr. J. T. Oden, Dr. R. K. Munshi, Dr. J. H. Talaba, Dr. E. Citipitioglu, Dr. F. A. Frusti, Dr. M. N. Reddy, Dr. M. E. Kamel, Dr. P. L. Koepsell, Dr. M. M. Douglas, Dr. E. P. Dallam, Dr. J. Ramey, Dr. G. Alberti, Dr. A. Lasker, and Dr. S. M. Aljaweini for their help in the development of this material.

Particular thanks are extended to Dr. C. G. Date for the preparation of graphs in Chapter 10 and for the calculation of tables in Chapter 31, to Dr. N. A. Seyedmadani for the preparation of graphs and tables in Chapters 6, 8, 23, 32, and to Mr. A. A. Mages for the permission to use his tables and graphs in Chapter 12. Although every effort was made to avoid errors, it would be presumptuous to assume that none had escaped detection in a work of this scope. The author earnestly solicits comments and recommendations for improvements and future additions.

Finally, but not least, gratitude is expressed to my wife Hana for her patience, understanding, and encouragement during the preparation of this book.

Tempe, Arizona *Jan J. Tuma*

HANDBOOK OF STRUCTURAL AND MECHANICAL MATRICES

Part I

Static Analysis of Bars

1

Analysis of Bars Notation, Signs, Basic Relations

STATIC STATE

1.01 INTRODUCTION

(1) Systems considered in Chaps. 1–16 are finite bars of given geometric shapes. Their cross-sectional dimensions are small in comparison to their length ($< 1/10$), and if curved, the radius of curvature of the undeformed bar axis is large in comparison to the largest dimension of the cross section ($> 10/1$). Each bar and its loads are in a state of static equilibrium and a state of time-independent elastic deformation.

(2) Assumptions used in the derivation of analytical relations are

(*a*) The material of the bar is homogeneous, isotropic, continuous, and follows Hooke's law.
(*b*) All deformations are small and do not alter (significantly) the initial geometry of the bar and of the load applications.
(*c*) The initial plane cross section of the bar remains plane during the deformation of the bar.
(*d*) The material constants (modulus of elasticity and of rigidity, spring constants of connections and of foundations, and volume change coefficients) are known from experiments and are independent of time.

(3) Symbols are defined where they appear first and are all summarized in Appendix A.

1.02 POSITION AND LOAD VECTORS

(1) Position of a point j is given by the position vector s_{0j}, represented symbolically by three coordinates x_{0j}, y_{0j}, z_{0j} and related to the right-handed orthogonal axes x, y, z. In column matrix form,

$$s_{0j} = \{x_{0j}, y_{0j}, z_{0j}\} \qquad (s_{0j} \text{ in m})$$

All coordinates are positive if measured in the positive direction of coordinate axes (Fig. 1.02-1).

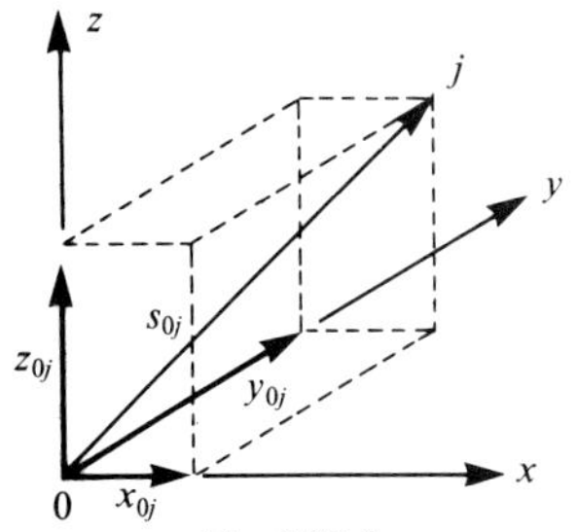

Fig. 1.02-1

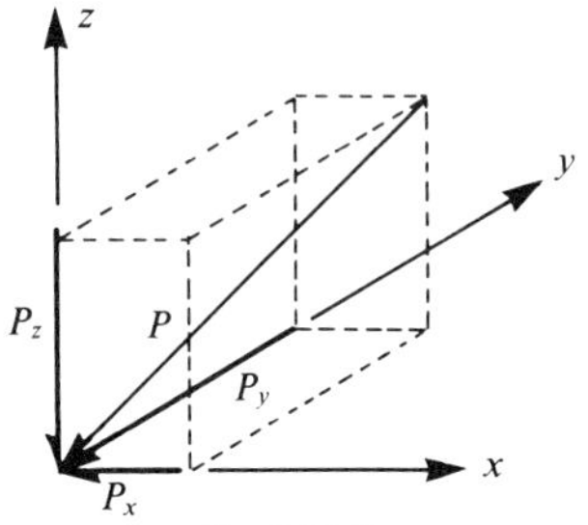

Fig. 1.02-2

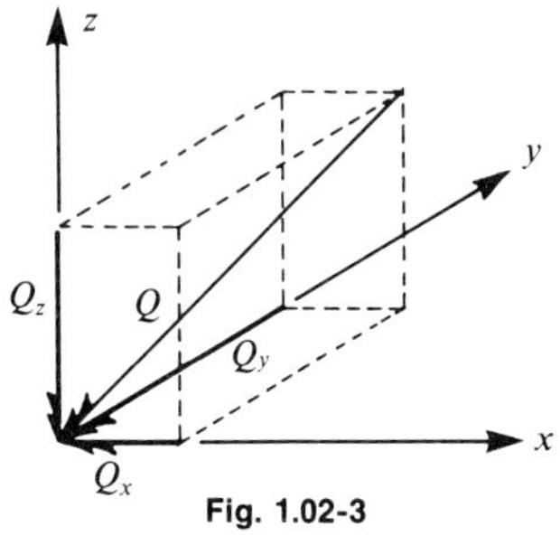

Fig. 1.02-3

(2) Concentrated load and applied moment are single force P and single couple Q represented symbolically by three orthogonal single- and double-headed vector components P_x, P_y, P_z and Q_x, Q_y, and Q_z respectively. In column matrix form,

$$P = \{P_x, P_y, P_z\} \qquad (P \text{ in N}) \qquad \text{and} \qquad Q = \{Q_x, Q_y, Q_z\} \qquad (Q \text{ in N}\cdot\text{m})$$

All concentrated-load and applied-moment components are positive if acting in the negative direction of coordinate axes (Figs. 1.02-2, 1.02-3).

(3) Intensity of distributed force and distributed moment are force and couple per unit length p and q represented symbolically by three orthogonal single- and double-headed vector components p_x, p_y, p_z and q_x, q_y, q_z respectively. In column matrix form,

$$p = \{p_x, p_y, p_z\} \qquad (p \text{ in N/m}) \qquad \text{and} \qquad q = \{q_x, q_y, q_z\} \qquad (q \text{ in N}\cdot\text{m/m})$$

All intensities of distributed-force and distributed-moment components are positive if acting in the negative direction of coordinate axes (Figs. 1.02-4, 1.02-5).

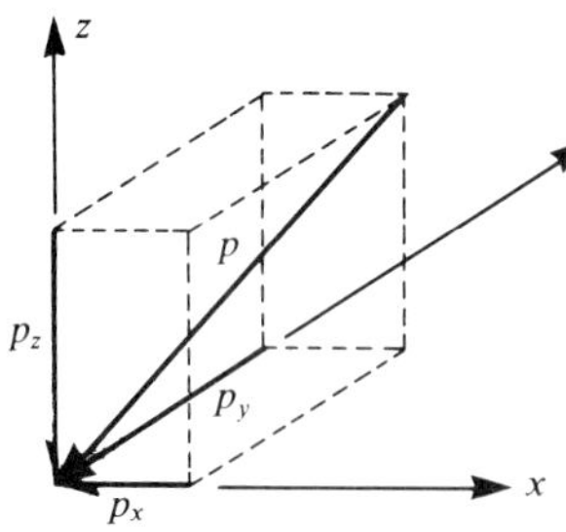

Fig. 1.02-4

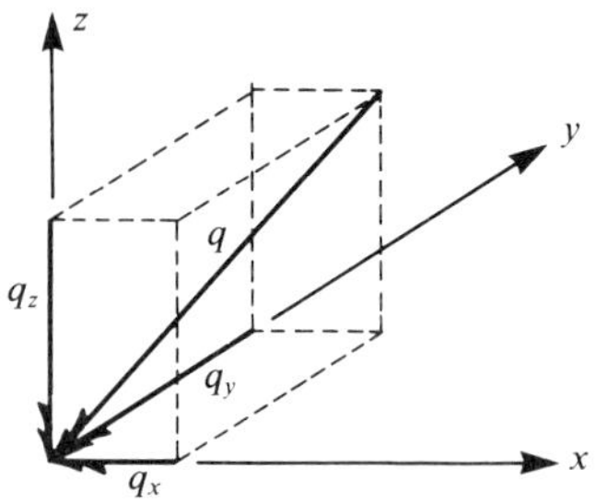

Fig. 1.02-5

(4) Sign convention of loads introduced above offers simplification in numerical calculations since many physical situations call for loads acting in these directions.

1.03 STRESS AND REACTIVE VECTORS

(1) Stress-resultant force vector is a single force acting at the centroid of the section and represented symbolically by three orthogonal single-headed vector components along the principal axes of the section. They are

U = normal force (N) $\quad V, W$ = shearing forces (N)

(2) Stress-resultant moment vector is a single moment acting at the centroid of the section and represented symbolically by three orthogonal double-headed vector components along the principal axes of the section. They are

X = torsional moment (N·m) $\quad Y, Z$ = bending moments (N·m)

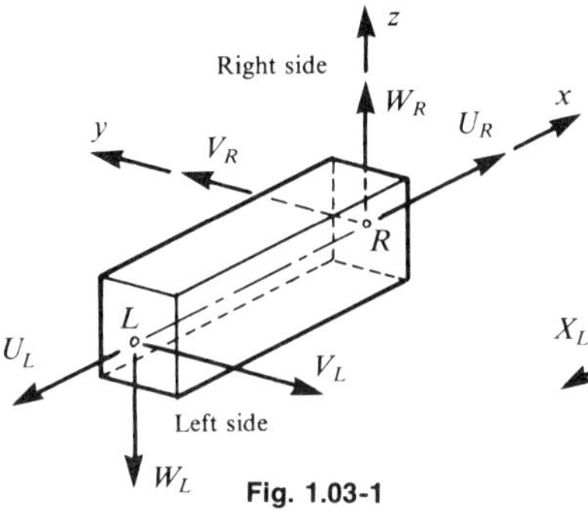

Fig. 1.03-1

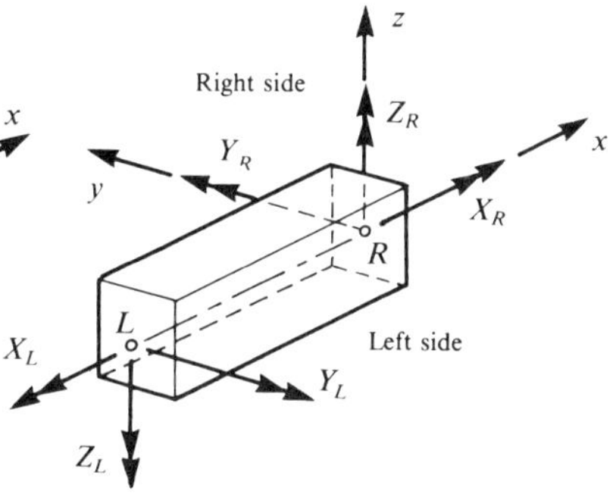

Fig. 1.03-2

(3) Stress-resultant vectors in column matrix form are

$S_L = \{U_L, V_L, W_L, X_L, Y_L, Z_L\} \qquad S_R = \{U_R, V_R, W_R, X_R, Y_R, Z_R\}$

where the subscripts L and R identify the left and right ends, respectively. All stress-resultant components acting on the right side (far side) of the section are positive if acting in the positive direction of the principal axes. For their left-side (near-side) counterparts, the opposite is true (Figs. 1.03-1, 1.03-2).

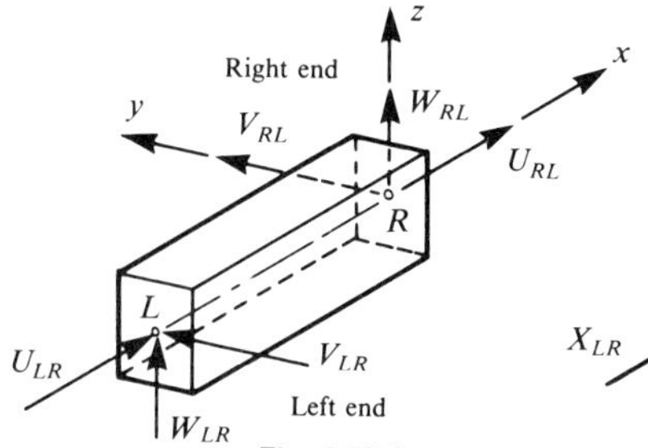

Fig. 1.03-3

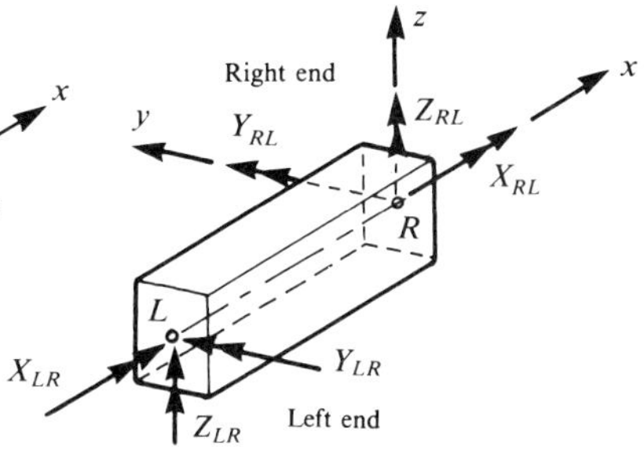

Fig. 1.03-4

(4) Reactions are force and moment vectors developed by loads and/or other causes at the points of support. In column matrix form they are

$S_{LR} = \{U_{LR}, V_{LR}, W_{LR}, X_{LR}, Y_{LR}, Z_{LR}\} \qquad S_{RL} = \{U_{RL}, V_{RL}, W_{RL}, X_{RL}, Y_{RL}, Z_{RL}\}$

where the first and second subscripts identify the near and far ends, respectively. All reactions are positive if acting in the positive direction of the respective axes (Figs. 1.03-3, 1.03-4).

1.04 DISPLACEMENTS

(1) Elastic curve. The deformation of a bar is defined as the change in its shape caused by loads and/or volume changes. As the bar deforms, its centroidal axis takes on a new shape, called the elastic curve. The coordinates and the slopes of this curve measured from the initial axis of the undeformed bar are designated as the linear displacements and angular displacements, respectively.

(2) Linear displacement is a directed segment represented symbolically by a single-headed vector resolved into three mutually perpendicular components u, v, w (m) measured along x, y, z axes, respectively (Fig. 1.04-1).

(3) Angular displacement is a directed segment represented symbolically by a double-headed vector resolved into three mutually perpendicular components ϕ, ψ, θ (rad) measured in the right-handed direction about x, y, z axes, respectively (Fig. 1.04-2).

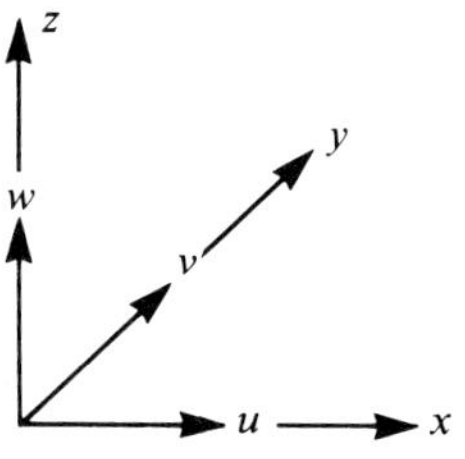

Fig. 1.04-1

(4) Displacement vector in column matrix form is

$$\Delta = \{u, v, w, \phi, \psi, \theta\}$$

All displacements are positive if acting in the positive direction of the respective coordinate axis.

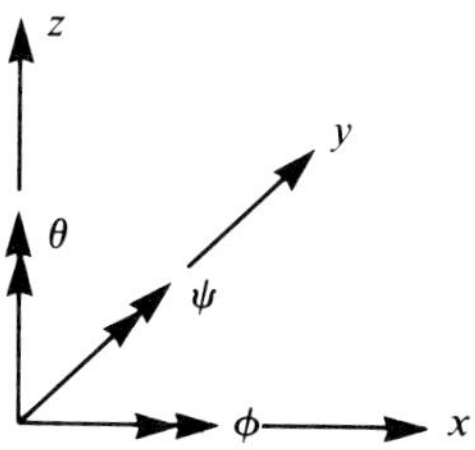

Fig. 1.04-2

(5) Geometry of small displacements allows the following simplifications:

$$ds \cong dx \qquad \begin{array}{lll} \sin\phi \cong \tan\phi \cong \phi & \quad & \cos\phi \cong 1 \\ \sin\psi \cong \tan\psi \cong \psi & & \cos\psi \cong 1 \\ \sin\theta \cong \tan\theta \cong \theta & & \cos\theta \cong 1 \end{array}$$

where ds is the elemental length of the elastic curve and dx is the corresponding elemental length of undeformed bar axis.

1.05 STATIC EQUILIBRIUM

(1) Definition. A system is in a state of static equilibrium when the resultant of all forces and moments is equal to zero. If a system is in a state of static equilibrium, any part of it is also in the same state.

(2) Equations. For such a state to exist, six conditions must be fulfilled simultaneously:

$$\begin{array}{lll} \Sigma F_x = 0 & \Sigma F_y = 0 & \Sigma F_z = 0 \\ \Sigma M_x = 0 & \Sigma M_y = 0 & \Sigma M_z = 0 \end{array}$$

where F and M designate the forces and moments, respectively. The subscripts x, y, z represent the axes along which force and moment components are added algebraically.

(3) Characteristics. A system is statically determinate if its reactions and stress resultants can be computed from the conditions of static equilibrium alone. A system is statically indeterminate if its reactions and stress resultants cannot be computed from the conditions of static equilibrium alone and deformation conditions must be considered. The superfluous forces and moments (which are not necessary for static equilibrium) are called redundants, and their number defines the degree of static indeterminacy of the system.

1.06 STATIC DETERMINACY AND INDETERMINACY

(1) Beams, arches, rings, and frames. For a system of bars connected together by joints, some or all of which are rigid, the total number of independent conditions of static equilibrium is

$$e = 6b + 6j + s$$

where b = number of bars, j = number of joints, and s = number of releases (number of zero forces and/or moments at supports and in connections). Since there are 12 unknown end stress resultants in each bar and r unknown reactions at supports, the number of redundants is

$$n = 12b + r - e = 6b + r - 6j - s > 0$$

If $n = 0$, the system is statically determinate, and if $n < 0$, the system is geometrically unstable. In these equations the number of joints j includes all internal joints and all points of support.

(2) Trusses. For a space system of bars connected together by frictionless hinges, the total number of independent conditions of static equilibrium is

$$e = b + 3j + s$$

where b, j, and s have the same meaning as in (1). Since there are only two unknown end stress resultants in each bar and r unknown reactions at supports, the number of redundants reduces to

$$n = 2b + r - e = b - 3j - s > 0$$

If $n = 0$, the truss is statically determinate, and if $n < 0$, the truss is geometrically unstable. As in (1), the number of joints j in both equations includes all internal joints and all points of support.

(3) Internal releases included in (1) and (2) are results of special conditions imposed on the system. Five typical releases at point j are:

(*a*) *Free end*

$$U_j = 0 \quad V_j = 0 \quad W_j = 0 \quad X_j = 0 \quad Y_j = 0 \quad Z_j = 0$$
$$u_j \neq 0 \quad v_j \neq 0 \quad w_j \neq 0 \quad \phi_j \neq 0 \quad \psi_j \neq 0 \quad \theta_j \neq 0$$

(*b*) *Spherical roller normal to x, y plane*

$$U_j = 0 \quad V_j = 0 \quad W_j \neq 0 \quad X_j = 0 \quad Y_j = 0 \quad Z_j = 0$$
$$u_j \neq 0 \quad v_j \neq 0 \quad w_j = 0 \quad \phi_j \neq 0 \quad \psi_j \neq 0 \quad \theta_j \neq 0$$

(*c*) *Spherical hinge*

$$U_j \neq 0 \quad V_j \neq 0 \quad W_j \neq 0 \quad X_j = 0 \quad Y_j = 0 \quad Z_j = 0$$
$$u_j = 0 \quad v_j = 0 \quad w_j = 0 \quad \phi_j \neq 0 \quad \psi_j \neq 0 \quad \theta_j \neq 0$$

(*d*) *Cylindrical hinge along x axis*

$$U_j \neq 0 \quad V_j \neq 0 \quad W_j \neq 0 \quad X_j = 0 \quad Y_j \neq 0 \quad Z_j \neq 0$$
$$u_j = 0 \quad v_j = 0 \quad w_j = 0 \quad \phi_j \neq 0 \quad \psi_j = 0 \quad \theta_j = 0$$

1.06 STATIC DETERMINACY AND INDETERMINACY
(Continued)

(*e*) *Linear guide along x axes*

$U_j = 0$	$V_j \neq 0$	$W_j \neq 0$	$X_j = 0$	$Y_j \neq 0$	$Z_j \neq 0$
$u_j \neq 0$	$v_j = 0$	$w_j = 0$	$\phi_j \neq 0$	$\psi_j = 0$	$\theta_j = 0$

This summary shows that the number of releases equals six in all cases. The conditions stated in the first row are used in (1.06-1, 2), and the conditions stated in the second row are used in (1.07-2, 3).

1.07 KINEMATIC DETERMINACY AND INDETERMINACY

(1) Definitions. Any system of bars can be always visualized as a system of joints connected together by elastic springs. Since each joint may have as many as six degrees of freedom (three linear and three angular), the kinematics of the system is defined by the displacements of each joint, which may be introduced as the kinematic redundants. The degree of kinematic indeterminacy of the system is then equal to the number of admissible, independent, and unknown joint displacements.

(2) Beams, arches, rings, and frames. For a system of b bars connected together by j joints (some or all of which are rigid), with s internal releases, g internal constraints, and r reactive constraints, the total number of admissible and independent joint displacements is

$$d = 6j + s - g - r$$

where the internal releases are those defined in (1.06-3). The total number of joints includes all internal joints and all points of support. The internal constraints are given by the type of connections, and the reactive constraints equal the number of reactions.

(3) Trusses. In a space truss with b bars and j frictionless hinges connecting these bars, the number of admissible and independent joint linear displacements is

$$d = 3j + s - g - r$$

where s, g, r have the same meaning as in (2) above.

1.08 CLASSIFICATION OF SYSTEMS

(1) System of order one is a coplanar system of bars acted upon by loads in the system plane. Typical examples of systems of order one are planar trusses, beams loaded in the plane of symmetry of the cross section, planar frames, and arches loaded in their plane. Chapters 4–10 show the matrices of straight and curved bars of order one.

(2) System of order two is a coplanar system of bars acted upon by loads normal to the system plane. The most typical systems in this category are arches and rings loaded normal to their plane, grids, planar frames loaded normal to their plane, and plane curved (bent) bars loaded normal to their plane. Chapters 11–15 show the matrices of straight and curved bars of order two.

(3) System of order three is a nonplanar system of bars acted upon by loads of arbitrary directions. A space truss, space frame, and circular helix girder are typical systems of order three. Chapters 2 and 3 show the matrices of straight and curved bars of order three.

(4) Resolution. A planar system of bars acted on by loads of arbitrary direction can always be resolved into a system of order one by taking the in-plane load components and into a system of order two by taking the normal-to-plane load components.

1.09 GEOMETRIC TRANSFORMATIONS

(1) Two coordinate systems are introduced in the analysis:

(*a*) *Reference system* (datum, global system, 0 system) is an arbitrarily selected set of orthogonal axes x^o, y^o, z^o whose direction is fixed and common for all parts of the structure.

(*b*) *Member system* (local system, S system) is given by the principal axes x^s, y^s, z^s at the station of investigation. The position coordinates of a joint, support, or cross section j related to these systems are directed segments, so that

$$x_{0j}^o = -x_{j0}^o \qquad y_{0j}^o = -y_{j0}^o \qquad z_{0j}^o = -z_{j0}^o$$
$$x_{0j}^s = -x_{j0}^s \qquad y_{0j}^s = -y_{j0}^s \qquad z_{0j}^s = -z_{j0}^s$$

where the superscript and first and second subscripts identify the system, origin, and position, respectively. They form the position vectors defined in (1.02-1) and are related to each other by the angular transformation matrices.

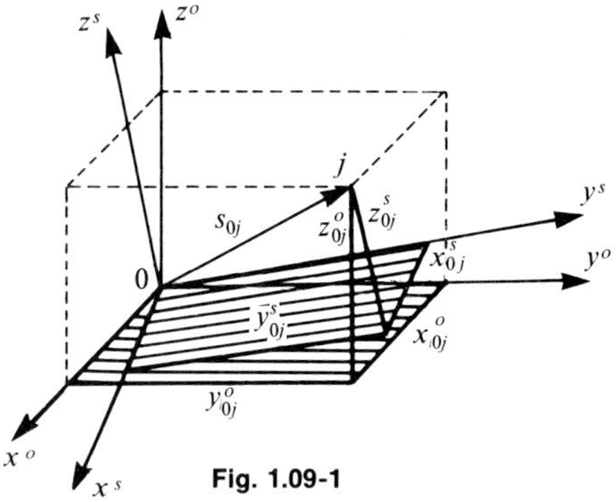

Fig. 1.09-1

(2) Angular transformation matrix equations are

$$\underbrace{\begin{bmatrix} x_{0j}^o \\ y_{0j}^o \\ z_{0j}^o \end{bmatrix}}_{s_{0j}^o} = \underbrace{\begin{bmatrix} \alpha_x & \alpha_y & \alpha_z \\ \beta_x & \beta_y & \beta_z \\ \gamma_x & \gamma_y & \gamma_z \end{bmatrix}}_{R^{os}} \underbrace{\begin{bmatrix} x_{0j}^s \\ y_{0j}^s \\ z_{0j}^s \end{bmatrix}}_{s_{0j}^s} \qquad \underbrace{\begin{bmatrix} x_{0j}^s \\ y_{0j}^s \\ z_{0j}^s \end{bmatrix}}_{s_{0j}^s} = \underbrace{\begin{bmatrix} \alpha_x & \beta_x & \gamma_x \\ \alpha_y & \beta_y & \gamma_y \\ \alpha_z & \beta_z & \gamma_z \end{bmatrix}}_{R^{so}} \underbrace{\begin{bmatrix} x_{0j}^o \\ y_{0j}^o \\ z_{0j}^o \end{bmatrix}}_{s_{0j}^o}$$

where s_{0j}^o and s_{0j}^s are the position vectors of j in the 0 and S systems, respectively, and R^{os}, R^{so} are the angular transformation matrices whose coefficients are the direction cosines of angles between the respective axes (Fig. 1.09-1).

(3) Relations of angular transformation matrices are

$$R^{so} = R^{os)T} = R^{os)-1} \qquad R^{os} = R^{so)T} = R^{so)-1}$$

where the superscripts $)T$ and $)-1$ identify the transpose and inverse of the matrix, respectively.

(4) Relations of direction cosines are summarized below (Ref. 5, p. 17).

Diagonal terms of matrix product $R^{os}R^{so}$	Off-diagonal terms of matrix product $R^{os}R^{so}$	Off-diagonal terms of matrix product $R^{so}R^{os}$	Diagonal terms of matrix product $R^{so}R^{os}$
$\alpha_x^2 + \alpha_y^2 + \alpha_z^2 = +1$	$\alpha_x\beta_x + \alpha_y\beta_y + \alpha_z\beta_z = 0$	$\alpha_x\alpha_y + \beta_x\beta_y + \gamma_x\gamma_y = 0$	$\alpha_x^2 + \beta_x^2 + \gamma_x^2 = +1$
$\beta_x^2 + \beta_y^2 + \beta_z^2 = +1$	$\beta_x\gamma_x + \beta_y\gamma_y + \beta_z\gamma_z = 0$	$\alpha_y\alpha_z + \beta_y\beta_z + \gamma_y\gamma_z = 0$	$\alpha_y^2 + \beta_y^2 + \gamma_y^2 = +1$
$\gamma_x^2 + \gamma_y^2 + \gamma_z^2 = +1$	$\gamma_x\alpha_x + \gamma_y\alpha_y + \gamma_z\alpha_z = 0$	$\alpha_z\alpha_x + \beta_z\beta_x + \gamma_z\gamma_x = 0$	$\alpha_z^2 + \beta_z^2 + \gamma_z^2 = +1$

1.10 DIRECTION COSINES BY ROTATION

(1) Procedure. The numerical values of the direction cosines in (1.09-2) can be obtained by successive rotation of coordinate axes. Symbols used in this procedure are

$$a,\ b,\ c = \text{right-handed angles (rad)}$$
$$S_a = \sin a,\ S_b = \sin b,\ S_c = \sin c,\ C_a = \cos a,\ C_b = \cos b,\ C_c = \cos c$$

This procedure requires a, b, c to be given angles.

(2) Successive rotations (subscript $0j$ omitted in x, y, z)

Rotation c about z^o	Rotation b about y^c	Rotation a about x^b
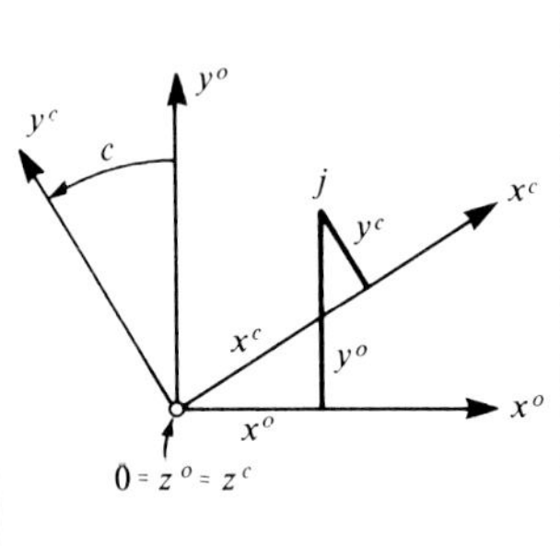	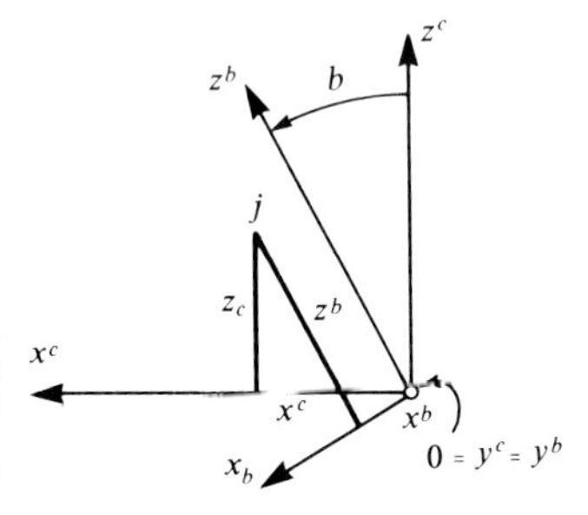	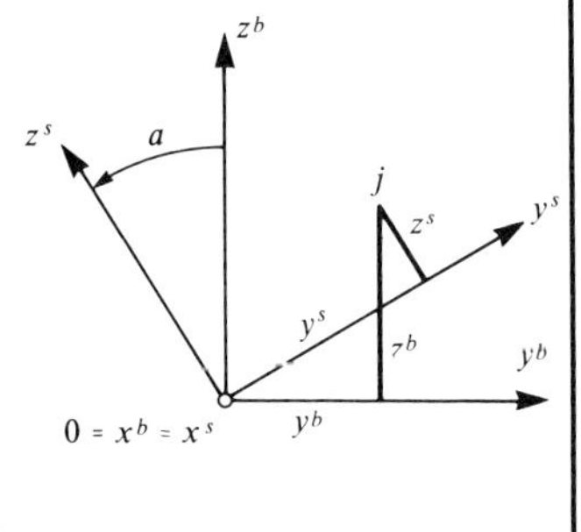
$R^{oc} = \begin{bmatrix} C_c & -S_c & 0 \\ S_c & C_c & 0 \\ 0 & 0 & 1 \end{bmatrix}$	$R^{cb} = \begin{bmatrix} C_b & 0 & S_b \\ 0 & 1 & 0 \\ -S_b & 0 & C_b \end{bmatrix}$	$R^{bs} = \begin{bmatrix} 1 & 0 & 0 \\ 0 & C_a & -S_a \\ 0 & S_a & C_a \end{bmatrix}$
$R^{co} = \begin{bmatrix} C_c & S_c & 0 \\ -S_c & C_c & 0 \\ 0 & 0 & 1 \end{bmatrix}$	$R^{bc} = \begin{bmatrix} C_b & 0 & -S_b \\ 0 & 1 & 0 \\ S_b & 0 & C_b \end{bmatrix}$	$R^{sb} = \begin{bmatrix} 1 & 0 & 0 \\ 0 & C_a & S_a \\ 0 & -S_a & C_a \end{bmatrix}$

(3) Space angular transformation matrices introduced in (1.09-2) are equal to the chain product of the component matrices (2) executed in the order of rotation. Thus

$$R^{os} = \begin{bmatrix} \alpha_x & \alpha_y & \alpha_z \\ \beta_x & \beta_y & \beta_z \\ \gamma_x & \gamma_y & \gamma_z \end{bmatrix} = \underbrace{\begin{bmatrix} C_bC_c & -C_aS_c + S_aS_bC_c & S_aS_c + C_aS_bC_c \\ S_bS_c & C_aC_c + S_aS_bS_c & -S_aC_c + C_aS_bS_c \\ -S_b & S_aC_b & C_aC_b \end{bmatrix}}_{R^{oc}R^{cb}R^{bs}}$$

By (1.09-2),

$$R^{so} = R^{os)T} = R^{sb}R^{bc}R^{co}$$

(4) Planar angular transformation matrices are the component matrices in (2) above.

1.11 DIRECTION COSINES BY GEOMETRY OF POSITION

(1) Equivalents used below are

$$x^o_{LR} = x^o_{0R} - y^o_{0L} \qquad S_\tau = \sin\tau$$
$$y^o_{LR} = y^o_{0R} - y^o_{0L} \qquad C_\tau = \cos\tau$$
$$z^o_{LR} = z^o_{0R} - z^o_{0L} \qquad e = \sqrt{(x^o_{LR})^2 + (y^o_{LR})^2}$$
$$f = \sqrt{(x^o_{LR})^2 + (y^o_{LR})^2 + (z^o_{LR})^2}$$

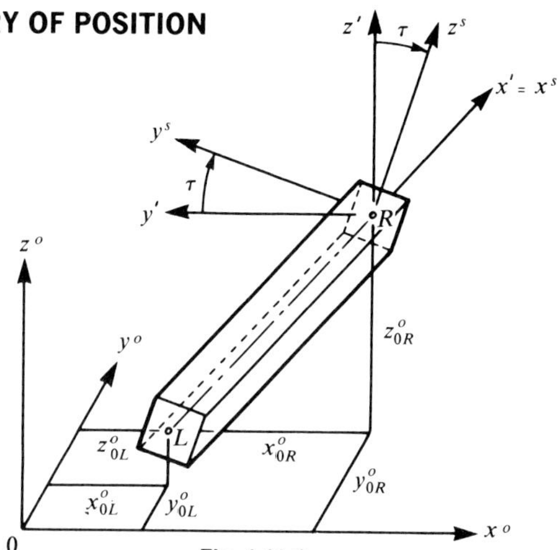

Fig. 1.11-1

(2) Procedure. If the position of the bar axis in Fig. 1.11-1 is given by the end points $L[x^o_{0L}, y^o_{0L}, z^o_{0L}]$, $R[x^o_{0R}, y^o_{0R}, z^o_{0R}]$ and by the position angle τ in the right-handed direction from y' which is parallel to the $x^o y^o$ plane, then the angular transformation matrix R^{os} and its transpose R^{so} can be expressed by equivalents shown in (1) above.

(3) Angular transformation matrices of order three corresponding to (1.10-3) become

$$R^{os} = \begin{bmatrix} \dfrac{x^o_{LR}}{f} & -\dfrac{y^o_{LR}}{e} & -\dfrac{x^o_{LR}z^o_{LR}}{ef} \\ \dfrac{y^o_{LR}}{f} & \dfrac{x^o_{LR}}{e} & -\dfrac{y^o_{LR}z^o_{LR}}{ef} \\ \dfrac{z^o_{LR}}{f} & 0 & \dfrac{e}{f} \end{bmatrix} \begin{bmatrix} 1 & 0 & 0 \\ 0 & C_\tau & -S_\tau \\ 0 & S_\tau & C_\tau \end{bmatrix}$$

$$R^{so} = \begin{bmatrix} 1 & 0 & 0 \\ 0 & C_\tau & S_\tau \\ 0 & -S_\tau & C_\tau \end{bmatrix} \begin{bmatrix} \dfrac{x^o_{LR}}{f} & \dfrac{y^o_{LR}}{f} & \dfrac{z^o_{LR}}{f} \\ -\dfrac{y^o_{LR}}{e} & \dfrac{x^o_{LR}}{e} & 0 \\ -\dfrac{x^o_{LR}z^o_{LR}}{ef} & -\dfrac{y^o_{LR}z^o_{LR}}{ef} & \dfrac{e}{f} \end{bmatrix}$$

where τ is frequently equal to zero.

(4) Angular transformation matrices of order two with $z^o_{LR} = 0$ and $\tau = 0$ are

$$R^{oc} = \begin{bmatrix} 1 & 0 & 0 \\ 0 & \dfrac{x^o_{LR}}{e} & -\dfrac{y_{LR}}{e} \\ 0 & \dfrac{y^o_{LR}}{e} & \dfrac{x^o_{LR}}{e} \end{bmatrix} \qquad R^{co} = \begin{bmatrix} 1 & 0 & 0 \\ 0 & \dfrac{x^o_{LR}}{e} & \dfrac{y^o_{LR}}{e} \\ 0 & -\dfrac{y^o_{LR}}{e} & \dfrac{x^o_{LR}}{e} \end{bmatrix}$$

(5) Angular transformation matrices of order one with $z^o_{LR} = 0$ and $\tau = 0$ are

$$R^{oc} = \begin{bmatrix} \dfrac{x^o_{LR}}{e} & -\dfrac{y^o_{LR}}{e} & 0 \\ \dfrac{y^o_{LR}}{e} & \dfrac{x^o_{LR}}{e} & 0 \\ 0 & 0 & 1 \end{bmatrix} \qquad R^{co} = \begin{bmatrix} \dfrac{x^o_{LR}}{e} & \dfrac{y^o_{LR}}{e} & 0 \\ -\dfrac{y^o_{LR}}{e} & \dfrac{x^o_{LR}}{e} & 0 \\ 0 & 0 & 1 \end{bmatrix}$$

(6) Applications of angular transformation matrices and their physical interpretations are shown in (1.12).

1.12 ANGULAR TRANSPORT AT A POINT

(1) Given vector at a point j in the 0 system is angularly transported to the C or S system by R^{co} or R^{so}, respectively, and inversely, the same vector given in the C or S system is angularly transported to the 0 system by R^{oc} or R^{os}, respectively (1.10–1.11). Typical operations at a point with a load vector are shown below. The same operations apply in transport of stress resultants, reactions, and displacements (Ref. 4, p. 8).

(2) Bar of order one In $x^o y^o$ plane is shown as a free body in Fig. 1.12-1. The load-vector angular transports are

$$\underbrace{\begin{bmatrix} P^c_{jx} \\ P^c_{jy} \\ Q^c_{jz} \end{bmatrix}}_{L^c_j} = \underbrace{\begin{bmatrix} C_c & S_c & 0 \\ -S_c & C_c & 0 \\ 0 & 0 & 1 \end{bmatrix}}_{R^{co}} \underbrace{\begin{bmatrix} P^o_{jx} \\ P^o_{jy} \\ Q^o_{jz} \end{bmatrix}}_{L^o_j} \qquad \underbrace{\begin{bmatrix} P^o_{jx} \\ P^o_{jy} \\ Q^o_{jz} \end{bmatrix}}_{L^o_j} = \underbrace{\begin{bmatrix} C_c & -S_c & 0 \\ S_c & C_c & 0 \\ 0 & 0 & 1 \end{bmatrix}}_{R^{oc}} \underbrace{\begin{bmatrix} P^c_{jx} \\ P^c_{jy} \\ Q^c_{jz} \end{bmatrix}}_{L^c_j}$$

where $C_c = \cos c$ and $S_c = \sin c$.

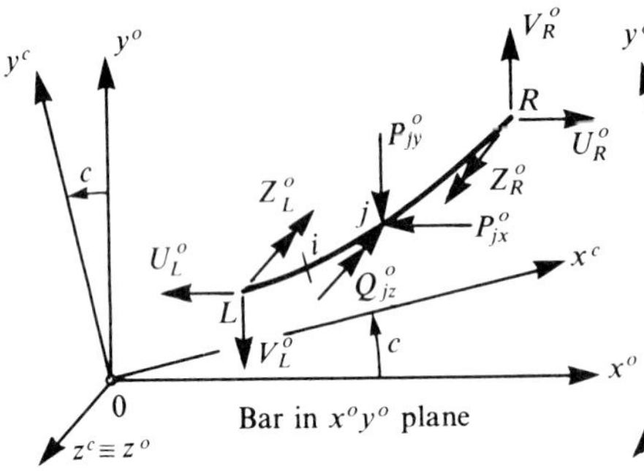

Fig. 1.12-1

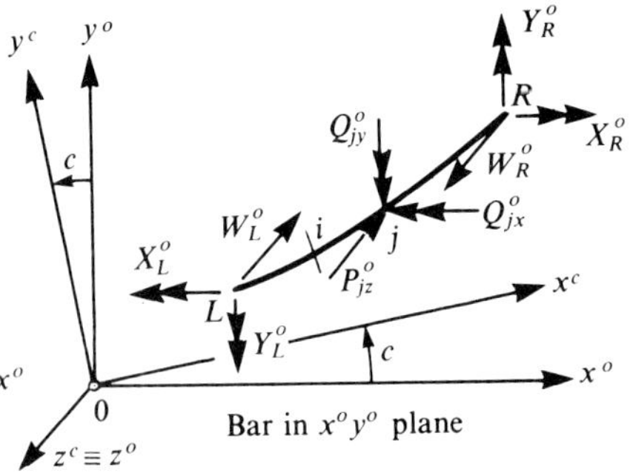

Fig. 1.12-2

(3) Bar of order two in $x^o y^o$ plane is shown as a free body in Fig. 1.12-2. The load-vector angular transports are

$$\underbrace{\begin{bmatrix} P^c_{jz} \\ Q^c_{jx} \\ Q^c_{jy} \end{bmatrix}}_{L^c_j} = \underbrace{\begin{bmatrix} 1 & 0 & 0 \\ 0 & C_c & S_c \\ 0 & -S_c & C_c \end{bmatrix}}_{R^{co}} \underbrace{\begin{bmatrix} P^o_{jz} \\ Q^o_{jx} \\ Q^o_{jy} \end{bmatrix}}_{L^o_j} \qquad \underbrace{\begin{bmatrix} P^o_{jz} \\ Q^o_{jx} \\ Q^o_{jy} \end{bmatrix}}_{L^o_j} = \underbrace{\begin{bmatrix} 1 & 0 & 0 \\ 0 & C_c & -S_c \\ 0 & S_c & C_c \end{bmatrix}}_{R^{oc}} \underbrace{\begin{bmatrix} P^c_{jz} \\ Q^c_{jx} \\ Q^c_{jz} \end{bmatrix}}_{L^c_j}$$

where C_c and S_c are defined in (2) above.

(4) Bar of order three in $x^o y^o z^o$ space with points L, j, R out of plane and with a load vector, and stress-resultant vectors consisting of six components, is not shown but implied. The load-vector angular transports are

$$\underbrace{\begin{bmatrix} P^s_j \\ Q^s_j \end{bmatrix}}_{L^s_j} = \underbrace{\begin{bmatrix} R^{so} & 0 \\ 0 & R^{so} \end{bmatrix}}_{\hat{R}^{so}} \underbrace{\begin{bmatrix} P^o_j \\ Q^o_j \end{bmatrix}}_{L^o_j} \qquad \underbrace{\begin{bmatrix} P^o_j \\ Q^o_j \end{bmatrix}}_{L^o} = \underbrace{\begin{bmatrix} R^{os} & 0 \\ 0 & R^{os} \end{bmatrix}}_{\hat{R}^{os}} \underbrace{\begin{bmatrix} P^s_j \\ Q^s_j \end{bmatrix}}_{L^s_j}$$

where R^{so} and R^{os} are the space angular transformation matrices introduced in (1.10-3) and (1.11-3).

1.13 LINEAR TRANSPORT OF STATIC VECTORS BETWEEN TWO POINTS

(1) Conditions of static equilibrium of bars in (1.12-2, 3, 4) yield two linear transport relations in the 0 system:

$$S_L^o = -t_{Lj}^o L_j^o + t_{LR}^o S_R^o \qquad S_R^o = t_{Rj}^o L_j^o + t_{RL}^o S_L^o$$

where S_L^o, S_R^o are the end stress-resultant vectors defined in (1.03-3) and L_j^o is the load vector introduced in (1.12-4). According to the order of the problem, each vector takes on the corresponding size. The linear transport matrices t_{LR}^o, t_{RL}^o and t_{Lj}^o, t_{Rj}^o are given for each particular order below.

(2) Linear transport matrices of order one in the bar of Fig. 1.12-1 are

$$t_{LR}^o = \begin{bmatrix} 1 & 0 & 0 \\ 0 & 1 & 0 \\ -y_{LR}^o & x_{LR}^o & 1 \end{bmatrix} \qquad t_{RL}^o = \begin{bmatrix} 1 & 0 & 0 \\ 0 & 1 & 0 \\ -y_{RL}^o & x_{RL}^o & 1 \end{bmatrix}$$

where $x_{LR}^o, y_{LR}^o, x_{RL}^o, y_{RL}^o$ are the position coordinates (1.11-1). The remaining matrices t_{Lj}^o and t_{Rj}^o have a similar form in terms of x_{Lj}^o, y_{Lj}^o and x_{Rj}^o, y_{Rj}^o, respectively (Ref. 4, p. 10).

(3) Linear transport matrices of order two in the bar of Fig. 1.12-2 are

$$t_{LR}^o = \begin{bmatrix} 1 & 0 & 0 \\ y_{LR}^o & 1 & 0 \\ -x_{LR}^o & 0 & 1 \end{bmatrix} \qquad t_{RL}^o = \begin{bmatrix} 1 & 0 & 0 \\ y_{RL}^o & 1 & 0 \\ -x_{RL}^o & 0 & 1 \end{bmatrix}$$

and t_{Lj}^o, t_{Rj}^o are constructed analogically (Ref. 5, p. 32).

(4) Linear transport matrices of order three, corresponding to the bar of (1.12-4), are

$$t_{LR}^o = \left[\begin{array}{ccc|ccc} 1 & 0 & 0 & 0 & 0 & 0 \\ 0 & 1 & 0 & 0 & 0 & 0 \\ 0 & 0 & 1 & 0 & 0 & 0 \\ \hline 0 & -z_{LR}^o & y_{LR}^o & 1 & 0 & 0 \\ z_{LR}^o & 0 & -x_{LR}^o & 0 & 1 & 0 \\ -y_{LR}^o & x_{LR}^o & 0 & 0 & 0 & 1 \end{array}\right]$$

$$t_{RL}^o = \left[\begin{array}{ccc|ccc} 1 & 0 & 0 & 0 & 0 & 0 \\ 0 & 1 & 0 & 0 & 0 & 0 \\ 0 & 0 & 1 & 0 & 0 & 0 \\ \hline 0 & -z_{RL}^o & y_{RL}^o & 1 & 0 & 0 \\ z_{RL}^o & 0 & -x_{RL}^o & 0 & 1 & 0 \\ -y_{RL}^o & x_{RL}^o & 0 & 0 & 0 & 1 \end{array}\right]$$

and t_{Lj}^o, t_{Rj}^o are constructed analogically (Ref. 5, p. 22).

(5) Properties. In all three cases,

$$\begin{aligned} t_{RL}^o t_{LR}^o &= I & t_{LR}^o t_{RL}^o &= I \\ t_{Lj}^o t_{jL}^o &= I & t_{Lj}^o t_{jR}^o &= t_{LR}^o \\ t_{Rj}^o t_{jR}^o &= I & t_{Rj}^o t_{jL}^o &= t_{RL}^o \end{aligned}$$

where I is the unit matrix. These relations hold in the 0 system and any other system.

1.14 ELEMENTAL DEFORMATIONS IN MEMBER SYSTEM

(1) Basic unit of the deformation analysis of bars is the elemental flexibility, defined as the deformation of the bar element ds due to a unit cause. Since the cause may be a force or a moment, the elemental flexibilities are classified as linear and angular, respectively (Ref. 5, p. 35).

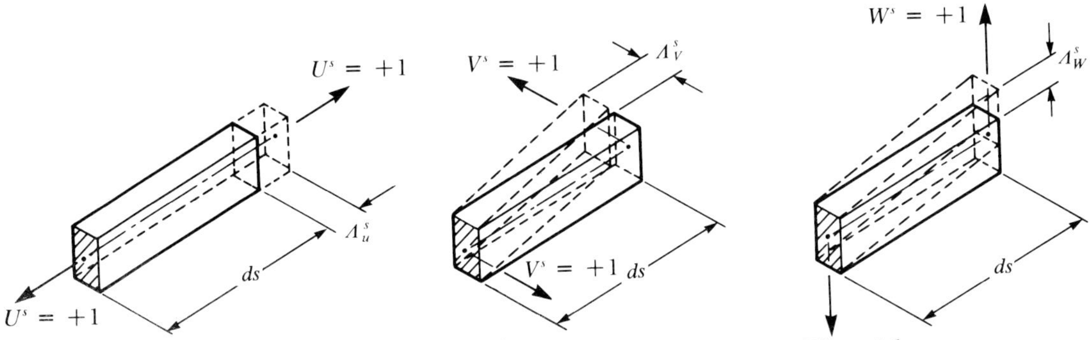

Fig. 1.14-1

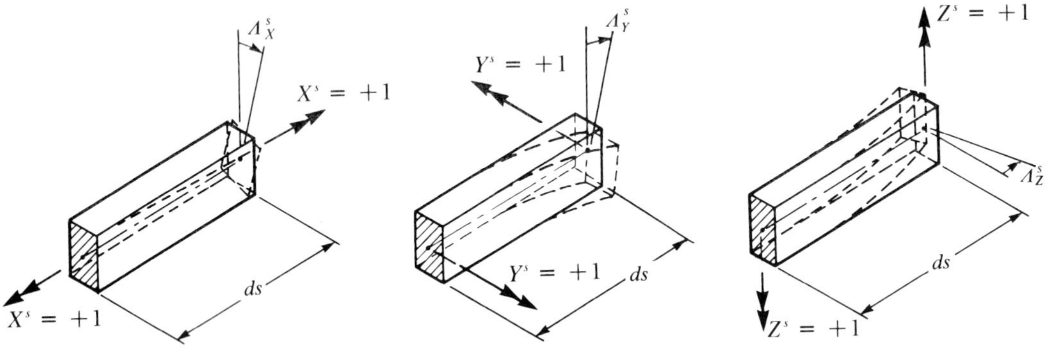

Fig. 1.14-2

(2) Linear elemental flexibilities in the principal axes of the normal section of the element ds are the linear deformations due to $U^s = +1$, $V^s = +1$, $W^s = +1$, respectively (Fig. 1.14-1). Analytically, they are

$$\Lambda_U^s = \frac{ds}{EA_x} \qquad \Lambda_V^s = \frac{ds}{GA_y} \qquad \Lambda_W^s = \frac{ds}{GA_z}$$

where ds = differential element of bar axis (m)

$A_x = A$ = area of normal section (m^2)

$A_y = A \cdot B_y$ = A modified by the shear shape factor in the y^s direction (m^2)

$A_z = A \cdot B_z$ = A modified by the shear shape factor in the z^s direction (m^2)

E = modulus of elasticity (Pa)

G = modulus of rigidity (Pa)

The analytical expressions for B_y and B_z are tabulated in Appendix B.

(3) Angular elemental flexibilities in the principal axes of the normal section of the bar element ds are the angular deformations (slopes) due to $X^s = +1$, $Y^s = +1$, $Z^s = +1$, respectively (Fig. 1.14-2). Analytically, they are

$$\Lambda_X^s = \frac{ds}{GI_x} \qquad \Lambda_Y^s = \frac{ds}{EI_y} \qquad \Lambda_Z^s = \frac{ds}{EI_z}$$

where I_x = torsional constant of A about x^s (m^4)

I_y, I_z = moments of inertia of A about y^s, z^s, respectively (m^4)

The analytical expressions for I_x, I_y, and I_z are tabulated in Appendix B.

(4) Elemental deformations of order one in the C system (Fig. 1.12-1) are

$$\underbrace{\begin{bmatrix} du^c \\ dv^c \\ d\theta^c \end{bmatrix}}_{d\Delta^c} = \underbrace{\begin{bmatrix} \Lambda_U^c & & \\ & \Lambda_V^c & \\ & & \Lambda_Z^c \end{bmatrix}}_{\Lambda^c} \underbrace{\begin{bmatrix} U^c \\ V^c \\ Z^c \end{bmatrix}}_{S^c}$$

where du^c, dv^c = elemental linear deformations along x^c, y^c, respectively (m)

$d\theta^c$ = elemental angular deformation about z^c (rad)

U^c, V^c, Z^c = components of stress-resultant vector (1.03) acting on the element

(5) Elemental deformations of order two in the C system (Fig. 1.12-2) are

$$\underbrace{\begin{bmatrix} dw^c \\ d\phi^c \\ d\psi^c \end{bmatrix}}_{d\Delta^c} = \underbrace{\begin{bmatrix} \Lambda_W^c & & \\ & \Lambda_X^c & \\ & & \Lambda_Y^c \end{bmatrix}}_{\Delta^c} \underbrace{\begin{bmatrix} W^c \\ X^c \\ Y^c \end{bmatrix}}_{S^c}$$

where dw^c = elemental linear deformation along z^c (m)

$d\phi^c$ = elemental angular deformation about x^c (rad)

$d\psi^c$ = elemental angular deformation about y^c (rad)

W^c, X^c, Y^c = components of stress-resultant vector (1.03) acting on the element

(6) Elemental deformations of order three in the S system are

$$\underbrace{\begin{bmatrix} du^s \\ dv^s \\ dw^s \\ \hline d\phi^s \\ d\psi^s \\ d\theta^s \end{bmatrix}}_{d\Delta^s} = \underbrace{\left[\begin{array}{ccc|ccc} \Lambda_U^s & & & & & \\ & \Lambda_V^s & & & & \\ & & \Lambda_W^s & & & \\ \hline & & & \Lambda_X^s & & \\ & & & & \Lambda_Y^s & \\ & & & & & \Lambda_Z^s \end{array}\right]}_{\Lambda^s} \underbrace{\begin{bmatrix} U^s \\ V^s \\ W^s \\ \hline X^s \\ Y^s \\ Z^s \end{bmatrix}}_{S^s}$$

(7) Angular transport of Λ^s in the S system to Λ^o in the 0 system is given by the following congruent transformation:

$$d\Delta^o = R^{os}\Lambda^s R^{so} S^o = \Lambda^o S^o$$

where R^{os}, R^{so} are the angular transformation matrices introduced for particular orders in (1.09–1.11). The congruent transformation given above preserves the symmetry of the matrix and applies in (4–6).

1.15 ELEMENTAL DEFORMATIONS IN 0 SYSTEM

(1) Elemental deformations of order one in the 0 system are, by (1.14-7),

$$\underbrace{\begin{bmatrix} du^o \\ dv^o \\ d\theta^o \end{bmatrix}}_{d\Delta^o} = \underbrace{\begin{bmatrix} C_c & -S_c & 0 \\ S_c & C_c & 0 \\ 0 & 0 & 1 \end{bmatrix} \begin{bmatrix} \Lambda_U^c & & \\ & \Lambda_V^c & \\ & & \Lambda_Z^c \end{bmatrix} \begin{bmatrix} C_c & S_c & 0 \\ -S_c & C_c & 0 \\ 0 & 0 & 1 \end{bmatrix}}_{\underbrace{\begin{bmatrix} \Lambda_{UU}^o & \Lambda_{UV}^o & 0 \\ \Lambda_{VU}^o & \Lambda_{VV}^o & 0 \\ 0 & 0 & \Lambda_{ZZ}^o \end{bmatrix}}_{\Lambda^o}} \underbrace{\begin{bmatrix} U^o \\ V^o \\ Z^o \end{bmatrix}}_{S^o}$$

where Λ^o is the elemental flexibility matrix of order one in the 0 system in terms of the elemental flexibilities Λ_U^c, Λ_V^c, Λ_Z^c (1.14-2, 3) and the direction functions

$$C_c = \cos c \qquad S_c = \sin c$$

The elements of Λ^o are then

$$\Lambda_{UU}^o = C_c^2\Lambda_U^c + S_c^2\Lambda_V^c \qquad \Lambda_{UV}^c = \Lambda_{VU}^c = C_cS_c(\Lambda_U^c - \Lambda_V^c)$$
$$\Lambda_{VV}^o = S_c^2\Lambda_U^c + C_c^2\Lambda_V^c \qquad \Lambda_{ZZ}^o = \Lambda_Z^c$$

and U^o, V^o, Z^o are the components of the stress-resultant vector (1.03) acting on the element in the 0 system.

(2) Elemental deformations of order two in the 0 system are, by (1.14-7),

$$\underbrace{\begin{bmatrix} dw^o \\ d\phi^o \\ d\psi^o \end{bmatrix}}_{d\Delta^o} = \underbrace{\begin{bmatrix} 1 & 0 & 0 \\ 0 & C_c & -S_c \\ 0 & S_c & C_c \end{bmatrix} \begin{bmatrix} \Lambda_W^c & & \\ & \Lambda_X^c & \\ & & \Lambda_Y^c \end{bmatrix} \begin{bmatrix} 1 & 0 & 0 \\ 0 & C_c & S_c \\ 0 & -S_c & C_c \end{bmatrix}}_{\underbrace{\begin{bmatrix} \Lambda_{WW}^o & 0 & 0 \\ 0 & \Lambda_{XX}^o & \Lambda_{XY}^o \\ 0 & \Lambda_{YX}^o & \Lambda_{YY}^o \end{bmatrix}}_{\Lambda^o}} \underbrace{\begin{bmatrix} W^o \\ X^o \\ Z^o \end{bmatrix}}_{S^o}$$

where Λ^o is the elemental flexibility matrix of order two in the 0 system in terms of the elemental flexibilities Λ_W^c, Λ_X^c, Λ_Y^c (1.14-2, 3) and the direction functions

$$C_c = \cos c \qquad S_c = \sin c$$

The elements of Λ^o are then

$$\Lambda_{WW}^o = \Lambda_W^c \qquad \Lambda_{XX}^o = C_c^2\Lambda_X^c + S_c^2\Lambda_Y^c$$
$$\Lambda_{XY}^o = \Lambda_{YX}^o = C_cS_c(\Lambda_X^c - \Lambda_Y^c) \qquad \Lambda_{YY}^o = S_c^2\Lambda_X^c + C_c^2\Lambda_Y^c$$

and W^o, X^o, Y^o are the components of the stress-resultant vector (1.03) acting on the element in the 0 system.

(3) Elemental deformations of order three in the 0 system are, by (1.14-7),

$$\underbrace{\left[\begin{array}{c} du^o\\ dv^o\\ dw^o\\ \hline d\phi^o\\ d\psi^o\\ d\theta^o\end{array}\right]}_{d\Delta^o} = \underbrace{\left[\begin{array}{c|c} R^{os} & 0\\ \hline 0 & R^{os}\end{array}\right]\left[\begin{array}{ccc|ccc}\Lambda^s_U &&&&&\\ &\Lambda^s_V&&&&\\ &&\Lambda^s_W&&&\\ \hline &&&\Lambda^s_X&&\\&&&&\Lambda^s_Y&\\&&&&&\Lambda^s_Z\end{array}\right]\left[\begin{array}{c|c}R^{so}&0\\ \hline 0&R^{so}\end{array}\right]}_{\left[\begin{array}{ccc|ccc}\Lambda^o_{UU} & \Lambda^o_{UV} & \Lambda^o_{UW} & & & \\ \Lambda^o_{VU} & \Lambda^o_{VV} & \Lambda^o_{VW} & & 0 & \\ \Lambda^o_{WU} & \Lambda^o_{WV} & \Lambda^o_{WW} & & & \\ \hline & & & \Lambda^o_{XX} & \Lambda^o_{XY} & \Lambda^o_{XZ}\\ & 0 & & \Lambda^o_{YX} & \Lambda^o_{YY} & \Lambda^o_{YZ}\\ & & & \Lambda^o_{ZX} & \Lambda^o_{ZY} & \Lambda^o_{ZZ}\end{array}\right] = \Lambda^o}\underbrace{\left[\begin{array}{c}U^o\\V^o\\W^o\\ \hline X^o\\Y^o\\Z^o\end{array}\right]}_{S^o}$$

where Λ^o is the elemental flexibility matrix of order three in the 0 system in terms of the elemental flexibilities (1.14-2, 3) in the S system, and the direction cosines $\alpha_x, \alpha_y, \ldots, \gamma_y, \gamma_z$ in R^{os} and R^{so} are given in (1.10, 1.11). The elements of Λ^o are then

$$\begin{aligned}
\Lambda^o_{UU} &= \alpha_x^2\Lambda^s_U + \alpha_y^2\Lambda^s_V + \alpha_z^2\Lambda^s_W\\
\Lambda^o_{VV} &= \beta_x^2\Lambda^s_U + \beta_y^2\Lambda^s_V + \beta_z^2\Lambda^s_W\\
\Lambda^o_{WW} &= \gamma_x^2\Lambda^s_U + \gamma_y^2\Lambda^s_V + \gamma_z^2\Lambda^s_W
\end{aligned}$$

$$\begin{aligned}
\Lambda^o_{UV} &= \Lambda^o_{VU} = \alpha_x\beta_x\Lambda^s_U + \alpha_y\beta_y\Lambda^s_V + \alpha_z\beta_z\Lambda^s_W\\
\Lambda^o_{UW} &= \Lambda^o_{WU} = \alpha_x\gamma_x\Lambda^s_U + \alpha_y\gamma_y\Lambda^s_V + \alpha_z\gamma_z\Lambda^s_W\\
\Lambda^o_{VW} &= \Lambda^o_{WV} = \beta_x\gamma_x\Lambda^s_U + \beta_y\gamma_y\Lambda^s_V + \beta_z\gamma_z\Lambda^s_W
\end{aligned}$$

$$\begin{aligned}
\Lambda^o_{XX} &= \alpha_x^2\Lambda^s_X + \alpha_y^2\Lambda^s_Y + \alpha_z^2\Lambda^s_Z\\
\Lambda^o_{YY} &= \beta_x^2\Lambda^s_X + \beta_y^2\Lambda^s_y + \beta_z^2\Lambda^s_Z\\
\Lambda^o_{ZZ} &= \gamma_x^2\Lambda^s_X + \gamma_y^2\Lambda^s_Y + \gamma_z^2\Lambda^s_Z
\end{aligned}$$

$$\begin{aligned}
\Lambda^o_{XY} &= \Lambda^o_{YX} = \alpha_x\beta_x\Lambda^s_X + \alpha_y\beta_y\Lambda^s_Y + \alpha_z\beta_z\Lambda^s_Z\\
\Lambda^o_{XZ} &= \Lambda^o_{ZX} = \alpha_x\gamma_x\Lambda^s_X + \alpha_y\gamma_y\Lambda^s_Y + \alpha_z\gamma_z\Lambda^s_Z\\
\Lambda^o_{YZ} &= \Lambda^o_{ZY} = \beta_x\gamma_x\Lambda^s_X + \beta_y\gamma_y\Lambda^s_Y + \beta_z\gamma_z\Lambda^s_Z
\end{aligned}$$

and $U^o, V^o, \ldots, Y^o, Z^o$ are the components of the stress-resultant vector (1.03) acting on the element in the 0 system.

1.16 LINEAR TRANSPORT OF DISPLACEMENT VECTORS BETWEEN TWO POINTS

(1) Conditions of consistent deformation of bars in (1.12-2, 3, 4) yield two linear transport relations in the 0 system:

$$\Delta_L^o = -\int_L^R [t_{iL}^{o)T}\Lambda_i^o S_i^o] + t_{RL}^{o)T}\Delta_R^o \qquad \Delta_R^o = \int_L^R [t_{iR}^{o)T}\Lambda_i^o S_i^o] + t_{LR}^{o)T}\Lambda_L^o$$

where Δ_L^o, Δ_R^o are the end displacement vectors (1.04-4), Λ_i^o and S_i^o are the elemental flexibility matrix and the stress-resultant matrix at an arbitrary section i (1.15-3), and $t_{LR}^{o)T}$, $t_{RL}^{o)T}$ are the transposes of t_{LR}^o, t_{RL}^o (1.13-4). The remaining matrices $t_{iL}^{o)T}$ and $t_{iR}^{o)T}$ are given for each particular order below in (2–4).

(2) Linear transport equations of order one in the bar of Fig. 1.12-1 are

$$\begin{bmatrix} u_L^o \\ v_L^o \\ \phi_L^o \end{bmatrix} = -\int_L^R \begin{bmatrix} 1 & 0 & -y_{iL}^o \\ 0 & 1 & x_{iL}^o \\ 0 & 0 & 1 \end{bmatrix} \begin{bmatrix} \Lambda_{UU}^o & \Lambda_{UV}^o & 0 \\ \Lambda_{VU}^o & \Lambda_{VV}^o & 0 \\ 0 & 0 & \Lambda_{ZZ}^o \end{bmatrix} \begin{bmatrix} U_i^o \\ V_i^o \\ Z_i^o \end{bmatrix} + \begin{bmatrix} 1 & 0 & -y_{RL}^o \\ 0 & 1 & x_{RL}^o \\ 0 & 0 & 1 \end{bmatrix} \begin{bmatrix} u_R^o \\ v_R^o \\ \theta_R^o \end{bmatrix}$$

$$\begin{bmatrix} u_R^o \\ v_R^o \\ \theta_R^o \end{bmatrix} = \int_L^R \begin{bmatrix} 1 & 0 & -y_{iR}^o \\ 0 & 1 & x_{iR}^o \\ 0 & 0 & 1 \end{bmatrix} \begin{bmatrix} \Lambda_{UU}^o & \Lambda_{UV}^o & 0 \\ \Lambda_{VU}^o & \Lambda_{VV}^o & 0 \\ 0 & 0 & \Lambda_{ZZ}^o \end{bmatrix} \begin{bmatrix} U_i^o \\ V_i^o \\ Z_i^o \end{bmatrix} + \begin{bmatrix} 1 & 0 & -y_{LR}^o \\ 0 & 1 & x_{LR}^o \\ 0 & 0 & 1 \end{bmatrix} \begin{bmatrix} u_L^o \\ v_L^o \\ \theta_L^o \end{bmatrix}$$

where x_{iL}^o, y_{iL}^o, x_{iR}^o, y_{iR}^o and U_i^o, V_i^o, Z_i^o are variables (Ref. 4, p. 24).

(3) Linear transport equations of order two in the bar of Fig. 1.12-2 are

$$\begin{bmatrix} w_L^o \\ \phi_L^o \\ \psi_L^o \end{bmatrix} = -\int_L^R \begin{bmatrix} 1 & y_{iL}^o & -x_{iL}^o \\ 0 & 1 & 0 \\ 0 & 0 & 1 \end{bmatrix} \begin{bmatrix} \Lambda_{WW}^o & 0 & 0 \\ 0 & \Lambda_{XX}^o & \Lambda_{XY}^o \\ 0 & \Lambda_{YX}^o & \Lambda_{YY}^o \end{bmatrix} \begin{bmatrix} W_i^o \\ X_i^o \\ Y_i^o \end{bmatrix} + \begin{bmatrix} 1 & y_{RL}^o & -x_{RL}^o \\ 0 & 1 & 0 \\ 0 & 0 & 1 \end{bmatrix} \begin{bmatrix} w_R^o \\ \phi_R^o \\ \psi_R^o \end{bmatrix}$$

$$\begin{bmatrix} w_R^o \\ \phi_R^o \\ \psi_R^o \end{bmatrix} = \int_L^R \begin{bmatrix} 1 & y_{iR}^o & -x_{iR}^o \\ 0 & 1 & 0 \\ 0 & 0 & 1 \end{bmatrix} \begin{bmatrix} \Lambda_{WW}^o & 0 & 0 \\ 0 & \Lambda_{XX}^o & \Lambda_{XY}^o \\ 0 & \Lambda_{YX}^o & \Lambda_{YY}^o \end{bmatrix} \begin{bmatrix} W_i^o \\ X_i^o \\ Y_i^o \end{bmatrix} + \begin{bmatrix} 1 & y_{LR}^o & -x_{LR}^o \\ 0 & 1 & 0 \\ 0 & 0 & 1 \end{bmatrix} \begin{bmatrix} w_L^o \\ \phi_L^o \\ \psi_L^o \end{bmatrix}$$

where x_{iL}^o, y_{iL}^o, x_{iR}^o, y_{iR}^o and W_i^o, X_i^o, Y_i^o are variables (Ref. 5, p. 40).

(4) Linear transport equations of order three in the bar of (1.12-3) are expressed in the forms given in (1) above and

$$t_{iL}^{o)T} = \left[\begin{array}{ccc|ccc} 1 & 0 & 0 & 0 & z_{iL}^o & -y_{iL}^o \\ 0 & 1 & 0 & -z_{iL}^o & 0 & x_{iL}^o \\ 0 & 0 & 1 & -y_{iL}^o & -x_{iL}^o & 0 \\ \hline 0 & 0 & 0 & 1 & 0 & 0 \\ 0 & 0 & 0 & 0 & 1 & 0 \\ 0 & 0 & 0 & 0 & 0 & 1 \end{array}\right] \qquad t_{iR}^{o)T} = \left[\begin{array}{ccc|ccc} 1 & 0 & 0 & 0 & z_{iR}^o & -y_{iR}^o \\ 0 & 1 & 0 & -z_{iR}^o & 0 & x_{iR}^o \\ 0 & 0 & 1 & y_{iR}^o & -x_{iR}^o & 0 \\ \hline 0 & 0 & 0 & 1 & 0 & 0 \\ 0 & 0 & 0 & 0 & 1 & 0 \\ 0 & 0 & 0 & 0 & 0 & 1 \end{array}\right]$$

where x_{iL}^o, y_{iL}^o, z_{iL}^o and x_{iR}^o, y_{iR}^o, z_{iR}^o are variables (Ref. 5, p. 40).

1.17 TEMPERATURE AND IMPOSED DEFORMATIONS

(1) Temperature-effect analysis assumes (in general) a linear variation of temperature along the y and z axes of the element of the bar (Figs. 1.17-1, 1.17-2).

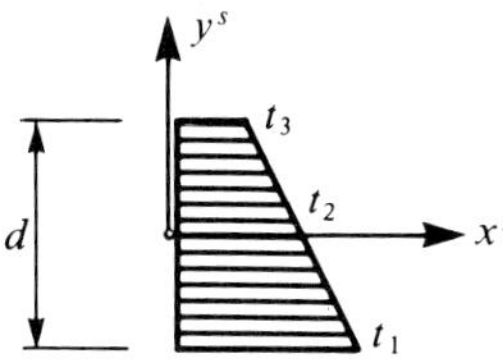

Fig. 1.17-1

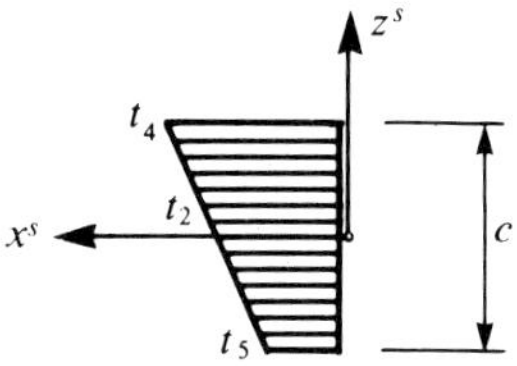

Fig. 1.17-2

The thermal strains in the S system are

$$e_t^s = \alpha_t(t_2 - t_0) \qquad f_t^s = \alpha_t(t_1 - t_3)/d \qquad g_t^s = \alpha_t(t_4 - t_5)/c$$

where c, d = width, depth, of normal section (m)

α_t = thermal coefficient (1/°C)

and the changes in temperature indicated by the diagrams are $t_1 - t_0, t_2 - t_0, \ldots, t_5 - t_0$, with $t_2 - t_0$ = change in temperature at the centroid (°C).

(2) Elemental temperature deformations in the 0 system are

$$\begin{bmatrix} du_t^o \\ dv_t^o \\ dw_t^o \\ \hline d\phi_t^o \\ d\psi_t^o \\ d\theta_t^o \end{bmatrix} = \left[\begin{array}{c|c} R^{os} & 0 \\ \hline 0 & R^{os} \end{array}\right] \begin{bmatrix} e_t^s \\ 0 \\ 0 \\ \hline 0 \\ f_t^s \\ g_t^s \end{bmatrix} ds = d\Delta_t^o$$

where R^{os} is the angular transformation matrix (1.10-3) or (1.11-3) of the element ds.

(3) Imposed deformation analysis assumes (in general) six abrupt changes at j in the bar axis, given in column matrix form in the 0 system as

$$\begin{bmatrix} u_j^o \\ v_j^o \\ w_j^o \\ \hline \phi_j^o \\ \psi_j^o \\ \theta_j^o \end{bmatrix} = \left[\begin{array}{c|c} R^{os} & 0 \\ \hline 0 & R^{os} \end{array}\right] \begin{bmatrix} u_j^s \\ v_j^s \\ w_j^s \\ \hline \phi_j^s \\ \psi_j^s \\ \theta_j^s \end{bmatrix} = d\Delta_j^o$$

where R^{os} is the angular transformation matrix (1.10-3) or (1.11-3) at j (arbitrary point).

(4) Conditions of consistent deformation due to causes (2) and (3) yield

$$\Delta_L^o = -\int_L^R [t_{iL}^{o)T}\, d\Delta_t^o] - t_{jL}^{o)T}\,\Delta_j^o + t_{RL}^{o)T}\,\Delta_R^o$$

$$\Delta_R^o = \int_L^R [t_{iR}^{o)T}\, d\Delta_t^o] + t_{jR}^{o)T}\,\Delta_j^o + t_{LR}^{o)T}\,\Delta_L^o$$

where $\Delta_L^o, \Delta_R^o, t_{iL}^{o)T}, t_{iR}^{o)T}, t_{jL}^{o)T}, t_{jR}^{o)T}$ are defined in (1.16-1) and $d\Delta_t^o, \Delta_j^o$ are given above.

1.18 INTERACTIVE SYSTEMS

(1) Analysis of free bars introduced in (1.13–1.17) assumes that the surrounding medium of the bar offers no resistance to the formation of the elastic curve and the effect of the axial forces on the bending moments is insignificant. If these two simplifications are not justified, the effect of the elastic medium and of the magnification effect of axial force on the bending moment must be included in the analysis.

(2) Analysis of interactive bars (such as bars encased in elastic foundation and beam-columns) cannot be performed by the simple operations described in (1.13–1.17). For their analysis, the governing differential equations must be derived first. For prismatic bars of order three,

$$EA_x \frac{d^2u(x)}{dx^2} - k_u u(x) = p_x \qquad GI_x \frac{d^2\phi(x)}{dx^2} - k_\phi \phi(x) = q_x$$

$$EI_z \frac{d^4v(x)}{dx^4} - \eta N \frac{d^2v(x)}{dx^2} + k_v v(x) = p_y$$

$$EI_y \frac{d^4w(x)}{dx^4} - \eta N \frac{d^2w(x)}{dx^2} + k_w w(x) = p_z$$

where k_u, k_v, k_w, k_ϕ are the foundation moduli (8.01, 13.01), p_x, p_y, p_z, q_x are the intensities of loads at x,

$$\eta = \begin{cases} +1 & \text{if } N = \text{axial tensile force (N)} \\ -1 & \text{if } N = \text{axial compressive force (N)} \end{cases}$$

and the remaining symbols are those defined in (1.14).

(3) Solutions of these differential equations are

$$u(x) = T_{uu}(x)u_L + T_{uU}(x)U_L + G_u(x)$$
$$v(x) = T_{vv}(x)v_L + T_{v\theta}(x)\theta_L + T_{vZ}(x)Z_L - T_{vV}(x)V_L + G_v(x)$$
$$w(x) = T_{ww}(x)w_L - T_{w\psi}(x)\psi_L - T_{wY}(x)Y_L - T_{wW}(x)W_L + G_w(x)$$
$$\phi(x) = T_{\phi\phi}(x)\phi_L + T_{\phi X}(x)X_L + G_\phi(x)$$

where $u(x)$, $v(x)$, $w(x)$ are the coordinates of the elastic curve at x, $\phi(x)$ is the angle of twist at the same point, $T_{uu}(x)$, $T_{uU}(x)$, . . . , $T_{\phi X}(x)$ are the transport coefficients in x obtained by the Laplace transform method, $G_u(x)$, $G_v(x)$, $G_w(x)$, $G_\phi(x)$ are the load functions obtained as the convolution integrals of the Laplace transform solution. The remaining constants u_L, v_L, w_L, ϕ_L, ψ_L, θ_L, U_L, V_L, W_L, X_L, Y_L, Z_L are the displacements and stress resultants of the left end. Particular cases of these symbolic solutions are given in Chaps. 8–10 and 13–15.

(4) Bar functions. Once the equations of the elastic curve are available, the remaining bar functions are to some scale the derivatives of these equations, so that

$$u(x) = u \qquad \frac{du(x)}{dx} = \frac{U}{EA_x} \qquad \phi(x) = \phi \qquad \frac{d\phi(x)}{dx} = \frac{X}{GI_x}$$

$$v(x) = v \qquad \frac{dv(x)}{dx} = \theta \qquad w(x) = w \qquad \frac{dw(x)}{dx} = -\psi$$

$$\frac{d^2v(x)}{dx^2} = \frac{Z}{EI_z} \qquad \frac{d^2w(x)}{dx^2} = -\frac{Y}{EI_y}$$

$$\frac{d^3v(x)}{dx^3} = -\frac{V}{EI_z} + \frac{\eta N\, dv(x)}{dx} \qquad \frac{d^3w(x)}{dx^3} = -\frac{W}{EI_y} + \frac{\eta N\, dw(x)}{dx}$$

where U, V, Z, X, Y, Z, E, G, A_x, I_x, I_y, I_z are symbols used in the elemental deformation analysis (1.14).

1.19 TRANSPORT METHOD

(1) Concepts presented in the preceding sections form the foundation of the analysis of bar systems, and their applications lead to a large variety of methods. Regardless of their names, each of these methods falls into one of the following categories:

(*a*) *The transport method*
(*b*) *The flexibility method*
(*c*) *The stiffness method*

The characteristics of these methods are summarized in this section and in (1.20–1.21), and their sign conventions are depicted in (1.22).

(2) Transport method, also called the transfer method or the mixed method, is based on the governing matrix equations

$$H_L = T_{LR}H_R \qquad H_R = T_{RL}H_L$$

where H_L, H_R are the *end state vectors* of the topological chain and T_{LR}, T_{RL} are their *transport matrices*. The state vectors define the end stress resultant and the end displacement at the left end L and the right end R, respectively. The relationships between H_L and H_R, and vice versa, are provided by the transport matrices T_{LR} and T_{RL}.

(3) Transport matrices give a clear, systematic, and complete record of the static and deformation properties of the selected segment; they are specific for the given segment and the selected loads, and they are independent of the end conditions. General forms of these matrices for free and interactive bars are introduced in Chap. 2, and their particular forms are given in Chaps. 4–15.

(4) Direct transport chain (without intermediate interferences) consists of a chain product of transport matrices, each corresponding to one segment of the system, so that

$$H_L = \underbrace{T_{Li}T_{ij}T_{jk}T_{kR}}_{T_{LR}}H_R \qquad H_R = \underbrace{T_{Rk}T_{kj}T_{ji}T_{iL}}_{T_{RL}}H_L$$

The result of the chain product is a new transport matrix, which is again characteristic for the given system and independent of the end conditions.

(5) Indirect transport chain accommodates intermediate interferences such as intermediate rigid and/or elastic supports and intermediate rigid and/or elastic releases. The inclusion of these special conditions is accomplished by using the *joint matrices,* so that

$$H_L = \underbrace{J_L T_{Li} J_i T_{ij} J_j T_{jk} J_k T_{kR} J_R}_{T_{LR}} H_R \qquad H_R = \underbrace{J_R^{)-1} T_{Rk} J_k^{)-1} T_{kj} J_j^{)-1} T_{ji} J_i^{)-1} T_{iL} J_L^{)-1}}_{T_{RL}} H_L$$

where J, $J^{)-1}$ is the joint matrix and its inverse, respectively. Particular forms of joint matrices are listed in (2.20).

(6) Practical applications of this method are restricted to the analysis of systems with distinct linear topology, such as single and continuous beams, arches, rings, multibay and multistory frames. The advantage of the transport method is the small number of unknowns involved in the solution of a particular problem. The disadvantage is the extensive sequence of multiplications, some of which may involve very small and very large numbers (Ref. 3, pp. 138–155; Ref. 4, pp. 32–47; Ref. 5, pp. 50–69).

1.20 FLEXIBILITY METHOD

(1) Concepts. The flexibility method, frequently called the force method, is the oldest and most direct method for analyzing statically indeterminate structures. For such structures, the conditions of static equilibrium and the special conditions generated by reactive and intermediate releases must be supplemented by the conditions of deformation compatibility whose number must equal the number of redundants (1.06).

(2) Component systems. To meet the requirements of statics and deformation, the given structure is resolved into two component systems:

(*a*) *Basic structure,* obtained from the initial system by removing the redundants (release of redundants) but retaining other causes such as the applied loads, change in volume, etc.

(*b*) *Complementary structure,* obtained from the initial system by retaining the redundants as loads of unknown magnitudes and removing the applied loads, changes in volume, etc.

The selection of these component systems is arbitrary, provided that each one independently is in a state of static equilibrium and is geometrically stable. The requirement of static determinacy is customary, but it is not mandatory. The selection of a statically indeterminate structure as a complementary and/or basic structure is useful in many special situations.

(3) Flexibility matrices. The displacement vectors at the sections of release in the basic structure and in the complementary structure are in matrix form

$$\underbrace{\begin{bmatrix} \Delta_{i0} \\ \Delta_{j0} \\ \cdots \\ \Delta_{n0} \end{bmatrix}}_{\Delta_0} = \underbrace{\begin{bmatrix} f_{i1} & f_{i2} & \cdots & f_{1n} \\ f_{j1} & f_{j2} & \cdots & f_{jn} \\ \cdots & \cdots & \cdots & \cdots \\ f_{n1} & f_{n2} & \cdots & f_{nn} \end{bmatrix}}_{F_L} \underbrace{\begin{bmatrix} L_1 \\ L_2 \\ \cdots \\ L_n \end{bmatrix}}_{L} + \underbrace{\begin{bmatrix} \Delta_{it} \\ \Delta_{jt} \\ \cdots \\ \Delta_{nt} \end{bmatrix}}_{\Delta_t}$$

$$\underbrace{\begin{bmatrix} \Delta_{iX} \\ \Delta_{jX} \\ \cdots \\ \Delta_{nX} \end{bmatrix}}_{\Delta_X} = \underbrace{\begin{bmatrix} f_{ii} & f_{ij} & \cdots & f_{in} \\ f_{ji} & f_{jj} & \cdots & f_{jn} \\ \cdots & \cdots & \cdots & \cdots \\ f_{ni} & f_{nj} & \cdots & f_{nn} \end{bmatrix}}_{F_X} \underbrace{\begin{bmatrix} X_i \\ X_j \\ \cdots \\ X_n \end{bmatrix}}_{X}$$

where L = load vector, X = redundant vector, Δ_t = temperature-displacement vector, Δ_0 = basic-structure displacement vector, Δ_X = complementary-structure displacement vector, F_L = unit-load flexibility matrix, and F_X = unit-redundant flexibility matrix.

(4) Conditions of compatibility require that

$$F_L L + \Delta_t + F_X X = \Delta$$

where Δ is the prescribed (given) displacement vector at the sections of release (which is zero in most cases). Since there are n redundants and n compatibility conditions,

$$X = -F_X^{-1}(F_L L + \Delta_t - \Delta)$$

which is the true value of the redundant vector.

(5) Practical applications of the flexibility method are presently restricted to the analysis of continuous beams, rings, and single-bay frames. The advantage of this method is the smaller number of unknowns as compared to the number of unknowns involved in the stiffness method described in the following section (Ref. 3, pp. 180–201; Ref. 4, pp. 48–187; Ref. 5, pp. 70–114, 157–183).

1.21 STIFFNESS METHOD

(1) Concepts. The stiffness method, frequently called the displacement method, is the third method introduced in (1.19). Instead of working with unknown stress resultants, this method introduces the joint displacements as the unknown quantities. Since these joints may be real joints or selected joints, the number of kinematic redundants (1.07) is governed by the number of admissible and independent displacements.

(2) Component systems. Although the formulation is inverted, the conditions to be satisfied (static equilibrium and deformation compatibility) remain the same and are met by the resolution of the given structure into two component systems:

(*a*) *Basic structure,* conceived by locking the ends of each member in the initial structure, preventing their displacements, but retaining all loads, volume changes, etc., thus producing fixed end forces and moments at the selected joints.

(*b*) *Complementary structure,* conceived from the initial system by removing the applied loads, volume changes, etc., and introducing unknown joint displacements, which satisfy the natural constraints of the system.

The selection of these component systems is arbitrary, provided that each one independently is in a state of compatible deformation (consistent with the remaining parts of the same system) and is also geometrically stable.

(3) Stiffness matrices. The joint forces and moments in the basic and complementary structure required to maintain the static equilibrium of the joints are in matrix form

$$\underbrace{\begin{bmatrix} S_{i0} \\ S_{j0} \\ \cdots \\ S_{n0} \end{bmatrix}}_{S_0} = \underbrace{\begin{bmatrix} k_{i1} & k_{i2} & \cdots & k_{in} \\ k_{j1} & k_{j2} & \cdots & k_{jn} \\ \cdots & \cdots & \cdots & \cdots \\ k_{n1} & k_{n2} & \cdots & k_{nn} \end{bmatrix}}_{K_L} \underbrace{\begin{bmatrix} L_1 \\ L_2 \\ \cdots \\ L_n \end{bmatrix}}_{L} + \underbrace{\begin{bmatrix} S_{it} \\ S_{jt} \\ \cdots \\ S_{nt} \end{bmatrix}}_{S_t}$$

$$\underbrace{\begin{bmatrix} S_i \\ S_j \\ \cdots \\ S_n \end{bmatrix}}_{S_\Delta} = \underbrace{\begin{bmatrix} k_{ii} & k_{ij} & \cdots & k_{in} \\ k_{ji} & k_{jj} & \cdots & k_{jn} \\ \cdots & \cdots & \cdots & \cdots \\ k_{ni} & k_{nj} & \cdots & k_{nn} \end{bmatrix}}_{K_\Delta} \underbrace{\begin{bmatrix} \Delta_i \\ \Delta_j \\ \cdots \\ \Delta_n \end{bmatrix}}_{\Delta}$$

where L = load vector, Δ = joint displacement vector, S_0 = fixed-joint load-effect vector, S_t = fixed-joint temperature-effect vector, K_L = unit-load joint-effect matrix, K_Δ = unit-joint displacement matrix.

(4) Conditions of equilibrium require that

$$K_L L + S_t + K_\Delta \Delta = S$$

where S is the prescribed (given) joint load vector (which is zero in most cases). Since there are n joint displacements and n joint equilibrium conditions,

$$\Delta = -K_\Delta^{-1}(K_L L + S_t - S)$$

which is the true value of the joint displacement vector.

(5) Practical applications of the stiffness method cover structures of all kinds, and the stiffness method is the most widely used tool of structural analysis. The advantage of this method is the simplicity of matrix constructions (Ref. 3, pp. 226–247; Ref. 4, pp. 188–247; Ref. 5, pp. 115–156, 184–211).

1.22 SIGN CONVENTIONS OF METHODS OF ANALYSIS

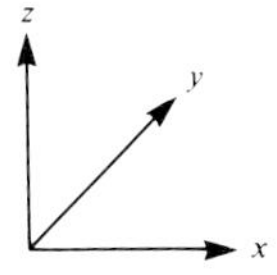

Fig. 1.22-1

(1) Positive directions of end forces, end moments, and end displacements in the transport method, flexibility method, and stiffness method are shown in the member system (Fig. 1.22-1). Although a straight bar is depicted in each case, the sign convention defined by each sketch is valid for bars of any shape (straight, bent, curved, or their combinations).

(2) Transport-method sign convention (Fig. 1.22-2)

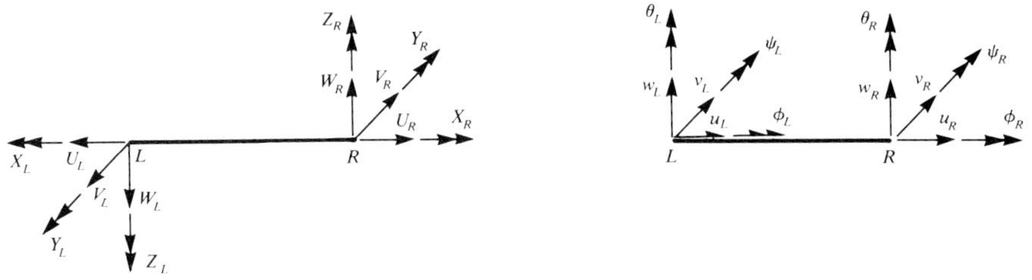

Fig. 1.22-2

(3) Flexibility-method sign convention (Fig. 1.22-3)

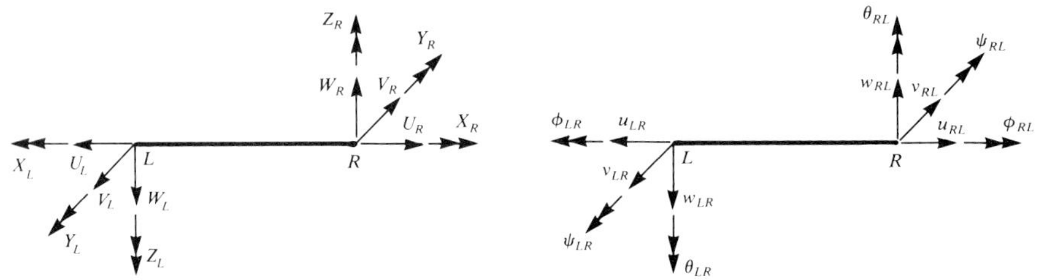

Fig. 1.22-3

(4) Stiffness-method sign convention (Fig. 1.22-4)

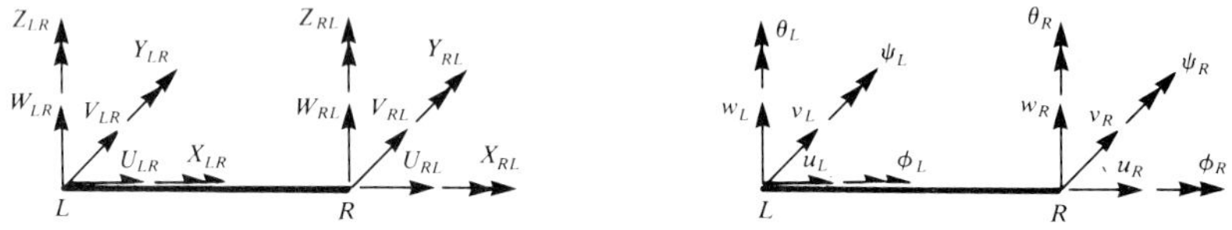

Fig. 1.22-4

(5) Load sign convention is defined in (1.02-2, 3, 4, 5) and is common for all methods.

2

Transport Matrices Free and Interactive Bars

STATIC STATE

2.01 DEFINITION OF STATE

(1) Systems considered are free and interactive bars of order one, two, and three defined in (1.01, 1.08). Geometric restrictions given by the shape of the bar axis and the shape of its normal section are stated in each case. Each bar is in a state of static equilibrium and a state of time-independent elastic deformation.

(2) Transport matrices and state vectors of these bars are presented in symbolic forms in (2.02). Particular cases of these matrices in free straight and curved bars are shown in (2.03–2.15). A general case of the same matrices in straight interactive bars is displayed in (2.16–2.18). Finally, the joint matrices, representing the effects of joint conditions, are discussed in (2.20). In all cases, the sign convention of the transport method is used (1.22-2).

(3) Assumptions used in the derivation of analytical relations are those stated in (1.01). In (2.03–2.15), the effects of normal, shearing, flexural, and torsional deformations are included, but the magnification effect of the axial force on the flexural moments is disregarded. In (2.16–2.18), this magnification effect with the effect of the surrounding medium are included.

(4) Symbols are defined where they appear first and are all summarized in Appendices A.1 and A.2.

2.02 TRANSPORT MATRIX EQUATIONS, BASIC RELATIONS IN S SYSTEM

(1) Concept. The stress-resultant equations and the displacement equations can be assembled into two transport matrix equations as

$$\underbrace{\begin{bmatrix} 1 \\ S_L \\ \Delta_L \end{bmatrix}}_{H_L} = \underbrace{\begin{bmatrix} 1 & 0 & 0 \\ F_S & t_{LR} & c_{LR} \\ F_\Delta & d_{LR} & t_{RL}^{)T} \end{bmatrix}}_{T_{LR}} \underbrace{\begin{bmatrix} 1 \\ S_R \\ \Delta_R \end{bmatrix}}_{H_R} \qquad \underbrace{\begin{bmatrix} 1 \\ S_R \\ \Delta_R \end{bmatrix}}_{H_R} = \underbrace{\begin{bmatrix} 1 & 0 & 0 \\ G_S & t_{RL} & c_{RL} \\ G_\Delta & d_{RL} & t_{LR}^{)T} \end{bmatrix}}_{T_{RL}} \underbrace{\begin{bmatrix} 1 \\ S_L \\ \Delta_L \end{bmatrix}}_{H_L}$$

where the *state vectors* of the left and right ends are, respectively,

$$H_L = \{1, U_L, V_L, W_L, X_L, Y_L\ Z_L, u_L, v_L, w_L, \phi_L, \psi_L, \theta_L\}$$
$$H_R = \{1, U_R, V_R, W_R, X_R, Y_R, Z_R, u_R, v_R, w_R, \phi_R, \psi_R, \theta_R\}$$

and T_{LR}, T_{RL} are their *transport matrices,* [13 × 13]. The superscript s for the S system is omitted in all vector terms (Ref. 5, p. 50).

(2) Submatrices of T_{LR}, T_{RL} are

t_{LR}, t_{RL} = static transport matrices, [6 × 6]
$t_{RL}^{)T}, t_{LR}^{)T}$ = kinematic transport matrices, [6 × 6]
c_{LR}, c_{RL} = displacement deviation matrices, [6 × 6]
d_{LR}, d_{RL} = stress deviation matrices, [6 × 6]

The *load functions* forming the first column in T_{LR}, T_{RL} are

$$F_S = \{F_U, F_V, F_W, F_X, F_Y, F_Z\} \qquad G_S = \{G_U, G_V, G_W, G_X, G_Y, G_Z\}$$
$$F_\Delta = \{F_u, F_v, F_w, F_\phi, F_\psi, F_\theta\} \qquad G_\Delta = \{G_u, G_v, G_w, G_\phi, G_\psi, G_\theta\}$$

Their analytical forms are given in (2.03).

(3) Properties. The transport matrices T_{LR}, T_{RL} give a clear, systematic, and complete record of the stress and deformation properties of the bar; they are specific for a given bar and independent of the end conditions. These matrices and their submatrices possess the following characteristics:

(*a*) *Inverse relations*

$$T_{LR}T_{RL} = I \qquad T_{RL}T_{LR} = I$$

where I = unit matrix, [13 × 13].

$$t_{LR}t_{RL} + c_{LR}d_{RL} = I \qquad t_{RL}t_{LR} + c_{RL}d_{LR} = I$$

where I = unit matrix, [6 × 6].
If $c_{LR} = c_{RL} = 0$, then $t_{LR}t_{RL} = t_{RL}t_{LR} = I$.

(*b*) *Shift relations*

$$d_{LR} = -d_{RL}^{)T} \qquad d_{RL} = -d_{LR}^{)T}$$
$$c_{LR} = -c_{RL}^{)T} \qquad c_{RL} = -c_{LR}^{)T}$$

$$\begin{bmatrix} F_S \\ F_\Delta \end{bmatrix} = -\begin{bmatrix} t_{LR} & c_{LR} \\ d_{LR} & t_{RL}^{)T} \end{bmatrix} \begin{bmatrix} G_S \\ G_\Delta \end{bmatrix}$$

$$\begin{bmatrix} G_S \\ G_\Delta \end{bmatrix} = -\begin{bmatrix} t_{RL} & c_{RL} \\ d_{RL} & t_{LR}^{)T} \end{bmatrix} \begin{bmatrix} F_S \\ F_\Delta \end{bmatrix}$$

(4) Sign convention. Stress-resultant components at the right end R (far end) are positive if acting in the positive direction of reference axes. For their left-end counterparts, the opposite is true (1.03-3). Displacements of either end are positive if acting in the positive direction of reference axes (1.22-2).

2.03 TRANSPORT MATRIX EQUATIONS, LOAD FUNCTIONS IN S SYSTEM

(1) Load functions due to singular loads are

$$F_S = -t_{Lj}L_j \qquad G_S = t_{Rj}L_j$$
$$F_\Delta = -d_{Lj}L_j \qquad G_\Delta = d_{Rj}L_j$$

where t_{Lj}, t_{Rj} = static transport matrices, [6 × 6]
d_{Lj}, d_{Rj} = stress deviation matrices, [6 × 6]

written for the segments Lj, Rj, respectively, and

$$L_j = \{P_{jx}, P_{jy}, P_{jz}, Q_{jx}, Q_{jy}, Q_{jz}\}$$

is the singular load vector at j (1.02-2).

(2) Load functions due to distributed load are

$$F_S = -\int_L^R t_{Li}\, dL_i \qquad G_S = \int_L^R t_{Ri}\, dL_i$$
$$F_\Delta = -\int_L^R d_{Li}\, dL_i \qquad G_\Delta = \int_L^R d_{Ri}\, dL_i$$

where t_{Li} , t_{Ri}, d_{Li}, d_{Ri} have the same meaning as their counterparts in (1) above but are written for the variable segments Li, Ri, respectively, and $dL_i = \{p_x, p_y, p_z, q_x, q_y, q_z\}\, ds$ is the load-intensity vector at i (1.02-4). The limits of integration are governed by the load geometry.

(3) Load functions due to abrupt displacements are

$$F_S = -c_{Lj}\Delta_j \qquad G_S = c_{Rj}\Delta_j$$
$$F_\Delta = -t_{jL}^{)T}\Delta_j \qquad G_\Delta = t_{jR}^{)T}\Delta_j$$

where c_{Lj}, c_{Rj} = displacement deviation matrices, [6 × 6]
$t_{jL}^{)T}, t_{jR}^{)T}$ = kinematic transport matrices, [6 × 6]

written for the segment Lj, Rj, respectively, and

$$\Delta_j = \{u_j, v_j, w_j, \phi_j, \psi_j, \phi_j\}$$

is the imposed displacement vector at j (1.17-3).

(4) Load functions due to temperature change are

$$F_S = -\int_L^R c_{Li}\, d\Delta_t \qquad G_S = \int_L^R c_{Ri}\, d\Delta_t$$
$$F_\Delta = -\int_L^R t_{iL}^{)T}\, d\Delta_t \qquad G_\Delta = \int_L^R t_{iR}^{)T}\, d\Delta_t$$

where c_{Li}, c_{Ri}, $t_{iL}^{)T}$, $t_{iR}^{)T}$ have the same meaning as their counterparts in (3) but are written for the variable segments Li, Ri, respectively, and $d\Delta_t$ is the elemental thermal deformation vector at i (1.17-2).

(5) Load-function integrals introduced in (2) and (4) above may be evaluated in many practical situations in a closed algebraic and/or transcendental form as shown in Chaps. 4–15. Whenever the closed-form integration is not feasible or practical, the numerical value of these functions can be found to any degree of accuracy by numerical quadrature formulas.

2.04 FREE STRAIGHT BAR OF ORDER THREE, COMPONENT MATRICES IN S SYSTEM

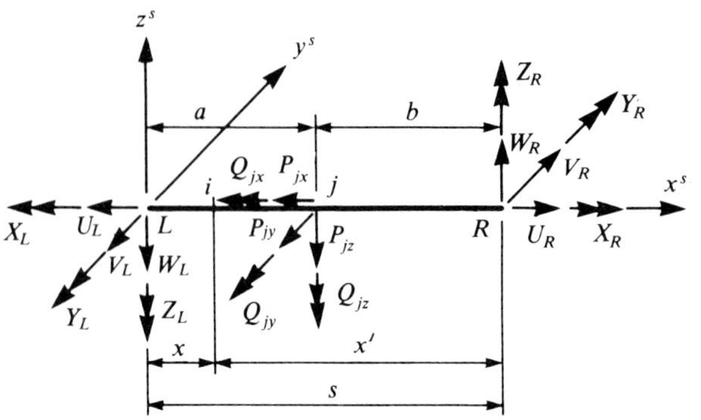

Fig. 2.04-1

(1) System considered in Fig. 2.04-1 is a free straight bar with constant normal section symmetrical about y^s and z^s centroidal axes and loaded along the centroidal x^s axis and in the planes of symmetry of the section. The properties of its normal section A_x, A_y, A_z, I_x, I_y, I_z and the elastic constants E, G are defined in (1.14).

s = span (m)
x, x' = coordinates (m)
a, b = segments (m)

The superscript s for the S system is omitted in all vector terms.

(2) Static and kinematic submatrices used in the construction of T_{LR}, T_{RL} in (2.02.1) are

$$t_{LR} = \begin{bmatrix} I & 0 \\ r_{LR} & I \end{bmatrix}_{[6\times6]} \qquad t_{RL} = \begin{bmatrix} I & 0 \\ r_{RL} & \mathrm{I} \end{bmatrix}_{[6\times6]}$$

where

$$r_{LR} = \begin{bmatrix} 0 & 0 & 0 \\ 0 & 0 & -s \\ 0 & s & 0 \end{bmatrix} \qquad r_{RL} = \begin{bmatrix} 0 & 0 & 0 \\ 0 & 0 & s \\ 0 & -s & 0 \end{bmatrix}$$

The r matrices in t_{Li}, t_{iL}, t_{Lj}, t_{jL}, t_{Ri}, t_{iR}, t_{Rj}, t_{jR} are, respectively,

$$r_{Li} = \begin{bmatrix} 0 & 0 & 0 \\ 0 & 0 & -x \\ 0 & x & 0 \end{bmatrix} \qquad r_{iL} = \begin{bmatrix} 0 & 0 & 0 \\ 0 & 0 & x \\ 0 & -x & 0 \end{bmatrix}$$

$$r_{Lj} = \begin{bmatrix} 0 & 0 & 0 \\ 0 & 0 & -a \\ 0 & a & 0 \end{bmatrix} \qquad r_{jL} = \begin{bmatrix} 0 & 0 & 0 \\ 0 & 0 & a \\ 0 & -a & 0 \end{bmatrix}$$

$$r_{Ri} = \begin{bmatrix} 0 & 0 & 0 \\ 0 & 0 & x' \\ 0 & -x' & 0 \end{bmatrix} \qquad r_{iR} = \begin{bmatrix} 0 & 0 & 0 \\ 0 & 0 & -x' \\ 0 & x' & 0 \end{bmatrix}$$

$$r_{Rj} = \begin{bmatrix} 0 & 0 & 0 \\ 0 & 0 & b \\ 0 & -b & 0 \end{bmatrix} \qquad r_{jR} = \begin{bmatrix} 0 & 0 & 0 \\ 0 & 0 & -b \\ 0 & b & 0 \end{bmatrix}$$

(3) Deviation submatrices in terms of t matrices given in (2) are

$$d_{LR} = -\int_L^R t_{iL}^{)T} \Lambda_i t_{iR} \qquad \Lambda_i = \left[\begin{array}{ccc|ccc} \Lambda_U & & & & & \\ & \Lambda_V & & & & \\ & & \Lambda_W & & & \\ \hline & & & \Lambda_X & & \\ & & & & \Lambda_Y & \\ & & & & & \Lambda_Z \end{array}\right]_{[6\times6]} \qquad c_{LR} = \begin{bmatrix} 0 & 0 \\ 0 & 0 \end{bmatrix}_{[6\times6]}$$

$$d_{RL} = \int_L^R t_{iR}^{)T} \Lambda_i t_{iL} \qquad c_{RL} = \begin{bmatrix} 0 & 0 \\ 0 & 0 \end{bmatrix}_{[6\times6]}$$

where Λ_i is the elemental flexibility matrix in the S system (1.14-6).

2.05 FREE STRAIGHT BAR OF ORDER THREE, TRANSPORT COEFFICIENTS AND LOAD FUNCTIONS IN S SYSTEM

(1) Transport coefficients which form the submatrices d_{LR}, d_{RL} in the transport matrices (2.02-1) are

$T_{1n} = T_{1n}(s) = \dfrac{s}{EA_x}$		$T_{1x} = T_{1x}(s) = \dfrac{s}{GI_x}$
$T_{1y} = T_{1y}(s) = \dfrac{s}{EI_y}$	$T_{2y} = T_{2y}(s) = \dfrac{s^2}{2EI_y}$	$T_{3y} = T_{3y}(s) = \dfrac{s^3}{6EI_y} - \dfrac{s}{GA_z}$
$T_{1z} = T_{1z}(s) = \dfrac{s}{EI_z}$	$T_{2z} = T_{2z}(s) = \dfrac{s^2}{2EI_z}$	$T_{3z} = T_{3z}(s) = \dfrac{s^3}{6EI_z} - \dfrac{s}{GA_y}$

For $s = a$, $T(a) = A$, and for $s = b$, $T(b) = B$.

(2) Load functions in T_{LR} (2.06) due to singular loads (2.03-1) are

$F_U = -P_{jx}$	$F_V = -P_{jy}$	$F_W = -P_{jz}$
$F_X = -Q_{jx}$	$F_Y = -Q_{jy} + aP_{jz}$	$F_Z = -Q_{jz} - aP_{jy}$
$F_u = A_{1n}P_{jx}$	$F_v = -A_{2z}Q_{jz} - A_{3z}P_{jy}$	$F_w = A_{2y}Q_{jy} - A_{3y}P_{jz}$
$F_\phi = A_{1x}Q_{jx}$	$F_\psi = A_{1y}Q_{jy} - A_{2y}P_{jz}$	$F_\theta = A_{1z}Q_{jz} + A_{2z}P_{jy}$

and in T_{RL} (2.06) are

$G_U = P_{jx}$	$G_V = P_{jy}$	$G_W = P_{jz}$
$G_X = Q_{jx}$	$G_Y = Q_{jy} + bP_{jz}$	$G_Z = Q_{jz} - bP_{jy}$
$G_u = B_{jn}P_{jx}$	$G_v = B_{2z}Q_{jz} - B_{3z}P_{jy}$	$G_w = -B_{zy}Q_{jy} - B_{3y}P_{jz}$
$G_\phi = B_{jx}Q_{jx}$	$G_\psi = B_{1y}Q_{jy} + B_{2y}P_{jz}$	$G_\theta = B_{1z}Q_{jz} - B_{2z}P_{jy}$

Other load functions are derived by the relations summarized in (2.03-2, 3, 4).

(3) Transport matrix equations constructed by (2.02-1) in terms of the transport coefficients and load functions introduced above are given in (2.06-1, 2). If the state vector is required at an intermediate section i of position coordinate x measured from L or x' measured from R, then the transport matrix T_{LR} or T_{RL}, respectively, must be reconstructed for the corresponding segment Li or Ri.

2.06 FREE STRAIGHT BAR OF ORDER THREE, TRANSPORT MATRIX EQUATIONS IN S SYSTEM

Notation (2.02, 2.05) Signs (2.02) Load functions (2.05)

(1) Left-end equation

$$
\underbrace{\begin{bmatrix} 1 \\ \hline U_L \\ V_L \\ W_L \\ \hline X_L \\ Y_L \\ Z_L \\ \hline u_L \\ v_L \\ w_L \\ \hline \phi_L \\ \psi_L \\ \theta_L \end{bmatrix}}_{H_L}
=
\underbrace{\left[\begin{array}{c|ccc|ccc|ccc|ccc}
1 & 0 & 0 & 0 & 0 & 0 & 0 & 0 & 0 & 0 & 0 & 0 & 0 \\ \hline
F_U & & & & & & & & & & & & \\
F_V & & I & & & 0 & & & 0 & & & 0 & \\
F_W & & & & & & & & & & & & \\ \hline
F_X & 0 & 0 & 0 & & & & & & & & & \\
F_Y & 0 & 0 & -s & & I & & & 0 & & & 0 & \\
F_Z & 0 & s & 0 & & & & & & & & & \\ \hline
F_u & -T_{1n} & 0 & 0 & 0 & 0 & 0 & & & & 0 & 0 & 0 \\
F_v & 0 & T_{3z} & 0 & 0 & 0 & T_{2z} & & I & & 0 & 0 & -s \\
F_w & 0 & 0 & T_{3y} & 0 & -T_{2y} & 0 & & & & 0 & s & 0 \\ \hline
F_\phi & 0 & 0 & 0 & -T_{1x} & 0 & 0 & & & & & & \\
F_\psi & 0 & 0 & T_{2y} & 0 & -T_{1y} & 0 & & 0 & & & I & \\
F_\phi & 0 & -T_{2z} & 0 & 0 & 0 & -T_{1z} & & & & & &
\end{array}\right]}_{T_{LR}}
\underbrace{\begin{bmatrix} 1 \\ \hline U_R \\ V_R \\ W_R \\ \hline X_R \\ Y_R \\ Z_R \\ \hline u_R \\ v_R \\ w_R \\ \hline \phi_R \\ \psi_R \\ \theta_R \end{bmatrix}}_{H_R}
$$

(2) Right-end equation

$$
\underbrace{\left[\begin{array}{c} 1 \\ \hline U_R \\ V_R \\ W_R \\ \hline X_R \\ Y_R \\ Z_R \\ \hline u_R \\ v_R \\ w_R \\ \hline \phi_R \\ \psi_R \\ \theta_R \end{array}\right]}_{H_R}
=
\underbrace{\left[\begin{array}{c|ccc|ccc|ccc|ccc}
1 & 0 & 0 & 0 & 0 & 0 & 0 & 0 & 0 & 0 & 0 & 0 & 0 \\ \hline
G_U & & & & & & & & & & & & \\
G_V & & I & & & 0 & & & 0 & & & 0 & \\
G_W & & & & & & & & & & & & \\ \hline
G_X & 0 & 0 & 0 & & & & & & & & & \\
G_Y & 0 & 0 & s & & I & & & 0 & & & 0 & \\
G_Z & 0 & -s & 0 & & & & & & & & & \\ \hline
G_u & T_{1n} & 0 & 0 & 0 & 0 & 0 & & & & 0 & 0 & 0 \\
G_v & 0 & -T_{3z} & 0 & 0 & 0 & T_{2z} & & I & & 0 & 0 & s \\
G_w & 0 & 0 & -T_{3y} & 0 & -T_{2y} & 0 & & & & 0 & -s & 0 \\ \hline
G_\phi & 0 & 0 & 0 & T_{1x} & 0 & 0 & & & & & & \\
G_\psi & 0 & 0 & T_{2y} & 0 & T_{1y} & 0 & & 0 & & & I & \\
G_\theta & 0 & -T_{2z} & 0 & 0 & 0 & T_{1z} & & & & & &
\end{array}\right]}_{T_{RL}}
\underbrace{\left[\begin{array}{c} 1 \\ \hline U_L \\ V_L \\ W_L \\ \hline X_L \\ Y_L \\ Z_L \\ \hline u_L \\ v_L \\ w_L \\ \hline \phi_L \\ \psi_L \\ \theta_L \end{array}\right]}_{H_L}
$$

(3) Scaling. The matrix equations given above are shown in their absolute form. For convenience of numerical work, the same matrices will occur later in scaled form (4.02–4.03, 11.02–11.03).

2.07 FREE CURVED BAR OF ORDER THREE, COMPONENT MATRICES IN 0 SYSTEM

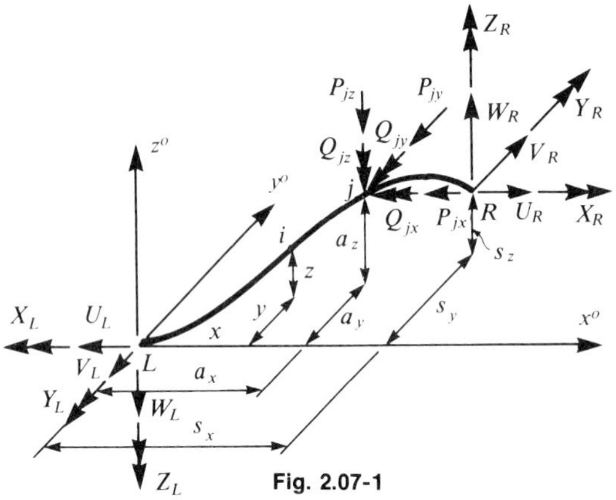

Fig. 2.07-1

(1) System considered in Fig. 2.07-1 is a free curved bar with normal section symmetrical about its principal axes y^s, z^s and acted on by loads of arbitrary type and direction. The properties of the section and the elastic constants of bar material are those defined in (1.14). Also, s_x, s_y, s_z = span projections (m), x, y, z, x', y', z' = coordinates of bar axis (m), and

$$x' = s_x - x \qquad y' = s_y - y \qquad z' = s_z - z$$
$$b_x = x_x - a_x \qquad b_y = s_y - a_y \qquad b_z = s_z - a_z$$

The superscript o for the 0 system is omitted in all vector terms.

(2) Submatrices used in construction of T_{LR}, T_{RL} in (2.02-1) are

$$t_{LR} = \begin{bmatrix} I & 0 \\ r_{LR} & I \end{bmatrix}_{[6\times 6]} \qquad c_{LR} = \begin{bmatrix} 0 & 0 \\ 0 & 0 \end{bmatrix}_{[6\times 6]} \qquad t_{RL} = \begin{bmatrix} I & 0 \\ r_{RL} & I \end{bmatrix}_{[6\times 6]} \qquad c_{RL} = \begin{bmatrix} 0 & 0 \\ 0 & 0 \end{bmatrix}_{[6\times 6]}$$

where

$$r_{LR} = \begin{bmatrix} 0 & -s_z & s_y \\ s_z & 0 & -s_x \\ -s_y & s_x & 0 \end{bmatrix} \qquad r_{RL} = \begin{bmatrix} 0 & s_z & -s_y \\ -s_z & 0 & s_x \\ s_y & -s_x & 0 \end{bmatrix}$$

The r matrices in t_{Li}, t_{Lj}, t_{Ri}, t_{Rj} are, respectively,

$$r_{Li} = \begin{bmatrix} 0 & -z & y \\ z & 0 & -x \\ -y & x & 0 \end{bmatrix} \qquad r_{Lj} = \begin{bmatrix} 0 & -a_z & a_y \\ a_z & 0 & -a_x \\ -a_y & a_x & 0 \end{bmatrix}$$

$$r_{Ri} = \begin{bmatrix} 0 & z' & -y' \\ -z' & 0 & x' \\ y' & -x' & 0 \end{bmatrix} \qquad r_{Rj} = \begin{bmatrix} 0 & b_z & -b_y \\ -b_z & 0 & b_x \\ b_y & -b_x & 0 \end{bmatrix}$$

and in terms of t matrices given above,

$$d_{LR} = -\int_L^R t_{iL}^{)T} \Lambda_i t_{iR} \qquad d_{RL} = \int_L^R t_{iR}^{)T} \Lambda_i t_{iL}$$

$$\Lambda_i = \left[\begin{array}{ccc|ccc} \Lambda_{UU} & \Lambda_{UV} & \Lambda_{UW} & & & \\ \Lambda_{VU} & \Lambda_{VV} & \Lambda_{VW} & & 0 & \\ \Lambda_{WU} & \Lambda_{WV} & \Lambda_{WW} & & & \\ \hline & & & \Lambda_{XX} & \Lambda_{XY} & \Lambda_{XZ} \\ & 0 & & \Lambda_{YX} & \Lambda_{YY} & \Lambda_{YZ} \\ & & & \Lambda_{ZX} & \Lambda_{ZY} & \Lambda_{ZZ} \end{array}\right]$$

where Λ_i is the elemental flexibility matrix (1.15-3) in the 0 system

2.08 FREE CURVED BAR OF ORDER THREE, ANGULAR TRANSFORMATION MATRICES

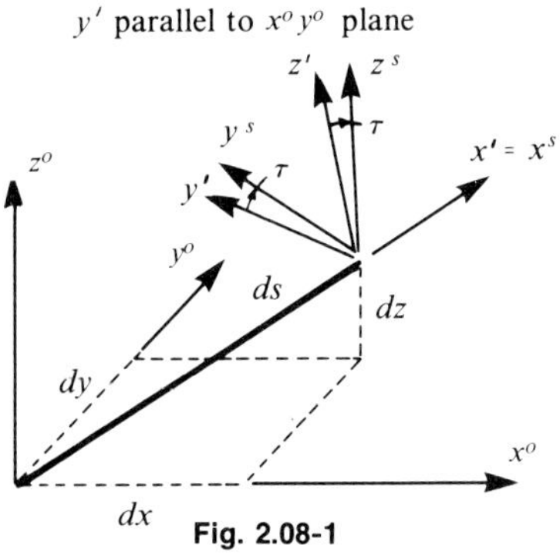

Fig. 2.08-1

(1) Geometry of the bar axis in Fig. 2.07-1 is given by

$$x = f_x(s) \qquad y = f_y(s) \qquad z = f_z(s)$$

where s is a curvilinear coordinate measured from L along the bar axis and by the angle τ measured about the x^s axis of the element ds from the y' axis, which is parallel to the $x^o y^o$ plane as shown in Fig. 2.08-1.

$$de = \sqrt{(dx)^2 + (dy)^2}$$

$$df = \sqrt{(dx)^2 + (dy)^2 + (dz)^2} = ds$$

(2) Angular transformation matrices of order three ($z \neq 0, \tau \neq 0$) are

$$R^{os} = \begin{bmatrix} \dfrac{dx}{df} & -\dfrac{dy}{de} & -\dfrac{dx\,dz}{de\,df} \\ \dfrac{dy}{df} & \dfrac{dx}{de} & -\dfrac{dy\,dz}{de\,df} \\ \dfrac{dz}{df} & 0 & \dfrac{de}{df} \end{bmatrix} \begin{bmatrix} 1 & 0 & 0 \\ 0 & \cos\tau & -\sin\tau \\ 0 & \sin\tau & \cos\tau \end{bmatrix}$$

$$R^{so} = \begin{bmatrix} 1 & 0 & 0 \\ 0 & \cos\tau & \sin\tau \\ 0 & -\sin\tau & \cos\tau \end{bmatrix} \begin{bmatrix} \dfrac{dx}{df} & \dfrac{dy}{df} & \dfrac{dz}{df} \\ -\dfrac{dy}{de} & \dfrac{dx}{de} & 0 \\ -\dfrac{dx\,dz}{de\,df} & -\dfrac{dy\,dz}{de\,df} & \dfrac{de}{df} \end{bmatrix}$$

which are the differential forms of (1.11-3).

(3) Angular transformation matrices of order two ($z = 0, \tau = 0$) are

$$R^{os} = \begin{bmatrix} 1 & 0 & 0 \\ 0 & \dfrac{dx}{ds} & -\dfrac{dy}{ds} \\ 0 & \dfrac{dy}{ds} & \dfrac{dx}{ds} \end{bmatrix} \qquad R^{so} = \begin{bmatrix} 1 & 0 & 0 \\ 0 & \dfrac{dx}{ds} & \dfrac{dy}{ds} \\ 0 & -\dfrac{dy}{ds} & \dfrac{dx}{dx} \end{bmatrix}$$

which are the differential forms of (1.11-4).

(4) Angular transformation matrices of order one ($z = 0, \tau = 0$) are

$$R^{os} = \begin{bmatrix} \dfrac{dx}{ds} & -\dfrac{dy}{ds} & 0 \\ \dfrac{dy}{ds} & \dfrac{dx}{ds} & 0 \\ 0 & 0 & 1 \end{bmatrix} \qquad R^{so} = \begin{bmatrix} \dfrac{dx}{ds} & \dfrac{dy}{ds} & 0 \\ -\dfrac{dy}{ds} & \dfrac{dx}{ds} & 0 \\ 0 & 0 & 1 \end{bmatrix}$$

which are the differential forms of (1.11-5).

2.09 FREE CURVED BAR OF ORDER THREE, TRANSPORT COEFFICIENTS IN 0 SYSTEM

(1) Transport coefficients which form the submatrices d_{LR}, d_{RL} in the transport matrices (2.02-1) are

$$T_{11} = \int_L^R [y'(y\Lambda_{ZZ} - a\Lambda_{ZY}) - z'(y\Lambda_{YZ} - z\Lambda_{YY}) - \Lambda_{UU}] \qquad T_{14} = \int_L^R (z\Lambda_{YX} - y\Lambda_{ZX})$$

$$T_{12} = \int_L^R [z'(y\Lambda_{XZ} - z\Lambda_{XY}) - x'(y\Lambda_{ZZ} - z\Lambda_{ZY}) - \Lambda_{UV}] \qquad T_{15} = \int_L^R (z\Lambda_{YY} - y\Lambda_{ZY})$$

$$T_{13} = \int_L^R [x'(y\Lambda_{YZ} - z\Lambda_{YY}) - y'(y\Lambda_{XZ} - z\Lambda_{XY}) - \Lambda_{UW}] \qquad T_{16} = \int_L^R (z\Lambda_{YZ} - y\Lambda_{ZZ})$$

$$T_{21} = \int_L^R [y'(z\Lambda_{ZX} - x\Lambda_{ZZ}) - z'(z\Lambda_{YX} - x\Lambda_{YZ}) - \Lambda_{VU}] \qquad T_{24} = \int_L^R (x\Lambda_{ZX} - z\Lambda_{XX})$$

$$T_{22} = \int_L^R [z'(z\Lambda_{XX} - x\Lambda_{XZ}) - x'(z\Lambda_{ZX} - x\Lambda_{ZZ}) - \Lambda_{VV}] \qquad T_{25} = \int_L^R (x\Lambda_{ZY} - z\Lambda_{XY})$$

$$T_{23} = \int_L^R [x'(z\Lambda_{YX} - x\Lambda_{YZ}) - y'(z\Lambda_{XX} - x\Lambda_{XZ}) - \Lambda_{VW}] \qquad T_{26} = \int_L^R (x\Lambda_{ZZ} - z\Lambda_{XZ})$$

$$T_{31} = \int_L^R [y'(x\Lambda_{ZY} - y\Lambda_{ZX}) - z'(x\Lambda_{YY} - y\Lambda_{YX}) - \Lambda_{WU}] \qquad T_{34} = \int_L^R (y\Lambda_{XX} - x\Lambda_{YX})$$

$$T_{32} = \int_L^R [z'(x\Lambda_{XY} - y\Lambda_{XX}) - x'(x\Lambda_{ZY} - y\Lambda_{ZX}) - \Lambda_{WV}] \qquad T_{35} = \int_L^R (y\Lambda_{XY} - x\Lambda_{YY})$$

$$T_{33} = \int_L^R [x'(x\Lambda_{YY} - y\Lambda_{YX}) - y'(x\Lambda_{XY} - y\Lambda_{XX}) - \Lambda_{WW}] \qquad T_{36} = \int_L^R (y\Lambda_{XZ} - x\Lambda_{YZ})$$

$$T_{41} = \int_L^R (y'\Lambda_{XZ} - z'\Lambda_{XY}) \qquad T_{42} = \int_L^R (z'\Lambda_{XX} - x'\Lambda_{XZ}) \qquad T_{43} = \int_L^R (x'\Lambda_{XY} - y'\Lambda_{XX})$$

$$T_{51} = \int_L^R (y'\Lambda_{YZ} - z'\Lambda_{YY}) \qquad T_{52} = \int_L^R (z'\Lambda_{YX} - x'\Lambda_{YZ}) \qquad T_{53} = \int_L^R (x'\Lambda_{YY} - y'\Lambda_{YX})$$

$$T_{61} = \int_L^R (y'\Lambda_{ZZ} - z'\Lambda_{ZY}) \qquad T_{62} = \int_L^R (z'\Lambda_{ZX} - x'\Lambda_{ZZ}) \qquad T_{63} = \int_L^R (x'\Lambda_{ZY} - y'\Lambda_{ZX})$$

$$T_{44} = \int_L^R \Lambda_{XX} \qquad T_{45} = \int_L^R \Lambda_{XY} \qquad T_{46} = \int_L^R \Lambda_{XZ}$$

$$T_{54} = \int_L^R \Lambda_{YX} \qquad T_{55} = \int_L^R \Lambda_{YY} \qquad T_{56} = \int_L^R \Lambda_{YZ}$$

$$T_{64} = \int_L^R \Lambda_{ZX} \qquad T_{65} = \int_L^R \Lambda_{ZY} \qquad T_{66} = \int_L^R \Lambda_{ZZ}$$

Symbols used above are defined in (2.08). For $s_x = a_x$, $T(a_x) = A$, and for $s_x = b_x$, $T(b_x) = B$.

(2) Transport matrix equations of the bar in Fig. 2.07-1 are displayed in (2.11) in terms of the load functions of (2.10) and the coefficients of (2.09). As before, if the state vector is required at an intermediate section i given by the position coordinate x measured from L, the transport matrix T_{RL} becomes T_{iL}, with $s_x = x$, $b_x = x - a_x$.

2.10 FREE CURVED BAR OF ORDER THREE, LOAD FUNCTIONS IN 0 SYSTEM

(1) Left-end load functions (2.03-1) due to singular loads in Fig. 2.07-1 are

$$
\begin{bmatrix} F_U \\ F_V \\ F_W \\ \hline F_X \\ F_Y \\ F_Z \\ \hline F_u \\ F_v \\ F_w \\ \hline F_\phi \\ F_\psi \\ F_\phi \end{bmatrix}
=
\left[\begin{array}{ccc|ccc}
-1 & & & & & \\
 & -1 & & & 0 & \\
 & & -1 & & & \\
\hline
0 & a_z & a_y & -1 & & \\
-a_z & 0 & a_x & & -1 & \\
a_y & -a_x & 0 & & & -1 \\
\hline
-A_{11} & -A_{12} & -A_{13} & -A_{14} & -A_{15} & -A_{16} \\
-A_{21} & -A_{22} & -A_{23} & -A_{24} & -A_{25} & -A_{26} \\
-A_{31} & -A_{32} & -A_{33} & -A_{34} & -A_{35} & -A_{36} \\
\hline
-A_{41} & -A_{42} & -A_{43} & -A_{44} & -A_{45} & -A_{46} \\
-A_{51} & -A_{52} & -A_{53} & -A_{54} & -A_{55} & -A_{56} \\
-A_{61} & -A_{62} & -A_{63} & -A_{64} & -A_{65} & -A_{66}
\end{array}\right]
\begin{bmatrix} P_{jx} \\ P_{jy} \\ P_{jz} \\ \hline Q_{jx} \\ Q_{jy} \\ Q_{jz} \end{bmatrix}
$$

where $A_{11}, A_{12}, \ldots, A_{66}$ are $T_{11}, T_{12}, \ldots, T_{66}$ (2.09-1) for $s_x = a_x$, $s_y = a_y$, $s_z = a_z$.

(2) Right-end load functions (2.03-1) due to the same loads are

$$
\begin{bmatrix} G_U \\ G_V \\ G_W \\ \hline G_X \\ G_Y \\ G_Z \\ \hline G_u \\ G_v \\ G_w \\ \hline G_\phi \\ G_\psi \\ G_\theta \end{bmatrix}
=
\left[\begin{array}{ccc|ccc}
1 & & & & & \\
 & 1 & & 0 & & \\
 & & 1 & & & \\
\hline
0 & b_z & -b_y & 1 & & \\
-b_z & 0 & b_x & & 1 & \\
b_y & -b_x & 0 & & & 1 \\
\hline
-B_{11} & -B_{21} & -B_{31} & -B_{41} & -B_{51} & -B_{61} \\
-B_{12} & -B_{22} & -B_{32} & -B_{42} & -B_{52} & -B_{62} \\
-B_{13} & -B_{23} & -B_{33} & -B_{43} & -B_{53} & -B_{63} \\
\hline
-B_{14} & -B_{24} & -B_{34} & -B_{44} & -B_{54} & -B_{64} \\
-B_{15} & -B_{25} & -B_{35} & -B_{45} & -B_{55} & -B_{65} \\
-B_{16} & -B_{26} & -B_{36} & -B_{46} & -B_{56} & -B_{66}
\end{array}\right]
\begin{bmatrix} P_{jx} \\ P_{jy} \\ P_{jz} \\ \hline Q_{jx} \\ Q_{jy} \\ Q_{jz} \end{bmatrix}
$$

where $B_{11}, B_{12}, \ldots, B_{66}$ are $T_{11}, T_{12}, \ldots, T_{66}$ (2.09-1) for $s_x = b_x$, $s_y = b_y$, $s_z = b_z$.

(3) Other load functions are derived by the relations summarized in (2.03-2, 3, 4).

2.11 FREE CURVED BAR OF ORDER THREE, TRANSPORT MATRIX EQUATIONS IN 0 SYSTEM

Notation (2.02, 2.08, 2.09) Signs (2.02) Load functions (2.10)

(1) Left-end equation

$$
\underbrace{\begin{bmatrix} 1 \\ \hline U_L \\ V_L \\ W_L \\ \hline X_L \\ Y_L \\ Z_L \\ \hline u_L \\ v_L \\ w_L \\ \hline \phi_L \\ \psi_L \\ \theta_L \end{bmatrix}}_{H_L}
=
\underbrace{\left[\begin{array}{c|ccc|ccc|ccc|ccc}
1 & 0 & 0 & 0 & 0 & 0 & 0 & 0 & 0 & 0 & 0 & 0 & 0 \\ \hline
F_U & & & & & & & & & & & & \\
F_V & & I & & & 0 & & & 0 & & & 0 & \\
F_W & & & & & & & & & & & & \\ \hline
F_X & 0 & -s_z & s_y & & & & & & & & & \\
F_Y & s_z & 0 & -s_x & & I & & & 0 & & & 0 & \\
F_Z & -s_y & s_x & 0 & & & & & & & & & \\ \hline
F_u & T_{11} & T_{12} & T_{13} & T_{14} & T_{15} & T_{16} & & & & 0 & -s_z & s_y \\
F_v & T_{21} & T_{22} & T_{23} & T_{24} & T_{25} & T_{26} & & I & & s_z & 0 & -s_x \\
F_w & T_{31} & T_{32} & T_{33} & T_{34} & T_{35} & T_{36} & & & & -s_y & s_x & 0 \\ \hline
F_\phi & T_{41} & T_{42} & T_{43} & T_{44} & T_{45} & T_{46} & & & & & & \\
F_\psi & T_{51} & T_{52} & T_{53} & T_{54} & T_{55} & T_{56} & & 0 & & & I & \\
F_\theta & T_{61} & T_{62} & T_{63} & T_{64} & T_{65} & T_{66} & & & & & &
\end{array}\right]}_{T_{LR}}
\underbrace{\begin{bmatrix} 1 \\ \hline U_R \\ V_R \\ W_R \\ \hline X_R \\ Y_R \\ Z_R \\ \hline u_R \\ v_R \\ w_R \\ \hline \phi_R \\ \psi_R \\ \theta_R \end{bmatrix}}_{H_R}
$$

(2) Right-end equation

$$
\underbrace{\left[\begin{array}{c} 1 \\ \hline U_R \\ V_R \\ W_R \\ \hline X_R \\ Y_R \\ Z_R \\ \hline u_R \\ v_R \\ w_R \\ \hline \phi_R \\ \psi_R \\ \theta_R \end{array}\right]}_{H_R}
=
\underbrace{\left[\begin{array}{c|ccc|ccc|ccc|ccc}
1 & 0 & 0 & 0 & 0 & 0 & 0 & 0 & 0 & 0 & 0 & 0 & 0 \\ \hline
G_U & & & & & & & & & & & & \\
G_V & & I & & & 0 & & & 0 & & & 0 & \\
G_W & & & & & & & & & & & & \\ \hline
G_X & 0 & s_z & -s_y & & & & & & & & & \\
G_Y & -s_z & 0 & s_x & & I & & & 0 & & & 0 & \\
G_Z & s_y & -s_x & 0 & & & & & & & & & \\ \hline
G_u & -T_{11} & -T_{21} & -T_{31} & -T_{41} & -T_{51} & -T_{61} & & & & 0 & s_z & -s_y \\
G_v & -T_{12} & -T_{22} & -T_{32} & -T_{42} & -T_{52} & -T_{62} & & I & & -s_z & 0 & s_x \\
G_w & -T_{13} & -T_{23} & -T_{33} & -T_{43} & -T_{53} & -T_{63} & & & & s_y & -s_x & 0 \\ \hline
G_\phi & -T_{14} & -T_{24} & -T_{34} & -T_{44} & -T_{54} & -T_{64} & & & & & & \\
G_\psi & -T_{15} & -T_{25} & -T_{35} & -T_{45} & -T_{55} & -T_{65} & & 0 & & & I & \\
G_\theta & -T_{16} & -T_{26} & -T_{36} & -T_{46} & -T_{56} & -T_{66} & & & & & &
\end{array}\right]}_{T_{RL}}
\underbrace{\left[\begin{array}{c} 1 \\ \hline U_L \\ V_L \\ W_L \\ \hline X_L \\ Y_L \\ Z_L \\ \hline u_L \\ v_L \\ w_L \\ \hline \phi_L \\ \psi_L \\ \theta_L \end{array}\right]}_{H_L}
$$

2.12 FREE CURVED BAR OF ORDER ONE, COMPONENT MATRICES AND COEFFICIENTS IN 0 SYSTEM

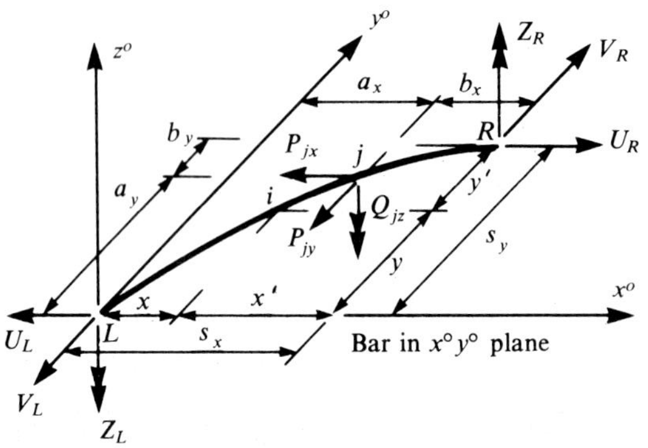

Fig. 2.12-1

(1) System considered in Fig. 2.12-1 is a free curved planar bar with normal section symmetrical about the centroidal axis y^s and acted on by loads in the bar plane. The properties of the bar section and the elastic constants of bar material are those defined in (1.14).

s_x, s_y = span projections (m)
x, y, x', y' = coordinates (m)
a_x, a_y, b_x, b_y = segments (m)
$x' = s_x - x \qquad y' = s_y - y$

The superscript o for the 0 system is omitted in all vector terms.

(2) Submatrices used in the construction of T_{LR}, T_{RL} (2.02-1) are

$$t_{LR} = \begin{bmatrix} 1 & -0 & 0 \\ 0 & 1 & 0 \\ -s_y & s_x & 1 \end{bmatrix} \qquad t_{RL} = \begin{bmatrix} 1 & 0 & 0 \\ 0 & 1 & 0 \\ s_y & -s_x & 1 \end{bmatrix} \qquad c_{LR} = \begin{bmatrix} 0 & 0 & 0 \\ 0 & 0 & 0 \\ 0 & 0 & 0 \end{bmatrix} = c_{RL}$$

$$t_{Li} = \begin{bmatrix} 1 & -0 & 0 \\ 0 & 1 & 0 \\ -y & x & 0 \end{bmatrix} \qquad t_{Lj} = \begin{bmatrix} 1 & -0 & 0 \\ 0 & 1 & 0 \\ -a_y & a_x & 1 \end{bmatrix}$$

$$t_{Ri} = \begin{bmatrix} 1 & 0 & 0 \\ -0 & 1 & 0 \\ y' & -x' & 1 \end{bmatrix} \qquad t_{Rj} = \begin{bmatrix} 1 & 0 & 0 \\ -0 & 1 & 0 \\ b_y & -b_x & 1 \end{bmatrix}$$

and in terms of t matrices given above,

$$d_{LR} = -\int_L^R t_{iL}^{)T} \Lambda_i t_{iR} \qquad \Lambda_i = \begin{bmatrix} \Lambda_{UU} & \Lambda_{UV} & 0 \\ \Lambda_{VU} & \Lambda_{VV} & 0 \\ 0 & 0 & \Lambda_{ZZ} \end{bmatrix}$$

$$d_{RL} = \int_L^R t_{iR}^{)T} \Lambda_i t_{iL}$$

where Λ_i is the elemental flexibility matrix (1.15-1) in the 0 system.

(3) Transport coefficients which form the submatrices d_{LR}, d_{RL} above are

$$T_{11} = \int_L^R (yy'\Lambda_{ZZ} - \Lambda_{UU}) \qquad T_{12} = \int_L^R (x'y\Lambda_{ZZ} + \Lambda_{UV}) \qquad T_{16} = \int_L^R y\Lambda_{ZZ}$$

$$T_{21} = \int_L^R (xy'\Lambda_{ZZ} + \Lambda_{VU}) \qquad T_{22} = \int_L^R (xx'\Lambda_{ZZ} - \Lambda_{VV}) \qquad T_{26} = \int_L^R x\Lambda_{ZZ}$$

$$T_{61} = \int_L^R y'\Lambda_{ZZ} \qquad T_{62} = \int_L^R x'\Lambda_{ZZ} \qquad T_{66} = \int_L^R \Lambda_{ZZ}$$

For $s_x = a$, $T(a_x) = A$, and for $s_x = b$, $T(b_x) = B$.

2.13 FREE CURVED BAR OF ORDER ONE, TRANSPORT MATRIX EQUATIONS IN 0 SYSTEM

Notation (2.02, 2.12) Signs (2.02) Transport coefficients (2.12)

(1) Transport matrix equations constructed by (2.02-1) are introduced below. If the state vector is required for any intermediate section given by the position coordinates x, y measured from L, the transport matrix must be reconstructed for $s_x = x$, $b_x = x - a_x$, $s_y = y$, $b_y = y - a$.

$$
\begin{bmatrix} 1 \\ \hline U_L \\ V_L \\ Z_L \\ \hline u_L \\ v_L \\ \theta_L \end{bmatrix}
=
\left[\begin{array}{c|ccc|ccc}
1 & 0 & 0 & 0 & 0 & 0 & 0 \\ \hline
F_U & 1 & 0 & 0 & & & \\
F_V & 0 & 1 & 0 & & 0 & \\
F_Z & -s_y & s_x & 1 & & & \\ \hline
F_u & T_{11} & -T_{12} & -T_{16} & 1 & 0 & -s_y \\
F_v & -T_{21} & T_{22} & T_{26} & 0 & 1 & s_x \\
F_\theta & T_{61} & -T_{62} & -T_{66} & 0 & 0 & 1
\end{array}\right]
\begin{bmatrix} 1 \\ \hline U_R \\ V_R \\ Z_R \\ \hline u_R \\ v_R \\ \theta_R \end{bmatrix}
$$

$$
\begin{bmatrix} 1 \\ \hline U_R \\ V_R \\ Z_R \\ \hline u_R \\ v_R \\ \theta_R \end{bmatrix}
=
\left[\begin{array}{c|ccc|ccc}
1 & 0 & 0 & 0 & 0 & 0 & 0 \\ \hline
G_U & 1 & 0 & 0 & & & \\
G_V & 0 & 1 & 0 & & 0 & \\
G_Z & s_y & -s_x & 1 & & & \\ \hline
G_u & -T_{11} & T_{21} & -T_{61} & 1 & 0 & -s_y \\
G_v & T_{12} & -T_{22} & T_{62} & 0 & 1 & s_x \\
G_\theta & T_{16} & -T_{26} & T_{66} & 0 & 0 & 1
\end{array}\right]
\begin{bmatrix} 1 \\ \hline U_L \\ V_L \\ Z_L \\ \hline u_L \\ v_L \\ \theta_L \end{bmatrix}
$$

Transport matrix equations of circular and parabolic bars of order one are shown in scaled form in (6.02–6.03, 7.02–7.03).

(2) Load functions (2.03-1) due to singular loads (Fig. 2.12-1) are

$$
\begin{bmatrix} F_U \\ F_V \\ F_Z \\ \hline F_u \\ F_v \\ F_\theta \end{bmatrix}
=
\left[\begin{array}{ccc}
-1 & 0 & 0 \\
0 & -1 & 0 \\
a_y & -a_x & -1 \\ \hline
-A_{11} & A_{12} & A_{16} \\
A_{21} & -A_{22} & -A_{26} \\
-A_{61} & A_{62} & A_{66}
\end{array}\right]
\begin{bmatrix} P_{jx} \\ P_{jy} \\ Q_{jz} \end{bmatrix}
\qquad
\begin{bmatrix} G_U \\ G_V \\ G_Z \\ \hline G_u \\ G_v \\ G_\theta \end{bmatrix}
=
\left[\begin{array}{ccc}
1 & 0 & 0 \\
0 & 1 & 0 \\
b_y & -b_x & 1 \\ \hline
-B_{11} & B_{21} & -B_{61} \\
B_{12} & -B_{22} & B_{62} \\
B_{16} & -B_{26} & B_{66}
\end{array}\right]
\begin{bmatrix} P_{jx} \\ P_{jy} \\ Q_{jz} \end{bmatrix}
$$

Other load functions are derived by the relations summarized in (2.03-2, 3, 4). Load functions of circular and parabolic bars of order one are shown in scaled form in (6.04–6.07, 7.04–7.07).

2.14 FREE CURVED BAR OF ORDER TWO, COMPONENT MATRICES AND COEFFICIENTS IN 0 SYSTEM

(1) System considered in Fig. 2.14-1 is a free curved planar bar with normal section symmetrical about the centroidal axis z^s and acted on by loads normal to the bar plane. The properties of the section and the elastic constants of the bar material are those defined in (1.14)

s_x, s_y = span projections (m)
x, y, x', y' = coordinates (m)
a_x, a_y, b_x, b_y = segments (m)
$x' = s_x - x \qquad y' = s_y - y$

The superscript o for the 0 system is omitted in all vector terms.

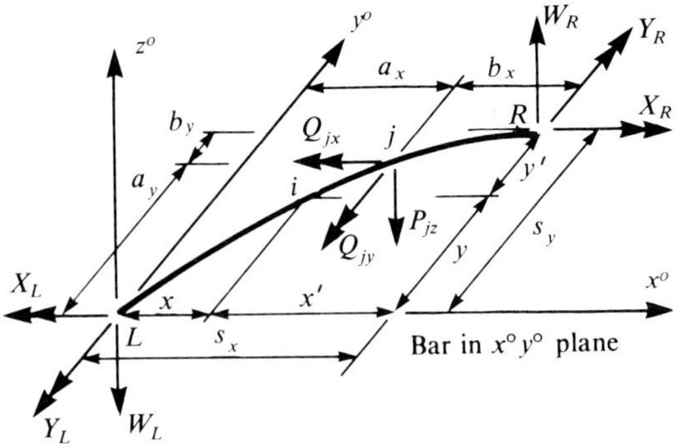

Fig. 2.14-1

(2) Submatrices used in the construction of T_{LR}, T_{RL} (2.02-1) are

$$t_{LR} = \begin{bmatrix} 1 & 0 & 0 \\ s_y & 1 & 0 \\ -s_x & 0 & 1 \end{bmatrix} \qquad t_{RL} = \begin{bmatrix} 1 & 0 & 0 \\ -s_y & 1 & 0 \\ s_x & 0 & 1 \end{bmatrix} \qquad c_{LR} = \begin{bmatrix} 0 & 0 & 0 \\ 0 & 0 & 0 \\ 0 & 0 & 0 \end{bmatrix} = c_{RL}$$

$$t_{Li} = \begin{bmatrix} 1 & 0 & 0 \\ y & 1 & 0 \\ -x & 0 & 1 \end{bmatrix} \qquad t_{Lj} = \begin{bmatrix} 1 & 0 & 0 \\ a_y & 1 & 0 \\ -a_x & 0 & 1 \end{bmatrix}$$

$$t_{Ri} = \begin{bmatrix} 1 & 0 & 0 \\ -y' & 1 & 0 \\ x' & 0 & 1 \end{bmatrix} \qquad t_{Rj} = \begin{bmatrix} 1 & 0 & 0 \\ -b_y & 1 & 0 \\ b_x & 0 & 1 \end{bmatrix}$$

and in terms of t matrices given above,

$$d_{LR} = -\int_0^{s_x} t_{iL}^{)T}\Lambda_i t_{iR} \qquad d_{RL} = \int_0^{s_x} t_{iR}^{)T}\Lambda_i t_{iL} \qquad \Lambda_i = \begin{bmatrix} \Lambda_{WW} & 0 & 0 \\ 0 & \Lambda_{XX} & \Lambda_{XY} \\ 0 & \Lambda_{YX} & \Lambda_{YY} \end{bmatrix}$$

where Λ_i is the elemental flexibility matrix (1.15-2) in the 0 system.

(3) Transport coefficients which form the submatrices d_{LR}, d_{RL} above are

$$T_{33} = \int_L^R [y'(y\Lambda_{XX} - x\Lambda_{XY}) - x'(y\Lambda_{XY} - x\Lambda_{YY}) - \Lambda_{WW}] \qquad T_{34} = \int_L^R (y\Lambda_{XX} - x\Lambda_{XY})$$

$$T_{35} = \int_L^R (y\Lambda_{XY} - \Lambda_{YY}) \qquad T_{43} = \int_L^R (y'\Lambda_{XX} - x'\Lambda_{XY}) \qquad T_{44} = \int_L^R \Lambda_{XX}$$

$$T_{45} = \int_L^R \Lambda_{XY} = T_{54} \qquad T_{53} = \int_L^R (y'\Lambda_{XY} - x'\Lambda_{YY}) \qquad T_{55} = \int_L^R \Lambda_{YY}$$

For $s_x = a_x$, $T(a_x) = A$, and for $s_x = b_x$, $T(b_x) = B$.

2.15 FREE CURVED BAR OF ORDER TWO, TRANSPORT MATRIX EQUATIONS IN 0 SYSTEM

Notation (2.02, 2.14) Signs (2.02) Transport coefficients (2.14)

(1) Transport matrix equations constructed by (2.02-1) are introduced below. If the state vector is required for any intermediate section given by the position coordinates x, y measured from L, the transport matrix must be reconstructed for $s_x = x$, $b_x = x - a_x$, $s_y = y$, $b_y = y - a_y$.

$$
\begin{bmatrix} 1 \\ \hline W_L \\ X_L \\ Y_L \\ \hline w_L \\ \phi_L \\ \psi_L \end{bmatrix}
=
\left[\begin{array}{c|ccc|ccc}
1 & 0 & 0 & 0 & 0 & 0 & 0 \\ \hline
F_W & 1 & 0 & 0 & & & \\
F_X & s_y & 1 & 0 & & 0 & \\
F_Y & -s_x & 0 & 1 & & & \\ \hline
F_w & T_{33} & T_{34} & T_{35} & 1 & -s_y & s_x \\
F_\phi & -T_{43} & -T_{44} & -T_{45} & 0 & 1 & 0 \\
F_\psi & -T_{53} & -T_{54} & -T_{55} & 0 & 0 & 1
\end{array}\right]
\begin{bmatrix} 1 \\ \hline W_R \\ X_R \\ Y_R \\ \hline w_R \\ \phi_R \\ \psi_R \end{bmatrix}
$$

$$
\begin{bmatrix} 1 \\ \hline W_R \\ X_R \\ Y_R \\ \hline w_R \\ \phi_R \\ \psi_R \end{bmatrix}
=
\left[\begin{array}{c|ccc|ccc}
1 & 0 & 0 & 0 & 0 & 0 & 0 \\ \hline
G_W & 1 & 0 & 0 & & & \\
G_X & -s_y & 1 & 0 & & 0 & \\
G_Y & s_x & 0 & 1 & & & \\ \hline
G_w & -T_{33} & T_{43} & T_{53} & 1 & s_y & -s_x \\
G_\phi & -T_{34} & T_{44} & T_{54} & 0 & 1 & 0 \\
G_\psi & -T_{35} & T_{45} & T_{55} & 0 & 0 & 1
\end{array}\right]
\begin{bmatrix} 1 \\ \hline W_L \\ X_L \\ Y_L \\ \hline w_L \\ \phi_L \\ \psi_L \end{bmatrix}
$$

Transport matrix equations of circular bars of order two are shown in scaled form in (12.02–12.03).

(2) Load functions (2.03-1) due to singular loads in Fig. 2.14-1 are

$$
\begin{bmatrix} F_W \\ F_X \\ F_Y \\ \hline F_w \\ F_\phi \\ F_\psi \end{bmatrix}
=
\left[\begin{array}{ccc}
-1 & 0 & 0 \\
-a_y & -1 & 0 \\
a_x & 0 & -1 \\ \hline
-A_{33} & -A_{34} & -A_{35} \\
A_{43} & A_{44} & A_{45} \\
A_{53} & A_{54} & A_{55}
\end{array}\right]
\begin{bmatrix} P_{jz} \\ Q_{jx} \\ Q_{jy} \end{bmatrix}
\qquad
\begin{bmatrix} G_W \\ G_X \\ G_Y \\ \hline G_w \\ G_\phi \\ G_\psi \end{bmatrix}
=
\left[\begin{array}{ccc}
1 & 0 & 0 \\
-b_y & 1 & 0 \\
b_x & 0 & 1 \\ \hline
-B_{33} & B_{43} & B_{53} \\
-B_{34} & B_{44} & B_{54} \\
-B_{35} & B_{45} & B_{55}
\end{array}\right]
\begin{bmatrix} P_{jz} \\ Q_{jx} \\ Q_{jy} \end{bmatrix}
$$

Other load functions are derived by the relations summarized in (2.03-2, 3, 4). Load functions of circular bars of order two are shown in scaled form in (12.04–12.06).

2.16 INTERACTIVE SYSTEMS OF ORDER THREE, CLASSIFICATION

(1) Systems considered are prismatic bars with normal section symmetrical about the centroidal axes y^s and z^s and acted on by loads applied along and normal to the centroidal axis x^s in the S system. The properties of their normal sections and the elastic constants of their material are those defined in (1.14). According to the nature of the surrounding medium and the type of axial force involved, they are classified as *bars encased in elastic foundation* (Fig. 2.16-1), *beam-columns* (Fig. 2.16-2), and *beam-columns encased in elastic foundation* (Fig. 2.16-3).

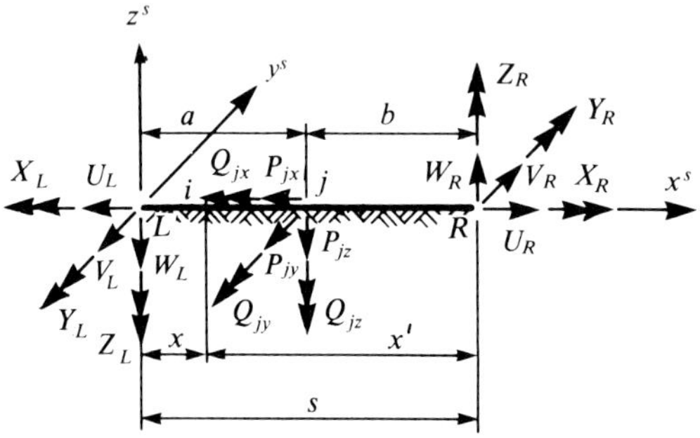

Fig. 2.16-1

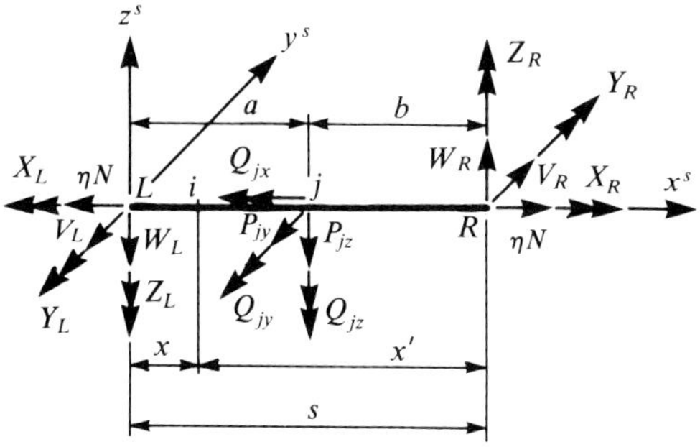

Fig. 2.16-2

(2) Elastic foundation fully surrounds the bar and offers resistance to its motion which is proportional to the displacement of the bar (Winkler's foundation). The constants of proportionality, called foundation moduli, are

k_u = longitudinal modulus along x^s (N/m^2)

k_v, k_w = transverse modulus along y^s, z^s (N/m^2)

k_ϕ = torsional modulus about x^s (N·m/m)

Each modulus corresponds to a unit length of the beam.

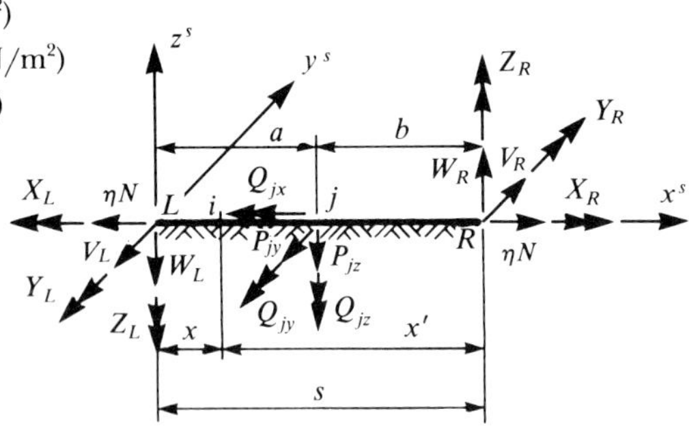

Fig. 2.16-3

(3) Axial force U in Fig. 2.16-1 may vary along the span of the bar, whereas the axial force in Figs. 2.16-2, 2.16-3 is constant, so that

$$U = \eta N = \begin{cases} +N = \text{axial tensile force (N)} \\ -N = \text{axial compressive force (N)} \end{cases}$$

(4) Transport coefficients and their derivatives introduced in (1.18-3) form for $x = s$ the elements of submatrices t_{RL}, c_{RL}, d_{RL}, $t_{LR}^{)T}$ in T_{RL} and for $x = -s$ the elements of t_{LR}, c_{LR}, d_{LR}, $t_{RL}^{)T}$ in T_{LR} (2.01-1). Since the Laplace transform method yields the coefficients of T_{RL}, these coefficients are derived first and their left-side counterparts are obtained by the relations shown in (2.02-3) (Ref. 17).

2.17 INTERACTIVE SYSTEMS OF ORDER THREE, LOAD FUNCTIONS IN *S* SYSTEM

(1) Left-end load functions (2.03-1) due to singular loads in Figs. 2.16-1 to 2.16-3 are

$$
\begin{bmatrix} F_U \\ F_V \\ F_Z \\ \hline F_X \\ F_Y \\ F_Z \\ \hline F_u \\ F_v \\ F_w \\ \hline F_\phi \\ F_\psi \\ F_\theta \end{bmatrix}
=
\left[\begin{array}{ccc|ccc}
-A_{UU} & 0 & 0 & 0 & 0 & 0 \\
0 & -A_{VV} & 0 & 0 & 0 & -A_{VZ} \\
0 & 0 & -A_{WW} & 0 & A_{WY} & 0 \\ \hline
0 & 0 & 0 & -A_{XX} & 0 & 0 \\
0 & 0 & A_{YW} & 0 & -A_{YY} & 0 \\
0 & -A_{ZV} & 0 & 0 & 0 & -A_{ZZ} \\ \hline
A_{uU} & 0 & 0 & 0 & 0 & 0 \\
0 & -A_{vV} & 0 & 0 & 0 & -A_{wZ} \\
0 & 0 & -A_{wW} & 0 & A_{wY} & 0 \\ \hline
0 & 0 & 0 & A_{\phi X} & 0 & 0 \\
0 & 0 & -A_{\psi W} & 0 & A_{\psi Y} & 0 \\
0 & A_{\theta V} & 0 & 0 & 0 & A_{\theta Z}
\end{array}\right]
\begin{bmatrix} P_{jx} \\ P_{jy} \\ P_{jz} \\ \hline Q_{jx} \\ Q_{jy} \\ Q_{jz} \end{bmatrix}
$$

where $A_{UU}, A_{VV}, \ldots, A_{\theta Z}$ are $T_{UU}, T_{VV}, \ldots, T_{\theta Z}$ (1.18-3) for $s = a$.

(2) Right-end load functions (2.03-1) due to the same loads are

$$
\begin{bmatrix} G_U \\ G_V \\ G_W \\ \hline G_X \\ G_Y \\ G_Z \\ \hline G_u \\ G_v \\ G_w \\ \hline G_\phi \\ G_\psi \\ G_\theta \end{bmatrix}
=
\left[\begin{array}{ccc|ccc}
B_{UU} & 0 & 0 & 0 & 0 & 0 \\
0 & B_{VV} & 0 & 0 & 0 & -B_{VZ} \\
0 & 0 & B_{WW} & 0 & B_{WY} & 0 \\ \hline
0 & 0 & 0 & B_{XX} & 0 & 0 \\
0 & 0 & B_{YW} & 0 & B_{YY} & 0 \\
0 & -B_{ZV} & 0 & 0 & 0 & B_{ZZ} \\ \hline
B_{uU} & 0 & 0 & 0 & 0 & 0 \\
0 & -B_{vV} & 0 & 0 & 0 & B_{vZ} \\
0 & 0 & -B_{wW} & 0 & -B_{wY} & 0 \\ \hline
0 & 0 & 0 & B_{\phi X} & 0 & 0 \\
0 & 0 & B_{\psi W} & 0 & B_{\psi Y} & 0 \\
0 & -B_{\theta V} & 0 & 0 & 0 & B_{\theta Z}
\end{array}\right]
\begin{bmatrix} P_{jx} \\ P_{jy} \\ P_{jz} \\ \hline Q_{jx} \\ Q_{jy} \\ Q_{jz} \end{bmatrix}
$$

where $B_{UU}, B_{VV}, \ldots, B_{\theta Z}$ are $T_{UU}, T_{VV}, \ldots, T_{\theta Z}$ (1.18-3) for $s = b$.

(3) Other load functions are derived by the relations summarized in (2.03-2, 3, 4). Extensive lists of scaled load functions for interactive bars of order one and two are given in (8.04–8.06, 9.06–9.08, 10.05–10.07, 13.04–13.06, 14.06–14.08, 15.05–15.07).

2.18 INTERACTIVE SYSTEMS OF ORDER THREE, TRANSPORT MATRIX EQUATIONS IN S SYSTEM

Notation (2.02, 2.16) Signs (2.02) Load functions (2.17)

(1) Left-end equation

$$
\underbrace{\left[\begin{array}{c} 1 \\ \hline U_L \\ V_L \\ W_L \\ \hline X_L \\ Y_L \\ Z_L \\ \hline u_L \\ v_L \\ w_L \\ \hline \phi_L \\ \psi_L \\ \theta_L \end{array}\right]}_{H_L}
=
\underbrace{\left[\begin{array}{c|ccc|ccc|ccc|ccc}
1 & 0 & 0 & 0 & 0 & 0 & 0 & 0 & 0 & 0 & 0 & 0 & 0 \\ \hline
F_U & T_{UU} & 0 & 0 & 0 & 0 & 0 & -T_{Uu} & 0 & 0 & 0 & 0 & 0 \\
F_V & 0 & T_{VV} & 0 & 0 & 0 & T_{VZ} & 0 & T_{Vv} & 0 & 0 & 0 & -T_{V\theta} \\
F_W & 0 & 0 & T_{WW} & 0 & -T_{WY} & 0 & 0 & 0 & T_{Ww} & 0 & T_{W\psi} & 0 \\ \hline
F_X & 0 & 0 & 0 & T_{XX} & 0 & 0 & 0 & 0 & 0 & -T_{X\phi} & 0 & 0 \\
F_Y & 0 & 0 & -T_{YW} & 0 & T_{YY} & 0 & 0 & 0 & -T_{Yw} & 0 & -T_{Y\psi} & 0 \\
F_Z & 0 & T_{ZV} & 0 & 0 & 0 & T_{ZZ} & 0 & T_{Zv} & 0 & 0 & 0 & -T_{Z\theta} \\ \hline
F_u & -T_{uU} & 0 & 0 & 0 & 0 & 0 & T_{uu} & 0 & 0 & 0 & 0 & 0 \\
F_v & 0 & T_{vV} & 0 & 0 & 0 & T_{vZ} & 0 & T_{vv} & 0 & 0 & 0 & -T_{v\theta} \\
F_w & 0 & 0 & T_{wW} & 0 & -T_{wY} & 0 & 0 & 0 & T_{ww} & 0 & T_{w\psi} & 0 \\ \hline
F_\phi & 0 & 0 & 0 & -T_{\phi X} & 0 & 0 & 0 & 0 & 0 & T_{\phi\phi} & 0 & 0 \\
F_\psi & 0 & 0 & T_{\psi W} & 0 & -T_{\psi Y} & 0 & 0 & 0 & T_{\psi w} & 0 & T_{\psi\psi} & 0 \\
F_\theta & 0 & -T_{\theta V} & 0 & 0 & 0 & -T_{\theta Z} & 0 & -T_{\theta v} & 0 & 0 & 0 & T_{\theta\theta}
\end{array}\right]}_{T_{LR}}
\underbrace{\left[\begin{array}{c} 1 \\ \hline U_R \\ V_R \\ W_R \\ \hline X_R \\ Y_R \\ Z_R \\ \hline u_R \\ v_R \\ w_R \\ \hline \phi_R \\ \psi_R \\ \theta_R \end{array}\right]}_{H_R}
$$

(2) Right-end equation

$$
\underbrace{\begin{bmatrix} 1 \\ \hline U_R \\ V_R \\ W_R \\ \hline X_R \\ Y_R \\ Z_R \\ \hline u_R \\ v_R \\ w_R \\ \hline \phi_R \\ \psi_R \\ \theta_R \end{bmatrix}}_{H_R}
=
\underbrace{\left[\begin{array}{c|ccc|ccc|ccc|ccc}
1 & 0 & 0 & 0 & 0 & 0 & 0 & 0 & 0 & 0 & 0 & 0 & 0 \\ \hline
G_U & T_{UU} & 0 & 0 & 0 & 0 & 0 & T_{Uu} & 0 & 0 & 0 & 0 & 0 \\
G_V & 0 & T_{VV} & 0 & 0 & 0 & -T_{VZ} & 0 & -T_{Vv} & 0 & 0 & 0 & -T_{V\theta} \\
G_W & 0 & 0 & T_{WW} & 0 & T_{WY} & 0 & 0 & 0 & -T_{Ww} & 0 & T_{W\psi} & 0 \\ \hline
G_X & 0 & 0 & 0 & T_{XX} & 0 & 0 & 0 & 0 & 0 & T_{X\phi} & 0 & 0 \\
G_T & 0 & 0 & T_{YW} & 0 & T_{YY} & 0 & 0 & 0 & -T_{Yw} & 0 & T_{Y\psi} & 0 \\
G_Z & 0 & -T_{ZV} & 0 & 0 & 0 & T_{ZZ} & 0 & T_{Zv} & 0 & 0 & 0 & T_{Z\theta} \\ \hline
G_u & T_{uU} & 0 & 0 & 0 & 0 & 0 & T_{uu} & 0 & 0 & 0 & 0 & 0 \\
G_v & 0 & -T_{vV} & 0 & 0 & 0 & T_{vZ} & 0 & T_{vv} & 0 & 0 & 0 & T_{v\theta} \\
G_w & 0 & 0 & -T_{wW} & 0 & -T_{wY} & 0 & 0 & 0 & T_{ww} & 0 & -T_{w\psi} & 0 \\ \hline
G_\phi & 0 & 0 & 0 & T_{\phi X} & 0 & 0 & 0 & 0 & 0 & T_{\phi\phi} & 0 & 0 \\
G_\psi & 0 & 0 & T_{\psi W} & 0 & T_{\psi Y} & 0 & 0 & 0 & -T_{\psi w} & 0 & T_{\psi\psi} & 0 \\
G_\theta & 0 & -T_{\theta V} & 0 & 0 & 0 & T_{\theta Z} & 0 & T_{\theta v} & 0 & 0 & 0 & T_{\theta\theta}
\end{array}\right]}_{T_{RL}}
\underbrace{\begin{bmatrix} 1 \\ \hline U_L \\ V_L \\ W_L \\ \hline X_L \\ Y_L \\ Z_L \\ \hline u_L \\ v_L \\ w_L \\ \hline \phi_L \\ \psi_L \\ \theta_L \end{bmatrix}}_{H_L}
$$

(3) Scaling. The matrix equations given above are shown in their absolute form. The same matrices will occur later in a scaled form (8.02–8.03, 9.02–9.03, 10.02–10.04, 13.02–13.03, 14.02–14.03, 15.02–15.04).

2.19 TRANSPORT METHOD, BOUNDARY VALUES, SCALING, ANGULAR TRANSFORMATIONS

(1) Application of transport matrix equations in the analysis of beams and frames is possible, and numerous examples of such applications are available in the technical literature (Refs. 1, 3–7, 9–11). In this book, the applications are restricted to the analysis of single segments and to the use of these equations in the development of stiffness matrices and their load functions. First, two distinct characteristics of these equations must be noticed:

(*a*) The *state vector* of one end of the segment (bar) is expressed in terms of the state vector of the opposite end and in terms of loads and other causes acting between them.

(*b*) This *relationship* makes no distinction between statically determinate and indeterminate systems (the number of unknowns to be solved is defined by the order of the system and not by the number of redundants).

(2) Scaling in free bars. A close inspection of the transport coefficients T and the load functions F, G reveals the involvement of such terms as EA_x, GA_y, GA_z, GI_x, EI_y, EI_z, which represent large numbers in the denominator of the respective expressions. Thus, for example, the horizontal displacement of the right end of the straight bar in (4.02-4), due to P_x applied in the span at the distance b from R, is in the true form

$$u_R = \frac{b}{EA_x} P_x + \frac{s}{EA_x} U_L + u_L$$

After multiplication by EA_x and in terms of equivalents

$$\bar{u}_R = EA_x u_R \qquad \text{and} \qquad \bar{u}_L = EA_x u_L$$

the scaled form of the same equations is

$$\bar{u}_R = bP_x + sU_L + \bar{u}_L$$

which is a more convenient form for numerical operations. The same scaling is done for the remaining displacements in (4.02-4) by means of EI_z. Algebraic examples of scaling are shown in (4.12–4.21).

(3) Scaling in interactive bars involves not only EA_x, GA_y, . . . , EI_z but also the shape parameters α, β, or λ. Thus, for example, the horizontal displacement of the right end of the straight bar encased in elastic foundation in (8.03-2), due to P_x applied in the span at the distance b from R, is in the true form

$$u_R = \frac{\sinh \alpha b}{\alpha EA_x} P_x + \frac{\sinh \alpha s}{\alpha EA_x} U_L + (\cosh \alpha s)\, u_L$$

After multiplication by αEA_x and in terms of equivalents

$$\bar{u}_R = \alpha EA_x u_R \qquad \text{and} \qquad \bar{u}_L = \alpha EA_x u_L$$

the scaled form of the same equation is

$$\bar{u}_R = (\sinh \alpha b)\, P_x + (\sinh \alpha s)\, U_L + (\cosh \alpha s)\, \bar{u}_L$$

which is again more convenient than the true form. The scaling of the remaining equations of the transport matrix equation of the straight bar encased in elastic foundation is shown in (8.03-2). All transport matrices in Chaps. 4–15 are given in the scaled form.

(4) In bars of order one, 12 boundary values are involved; they are

$$H_L = \{1, U_L, V_L, Z_L, u_L, v_L, \theta_L\} \qquad H_R = \{1, U_R, V_R, Z_R, u_R, v_R, \theta_R\}$$

of which six are always zero and six are unknown. Thus six equations, in general, are necessary for the solution of a given problem. The transport matrix equation provides these equations. In straight bars only two equations and in curved bars only three equations must be solved simultaneously. The remaining unknowns are then calculated by the substitution of the obtained values in the remaining equations of the matrix equation. The left- or the right-end equation may be used for this purpose (but not both). Six typical cases are outlined symbolically in (4.07-2), and numerical examples are shown in Ref. 3, pp. 148–155, and Ref. 4, pp. 44–47.

(5) In bars of order two, again only 12 boundary values are involved. They are

$$H_L = \{1, W_L, X_L, Y_L, w_L, \phi_L, \psi_L\} \qquad H_R = \{1, W_R, X_R, Y_R, w_R, \phi_R, \psi_R\}$$

of which six are always zero and six are unknown. Thus again six equations, in general, are necessary for the solution. The procedure of calculation of the remaining unknowns is the same as in (4) above. Numerical examples are shown in Ref. 5, pp. 58–67.

(6) In bars of order three, 24 boundary values are involved. They are

$$H_L = \{1, U_L, V_L, W_L, X_L, Y_L, Z_L, u_L, v_L, w_L, \phi_L, \psi_L, \theta_L\}$$
$$H_R = \{1, U_R, V_R, W_R, X_R, Y_R, Z_R, u_R, v_R, w_R, \phi_R, \psi_R, \theta_R\}$$

of which 12 are always zero and 12 are unknown. Thus 12 equations, in general, are necessary for the solution. In straight bars only two sets of two equations and in curved bars only two sets of three equations must be solved simultaneously. The calculation of the remaining unknowns is the same as in (4) above. Numerical examples are shown in Ref. 5, pp. 58–67.

(7) Angular transformations. For curved bars such as the bar shown in Fig. 2.07-1, it is necessary to calculate the state vectors at each end in the principal axes of the respective normal sections, called the member axes and designated by the superscripts of the respective ends as X^L, Y^L, Z^L and X^R, Y^R, Z^R. The transport matrix equations (2.02-1) in these systems become

$$\underbrace{\begin{bmatrix} 1 \\ S_L^L \\ \Delta_L^L \end{bmatrix}}_{H_L^L} = \underbrace{\begin{bmatrix} 1 & 0 & 0 \\ 0 & \hat{R}^{L0} & 0 \\ 0 & 0 & \hat{R}^{L0} \end{bmatrix} \underbrace{\begin{bmatrix} 1 & 0 & 0 \\ F_S^0 & t_{LR}^0 & 0 \\ F_\Delta^0 & d_{LR}^0 & t_{RL}^{0)T} \end{bmatrix}}_{T_{LR}^0} \begin{bmatrix} 1 & 0 & 0 \\ 0 & \hat{R}^{0R} & 0 \\ 0 & 0 & \hat{R}^{0R} \end{bmatrix}}_{T_{LR}^{LR}} \underbrace{\begin{bmatrix} 1 \\ S_R^R \\ \Delta_R^R \end{bmatrix}}_{H_R^R}$$

$$\underbrace{\begin{bmatrix} 1 \\ S_R^R \\ \Delta_R^R \end{bmatrix}}_{H_R^R} = \underbrace{\begin{bmatrix} 1 & 0 & 0 \\ 0 & \hat{R}^{R0} & 0 \\ 0 & 0 & \hat{R}^{R0} \end{bmatrix} \underbrace{\begin{bmatrix} 1 & 0 & 0 \\ G_S^0 & t_{RL}^0 & 0 \\ G_\Delta^R & d_{RL}^0 & t_{LR}^{0)T} \end{bmatrix}}_{T_{RL}^0} \begin{bmatrix} 1 & 0 & 0 \\ 0 & \hat{R}^{0L} & 0 \\ 0 & 0 & \hat{R}^{0L} \end{bmatrix}}_{T_{RL}^{RL}} \underbrace{\begin{bmatrix} 1 \\ S_L^L \\ \Delta_L^L \end{bmatrix}}_{H_L^L}$$

where $\hat{R}^{0L}$, $\hat{R}^{L0}$ = angular transformation matrices (1.12-4) of the left end, [6 × 6]
$\hat{R}^{0R}$, $\hat{R}^{R0}$ = angular transformation matrices (1.12-4) of the right end, [6 × 6]
H_L^L, H_R^R = state vectors of the respective ends in the end system, [13 × 1]

2.20 TRANSPORT METHOD, JOINT-EFFECT MATRICES

(1) Direct transport chain, given by (1.19-4), must be frequently modified to accommodate intermediate interferences such as:

(*a*) Intermediate rigid constraints (rigid supports)
(*b*) Intermediate elastic constraints (elastic supports)
(*c*) Intermediate rigid releases (hinges, guides, etc.)
(*d*) Intermediate elastic releases (elastic connections)

Their effects can always be expressed analytically by the joint-effect matrices and included in the transport chain, whose uniformity when interrupted by this inclusion is designated as the *indirect transport chain* (1.19-5).

(2) Joint-effect matrices at j (any joint or section) are

$$J_j = J_{LjR} = \begin{bmatrix} 1 & 0 & 0 \\ S_{LjR} & t_{LjR} & c_{LjR} \\ \Delta_{LjR} & d_{LjR} & s_{LjR} \end{bmatrix} \qquad J_j^{-1} = J_{RjL} = \begin{bmatrix} 1 & 0 & 0 \\ S_{RjL} & t_{RjL} & c_{RjL} \\ \Delta_{RjL} & d_{RjL} & s_{RjL} \end{bmatrix}$$

where $S_{LjR} = -S_{RjL}$ = joint stress-resultant gap matrix, $[6 \times 1]$
$\Delta_{LjR} = -\Delta_{RjL}$ = joint displacement gap matrix, $[6 \times 1]$
$t_{LjR} = t_{RjL}$ = joint stress-resultant transport matrix, $[6 \times 6]$
$s_{LjR} = s_{RjL}$ = joint displacement transport matrix, $[6 \times 6]$
$c_{LjR} = -c_{RjL}$ = joint elastic constraint matrix, $[6 \times 6]$
$d_{LjR} = -d_{RjL}$ = joint elastic release matrix, $[6 \times 6]$

The joint stress-resultant gap matrix represents the unknown intermediate reactions at j, treated as loads of unknown magnitude. Similarly, the joint displacement gap matrix represents the sudden change in the continuity of the elastic curve caused by mechanical releases, such as hinges, guides, etc., and treated as displacements of unknown magnitude. The joint stress-resultant and displacement matrices are unit matrices with as many zeros on the diagonal as there are zero transports of stress resultants and displacements. Finally, the joint elastic constraint and release matrices consist of the spring coefficients of supports and connections, respectively.

(3) Special conditions. In addition to the 12 unknown boundary values, intermediate interferences generate additional unknowns. Since each interference is always accompanied by the corresponding condition (rigid support by zero displacement, rigid release by zero stress resultant), allowing the elimination of the interference by the modification of the transport matrix, the final number of unknowns again cannot exceed 12, of which not more than six must be solved simultaneously (Ref. 5, pp. 54–55, 60–69).

3

Flexibility and Stiffness Matrices Free and Interactive Bars

STATIC STATE

3.01 DEFINITION OF STATE

(1) Systems considered are free straight and curved bars introduced in (2.04, 2.07, 2.12, 2.14) and interactive straight bars classified in (2.16). Each bar is in a state of static equilibrium and a state of time-independent elastic deformation.

(2) Flexibility and stiffness matrices and the respective load functions of the bars mentioned above are presented in symbolic form in (3.02, 3.14). Particular cases of these matrices in free straight and curved bars are shown in (3.03–3.08, 3.15–3.17). General cases of the same matrices in interactive straight bars are displayed in (3.09–3.10, 3.18–3.19). In all cases, the sign convention of the respective method is used (1.22).

(3) Assumptions used in the derivation of analytical relations are those adapted in (1.01). The restrictive assumptions introduced in (2.01) remain valid in this chapter.

(4) Symbols used in this chapter are defined where they appear first and are all summarized in Appendices A.1, A.3, A.4.

3.02 FLEXIBILITY MATRIX EQUATIONS, BASIC RELATIONS IN S SYSTEM

(1) End flexibility of a bar is defined as the end displacement produced by a unit cause (force or moment). Since the unit causes can be applied along three orthogonal axes at each end and each unit cause can (in general) produce six mutually perpendicular displacements at each end, the derivation of the flexibility matrix requires the calculation of 144 coefficients. If the derivation is carried out in the member system (principal flexibilities) and the reciprocal relations are observed, the total number of distinct coefficients reduces to a maximum of 32. These coefficients can be obtained from the equations of the elastic curve or directly from the transport matrix equations introduced in (2.02-1). The latter approach is used below (Ref. 5, p. 157).

(2) Free-bar flexibility matrix equations are derived by their definition from the basic segments, of which the most typical ones are the cantilever bar and the simple bar. The *cantilever-bar flexibility matrix equations* stating the free end displacements as functions of the causes acting on the bar are

$$\Delta_{LR} = f_{LL}S_L + \Delta_{L0} \qquad \Delta_{RL} = f_{RR}S_R + \Delta_{R0}$$

where

$$\Delta_{LR} = \{u_{LR}, v_{LR}, w_{LR}, \phi_{LR}, \psi_{LR}, \theta_{LR}\} \qquad \Delta_{RL} = \{u_{RL}, v_{RL}, w_{RL}, \phi_{RL}, \psi_{RL}, \theta_{RL}\}$$

are the end displacement vectors of the cantilever bar free at L and at R, respectively. The causes of these end displacements are the end stress resultants

$$S_L = \{U_L, V_L, W_L, X_L, Y_L, Z_L\} \qquad S_R = \{U_R, V_R, W_R, X_R, Y_R, Z_R\}$$

and the loads and/or volume changes.

(3) Free-bar flexibility matrices in (2) above are

$$f_{LL} = -d_{LR}t_{RL} \qquad f_{RR} = d_{RL}t_{LR}$$

$$= \left(\int_L^R t_{iL}^{)T}\Lambda_i t_{iR}\right) t_{RL} \qquad = \left(\int_L^R t_{iR}^{)T}\Lambda_i t_{iL}\right) t_{LR}$$

$$= \int_L^R t_{iL}^{)T}\Lambda_i t_{iL} \qquad = \int_L^R t_{iR}^{)T}\Lambda_i t_{iR}$$

where d_{LR}, t_{LR}, d_{RL}, t_{RL}, are the submatrices of the transport matrices T_{LR}, T_{RL} in (2.02-1).

(4) Free-bar flexibility load functions in (2) above are

$$\Delta_{L0} = t_{RL}^{)T}G_\Delta \qquad \Delta_{R0} = -t_{LR}^{)T}F_\Delta$$

where

$$\Delta_{L0} = \{u_{L0}, v_{L0}, w_{L0}, \phi_{L0}, \psi_{L0}, \theta_{L0}\} \qquad \Delta_{R0} = \{u_{R0}, v_{R0}, w_{R0}, \phi_{R0}, \psi_{R0}, \theta_{R0}\}$$

are the end displacements of the respective cantilever beam due to loads (or other causes) applied to its span and G_Δ, F_Δ are the transport load functions introduced in (2.02-1).

(5) Beam-column flexibility matrix equations can also be obtained from the cantilever-bar transport matrix equations.

(6) Sign conventions of the transport method and the flexibility method are introduced in (1.22), and their comparison shows that the end stress-resultant vectors of both methods are identical, whereas the end displacement vectors are related as

$$\Delta_{LR} = -\Delta_L \qquad \Delta_{RL} = \Delta_R$$

(7) Interactive-bar flexibility matrix equations are derived by their definition from the transport matrix equations (2.02-1) by choosing the submatrix rows involving two stress-resultant vectors and one displacement vector. In terms of the sign convention discussed in (6) above,

$$S_L = F_S + t_{LR}S_R + c_{LR}\Delta_{RL} \qquad S_R = G_S + t_{RL}S_L - c_{RL}\Delta_{LR}$$

and yield in reversed order,

$$\Delta_{LR} = f_{LL}S_L + f_{LR}S_R + \Delta_{L0} \qquad \Delta_{RL} = f_{RR}S_R + f_{RL}S_L + \Delta_{R0}$$

where Δ_{LR}, Δ_{RL} are the end displacements of the bar produced by the end stress resultants S_L, S_R and the loads.

(8) Interactive-bar flexibility matrices in (7) above are

$$f_{LL} = c_{RL}^{-1}t_{RL} \qquad f_{LR} = -c_{RL}^{-1} \qquad f_{RL} = c_{LR}^{-1} \qquad f_{RR} = -c_{LR}^{-1}t_{LR}$$

where t_{LR}, t_{RL}, c_{RL}, c_{LR} are the submatrices of the transport matrices T_{LR}, T_{RL} (2.02-1).

(9) Interactive-bar load functions in (7) above are

$$\Delta_{L0} = c_{RL}^{-1}G_S = -f_{LR}G_S \qquad \Delta_{R0} = -c_{LR}^{-1}F_S = -f_{RL}F_S$$

where Δ_{L0}, Δ_{R0} are the end displacements due to loads (or other causes) applied in the span and F_S, G_S are the transport load functions in (2.02-1).

(10) Properties. The free- and interactive-bar flexibility matrices and their load functions possess the following characteristics:

(*a*) Transpose relations

$$f_{LL} = f_{LL}^T \qquad {}^\dagger f_{LR} = f_{RL}^T \qquad {}^\dagger f_{RL} = f_{LR}^T \qquad f_{RR} = f_{RR}^T$$

which shows that the bar flexibility matrix

$$f = \begin{bmatrix} f_{LL} & {}^\dagger f_{LR} \\ {}^\dagger f_{RL} & f_{RR} \end{bmatrix}$$

is a symmetrical matrix.

(*b*) Shift relations

$$f_{LL} = t_{RL}^T f_{RR} t_{RL} \qquad f_{RR} = t_{LR}^T f_{LL} t_{LR}$$

$${}^\dagger f_{LR} = -t_{RL}^T f_{RR} \qquad {}^\dagger f_{RL} = -t_{LR}^T f_{LL}$$

$$\Delta_{L0} = -t_{RL}^T \Delta_{R0} - F_\Delta \qquad \Delta_{R0} = -t_{LR}^T \Delta_{L0} + G_\Delta$$

where F_Δ, G_Δ are the transport load functions in (2.02-1).

† Valid only for interactive systems; for free bars, $f_{LR} = f_{RL} = 0$.

3.03 FREE STRAIGHT BAR OF ORDER THREE, END FLEXIBILITY LOAD FUNCTIONS IN S SYSTEM

(1) Systems considered in Figs. 3.03-1, 3.03-2 are free straight cantilever bars of constant section. Their cross-sectional properties and elastic constants are defined in (2.04-1). The superscript s for the S system is omitted in all vector terms.

s = span (m) $\quad a, b$ = segments (m)

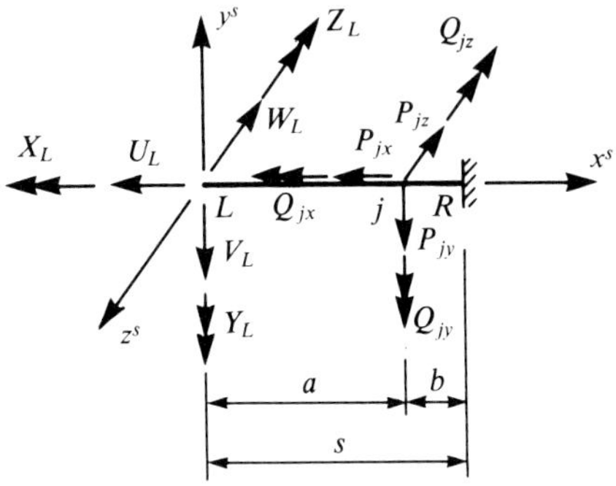

Fig. 3.03-1

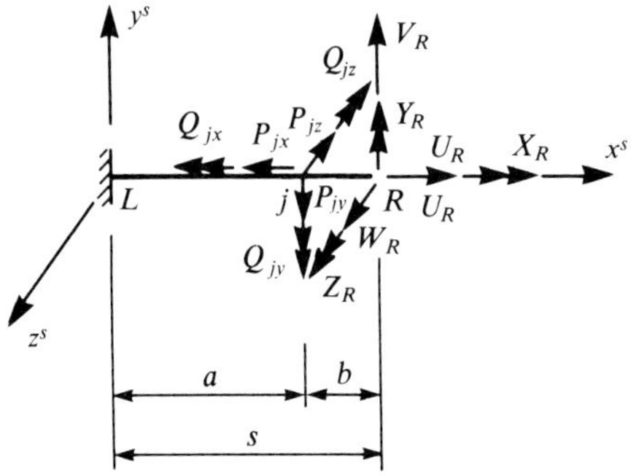

Fig. 3.03-2

(2) Flexibility load functions of these bars are, by (3.02-4),

$$
\begin{bmatrix} u_{L0} \\ v_{L0} \\ w_{L0} \\ \hline \phi_{L0} \\ \psi_{L0} \\ \theta_{L0} \end{bmatrix}
=
\left[\begin{array}{ccc|ccc}
1 & 0 & 0 & 0 & 0 & 0 \\
0 & 1 & 0 & 0 & 0 & -s \\
0 & 0 & 1 & 0 & s & 0 \\
\hline
0 & 0 & 0 & 1 & 0 & 0 \\
0 & 0 & 0 & 0 & 1 & 0 \\
0 & 0 & 0 & 0 & 0 & 1
\end{array}\right]
\begin{bmatrix} G_u \\ G_v \\ G_w \\ \hline G_\phi \\ G_\psi \\ G_\theta \end{bmatrix}
$$

$$
\begin{bmatrix} u_{R0} \\ v_{R0} \\ w_{R0} \\ \hline \phi_{R0} \\ \psi_{R0} \\ \theta_{R0} \end{bmatrix}
=
\left[\begin{array}{ccc|ccc}
-1 & 0 & 0 & 0 & 0 & 0 \\
0 & -1 & 0 & 0 & 0 & -s \\
0 & 0 & -1 & 0 & s & 0 \\
\hline
0 & 0 & 0 & -1 & 0 & 0 \\
0 & 0 & 0 & 0 & -1 & 0 \\
0 & 0 & 0 & 0 & 0 & -1
\end{array}\right]
\begin{bmatrix} F_u \\ F_v \\ F_w \\ \hline F_\phi \\ F_\psi \\ F_\theta \end{bmatrix}
$$

where $F_u, F_v, \ldots, F_\theta$ and $G_u, G_v, \ldots, G_\theta$ are the transport load functions introduced in (2.03) and shown for singular loads in (2.05-2).

3.04 FREE STRAIGHT BAR OF ORDER THREE, END FLEXIBILITY MATRIX EQUATIONS IN S SYSTEM

Notation (3.02) Signs (3.02) Load functions (3.03)

(1) Flexibility coefficients used in the construction of f_{LL} and f_{RR} in (2, 3) below are

$F_{1n} = \dfrac{s}{EA_x}$		$F_{1x} = \dfrac{s}{GI_x}$
$F_{1y} = \dfrac{s}{EI_y}$	$F_{2y} = \dfrac{s^2}{2EI_y}$	$F_{3y} = \dfrac{s^3}{3EI_y} + \dfrac{s}{GA_z}$
$F_{1z} = \dfrac{s}{EI_z}$	$F_{2z} = \dfrac{s^2}{2EI_z}$	$F_{3z} = \dfrac{s^3}{3EI_z} + \dfrac{s}{GA_y}$

(2) Left-end flexibility matrix equation (3.02-2) is

$$\underbrace{\begin{bmatrix} u_{LR} \\ v_{LR} \\ w_{LR} \\ \hline \phi_{LR} \\ \psi_{LR} \\ \theta_{LR} \end{bmatrix}}_{\Delta_{LR}} = \underbrace{\left[\begin{array}{ccc|ccc} F_{1n} & 0 & 0 & 0 & 0 & 0 \\ 0 & F_{3z} & 0 & 0 & 0 & -F_{2z} \\ 0 & 0 & F_{3y} & 0 & F_{2y} & 0 \\ \hline 0 & 0 & 0 & F_{1x} & 0 & 0 \\ 0 & 0 & F_{2y} & 0 & F_{1y} & 0 \\ 0 & -F_{2z} & 0 & 0 & 0 & F_{1z} \end{array}\right]}_{f_{LL}} \underbrace{\begin{bmatrix} U_L \\ V_L \\ W_L \\ \hline X_L \\ Y_L \\ Z_L \end{bmatrix}}_{S_L} + \underbrace{\begin{bmatrix} u_{L0} \\ v_{L0} \\ w_{L0} \\ \hline \phi_{L0} \\ \psi_{L0} \\ \theta_{L0} \end{bmatrix}}_{\Delta_{L0}}$$

where Δ_{LR} is the displacement vector at L of the cantilever beam of Fig. 3.03-1.

(3) Right-end flexibility matrix equation (3.02-2) is

$$\underbrace{\begin{bmatrix} u_{RL} \\ v_{RL} \\ w_{RL} \\ \hline \phi_{RL} \\ \psi_{RL} \\ \theta_{RL} \end{bmatrix}}_{\Delta_{RL}} = \underbrace{\left[\begin{array}{ccc|ccc} F_{1n} & 0 & 0 & 0 & 0 & 0 \\ 0 & F_{3z} & 0 & 0 & 0 & F_{2z} \\ 0 & 0 & F_{3y} & 0 & -F_{2y} & 0 \\ \hline 0 & 0 & 0 & F_{1x} & 0 & 0 \\ 0 & 0 & -F_{2y} & 0 & F_{1y} & 0 \\ 0 & F_{2z} & 0 & 0 & 0 & F_{1z} \end{array}\right]}_{f_{RR}} \underbrace{\begin{bmatrix} U_R \\ V_R \\ W_R \\ \hline X_R \\ Y_R \\ Z_R \end{bmatrix}}_{S_R} = \underbrace{\begin{bmatrix} u_{R0} \\ v_{R0} \\ w_{R0} \\ \hline \phi_{R0} \\ \psi_{R0} \\ \theta_{R0} \end{bmatrix}}_{\Delta_{R0}}$$

where Δ_{RL} is the displacement vector at R of the cantilever beam of Fig. 3.03-2.

(4) Curved bar of order three. The flexibility matrix equations for curved bars of order three in the 0 system can be derived from the submatrices of T_{LR}, T_{RL} in (2.11) by the relations introduced in (3.02).

3.05 FREE CURVED BAR OF ORDER ONE, END FLEXIBILITY MATRIX EQUATIONS IN 0 SYSTEM

(1) Systems considered in Figs. 3.05-1, 3.05-2 are free curved cantilever bars of constant section. Their cross-sectional properties and elastic constants are defined in (2.12-1). Also, s_x, s_y = span projections (m), x, x', y, y' = coordinates (m), $x' = s_x - x$, $y' = s_y - y$. The symbols and signs are given in (3.02).

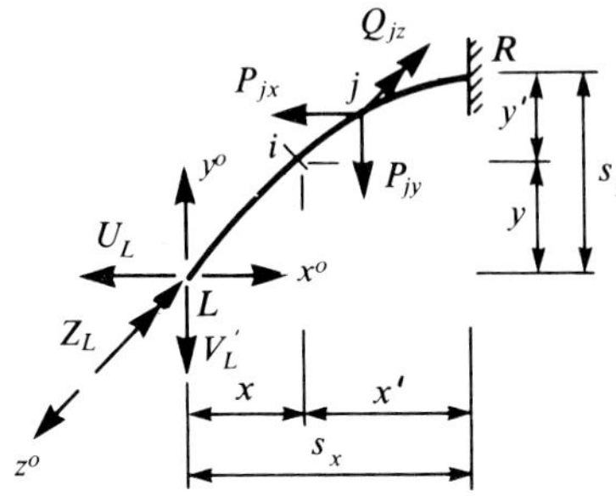

Fig. 3.05-1

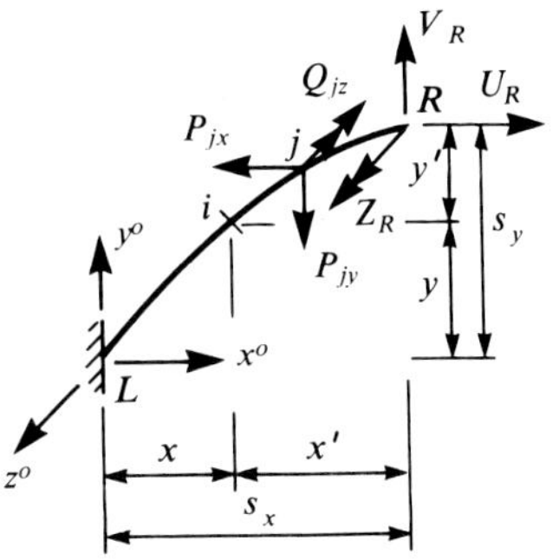

Fig. 3.05-2

(2) Flexibility coefficients used in the construction of f_{LL} and f_{RR} below are

$$F_{11} = \int_L^R (y^2\Lambda_{ZZ} + \Lambda_{UU}) \qquad F_{12} = F_{21} = \int_L^R (xy\Lambda_{ZZ} - \Lambda_{UV}) \qquad F_{16} = F_{61} = \int_L^R y\Lambda_{ZZ}$$

$$F_{22} = \int_L^R (x^2\Lambda_{ZZ} + \Lambda_{VV}) \qquad F_{26} = F_{62} = \int_L^R x\Lambda_{ZZ} \qquad F_{66} = \int_L^R \Lambda_{ZZ}$$

$$G_{11} = \int_L^R (y'^2\Lambda_{ZZ} + \Lambda_{UU}) \qquad G_{12} = G_{21} = \int_L^R (x'y'\Lambda_{ZZ} - \Lambda_{UV}) \qquad G_{16} = G_{61} = \int_L^R y'\Lambda_{ZZ}$$

$$G_{22} = \int_L^R (x'^2\Lambda_{ZZ} + \Lambda_{VV}) \qquad G_{26} = G_{62} = \int_L^R x'\Lambda_{ZZ} \qquad G_{66} = \int_L^R \Lambda_{ZZ}$$

where Λ_{UU}, Λ_{UV}, Λ_{VV}, Λ_{ZZ} are the elemental flexibilities (1.15-3). For symmetrical curves, $F_{11} = G_{11}$, $F_{12} = G_{12}, \ldots, s_y = 0$.

(3) Flexibility matrix equations (3.02-2) are

$$\underbrace{\begin{bmatrix} u_{LR} \\ v_{LR} \\ \theta_{LR} \end{bmatrix}}_{\Delta_{LR}} = \underbrace{\begin{bmatrix} F_{11} & -F_{12} & F_{16} \\ -F_{21} & F_{22} & -F_{26} \\ F_{61} & -F_{62} & F_{66} \end{bmatrix}}_{f_{LL}} \underbrace{\begin{bmatrix} U_L \\ V_L \\ Z_L \end{bmatrix}}_{S_L} + \underbrace{\begin{bmatrix} 1 & 0 & s_y \\ 0 & 1 & -s_x \\ 0 & 0 & 1 \end{bmatrix} \begin{bmatrix} G_u \\ G_v \\ G_\theta \end{bmatrix}}_{\Delta_{L0}}$$

$$\underbrace{\begin{bmatrix} u_{RL} \\ v_{RL} \\ \theta_{RL} \end{bmatrix}}_{\Delta_{RL}} = \underbrace{\begin{bmatrix} G_{11} & -G_{12} & -G_{16} \\ -G_{21} & G_{22} & G_{26} \\ -G_{61} & G_{62} & G_{66} \end{bmatrix}}_{f_{RR}} \underbrace{\begin{bmatrix} U_R \\ V_R \\ Z_R \end{bmatrix}}_{S_R} + \underbrace{\begin{bmatrix} -1 & 0 & s_y \\ 0 & -1 & -s_x \\ 0 & 0 & -1 \end{bmatrix} \begin{bmatrix} F_u \\ F_v \\ F_\theta \end{bmatrix}}_{\Delta_{R0}}$$

where Δ_{LR}, Δ_{RL} are the displacement vectors at L, R, respectively, and Δ_{L0}, Δ_{R0} are the respective flexibility load functions in terms of F, G defined in (2.02-1).

3.06 FREE CURVED BAR OF ORDER TWO, END FLEXIBILITY MATRIX EQUATIONS IN 0 SYSTEM

(1) Systems considered in Figs. 3.06-1, 3.06-2 are free curved cantilever bars of constant section. Their cross-sectional properties and elastic constants are defined in (2.13-1). Also, s_x, s_y = segments (m), x, x', y, y' = coordinates (m), a_x, a_y, b_x, b_y = segments (m), $x' = s_x - x$, $y' = s_y - y$. The symbols and signs are given in (3.02).

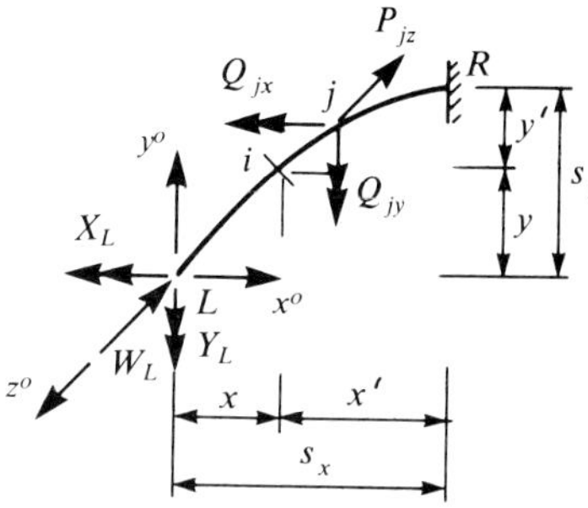

Fig. 3.06-1

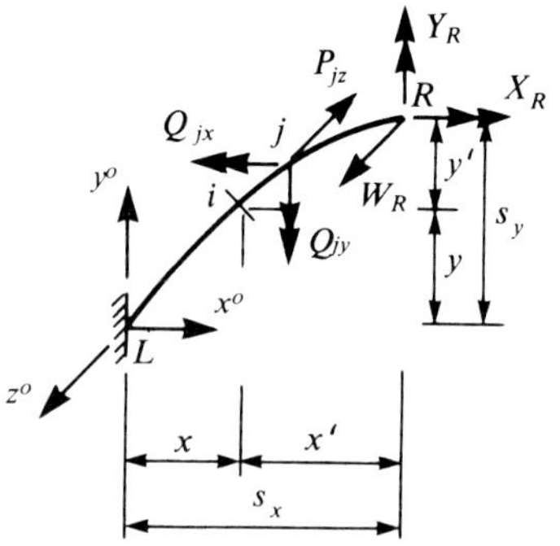

Fig. 3.06-2

(2) Flexibility coefficients used in the construction of f_{LL} and f_{RR} below are

$$F_{33} = \int_L^R (y^2\Lambda_{XX} - 2_{xy}\Lambda_{XY} + x^2\Lambda_{YY} + \Lambda_{WW}) \qquad F_{44} = \int_L^R \Lambda_{XX} \qquad F_{55} = \int_L^R \Lambda_{YY}$$

$$F_{34} = F_{43} = \int_L^R (y\Lambda_{XX} - x\Lambda_{XY}) \qquad F_{35} = F_{53} = \int_L^R (x\Lambda_{YY} - y\Lambda_{XY}) \qquad F_{45} = F_{54} = \int_L^R \Lambda_{XY}$$

$$G_{33} = \int_L^R (y'^2\Lambda_{XX} - 2x'y'\Lambda_{XY} + x'^2\Lambda_{YY} + \Lambda_{WW}) \qquad G_{44} = F_{44} \qquad G_{55} = F_{55}$$

$$G_{34} = G_{43} = \int_L^R (y'\Lambda_{XX} - x'\Lambda_{XY}) \qquad G_{35} = G_{53} = \int_L^R (x'\Lambda_{YY} - y'\Lambda_{XY}) \qquad G_{45} = G_{54} = F_{45}$$

where Λ_{WW}, Λ_{XX}, Λ_{XY}, Λ_{YY} are the elemental flexibilities (1.15-2). For symmetrical curves, $F_{33} = G_{33}$, $F_{34} = G_{34}, \ldots, s_y = 0$.

(3) Flexibility matrix equations (3.02-2) are

$$\underbrace{\begin{bmatrix} w_{LR} \\ \phi_{LR} \\ \psi_{LR} \end{bmatrix}}_{\Delta_{LR}} = \underbrace{\begin{bmatrix} F_{33} & -F_{34} & F_{35} \\ -F_{43} & F_{44} & F_{45} \\ F_{53} & F_{53} & F_{55} \end{bmatrix}}_{f_{LL}} \underbrace{\begin{bmatrix} W_L \\ X_L \\ Y_L \end{bmatrix}}_{S_L} + \underbrace{\begin{bmatrix} 1 & -s_y & s_x \\ 0 & 1 & 0 \\ 0 & 0 & 1 \end{bmatrix} \begin{bmatrix} G_w \\ G_\phi \\ G_\psi \end{bmatrix}}_{\Delta L_0}$$

$$\underbrace{\begin{bmatrix} w_{RL} \\ \phi_{RL} \\ \psi_{RL} \end{bmatrix}}_{\Delta_{RL}} = \underbrace{\begin{bmatrix} G_{33} & G_{34} & -G_{35} \\ G_{43} & G_{44} & G_{45} \\ -G_{53} & G_{54} & G_{55} \end{bmatrix}}_{f_{RR}} \underbrace{\begin{bmatrix} W_R \\ X_R \\ Y_R \end{bmatrix}}_{S_R} + \underbrace{\begin{bmatrix} -1 & -s_y & s_x \\ 0 & -1 & 0 \\ 0 & 0 & -1 \end{bmatrix} \begin{bmatrix} F_w \\ F_\phi \\ F_\psi \end{bmatrix}}_{\Delta_{R0}}$$

where Δ_{LR}, Δ_{RL} are the displacement vectors at L, R, respectively, and Δ_{L0}, Δ_{R0} are the respective flexibility load functions in terms of F, G defined in (2.02-1).

3.07 FREE CURVED BAR OF ORDER ONE, NEUTRAL-POINT FLEXIBILITY MATRIX EQUATIONS IN 0 SYSTEM

(1) Concept. The transfer of the end stress-resultant vector from its point of application (L or R) to the neutral point C leads to the diagonalization of the end flexibility matrix. The transfer is accomplished by means of a rigid arm (LC or RC), and the neutral point (elastic center) is defined as the point of application of the stress-resultant vector S_C, each component of which produces one displacement in its own direction only.

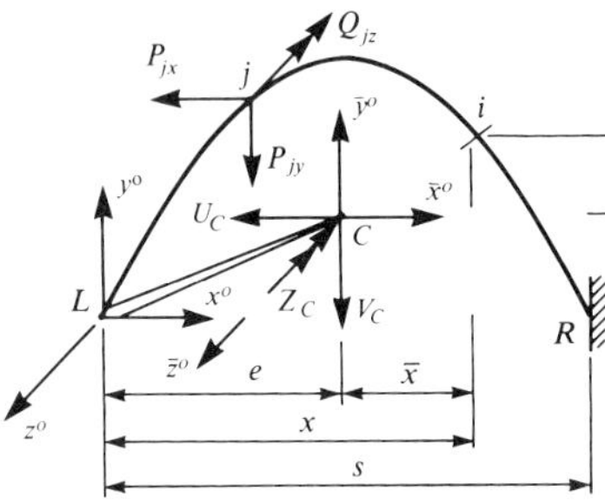

Fig. 3.07-1

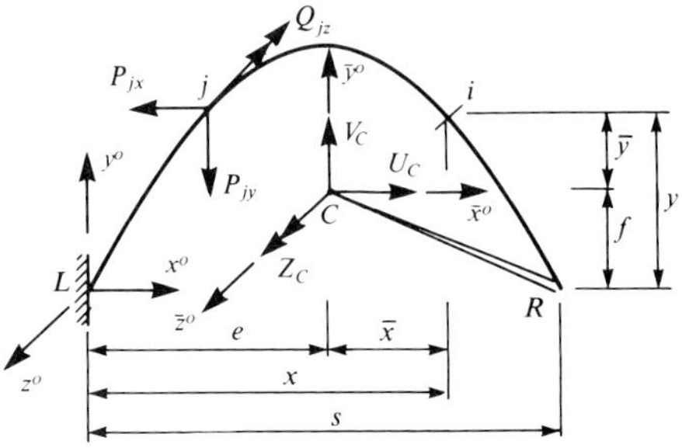

Fig. 3.07-2

(2) For a symmetrical bar such as those shown in Figs. 3.07-1, 3.07-2, the coordinates of the neutral point C are

$$e = \tfrac{1}{2}s \qquad f = \frac{\int_L^R y\Lambda_{ZZ}}{\int_L^R \Lambda_{ZZ}}$$

where s = span (m), x, y = coordinates measured from L (m), and the remaining symbols are those of (3.05).

(3) Neutral-point flexibility coefficients by (3.02-3) are

$$F_1 = \int_L^R (\bar{y}^2\Lambda_{ZZ} + \Lambda_{UU})$$

$$F_2 = \int_L^R (\bar{x}^2\Lambda_{ZZ} + \Lambda_{VV}) \qquad F_6 = \int_L^R \Lambda_{ZZ}$$

where $\bar{x}$, $\bar{y}$ = coordinates measured from C (m) and, again, the remaining symbols are those of (3.05).

(4) Neutral-point flexibility matrix equations (3.02-2) are then

$$\underbrace{\begin{bmatrix} u_{CR} \\ v_{CR} \\ \theta_{CR} \end{bmatrix}}_{\Delta_{CR}} = \underbrace{\begin{bmatrix} F_1 & 0 & 0 \\ 0 & F_2 & 0 \\ 0 & 0 & F_6 \end{bmatrix}}_{f_{CC}} \underbrace{\begin{bmatrix} U_C \\ V_C \\ Z_C \end{bmatrix}}_{S_C} + \underbrace{\begin{bmatrix} 1 & 0 & f \\ 0 & 1 & -e \\ 0 & 0 & 1 \end{bmatrix} \begin{bmatrix} G_u \\ G_v \\ G_\theta \end{bmatrix}}_{\Delta_{C0,R}}$$

$$\underbrace{\begin{bmatrix} u_{CL} \\ v_{CL} \\ \theta_{CL} \end{bmatrix}}_{\Delta_{CL}} = \underbrace{\begin{bmatrix} F_1 & 0 & 0 \\ 0 & F_2 & 0 \\ 0 & 0 & F_6 \end{bmatrix}}_{f_{CC}} \underbrace{\begin{bmatrix} U_C \\ V_C \\ Z_C \end{bmatrix}}_{S_C} + \underbrace{\begin{bmatrix} -1 & 0 & f \\ 0 & -1 & -e \\ 0 & 0 & -1 \end{bmatrix} \begin{bmatrix} F_u \\ F_v \\ F_\theta \end{bmatrix}}_{\Delta_{C0,L}}$$

where Δ_{CR}, Δ_{CL} are the displacement vectors at the neutral point C in Figs. 3.07-1, 3.07-2, respectively, and $\Delta_{C0,R}$, $\Delta_{C0,L}$ are the respective neutral-point flexibility load functions in terms of F, G defined in (2.02-1).

3.08 FREE CURVED BAR OF ORDER TWO, NEUTRAL-POINT FLEXIBILITY MATRIX EQUATIONS IN 0 SYSTEM

(1) Concept. As in (3.07), the transfer of the stress-resultant vector from its point of application (L or R) to the neutral point D leads again to the diagonalization of the end flexibility matrix. The definition of the neutral point D is the same as in (3.07), but the neutral point C of order one and the neutral point D of order two are, in general, two different points in the plane $\bar{x}\bar{y}$.

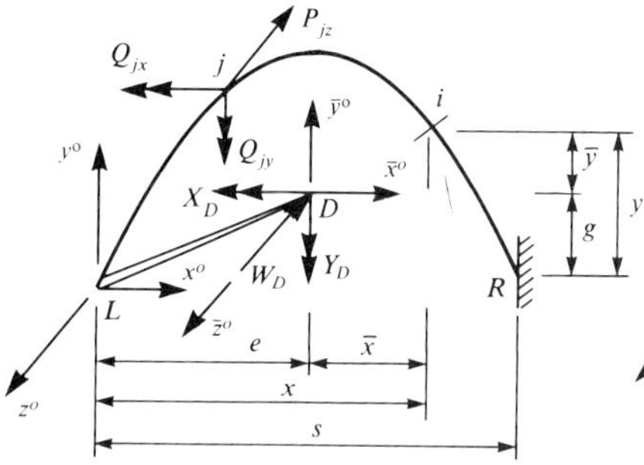

Fig. 3.08-1

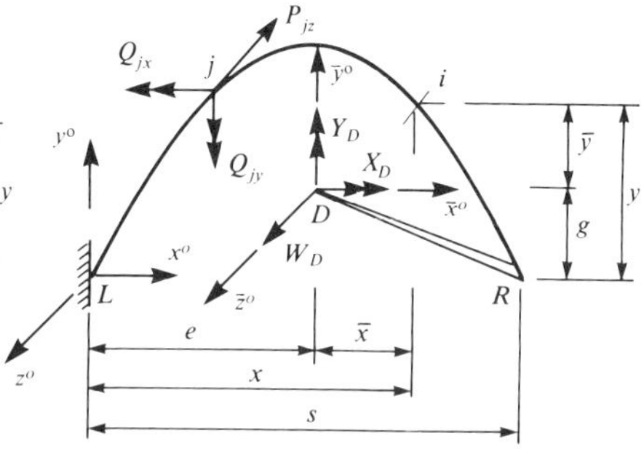

Fig. 3.08-2

(2) For a symmetrical bar such as those in Figs. 3.08-1, 3.08-2, the coordinates of the neutral point D are

$$e = \tfrac{1}{2}s \qquad g = \frac{\int_L^R (y\Lambda_{XX} - x\Lambda_{XY})}{\int_L^R \Lambda_{XX}}$$

where s = span (m), x, y = coordinates measured from L (m), and the remaining symbols are those of (3.06).

(3) Neutral-point flexibility coefficients by (3.02-3) are

$$F_3 = \int_L^R (\bar{y}^2\Lambda_{XX} - 2\bar{x}\bar{y}\Lambda_{XY} + \bar{x}^2\Lambda_{YY} + \Lambda_{WW})$$

$$F_4 = \int_L^R \Lambda_{XX} \qquad F_5 = \int_L^R \Lambda_{YY}$$

where $\bar{x}$, $\bar{y}$ = coordinates measured from D (m) and, again, the remaining symbols are those of (3.06).

(4) Neutral-point flexibility matrix equations (3.02-2) are then

$$\underbrace{\begin{bmatrix} w_{DR} \\ \phi_{DR} \\ \psi_{DR} \end{bmatrix}}_{\Delta_{DR}} = \underbrace{\begin{bmatrix} F_3 & 0 & 0 \\ 0 & F_4 & 0 \\ 0 & 0 & F_5 \end{bmatrix}}_{f_{DD}} \underbrace{\begin{bmatrix} W_D \\ X_D \\ Y_D \end{bmatrix}}_{S_D} + \underbrace{\begin{bmatrix} 1 & -g & e \\ 0 & 1 & 0 \\ 0 & 0 & 1 \end{bmatrix}\begin{bmatrix} G_w \\ G_\phi \\ G_\psi \end{bmatrix}}_{\Delta_{D0,R}}$$

$$\underbrace{\begin{bmatrix} w_{DL} \\ \phi_{DL} \\ \psi_{DL} \end{bmatrix}}_{\Delta_{DL}} = \underbrace{\begin{bmatrix} F_3 & 0 & 0 \\ 0 & F_4 & 0 \\ 0 & 0 & F_5 \end{bmatrix}}_{f_{DD}} \underbrace{\begin{bmatrix} W_D \\ X_D \\ Y_D \end{bmatrix}}_{S_D} + \underbrace{\begin{bmatrix} -1 & -g & e \\ 0 & -1 & 0 \\ 0 & 0 & -1 \end{bmatrix}\begin{bmatrix} F_w \\ F_\phi \\ F_\psi \end{bmatrix}}_{\Delta_{D0,L}}$$

where Δ_{DR}, Δ_{DL} are the displacement vectors at D in Figs. 3.08-1, 3.08-2, respectively, and $\Delta_{D0,R}$, $\Delta_{D0,L}$ are the respective neutral-point flexibility load functions in terms of F, G defined in (2.02-1).

3.09 INTERACTIVE SYSTEMS OF ORDER THREE, FLEXIBILITY COEFFICIENTS IN S SYSTEM

(1) Systems considered are the bars encased in elastic foundation (Fig. 2.16-1) and the beam-columns encased in the same foundation (Fig. 2.16-3). Their definitions are given in (2.16), and other properties are described in (3.01-1).

s = span (m) x, x' = coordinates (m) a, b = segments (m)

(2) Relations. According to (3.02-8), the flexibility coefficients which form the submatrices f_{LL}, f_{LR}, f_{RL}, f_{RR} are derived from the submatrices t_{LR}, c_{LR}, t_{RL}, c_{RL} of the transport matrices in (2.18). A close inspection of the transport coefficients in (2.18) reveals the following identities:

$$T_{Uu} = T_{uU} \qquad T_{X\phi} = T_{\phi X}$$
$$T_{V\theta} = T_{Zv} \qquad T_{W\psi} = T_{Yw}$$

where the first row of identities is based on the transpose relations and the last row is apparent from the reciprocal theorem (Ref. 17).

(3) Flexibility coefficients derived as indicated above are

$F_{1x} = \dfrac{T_{XX}}{T_{X\phi}}$	$F_{2x} = \dfrac{1}{T_{X\phi}}$
$F_{1y} = (T_{YW}T_{W\psi} - T_{WW}T_{Y\psi})D_y$ $F_{2y} = (T_{YY}T_{W\psi} - T_{WY}T_{Y\psi})D_y$ $F_{3y} = (T_{YY}T_{Ww} - T_{WY}T_{W\psi})D_y$	$F_{4y} = T_{Y\psi}D_y$ $F_{5y} = T_{W\psi}D_y$ $F_{6y} = T_{Ww}D_v$

$F_{1n} = \dfrac{T_{UU}}{T_{Uu}}$	$F_{2n} = \dfrac{1}{T_{Uu}}$
$F_{1z} = (T_{ZV}T_{V\theta} - T_{VV}T_{Z\theta})D_z$ $F_{2z} = (T_{ZZ}T_{V\theta} - T_{VZ}T_{Z\theta})D_z$ $F_{3z} = (T_{ZZ}T_{Vv} - T_{VZ}T_{V\theta})D_z$	$F_{4z} = T_{Z\theta}D_z$ $F_{5z} = T_{V\theta}D_z$ $F_{6z} = T_{Vv}D_z$

where the T coefficients are the transport coefficients introduced in symbolic form in (2.16-4) and given in particular form in Chaps. 8–10 and 13–15. Also,

$$D_y = \frac{1}{T_{Ww}T_{Y\psi} - T_{W\psi}^2} \qquad D_z = \frac{1}{T_{Vv}T_{Z\theta} - T_{V\theta}^2}$$

The subscripts y and z identify the normal to the plane of action, so that $F_{1y}, F_{2y}, \ldots, F_{6y}$ are the end flexibilities of order two and $F_{1z}, F_{2z}, \ldots, F_{6z}$ are the end flexibilities of order one, which are used for the construction of the flexibility matrices of order three in (3.10).

(4) End flexibility matrix equations, derived by means of relations (3.02-7, 8, 9), are shown in (3.10) in terms of the flexibility coefficients listed above and the transport load functions (2.17-1, 2). The physical interpretation of the flexibility coefficients is shown in Fig. 3.09-1.

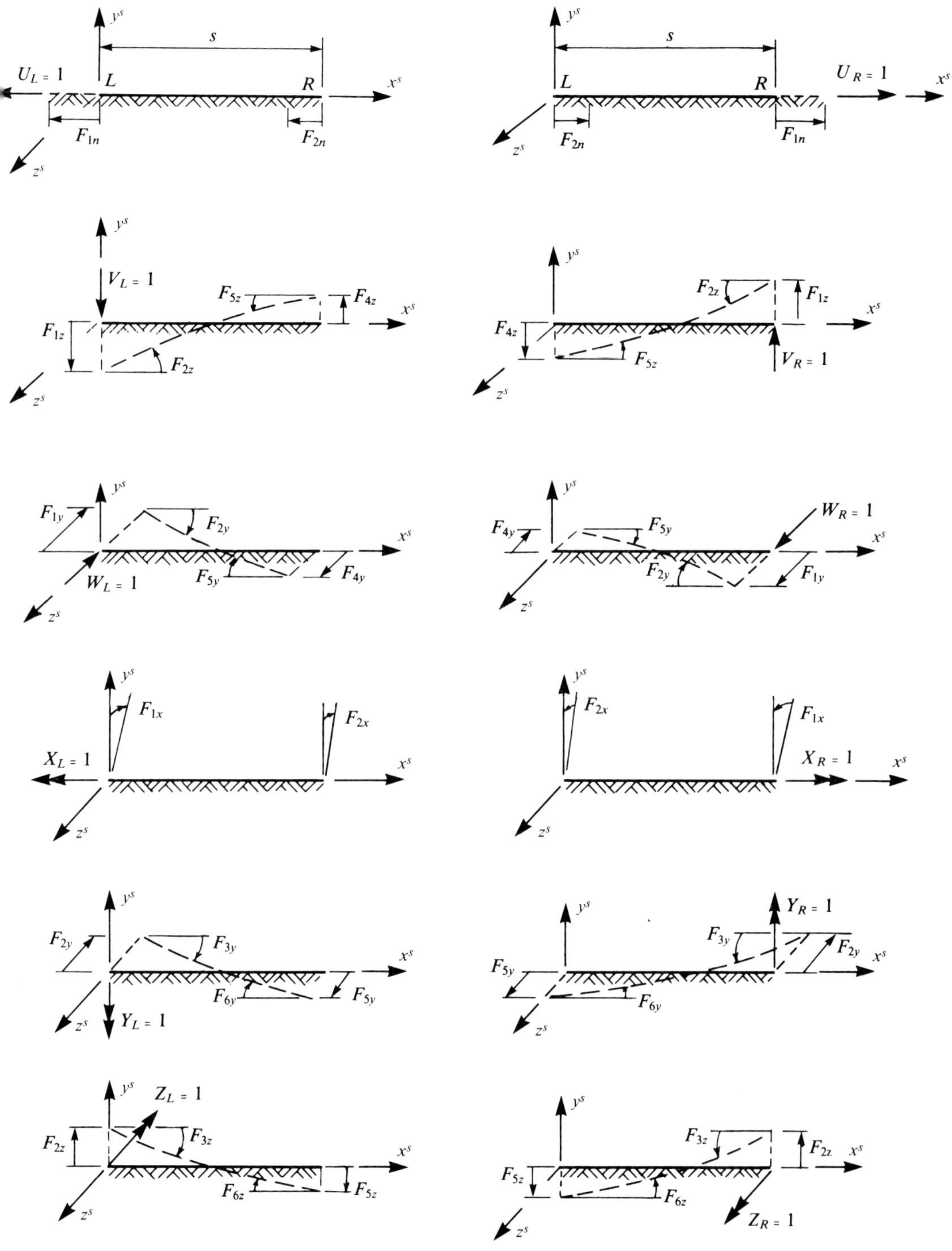
ys
s
UL = 1
L
R
xs
F1n
F2n
zs
ys
s
L
R
UR = 1
xs
zs
F2n
F1n
ys
VL = 1
F5z
F4z
xs
F1z
F2z
zs
ys
F2z
F1z
xs
F4z
F5z
VR = 1
zs
ys
F1y
F2y
xs
WL = 1
F5y
F4y
zs
ys
WR = 1
F4y
F5y
xs
F2y
F1y
zs
ys
F1x
F2x
XL = 1
xs
zs
ys
F2x
F1x
XR = 1
xs
zs
ys
F2y
F3y
xs
F6y
F5y
zs
YL = 1
ys
YR = 1
F3y
F2y
xs
F5y
F6y
zs
ys
ZL = 1
F2z
F3z
xs
F6z
F5z
zs
ys
F3z
F2z
xs
F5z
F6z
ZR = 1
zs

Fig. 3.09-1

3.10 INTERACTIVE SYSTEMS OF ORDER THREE, END FLEXIBILITY MATRIX EQUATIONS IN S SYSTEM

Notation (3.02, 3.09) Signs (3.02) Coefficients (3.09)

(1) End flexibility matrix equations in (3.02-7) can be assembled into one matrix equation as

$$\begin{bmatrix} \Delta_{LR} \\ \Delta_{RL} \end{bmatrix} = \underbrace{\begin{bmatrix} f_{LL} & f_{LR} \\ f_{RL} & f_{RR} \end{bmatrix}}_{f} \begin{bmatrix} S_L \\ S_R \end{bmatrix} + \begin{bmatrix} \Delta_{L0} \\ \Delta_{R0} \end{bmatrix}$$

in which

$$f = \left[\begin{array}{ccc|ccc|ccc|ccc}
F_{1n} & 0 & 0 & 0 & 0 & 0 & -F_{2n} & 0 & 0 & 0 & 0 & 0 \\
0 & F_{1z} & 0 & 0 & 0 & -F_{2z} & 0 & F_{4z} & 0 & 0 & 0 & F_{5z} \\
0 & 0 & F_{1y} & 0 & F_{2y} & 0 & 0 & 0 & F_{4y} & 0 & -F_{5y} & 0 \\
\hline
0 & 0 & 0 & F_{1x} & 0 & 0 & 0 & 0 & 0 & -F_{2x} & 0 & 0 \\
0 & 0 & F_{2y} & 0 & F_{3y} & 0 & 0 & 0 & F_{5y} & 0 & -F_{6y} & 0 \\
0 & -F_{2z} & 0 & 0 & 0 & F_{3z} & 0 & -F_{5z} & 0 & 0 & 0 & -F_{6z} \\
\hline
-F_{2n} & 0 & 0 & 0 & 0 & 0 & F_{1n} & 0 & 0 & 0 & 0 & 0 \\
0 & F_{4z} & 0 & 0 & 0 & -F_{5z} & 0 & F_{1z} & 0 & 0 & 0 & F_{2z} \\
0 & 0 & F_{4y} & 0 & F_{5y} & 0 & 0 & 0 & F_{1y} & 0 & -F_{2y} & 0 \\
\hline
0 & 0 & 0 & -F_{2x} & 0 & 0 & 0 & 0 & 0 & F_{1x} & 0 & 0 \\
0 & 0 & -F_{5y} & 0 & -F_{6y} & 0 & 0 & 0 & -F_{2y} & 0 & F_{3y} & 0 \\
0 & F_{5z} & 0 & 0 & 0 & -F_{6z} & 0 & F_{2z} & 0 & 0 & 0 & F_{3z}
\end{array}\right]$$

where Δ_{LR}, Δ_{RL} and S_L, S_R are the end displacement vectors and the end stress-resultant vectors defined in (3.02-2).

(2) Flexibility load functions forming the last column matrix in (1) above are derived by (3.02-9).

$$\Delta_{L0} = \begin{bmatrix} u_{L0} \\ v_{L0} \\ \theta_{L0} \\ \hline \phi_{L0} \\ \psi_{L0} \\ \theta_{L0} \end{bmatrix} = \left[\begin{array}{ccc|ccc} F_{2n} & 0 & 0 & 0 & 0 & 0 \\ 0 & -F_{4z} & 0 & 0 & 0 & -F_{5z} \\ 0 & 0 & -F_{4y} & 0 & F_{5y} & 0 \\ \hline 0 & 0 & 0 & F_{2x} & 0 & 0 \\ 0 & 0 & -F_{5y} & 0 & F_{6y} & 0 \\ 0 & F_{5z} & 0 & 0 & 0 & F_{6z} \end{array}\right] \begin{bmatrix} G_U \\ G_V \\ G_W \\ \hline G_X \\ G_Y \\ G_Z \end{bmatrix} = -f_{LR} G_S$$

$$\Delta_{R0} = \begin{bmatrix} u_{R0} \\ v_{R0} \\ w_{R0} \\ \hline \phi_{R0} \\ \psi_{R0} \\ \theta_{R0} \end{bmatrix} = \left[\begin{array}{ccc|ccc} F_{2n} & 0 & 0 & 0 & 0 & 0 \\ 0 & -F_{4z} & 0 & 0 & 0 & F_{5z} \\ 0 & 0 & -F_{4y} & 0 & -F_{5y} & 0 \\ \hline 0 & 0 & 0 & F_{2x} & 0 & 0 \\ 0 & 0 & F_{5y} & 0 & F_{6y} & 0 \\ 0 & -F_{5z} & 0 & 0 & 0 & F_{6z} \end{array}\right] \begin{bmatrix} F_U \\ F_V \\ F_W \\ \hline F_Z \\ F_X \\ F_Y \end{bmatrix} = -f_{RL} F_S$$

where G_U, G_V, . . . , G_Z and F_U, F_V, . . . , F_Z are the transport load functions introduced in symbolic form in (2.18) and given in particular forms in Chaps. 8–10 and 13–15.

3.11 STIFFNESS MATRIX EQUATIONS, BASIC RELATIONS IN S SYSTEM

(1) End stiffness of a bar is defined as the end reaction produced by a unit displacement (linear or angular). Since the unit displacement can be applied along three orthogonal axes at each end and each unit displacement can (in general) produce six reactions at each end, the derivation of the stiffness matrix requires the calculation of 144 coefficients. If the derivation is carried out in the member system (principal stiffnesses) and the reciprocal relations are observed, the total number of distinct coefficients reduces to a maximum of 32. These coefficients can be obtained from the equations of the elastic curve or directly from the transport matrix equations introduced in (2.02-1). The latter approach is used below (Ref. 5, p.184).

(2) Sign conventions of the transport method and the stiffness method are given in (1.22), and their comparison shows that the end displacement vectors are identical, whereas the end forces and moments are related as

$$S_{LR} = -S_L \qquad S_{RL} = S_R$$

(3) Stiffness matrix equations of free and interactive bars are derived by their definition from the basic segment, which is the fixed-end bar. The lower submatrix equations in (2.02-1) in terms of the sign convention discussed in (2) above are

$$\Delta_L = F_\Delta + d_{LR} S_{RL} + t_{RL}^{)T} \Delta_R \qquad \Delta_R = G_\Delta - d_{RL} S_{LR} + t_{LR}^{)T} \Delta_L$$

and yield in reversed order

$$S_{LR} = k_{LL}\Delta_L + k_{LR}\Delta_R + S_{L0} \qquad S_{RL} = k_{RL}\Delta_L + k_{RR}\Delta_R + S_{R0}$$

where

$$S_{LR} = \{U_{LR}, V_{LR}, W_{LR}, X_{LR}, Y_{LR}, Z_{LR}\} \qquad S_{RL} = \{U_{RL}, V_{RL}, W_{RL}, X_{RL}, Y_{RL}, Z_{RL}\}$$

$$\Delta_L = \{u_L, v_L, w_L, \phi_L, \psi_L, \theta_L\} \qquad \Delta_R = \{u_R, v_R, w_R, \phi_R, \psi_R, \theta_R\}$$

are the reactions and displacements of the respective ends.

(4) Stiffness submatrices in (3) above are

$$k_{LL} = d_{RL}^{)-1} t_{LR}^{)T} \qquad k_{LR} = -d_{RL}^{)-1} \qquad k_{RL} = d_{LR}^{)-1} \qquad k_{RR} = -d_{LR}^{)-1} t_{RL}^{)T}$$

where t_{LR}, t_{RL}, d_{LR}, d_{RL} are the submatrices of the transport matrices T_{LR}, T_{RL} in (2.02-1).

(5) Load functions in (3) above are

$$S_{L0} = d_{RL}^{)-1} G_\Delta = -k_{LR} G_\Delta \qquad S_{R0} = -d_{LR}^{)-1} F_\Delta = -k_{RL} F_\Delta$$

where

$$S_{L0} = \{U_{L0}, V_{L0}, W_{L0}, X_{L0}, Y_{L0}, Z_{L0}\} \qquad S_{R0} = \{U_{R0}, V_{R0}, W_{R0}, X_{R0}, Y_{R0}, Z_{R0}\}$$

are the end reactions of a fixed-end bar due to loads and/or other causes and

$$F_\Delta = \{F_u, F_v, F_w, F_\phi, F_y, F_\theta\} \qquad G_\Delta = \{G_u, G_v, G_w, G_\phi, G_y, G_\theta\}$$

are the load functions of the transport matrices T_{LR}, T_{RL} in (2.02-1).

(6) Evaluation of stiffness integrals in (3) depends on the geometry of the bar. In straight bars of constant sections and in planar circular and parabolic bars of constant sections, the stiffness coefficients can be expressed in closed form. In other cases, numerical methods of integration must be employed.

(7) Symmetrical case. If the bar is symmetrical with respect to a plane normal to the axis LR in the S system and the displacement vectors are symmetrical with respect to the same plane, so that

$$u_L = -u_R \qquad v_L = v_R \qquad w_L = w_R \qquad \phi_L = \phi_R \qquad \psi_L = -\psi_R \qquad \theta_L = -\theta_R$$

then the stiffness matrix equation in (3) of the left end reduces to

$$S_{LR} = (k_{LL} - k_{LR}i)\Delta_L + S_{L0}$$

where $i = \text{Diag}\,[1, -1, -1, -1, +1, +1]$.

(8) Antisymmetrical case. If the bar is symmetrical with respect to the same plane and the end displacement vectors are antisymmetrical with respect to the same plane, so that

$$u_L = u_R \qquad v_L = -v_R \qquad w_L = -w_R \qquad \phi_L = -\phi_R \qquad \psi_L = \psi_R \qquad \theta_L = \theta_R$$

then the stiffness matrix equation in (3) of the left end reduces to

$$S_{LR} = (k_{LL} + k_{LR}i)\Delta_L + S_{L0}$$

where i is the same as in (7) above.

(9) Properties. The free- and interactive-bar stiffness matrices and load functions possess the following characteristics:

(*a*) Transpose relations

$$k_{LL} = k_{LL}^{T} \qquad k_{LR} = k_{RL}^{T} \qquad k_{RL} = k_{LR}^{T} \qquad k_{RR} = k_{RR}^{T}$$

which show that the bar stiffness matrix

$$k = \begin{bmatrix} k_{LL} & k_{LR} \\ k_{RL} & k_{RR} \end{bmatrix}$$

is a symmetrical matrix.

(*b*) Shift relations

$$k_{LL} = t_{LR}k_{RR}t_{LR}^{T} \qquad k_{RR} = t_{RL}k_{LL}t_{RL}^{T}$$

$$k_{LR} = -k_{LL}t_{RL}^{T} = -t_{LR}k_{RR} \qquad k_{RL} = -k_{RR}t_{LR}^{T} = -t_{RL}k_{LL}$$

$$S_{L0} = -t_{LR}S_{R0} - F_S \qquad S_{R0} = -t_{RL}S_{L0} + G_S$$

where F_S and G_S are the transport load functions in (2.02-1).

(*c*) Conversions

$$k_{LL} = t_{LR}f_{RR}^{-1}t_{LR}^{T} = f_{LL}^{-1} \qquad k_{RR} = t_{RL}f_{LL}^{-1}t_{RL}^{T} = f_{RR}^{-1}$$

$$k_{LR} = -f_{LL}^{-1}t_{RL}^{T} = -t_{LR}f_{RR}^{-1} \qquad k_{RL} = -f_{RR}^{-1}t_{LR}^{T} = -t_{RL}f_{LL}^{-1}$$

$$S_{L0} = k_{LL}\Delta_{L0} \qquad S_{R0} = -k_{RR}\Delta_{R0}$$

$$f_{LL} = t_{RL}^{T}k_{RR}^{-1}t_{RL} = k_{LL}^{-1} \qquad f_{RR} = t_{LR}^{T}k_{LL}^{-1}t_{RL} = k_{RR}^{-1}$$

$$\Delta_{L0} = f_{LL}S_{L0} \qquad \Delta_{R0} = f_{RR}S_{R0}$$

which show the relations between the stiffness coefficients and stiffness load functions (3.11-4, 5) and between the flexibility coefficients and flexibility load functions (3.02-3, 4).

3.12 FREE STRAIGHT BAR OF ORDER THREE, STIFFNESS COEFFICIENTS IN S SYSTEM

(1) System considered in Fig. 3.12-1 is a free straight bar of constant cross section. The section properties and elastic constants are defined in (1.14). Also, s = span (m), x, x' = coordinates (m), a, b = segments (m). Positive forces and moments are shown in Fig. 3.12-1, and positive displacements are shown in Fig. 3.12-2.

(2) Free-body diagram

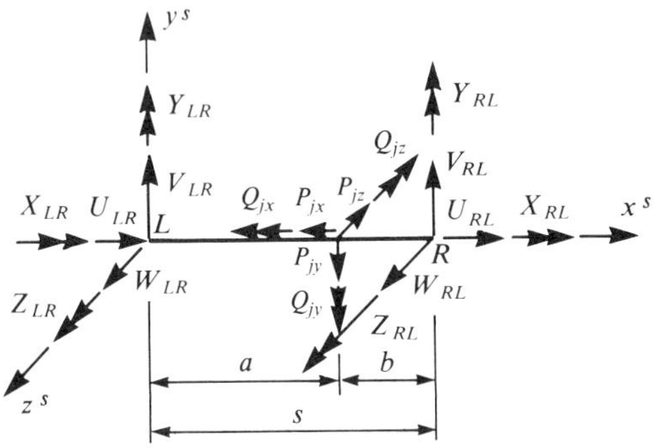

Fig. 3.12-1

(3) Elastic curve

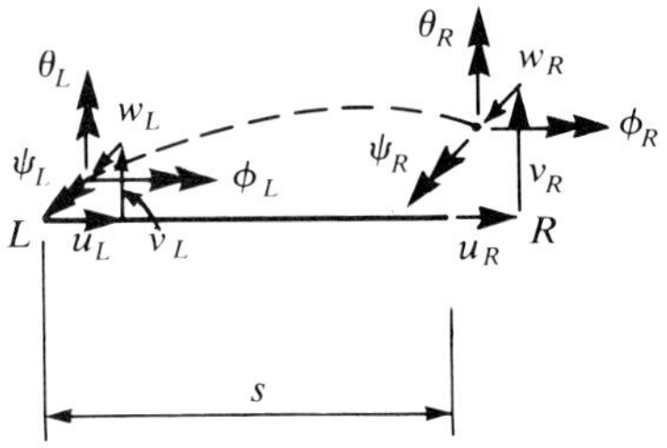

Fig. 3.12-2

(4) Stiffness coefficients which form the submatrices k_{LL}, k_{LR}, k_{LR}, k_{RR} derived from the submatrices d_{LR}, d_{RL}, $t_{LR}^{)T}$, $t_{RL}^{)T}$ of the transport matrices in (2.04) are

$K_{1n} = \dfrac{EA_x}{s}$	$K_{1x} = \dfrac{GI_x}{s}$
$K_{1y} = \dfrac{12}{\alpha s^3} EI_y$	$K_{1z} = \dfrac{12}{\beta s^3} EI_z$
$K_{2y} = \dfrac{6}{\alpha s^2} EI_y$	$K_{2z} = \dfrac{6}{\beta s^2} EI_z$
$K_{3y} = \dfrac{3+\alpha}{\alpha s} EI_y$	$K_{3z} = \dfrac{3+\beta}{\beta s} EI_z$
$K_{4y} = \dfrac{3-\alpha}{\alpha s} EI_y$	$K_{4z} = \dfrac{3-\beta}{\beta s} EI_z$

where $\alpha = 1 + \dfrac{12EI_y}{GA_z s^2}$ $\qquad \beta = 1 + \dfrac{12EI_z}{GA_y s^2}$

and in slender bars $\alpha \cong \beta \cong 1$.

(3) End stiffness matrix equations derived by means of relations (3.11-2, 3, 4, 5) are shown in symbolic form below.

$$\begin{bmatrix} S_{LR} \\ S_{RL} \end{bmatrix} = \begin{bmatrix} k_{LL} & k_{LR} \\ k_{RL} & k_{RR} \end{bmatrix} \begin{bmatrix} \Delta_L \\ \Delta_R \end{bmatrix} + \begin{bmatrix} S_{L0} \\ S_{R0} \end{bmatrix}$$

Their full form is given in (3.13). The physical interpretation of the stiffness coefficients is shown in Fig. 3.12-3.

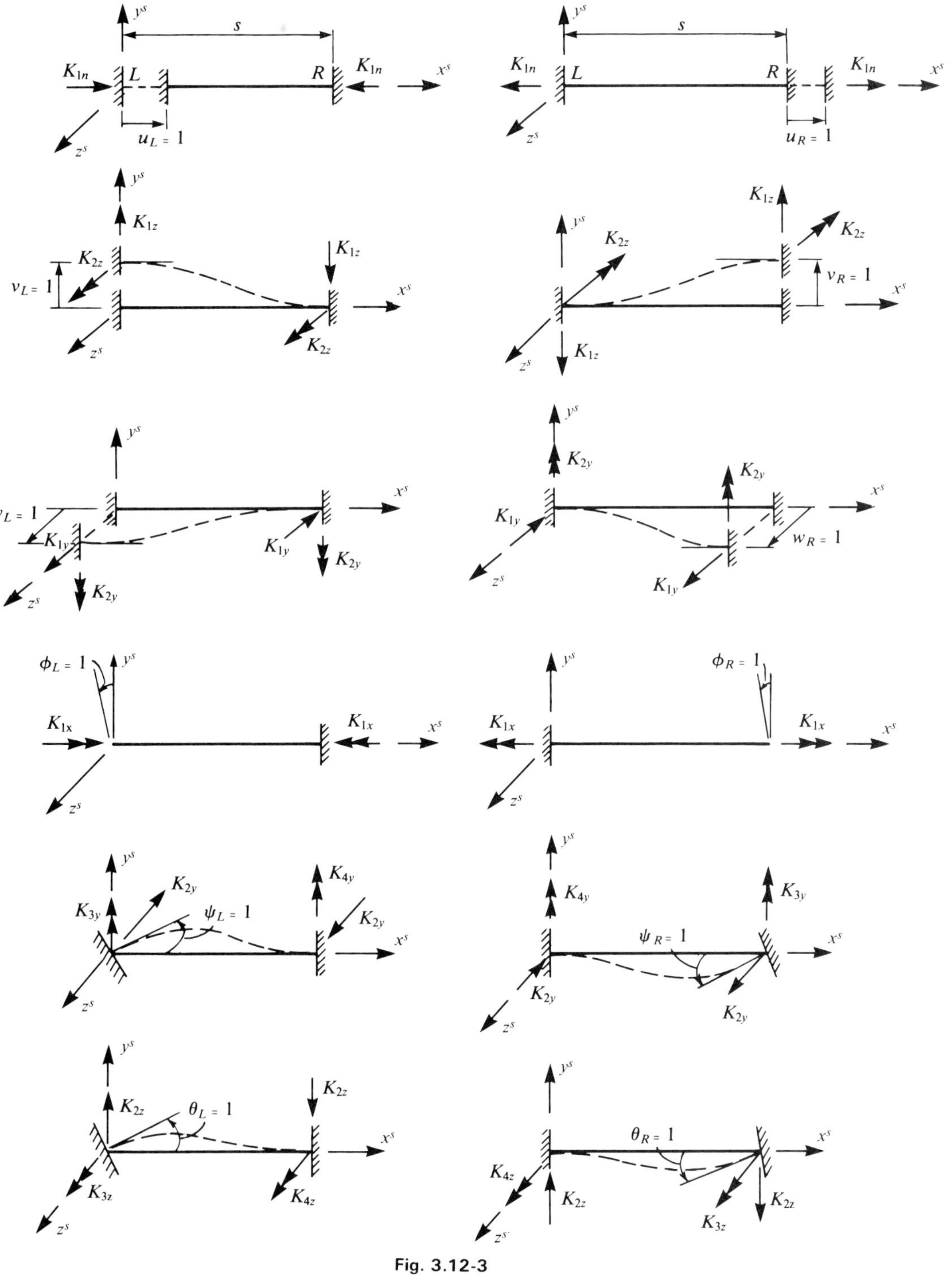
s
L
R
K1n
xs
ys
zs
uL = 1
uR = 1
K1z
K2z
vL = 1
vR = 1
K1y
K2y
wL = 1
wR = 1
φL = 1
φR = 1
K1x
K3y
K4y
ψL = 1
ψR = 1
θL = 1
θR = 1
K3z
K4z

Fig. 3.12-3

3.13 FREE STRAIGHT BAR OF ORDER THREE, END STIFFNESS MATRIX EQUATIONS IN S SYSTEM

Notation (3.11, 3.12) Signs (3.11) Coefficients (3.12)

(1) Stiffness matrix equations in (3.11-3) are in full form

$$
\left[\begin{array}{c} U_{LR} \\ V_{LR} \\ W_{LR} \\ \hline X_{LR} \\ Y_{LR} \\ Z_{LR} \\ \hline U_{RL} \\ V_{RL} \\ W_{RL} \\ \hline X_{RL} \\ Y_{RL} \\ Z_{RL} \end{array}\right]
=
\left[\begin{array}{ccc|ccc|ccc|ccc}
K_{1n} & 0 & 0 & 0 & 0 & 0 & -K_{1n} & 0 & 0 & 0 & 0 & 0 \\
0 & K_{1z} & 0 & 0 & 0 & K_{2z} & 0 & -K_{1z} & 0 & 0 & 0 & K_{2z} \\
0 & 0 & K_{1y} & 0 & -K_{2y} & 0 & 0 & 0 & -K_{1y} & 0 & -K_{2y} & 0 \\
\hline
0 & 0 & 0 & K_{1x} & 0 & 0 & 0 & 0 & 0 & -K_{1x} & 0 & 0 \\
0 & 0 & -K_{2y} & 0 & K_{3y} & 0 & 0 & 0 & K_{2y} & 0 & K_{4y} & 0 \\
0 & K_{2z} & 0 & 0 & 0 & K_{3z} & 0 & -K_{2z} & 0 & 0 & 0 & K_{4z} \\
\hline
-K_{1n} & 0 & 0 & 0 & 0 & 0 & K_{1n} & 0 & 0 & 0 & 0 & 0 \\
0 & -K_{1z} & 0 & 0 & 0 & -K_{2z} & 0 & K_{1z} & 0 & 0 & 0 & -K_{2z} \\
0 & 0 & -K_{1y} & 0 & K_{2y} & 0 & 0 & 0 & K_{1y} & 0 & K_{2y} & 0 \\
\hline
0 & 0 & 0 & -K_{1x} & 0 & 0 & 0 & 0 & 0 & K_{1x} & 0 & 0 \\
0 & 0 & -K_{2y} & 0 & K_{4y} & 0 & 0 & 0 & K_{2y} & 0 & K_{3y} & 0 \\
0 & K_{2z} & 0 & 0 & 0 & K_{4z} & 0 & -K_{2z} & 0 & 0 & 0 & K_{3z}
\end{array}\right]
\left[\begin{array}{c} u_L \\ v_L \\ w_L \\ \hline \phi_L \\ \psi_L \\ \theta_L \\ \hline u_R \\ v_R \\ w_R \\ \hline \phi_R \\ \psi_R \\ \theta_R \end{array}\right]
+
\left[\begin{array}{c} U_{L0} \\ V_{L0} \\ W_{L0} \\ \hline X_{L0} \\ Y_{L0} \\ Z_{L0} \\ \hline U_{R0} \\ V_{R0} \\ W_{R0} \\ \hline X_{R0} \\ Y_{R0} \\ Z_{R0} \end{array}\right]
$$

(2) Stiffness load functions forming the last column in (1) above are derived by (3.11-5) as

$$S_{L0} = \begin{bmatrix} U_{L0} \\ V_{L0} \\ W_{L0} \\ \hline X_{L0} \\ Y_{L0} \\ Z_{L0} \end{bmatrix} = \left[\begin{array}{ccc|ccc} K_{1n} & 0 & 0 & 0 & 0 & 0 \\ 0 & K_{1z} & 0 & 0 & 0 & -K_{2z} \\ 0 & 0 & K_{1y} & 0 & K_{2y} & 0 \\ \hline 0 & 0 & 0 & K_{1x} & 0 & 0 \\ 0 & 0 & -K_{2y} & 0 & -K_{4y} & 0 \\ 0 & K_{2z} & 0 & 0 & 0 & -K_{4z} \end{array}\right] \begin{bmatrix} G_u \\ G_v \\ G_w \\ \hline G_\phi \\ G_\psi \\ G_\theta \end{bmatrix} = -k_{LR} G_\Delta$$

$$S_{R0} = \begin{bmatrix} U_{R0} \\ V_{R0} \\ W_{R0} \\ \hline X_{R0} \\ Y_{R0} \\ Z_{R0} \end{bmatrix} = \left[\begin{array}{ccc|ccc} K_{1n} & 0 & 0 & 0 & 0 & 0 \\ 0 & K_{1z} & 0 & 0 & 0 & K_{2z} \\ 0 & 0 & K_{1y} & 0 & -K_{2y} & 0 \\ \hline 0 & 0 & 0 & K_{1x} & 0 & 0 \\ 0 & 0 & K_{2y} & 0 & -K_{4y} & 0 \\ 0 & -K_{2z} & 0 & 0 & 0 & -K_{4z} \end{array}\right] \begin{bmatrix} F_u \\ F_v \\ F_w \\ \hline F_\phi \\ F_\psi \\ F_\theta \end{bmatrix} = -k_{RL} F_\Delta$$

where $G_u, G_v, \ldots, G_\theta$ and $F_u, F_v, \ldots, F_\theta$ are transport load functions introduced in (2.04) and given in particular form in Chaps. 4 and 11.

3.14 FREE SYMMETRICAL CURVED BAR OF ORDER ONE, STIFFNESS COEFFICIENTS IN 0 SYSTEM

(1) System considered in Fig. 3.14-1 is a free symmetrical planar curved bar acted on by loads in the bar plane. The parameters of the bar section and the constants of bar material are those defined in (1.14). The positive end forces, end moments, and loads are shown in Fig. 3.14-1, and the positive displacements are shown in Fig. 3.14-2. The superscript o for the 0 system is omitted in all vectors.

(2) Free-body diagram

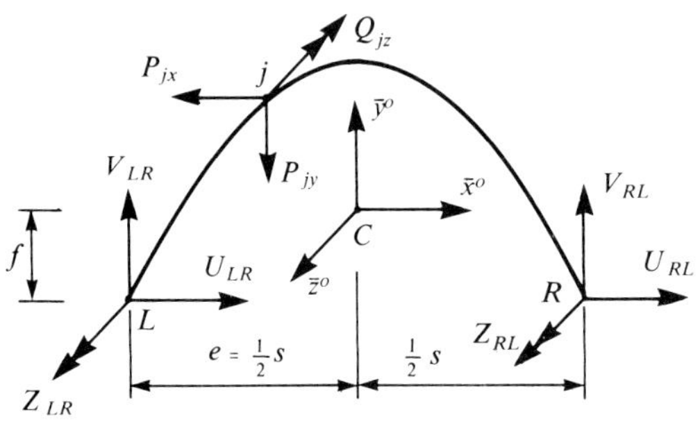

Fig. 3.14-1

(3) Elastic curve

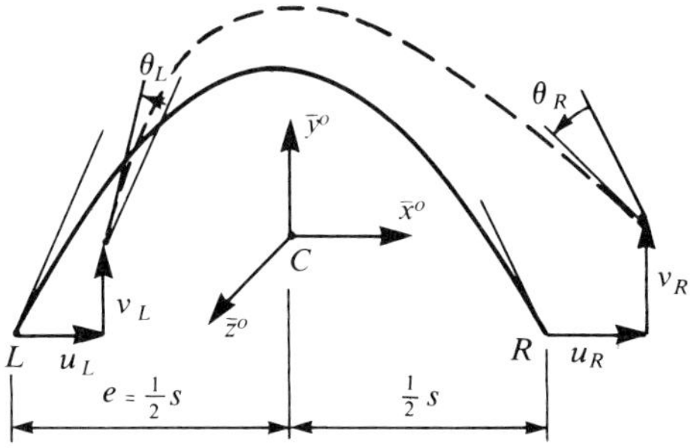

Fig. 3.14-2

(4) Coordinates of neutral point C are

$$e = \tfrac{1}{2}s \qquad f = \frac{\int_L^R y\Lambda_{ZZ}}{\int_L^R \Lambda_{ZZ}}$$

where s = span (m), x, y = coordinates measured from L (m), and the remaining symbols are those of (3.05).

(5) Neutral-point flexibility coefficients (3.07-3) are

$$F_1 = \int_L^R (\bar{y}^2\Lambda_{ZZ} + \Lambda_{UU}) \qquad F_2 = \int_L^R (\bar{x}^2\Lambda_{ZZ} + \Lambda_{VV}) \qquad F_6 = \int_L^R \Lambda_{ZZ}$$

where $\bar{x}$, $\bar{y}$ = coordinates measured from C (m) and, again, the remaining symbols are those of (3.05).

(6) Stiffness coefficients derived in terms of F_1, F_2, F_6, e, and f introduced above are

$$K_{1z} = \frac{1}{F_1} \qquad K_{3z} = \frac{1}{F_2} \qquad K_{5z} = \frac{f^2}{F_1} + \frac{e^2}{F_2} + \frac{1}{F_6}$$

$$K_{2z} = \frac{f}{F_1} \qquad K_{4z} = \frac{e}{F_2} \qquad K_{6z} = \frac{f^2}{F_1} - \frac{e^2}{F_2} + \frac{1}{F_6}$$

The closed-form evaluations of these coefficients for circular and parabolic bars are given in scaled form in (6.09, 7.07). For bars of other forms or for bars of variable section, the numerical integration must be used. The construction of the stiffness matrix equations is shown in (3.15-1).

3.15 FREE SYMMETRICAL CURVED BAR OF ORDER ONE, STIFFNESS MATRIX EQUATIONS IN 0 SYSTEM

Notation (3.11, 3.14) Signs (3.11) Coefficients (3.14)

(1) Stiffness matrix equations defined in (3.11-3) are

$$\begin{bmatrix} U_{LR} \\ V_{LR} \\ Z_{LR} \\ \hline U_{RL} \\ V_{RL} \\ Z_{RL} \end{bmatrix} = \left[\begin{array}{ccc|ccc} K_{1z} & 0 & -K_{2z} & -K_{1z} & 0 & K_{2z} \\ 0 & K_{3z} & K_{4z} & 0 & -K_{3z} & K_{4z} \\ -K_{2z} & K_{4z} & K_{5z} & K_{2z} & -K_{4z} & -K_{6z} \\ \hline -K_{1z} & 0 & K_{2z} & K_{1z} & 0 & -K_{2z} \\ 0 & -K_{3z} & -K_{4z} & 0 & K_{3z} & -K_{4z} \\ K_{2z} & K_{4z} & -K_{6z} & -K_{2z} & -K_{4z} & K_{5z} \end{array}\right] \begin{bmatrix} u_L \\ v_L \\ \theta_L \\ \hline u_R \\ v_R \\ \theta_R \end{bmatrix} + \begin{bmatrix} U_{L0} \\ V_{L0} \\ Z_{L0} \\ \hline U_{R0} \\ V_{R0} \\ Z_{R0} \end{bmatrix}$$

where $U_{L0}, V_{L0}, \ldots, Z_{R0}$ are the stiffness load functions given below.

(2) Stiffness load functions defined in (3.11-5) are

$$S_{L0} = \begin{bmatrix} U_{L0} \\ V_{L0} \\ Z_{L0} \end{bmatrix} = \begin{bmatrix} K_{1z} & 0 & -K_{2z} \\ 0 & K_{3z} & -K_{4z} \\ -K_{2z} & K_{4z} & K_{6z} \end{bmatrix} \begin{bmatrix} G_u \\ G_v \\ G_\theta \end{bmatrix} = -k_{LR} G_\Delta$$

$$S_{R0} = \begin{bmatrix} U_{R0} \\ V_{R0} \\ Z_{R0} \end{bmatrix} = \begin{bmatrix} K_{1z} & 0 & -K_{2z} \\ 0 & K_{3z} & K_{4z} \\ -K_{2z} & -K_{4z} & K_{6z} \end{bmatrix} \begin{bmatrix} F_u \\ F_v \\ F_\theta \end{bmatrix} = -k_{RL} F_\Delta$$

where $F_u, F_v, \ldots, G_\theta$ are the load functions of the transport matrices T_{LR}, T_{RL} in (2.14-2). Stiffness load functions of circular and parabolic bars are listed in (6.12–6.17, 7.08–7.09).

(3) Symmetrical end displacements If $u_L = -u_R$, $v_L = v_R$, $\theta_L = -\theta_R$, and *symmetrical causes* are applied, the stiffness matrix equation in (1) above reduces to

$$\begin{bmatrix} U_{LR} \\ V_{LR} \\ Z_{LR} \end{bmatrix} = \begin{bmatrix} 2K_{1z} & 0 & -2K_{2z} \\ 0 & 0 & 0 \\ -2K_{2z} & 0 & K_{5z} + K_{6z} \end{bmatrix} \begin{bmatrix} u_L \\ v_L \\ \theta_L \end{bmatrix} + \begin{bmatrix} U_{L0} \\ V_{L0} \\ Z_{L0} \end{bmatrix}$$

(4) Antisymmetrical end displacements If $u_L = u_R$, $v_L = -v_R$, $\theta_L = \theta_R$, and *antisymmetrical causes* are applied, the stiffness matrix equation in (1) above reduces to

$$\begin{bmatrix} U_{LR} \\ V_{LR} \\ Z_{LR} \end{bmatrix} = \begin{bmatrix} -0 & 0 & 0 \\ 0 & 2K_{3z} & 2K_{4z} \\ 0 & 2K_{4z} & K_{5z} - K_{6z} \end{bmatrix} \begin{bmatrix} u_L \\ v_L \\ \theta_L \end{bmatrix} + \begin{bmatrix} U_{L0} \\ V_{L0} \\ Z_{L0} \end{bmatrix}$$

3.16 FREE SYMMETRICAL CURVED BAR OF ORDER TWO, STIFFNESS COEFFICIENTS IN 0 SYSTEM

(1) System considered in Fig. 3.16-1 is a free symmetrical planar curved bar acted on by loads normal to the bar plane. The parameters of the bar section and the constants of the bar material are those defined in (1.14). The positive end forces, end moments, and loads are shown in Fig. 3.16-1, and the positive displacements are shown in Fig. 3.16-2. The superscript o for the 0 system is omitted in all vectors.

(2) Free-body diagram **(3) Elastic curve**

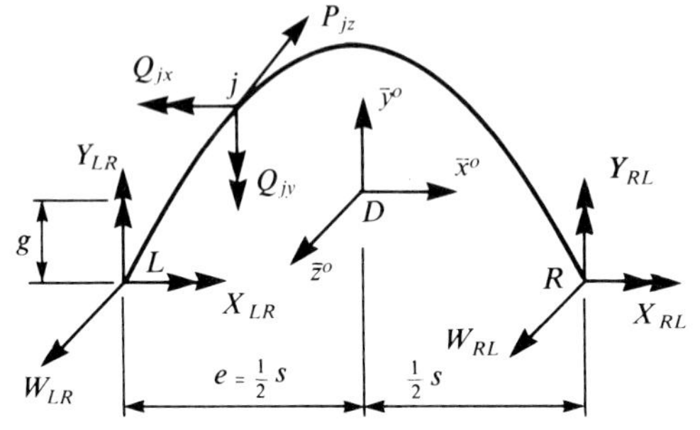

Fig. 3.16-1

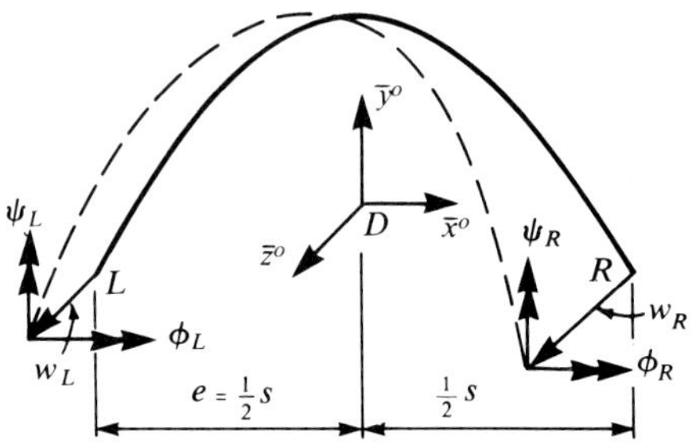

Fig. 3.16-2

(4) Coordinates of neutral point D are

$$e = \tfrac{1}{2}s \qquad g = \frac{\int_L^R (y\Lambda_{XX} - x\Lambda_{XY})}{\int_L^R \Lambda_{XX}}$$

where s = span (m), x, y = coordinates measured from L (m), and the remaining symbols are those of (3.06).

(5) Neutral-point flexibility coefficients (3.08-3) are

$$F_3 = \int_L^R (\bar{y}^2\Lambda_{XX} - 2\bar{x}\bar{y}\Lambda_{XY} + \bar{x}^2\Lambda_{YY} + \Lambda_{WW}) \qquad F_4 = \int_L^R \Lambda_{XX} \qquad F_5 = \int_L^R \Lambda_{YY}$$

where $\bar{x}$, $\bar{y}$ = coordinates measured from D (m) and, again, the remaining symbols are those of (3.06).

(6) Stiffness coefficients derived in terms of F_3, F_4, F_5, e, and g introduced above are

$$K_{1y} = \frac{1}{F_3} \qquad K_{2y} = \frac{g}{F_3} \qquad K_{5y} = \frac{g^2}{F_3} + \frac{1}{F_4}$$

$$K_{3y} = \frac{e}{F_3} \qquad K_{4y} = \frac{eg}{F_3} \qquad K_{6y} = \frac{e^2}{F_3} + \frac{1}{F_5}$$

$$K_{7y} = \frac{e^2}{F_3} - \frac{1}{F_5}$$

The closed-form evaluations of these coefficients for circular bars are shown in Chap. 12. For bars of other forms, the numerical integration must be used. The construction of the stiffness matrix equations is shown in (3.17-1).

3.17 FREE SYMMETRICAL CURVED BAR OF ORDER TWO, STIFFNESS MATRIX EQUATIONS IN 0 SYSTEM

Notation (3.11, 3.16) Signs (3.11) Coefficients (3.16)

(1) Stiffness matrix equations defined in (3.11-3) are

$$\begin{bmatrix} W_{LR} \\ X_{LR} \\ Y_{LR} \\ \hline W_{RL} \\ X_{RL} \\ Y_{RL} \end{bmatrix} = \left[\begin{array}{ccc|ccc} K_{1y} & K_{2y} & -K_{3y} & -K_{1y} & -K_{2y} & -K_{3y} \\ K_{2y} & K_{5y} & -K_{4y} & -K_{2y} & -K_{5y} & -K_{4y} \\ -K_{3y} & -K_{4y} & K_{6y} & K_{3y} & K_{4y} & K_{7y} \\ \hline -K_{1y} & -K_{2y} & K_{3y} & K_{1y} & K_{2y} & K_{3y} \\ -K_{2y} & -K_{5y} & K_{4y} & K_{2y} & K_{5y} & K_{4y} \\ -K_{3y} & -K_{4y} & K_{7y} & K_{3y} & K_{4y} & K_{6y} \end{array}\right] \begin{bmatrix} w_L \\ \phi_L \\ \psi_L \\ \hline w_R \\ \phi_R \\ \psi_R \end{bmatrix} + \begin{bmatrix} W_{L0} \\ X_{L0} \\ Y_{L0} \\ \hline W_{R0} \\ X_{R0} \\ Y_{R0} \end{bmatrix}$$

where $W_{L0}, X_{L0}, \ldots, Y_{R0}$ are the stiffness load functions given below.

(2) Stiffness load functions defined in (3.11-5) are

$$S_{L0} = \begin{bmatrix} W_{L0} \\ X_{L0} \\ Y_{L0} \end{bmatrix} = \begin{bmatrix} K_{1y} & K_{2y} & K_{3y} \\ K_{2y} & K_{5y} & K_{4y} \\ -K_{3y} & -K_{4y} & -K_{7y} \end{bmatrix} \begin{bmatrix} G_w \\ G_\phi \\ G_\psi \end{bmatrix} = -k_{LR} G_\Delta$$

$$S_{R0} = \begin{bmatrix} W_{R0} \\ X_{R0} \\ Y_{R0} \end{bmatrix} = \begin{bmatrix} K_{1y} & K_{2y} & -K_{3y} \\ K_{2y} & K_{5y} & -K_{4y} \\ K_{3y} & K_{4y} & -K_{7y} \end{bmatrix} \begin{bmatrix} F_w \\ F_\phi \\ F_\psi \end{bmatrix} = -k_{RL} F_\Delta$$

where $F_w, F_\phi, \ldots, G_\psi$ are the load functions of the transport matrices T_{LR}, T_{RL} in (2.14-2). Stiffness load functions of circular bars are listed in Chap. 12.

(3) Symmetrical end displacements If $w_L = w_R$, $\phi_L = \phi_R$, $\psi_L = -\psi_R$, and *symmetrical causes* are applied, the stiffness matrix equation in (1) above reduces to

$$\begin{bmatrix} W_{LR} \\ X_{LR} \\ Y_{LR} \end{bmatrix} = \begin{bmatrix} -0 & -0 & 0 \\ 0 & 0 & 0 \\ 0 & 0 & K_{6y} - K_{7y} \end{bmatrix} \begin{bmatrix} w_L \\ \phi_L \\ \psi_L \end{bmatrix} + \begin{bmatrix} W_{L0} \\ X_{L0} \\ Y_{L0} \end{bmatrix}$$

(4) Antisymmetrical end displacements If $w_L = -w_R$, $\phi_L = -\phi_R$, $\psi_L = \psi_R$, and *antisymmetrical causes* are applied, the stiffness matrix in (1) above reduces to

$$\begin{bmatrix} W_{LR} \\ X_{LR} \\ Y_{LR} \end{bmatrix} = \begin{bmatrix} 2K_{1y} & 2K_{2y} & -2K_{3y} \\ 2K_{2y} & 2K_{5y} & -2K_{4y} \\ -2K_{3y} & -2K_{4y} & K_{6y} + K_{7y} \end{bmatrix} \begin{bmatrix} w_L \\ \phi_L \\ \psi_L \end{bmatrix} + \begin{bmatrix} W_{L0} \\ X_{L0} \\ Y_{L0} \end{bmatrix}$$

3.18 INTERACTIVE SYSTEMS OF ORDER THREE, STIFFNESS COEFFICIENTS IN *S* SYSTEM

(1) Systems considered are the bars encased in elastic foundation (Fig. 2.16-1), free beam-columns (Fig. 2.16-2), and beam-columns encased in elastic foundation (Fig. 2.16-3). Their definitions are given in (2.16), and other properties are described in (3.01-1). As before, s = span (m), a, b = segments (m).

(2) Relations. According to (3.11-4), the stiffness coefficients which form the submatrices k_{LL}, k_{LE}, k_{RL}, k_{RR} are derived from the submatrices t_{LR}, d_{LR}, t_{RL}, d_{RL} of the transport matrices in (2.18). A close inspection of the transport coefficients in (2.18) reveals the following identities:

$$T_{Uu} = T_{uU} \qquad T_{X\phi} = T_{\phi X}$$
$$T_{vZ} = T_{\theta V} \qquad T_{wY} = T_{\psi W}$$

where the first row of identities is based on the transpose relations and the second row is apparent from the reciprocal theorem (Ref. 17).

(3) Stiffness coefficients derived as indicated above are

$K_{1x} = \dfrac{T_{\phi\phi}}{T_{\phi X}}$	$K_{2x} = \dfrac{1}{T_{\phi X}}$
$K_{1y} = D_y(T_{ww}T_{\psi Y} - T_{\psi w}T_{wY})$ $K_{2y} = D_y(T_{\psi\psi}T_{wY} - T_{w\psi}T_{\psi Y})$ $K_{3y} = D_y(T_{w\psi}T_{wY} - T_{\psi\psi}T_{wW})$	$K_{4y} = D_yT_{\psi Y}$ $K_{5y} = D_yT_{wY}$ $K_{6y} = D_yT_{wW}$

$K_{1n} = \dfrac{T_{uu}}{T_{uU}}$	$K_{2n} = \dfrac{1}{T_{uU}}$
$K_{1z} = D_z(T_{vv}T_{\theta Z} - T_{\theta v}T_{vZ})$ $K_{2z} = D_z(T_{\theta\theta}T_{vZ} - T_{v\theta}T_{\theta Z})$ $K_{3z} = D_z(T_{v\theta}T_{\theta V} - T_{\theta\theta}T_{vV})$	$K_{4z} = D_zT_{\theta Z}$ $K_{5z} = D_zT_{vZ}$ $K_{6z} = D_zT_{vV}$

where the T coefficients are the transport coefficients introduced symbolically in (2.16-4) and given in particular form in Chaps. 8–10 and 14–15. Also,

$$D_y = \frac{1}{T_{wY}^2 - T_{wW}T_{\psi Y}} \qquad D_z = \frac{1}{T_{vZ}^2 - T_{vV}T_{\theta Z}}$$

The subscripts y and z identify the normal to the plane action, so that $K_{1y}, K_{2y}, \ldots, K_{6y}$ are the end stiffnesses of order two and $K_{1z}, K_{2z}, \ldots, K_{6z}$ are the end stiffnesses of order one, which are used for the construction of the stiffness matrices of order three in (3.19).

(4) End stiffness matrix equations, derived by means of relations (3.12-2, 3, 4, 5) in terms of the stiffness coefficients given above and the load functions of the transport matrices in (2.18), are displayed in (3.19). Particular forms of these equations are given in Chaps. 8–10 and 13–15. The physical interpretation of the stiffness coefficients is shown in Fig. 3.18-1.

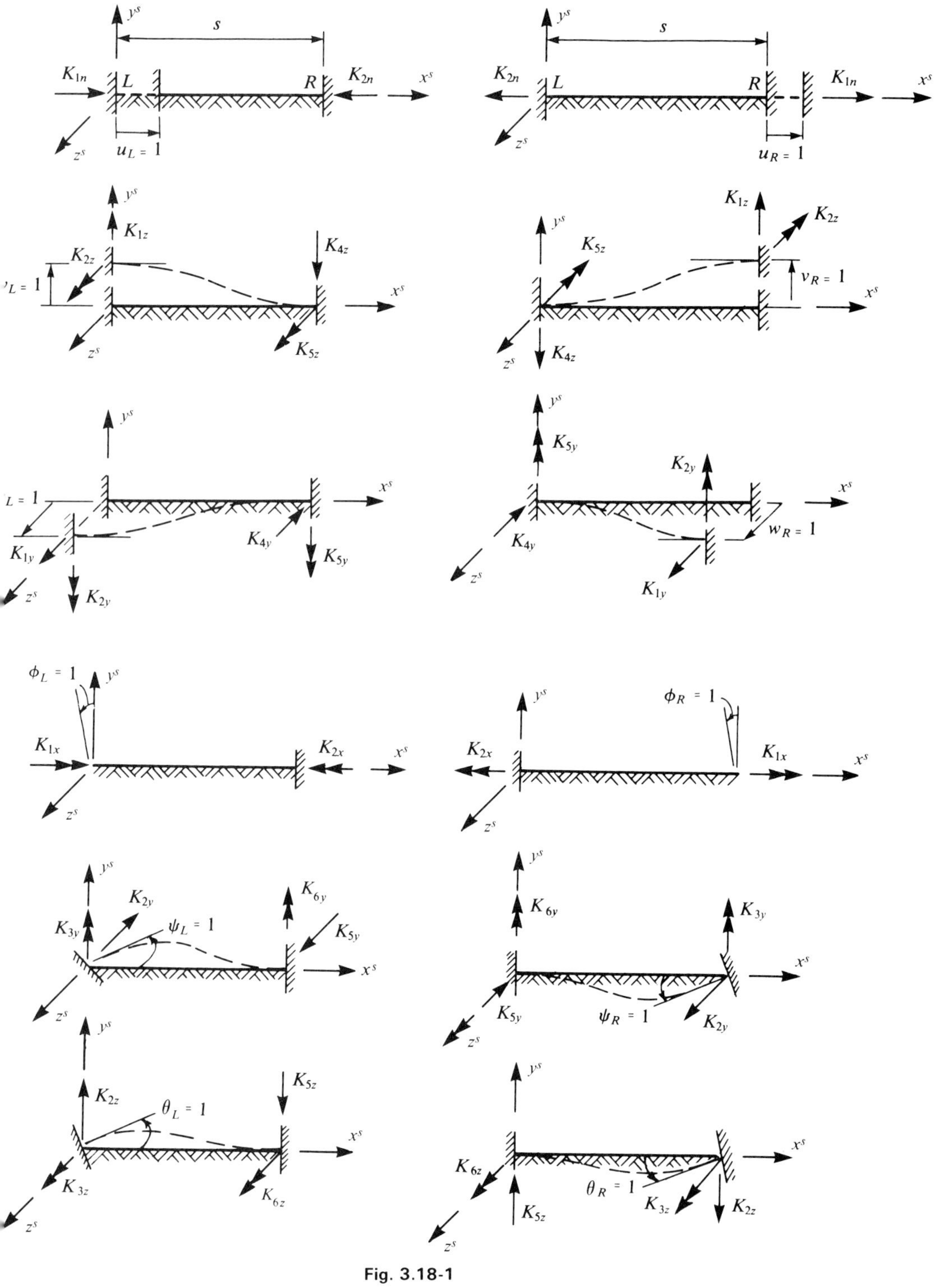

y^s
s
K_{1n}
L
R
K_{2n}
x^s
z^s
$u_L = 1$
$u_R = 1$
K_{1z}
K_{2z}
K_{4z}
K_{5z}
$v_R = 1$
K_{5y}
K_{2y}
K_{4y}
K_{1y}
$w_R = 1$
$\phi_L = 1$
$\phi_R = 1$
K_{1x}
K_{2x}
K_{3y}
K_{6y}
$\psi_L = 1$
$\psi_R = 1$
$\theta_L = 1$
$\theta_R = 1$
K_{3z}
K_{6z}

Fig. 3.18-1

3.19 INTERACTIVE SYSTEM OF ORDER THREE, END STIFFNESS MATRIX EQUATIONS IN S SYSTEM

Notation (3.11, 3.18) Signs (3.11) Coefficients (3.18)

(1) Stiffness matrix equations in (3.11-3) are in full form

$$
\begin{bmatrix} U_{LR} \\ V_{LR} \\ W_{LR} \\ \hline X_{LR} \\ Y_{LR} \\ Z_{LR} \\ \hline U_{RL} \\ V_{RL} \\ W_{RL} \\ \hline X_{RL} \\ Y_{RL} \\ Z_{RL} \end{bmatrix}
=
\left[\begin{array}{ccc|ccc|ccc|ccc}
K_{1n} & 0 & 0 & 0 & 0 & 0 & -K_{2n} & 0 & 0 & 0 & 0 & 0 \\
0 & K_{1z} & 0 & 0 & 0 & K_{2z} & 0 & -K_{4z} & 0 & 0 & 0 & K_{5z} \\
0 & 0 & K_{1y} & 0 & -K_{2y} & 0 & 0 & 0 & -K_{4y} & 0 & -K_{5y} & 0 \\
\hline
0 & 0 & 0 & K_{1x} & 0 & 0 & 0 & 0 & 0 & -K_{2x} & 0 & 0 \\
0 & 0 & -K_{2y} & 0 & K_{3y} & 0 & 0 & 0 & K_{5y} & 0 & K_{6y} & 0 \\
0 & K_{2z} & 0 & 0 & 0 & K_{3z} & 0 & -K_{5z} & 0 & 0 & 0 & K_{6z} \\
\hline
-K_{2n} & 0 & 0 & 0 & 0 & 0 & K_{1n} & 0 & 0 & 0 & 0 & 0 \\
0 & -K_{4z} & 0 & 0 & 0 & -K_{5z} & 0 & K_{1z} & 0 & 0 & 0 & -K_{2z} \\
0 & 0 & -K_{4y} & 0 & K_{5y} & 0 & 0 & 0 & K_{1y} & 0 & K_{2y} & 0 \\
\hline
0 & 0 & 0 & -K_{2x} & 0 & 0 & 0 & 0 & 0 & K_{1x} & 0 & 0 \\
0 & 0 & -K_{5y} & 0 & K_{6y} & 0 & 0 & 0 & K_{2y} & 0 & K_{3y} & 0 \\
0 & K_{5z} & 0 & 0 & 0 & K_{6z} & 0 & -K_{2z} & 0 & 0 & 0 & K_{3z}
\end{array}\right]
\begin{bmatrix} u_L \\ v_L \\ w_L \\ \hline \phi_L \\ \psi_L \\ \theta_L \\ \hline u_R \\ v_R \\ w_R \\ \hline \phi_R \\ \psi_R \\ \theta_R \end{bmatrix}
+
\begin{bmatrix} U_{L0} \\ V_{L0} \\ W_{L0} \\ \hline X_{L0} \\ Y_{L0} \\ Z_{L0} \\ \hline U_{R0} \\ V_{R0} \\ W_{R0} \\ \hline X_{R0} \\ Y_{R0} \\ Z_{R0} \end{bmatrix}
$$

(2) Stiffness load functions forming the last column in (1) above are derived by (3.12-5) as

$$S_{L0} = \begin{bmatrix} U_{L0} \\ V_{L0} \\ W_{L0} \\ \hline X_{L0} \\ Y_{L0} \\ Z_{L0} \end{bmatrix} = \left[\begin{array}{ccc|ccc} K_{2n} & 0 & 0 & 0 & 0 & 0 \\ 0 & K_{4z} & 0 & 0 & 0 & -K_{5z} \\ 0 & 0 & K_{4y} & 0 & K_{5y} & 0 \\ \hline 0 & 0 & 0 & K_{2\phi} & 0 & 0 \\ 0 & 0 & -K_{5y} & 0 & -K_{6y} & 0 \\ 0 & K_{5z} & 0 & 0 & 0 & -K_{6z} \end{array}\right] \begin{bmatrix} G_u \\ G_v \\ G_w \\ \hline G_\phi \\ G_\psi \\ G_\theta \end{bmatrix} = -k_{LR} G_\Delta$$

$$S_{R0} = \begin{bmatrix} U_{R0} \\ V_{R0} \\ W_{R0} \\ \hline X_{R0} \\ Y_{R0} \\ Z_{R0} \end{bmatrix} = \left[\begin{array}{ccc|ccc} K_{2n} & 0 & 0 & 0 & 0 & 0 \\ 0 & K_{4z} & 0 & 0 & 0 & K_{5z} \\ 0 & 0 & K_{4y} & 0 & -K_{5y} & 0 \\ \hline 0 & 0 & & K_{2\phi} & 0 & 0 \\ 0 & 0 & K_{5y} & 0 & -K_{6y} & 0 \\ 0 & -K_{5z} & 0 & 0 & 0 & -K_{6z} \end{array}\right] \begin{bmatrix} F_u \\ F_v \\ F_w \\ \hline F_\phi \\ F_\psi \\ F_\theta \end{bmatrix} = -k_{RL} F_\Delta$$

where G_u, G_v, . . . , G_θ and F_u, F_v, . . . , F_θ are the transport load functions introduced in (2.04) and given in particular forms in Chaps. 8–10 and 13–15.

3.20 GENERAL RELATIONS OF TRANSPORT, FLEXIBILITY AND STIFFNESS MATRICES

(1) Transport-flexibility relations. The transport coefficients in t_{LR}, d_{LR}, t_{RL}, d_{RL} and the transport load functions in F_S, F_Δ, G_S, G_Δ introduced in (2.02) are related to their flexibility counterparts in (3.02) as

$$\dagger c_{LR} = -t_{LR} f_{RR}^{-1} \qquad \dagger c_{RL} = t_{RL} f_{LL}^{-1}$$

$$d_{LR} = -f_{LL} t_{LR} \qquad d_{RL} = f_{RR} t_{RL}$$

$$\dagger F_S = -f_{RL}^{-1} \Delta_{R0} \qquad \dagger G_S = -f_{LR}^{-1} \Delta_{L0}$$

$$F_\Delta = -t_{RL}^{T} \Delta_{R0} - \Delta_{L0} \qquad G_\Delta = t_{LR}^{T} \Delta_{L0} + \Delta_{R0}$$

where the relations designated by a dagger are valid for interactive bars only.

(2) Transport-stiffness relations. The same coefficients are related to their stiffness counterparts in (3.11) as

$$\dagger c_{LR} = k_{LL} t_{LR} \qquad \dagger c_{RL} = k_{RR} t_{RL}$$

$$d_{LR} = k_{RL}^{-1} \qquad d_{RL} = -k_{RL}^{-1}$$

$$F_S = -S_{L0} - t_{LR} S_{R0} \qquad G_S = S_{R0} + t_{RL} S_{L0}$$

$$F_\Delta = -k_{RL}^{-1} S_{R0} \qquad G_\Delta = -k_{LR}^{-1} S_{L0}$$

where again the relations designated by a dagger are valid for interactive bars only.

4

Free Straight Bar of Order One Constant Section

STATIC STATE

4.01 DEFINITION OF STATE

(1) System considered is a free finite straight bar of constant and symmetrical section. The centroidal axis x of the bar and the vertical axis y define the plane of symmetry of this section. The bar is part of a planar system acted on by loads in the system plane (system of order one).

(2) Notation (A.1, A.2, A.4)

s = span (m)	A_x = area of normal section (m^2)
a, b = segments (m)	I_z = moment of inertia of A_x about z (m^4)
x, x' = coordinates (m)	E = modulus of elasticity (Pa)
U = normal force (N)	u = linear displacement along x (m)
V = shearing force (N)	v = linear displacement along y (m)
Z = flexural moment (N·m)	θ = flexural slope about z (rad)

(3) Assumptions of analysis are stated in (1.01). In this chapter, the effects of axial and flexural deformations are included, but the effects of shearing deformations and the effect of axial force on the magnitude of the flexural moment are neglected. The beam-column effect is considered in Chap. 9. For the inclusion of shearing deformations, reference is made to Chaps. 2 and 3.

4.02 TRANSPORT MATRIX EQUATIONS

(1) Equivalents and signs

$\bar{u} = uEA_x =$ scaled u (N·m) $\qquad \bar{v} = vEI_z =$ scaled v (N·m^3)

$\bar{\theta} = \theta EI_z =$ scaled θ (N·m^2)

All forces, moments, and displacements are in the principal axes of the normal section, and their positive directions are shown in (2, 3) below.

(2) Free-body diagram

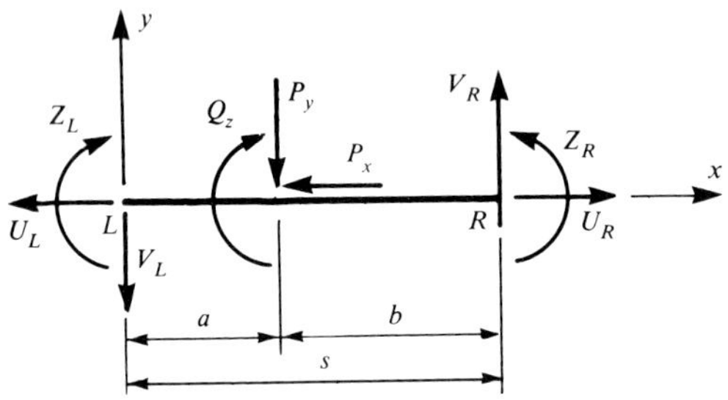

(3) Elastic curve

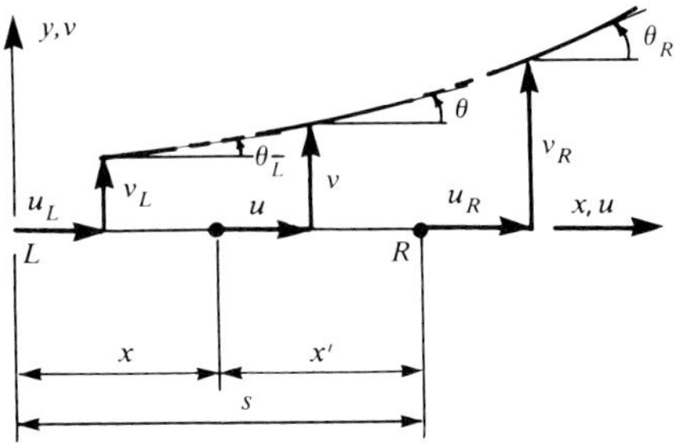

(4) Transport matrix equations

$$
\underbrace{\begin{bmatrix} 1 \\ \hline U_L \\ V_L \\ Z_L \\ \hline \bar{u}_L \\ \bar{v}_L \\ \bar{\theta}_L \end{bmatrix}}_{H_L}
=
\underbrace{\left[\begin{array}{c|ccc|ccc}
1 & 0 & 0 & 0 & 0 & 0 & 0 \\ \hline
F_U & 1 & 0 & 0 & 0 & 0 & 0 \\
F_V & 0 & 1 & 0 & 0 & 0 & 0 \\
F_Z & 0 & s & 1 & 0 & 0 & 0 \\ \hline
F_u & -s & 0 & 0 & 1 & 0 & 0 \\
F_v & 0 & \frac{1}{6}s^3 & \frac{1}{2}s^2 & 0 & 1 & -s \\
F_\theta & 0 & -\frac{1}{2}s^2 & -s & 0 & 0 & 1
\end{array}\right]}_{T_{LR}}
\underbrace{\begin{bmatrix} 1 \\ \hline U_R \\ V_R \\ Z_R \\ \hline \bar{u}_R \\ \bar{v}_R \\ \bar{\theta}_R \end{bmatrix}}_{H_R}
$$

$$
\underbrace{\begin{bmatrix} 1 \\ \hline U_R \\ V_R \\ Z_R \\ \hline \bar{u}_R \\ \bar{v}_R \\ \bar{\theta}_R \end{bmatrix}}_{H_R}
=
\underbrace{\left[\begin{array}{c|ccc|ccc}
1 & 0 & 0 & 0 & 0 & 0 & 0 \\ \hline
G_U & 1 & 0 & 0 & 0 & 0 & 0 \\
G_V & 0 & 1 & 0 & 0 & 0 & 0 \\
G_Z & 0 & -s & 1 & 0 & 0 & 0 \\ \hline
G_u & s & 0 & 0 & 1 & 0 & 0 \\
G_v & 0 & -\frac{1}{6}s^3 & \frac{1}{2}s^2 & 0 & 1 & s \\
G_\theta & 0 & -\frac{1}{2}s^2 & s & 0 & 0 & 1
\end{array}\right]}_{T_{RL}}
\underbrace{\begin{bmatrix} 1 \\ \hline U_L \\ V_L \\ Z_L \\ \hline \bar{u}_L \\ \bar{v}_L \\ \bar{\theta}_L \end{bmatrix}}_{H_L}
$$

where H_L, H_R are the *end state vectors* and T_{LR}, T_{RL} are the respective *transport matrices*. The elements of these matrices are the *transport coefficients* shown above and the *load functions F, G* listed for particular cases in (4.04–4.06).

4.03 TRANSPORT RELATIONS

(1) Load functions are expressed in terms of the following symbols:

P_x, P_y = concentrated loads (N) $\quad p_x, p_y$ = intensities of distributed load (N/m)

Q_z = applied couple (N·m) $\quad q_z$ = intensity of distributed couple (N·m/m)

In cases of variable load distribution, p_x, p_y, q_z are the maximum intensities. The effects of abrupt changes in bar axis at a particular point j are expressed in terms of u_j, v_j, θ_j = abrupt displacements (m, m, rad). The temperature-effect analysis assumes (in general) a linear variation of temperature change along the depth d (m) of the bar in a selected segment. The symbols used in the temperature analysis are

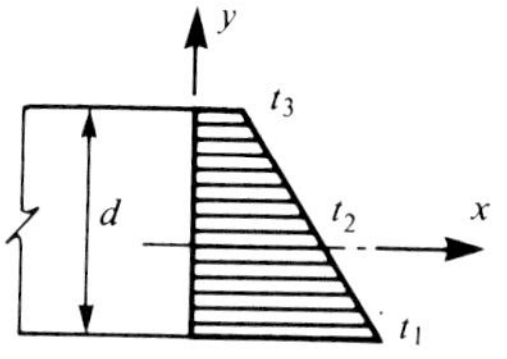

$t_1 - t_0$ = temperature change at the bottom (°C)

$t_2 - t_0$ = temperature change at the centroid (°C)

$t_3 - t_0$ = temperature change at the top (°C)

α_t = thermal coefficient (1/°C)

$e_t = \alpha_t(t_2 - t_0)$ = linear thermal strain

$f_t = \alpha_t(t_1 - t_3)/d$ = angular thermal strain per unit length (rad/m)

The load functions have a physical meaning and can be interpreted as the end stress resultants of the respective cantilever beam and as the linear and angular deviations of the tangent drawn to the elastic curve at the free end and measured at the opposite end. Case (4.04-1) is interpreted graphically below.

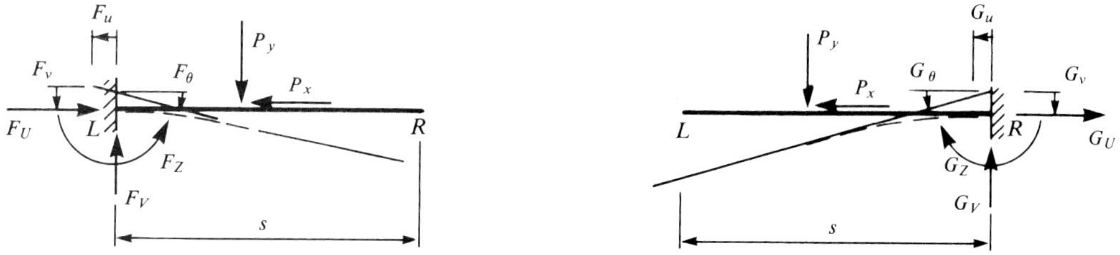

(2) Differential relations

$\dfrac{du}{dx} = \dfrac{U}{EA_x}$	$\dfrac{dv}{dx} = \theta$	$\dfrac{d^3v}{dx^3} = \dfrac{d^2\theta}{dx^2} = \dfrac{dZ}{EI_z\,dx} = -\dfrac{V}{EI_z}$
$\dfrac{d^2u}{dx^2} = \dfrac{p_x}{EA_x}$	$\dfrac{d^2v}{dx^2} = \dfrac{d\theta}{dx} = \dfrac{Z}{EI_z}$	$\dfrac{d^4v}{dx^4} = \dfrac{d^3\theta}{dx^3} = \dfrac{d^2Z}{EI_z\,dx^2} = -\dfrac{dV}{EI_z\,dx} = -\dfrac{p_y}{EI_z}$

(3) Transport matrices T_{LR} and T_{RL} give a clear, systematic, and complete record of the static and deformation properties of the segment s; they are specific for the selected segment and for the given loads and independent of the end conditions. They apply to the whole bar or to any part of it. In the latter case, s becomes x and the load function G becomes a function of x, which is the position coordinate of an intermediate point D measured from L, or s becomes x' and the load function F becomes a function of x', which is the position coordinate of D measured from R, as shown in (4.02-3). Single-segment applications of the transport matrix equations are discussed in (4.07), and the results produced by particular loads and end conditions are summarized in (4.08–4.21).

(4) General transport matrix equations for free bars of order three are obtained by matrix composition of (4.02–4.06) and (11.02–11.06) as shown in (2.04–2.06). This general case includes the effects of all deformations.

4.04 TRANSPORT LOAD FUNCTIONS

Notation (4.02–4.03) P_x, P_y = concentrated loads (N) Q_z = applied couple (N·m) q_z = intensity of distributed couple (N·m/m)	Signs (4.02)	Matrix (4.02)
$F_U = -P_x$ $F_V = -P_y$ $F_Z = -aP_y$ $F_u = aP_x$ $F_v = -\frac{1}{6}a^3P_y$ $F_\theta = \frac{1}{2}a^2P_y$	**1.** 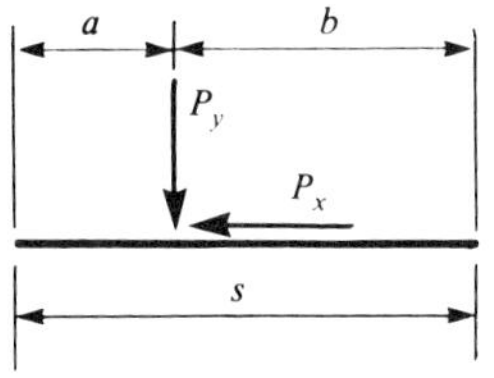 	$G_U = P_x$ $G_V = P_y$ $G_Z = -bP_y$ $G_u = bP_x$ $G_v = -\frac{1}{6}b^3P_y$ $G_\theta = -\frac{1}{2}b^2P_y$
$F_U = 0$ $F_V = 0$ $F_Z = -Q_z$ $F_u = 0$ $F_v = -\frac{1}{2}a^2Q_z$ $F_\theta = aQ_z$	**2.** 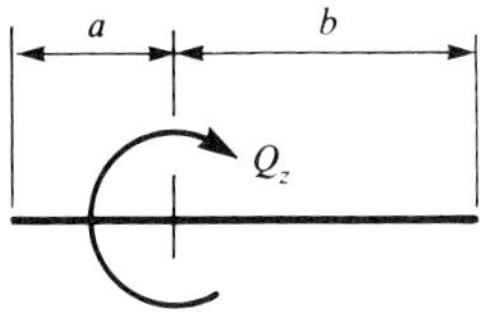 	$G_U = 0$ $G_V = 0$ $G_Z = Q_z$ $G_u = 0$ $G_v = \frac{1}{2}b^2Q_z$ $G_\theta = bQ_z$
$F_U = 0$ $F_V = 0$ $F_Z = -sq_z$ $F_u = 0$ $F_v = -\frac{1}{6}s^3q_z$ $F_\theta = \frac{1}{2}s^2q_z$	**3.** 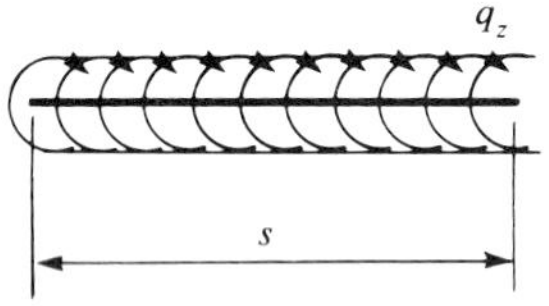 	$G_U = 0$ $G_V = 0$ $G_Z = sq_z$ $G_u = 0$ $G_v = \frac{1}{6}s^3q_z$ $G_\theta = \frac{1}{2}s^2q_z$
$F_U = 0$ $F_V = 0$ $F_Z = -bq_z$ $F_u = 0$ $F_v = -\frac{1}{6}(s^3 - a^3)q_z$ $F_\theta = \frac{1}{2}(s^2 - a^2)q_z$	**4.** 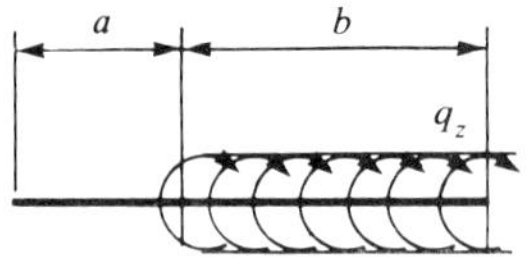 	$G_U = 0$ $G_V = 0$ $G_Z = bq_z$ $G_u = 0$ $G_v = \frac{1}{6}b^3q_z$ $G_\theta = \frac{1}{2}b^2q_z$

4.05 TRANSPORT LOAD FUNCTIONS

Notation (4.02–4.03) p_x, p_y = intensities of distributed load (N/m)	Signs (4.02)	Matrix (4.02)
$F_U = -sp_x$ $F_V = -sp_y$ $F_Z = -\frac{1}{2}s^2p_y$	**1.** 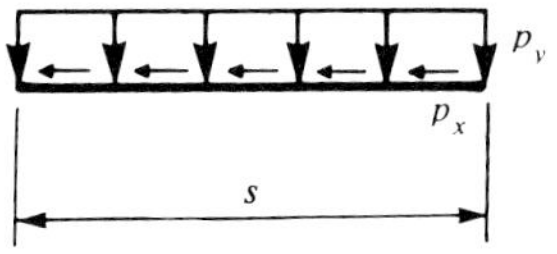	$G_U = sp_x$ $G_V = sp_y$ $G_Z = -\frac{1}{2}s^2p_y$
$F_u = \frac{1}{2}s^2p_x$ $F_v = -\frac{1}{24}s^4p_y$ $F_\theta = \frac{1}{6}s^3p_y$		$G_u = \frac{1}{2}s^2p_x$ $G_v = -\frac{1}{24}s^4p_y$ $G_\theta = -\frac{1}{6}s^3p_y$
$F_U = -bp_x$ $F_V = -bp_y$ $F_Z = -\frac{1}{2}(s^2 - a^2)p_y$	**2.** 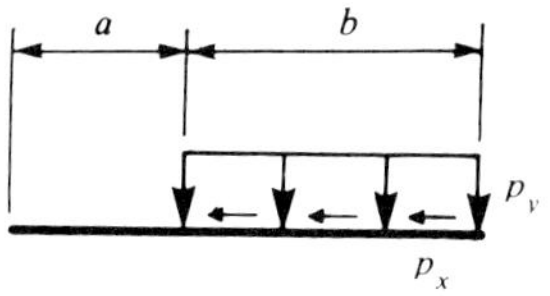	$G_U = bp_x$ $G_V = bp_y$ $G_Z = -\frac{1}{2}b^2p_y$
$F_u = \frac{1}{2}(s^2 - a^2)p_x$ $F_v = -\frac{1}{24}(s^4 - a^4)p_y$ $F_\theta = \frac{1}{6}(s^3 - a^3)p_y$		$G_u = \frac{1}{2}b^2p_x$ $G_v = -\frac{1}{24}b^4p_y$ $G_\theta = -\frac{1}{6}b^3p_y$
$F_U = -\frac{1}{2}sp_x$ $F_V = -\frac{1}{2}sp_y$ $F_Z = -\frac{1}{3}s^2p_y$	**3.** 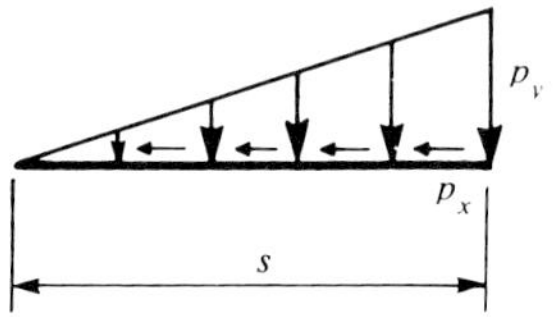	$G_U = \frac{1}{2}sp_x$ $G_V = \frac{1}{2}sp_y$ $G_Z = -\frac{1}{6}s^2p_y$
$F_u = \frac{1}{3}s^2p_x$ $F_v = -\frac{1}{30}s^4p_y$ $F_\theta = \frac{1}{8}s^3p_y$		$G_u = \frac{1}{6}s^2p_x$ $G_v = -\frac{1}{120}s^4p_y$ $G_\theta = -\frac{1}{24}s^3p_y$
$F_U = -\frac{1}{2}bp_x$ $F_V = -\frac{1}{2}bp_y$ $F_Z = -\frac{1}{6}(2s + a)bp_y$	**4.** 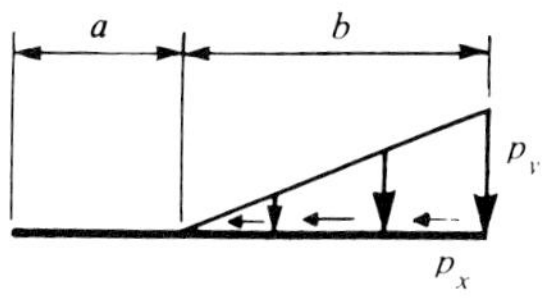	$G_U = \frac{1}{2}bp_x$ $G_V = \frac{1}{2}bp_y$ $G_Z = -\frac{1}{6}b^2p_y$
$F_U = \frac{1}{6}(2s + a)bp_x$ $F_v = -\frac{1}{120}b^{-1}(a^5 - 5bs^4 - s^5)p_y$ $F_\theta = \frac{1}{24}b^{-1}(a^4 - 4bs^3 - s^4)p_y$		$G_u = \frac{1}{6}b^2p_x$ $G_v = -\frac{1}{120}b^4p_y$ $G_\theta = -\frac{1}{24}b^3p_y$

4.06 TRANSPORT LOAD FUNCTIONS

Notation (4.02–4.03) | Signs (4.02) | Matrix (4.02)

u_j, v_j, θ_j = abrupt displacements (m, m, rad)

e_t, f_t = thermal strains (4.03)

1. Abrupt displacements

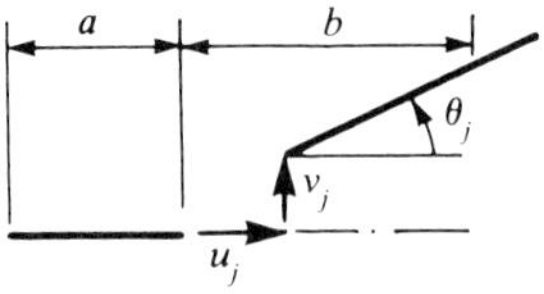

$F_U = 0$

$F_V = 0$

$F_Z = 0$

$F_u = -EA_x u_j$

$F_v = -EI_z(v_j - a\theta j)$

$F_\theta = -EI_z\theta_j$

$G_U = 0$

$G_V = 0$

$G_Z = 0$

$G_u = EA_x u_j$

$G_v = EI_z(v_j + b\theta_j)$

$G_\theta = EI_z\theta_j$

2. Temperature deformation of segment *b*

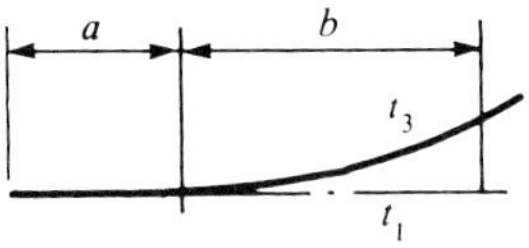

$F_U = 0$

$F_V = 0$

$F_Z = 0$

$F_u = -EA_x b e_t$

$F_v = \frac{1}{2}EI_z(s^2 - a^2)f_t$

$F_\theta = -EI_z b f_t$

$G_U = 0$

$G_V = 0$

$G_Z = 0$

$G_u = EA_x b e_t$

$G_v = \frac{1}{2}EI_z b^2 f_t$

$G_\theta = EI_z b f_t$

3.

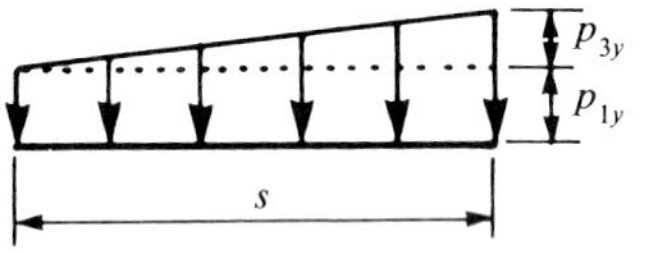

Sec. 4.05 (case 1 + case 3)

4.

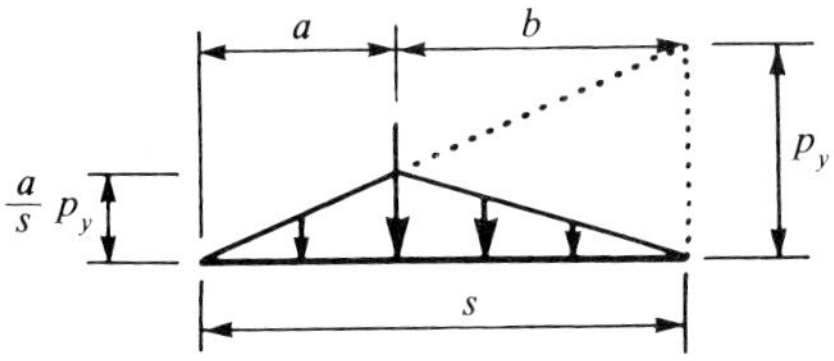

Sec. 4.05 (case 3 − case 4)

5.

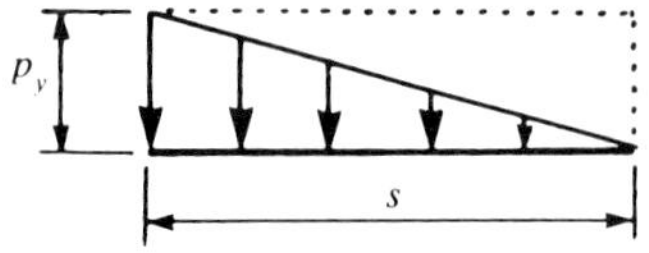

Sec. 4.05 (case 1 − case 3)

6.

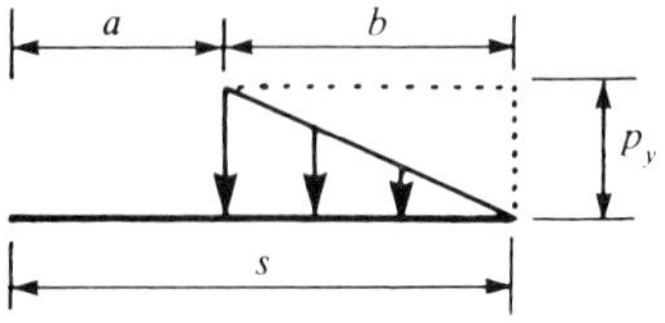

Sec. 4.05 (case 2 − case 4)

4.07 TRANSPORT METHOD

(1) Application of the transport method is restricted in this book to the calculation of forces, moments, and displacements in single-span beams. The transport method makes no distinctions between the statically determinate and indeterminate beams. There are always twelve boundary values involved, of which six are always known and six are unknown. Thus six equations, in general, are necessary for the solution of a given problem. The transport matrix equation provides these equations, of which two and only two must be solved simultaneously. Six typical cases are symbolically outlined in (2) below. This table shows that in all cases, regardless of the type of end conditions, the transport matrix remains the same.

(2) Typical cases (Ref. 3, p. 144)

1. Fixed-free beam	2. Free-fixed beam	3. Fixed-fixed beam
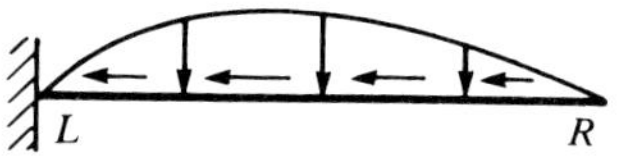	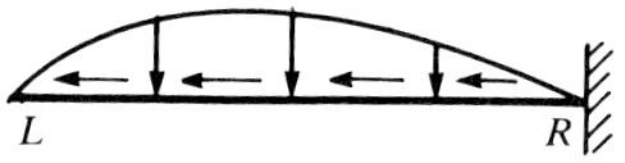	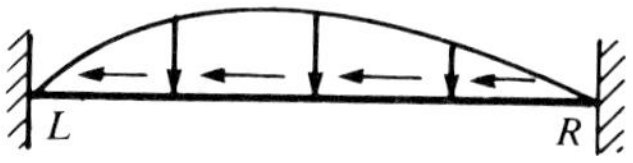
$\begin{bmatrix}1\\ \hline U_L\\ V_L\\ Z_L\\ \hline 0\\ 0\\ 0\end{bmatrix} = \begin{bmatrix} & T_{LR} & \end{bmatrix}\begin{bmatrix}1\\ \hline 0\\ 0\\ 0\\ \hline \bar{u}_R\\ \bar{v}_R\\ \bar{\theta}_R\end{bmatrix}$	$\begin{bmatrix}1\\ \hline 0\\ 0\\ 0\\ \hline \bar{u}_L\\ \bar{v}_L\\ \bar{\theta}_L\end{bmatrix} = \begin{bmatrix} & T_{LR} & \end{bmatrix}\begin{bmatrix}1\\ \hline U_R\\ V_R\\ Z_R\\ \hline 0\\ 0\\ 0\end{bmatrix}$	$\begin{bmatrix}1\\ \hline U_L\\ V_L\\ Z_L\\ \hline 0\\ 0\\ 0\end{bmatrix} = \begin{bmatrix} & T_{LR} & \end{bmatrix}\begin{bmatrix}1\\ \hline U_R\\ V_R\\ Z_R\\ \hline 0\\ 0\\ 0\end{bmatrix}$
4. Roller-hinge beam	5. Roller-fixed beam	6. Guide-fixed beam
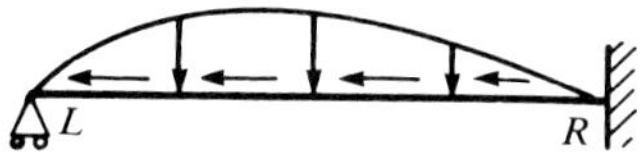	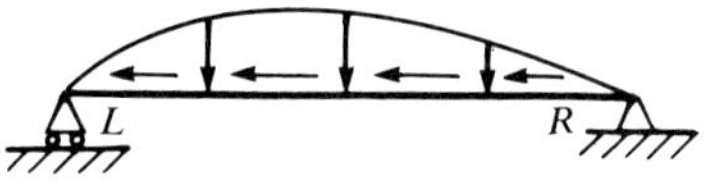	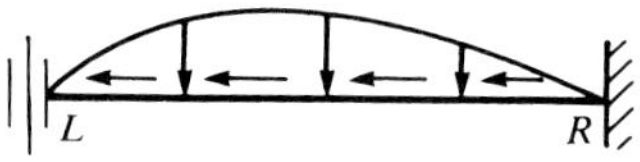
$\begin{bmatrix}1\\ \hline 0\\ V_L\\ 0\\ \hline \bar{u}_L\\ 0\\ \bar{\theta}_L\end{bmatrix} = \begin{bmatrix} & T_{LR} & \end{bmatrix}\begin{bmatrix}1\\ \hline U_R\\ V_R\\ 0\\ \hline 0\\ 0\\ \bar{\theta}_R\end{bmatrix}$	$\begin{bmatrix}1\\ \hline 0\\ V_L\\ 0\\ \hline \bar{u}_L\\ 0\\ \bar{\theta}_R\end{bmatrix} = \begin{bmatrix} & T_{LR} & \end{bmatrix}\begin{bmatrix}1\\ \hline U_R\\ V_R\\ Z_R\\ \hline 0\\ 0\\ 0\end{bmatrix}$	$\begin{bmatrix}1\\ \hline U_L\\ 0\\ Z_L\\ \hline 0\\ \bar{v}_L\\ 0\end{bmatrix} = \begin{bmatrix} & T_{LR} & \end{bmatrix}\begin{bmatrix}1\\ \hline U_R\\ V_R\\ Z_R\\ \hline 0\\ 0\\ 0\end{bmatrix}$

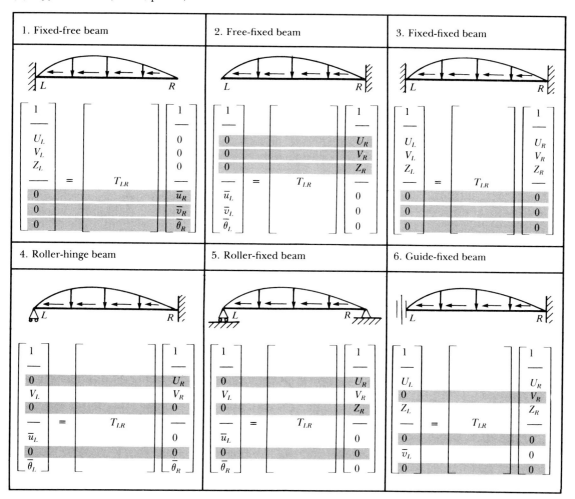

(3) Unit step function used in (4.12, 4.14, 4.16, 4.18, 4.20) is defined for $n = 0, 1, 2, 3, \ldots$ as

$$\langle x - a\rangle^n = \begin{cases} 0 & \text{if } x < a \\ (x-a)^n & \text{if } x \geq a \end{cases}$$

This function is used in the evaluation of load functions of the transport method.

4.08 CANTILEVER BEAM, DIAGRAMS†

Notation (4.02)	Signs (4.02)	Matrix (4.02)

Loaded beam

Concentrated load ① Pa , P , a , b , P , ℓ

Applied couple ② Q , a , b , Q , ℓ

Uniformly distributed load ③ $\dfrac{p\ell^2}{2}$, p , $p\ell$, ℓ

Shear

Concentrated: $-$, P , a , b

Applied couple: Zero shear , ℓ

Uniformly distributed: $-$, $p\ell$, ℓ

Bending moment

Concentrated: $-$, Pa , a , b

Applied couple: $-$, Q , a , b

Uniformly distributed: 2° Parabola , $-$, $\dfrac{p\ell^2}{2}$, ℓ

Loaded beam

Triangular load ④ $\dfrac{p\ell^2}{6}$, $\dfrac{px}{\ell}$, x , p , $\dfrac{p\ell}{2}$, ℓ

Parabolic load ⑤ $\dfrac{p\ell^2}{12}$, $\dfrac{px^2}{\ell^2}$, 2° Parabola , x , p , $\dfrac{p\ell}{3}$, ℓ

Sinusoidal load ⑥ $\dfrac{p\ell^2}{\pi}$, x , p , $\dfrac{2p\ell}{\pi}$, $p\sin \pi x/\ell$, ℓ

Shear

Triangular: 2° Parabola , $-$, $\dfrac{p\ell}{2}$, ℓ

Parabolic: 3° Parabola , $-$, $\dfrac{p\ell}{3}$, ℓ

Sinusoidal: $\dfrac{p\ell}{\pi}$, Cosine curve , $-$, $\dfrac{2p\ell}{\pi}$, $\ell/2$, $\ell/2$

Bending moment

Triangular: 3° Parabola , $-$, 3° , $\dfrac{p\ell^2}{6}$, ℓ

Parabolic: 4° Parabola , $-$, 4° , $\dfrac{p\ell^2}{12}$, ℓ

Sinusoidal: Declining sine curve , $\dfrac{\pi p\ell^2}{2}$, $-$, $\dfrac{p\ell^2}{\pi}$, $\ell/2$, $\ell/2$

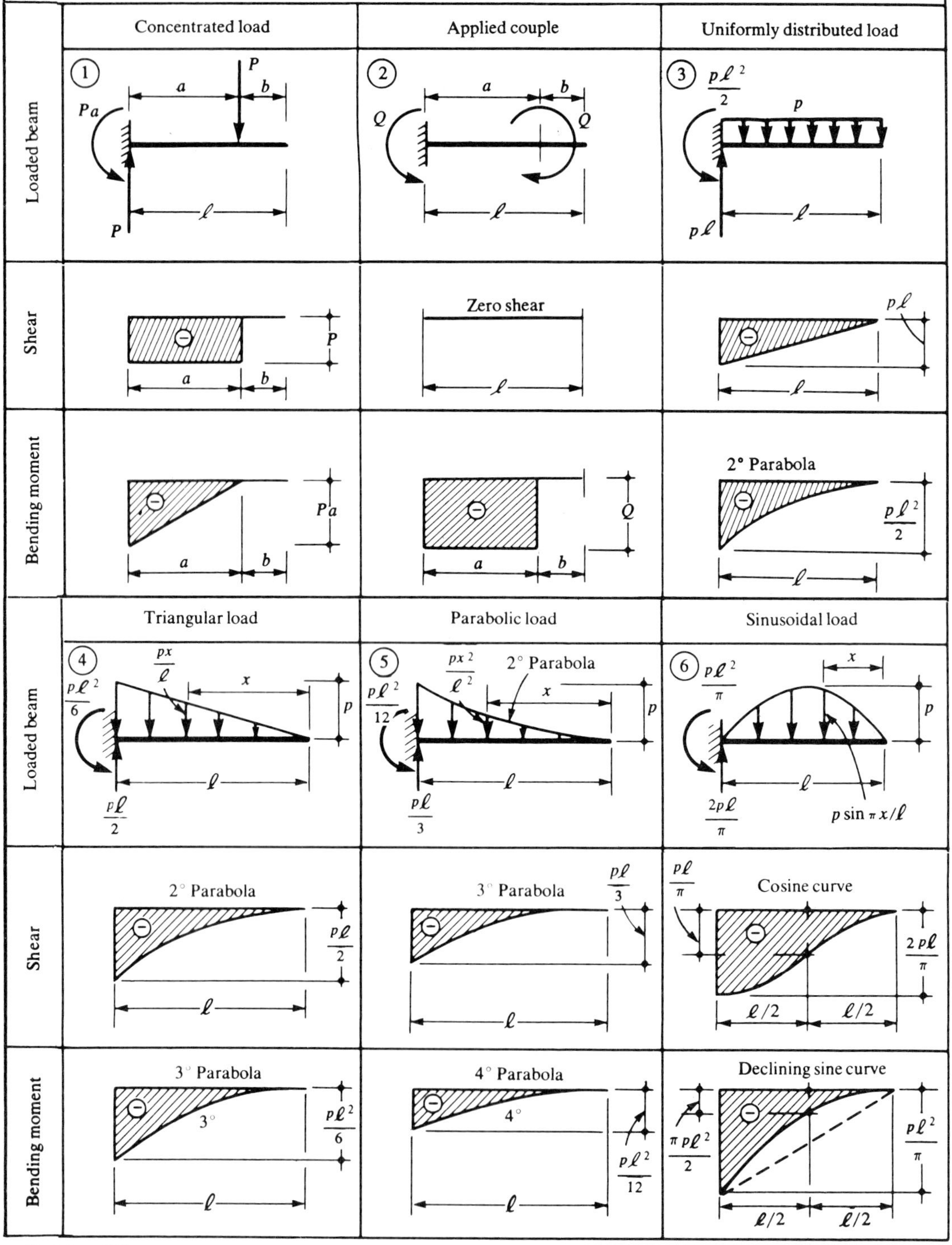

†Ref. 7, p. 119.

4.09 SIMPLE BEAM, DIAGRAMS†

Notation (4.02)	Signs (4.02)	Matrix (4.02)

	Concentrated load	Applied couple	Uniformly distributed load
Loaded beam	① $\dfrac{Pb}{\ell}$ $\dfrac{Pa}{\ell}$	② $\dfrac{Q}{\ell}$ $\dfrac{Q}{\ell}$	③ $\dfrac{p\ell}{2}$ $\dfrac{p\ell}{2}$
Shear	$\dfrac{Pa}{\ell}$ $\dfrac{Pb}{\ell}$ a b P	$\dfrac{Q}{\ell}$ ℓ	$\dfrac{p\ell}{2}$ $\ell/2$ $\ell/2$ $p\ell$
Bending moment	$\dfrac{Pab}{\ell}$ a b	$\dfrac{Qb}{\ell}$ $\dfrac{Qa}{\ell}$ a b Q	2° Parabola $\dfrac{p\ell^2}{8}$ $\ell/2$ $\ell/2$

	Triangular load	Parabolic load	Sinusoidal load
Loaded beam	④ x $\dfrac{px}{\ell}$ p ℓ $\dfrac{p\ell}{6}$ $\dfrac{p\ell}{3}$	⑤ x $\dfrac{px^2}{\ell^2}$ p 2° Parabola ℓ $\dfrac{p\ell}{12}$ $\dfrac{p\ell}{4}$	⑥ x $p\sin\dfrac{\pi x}{\ell}$ p ℓ $\dfrac{p\ell}{\pi}$ $\dfrac{p\ell}{\pi}$
Shear	$\ell\sqrt3/3 \cong 0.577\ell$ $\dfrac{p\ell}{6}$ $\dfrac{p\ell}{3}$ 2° Parabola	$\ell/\sqrt[3]{4} \cong 0.630\,\ell$ $\dfrac{p\ell}{12}$ $\dfrac{p\ell}{4}$ 3° Parabola	$\ell/2$ $\dfrac{p\ell}{\pi}$ $\dfrac{p\ell}{\pi}$ Cosine curve
Bending moment	$\dfrac{p\ell^2\sqrt3}{27}$ $\ell\sqrt3/3$ 3° Parabola	$\dfrac{p\ell^2}{16\sqrt[3]{4}}$ $\ell/\sqrt[3]{4}$ 4° Parabola	$\dfrac{p\ell^2}{\pi^2}$ $\ell/2$ Sine curve

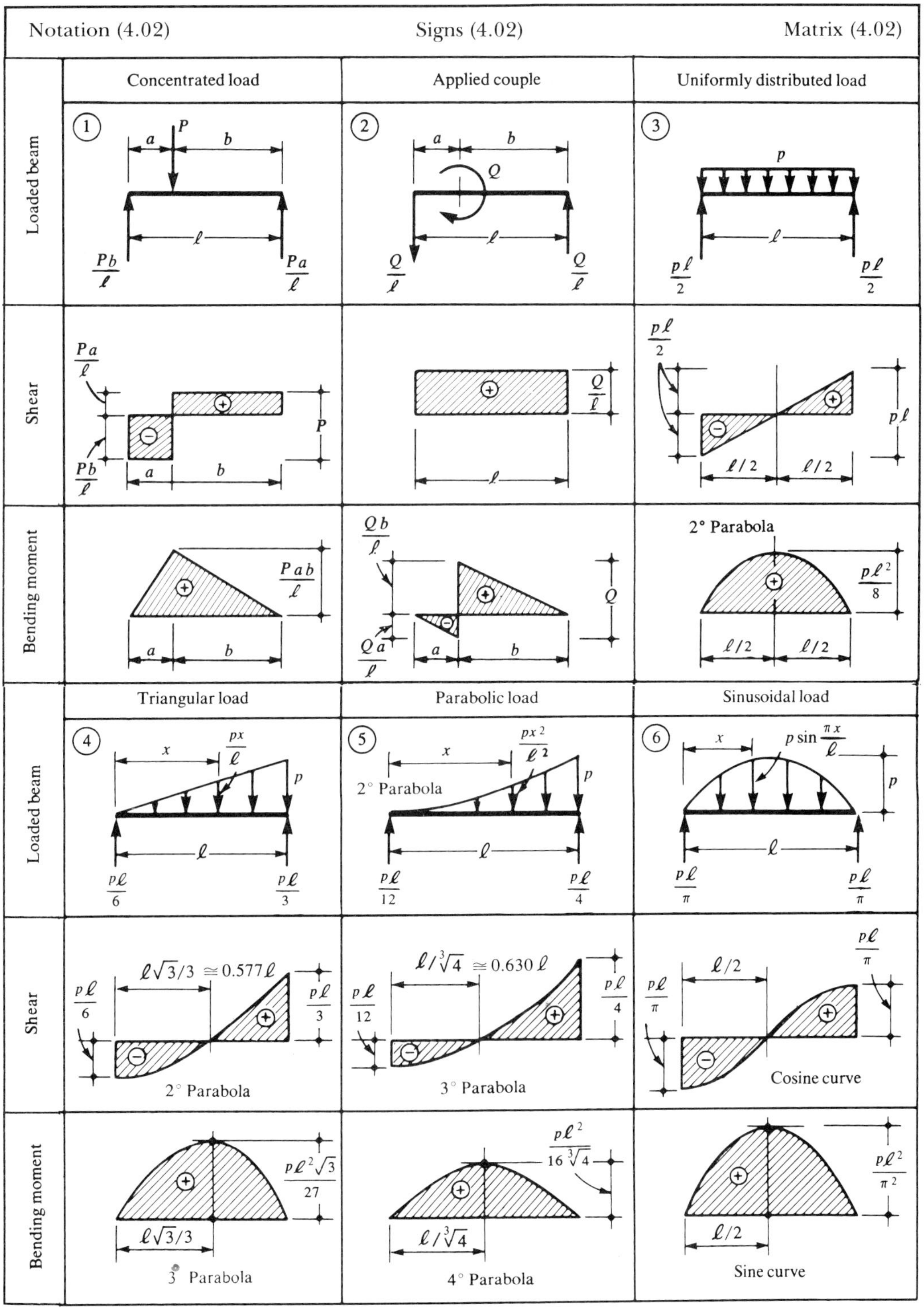

†Ref. 7, p. 118.

4.10 PROPPED-END BEAM, DIAGRAMS†

Notation (4.02) Signs (4.02) Matrix (4.02)

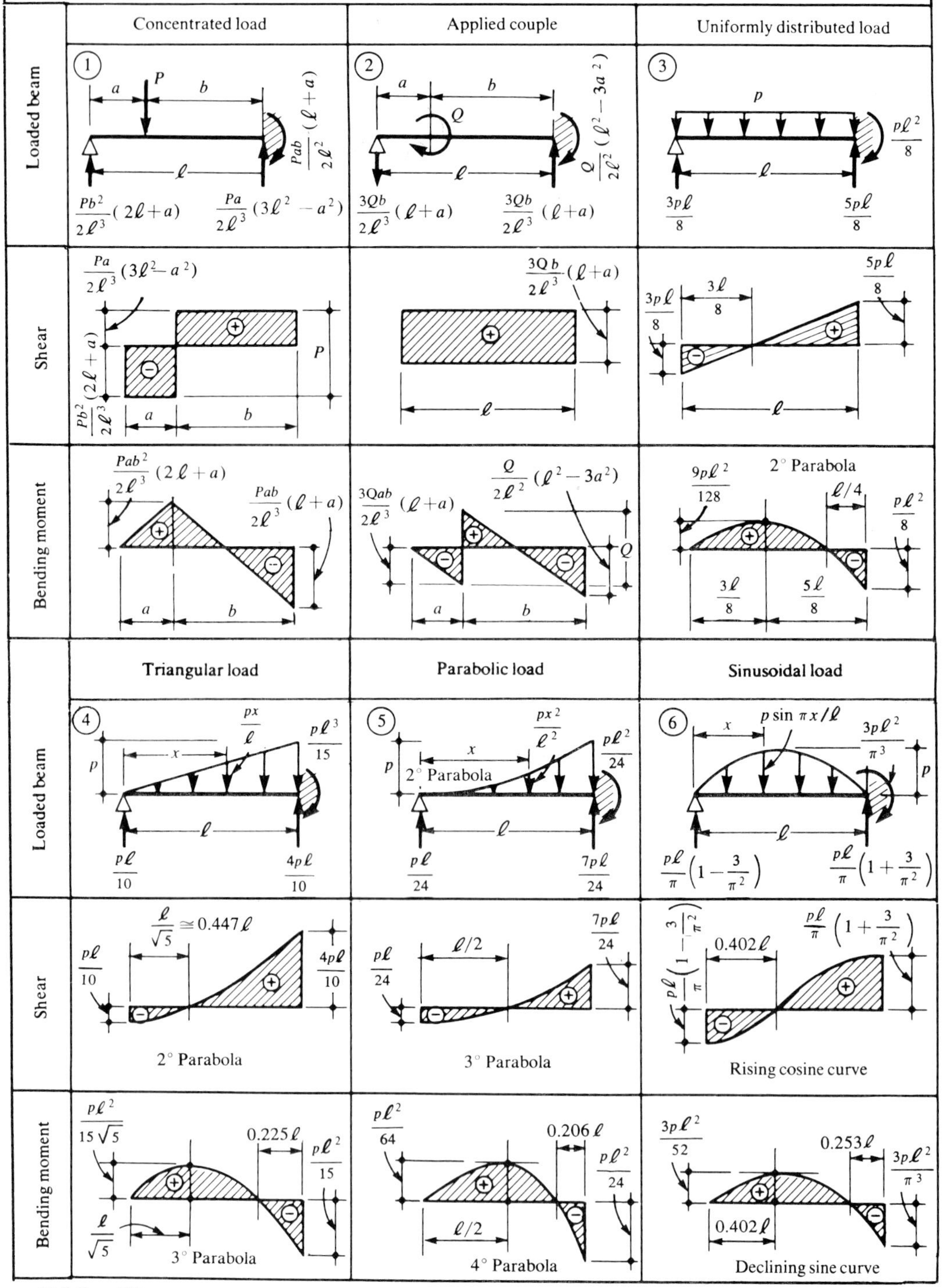

†Ref. 7, p. 120.

4.11 FIXED-END BEAM, DIAGRAMS†

Notation (4.02) | Signs (4.02) | Matrix (4.02)

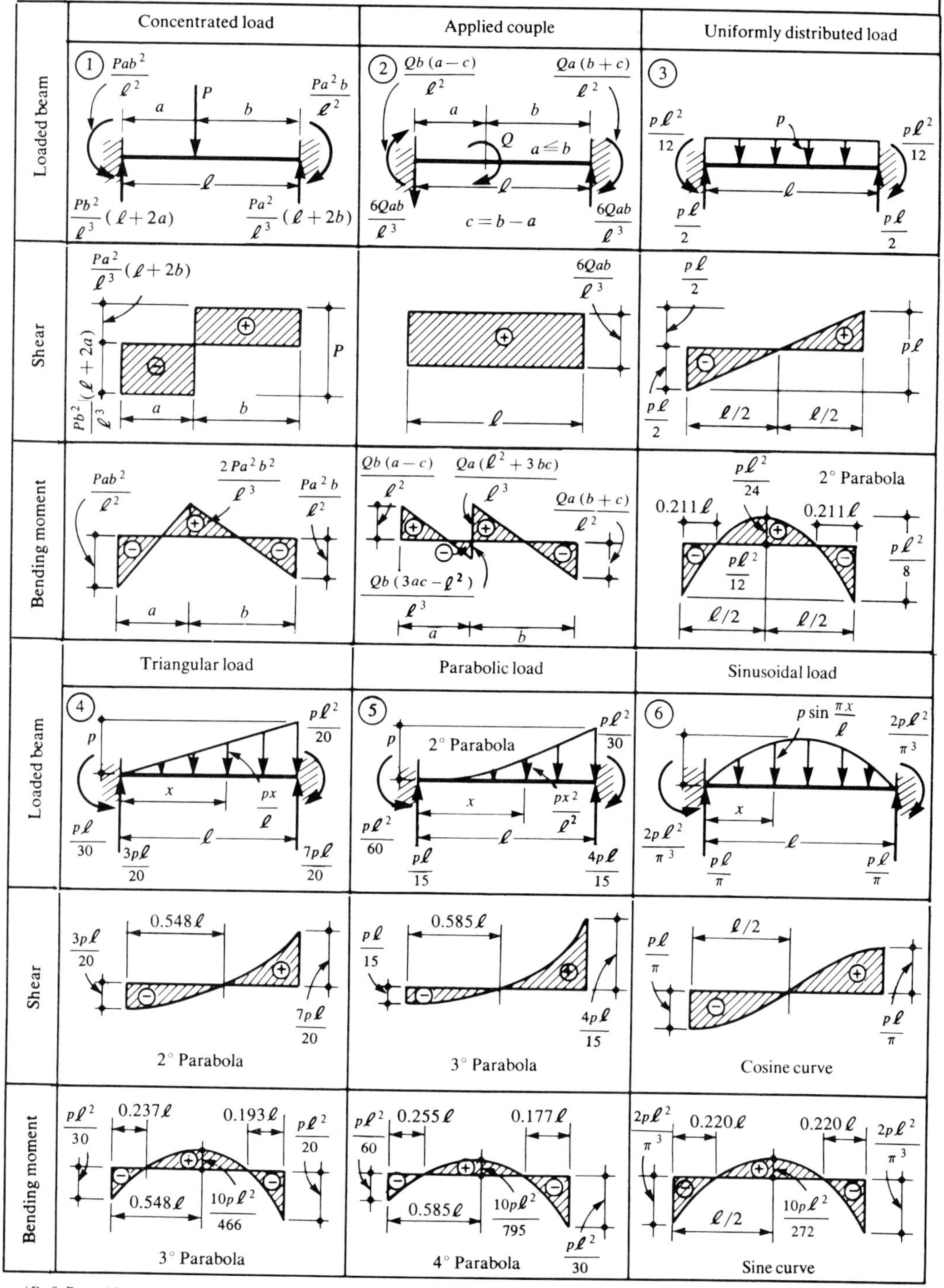

†Ref. 7, p. 121.

4.12 ELASTIC CURVE, CONCENTRATED LOAD

Matrix (4.02, 4.03) $\quad \bar{v} = EI_z v \quad \bar{\theta} = EI_z\theta \quad m = \dfrac{a}{s} \quad n = \dfrac{b}{s}$

v_{max} = maximum deflection (m) $\quad v_C$ = central deflection (m)

x_1, x_2 = position coordinate of v_{max} measured from L, R, respectively

Equation of elastic curve: $$\bar{v} = \bar{v}_L + \bar{\theta}_L x + Z_L \frac{x^2}{2} - V_L \frac{x^3}{6} - P\frac{\langle x - a\rangle^3}{6} \qquad (4.07\text{-}3)$$

(1) Cantilever beam

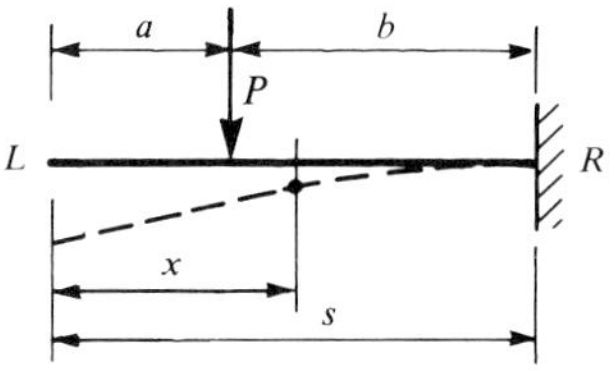

$V_L = 0$

$Z_L = 0$

$\bar{v}_L = -\dfrac{Psb^2}{6}(3 - n)$

$\bar{\theta}_L = \dfrac{Pb^2}{2}$

$V_R = P$

$Z_R = -Pb$

$\bar{v}_R = 0$

$\bar{\theta}_R = 0$

$a = 0 \qquad \bar{v}_{max} = -\dfrac{Ps^3}{3} \quad$ at L

(2) Propped-end beam

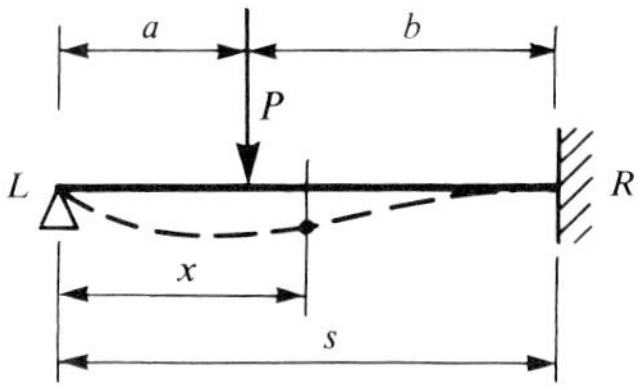

$V_L = -\dfrac{P}{2}(2 + m)n^2$

$Z_L = 0$

$\bar{v}_L = 0$

$\bar{\theta}_L = -\dfrac{P}{4}b^2 m$

$V_R = \dfrac{P}{2}(3 - m^2)m$

$Z_R = -\dfrac{Pa}{2}(1 - m^2)$

$\bar{v}_R = 0$

$\bar{\theta}_R = 0$

$$\bar{v}_{max} = \begin{cases} -\dfrac{Ps^3 m(1 - m^2)^3}{3(3 - m^2)^2} & \text{at } x_1 = s\dfrac{1 + m^2}{3 - m^2} \quad a < 0.4142s \\ -\dfrac{Ps^3}{48\sqrt{5}} & \text{at } x_1 = \dfrac{s}{\sqrt{5}} \quad a = b \\ -\dfrac{Ps^2 mn^2 x_1}{6} & \text{at } x_1 = s\sqrt{\dfrac{m}{2 + m}} \quad a \geq 0.4142s \end{cases}$$

(3) Fixed-end beam

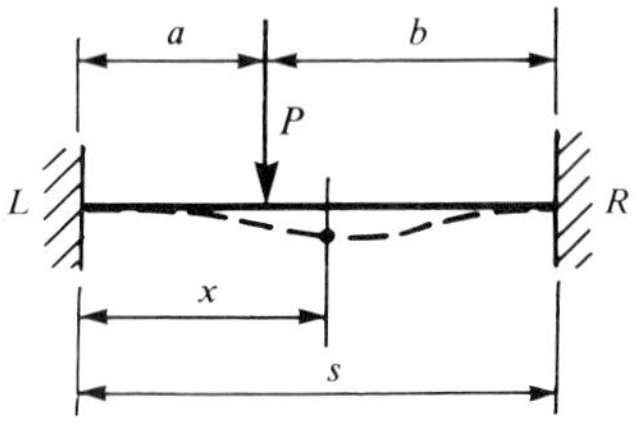

$V_L = -P(1 + 2m)n^2$

$Z_L = -Pan^2$

$\bar{v}_L = 0$

$\bar{\theta}_L = 0$

$V_R = P(1 + 2n)m^2$

$Z_R = -Pbm^2$

$\bar{v}_R = 0$

$\bar{\theta}_R = 0$

$$\bar{v}_{max} = \begin{cases} \dfrac{2Z_R^3}{3V_R^2} & \text{at } x_2 = -\dfrac{2Z_R}{V_R} \quad a < b \\ -\dfrac{Ps^3}{192} & \text{at } x_1 = \dfrac{s}{2} \quad a = b \\ \dfrac{2Z_L^3}{3V_L^2} & \text{at } x_1 = \dfrac{2Z_L}{V_L} \quad a > b \end{cases}$$

4.13 DEFLECTION OF SIMPLE BEAM, CONCENTRATED LOAD

For definition of symbols and equation of elastic curve refer to (4.02, 4.03, 4.12).

(1) Simple beam

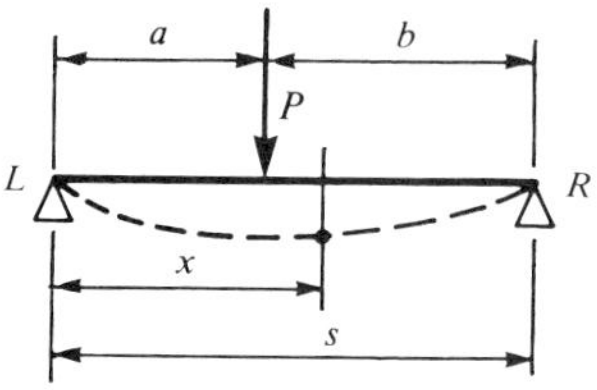

$$V_L = -Pn \qquad V_R = Pm$$

$$Z_L = 0 \qquad Z_R = 0$$

$$\bar{v}_L = 0 \qquad \bar{v}_R = 0$$

$$\bar{\theta}_L = -\frac{Pab(1+n)}{6} \qquad \bar{\theta}_R = \frac{Pab(1+m)}{6}$$

$$\bar{v}_{max} = \begin{cases} -\dfrac{Pmx_2^3}{3} & \text{at } x_2 = s\sqrt{\dfrac{1-m^2}{3}} & a < b \\[2ex] -\dfrac{Ps^3}{48} & \text{at } x_1 = \dfrac{s}{2} & a = b \\[2ex] -\dfrac{Pnx_1^3}{3} & \text{at } x_1 = s\sqrt{\dfrac{1-n^2}{3}} & a > b \end{cases}$$

(2) Maximum deflection v_{max} of a simple beam produced by transverse loads acting in the same direction can be always approximated by the central deflection v_C. The error of this approximation cannot be greater than 2.57 percent (Ref. 3, pp. 121–122).

(3) Central deflection coefficients

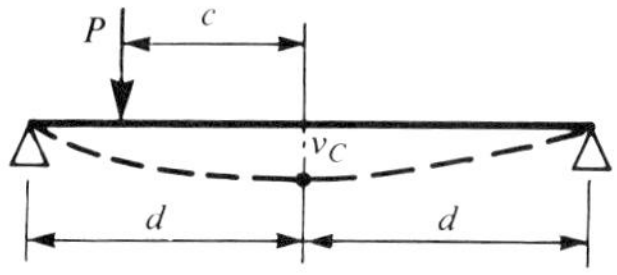

$$v_C = -\frac{Pd^3}{10EI_z}A_k \qquad A_k = \frac{10(2 - 3k^2 + k^3)}{12}$$

$$k = \frac{c}{d} \qquad A_{max} = \frac{10}{6} \quad \text{at } k = 0.00$$

100-point table of coefficients A_k (Ref. 16)

k	0	1	2	3	4	5	6	7	8	9
0.0	1.6667	1.6664	1.6657	1.6644	1.6627	1.6605	1.6579	1.6547	1.6511	1.6469
0.1	1.6425	1.6375	1.6321	1.6263	1.6200	1.6132	1.6061	1.5985	1.5901	1.5821
0.2	1.5733	1.5641	1.5545	1.5446	1.5342	1.5226	1.5123	1.5001	1.4890	1.4767
0.3	1.4642	1.4513	1.4380	1.4244	1.4104	1.3961	1.3815	1.3666	1.3514	1.3359
0.4	1.3200	1.3039	1.2874	1.2707	1.2537	1.2364	1.2188	1.2009	1.1828	1.1645
0.5	1.1458	1.1270	1.1078	1.0885	1.0689	1.0491	1.0290	1.0095	0.9883	0.9676
0.6	0.9467	0.9256	0.9043	0.8828	0.8611	0.8393	0.8172	0.7951	0.7727	0.7502
0.7	0.7275	0.7047	0.6817	0.6586	0.6354	0.6120	0.5885	0.5649	0.5411	0.5173
0.8	0.4933	0.4393	0.4451	0.4209	0.3966	0.3722	0.3477	0.3232	0.2986	0.2739
0.9	0.2492	0.2244	0.1996	0.1747	0.1498	0.1249	0.0999	0.0750	0.0500	0.0250

(4) Superposition

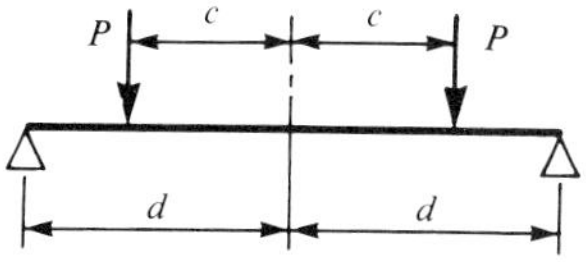

$$v_C = -\frac{2Pd^3}{10EI_z}A_k$$

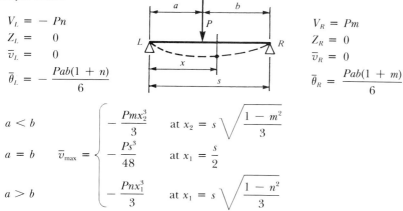

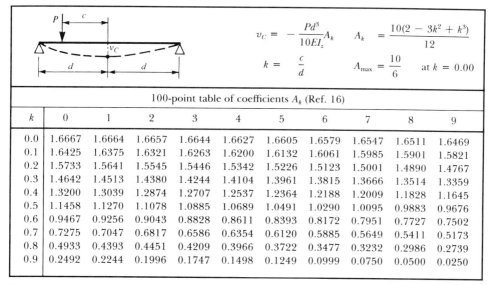

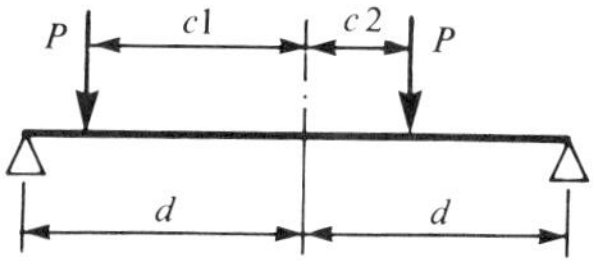

$$v_C = -\frac{Pd^3}{10EI_z}(A_{k1} + A_{k2})$$

4.14 ELASTIC CURVE, UNIFORMLY DISTRIBUTED LOAD

Matrix (4.02, 4.03) $\quad \bar{v} = EI_z v \quad \bar{\theta} = EI_z\theta \quad m = \frac{a}{s} \quad n = \frac{b}{s}$

v_{max} = maximum deflection (m) $\quad v_C$ = central deflection (m)

$$e = \frac{1}{2p}\left(3V_R - \sqrt{9V_R^2 + 24pM_R}\right)$$

Equation of elastic curve: $\bar{v} = \bar{v}_L + \bar{\theta}_L x + Z_L \frac{x^2}{2} - V_L \frac{x^3}{6} - \frac{p\langle x - a\rangle^4}{24}$ (4.07-3)

(1) Cantilever beam

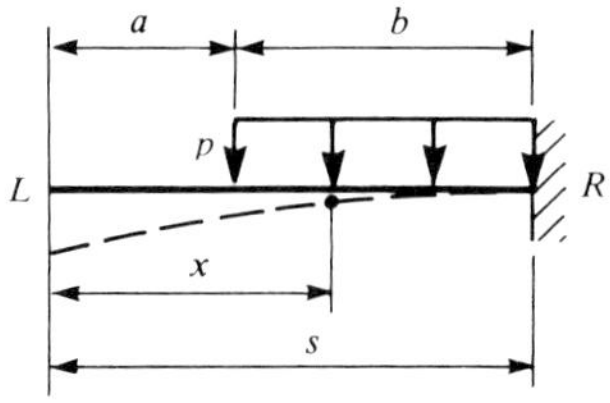

$V_L = 0$

$Z_L = 0$

$\bar{v}_L = -\frac{pb^3(4-n)}{24n}$

$\bar{\theta}_L = \frac{pb^3}{6}$

$V_R = pb$

$Z_R = -\frac{pb^2}{2}$

$\bar{v}_R = 0$

$\bar{\theta}_R = 0$

$a = 0 \quad \bar{v}_{max} = -\frac{ps^4}{8}$ at L

(2) Propped-end beam

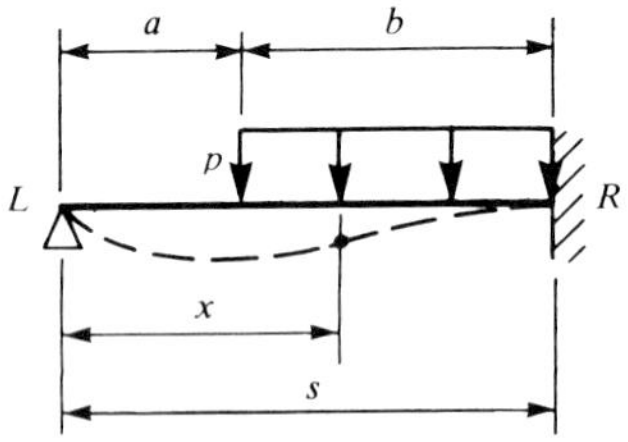

$V_L = -\frac{pb}{8} n^2(4 - n)$

$Z_L = 0$

$\bar{v}_L = 0$

$\bar{\theta}_L = -\frac{pb^3}{48}(4 - 3n)$

$V_R = \frac{pb}{8}(8 - 4n^2 + n^3)$

$Z_R = -\frac{pb^2}{8}(2 - n)^2$

$\bar{v}_R = 0$

$\bar{\theta}_R = 0$

$$\bar{v}_{max} = \begin{cases} -\dfrac{ps^4}{185} & a = 0, \text{ at } x_1 = 0.4215s \\ V_R \dfrac{e^3}{6} + Z_R \dfrac{e^2}{2} + \dfrac{pe^4}{24} & a < 0.4845s, \text{ at } x_1 = s - e \\ \dfrac{2\bar{\theta}_L}{3} x_1 & a \geq 0.4845s, \text{ at } x_1 = \sqrt{\dfrac{2\bar{\theta}_L}{V_L}} \end{cases}$$

(3) Fixed-end beam

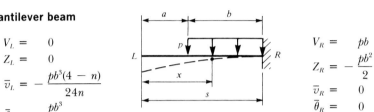

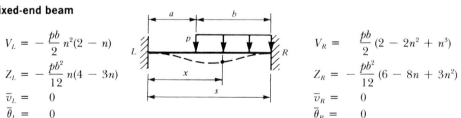

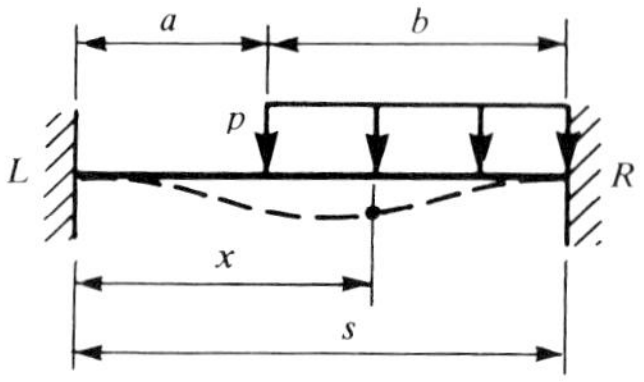

$V_L = -\frac{pb}{2} n^2(2 - n)$

$Z_L = -\frac{pb^2}{12} n(4 - 3n)$

$\bar{v}_L = 0$

$\bar{\theta}_L = 0$

$V_R = \frac{pb}{2}(2 - 2n^2 + n^3)$

$Z_R = -\frac{pb^2}{12}(6 - 8n + 3n^2)$

$\bar{v}_R = 0$

$\bar{\theta}_R = 0$

$$\bar{v}_{max} = \begin{cases} -\dfrac{ps^4}{384} & a = 0, \text{ at } x_1 = \dfrac{s}{2} \\ V_R \dfrac{e^3}{6} + Z_R \dfrac{e^2}{2} + \dfrac{pe^4}{24} & a < 0.5774s, \text{ at } x_1 = s - e \\ \dfrac{2Z_L^3}{3V_L^2} & a \geq 0.5774s, \text{ at } x_1 = \dfrac{2Z_L}{V_L} \end{cases}$$

4.15 DEFLECTION OF SIMPLE BEAM, UNIFORMLY DISTRIBUTED LOAD

For definition of symbols and equation of elastic curve refer to (4.02, 4.03, 4.14).

(1) Simple beam

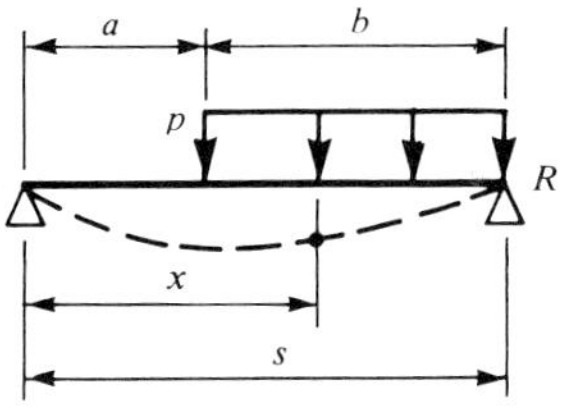

$V_L = -\dfrac{pbn}{2}$ $\qquad$ $V_R = \dfrac{pb}{2}(2 - n)$

$Z_L = 0$ $\qquad$ $Z_R = 0$

$\bar{v}_L = 0$ $\qquad$ $\bar{v}_R = 0$

$\bar{\theta}_L = -\dfrac{pb^3}{24n}(2 - n^2)$ $\qquad$ $\bar{\theta}_R = \dfrac{pb^3}{24n}(2 - n)^2$

$a = 0 \qquad \bar{v}_{max} = -\dfrac{5ps^4}{384} \qquad$ at $x_1 = \dfrac{s}{2}$

$a < b \qquad \bar{v}_{max} \cong \bar{v}_C = -\dfrac{ps^4}{384}(5 - 12m^2 + 8m^4) \qquad$ at $x_1 = \dfrac{s}{2}$

$a \geq b \qquad \bar{v}_{max} \cong \bar{v}_C = -\dfrac{ps^4}{96}(3 - 2n^2)n^2 \qquad$ at $x_1 = \dfrac{s}{2}$

(2) Central deflection coefficients†

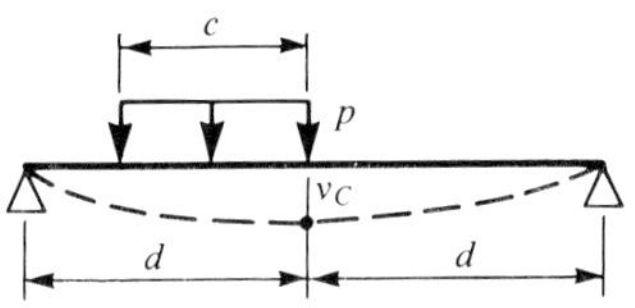

$v_C = -\dfrac{pd^4}{10EI_z}B_k \qquad B_k = \dfrac{10(8k - 4k^3 + k^4)}{48}$

$k = \dfrac{c}{d} \qquad B_{max} = \dfrac{50}{48} \qquad$ at $k = 1.00$

100-point table of coefficients B_k (Ref. 16)

k	0	1	2	3	4	5	6	7	8	9
0.0	0.0000	0.0167	0.0333	0.0500	0.0667	0.0832	0.0982	0.1164	0.1329	0.1494
0.1	0.1659	0.1823	0.1986	0.2148	0.2311	0.2473	0.2634	0.2794	0.2912	0.3112
0.2	0.3270	0.3427	0.3583	0.3738	0.3892	0.4045	0.4196	0.4347	0.4497	0.4645
0.3	0.4792	0.4938	0.5082	0.5225	0.5367	0.5507	0.5646	0.5784	0.5920	0.6054
0.4	0.6187	0.6318	0.6447	0.6575	0.6702	0.6826	0.6949	0.7070	0.7190	0.7304
0.5	0.7422	0.7536	0.7647	0.7757	0.7865	0.7971	0.8075	0.8177	0.8277	0.8374
0.6	0.8470	0.8564	0.8655	0.8745	0.8832	0.8917	0.9000	0.9080	0.9159	0.9235
0.7	0.9309	0.9380	0.9450	0.9517	0.9581	0.9644	0.9704	0.9761	0.9817	0.9869
0.8	0.9920	0.9968	1.0014	1.0057	1.0098	1.0136	1.0173	1.0206	1.0237	1.0266
0.9	1.0292	1.0316	1.0337	1.0356	1.0372	1.0385	1.0397	1.0405	1.0412	1.0415

†For approximation of v_{max} by v_C, refer to (4.13-2).

(3) Superposition

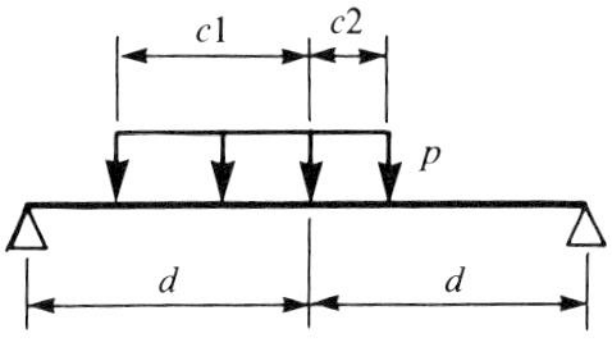

$v_C = -\dfrac{pd^4}{10EI_z}(B_{k1} + B_{k2})$

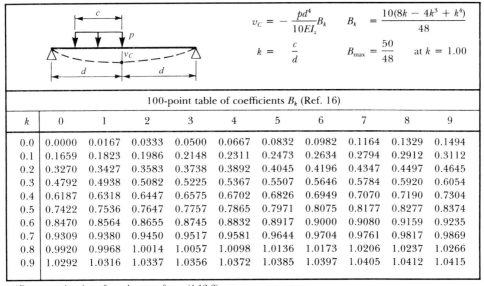

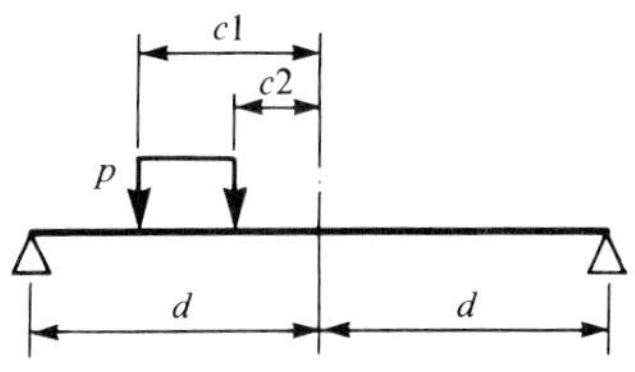

$v_L = -\dfrac{pd^4}{10EI_z}(B_{k1} - B_{k2})$

4.16 ELASTIC CURVE, INCREASING TRIANGULAR LOAD

Matrix (4.02, 4.03) $\quad \bar{v} = EI_z v \quad \bar{\theta} = EI_z \theta \quad m = \dfrac{a}{s} \quad n = \dfrac{b}{s}$

$v_{\max}$ = maximum deflection (m) $\quad v_C$ = central deflection (m)

e, f = roots of quartic equations given below to be used as position coordinates of $v_{\max}$

$$0 = \bar{\theta}_L - V_L \frac{e^2}{2} - \frac{p(e-a)^4}{24b} \qquad 0 = Z_L f - V_L \frac{f^2}{2} - \frac{p(f-a)^4}{24b}$$

Equation of elastic curve: $\quad \bar{v} = \bar{v}_L + \bar{\theta}_L x + Z_L \dfrac{x^2}{2} - V_L \dfrac{x^3}{6} - \dfrac{p\langle x-a\rangle^5}{120b} \qquad$ (4.07-3)

(1) Cantilever beam

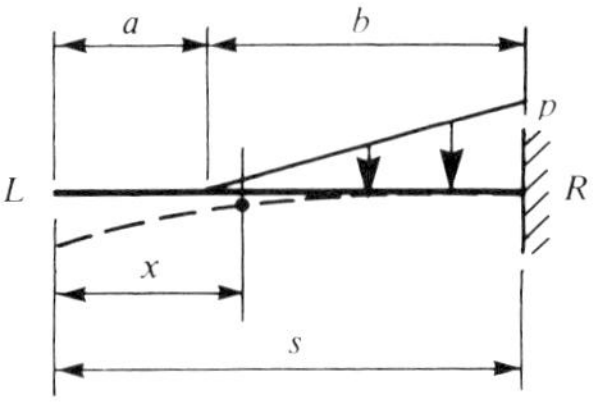

$V_L = 0$

$Z_L = 0$

$\bar{v}_L = -\dfrac{pb^4(5-n)}{120n}$

$\bar{\theta}_L = \dfrac{pb^3}{24}$

$V_R = \dfrac{pb}{2}$

$Z_R = -\dfrac{pb^2}{6}$

$\bar{v}_R = 0$

$\bar{\theta}_R = 0$

$a = 0 \qquad \bar{v}_{\max} = -\dfrac{ps^4}{30} \qquad$ at L

(2) Propped-end beam

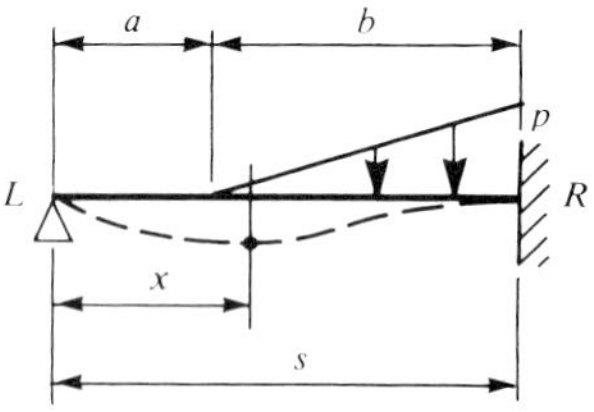

$V_L = -\dfrac{pb}{40}(5-n)n^2$

$Z_L = 0$

$\bar{v}_L = 0$

$\bar{\theta}_L = -\dfrac{pb^3}{240}(5-3n)$

$V_R = \dfrac{pb}{40}(20 - 5n + n^3)$

$Z_R = -\dfrac{pb^2}{120}(20 - 15n + 3n^2)$

$\bar{v}_R = 0$

$\bar{\theta}_R = 0$

$$\bar{v}_{\max} = \begin{cases} -\dfrac{4ps^4}{1677} & a = 0, \text{ at } x_1 = 0.4472s \\ \bar{\theta}_L e - V_L \dfrac{e^3}{6} - \dfrac{p\langle e-a\rangle^5}{120b} & a < 0.4472s, \text{ at } x_1 = e \\ -\dfrac{2\bar{\theta}_L}{3} x_1 & a \geq 0.4472s, \text{ at } x_1 = \sqrt{\dfrac{2\bar{\theta}_L}{V_L}} \end{cases}$$

(3) Fixed-end beam

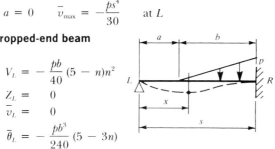

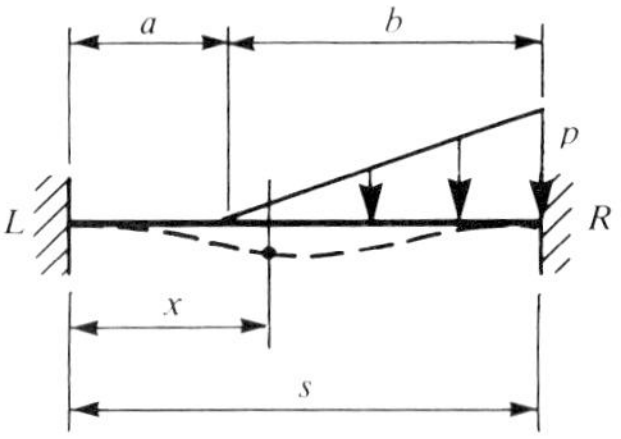

$V_L = -\dfrac{pb}{24}(5-2n)n^2$

$Z_L = -\dfrac{pb^2}{60}(5-3n)n$

$\bar{v}_L = 0$

$\bar{\theta}_L = 0$

$V_R = \dfrac{pb}{20}(10 - 5n^2 + 2n^3)$

$Z_R = -\dfrac{pb^2}{60}(10 - 10n + 3n^2)$

$\bar{v}_R = 0$

$\bar{\theta}_R = 0$

$a = 0 \qquad \bar{v}_{\max} \cong -\dfrac{ps^4}{764} \qquad$ at $x_1 = 0.5247s$

$a < 0.6039s \qquad \bar{v}_{\max} = Z_L \dfrac{f^2}{2} - V_L \dfrac{f^3}{6} - \dfrac{p\langle f-a\rangle^5}{120b} \qquad$ at $x_1 = f$

$a \geq 0.6039s \qquad \bar{v}_{\max} = \dfrac{V_L}{12} x_1^3 \qquad$ at $x_1 = \dfrac{2Z_L}{V_L}$

4.17 DEFLECTION OF SIMPLE BEAM, INCREASING TRIANGULAR LOAD

For definition of symbols and equation of elastic curve refer to (4.02, 4.03, 4.16).

(1) Simple beam

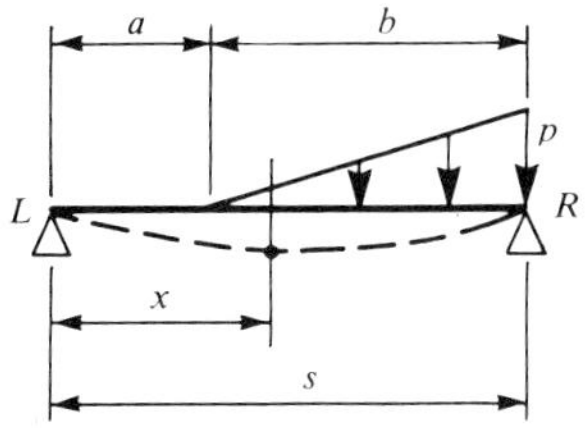

$$V_L = -\frac{pbn}{6} \qquad V_R = \frac{pb}{6}(3 - n)$$

$$Z_L = 0 \qquad Z_R = 0$$

$$\bar{v}_L = 0 \qquad \bar{v}_R = 0$$

$$\bar{\theta}_L = -\frac{pb^3}{360n}(10 - 3n^2) \qquad \bar{\theta}_R = \frac{pb^3}{360n}(20 - 15n + 3n^2)$$

$a = 0$ $\quad \bar{v}_{\max} \cong -\dfrac{3ps^4}{460}$ $\quad$ at $x_1 \cong 0.5193s$

$a < b$ $\quad \bar{v}_{\max} \cong \bar{v}_C = -\dfrac{ps^4}{3840n}(25 - 50m + 40m^3 - 16m^5)$ $\quad$ at $x_1 = \dfrac{s}{2}$

$a = b$ $\quad \bar{v}_{\max} \cong \bar{v}_C = -\dfrac{3ps^4}{1280}$ $\quad$ at $x_1 = \dfrac{s}{2}$

$a \geq b$ $\quad \bar{v}_{\max} \cong \bar{v}_C = -\dfrac{ps^4}{480}(5 - 2n^2)n^2$ $\quad$ at $x_1 = \dfrac{s}{2}$

(2) Central deflection coefficients†

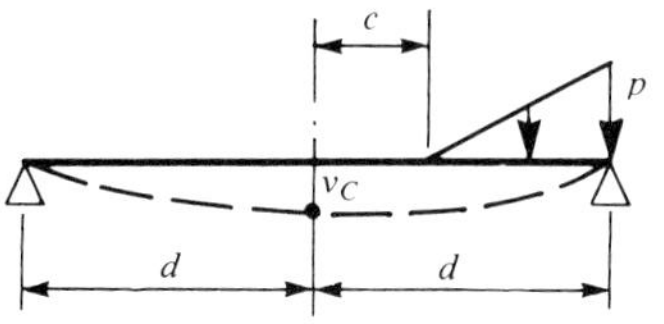

$$v_C = -\frac{pd^4}{100EI_z}C_k \qquad C_k = \frac{10(1 - k)^2(9 + 2k - k^2)}{24}$$

$$k = \frac{c}{d} \qquad C_{\max} = \frac{90}{24} \quad \text{for } k = 0.00$$

100-point table of coefficients C_k (Ref. 16)

k	0	1	2	3	4	5	6	7	8	9
0.0	3.7500	3.6835	3.6173	3.5513	3.4861	3.4210	3.3563	3.2920	3.2282	3.1647
0.1	3.1016	3.0390	2.9768	2.9087	2.8537	2.7929	2.7325	2.6727	2.6133	2.5544
0.2	2.4960	2.4381	2.3808	2.3239	2.2676	2.2119	2.1567	2.1021	2.0480	1.9945
0.3	1.9416	1.8593	1.8376	1.7864	1.7359	1.6860	1.6367	1.5881	1.5401	1.4927
0.4	1.4460	1.4000	1.3545	1.3098	1.2657	1.2223	1.1796	1.1375	1.0962	1.0596
0.5	1.0156	0.9764	0.9379	0.9001	0.8630	0.8267	0.7910	0.7562	0.7220	0.6886
0.6	0.6560	0.6242	0.5930	0.5626	0.5330	0.5042	0.4761	0.4488	0.4223	0.3966
0.7	0.3716	0.3475	0.3241	0.3015	0.2790	0.2588	0.2386	0.2192	0.2007	0.1829
0.8	0.1660	0.1500	0.1346	0.1201	0.1064	0.0935	0.0815	0.0703	0.0600	0.0503
0.9	0.0416	0.0337	0.0266	0.0204	0.0150	0.0104	0.0067	0.0037	0.0017	0.0004

†For approximation of $v_{\max}$ by v_C, refer to (4.13-2).

(3) Superposition [for $B_{1.00}$ see (4.15-2)]

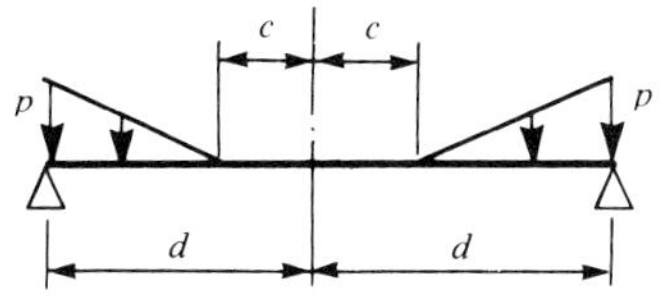

$$v_C = -\frac{2pd^4}{100EI_z}C_k$$

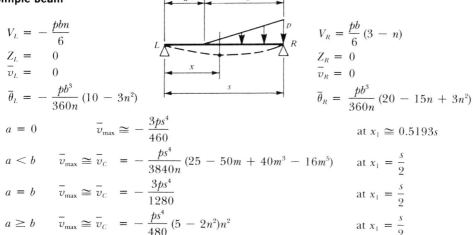

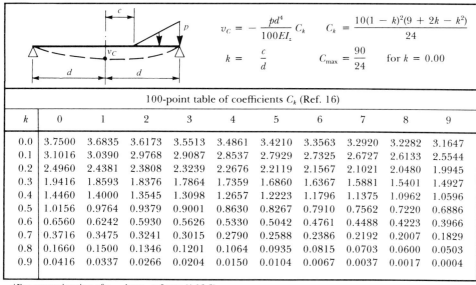

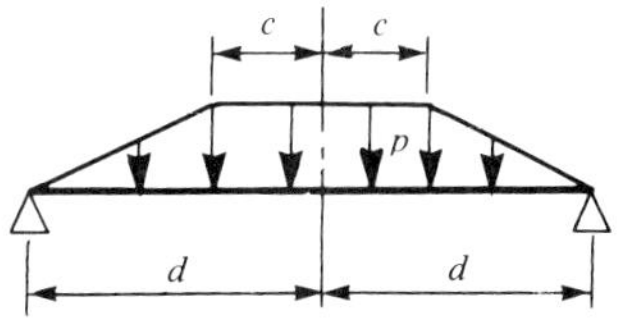

$$v_C = -\frac{pd^4}{10EI_z}B_{1.00} + \frac{2pd^4}{100EI_z}C_k$$

4.18 ELASTIC CURVE, DECREASING TRIANGULAR LOAD

Matrix (4.02, 4.03) $\bar{v} = EI_z v$ $\bar{\theta} = EI_z\theta$ $m = \dfrac{a}{s}$ $n = \dfrac{b}{s}$

v_{max} = maximum deflection (m) v_C = central deflection (m)

e = root of quartic equation given below to be used as the position coordinate of v_{max}

$$0 = V_R\frac{e^2}{2} + Z_R e + \frac{pe^4}{24n}$$

Equation of elastic curve: $$\bar{v} = \bar{v}_L + \bar{\theta}_L x + Z_L\frac{x^2}{2} - V_L\frac{x^3}{6} - \frac{p\langle x-a\rangle^4}{24} + \frac{p\langle x-a\rangle^5}{120b} \qquad (4.07\text{-}3)$$

(1) Cantilever beam

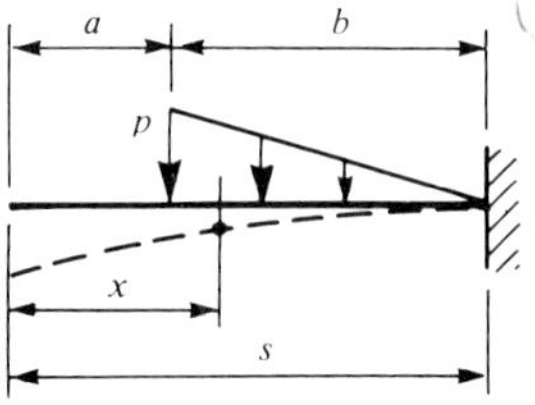

$V_L = 0$

$Z_L = 0$

$\bar{v}_L = -\dfrac{pb^4}{120n}(15 - 4n)$

$\bar{\theta}_L = \dfrac{pb^3}{8}$

$V_R = \dfrac{pb}{2}$

$Z_R = -\dfrac{pb^2}{3}$

$\bar{v}_R = 0$

$\bar{\theta}_R = 0$

$a = 0$ $\bar{v}_{max} = -\dfrac{11pb^4}{120}$ at L

(2) Propped-end beam

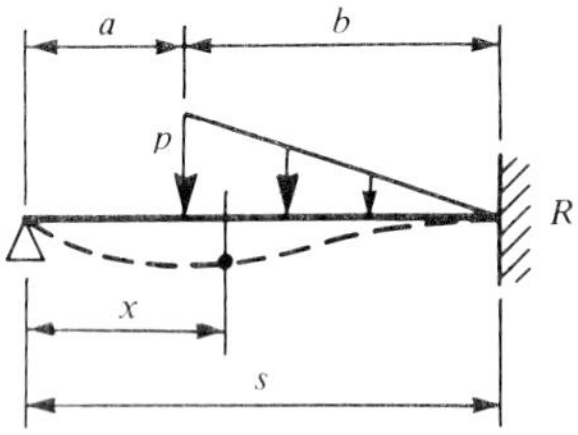

$V_L = -\dfrac{pb}{40}(15 - 4n)n^2$

$Z_L = 0$

$\bar{v}_L = 0$

$\bar{\theta}_L = -\dfrac{pb^3}{80}(5 - 4n)$

$V_R = \dfrac{pb}{40}(20 - 15n^2 + 4n^3)$

$Z_R = -\dfrac{pb^2}{120}(40 - 45n + 12n^2)$

$\bar{v}_R = 0$

$\bar{\theta}_R = 0$

$a = 0$ $\bar{v}_{max} \cong -\dfrac{3ps^4}{329}$ at $x_1 = 0.4025s$

$a < 0.4738s$ $\bar{v}_{max} = V_R\dfrac{e^3}{6} + Z_R\dfrac{e^2}{2} + \dfrac{pe^5}{120b}$ at $x_1 = s - e$

$a \geq 0.4738s$ $\bar{v}_{max} = \dfrac{2\bar{\theta}_L}{3}x_1$ at $x_1 = \sqrt{\dfrac{2\bar{\theta}_L}{V_L}}$

(3) Fixed-end beam

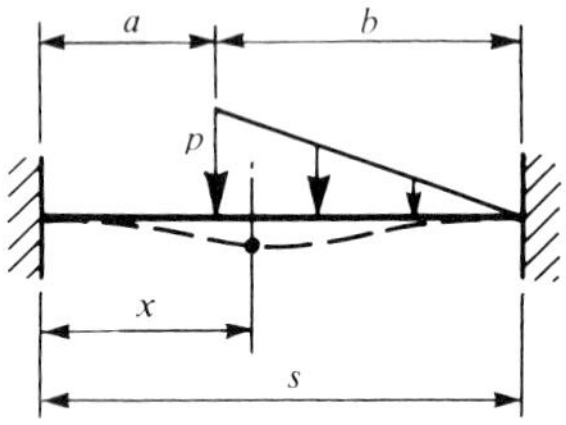

$V_L = -\dfrac{pb}{20}(15 - 8n)n^2$

$Z_L = -\dfrac{pb^2}{20}(5 - 4n)n$

$\bar{v}_L = 0$

$\bar{\theta}_L = 0$

$V_R = \dfrac{pb}{20}(10 - 15n^2 + 8n^3)$

$Z_R = -\dfrac{pb^2}{30}(10 - 15n + 6n^2)$

$\bar{v}_R = 0$

$\bar{\theta}_R = 0$

$a = 0$ $\bar{v}_{max} \cong -\dfrac{ps^4}{764}$ at $x_1 = 0.4753s$

$a < 0.5664s$ $\bar{v}_{max} = V_R\dfrac{e^3}{6} + Z_R\dfrac{e^2}{2} + \dfrac{pe^5}{120b}$ at $x_1 = s - e$

$a \geq 0.5664s$ $\bar{v}_{max} = \dfrac{V_L}{12}x_1^3$ at $x_1 = \dfrac{2Z_L}{V_L}$

4.19 DEFLECTION OF SIMPLE BEAM, DECREASING TRIANGULAR LOAD

For definition of symbols and equation of elastic curve refer to (4.02, 4.03, 4.18).

(1) Simple beam

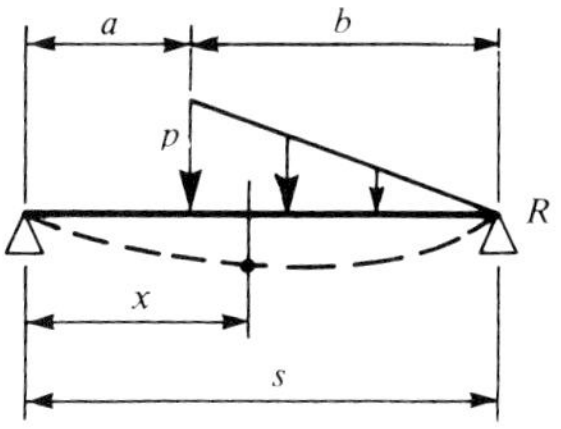

$$V_L = -\frac{pbn}{3} \qquad V_R = \frac{pb}{6}(3 - 2n)$$

$$Z_L = 0 \qquad Z_R = 0$$

$$\bar{v}_L = 0 \qquad \bar{v}_R = 0$$

$$\bar{\theta}_L = -\frac{pb^3}{90n}(5 - 3n^2) \qquad \bar{\theta}_R = \frac{pb^3}{360n}(40 - 45n + 12n^2)$$

$a = 0 \qquad \bar{v}_{\max} \cong -\dfrac{3ps^4}{460} \qquad$ at $x_1 = 0.4807s$

$a < b \qquad \bar{v}_{\max} \cong \bar{v}_C = -\dfrac{ps^4}{3840n}(25 - 120m^2 + 80m^3 + 80m^4 - 64m^5) \qquad$ at $x_1 = \dfrac{s}{2}$

$a \geq b \qquad \bar{v}_{\max} \cong \bar{v}_C = -\dfrac{ps^4}{240}(5 - 4n^2)n^2 \qquad$ at $x_1 = \dfrac{s}{2}$

(2) Central deflection coefficients†

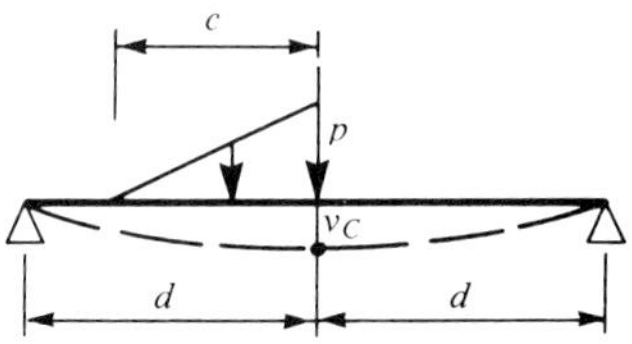

$$v_C = -\frac{pd^4}{100EI_z}D_k \qquad D_k = \frac{10(20k - 5k^2 + k^3)}{24}$$

$$k = \frac{c}{d} \qquad D_{\max} = \frac{20}{3} \quad \text{for } k = 1.00$$

100-point table of coefficients D_k (Ref. 16)

k	0	1	2	3	4	5	6	7	8	9
0.0	0.0000	0.0833	0.1667	0.2499	0.3332	0.4164	0.4995	0.5828	0.6656	0.7485
0.1	0.8313	0.9139	0.9965	1.0789	1.1611	1.2432	1.3251	1.4068	1.4883	1.5696
0.2	1.6507	1.7315	1.8121	1.8925	1.9726	2.0524	2.1315	2.2112	2.2902	2.3688
0.3	2.4471	2.5251	2.6028	2.6800	2.7570	2.8336	2.9098	2.9856	3.0610	3.1360
0.4	3.2107	3.2848	3.3511	3.4319	3.5048	3.5772	3.6492	3.7207	3.7917	3.8622
0.5	3.9323	4.0018	4.0709	4.1394	4.2074	4.2748	4.3417	4.4081	4.4740	4.5393
0.6	4.6040	4.6681	4.7317	4.7947	4.8571	4.9189	4.9801	5.0407	5.1007	5.1600
0.7	5.2188	5.2769	5.3343	5.3912	5.4474	5.5029	5.5578	5.6120	5.6655	5.7184
0.8	5.7706	5.8222	5.8730	5.9232	5.9726	6.0214	6.0694	6.1168	6.1635	6.2094
0.9	6.2546	6.2991	6.3429	6.3859	6.4282	6.4698	6.5108	6.5508	6.5901	6.6288

†For approximation of $v_{\max}$ by v_C, refer to (4.13-2)

(3) Superposition [for C_k refer to (4.17-2)]

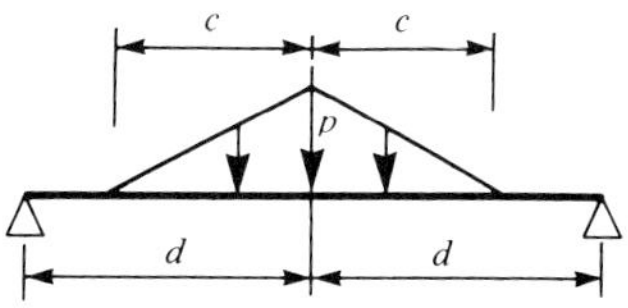

$$v_C = -\frac{2pd^4}{100EI_z}D_k$$

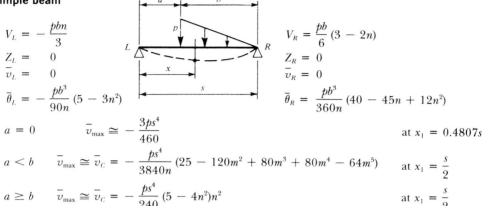

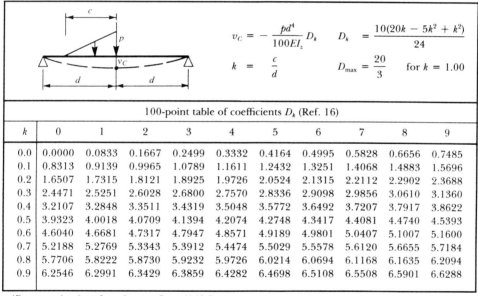

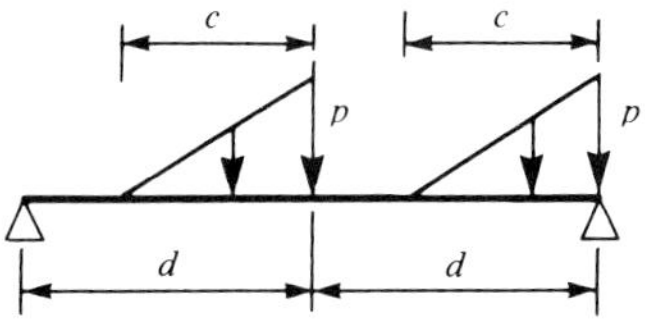

$$v_C = -\frac{pd^4}{100EI_z}(C_k + D_k)$$

4.20 ELASTIC CURVE, APPLIED COUPLE

Matrix (4.02, 4.03) $\bar{v} = EI_z v$ $\bar{\theta} = EI_z\theta$ $m = \dfrac{a}{s}$ $n = \dfrac{b}{s}$

v_{max} = maximum deflection (m)

Equation of elastic curve: $\bar{v} = \bar{v}_L + \bar{\theta}_L x + Z_L \dfrac{x^2}{2} - V_L \dfrac{x^3}{6} + \dfrac{Q\langle x - a\rangle^2}{2}$ (4.07-3)

(1) Cantilever beam

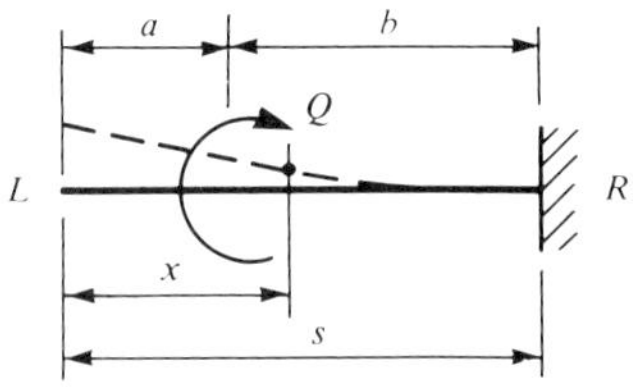

$V_L = 0$

$Z_L = 0$

$\bar{v}_L = \dfrac{Qb^2}{2n}(2 - n)$

$\bar{\theta}_L = -Qb$

$a = 0$ $\bar{v}_{max} = \dfrac{Qs^2}{2}$ at L

$V_R = 0$

$Z_R = Q$

$\bar{v}_R = 0$

$\bar{\theta}_R = 0$

(2) Propped-end beam

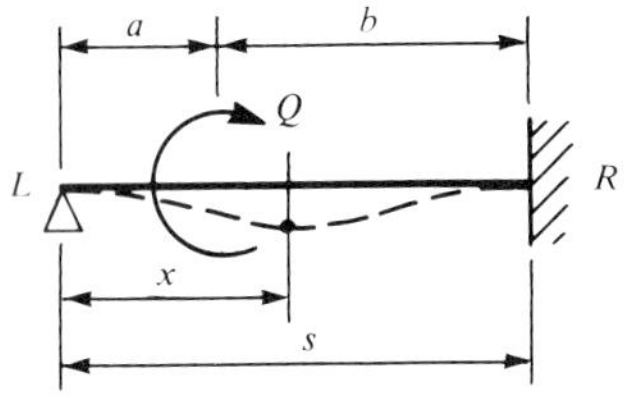

$V_L = \dfrac{3Q}{2s}(1 - m^2)$

$Z_L = 0$

$\bar{v}_L = 0$

$\bar{\theta}_L = \dfrac{Qb}{4}(3m - 1)$

$V_R = \dfrac{3Q}{2s}(1 - m^2)$

$Z_R = -\dfrac{Q}{2}(1 - 3m^2)$

$\bar{v}_R = 0$

$\bar{\theta}_R = 0$

$$\bar{v}_{max} = \begin{cases} -\dfrac{Qs^2}{27} & a = 0, \quad x_1 = \dfrac{s}{3} \\ -\dfrac{2Z_R^3}{3V_R^2} & a \le \dfrac{s}{3}, \quad x_1 = s + \dfrac{2Z_R}{V_R} \\ \dfrac{Q}{2s}(1 - m^2)x_1^3 & a > \dfrac{s}{3}, \quad x_1 = s\sqrt{\dfrac{3m - 1}{3m + 3}} \\ \dfrac{Qs^2}{72} & a = \dfrac{s}{2}, \quad x_1 = \dfrac{s}{3} \end{cases}$$

(3) Fixed-end beam

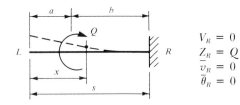

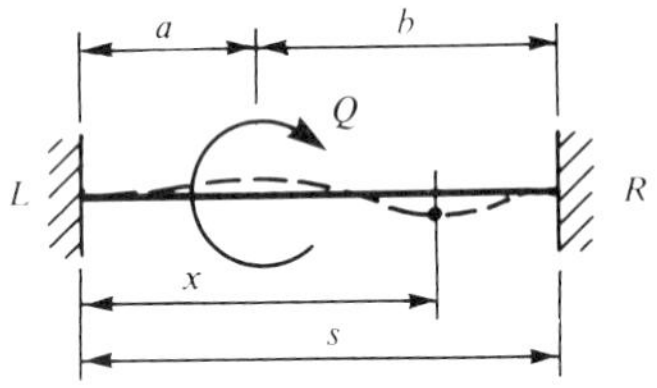

$V_L = \dfrac{6Qmn}{s}$

$Z_L = Qn(2 - 3n)$

$\bar{v}_L = 0$

$\bar{\theta}_L = 0$

$V_R = \dfrac{6Qmn}{s}$

$Z_R = -Qm(2 - 3m)$

$\bar{v}_R = 0$

$\bar{\theta}_R = 0$

$$\bar{v}_{max} = \begin{cases} \dfrac{2Z_R^3}{3V_R^2} & a < \dfrac{s}{3}, \quad x_1 = s + \dfrac{2Z_R}{V_R} \\ \pm\dfrac{Qs^2}{216} & a = b, \quad x_1 = \dfrac{s}{3}, \dfrac{2s}{3} \\ \dfrac{2Z_L^3}{3V_L^2} & a > \dfrac{s}{3}, \quad x_1 = \dfrac{2Z_L}{V_L} \end{cases}$$

4.21 DEFLECTION OF SIMPLE BEAM, APPLIED COUPLE

For definition of symbols refer to (4.02, 4.03, 4.20). For the equation of elastic curve refer to (4.20). x_1, x_2 = position coordinate of $v_{\max}$ measured from L, R, respectively.

(1) Single couple

$$V_L = \frac{Q}{s} \qquad Z_L = 0 \qquad \bar{v}_L = 0 \qquad \bar{\theta}_L = \frac{Qs}{6}(1 - 3n^2)$$

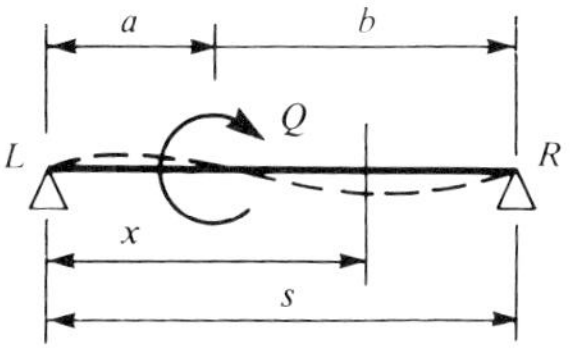

$$V_R = \frac{Q}{s} \qquad Z_R = 0 \qquad \bar{v}_R = 0 \qquad \bar{\theta}_R = \frac{Qs}{6}(1 - 3m^2)$$

(*a*) *Negative elastic curve* ($0 \le a \le 0.4227s$)

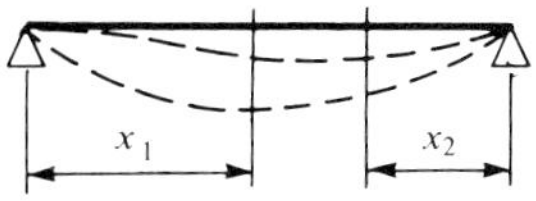

$a = 0$ $\qquad \bar{v}_{\max} = -\dfrac{Qs^2}{27}\sqrt{3}$ at $x_1 = 0.4227s$

$a \le 0.4227s$ $\qquad \bar{v}_{\max} = -\dfrac{Q}{3s}x_2^3$ at $x_2 = s\sqrt{\dfrac{1 - 3m^2}{3}}$

(*b*) *Alternating elastic curve* ($0.4227s < a < 0.5773s$)

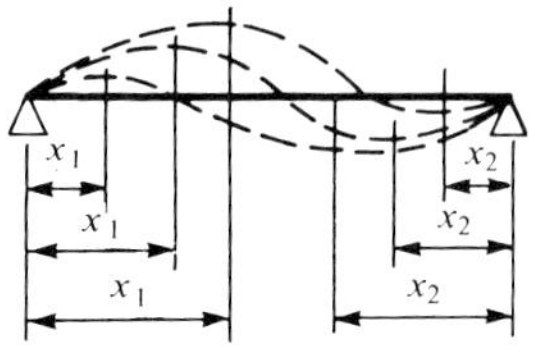

$$0.4227s < a < \frac{s}{2} \qquad \bar{v}_{\max} = \begin{cases} \dfrac{Q}{3s}x_1^3 & \text{at } x_1 = s\sqrt{\dfrac{1-3m^2}{3}} \\ -\dfrac{Q}{3s}x_2^3 & \text{at } x_2 = s\sqrt{\dfrac{1-3n^2}{3}} \end{cases}$$

$$a = \frac{s}{2} \qquad \bar{v}_{\max} = \pm\frac{Qs^2\sqrt{3}}{54} \qquad \begin{array}{l} \text{at } x_1 = 0.2887s \\ \text{at } x_2 = 0.2887s \end{array}$$

$$0.5773s > a > \frac{s}{2} \qquad \bar{v}_{\max} = \begin{cases} \dfrac{Q}{3s}x_1^3 & \text{at } x_1 = s\sqrt{\dfrac{1-3n^2}{3}} \\ -\dfrac{Q}{3s}x_2^3 & \text{at } x_2 = s\sqrt{\dfrac{1-3m^2}{3}} \end{cases}$$

(*c*) *Positive elastic curve* ($0.5773 \le a \le s$)

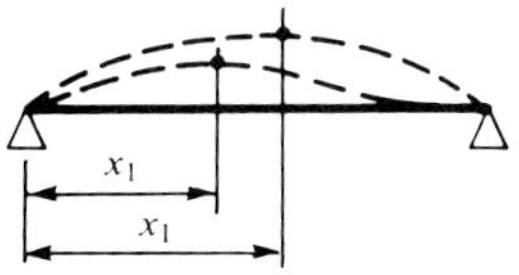

$$\begin{array}{l} a \ge 0.5773s \\ a = s \end{array} \qquad \bar{v}_{\max} = \begin{cases} \dfrac{Qs^2}{3}x_1^3 & \text{at } x_1 = s\sqrt{\dfrac{1-3n^2}{3}} \\ \dfrac{Qs^2}{27}\sqrt{3} & \text{at } x_1 = 0.5773s \end{cases}$$

(2) Two couples

(*a*) *Symmetrical elastic curve*

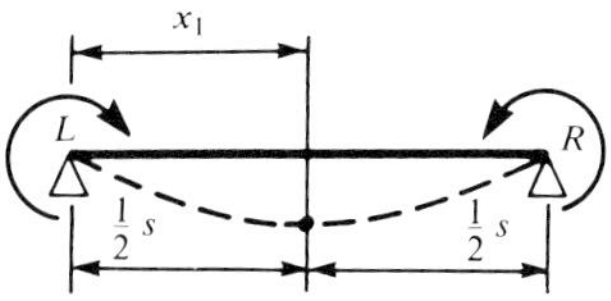

$$\bar{\theta}_L = -\tfrac{1}{2}Qs \qquad \bar{\theta}_R = \tfrac{1}{2}Qs$$

$$\bar{v}_{\max} = -\frac{Qs^2}{8} \quad \text{at } x_1 = \tfrac{1}{2}s$$

(*b*) *Antisymmetrical elastic curve*

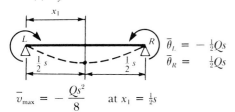

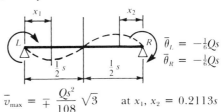

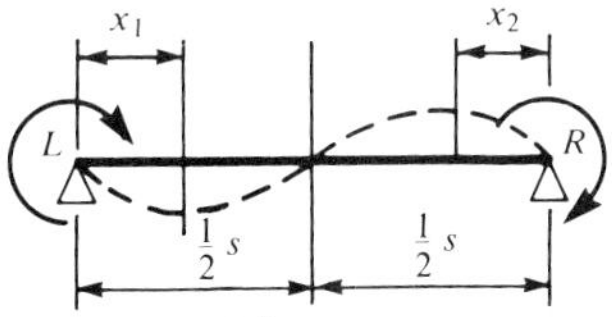

$$\bar{\theta}_L = -\tfrac{1}{6}Qs \qquad \bar{\theta}_R = -\tfrac{1}{6}Qs$$

$$\bar{v}_{\max} = \mp\frac{Qs^2}{108}\sqrt{3} \quad \text{at } x_1, x_2 = 0.2113s$$

4.22 FLEXIBILITY MATRIX EQUATIONS, SIMPLE BEAM

(1) Notation

A_x = area of section normal to x (m^2)
I_z = moment of inertia of A_x about z (m^4)
E = modulus of elasticity (Pa)
s = span (m)
a, b = segments (m)
r = I_z/A_x = ratio (m^2)

Positive forces, moments, and displacements are shown in (2) and (3).

(2) Free-body diagram

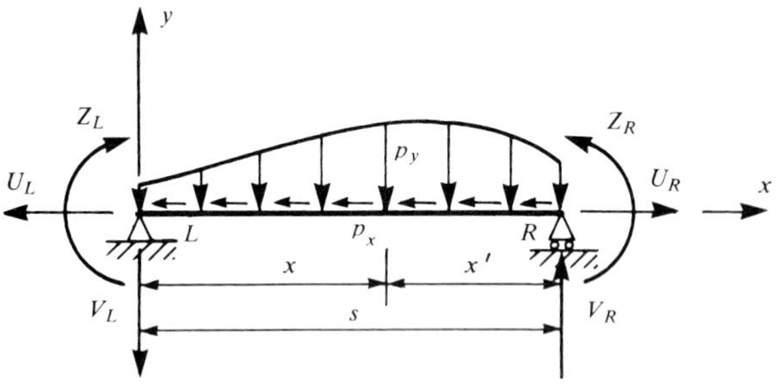

(3) Elastic curve

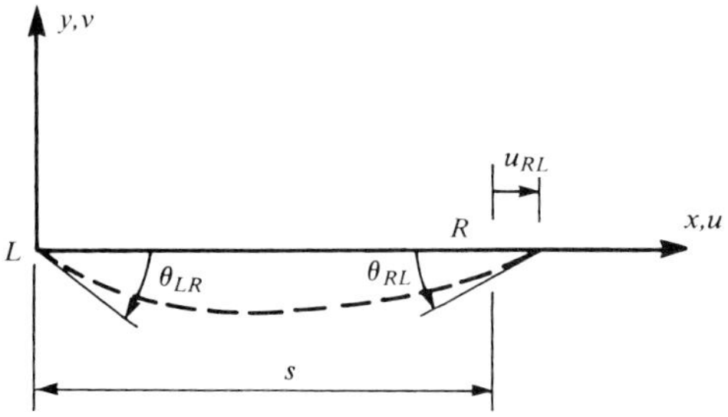

(4) Matrix equation defining the end displacement vector of a simple beam as a function of the end stress vectors and loads is the flexibility matrix equation (Ref. 4, p. 52).

$$\begin{bmatrix} u_{RL} \\ \hline \theta_{LR} \\ \theta_{RL} \end{bmatrix} = \frac{s}{EI_z} \left[\begin{array}{c|cc} r & 0 & 0 \\ \hline 0 & \frac{1}{3} & \frac{1}{6} \\ 0 & \frac{1}{6} & \frac{1}{3} \end{array}\right] \begin{bmatrix} U_R \\ \hline Z_L \\ Z_R \end{bmatrix} + \begin{bmatrix} u_{R0} \\ \hline \theta_{L0} \\ \theta_{R0} \end{bmatrix}$$

where u_{RL}, θ_{LR}, θ_{RL} are the end displacements shown in (3); U_R, Z_L, Z_R are the normal force at R, the bending moment at L, and the bending moment at R, respectively; and u_{R0}, θ_{L0}, θ_{R0} are the flexibility load functions defined analytically below.

(5) Flexibility load functions are the end displacements of the simple beam due to loads and other causes. In general,

$$u_{R0} = \int_0^s \frac{U\,dx}{EA_x} + \int_0^s e_t\,dx$$

$$\theta_{L0} = \int_0^s \frac{Zx'\,dx}{sEI_z} + \frac{1}{s}\int_0^s f_t x'\,dx$$

$$\theta_{R0} = \int_0^s \frac{Zx\,dx}{sEI_z} + \frac{1}{s}\int_0^s f_t x\,dx$$

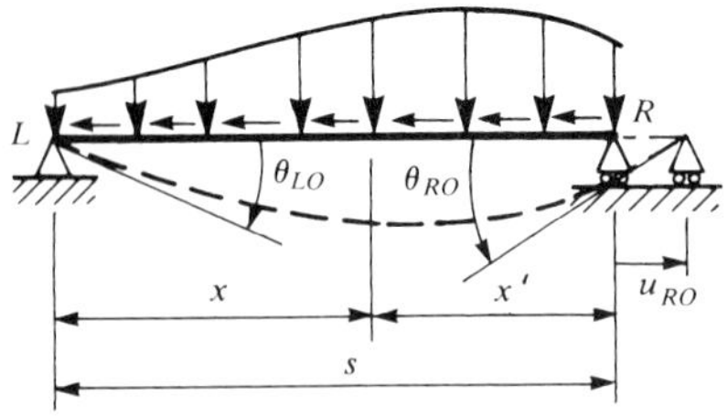

where U = normal force at x (N) Z = bending moment at x (N·m)
e_t, f_t = thermal strains defined in (4.03)

Particular cases of load functions are tabulated in (4.24–4.29).

4.23 THREE-MOMENT RELATIONS

(1) Continuous beam is defined as a slender bar supported at more than two points and acted upon by causes producing primarily bending. According to the number of spans, continuous beams are classified as two-, three-, . . . , and N-span beams. Their degree of statical indeterminacy is defined by the number of redundant moments over supports as

$$N = R - S - 3$$

where N = number of redundants, R = number of independent reactions, and S = number of releases (such as internal hinges and/or guides).

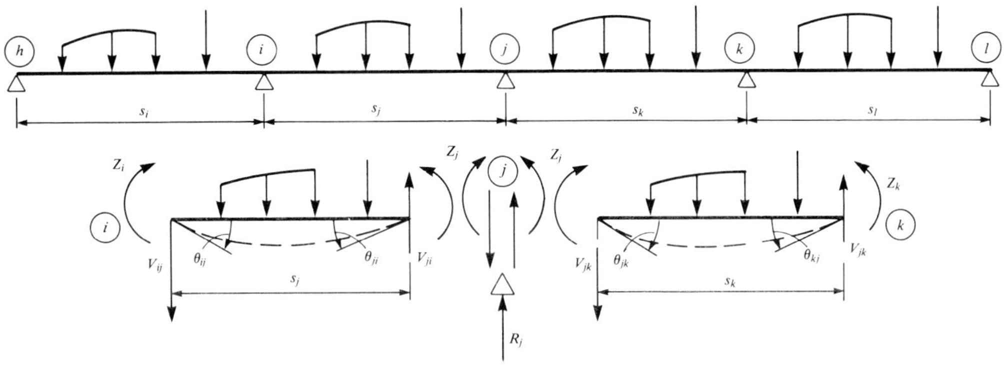

(2) Compatibility condition written for the intermediate support j (any support) in the continuous beam shown above requires that the sum of end slopes θ_{ji} and θ_{jk} equals zero, which in terms of the parameters introduced in (4.22) becomes the *three-moment equation*

$$\frac{s_j}{6EI_{jz}} Z_i + \left(\frac{s_j}{3EI_{jz}} + \frac{s_k}{3EI_{kz}} \right) Z_j + \frac{s_k}{6EI_{kz}} + \theta_{j0,i} + \theta_{j0,k} = 0$$

where Z_i, Z_j, Z_k = unknown bending moments over the respective supports (N·m)
s_j, s_k = spans ij, jk, respectively (m)
I_{jz}, I_{kz} = moments of inertia of the normal sections in ij, jk, respectively (m^4)
$\theta_{j0,i}$ = right-end load function in span ij (rad)
$\theta_{j0,k}$ = left-end load function in span jk (rad)

(3) Three-moment equation can be used as a recurrent formula for the construction of the system flexibility matrix equation in a given continuous beam, where the number of unknowns equals the number of redundant bending moments over supports. When the spans are equal, so that $s_j = s_k$ and $I_{jz} = I_{kz}$, the three-moment equation in (2) reduces to

$$Z_i + 4Z_j + Z_k + \frac{6EI}{s} z(\theta_{j0,i} + \theta_{j0,k}) = 0$$

(4) Fixed-end. If the exterior end h in (1) is fixed, the three-moment equation at h reduces to

$$\frac{s_i}{3EI_{iz}} Z_h + \frac{s_i}{6EI_{iz}} Z_i + \theta_{h0,i} = 0$$

(5) Overhang. If the end h in (1) is free and the span s_i overhangs the support i, the bending moment at i is the bending moment Z_i in the cantilever beam hi, which is a statically determinate quantity. The three-moment equation written for the compatibility at j includes this moment as a known quantity.

4.24 FLEXIBILITY LOAD FUNCTIONS, SIMPLE BEAM

Notation (4.22)	Signs (4.22)		Matrix (4.22)
For concentrated load P:	$\theta_{L0} = \dfrac{Ps^2}{EI_z} C_L$	$\theta_{R0} = \dfrac{Ps^2}{EI_z} C_R$	$m = \dfrac{a}{s}$
For applied couple Q:	$\theta_{L0} = \dfrac{Qs}{EI_z} C_L$	$\theta_{R0} = \dfrac{Qs}{EI_z} C_R$	$n = \dfrac{b}{s}$

C_L	Case	C_R
$\dfrac{mn(1 + n)}{6}$	**1.** 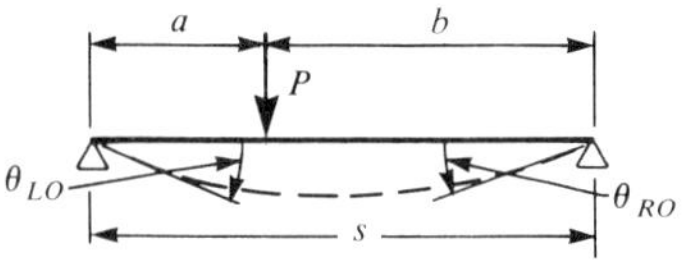	$\dfrac{mn(1 + m)}{6}$
$\dfrac{1}{16}$	**2.** 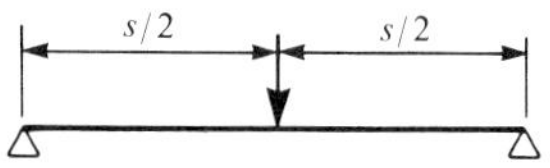	$\dfrac{1}{16}$
$\dfrac{m(1 - m)}{2}$	**3.** 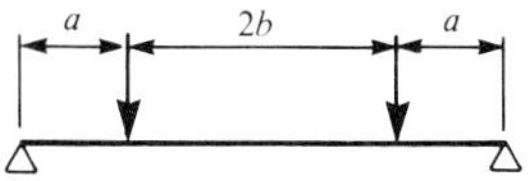	$\dfrac{m(1 - m)}{2}$
$\dfrac{c^2 - 1}{24c}$	**4.** 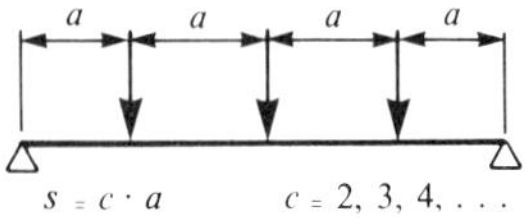	$\dfrac{c^2 - 1}{24c}$
$\dfrac{2c^2 + 1}{48c}$	**5.** 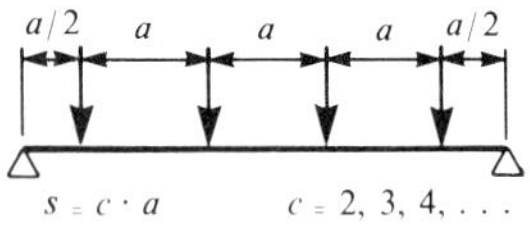	$\dfrac{2c^2 + 1}{48c}$
$-\dfrac{1 - 3n^2}{6}$	**6.** 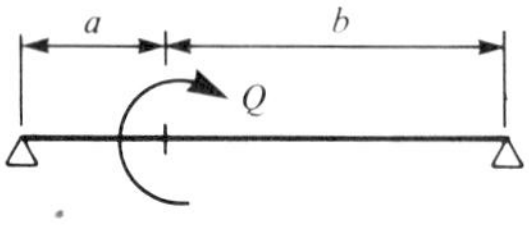	$\dfrac{1 - 3m^2}{6}$
n	**7.** 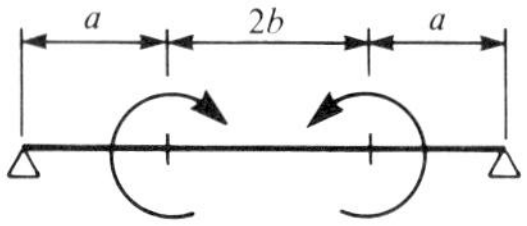	n
$-\dfrac{1}{24}$	**8.** 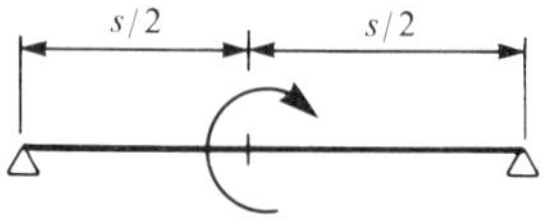	$\dfrac{1}{24}$

4.25 FLEXIBILITY LOAD FUNCTIONS, SIMPLE BEAM

Notation (4.22) Signs (4.22) Matrix (4.22)

For uniformly distributed load of intensity p: $\theta_{LO} = \frac{ps^3}{EI_z} C_L$ $\theta_{RO} = \frac{ps^3}{EI_z} C_R$ $m = \frac{a}{s}$

For uniformly distributed couple of intensity q: $\theta_{LO} = \frac{qs^2}{EI_z} C_L$ $\theta_{RO} = \frac{qs^2}{EI_z} C_R$ $n = \frac{b}{s}$

C_L	Case	C_R
$\frac{1}{24}$	**1.** 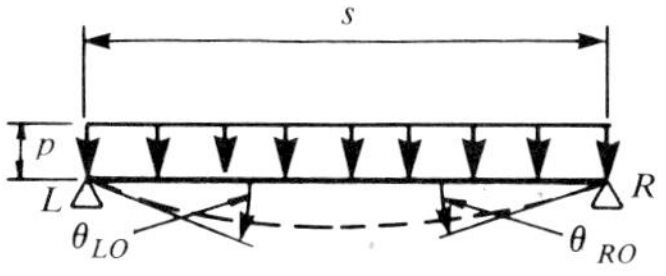	$\frac{1}{24}$
$\frac{n^2(2 - n^2)}{24}$	**2.** 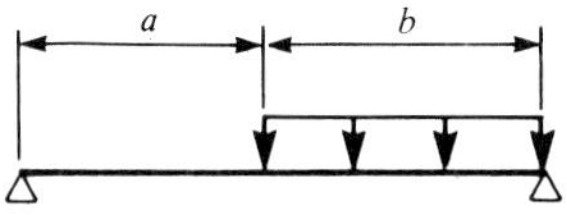	$\frac{n^2(2 - n)^2}{24}$
$\frac{m^2(1 + n)}{6}$	**3.** 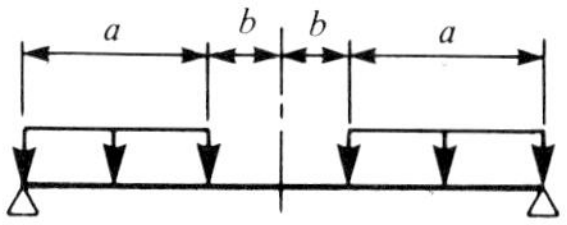	$\frac{m^2(1 + n)}{6}$
$\frac{n(3 - 4n^2)}{24}$	**4.** 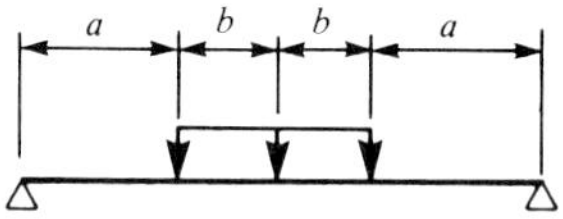	$\frac{n(3 - 4n^2)}{24}$
$\frac{m(3 + 6m - 8m^2)}{24}$	**5.** 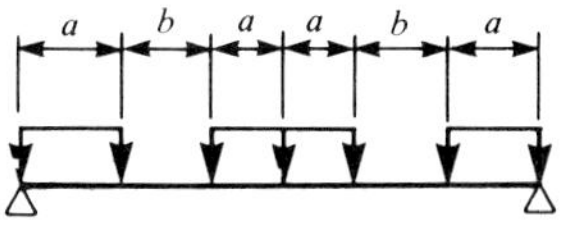	$\frac{m(3 + 6m - 8m^2)}{24}$
0	**6.** 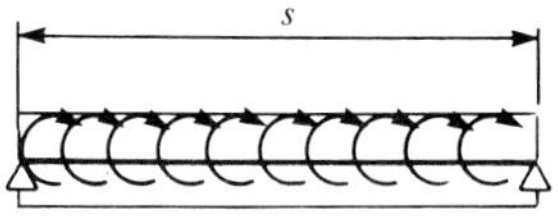	0
$-\frac{n}{6}(1 - n^2)$	**7.** 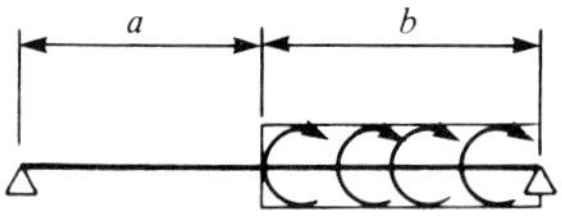	$-\frac{m}{6}(1 - m^2)$
$-\frac{m}{2}(1 - m)$	**8.** 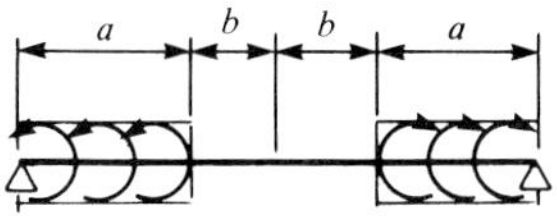	$-\frac{m}{2}(1 - m)$

4.26 FLEXIBILITY LOAD FUNCTIONS, SIMPLE BEAM

Notation (4.22) Signs (4.22) Matrix (4.22)

For triangular load of maximum intensity p:

$$\begin{cases} \theta_{L0} = \dfrac{ps^3}{EI_z} C_L & a = \dfrac{a}{s} \\ \theta_{R0} = \dfrac{ps^3}{EI_z} C_R & b = \dfrac{b}{s} \end{cases}$$

C_L	Case	C_R
$\dfrac{7}{360}$	**1.** 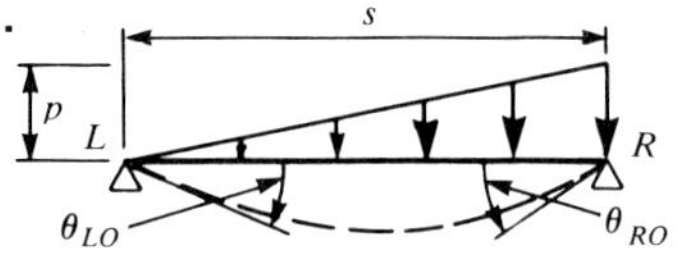	$\dfrac{8}{360}$
$\dfrac{n^2(10-3n^2)}{360}$	**2.** 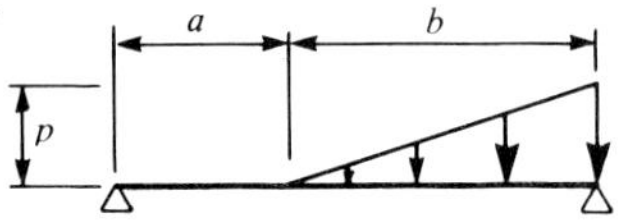	$\dfrac{n^2(20-15n+3n^2)}{360}$
$\dfrac{n^2(5-3n^2)}{90}$	**3.** 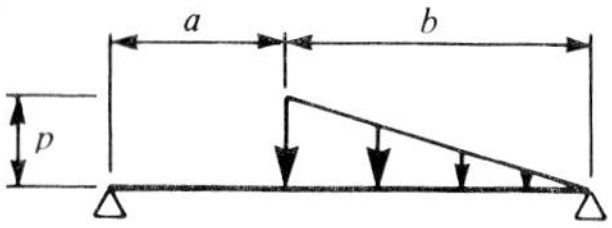	$\dfrac{n^2(40-45n+12n^2)}{360}$
$\dfrac{(1+n)(7-3n^2)}{360}$	**4.** 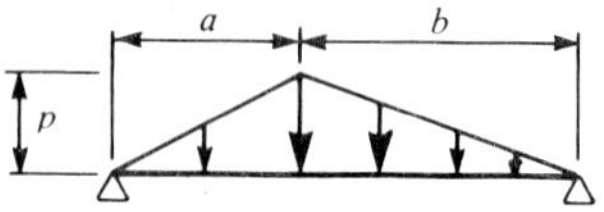	$\dfrac{(1+m)(7-3m^2)}{360}$
$\dfrac{119}{5760}$	**5.** 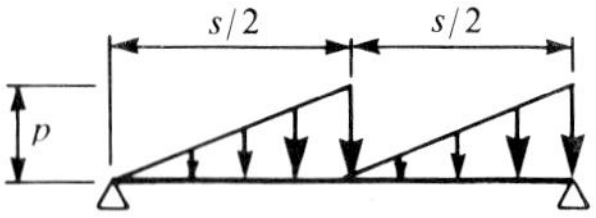	$\dfrac{121}{5760}$
$\dfrac{15-(1+n)(7-3n^2)}{360}$	**6.** 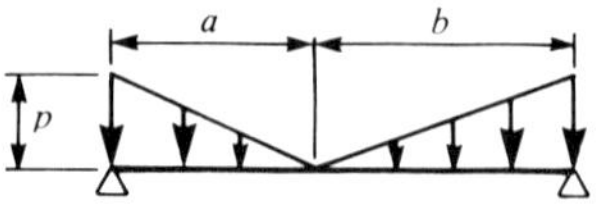	$\dfrac{15-(1+m)(7-3m^2)}{360}$
$-\dfrac{1}{360}$	**7.** 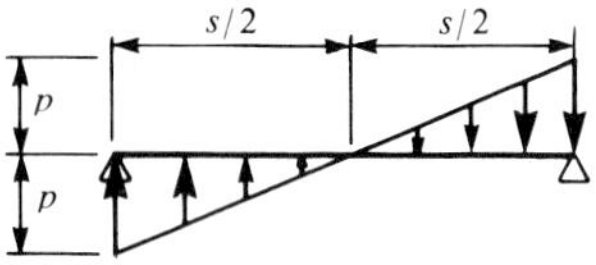	$\dfrac{1}{360}$
$\dfrac{8p+7p_1}{360p}$	**8.** 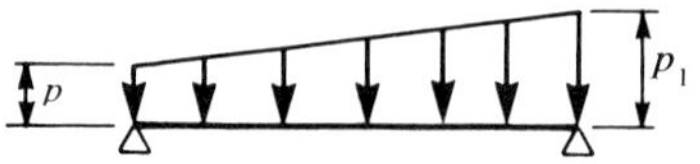	$\dfrac{7p+8p_1}{360p}$

4.27 FLEXIBILITY LOAD FUNCTIONS, SIMPLE BEAM

Notation (4.22) Signs (4.22) Matrix (4.22)

For triangular load of maximum intensity p:

$$\begin{cases} \theta_{L0} = \dfrac{ps^3}{EI_z} C_L & m = \dfrac{a}{s} \\ \theta_{R0} = \dfrac{ps^3}{EI_z} C_R & n = \dfrac{b}{s} \end{cases}$$

C_L	Case	C_R
$\dfrac{1}{64}$	1. 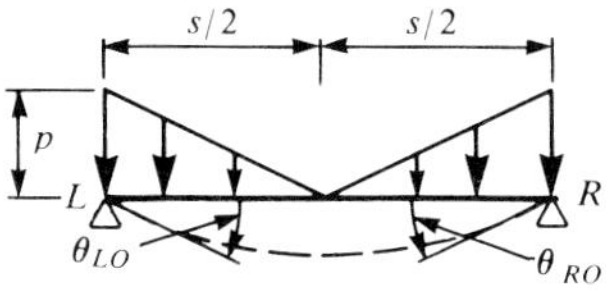	$\dfrac{1}{64}$
$\dfrac{m^2(2-m)}{24}$	2. 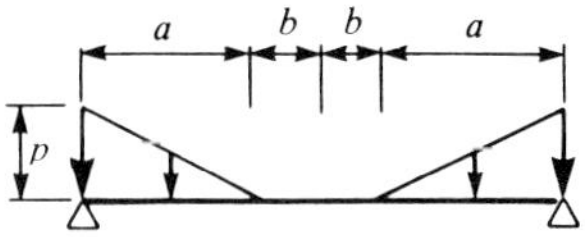	$\dfrac{m^2(2-m)}{24}$
$\dfrac{5}{192}$	3. 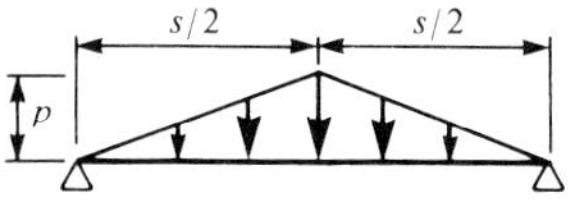	$\dfrac{5}{192}$
$\dfrac{n(3-2n^2)}{48}$	4. 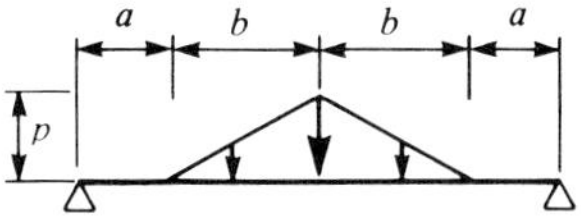	$\dfrac{n(3-2n^2)}{48}$
$\dfrac{n^2(5-6n^2)}{360}$	5. 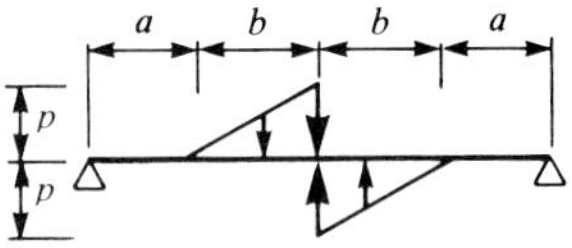	$-\dfrac{n^2(5-6n^2)}{360}$
$\dfrac{17}{768}$	6. 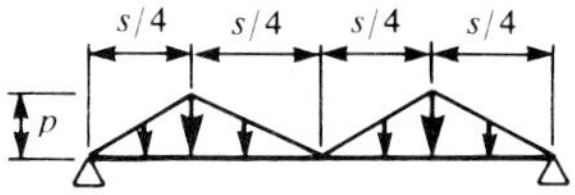	$\dfrac{17}{768}$
$\dfrac{5}{256}$	7. 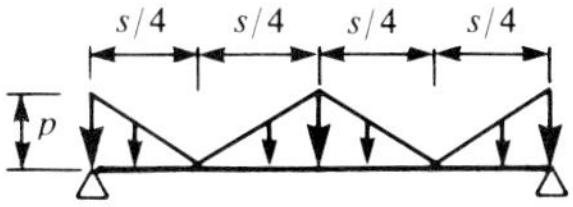	$\dfrac{5}{256}$
$\dfrac{1-2m^2+m^3}{24}$	8. 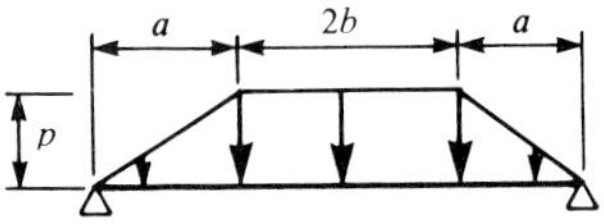	$\dfrac{1-2m^2+m^3}{24}$

4.28 FLEXIBILITY LOAD FUNCTIONS, SIMPLE BEAM

Notation (4.22) Signs (4.22) Matrix (4.22)

For harmonic load of maximum intensity p:

$$\begin{cases} \theta_{L0} = \dfrac{ps^3}{EI_z} C_L \\ \theta_{R0} = \dfrac{ps^3}{EI_z} C_R \end{cases} \qquad c = \frac{s}{m}$$

C_L	Case	C_R
$\dfrac{1}{\pi^3}$	**1.** 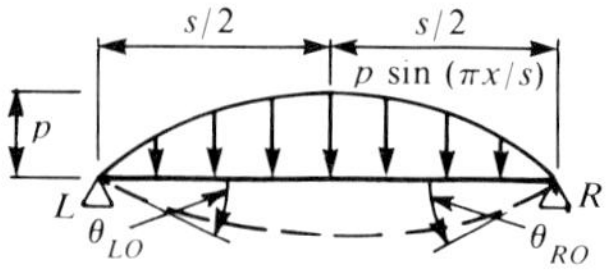	$\dfrac{1}{\pi^3}$
$\dfrac{1}{(2\pi)^3}$	**2.** 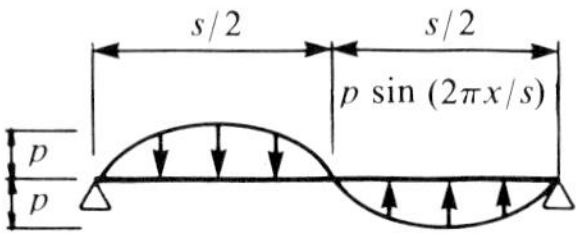	$-\dfrac{1}{(2\pi)^3}$
$\dfrac{1}{(m\pi)^3}$	**3.** 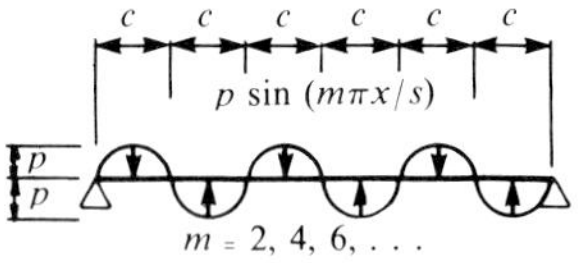	$-\dfrac{1}{(m\pi)^3}$
$\dfrac{1}{(m\pi)^3}$	**4.** 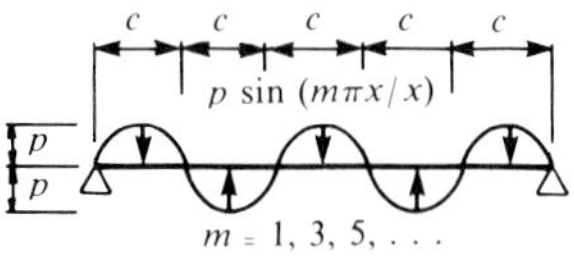	$\dfrac{1}{(m\pi)^3}$
$\dfrac{12-\pi^2}{6\pi^4}$	**5.** 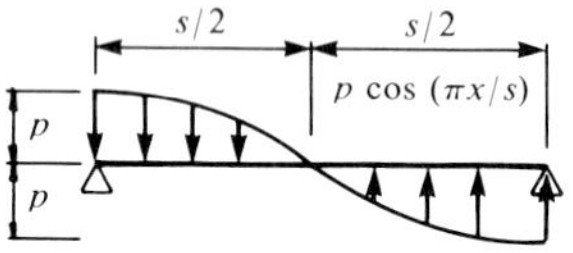	$-\dfrac{12-\pi^2}{6\pi^4}$
$\dfrac{1}{8\pi^2}$	**6.** 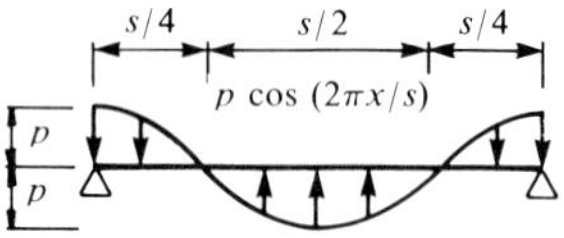	$\dfrac{1}{8\pi^2}$
$\dfrac{1}{2(m\pi)^2}$	**7.** 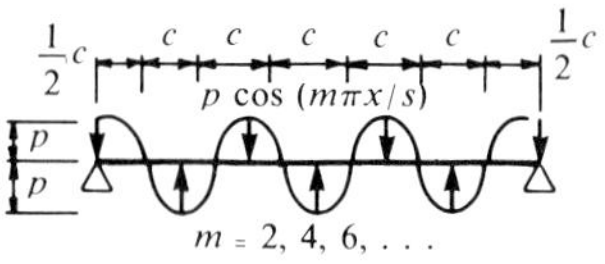	$\dfrac{1}{2(m\pi)^2}$
$\dfrac{12-(m\pi)^2}{6(m\pi)^4}$	**8.** 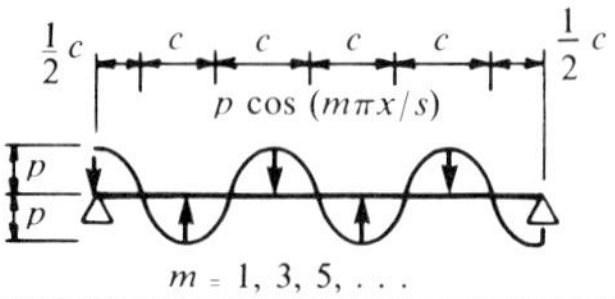	$-\dfrac{12-(m\pi)^2}{6(m\pi)^4}$

4.29 FLEXIBILITY LOAD FUNCTIONS, SIMPLE BEAM

Notation (4.22)	Signs (4.22)	Matrix (4.22)
For parabolic load of maximum intensity p:		$\theta_{L0} = \dfrac{ps^3}{EI_z} C_L \qquad m = \dfrac{a}{s}$ $\theta_{R0} = \dfrac{ps^3}{EI_z} C_R \qquad n = \dfrac{b}{s}$

C_L	Case	C_R
$\dfrac{k(k+4)}{24(k+1)(k+3)}$	**1.** 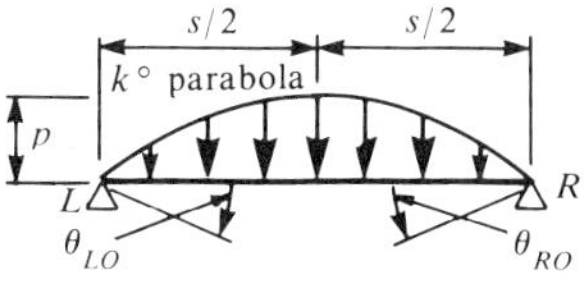	$\dfrac{k(k+4)}{24(k+1)(k+3)}$
$\dfrac{6+7k+k^2}{6(k+1)(k+2)(k+3)(k+4)}$	**2.** 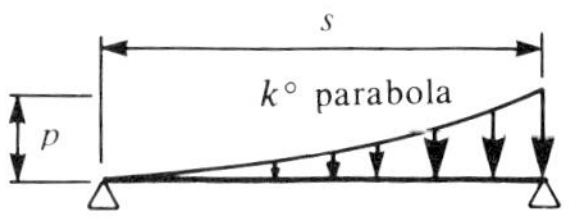	$\dfrac{1}{3(k+2)(k+4)}$
$\dfrac{n^2[(k+3)(k+4)-6n^2]}{6(k+1)(k+2)(k+3)(k+4)}$	**3.** 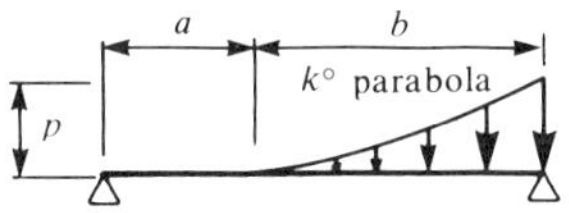	$\dfrac{n^2[(k+4)(k+3m)+3n^2]}{3(k+1)(k+2)(k+3)(k+4)}$

v_L, v_R = vertical displacements of supports (m)

v_j, θ_j = externally applied displacements (m, rad)

f_t = thermal strain (4.03)

θ_{L0}	Case	θ_{R0}
$-\dfrac{v_L - v_R}{s}$	**4. Displacement of supports** 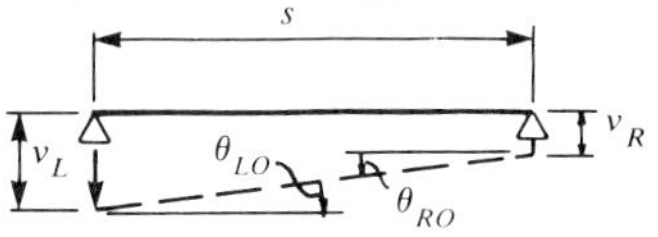	$\dfrac{v_L - v_R}{s}$
$\dfrac{v_j}{s}$	**5. Abrupt linear displacement** 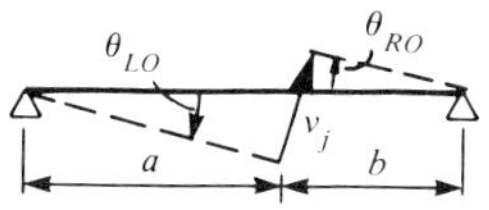	$-\dfrac{v_j}{s}$
$n\theta_j$	**6. Abrupt angular displacement** 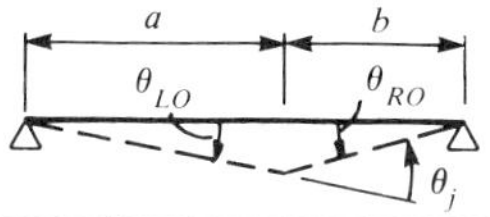	$m\theta_j$
$\frac{1}{2}bnf_t$	**7. Angular temperature deformation of segment *b*** 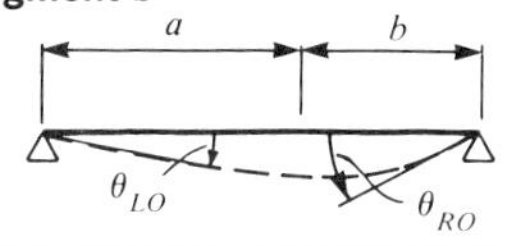	$\frac{1}{2}(1-m^2)sf_t$

4.30 EQUAL-SPAN CONTINUOUS BEAMS, CONCENTRATED LOAD

(1) Notation is simplified to facilitate the tabulation of large number of coefficients. All spans have the same length and the same cross section The concentrated load is applied at the center of the respective span. For sign convention refer to (4.02).

s = span length (m)
E = modulus of elasticity (Pa)
V_{jL} = left-side shear (N)
Z_i = max. bending moment in span i (N·m)
Z_j = bending moment over j (N·m)
i = a, b, c, . . . = subscript of span

P = concentrated load (N)
I = moment of inertia (m^4)
V_{jR} = right-side shear (N)
v_i = central deflection in span i (m)
R_j = reaction at j (N)
j = 0, 1, 2, . . . = subscript of support

(2) Relations

$$v_i = \frac{k_i s^3 P}{10^5 EI} \qquad Z_j = \frac{C_j s P}{10} \qquad Z_i = \frac{C_i s P}{10} \qquad V_{jL} = \frac{D_{jL} P}{10} \qquad V_{jR} = \frac{D_{jR} P}{10} \qquad R_j = V_{jL} - V_{jR}$$

(3) Table of coefficients to be used in relations stated above.†

Loading					C_1	C_2	C_3			D_{1L}	D_{2L}	D_{3L}	D_{4L}
	k_a	k_b	k_c	k_d	C_a	C_b	C_c	C_d	D_{0R}	D_{1R}	D_{2R}	D_{3R}	
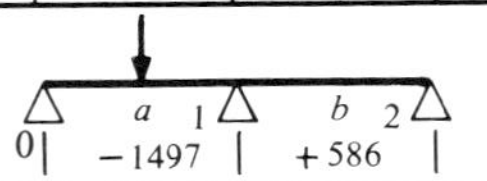	−1497	+586			−0.938					+5.938	−0.938		
					+2.031	−0.938			−4.062	−0.938			
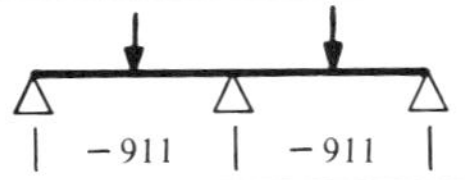	−911	−911			−1.875					+6.875	+3.125		
					+1.562	+1.562			−3.125	−6.875			
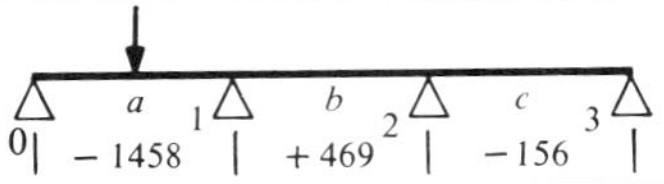	−1458	+469	−156		−1.000	+0.250				+6.000	−1.250	+0.250	
					+2.000	−1.000	+0.250		−4.000	−1.250	+0.250		
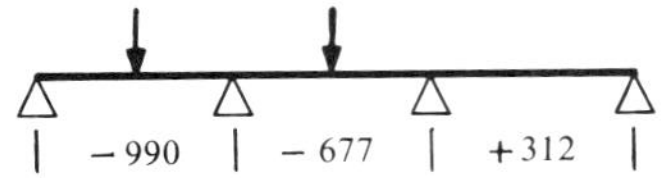	−990	−677	+312		−1.750	−0.500				+6.750	+3.750	−0.500	
					+1.625	+1.375	−0.500		−3.250	−6.250	−0.500		

Loads at centers of spans 1 and 2 and 3; moments: −1146, −208, −1146		−1.500	−1.500			+6.500	+5.000	+3.500	
	+1.750	+1.000	+1.750		−3.500	−5.000	−6.500		
Loads at centers of spans 1 and 3; moments: −1146, +208, −1146		−0.750	−0.750			+5.750	0	+4.250	
	+2.125	−0.750	+2.125		−4.250	0	−5.750		
Load at center of span 2; moments: +469, −1146, +469		−0.750	−0.750			+0.750	+5.000	−0.750	
	−0.750	+1.750	−0.750		+0.750	−5.000	−0.750		
Supports 0, 1, 2, 3, 4; spans a, b, c, d; load at center of span a; moments: −1456, +460, −126, +42		−1.004	+0.268	−0.067		+6.004	−1.272	+0.335	−0.067
	+1.998	−1.004	+0.268	−0.067	−3.996	−1.272	+0.335	−0.067	
Load at center of span b; moments: +460, −1121, +377, −126		−0.737	−0.804	+0.201		+0.737	+5.067	−1.005	+0.201
	−0.737	+1.730	−0.804	+0.201	+0.737	−4.933	−1.005	+0.201	
Loads at centers of spans a and c; moments: −1581, +837, −1246, +502		−0.804	−0.536	−0.804		+5.804	−0.268	+5.268	−0.804
	+2.098	−0.804	+1.830	−0.804	−4.196	−0.268	−4.732	−0.804	
Loads at centers of spans a, b, c, d; moments: −1079, −409, −409, −1079		−1.607	−1.071	−1.607		+6.607	+4.464	+5.536	+3.393
	+1.697	+1.161	1.161	+1.697	−3.393	−5.536	−4.464	−6.607	
Loads at centers of spans b and c; moments: +335, −744, −744, +335		−0.536	−1.607	−0.536		+0.536	+6.071	+3.929	−0.530
	−0.536	+1.428	+1.428	−0.536	+0.536	−3.929	−6.071	−0.530	
Loads at centers of spans a, b, d; moments: −953, −786, +711, −1539		−1.808	−0.268	−0.871		+6.808	+3.460	+0.603	+4.129
	+1.596	+1.462	−0.871	+2.065	−3.192	−6.540	+0.603	−5.871	

†Ref. 8, pp. 778–779.

4.31 EQUAL-SPAN CONTINUOUS BEAMS, UNIFORM LOAD

(1) Notation is simplified to facilitate the tabulation of large numbers of coefficients. All spans have the same length and the same cross section For sign convention refer to (4.02)

s = span length (m)
E = modulus of elasticity (Pa)
V_{jL} = left-side shear (N)
Z_i = max. bending moment in span i (N·m)
Z_j = bending moment over j (N·m)
i = $a, b, c, \ldots$ = subscript of span

p = intensity of load (N/m)
I = moment of inertia (m^4)
V_{jR} = right-side shear (N)
v_i = central deflection in span i (m)
R_j = reaction at j (N)
j = $0, 1, 2, \ldots$ = subscript of support

(2) Relations

$$v_i = \frac{k_i s^4 p}{10^5 EI} \qquad Z_j = \frac{C_j s^2 p}{10} \qquad Z_i = \frac{C_i s^2 p}{10} \qquad V_{jL} = \frac{D_{jL} s p}{10} \qquad V_{jR} = \frac{D_{jR} s p}{10} \qquad R_j = V_{jL} - V_{jR}$$

(3) Table of coefficients to be used in relations stated above.†

Loading	k_a	k_b	k_c	k_d	C_a	C_1	C_b	C_2	C_c	C_3	C_d	D_{0R}	D_{1L}	D_{1R}	D_{2L}	D_{2R}	D_{3L}	D_{3R}	D_{4L}
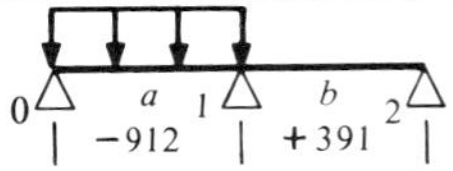	−912	+391			+0.957	−0.625	−0.625					−4.375	+5.625	−0.625	−0.625				
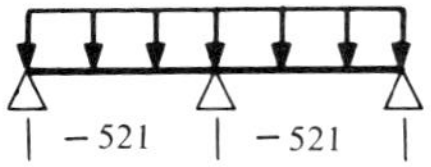	−521	−521			+0.703	−1.250	+0.703					−3.750	+6.250	−6.250	+3.750				
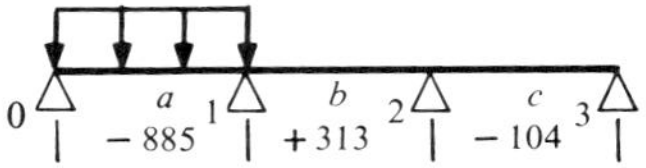	−885	+313	−104		+0.959	−0.667	−0.667	+0.167	+0.167			−4.333	+5.667	−0.834	−0.834	+0.167	+0.167		
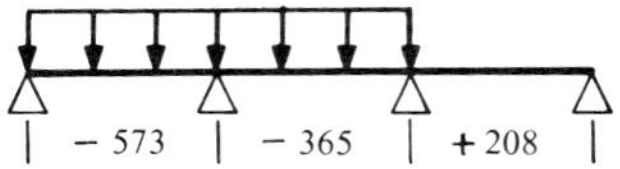	−573	−365	+208		+0.735	−1.167	+0.535	−0.333	−0.333			−3.833	+6.167	−5.834	+4.166	−0.333	−0.333		

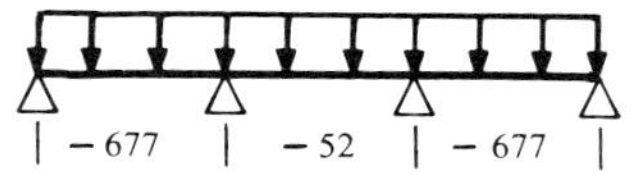		−1.000	−1.000				+6.000	+5.000	+4.000	
	+0.800	+0.250	+0.800			−4.000	−5.000	−6.000		
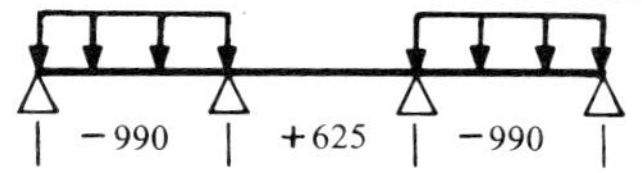		−0.500	−0.500				+5.500	+0.000	+4.500	
	+1.013	−0.500	+1.013			−4.500	−0.000	−5.550		
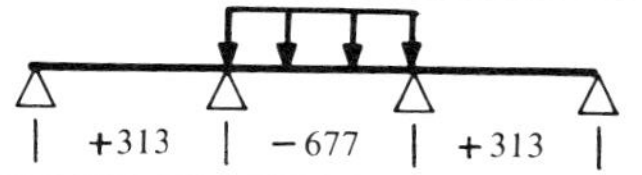		−0.500	−0.500				+0.500	+5.000	−0.500	
	−0.500	+0.750	−0.500			+0.500	−5.000	−0.500		
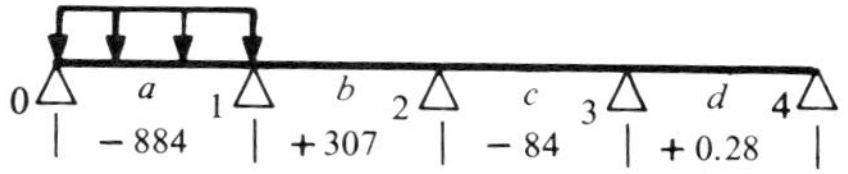		−0.665	+0.179	−0.045			+5.665	−0.844	+0.224	−0.045
	+0.940	−0.665	+0.179	−0.045		−4.335	−0.844	+0.224	−0.045	
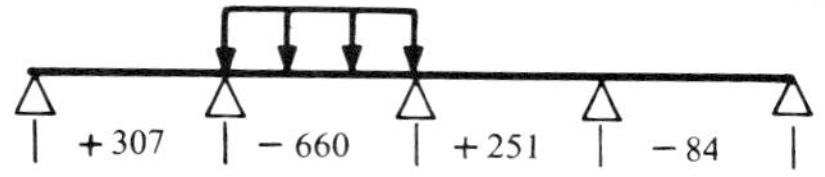		−0.491	−0.536	+0.134			+0.491	+5.045	−0.670	+0.134
	−0.491	+0.737	−0.536	+0.134		+0.491	−4.955	−0.670	+0.134	
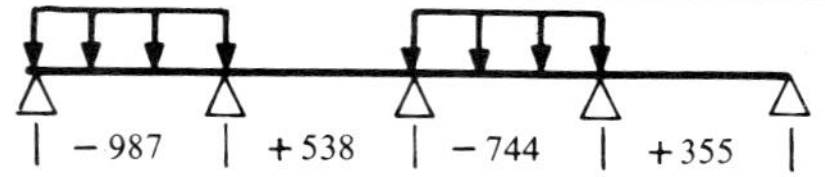		−0.536	−0.357	−0.536			+5.536	−0.179	+5.179	−0.536
	+0.996	−0.536	+0.805	−0.536		−4.464	−0.179	−4.821	−0.536	
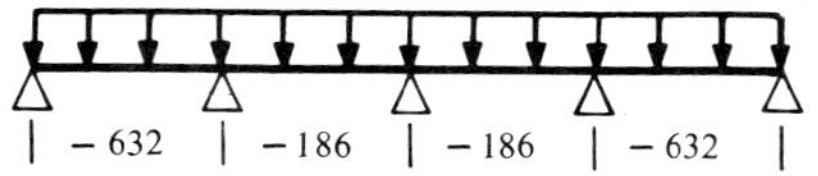		−1.071	−0.714	−1.071			+6.071	+4.643	+5.357	+3.929
	+0.772	+0.364	+0.364	+0.772		−3.929	−5.357	−4.643	−6.071	
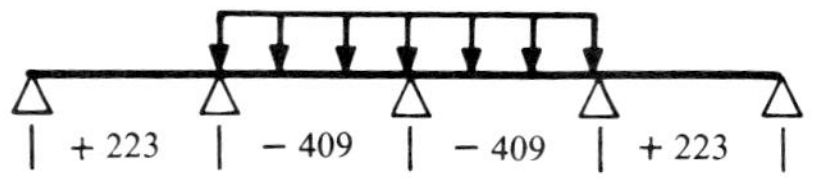		−0.357	−1.072	−0.357			−0.357	+5.715	+4.285	−0.357
	−0.357	+0.561	+0.561	−0.357		+0.357	−4.285	−5.717	−0.357	
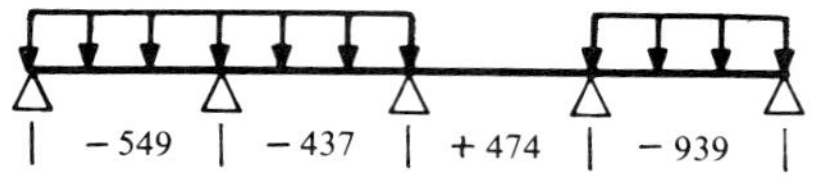		−1.205	−0.179	−0.580			+6.205	+3.974	+0.401	+4.420
	+0.720	+0.610	−0.580	+0.977		−3.795	−6.026	+0.401	−5.580	

†Ref. 8, pp. 776–777

4.32 STIFFNESS MATRIX EQUATIONS

(1) Notation

A_x = area of section normal to x (m^2)
I_z = moment of inertia of A_x about z (m^4)
E = modulus of elasticity (Pa)

s = span (m)
a, b = segments (m)

$$r = \frac{s^2 A_x}{I_z}$$

Positive forces, moments, and displacements are shown in (2, 3) below.

(2) Free-body diagram

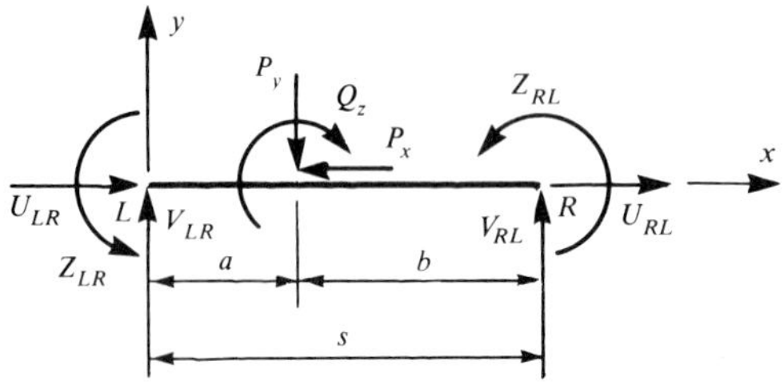

(3) Elastic curve

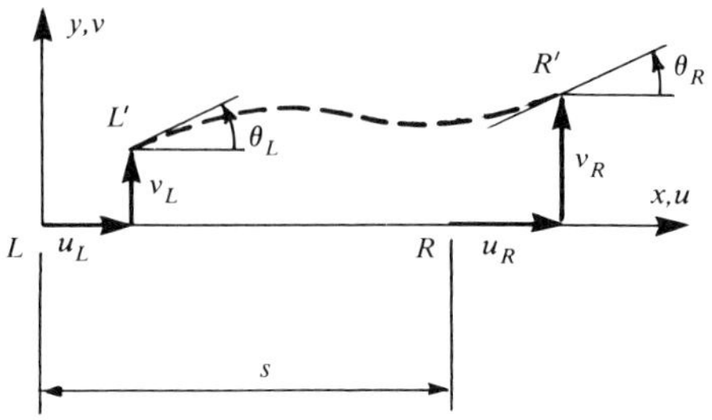

(4) Matrix equation, defining the end reaction vectors as functions of end displacements and causes acting on the bar, is the stiffness matrix equation (Ref. 4, p. 193)

$$\begin{bmatrix} U_{LR} \\ V_{LR} \\ Z_{LR} \\ \hline U_{RL} \\ V_{RL} \\ Z_{RL} \end{bmatrix} = \frac{EI_z}{s^3} \left[\begin{array}{ccc|ccc} r & 0 & 0 & -r & 0 & 0 \\ 0 & 12 & 6s & 0 & -12 & 6s \\ 0 & 6s & 4s^2 & 0 & -6s & 2s^2 \\ \hline -r & 0 & 0 & r & 0 & 0 \\ 0 & -12 & -6s & 0 & 12 & -6s \\ 0 & 6s & 2s^2 & 0 & -6s & 4s^2 \end{array}\right] \begin{bmatrix} u_L \\ v_L \\ \theta_L \\ \hline u_R \\ v_R \\ \theta_R \end{bmatrix} + \begin{bmatrix} U_{L0} \\ V_{L0} \\ Z_{L0} \\ \hline U_{R0} \\ V_{R0} \\ Z_{R0} \end{bmatrix}$$

(5) Load functions $U_{L0}, V_{L0}, \ldots, Z_{R0}$ shown as the last column matrix in (4) above are the reactions of fixed-end bars due to loads and other causes (4.34–4.41).

(6) Symmetrical end displacements If $u_L = -u_R$, $v_L = v_R$, $\theta_L = -\theta_R$, and *symmetrical causes* are applied, the stiffness matrix equation in (4) above reduces to

$$\begin{bmatrix} U_{LR} \\ V_{LR} \\ Z_{LR} \end{bmatrix} = \frac{EI_z}{s^3} \begin{bmatrix} 2r & 0 & 0 \\ 0 & 0 & 0 \\ 0 & 0 & 2s^2 \end{bmatrix} \begin{bmatrix} u_L \\ v_L \\ \theta_L \end{bmatrix} + \begin{bmatrix} U_{L0} \\ V_{L0} \\ Z_{L0} \end{bmatrix}$$

(7) Antisymmetrical end displacements If $u_L = u_R$, $v_L = -v_R$, $\theta_L = \theta_R$, and *antisymmetrical causes* are applied, the stiffness matrix equation in (4) above reduces to

$$\begin{bmatrix} U_{LR} \\ V_{LR} \\ Z_{LR} \end{bmatrix} = \frac{EI_z}{s^3} \begin{bmatrix} 0 & 0 & 0 \\ 0 & 24 & 12s \\ 0 & 12s & 6s^2 \end{bmatrix} \begin{bmatrix} u_L \\ v_L \\ \theta_L \end{bmatrix} + \begin{bmatrix} U_{L0} \\ V_{L0} \\ Z_{L0} \end{bmatrix}$$

4.33 MODIFIED STIFFNESS MATRIX EQUATIONS

(1) Left end hinged. If the left end moment $Z_{LR} = 0$, the stiffness matrix equation in (4.32-4) reduces to

$$\begin{bmatrix} U_{LR} \\ V_{LR} \\ \hline U_{RL} \\ V_{RL} \\ Z_{RL} \end{bmatrix} = \frac{EI_z}{s^3} \left[\begin{array}{cc|ccc} r & 0 & -r & 0 & 0 \\ 0 & 3 & 0 & -3 & 3s \\ \hline -r & 0 & r & 0 & 0 \\ 0 & -3 & 0 & 3 & -3s \\ 0 & 3s & 0 & -3s & 3s^2 \end{array}\right] \begin{bmatrix} u_L \\ v_L \\ \hline u_R \\ v_R \\ \theta_R \end{bmatrix} + \begin{bmatrix} U^*_{L0} \\ V^*_{L0} \\ \hline U^*_{R0} \\ V^*_{R0} \\ Z^*_{R0} \end{bmatrix}$$

The load functions shown as the last column matrix above are the reactions of the respective propped-end beam (4.42–4.43).

(2) Right end hinged. If the right end moment $Z_{RL} = 0$, the stiffness matrix equation in (4.32-4) reduces to

$$\begin{bmatrix} U_{LR} \\ V_{LR} \\ Z_{LR} \\ \hline U_{RL} \\ V_{RL} \end{bmatrix} = \frac{EI_z}{s^3} \left[\begin{array}{ccc|cc} r & 0 & 0 & -r & 0 \\ 0 & 3 & 3s & 0 & -3 \\ 0 & 3s & 3s^2 & 0 & -3s \\ \hline -r & 0 & 0 & r & 0 \\ 0 & -3 & -3s & 0 & 3 \end{array}\right] \begin{bmatrix} u_L \\ v_L \\ \theta_L \\ \hline u_R \\ v_R \end{bmatrix} + \begin{bmatrix} U^*_{L0} \\ V^*_{L0} \\ Z^*_{L0} \\ \hline U^*_{R0} \\ V^*_{R0} \end{bmatrix}$$

The load functions shown as the last column matrix above are the reactions of the respective propped-end beam.

(3) Left end guided. If the left-end vertical force $V_{LR} = 0$, the stiffness matrix equation in (4.32-4) reduces to

$$\begin{bmatrix} U_{LR} \\ Z_{LR} \\ \hline U_{RL} \\ V_{RL} \\ Z_{RL} \end{bmatrix} = \frac{EI_z}{s^3} \left[\begin{array}{cc|ccc} r & 0 & -r & 0 & 0 \\ 0 & s^2 & 0 & 0 & -s^2 \\ \hline -r & 0 & r & 0 & 0 \\ 0 & 0 & 0 & 0 & 0 \\ 0 & -s^2 & 0 & 0 & s^2 \end{array}\right] \begin{bmatrix} u_L \\ \theta_L \\ \hline u_R \\ v_R \\ \theta_R \end{bmatrix} + \begin{bmatrix} U^*_{L0} \\ Z^*_{L0} \\ \hline U^*_{R0} \\ V^*_{R0} \\ Z^*_{R0} \end{bmatrix}$$

The load functions shown as the last column matrix above are the reactions of the respective guided beam (4.44–4.45).

(4) Right end guided. If the right-end vertical force $V_{RL} = 0$, the stiffness matrix equation in (4.32-4) reduces to

$$\begin{bmatrix} U_{LR} \\ V_{LR} \\ Z_{LR} \\ \hline U_{RL} \\ Z_{RL} \end{bmatrix} = \frac{EI_z}{s^3} \left[\begin{array}{ccc|cc} r & 0 & 0 & -r & 0 \\ 0 & 0 & 0 & 0 & 0 \\ 0 & 0 & s^2 & 0 & -s^2 \\ \hline -r & 0 & 0 & r & 0 \\ 0 & 0 & -s^2 & 0 & s^2 \end{array}\right] \begin{bmatrix} u_L \\ v_L \\ \theta_L \\ \hline u_R \\ \theta_R \end{bmatrix} + \begin{bmatrix} U^*_{L0} \\ V^*_{L0} \\ Z^*_{L0} \\ \hline U^*_{R0} \\ Z^*_{R0} \end{bmatrix}$$

The load functions shown as the last column matrix above are the reactions of the respective guided beam.

4.34 STIFFNESS LOAD FUNCTIONS, FIXED-END BEAM

Notation (4.32)	Signs (4.32)	Matrix (4.32)
P_x, P_y = concentrated loads (N)	$m = \frac{a}{s}$ $\quad n = \frac{b}{s}$	$c = \frac{s}{a}$
$U_{LO} = nP_x$ $V_{LO} = (1 + 2m)n^2 P_y$ $Z_{LO} = mnbP_y$	**1.** 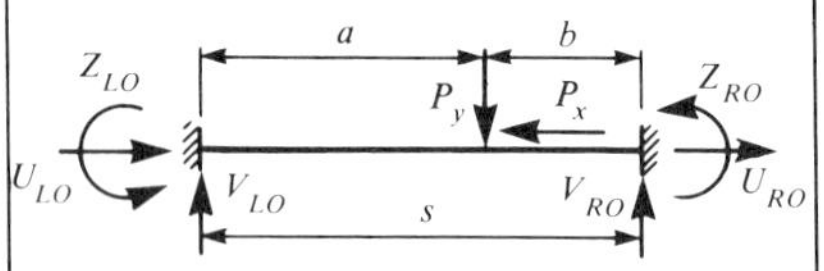	$U_{RO} = mP_x$ $V_{RO} = (1 + 2n)m^2 P_y$ $Z_{RO} = -mnaP_y$
$U_{LO} = \frac{1}{2}P_x$ $V_{LO} = \frac{1}{2}P_y$ $Z_{LO} = \frac{1}{8}sP_y$	**2.** 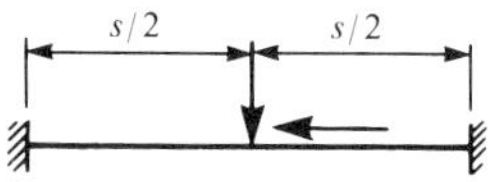	$U_{RO} = \frac{1}{2}P_x$ $V_{RO} = \frac{1}{2}P_y$ $Z_{RO} = -\frac{1}{8}sP_y$
$U_{LO} = P_x$ $V_{LO} = P_y$ $Z_{LO} = (1 - m)aP_y$	**3.** 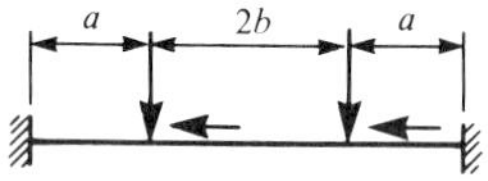	$U_{RO} = P_x$ $V_{RO} = P_y$ $Z_{RO} = -(1 - m)aP_y$
$U_{LO} = 2nP_x$ $V_{LO} = 2n(1 + 2m - 2m^2)P_y$ $Z_{LO} = 2n(1 - m)aP_y$	**4.** 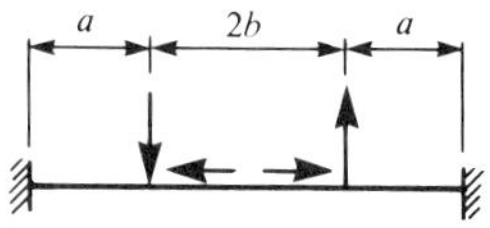	$U_{RO} = -2nP_x$ $V_{RO} = -2n(1 + 2m - 2m^2)P_y$ $Z_{RO} = 2n(1 - m)aP_y$
$U_{LO} = \frac{1}{2}(c - 1)P_x$ $V_{LO} = \frac{1}{2}(c - 1)P_y$ $Z_{LO} = \frac{c^2 - 1}{12} aP_y$	**5.** 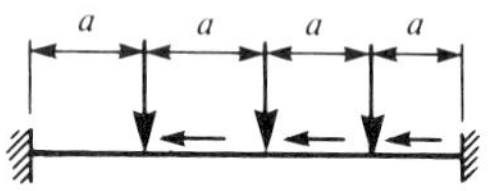 $s = c \cdot a \quad c = 2, 3, 4, \ldots$	$U_{RO} = \frac{1}{2}(c - 1)P_x$ $V_{RO} = \frac{1}{2}(c - 1)P_y$ $Z_{RO} = -\frac{c^2 - 1}{12} aP_y$
$U_{LO} = \frac{1}{2}cP_x$ $V_{LO} = \frac{1}{2}cP_y$ $Z_{LO} = \frac{2c^2 + 1}{24} aP_y$	**6.** 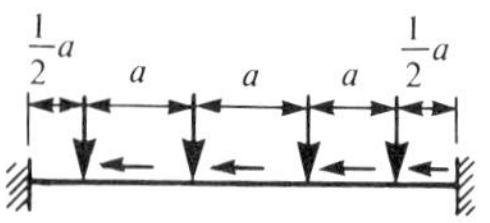 $s = c \cdot a \quad c = 2, 3, 4, \ldots$	$U_{RO} = \frac{1}{2}cP_x$ $V_{RO} = \frac{1}{2}cP_y$ $Z_{RO} = -\frac{2c^2 + 1}{24} aP_y$

4.35 STIFFNESS LOAD FUNCTIONS, FIXED-END BEAM

Notation (4.32) — Signs (4.32) — Matrix (4.32)

p_x, p_y = intensities of distributed load (N/m) $\quad m = \dfrac{a}{s} \quad n = \dfrac{b}{s}$

1.

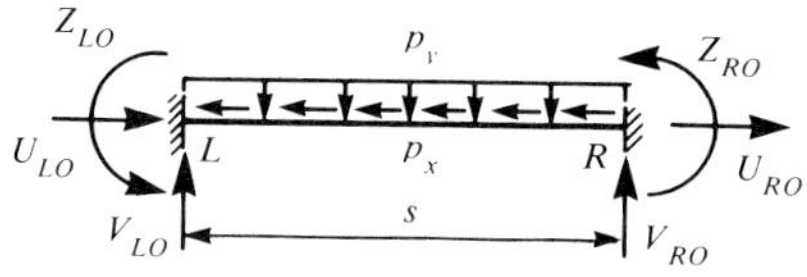

$$U_{LO} = \tfrac{1}{2} s p_x \qquad U_{RO} = \tfrac{1}{2} s p_x$$
$$V_{LO} = \tfrac{1}{2} s p_y \qquad V_{RO} = \tfrac{1}{2} s p_y$$
$$Z_{LO} = \frac{s^2}{12} p_y \qquad Z_{RO} = -\frac{s^2}{12} p_y$$

2.

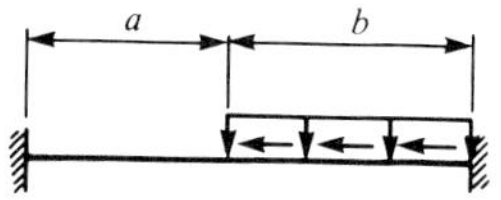

$$U_{LO} = \tfrac{1}{2} n b p_x \qquad U_{RO} = \tfrac{1}{2}(2 - n) b p_x$$
$$V_{LO} = \tfrac{1}{2}(2 - n) n^2 b p_y \qquad V_{RO} = \tfrac{1}{2}(1 + m + m^2 n) b p_y$$
$$Z_{LO} = \frac{(4 - 3n)}{12} n b^2 p_y \qquad Z_{RO} = -\frac{6 - 8n + 3n^2}{12} b^2 p_y$$

3.

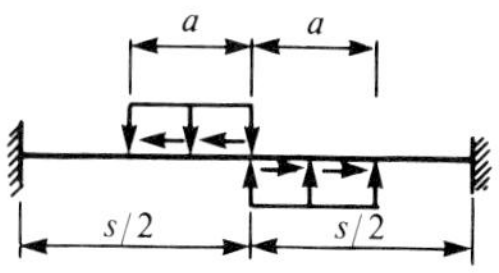

$$U_{LO} = m a p_x \qquad U_{RO} = -m a p_x$$
$$V_{LO} = \frac{3 - 2m^2}{2} m a p_y \qquad V_{RO} = -\frac{3 - 2m^2}{2} m a p_y$$
$$Z_{LO} = \frac{1 - 2m^2}{4} a^2 p_y \qquad Z_{RO} = \frac{1 - 2m^2}{4} a^2 p_y$$

4.

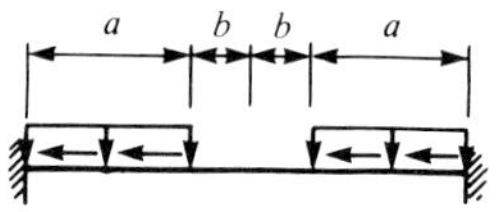

$$U_{LO} = a p_x \qquad U_{RO} = a p_x$$
$$V_{LO} = a p_y \qquad V_{RO} = a p_y$$
$$Z_{LO} = \frac{1 + n}{3} a^2 p_y \qquad Z_{RO} = -\frac{1 + n}{3} a^2 p_y$$

5.

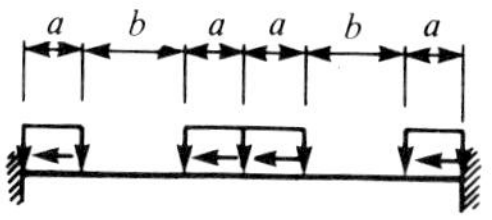

$$U_{LO} = 2 a p_x \qquad U_{RO} = 2 a p_x$$
$$V_{LO} = 2 a p_y \qquad V_{RO} = 2 a p_y$$
$$Z_{LO} = \frac{3 + 6m - 8m^2}{12} a s p_y \qquad Z_{RO} = -\frac{3 + 6m - 8m^2}{12} a s p_y$$

6.

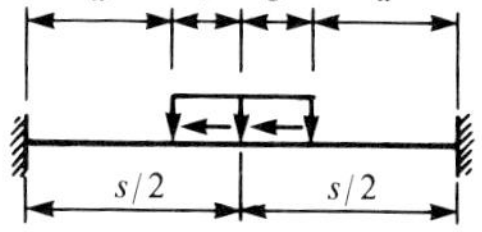

Case 1 − case 4

7.

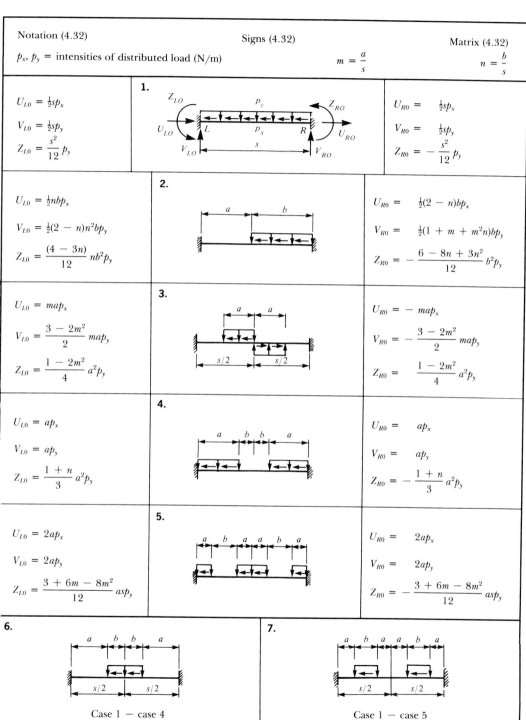

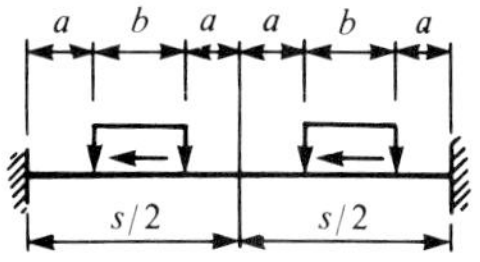

Case 1 − case 5

4.36 STIFFNESS LOAD FUNCTIONS, FIXED-END BEAM

Notation (4.32) Q_z = applied couple (N·m)	Signs (4.32) $m = \frac{a}{s}$ $\quad\quad$ $n = \frac{b}{s}$	Matrix (4.32) $c = \frac{s}{a}$
$U_{L0} = 0$ $V_{L0} = -\frac{6}{s} mnQ_z$ $Z_{L0} = -(2 - 3n)nQ_z$	**1.** 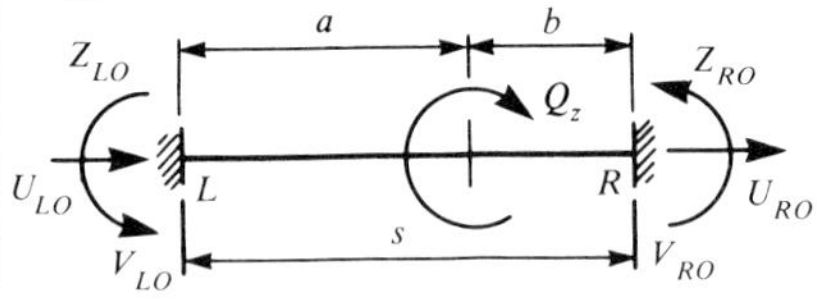	$U_{R0} = 0$ $V_{R0} = \frac{6}{s} mnQ_z$ $Z_{R0} = -(2 - 3m)mQ_z$
$U_{L0} = 0$ $V_{L0} = -\frac{3}{2s} Q_z$ $Z_{L0} = -\frac{1}{4}Q_z$	**2.** 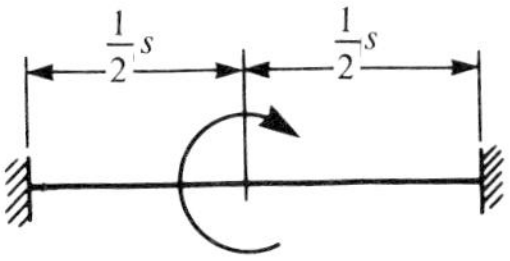	$U_{R0} = 0$ $V_{R0} = \frac{3}{2s} Q_z$ $Z_{R0} = -\frac{1}{4}Q_z$
$U_{L0} = 0$ $V_{L0} = 0$ $Z_{L0} = 2nQ_z$	**3.** 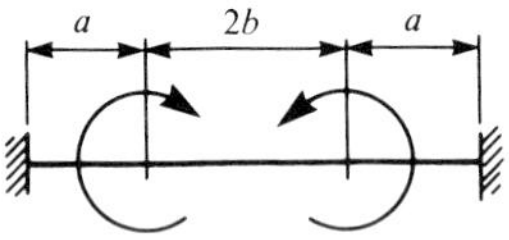	$U_{R0} = 0$ $V_{R0} = 0$ $Z_{R0} = -2nQ_z$
$U_{L0} = 0$ $V_{L0} = -\frac{12}{s} m(1 - m)Q_z$ $Z_{L0} = (1 - 6m + 6m^2)Q_z$	**4.** 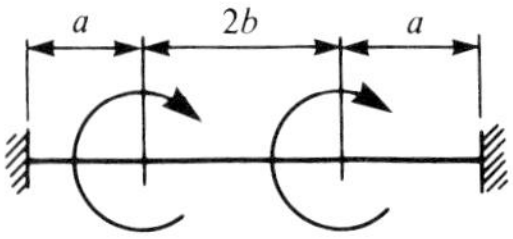	$U_{R0} = 0$ $V_{R0} = \frac{12}{s} m(1 - m)Q_z$ $Z_{R0} = (1 - 6m + 6m^2)Q_z$
U_{L0} $\quad 0$ $V_{L0} = -\frac{c^2 - 1}{cs} Q_z$ $Z_{L0} = -\frac{c - 1}{2c} Q_z$	**5.** 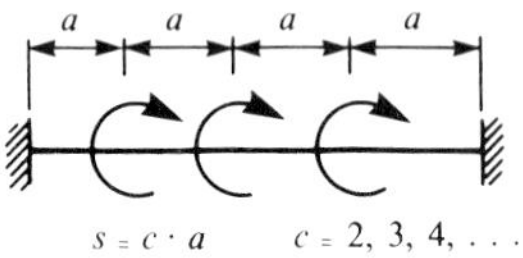 $s = c \cdot a$ $\quad$ $c = 2, 3, 4, \ldots$	$U_{R0} = 0$ $V_{R0} = \frac{c^2 - 1}{cs} Q_z$ $Z_{R0} = -\frac{c - 1}{2c} Q_z$
$U_{L0} = 0$ $V_{L0} = -\frac{3c}{2s} Q_z$ $Z_{L0} = -\frac{1}{4}cQ_z$	**6.** 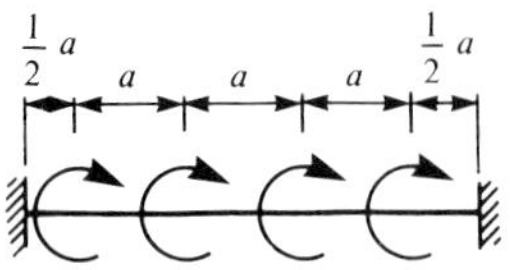 $s = c \cdot a$ $\quad$ $c = 2, 3, 4, \ldots$	$U_{R0} = 0$ $V_{R0} = \frac{3c}{2s} Q_z$ $Z_{R0} = -\frac{1}{4}cQ_z$

4.37 STIFFNESS LOAD FUNCTIONS, FIXED-END BEAM

Notation (4.32)	Signs (4.32)	Matrix (4.32)
q_z = intensity of distributed moment (N·m/m)	$m = \dfrac{a}{s}$	$n = \dfrac{b}{s}$
$U_{L0} = 0$ $V_{L0} = -q_z$ $Z_{L0} = 0$	**1.** 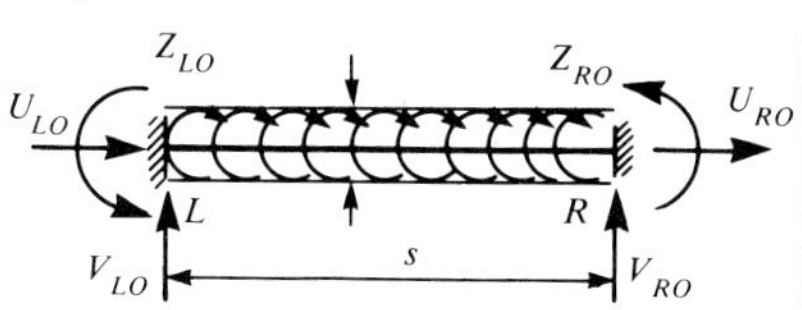	$U_{L0} = 0$ $V_{L0} = q_z$ $Z_{L0} = 0$
$U_{L0} = 0$ $V_{L0} = -(3 - 2n)n^2 q_z$ $Z_{L0} = -mnbq_z$	**2.** 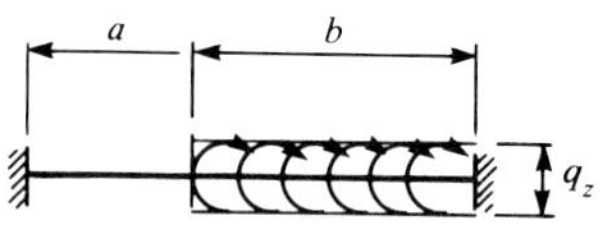	$U_{L0} = 0$ $V_{L0} = (3 - 2n)n^2 q_z$ $Z_{L0} = mnaq_z$
$U_{L0} = 0$ $V_{L0} = -\frac{1}{2}q_z$ $Z_{L0} = -\frac{1}{12}sq_z$	**3.** 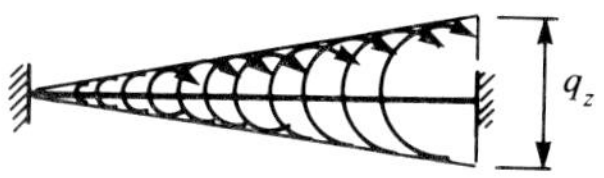	$U_{L0} = 0$ $V_{L0} = \frac{1}{2}q_z$ $Z_{L0} = \frac{1}{12}sq_z$
$U_{L0} = 0$ $V_{L0} = -\frac{1}{2}(2 - n)n^2 q_z$ $Z_{L0} = -\dfrac{1 + 2m - 3m^2}{12} bq_z$	**4.** 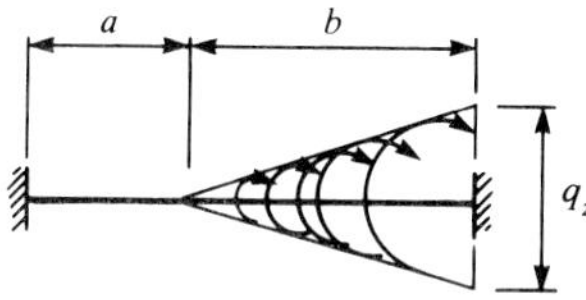	$U_{L0} = 0$ $V_{L0} = \frac{1}{2}(2 - n)n^2 q_z$ $Z_{L0} = \dfrac{1 + 2m + 3m^2}{12} bq_z$
$U_{L0} = 0$ $V_{L0} = 0$ $Z_{L0} = -\frac{1}{4}sq_z$	**5.** 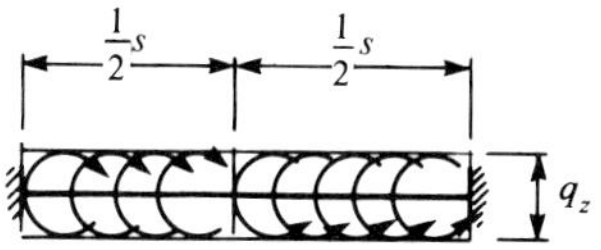	$U_{L0} = 0$ $V_{L0} = 0$ $Z_{L0} = \frac{1}{4}sq_z$
6. 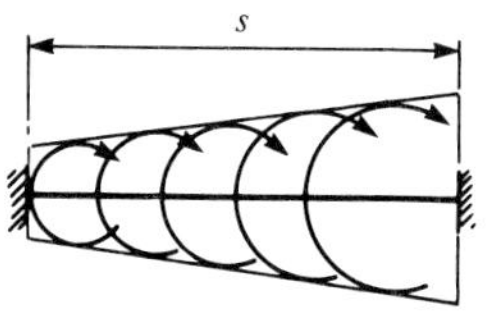 Case 1 + case 3		**7.** 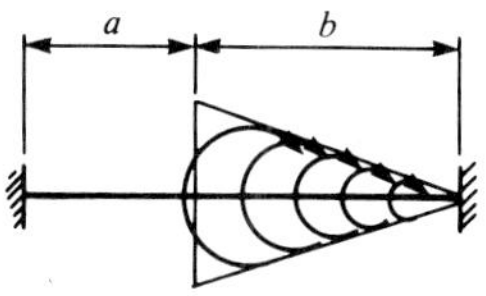 Case 2 − case 4

4.38 STIFFNESS LOAD FUNCTIONS, FIXED-END BEAM

Notation (4.32) — Signs (4.32) — Matrix (4.32)

p_x, p_y = maximum intensities of triangular loads (N/m) $\qquad m = \dfrac{a}{s} \qquad n = \dfrac{b}{s}$

$U_{LO} = \frac{1}{6} s p_x$ $V_{LO} = \dfrac{3s}{20} p_y$ $Z_{LO} = \dfrac{s^2}{30} p_y$	**1.** 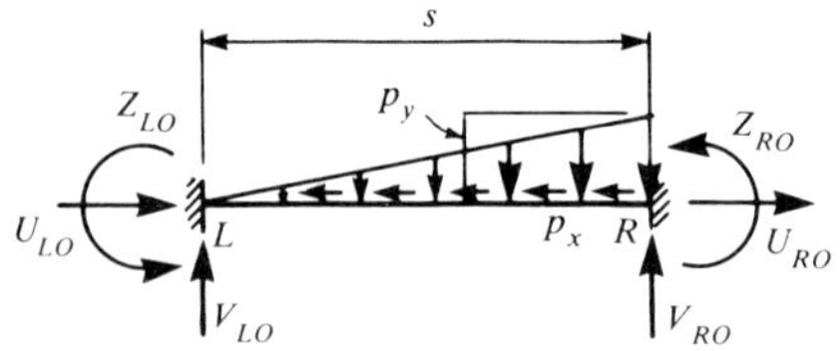 	$U_{RO} = \frac{1}{3} s p_x$ $V_{RO} = \dfrac{7s}{20} p_y$ $Z_{RO} = -\dfrac{s^2}{20} p_y$
$U_{LO} = \frac{1}{6} n b p_x$ $V_{LO} = \dfrac{n^2(5 - 2n)}{20} b p_y$ $Z_{LO} = \dfrac{5n - 3n^2}{60} b^2 p_y$	**2.** 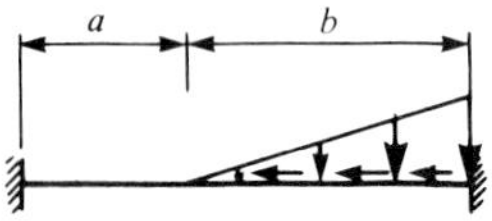 	$U_{RO} = \frac{1}{6}(3 - n) b p_x$ $V_{RO} = \dfrac{10 - 5n^2 + 2n^3}{20} b p_y$ $Z_{RO} = -\dfrac{10 - 10n + 3n^2}{60} b^2 p_y$
$U_{LO} = \frac{1}{2} a p_x$ $V_{LO} = \frac{1}{2} a p_y$ $Z_{LO} = \dfrac{2 - m}{12} a^2 p_y$	**3.** 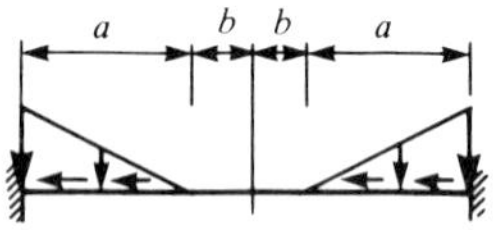 	$U_{RO} = \frac{1}{2} a p_x$ $V_{RO} = \frac{1}{2} a p_y$ $Z_{RO} = -\dfrac{2 - m}{12} a^2 p_y$
$U_{LO} = \dfrac{n}{3} b p_x$ $V_{LO} = \dfrac{(15 - 8n)n^2}{20} b p_y$ $Z_{LO} = \dfrac{5n - 4n^2}{20} b^2 p_y$	**4.** 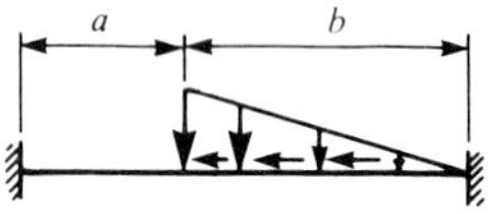 	$U_{RO} = \frac{1}{6}(1 + 2m) b p_x$ $V_{RO} = \dfrac{10 - 15n^2 + 8n^3}{20} b p_y$ $Z_{RO} = -\dfrac{10 - 15n + 6n^2}{30} b^2 p_y$
$U_{LO} = \frac{1}{2} b p_x$ $V_{LO} = \frac{1}{2} b p_y$ $Z_{LO} = \dfrac{3 - 2n^2}{24} b s p_y$	**5.** 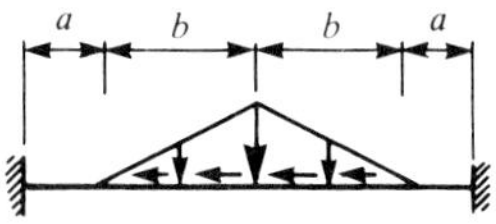 	$U_{RO} = \frac{1}{2} b p_x$ $V_{RO} = \frac{1}{2} b p_y$ $Z_{RO} = -\dfrac{3 - 2n^2}{24} b s p_y$
$U_{LO} = \frac{1}{3} n b p_x$ $V_{LO} = \dfrac{(5 - 2n^2) n b p_y}{10}$ $Z_{LO} = \dfrac{5 - 6n^2}{60} b^2 p_y$	**6.** 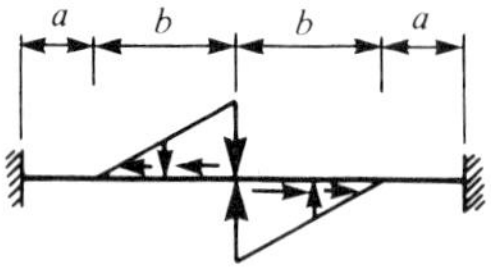 	$U_{RO} = -\frac{1}{3} n b p_x$ $V_{RO} = -\dfrac{(5 - 2n^2) n b p_y}{10}$ $Z_{RO} = \dfrac{5 - 6n^2}{60} b^2 p_y$

4.39 STIFFNESS LOAD FUNCTIONS, FIXED-END BEAM

Notation (4.32) Signs (4.32) Matrix (4.32)

p_x, p_y = maximum intensities of triangular loads (N/m) $m = \frac{a}{s}$ $n = \frac{b}{s}$

Left end	Case	Right end
$U_{L0} = \frac{1}{4}sp_x$ $V_{L0} = \frac{1}{4}sp_y$ $Z_{L0} = \frac{5}{96}s^2p_y$	1. 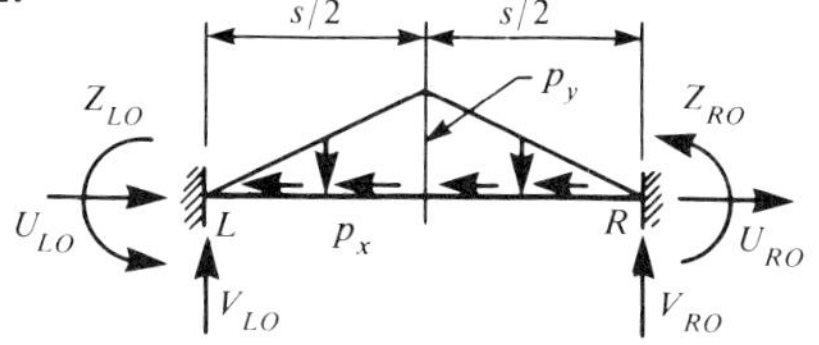	$U_{R0} = \frac{1}{4}sp_x$ $V_{R0} = \frac{1}{4}sp_y$ $Z_{R0} = -\frac{5}{96}s^2p_y$
$U_{L0} = \frac{1}{4}sp_x$ $V_{L0} = \frac{1}{4}sp_y$ $Z_{L0} = \frac{s^2}{32}p_y$	2. 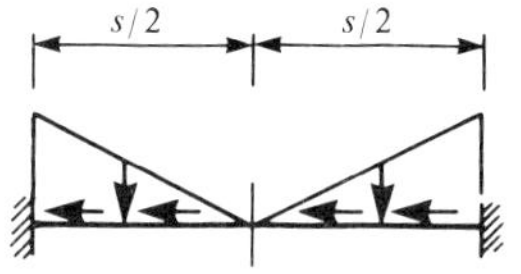	$U_{R0} = \frac{1}{4}sp_x$ $V_{R0} = \frac{1}{4}sp_y$ $Z_{R0} = -\frac{s^2}{32}p_y$
$U_{L0} = \frac{1}{4}sp_x$ $V_{L0} = \frac{1}{4}sp_y$ $Z_{L0} = \frac{17}{384}s^2p_y$	3. 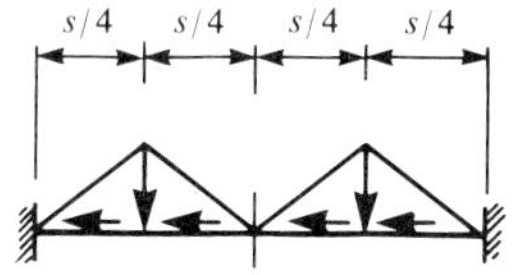	$U_{R0} = \frac{1}{4}sp_x$ $V_{R0} = \frac{1}{4}sp_y$ $Z_{R0} = -\frac{17}{384}s^2p_y$
$U_{L0} = \frac{1}{4}sp_x$ $V_{L0} = \frac{1}{4}sp_y$ $Z_{L0} = \frac{5}{128}s^2p_y$	4. 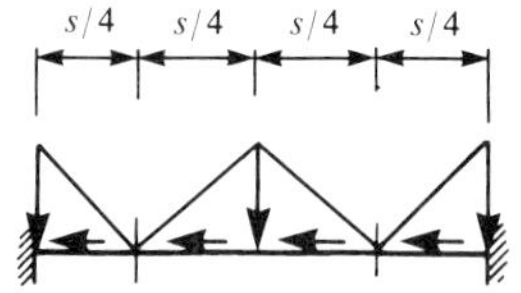	$U_{R0} = \frac{1}{4}sp_x$ $V_{R0} = \frac{1}{4}sp_y$ $Z_{R0} = -\frac{5}{128}s^2p_y$
$U_{L0} = \frac{1}{6}(2+\alpha)sp_x$ $V_{L0} = \frac{7+3\alpha}{20}sp_y$ $Z_{L0} = \frac{3+2\alpha}{60}s^2p_y$	5. 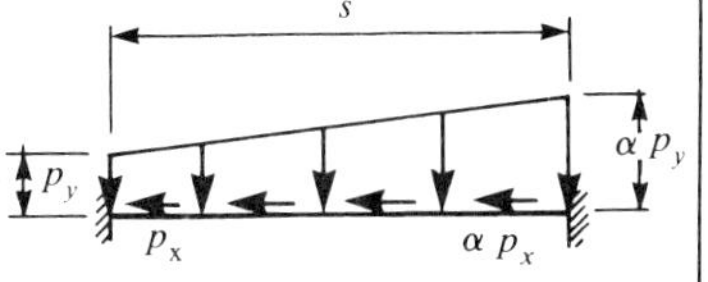	$U_{R0} = \frac{1}{6}(1+2\alpha)sp_x$ $V_{R0} = \frac{3+7\alpha}{20}sp_y$ $Z_{R0} = -\frac{2+3\alpha}{60}s^2p_y$
$U_{L0} = \frac{1}{2}(1-m)sp_x$ $V_{L0} = \frac{1}{2}(1-m)sp_y$ $Z_{L0} = \frac{1-2m^2+m^3}{12}s^2p_y$	6. 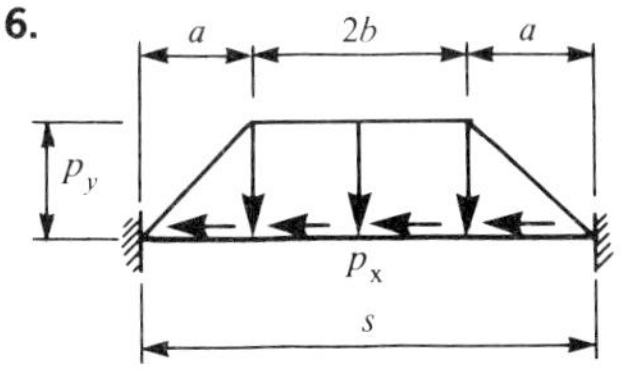	$U_{R0} = \frac{1}{2}(1-m)sp_x$ $V_{R0} = \frac{1}{2}(1-m)sp_y$ $Z_{R0} = -\frac{1-2m^2+m^3}{12}s^2p_y$

4.40 STIFFNESS LOAD FUNCTIONS, FIXED-END BEAM

Notation (4.32) Signs (4.32) Matrix (4.32)

p_x, p_y = maximum intensities of harmonic loads (N/m) $c = \dfrac{s}{m}$ m = integer

$U_{L0} = \dfrac{s}{\pi} p_x$ $V_{L0} = \dfrac{s}{\pi} p_y$ $Z_{R0} = \dfrac{2s^2}{\pi^3} p_y$	**1.** 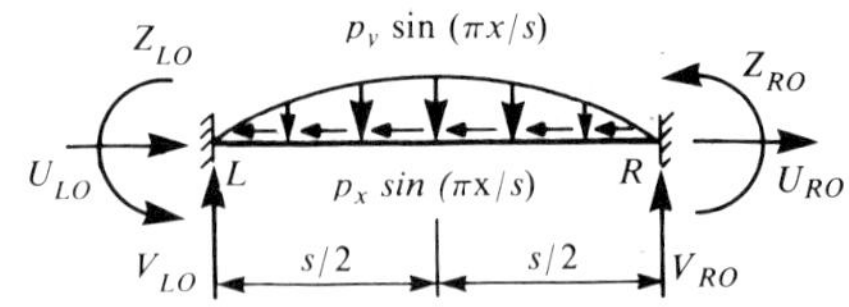 	$U_{R0} = \dfrac{s}{\pi} p_x$ $V_{R0} = \dfrac{s}{\pi} p_y$ $Z_{R0} = -\dfrac{2s^2}{\pi^3} p_y$
$U_{L0} = \dfrac{s}{m\pi} p_x$ $V_{L0} = \dfrac{12 + (m\pi)^2}{(m\pi)^3} s p_y$ $Z_{L0} = \dfrac{6s^2}{(m\pi)^3} p_y$	**2.** 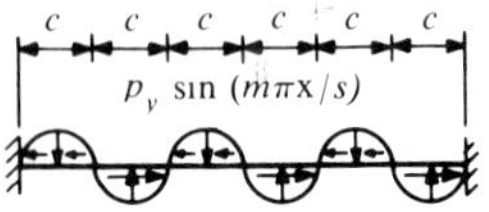 $m = 2, 4, 6, \ldots$	$U_{R0} = -\dfrac{s}{m\pi} p_x$ $V_{R0} = -\dfrac{12 + (m\pi)^2}{(m\pi)^3} s p_y$ $Z_{R0} = \dfrac{6s^2}{(m\pi)^3} p_y$
$U_{L0} = \dfrac{s}{m\pi} p_x$ $V_{L0} = \dfrac{s}{m\pi} p_y$ $Z_{L0} = \dfrac{2s^2}{(m\pi)^3} p_y$	**3.** 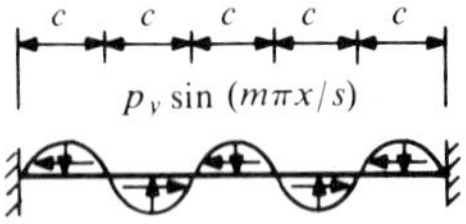 $m = 1, 3, 5, \ldots$	$U_{R0} = \dfrac{s}{m\pi} p_x$ $V_{R0} = \dfrac{s}{m\pi} p_y$ $Z_{R0} = -\dfrac{2s^2}{(m\pi)^3} p_y$
$U_{L0} = \dfrac{2s}{\pi^2} p_x$ $V_{L0} = \dfrac{24s}{\pi^2} p_y$ $Z_{L0} = \dfrac{12 - \pi^2}{\pi^4} s^2 p_y$	**4.** 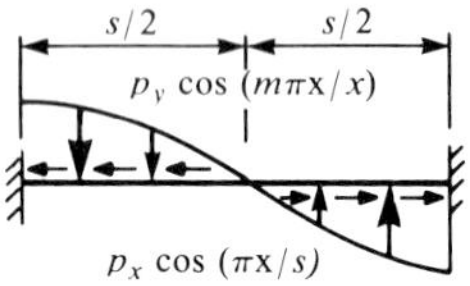 	$U_{R0} = -\dfrac{2s}{\pi^2} p_x$ $V_{R0} = -\dfrac{24s}{\pi^2} p_y$ $Z_{R0} = \dfrac{12 - \pi^2}{\pi^4} s^2 p_y$
$U_{L0} = 0$ $V_{L0} = 0$ $Z_{L0} = -\dfrac{s^2}{(m\pi)^2} p_y$	**5.** 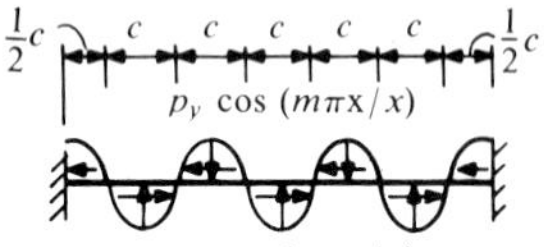 $m = 2, 4, 6, \ldots$	$U_{R0} = 0$ $V_{R0} = 0$ $Z_{R0} = \dfrac{s^2}{(m\pi)^2} p_y$
$U_{L0} = \dfrac{2s}{(m\pi)^2} p_y$ $V_{L0} = \dfrac{24s}{(m\pi)^2} p_y$ $Z_{L0} = \dfrac{12 - (m\pi)^2}{(m\pi)^4} s^2 p_y$	**6.** 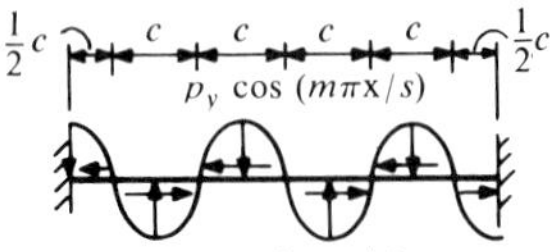 $m = 1, 3, 5, \ldots$	$U_{R0} = -\dfrac{2s}{(m\pi)^2} p_x$ $V_{R0} = -\dfrac{24s}{(m\pi)^2} p_y$ $Z_{R0} = \dfrac{12 - (m\pi)^2}{(m\pi)^4} s^2 p_y$

4.41 STIFFNESS LOAD FUNCTIONS, FIXED-END BEAM

Notation (4.32)	Signs (4.32)	Matrix (4.32)
p_x, p_y = maximum intensities of parabolic loads (N/m)	$m = \frac{a}{s}$	$n = \frac{b}{s}$

$U_{LO} = \frac{1}{3} s p_x$ $V_{LO} = \frac{1}{3} s p_y$ $Z_{LO} = \frac{s^2}{15} p_y$	**1.** 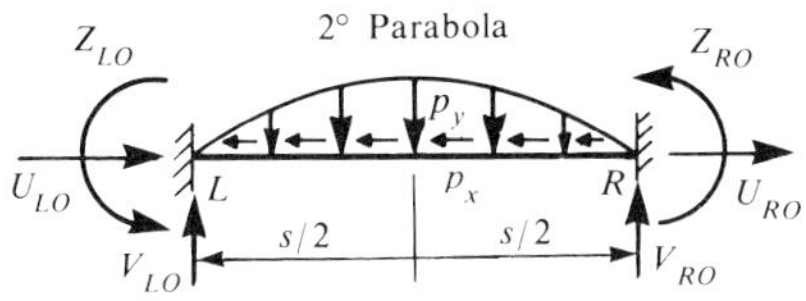	$U_{RO} = \frac{1}{3} s p_x$ $V_{RO} = \frac{1}{3} s p_y$ $Z_{RO} = -\frac{s^2}{15} p_y$
$U_{LO} = \frac{1}{6} s p_x$ $V_{LO} = \frac{1}{6} s p_y$ $Z_{LO} = \frac{s^2}{60} p_y$	**2.** 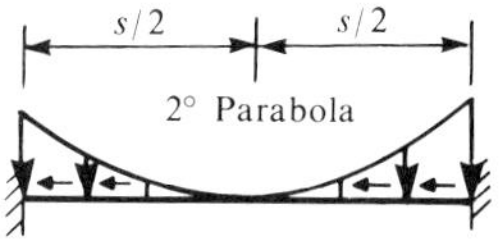	$U_{RO} = \frac{1}{6} s p_x$ $V_{RO} = \frac{1}{6} s p_y$ $Z_{RO} = -\frac{s^2}{60} p_y$
$U_{LO} = \frac{nb}{12} p_x$ $V_{LO} = \frac{3-n}{30} n^2 b p_y$ $Z_{LO} = \frac{2-n}{60} n b^2 p_y$	**3.** 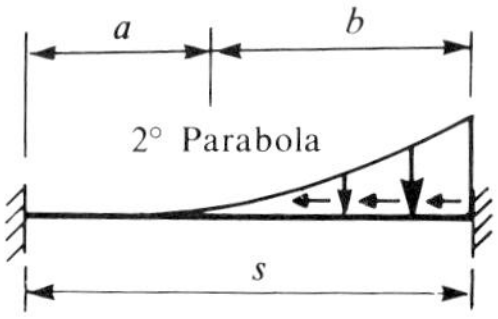	$U_{RO} = \frac{4-n}{12} bp$ $V_{RO} = \frac{10 - 3n^2 + n^3}{30} b p_y$ $Z_{RO} = -\frac{5 - 4n + n^2}{60} b^2 p_y$
$U_{LO} = E A_x s^{-1} u_j$ $V_{LO} = \frac{12}{s^3} E I_z v_j$ $Z_{LO} = \frac{6}{s^2} E I_z v_j$	**4. Abrupt linear displacement** 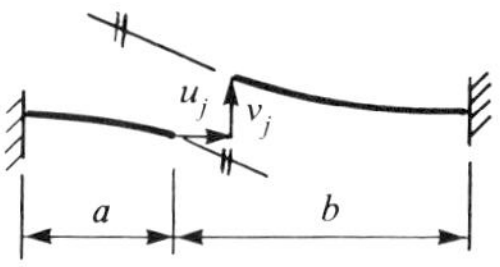	$U_{RO} = -E A_x s^{-1} u_j$ $V_{RO} = -\frac{12}{s^3} E I_z v_j$ $Z_{RO} = \frac{6}{s^2} E I_z v_j$
$U_{LO} = 0$ $V_{LO} = \frac{6(1-2m)}{s^2} E I_z \theta_j$ $Z_{LO} = \frac{2(2-3m)}{s} E I_z \theta_j$	**5. Abrupt angular displacement** 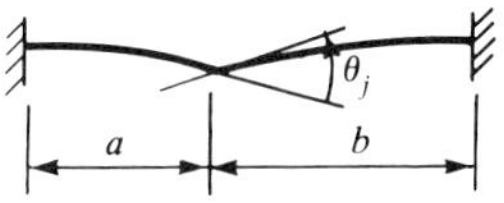	$U_{RO} = 0$ $V_{RO} = -\frac{6(1-2m)}{s^2} E I_z \theta_j$ $Z_{RO} = \frac{2(1-3m)}{s} E I_z \theta_j$
$U_{LO} = n E A_x e_t$ $V_{LO} = -\frac{6mn}{s} E I_z f_t$ $Z_{LO} = (1 - 3m) n E I_z f_t$	**6. Temperature deformation of segment *b*** 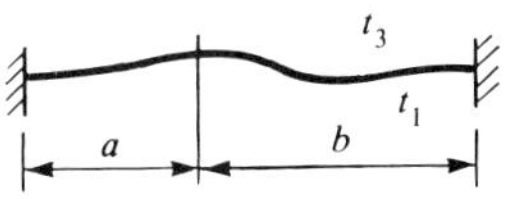 e_t, f_t (4.03)	$U_{RO} = -n E A_x e_t$ $V_{RO} = \frac{6mn}{s} E I_z f_t$ $Z_{RO} = -(1 + 3m) n E I_z f_t$

4.42 STIFFNESS LOAD FUNCTIONS, PROPPED-END BEAM

Notation (4.33) Signs (4.33) Matrix (4.33)

$m = \frac{a}{s}$ $n = \frac{b}{s}$ P_x, P_y, Q_z, p_x, p_y = loads defined in (4.34–4.39)

1.

$U^*_{LO} = 0$

$V^*_{LO} = -\frac{3(1-m^2)}{2s}Q_z$

$Z^*_{LO} = 0$

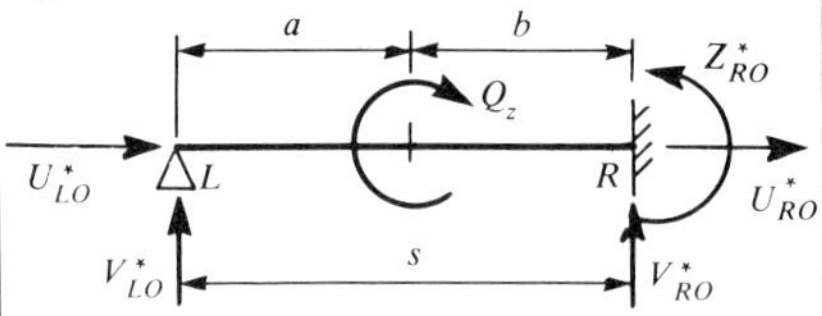

$U^*_{RO} = 0$

$V^*_{RO} = \frac{3(1-m^2)}{2s}Q_z$

$Z^*_{RO} = -\frac{1}{2}(1-3m^2)Q_z$

2.

$U^*_{LO} = nP_x$

$V^*_{LO} = \frac{1}{2}(3-n)n^2P_y$

$Z^*_{LO} = 0$

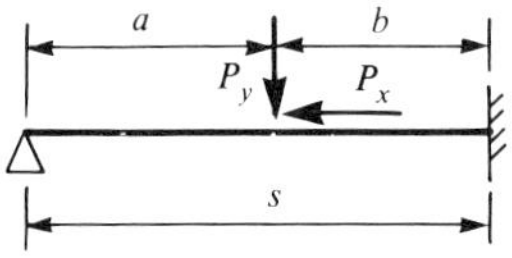

$U^*_{RO} = mP_x$

$V^*_{RO} = \frac{1}{2}(3-m^2)mP_y$

$Z^*_{RO} = -\frac{1}{2}(1+m)mbP_y$

3.

$U^*_{LO} = \frac{1}{2}sp_x$

$V^*_{LO} = \frac{3}{8}sp_y$

$Z^*_{LO} = 0$

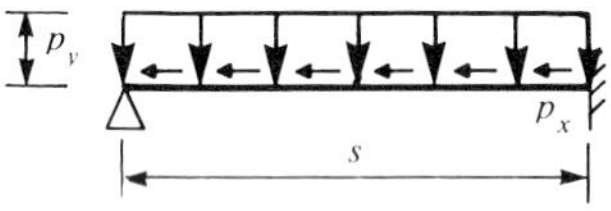

$U^*_{RO} = \frac{1}{2}sp_x$

$V^*_{RO} = \frac{5}{8}sp_y$

$Z^*_{RO} = -\frac{1}{8}s^2p_y$

4.

$U^*_{LO} = \frac{1}{6}sp_x$

$V^*_{LO} = \frac{s}{10}p_y$

$Z^*_{LO} = 0$

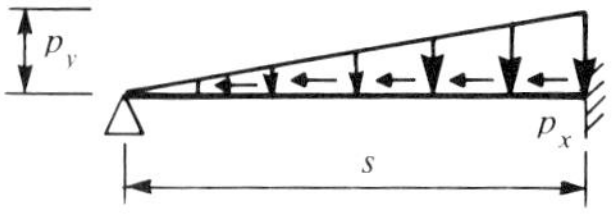

$U^*_{RO} = \frac{1}{3}sp_x$

$V^*_{RO} = \frac{4s}{10}p_y$

$Z^*_{RO} = -\frac{s^2}{15}p_y$

5.

$U^*_{LO} = \frac{s}{12}p_x$

$V^*_{LO} = \frac{s}{24}p_y$

$Z^*_{LO} = 0$

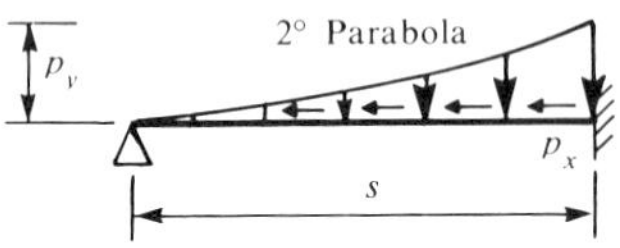

$U^*_{RO} = \frac{s}{4}p_x$

$V^*_{RO} = \frac{7s}{24}p_y$

$Z^*_{RO} = -\frac{s^2}{24}p_y$

6.

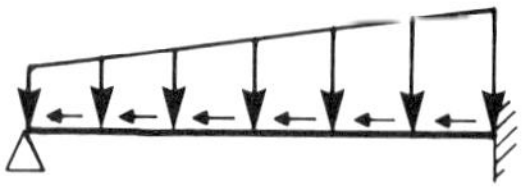

Case 3 + case 4

7.

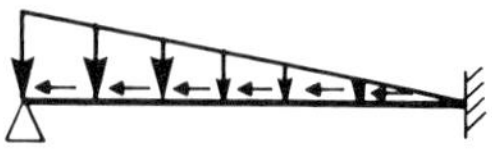

Case 3 − case 4

4.43 STIFFNESS LOAD FUNCTIONS, PROPPED-END BEAM

Notation (4.33) Signs (4.32) Matrix (4.33)

$m = \dfrac{a}{s}$ $\qquad n = \dfrac{b}{s}$ $\qquad p_x, p_y, \theta_j, v_j, e_t, f_t$ = loads and causes defined in (4.03)

1.

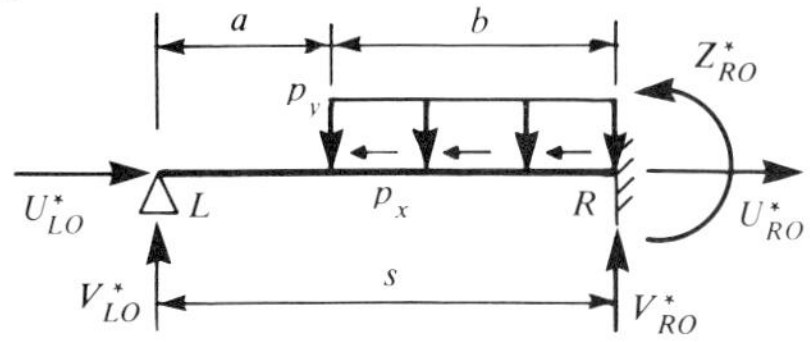

$U^*_{LO} = \frac{1}{2} n b p_x$

$V^*_{LO} = \frac{1}{8}(4 - n)n^2 b p_y$

$Z^*_{LO} = 0$

$U^*_{RO} = \frac{1}{2}(2 - n) b p_x$

$V^*_{RO} = \frac{1}{8}(8 - 4n^2 + n^3) b p_y$

$Z^*_{RO} = -\frac{1}{8}(2 - n)^2 b^2 p_y$

2.

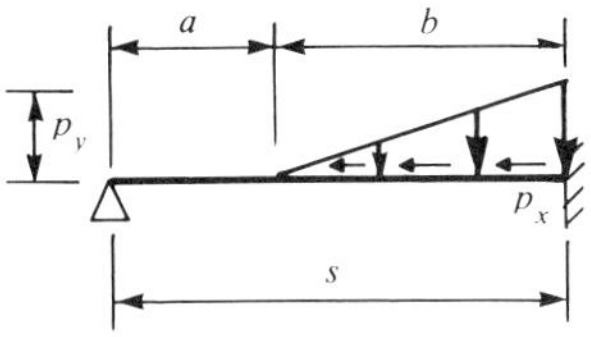

$U^*_{LO} = \frac{1}{6} n b p_x$

$V^*_{LO} = \dfrac{5n^2 - n^3}{40} b p_y$

$Z^*_{LO} = 0$

$U^*_{RO} = \frac{1}{6}(3 - n) b p_x$

$V^*_{RO} = \dfrac{20 - 5n^2 + n^3}{40} b p_y$

$Z^*_{RO} = -\dfrac{20 - 15n + 3n^2}{120} b^2 p_y$

3. Abrupt linear displacement

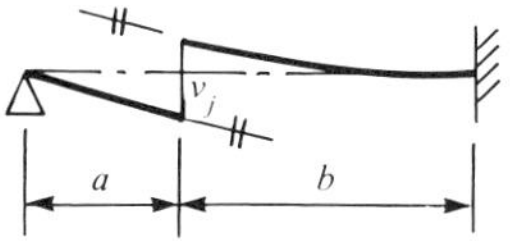

$U^*_{LO} = 0$

$V^*_{LO} = \dfrac{3}{s^3} EI_z v_j$

$Z^*_{LO} = 0$

$U^*_{RO} = 0$

$V^*_{RO} = -\dfrac{3}{s^3} EI_z v_j$

$Z^*_{RO} = \dfrac{3}{s^2} EI_z v_j$

4. Abrupt angular displacement

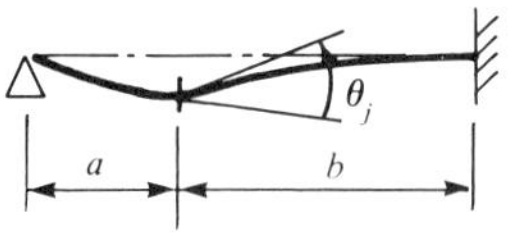

$U^*_{LO} = 0$

$V^*_{LO} = -\dfrac{3mEI_z}{s^2} \theta_j$

$Z^*_{LO} = 0$

$U^*_{RO} = 0$

$V^*_{RO} = \dfrac{3mEI_z}{s^2} \theta_j$

$Z^*_{RO} = -\dfrac{3mEI_z}{s} \theta_j$

5. Temperature deformation of segment *b*

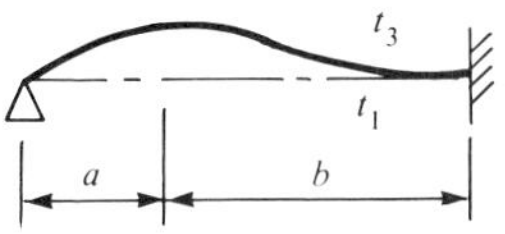

$U^*_{LO} = nEA_x e_t$

$V^*_{LO} = -\dfrac{3}{2s}(1 + m) nEI_z f_t$

$Z^*_{LO} = 0$

$U^*_{RO} = -nEA_x e_t$

$V^*_{RO} = \dfrac{3}{2s}(1 + m) nEI_z f_t$

$Z^*_{RO} = -\frac{3}{2}(1 + m) nEI_z f_t$

6.

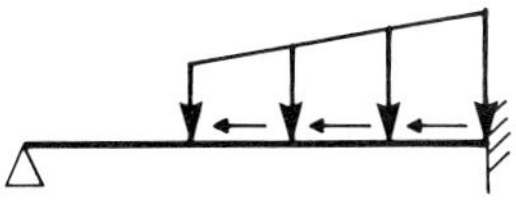

Case 1 + case 2

7.

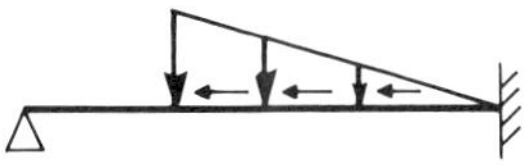

Case 1 − case 2

4.44 STIFFNESS LOAD FUNCTIONS, GUIDED BEAM

Notation (4.33)	Signs (4.32)	Matrix (4.33)
$m = \frac{a}{s}$ $\quad n = \frac{b}{s}$	P_x, P_y, Q_z, p_x, p_y = loads defined in (4.34–4.39)	

$U^*_{L0} = 0$ $V^*_{L0} = 0$ $Z^*_{L0} = nQ_z$	**1.** 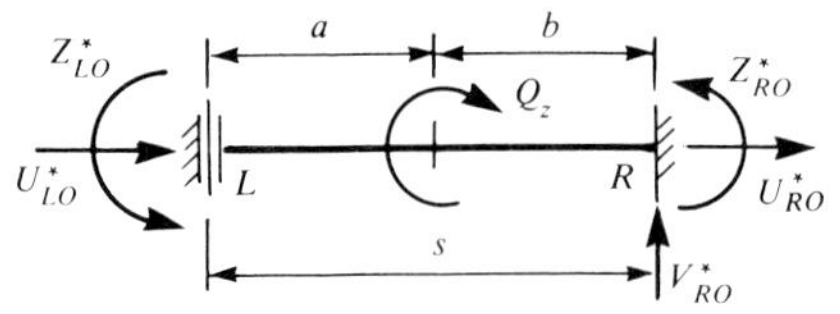 	$U^*_{R0} = 0$ $V^*_{R0} = 0$ $Z^*_{R0} = mQ_z$
$U^*_{L0} = nP_x$ $V^*_{L0} = 0$ $Z^*_{L0} = -\frac{1}{2}nbP_y$	**2.** 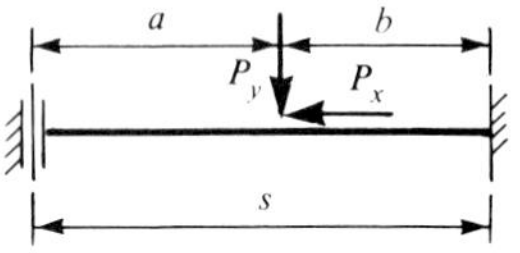 	$U^*_{R0} = mP_x$ $V^*_{R0} = P_y$ $Z^*_{R0} = -\frac{1}{2}(1 + m)bP_y$
$U^*_{L0} = \frac{1}{2}sp_x$ $Z^*_{L0} = -\frac{1}{6}s^2p_y$ $V^*_{L0} = 0$	**3.** 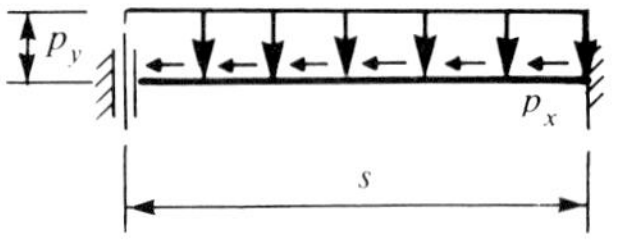 	$U^*_{R0} = \frac{1}{2}sp_x$ $V^*_{R0} = sp_y$ $Z^*_{R0} = -\frac{1}{3}s^2p_y$
$U^*_{L0} = \frac{1}{6}sp_x$ $V^*_{L0} = 0$ $Z^*_{L0} = -\frac{s^2}{24}p_y$	**4.** 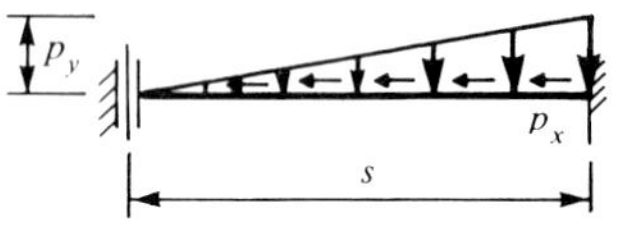 	$U^*_{R0} = \frac{1}{3}sp_x$ $V^*_{R0} = \frac{1}{2}sp_y$ $Z^*_{R0} = -\frac{s^2}{8}p_y$
$U^*_{L0} = \frac{s}{12}p_x$ $V^*_{L0} = 0$ $Z^*_{L0} = -\frac{s^2}{60}p_y$	**5.** 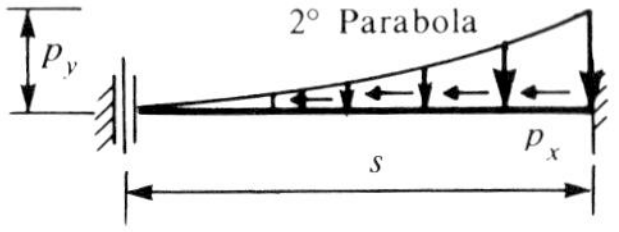 	$U^*_{R0} = \frac{1}{4}sp_x$ $V^*_{R0} = \frac{1}{3}sp_y$ $Z^*_{R0} = -\frac{s^2}{15}p_y$

6. 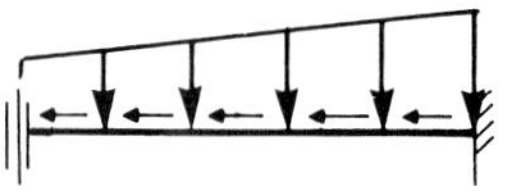 Case 3 + case 4	**7.** 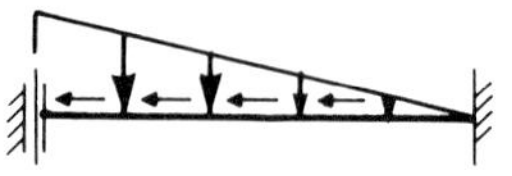 Case 3 − case 4

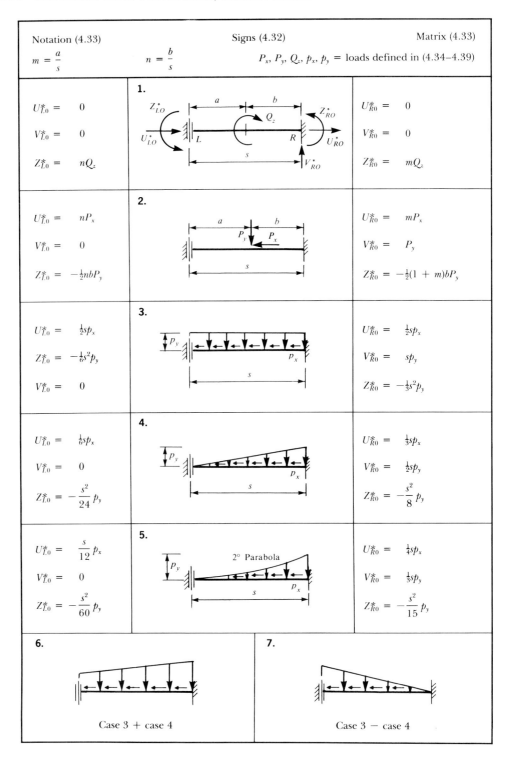

4.45 STIFFNESS LOAD FUNCTIONS, GUIDED BEAM

Notation (4.33) $m = \frac{a}{s}$ $\quad n = \frac{b}{s}$	Signs (4.32) $p_x, p_y, \theta_j, v_j, e_t, f_t$ = loads and causes defined in (4.03)	Matrix (4.33)
$U^*_{L0} = \frac{1}{2}nbp_x$ $V^*_{L0} = 0$ $Z^*_{L0} = -\frac{1}{6}nb^2p_y$	**1.** 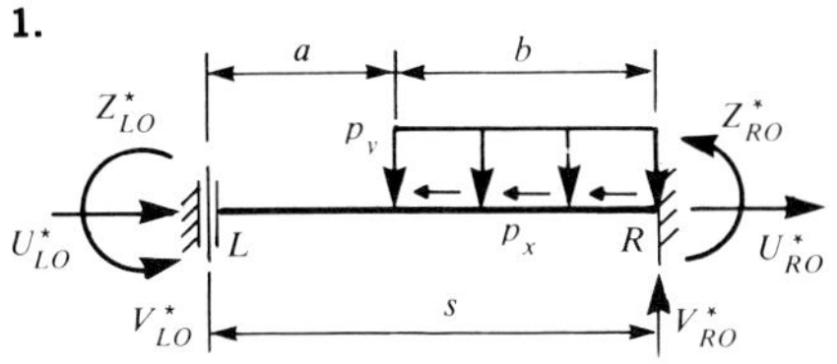 	$U^*_{R0} = \frac{1}{2}(2 - n)bp_x$ $V^*_{R0} = bp_y$ $Z^*_{R0} = -\frac{1}{6}(3 - n)b^2p_y$
$U^*_{L0} = \frac{1}{6}nbp_x$ $V^*_{L0} = 0$ $Z^*_{L0} = -\frac{n}{24}b^2p_y$	**2.** 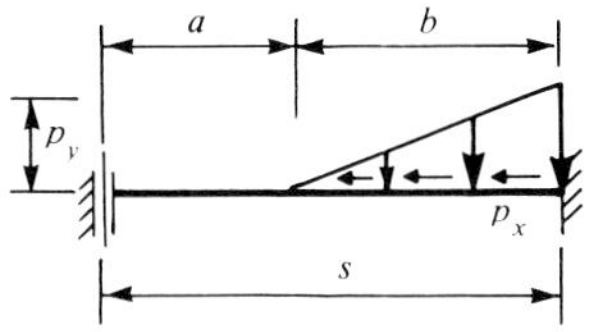 	$U^*_{R0} = \frac{1}{6}(3 - n)bp_x$ $V^*_{R0} = \frac{1}{2}bp_y$ $Z^*_{R0} = -\frac{4-n}{24}b^2p_y$
$U^*_{L0} = 0$ $V^*_{L0} = 0$ $Z^*_{L0} = 0$	**3. Abrupt linear displacement** 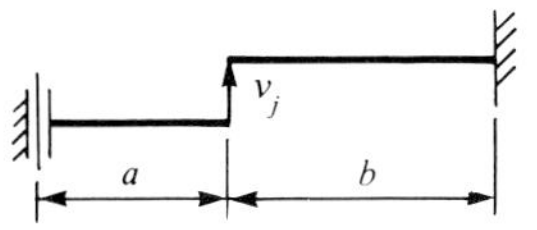 	$U^*_{R0} = 0$ $V^*_{R0} = 0$ $Z^*_{R0} = 0$
$U^*_{L0} = 0$ $V^*_{L0} = 0$ $Z^*_{L0} = \frac{1}{s}EI_z\theta_j$	**4. Abrupt angular displacement** 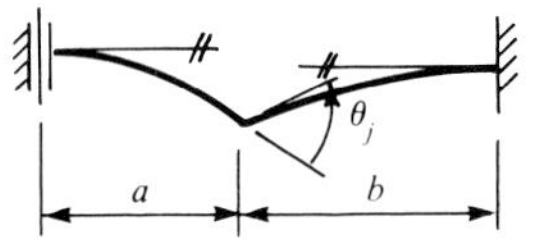 	$U^*_{R0} = 0$ $V^*_{R0} = 0$ $Z^*_{R0} = -\frac{1}{s}EI_z\theta_j$
$U^*_{L0} = nEA_xe_t$ $V^*_{L0} = 0$ $Z^*_{L0} = nEI_zf_t$	**5. Temperature deformation of segment *b*** 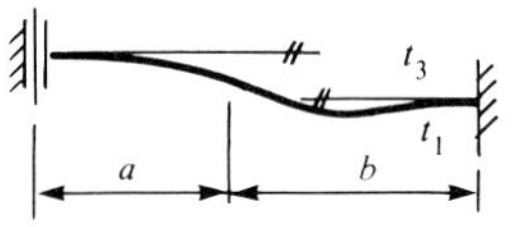 	$U^*_{R0} = -nEA_xe_t$ $V^*_{R0} = 0$ $Z^*_{R0} = -nEI_zf_t$

6. 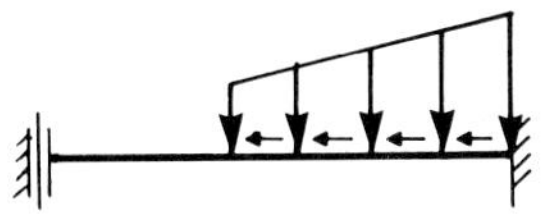 Case 1 + case 2	**7.** 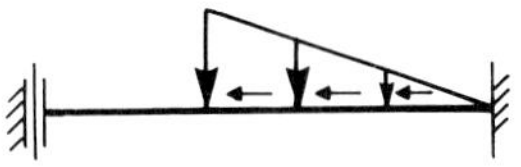 Case 1 − case 2

4.46 STIFFNESS MATRIX EQUATIONS, BAR WITH FLEXIBLE CONNECTIONS

(1) Bar with flexible connections is shown in (2) below. The spring constants of these connections are K_L, K_R, defined as the moments required to produce a unit rotation of the respective connection (N·m/rad).

(2) Free-body diagram

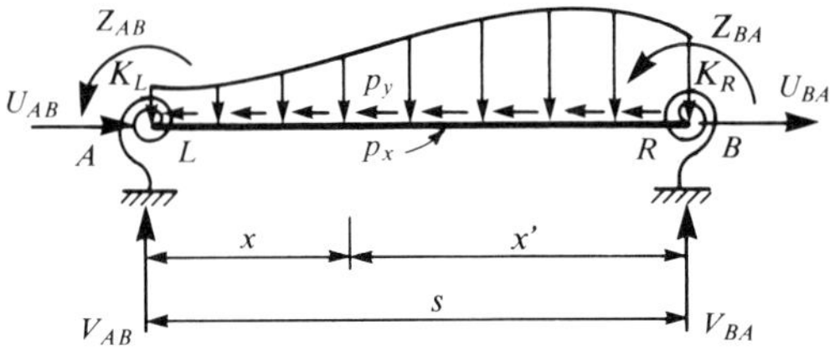

(3) Notation and signs are given in (4.32).

$$\alpha = 2 + \frac{6EI_z}{sK_R} \qquad \beta = 2 + \frac{6EI_z}{sK_L}$$

$$\lambda_1 = \frac{2\alpha - 1}{\alpha\beta - 1} \qquad \lambda_2 = \frac{\alpha - 2}{\alpha\beta - 1}$$

$$\lambda_3 = \frac{\beta - 2}{\alpha\beta - 1} \qquad \lambda_4 = \frac{2\beta - 1}{\alpha\beta - 1}$$

$$\lambda_5 = \lambda_1 - \lambda_3 \qquad \lambda_6 = \lambda_2 - \lambda_4$$

(4) Dimensionless stiffness coefficients

$$K_1 = \frac{6\alpha}{\alpha\beta - 1} \qquad K_2 = \frac{6}{\alpha\beta - 1} \qquad K_3 = \frac{6\beta}{\alpha\beta - 1}$$

$$K_4 = K_1 + K_2 \qquad K_5 = K_2 + K_3 \qquad K_6 = K_4 + K_5$$

(5) Stiffness matrix equations ($r = s^2 A_x / I_z$)

$$\begin{bmatrix} U_{AB} \\ V_{AB} \\ Z_{AB} \\ \hline U_{BA} \\ V_{BA} \\ Z_{BA} \end{bmatrix} = \frac{EI_z}{s^3} \left[\begin{array}{ccc|ccc} r & 0 & 0 & -r & 0 & 0 \\ 0 & K_6 & sK_4 & 0 & -K_6 & sK_5 \\ 0 & sK_4 & s^2K_1 & 0 & -sK_4 & s^2K_2 \\ \hline -r & 0 & 0 & r & 0 & 0 \\ 0 & -K_6 & -sK_4 & 0 & K_6 & -sK_5 \\ 0 & sK_5 & s^2K_2 & 0 & -sK_5 & s^2K_3 \end{array}\right] \begin{bmatrix} u_A \\ v_A \\ \theta_A \\ \hline u_B \\ v_B \\ \theta_B \end{bmatrix} + \begin{bmatrix} U_{A0} \\ V_{A0} \\ Z_{A0} \\ \hline U_{B0} \\ V_{B0} \\ Z_{B0} \end{bmatrix}$$

where A, B are the exterior ends of the segment $\overline{AB}$ which includes the springs and the bar $\overline{LR}$.

(6) Load functions $U_{A0}, V_{A0}, \ldots, Z_{B0}$ given as the last column matrix in (5) above are functions of the reactions U_{L0}, Z_{L0}, U_{R0}, Z_{R0} of the fixed-end bar $\overline{LR}$ introduced in (4.34–4.41).

$$U_{A0} = U_{L0} \qquad U_{B0} = U_{R0}$$

$$Z_{A0} = \lambda_1 Z_{L0} - \lambda_2 Z_{R0} \qquad Z_{B0} = -\lambda_3 Z_{L0} + \lambda_4 Z_{R0}$$

The vertical load functions V_{A0}, V_{B0} are obtained by statics in terms of Z_{L0}, Z_{R0} and the applied causes. Particular cases are given in (4.47) for the left end of the bar. The right-end load functions may be calculated by statics.

4.47 STIFFNESS LOAD FUNCTIONS, BAR WITH FLEXIBLE CONNECTIONS

Notation (4.32, 4.46) Signs (4.32) Matrix (4.46)

P_x, P_y = concentrated loads (N)

p_x, p_y = intensities of distributed loads (N/m)

e_t, f_t = thermal strains (4.03)

Q_z = applied couple (N·m)

$m = \dfrac{a}{s}$

$n = \dfrac{b}{s}$

1.

$$U_{A0} = nP_x$$

$$V_{A0} = n(1 + mn\lambda_5 + m^2\lambda_6)P_y$$

$$Z_{A0} = mn(n\lambda_1 + m\lambda_2)sP_y$$

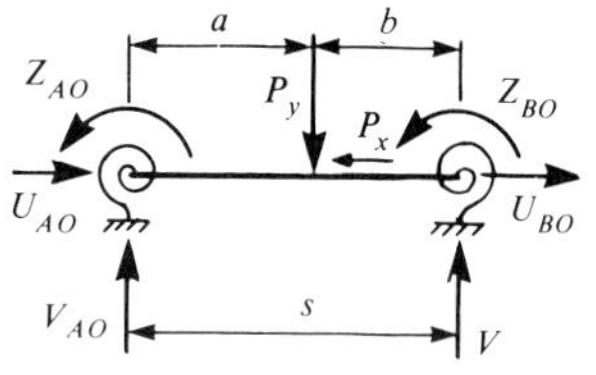

2.

$$U_{A0} = 0$$

$$V_{A0} = -\frac{1}{s}[1 + n(2 - 3n)\lambda_5 - m(3 - 3m)\lambda_6]Q_z$$

$$Z_{A0} = -[n(2 - 3n)\lambda_1 - m(2 - 3m)\lambda_2]Q_z$$

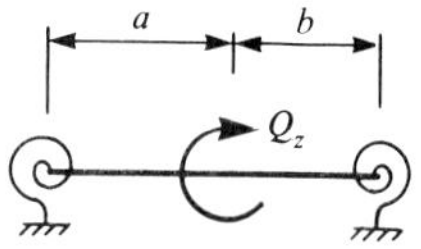

3.

$$U_{A0} = \tfrac{1}{2}sp_x$$

$$V_{A0} = \tfrac{1}{12}(6 + \lambda_5 + \lambda_6)sp_y$$

$$Z_{A0} = \tfrac{1}{12}(\lambda_1 + \lambda_2)s^2p_y$$

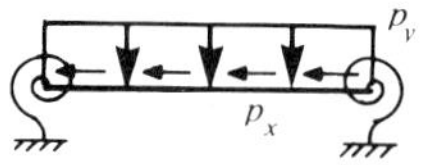

4.

$$U_{A0} = \tfrac{1}{6}sp_x$$

$$V_{A0} = \tfrac{1}{60}[10 + 2\lambda_5 + 3\lambda_6]sp_y$$

$$Z_{A0} = \tfrac{1}{6}(2\lambda_1 + 3\lambda_2)s^2p_y$$

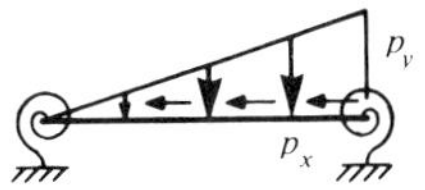

5. Temperature deformation of segment *b*

$$U_{A0} = nEA_xe_t$$

$$V_{A0} = \frac{n}{s}[(1 - 3m)\lambda_5 + (1 + 3m)\lambda_6]EI_zf_t$$

$$Z_{A0} = n[(1 - 3m)\lambda_1 + (1 + 3m)\lambda_2]EI_zf_t$$

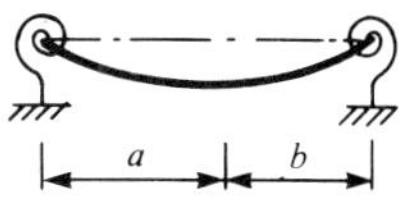

4.48 STIFFNESS MATRIX EQUATIONS, BAR WITH RIGID ENDS

(1) Bar with rigid ends shown in (2) below consists of two rigid segments and one elastic segment.

(2) Free-body diagram

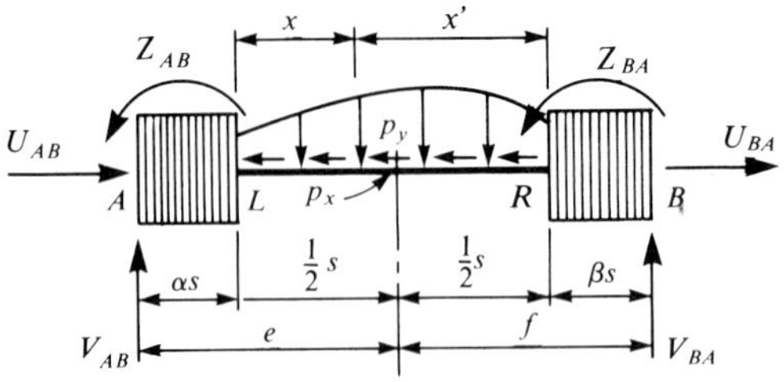

(3) Notation and signs are given in (4.32).

$$\alpha s,\ \beta s = \text{rigid segments (m)}$$

$$s = \text{elastic segment (m)}$$

$$r = \frac{s^2 A_x}{12 I_z} \qquad e = s\left(\alpha + \tfrac{1}{2}\right)$$

$$g = \frac{s^2}{12} \qquad f = s\left(\beta + \tfrac{1}{2}\right)$$

(4) Stiffness matrix equations

$$\begin{bmatrix} U_{AB} \\ V_{AB} \\ Z_{AB} \\ \hline U_{BA} \\ V_{BA} \\ Z_{BA} \end{bmatrix} = \frac{12EI_z}{s^3} \left[\begin{array}{ccc|ccc} r & 0 & 0 & -r & 0 & 0 \\ 0 & 1 & e & 0 & -1 & f \\ 0 & e & e^2 + g & 0 & -e & ef - g \\ \hline -r & 0 & 0 & r & 0 & 0 \\ 0 & -1 & -e & 0 & 1 & -f \\ 0 & f & ef - g & 0 & -f & f^2 + g \end{array}\right] \begin{bmatrix} u_A \\ v_A \\ \theta_A \\ \hline u_B \\ v_B \\ \theta_B \end{bmatrix} + \begin{bmatrix} U_{A0} \\ V_{A0} \\ Z_{A0} \\ \hline U_{B0} \\ V_{B0} \\ Z_{B0} \end{bmatrix}$$

where A, B are the exterior points of the bar and L, R are the end points of the elastic segment. If only one segment is rigid, then $\alpha = 0$ or $\beta = 0$ in all expressions.

(5) Load functions $U_{A0}, V_{A0}, \ldots, Z_{B0}$ given as the last column matrix in (4) above are functions of the reactions $U_{L0}, V_{L0}, Z_{L0}, U_{R0}, V_{R0}, Z_{R0}$ of the fixed-end bar $\overline{LR}$ introduced in (4.34–4.41).

$$\begin{bmatrix} U_{A0} \\ V_{A0} \\ Z_{A0} \end{bmatrix} = \begin{bmatrix} 1 & 0 & 0 \\ 0 & 1 & 0 \\ 0 & \alpha s & 1 \end{bmatrix} \begin{bmatrix} U_{L0} \\ V_{L0} \\ Z_{L0} \end{bmatrix} \qquad \begin{bmatrix} U_{B0} \\ V_{B0} \\ Z_{B0} \end{bmatrix} = \begin{bmatrix} 1 & 0 & 0 \\ 0 & 1 & 0 \\ 0 & -\beta s & 1 \end{bmatrix} \begin{bmatrix} U_{R0} \\ V_{R0} \\ Z_{R0} \end{bmatrix}$$

Particular cases are tabulated in (4.49).

4.49 STIFFNESS LOAD FUNCTIONS, BAR WITH RIGID ENDS

Notation (4.32, 4.48) Signs (4.32) Matrix (4.48)

P_x, P_y = concentrated loads (N)

p_x, p_y = intensities of distributed loads (N/m)

e_t, f_t = thermal strains (4.03)

Q_z = applied couple (N·m)

$m = \frac{a}{s}$ $n = \frac{b}{s}$

$\alpha = \frac{\alpha s}{s}$ $\beta = \frac{\beta s}{s}$

$U_{A0} = nP_x$ $V_{A0} = (1 + 2m)n^2P_y$ $Z_{A0} = n^2(m + \alpha + 2m\alpha)sP_y$	**1.** 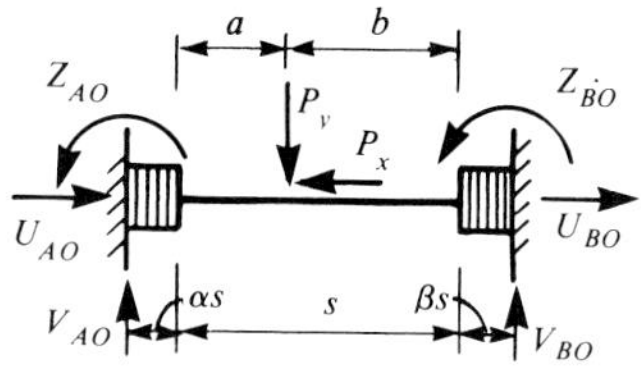	$U_{B0} = mP_x$ $V_{B0} = (1 + 2n)m^2P_y$ $Z_{B0} = -m^2(n + \beta + 2n\beta)sP_y$
$U_{A0} = 0$ $V_{A0} = -\frac{6}{s}mnQ_z$ $Z_{A0} = -n(2 - 3n + 6m\alpha)Q_z$	**2.** 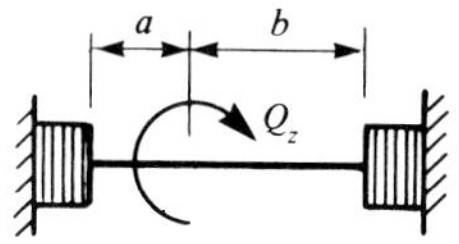	$U_{B0} = 0$ $V_{B0} = \frac{6}{s}mnQ_z$ $Z_{B0} = -m(2 - 3m + 6n\beta)Q_z$
$U_{A0} = \frac{1}{2}sp_x$ $V_{A0} = \frac{1}{2}sp_y$ $Z_{A0} = \frac{1 + 6\alpha}{12}s^2p_y$	**3.** 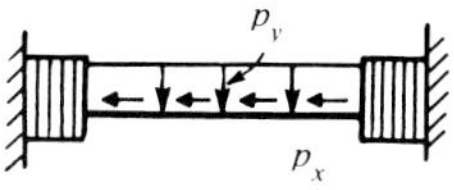	$U_{B0} = \frac{1}{2}sp_x$ $V_{B0} = \frac{1}{2}sp_y$ $Z_{B0} = -\frac{1 + 6\beta}{12}s^2p_y$
$U_{A0} = \frac{1}{6}sp_x$ $V_{A0} = \frac{3}{20}sp_y$ $Z_{A0} = \frac{2 + 9\alpha}{60}s^2p_y$	**4.** 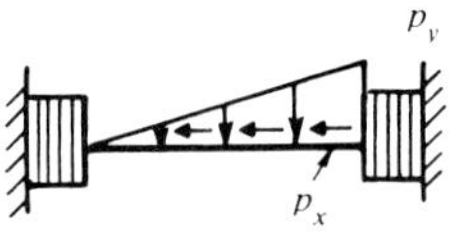	$U_{B0} = \frac{1}{3}sp_x$ $V_{B0} = \frac{7s}{20}p_y$ $Z_{B0} = -\frac{1 + 7\beta}{20}s^2p_y$
$U_{A0} = nEA_xe_t$ $V_{A0} = -\frac{6mn}{s}EI_zf_t$ $Z_{A0} = n(1 - 3m - 6m\alpha)EI_zf_t$	**5. Temperature deformation of segment *b*** 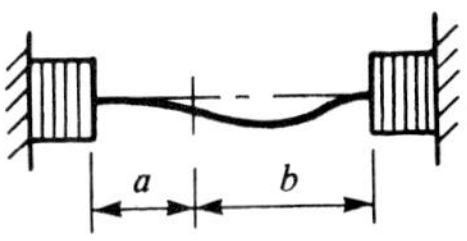	$U_{B0} = -nEA_xe_t$ $V_{B0} = \frac{6mn}{s}EI_zf_t$ $Z_{B0} = -n(1 + 3m + 6m\beta)EI_zf_t$

4.50 STIFFNESS MATRIX EQUATIONS, COMPOUND BAR

(1) Bar with intermediate hinge is called a compound bar. The position of the hinge is given by a, b measured from the respective end as shown in (2) below.

(2) Free-body diagram

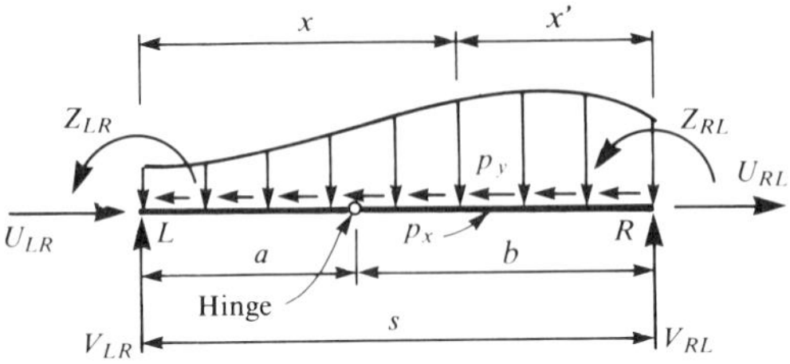

(3) Notation and signs are defined in (4.32).

$$\lambda = \frac{3}{a^3 + b^3}$$

$$r_1 = \frac{A_x}{\lambda s I_z}$$

$$r_2 = \frac{s^2 A_x}{3 I_z}$$

(4) Stiffness matrix equations $(a \neq b)$

$$\begin{bmatrix} U_{LR} \\ V_{LR} \\ Z_{LR} \\ \hline U_{RL} \\ V_{RL} \\ Z_{RL} \end{bmatrix} = \lambda E I_z \left[\begin{array}{ccc|ccc} r_1 & 0 & 0 & -r_1 & 0 & 0 \\ 0 & 1 & a & 0 & -1 & b \\ 0 & a & a^2 & 0 & -a & ab \\ \hline -r_1 & 0 & 0 & r_1 & 0 & 0 \\ 0 & -1 & -a & 0 & 1 & -b \\ 0 & b & ab & 0 & -b & b^2 \end{array}\right] \begin{bmatrix} u_L \\ v_L \\ \theta_L \\ \hline u_R \\ v_R \\ \theta_R \end{bmatrix} + \begin{bmatrix} U_{L0} \\ V_{L0} \\ Z_{L0} \\ \hline U_{R0} \\ V_{R0} \\ Z_{R0} \end{bmatrix}$$

(5) Stiffness matrix equations $(a = b)$

$$\begin{bmatrix} U_{LR} \\ V_{LR} \\ Z_{LR} \\ \hline U_{RL} \\ V_{RL} \\ Z_{RL} \end{bmatrix} = \frac{3EI_z}{s^3} \left[\begin{array}{ccc|ccc} r_2 & 0 & 0 & -r_2 & 0 & 0 \\ 0 & 4 & 2s & 0 & -4 & 2s \\ 0 & 2s & s^2 & 0 & -2s & s^2 \\ \hline -r_2 & 0 & 0 & r_2 & 0 & 0 \\ 0 & -4 & -2s & 0 & 4 & -2s \\ 0 & 2s & s^2 & 0 & -2s & s^2 \end{array}\right] \begin{bmatrix} u_L \\ v_L \\ \theta_L \\ \hline u_R \\ v_R \\ \theta_R \end{bmatrix} + \begin{bmatrix} U_{L0} \\ V_{L0} \\ Z_{L0} \\ \hline U_{R0} \\ V_{R0} \\ Z_{R0} \end{bmatrix}$$

(6) Load functions $U_{L0}, V_{L0}, \ldots, Z_{R0}$ shown in the last column matrix in (4, 5) above are the reactions of the fixed-end compound bar due to loads and other causes listed in (4.51–4.52).

4.51 STIFFNESS LOAD FUNCTIONS, FIXED-END COMPOUND BEAM

Notation (4.32, 4.50) Signs (4.32) Matrix (4.50)

P_x, P_y = concentrated loads (N) c = segment (m)

p_x, p_y = intensities of uniformly distributed load (N/m)

$$\lambda = \frac{3}{a^3 + b^3}$$

1.

$U_{L0} = \frac{c}{s} P_x$

$V_{L0} = \frac{\lambda}{6}(3b - c)c^2 P_y$

$Z_{L0} = aV_{L0}$

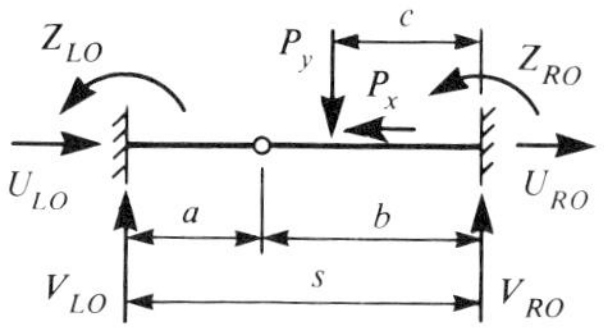

$U_{R0} = P_x - U_{L0}$

$V_{R0} = P_y - V_{L0}$

$Z_{R0} = bV_{L0} - cP_y$

2.

$U_{L0} = \frac{c^2}{2s} p_x$

$V_{L0} = \frac{\lambda}{24}(4b - c)c^3 p_y$

$Z_{L0} = aV_{L0}$

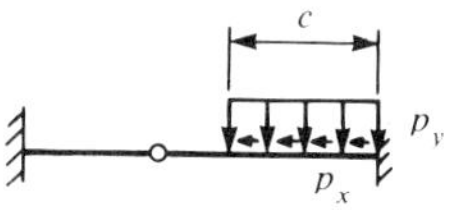

$U_{R0} = cp_x - U_{L0}$

$V_{R0} = cp_y - V_{L0}$

$Z_{R0} = bV_{L0} - \frac{c^2}{2} p_y$

3.

$U_{L0} = \frac{b^2}{2s} p_x$

$V_{L0} = \frac{\lambda}{8} b^4 p_y$

$Z_{L0} = aV_{L0}$

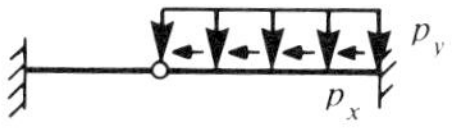

$U_{R0} = bp_x - U_{L0}$

$V_{R0} = bp_y - V_{L0}$

$Z_{R0} = bV_{L0} - \frac{b^2}{2} p_y$

4.

$U_{L0} = \frac{1}{2} s p_x$

$V_{L0} = (a + \lambda\Delta) p_y$

$Z_{L0} = aV_{L0} - \frac{1}{2} a^2 p_y$

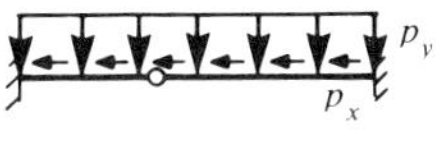

$\Delta = \frac{1}{8}(b^4 - a^4)$

$U_{R0} = \frac{1}{2} s p_x$

$V_{R0} = (b - \lambda\Delta) p_y$

$Z_{R0} = -bV_{R0} + \frac{1}{2} b^2 p_y$

5.

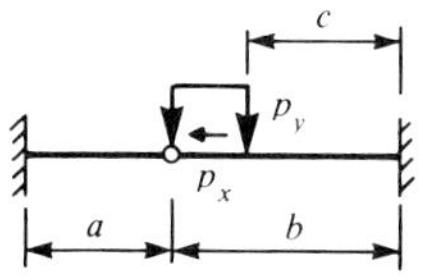

Case 3 − case 2

6.

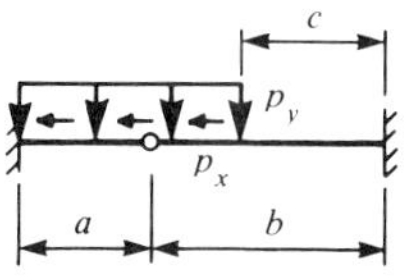

Case 4 − case 2

4.52 STIFFNESS LOAD FUNCTIONS, FIXED-END COMPOUND BEAM

Notation (4.32, 4.50)	Signs (4.32)	Matrix (4.50)
$Q_z =$ applied couple (N·m) $c =$ segment (m) $q_z =$ intensity of distributed couple (N·m/m) $p_x, p_y =$ maximum intensities of distributed loads (N/m) $e_t, f_t =$ thermal strains (4.03)		$\lambda = \dfrac{3}{a^3 + b^3}$

$U_{L0} = 0$ $V_{L0} = -\dfrac{\lambda}{2}(2b - c)cQ_z$ $Z_{L0} = aV_{L0}$	**1.** 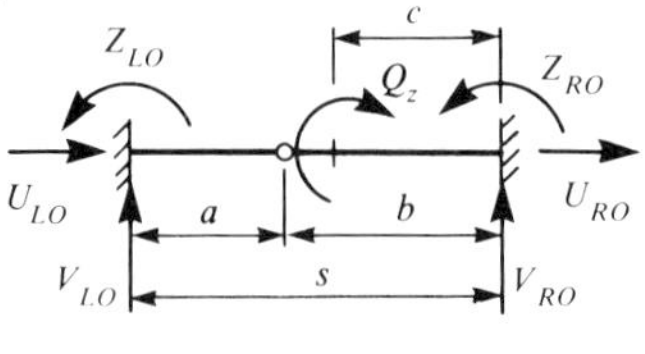	$U_{R0} = 0$ $V_{R0} = -V_{L0}$ $Z_{R0} = bV_{L0} + Q_z$
$U_{L0} = 0$ $V_{L0} = -\dfrac{\lambda}{6}(3b - c)c^2q_z$ $Z_{L0} = -\dfrac{\lambda}{6}(3b - c)ac^2q_z$	**2.** 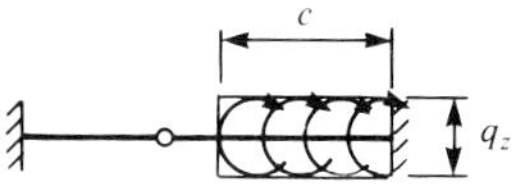	$U_{R0} = 0$ $V_{R0} = -V_{L0}$ $Z_{R0} = cq_z + bV_{L0}$
$U_{L0} = \dfrac{c^2}{6s}p_x$ $V_{L0} = \dfrac{\lambda}{120}(5b - c)c^3p_y$ $Z_{L0} = aV_{L0}$	**3.** 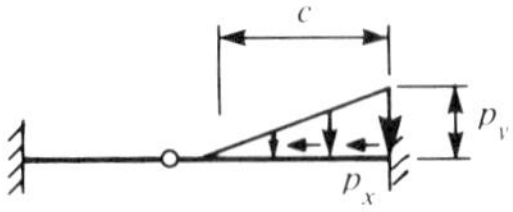	$U_{R0} = \dfrac{(3s - c)c}{6s}p_x$ $V_{R0} = \dfrac{c}{2}p_y - V_{L0}$ $Z_{R0} = bV_{L0} - \dfrac{c^2}{6}p_y$
$U_{L0} = \dfrac{s}{6}p_x$ $V_{L0} = \dfrac{1}{2s}(a^2 + \lambda\Delta)p_y$ $Z_{L0} = aV_{L0} - \dfrac{a^3}{6s}p_y$	**4.**  $\Delta = \frac{1}{60}(4b^5 + 15ab^4 - 11a^5)$	$U_{R0} = \dfrac{s}{3}p_x$ $V_{R0} = \dfrac{1}{2s}(2ab + b^2 - \lambda\Delta)p_y$ $Z_{R0} = bV_{L0} - \dfrac{1}{6s}(s^3 - a^3)p_y$
$U_{L0} = EA_xe_t$ $V_{L0} = -\dfrac{\lambda}{2}(b - a)sEI_zf_t$ $Z_{L0} = aV_{L0}$	**5. Temperature deformation of segment *s*** 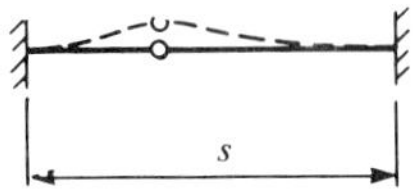	$U_{R0} = -EA_xe_t$ $V_{R0} = -V_{L0}$ $Z_{R0} = bV_{L0}$

5

Free Straight Bar of Order One Variable Section

STATIC STATE

5.01 DEFINITION OF STATE

(1) System considered is a free finite straight bar of variable and symmetrical section. The centroidal axis of the bar and the vertical axis y define the plane of symmetry of this section. The width of the section remains constant for the length of the bar, whereas the depth is variable along the entire length or along one or more segments of the bar. The bar is a part of a planar system acted on by loads in the system plane (system of order one).

(2) Notation and sign convention of the transport method for bars of variable section are the same as in (4.02–4.03) with the following exceptions:

A_{Lx}, A_{Rx} = area of section normal to x at L, R, respectively (m^2)

A_x = variable area of section normal to x at the distance x from L (m^2)

d_L, d_R = depth of A_x at L, R, respectively (m) c = constant width of A_x (m)

d = variable depth of A_x at the distance x from L (m)

I_{Lz}, I_{Rz} = moment of inertia of the normal section about z at L, R, respectively (m^4)

I_z = variable moment of inertia of A_x about z at the distance x from L (m^4)

(3) Assumptions of analysis are stated in (1.01). In this chapter, the effects of axial and flexural deformations are included, but the effects of shearing deformations and the effect of axial force on the magnitude of the flexural moment are neglected. The beam-column effect is considered in Chap. 9. For inclusion of shearing deformations, reference is made to Chaps. 2 and 3.

5.02 TRANSPORT MATRIX EQUATIONS

(1) Geometry. If the cross section of the bar varies along the entire length or along one or more segments of the bar, the area of the section and its moment of inertia become functions of the position coordinate and add considerably to the difficulty in evaluating the integrals of the transport coefficients. In a few cases, it is possible to carry out the exact evaluation of the respective integrals (5.04–5.05). In the majority of cases, it is not possible to proceed with the given function, and a *substitute function* or a *substitute model* must be introduced (5.06–5.07).

(2) Equivalents used in this chapter take on the following forms:

$$\bar{u} = EA_{Lx}u = \text{scaled } u \ (\text{N}\cdot\text{m}) \qquad A_x = A_{Lx}R_x \ (\text{m}^2)$$

$$\bar{v} = EI_{Lz}v = \text{scaled } v \ (\text{N}\cdot\text{m}^3) \qquad I_z = I_{Lz}R_z \ (\text{m}^4)$$

$$\bar{\theta} = EI_{Lz}\theta = \text{scaled } \theta \ (\text{N}\cdot\text{m}^2) \qquad \bar{x} = x/s = \text{dimensionless coordinate}$$

where s is the span length (m) and R_x, R_z are the dimensionless shape factors (5.04–5.07) defining the variations of A_x, I_z, respectively. The components of stress resultants, loads, and displacements are the same as those depicted in (4.02-2, 3).

(3) Transport matrix equations

$$\underbrace{\begin{bmatrix} 1 \\ \hline U_L \\ V_L \\ Z_L \\ \hline \bar{u}_L \\ \bar{v}_L \\ \bar{\theta}_L \end{bmatrix}}_{H_L} = \underbrace{\left[\begin{array}{c|ccc|ccc} 1 & 0 & 0 & 0 & 0 & 0 & 0 \\ \hline F_U & 1 & 0 & 0 & 0 & 0 & 0 \\ F_V & 0 & 1 & 0 & 0 & 0 & 0 \\ F_Z & 0 & s & 1 & 0 & 0 & 0 \\ \hline F_u & -sT_{0n} & 0 & 0 & 1 & 0 & 0 \\ F_v & 0 & s^3(T_{1z} - T_{2z}) & s^2T_{1z} & 0 & 1 & -s \\ F_\theta & 0 & -s^2(T_{0z} - T_{1z}) & -sT_{0z} & 0 & 0 & 1 \end{array}\right]}_{T_{LR}} \underbrace{\begin{bmatrix} 1 \\ \hline U_R \\ V_R \\ Z_R \\ \hline \bar{u}_R \\ \bar{v}_R \\ \bar{\theta}_R \end{bmatrix}}_{H_R}$$

$$\underbrace{\begin{bmatrix} 1 \\ \hline U_R \\ V_R \\ Z_R \\ \hline \bar{u}_R \\ \bar{v}_R \\ \bar{\theta}_R \end{bmatrix}}_{H_R} = \underbrace{\left[\begin{array}{c|ccc|ccc} 1 & 0 & 0 & 0 & 0 & 0 & 0 \\ \hline G_U & 1 & 0 & 0 & 0 & 0 & 0 \\ G_V & 0 & 1 & 0 & 0 & 0 & 0 \\ G_Z & 0 & -s & 1 & 0 & 0 & 0 \\ \hline G_u & sT_{0n} & 0 & 0 & 1 & 0 & 0 \\ G_v & 0 & -s^3(T_{1z} - T_{2z}) & s^2(T_{0z} - T_{1z}) & 0 & 1 & s \\ G_\theta & 0 & -s^2T_{1z} & sT_{0z} & 0 & 0 & 1 \end{array}\right]}_{T_{RL}} \underbrace{\begin{bmatrix} 1 \\ \hline U_L \\ V_L \\ Z_L \\ \hline \bar{u}_L \\ \bar{v}_L \\ \bar{\theta}_L \end{bmatrix}}_{H_L}$$

where H_L, H_R are the *end state vectors* and T_{LR}, T_{RL} are the respective *transport matrices*. The elements of these matrices are the *transport coefficients* defined in (5.03) and the *load functions* F, G listed for particular loadings and deformation conditions in (5.08–5.09).

5.03 TRANSPORT RELATIONS

(1) Dimensionless transport coefficients in (5.02-3) and in (5.08–5.09) are

$T_{jn} = \int_0^1 \frac{\bar{x}^j\,d\bar{x}}{R_x}$	$T_{jz} = \int_0^1 \frac{\bar{x}^j\,d\bar{x}}{R_z}$
$A_{jn} = \int_0^m \frac{\bar{x}^j\,d\bar{x}}{R_x}$	$A_{jz} = \int_0^m \frac{\bar{x}^j\,d\bar{x}}{R_z}$
$C_{jn} = \int_m^1 \frac{\bar{x}^j\,d\bar{x}}{R_x} = T_{jn} - A_{jn}$	$C_{jz} = \int_m^1 \frac{\bar{x}^j\,d\bar{x}}{R_z} = T_{jz} - A_{jz}$

where $j = 0, 1, 2, \ldots, 3$ and $m = a/s$. The closed-form evaluations of these integrals for particular variations of cross section are tabulated in (5.04–5.07)

(2) Load functions in bars of variable cross section are complex expressions, and for lack of space only the left-side functions are tabulated in (5.08–5.09). The symbols used in these tables are those defined in (4.03). The right-side load functions (if needed) can be derived by the transport-shift theorem as

$$G_U = -F_U \qquad G_u = -sT_{0n}F_U - F_u$$

$$G_V = -F_V \qquad G_v = s^3(T_{1z} - T_{2z})T_v - s^2(T_{0z} - T_{1z})F_Z - F_v - sF_\theta$$

$$G_Z = sF_V - F_Z \qquad G_\theta = s^2T_{1z}F_V - sT_{0z}F_Z - F_\theta$$

where F_U, F_V, F_Z, F_u, F_v, F_θ are the left-side load functions in the respective tables.

(3) Relationships of the state-vector components are given by the following differential equations.

$U = EA_x \frac{du}{dx}$	$\theta = \frac{dv}{dx}$	$-V = \frac{d}{dx}\left(EI_z \frac{d^2v}{dx^2}\right)$
$p_x = \frac{d}{dx}\left(EA_x \frac{du}{dx}\right)$	$Z = EI_z \frac{d^2v}{dx^2}$	$-p_y = \frac{d^2}{dx^2}\left(EI_z \frac{d^2v}{dx^2}\right)$

where x is the position coordinate measured from L and p_x, p_y are the intensities of the distributed load.

(4) Elemental temperature deformations of a bar of variable section are

$$du_t = e_t\,dx \qquad d\theta_t = f_t\,dx$$

where e_t and f_t are the thermal strains defined in (4.03).

(5) Applications of the transport matrices in bars of variable section is formally indentical to the procedure described in (4.07).

5.04 TRANSPORT COEFFICIENTS, RECTANGULAR SECTION, LINEAR VARIATION OF DEPTH

(1) Geometry. When the section of the bar is rectangular of constant width c and linearly varying depth d, the closed-form evaluation of the integrals of the transport coefficients is possible and practical (Ref. 4, p. 77). The equivalents used are

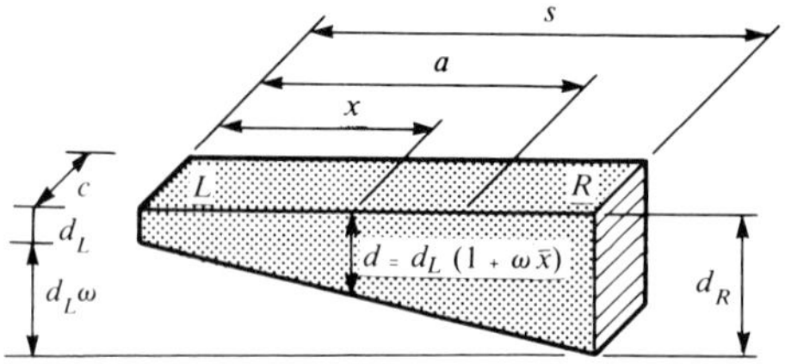

$$\bar{x} = \frac{x}{s} \qquad m = \frac{a}{s} \qquad \omega = \frac{d_R - d_L}{d_L}$$

$$A_x = A_{Lx}(1 + \omega\bar{x}) = A_{Lx}R_x$$

$$I_z = I_{Lz}(1 + \omega\bar{x})^3 = I_{Lz}R_z$$

$$M = 1 + \omega m \qquad N = 1 + \omega$$

(2) Coefficients T_{jn}, T_{jz}, A_{jn}, A_{jz} (5.03, 5.08–5.09)

$T_{0n} = \dfrac{1}{\omega}\ln N$ $T_{1n} = \dfrac{1}{\omega^2}(\omega - \ln N)$ $T_{2n} = \dfrac{1}{\omega^3}\left(\dfrac{\omega^2}{2} - \omega + \ln N\right)$	$T_{3n} = \dfrac{1}{\omega^4}\left(\dfrac{\omega^3}{3} - \dfrac{\omega^2}{2} + \omega - \ln N\right)$ $T_{4n} = \dfrac{1}{\omega^5}\left(\dfrac{\omega^4}{4} - \dfrac{\omega^3}{3} + \dfrac{\omega^2}{2} - \omega + \ln N\right)$
$T_{0z} = \dfrac{N+1}{2N^2}$ $T_{1z} = \dfrac{1}{2N^2}$ $T_{2z} = \dfrac{1}{\omega^3}\left[\dfrac{\omega(1-3N)}{2N^2} + \ln N\right]$	$T_{3z} = \dfrac{1}{\omega^4}\left[\dfrac{\omega(2N^2 + 5N - 1)}{2N^2} - 3\ln N\right]$ $T_{4z} = \dfrac{1}{\omega^5}\left[\dfrac{\omega(N^3 + 7N^2 - 7N + 1)}{2N^2} + 6\ln N\right]$
$A_{0n} = \dfrac{1}{\omega}\ln M$ $A_{1n} = \dfrac{1}{\omega^2}(\omega m - \ln M)$ $A_{2n} = \dfrac{1}{\omega^3}\left(\dfrac{\omega^2 m^2}{2} - \omega m + \ln M\right)$	$A_{3n} = \dfrac{1}{\omega^4}\left(\dfrac{\omega^3 m^3}{3} - \dfrac{\omega^2 m^2}{2} + \omega m - \ln M\right)$ $A_{4n} = \dfrac{1}{\omega^5}\left(\dfrac{\omega^4 m^4}{4} - \dfrac{\omega^3 m^3}{3} + \dfrac{\omega^2 m^2}{2} - \omega m + \ln M\right)$
$A_{0z} = \dfrac{m(1+M)}{2M^2}$ $A_{1z} = \dfrac{m^2}{2M^2}$ $A_{2z} = \dfrac{1}{\omega^3}\left[\dfrac{\omega m(1-3M)}{2M^2} + \ln M\right]$	$A_{3z} = \dfrac{1}{\omega^4}\left[\dfrac{\omega m(2M^2 + 5M - 1)}{2M^2} - 3\ln M)\right]$ $A_{4z} = \dfrac{1}{\omega^5}\left[\dfrac{\omega m(M^3 + 7M^2 - 7M + 1)}{2M^2} + 6\ln M)\right]$

5.05 TRANSPORT COEFFICIENTS, RECTANGULAR SECTION, PARABOLIC VARIATION OF DEPTH

(1) Geometry. When the section of the bar is rectangular of constant width c and parabolically varying depth d, the closed-form evaluation of the integrals of the transport coefficients is again possible and practical (Ref. 4, p. 78). The equivalents used are

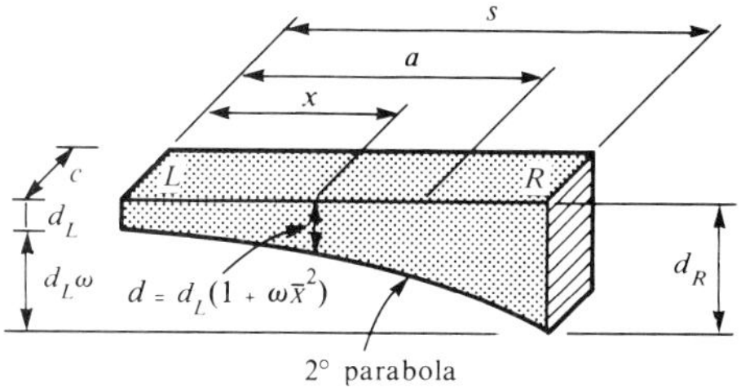

$$\bar{x} = \frac{x}{s} \qquad m = \frac{a}{s} \qquad \omega = \frac{d_R - d_L}{d_L}$$

$$A_x = A_{Lx}(1 + \omega\bar{x}^2) = A_{Lx}R_x$$

$$I_z = I_{Lz}(1 + \omega\bar{x}^2)^3 = I_{Lz}R_z$$

$$M = 1 + \omega m^2 \qquad N = 1 + \omega$$

$$K = \frac{\tan^{-1} m\sqrt{\omega}}{m\sqrt{\omega}} \qquad L = \frac{\tan^{-1}\sqrt{\omega}}{\sqrt{\omega}}$$

(2) Coefficients T_{jn}, T_{jz}, A_{jn}, A_{jz} (5.03, 5.08–5.09)

$T_{0n} = L$ $T_{1n} = \dfrac{\ln N}{2\omega}$ $T_{2n} = \dfrac{1}{\omega}(1 - L)$	$T_{3n} = \dfrac{1}{2\omega}\left(1 - \dfrac{\ln N}{\omega}\right)$ $T_{4n} = \dfrac{1}{\omega^2}\left(\dfrac{N-4}{3} + L\right)$ $T_{5n} = \dfrac{1}{2\omega^2}\left(\dfrac{N-3}{2} + \dfrac{\ln N}{\omega}\right)$
$T_{0z} = \dfrac{1}{\omega}\left(\dfrac{3N+2}{N^2} + 3L\right)$ $T_{1z} = \dfrac{N+1}{4N^2}$ $T_{2z} = \dfrac{1}{8\omega}\left(\dfrac{N-2}{N^2} + L\right)$	$T_{3z} = \dfrac{1}{4N^2}$ $T_{4z} = \dfrac{1}{8\omega^2}\left(3L - \dfrac{5N-2}{N^2}\right)$ $T_{5z} = \dfrac{1}{2\omega^3}\left[\ln N - \dfrac{\omega(3N-1)}{2N^2}\right]$
$A_{0n} = K$ $A_{1n} = \dfrac{\ln M}{2\omega}$ $A_{2n} = \dfrac{m}{\omega}(1 - K)$	$A_{3n} = \dfrac{m^2}{2\omega}\left(1 - \dfrac{\ln M}{\omega m^2}\right)$ $A_{4n} = \dfrac{m}{\omega^2}\left(\dfrac{M-4}{3} + K\right)$ $A_{5n} = \dfrac{m^2}{2\omega^2}\left(\dfrac{M-3}{2} + \dfrac{\ln M}{\omega m^2}\right)$
$A_{0z} = \dfrac{m}{\omega}\left(\dfrac{3M+2}{M^2} + 3K\right)$ $A_{1z} = \dfrac{m^2(M+1)}{4M^2}$ $A_{2z} = \dfrac{m}{8\omega}\left(\dfrac{M-2}{M^2} + K\right)$	$A_{3z} = \dfrac{m^4}{4M^2}$ $A_{4z} = \dfrac{m}{8\omega^2}\left(3K - \dfrac{5M-2}{M^2}\right)$ $A_{5z} = \dfrac{1}{2\omega^3}\left[\ln M - \dfrac{\omega m^2(3M-1)}{2M^2}\right]$

5.06 TRANSPORT COEFFICIENTS, SYMMETRICAL SECTION, LINEAR VARIATION OF DEPTH

(1) Geometry. When the symmetrical section of the bar has a constant width c and a linearly varying depth d but otherwise has any arbitrary shape, such as those shown below, the closed form of the integrals of the transport coefficients is not feasible. To surmount this difficulty, the variation of A_x and I_z is represented with a sufficient accuracy by the substitute functions shown below (Ref. 4, p. 80).

$$A_x = A_{Lx}(1 + \omega\bar{x})^{\mu} = A_{Lx}R_x \qquad\qquad I_z = I_{Lz}(1 + \omega\bar{x})^{\rho} = I_{Lz}R_z$$

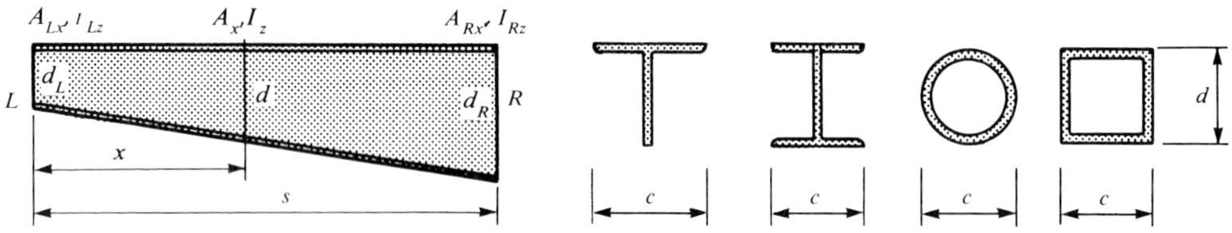

(2) Equivalents used in the substitute functions and in their integrals are

$$\bar{x} = x/s \qquad m = a/s \qquad \omega = (d_R - d_L)/d_L \qquad j = 0, 1, 2, \ldots$$

$$\mu = \frac{\ln(A_{Rx} - A_{Lx})}{\ln(d_R - d_L)} \qquad e_j = \frac{(1+\omega)^{1+j-\mu} - 1}{1 + j - \mu} \qquad f_j = \frac{(1+\omega)^{1+j-\rho} - 1}{1 + j - \rho}$$

$$\rho = \frac{\ln(I_{Rz} - I_{Lz})}{\ln(d_R - d_L)} \qquad e_{jm} = \frac{(1+\omega m)^{1+j-\mu} - 1}{1 + j - \mu} \qquad f_{jm} = \frac{(1+\omega m)^{1+j-\rho} - 1}{1 + j - \rho}$$

(3) Coefficients T_{jn}, T_{jz}, A_{jn}, A_{jz} (5.03, 5.08–5.09)

$T_{0n} = \frac{1}{\omega} e_0$	$T_{0z} = \frac{1}{\omega} f_0$
$T_{1n} = \frac{1}{\omega^2}(e_1 - e_0)$	$T_{1z} = \frac{1}{\omega^2}(f_1 - f_0)$
$T_{2n} = \frac{1}{\omega^3}(e_2 - 2e_1 + e_0)$	$T_{2z} = \frac{1}{\omega^3}(f_2 - 2f_1 + f_0)$
$T_{3n} = \frac{1}{\omega^4}(e_3 - 3e_2 + 3e_1 - e_0)$	$T_{3z} = \frac{1}{\omega^4}(f_3 - 3f_2 + 3f_1 - f_0)$
$T_{jn} = \frac{1}{\omega^{j+1}} \sum_{k=0}^{j} (-1)^k \binom{j}{k} e_{j-k}$	$T_{jz} = \frac{1}{\omega^{j+1}} \sum_{k=0}^{j} (-1)^k \binom{j}{k} f_{j-k}$
$A_{0n} = \frac{1}{\omega} e_{0m}$	$A_{0z} = \frac{1}{\omega} f_{0m}$
$A_{1n} = \frac{1}{\omega^2}(e_{1m} - e_{0m})$	$A_{1z} = \frac{1}{\omega^2}(f_{1m} - f_{0m})$
$A_{2n} = \frac{1}{\omega^3}(e_{2m} - 2e_{1m} + e_{0m})$	$A_{2z} = \frac{1}{\omega^3}(f_{2m} - 2f_{1m} + f_{0m})$
$A_{3n} = \frac{}{\omega^4}(e_{3m} - 3e_{2m} + 3e_{1m} - e_{0m})$	$A_{3z} = \frac{1}{\omega^4}(f_{3m} - 3f_{2m} + 3f_{1m} - f_{0m})$
$A_{jn} = \frac{1}{\omega^{j+1}} \sum_{k=0}^{j} (-1)^k \binom{j}{k} e_{j-k,m}$	$A_{jz} = \frac{1}{\omega^{j+1}} \sum_{k=0}^{j} (-1)^k \binom{j}{k} f_{j-k,m}$

5.07 TRANSPORT COEFFICIENTS, SYMMETRICAL SECTION, NONLINEAR VARIATION OF DEPTH

(1) Geometry. When the variation of depth d in (5.06) is nonlinear but continuous, another set of substitute functions may be used (Ref. 4, p. 81).

$$A_x = \frac{A_{Lx}}{1 - e\bar{x}^{\mu}} = A_{Lx}R_x \qquad I_z = \frac{I_{Lz}}{1 - g\bar{x}^{\rho}} = I_{Lz}R_z$$

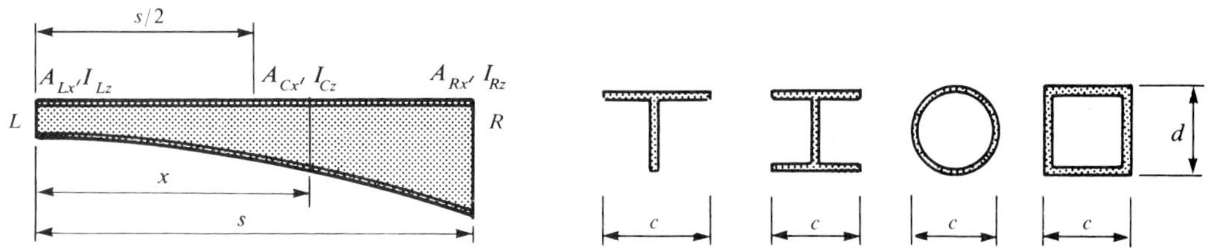

(2) Equivalents used in the substitute functions and in their integrals are

$$\bar{x} = x/s \qquad e = 1 - A_{Lx}/A_{Rx} \qquad g = 1 - I_{Lz}/I_{Rz} \qquad \mu = \frac{\ln(e/f)}{\ln 2} \qquad \rho = \frac{\ln(g/h)}{\ln 2}$$

$$m = a/s \qquad f = 1 - A_{Lx}/A_{Cx} \qquad h = 1 - I_{Lz}/I_{Cz}$$

and A_{Cx}, I_{Cz} are the area and the moment of inertia of the section at the center of the span, respectively.

(3) Coefficients T_{jn}, T_{jz}, A_{jn}, A_{jz} (5.03, 5.08–5.09)

$T_{0n} = 1 - \dfrac{e}{\mu + 1}$ $T_{1n} = \dfrac{1}{2} - \dfrac{e}{\mu + 2}$ $T_{2n} = \dfrac{1}{3} - \dfrac{e}{\mu + 3}$ $T_{3n} = \dfrac{1}{4} - \dfrac{e}{\mu + 4}$ $T_{jn} = \dfrac{1}{j+1} - \dfrac{e}{\mu + j + 1}$	$T_{0z} = 1 - \dfrac{g}{\rho + 1}$ $T_{1z} = \dfrac{1}{2} - \dfrac{g}{\rho + 2}$ $T_{2z} = \dfrac{1}{3} - \dfrac{g}{\rho + 3}$ $T_{3z} = \dfrac{1}{4} - \dfrac{g}{\rho + 4}$ $T_{jz} = \dfrac{1}{j+1} - \dfrac{g}{\rho + j + 1}$
$A_{0n} = m\left(1 - \dfrac{em^{\mu}}{\mu + 1}\right)$ $A_{1n} = m^2\left(\dfrac{1}{2} - \dfrac{em^{\mu}}{\mu + 2}\right)$ $A_{2n} = m^3\left(\dfrac{1}{3} - \dfrac{em^{\mu}}{\mu + 3}\right)$ $A_{3n} = m^4\left(\dfrac{1}{4} - \dfrac{em^{\mu}}{\mu + 4}\right)$ $A_{jn} = m^{j+1}\left(\dfrac{1}{j+1} - \dfrac{em^{\mu}}{\mu + j + 1}\right)$	$A_{0z} = m\left(1 - \dfrac{gm^{\rho}}{\rho + 1}\right)$ $A_{1z} = m^2\left(\dfrac{1}{2} - \dfrac{gm^{\rho}}{\rho + 2}\right)$ $A_{2z} = m^3\left(\dfrac{1}{3} - \dfrac{gm^{\rho}}{\rho + 3}\right)$ $A_{3z} = m^4\left(\dfrac{1}{4} - \dfrac{gm^{\rho}}{\rho + 4}\right)$ $A_{jz} = m^{j+1}\left(\dfrac{1}{j+1} - \dfrac{gm^{\rho}}{\rho + j + 1}\right)$

5.08 TRANSPORT LOAD FUNCTIONS

Notation (5.02) Signs (5.02) Matrix (5.02)

P_x, P_y = concentrated loads (N)

p_x, p_y = intensities of distributed loads (N/m)

q_z = intensity of distributed couple (N·m/m)

A_j, C_j, T_j = transport coefficients (5.03–5.07)

Q_z = applied couple (N·m)

$m = a/s$

$n = b/s$

1.

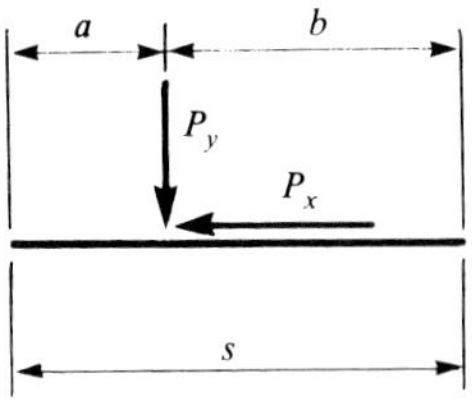

$$F_U = -P_x$$
$$F_V = -P_y$$
$$F_Z = -aP_y$$
$$F_u = A_{0n}sP_x$$
$$F_v = -(A_{1z} - A_{2z})s^3P_y$$
$$F_\theta = (A_{0z} - A_{1z})s^2P_y$$

2.

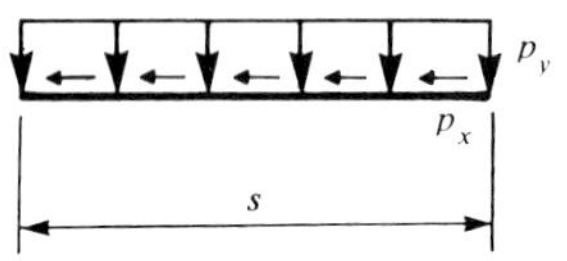

$$F_U = -sp_x$$
$$F_V = -sp_y$$
$$F_Z = -\tfrac{1}{2}s^2p_y$$
$$F_u = (T_{0n} - T_{1n})s^2p_x$$
$$F_v = -\tfrac{1}{2}(T_{1z} - 2T_{2z} + T_{3z})s^4p_y$$
$$F_\theta = \tfrac{1}{2}(T_{0z} - 2T_{1z} + T_{2z})s^3p_y$$

3.

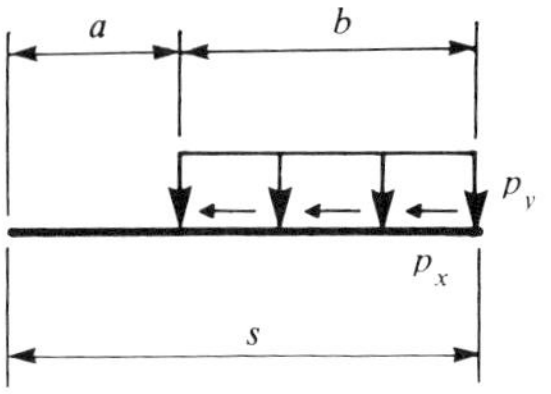

$$F_U = -bp_x$$
$$F_V = -bp_y$$
$$F_Z = -\tfrac{1}{2}(s^2 - a^2)p_y$$
$$F_u = (C_{0n} - C_{1n})s^2p_x$$
$$F_v = -\tfrac{1}{2}(C_{1z} - 2C_{2z} + C_{3z})s^4p_y$$
$$F_\theta = \tfrac{1}{2}(C_{0z} - 2C_{1z} + C_{2z})s^3p_y$$

4.

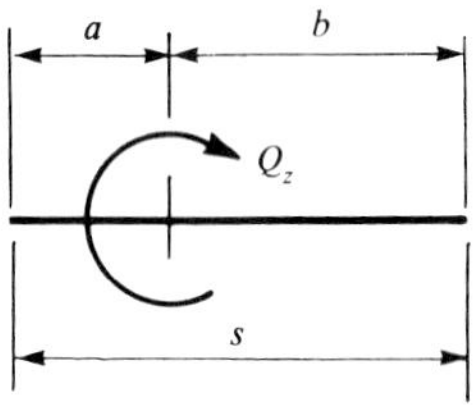

$$F_U = 0$$
$$F_V = 0$$
$$F_Z = -Q_z$$
$$F_u = 0$$
$$F_v = -A_{1z}s^2Q_z$$
$$F_\theta = A_{0z}sQ_z$$

5.

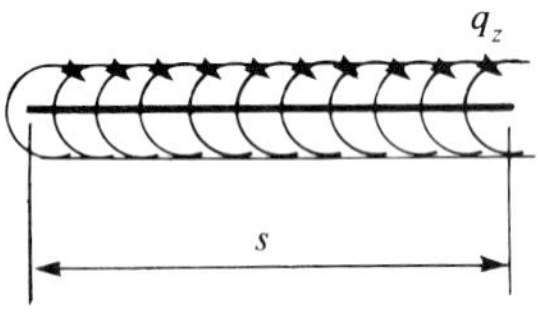

$$F_U = 0$$
$$F_V = 0$$
$$F_Z = -sq_z$$
$$F_u = 0$$
$$F_v = -(T_{1z} - T_{2z})s^3q_z$$
$$F_\theta = (T_{0z} - T_{1z})s^2q_z$$

6.

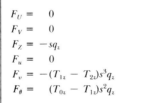

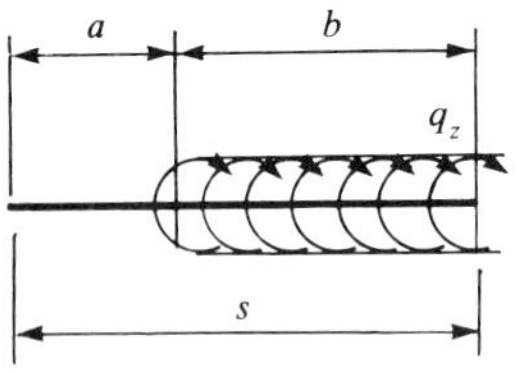

$$F_U = 0$$
$$F_V = 0$$
$$F_Z = -bq_z$$
$$F_u = 0$$
$$F_v = -(C_{1z} - C_{2z})s^3q_z$$
$$F_\theta = (C_{0z} - C_{1z})s^2q_z$$

5.09 TRANSPORT LOAD FUNCTIONS

Notation (5.02) Signs (5.02) Matrix (5.02)

p_x, p_y = intensities of distributed loads (N/m)

u_j, v_j, θ_j = externally applied displacements (m, m, rad)

e_t, f_t = thermal strains (5.03, 4.03)

A_j, C_j, T_j = transport coefficients (5.03–5.07)

$m = a/s$

$n = b/s$

1.

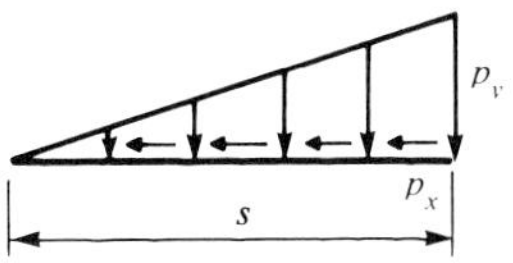

$$F_U = -\tfrac{1}{2} s p_x$$
$$F_V = -\tfrac{1}{2} s p_y$$
$$F_Z = -\tfrac{1}{3} s^2 p_y$$
$$F_u = \tfrac{1}{2}(T_{0n} - T_{2n}) s^2 p_x$$
$$F_v = -\tfrac{1}{6}(2T_{1z} - 3T_{2z} + T_{4z}) s^4 p_y$$
$$F_\theta = \tfrac{1}{6}(2T_{0z} - 3T_{1z} + T_{3z}) s^3 p_y$$

2.

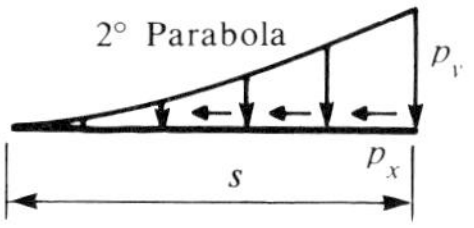

$$F_U = -\tfrac{1}{3} s p_x$$
$$F_V = -\tfrac{1}{3} s F_y$$
$$F_Z = -\tfrac{1}{4} s^2 p_y$$
$$F_u = \tfrac{1}{3}(T_{0n} - T_{3n}) s^2 p_x$$
$$F_v = -\tfrac{1}{12}(3T_{1z} - 4T_{2z} + T_{4z}) s^4 p_y$$
$$F_\theta = \tfrac{1}{12}(3T_{0z} - 4T_{1z} + T_{3z}) s^3 p_y$$

3.

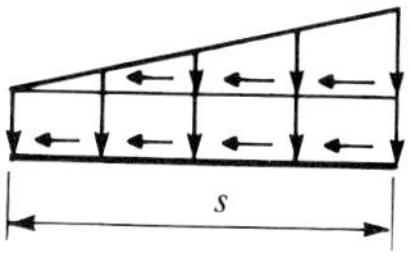

Sec. 5.08 (Case 2)
+ Sec. 5.09 (case 1)

4.

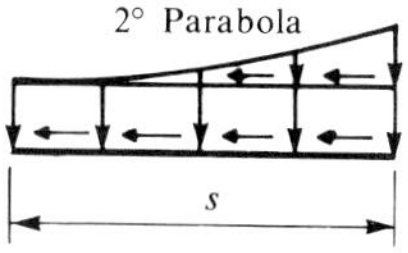

Sec. 5.08 (case 2)
+ Sec. 5.09 (case 2)

5. Externally applied displacements at j

$$F_U = 0$$
$$F_V = 0$$
$$F_Z = 0$$
$$F_u = -EA_{Lx} u_j$$
$$F_v = -EI_{Lz}(v_i - a\theta_j)$$
$$F_\theta = -EI_{Lz}\theta_j$$

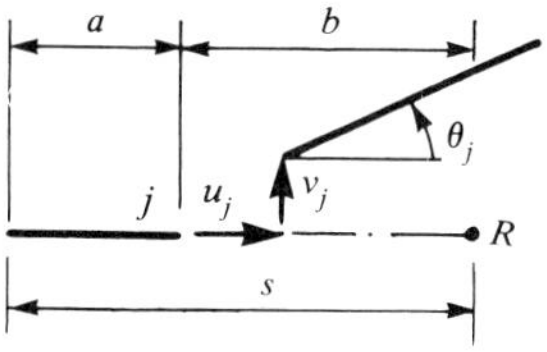

$$G_U = 0$$
$$G_V = 0$$
$$G_Z = 0$$
$$G_u = EA_{Lx} u_j$$
$$G_v = EI_{Lz}(v_j + b\theta_j)$$
$$G_\theta = EI_{Lz}\theta_j$$

6. Temperature deformation of segment b

$$F_U = 0$$
$$F_V = 0$$
$$F_Z = 0$$
$$F_u = -EA_{Lx} b e_t$$
$$F_v = EI_{Lz} C_{1z} s^2 f_t$$
$$F_\theta = -EI_{Lz} C_{0z} s f_t$$

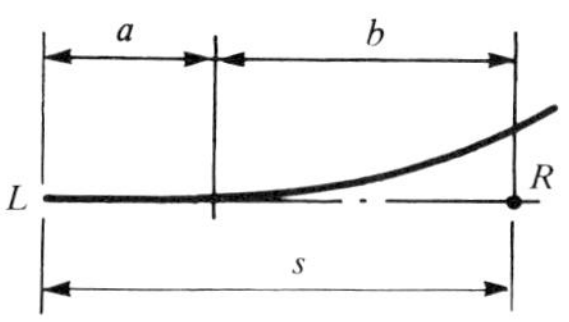

$$G_U = 0$$
$$G_V = 0$$
$$G_Z = 0$$
$$G_u = EA_{Lx} b e_t$$
$$G_v = EI_{Lz}(C_{0z} - C_{1z}) s^2 f_t$$
$$G_\theta = EI_{Lz} C_{0z} s f_t$$

5.10 FLEXIBILITY MATRIX EQUATIONS

(1) Notation used in the flexibility method (4.22) is summarized in Appendices A.1, A.3. Positive forces, moments, and displacements are shown in (4.22-2, 3). The flexibility coefficients introduced below for bars of variable section are expressed in terms of

$A_x, I_z, A_{Lx}, I_{Lz}, R_x, R_z$ = parameters of normal section (5.02) $\qquad r = I_{Lz}/A_{Lx}$

$T_{0n}, T_{0z}, T_{1z}, T_{2z}$ = transport coefficients (5.03, 5.04–5.07) $\qquad \bar{x} = x/s$

(2) Flexibility coefficients

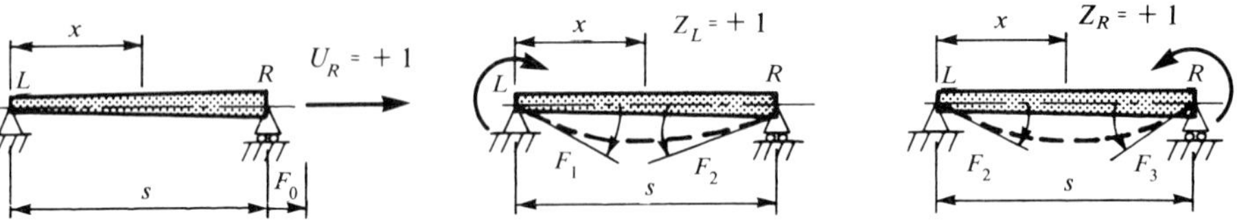

$$F_0 = \int_L^R \frac{dx}{EA_x} = \frac{s}{EA_{Lx}} \int_0^1 \frac{d\bar{x}}{R_x} = \frac{s}{EA_{Lx}} T_{0n}$$

$$F_1 = \int_L^R \frac{(s-x)^2\,dx}{EI_z s^2} = \frac{s}{EI_{Lz}} \int_0^1 \frac{(1-\bar{x})^2\,d\bar{x}}{R_z} = \frac{s}{EI_{Lz}} (T_{0z} - 2T_{1z} + T_{2z})$$

$$F_2 = \int_L^R \frac{x(s-x)\,dx}{EI_z s^2} = \frac{s}{EI_{Lz}} \int_0^1 \frac{\bar{x}(1-\bar{x})\,d\bar{x}}{R_z} = \frac{s}{EI_{Lz}} (T_{1z} - T_{2z})$$

$$F_3 = \int_L^R \frac{x^2\,dx}{EI_z s^2} = \frac{s}{EI_{Lz}} \int_0^1 \frac{\bar{x}^2\,d\bar{x}}{R_z} = \frac{s}{EI_{Lz}} T_{2z}$$

In cases of monotonic variation of section, such as those shown in (5.04–5.07), the closed-form formulas given in the last column above lead to a rapid evaluation of F_0, F_1, F_2, F_3. If, however, haunches are involved, the integrals given in the second column of the above table must be evaluated in parts, each corresponding to a particular segment of the bar.

(3) Matrix equations (4.22-4) in the case of a variable section become

$$\begin{bmatrix} u_{RL} \\ \theta_{LR} \\ \theta_{RL} \end{bmatrix} = \begin{bmatrix} F_0 & 0 & 0 \\ 0 & F_1 & F_2 \\ 0 & F_2 & F_3 \end{bmatrix} \begin{bmatrix} U_R \\ Z_L \\ Z_R \end{bmatrix} + \begin{bmatrix} u_{R0} \\ \theta_{L0} \\ \theta_{R0} \end{bmatrix}$$

(4) Load functions u_{R0}, θ_{L0}, θ_{R0} shown as the last column matrix above are the end displacements of the simple beam due to loads and other causes. In symbolic form,

$$\begin{bmatrix} u_{R0} \\ \theta_{L0} \\ \theta_{R0} \end{bmatrix} = \frac{1}{EI_{Lz}} \begin{bmatrix} r & 0 & 0 \\ 0 & s^{-1} & 0 \\ 0 & 0 & s^{-1} \end{bmatrix} \begin{bmatrix} G_u \\ G_v \\ F_v \end{bmatrix} + \begin{bmatrix} F_0 & 0 & 0 \\ 0 & F_2 & 0 \\ 0 & 0 & F_2 \end{bmatrix} \begin{bmatrix} F_U \\ F_Z \\ G_Z \end{bmatrix}$$

where F_U, F_Z, G_Z, G_u, G_v, F_v are the *transport load functions* (5.08–5.09). The same functions may be calculated by the relations defined in (4.22-5) with the help of quadrature formulas (Ref. 4, pp. 81–82).

5.11 STIFFNESS MATRIX EQUATIONS

(1) Notation used in the stiffness method (4.32) is summarized in (A.1, A.4). Positive forces, moments, and displacements are shown in (4.32-2, 3). The stiffness coefficients introduced below for bars of variable section are expressed in terms of the flexibility coefficients F_0, F_1, F_2, F_3 defined in (5.10-2).

(2) Stiffness coefficients

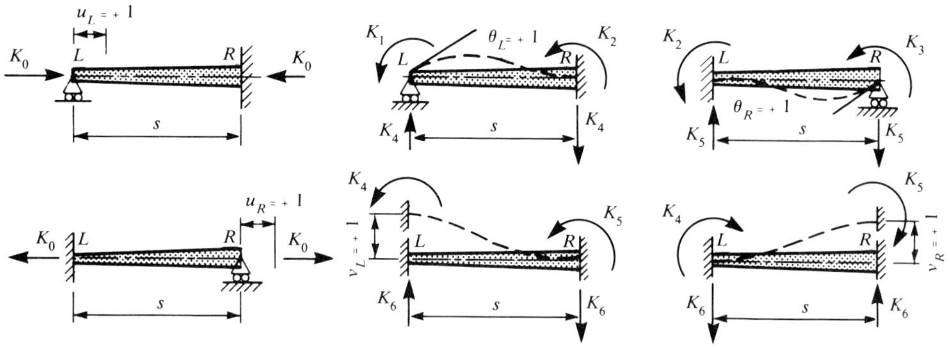

$$D_0 = \frac{1}{F_1F_3 - F_2^2} \qquad K_0 = \frac{1}{F_0} \qquad K_1 = F_1D_0 \qquad K_2 = F_2D_0 \qquad K_3 = F_3D_0$$

$$K_4 = \frac{K_1 + K_2}{s} \qquad K_5 = \frac{K_2 + K_3}{s} \qquad K_6 = \frac{K_4 + K_5}{s}$$

The numerical values of K_0, K_1, K_2, K_3 for particular bars are tabulated in (5.13–5.17, 5.19–5.23).

(3) Matrix equations (4.32-4) in the case of a variable section become

$$\begin{bmatrix} U_{LR} \\ V_{LR} \\ Z_{LR} \\ \hline U_{RL} \\ V_{RL} \\ Z_{RL} \end{bmatrix} = \left[\begin{array}{ccc|ccc} K_0 & 0 & 0 & -K_0 & 0 & 0 \\ 0 & K_6 & K_4 & 0 & -K_6 & K_5 \\ 0 & K_4 & K_1 & 0 & -K_4 & K_2 \\ \hline -K_0 & 0 & 0 & K_0 & 0 & 0 \\ 0 & -K_6 & -K_4 & 0 & K_6 & -K_5 \\ 0 & K_5 & K_2 & 0 & -K_5 & K_3 \end{array}\right] \begin{bmatrix} u_L \\ v_L \\ \theta_L \\ \hline u_R \\ v_R \\ \theta_R \end{bmatrix} + \begin{bmatrix} U_{L0} \\ V_{L0} \\ Z_{L0} \\ \hline U_{R0} \\ V_{R0} \\ Z_{R0} \end{bmatrix}$$

(4) Load functions U_{L0}, V_{L0}, . . . , Z_{R0} shown as the last column matrix above are the reactions of the fixed-end bar due to loads and other causes. In symbolic form,

$$\begin{bmatrix} U_{L0} \\ V_{L0} \\ Z_{L0} \end{bmatrix} = \frac{1}{EI_{Lz}} \begin{bmatrix} rK_0 & 0 & 0 \\ 0 & K_6 & -K_5 \\ 0 & K_4 & -K_2 \end{bmatrix} \begin{bmatrix} G_u \\ G_v \\ G_\theta \end{bmatrix} \qquad \begin{bmatrix} U_{R0} \\ V_{R0} \\ Z_{R0} \end{bmatrix} = \frac{1}{EI_{Lz}} \begin{bmatrix} rK_0 & 0 & 0 \\ 0 & K_6 & K_4 \\ 0 & -K_5 & -K_2 \end{bmatrix} \begin{bmatrix} F_u \\ F_v \\ F_\theta \end{bmatrix}$$

where F_u, F_v, F_θ, G_u, G_v, G_θ are the transport load functions (5.08–5.09). The numerical values of U_{L0}, Z_{L0}, U_{R0}, Z_{R0} due to selected loads are tabulated in (5.13–5.17, 5.19–5.23).

5.12 PROPERTIES OF PRISMATIC BAR WITH STRAIGHT HAUNCHES

(1) Geometry of bars considered in (5.13–5.17) is shown below. The width of the section is constant, and the depth varies linearly in the haunch. The geometry of the haunch is defined in (5.04), but the length of the haunch is βs.

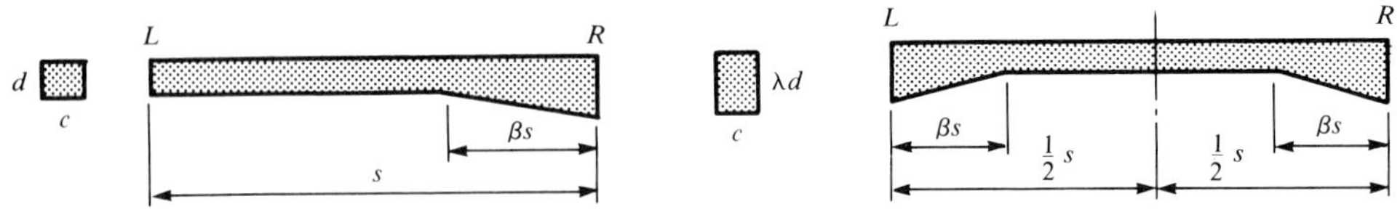

(2) Notation introduced in (5.10–5.11) is supplemented by the following symbols:

d = minimum depth (m) $\quad$ λd = maximum depth (m)
s = span (m) $\quad$ c = constant width (m)
sm = variable coordinate (m) $\quad$ βs = length of haunch (m)
$A = cd$ = minimum area of normal section (m²)
$I = \frac{1}{12} cd^3$ = minimum moment of inertia of A about z axis (m⁴)
P_x, P_y = components of concentrated load (N)
p_x, p_y = intensities of uniformly distributed load (N/m)
g_x, g_y = maximum intensities of haunch load (N/m)

(3) Stiffness coefficients (5.11-2)

$$K_0 = \frac{EA}{100s}\bar{K}_0 \qquad K_1 = \frac{EI}{100s}\bar{K}_1 \qquad K_2 = \frac{EI}{100s}\bar{K}_2 \qquad K_3 = \frac{EI}{100s}\bar{K}_3$$

where $\bar{K}_0, \bar{K}_1, \bar{K}_2, \bar{K}_3$ are dimensionless factors listed in the respective tables (5.13–5.17). The remaining coefficients K_4, K_5, K_6 (which are required for the construction of the stiffness matrix) are computed by the relations given in (5.11-2).

(4) Load functions (5.11-3)

$U_{L0}^{(P)} = \frac{1}{100} P_x \bar{L}_0$ $Z_{L0}^{(P)} = \frac{s}{100} P_y \bar{L}_1$	$U_{R0}^{(P)} = \frac{1}{100} P_x(100 - \bar{L}_0)$ $Z_{R0}^{(P)} = -\frac{s}{100} P_y \bar{L}_2$	Fixed-end beam reactions due to concentrated load at sm
$U_{L0}^{(U)} = \frac{s}{100} p_y \bar{L}_3$ $Z_{L0}^{(U)} = \frac{s^2}{100} p_y \bar{L}_4$	$U_{R0}^{(U)} = \frac{s}{100} p_y(100 - \bar{L}_3)$ $Z_{R0}^{(U)} = -\frac{s^2}{100} p_y \bar{L}_5$	Fixed-end beam reactions due to uniformly distributed load over s
$U_{L0}^{(H)} = \frac{s}{100} g_y \bar{L}_6$ $Z_{L0}^{(H)} = \frac{s^2}{100} g_y \bar{L}_8$	$U_{R0}^{(H)} = \frac{s}{100} g_y \bar{L}_7$ $Z_{R0}^{(H)} = -\frac{s^2}{100} g_y \bar{L}_9$	Fixed-end beam reactions due to linearly varying haunch load in βs

$\bar{L}_0, \bar{L}_1, \ldots, \bar{L}_9$ are dimensionless factors listed in the respective tables (5.13–5.17). The vertical reactions V_{L0}, V_{R0} are obtained by statics in terms of loads and Z_{L0}, Z_{R0}.

(5) Tables of dimensionless factors in (5.13–5.17) are independently calculated and compared with those published elsewhere (Refs. 24–25).

5.13 AXIAL STIFFNESS COEFFICIENTS AND LOAD FUNCTIONS BAR WITH ONE STRAIGHT HAUNCH

Notation (5.12) Signs (4.38) Matrix (5.11)

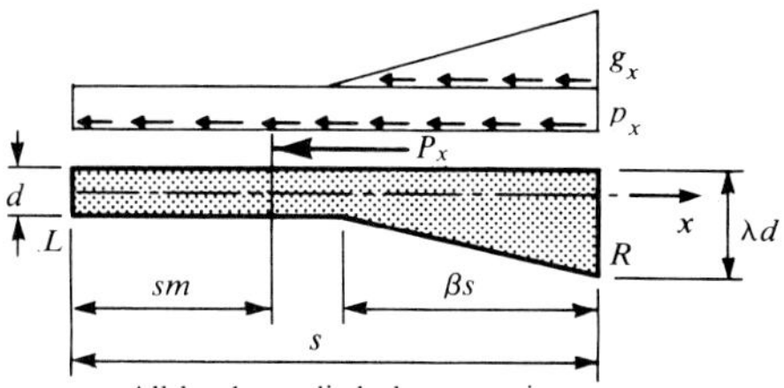

All loads applied along x axis

(1) Stiffness coefficient and load functions

$$K_0 = \frac{EA}{100s}\bar{K}_0$$

$$U_{L0}^{(P)} = \frac{1}{100}P_x\bar{L}_0 \qquad U_{L0}^{(U)} = \frac{s}{100}p_x\bar{L}_3 \qquad U_{L0}^{(H)} = \frac{s}{100}g_x\bar{L}_6 \qquad U_{R0}^{(H)} = \frac{s}{100}g_x\bar{L}_7$$

(2) Table of dimensionless factors

β	λ	$\bar{K}_0$	$\bar{L}_0$ $m = 0.1$	$m = 0.3$	$m = 0.5$	$m = 0.7$	$m = 0.9$	$\bar{L}_3$	$\bar{L}_6$	$\bar{L}_7$
0.1	1.4	102	89.8	69.5	49.2	28.9	8.5	49.2	0.13	4.87
	1.6	102	89.7	69.3	48.9	28.3	8.0	49.0	0.12	4.88
	2.0	103	38.6	69.0	48.4	27.8	7.2	48.5	0.10	4.90
	2.5	104	89.6	68.8	48.0	27.2	6.4	48.1	0.08	4.92
	3.0	105	89.5	68.6	47.6	26.7	5.8	47.8	0.07	4.93
0.2	1.4	103	89.8	69.0	48.4	27.7	8.0	48.6	0.53	9.47
	1.6	105	89.7	68.6	47.7	26.8	7.2	48.0	0.48	9.52
	2.0	107	89.7	68.0	46.7	25.4	6.1	47.2	0.41	9.59
	2.5	108	89.6	67.5	45.8	24.1	5.2	46.4	0.35	9.65
	3.0	110	89.5	67.0	45.0	23.1	4.5	45.8	0.30	9.70
0.3	1.4	105	89.7	68.5	47.5	26.5	7.9	48.0	1.22	13.78
	1.6	107	89.5	67.9	46.5	25.1	7.1	47.3	1.11	13.89
	2.0	110	89.4	67.0	44.9	23.9	6.0	46.1	0.96	14.04
	2.5	113	89.2	66.0	43.4	20.8	5.0	44.9	0.82	14.18
	3.0	116	89.0	65.3	42.2	19.0	4.4	44.0	0.71	14.29
0.4	1.4	107	89.5	68.5	46.6	25.8	7.9	47.6	2.20	17.80
	1.6	109	89.3	67.9	45.3	24.1	7.2	46.6	2.03	17.97
	2.0	114	89.0	67.0	43.0	21.4	6.1	45.1	1.76	18.24
	2.5	118	88.7	66.0	40.8	18.9	5.3	43.6	1.52	18.48
	3.0	122	88.4	65.3	39.0	16.9	4.4	42.4	1.34	18.66
0.5	1.4	109	89.3	68.0	45.7	25.5	8.0	47.2	3.49	21.51
	1.6	112	89.0	67.2	43.9	23.8	7.3	46.1	3.25	21.75
	2.0	118	88.6	65.8	40.9	21.1	6.2	44.3	2.85	22.15
	2.5	124	88.2	64.5	37.9	18.5	5.3	42.5	2.49	22.51
	3.0	129	87.8	63.4	35.4	16.5	4.6	41.1	2.22	22.78
0.75	1.4	114	88.8	66.4	45.0	25.8	8.3	46.8	8.26	29.24
	1.6	119	88.4	64.8	42.5	24.3	7.7	45.5	7.82	29.68
	2.0	130	87.5	62.1	39.6	21.8	6.7	43.4	7.11	30.39
	2.5	141	86.6	58.8	36.2	19.4	5.9	41.2	6.42	31.08
	3.0	149	85.8	56.8	33.4	17.6	5.3	39.5	5.90	31.60
1.0	1.4	119	88.3	66.3	45.8	26.6	8.6	47.2	15.30	34.70
	1.6	128	87.6	64.8	44.2	25.4	8.2	46.1	14.78	35.22
	2.0	144	86.2	62.1	41.5	23.4	7.4	44.3	13.93	36.07
	2.5	164	84.7	59.4	38.9	21.7	6.7	42.5	13.13	36.87
	3.0	182	83.4	57.2	36.9	20.3	6.3	41.4	12.50	37.50

5.14 FLEXURAL STIFFNESS COEFFICIENTS AND LOAD FUNCTIONS BAR WITH ONE STRAIGHT HAUNCH

Notation (5.12) Signs (4.38) Matrix (5.11)

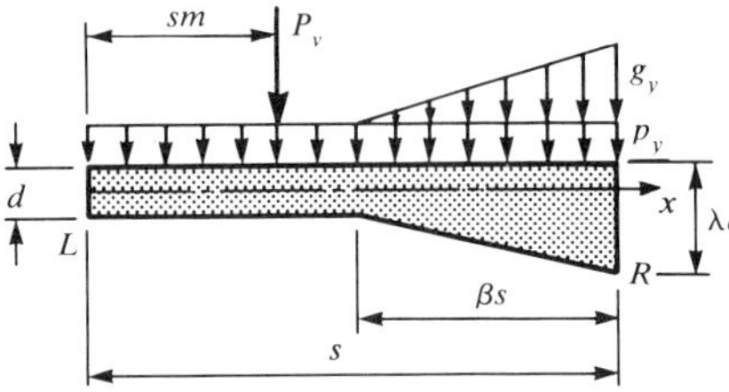

All loads applied along x axis

(1) Stiffness coefficients and load functions

$$K_1 = \frac{EI}{100s}\bar{K}_1 \qquad Z_{L0}^{(P)} = \frac{s}{100}P_y\bar{L}_1$$

$$K_2 = \frac{EI}{100s}\bar{K}_2 \qquad Z_{L0}^{(U)} = \frac{s^2}{100}p_y\bar{L}_4 \qquad Z_{L0}^{(H)} = \frac{s^2}{100}g_y\bar{L}_8$$

(2) Table of dimensionless factors

β	λ	$\bar{K}_1$	$\bar{K}_2$	$\bar{L}_1$					$\bar{L}_4$	$\bar{L}_8$
				$m = 0.1$	$m = 0.3$	$m = 0.5$	$m = 0.7$	$m = 0.9$		
0.1	1.4	414	230	8.04	14.26	11.64	5.34	0.52	7.80	0.00
	1.6	419	240	8.02	14.12	11.37	5.05	0.42	7.63	0.00
	2.0	425	253	7.99	13.93	11.00	4.64	0.28	7.41	0.00
	2.5	430	264	7.97	13.78	10.72	4.34	0.18	7.24	0.00
	3.0	433	270	7.95	13.69	10.55	4.15	0.13	7.14	0.00
0.2	1.4	426	258	7.98	13.91	11.02	4.76	0.50	7.47	0.03
	1.6	435	279	7.94	13.63	10.49	4.23	0.38	7.17	0.03
	2.0	449	312	7.88	13.21	9.71	3.46	0.23	6.73	0.02
	2.5	461	339	7.83	12.87	9.08	2.85	0.14	6.38	0.01
	3.0	468	358	7.80	12.64	8.66	2.46	0.09	6.16	0.01
0.3	1.4	434	281	7.95	13.68	10.64	4.53	0.51	7.30	0.12
	1.6	448	313	7.89	13.27	9.90	3.87	0.39	6.90	0.09
	2.0	471	354	7.79	12.62	8.74	2.89	0.24	6.30	0.06
	2.5	491	425	7.70	12.03	7.71	2.08	0.15	5.77	0.04
	3.0	506	465	7.64	11.61	7.00	1.55	0.10	5.42	0.03
0.4	1.4	439	298	7.93	13.55	10.44	4.49	0.53	7.22	0.27
	1.6	457	345	7.85	13.05	9.58	3.83	0.42	6.78	0.23
	2.0	418	428	7.72	12.19	8.15	2.83	0.28	6.07	0.16
	2.5	518	516	7.58	11.34	6.79	1.99	0.18	5.47	0.11
	3.0	542	586	7.47	10.70	5.78	1.44	0.12	4.94	0.09
0.5	1.4	443	308	7.91	13.46	10.32	4.48	0.55	7.18	0.51
	1.6	462	364	7.83	12.91	9.41	3.84	0.44	6.72	0.44
	2.0	499	473	7.67	11.92	7.88	2.88	0.30	5.97	0.33
	2.5	539	600	7.49	10.87	6.36	2.07	0.20	5.24	0.24
	3.0	573	713	7.35	10.01	5.19	1.53	0.14	4.58	0.17
0.75	1.4	456	315	7.85	12.97	9.72	4.30	0.54	6.92	1.43
	1.6	480	378	7.74	12.27	8.71	3.69	0.45	6.41	1.26
	2.0	525	516	7.55	11.08	7.17	2.82	0.32	5.62	1.01
	2.5	576	706	7.33	9.86	5.80	2.12	0.22	4.91	0.81
	3.0	624	912	7.13	8.85	4.81	1.67	0.17	4.37	0.66
1.0	1.4	517	332	7.66	12.43	9.53	4.34	0.55	6.75	2.56
	1.6	574	407	7.44	11.54	8.50	3.75	0.48	6.18	2.30
	2.0	685	571	7.06	10.05	6.91	2.89	0.35	5.29	1.90
	2.5	823	807	6.64	8.60	5.55	2.21	0.26	4.50	1.56
	3.0	957	1071	6.27	7.52	4.60	1.75	0.20	3.92	1.31

5.15 FLEXURAL STIFFNESS COEFFICIENTS AND LOAD FUNCTIONS BAR WITH ONE STRAIGHT HAUNCH

Notation (5.12) Signs (4.38) Matrix (5.11)

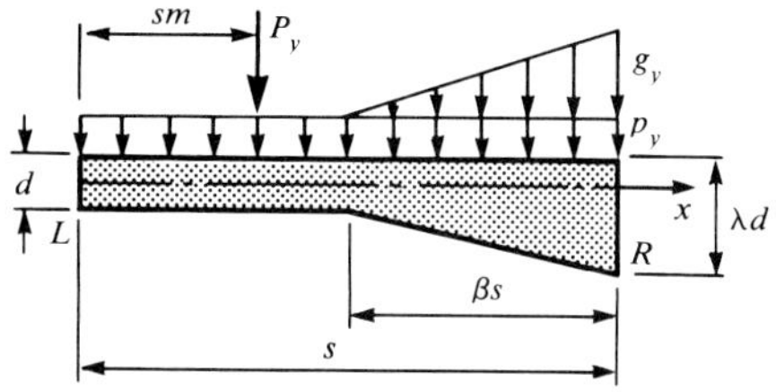

All loads applied along x axis

(1) Stiffness coefficients and load functions

$$K_3 = \frac{EI}{100s}\bar{K}_3 \qquad Z_{R0}^{(P)} = -\frac{s}{100}P_y\bar{L}_2$$

$$K_2 = \frac{EI}{100s}\bar{K}_2 \qquad Z_{R0}^{(U)} = -\frac{s^2}{100}p_y\bar{L}_5 \qquad Z_{R0}^{(H)} = -\frac{s^2}{100}g_y\bar{L}_9$$

(2) Table of dimensionless factors

β	λ	$\bar{K}_3$	$\bar{K}_2$	$\bar{L}_2$, $m = 0.1$	$m = 0.3$	$m = 0.5$	$m = 0.7$	$m = 0.9$	$\bar{L}_5$	$\bar{L}_9$
0.1	1.4	474	230	1.03	7.24	14.32	16.72	8.89	9.46	0.16
	1.6	485	240	1.08	7.54	14.90	17.38	9.11	9.81	0.16
	2.0	514	253	1.14	7.95	15.68	18.22	9.40	10.29	0.16
	2.5	536	264	1.18	8.26	16.29	18.87	9.60	10.66	0.16
	3.0	550	270	1.21	8.46	16.67	19.28	9.72	10.88	0.16
0.2	1.4	531	258	1.16	8.06	15.81	18.09	8.93	10.25	0.59
	1.6	581	279	1.25	8.71	17.01	19.29	9.17	10.93	0.61
	2.0	657	312	1.40	9.68	18.81	21.05	9.48	11.92	0.62
	2.5	722	339	1.52	10.41	20.30	22.47	9.69	12.74	0.64
	3.0	766	358	1.64	11.04	21.29	23.39	9.80	13.27	0.65
0.3	1.4	598	281	1.27	8.70	16.84	18.65	8.87	10.69	1.23
	1.6	684	313	1.41	9.70	18.62	20.17	9.11	11.62	1.28
	2.0	829	354	1.66	11.34	21.50	22.52	9.43	13.11	1.35
	2.5	958	425	1.90	12.86	24.12	24.52	9.65	14.42	1.40
	3.0	1072	465	2.10	13.97	25.94	25.85	9.77	15.34	1.43
0.4	1.4	659	298	1.33	9.11	17.34	18.49	8.77	10.84	1.99
	1.6	786	345	1.53	10.41	19.50	20.01	9.00	11.92	2.09
	2.0	1024	428	1.90	12.73	23.27	22.41	9.31	13.76	2.24
	2.5	1282	516	2.28	15.13	27.04	24.53	9.54	15.54	2.37
	3.0	1494	586	2.58	17.09	29.93	25.97	9.68	16.88	2.44
0.5	1.4	712	308	1.37	9.30	17.33	18.12	8.69	10.79	2.80
	1.6	881	364	1.61	10.79	19.58	19.50	8.90	11.91	2.95
	2.0	1228	473	2.08	13.64	23.71	21.75	9.19	13.90	3.21
	2.5	1652	600	2.63	16.88	28.12	23.82	9.43	15.99	3.44
	3.0	2042	713	3.11	19.69	31.74	25.29	9.58	17.00	3.65
0.75	1.4	802	315	1.39	9.15	16.48	17.37	8.56	10.41	4.62
	1.6	1055	378	1.65	10.58	18.24	18.37	8.71	11.34	4.91
	2.0	1660	516	2.22	13.43	21.37	20.02	8.94	13.02	5.40
	2.5	2592	706	2.98	16.94	24.71	21.60	9.15	14.90	5.89
	3.0	3695	912	3.80	20.37	27.56	22.81	9.28	16.59	6.29
1.0	1.4	857	332	1.39	8.85	15.83	16.89	8.50	10.11	5.86
	1.6	1163	407	1.68	10.01	17.17	17.86	8.58	10.86	6.21
	2.0	1946	571	2.24	12.21	19.51	18.93	8.77	12.16	6.80
	2.5	3269	807	2.96	14.75	21.84	20.10	8.93	13.52	7.38
	3.0	5013	1071	3.70	16.82	23.74	20.97	9.05	14.66	7.87

5.16 AXIAL STIFFNESS COEFFICIENTS AND LOAD FUNCTIONS BAR WITH TWO SYMMETRICAL STRAIGHT HAUNCHES

Notation (5.12) Signs (4.38) Matrix (5.11)

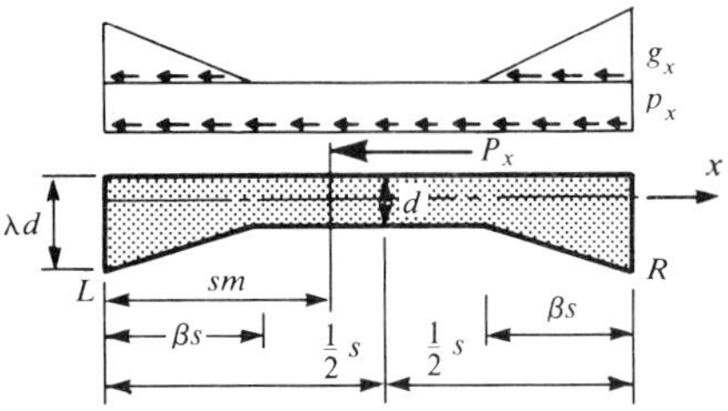

All loads applied along x axis

(1) Stiffness coefficient and load functions

$$K_0 = \frac{EA}{100s}\bar{K}_0 \qquad U_{L0}^{(P)} = \frac{1}{100}P_x\bar{L}_0 \qquad U_{L0}^{(U)} = \frac{s}{100}p_x\bar{L}_3 = U_{R0}^{(U)} \qquad U_{L0}^{(H)} = \frac{s}{100}g_x\bar{L}_6$$

(2) Table of dimensionless factors

β	λ	$\bar{K}_0$	$\bar{L}_0$: $m = 0.1$	$m = 0.3$	$m = 0.5$	$m = 0.7$	$m = 0.9$	$\bar{L}_3$	$\bar{L}_6$
0.1	1.4	103	91.3	70.7	50.0	29.3	8.7	50.0	5.0
	1.6	104	91.8	70.9	50.0	29.1	8.2	50.0	5.0
	2.0	106	92.6	71.3	50.0	28.7	7.4	50.0	5.0
	2.5	108	93.4	71.7	50.0	28.3	6.6	50.0	5.0
	3.0	110	94.0	72.0	50.0	28.0	6.0	50.0	5.0
0.2	1.4	107	91.8	71.4	50.0	28.6	8.2	50.0	10.0
	1.6	110	92.4	71.9	50.0	28.1	7.6	50.0	10.0
	2.0	114	93.4	72.8	50.0	27.2	6.6	50.0	10.0
	2.5	118	94.4	73.7	50.0	26.3	5.6	50.0	10.0
	3.0	122	95.1	74.4	50.0	25.6	4.9	50.0	10.0
0.3	1.4	111	91.7	72.1	50.0	27.9	8.3	50.0	15.0
	1.6	115	92.3	73.0	50.0	27.0	7.7	50.0	15.0
	2.0	123	93.3	74.5	50.0	25.5	6.7	50.0	15.0
	2.5	131	94.2	76.1	50.0	23.9	5.8	50.0	15.0
	3.0	137	94.8	74.4	50.0	22.6	5.2	50.0	15.0
0.4	1.4	115	91.5	72.4	50.0	27.6	8.5	50.0	20.0
	1.6	121	92.1	73.4	50.0	26.6	7.9	50.0	20.0
	2.0	133	92.9	75.1	50.0	24.9	7.1	50.0	20.0
	2.5	145	93.7	76.9	50.0	23.2	6.3	50.0	20.0
	3.0	156	94.3	78.3	50.0	21.7	5.7	50.0	20.0
0.5	1.4	119	91.3	72.1	50.0	27.9	8.9	50.0	25.0
	1.6	128	91.7	72.9	50.0	27.1	8.3	50.0	25.0
	2.0	144	92.4	74.3	50.0	25.7	7.6	50.0	25.0
	2.5	168	93.0	75.7	50.0	24.3	7.0	50.0	25.0
	3.0	182	93.5	76.8	50.0	23.2	6.5	50.0	25.0
β	λ	$\bar{K}_0$	$m = 0.9$	$m = 0.7$	$m = 0.5$	$m = 0.3$	$m = 0.1$	$\bar{L}_3$	$\bar{L}_6$
			$100 - \bar{L}_0$						

5.17 FLEXURAL STIFFNESS COEFFICIENTS AND LOAD FUNCTIONS BAR WITH TWO SYMMETRICAL STRAIGHT HAUNCHES

Notation (5.12)   Signs (4.38)   Matrix(5.11)

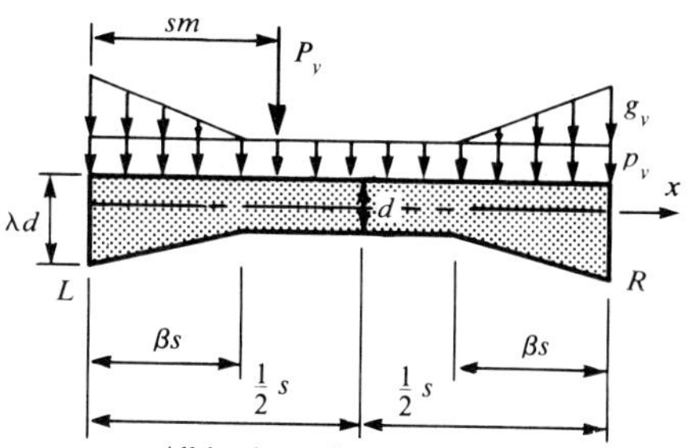

All loads applied along x axis

(1) Stiffness coefficients and load functions

$$K_1 = K_3 = \frac{EI}{100s}\bar{K}_1 \qquad K_2 = \frac{EI}{100s}\bar{K}_2$$

$$Z_{L0}^{(P)} = \frac{s}{100}P_y\bar{L}_1 \qquad Z_{R0}^{(P)} = -\frac{s}{100}P_y\bar{L}_2 \qquad Z_{L0}^{(U)} = \frac{s^2}{100}p_y\bar{L}_4 = -Z_{R0}^{(U)} \qquad Z_{L0}^{(H)} = \frac{s^2}{100}g_y\bar{L}_8 = -Z_{R0}^{(H)}$$

(2) Table of dimensionless factors

β	λ	$\bar{K}_1$	$\bar{K}_2$	$\bar{L}_1$ $m = 0.1$	$m = 0.3$	$m = 0.5$	$m = 0.7$	$m = 0.9$	$\bar{L}_4$	$\bar{L}_8$
0.1	1.4	483	267	8.84	16.29	13.40	6.17	0.60	8.89	0.16
	1.6	512	290	9.06	16.79	13.66	6.09	0.50	9.05	0.16
	2.0	554	326	9.36	17.49	14.00	5.94	0.36	9.25	0.16
	2.5	589	355	9.57	18.02	14.25	5.81	0.25	9.41	0.16
	3.0	611	375	9.69	18.39	14.41	5.72	0.18	9.50	0.17
0.2	1.4	575	338	8.85	17.32	14.12	6.13	0.65	9.25	0.63
	1.6	651	402	9.08	18.28	14.63	6.00	0.55	9.54	0.64
	2.0	781	515	9.39	19.73	15.33	5.61	0.39	9.93	0.65
	2.5	908	627	9.62	20.97	15.87	5.15	0.25	10.21	0.66
	3.0	1005	715	9.74	21.84	16.21	4.78	0.18	10.39	0.66
0.3	1.4	665	404	8.75	17.62	14.61	6.40	0.73	9.45	1.37
	1.6	804	520	8.97	18.76	15.34	6.25	0.66	9.82	1.40
	2.0	1085	765	9.24	20.63	16.40	5.77	0.52	10.34	1.47
	2.5	1427	1075	9.45	22.41	17.25	5.06	0.39	10.74	1.45
	3.0	1742	1367	9.62	23.75	17.81	4.38	0.31	10.99	1.48
0.4	1.4	744	454	8.62	17.29	14.81	6.66	0.81	9.47	2.33
	1.6	950	620	8.80	18.29	15.67	6.67	0.76	9.87	2.39
	2.0	1426	1027	9.05	19.91	17.00	6.49	0.67	10.46	2.48
	2.5	2131	1660	9.25	21.45	18.16	6.08	0.57	10.97	2.54
	3.0	2935	2407	9.39	22.64	18.97	5.59	0.49	11.25	2.58
0.5	1.4	807	480	8.52	16.82	14.58	6.72	0.85	9.33	3.42
	1.6	1072	679	8.67	17.61	15.38	6.80	0.81	9.69	3.52
	2.0	1734	1200	8.87	18.86	16.67	6.86	0.87	10.23	3.66
	2.5	2832	2118	9.03	19.99	17.86	6.79	0.71	10.70	3.79
	3.0	4261	3362	9.14	20.83	18.75	6.67	0.67	11.03	3.85
β	λ	$\bar{K}_1$	$\bar{K}_2$	$m = 0.9$ $\bar{L}_2$	$m = 0.7$	$m = 0.5$	$m = 0.3$	$m = 0.1$	$\bar{L}_4$	$\bar{L}_8$

5.18 PROPERTIES OF PRISMATIC BARS WITH PARABOLIC HAUNCHES

(1) Geometry of bars considered in (5.19–5.23) is shown below. The width of the section is constant, and the depth varies parabolically in the haunch. The geometry of the haunch is defined in (5.05), but the length of the haunch is βs.

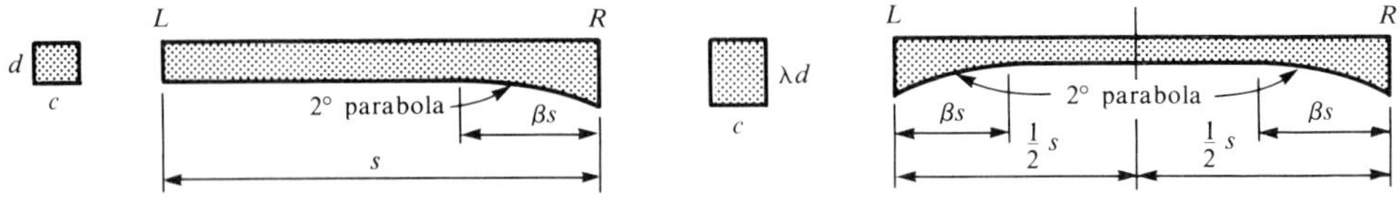

(2) Notation introduced in (5.10–5.11) is supplemented by the following symbols:

d = minimum depth (m) | λd = maximum depth (m)
s = span (m) | c = constant width (m)
sm = variable coordinate (m) | βs = length of haunch
$A = cd$ = minimum area of right section (m^2)
$I = \frac{1}{12}cd^3$ = minimum moment of inertia of A about z axis (m^4)
P_x, P_y = components of concentrated load (N)
p_x, p_y = intensities of uniformly distributed load (N/m)
g_x, g_y = maximum intensities of parabolic haunch load (N/m)

(3) Stiffness coefficients (5.11-2)

$$K_0 = \frac{EA}{100s}\overline{K}_0 \qquad K_1 = \frac{EI}{100s}\overline{K}_1 \qquad K_2 = \frac{EI}{100s}\overline{K}_2 \qquad K_3 = \frac{EI}{100s}\overline{K}_3$$

where $\overline{K}_0$, $\overline{K}_1$, $\overline{K}_2$, $\overline{K}_3$ are dimensionless factors listed in the respective tables (5.19–5.23). The remaining stiffness coefficients K_4, K_5, K_6 (which are required for the constructions of the stiffness matrix) are computed by the relations given in (5.11-2).

(4) Load functions (5.11-3)

$U_{L0}^{(P)} = \frac{1}{100}P_x\overline{L}_0$ $Z_{L0}^{(P)} = \frac{s}{100}P_y\overline{L}_1$	$U_{R0}^{(P)} = \frac{1}{100}P_x(100 - \overline{L}_0)$ $Z_{R0}^{(P)} = -\frac{2}{100}P_y\overline{L}_2$	Fixed-end beam reactions due to concentrated loads at sm
$U_{L0}^{(U)} = \frac{s}{100}p_y\overline{L}_3$ $Z_{L0}^{(U)} = \frac{s^2}{100}p_y\overline{L}_4$	$U_{R0}^{(U)} = \frac{s}{100}p_y(100 - \overline{L}_3)$ $Z_{R0}^{(U)} = -\frac{s^2}{100}p_y\overline{L}_5$	Fixed-end beam reactions due to uniformly distributed load over s
$U_{L0}^{(H)} = \frac{s}{100}g_x\overline{L}_6$ $Z_{L0}^{(H)} = \frac{s^2}{100}g_y\overline{L}_8$	$U_{R0}^{(H)} = \frac{s}{100}g_x\overline{L}_7$ $Z_{R0}^{(H)} = -\frac{s^2}{100}g_y\overline{L}_9$	Fixed end beam reactions due to parabolically varying load in βs

$\overline{L}_0$, $\overline{L}_1$, . . . , $\overline{L}_9$ are the dimensionless factors listed in the respective tables (5.19–5.23). The vertical reactions V_{L0}, V_{R0} are obtained by statics in terms of the loads and the respective Z_{L0}, Z_{R0}.

(5) Tables of dimensionless factors (5.19–5.23) are independently calculated and compared with those published elsewhere (Refs. 24, 25).

5.19 AXIAL STIFFNESS COEFFICIENTS AND LOAD FUNCTIONS BAR WITH ONE PARABOLIC HAUNCH

Notation (5.18)   Signs (4.38)   Matrix (5.11)

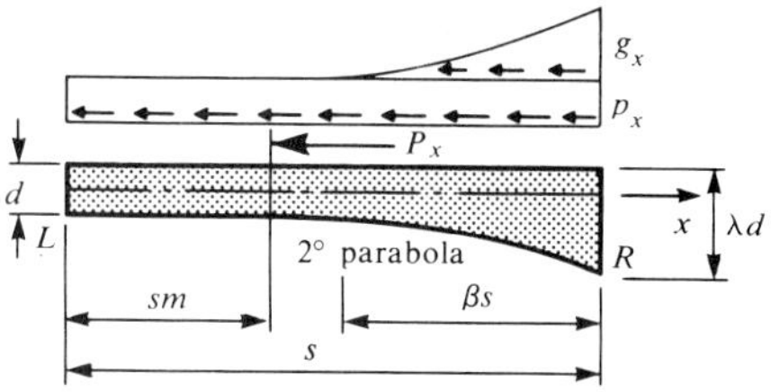

All loads applied along x axis

(1) Stiffness coefficient and load functions

$$K_0 = \frac{EA}{100s}\bar{K}_0$$

$$U_{L0}^{(P)} = \frac{1}{100}P_x\bar{L}_0 \qquad U_{L0}^{(U)} = \frac{s}{100}p_x\bar{L}_3 \qquad U_{L0}^{(H)} = \frac{s}{100}g_x\bar{L}_6 \qquad U_{R0}^{(H)} = \frac{s}{100}g_x\bar{L}_7$$

(2) Table of dimensionless factors

β	λ	$\bar{K}_0$	$\bar{L}_0$					$\bar{L}_3$	$\bar{L}_6$	$\bar{L}_7$
			$m = 0.1$	$m = 0.3$	$m = 0.5$	$m = 0.7$	$m = 0.9$			
0.1	1.4	101	89.9	69.7	49.4	29.2	9.0	49.5	0.07	3.27
	1.6	102	89.8	69.4	49.2	28.9	8.6	49.3	0.06	3.27
	2.0	102	89.8	69.3	48.9	28.5	8.0	49.0	0.05	3.28
	2.5	103	89.7	69.1	48.6	28.0	7.4	48.7	0.04	3.29
	3.0	103	89.7	69.0	48.3	27.6	7.0	48.4	0.04	3.29
0.2	1.4	102	89.8	69.3	48.9	28.4	8.3	49.3	0.27	6.40
	1.6	103	89.7	69.0	48.5	27.8	7.7	48.6	0.25	6.42
	2.0	104	89.6	68.6	47.8	26.9	6.7	48.0	0.21	6.46
	2.5	106	89.4	68.2	47.0	25.9	5.8	47.4	0.18	6.49
	3.0	107	89.3	67.9	46.1	25.2	5.1	47.0	0.16	6.51
0.3	1.4	103	89.7	69.0	48.3	27.6	8.2	48.6	0.62	9.38
	1.6	105	89.5	68.6	47.7	26.7	7.4	48.0	0.57	9.43
	2.0	107	89.3	67.9	46.6	25.2	6.3	47.1	0.49	9.51
	2.5	109	89.1	67.3	45.5	23.7	5.4	46.3	0.42	9.58
	3.0	111	88.9	66.8	44.6	22.4	4.7	45.6	0.37	9.63
0.4	1.4	105	89.6	68.6	47.8	27.0	8.1	48.2	1.11	12.22
	1.6	106	89.4	68.1	46.8	25.7	7.3	47.5	1.02	12.31
	2.0	109	89.1	67.2	45.3	23.4	6.2	46.4	0.90	12.43
	2.5	112	88.8	66.3	43.8	21.6	5.2	45.3	0.78	12.55
	3.0	115	88.5	65.5	42.5	20.0	4.6	44.4	0.69	12.65
0.5	1.4	106	89.4	68.3	47.1	26.4	8.1	47.9	1.75	14.82
	1.6	108	89.2	67.6	46.0	25.0	7.3	47.1	1.63	15.04
	2.0	112	88.8	66.4	44.0	22.7	6.2	45.7	1.43	15.24
	2.5	116	88.4	65.2	42.0	20.4	5.3	44.4	1.25	15.42
	3.0	119	88.1	64.2	40.3	18.6	4.6	43.3	1.12	15.55
0.75	1.4	109	89.1	67.4	46.0	26.2	8.0	47.4	4.05	20.95
	1.6	113	88.7	66.2	44.3	24.4	7.4	46.3	3.91	21.08
	2.0	119	88.1	64.8	41.4	21.9	6.4	44.5	3.43	21.57
	2.5	126	87.4	62.2	38.6	19.5	5.5	43.2	3.07	21.93
	3.0	134	86.8	60.4	36.2	17.7	4.8	41.4	2.79	22.21
1.0	1.4	112	88.8	66.8	45.7	26.1	8.2	47.2	7.42	25.91
	1.6	118	88.3	65.4	44.0	24.6	7.6	46.0	7.07	26.26
	2.0	127	87.4	62.9	41.0	22.2	6.7	44.1	6.51	26.82
	2.5	138	86.2	60.3	38.0	20.0	5.9	42.2	5.98	27.35
	3.0	148	85.3	58.0	35.6	18.3	5.3	40.7	5.56	27.77

5.20 FLEXURAL STIFFNESS COEFFICIENTS AND LOAD FUNCTIONS BAR WITH ONE PARABOLIC HAUNCH

Notation (5.18) Signs (4.38) Matrix (5.11)

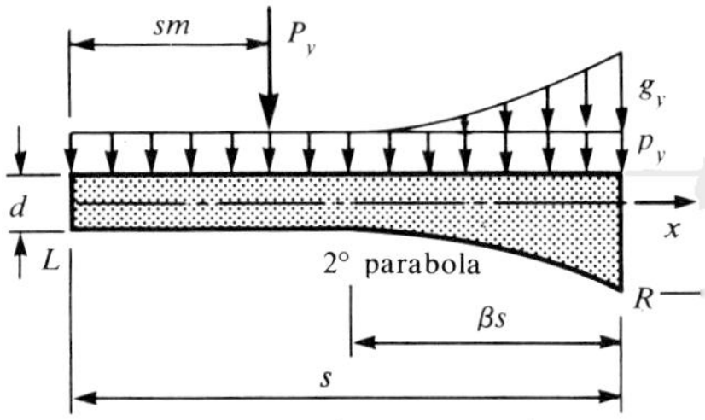

All loads applied along x axis

(1) Stiffness of coefficients and load functions

$$K_1 = \frac{EI}{100s}\bar{K}_1 \qquad Z_{L0}^{(P)} = \frac{s}{100}P_y\bar{L}_1$$

$$K_2 = \frac{EI}{100s}\bar{K}_2 \qquad Z_{L0}^{(U)} = \frac{s^2}{100}p_y\bar{L}_4 \qquad Z_{L0}^{(H)} = \frac{s^2}{100}g_y\bar{L}_8$$

(2) Table of dimensionless factors

β	λ	$\bar{K}_1$	$\bar{L}_2$	$\bar{L}_1$ $m=0.1$	$m=0.3$	$m=0.5$	$m=0.7$	$m=0.9$	$\bar{L}_4$	$\bar{L}_8$
0.1	1.4	410	220	8.06	14.39	11.89	5.61	0.60	7.95	0.00
	1.6	413	228	8.04	14.29	11.69	5.39	0.51	7.82	0.00
	2.0	418	238	8.02	14.14	11.41	5.08	0.39	7.65	0.00
	2.5	422	246	8.00	14.02	11.19	4.83	0.30	7.51	0.00
	3.0	424	251	7.99	13.64	11.03	4.66	0.24	7.42	0.00
0.2	1.4	419	241	8.02	14.13	11.41	5.13	0.52	7.68	0.01
	1.6	425	255	7.99	13.93	11.03	4.72	0.41	7.45	0.01
	2.0	435	278	7.54	13.63	10.46	4.13	0.27	7.12	0.01
	2.5	444	297	7.91	13.37	9.97	3.64	0.17	6.84	0.00
	3.0	450	311	7.88	13.18	9.63	3.31	0.12	6.65	0.00
0.3	1.4	425	258	7.99	13.63	11.06	4.83	0.51	7.50	0.05
	1.6	435	281	7.94	13.64	10.52	4.30	0.40	7.20	0.04
	2.0	451	318	7.87	12.18	9.67	3.51	0.27	6.73	0.03
	2.5	465	352	7.81	12.76	8.92	2.83	0.16	6.33	0.02
	3.0	475	379	7.77	12.45	8.37	2.36	0.11	6.05	0.01
0.4	1.4	431	272	7.97	13.79	10.82	4.68	0.52	7.39	0.11
	1.6	443	303	7.91	13.42	10.16	4.09	0.41	7.03	0.09
	2.0	464	356	7.81	12.80	9.07	3.19	0.27	6.47	0.06
	2.5	484	410	7.73	12.23	8.08	2.42	0.17	5.96	0.04
	3.0	500	453	7.66	11.78	7.33	1.90	0.11	5.59	0.03
0.5	1.4	434	283	7.94	13.68	10.66	4.63	0.53	7.32	0.21
	1.6	448	321	7.88	13.25	9.91	3.99	0.42	6.93	0.18
	2.0	475	391	7.77	12.52	8.66	3.06	0.27	6.28	0.13
	2.5	502	466	7.65	11.79	7.48	2.27	0.18	5.71	0.10
	3.0	523	530	7.56	11.20	6.60	1.73	0.12	5.27	0.07
0.75	1.4	442	299	7.91	13.44	10.29	4.49	0.54	7.17	0.63
	1.6	461	351	7.81	12.91	9.41	3.86	0.44	6.72	0.57
	2.0	496	452	7.68	11.95	7.97	2.94	0.29	6.00	0.45
	2.5	534	577	7.51	10.95	6.60	2.18	0.20	5.24	0.35
	3.0	568	698	7.37	10.04	5.54	1.67	0.14	4.81	0.28
1.0	1.4	459	309	7.64	13.04	9.93	4.40	0.55	6.98	1.27
	1.6	486	367	7.72	12.34	8.96	3.77	0.44	6.48	1.13
	2.0	536	488	7.50	11.14	7.42	2.87	0.31	5.68	0.92
	2.5	593	647	7.25	9.91	6.03	2.14	0.22	4.92	0.74
	3.0	645	813	7.04	8.91	5.00	1.63	0.16	4.39	0.62

5.21 FLEXURAL STIFFNESS COEFFICIENTS AND LOAD FUNCTIONS BAR WITH ONE PARABOLIC HAUNCH

Notation (5.18) Signs (4.38) Matrix (5.11)

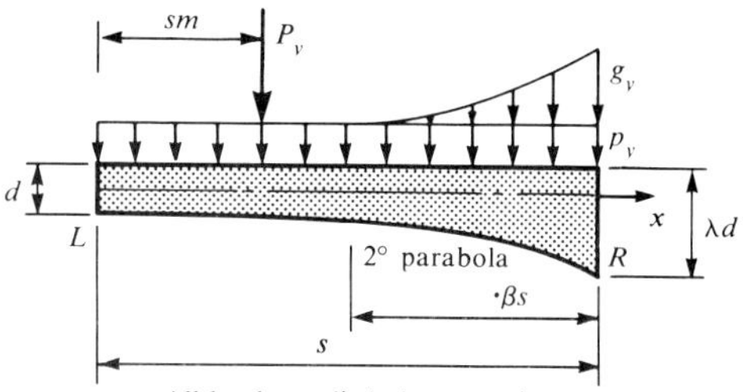

All loads applied along x axis

(1) Stiffness coefficients and load functions

$$K_3 = \frac{EI}{100s}\bar{K}_3 \qquad Z_{R0}^{(P)} = -\frac{s}{100}P_y\bar{L}_2$$

$$K_2 = \frac{EI}{100s}\bar{K}_2 \qquad Z_{R0}^{(U)} = -\frac{s^2}{100}p_y\bar{L}_5 \qquad Z_{R0}^{(H)} = -\frac{s^2}{100}g_y\bar{L}_9$$

(2) Table of dimensionless factors

β	λ	$\bar{K}_3$	$\bar{K}_2$	$\bar{L}_2$					$\bar{L}_5$	$\bar{L}_9$
				$m = 0.1$	$m = 0.3$	$m = 0.5$	$m = 0.7$	$m = 0.9$		
0.1	1.4	444	220	1.00	6.96	13.78	16.14	8.72	9.14	0.08
	1.6	459	228	1.03	7.17	14.19	16.60	8.91	9.40	0.08
	2.0	479	238	1.07	7.47	14.78	17.26	9.17	9.76	0.08
	2.5	497	246	1.11	7.72	15.27	17.79	9.36	10.07	0.08
	3.0	509	251	1.13	7.90	15.60	18.16	9.48	10.26	0.08
0.2	1.4	489	241	1.08	7.54	14.87	17.24	8.88	9.75	0.30
	1.6	522	255	1.15	7.99	15.72	18.12	9.12	10.25	0.31
	2.0	572	278	1.25	8.67	16.99	19.43	9.42	10.98	0.32
	2.5	617	297	1.33	9.25	18.10	20.54	9.63	11.62	0.32
	3.0	650	311	1.40	9.68	18.89	21.32	9.75	12.05	0.33
0.3	1.4	533	258	1.16	8.04	15.74	17.93	8.89	10.18	0.64
	1.6	588	281	1.26	8.72	17.00	19.13	9.12	10.88	0.66
	2.0	677	318	1.42	9.82	18.59	20.97	9.42	11.96	0.69
	2.5	764	352	1.58	10.85	20.82	22.58	9.64	12.93	0.71
	3.0	830	379	1.69	11.01	22.18	23.72	9.76	13.62	0.72
0.4	1.4	576	272	1.20	8.44	16.36	18.23	8.85	10.45	1.07
	1.6	654	303	1.36	9.34	17.95	19.58	9.08	11.29	1.11
	2.0	793	356	1.59	10.86	20.60	21.69	9.39	12.66	1.19
	2.5	937	410	1.83	12.39	23.19	23.57	9.61	13.95	1.22
	3.0	1055	453	2.01	13.59	25.19	24.89	9.73	14.91	1.25
0.5	1.4	616	283	1.27	8.73	16.72	18.23	8.77	10.58	1.55
	1.6	721	323	1.44	9.80	18.55	19.63	9.03	11.51	1.62
	2.0	912	390	1.74	11.73	21.71	21.81	9.33	13.10	1.73
	2.5	1135	466	2.06	13.77	24.02	23.76	9.56	14.65	1.82
	3.0	1329	530	2.34	15.47	27.50	25.14	9.69	15.86	1.88
0.75	1.4	697	299	1.33	9.02	16.81	17.92	8.70	10.59	2.83
	1.6	862	351	1.55	10.31	18.73	19.18	8.89	11.57	2.98
	2.0	1213	452	1.98	12.87	22.17	21.19	9.17	13.31	3.25
	2.5	1682	577	2.51	15.82	25.80	23.04	9.40	15.38	3.50
	3.0	2169	698	3.01	18.54	28.82	24.38	9.54	16.68	3.67
1.0	1.4	755	309	1.34	9.03	16.52	17.58	8.61	10.45	3.94
	1.6	968	367	1.61	10.36	18.28	18.65	8.79	11.35	4.19
	2.0	1462	488	2.11	12.92	21.38	20.40	9.03	13.07	4.61
	2.5	2199	647	2.74	15.01	24.00	22.02	9.23	14.80	5.01
	3.0	3058	813	3.39	18.68	27.27	23.18	9.38	16.22	5.33

5.22 AXIAL STIFFNESS COEFFICIENTS AND LOAD FUNCTIONS BAR WITH TWO SYMMETRICAL HAUNCHES

Notation (5.18) Signs (4.38) Matrix (5.11)

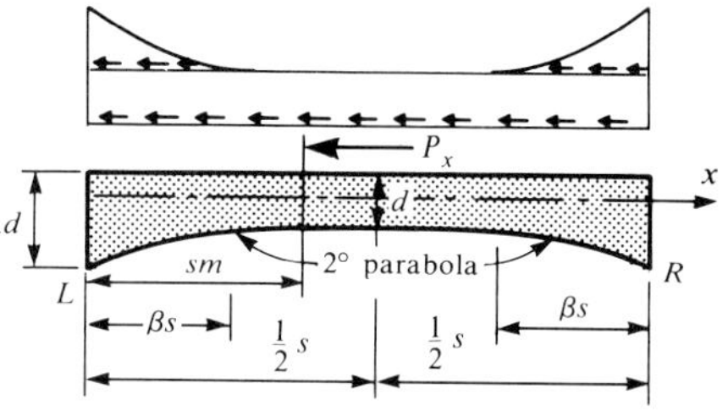

All loads applied along x axis

(1) Stiffness coefficient and load functions

$$K_0 = \frac{EA}{100s}\bar{K}_0 \qquad U_{L0}^{(P)} = \frac{1}{100}P_x\bar{L}_0 \qquad U_{L0}^{(U)} = \frac{s}{100}p_x\bar{L}_3 \qquad U_{L0}^{(H)} = \frac{s}{100}g_x\bar{L}_6$$

(2) Table of dimensionless factors

β	λ	$\bar{K}_0$	$\bar{L}_0$					$\bar{L}_3$	$\bar{L}_6$
			$m = 0.1$	$m = 0.3$	$m = 0.5$	$m = 0.7$	$m = 0.9$		
0.1	1.4	102	90.9	70.4	50.0	29.6	9.1	50.0	3.33
	1.6	103	91.2	70.6	50.0	29.4	8.8	50.0	3.33
	2.0	104	91.8	70.9	50.0	29.1	8.2	50.0	3.33
	2.5	106	92.3	71.2	50.0	28.8	7.7	50.0	3.33
	3.0	107	92.8	71.4	50.0	28.6	7.2	50.0	3.33
0.2	1.4	105	91.5	70.9	50.0	29.1	8.5	50.0	6.67
	1.6	106	92.0	71.3	50.0	28.7	8.0	50.0	6.67
	2.0	109	93.0	71.9	50.0	28.1	7.0	50.0	6.67
	2.5	112	93.8	72.5	50.0	27.5	6.2	50.0	6.67
	3.0	115	94.5	73.0	50.0	27.0	5.5	50.0	6.67
0.3	1.4	107	91.6	71.4	50.0	28.6	8.4	50.0	10.00
	1.6	110	92.2	72.0	50.0	28.0	7.8	50.0	10.00
	2.0	115	93.0	72.7	50.0	27.3	7.0	50.0	10.00
	2.5	120	94.1	74.0	50.0	26.0	5.9	50,0	10.00
	3.0	124	94.7	74.8	50.0	25.2	5.3	50.0	10.00
0.4	1.4	109	91.6	7.18	50.0	28.2	8.4	50.0	13.33
	1.6	114	92.2	72.6	50.0	27.4	7.8	50.0	13.33
	2.0	121	93.1	73.9	50.0	26.1	6.9	50.0	13.33
	2.5	128	94.0	75.3	50.0	24.7	6.4	50.0	13.33
	3.0	135	94.6	76.5	50.0	23.5	5.4	50.0	13.33
0.5	1.4	112	91.5	72.0	50.0	28.0	8.5	50.0	16.67
	1.6	118	92.1	72.8	50.0	27.2	7.9	50.0	16.67
	2.0	127	93.0	74.2	50.0	25.8	7.0	50.0	16.67
	2.5	138	93.7	75.7	50,0	24.1	6.3	50.0	16.67
	3.0	148	94.3	76.9	50.0	23.1	5.7	50.0	16.67
β	λ	$\bar{K}_0$	$m = 0.9$	$m = 0.7$	$m = 0.5$	$m = 0.3$	$m = 0.1$	$\bar{L}_3$	$\bar{l}_6$
			$1 - \bar{L}_0$						

5.23 FLEXURAL STIFFNESS COEFFICIENTS AND LOAD FUNCTIONS BAR WITH TWO SYMMETRICAL HAUNCHES

Notation (5.18) Signs (4.38) Matrix (5.11)

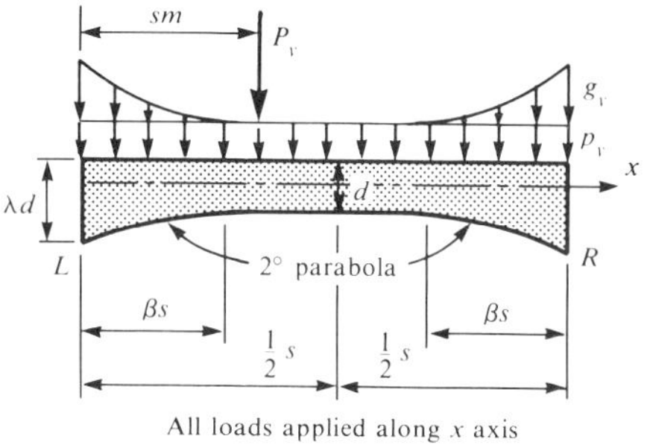

(1) Stiffness coefficients and load functions

$$K_1 = K_3 = \frac{EI}{100s}\bar{K}_1 \qquad K_2 = \frac{EI}{100s}\bar{K}_2$$

$$Z_{L0}^{(P)} = \frac{s}{100}P_y\bar{L}_1 \qquad Z_{R0}^{(P)} = -\frac{s}{100}P_y\bar{L}_2 \qquad Z_{L0}^{(U)} = \frac{s^2}{100}p_y\bar{L}_4 \qquad Z_{L0}^{(H)} = \frac{s^2}{100}g_y\bar{L}_8$$

(2) Table of dimensionless factors

				$\bar{L}_1$						
β	λ	$\bar{K}_1$	$\bar{K}_2$	$m = 0.1$	m 0.3	$m = 0.5$	$m = 0.7$	$m = 0.9$	$\bar{L}_4$	$\bar{L}_8$
0.1	1.4	456	245	8.69	15.83	13.13	6.21	0.66	8.73	0.08
	1.6	476	261	8.87	16.20	13.53	6.76	0.58	8.85	0.08
	2.0	505	285	9.12	16.72	13.60	6.07	0.46	9.02	0.08
	2.5	530	306	9.32	17.15	13.81	5.99	0.37	9.15	0.08
	3.0	548	321	9.45	17.44	13.95	5.92	0.30	9.24	0.08
0.2	1.4	516	292	8.83	16.68	13.67	6.18	0.63	9.03	0.30
	1.6	563	330	9.05	17.38	14.05	6.07	0.53	9.25	0.32
	2.0	641	397	9.35	18.44	14.59	5.84	0.38	9.56	0.33
	2.5	717	462	9.57	19.36	15.02	5.58	0.27	9.80	0.33
	3.0	777	514	9.70	20.03	15.32	5.35	0.19	9.96	0.33
0.3	1.4	576	337	8.80	17.17	14.10	6.25	0.67	9.23	0.70
	1.6	658	404	9.02	18.10	14.64	6.12	0.58	9.52	0.71
	2.0	810	535	9.32	19.58	15.43	5.78	0.43	9.94	0.72
	2.5	978	682	9.54	20.95	16.09	5.33	0.31	10.28	0.73
	3.0	1127	815	9.68	21.98	16.55	4.91	0.23	10.50	0.74
0.4	1.4	632	377	8.75	17.30	14.39	6.30	0.72	9.34	1.20
	1.6	754	475	8.95	18.31	15.05	6.29	0.54	9.68	1.22
	2.0	1002	687	9.24	19.92	16.07	5.97	0.51	10.17	1.23
	2.5	1318	966	9.46	21.45	19.64	5.48	0.39	10.58	1.29
	3.0	1634	1253	9.59	22.62	17.57	5.00	0.31	10.85	1.30
0.5	1.4	684	410	8.66	17.18	14.51	6.50	0.76	9.37	1.80
	1.6	842	536	8.87	18.15	15.25	6.47	0.69	9.72	1.86
	2.0	1203	835	9.13	19.70	16.39	6.26	0.57	10.25	1.92
	2.5	1713	1280	9.34	21.14	17.41	5.89	0.48	10.69	1.96
	3.0	2283	1790	9.47	22.24	18.16	5.49	0.40	10.90	1.99
β	λ	$\bar{K}_1$	$\bar{K}_2$	$m = 0.9$	$m = 0.7$	$m = 0.5$	$m = 0.3$	$m = 0.1$	$\bar{L}_4$	$\bar{L}_8$
				$\bar{L}_2$						

5.24 MODIFIED STIFFNESS MATRIX EQUATIONS

(1) Modified stiffness coefficients are

$$K_7 = \frac{1 - \alpha_1\alpha_3}{s^2} K_3 \qquad K_8 = \frac{1 - \alpha_1\alpha_3}{s^2} K_1 \qquad K_9 = K_1 - \alpha_4 K_4$$

where

$$\alpha_1 = \frac{K_2}{K_1} \qquad \alpha_3 = \frac{K_2}{K_3} \qquad \alpha_4 = \frac{K_4}{K_6} \qquad \alpha_5 = \frac{K_5}{K_6}$$

and K_1, K_2, K_3, K_4, K_5, K_6 are the stiffness coefficients (5.11-2).

(2) Modifications of the stiffness matrix equations (5.11-3) due to hinged and guided ends are shown in (3, 4) and (5, 6) below. Because the hinged end and the guided end do not affect the expressions for U_{LR} and U_{RL}, only the affected part of the respective matrix equation is presented.

(3) Left end hinged ($Z_{LR} = 0$)

$$\begin{bmatrix} V_{LR} \\ V_{RL} \\ Z_{RL} \end{bmatrix} = K_7 \begin{bmatrix} 1 & -1 & s \\ -1 & 1 & -s \\ s & s & s^2 \end{bmatrix} \begin{bmatrix} v_L \\ v_R \\ \theta_R \end{bmatrix} + \begin{bmatrix} V_{L0} - (1 + \alpha_1)\, s^{-1} Z_{L0} \\ V_{R0} + (1 + \alpha_1) s^{-1} Z_{L0} \\ Z_{R0} - \alpha_1 Z_{L0} \end{bmatrix}$$

(4) Right end hinged ($Z_{RL} = 0$)

$$\begin{bmatrix} V_{LR} \\ Z_{LR} \\ V_{RL} \end{bmatrix} = K_8 \begin{bmatrix} 1 & x & -1 \\ s & s^2 & -s \\ -1 & -s & 1 \end{bmatrix} \begin{bmatrix} v_L \\ \theta_L \\ v_R \end{bmatrix} + \begin{bmatrix} V_{L0} + (1 + \alpha_3) s^{-1} Z_{R0} \\ Z_{L0} - \alpha_3 Z_{R0} \\ V_{R0} - (1 + \alpha_3) s^{-1} Z_{R0} \end{bmatrix}$$

(5) Left end guided ($V_{LR} = 0$)

$$\begin{bmatrix} Z_{LR} \\ V_{RL} \\ Z_{RL} \end{bmatrix} = K_9 \begin{bmatrix} 1 & 0 & -1 \\ 0 & 0 & 0 \\ -1 & 0 & 1 \end{bmatrix} \begin{bmatrix} \theta_L \\ v_R \\ \theta_L \end{bmatrix} + \begin{bmatrix} Z_{L0} - \alpha_4 V_{L0} \\ V_{R0} + V_{L0} \\ Z_{R0} - \alpha_5 V_{L0} \end{bmatrix}$$

(6) Right end guided ($V_{RL} = 0$)

$$\begin{bmatrix} V_{LR} \\ Z_{LR} \\ Z_{RL} \end{bmatrix} = K_9 \begin{bmatrix} 0 & 0 & 0 \\ 1 & 1 & -1 \\ -1 & -1 & 1 \end{bmatrix} \begin{bmatrix} v_L \\ \theta_L \\ \theta_R \end{bmatrix} + \begin{bmatrix} V_{L0} + V_{R0} \\ Z_{L0} + \alpha_5 V_{R0} \\ Z_{R0} - \alpha_4 V_{R0} \end{bmatrix}$$

(7) Modified load functions shown in the last column matrix in (3–6) above are the reactions of the respective beams (propped-end beam or guided beam) expressed in terms of the reactions of the fixed-end beam V_{L0}, Z_{L0}, V_{R0}, Z_{R0} (5.11-4) and the respective carry-over ratio α_1, α_3, α_4, or α_5.

6

Free Circular Bar of Order One Constant Section

STATIC STATE

6.01 DEFINITION OF STATE

(1) System considered is a free finite circular bar of constant and symmetrical section. The bar and its loads form a coplanar system (system of order one). The system plane is the plane of symmetry of the bar section. Two coordinate systems are used in this chapter: (*a*) the local system given by the principal axes of the normal section at each point of the centroidal axis and (*b*) the 0 system related to the center 0 of the circle with x axis parallel to the segment LR and y axis bisecting the central angle $a_0 = 2b_0$ (rad).

(2) Notation (A.1, A.2, A.4)

s = span (m)	A_x = area of normal section (m^2)
a, b = angles (rad)	I_z = moment of intertia of A_x about z (m^4)
R = radius (m)	E = modulus of elasticity (Pa)
U = normal force (N)	u = linear displacement along x (m)
V = shearing force (N)	v = linear displacement along y (m)
Z = flexural moment (N·m)	θ = flexural slope about z (rad)

$r = \dfrac{I_z}{A_x R^2}$	$\alpha = \frac{1}{2}(1 + r)$	$\beta = \frac{1}{2}(1 - r)$

(3) Assumptions of analysis are stated in (1.01). In this chapter, the effects of axial and flexural deformations are included. Since only bars of large radii are considered, the shearing deformation effects are too small and as such are neglected.

6.02 TRANSPORT MATRIX EQUATIONS IN LOCAL SYSTEMS

(1) Equivalents and signs

$$\bar{u} = uEI_z = \text{scaled } u \ (\text{N}\cdot\text{m}^3)$$
$$\bar{v} = vEI_z = \text{scaled } v \ (\text{N}\cdot\text{m}^3)$$
$$\bar{\theta} = \theta EI_z = \text{scaled } \theta \ (\text{N}\cdot\text{m}^2)$$

$$r = \frac{I_z}{A_x R^2}$$
$$\alpha = \tfrac{1}{2}(1 + r) \qquad \beta = \tfrac{1}{2}(1 - r)$$

All forces, moments, and displacements are in the principal axes of the normal section, and their positive directions are shown in (2, 3) below.

(2) Free-body diagram

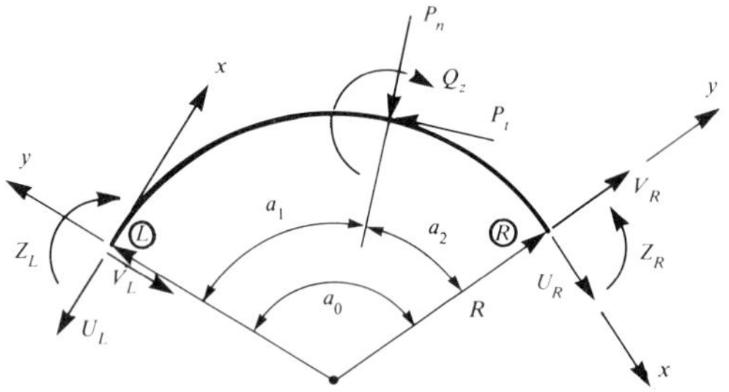

(3) Elastic curve

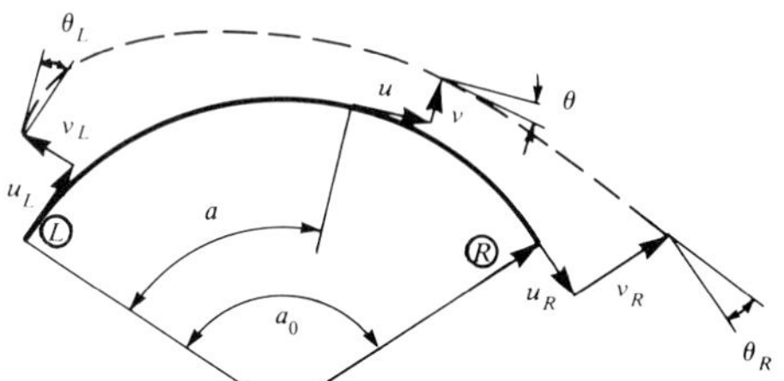

(4) Transport matrix equations

$$\underbrace{\begin{bmatrix} 1 \\ \hline U_L \\ V_L \\ Z_L \\ \hline \bar{u}_L \\ \bar{v}_L \\ \bar{\theta}_L \end{bmatrix}}_{H_L} = \underbrace{\left[\begin{array}{c|ccc|ccc} 1 & 0 & 0 & 0 & 0 & 0 & 0 \\ \hline F_U & T_1 & T_2 & 0 & 0 & 0 & 0 \\ F_V & -T_2 & T_1 & 0 & 0 & 0 & 0 \\ F_Z & -T_3R & T_2R & 1 & 0 & 0 & 0 \\ \hline F_u & -T_7R^3 & T_6R^3 & T_4R^2 & T_1 & T_2 & -T_3R \\ F_v & -T_6R^3 & T_5R^3 & T_3R^2 & -T_2 & T_1 & -T_2R \\ F_\theta & T_4R^2 & -T_3R^2 & -T_0R & 0 & 0 & 1 \end{array}\right]}_{T_{LR}} \underbrace{\begin{bmatrix} 1 \\ \hline U_R \\ V_R \\ Z_R \\ \hline \bar{u}_R \\ \bar{v}_R \\ \bar{\theta}_R \end{bmatrix}}_{H_R}$$

$$\underbrace{\begin{bmatrix} 1 \\ \hline U_R \\ V_R \\ Z_R \\ \hline \bar{u}_R \\ \bar{v}_R \\ \bar{\theta}_R \end{bmatrix}}_{H_R} = \underbrace{\left[\begin{array}{c|ccc|ccc} 1 & 0 & 0 & 0 & 0 & 0 & 0 \\ \hline G_U & T_1 & -T_2 & 0 & 0 & 0 & 0 \\ G_V & T_2 & T_1 & 0 & 0 & 0 & 0 \\ G_Z & -T_3R & -T_2R & 1 & 0 & 0 & 0 \\ \hline G_u & T_7R^3 & T_6R^3 & -T_4R^2 & T_1 & -T_2 & -T_3R \\ G_v & -T_6R^3 & -T_5R^3 & T_3R^2 & T_2 & T_1 & T_2R \\ G_\theta & -T_4R^2 & -T_3R^2 & T_0R & 0 & 0 & 1 \end{array}\right]}_{T_{RL}} \underbrace{\begin{bmatrix} 1 \\ \hline U_L \\ V_L \\ Z_L \\ \hline \bar{u}_L \\ \bar{v}_L \\ \bar{\theta}_L \end{bmatrix}}_{H_L}$$

where H_L, H_R are the end *state vectors* in their respective local systems and T_{LR}, T_{RL} are the *transport matrices*. The elements of the transport matrices are the *transport coefficients* T_j defined in (6.03) and the *load functions* F, G listed in (6.06–6.08) (Ref. 5, p. 68).

6.03 TRANSPORT RELATIONS

(1) Transport functions of the variable-position angle a are

$T_0(a) = a$ $T_1(a) = \cos a$ $T_2(a) = \sin a$ $T_3(a) = 1 - \cos a$	$T_4(a) = a - \sin a$ $T_5(a) = \alpha(\sin a - a\cos a)$ $T_6(a) = 1 - \cos a - \alpha a \sin a$ $T_7(a) = a - \sin a + \alpha a \cos a - \beta \sin a$
$T_{10}(a) = \frac{1}{2}a^2$ $T_{11}(a) = \sin a$ $T_{12}(a) = 1 - \cos a$ $T_{13}(a) = a - \sin a$	$T_{14}(a) = \frac{1}{2}a^2 - 1 + \cos a$ $T_{15}(a) = \alpha(2\cos a + a\sin a - 2)$ $T_{16}(a) = a - \sin a - \alpha(\sin a - a\cos a)$ $T_{17}(a) = \frac{1}{2}a^2 - 2(1 - \cos a) + \alpha a \sin a$

where $T_0(a), T_1(a), \ldots, T_7(a)$ are the basic functions and $T_{10}(a), T_{11}(a), \ldots, T_{17}(a)$ are their definite integrals in $[0, a]$. Table, graphs, and series of these functions are shown in (6.04–6.07).

(2) Coefficients T_j, A_j, B_j are values of $T_j(a)$ for a particular argument a.

For $a = a_0$, $\quad T_0(a_0) = T_0,\ T_1(a_0) = T_1, \ldots, T_{17}(a_0) = T_{17}$
For $a = a_1$, $\quad T_0(a_1) = A_0,\ T_1(a_1) = A_1, \ldots, T_{17}(a_1) = A_{17}$
For $a = a_2$, $\quad T_0(a_2) = B_0,\ T_1(a_2) = B_1, \ldots, T_{17}(a_2) = B_{17}$

(3) Load functions F, G are expressed in terms of the same loads and causes as in (4.03). Since the local systems are involved, the concentrated loads and the distributed loads are applied in the respective local systems. The abrupt displacements and the temperature strains are also related to the local systems (6.06–6.08).

(4) Relations of the state-vector components are given by the following differential equations:

$\dfrac{dU}{da} = p_t R - V$ $\dfrac{dV}{da} = p_n R + U$ $\dfrac{dZ}{da} = q_z R - VR$	$\dfrac{du}{da} = \dfrac{RU}{EA_x} - v$ $\dfrac{dv}{da} = R\theta + u$ $\dfrac{d\theta}{da} = \dfrac{RZ}{EI_z}$

where U = normal force (N), V = shearing force (N), Z = bending moment (N·m), u = axial displacement (m), v = transverse displacement (m), θ = slope (rad), p_n, p_t = intensities of normal and tangential distributed loads (N/m), q_z = intensity of distributed moment (N·m/m), and a = variable-position angle (rad) measured from the left end L.

(5) Transport matrices of this bar have analogical properties to those of (4.02) and can be used in similar ways. If the state vector is required in the segmental system, the angular transformation matrices introduced in (2.19-7) must be used.

(6) General transport matrix equations for free bars of order three are obtained by matrix composition of (6.02) and (12.02).

6.04 TABLE OF TRANSPORT COEFFICIENTS

(1) Reduced coefficients. When $a \geq \pi/4$ (45°), the effect of the normal force deformations on the transport coefficients becomes very small and may be neglected (Ref. 18). In such cases $\alpha \cong \frac{1}{2}$, and the transport coefficients become independent of the section size and can be tabulated as functions of a.

(2) Table of coefficients (a in degrees)†

a	$T_4(a)$	$T_5(a)$	$T_6(a)$	$T_7(a)$
45	0.07829	0.07587	0.01521	0.00242
51	0.11297	0.10849	0.02480	0.00448
57	0.15617	0.14842	0.03819	0.00775
63	0.20855	0.19591	0.05615	0.01264
69	0.27070	0.25100	0.07949	0.01969
75	0.34307	0.31357	0.10898	0.02950
78	0.38321	0.34755	0.12628	0.03566
81	0.42603	0.38327	0.14541	0.04276
84	0.47155	0.42064	0.16645	0.05092
87	0.51981	0.45958	0.18949	0.06023
90	0.57080	0.50000	0.21460	0.07080
93	0.62453	0.54179	0.24187	0.08274
96	0.68099	0.58483	0.27136	0.09616
99	0.74019	0.62899	0.30313	0.11119
102	0.80209	0.67414	0.33724	0.12795
105	0.86667	0.72012	0.37374	0.14655
108	0.93390	0.76677	0.41267	0.16713
111	1.00374	0.81393	0.45405	0.18981
114	1.07613	0.86141	0.49791	0.21472
117	1.15103	0.90904	0.54426	0.24199
120	1.22837	0.95661	0.59310	0.27176
123	1.30808	1.00394	0.64443	0.30415
126	1.39010	1.05081	0.69822	0.33929
129	1.47433	1.09702	0.75446	0.37731
132	1.56069	1.14236	0.81309	0.41833
135	1.64909	1.18659	0.87407	0.46249
138	1.73942	1.22952	0.93733	0.50991
141	1.83159	1.27090	1.00279	0.56069
144	1.92549	1.31053	1.07038	0.61496
147	2.02099	1.34818	1.14000	0.67281
150	2.11799	1.38362	1.21153	0.73437
153	2.21636	1.41665	1.28485	0.79972
156	2.31598	1.44703	1.35983	0.86895
159	2.41671	1.47456	1.43633	0.94214
162	2.51842	1.49903	1.51419	1.01938
165	2.62097	1.52024	1.59325	1.10073
168	2.72424	1.53800	1.67333	1.18625
171	2.82808	1.55210	1.75425	1.27598
174	2.93234	1.56238	1.83580	1.36996
177	3.03690	1.56867	1.91779	1.46823
180	3.14159	1.57080	2.00000	1.57080

†Values of $T_1(a)$, $T_2(a)$, $T_3(a)$ are available in standard mathematical tables.

6.05 GRAPHS AND SERIES OF TRANSPORT FUNCTIONS

(1) Graphs of $T_j(a)$ ($j = 4, 5, 6, 7;\ a \geq 45°$)

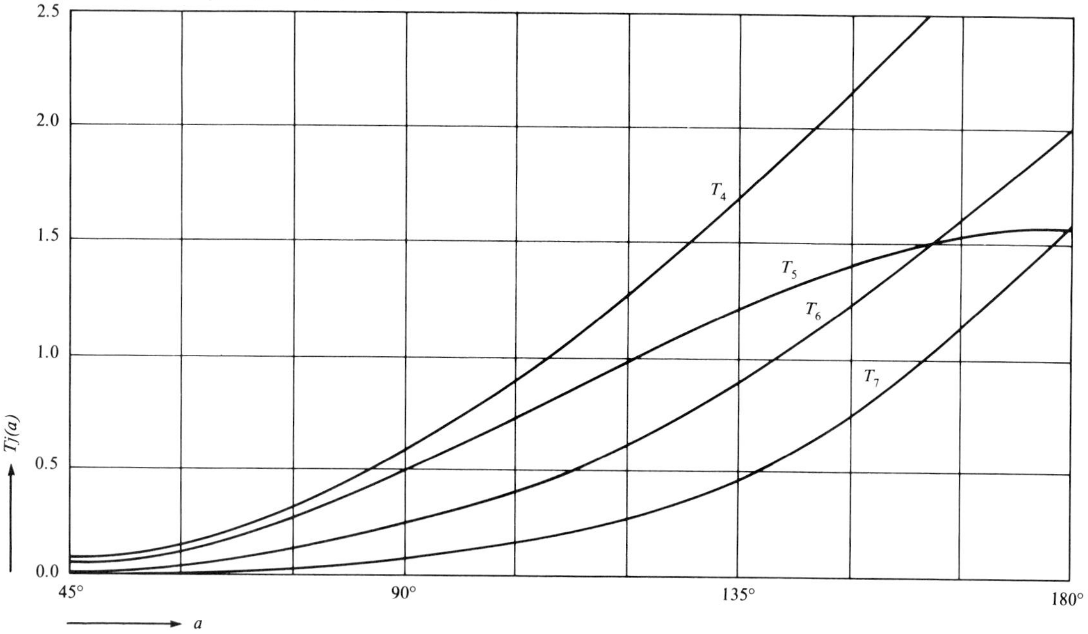

(2) Series expansion of $T_j(a)$ ($j = 1, 2, \ldots, 7);\ a \geq \pi/4$)

$$T_1(a) = 1 - \frac{a^2}{2!} + \frac{a^4}{4!} - \cdots = \sum_{k=0}^{\infty} (-1)^k \frac{a^{2k}}{(2k)!}$$

$$T_2(a) = \frac{a}{1!} - \frac{a^3}{3!} + \frac{a^5}{5!} - \cdots = \sum_{k=0}^{\infty} (-1)^k \frac{a^{2k+1}}{(2k+1)!}$$

$$T_3(a) = \frac{a^2}{2!} - \frac{a^4}{4!} + \frac{a^6}{6!} - \cdots = \sum_{k=0}^{\infty} (-1)^k \frac{a^{2k+2}}{(2k+2)!}$$

$$T_4(a) = \frac{a^3}{3!} - \frac{a^5}{5!} + \frac{a^7}{7!} - \cdots = \sum_{k=0}^{\infty} (-1)^k \frac{a^{2k+3}}{(2k+3)!}$$

$$T_5(a) = \frac{a^3}{3!} - \frac{2a^5}{5!} + \frac{3a^7}{7!} - \cdots = \sum_{k=0}^{\infty} (-1)^k \frac{(k+1)a^{2k+3}}{(2k+3)!}$$

$$T_6(a) = \frac{a^4}{4!} - \frac{2a^6}{6!} + \frac{3a^8}{8!} - \cdots = \sum_{k=0}^{\infty} (-1)^k \frac{(k+1)a^{2k+4}}{(2k+4)!}$$

$$T_7(a) = \frac{a^5}{5!} - \frac{2a^7}{7!} + \frac{3a^9}{9!} - \cdots = \sum_{k=0}^{\infty} (-1)^k \frac{(k+1)a^{2k+5}}{(2k+5)!}$$

6.06 TRANSPORT LOAD FUNCTIONS IN LOCAL SYSTEMS

Notation (6.02–6.03) P_t, P_n = concentrated loads (N) Q_z = applied couple (N·m)	Signs (6.02) p_t, p_n = intensities of uniform load (N/m) T_j, A_j, B_j = coefficents (6.03)	Matrix (6.02)
$F_U = -A_1P_t - A_2P_n$ $F_V = A_2P_t - A_1P_n$ $F_Z = (A_3P_t - A_2P_n)R$ $F_u = (A_7P_t - A_6P_n)R^3$ $F_v = (A_6P_t - A_5P_n)R^3$ $F_\theta = -(A_4P_t - A_3P_n)R^2$	1. 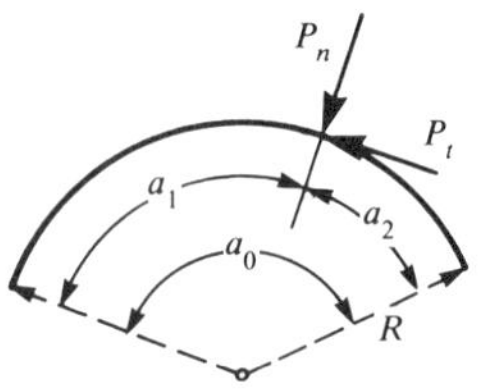	$G_U = B_1P_t - B_2P_n$ $G_V = B_2P_t + B_1P_n$ $G_Z = -(B_3P_t + B_2P_n)R$ $G_u = (B_7P_t + B_6P_n)R^3$ $G_v = -(B_6P_t + B_5P_n)R^3$ $G_\theta = -(B_4P_t + B_3P_n)R^2$
$F_U = -(T_{11} - A_{11})Rp_t$ $F_V = (T_{12} - A_{12})Rp_t$ $F_Z = (T_{13} - A_{13})R^2p_t$ $F_u = (T_{17} - A_{17})R^4P_t$ $F_v = (T_{16} - A_{16})R^4p_t$ $F_\theta = -(T_{14} - A_{14})R^3p_t$	2. 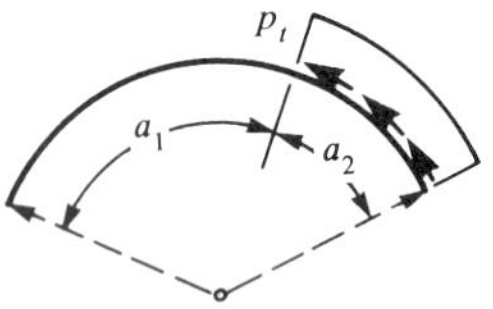	$G_U = B_{11}Rp_t$ $G_V = B_{12}Rp_t$ $G_Z = -B_{13}R^2p_t$ $G_u = B_{17}R^4p_t$ $G_v = -B_{16}R^4p_t$ $G_\theta = -B_{14}R^3p_t$
$F_U = -(T_{12} - A_{12})Rp_n$ $F_V = -(T_{11} - A_{11})Rp_n$ $F_Z = -(T_{12} - A_{12})R^2p_n$ $F_u = -(T_{16} - A_{16})R^4p_n$ $F_v = -(T_{15} - A_{15})R^4p_n$ $F_\theta = (T_{13} - A_{13})R^3p_n$	3. 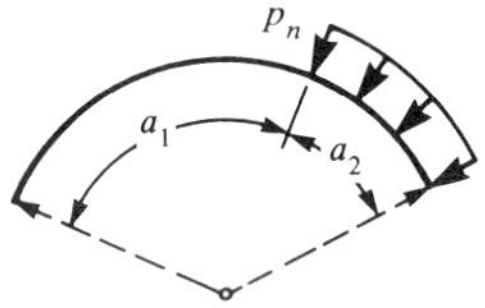	$G_U = -B_{12}Rp_n$ $G_v = B_{11}Rp_n$ $G_Z = -B_{12}R^2p_n$ $G_u = B_{16}R^4p_n$ $G_v = -B_{15}R^4p_n$ $G_\theta = -B_{13}R^3p_n$
$F_U = 0$ $F_V = 0$ $F_Z = -Q_z$ $F_u = -A_4R^2Q_z$ $F_v = -A_3R^2Q_z$ $F_\theta = A_0RQ_z$	4. 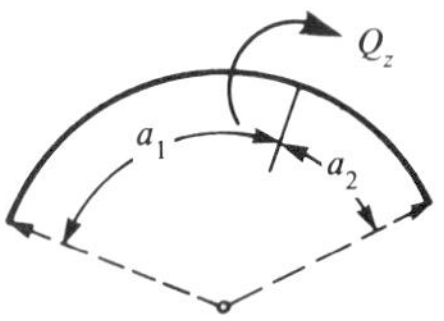	$G_U = -0$ $G_V = 0$ $G_Z = Q_z$ $G_u = -B_4R^2Q_z$ $G_v = B_3R^2Q_z$ $G_\theta = B_0RQ_z$

6.07 TRANSPORT LOAD FUNCTIONS IN LOCAL SYSTEMS

Notation (6.02–6.03)	Signs (6.02)	Matrix (6.02)
P_x, P_y = concentrated loads (N)		p_t, p_n = intensities of uniform load (N/m)
T_j, A_j, B_j = coefficients (6.03)		$S_3 = \sin a_3$ $C_3 = \cos a_3$

$F_U = -(A_1C_3 + A_2S_3)P_x$	**1.** 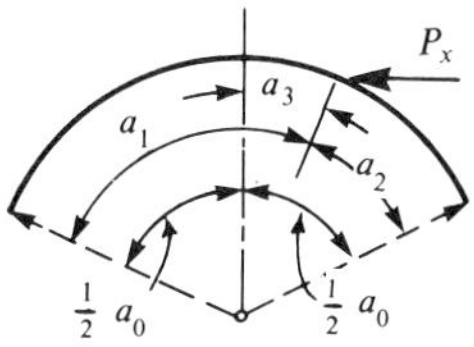	$G_U = (B_1C_3 - B_2S_3)P_x$
$F_V = (A_2C_3 - A_1S_3)P_x$		$G_V = (B_2C_3 + B_1S_3)P_x$
$F_Z = (A_3C_3 - A_2S_3)RP_x$		$G_Z = -(B_3C_3 + B_2S_3)RP_x$
$F_u = (A_7C_3 - A_6S_3)R^3P_x$		$G_u = (B_7C_3 + B_6C_3)R^3P_x$
$F_v = (A_6C_3 - A_5S_3)R^3P_x$		$G_v = -(B_6C_3 + B_5S_3)R^3P_x$
$F_\theta = -(A_4C_3 - A_3S_3)R^2P_x$		$G_\theta = -(B_4C_3 + B_3S_3)R^2P_x$
$F_U = (A_1S_3 - A_2C_3)P_y$	**2.** 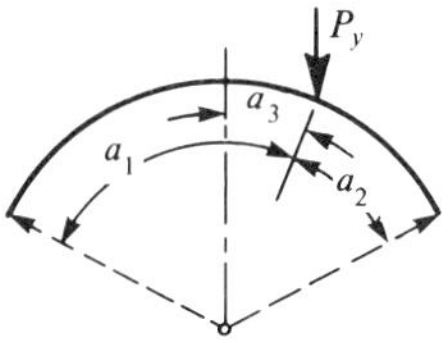	$G_U = -(B_1S_3 + B_2C_3)P_y$
$F_V = (A_2S_3 - A_1C_3)P_y$		$G_V = -(B_2S_3 - B_1C_3)P_y$
$F_Z = -(A_3S_3 + A_2C_3)RP_y$		$G_Z = (B_3S_3 - B_2C_3)RP_y$
$F_u = -(A_7S_3 + A_6C_3)R^3P_y$		$G_u = -(B_7S_3 - B_6C_3)R^3P_y$
$F_v = -(A_6S_3 + A_5C_3)R^3P_y$		$G_v = -(B_6S_3 - B_5C_3)R^3P_y$
$F_\theta = (A_4S_3 + A_3C_3)R^2P_y$		$G_\theta = (B_4S_3 - B_3C_3)R^2P_y$
$F_U = -T_{12}Rp_n$	**3.** 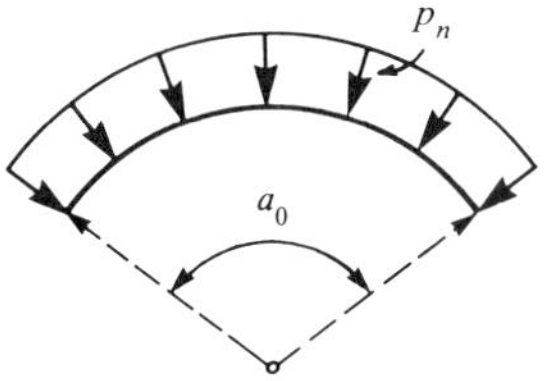	$G_U = -T_{12}Rp_n$
$F_V = -T_{11}Rp_n$		$G_V = T_{11}Rp_n$
$F_Z = -T_{12}R^2p_n$		$G_Z = -T_{12}R^2p_n$
$F_u = -T_{16}R^4p_n$		$G_u = T_{16}R^4p_n$
$F_v = -T_{15}R^4p_n$		$G_v = -T_{15}R^4p_n$
$F_\theta = T_{13}R^3p_n$		$G_\theta = -T_{13}R^3p_n$
$F_U = -T_{11}Rp_t$	**4.** 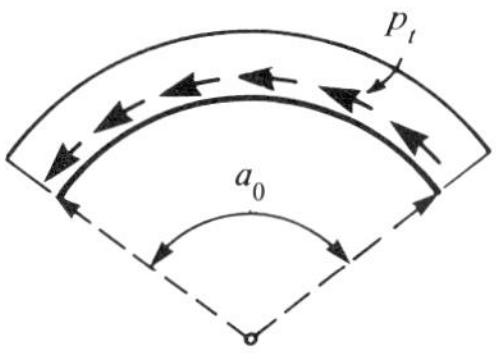	$G_U = T_{11}Rp_t$
$F_V = T_{12}Rp_t$		$G_V = T_{12}Rp_t$
$F_Z = T_{13}R^2p_t$		$G_Z = -T_{13}R^2p_t$
$F_u = T_{17}R^4p_t$		$G_u = T_{17}R^4p_t$
$F_v = T_{16}R^4p_t$		$G_v = -T_{16}R^4p_t$
$F_\theta = -T_{14}R^3p_t$		$G_\theta = -T_{14}R^3p_t$

6.08 TRANSPORT LOAD FUNCTIONS IN LOCAL SYSTEMS

<table>
<tr><td colspan="3">Notation (6.02–6.03)
u_j, v_j θ_j = abrupt displacements (4.03)
e_t, f_t = thermal strains (4.03)

Signs (6.02)

Matrix (6.02)
T_j, A_j, B_j = coefficients (6.03)</td></tr>
<tr>
<td>$F_U = -0$
$F_V = 0$
$F_Z = 0$

$F_u = -(A_1 u_j + A_2 v_j)EI_z$
$F_v = (A_2 u_j - A_1 v_j)EI_z$
$F_\theta = 0$</td>
<td>1. Abrupt linear displacements u_j, v_j

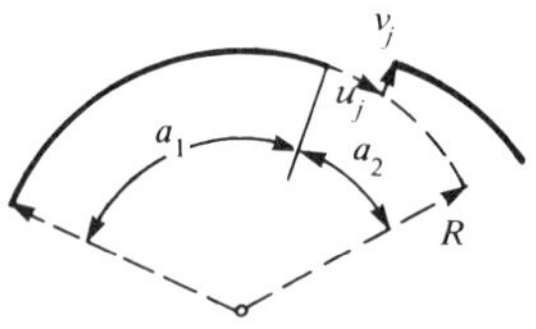

</td>
<td>$G_U = -0$
$G_V = 0$
$G_Z = 0$

$G_u = -(B_1 u_j - B_2 v_j)EI_z$
$G_v = (B_2 u_j + B_1 v_j)EI_z$
$G_\theta = 0$</td>
</tr>
<tr>
<td>$F_U = -0$
$F_V = 0$
$F_Z = 0$

$F_u = A_3 R\theta_j EI_z$
$F_v = A_2 R\theta_j EI_z$
$F_\theta = -\theta_j EI_z$</td>
<td>2. Abrupt angular displacement θ_j

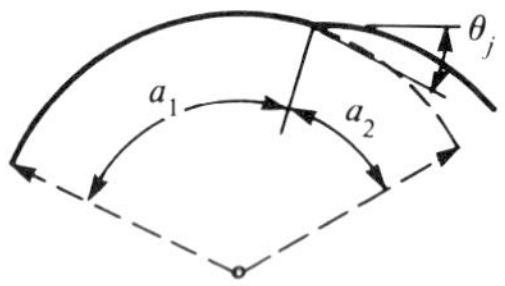

</td>
<td>$G_U = -0$
$G_V = 0$
$G_Z = 0$

$G_u = -B_3 R\theta_j EI_z$
$G_v = B_2 R\theta_j EI_z$
$G_\theta = \theta_j EI_z$</td>
</tr>
<tr>
<td>$F_U = -0$
$F_V = 0$
$F_Z = 0$

$F_u = -(T_{11} - A_{11})Re_t EI_z$
$F_v = (T_{12} - A_{12})Re_t EI_z$
$F_\theta = 0$</td>
<td>3. Linear temperature deformation of Ra_2

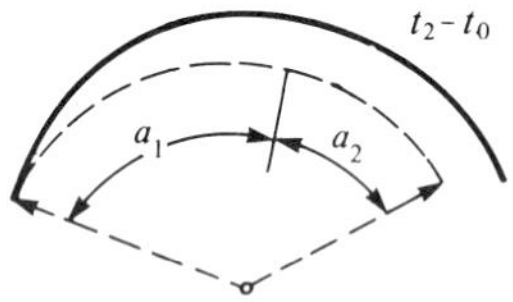

</td>
<td>$G_U = -0$
$G_V = 0$
$G_Z = 0$

$G_u = -B_{11} Re_t EI_z$
$G_v = B_{12} Re_t EI_z$
$G_\theta = 0$</td>
</tr>
<tr>
<td>$F_U = -0$
$F_V = 0$
$F_Z = 0$

$F_u = (T_{13} - A_{13})R^2 f_t EI_z$
$F_v = (T_{12} - A_{12})R^2 f_t EI_z$
$F_\theta = -Rf_t EI_z$</td>
<td>4. Angular temperature deformation of Ra_2

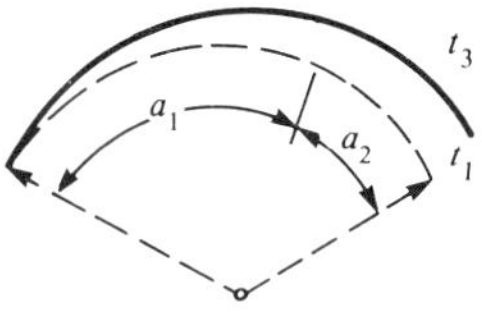

</td>
<td>$G_U = -0$
$G_V = 0$
$G_Z = 0$

$G_u = -B_{13} R^2 f_t EI_z$
$G_v = B_{12} R^2 f_t EI_z$
$G_\theta = Rf_t EI_z$</td>
</tr>
</table>

6.09 STIFFNESS MATRIX EQUATIONS IN 0 SYSTEM

(1) Equivalents and signs

$$F_1 = 2\alpha(b_0 + \sin b_0 \cos b_0) - \frac{2\sin^2 b_0}{b_0}$$

$$F_2 = 2\alpha(b_0 - \sin b_0 \cos b_0) \qquad F_6 = 2b_0$$

$$e = \sin b_0 \qquad f = \frac{\sin b_0}{b_0} - \cos b_0$$

$$b_0 = \tfrac{1}{2} a_0 = 2\tan^{-1}\frac{2h}{s}$$

$$r = \frac{I_z}{R^2 A_x}$$

$$\alpha = \tfrac{1}{2}(1+r) \qquad \beta = \tfrac{1}{2}(1-r)$$

All forces, moments, and displacements are in the 0 system, and their positive directions are shown in (2, 3) below.

(2) Free-body diagram

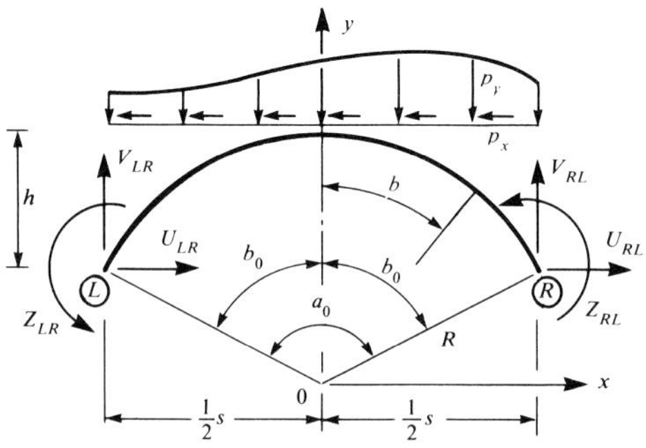

(3) Elastic curve

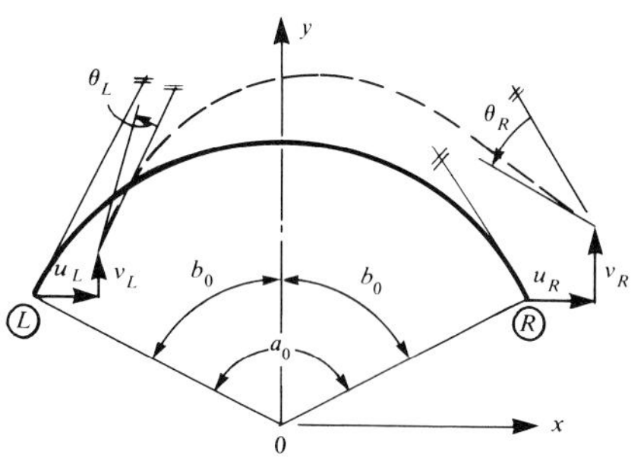

(4) Dimensionless stiffness coefficents

$$K_1 = \frac{1}{F_1} \qquad K_2 = \frac{f}{F_1} \qquad K_5 = \frac{f^2}{F_1} + \frac{e^2}{F_2} + \frac{1}{F_6}$$

$$K_3 = \frac{1}{F_2} \qquad K_4 = \frac{e}{F_2} \qquad K_6 = \frac{f^2}{F_1} - \frac{e^2}{F_2} + \frac{1}{F_6}$$

(5) Stiffness matrix equations

$$\begin{bmatrix} U_{LR} \\ V_{LR} \\ Z_{LR} \\ \hline U_{RL} \\ V_{RL} \\ Z_{RL} \end{bmatrix} = \frac{EI_z}{R^3} \left[\begin{array}{ccc|ccc} K_1 & 0 & -K_2R & -K_1 & 0 & K_2R \\ 0 & K_3 & K_4R & 0 & -K_3 & K_4R \\ -K_2R & K_4R & K_5R^2 & K_2R & -K_4R & -K_6R^2 \\ \hline -K_1 & 0 & K_2R & K_1 & 0 & -K_2R \\ 0 & -K_3 & -K_4R & 0 & K_3 & -K_4R \\ K_2R & K_4R & -K_6R^2 & -K_2R & -K_4R & K_5R^2 \end{array}\right] \begin{bmatrix} u_L \\ v_L \\ \theta_L \\ \hline u_R \\ v_R \\ \theta_R \end{bmatrix} + \begin{bmatrix} U_{L0} \\ V_{L0} \\ Z_{L0} \\ \hline U_{R0} \\ V_{R0} \\ Z_{R0} \end{bmatrix}$$

where U_{L0}, V_{L0}, Z_{L0} and U_{R0}, V_{R0}, Z_{R0} are the *reactions of the fixed-end bar* due to loads and other causes. Because of the complexity of their analytical form, only the left-side functions are tabulated in (6.12–6.17). The right-side functions can be calculated by statics. The numerical values of the stiffness coefficients $K_1, K_2, \ldots, K_6$ for bars of $2b_0 \geq \pi/4$ (45°) are tabulated in (6.10) (Ref. 4, p. 206).

(6) General stiffness matrix equations for free bars of order three are obtained by matrix composition of (6.09) and (12.11).

6.10 TABLE OF DIMENSIONLESS STIFFNESS COEFFICIENTS IN 0 SYSTEM

(1) Reduced coefficients. When $a_0 \geq \pi/4$ (45°), the effect of the normal force deformations on the stiffness coefficients becomes very small and may be neglected. In such cases $\alpha \cong \frac{1}{2}$, and the dimensionless stiffness coefficients become independent of the section size and can be tabulated as functions of a_0 (Ref. 18).

(2) Table of stiffness coefficients (a_0 in degrees)

a_0	K_1	K_2	K_3	K_4	K_5	K_6
45	2462.96237	124.66487	25.54560	9.77588	11.32432	3.84219
51	1325.54810	85.79904	17.70351	7.62156	9.95815	3.39582
57	765.49125	61.58566	12.80679	6.11087	8.87576	3.04405
63	467.75935	45.71861	9.58998	5.01075	7.99609	2.75986
69	299.38746	34.88804	7.38834	4.18480	7.26622	2.52563
75	199.19125	27.24251	5.82970	3.54889	6.65021	2.32935
78	164.54376	24.25421	5.21908	3.28447	6.37669	2.24271
81	136.95987	21.69103	4.69452	3.04885	6.12274	2.16260
84	114.80803	19.47985	4.24129	2.83798	5.88628	2.08833
87	96.87440	17.56216	3.84758	2.64850	5.66549	2.01927
90	82.24618	15.89075	3.50388	2.47762	5.45880	1.95493
93	70.23007	14.42722	3.20243	2.32296	5.26485	1.89483
96	60.29453	13.14018	2.93688	2.18253	5.08245	1.83858
99	52.02846	12.00373	2.70202	2.05463	4.91053	1.78583
102	45.11136	10.99642	2.49349	1.93781	4.74819	1.73627
105	39.29135	10.10036	2.30768	1.83081	4.59458	1.68963
108	34.36912	9.30057	2.14156	1.73256	4.44900	1.64566
111	30.18584	8.58442	1.99256	1.64212	4.31078	1.60415
114	26.61412	7.94121	1.85851	1.55868	4.17934	1.56490
117	23.55118	7.36186	1.73758	1.48153	4.05416	1.52774
120	20.91360	6.83861	1.62817	1.41004	3.93478	1.49251
123	18.63333	6.36479	1.52895	1.34367	3.82075	1.45907
126	16.65451	5.93469	1.43875	1.28193	3.71172	1.42729
129	14.93112	5.54335	1.35655	1.22440	3.60731	1.39706
132	13.42505	5.18648	1.28148	1.17079	3.50723	1.36826
135	12.10457	4.86035	1.21279	1.12047	3.41117	1.34081
138	10.94319	4.56170	1.14981	1.07344	3.31888	1.31460
141	9.91867	4.28768	1.09195	1.02931	3.23012	1.28957
144	9.01228	4.03579	1.03870	0.98786	3.14466	1.26564
147	8.20819	3.80382	0.98961	0.94886	3.06231	1.24274
150	7.49296	3.58983	0.94429	0.91211	2.98287	1.22080
153	6.85516	3.39211	0.90238	0.87745	2.90618	1.19978
156	6.28501	3.20911	0.86357	0.84470	2.83208	1.17961
159	5.77413	3.03949	0.82757	0.81372	2.76042	1.16024
162	5.31533	2.88204	0.79415	0.78437	2.69107	1.14164
165	4.90240	2.73567	0.76308	0.75655	2.62390	1.12375
168	4.52996	2.59942	0.73415	0.73013	2.55879	1.10653
171	4.19337	2.47241	0.70719	0.70501	2.49564	1.08996
174	3.88857	2.35388	0.68205	0.68111	2.43435	1.07399
177	3.61204	2.24312	0.65857	0.65834	2.37482	1.05859
180	3.36070	2.13949	0.63662	0.63662	2.31697	1.04373

6.11 UNIFORM-LOAD STIFFNESS LOAD FUNCTIONS IN 0 SYSTEM

(1) Two special cases of uniform-load stiffness load functions are given below. In both cases $\alpha \cong \frac{1}{2}$, $\beta \cong \frac{1}{2}$, and $a_0 = 2b_0$ (Ref. 18).

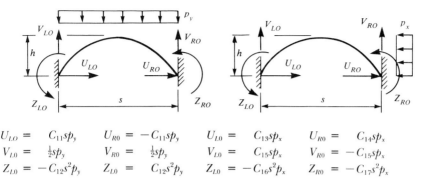

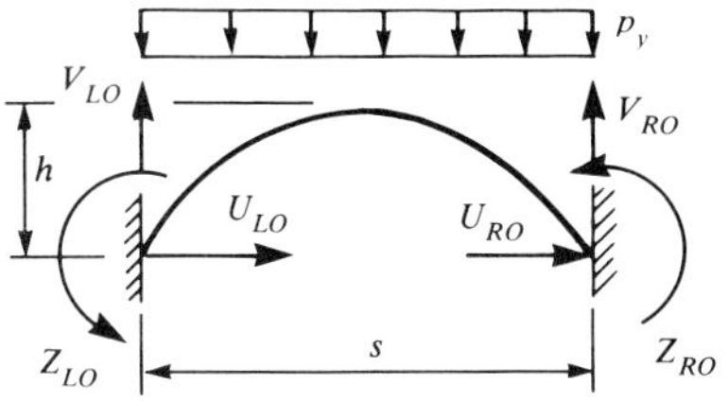

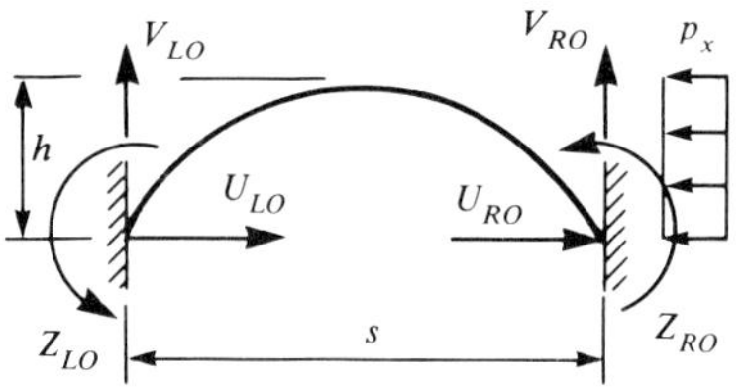

$$U_{LO} = C_{11} s p_y \qquad U_{R0} = -C_{11} s p_y \qquad U_{L0} = C_{13} s p_x \qquad U_{R0} = C_{14} s p_x$$
$$V_{LO} = \tfrac{1}{2} s p_y \qquad V_{R0} = \tfrac{1}{2} s p_y \qquad V_{L0} = C_{15} s p_x \qquad V_{R0} = -C_{15} s p_x$$
$$Z_{L0} = -C_{12} s^2 p_y \qquad Z_{L0} = C_{12} s^2 p_y \qquad Z_{L0} = -C_{16} s^2 p_x \qquad Z_{R0} = -C_{17} s^2 p_x$$

(2) Table of uniform-load stiffness coefficients (a_0 in degrees)

a_0	C_{11}	C_{12}	C_{13}	C_{14}	C_{15}	C_{16}	C_{17}
45	1.26388	0.00113	0.02134	0.07811	0.00245	0.00068	0.00181
51	1.11283	0.00146	0.02429	0.08885	0.00317	0.00089	0.00234
57	0.99332	0.00183	0.02728	0.09971	0.00398	0.00113	0.00296
63	0.89633	0.00226	0.03031	0.11071	0.00489	0.00140	0.00365
69	0.81599	0.00273	0.03339	0.12187	0.00591	0.00170	0.00443
75	0.74830	0.00326	0.03652	0.13321	0.00704	0.00205	0.00531
81	0.69045	0.00384	0.03972	0.14474	0.00829	0.00244	0.00628
87	0.64040	0.00448	0.04299	0.15649	0.00965	0.00288	0.00737
93	0.59664	0.00519	0.04633	0.16849	0.01114	0.00337	0.00856
99	0.55803	0.00596	0.04976	0.18075	0.01276	0.00392	0.00988
105	0.52368	0.00680	0.05328	0.19330	0.01452	0.00454	0.01134
111	0.49291	0.00772	0.05690	0.20617	0.01642	0.00523	0.01295
117	0.46516	0.00873	0.06063	0.21939	0.01849	0.00599	0.01472
123	0.43999	0.00982	0.06448	0.23299	0.02072	0.00685	0.01667
129	0.41703	0.01101	0.06847	0.24701	0.02313	0.00781	0.01882
135	0.39600	0.01230	0.07260	0.26149	0.02573	0.00889	0.02119
138	0.38612	0.01299	0.07472	0.26892	0.02711	0.00947	0.02246
141	0.37664	0.01371	0.07689	0.27647	0.02854	0.01009	0.02380
144	0.36753	0.01446	0.07910	0.28417	0.03003	0.01075	0.02521
147	0.35875	0.01525	0.08136	0.29201	0.03157	0.01144	0.02669
150	0.35031	0.01607	0.08367	0.30000	0.03317	0.01218	0.02825
153	0.34216	0.01693	0.08602	0.30815	0.03483	0.01296	0.02989
156	0.33431	0.01782	0.08843	0.31646	0.03656	0.01379	0.03161
159	0.32672	0.01876	0.09090	0.32495	0.03836	0.01468	0.03343
162	0.31939	0.01973	0.09342	0.33362	0.04022	0.01561	0.03535
165	0.31230	0.02076	0.09601	0.34248	0.04216	0.01661	0.03737
168	0.30544	0.02183	0.09866	0.35155	0.04417	0.01767	0.03950
171	0.29880	0.02295	0.10137	0.36082	0.04626	0.01880	0.04175
174	0.29236	0.02412	0.10416	0.37032	0.04844	0.02000	0.04413
177	0.28612	0.02535	0.10703	0.38005	0.05070	0.02128	0.04664
180	0.28006	0.02665	0.10997	0.39003	0.05305	0.02265	0.04930

6.12 STIFFNESS LOAD FUNCTIONS, FIXED-END BAR IN 0 SYSTEM

Notation (6.01–6.03, 6.09) Signs (6.09) Matrix (6.09)

P_x = concentrated load (N)

p_x = intensity of distributed load (N/m)

$K_1, K_2, \ldots$ = stiffness coefficients (6.09)

b_0, b_3 = angles (rad)

$$r = \frac{I_z}{A_x R^2}$$

$$\alpha = \tfrac{1}{2}(1 + r)$$

$$\beta = \tfrac{1}{2}(1 - r)$$

$$D_{11} = (b_0 - b_3)C_0C_3 - (S_0 - S_3)(C_0 + C_3) + (b_0 - b_3 + S_0C_0 - S_3C_3)\alpha$$

$$D_{12} = (b_0 - b_3)S_0C_3 - (S_0 - S_3)S_0 - (C_0 - C_3)C_3 + (C_0^2 - C_3^2)\alpha$$

$$D_{13} = S_0 - S_3 - (b_0 - b_3)C_3$$

$$D_{14} = b_0C_0 - (1 + C_0)S_0 + (b_0 + S_0C_0)\alpha$$

$$D_{15} = b_0S_0 + 1 - C_0 - (1 + \alpha)S_0^2$$

$$D_{16} = S_0 - b_0$$

$$D_{17} = \tfrac{1}{4}(3b_0C_0 - 3S_0 + S_0^3) - (S_0 - b_0)\alpha - (S_0C_0 - S_0 + \tfrac{2}{3}S_0^3)\beta$$

$$D_{18} = C_0^2 - S_0^2 - \tfrac{3}{4}(C_0^3 - b_0S_0) + \tfrac{1}{3}(2C_0^3 - 3C_0^2 + 1)\beta$$

$$D_{19} = -\tfrac{1}{4}(3b_0 - 4S_0 + S_0C_0)$$

$S_0 = \sin b_0$

$C_0 = \cos b_0$

$S_3 = \sin b_3$

$C_3 = \cos b_3$

1.

$$U_{L0} = (K_1D_{11} - K_2D_{13})P_x$$

$$V_{L0} = (K_3D_{12} + K_4D_{13})P_x$$

$$Z_{L0} = -(K_2D_{11} - K_4D_{12} - K_5D_{13})RP_x$$

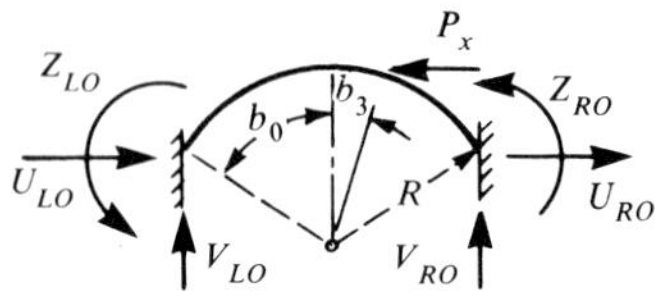

2.

$$U_{L0} = \tfrac{1}{2}P_x$$

$$V_{L0} = (K_3D_{15} + K_4D_{16})P_x$$

$$Z_{L0} = -(K_2D_{14} - K_4D_{15} - K_5D_{16})RP_x$$

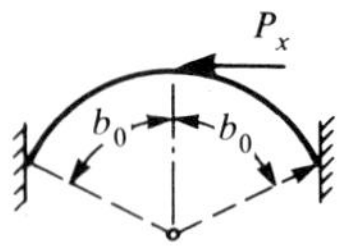

3.

$$U_{L0} = (K_1D_{17} - K_2D_{19})Rp_x$$

$$V_{L0} = (K_3D_{18} + K_4D_{19})Rp_x$$

$$Z_{L0} = -(K_2D_{17} - K_4D_{18} - K_5D_{19})R^2p_x$$

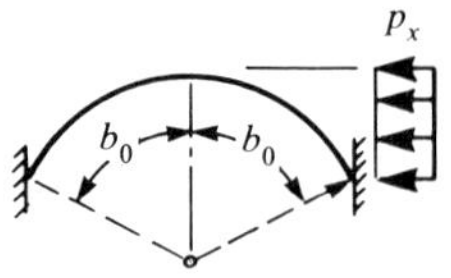

6.13 STIFFNESS LOAD FUNCTIONS, FIXED-END BAR IN 0 SYSTEM

Notation (6.01–6.03, 6.09) Signs (6.09) Matrix (6.09)

P_y = concentrated load (N) b_0, b_3 = angles (rad)

p_y = intensity of distributed load (N/m)

$K_1, K_2, \ldots$ = stiffness coefficients (6.09)

$$r = \frac{I_z}{A_x R^2}$$

$$\alpha = \tfrac{1}{2}(1 + r)$$

$$\beta = \tfrac{1}{2}(1 - r)$$

$$D_{21} = (S_0 - S_3)S_3 - (C_0 - C_3)C_0 - (b_0 - b_3)C_0S_3 + (C_0^2 - C_3^2)\alpha$$

$$D_{22} = -(S_0 - S_3)(C_0 - C_3) - (b_0 - b_3)S_0S_3 + (b_0 - b_3 - S_0C_0 + S_3C_3)\alpha$$

$$D_{23} = C_0 - C_3 + (b_0 - b_3)S_3$$

$$D_{24} = C_0 - \alpha - \beta C_0^2$$

$$D_{25} = (1 - C_0)S_0 + (b_0 - S_0C_0)\alpha$$

$$D_{26} = C_0 - 1$$

$$D_{27} = \tfrac{1}{4}(b_0 - S_0C_0)C_0 - \tfrac{1}{6}(4\alpha - 1)S_0^3$$

$$D_{28} = \tfrac{1}{4}(b_0 - S_0C_0)S_0 + \tfrac{1}{6}(4\alpha - 1)(2 - 3C_0 + C_0^3)$$

$$D_{29} = -\tfrac{1}{4}(b_0 - S_0C_0)$$

$S_0 = \sin b_0$

$C_0 = \cos b_0$

$S_3 = \sin b_3$

$C_3 = \cos b_3$

1.

$$U_{L0} = (K_1D_{21} - K_2D_{23})P_y$$

$$V_{L0} = -(K_3D_{22} + K_4D_{23})P_y$$

$$Z_{L0} = -(K_2D_{21} - K_4D_{22} - K_5D_{23})RP_y$$

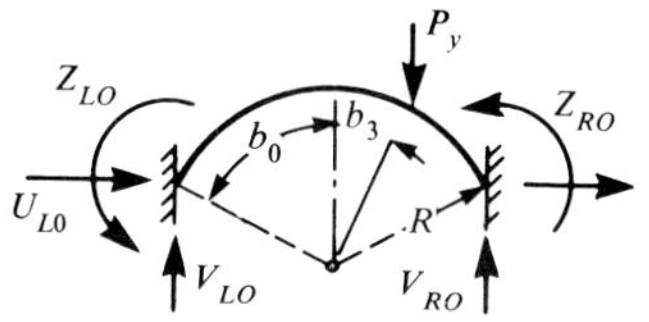

2.

$$U_{L0} = (K_1D_{24} - K_2D_{26})P_y$$

$$V_{L0} = \tfrac{1}{2}P_y$$

$$Z_{L0} = -(K_2D_{24} - K_4D_{25} - K_5D_{26})RP_y$$

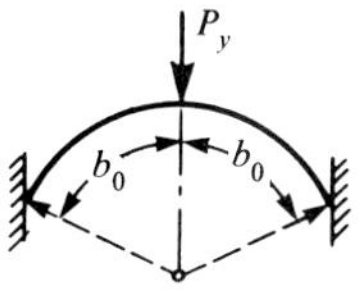

3.

$$U_{L0} = (K_1D_{27} - K_2D_{29})Rp_y$$

$$V_{L0} = (K_3D_{28} + K_4D_{29})Rp_y$$

$$Z_{L0} = -(K_2D_{27} - K_4D_{28} - K_5D_{29})R^2p_y$$

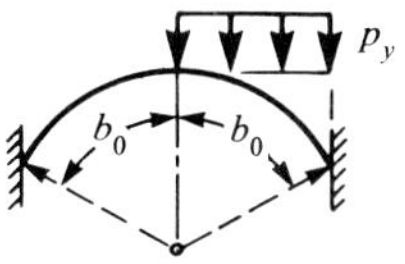

6.14 STIFFNESS LOAD FUNCTIONS, FIXED-END BAR IN 0 SYSTEM

Notation (6.02–6.03, 6.09) Signs (6.09) Matrix (6.09)

Pt = concentrated load (N)

p_t = intensity of distributed load (N/m)

$K_1, K_2, \ldots$ = stiffness coefficients (6.09)

b_0, b_2 = angles (rad)

$$r = \frac{I_z}{A_x R^2}$$

$$\alpha = \tfrac{1}{2}(1 + r)$$

$$\beta = \tfrac{1}{2}(1 - r)$$

$$D_{31} = b_2 C_0 - S_0 + S_0 C_2 - C_0 S_2 + b_2(C_0 C_2 + S_0 S_2)\alpha - C_0 S_2 \beta$$

$$D_{32} = -b_2 S_0 + C_0 - C_0 C_2 - S_0 S_2 - b_2(S_0 C_2 - C_0 S_2)\alpha + S_0 S_2 \beta$$

$$D_{33} = -(b_2 - S_2)$$

$$D_{34} = 2C_0(b_0^2 - 2S_0^2) - 2S_0(b_0 - S_0 C_0)\beta$$

$$D_{35} = -2S_0(b_0^2 - 2S_0^2) + 2C_0(b_0 - S_0 C_0)\beta$$

$$D_{36} = -2(b_0^2 - S_0^2)$$

$S_0 = \sin b_0$

$C_0 = \cos b_0$

$S_2 = \sin b_2$

$C_2 = \cos b_2$

$$D_{37} = C_0(\tfrac{1}{2}b_2^2 + 2C_2 - 2) - S_0(b_2 - S_2) - [b_2(S_0 C_2 - C_0 S_2) + S_0 S_2]\alpha$$

$$D_{38} = -S_0(\tfrac{1}{2}b_2 + 2C_2 - 2) - C_0(b_2 - S_2) - [b_2(S_0 C_2 - C_0 S_2) - C_0 S_2]\alpha$$

$$D_{39} = -(\tfrac{1}{2}b_2^2 - 1 + C_2)$$

1.

$$U_{L0} = (K_1 D_{31} - K_2 D_{33})P_t$$

$$V_{L0} = (K_3 D_{32} - K_4 D_{33})P_t$$

$$Z_{L0} = -(K_2 D_{31} - K_4 D_{32} - K_6 D_{33})RP_t$$

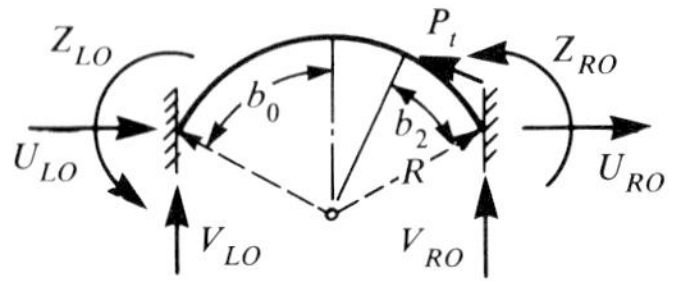

2.

$$U_{L0} = S_0 R p_t$$

$$V_{L0} = (K_3 D_{35} - K_4 D_{36})Rp_t$$

$$Z_{L0} = -(K_2 D_{34} - K_4 D_{35} - K_6 D_{36})R^2 p_t$$

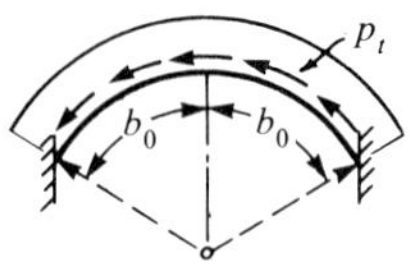

3.

$$U_{L0} = (K_1 D_{37} - K_2 D_{39})Rp_t$$

$$V_{L0} = (K_3 D_{38} - K_4 D_{39})Rp_t$$

$$Z_{L0} = -(K_2 D_{37} - K_4 D_{38} - K_6 D_{39})R^2 p_t$$

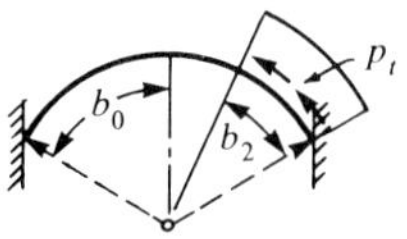

6.15 STIFFNESS LOAD FUNCTIONS, FIXED-END BAR IN 0 SYSTEM

Notation (6.01–6.03, 6.09) Signs (6.09) Matrix (6.09)

P_n = concentrated load (N)

p_n = intensity of distributed load (N/m)

$K_1, K_2, \ldots$ = stiffness coefficients (6.09)

b_0, b_2 = angles (rad)

$$r = \frac{I_z}{A_x R^2}$$

$$\alpha = \tfrac{1}{2}(1 + r)$$

$$\beta = \tfrac{1}{2}(1 - r)$$

$$D_{41} = C_0(1 - C_2) - [b_2(S_0C_2 - C_0S_2) - S_0S_2]\alpha$$

$$D_{42} = -S_0(1 - C_2) + [b_2(C_0C_2 + S_0S_2) - C_0S_2]\alpha$$

$$D_{43} = -(1 - C_2)$$

$$D_{44} = 2C_0(b_0 - S_0C_0)(1 + \alpha)$$

$$D_{45} = -2S_0(b_0 - S_0C_0) + 2[S_0C_0(S_0 - 2b_0C_0) + \alpha(C_0^2 - S_0^2)(C_0 + 2b_0S_0) + 2C_0]$$

$$D_{46} = -2(b_0 - S_0C_0)$$

$$D_{47} = C_0(b_2 - S_2) + [b_2(C_0C_2 + S_0S_2) - 2S_0C_2 - C_0S_2 + 2S_0]\alpha$$

$$D_{48} = -S_0(b_2 - S_2) - [b_2(S_0C_2 + C_0S_2) + 2C_0C_2 - S_0S_2 - 2C_0]\alpha$$

$$D_{49} = -(b_2 - S_2)$$

$S_0 = \sin b_0$

$C_0 = \cos b_0$

$S_2 = \sin b_2$

$C_2 = \cos b_2$

1.

$$U_{LO} = (K_1D_{41} - K_2D_{43})P_n$$

$$V_{LO} = (K_3D_{42} - K_4D_{43})P_n$$

$$Z_{LO} = -(K_2D_{41} - K_4D_{42} - K_6D_{43})RP_n$$

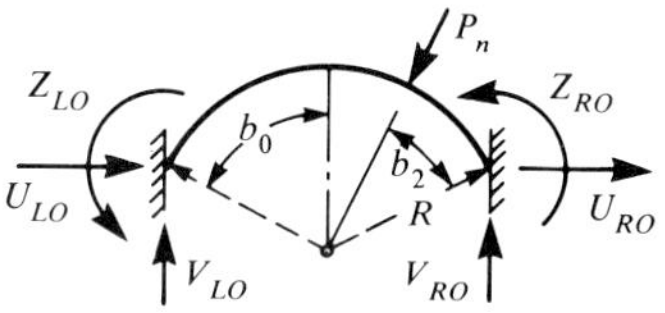

2.

$$U_{LO} = (K_1D_{44} - K_2D_{46})Rp_n$$

$$V_{LO} = S_0Rp_n$$

$$Z_{LO} = -(K_1D_{44} - K_4D_{45} - K_6D_{46})R^2p_n$$

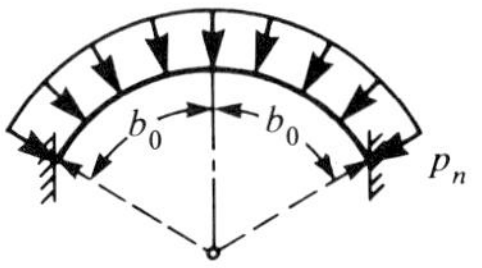

3.

$$U_{LO} = (K_1D_{47} - K_2D_{49})Rp_n$$

$$V_{LO} = (K_3D_{48} - K_4D_{49})Rp_n$$

$$Z_{LO} = -(K_1D_{47} - K_4D_{48} - K_6D_{49})R^2p_n$$

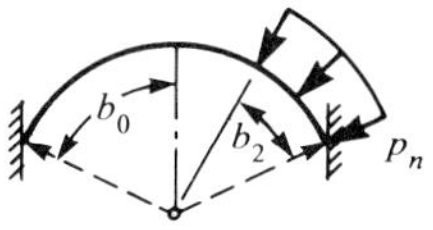

6.16 STIFFNESS LOAD FUNCTIONS, FIXED-END BAR IN 0 SYSTEM

Notation (6.01–6.03, 6.09) Signs (6.09) Matrix (6.09)

Q_z = applied couple (N·m) q_z = intensity of distributed couple (N·m/m)

$K_1, K_2, \ldots$ = stiffness coefficients (6.09) b_0, b_2, b_3 = angles (rad)

$S_0 = \sin b_0$ $C_0 = \cos b_0$ $S_2 = \sin b_2$ $C_2 = \cos b_2$ $S_3 = \sin b_3$ $C_3 = \cos b_3$

$D_{51} = S_0 - S_3 - b_2 C_0$

$D_{52} = C_0 - C_3 - b_2 S_0$

$D_{55} = C_0 - 1 + b_0 S_0 - \frac{1}{2} b_0^2 C_0$

$D_{56} = b_0 C_0 - S_0 - \frac{1}{2} b_0^2 S_0$

$D_{53} = 2b_0(S_0 - b_0 C_0)$

$D_{54} = 2(b_0 C_0 - S_0 - b_0^2 S_0)$

$D_{57} = b_2 S_0 + C_0 - C_3 - \frac{1}{2} b_2^2 C_0$

$D_{58} = b_0 C_0 - S_3 - \frac{1}{2} b_2^2 S_0$

1.

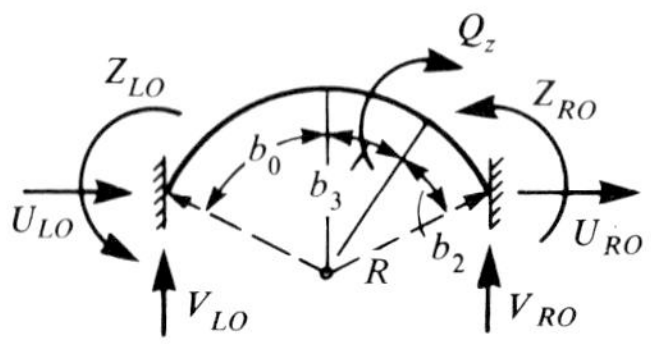

$$U_{L0} = (K_1 D_{51} - K_2 b_2) \frac{Q_z}{R}$$

$$V_{L0} = (K_3 D_{52} - K_4 b_2) \frac{Q_z}{R}$$

$$Z_{L0} = -(K_2 D_{51} - K_4 D_{52} - K_5 b_z) Q_z$$

3.

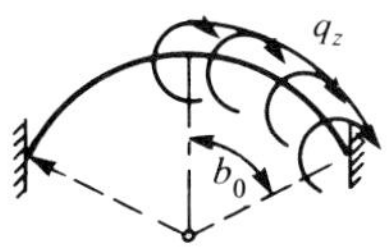

$$U_{L0} = (K_1 D_{55} - \tfrac{1}{2} K_2 b_0^2) q_z$$

$$V_{L0} = (K_3 D_{56} - \tfrac{1}{2} K_4 b_0^2) q_z$$

$$Z_{L0} = -(K_2 D_{55} - K_4 D_{56} - \tfrac{1}{2} K_5 b_0^2) R q_z$$

2.

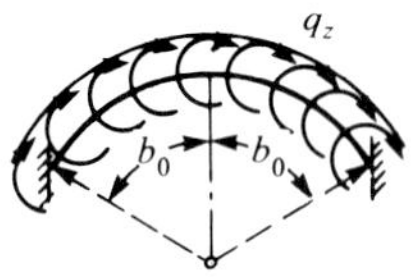

$$U_{L0} = (K_1 D_{53} - K_2 b_0^2) q_z$$

$$V_{L0} = (K_3 D_{54} - K_4 b_0^2) q_z$$

$$Z_{L0} = -(K_2 D_{53} - K_4 D_{54} - K_5 b_0^2) R q_z$$

4.

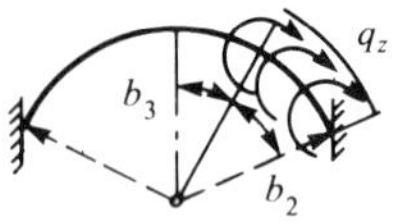

$$U_{L0} = (K_1 D_{57} - \tfrac{1}{2} K_2 b_2^2) q_z$$

$$V_{L0} = (K_3 D_{58} - \tfrac{1}{2} K_4 b_2^2) q_z$$

$$Z_{L0} = -(K_2 D_{57} - K_4 D_{58} - \tfrac{1}{2} K_5 b_2^2) R q_z$$

6.17 STIFFNESS LOAD FUNCTIONS, FIXED-END BAR IN 0 SYSTEM

Notation (6.01–6.03, 6.09) — Signs (6.09) — Matrix (6.09)

u_j, v_j, θ_j = abrupt displacements (4.03) — e_t, f_t = temperature strains (4.03)

$K_1, K_2, \ldots$ = stiffness coefficients (6.09) — b_0, b_2, b_3 = angles (rad)

$S_0 = \sin b_0$ $\quad C_0 = \cos b_0$ $\quad S_2 = \sin b_2$ $\quad C_2 = \cos b_2$ $\quad S_3 = \sin b_3$ $\quad C_3 = \cos b_3$

$$D_{61} = (C_3 u_j + S_3 v_j)R^{-3}$$

$$D_{62} = -(S_3 u_j - C_3 v_j)R^{-3}$$

$$D_{65} = (C_3 - C_0)\theta_j R^{-2}$$

$$D_{66} = (S_0 + S_3)\theta_j R^{-2}$$

$$D_{63} = (S_0 - S_3)e_t R^{-2}$$

$$D_{64} = (C_0 - C_3)e_t R^{-2}$$

$$D_{67} = (S_0 - S_3 - b_2 C_0)f_t R^{-1}$$

$$D_{68} = (C_0 - C_3 + b_2 S_0)f_t R^{-1}$$

1. Abrupt displacements u_j, v_j

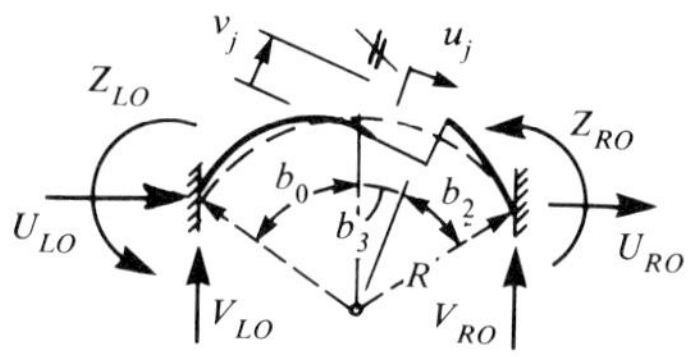

$$U_{L0} = EI_z K_1 D_{61}$$

$$V_{L0} = EI_z K_3 D_{62}$$

$$Z_{L0} = -EI_z(K_2 D_{61} - K_4 D_{62})R$$

2. Linear temperature deformation of segment Rb_2

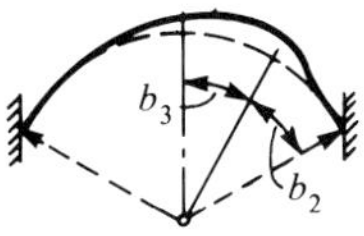

$$U_{L0} = EI_z K_1 D_{63}$$

$$V_{L0} = EI_z K_3 D_{64}$$

$$Z_{L0} = -EI_z(K_2 D_{63} - K_4 D_{64})R$$

3. Abrupt displacement θ_j

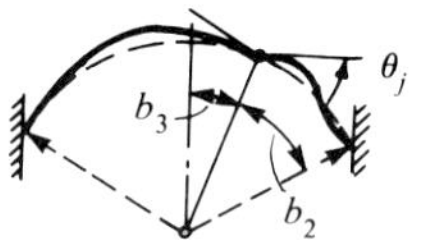

$$U_{L0} = EI_z(K_1 D_{65} - K_2 \theta_j R^{-2})$$

$$V_{L0} = EI_z(K_3 D_{66} - K_4 \theta_j R^{-2})$$

$$Z_{L0} = -EI_z(K_2 D_{65} - K_4 D_{66} - K_6 \theta_j R^{-2})R$$

4. Angular temperature deformation of segment Rb_2

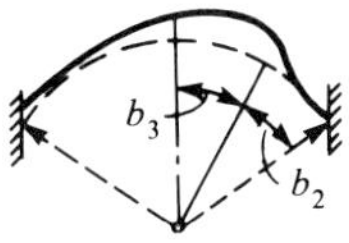

$$U_{L0} = EI_z(K_1 D_{67} - K_2 b_2 f_t R^{-1})$$

$$V_{L0} = EI_z(K_3 D_{68} - K_4 b_2 f_t R^{-1})$$

$$Z_{L0} = -EI_z(K_2 D_{67} - K_4 D_{68} - K_6 b_2 f_t R^{-1})R$$

6.18 STIFFNESS LOAD FUNCTIONS, TWO-HINGED BAR IN 0 SYSTEM

Notation (6.01–6.03, 6.09) Signs (6.09) r, α, β (6.09)

P_x, P_y = concentrated loads (N) p_x, p_y = intensities of distributed load (N/m)

b_0, b_3 = angles (rad) $S_0 = \sin b_0$ $C_0 = \cos b_0$ $S_3 = \sin b_3$ $C_3 = \cos b_3$

$D_{00} = 2[b_0 C_0^2 - 2S_0 C_0 + (b_0 + S_0 C_0)\alpha]$ $D_{10} = 2(S_0 - b_0 C_0)S_0$

$D_{11}, D_{17}, D_{21}, D_{24}, D_{27}$ = flexibility load functions (6.12–6.13)

1.

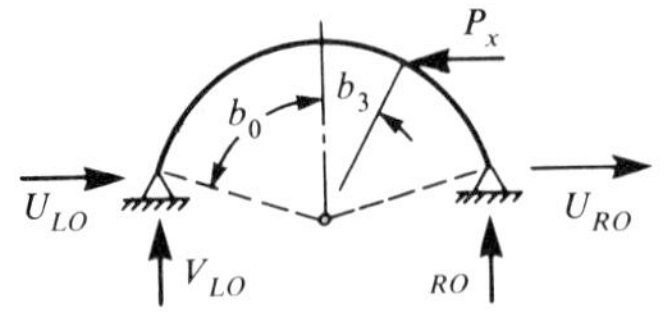

$V_{L0} = \dfrac{C_3 - C_0}{2S_0} P_x \qquad U_{L0} = \dfrac{D_{11}P_x + D_{10}V_{L0}}{D_{00}}$

2.

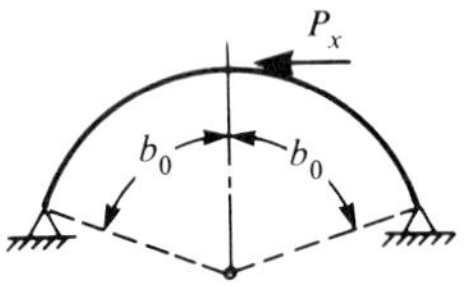

$V_{L0} = \dfrac{l - C_0}{2S_0} P_x \qquad U_{L0} = \tfrac{1}{2}P_x$

3.

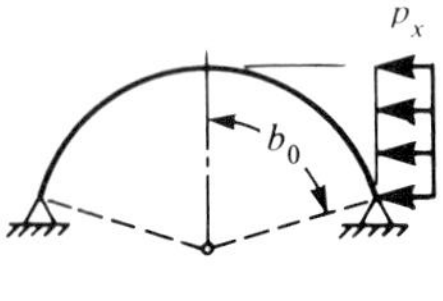

$V_{L0} = \dfrac{(1 - C_0)^2}{2S_0} Rp_x \qquad U_{L0} = \dfrac{D_{17}Rp_x + D_{10}V_{L0}}{D_{00}}$

4.

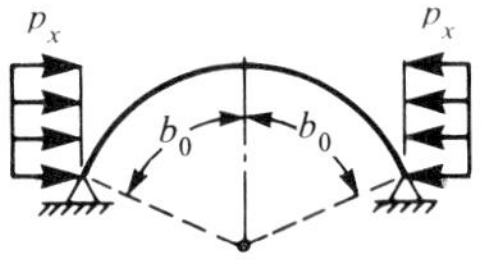

$V_{L0} = 0 \qquad U_{L0} = \left(\dfrac{2D_{17}}{D_{00}} - 1 + C_0\right) Rp_x$

5.

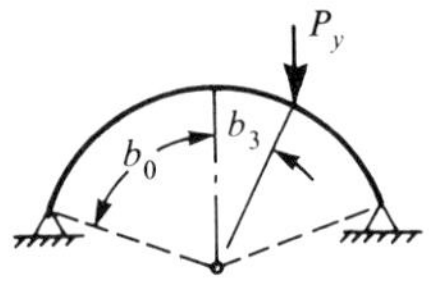

$V_{L0} = \dfrac{S_0 - S_3}{2S_0} P_y \qquad U_{L0} = \dfrac{D_{21}P_y + D_{10}V_{L0}}{D_{00}}$

6.

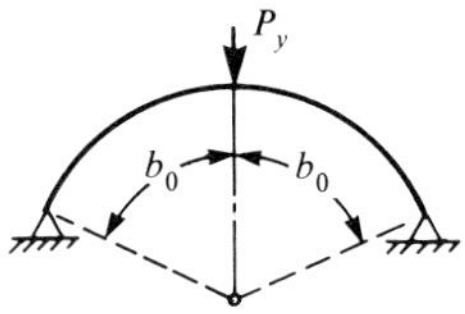

$V_{L0} = \tfrac{1}{2}P_y \qquad U_{L0} = \dfrac{D_{24} + \frac{1}{2}D_{10}}{D_{00}} P_y$

7.

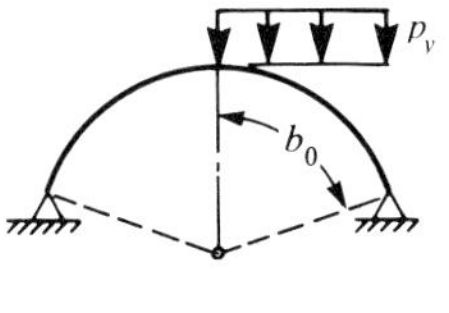

$V_{L0} = \dfrac{S_0}{4} Rp_y \qquad U_{L0} = \dfrac{D_{27}Rp_y + D_{10}V_{L0}}{D_{00}}$

8.

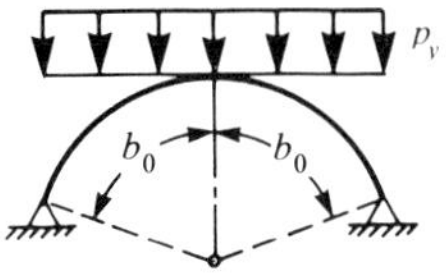

$V_{L0} = RS_0 p_y \qquad U_{L0} = \dfrac{2D_{27} + \frac{1}{2}D_{10}S_0}{D_{00}} Rp_y$

6.19 STIFFNESS LOAD FUNCTIONS, TWO-HINGED BAR IN 0 SYSTEM

Notation (6.01–6.03, 6.09) Signs (6.09) r, α, β (6.09)

P_t, P_n = concentrated loads (N) p_t, p_n = intensities of distributed load (N/m)

b_0, b_2 = angles (rad) $S_0 = \sin b_0$ $C_0 = \cos b_0$ $S_2 = \sin b_2$ $C_2 = \cos b_2$

$$D_{00} = 2[b_0C_0^2 - 2S_0C_0 + (b_0 + S_0C_0)\alpha] \qquad D_{10} = 2(S_0 - b_0C_0)S_0$$

$D_{31}, D_{34}, D_{37}, D_{41}, D_{44}, D_{47}$ = transport load functions (6.14–6.15)

1.

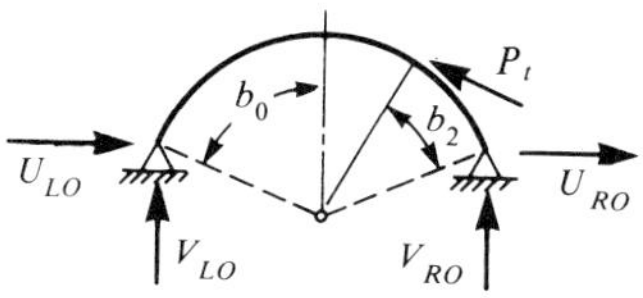

$$V_{L0} = \frac{1 - C_2}{2S_0} P_t \qquad U_{L0} = \frac{D_{31}P_t + D_{10}V_{L0}}{D_{00}}$$

2.

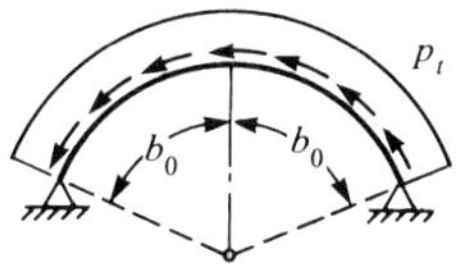

$$V_{L0} = \frac{b_0 - S_0C_0}{S_0} Rp_t \qquad U_{L0} = \frac{D_{34}Rp_t + D_{10}V_{L0}}{D_{00}}$$

3.

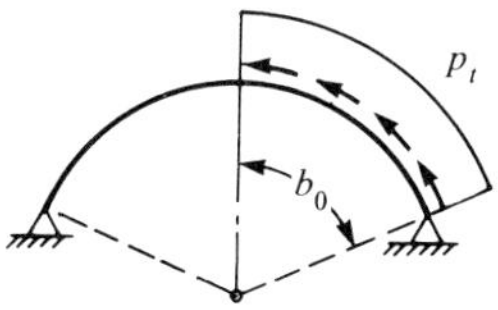

$$V_{L0} = \frac{b_0 - S_0}{2S_0} Rp_t \qquad U_{L0} = \frac{\frac{1}{2}D_{34}Rp_t + D_{10}V_{L0}}{D_{00}}$$

4.

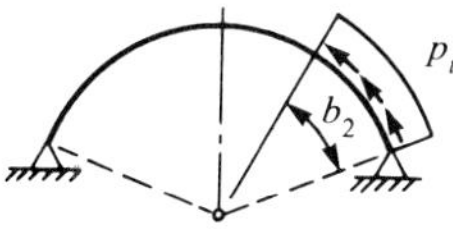

$$V_{L0} = \frac{b_2 - S_2}{2S_0} Rp_t \qquad U_{L0} = \frac{D_{37}Rp_t + D_{10}V_{L0}}{D_{00}}$$

5.

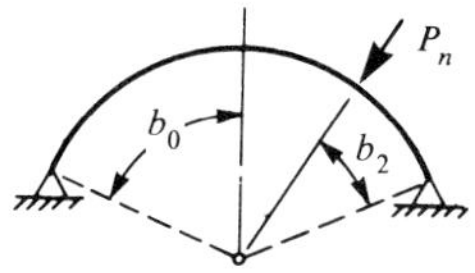

$$V_{L0} = \frac{S_2}{2S_0} P_n \qquad U_{L0} = \frac{D_{41}P_n + D_{10}V_{L0}}{D_{00}}$$

6.

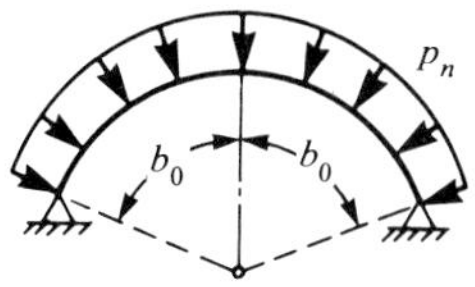

$$V_{L0} = RS_0p_n \qquad U_{L0} = \frac{D_{44} + D_{10}S_0}{D_{00}} Rp_n$$

7.

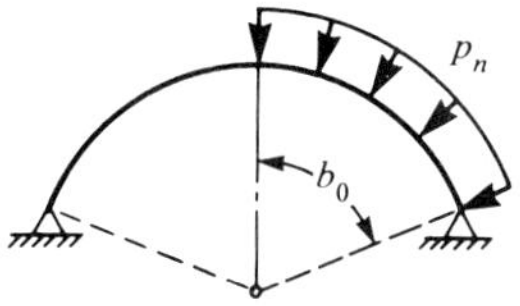

$$V_{L0} = \tfrac{1}{2}RS_0p_n \qquad U_{L0} = \frac{D_{44} + D_{10}S_0}{2D_{00}} Rp_n$$

8.

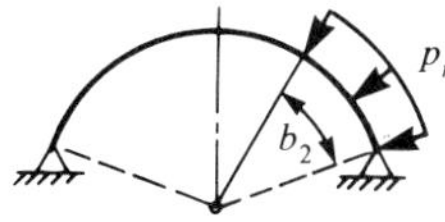

$$V_{L0} = \frac{S_2^2}{S_0} Rp_n \qquad U_{L0} = \frac{D_{47}Rp_n + D_{10}V_{L0}}{D_{00}}$$

6.20 STIFFNESS LOAD FUNCTIONS, TWO-HINGED BAR IN 0 SYSTEM

Notation (6.01–6.03, 6.09) Signs (6.09) r, α, β (6.09)

Q_z = applied moment (N·m) q_z = intensity of distributed moment (N·m/m)

u_j, v_j, θ_j = abrupt displacements (4.03) e_t, f_t = temperature strains (4.03)

For definition of $b_0, b_2, b_3, S_0, C_0, S_2, C_2, S_3, C_3$ refer to (6.18–6.19)

$D_{00} = 2[b_0 C_0^2 - 2S_0 C_0 + (b_0 + S_0 C_0)\alpha]$ $D_{10} = 2(S_0 - b_0 C_0) S_0$

$D_{51}, D_{55}, D_{57}, D_{61}, D_{63}, D_{65}, D_{67}$ = transport load functions (6.16–6.17)

1.

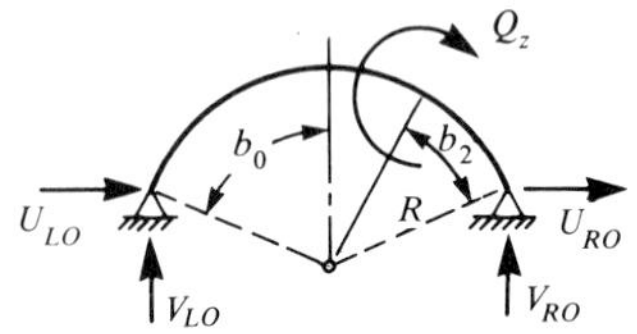

$$V_{L0} = -\frac{Q_z}{2RS_0} \qquad U_{L0} = \frac{D_{51}R^{-1}Q_z + D_{10}V_{L0}}{D_{00}}$$

2.

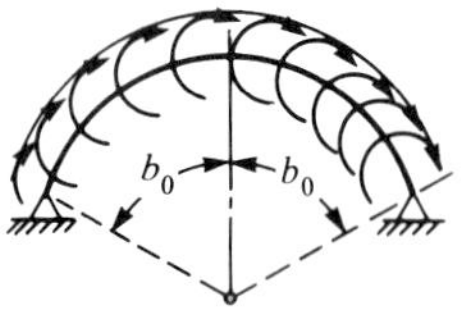

$$V_{L0} = -\frac{b_0 q_z}{S_0} \qquad U_{L0} = \frac{D_{53}q_z + D_{10}V_{L0}}{D_{00}}$$

3.

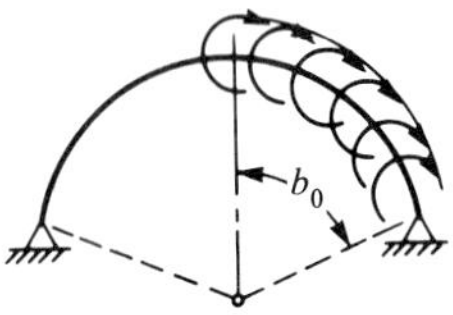

$$V_{L0} = -\frac{b_0 q_z}{2S_0} \qquad U_{L0} = \frac{D_{55}q_z + D_{10}V_{L0}}{D_{00}}$$

4.

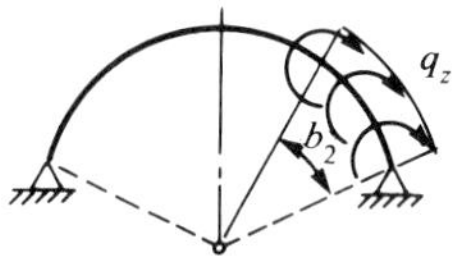

$$V_{L0} = -\frac{b_2 q_z}{2S_0} \qquad U_{L0} = \frac{D_{57}q_z + D_{10}V_{L0}}{D_{00}}$$

5. Abrupt displacements u_j, v_j

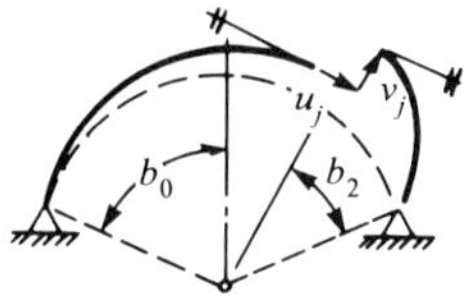

$$V_{L0} = 0 \qquad U_{L0} = \frac{EI_z D_{61}}{D_{00}}$$

6. Linear temperature deformation of segment Rb_2

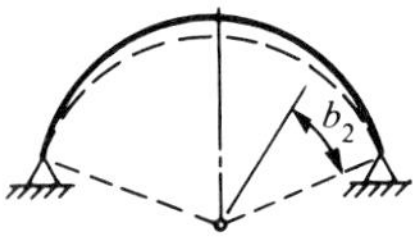

$$V_{L0} = 0 \qquad U_{L0} = \frac{EI_z D_{63}}{D_{00}}$$

7. Abrupt displacement θ_j

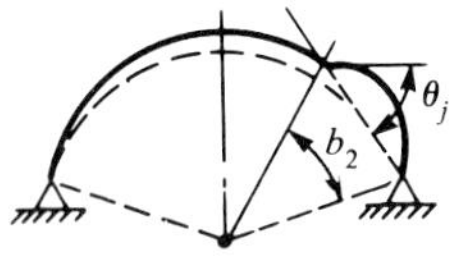

$$V_{L0} = 0 \qquad U_{L0} = \frac{EI_z D_{65}}{D_{00}}$$

8. Angular temperature deformation of segment Rb_2

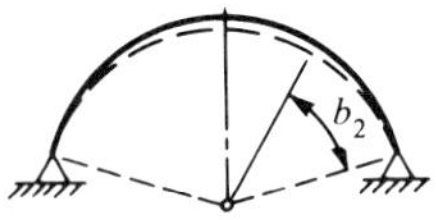

$$V_{L0} = 0 \qquad U_{L0} = \frac{EI_z D_{67}}{D_{00}}$$

7

Free Parabolic Bar of Order One Constant Section

STATIC STATE

7.01 DEFINITION OF STATE

(1) System considered is a finite symmetrical parabolic bar of constant and symmetrical section. The bar and its loads form a planar system (system of order one). The system plane is the plane of symmetry of the bar section, and the centroidal axis of the bar is a 2° parabola.

(2) Notation (A.1, A.2, A.4)

s	= span (m)	A_x	= area of normal section (m^2)
a, b, c	= segments (m)	I_z	= moment of inertia of A_y about z (m^4)
h	= height (m)	R	= radius of curvature at vertex (m)
x, y	= coordinates (m)	E	= modulus of elasticity (Pa)
U	= normal force (N)	u	= linear displacement along x (m)
V	= shearing force (N)	v	= linear displacement along y (m)
Z	= flexural moment (N·m)	θ	= flexural slope about z (rad)

$R = \dfrac{c^2}{2h}$	$s = 2c$	$e = \dfrac{2h}{c}$	$y = h\left[1 - \left(\dfrac{x}{c}\right)^2\right]$

(3) Assumptions of analysis are stated in (1.01). In this chapter, the effects of axial and flexural deformations are included. Since only bars of large radii of curvature are considered, the shearing deformation effects are too small and as such are neglected.

7.02 TRANSPORT MATRIX EQUATIONS IN 0 SYSTEM

(1) Equivalents and signs

$\bar{u} = uEI_z =$ scaled u (N·m³) $m = a/R$

$\bar{v} = vEI_z =$ scaled v (N·m³) $n = b/R$

$\bar{\theta} = \theta EI_z =$ scaled θ (N·m²·rad) $e = c/R = 2h/c$

$$R = \frac{c^2}{2h} \qquad r = \frac{I_z}{A_x R^2}$$

All forces, moments, and displacements are in the 0 system as shown in (2, 3) below. Their positive directions are indicated in the same figures (Ref. 19).

(2) Free-body diagram

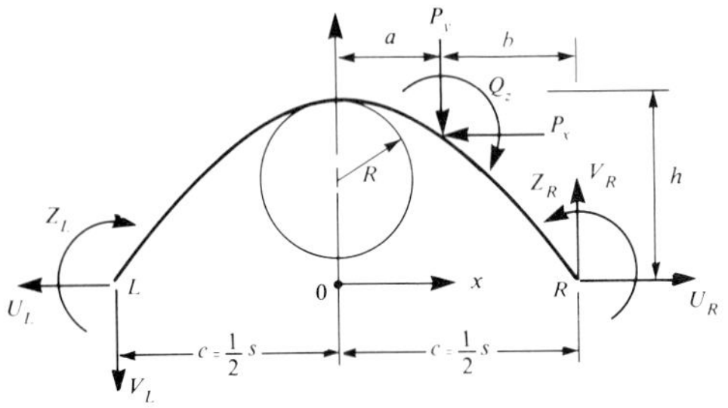

(3) Elastic curve

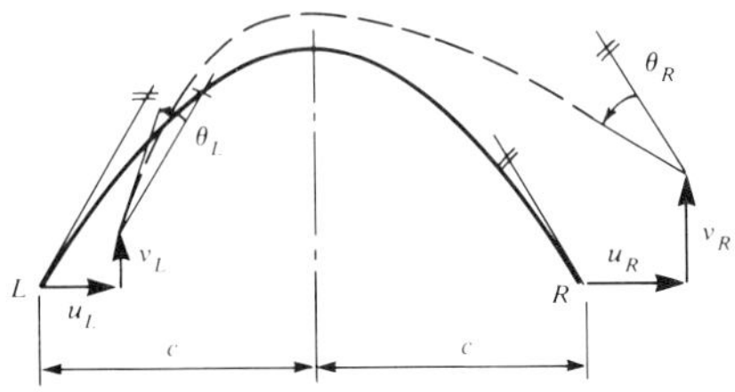

(4) Transport matrix equations

$$\underbrace{\begin{bmatrix} 1 \\ \hline U_L \\ V_L \\ Z_L \\ \hline \bar{u}_L \\ \bar{v}_L \\ \bar{\theta}_L \end{bmatrix}}_{H_L} = \underbrace{\left[\begin{array}{c|ccc|ccc} 1 & 0 & 0 & 0 & 0 & 0 & 0 \\ \hline F_U & 1 & 0 & 0 & 0 & 0 & 0 \\ F_V & 0 & 1 & 0 & 0 & 0 & 0 \\ F_Z & 0 & s & 1 & 0 & 0 & 0 \\ \hline F_u & -T_1R^3 & -T_2R^3 & -T_3R^2 & 1 & 0 & 0 \\ F_v & T_2R^3 & T_5R^3 & T_4R^2 & 0 & 1 & -s \\ F_\theta & -T_3R^2 & -T_4R^2 & -T_0R & 0 & 0 & 1 \end{array}\right]}_{T_{LR}} \underbrace{\begin{bmatrix} 1 \\ \hline U_R \\ V_R \\ Z_R \\ \hline \bar{u}_R \\ \bar{v}_R \\ \bar{\theta}_R \end{bmatrix}}_{H_R}$$

$$\underbrace{\begin{bmatrix} 1 \\ \hline U_R \\ V_R \\ Z_R \\ \hline \bar{u}_R \\ \bar{v}_R \\ \bar{\theta}_R \end{bmatrix}}_{H_R} = \underbrace{\left[\begin{array}{c|ccc|ccc} 1 & 0 & 0 & 0 & 0 & 0 & 0 \\ \hline G_U & 1 & 0 & 0 & 0 & 0 & 0 \\ G_V & 0 & 1 & 0 & 0 & 0 & 0 \\ G_Z & 0 & -s & 1 & 0 & 0 & 0 \\ \hline G_u & T_1R^3 & -T_2R^3 & T_3R^2 & 1 & 0 & 0 \\ G_v & T_2R^3 & -T_5R^3 & T_4R^2 & 0 & 1 & s \\ G_\theta & T_3R^2 & -T_4R^2 & T_0R & 0 & 0 & 1 \end{array}\right]}_{T_{RL}} \underbrace{\begin{bmatrix} 1 \\ \hline U_L \\ V_L \\ Z_L \\ \hline \bar{u}_L \\ \bar{v}_L \\ \bar{\theta}_L \end{bmatrix}}_{H_L}$$

where H_L, H_R are the *state vectors* of the respective ends and T_{LR}, T_{RL} are their *transport matrices*. The elements of the transport matrices are the *transport coefficients* defined in (7.03) and the *load functions* F, G discussed in (7.03). The right-side load functions G are tabulated for particular cases in (7.04–7.06).

7.03 TRANSPORT RELATIONS IN 0 SYSTEM

(1) Equivalents used in the transport coefficients and load functions are tabulated below.

$S_e = \sinh^{-1} e$	$L_e = \sqrt{1 + e^2}$ $S_m = \sinh^{-1} m$	$L_m = \sqrt{1 + m^2}$
$C_0 = \frac{1}{2}(L_e e + S_e)$	$A_0 = \frac{1}{2}(L_e e + L_m m + S_e + S_m)$	$B_0 = \frac{1}{2}(L_e e - L_m m + S_e - S_m)$
$C_1 = \frac{1}{3}(L_e^3 - 1)$	$A_1 = -\frac{1}{3}(L_e^3 - L_m^3)$	$B_1 = \frac{1}{3}(L_e^3 - L_m^3)$
$C_2 = \frac{1}{4}(L_e^3 e - C_0)$	$A_2 = \frac{1}{4}(L_e^3 e + L_m^3 m - A_0)$	$B_2 = \frac{1}{4}(L_e^3 - L_m^3 m - B_0)$
$C_3 = \frac{1}{5}(L_e^3 e^2 - 2C_1)$	$A_3 = \frac{1}{5}(L_e^3 e^2 - L_m^3 m^2 + 2A_1)$	$B_3 = \frac{1}{5}(L_e^3 e^2 - L_m^3 m^2 - 2B_1)$
$C_4 = \frac{1}{6}(L_e^3 e^3 - 3C_2)$	$A_4 = \frac{1}{6}(L_e^3 e^3 + L_m^3 m^4 + 4A_2)$	$B_4 = \frac{1}{6}(L_e^3 e^3 - L_m^3 m^3 - 3B_2)$
$C_5 = \frac{1}{7}(L_e^3 e^4 - 4C_3)$	$A_5 = \frac{1}{7}(L_e^3 e^4 - L_m^3 m^4 + 4A_3)$	$B_5 = \frac{1}{7}(L_e^3 e^4 - L_m^3 m^4 - 4B_3)$
$C_6 = \frac{1}{8}(L_e^3 e^5 - 5C_4)$	$A_6 = \frac{1}{8}(L_e^3 e^5 + L_m^3 m^5 - 5A_4)$	$B_6 = \frac{1}{8}(L_e^3 e^5 - L_m^3 m^5 - 5B_4)$
$C_{-1} = S_e$	$A_{-1} = S_e + S_m$	$B_{-1} = S_e - S_m$
$C_{-2} = L_e - 1$	$A_{-2} = -L_e + L_m$	$B_{-2} = L_e - L_m$
$C_{-3} = C_0 - C_{-1}$	$A_{-3} = A_0 - A_{-1}$	$B_{-3} = B_0 - B_{-1}$
$C_{-4} = C_1 - C_{-2}$	$A_{-4} = A_1 - A_{-2}$	$B_{-4} = B_1 - B_{-2}$
$C_{-5} = C_2 - C_{-3}$	$A_{-5} = A_2 - A_{-3}$	$B_{-5} = B_2 - B_{-3}$

(2) Transport coefficients T_j expressed in terms of the equivalents listed above are

$T_0 = 2C_0$	$T_3 = e^2 C_0 - C_2$
$T_1 = \frac{1}{2}(e^4 C_0 - 2e^2 C_2 + C_4) + 2rC_{-1}$	$T_4 = 2eC_0$
$T_2 = e(e^2 C_0 - C_2)$	$T_5 = 2e^2 C_0 - 2C_2 - 2r(C_0 - C_{-1})$

(3) Load functions in parabolic bars of constant section are complex expressions, and for lack of space only the right-side functions are tabulated in (7.04–7.06). The left-side load functions (if needed) can be derived by the transport-shift theorem as

$$F_U = -G_U \qquad F_u = R^3(T_1 G_U + T_2 G_V) + R^2 T_3 G_Z - G_u$$

$$F_V = -G_V \qquad F_v = -R^3(T_2 G_U + T_5 G_V) - R^2 T_4 G_Z - G_v + sG_\theta$$

$$F_Z = -sG_V - G_Z \qquad F_\theta = R^2(T_3 G_U + T_4 G_V) + RT_0 G_Z - G_\theta$$

where $G_U, G_V, \ldots, G_\theta$ are the right-side load functions in the respective tables. The temperature strains used in (7.06) have the same meaning as in (6.03).

(4) Transport matrices of this bar have analogical properties to those of (6.02) and can be used in similar ways.

7.04 TRANSPORT LOAD FUNCTIONS IN 0 SYSTEM

Notation (7.01–7.03) Signs (7.02) Matrix (7.02)

P_x = concentrated load (N) p_x = intensity of distributed load (N/m)

B_j, C_j = equivalents (7.03) $R = \dfrac{c^2}{2h}$ $m = \dfrac{a}{R}$ $n = \dfrac{b}{R}$ $e = \dfrac{c}{R}$ $r = \dfrac{I_z}{A_x R^2}$

$$G_{11} = -\tfrac{1}{4}[-e^2m^2B_0 + (e^2 + m^2)B_2 - B_4 - 4rB_{-1}]$$
$$G_{12} = -\tfrac{1}{2}(-em^2B_0 + m^2B_1 + eB_2 - B_3 - 2rB_{-2})$$
$$G_{13} = -\tfrac{1}{2}(B_2 - m^2B_0)$$

$$G_{14} = -\tfrac{1}{16}(e^2C_4 - C_6 - 8rC_{-3})$$
$$G_{15} = -\tfrac{1}{8}(C_4 - C_5 - 4rC_{-2})$$
$$G_{16} = -\tfrac{1}{8}C_4$$

$$G_{17} = -\tfrac{1}{16}[e^2m^2B_0 - m^2(2e^2 + m^2)B_2 + (2m^2 + e^2)B_4 - B_6 - 8r(B_{-3} - m^2B_{-1})]$$
$$G_{18} = -\tfrac{1}{8}[\text{em}^4B_0 - m^4B_1 - 2em^2B_3 + eB_4 - B_5 - 4r(B_{-4} - m^2B_{-2})]$$
$$G_{19} = -\tfrac{1}{8}(m^2B_0 - 2m^2B_2 + B_4)$$

1.

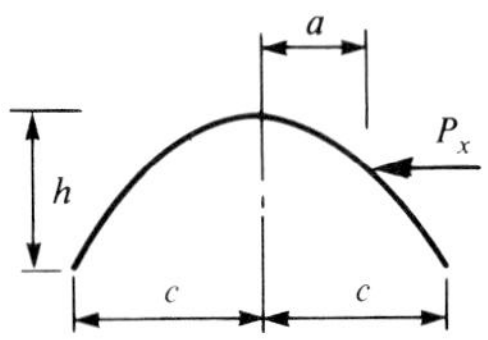

$G_U = P_x$ $G_u = G_{11}R^3P_x$

$G_V = 0$ $G_v = G_{12}R^3P_x$

$G_Z = -\tfrac{1}{2}(e^2 - m^2)RP_x$ $G_\theta = G_{13}R^2P_x$

2.

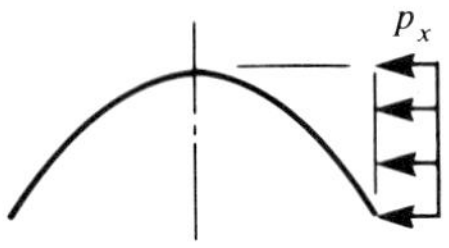

$G_U = \tfrac{1}{2}e^2Rp_x$ $G_u = G_{14}R^4p_x$

$G_V = 0$ $G_v = G_{15}R^4p_x$

$G_Z = -\tfrac{1}{8}e^4R^2p_x$ $G_\theta = G_{16}R^3p_x$

3.

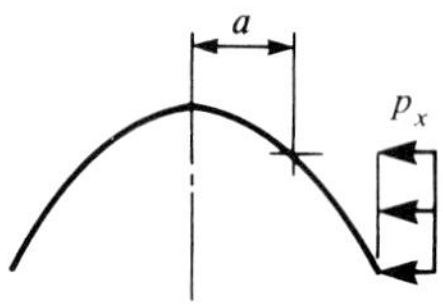

$G_U = \tfrac{1}{2}(e^2 - m^2)Rp_x$ $G_u = G_{17}R^4p_x$

$G_V = 0$ $G_v = G_{18}R^4p_x$

$G_Z = -\tfrac{1}{8}(e^2 - m^2)^2R^2p_x$ $G_\theta = G_{19}R^3p_x$

4.

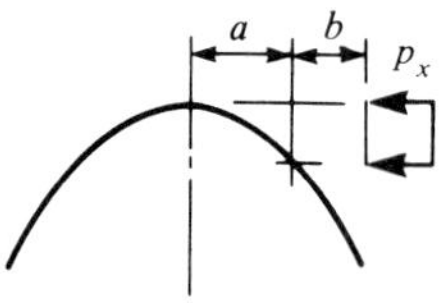

$G_U = \tfrac{1}{2}m^2Rp_x$ $G_u = (G_{14} - G_{17})R^4p_x$

$G_V = 0$ $G_v = (G_{15} - G_{18})R^4p_x$

$G_Z = -\tfrac{1}{8}m^2(2e^2 - m^2)R^2p_x$ $G_\theta = (G_{16} - G_{19})R^3p_x$

7.05 TRANSPORT LOAD FUNCTIONS IN 0 SYSTEM

Notation (7.02–7.03) Signs (7.02) Matrix (7.02)

P_y = concentrated load (N) p_y = intensity of distributed load (N/m)

B_j, C_j = equivalents (7.03) $R = \dfrac{c^2}{2h}$ $m = \dfrac{a}{R}$ $n = \dfrac{b}{R}$ $e = \dfrac{c}{R}$ $r = \dfrac{I_z}{A_x R^2}$

$$G_{21} = -\tfrac{1}{2}(-me^2B_0 + e^2B_1 + mB_2 - B_3 - 2rB_{-2})$$
$$G_{22} = -[-meB_0 + (e+m)B_1 - B_2 - rB_{-3}]$$
$$G_{23} = -(B_1 - mB_0)$$

$$G_{24} = -\tfrac{1}{2}(e^4C_0 - C_4 - 4rC_{-3})$$
$$G_{25} = -e(e^2C_0 - C_2 - 2rC_{-3}]$$
$$G_{26} = -(e^2C_0 + C_2)$$

$$G_{27} = -\tfrac{1}{4}[m^2e^2B_0 - 2e^2mB_1 + (e^2 - m^2)B_2 + 2mB_3 - B_4 - 4r(B_{-3} - mB_{-2})]$$
$$G_{28} = -\tfrac{1}{2}[em^2B_0 - m(m+2e)B_1 + (e+2m)B_2 - B_3 - 2r(B_{-4} - mB_{-3})]$$
$$G_{29} = -\tfrac{1}{2}(m^2B_0 - 2mB_1 + B_2)$$

1.

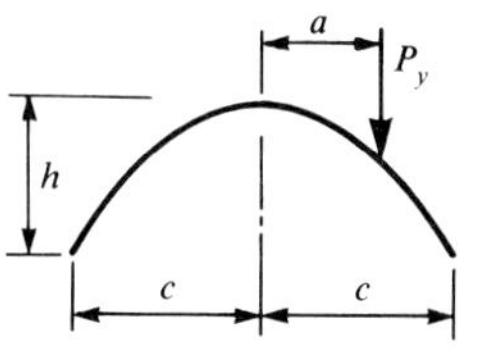

$G_U = 0$ $G_u = G_{21}R^3P_y$

$G_V = P_y$ $G_v = G_{22}R^3P_y$

$G_Z = -nRP_y$ $G_\theta = G_{23}R^2P_y$

2.

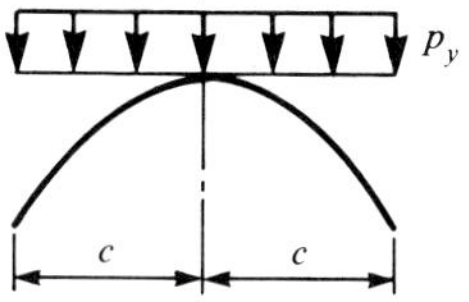

$G_U = 0$ $G_u = G_{24}R^4p_y$

$G_V = 2eRp_y$ $G_v = G_{25}R^4p_y$

$G_Z = -2e^2R^2p_y$ $G_\theta = G_{26}R^3p_y$

3.

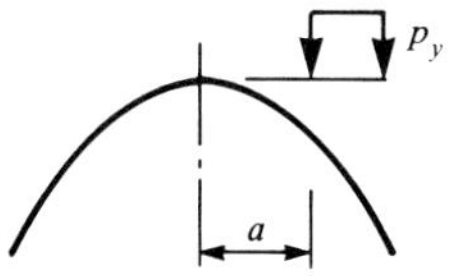

$G_U = 0$ $G_u = G_{27}R^4p_y$

$G_V = nRp_y$ $G_v = G_{28}R^4p_y$

$G_Z = -\tfrac{1}{2}n^2R^2p_y$ $G_\theta = G_{29}R^3p_y$

4.

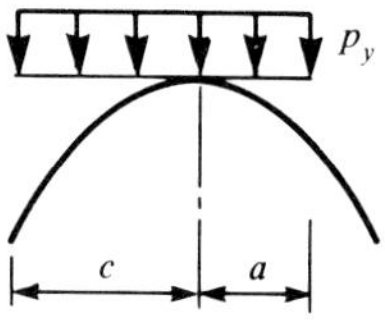

$G_U = 0$ $G_u = (G_{24} - G_{27})R^4p_y$

$G_V = (e+m)Rp_y$ $G_v = (G_{25} - G_{28})R^4p_y$

$G_Z = -\tfrac{1}{2}(4e^2 - m^2)R^2p_y$ $G_\theta = (G_{26} - G_{29})R^4p_y$

7.06 TRANSPORT LOAD FUNCTIONS IN 0 SYSTEM

Notation (7.01–7.03) Signs (7.02) Matrix (7.02)

P_x, P_y = concentrated loads (N) Q_z = applied couple (N·m)

u_i, v_i, θ_i = abrupt displacements (4.03) e_t, f_t = temperature strains (4.03)

B_j, C_j = equivalents (7.03) $R = \dfrac{c^2}{2h}$ $m = \dfrac{a}{R}$ $n = \dfrac{b}{R}$ $e = \dfrac{c}{R}$ $r = \dfrac{I_z}{A_x R^2}$

$G_{31} = \frac{1}{2}(e^2 B_0 - B_2)$ $G_{32} = eB_0 - B_1$

$G_{33} = -\frac{1}{4}(e^2 C_2 - C_4 - 4rC_{-1})$ $G_{34} = -\frac{1}{2}(e^2 C_1 - C_3 - 2rC_{-2})$

$G_{35} = -\frac{1}{2}(eC_2 - C_3 - 2rC_{-2})$ $G_{36} = -(eC_1 - C_2 - rC_{-3})$

1.

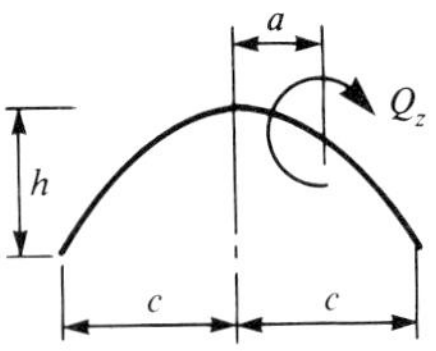

$G_U = 0$ $G_u = G_{31} R^2 Q_z$

$G_V = 0$ $G_v = G_{32} R^2 Q_z$

$G_Z = Q_z$ $G_\theta = B_0 R Q_z$

2.

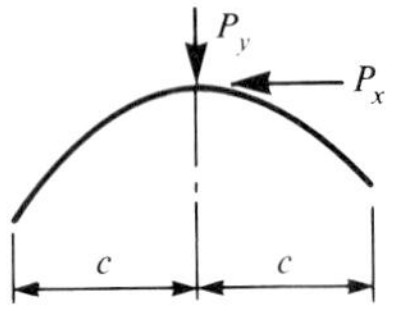

$G_U = P_x$ $G_u = (G_{33} P_x + G_{34} P_y) R^3$

$G_V = P_y$ $G_v = (G_{35} P_x + G_{36} P_y) R^3$

$G_Z = -hP_x - cP_y$ $G_\theta = (\frac{1}{2} C_2 P_x + C_1 P_y) R^2$

3. Abrupt displacements u_j, v_j, θ_j

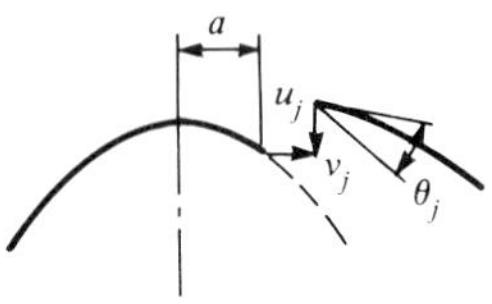

$G_U = 0$ $G_u = \left(u_j + \dfrac{e^2 - m^2}{2} R\theta_j\right) EI_z$

$G_V = 0$ $G_v = (v_i + nR\theta_j) EI_z$

$G_Z = 0$ $G_\theta = \theta_i EI_z$

4. Temperature deformations of segment b

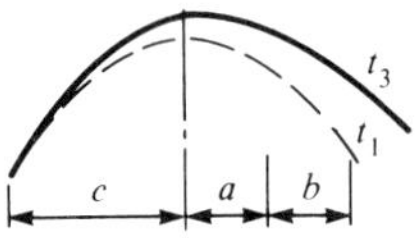

$G_U = 0$ $G_u = (be_t + G_{31} R^2 f_t) EI_z$

$G_V = 0$ $G_v = G_{32} R^2 f_t EI_z$

$G_Z = 0$ $G_\theta = B_0 R f_t EI_z$

7.07 STIFFNESS MATRIX EQUATIONS IN 0 SYSTEM

$$R = \frac{c^2}{2h} \qquad r = \frac{I}{A_x R^2}$$

$$e = \frac{c}{R} \qquad f = \frac{1}{2}\left(e^2 - \frac{C_2}{C_0}\right)$$

(1) Equivalents and signs. All forces, moments, and displacements are in the 0 system as shown in (2, 3) below. Their positive directions are indicated in the same figures. The superscript o indicating the 0 system is omitted but implied in all vector terms (Ref. 19).

$$F_1 = 2rC_{-1} + \frac{1}{2}\left(C_4 - \frac{C_2}{C_0}\right) \qquad F_2 = 2rC_{-3} + 2C_2 \qquad F_6 = 2C_0$$

where C_0, C_{-1}, C_2, C_{-3}, C_4 are the equivalents given in (7.03).

(2) Free-body diagram **(3) Elastic curve**

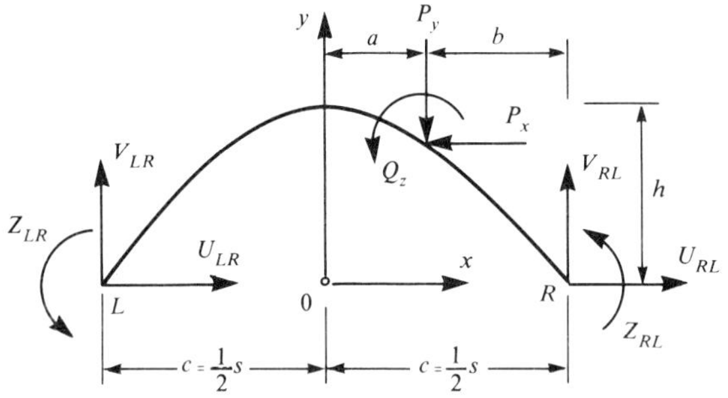

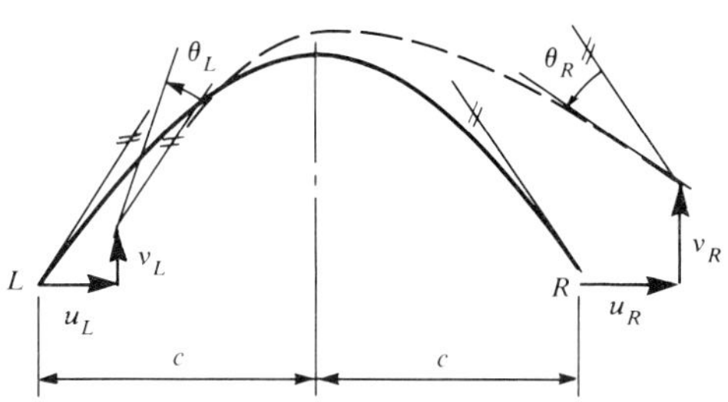

(4) Dimensionless stiffness coefficients

$$K_1 = \frac{1}{F_1} \qquad K_2 = \frac{f}{F_2} \qquad K_5 = \frac{f^2}{F_1} + \frac{e^2}{F_2} + \frac{1}{F_6}$$

$$K_3 = \frac{1}{F_2} \qquad K_4 = \frac{e}{F_2} \qquad K_6 = \frac{f^2}{F_1} - \frac{e^2}{F_2} + \frac{1}{F_6}$$

(5) Stiffness matrix equations

$$\begin{bmatrix} U_{LR} \\ V_{LR} \\ Z_{LR} \\ \hline U_{RL} \\ V_{RL} \\ Z_{RL} \end{bmatrix} = \frac{EI_z}{R^3} \left[\begin{array}{ccc|ccc} K_1 & 0 & -K_2R & -K_1 & 0 & K_2R \\ 0 & K_3 & K_4R & 0 & -K_3 & K_4R \\ -K_2R & K_4R & K_5R^2 & K_2R & -K_4R & -K_6R^2 \\ \hline -K_1 & 0 & K_2R & K_1 & 0 & -K_2R \\ 0 & -K_3 & -K_4R & 0 & K_3 & -K_4R \\ K_2R & K_4R & -K_6R^2 & -K_2R & -K_4R & K_5R^2 \end{array}\right] \begin{bmatrix} u_L \\ v_L \\ \theta_L \\ \hline u_R \\ v_R \\ \theta_R \end{bmatrix} + \begin{bmatrix} U_{L0} \\ V_{L0} \\ Z_{L0} \\ \hline U_{R0} \\ V_{R0} \\ Z_{R0} \end{bmatrix}$$

where U_{L0}, V_{L0}, Z_{L0} and U_{R0}, V_{R0}, Z_{R0} are the *reactions of the fixed-end parabolic bar* due to loads and other causes. Because of the complexity of their analytical forms, only the left-side reactions are given in (7.08). The right-side reactions can be calculated from the conditions of static equilibrium in terms of the left-side reactions and the applied loads.

7.08 STIFFNESS LOAD FUNCTIONS, FIXED-END BAR IN 0 SYSTEM

(1) General form of the stiffness load functions in (7.07) is given by the following matrix equations.

$$\begin{bmatrix} U_{L0} \\ V_{L0} \\ Z_{L0} \end{bmatrix} = \frac{1}{R^3} \begin{bmatrix} K_1 & 0 & -K_2R \\ 0 & K_3 & -K_4R \\ -K_3R & K_4R & K_6R^2 \end{bmatrix} \begin{bmatrix} G_u \\ G_v \\ G_\theta \end{bmatrix}$$

$$\begin{bmatrix} U_{R0} \\ V_{R0} \\ Z_{R0} \end{bmatrix} = \frac{1}{R^3} \begin{bmatrix} K_1 & 0 & -K_2R \\ 0 & K_3 & K_4R \\ -K_2R & -K_4R & K_6R \end{bmatrix} \begin{bmatrix} F_u \\ F_v \\ F_\theta \end{bmatrix}$$

where F_u, F_v, F_θ and G_u, G_v, G_θ are the transport load functions listed in (7.04–7.06) and K_1, K_2, K_3, K_4, K_5, K_6 are the stiffness coefficients (7.07).

(2) Particular cases

Notation (7.01–7.03, 7.07) Signs (7.07) $R = \dfrac{c^2}{2h}$

$B_0, B_1, \ldots, C_0, C_1, \ldots$ = equivalents (7.03)

$G_{11}, G_{12}, \ldots$ = equivalents (7.04–7.06)

1.

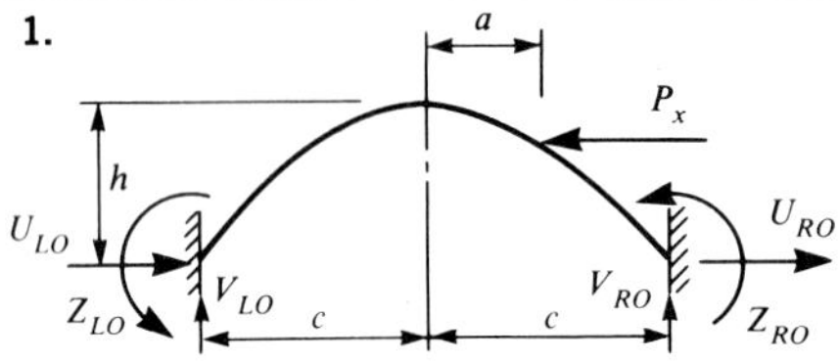

$$U_{L0} = (K_1G_{11} - K_2G_{13})P_x$$

$$V_{L0} = (K_3G_{12} - K_4G_{13})P_x$$

$$Z_{L0} = -(K_2G_{11} - K_4G_{12} - K_6G_{13})RP_x$$

2.

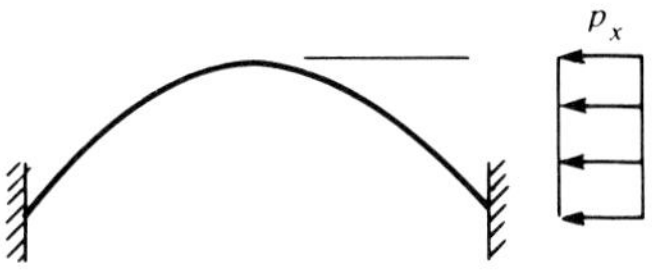

$$U_{L0} = (K_1G_{14} - K_2G_{16})Rp_x$$

$$V_{L0} = (K_3G_{15} - K_4G_{16})Rp_x$$

$$Z_{L0} = -(K_2G_{14} - K_4G_{15} - K_6G_{16})R^2p_x$$

3.

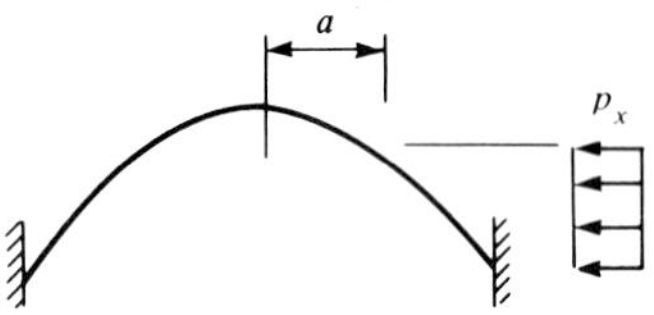

$$U_{L0} = (K_1G_{17} - K_2G_{19})Rp_x$$

$$V_{L0} = (K_3G_{18} - K_4G_{19})Rp_x$$

$$Z_{L0} = -(K_2G_{17} - K_4G_{18} - K_6G_{19})R^2p_x$$

4.

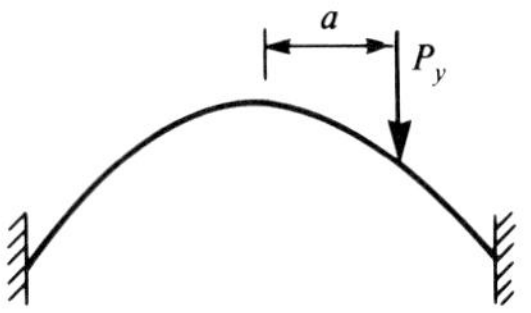

$$U_{L0} = (K_1G_{21} - K_2G_{23})P_y$$

$$V_{L0} = (K_3G_{22} - K_4G_{23})P_y$$

$$Z_{L0} = -(K_2G_{21} - K_4G_{22} - K_6G_{23})RP_y$$

5.

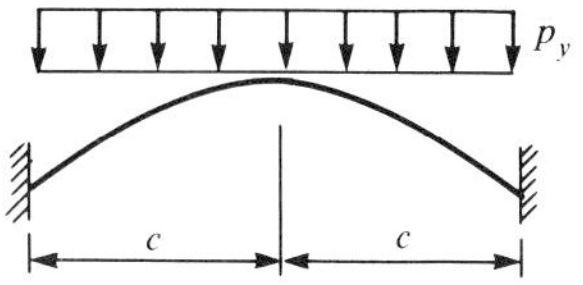

$$U_{L0} = (K_1G_{24} - K_2G_{26})Rp_y$$

$$V_{L0} = (K_3G_{25} - K_4G_{26})Rp_y$$

$$Z_{L0} = -(K_2G_{24} - K_4G_{25} - K_6G_{26})R^2p_y$$

6.

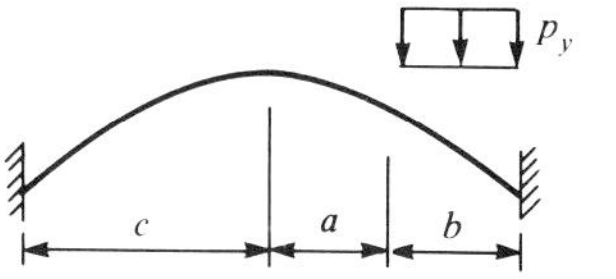

$$U_{L0} = (K_1G_{27} - K_2G_{29})Rp_y$$

$$V_{L0} = (K_3G_{28} - K_4G_{29})Rp_y$$

$$Z_{L0} = -(K_2G_{27} - K_4G_{28} - K_6G_{29})R^2p_y$$

7.

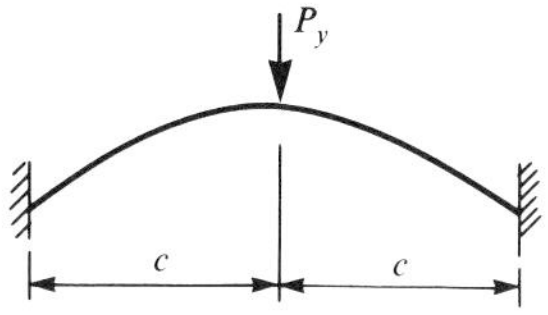

$$U_{L0} = (K_1G_{34} + K_2C_1)P_y$$

$$V_{L0} = (K_3G_{36} + K_4C_1)P_y$$

$$Z_{L0} = -(K_2G_{34} - K_4G_{36} + K_6C_1)RP_y$$

8.

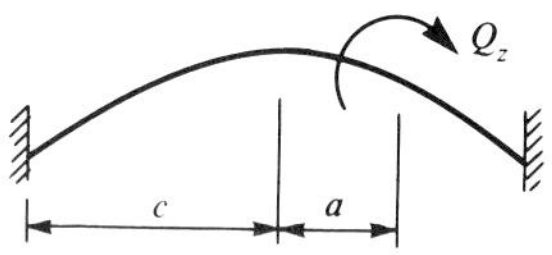

$$U_{L0} = (K_1G_{31} - K_2B_0)R^{-1}Q_z$$

$$V_{L0} = (K_3G_{32} - K_4B_0)R^{-1}Q_z$$

$$Z_{L0} = -(K_2G_{31} - K_4G_{32} - K_6B_0)Q_z$$

9. Abrupt displacement u_j, v_j

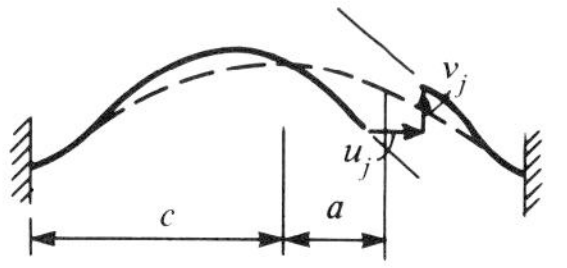

$$U_{L0} = EI_zK_1R^{-3}u_j$$

$$V_{L0} = EI_zK_3R^{-3}v_j$$

$$Z_{L0} = -EI_z(K_2u_j - K_4v_j)R^{-2}$$

10. Abrupt displacement θ_j

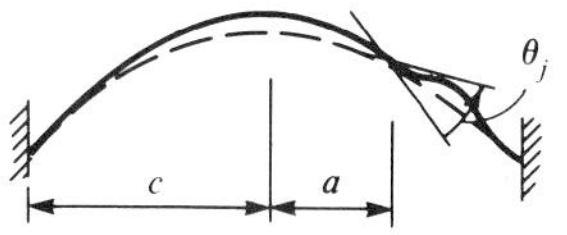

$$U_{L0} = EI_z(\tfrac{1}{2}(e^2 - m^2)K_1 - K_2)R^{-2}\theta_j$$

$$V_{L0} = EI_z(nK_3 - K_4)R^{-2}\theta_j$$

$$Z_{L0} = -EI_z[\tfrac{1}{2}(e^2 - m^2)K_2 - nK_4 - K_6]R^{-1}\theta_j$$

11. Linear temperature deformation of segment b

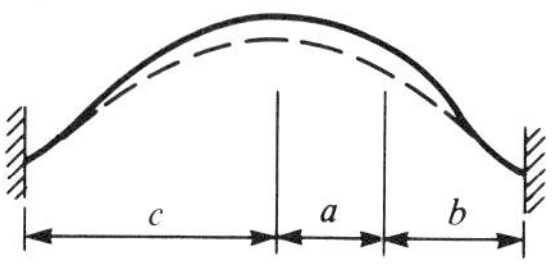

$$U_{L0} = EI_2nK_1R^{-2}e_t$$

$$V_{L0} = 0$$

$$Z_{L0} = -EI_znK_2R^{-1}e_t$$

12. Angular temperature deformation of segment b

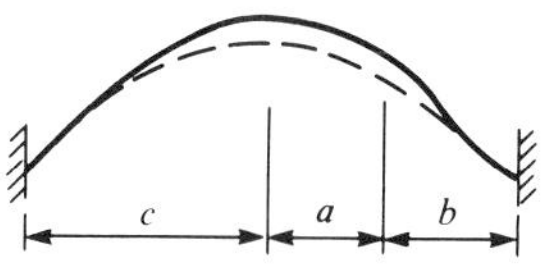

$$U_{L0} = EI_z(K_1G_{31} - K_2B_0)R^{-1}f_t$$

$$V_{L0} = EI_z(K_3G_{32} - K_4B_0)R^{-1}f_t$$

$$Z_{L0} = -EI_z(K_2G_{31} - K_4G_{32} - K_6B_0)f_t$$

8

Straight Bar of Order One Encased in Elastic Foundation Constant Section

STATIC STATE

8.01 DEFINITION OF STATE

(1) System considered is a finite straight bar of constant and symmetrical section. The centroidal axis x of the bar and the vertical axis y define the plane of symmetry of the section. The elastic foundation fully surrounds the bar and offers resistance, which is proportional to its displacement (Winkler's foundation). The constants of proportionality, called the foundation moduli, are

k_u = longitudinal modulus along x (N/m^2)	k_v = transverse modulus along y (N/m^2)

Each modulus corresponds to a unit length of the bar. The bar is a part of a planar system acted on by loads in the system plane (system of order one).

(2) Notation (A.1, A.2, A.4)

s = span (m)	A_x = area of normal section (m^2)
a, b = segments (m)	I_z = moment of inertia of A_x about z (m^4)
x, x' = coordinates (m)	E = modulus of elasticity (Pa)
U = normal force (N)	u = linear displacement along x (m)
V = shearing force (N)	v = linear displacement along y (m)
Z = flexural moment (N·m)	θ = flexural slope about z (rad)

$\alpha = \sqrt{\dfrac{k_u}{EA_x}}$ = shape parameter (m^{-1})	$\beta = \sqrt[4]{\dfrac{k_v}{4EI_z}}$ = shape parameter (m^{-1})

(3) Assumptions of analysis are stated in (1.01). In this chapter, the effects of axial and flexural deformations are included, but the effects of shearing deformations and of axial force on the magnitude of the bending moment are neglected. The beam-column effect is considered in Chap. 10.

8.02 TRANSPORT MATRIX EQUATIONS

(1) Equivalents and signs

$\bar{u} = uEA_x\alpha$ = scaled u (N) $\qquad \bar{v} = vEI\beta^3$ = scaled v (N)

$\bar{Z} = Z\beta$ = scaled Z (N) $\qquad \bar{\theta} = \theta EI_z\beta^2$ = scaled θ (N)

All forces, moments, and displacements are in the principal axes of the normal section, and their positive directions are shown in (2, 3) below (Ref. 17).

(2) Free-body diagram

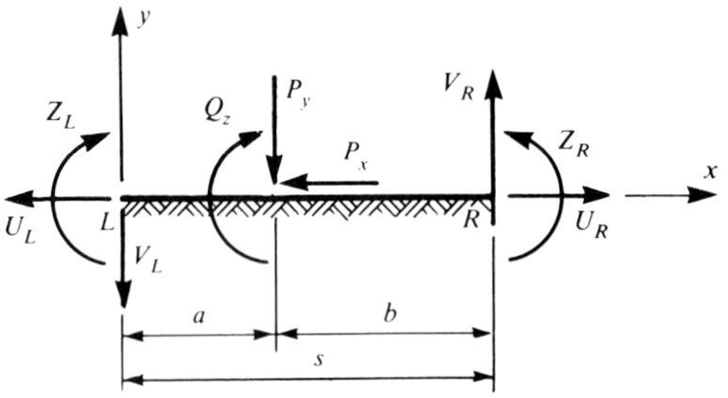

(3) Elastic curve

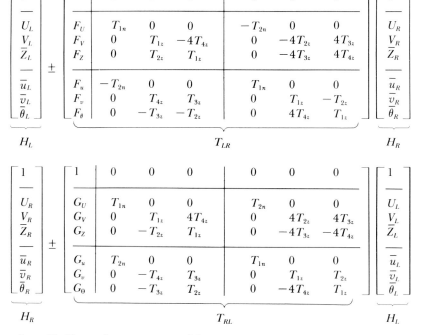

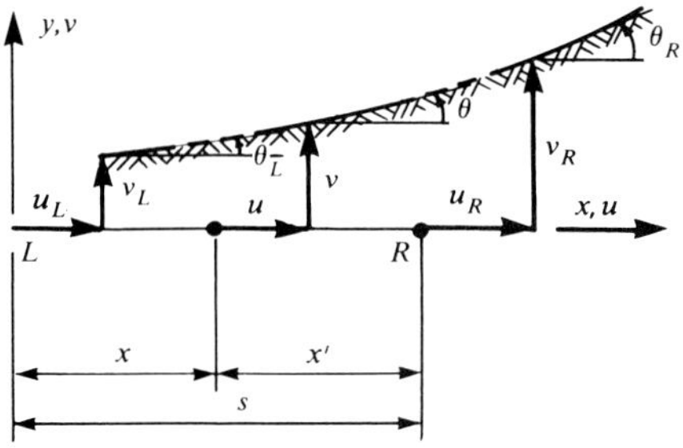

(4) Transport matrix equations

$$
\underbrace{\begin{bmatrix} 1 \\ \hline U_L \\ V_L \\ Z_L \\ \hline \bar{u}_L \\ \bar{v}_L \\ \bar{\theta}_L \end{bmatrix}}_{H_L}
\pm
\underbrace{\left[\begin{array}{c|ccc|ccc}
1 & 0 & 0 & 0 & 0 & 0 & 0 \\ \hline
F_U & T_{1n} & 0 & 0 & -T_{2n} & 0 & 0 \\
F_V & 0 & T_{1z} & -4T_{4z} & 0 & -4T_{2z} & 4T_{3z} \\
F_Z & 0 & T_{2z} & T_{1z} & 0 & -4T_{3z} & 4T_{4z} \\ \hline
F_u & -T_{2n} & 0 & 0 & T_{1n} & 0 & 0 \\
F_v & 0 & T_{4z} & T_{3z} & 0 & T_{1z} & -T_{2z} \\
F_\theta & 0 & -T_{3z} & -T_{2z} & 0 & 4T_{4z} & T_{1z}
\end{array}\right]}_{T_{LR}}
\underbrace{\begin{bmatrix} 1 \\ \hline U_R \\ V_R \\ Z_R \\ \hline \bar{u}_R \\ \bar{v}_R \\ \bar{\theta}_R \end{bmatrix}}_{H_R}
$$

$$
\underbrace{\begin{bmatrix} 1 \\ \hline U_R \\ V_R \\ Z_R \\ \hline \bar{u}_R \\ \bar{v}_R \\ \bar{\theta}_R \end{bmatrix}}_{H_R}
\pm
\underbrace{\left[\begin{array}{c|ccc|ccc}
1 & 0 & 0 & 0 & 0 & 0 & 0 \\ \hline
G_U & T_{1n} & 0 & 0 & T_{2n} & 0 & 0 \\
G_V & 0 & T_{1z} & 4T_{4z} & 0 & 4T_{2z} & 4T_{3z} \\
G_Z & 0 & -T_{2z} & T_{1z} & 0 & -4T_{3z} & -4T_{4z} \\ \hline
G_u & T_{2n} & 0 & 0 & T_{1n} & 0 & 0 \\
G_v & 0 & -T_{4z} & T_{3z} & 0 & T_{1z} & T_{2z} \\
G_\theta & 0 & -T_{3z} & T_{2z} & 0 & -4T_{4z} & T_{1z}
\end{array}\right]}_{T_{RL}}
\underbrace{\begin{bmatrix} 1 \\ \hline U_L \\ V_L \\ Z_L \\ \hline \bar{u}_L \\ \bar{v}_L \\ \bar{\theta}_L \end{bmatrix}}_{H_L}
$$

where H_L, H_R are the *state vectors* of the respective ends and T_{LR}, T_{RL} are the *transport matrices*. The elements of the transport matrices are the *transport coefficients* T_{jn}, T_{jz} defined in (8.03) and the *load functions* F, G displayed for particular cases in (8.06–8.08).

8.03 TRANSPORT RELATIONS

(1) Transport functions

$$\alpha = \sqrt{\frac{k_u}{EA_x}} \qquad \beta = \sqrt[4]{\frac{k_v}{4EI_z}}$$

$T_{1n}(x) = \cosh \alpha x$	$T_{1z}(x) = \cosh \beta x \cos \beta x$
$T_{2n}(x) = \sinh \alpha x$	$T_{2z}(x) = \frac{1}{2}(\cosh \beta x \sin \beta x + \sinh \beta x \cos \beta x)$
$T_{3n}(x) = \cosh \alpha x - 1$	$T_{3z}(x) = \frac{1}{2} \sinh \beta x \sin \beta x$
$T_{4n}(x) = \sinh \alpha x - \alpha x$	$T_{4z}(x) = \frac{1}{4}(\cosh \beta x \sin \beta x - \sinh \beta x \cos \beta x)$
$T_{5n}(x) = T_{3n}(x) - \dfrac{(\alpha x)^2}{2!}$	$T_{5z}(x) = \frac{1}{4}[1 - T_{1z}(x)]$
$T_{6n}(x) = T_{4n}(x) - \dfrac{(\alpha x)^3}{3!}$	$T_{6z}(x) = \frac{1}{4}[\beta x - T_{2z}(x)]$

The graphs and table of these functions are shown in (8.04–8.05).

(2) Transport coefficients T_j, A_j, B_j, C_j are the values of $T_j(x)$ for a particular argument x.

For $x = s$, $\quad T_{jn}(s) = T_{jn}$, $\quad T_{jz}(s) = T_{jz}$

For $x = a$, $\quad T_{jn}(a) = A_{jn}$, $\quad T_{jz}(a) = A_{jz}$ $\qquad j = 1, 2, \ldots, 6$

For $x = b$, $\quad T_{jn}(b) = B_{jn}$, $\quad T_{jz}(b) = B_{jz}$

and $T_{jn} - A_{jn} = C_{jn}$, $\quad T_{jz} - A_{jz} = C_{jz}$

(3) Transport matrices of this bar have analogical properties to those of (4.02) and can be used in similar ways.

(4) Differential relations

$\dfrac{du}{dx} = \dfrac{U}{EA_x}$	$\dfrac{dv}{dx} = \theta$
$\dfrac{d^2u}{dx^2} = \dfrac{uk_u}{EA_x} + \dfrac{p_x}{EA_x}$	$\dfrac{d^2v}{dx^2} = \dfrac{d\theta}{dx} = \dfrac{Z}{EI_z}$
	$\dfrac{d^3v}{dx^3} = \dfrac{d^2\theta}{dx^2} = \dfrac{dZ}{dx\, EI_z} = -\dfrac{V}{EI_z}$
	$\dfrac{d^4_v}{dx^4} = \dfrac{d^3\theta}{dx^3} = \dfrac{d^2Z}{dx^2\, EI_z} = -\dfrac{dV}{dx\, EI_z} = -\dfrac{vk_v}{EI_z} - \dfrac{p_y}{EI_z}$

(5) Bars encased in elastic medium are classified as *(a) short bars* ($\beta s < 0.6$), which can be considered as infinitely rigid, *(b) intermediate bars* ($0.6 < \beta s < 5.0$), which must be treated as finite and elastic and must be analyzed by the methods introduced in (8.02–8.13), *(c) long bars* ($\beta s > 5.0$), which may be treated as semi-infinite and must be analyzed by the method introduced in (8.14).

(6) General transport matrix equations for bars of order three are obtained by matrix composition of (8.02–8.08) and (13.02–13.06) as shown in (2.18).

8.04 TABLE OF TRANSPORT COEFFICIENTS†

βx	$T_{1z}(x)$	$T_{2z}(x)$	$T_{3z}(x)$	$T_{4z}(x)$	$T_{5z}(x)$	$T_{6z}(x)$
0.0	1.00000	0.00000	0.00000	0.00000	0.00000	0.00000
0.50	0.98958	0.49896	0.12491	0.02083	0.00260	0.00026
0.60	0.97841	0.59741	0.17974	0.03598	0.00540	0.00065
0.70	0.96001	0.69440	0.24435	0.05710	0.01000	0.00140
0.80	0.93180	0.78908	0.31854	0.08517	0.01705	0.00273
0.90	0.89082	0.88033	0.40205	0.12112	0.02729	0.00492
1.00	0.83373	0.96671	0.49445	0.16587	0.04157	0.00832
1.10	0.75683	1.04642	0.59517	0.22029	0.06079	0.01339
1.20	0.65611	1.11728	0.70344	0.28516	0.08597	0.02068
1.30	0.52722	1.17670	0.81825	0.36119	0.11820	0.03082
1.40	0.36558	1.22164	0.93830	0.44898	0.15860	0.04459
1.50	0.16640	1.24857	1.06197	0.54897	0.20840	0.06286
1.5708	0.00000	1.25459	1.15065	0.62729	0.25000	0.07905
1.70	−0.36441	1.23193	1.31179	0.78640	0.34110	0.11702
1.80	−0.70602	1.17887	1.43261	0.92367	0.42651	0.15528
1.90	−1.10492	1.08882	1.54633	1.07269	0.52623	0.20280
2.00	−1.56563	0.95582	1.64895	1.23257	0.64141	0.26104
2.10	−2.09224	0.77350	1.73585	1.40196	0.77306	0.33163
2.20	−2.68822	0.53506	1.80178	1.57904	0.92205	0.41623
2.30	−3.35618	0.23345	1.84076	1.76142	1.08904	0.51664
2.3651	−3.83049	0.00000	1.84852	1.88158	1.20762	0.59135
2.50	−4.91284	−0.58854	1.81044	2.12927	1.47821	0.77213
2.60	−5.80028	−1.12360	1.72557	2.30652	1.70007	0.93090
2.70	−6.75655	−1.75089	1.58264	2.47245	1.93914	1.11272
2.80	−7.77591	−2.47702	1.37210	2.62079	2.19398	1.31926
2.90	−8.84988	−3.30790	1.08375	2.74428	2.46247	1.55197
3.00	−9.96691	−4.24844	0.70686	2.83459	2.74173	1.81211
3.1416	−11.59195	−5.77437	0.00000	2.88718	3.14799	2.22899
3.20	−12.26569	−6.47111	−0.35742	2.87694	3.31642	2.41778
3.30	−13.40480	−7.75487	−1.06777	2.80676	3.60120	2.76372
3.40	−14.50075	−9.15064	−1.91213	2.65892	3.87519	3.13766
3.50	−15.51973	−10.65246	−2.90144	2.41950	4.12993	3.53812
3.60	−16.42214	−12.25071	−4.04584	2.07346	4.35553	3.96268
3.70	−17.16216	−13.93148	−5.35434	1.60486	4.54054	4.40787
3.80	−17.68744	−15.67599	−6.83427	0.99688	4.67186	4.86900
3.9266	−17.95122	−17.93724	−8.96163	0.00000	4.73781	5.46596
4.00	−17.84985	−19.25241	−10.32654	−0.70726	4.71246	5.81310
4.10	−17.34728	−21.01604	−12.34038	−1.83914	4.58682	6.27901
4.20	−16.35052	−22.70540	−14.52728	−3.18111	4.33763	6.72635
4.30	−14.77213	−24.26676	−16.87720	−4.75004	3.94303	7.14169
4.40	−12.51815	−25.63731	−19.37428	−6.56147	3.37954	7.50933
4.50	−9.48879	−26.74455	−21.99590	−8.62905	2.62220	7.81114
4.60	−5.57927	−27.50574	−24.71167	−10.96380	1.64482	8.02643
4.7124	0.00000	−27.83169	−27.82720	−13.91585	0.25000	8.13602
4.80	5.31638	−27.60531	−30.25904	−16.46049	−1.07909	8.10133
4.90	12.52405	−26.72384	−32.98151	−19.62325	−2.88101	7.90596
5.00	21.05056	−25.05654	−35.57763	−23.05259	−5.01264	7.51413
6.00	193.68136	68.65825	−28.18089	−62.51036	−48.17034	−15.66456

†Values of $T_{jn}(x)$ are available in standard mathematical tables.

8.05 GRAPHS AND SERIES OF TRANSPORT COEFFICIENTS

(1) Graphs of $T_{jz}(x)$ $(j = 1, 2, \ldots, 6)$

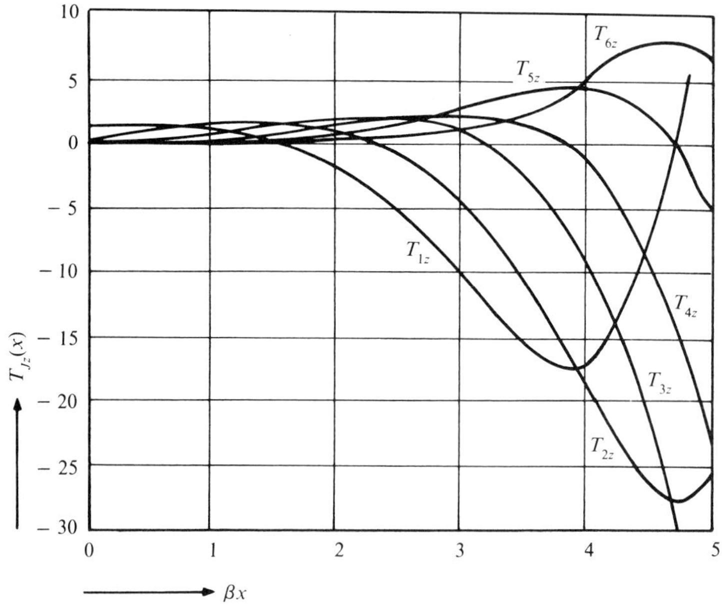

(2) Series expansion of $T_{jn}(x)$ and $T_{jz}(x)$

$$T_{1n}(x) = 1 + \frac{(\alpha x)^2}{2!} + \frac{(\alpha x)^4}{4!} + \frac{(\alpha x)^6}{6!} + \cdots = \sum_{k=0}^{\infty} \frac{(\alpha x)^{2k}}{(2k)!}$$

$$T_{2n}(x) = \frac{(\alpha x)}{1!} + \frac{(\alpha x)^3}{3!} + \frac{(\alpha x)^5}{5!} + \frac{(\alpha x)^7}{7!} + \cdots = \sum_{k=0}^{\infty} \frac{(\alpha x)^{2k+1}}{(2k+1)!}$$

$$T_{3n}(x) = \frac{(\alpha x)^2}{2!} + \frac{(\alpha x)^4}{4!} + \frac{(\alpha x)^6}{6!} + \frac{(\alpha x)^8}{8!} + \cdots = \sum_{k=0}^{\infty} \frac{(\alpha x)^{2k+2}}{(2k+2)!}$$

$$T_{jn}(x) = \frac{(\alpha x)^{j-1}}{(j-1)!} + \frac{(\alpha x)^{j+1}}{(j+1)!} + \frac{(\alpha x)^{j+3}}{(j+3)!} + \frac{(\alpha x)^{j+5}}{(j+5)!} + \cdots = \sum_{k=0}^{\infty} \frac{(\alpha x)^{j+2k-1}}{(j+2k-2)!}$$

$$T_{1z}(x) = 1 - 4\frac{(\beta x)^4}{4!} + 4^2\frac{(\beta x)^8}{8!} - 4^3\frac{(\beta x)^{12}}{12!} + \cdots = \sum_{k=0}^{\infty} (-4)^k \frac{(\beta x)^{4k}}{(4k)!}$$

$$T_{2z}(x) = \frac{(\beta x)}{1!} - 4\frac{(\beta x)^5}{5!} + 4^2\frac{(\beta x)^9}{9!} - 4^3\frac{(\beta x)^{13}}{13!} + \cdots = \sum_{k=0}^{\infty} (-4)^k \frac{(\beta x)^{4k+1}}{(4k+1)!}$$

$$T_{3z}(x) = \frac{(\beta x)^2}{2!} - 4\frac{(\beta x)^6}{6!} + 4^2\frac{(\beta x)^{10}}{10!} - 4^3\frac{(\beta x)^{14}}{14!} + \cdots = \sum_{k=0}^{\infty} (-4)^k \frac{(\beta x)^{4k+2}}{(4k+2)!}$$

$$T_{jz}(x) = \frac{(\beta x)^{j-1}}{(j-1)!} - 4\frac{(\beta x)^{j+3}}{(j+3)!} + 4^2\frac{(\beta x)^{j+7}}{(j+7)!} - 4^3\frac{(\beta x)^{j+11}}{(j+11)!} + \cdots = \sum_{k=0}^{\infty} (-4)^k \frac{(\beta x)^{j+4k-1}}{(j+4k-1)!}$$

8.06 TRANSPORT LOAD FUNCTIONS

Notation (8.01–8.03) Signs (8.02) Matrix (8.02) P_x, P_y = concentrated loads (N) Q_z = applied couple (N·m) q_z = intensity of distributed couple (N·m/m) A_j, B_j, C_j, T_j = transport coefficients (8.03–8.05)		$\alpha = \sqrt{\dfrac{k_u}{EA_x}}$ $\beta = \sqrt[4]{\dfrac{k_v}{4EI_z}}$
$F_U = -A_{1n}P_x$ $F_V = -A_{1z}P_y$ $F_Z = -A_{2z}P_y$ $F_u = A_{2n}P_x$ $F_v = -A_{4z}P_y$ $F_\theta = A_{3z}P_y$	**1.** 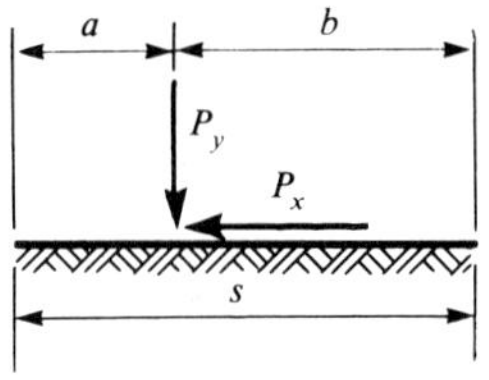	$G_U = B_{1n}P_x$ $G_V = B_{1z}P_y$ $G_Z = -B_{2z}P_y$ $G_u = B_{2n}P_x$ $G_v = -B_{4z}P_y$ $G_\theta = -B_{3z}P_y$
$F_U = 0$ $F_V = 4A_{4z}\beta Q_z$ $F_Z = -A_{1z}\beta Q_z$ $F_u = 0$ $F_v = -A_{3z}\beta Q_z$ $F_\theta = A_{2z}\beta Q_z$	**2.** 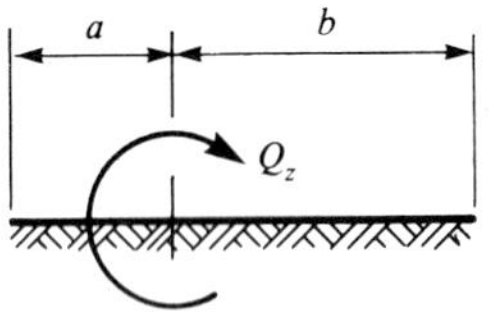	$G_U = 0$ $G_V = 4B_{4z}\beta Q_z$ $G_Z = B_{1z}\beta Q_z$ $G_u = 0$ $G_v = B_{3z}\beta Q_z$ $G_\theta = B_{2z}\beta Q_z$
$F_U = 0$ $F_V = 4T_{5z}q_z$ $F_Z = -T_{2z}q_z$ $F_u = 0$ $F_v = -T_{4z}q_z$ $F_\theta = T_{3z}q_z$	**3.** 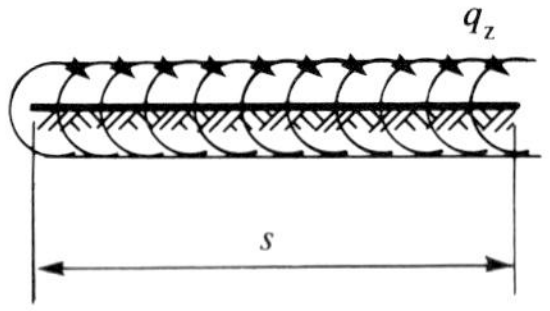	$G_U = 0$ $G_V = 4T_{5z}q_z$ $G_Z = T_{2z}q_z$ $G_u = 0$ $G_v = T_{4z}q_z$ $G_\theta = T_{3z}q_z$
$F_U = 0$ $F_V = 4C_{5z}q_z$ $F_Z = -C_{2z}q_z$ $F_u = 0$ $F_v = -C_{4z}q_z$ $F_\theta = C_{3z}q_z$	**4.** 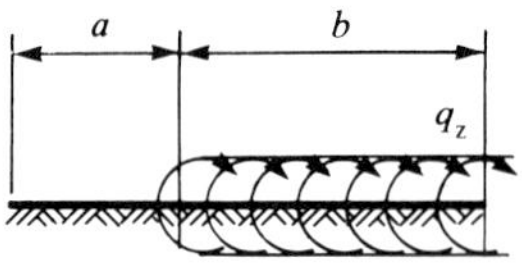	$G_U = 0$ $G_V = 4B_{5z}q_z$ $G_Z = B_{2z}q_z$ $G_u = 0$ $G_V = B_{4z}q_z$ $G_\theta = B_{3z}q_z$

8.07 TRANSPORT LOAD FUNCTIONS

Notation (8.01–8.03) p_x, p_y = intensities of distributed load (N/m) A_j, B_j, C_j, T_j = transport coefficients (8.03–8.05)	Signs (8.02) Matrix (8.02)	$\alpha = \sqrt{\dfrac{k_u}{EA_x}}$ $\beta = \sqrt[4]{\dfrac{k_v}{4EI_z}}$
$F_U = -T_{2n}\alpha^{-1}p_x$ $F_V = -T_{2z}\beta^{-1}p_y$ $F_Z = -T_{3z}\beta^{-1}p_y$	1. 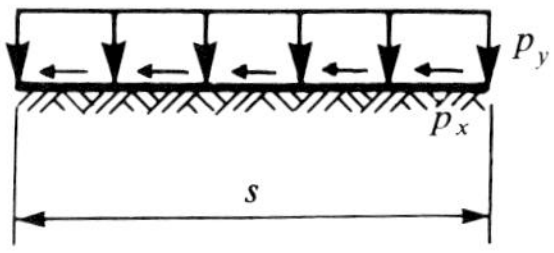	$G_U = T_{2n}\alpha^{-1}p_x$ $G_V = T_{2z}\beta^{-1}p_y$ $G_Z = -T_{3z}\beta^{-1}p_y$
$F_u = T_{3n}\alpha^{-1}p_x$ $F_v = -T_{5z}\beta^{-1}p_y$ $F_\theta = T_{4z}\beta^{-1}p_y$		$G_u = T_{3n}\alpha^{-1}p_x$ $G_v = -T_{5z}\beta^{-1}p_y$ $G_\theta = -T_{4z}\beta^{-1}p_y$
$F_U = -C_{2n}\alpha^{-1}p_x$ $F_V = -C_{2z}\beta^{-1}p_y$ $F_Z = -C_{3z}\beta^{-1}p_y$	2. 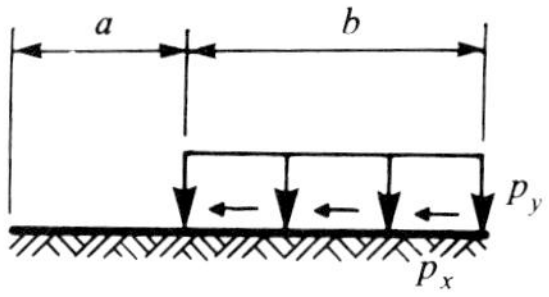	$G_U = B_{2n}\alpha^{-1}p_x$ $G_V = B_{2z}\beta^{-1}p_y$ $G_Z = -B_{3z}\beta^{-1}p_y$
$F_u = C_{3n}\alpha^{-1}p_x$ $F_v = -C_{5z}\beta^{-1}p_y$ $F_\theta = C_{4z}\beta^{-1}p_y$		$G_u = B_{3n}\alpha^{-1}p_x$ $G_v = -B_{5z}\beta^{-1}p_y$ $G_\theta = -B_{4z}\beta^{-1}p_y$
$F_U = -(\alpha s T_{2n} - T_{3n})\alpha^{-2}s^{-1}p_x$ $F_V = -(\beta s T_{2z} - T_{3z})\beta^{-2}s^{-1}p_y$ $F_Z = -(\beta s T_{3z} - T_{4z})\beta^{-2}s^{-1}p_y$	3. 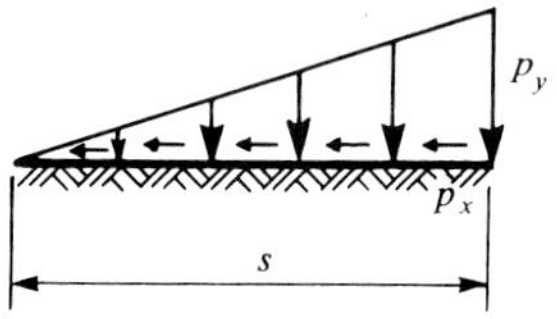	$G_U = T_{3n}\alpha^{-2}s^{-1}p_x$ $G_V = T_{3z}\beta^{-2}s^{-1}p_y$ $G_Z = -T_{4z}\beta^{-2}s^{-1}p_y$
$F_u = (\alpha s T_{3n} - T_{4n})\alpha^{-2}s^{-1}p_x$ $F_v = -(\beta s T_{5z} - T_{6z})\beta^{-2}s^{-1}p_y$ $F_\theta = (\beta s T_{4z} - T_{5z})\beta^{-2}s^{-1}p_y$		$G_u = T_{4n}\alpha^{-2}s^{-1}p_x$ $G_v = -T_{6z}\beta^{-2}s^{-1}p_y$ $G_\theta = -T_{5z}\beta^{-2}s^{-1}p_y$
$F_U = -(\alpha b T_{2n} - C_{3n})\alpha^{-2}b^{-1}p_x$ $F_V = -(\beta b T_{2z} - C_{3z})\beta^{-2}b^{-1}p_y$ $F_Z = -(\beta b T_{3z} - C_{4z})\beta^{-2}b^{-1}p_y$	4. 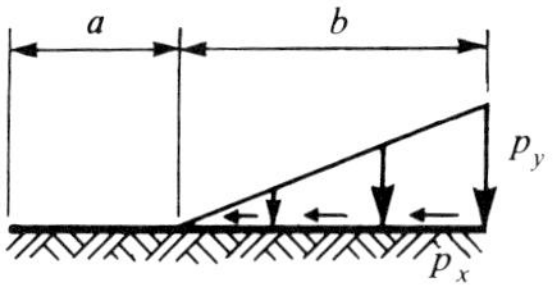	$G_U = B_{3n}\alpha^{-2}b^{-1}p_x$ $G_V = B_{3z}\beta^{-2}b^{-1}p_y$ $G_Z = -B_{4z}\beta^{-2}b^{-1}p_y$
$F_u = (\alpha b T_{3n} - C_{4n})\alpha^{-2}b^{-1}p_x$ $F_v = -(\beta b T_{5z} - C_{6z})\beta^{-2}b^{-1}p_y$ $F_\theta = (\beta b T_{4z} - C_{5z})\beta^{-2}b^{-1}p_y$		$G_u = B_{4n}\alpha^{-2}b^{-1}p_x$ $G_v = -B_{6z}\beta^{-2}b^{-1}p_y$ $G_\theta = -B_{5z}\beta^{-2}b^{-1}p_y$

8.08 TRANSPORT LOAD FUNCTIONS

Notation (8.01–8.03) Signs (8.02) Matrix (8.02) u_j, v_j, θ_j = abrupt displacements (m, m, rad) e_t, f_t = temperature strains (4.03) A_j, B_j, C_j, T_j = transport coefficients (8.03–8.05)	$\alpha = \sqrt{\dfrac{k_u}{EA_x}}$ $\beta = \sqrt[4]{\dfrac{k_v}{4EI_z}}$

$F_U = EA_x A_{2n} \alpha u_j$ $F_V = 4EI_z A_{2z} \beta^3 v_j$ $F_Z = 4EI_z A_{3z} \beta^3 v_j$	**1. Abrupt displacements** u_j, v_j 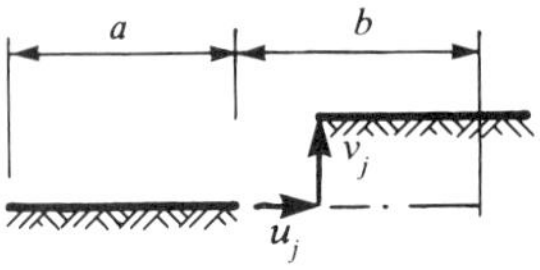	$G_U = EA_x B_{2n} \alpha u_j$ $G_V = 4EI_z B_{2z} \beta^3 v_j$ $G_Z = -4EI_z B_{3z} \beta^3 v_j$
$F_u = -EA_x A_{1n} \alpha u_j$ $F_v = -4EI_z A_{1z} \beta^3 v_j$ $F_\theta = -4EI_z A_{4z} \beta^3 v_j$		$G_u = EA_x B_{1n} \alpha u_j$ $G_v = EI_z B_{1z} \beta^3 v_j$ $G_\theta = -4EI_z B_{4z} \beta^3 v_j$
$F_U = 0$ $F_V = -4EI_z A_{3z} \beta^2 \theta_j$ $F_Z = -4EI_z A_{4z} \beta^2 \theta_j$	**2. Abrupt displacement θ_j** 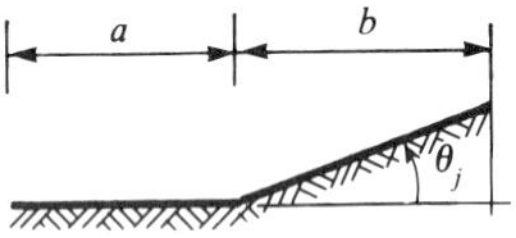	$G_U = 0$ $G_V = 4EI_z B_{3z} \beta^2 \theta_j$ $G_Z = -4EI_z B_{4z} \beta^2 \theta_j$
$F_u = 0$ $F_v = EI_z A_{2z} \beta^2 \theta_j$ $F_\theta = -EI_z A_{1z} \beta^2 \theta_j$		$G_u = 0$ $G_v = EI_z B_{2z} \beta^2 \theta_j$ $G_\theta = EI_z B_{1z} \beta^2 \theta_j$
$F_U = EA_x C_{3n} e_t$ $F_V = -4EI_z C_{4z} \beta f_t$ $F_Z = -4EI_z C_{5z} \beta f_t$	**3. Temperature deformation of segment** b 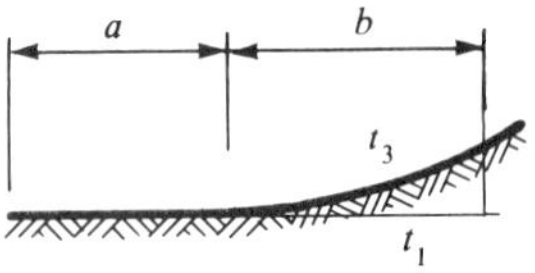	$G_U = EA_x B_{3n} e_t$ $G_V = 4EI_z B_{4z} \beta f_t$ $G_Z = -4EI_z B_{5z} \beta f_t$
$F_u = -EA_x C_{2n} e_t$ $F_v = EI_z C_{3z} \beta f_t$ $F_\theta = -EI_z C_{2z} \beta f_t$		$G_u = EA_x B_{2n} e_t$ $G_v = EI_z B_{3z} \beta f_t$ $G_\theta = EI_z B_{2z} \beta f_t$

4. 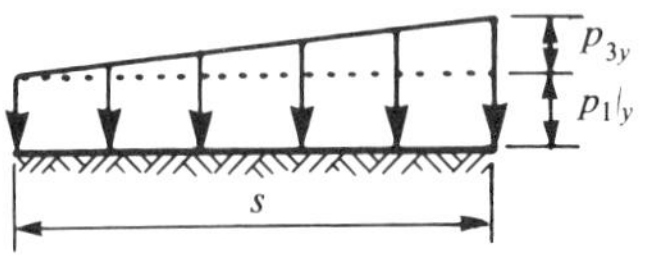 Sec. 8.07 (case 1 + case 3)	**5.** 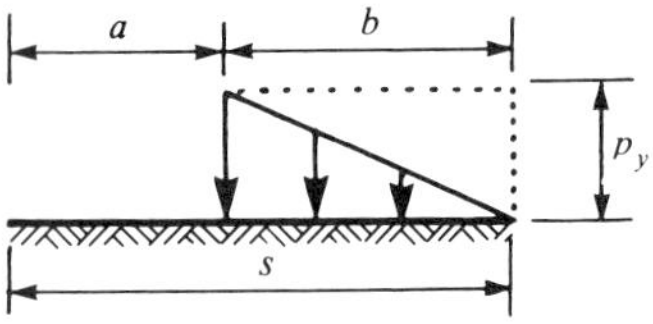 Sec. 8.07 (case 2 − case 4)

8.09 STIFFNESS MATRIX EQUATIONS

(1) Equivalents and signs. All forces, moments, and displacements are in the principal axes of the normal section, and their positive directions are shown in (2, 3) below (Ref. 17).

$$r = \frac{s^2 EA_x}{EI_z}$$

(2) Free-body diagram

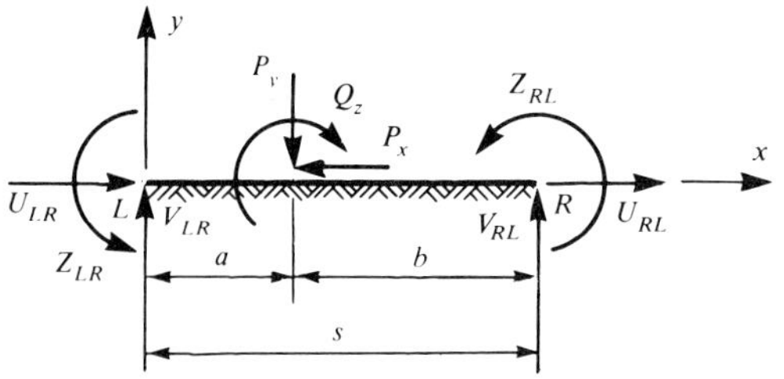

(3) Elastic curve

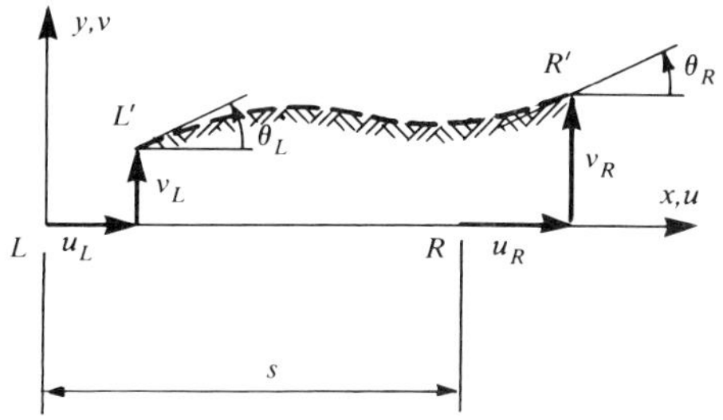

(4) Dimensionless stiffness coefficients

$$D_0 = \frac{1}{\sinh^2 \beta s - \sin^2 \beta s}$$

$$L_1 = \frac{\cosh \alpha s}{\sinh \alpha s} \qquad K_1 = (\alpha s) L_1$$

$$L_2 = 2D_0 (\sinh 2\beta s + \sin 2\beta s) \qquad K_2 = (\beta s)^3 L_2$$

$$L_3 = D_0 (\cosh 2\beta s - \cos 2\beta s) \qquad K_3 = (\beta s)^2 L_3$$

$$L_4 = D_0 (\sinh 2\beta s - \sin 2\beta s) \qquad K_4 = (\beta s) L_4$$

$$L_5 = \frac{1}{\sinh \alpha s} \qquad K_5 = (\alpha s) L_5$$

$$L_6 = 4D_0 (\cosh \beta s \sin \beta s + \sinh \beta s \cos \beta s) \qquad K_6 = (\beta s)^3 L_6$$

$$L_7 = 4D_0 \sinh \beta s \sin \beta s \qquad K_7 = (\beta s)^2 L_7$$

$$L_8 = 2D_0 (\cosh \beta s \sin \beta s - \sinh \beta s \cos \beta s) \qquad K_8 = (\beta s) L_8$$

$$\alpha = \sqrt{\frac{k_u}{EA_x}} \qquad \beta = \sqrt[4]{\frac{k_v}{4EI_z}}$$

(5) Stiffness matrix equations

$$\begin{bmatrix} U_{LR} \\ V_{LR} \\ Z_{LR} \\ \hline U_{RL} \\ V_{RL} \\ Z_{RL} \end{bmatrix} = \frac{EI_z}{s^3} \left[\begin{array}{ccc|ccc} K_1 r & 0 & 0 & -K_5 r & 0 & 0 \\ 0 & K_2 & K_3 s & 0 & -K_6 & K_2 s \\ 0 & K_3 s & K_4 s^2 & 0 & -K_7 s & K_8 s^2 \\ \hline -K_5 r & 0 & 0 & K_1 r & 0 & 0 \\ 0 & -K_6 & -K_7 s & 0 & K_2 & -K_3 s \\ 0 & K_7 s & K_8 s^2 & 0 & -K_3 s & K_4 s^2 \end{array} \right] \begin{bmatrix} u_L \\ v_L \\ \theta_L \\ \hline u_R \\ v_R \\ \theta_R \end{bmatrix} + \begin{bmatrix} U_{L0} \\ V_{L0} \\ Z_{L0} \\ \hline U_{R0} \\ V_{R0} \\ Z_{R0} \end{bmatrix}$$

where U_{L0}, V_{L0}, Z_{L0} and U_{R0}, V_{R0}, Z_{R0} are the reactions of a *fixed-end* beam *encased in elastic foundation* and acted on by loads and other causes shown in (8.12). The graphs and tables of stiffness coefficients are shown in (8.10–8.11).

(6) General stiffness matrix equations for beams of order three encased in elastic foundation are obtained by matrix composition of (8.09) and (13.07) as shown in (3.19).

8.10 TABLE OF DIMENSIONLESS STIFFNESS COEFFICIENTS†

βs	K_2	K_3	K_4	K_6	K_7	K_8
0.0	12.00000	6.00000	4.00000	12.00000	6.00000	2.00000
0.50	12.09283	6.01309	4.00238	11.96788	5.99227	1.99822
0.60	12.19245	6.02713	4.00493	11.93344	5.98397	1.99630
0.70	12.35638	6.05024	4.00913	11.87682	5.97034	1.99315
0.80	12.60757	6.08562	4.01556	11.79023	5.94948	1.98834
0.90	12.97228	6.13694	4.02488	11.66483	5.91926	1.98136
1.00	13.47992	6.20831	4.03784	11.49094	5.87734	1.97168
1.10	14.16286	6.30417	4.05522	11.25820	5.82118	1.95870
1.20	15.05606	6.42928	4.07789	10.95587	5.74815	1.94182
1.30	16.19674	6.58864	4.10673	10.57317	5.65559	1.92040
1.40	17.62381	6.78734	4.14262	10.09973	5.54090	1.89384
1.50	19.37734	7.03048	4.18644	9.52611	5.40164	1.86154
1.60	21.49791	7.32303	4.23903	8.84434	5.23569	1.82300
1.70	24.02584	7.66965	4.30114	8.04856	5.04135	1.77778
1.80	27.00048	8.07457	4.37342	7.13560	4.81750	1.72558
1.90	30.45950	8.54144	4.45637	6.10559	4.56374	1.66625
2.00	34.43820	9.07316	4.55034	4.96238	4.28045	1.59979
2.10	38.96903	9.67184	4.65549	3.71393	3.96890	1.52641
2.20	44.08117	10.33870	4.77178	2.37240	3.63129	1.44654
2.30	49.80045	11.07403	4.89895	0.95405	3.27074	1.36076
2.3650	54.27057	11.67805	5.02581	0.00000	3.04931	1.31229
2.50	63.14719	12.74703	5.18405	−2.02962	2.49738	1.17487
2.60	70.81093	13.68128	5.34061	−3.54601	2.09453	1.07680
2.70	79.15542	14.67742	5.50537	−5.04303	1.68831	0.97686
2.80	88.19430	15.73253	5.67740	−6.49306	1.28454	0.87627
2.90	97.94070	16.84348	5.85573	−7.86910	0.88900	0.77626
3.00	108.40791	18.00715	6.03937	−9.14572	0.50723	0.67801
3.1416	124.48919	19.73921	6.30670	−10.73928	0.00000	0.54406
3.20	131.56222	20.48093	6.41896	−11.31224	−0.19525	0.49114
3.30	144.28140	21.78592	6.61327	−12.16646	−0.50764	0.40436
3.40	157.78603	23.13349	6.80965	−12.85080	−0.78980	0.32302
3.50	172.09631	23.52204	7.00753	−13.35766	−1.03950	0.24767
3.60	187.23402	25.95038	7.20643	−13.68364	−1.25530	0.17870
3.70	203.22238	27.41766	7.40600	−13.82939	−1.43651	0.11637
3.80	220.08575	28.92336	7.60595	−13.79934	−1.58319	0.06077
3.9266	242.72937	30.88433	7.85930	−13.52164	−1.72046	0.00000
4.00	256.53915	32.04926	8.00621	−13.24600	−1.77621	−0.03041
4.10	276.18129	33.66953	8.20631	−12.74665	−1.82553	−0.06636
4.20	296.80215	35.32827	8.40630	−12.11841	−1.84608	−0.09625
4.30	318.42805	37.02576	8.60616	−11.37793	−1.84028	−0.12045
4.40	341.08512	38.76232	8.80591	−10.54283	−1.81075	−0.13937
4.50	364.79918	40.53824	9.00555	−9.63129	−1.76027	−0.15346
4.60	389.59574	42.35380	9.20512	−8.66161	−1.69169	−0.16316
4.70	415.49991	44.20925	9.40463	−7.65182	−1.60786	−0.16896
4.80	442.53647	46.10479	9.60411	−6.61934	−1.51160	−0.17131
490	470.72985	48.04057	9.80358	−5.58075	−1.40562	−0.17068
5.00	500.10422	50.01670	10.00307	−4.55143	−1.29251	−0.16750
5.4978	664.72402	60.45567	10.99671	0.00000	−0.70033	−0.12739

†Values of K_1 and K_5 are available in standard mathematical tables.

8.11 GRAPHS AND SERIES OF DIMENSIONLESS STIFFNESS COEFFICIENTS

(1) Graphs of K_j ($j = 2, 3, 4, 6, 7, 8$)

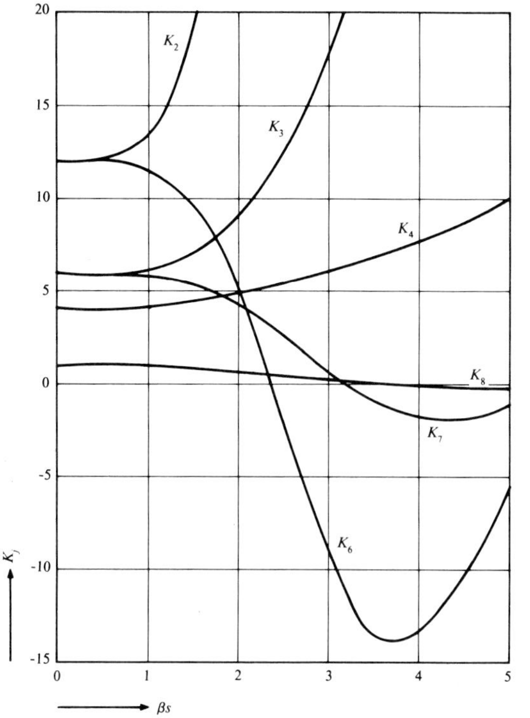

(2) Series expansion of K_j

$$K_1 = \frac{\displaystyle\sum_{k=0}^{\infty} \frac{(\alpha s)^{2k}}{(2k)!}}{\displaystyle\sum_{k=0}^{\infty} \frac{(\alpha s)^{2k}}{(2k+1)!}} = 1 + \sum_{k=1}^{\infty} C_{1k}(\alpha s)^{2k} \qquad K_5 = \frac{1}{\displaystyle\sum_{k=0}^{\infty} \frac{(\alpha s)^{2k}}{(2k+1)!}} = 1 + \sum_{k=1}^{\infty} C_{5k}(\alpha s)^{2k}$$

$$K_2 = \frac{1}{2}\,\frac{\displaystyle\sum_{k=0}^{\infty} \frac{(2\beta s)^{4k}}{(4k+1)!}}{\displaystyle\sum_{k=0}^{\infty} \frac{(2\beta s)^{4k}}{(4k+4)!}} = 12 + \sum_{k=1}^{\infty} C_{2k}(\beta s)^{4k} \qquad K_6 = 8\,\frac{\displaystyle\sum_{k=0}^{\infty} (-4)^k \frac{(\beta s)^{4k}}{(4k+1)!}}{\displaystyle\sum_{k=0}^{\infty} \frac{(2\beta s)^{4k}}{(4k+4)!}} = 12 + \sum_{k=1}^{\infty} C_{6k}(\beta s)^{4k}$$

$$K_3 = \frac{1}{2}\,\frac{\displaystyle\sum_{k=0}^{\infty} \frac{(2\beta s)^{4k}}{(4k+2)!}}{\displaystyle\sum_{k=0}^{\infty} \frac{(2\beta s)^{4k}}{(4k+4)!}} = 6 + \sum_{k=1}^{\infty} C_{3k}(\beta s)^{4k} \qquad K_7 = 8\,\frac{\displaystyle\sum_{k=0}^{\infty} (-4)^k \frac{(\beta s)^{4k}}{(4k+2)!}}{\displaystyle\sum_{k=0}^{\infty} \frac{(2\beta s)^{4k}}{(4k+4)!}} = 6 + \sum_{k=1}^{\infty} C_{7k}(\beta s)^{4k}$$

$$K_4 = \frac{\displaystyle\sum_{k=0}^{\infty} \frac{(2\beta s)^{4k}}{(4k+3)!}}{\displaystyle\sum_{k=0}^{\infty} \frac{(2\beta s)^{4k}}{(4k+4)!}} = 4 + \sum_{k=1}^{\infty} C_{4k}(\beta s)^{4k} \qquad K_8 = 8\,\frac{\displaystyle\sum_{k=0}^{\infty} (-4)^k \frac{(\beta s)^{4k}}{(4k+3)!}}{\displaystyle\sum_{k=0}^{\infty} \frac{(2\beta s)^{4k}}{(4k+4)!}} = 2 + \sum_{k=1}^{\infty} C_{8k}(\beta s)^{4k}$$

where $C_{1k}, C_{2k}, \ldots$ are the numerical constants of the respective series obtained by the method described in Ref. 1, p. 105.

8.12 STIFFNESS LOAD FUNCTIONS, FIXED-END BEAM

(1) General form of the stiffness load functions in (8.09) is given by the matrix equations

$$\begin{bmatrix} U_{LO} \\ V_{LO} \\ Z_{LO} \end{bmatrix} = \begin{bmatrix} L_5 & -0 & 0 \\ 0 & L_6 & -L_7 \\ 0 & L_7\beta^{-1} & -L_8\beta^{-1} \end{bmatrix} \begin{bmatrix} G_u \\ G_v \\ G_\theta \end{bmatrix}$$

$$\begin{bmatrix} U_{RO} \\ V_{RO} \\ Z_{RO} \end{bmatrix} = \begin{bmatrix} L_5 & 0 & 0 \\ 0 & L_6 & L_7 \\ 0 & -L_7\beta^{-1} & -L_8\beta^{-1} \end{bmatrix} \begin{bmatrix} F_u \\ F_v \\ F_\theta \end{bmatrix}$$

where F_u, F_v, F_θ are the left-end transport load functions and G_u, G_v, and G_θ are their right-end counterparts, given for particular loads and causes in (8.06–8.08), and L_5, L_6, L_7, L_8 are the stiffness coefficients (8.09). The load vectors and other causes are defined by symbols introduced in (8.06–8.08).

(2) Particular cases

Notation (8.02–8.03, 8.09) Signs (8.09) A_j, B_j, C_j, T_j = transport coefficients (8.03)		$\alpha = \sqrt{\dfrac{k_u}{EA_x}}$ $\beta = \sqrt[4]{\dfrac{k_v}{4EI_z}}$
$V_{LO} = L_5 B_{2n} P_x$ $V_{LO} = (L_7 B_{3z} - L_6 B_{4z}) P_y$ $Z_{LO} = (L_8 B_{3z} - L_7 B_{4z}) \beta^{-1} P_y$	**1.** 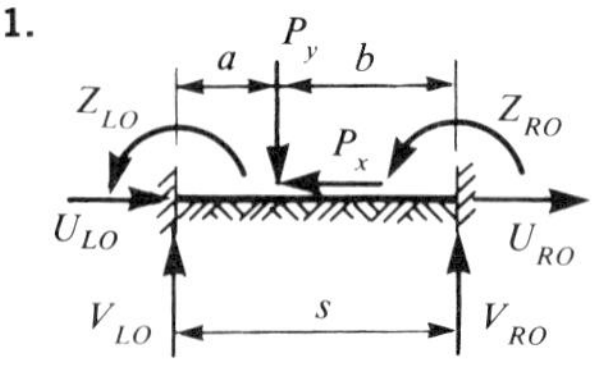	$U_{RO} = L_5 A_{2n} P_x$ $V_{RO} = (L_7 A_{3z} - L_6 A_{4z}) P_y$ $Z_{RO} = -(L_8 A_{3z} - L_7 A_{4z}) \beta^{-1} P_y$
$U_{LO} = 0$ $V_{LO} = -(L_7 B_{2z} - L_7 B_{3z}) \beta Q_z$ $Z_{LO} = -(L_8 B_{2z} - L_7 B_{3z}) Q_z$	**2.** 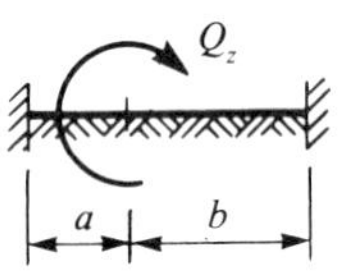	$U_{RO} = 0$ $V_{RO} = (L_7 A_{2z} - L_6 A_{3z}) \beta Q_z$ $Z_{RO} = -(L_8 A_{2z} - L_7 A_{3z}) Q_z$
$U_{LO} = \dfrac{\cosh \alpha s - 1}{\sinh \alpha s} \dfrac{p_x}{\alpha}$ $V_{LO} = \dfrac{\cosh \beta s + \cos \beta s}{\sinh \beta s + \sin \beta s} \dfrac{p_y}{\beta}$ $Z_{LO} = \dfrac{\sinh \beta s - \sin \beta s}{\sinh \beta s + \sin \beta s} \dfrac{p_y}{2\beta^2}$	**3.** 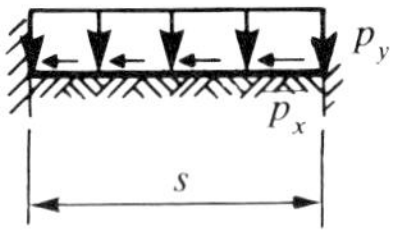	$U_{RO} = \dfrac{\cosh \alpha s - 1}{\sinh \alpha s} \dfrac{p_x}{\alpha}$ $V_{RO} = \dfrac{\cosh \beta s + \cos \beta s}{\sinh \beta s + \sin \beta s} \dfrac{p_y}{\beta}$ $Z_{RO} = -\dfrac{\sinh \beta s - \sin \beta s}{\sinh \beta s + \sin \beta s} \dfrac{p_y}{2\beta^2}$

$$U_{L0} = \frac{1}{2\cosh \frac{1}{2}\alpha s} P_x = U_{R0}$$

$$V_{L0} = \frac{\cosh \frac{1}{2}\beta s \sin \frac{1}{2}\beta s + \sinh \frac{1}{2}\beta s \cos \frac{1}{2}\beta s}{\sinh \beta s + \sin \beta s} P_y = V_{R0}$$

$$Z_{L0} = \frac{\sinh \frac{1}{2}\beta s \sin \frac{1}{2}\beta s}{\sinh \beta s + \sin \beta s} \frac{P_y}{\beta} = -Z_{R0}$$

4.

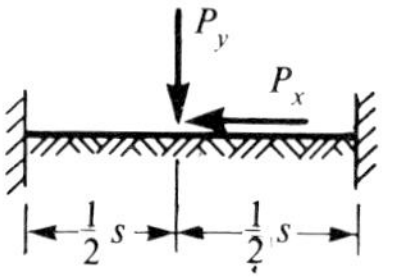

$$U_{L0} = 0 = U_{r0}$$

$$V_{L0} = -\frac{\sinh \frac{1}{2}\beta s \sin \frac{1}{2}\beta s}{\sinh \beta s + \sin \beta s} \beta Q_z = -V_{R0}$$

$$Z_{L0} = -\frac{\cosh \frac{1}{2}\beta s \sin \frac{1}{2}\beta s - \sinh \frac{1}{2}\beta s \cos \frac{1}{2}\beta s}{\sinh \beta s + \sin \beta s} Q_z = Z_{R0}$$

5.

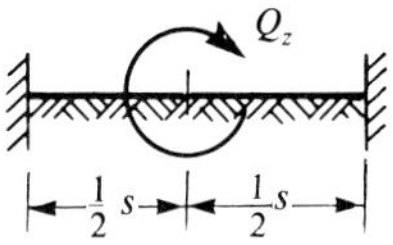

6.

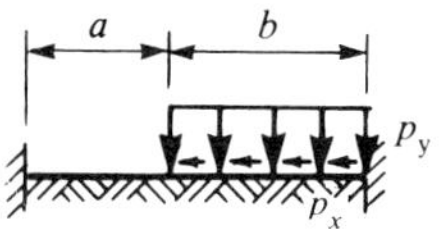

Use general form in terms of (8.07-2)

7.

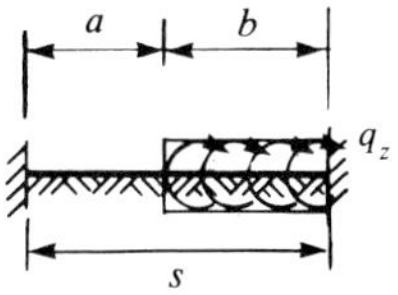

Use general form in terms of (8.06-4)

8.

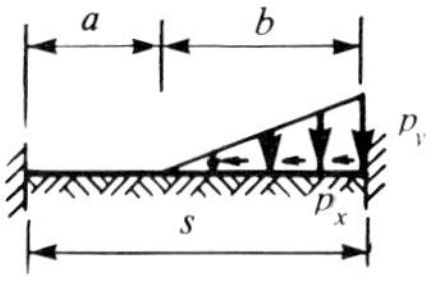

Use general form in terms of (8.07-4)

9. Abrupt displacements u_j, v_j

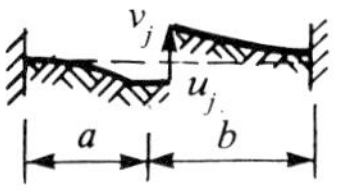

$$U_{L0} = EA_x L_5 B_{1n} \alpha u_j$$

$$V_{L0} = EI_z(L_6 B_{1z} + 4L_7 B_{4z})\beta^3 v_j$$

$$Z_{L0} = EI_z(L_7 B_{1z} + 4L_8 B_{4z})\beta^2 v_j$$

$$U_{R0} = -EA_x L_5 A_{1n} \alpha u_j$$

$$V_{R0} = -EI_z(L_6 A_{1z} + 4L_7 A_{4z})\beta^3 v_j$$

$$Z_{R0} = EI_z(L_7 A_{1z} + 4L_8 A_{4z})\beta^2 v_j$$

10. Abrupt displacement θ_j

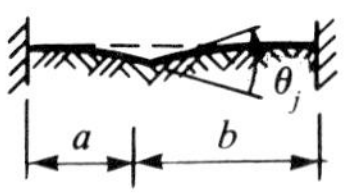

$$U_{L0} = 0$$

$$V_{L0} = EI_z(L_6 B_{2z} - L_7 B_{1z})\beta^2 \theta_j$$

$$Z_{L0} = EI_z(L_7 B_{2z} - L_8 B_{1z})\beta \theta_j$$

$$U_{R0} = 0$$

$$V_{R0} = EI_z(L_6 A_{2z} - L_7 A_{1z})\beta^2 \theta_j$$

$$Z_{R0} = -EI_z(L_7 A_{2z} - L_8 A_{1z})\beta \theta_j$$

11. Temperature deformation of segment b

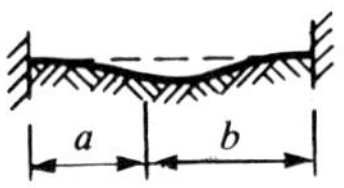

$$U_{L0} = EA_x L_5 B_{2n} e_t$$

$$V_{L0} = EI_z(L_6 B_{3z} - L_7 B_{2z})\beta f_t$$

$$Z_{L0} = EI_z(L_7 B_{3z} - L_8 B_{2z}) f_t$$

$$U_{R0} = -EA_x L_5 C_{2n} e_t$$

$$V_{R0} = EI_z(L_6 C_{3z} - L_7 C_{2z})\beta f_t$$

$$Z_{R0} = -EI_z(L_7 C_{3z} - L_8 C_{2z}) f_t$$

8.13 MODIFIED STIFFNESS MATRIX EQUATIONS

(1) Dimensionless stiffness coefficients

	$\beta = \sqrt[4]{\dfrac{k_v}{4EI_z}}$
$D_1 = \dfrac{1}{\cosh \beta s \sinh \beta s - \cos \beta s \sin \beta s}$	$D_2 = \dfrac{1}{\cosh^2 \beta s + \cos^2 \beta s}$
$L_9 = 2D_1 (\cosh^2 \beta s + \cos^2 \beta s)$	$K_9 = (\beta s)^3 L_9$
$L_{10} = 4D_1 \cosh \beta s \cos \beta s$	$K_{10} = (\beta s)^3 L_{10}$
$L_{11} = 2D_1 (\cosh \beta s \sin \beta s + \sinh \beta s \cos \beta s)$	$K_{11} = (\beta s)^3 L_{11}$
$L_{12} = 4D_1 (\cosh^2 \beta s - \sin^2 \beta s)$	$K_{12} = (\beta s)^2 L_{12}$
$L_{13} = 2D_1 (\cosh \beta s \sinh \beta s + \cos \beta s \sin \beta s)$	$K_{13} = (\beta s)^2 L_{13}$
$L_{14} = 2D_1 (\sinh^2 \beta s + \sin^2 \beta s)$	$K_{14} = (\beta s) L_{14}$
$L_{15} = 2D_1 \sinh \beta s \sin \beta s$	
$L_{16} = D_1 (\cosh \beta s \sin \beta s - \sinh \beta s \cos \beta s)$	
$L_{17} = 4D_2 (\cosh \beta s \sinh \beta s + \cos \beta s \sin \beta s)$	$K_{17} = (\beta s)^3 L_{17}$
$L_{18} = 2D_2 (\cosh^2 \beta s - \cos^2 \beta s)$	$K_{18} = (\beta s)^2 L_{18}$
$L_{19} = 2D_2 (\cosh \beta s \sinh \beta s - \cos \beta s \sin \beta s)$	$K_{19} = (\beta s) L_{19}$
$L_{20} = 2D_2 \cosh \beta s \cos \beta s$	
$L_{21} = 2D_2 (\cosh \beta s \sin \beta s - \sinh \beta s \cos \beta s)$	
$L_{22} = D_2 (\cosh \beta s \sin \beta s + \sinh \beta s \cos \beta s)$	

(2) Modifications of stiffness matrix equations (8.09) due to hinged end and to free end are shown in (3, 4) and (5, 6), respectively. Since these modifications do not affect the expressions for U_{LR} and U_{RL}, only the modified part of the respective stiffness matrix is presented. The modified load functions designated by the asterisk are the reactions of the respective propped-end beams in (3, 4) and of the respective cantilever beams in (5, 6). They are all linear combinations of the L coefficients and the transport load functions F or G introduced in (8.06–8.08).

(3) Left end hinged ($Z_{LR} = 0$)

$$\begin{bmatrix} V_{LR} \\ V_{RL} \\ Z_{RL} \end{bmatrix} = \frac{EI_z}{s^3} \begin{bmatrix} K_9 & -K_{10} & K_{11}s \\ -K_{10} & K_{12} & -K_{13}s \\ K_{11}s & -K_{13}s & K_{14}s^2 \end{bmatrix} \begin{bmatrix} v_L \\ v_R \\ \theta_R \end{bmatrix} + \begin{bmatrix} V_{L0}^* \\ V_{R0}^* \\ Z_{R0}^* \end{bmatrix}$$

$$\begin{bmatrix} V_{L0}^* \\ V_{R0}^* \\ Z_{R0}^* \end{bmatrix} = \begin{bmatrix} -1 & \frac{1}{2}L_{14} & -L_9 \\ 0 & -L_{15} & L_{10} \\ 0 & L_{16}\beta^{-1} & L_{11}\beta^{-1} \end{bmatrix} \begin{bmatrix} F_V \\ F_Z \\ F_v \end{bmatrix}$$

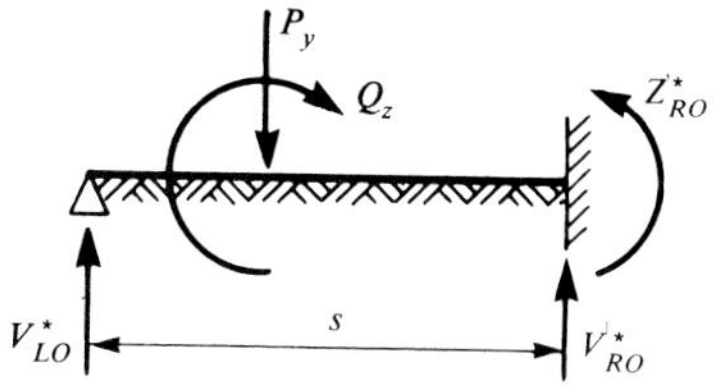

(4) Right end hinged ($Z_{RL} = 0$)

$$\begin{bmatrix} V_{LR} \\ Z_{LR} \\ V_{RL} \end{bmatrix} = \frac{EI_z}{s^3} \begin{bmatrix} K_{12} & K_{13}s & -K_{10} \\ K_{13}s & K_{14}s^2 & -K_{11}s \\ -K_{10} & -K_{11}s & K_9 \end{bmatrix} \begin{bmatrix} v_L \\ \theta_L \\ v_R \end{bmatrix} + \begin{bmatrix} V_{L0}^* \\ Z_{L0}^* \\ V_{R0}^* \end{bmatrix}$$

$$\begin{bmatrix} V_{L0}^* \\ Z_{L0}^* \\ V_{R0}^* \end{bmatrix} = \begin{bmatrix} 0 & -L_{15} & L_{10} \\ 0 & -L_{16}\beta^{-1} & L_{11}\beta^{-1} \\ 1 & \frac{1}{2}L_{14} & L_9 \end{bmatrix} \begin{bmatrix} G_V \\ G_Z \\ G_v \end{bmatrix}$$

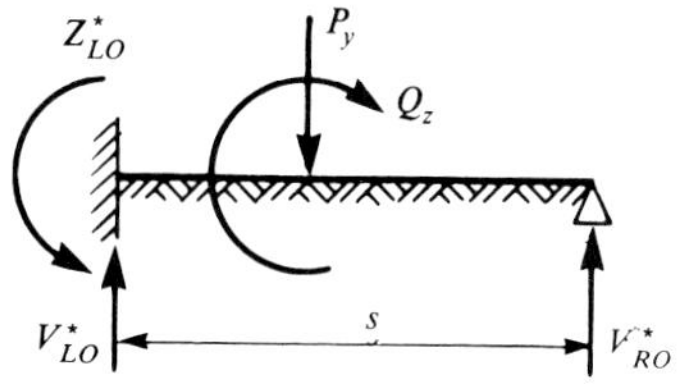

(5) Left end free ($V_{LR} = 0,\ Z_{LR} = 0$)

$$\begin{bmatrix} V_{RL} \\ Z_{RL} \end{bmatrix} = \frac{EI_z}{s^3} \begin{bmatrix} K_{17} & -K_{18}s \\ -K_{18}s & K_{19}s^2 \end{bmatrix} \begin{bmatrix} v_R \\ \theta_R \end{bmatrix} + \begin{bmatrix} V_{R0}^* \\ Z_{R0}^* \end{bmatrix}$$

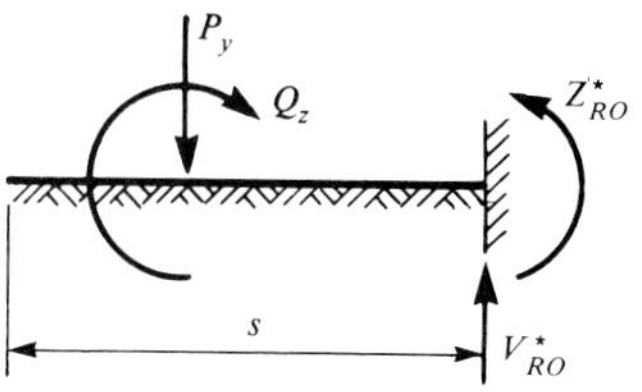

$$\begin{bmatrix} V_{R0}^* \\ Z_{R0}^* \end{bmatrix} = \begin{bmatrix} -L_{20} & -L_{21} \\ L_{22}\beta^{-1} & -L_{20}\beta^{-1} \end{bmatrix} \begin{bmatrix} F_V \\ F_Z \end{bmatrix}$$

(6) Right end free ($V_{RL} = 0,\ Z_{RL} = 0$)

$$\begin{bmatrix} V_{LR} \\ Z_{LR} \end{bmatrix} = \frac{EI_2}{s^3} \begin{bmatrix} K_{17} & K_{18}s \\ K_{18}s & K_{19}s^2 \end{bmatrix} \begin{bmatrix} v_L \\ \theta_L \end{bmatrix} + \begin{bmatrix} V_{L0}^* \\ Z_{L0}^* \end{bmatrix}$$

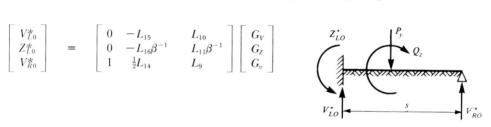

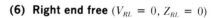

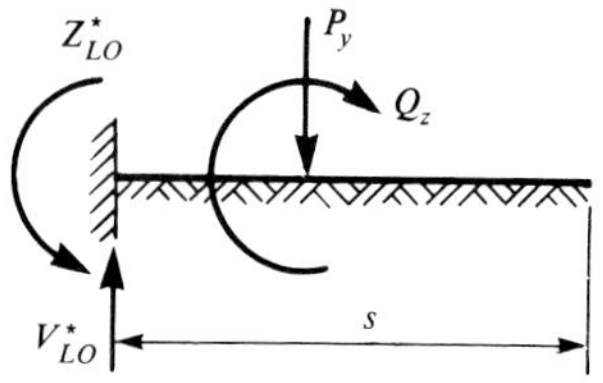

$$\begin{bmatrix} V_{L0}^* \\ Z_{L0}^* \end{bmatrix} = \begin{bmatrix} L_{20} & L_{21} \\ L_{22}\beta^{-1} & L_{20}\beta^{-1} \end{bmatrix} \begin{bmatrix} G_V \\ G_Z \end{bmatrix}$$

8.14 SEMI-INFINITE BAR ENCASED IN ELASTIC FOUNDATION

(1) System. If $\beta s > 5.0$, the effect of the far-end state vector in (8.02) is negligible and the bar can be assumed to be semi-infinite, which means the far-end state vector is assumed to be zero. In such cases the stiffness matrix equations reduce to two very simple forms shown below in terms of the symbols and signs of (8.09).

$$\alpha = \sqrt{\frac{k_u}{EA_x}} \qquad \beta = \sqrt[4]{\frac{k_v}{4EI_z}} \qquad r = \frac{s^2 EA_x}{EI_z}$$

(2) Left end finite

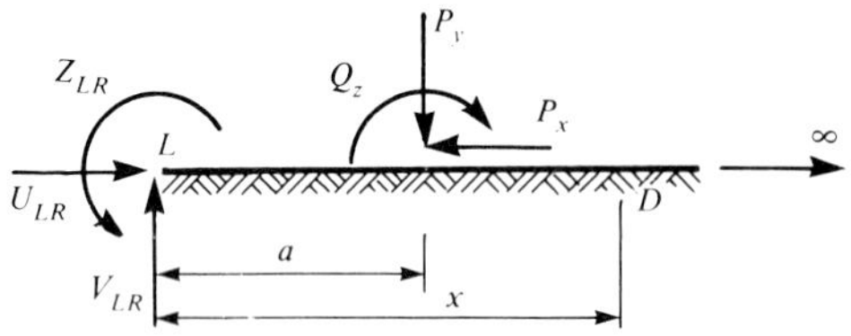

(3) Right end finite

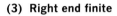

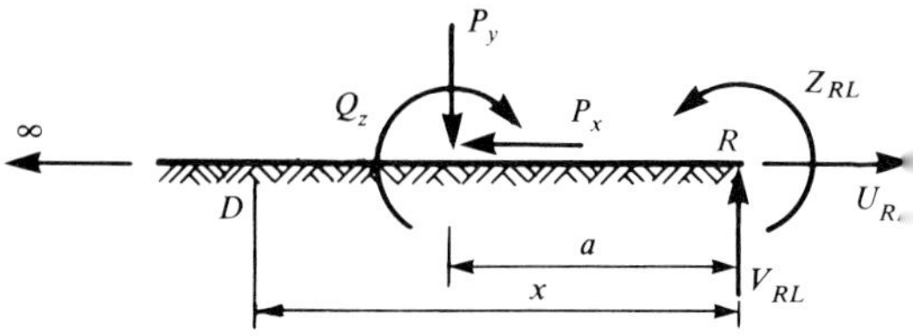

(4) Stiffness matrix equations

$$\begin{bmatrix} U_{LR} \\ V_{LR} \\ Z_{LR} \end{bmatrix} = EI_z \begin{bmatrix} ras^{-2} & -0 & -0 \\ 0 & 4\beta^3 & 2\beta^2 \\ 0 & 2\beta^2 & 2\beta \end{bmatrix} \begin{bmatrix} u_L \\ v_L \\ \theta_L \end{bmatrix} + \begin{bmatrix} U_{L0} \\ V_{L0} \\ Z_{L0} \end{bmatrix}$$

$$\begin{bmatrix} U_{RL} \\ V_{RL} \\ Z_{RL} \end{bmatrix} = EI_z \begin{bmatrix} ras^{-2} & 0 & 0 \\ 0 & 4\beta^3 & -2\beta^2 \\ 0 & -2\beta^2 & 2\beta \end{bmatrix} \begin{bmatrix} u_R \\ v_R \\ \theta_R \end{bmatrix} + \begin{bmatrix} U_{R0} \\ V_{R0} \\ Z_{R0} \end{bmatrix}$$

(5) Load functions $U_{L0}, V_{L0}, \ldots, Z_{R0}$ given as the last matrix columns in (4) are the *reactions due to loads and other causes* at the fixed end of the respective semi-infinite beams. Their analytical expressions are tabulated in (8.15–8.16). The shape functions used in these expressions are introduced in (6) below and in (8); they take on particular values given by the argument x.

$$\left.\begin{array}{lll} \text{For } x = a, & T_{ijn}(a) = A_{ijn}, & T_{ijz}(a) = A_{ijz} \\ \text{For } x = b, & T_{ijn}(b) = B_{ijn}, & T_{ijz}(b) = B_{ijz} \\ \text{and } T_{ijn}(b) - T_{ijn}(a) = C_{ijn}, & & T_{ijz}(b) - T_{ijz}(a) = C_{ijz} \end{array}\right\} \quad ij = 11, 12, \ldots$$

(6) Shape functions

$T_{11n}(x) = e^{-\alpha x}$	$T_{11z}(x) = e^{-\beta x}(\cos\beta x + \sin\beta x)$	$T_{12z}(x) = e^{-\beta x}\cos\beta x$
$T_{12n}(x) = 1 - e^{-\alpha x}$	$T_{13z}(x) = e^{-\beta x}(\cos\beta x - \sin\beta x)$	$T_{14z}(x) = e^{-\beta x}\sin\beta x$
$T_{13n}(x) = 1 - (1 + \alpha x)e^{-\alpha x}$	$T_{15z}(x) = \frac{1}{2}[1 - T_{12z}(x) - \beta x T_{11z}(x)]$	$T_{16z}(x) = \frac{1}{2}[1 - T_{11z}(x)]$
	$T_{17z}(x) = \frac{1}{4}[1 - T_{13z}(x) - 2\beta x T_{12z}(x)]$	$T_{18z}(x) = \frac{1}{2}[1 - T_{12z}(x)]$

(7) State vector at L **or at** R can be calculated by the standard application of the stiffness method, and it may include any prescribed end condition at that end. For isolated beams, which are not a part of a continuous system, the stiffness matrix equations in (4) provide directly the calculation of this vector. *State vector* at an arbitrary section D is calculated in terms of the state vector of the finite end by means of the transport matrix equations (8.02) in terms of their load functions (8.06–8.08).

(8) Table of shape coefficients

βx	$T_{11}(x)$	$T_{12}(x)$	$T_{13}(x)$	$T_{14}(x)$	$T_{15}(x)$	$T_{16}(x)$	$T_{17}(x)$	$T_{18}(x)$
0.0	1.00000	1.00000	1.00000	0.00000	0.00000	0.00000	0.00000	0.00000
0.10	0.99065	0.90032	0.80998	0.09033	0.00031	0.00467	0.00498	0.04984
0.20	0.96507	0.80241	0.63975	0.16266	0.00229	0.01747	0.01964	0.09879
0.30	0.92666	0.70773	0.48880	0.21893	0.00714	0.03667	0.04328	0.14613
0.40	0.87844	0.61741	0.35637	0.26103	0.01561	0.06078	0.07485	0.19130
0.50	0.82307	0.53228	0.24149	0.29079	0.02809	0.08847	0.11311	0.23386
0.60	0.76284	0.45295	0.14307	0.30988	0.04467	0.11858	0.15669	0.27352
0.70	0.69972	0.37981	0.05990	0.31991	0.06519	0.15014	0.20418	0.31010
0.25π	0.64479	0.32240	0.00000	0.32240	0.08559	0.17760	0.24679	0.33880
0.90	0.57120	0.25273	−0.06575	0.31848	0.11659	0.21440	0.30542	0.37364
1.00	0.50833	0.19877	−0.11079	0.30956	0.14645	0.24584	0.35663	0.40062
1.10	0.44765	0.15099	−0.14567	0.29666	0.17830	0.27618	0.40675	0.42451
1.20	0.38986	0.10914	−0.17158	0.28072	0.21151	0.30507	0.45482	0.44543
1.30	0.33550	0.07290	−0.18970	0.26260	0.24547	0.33225	0.50008	0.46355
1.40	0.28492	0.04191	−0.20110	0.24301	0.27960	0.35754	0.54187	0.47904
1.50	0.23835	0.01578	−0.20679	0.22257	0.31334	0.38082	0.57972	0.49211
0.5π	0.20788	0.00000	−0.20788	0.20788	0.33673	0.39606	0.60394	0.50000
1.70	0.15762	−0.02354	−0.20470	0.18116	0.37779	0.42119	0.64236	0.51177
1.80	0.12342	−0.03756	−0.19853	0.16098	0.40770	0.43829	0.66687	0.51878
1.90	0.09318	−0.04835	−0.18989	0.14154	0.43565	0.45341	0.68682	0.52418
2.00	0.06674	−0.05632	−0.17938	0.12306	0.46142	0.46663	0.70233	0.52816
2.10	0.04388	−0.06182	−0.16753	0.10571	0.48483	0.47806	0.71359	0.53091
2.20	0.02438	−0.06521	−0.15479	0.08958	0.50579	0.48781	0.72085	0.53260
2.30	0.00796	0.06680	−0.14156	0.07476	0.52424	0.49602	0.72442	0.53340
0.75π	0.00000	−0.06702	−0.13404	0.06702	0.53351	0.50000	0.72493	0.53351
2.50	−0.01664	−0.06576	−0.11489	0.04913	0.55368	0.50832	0.72185	0.53288
2.60	−0.02536	−0.06364	−0.10193	0.03829	0.56478	0.51268	0.71644	0.53182
2.70	−0.03204	−0.06076	−0.08948	0.02872	0.57363	0.51602	0.70879	0.53038
2.80	−0.03693	−0.05730	−0.07767	0.02037	0.58034	0.51846	0.69926	0.52865
2.90	−0.04026	−0.05343	−0.06659	0.01316	0.58509	0.52013	0.68823	0.52671
3.00	−0.04226	−0.04929	−0.05631	0.00703	0.58804	0.52113	0.67602	0.52464
1.00π	−0.04321	−0.04321	−0.04321	0.00000	0.58949	0.52161	0.65737	0.52161
3.20	−0.04307	−0.04069	−0.03831	−0.00238	0.58926	0.52154	0.64937	0.52035
3.30	−0.04224	−0.03642	−0.03060	−0.00582	0.58791	0.52112	0.63549	0.51821
3.40	−0.04079	−0.03227	−0.02374	−0.00853	0.58548	0.52040	0.62157	0.51613
3.50	−0.03887	−0.02828	−0.01769	−0.01059	0.58216	0.51944	0.60782	0.51414
3.60	−0.03659	−0.02450	−0.01241	−0.01209	0.57812	0.51830	0.59442	0.51225
3.70	−0.03407	−0.02097	−0.00787	−0.01310	0.57351	0.51703	0.58152	0.51048
3.80	−0.03138	−0.01769	−0.00401	−0.01369	0.56847	0.51569	0.56924	0.50885
1.25π	−0.02786	−0.01393	0.00000	−0.01393	0.56168	0.51393	0.55471	0.50697
1.50π	−0.00898	0.00000	0.00898	−0.00898	0.52117	0.50449	0.49551	0.50000
1.75π	0.00000	0.00290	0.00579	−0.00290	0.49855	0.50000	0.48118	0.49855
2.00π	0.00187	0.00187	0.00187	0.00000	0.49320	0.49907	0.48733	0.49907
3.00π	−0.00008	−0.00008	−0.00008	0.00000	0.50042	0.50004	0.50080	0.50004
4.00π	0.00000	0.00000	0.00000	0.00000	0.49998	0.50000	0.49995	0.50000

8.15 SEMI-INFINITE BAR ENCASED IN ELASTIC FOUNDATION GRAPHS OF SHAPE FUNCTIONS

$T_{11z}(x) = e^{-\beta x}(\cos \beta x + \sin \beta x)$

$T_{12z}(x) = e^{-\beta x} \cos \beta x$

$T_{13z}(x) = e^{-\beta x}(\cos \beta x - \sin \beta x)$

$T_{14z}(x) = e^{-\beta x} \sin \beta x$

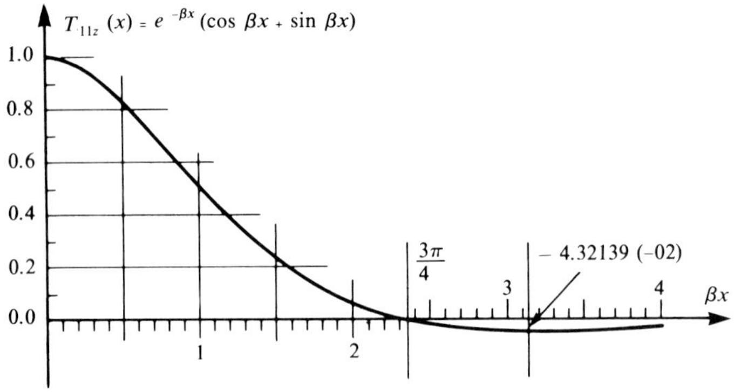

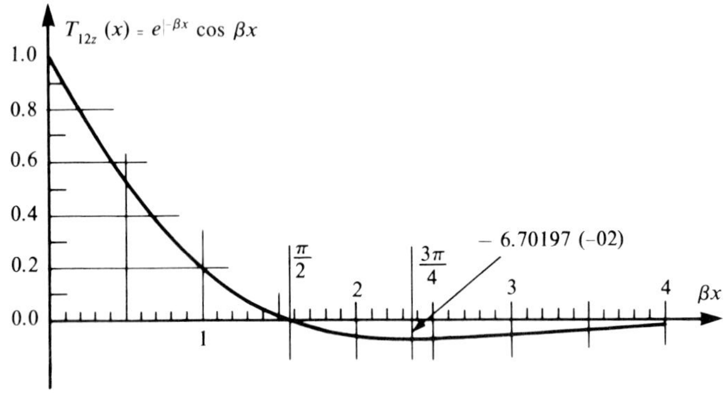

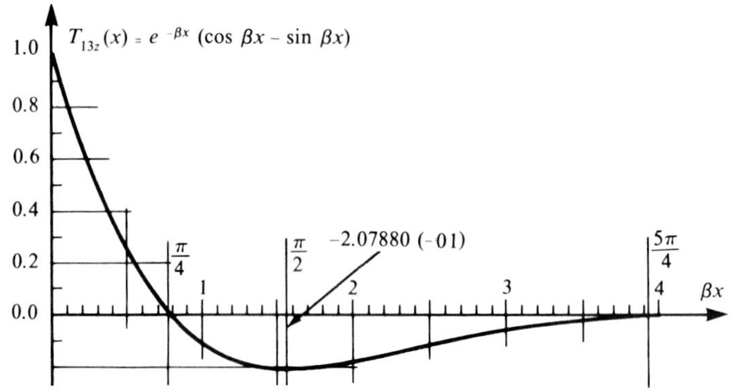

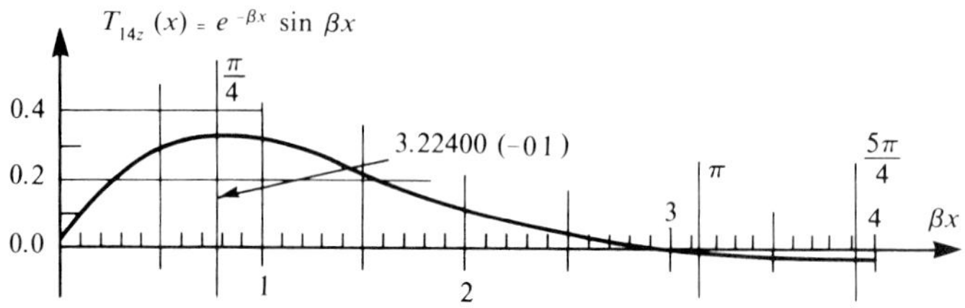

8.16 SEMI-INFINITE FIXED-END BEAM FUNCTIONS

Notation (8.09, 8.14) Signs (8.09) Matrix (8.14) P_x, P_y = concentrated loads (N) p_x, p_y = intensities of distributed loads (N/m) A_{ij}, C_{ij} = shape coefficients (8.14)			$\alpha = \sqrt{\dfrac{k_u}{EA_x}}$ $\beta = \sqrt[4]{\dfrac{k_v}{4EI_2}}$
$U_{L0} = A_{11n}P_x$ $V_{L0} = A_{11z}P_y$ $Z_{L0} = A_{14z}\beta^{-1}P_y$	**1.** 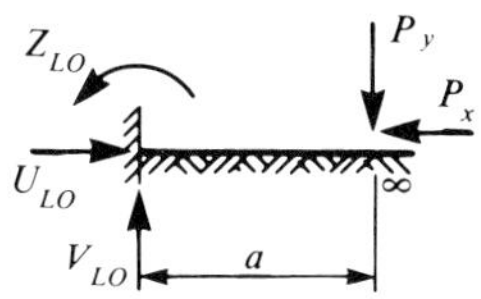	**2.** 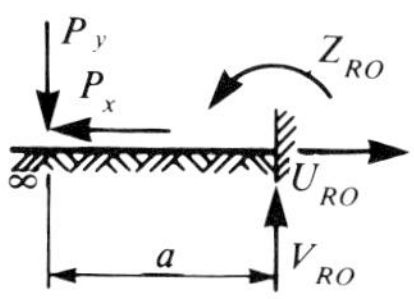	$U_{R0} = A_{11n}P_x$ $V_{R0} = A_{11z}P_y$ $Z_{R0} = -A_{14z}\beta^{-1}P_y$
$U_{L0} = \alpha^{-1}p_x$ $V_{L0} = \beta^{-1}p_y$ $Z_{L0} = \frac{1}{2}\beta^{-2}p_y$	**3.** 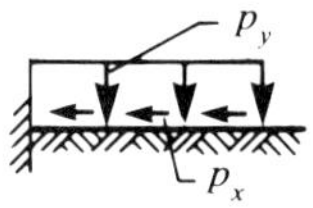	**4.** 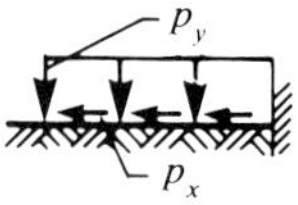	$U_{R0} = \alpha^{-1}p_x$ $V_{R0} = \beta^{-1}p_y$ $Z_{R0} = -\frac{1}{2}\beta^{-2}p_y$
$U_{L0} = A_{12n}\alpha^{-1}p_x$ $V_{L0} = 2A_{18z}\beta^{-1}p_y$ $Z_{L0} = A_{16z}\beta^{-2}p_y$	**5.** 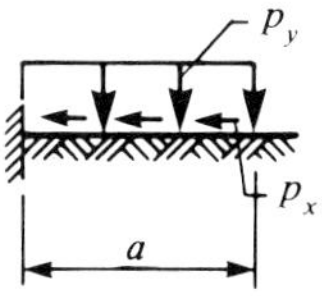	**6.** 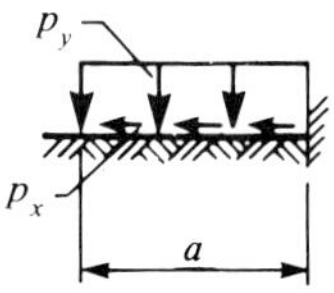	$U_{R0} = A_{12n}\alpha^{-1}p_x$ $V_{R0} = 2A_{18z}\beta^{-1}p_y$ $Z_{R0} = -A_{16z}\beta^{-2}p_y$
$U_{L0} = C_{12n}\alpha^{-1}p_x$ $V_{L0} = C_{12z}\beta^{-1}p_y$ $Z_{L0} = \frac{1}{2}C_{11z}\beta^{-2}p_y$	**7.** 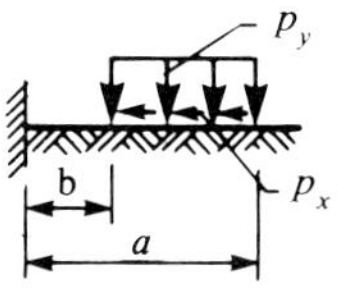	**8.** 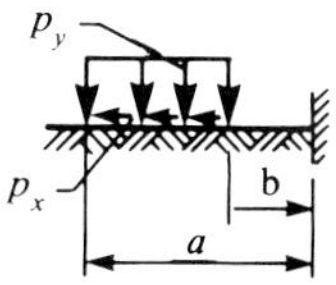	$U_{R0} = C_{12n}\alpha^{-1}p_x$ $V_{R0} = C_{12n}\beta^{-1}p_x$ $Z_{R0} = -\frac{1}{2}C_{11z}\beta^{-2}p_y$
$U_{L0} = A_{13n}\alpha^{-2}a^{-1}p_x$ $V_{L0} = A_{17z}\beta^{-2}a^{-1}p_y$ $Z_{L0} = A_{15z}\beta^{-3}a^{-1}p_y$	**9.** 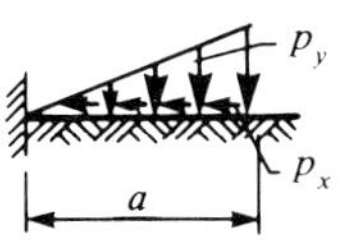	**10.** 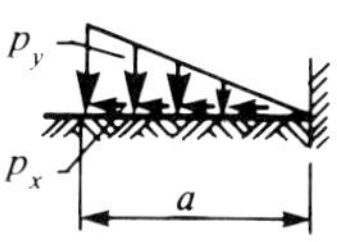	$U_{R0} = A_{13n}\alpha^{-2}a^{-1}p_x$ $V_{R0} = A_{17z}\beta^{-2}a^{-1}p_y$ $Z_{R0} = -A_{15z}\beta^{-3}a^{-1}p_y$

8.17 SEMI-INFINITE FIXED-END BEAM FUNCTIONS

Notation (8.09, 8.14) Signs (8.09) Matrix (8.14) Q_z = applied couple (N·m) e_f, f_t = temperature strains (4.03) q_z = intensity of distributed couple (N·m/m) A_{ij}, C_{ij} = shape coefficients (8.14)			$\alpha = \sqrt{\dfrac{k_u}{EA_x}}$ $\beta = \sqrt[4]{\dfrac{k_v}{4EI_z}}$
$U_{L0} = 0$ $V_{L0} = -2A_{14z}\beta Q_z$ $Z_{L0} = -A_{13z}Q_z$	**1.** 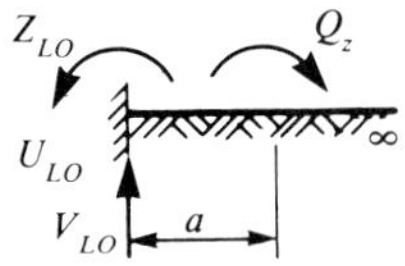 	**2.** 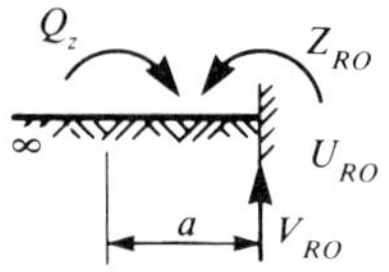 	$U_{R0} = 0$ $V_{R0} = 2A_{14z}\beta Q_z$ $Z_{R0} = A_{13z}Q_z$
$U_{L0} = 0$ $V_{L0} = -A_{16z}q_z$ $Z_{L0} = A_{14z}\beta^{-1}q_z$	**3.** 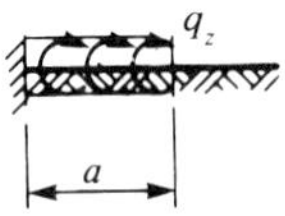 	**4.** 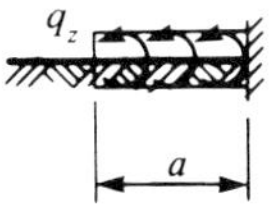 	$U_{R0} = 0$ $V_{R0} = A_{16z}q_z$ $Z_{R0} = A_{14z}\beta^{-1}q_z$
$U_{L0} = EA_xA_{11n}\alpha u_j$ $V_{L0} = 4EI_zA_{12z}\beta^3 v_j$ $Z_{L0} = 2EI_zA_{11z}\beta^2 v_j$	**5. Abrupt displacements** u_j, v_j 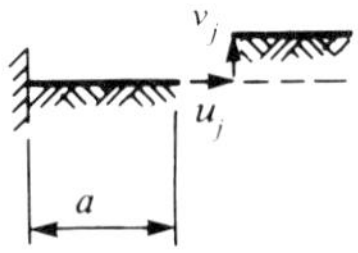 	**6. Abrupt displacements** u_j, v_j 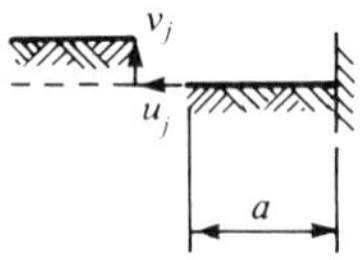 	$U_{R0} = -EA_xA_{11n}\alpha u_j$ $V_{R0} = 4EI_zA_{12z}\beta^3 v_j$ $Z_{R0} = -2EI_zA_{11z}\beta^2 v_j$
$U_{L0} = 0$ $V_{L0} = 2EI_zA_{13z}\beta^2\theta_j$ $Z_{L0} = 2EI_zA_{12z}\beta\theta_j$	**7. Abrupt displacement** θ_j 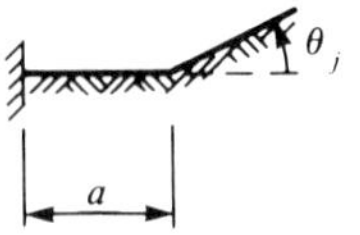 	**8. Abrupt displacement** θ_j 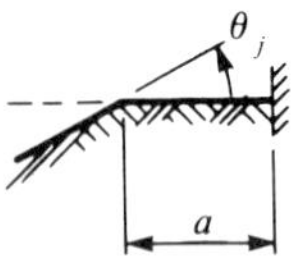 	$U_{R0} = 0$ $V_{R0} = -2EI_zA_{13z}\beta^2\theta_j$ $Z_{R0} = 2EI_zA_{12z}\beta\theta_j$
$U_{L0} = EA_xA_{12n}e_t$ $V_{L0} = EI_zA_{14z}\beta^{-1}f_t$ $Z_{L0} = EI_z(1 - A_{13z})f_t$	**9. Temperature deformation of segment** a 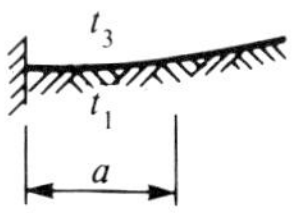 	**10. Temperature deformation of segment** a 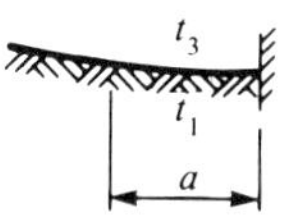 	$U_{R0} = EA_xA_{12n}e_t$ $V_{R0} = EI_zA_{14z}b^{-1}f_t$ $Z_{R0} = -EI_z(1 - A_{13z})f_t$

9

Free Beam-Column of Order One Constant Section

STATIC STATE

9.01 DEFINITION OF STATE

(1) System considered is a finite straight beam-column of constant and symmetrical section. The centroidal axis x of the bar and the vertical axis y define the plane of symmetry of this section. The axial force N is constant along the entire span s. The bar is a part of a planar system acted on by loads in the system plane (system of order one).

(2) Notation (A.1, A.2, A.4)

s = span (m)	A_x = area of normal section (m^2)
a, b = segments (m)	I_z = moment of inertia of A_x about z (m^4)
x, x' = coordinates (m)	E = modulus of elasticity (Pa)
U = normal force (N)	u = linear displacement along x (m)
V = shearing force (N)	v = linear displacement along y (m)
Z = flexural moment (N·m)	θ = flexural slope about z (rad)

$\eta = \begin{cases} +1 & \text{if } N = \text{axial tension (N)} \\ -1 & \text{if } N = \text{axial compression (N)} \end{cases}$	$\beta = \sqrt{\dfrac{N}{EI_z}}$ = shape parameter (m^{-1})

(3) Assumptions of analysis are stated in (1.01). In this chapter, the effects of axial and flexural deformations and the beam-column effects are included. Since only long and slender bars are capable of developing the beam-column action, the effects of shearing deformations are neglected.

9.02 TRANSPORT MATRIX EQUATIONS

(1) Equivalents and signs

$\bar{u} = uEA_x s^{-1}$ = scaled u (N) $\qquad \bar{v} = vEI_z\beta^3$ = scaled v (N)

$\bar{Z} = Z\beta$ = scaled Z (N) $\qquad \bar{\theta} = \theta EI_z\beta^2$ = scaled θ (N)

All forces, moments, and displacements are in the principal axes of the normal section, and their positive directions are shown in (2, 3) below.

(2) Free-body diagram

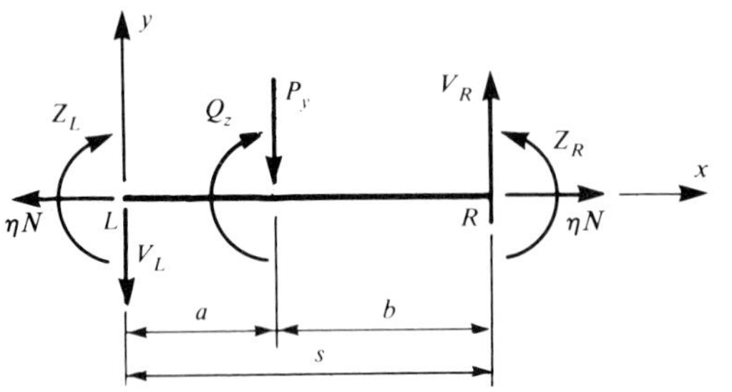

(3) Elastic curve

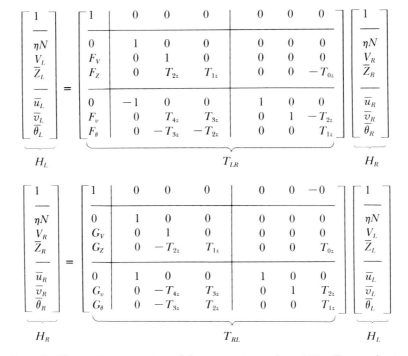

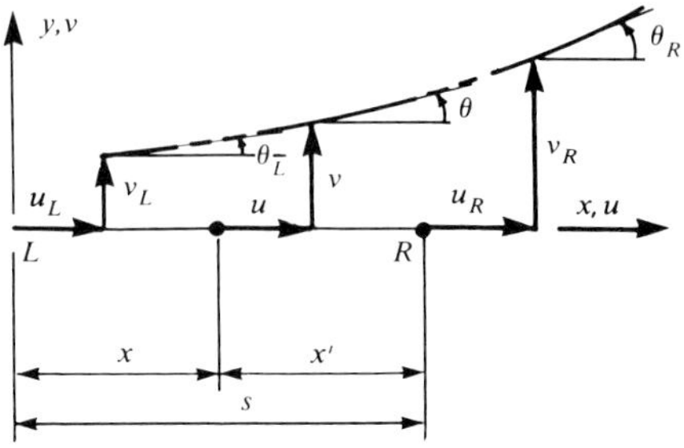

(4) Transport matrix equations

$$\underbrace{\left[\begin{array}{c} 1 \\ \hline \eta N \\ V_L \\ \bar{Z}_L \\ \hline \bar{u}_L \\ \bar{v}_L \\ \bar{\theta}_L \end{array}\right]}_{H_L} = \underbrace{\left[\begin{array}{c|ccc|ccc} 1 & 0 & 0 & 0 & 0 & 0 & 0 \\ \hline 0 & 1 & 0 & 0 & 0 & 0 & 0 \\ F_V & 0 & 1 & 0 & 0 & 0 & 0 \\ F_Z & 0 & T_{2z} & T_{1z} & 0 & 0 & -T_{0z} \\ \hline 0 & -1 & 0 & 0 & 1 & 0 & 0 \\ F_v & 0 & T_{4z} & T_{3z} & 0 & 1 & -T_{2z} \\ F_\theta & 0 & -T_{3z} & -T_{2z} & 0 & 0 & T_{1z} \end{array}\right]}_{T_{LR}} \underbrace{\left[\begin{array}{c} 1 \\ \hline \eta N \\ V_R \\ \bar{Z}_R \\ \hline \bar{u}_R \\ \bar{v}_R \\ \bar{\theta}_R \end{array}\right]}_{H_R}$$

$$\underbrace{\left[\begin{array}{c} 1 \\ \hline \eta N \\ V_R \\ \bar{Z}_R \\ \hline \bar{u}_R \\ \bar{v}_R \\ \bar{\theta}_R \end{array}\right]}_{H_R} = \underbrace{\left[\begin{array}{c|ccc|ccc} 1 & 0 & 0 & 0 & 0 & 0 & -0 \\ \hline 0 & 1 & 0 & 0 & 0 & 0 & 0 \\ G_V & 0 & 1 & 0 & 0 & 0 & 0 \\ G_Z & 0 & -T_{2z} & T_{1z} & 0 & 0 & T_{0z} \\ \hline 0 & 1 & 0 & 0 & 1 & 0 & 0 \\ G_v & 0 & -T_{4z} & T_{3z} & 0 & 1 & T_{2z} \\ G_\theta & 0 & -T_{3z} & T_{2z} & 0 & 0 & T_{1z} \end{array}\right]}_{T_{RL}} \underbrace{\left[\begin{array}{c} 1 \\ \hline \eta N \\ V_L \\ \bar{Z}_L \\ \hline \bar{u}_L \\ \bar{v}_L \\ \bar{\theta}_L \end{array}\right]}_{H_L}$$

where H_L, H_R are the *state vectors* of the respective ends and T_{LR}, T_{RL} are the *transport matrices*. The elements of the transport matrices are the *transport coefficients* T_{jz} defined in (9.03) and the *load functions* F, G displayed for particular cases in (9.06–9.08).

9.03 TRANSPORT RELATIONS

(1) Transport functions

$$\beta = \sqrt{\frac{N}{EI_z}}$$

Axial tension ($\eta = +1$)	Axial compression ($\eta = -1$)
$T_{0z}(x) = \sinh \beta x$	$T_{0z}(x) = -\sin \beta x$
$T_{1z}(x) = \cosh \beta x$	$T_{1z}(x) = \cos \beta x$
$T_{2z}(x) = \sinh \beta x$	$T_{2z}(x) = \sin \beta x$
$T_{3z}(x) = \cosh \beta x - 1$	$T_{3z}(x) = 1 - \cos \beta x$
$T_{4z}(x) = \sinh \beta x - \beta x$	$T_{4z}(x) = \beta x - \sin \beta x$
$T_{5z}(x) = T_{3z}(x) - \dfrac{(\beta x)^2}{2!}$	$T_{5z}(x) = \dfrac{(\beta x)^2}{2!} - T_{3z}(x)$
$T_{6z}(x) = T_{4z}(x) - \dfrac{(\beta x)^3}{3!}$	$T_{6z}(x) = \dfrac{(\beta x)^3}{3!} - T_{4z}(x)$

The graphs and series of these functions are shown in (9.04–9.05).

(2) Transport coefficients T_{jz}, A_{jz}, B_{jz}, C_{jz} are the values of $T_{jz}(x)$ for a particular argument x.

$$\left.\begin{array}{ll} \text{For } x = s, & T_{jz}(s) = T_{jz} \\ \text{For } x = a, & T_{jz}(a) = A_{jz} \\ \text{For } x = b, & T_{jz}(b) = B_{jz} \\ \text{and } T_{jz} - A_{jz} = C_{jz} & \end{array}\right\} \quad j = 0, 1, 2, \ldots, 6$$

(3) Transport matrices of this bar have analogical properties to those of (4.02) and can be used in similar ways.

(4) Differential relations

$$\frac{du}{dx} = \frac{U}{EA_x} \qquad \frac{d^2u}{dx^2} = 0$$

$$\frac{dv}{dx} = \theta$$

$$\frac{d^2v}{dx^2} = \frac{d\theta}{dx} = \frac{Z}{EI_z}$$

$$\frac{d^3v}{dx^3} = \frac{d^2\theta}{dx^2} = \frac{dZ}{dx\,EI_z} = -\frac{V}{EI_z} + \frac{\eta N\,dv}{EI_z\,dx}$$

$$\frac{d^4v}{dx^4} = \frac{d^3\theta}{dx^3} = \frac{d^2Z}{dx^2\,EI_z} = -\frac{p_y}{EI_z} + \frac{\eta N\,d^2v}{EI_z\,dx^2}$$

(5) Two forms of solution are shown in each matrix in this chapter: one corresponding to the axial tension with $\eta = +1$ (expressed in terms of hyperbolic functions) and the other one corresponding to the axial compression with $\eta = -1$ (expressed in terms of trigonometric functions). The general expressions for load functions in (9.06–9.08) and in (9.14) must be evaluated in terms of functions corresponding to the given state.

(6) General transport matrix equations for beam-columns of order three are obtained by matrix composition of (9.02) and (14.02) as shown in (2.18).

9.04 GRAPHS AND SERIES OF TRANSPORT COEFFICIENTS, AXIAL TENSION

(1) Graphs of $T_{jz}(x)$

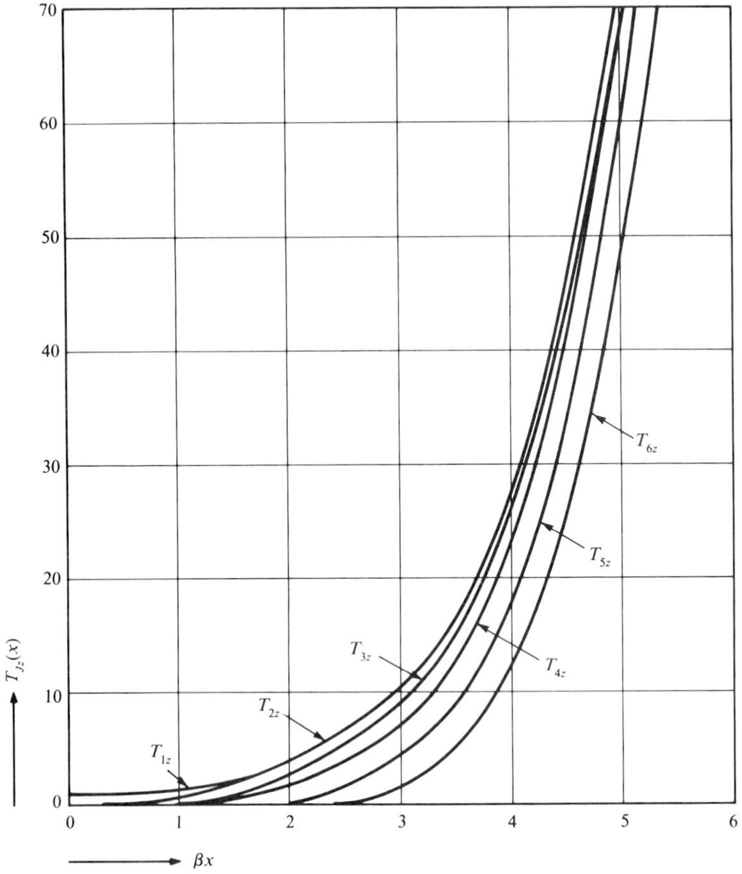

(2) Series expansion of $T_{jz}(x)$

$$T_{1z}(x) = 1 + \frac{(\beta x)^2}{2!} + \frac{(\beta x)^4}{4!} + \cdots = \sum_{k=0}^{\infty} \frac{(\beta x)^{2k}}{(2k)\,!}$$

$$T_{2z}(x) = \frac{(\beta x)}{1!} + \frac{(\beta x)^3}{3!} + \frac{(\beta x)^5}{5!} + \cdots = \sum_{k=0}^{\infty} \frac{(\beta x)^{2k+1}}{(2k+1)!}$$

$$T_{3z}(x) = \frac{(\beta x)^2}{2!} + \frac{(\beta x)^4}{4!} + \frac{(\beta x)^6}{6!} + \cdots = \sum_{k=0}^{\infty} \frac{(\beta x)^{2k+2}}{(2k+2)!}$$

$$T_{4z}(x) = \frac{(\beta x)^3}{3!} + \frac{(\beta x)^5}{5!} + \frac{(\beta x)^7}{7!} + \cdots = \sum_{k=0}^{\infty} \frac{(\beta x)^{2k+3}}{(2k+3)!}$$

$$T_{jz}(x) = \frac{(\beta x)^{j-1}}{(j-1)!} + \frac{(\beta x)^{j+1}}{(j+1)!} + \frac{(\beta x)^{j+3}}{(j+3)!} + \cdots = \sum_{k=0}^{\infty} \frac{(\beta x)^{j+2k-1}}{(j+2k-1)!}$$

9.05 GRAPHS AND SERIES OF TRANSPORT COEFFICIENTS, AXIAL COMPRESSION

(1) Graphs of $T_{jz}(x)$

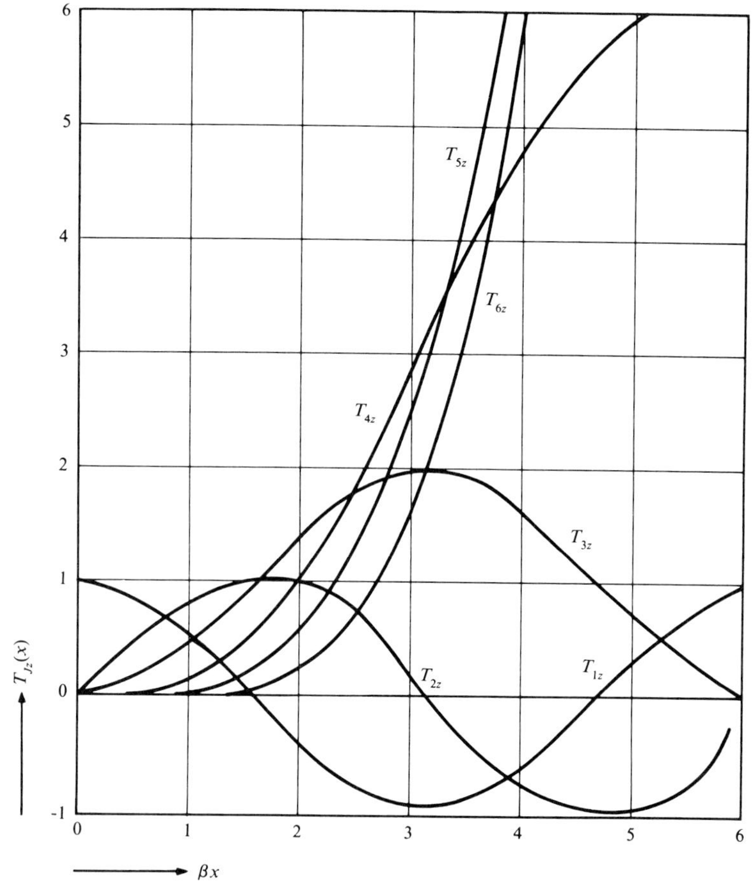

(2) Series expansion of $T_{jz}(x)$

$$T_{1x}(x) = 1 - \frac{(\beta x)^2}{2!} + \frac{(\beta x)^4}{4!} - \cdots = \sum_{k=0}^{\infty} (-1)^k \frac{(\beta x)^{2k}}{(2k)!}$$

$$T_{2z}(x) = \frac{(\beta x)}{1!} - \frac{(\beta x)^3}{3!} + \frac{(\beta x)^5}{5!} - \cdots = \sum_{k=0}^{\infty} (-1)^k \frac{(\beta x)^{2k+1}}{(2k+1)!}$$

$$T_{3z}(x) = \frac{(\beta x)^2}{2!} - \frac{(\beta x)^4}{4!} + \frac{(\beta x)^6}{6!} - \cdots = \sum_{k=0}^{\infty} (-1)^k \frac{(\beta x)^{2k+2}}{(2k+2)!}$$

$$T_{4z}(x) = \frac{(\beta x)^3}{3!} - \frac{(\beta x)^5}{5!} + \frac{(\beta x)^7}{7!} - \cdots = \sum_{k=0}^{\infty} (-1)^k \frac{(\beta x)^{2k+3}}{(2k+3)!}$$

$$T_{jz}(x) = \frac{(\beta x)^{j-1}}{(j-1)!} - \frac{(\beta x)^{j+1}}{(j+1)!} + \frac{(\beta x)^{j+3}}{(j+3)!} - \cdots = \sum_{k=0}^{\infty} (-1)^k \frac{(\beta x)^{j+2k-1}}{(j+2k-1)!}$$

9.06 TRANSPORT LOAD FUNCTIONS

Notation (9.02–9.03) Signs (9.02) Matrix (9.02)		$\beta = \sqrt{\dfrac{N}{EI_z}}$
P_y = concentrated load (N) Q_z = applied couple (N·m) q_z = intensity of distributed couple (N·m/m) A_j, B_j, C_j, T_j = transport coefficients (9.03–9.05)		
$F_U = 0$ $F_V = -P_y$ $F_Z = -A_{2z}P_y$	**1.** 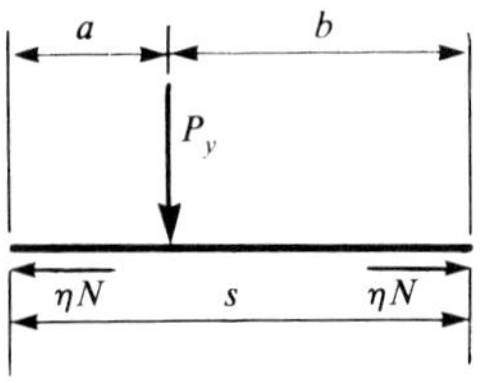	$G_U = 0$ $G_V = P_y$ $G_Z = -B_{2z}P_y$
$F_u = 0$ $F_v = -A_{4z}P_y$ $F_\theta = A_{3z}P_y$		$G_u = 0$ $G_v = -B_{4z}P_y$ $G_\theta = -B_{3z}P_y$
$F_U = 0$ $F_V = 0$ $F_Z = -A_{1z}\beta Q_z$	**2.** 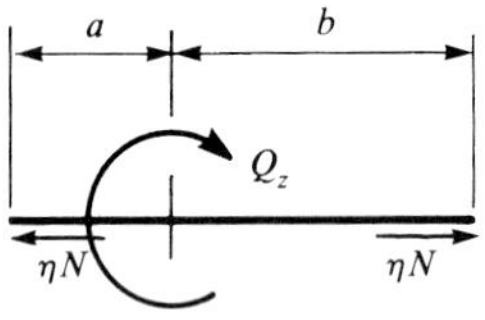	$G_U = 0$ $G_V = 0$ $G_Z = B_{1z}\beta Q_z$
$F_u =$ $F_v = -A_{3z}\beta Q_z$ $F_\theta = A_{2z}\beta Q_z$		$G_u = 0$ $G_v = B_{3z}\beta Q_z$ $G_\theta = B_{2z}\beta Q_z$
$F_U = 0$ $F_V = 0$ $F_Z = -T_{2z}q_z$	**3.** 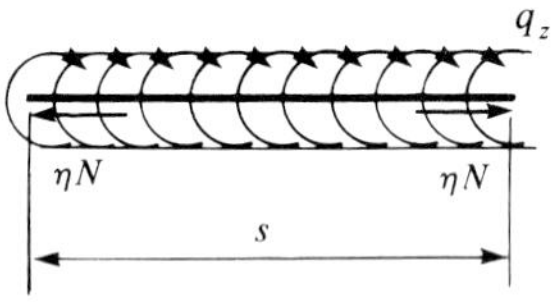	$G_U = 0$ $G_V = 0$ $G_Z = T_{2z}q_z$
$F_u = 0$ $F_v = -T_{4z}q_z$ $F_\theta = T_{3z}q_z$		$G_u = 0$ $G_v = T_{4z}q_z$ $G_\theta = T_{3z}q_z$
$F_U = 0$ $F_V = 0$ $F_Z = -C_{2z}q_z$	**4.** 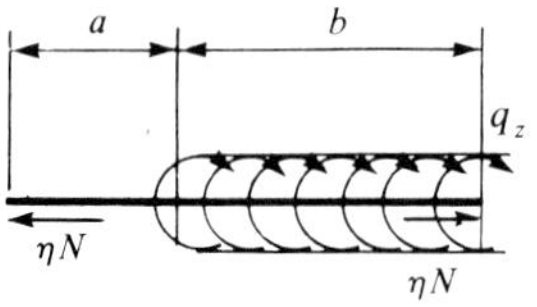	$G_U = 0$ $G_V = 0$ $G_Z = B_{2z}q_z$
$F_u = 0$ $F_v = -C_{4z}q_z$ $F_\theta = C_{3z}q_z$		$G_u = 0$ $G_v = B_{4z}q_z$ $G_\theta = B_{3z}q_z$

9.07 TRANSPORT LOAD FUNCTIONS

Notation (9.02–9.03) Signs (9.02) Matrix (9.02)	
p_y = intensity of distributed load (N/m) A_j, B_j, C_j, T_j = transport coefficients (9.03–9.05)	$\beta = \sqrt{\dfrac{N}{EI_z}}$

F	Case	G
$F_U = 0$ $F_V = -sp_y$ $F_Z = -T_{3z}\beta^{-1}p_y$	**1.** 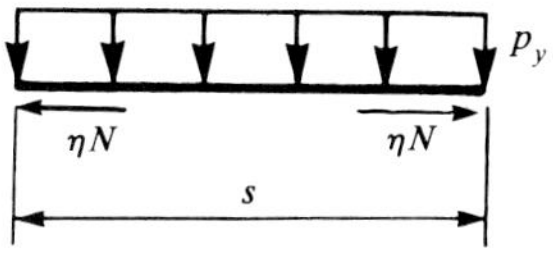	$G_U = 0$ $G_V = sp_y$ $G_Z = -T_{3z}\beta^{-1}p_y$
$F_u = 0$ $F_v = -T_{5z}\beta^{-1}p_y$ $F_\theta = T_{4z}\beta^{-1}p_y$		$G_u = 0$ $G_v = -T_{5z}\beta^{-1}p_y$ $G_\theta = -T_{4z}\beta^{-1}p_y$
$F_U = 0$ $F_V = -bp_y$ $F_Z = -C_{3z}\beta^{-1}p_y$	**2.** 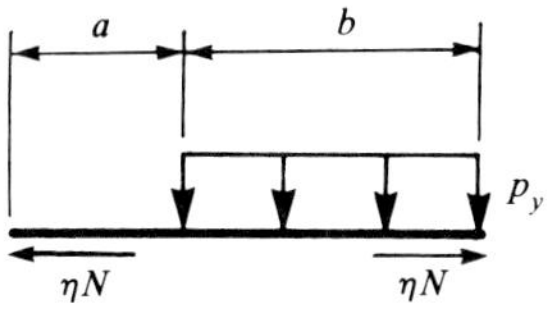	$G_U = 0$ $G_V = bp_y$ $G_\theta = -B_{3z}\beta^{-1}p_y$
$F_u = 0$ $F_v = -C_{5z}\beta^{-1}p_y$ $F_\theta = C_{4z}\beta^{-1}p_y$		$G_u = 0$ $G_v = -B_{5z}\beta^{-1}p_y$ $G_\theta = -B_{4z}\beta^{-1}p_y$
$F_U = 0$ $F_V = -\frac{1}{2}sp_y$ $F_Z = -(\beta s T_{3z} - T_{4z})\beta^{-2}s^{-1}p_y$	**3.** 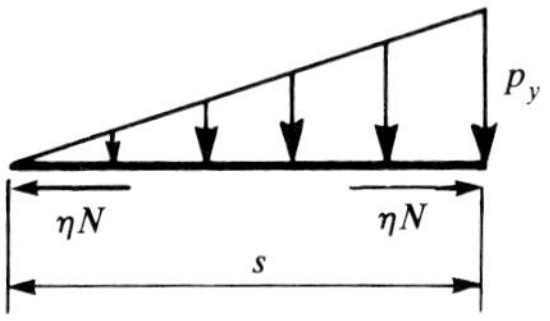	$G_U = 0$ $G_V = \frac{1}{2}sp_y$ $G_Z = -T_{4z}\beta^{-2}s^{-1}p_y$
$F_u = 0$ $F_v = -(\beta s T_{5z} - T_{6z})\beta^{-2}s^{-1}p_y$ $F_\theta = (\beta s T_{4z} - T_{5z})\beta^{-2}s^{-1}p_y$		$G_u = 0$ $G_v = -T_{6z}\beta^{-2}s^{-1}p_y$ $G_\theta = -T_{5z}\beta^{-2}s^{-1}p_y$
$F_U = 0$ $F_V = -\frac{1}{2}bp_y$ $F_Z = -(\beta b T_{3z} - C_{4z})\beta^{-2}b^{-1}p_y$	**4.** 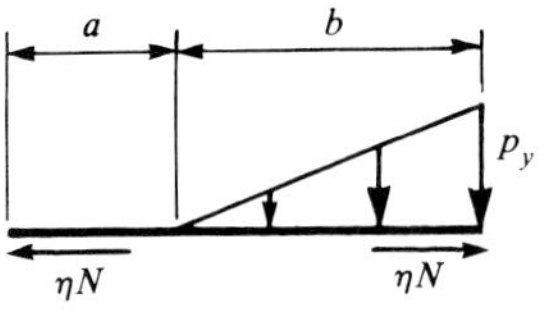	$G_U = 0$ $G_V = \frac{1}{2}bp_y$ $G_Z = -B_{4z}\beta^{-2}b^{-1}p_y$
$F_u = 0$ $F_v = -(\beta b T_{5z} - C_{6z})\beta^{-2}b^{-1}p_y$ $F_\theta = (\beta b T_{4z} - C_{5z})\beta^{-2}b^{-1}p_y$		$G_u = 0$ $G_v = -B_{6z}\beta^{-2}b^{-1}p_y$ $G_\theta = -B_{5z}\beta^{-2}b^{-1}p_y$

9.08 TRANSPORT LOAD FUNCTIONS

Notation (9.01–9.03) Signs (9.02) Matrix (9.02)

u_j, v_j, θ_j = abrupt displacements (m, m, rad)
e_g, f_t = temperature strains (4.03)
A_j, B_j, C_j = transport coefficients (8.03–8.5)

$$\beta = \sqrt{\frac{N}{EI_z}}$$

1. Abrupt displacements displacements u_j, v_j

$F_U = 0$
$F_V = 0$
$F_Z = -EI_z A_{1z} \beta^3 \eta v_j$

$F_u = -EA_x s^{-1} u_j$
$F_v = -EI_z (A_{3z} + \eta) \beta^3 v_j$
$F_\theta = EI_z A_{2z} \beta^3 \eta v_j$

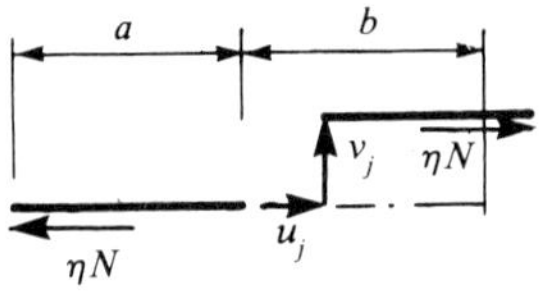

$G_U = 0$
$G_V = 0$
$G_Z = EI_z B_{1z} \beta^3 \eta v_j$

$G_u = EA_x s^{-1} u_j$
$G_v = EI_z (B_{3z} + \eta) \beta^3 v_j$
$G_\theta = EI_z B_{2z} \beta^3 \eta v_j$

2. Abrupt displacement θ_j

$F_U = 0$
$F_V = 0$
$F_Z = EI_z Z_{0z} \beta^2 \theta_j$

$F_u = 0$
$F_v = EI_z A_{2z} \beta^2 \theta_j$
$F_\theta = -EI_z A_{1z} \beta^2 \theta_j$

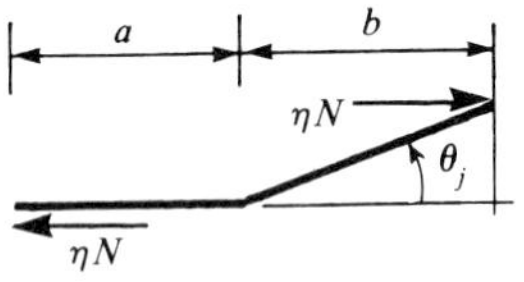

$G_U = 0$
$G_V = 0$
$G_Z = EI_z B_{0z} \beta^2 \theta_j$

$G_u = 0$
$G_v = EI_z B_{2z} \beta^2 \theta_j$
$G_\theta = EI_z B_{1z} \beta^2 \theta_j$

3. Temperature deformation of segment b

$F_U = 0$
$F_V = 0$
$F_Z = EI_z C_{1z} \beta f_t$

$F_u = -EA_x b s^{-1} e_t$
$F_v = EI_z C_{3z} \beta f_t$
$F_\theta = -EI_z C_{2z} \beta f_t$

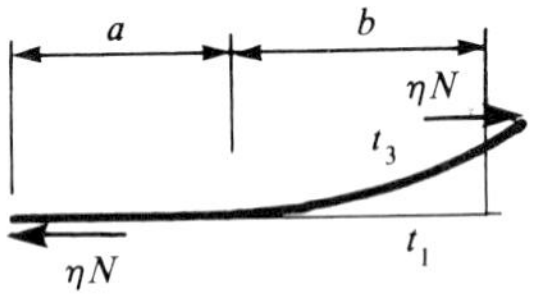

$G_U = 0$
$G_V = 0$
$G_Z = EI_z B_{1z} \beta f_t$

$G_u = EA_x b s^{-1} e_t$
$G_v = EI_z B_{3z} \beta f_t$
$G_\theta = EI_z B_{2z} \beta f_t$

4.

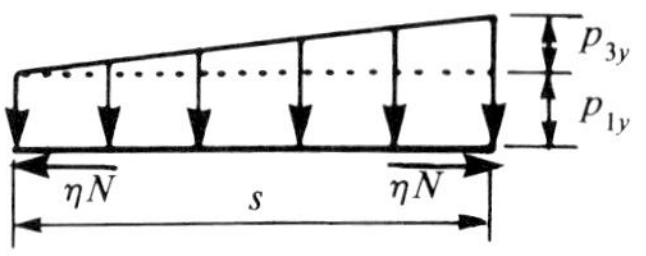

Sec. 9.07 (case 1 + case 3)

5.

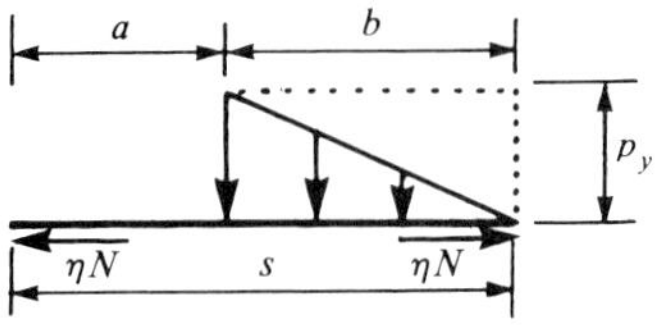

Sec. 9.07 (case 2 − case 4)

9.09 STIFFNESS MATRIX EQUATIONS

$$r = \frac{s^2 EA_x}{EI_z} \qquad \beta = \sqrt{\frac{N}{EI_z}}$$

(1) Equivalents and signs. All forces, moments, and displacements are in the principal axes of the normal section, and their positive directions are shown in (2, 3) below.

(2) Free-body diagram

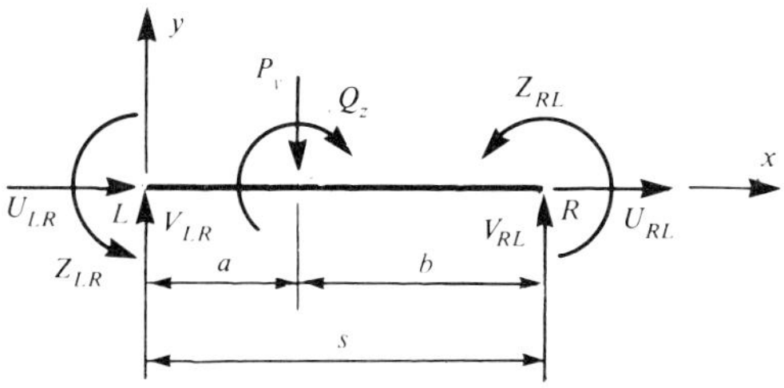

(3) Elastic curve

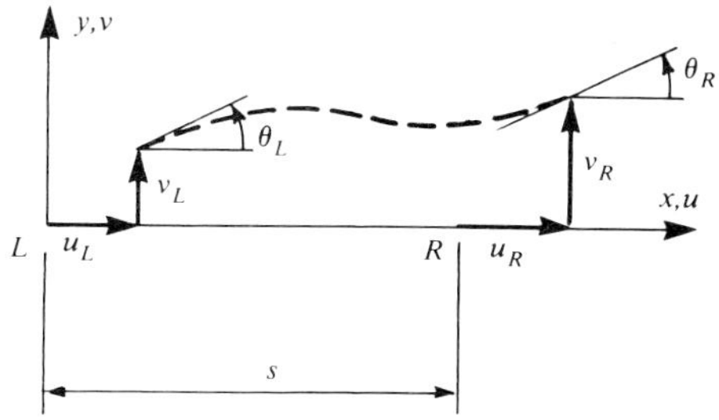

(4) Dimensionless stiffness coefficients

Axial tension ($\eta = +1$)	Axial compression ($\eta = -1$)
$D_0 = \dfrac{1}{2(1 - \cosh \beta s) + \beta s \sinh \beta s}$	$D_0 = \dfrac{1}{2(1 - \cos \beta s) - \beta s \sin \beta s}$
$K_1 = (\beta s)^3 D_0 \sinh \beta s$	$K_1 = (\beta s)^3 D_0 \sin \beta s$
$K_2 = (\beta s)^2 D_0 (\cosh \beta s - 1)$	$K_2 = (\beta s)^2 D_0 (1 - \cos \beta s)$
$K_3 = \beta s D_0 (\beta s \cosh \beta s - \sinh \beta s)$	$K_3 = \beta s D_0 (\sin \beta s - \beta s \cos \beta s)$
$K_4 = \beta s D_0 (\sinh \beta s - \beta s)$	$K_4 = \beta s D_0 (\beta s - \sin \beta s)$
$L_1 = D_0 \sinh \beta s$	$L_1 = D_0 \sin \beta s$
$L_2 = D_0 (\cosh \beta s - 1)$	$L_2 = D_0 (1 - \cos \beta s)$
$L_3 = D_0 (\beta s \cosh \beta s - \sinh \beta s)$	$L_3 = D_0 (\sin \beta s - \beta s \cos \beta s)$
$L_4 = D_0 (\sinh \beta s - \beta s)$	$L_4 = D_0 (\beta s - \sin \beta s)$

(5) Stiffness matrix equations

$$\begin{bmatrix} U_{LR} \\ V_{LR} \\ Z_{LR} \\ \hline U_{RL} \\ V_{RL} \\ Z_{RL} \end{bmatrix} = \frac{EI_z}{s^3} \left[\begin{array}{cccccc} r & 0 & 0 & -r & 0 & 0 \\ 0 & K_1 & K_2 s & 0 & -K_2 s & -K_2 s \\ 0 & K_2 s & K_3 s^2 & 0 & -K_4 s^2 & -K_4 s^2 \\ \hline -r & 0 & 0 & r & 0 & 0 \\ 0 & -K_1 & -K_2 s & 0 & K_2 s & -K_2 s \\ 0 & K_2 s & K_4 s^2 & 0 & -K_3 s^2 & K_3 s^2 \end{array} \right] \begin{bmatrix} u_L \\ v_L \\ \theta_L \\ \hline u_R \\ v_R \\ \theta_R \end{bmatrix} + \begin{bmatrix} -\eta N \\ V_{L0} \\ Z_{L0} \\ \hline \eta N \\ V_{R0} \\ Z_{R0} \end{bmatrix}$$

where V_{L0}, Z_{L0} and V_{R0}, Z_{R0} are the *reactions of fixed-end beam-columns* due to loads and other causes shown in (9.14). The graphs and tables of stiffness coefficients are shown in (9.10–9.13).

(6) General stiffness matrix equations for beam-columns of order three are obtained by matrix composition of (9.09) and (14.07) as shown in (3.19).

9.10 TABLE OF DIMENSIONLESS STIFFNESS COEFFICIENTS, AXIAL TENSION

ps	K_1	K_2	K_3	K_4	K_5	K_6
0.0	12.00000	6.00000	4.00000	2.00000	3.00000	3.00000
0.50	12.29991	6.02496	4.03322	1.99173	3.29965	3.04965
0.60	12.43182	6.03591	4.04778	1.98813	3.43127	3.07127
0.70	12.58766	6.04883	4.06492	1.98391	3.58666	3.09666
0.80	12.76742	6.06371	4.08463	1.97908	3.76572	3.12572
0.90	12.97107	6.08054	4.10687	1.97366	3.96838	3.15838
1.00	13.19859	6.09929	4.13162	1.96767	4.19453	3.19453
1.10	13.44994	6.11997	4.15884	1.96113	4.44406	3.23406
1.20	13.72508	6.14254	4.18849	1.95406	4.71686	3.27686
1.30	14.02399	6.16700	4.22052	1.94648	5.01281	3.32281
1.40	14.34663	6.19331	4.25489	1.93842	5.33180	3.37180
1.50	14.69294	6.22147	4.29156	1.92991	5.67368	3.42368
1.60	15.06290	6.25145	4.33048	1.92097	6.03835	3.47835
1.70	15.45644	6.28322	4.37159	1.91163	6.42566	3.53566
1.80	15.87352	6.31676	4.41484	1.90192	6.83549	3.59549
1.90	16.31410	6.35205	4.46018	1.89186	7.26772	3.65772
2.00	16.77811	6.38906	4.50756	1.88149	7.72221	3.72221
2.10	17.26551	6.42775	4.55692	1.87083	8.19885	3.78885
2.20	17.77624	6.46812	4.60820	1.85992	8.69752	3.85752
2.30	18.31023	6.51012	4.66135	1.84877	9.21810	3.92810
2.40	18.86744	6.55372	4.71631	1.83742	9.76047	4.00047
2.50	19.44780	6.59890	4.77301	1.82589	10.32453	4.07453
2.60	20.05125	6.64563	4.83142	1.81421	10.91018	4.15018
2.70	20.67774	6.69387	4.89146	1.80241	11.51731	4.22731
2.80	21.32718	6.74359	4.95309	1.79050	12.14584	4.30584
2.90	21.99954	6.79477	5.01624	1.77853	12.79566	4.38566
3.00	22.69473	6.84737	5.08087	1.76650	13.46670	4.46670
3.10	23.41270	6.90135	5.14692	1.75443	14.15888	4.54888
3.20	24.15338	6.95669	5.21433	1.74236	14.87212	4.63212
3.30	24.91672	7.01336	5.28305	1.73030	15.60635	4.71635
3.40	25.70263	7.07132	5.35304	1.71827	16.36150	4.80150
3.50	26.51107	7.13054	5.42425	1.70629	17.13751	4.88751
3.60	27.34196	7.19098	5.49662	1.69436	17.93432	4.97432
3.70	28.19525	7.25263	5.57011	1.68252	18.75189	5.06189
3.80	29.07087	7.31544	5.64467	1.67076	19.59014	5.15014
3.90	29.96876	7.37938	5.72026	1.65912	20.44905	5.23905
4.00	30.88885	7.44443	5.79684	1.64758	21.32856	5.32856
4.10	31.83109	7.51055	5.87437	1.63618	22.22864	5.41864
4.20	32.79542	7.57771	5.95279	1.62491	23.14925	5.50925
4.30	33.78177	7.64588	6.03209	1.61379	24.09034	5.60034
4.40	34.79008	7.71504	6.11221	1.60283	25.05190	5.69190
4.50	35.82031	7.78516	6.19313	1.59203	26.03388	5.78388
4.60	36.87239	7.85619	6.27481	1.58139	27.03626	5.87626
4.70	37.94627	7.92813	6.35721	1.57093	28.05902	5.96902
4.80	39.04188	8.00094	6.44030	1.56064	29.10212	6.06212
4.90	40.15919	8.07459	6.52406	1.55054	30.16555	6.15555
5.00	41.29813	8.14907	6.60845	1.54061	31.24929	6.24929
6.00	53.86681	8.93340	7.48161	1.45179	43.19989	7.19989

9.11 GRAPHS AND SERIES OF DIMENSIONLESS STIFFNESS COEFFICIENTS AXIAL TENSION

(1) Graphs of K_j

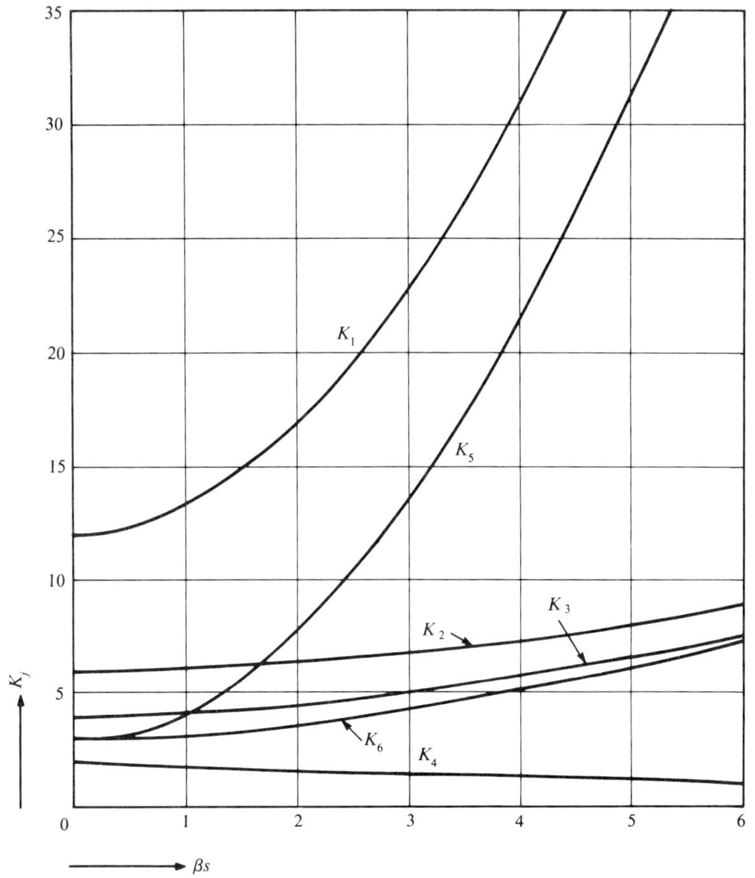

(2) Series expansion of K_j

$$K_1 = \frac{1}{2}\,\frac{\displaystyle\sum_{k=0}^{\infty}\frac{(\beta s)^{2k}}{(2k+1)!}}{\displaystyle\sum_{k=0}^{\infty}\frac{(k+1)(\beta s)^{2k}}{(2k+4)!}} \qquad K_4 = \frac{1}{2}\,\frac{\displaystyle\sum_{k=0}^{\infty}\frac{(\beta s)^{2k}}{(2k+3)!}}{\displaystyle\sum_{k=0}^{\infty}\frac{(k+1)(\beta s)^{2k}}{(2k+4)!}}$$

$$K_2 = \frac{1}{2}\,\frac{\displaystyle\sum_{k=0}^{\infty}\frac{(\beta s)^{2k}}{(2k+2)!}}{\displaystyle\sum_{k=0}^{\infty}\frac{(k+1)(\beta s)^{2k}}{(2k+4)!}} \qquad K_5 = \frac{1}{2}\,\frac{\displaystyle\sum_{k=0}^{\infty}\frac{(\beta s)^{2k}}{(2k)!}}{\displaystyle\sum_{k=0}^{\infty}\frac{(k+1)\,(\beta s)^{2k}}{(2k+3)!}}$$

$$K_3 = \frac{\displaystyle\sum_{k=0}^{\infty}\frac{(k+1)(\beta s)^{2k}}{(2k+3)!}}{\displaystyle\sum_{k=0}^{\infty}\frac{(k+1)(\beta s)^{2k}}{(2k+4)!}} \qquad K_6 = \frac{1}{2}\,\frac{\displaystyle\sum_{k=0}^{\infty}\frac{(\beta s)^{2k}}{(2k+1)!}}{\displaystyle\sum_{k=0}^{\infty}\frac{(k+1)(\beta s)^{2k}}{(2k+3)!}}$$

9.12 TABLE OF DIMENSIONLESS STIFFNESS COEFFICIENTS, AXIAL COMPRESSION

βs	K_1	K_2	K_3	K_4	K_5	K_6
0.0	12.00000	6.00000	4.00000	2.00000	3.00000	3.00000
0.50	11.69991	5.97496	3.96656	2.00840	2.69964	2.94964
0.60	11.56781	5.96391	3.95177	2.01214	2.56725	2.92725
0.70	11.41166	5.95083	3.93424	2.01658	2.41060	2.90060
0.80	11.23141	5.93571	3.91394	2.02176	2.22959	2.86959
0.90	11.02705	5.91853	3.89083	2.02769	2.02411	2.83411
1.00	10.79856	5.89928	3.86488	2.03439	1.79402	2.79402
1.10	10.54588	5.87794	3.83604	2.04190	1.53916	2.74916
1.20	10.26899	5.85449	3.80426	2.05023	1.25934	2.69934
1.30	9.96784	5.82892	3.76949	2.05943	0.95435	2.64435
1.40	9.64239	5.80119	3.73167	2.06953	0.62394	2.58394
1.50	9.29258	5.77129	3.69072	2.08058	0.26783	2.51783
1.60	8.91836	5.73918	3.64656	2.09262	−0.11431	2.44569
1.70	8.51967	5.70484	3.59912	2.10571	−0.52285	2.36715
1.80	8.09644	5.66822	3.54831	2.11991	−0.95822	2.28178
1.90	7.64860	5.62930	3.49401	2.13529	−1.42094	2.18906
2.00	7.17608	5.58804	3.43611	2.15193	−1.91157	2.08843
2.10	6.67878	5.54439	3.37450	2.16989	−2.43081	1.97919
2.20	6.15662	5.49831	3.30902	2.18929	−2.97944	1.86056
2.30	5.60951	5.44976	3.23954	2.21022	−3.55842	1.73158
2.40	5.03734	5.39867	3.16587	2.23280	−4.16886	1.59114
2.50	4.44001	5.34500	3.08784	2.25716	−4.81210	1.43790
2.60	3.81738	5.28869	3.00525	2.28344	−5.48976	1.27024
2.70	3.16934	5.22967	2.91785	2.31182	−6.20381	1.08619
2.80	2.49575	5.16787	2.82540	2.34247	−6.95668	0.88332
2.90	1.79645	5.10323	2.72763	2.37560	−7.75139	0.65861
3.00	1.07131	5.03565	2.62420	2.41145	−8.59176	0.40824
3.14159	0.00000	4.93480	2.46740	2.46740	−9.86940	0.00000
3.20	−0.45724	4.89138	2.39895	2.49243	−10.43060	−0.19060
3.30	−1.26101	4.81449	2.27629	2.53821	−11.44398	−0.55398
3.40	−2.09139	4.73430	2.14627	2.58803	−12.53443	−0.97443
3.50	−2.94861	4.65070	2.00834	2.64236	−13.71818	−1.46818
3.60	−3.83289	4.56355	1.86183	2.70172	−15.01867	−2.05867
3.70	−4.74451	4.47274	1.70599	2.76675	−16.47109	−2.78109
3.80	−5.68375	4.37813	1.53996	2.83817	−18.13085	−3.69085
3.90	−6.65089	4.27955	1.36272	2.91683	−20.09060	−4.88060
4.00	−7.64628	4.17686	1.17311	3.00374	−22.51794	−6.51794
4.10	−8.67027	4.06987	0.96976	3.10011	−25.75068	−8.94068
4.20	−9.72323	3.95839	0.75101	3.20737	−30.58681	−12.94681
4.30	−10.80558	3.84221	0.51494	3.32727	−39.47418	−20.98418
4.4934	−12.98394	3.60335	0.00000	3.60332	$\pm\infty$	$\pm\infty$
4.50	−13.06029	3.59485	−0.01910	3.61395	663.53773	683.78773
4.60	−14.23368	3.46316	−0.32343	3.78659	22.84789	44.00789
4.7124	−15.59001	3.30831	−0.70204	4.01035	0.00000	22.20661
4.80	−16.67544	3.18228	−1.02891	4.21119	−6.83305	16.20695
4.90	−17.94515	3.03243	−1.44266	4.47509	−11.57109	12.43891
5.00	−19.24840	2.87580	−1.90872	4.78452	−14.91555	10.08445
6.2832	−39.47842	0.00000	$\pm\infty$	$\pm\infty$	−39.47842	0.00000

9.13 GRAPHS AND SERIES OF DIMENSIONLESS STIFFNESS COEFFICIENTS AXIAL COMPRESSION

(1) Graphs of K_j

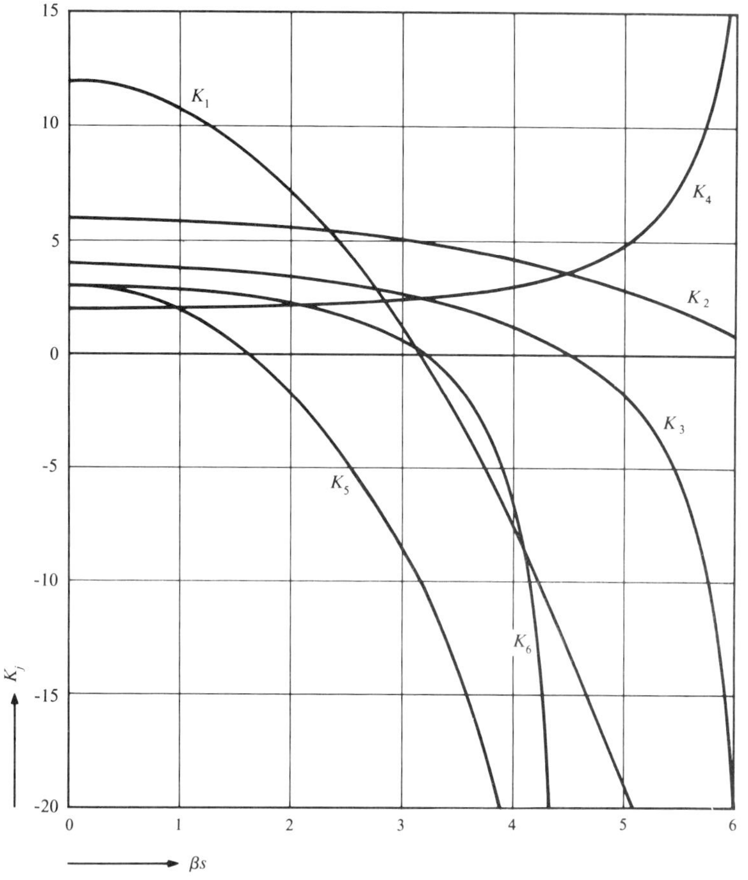

(2) Series of expansion of K_j

$$K_1 = \frac{1}{2}\,\frac{\displaystyle\sum_{k=0}^{\infty}(-1)^k\frac{(\beta s)^{2k}}{(2k+1)!}}{\displaystyle\sum_{k=0}^{\infty}(-1)^k(k+1)\frac{(\beta s)^{2k)}}{(2k+4)!}}$$

$$K_4 = \frac{1}{2}\,\frac{\displaystyle\sum_{k=0}^{\infty}(-1)^k\frac{(\beta s)^{2k}}{(2k+3)!}}{\displaystyle\sum_{k=0}^{\infty}(-1)^k\,(k+1)\frac{(\beta s)^{2k}}{(2k+4)!}}$$

$$K_2 = \frac{1}{2}\,\frac{\displaystyle\sum_{k=0}^{\infty}(-1)^k\frac{(\beta s)^{2k}}{(2k+2)!}}{\displaystyle\sum_{k=0}^{\infty}(-1)^k(k+1)\frac{(\beta s)^{2k}}{(2k+4)!}}$$

$$K_5 = \frac{1}{2}\,\frac{\displaystyle\sum_{k=0}^{\infty}(-1)^k\frac{(\beta s)^{2k}}{(2k)!}}{\displaystyle\sum_{k=0}^{\infty}\frac{(k+1)(\beta s)^{2k}}{(2k+3)!}}$$

$$K_3 = \frac{\displaystyle\sum_{k=0}^{\infty}(-1)^k(k+1)\frac{(\beta s)^{2k}}{(2k+3)!}}{\displaystyle\sum_{k=0}^{\infty}(-1)^k(k+1)\frac{(\beta s)^{2k}}{(2k+4)!}}$$

$$K_6 = \frac{1}{2}\,\frac{\displaystyle\sum_{k=0}^{\infty}(-1)^k\frac{(\beta s)^{2k}}{(2k+1)!}}{\displaystyle\sum_{k=0}^{\infty}\frac{(k+1)(\beta s)^{2k}}{(2k+3)!}}$$

9.14 STIFFNESS LOAD FUNCTIONS, FIXED-END BEAM-COLUMN

(1) General form of the stiffness load functions in (9.09-5) is given by the matrix equations

$$\begin{bmatrix} U_{L0} \\ V_{L0} \\ Z_{L0} \end{bmatrix} = \begin{bmatrix} 1 & 0 & 0 \\ 0 & L_1 & -L_2 \\ 0 & L_2\beta^{-1} & -L_4\beta^{-1} \end{bmatrix} \begin{bmatrix} -\eta N \\ G_v \\ G_\theta \end{bmatrix}$$

$$\begin{bmatrix} U_{R0} \\ V_{R0} \\ Z_{R0} \end{bmatrix} = \begin{bmatrix} 1 & 0 & 0 \\ 0 & L_1 & L_2 \\ 0 & -L_2\beta^{-1} & -L_4\beta^{-1} \end{bmatrix} \begin{bmatrix} \eta N \\ F_v \\ F_\theta \end{bmatrix}$$

where F_v, F_θ are the left-end transport load functions and G_v, G_θ are their right-end counterparts, given for particular loads and causes in (9.06–9.08), and L_1, L_2, L_4 are the stiffness coefficients (9.09). The load vectors and other causes are defined by symbols introduced in (9.06–9.08).

(2) Particular cases

Notation (9.02–9.03, 9.09) A_j, B_j, C_j, T_j = transport coefficients (9.03)	Signs (9.09)	$\beta = \sqrt{\dfrac{N}{EI_z}}$
$U_{L0} = -\eta N$ $V_{L0} = (L_2 B_{3z} - L_1 B_{4z}) P_y$ $Z_{L0} = (L_A B_{3z} - L_2 B_{4z}) \beta^{-1} P_y$	**1.** 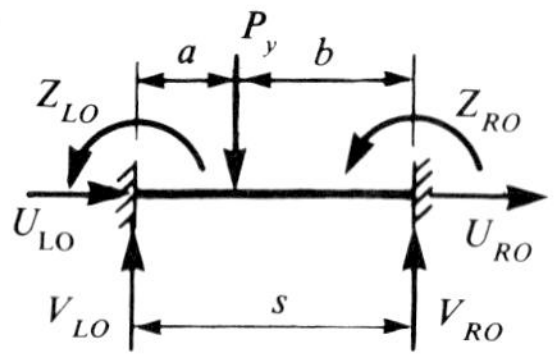	$U_{R0} = \eta N$ $V_{R0} = (L_2 A_{3z} - L_1 A_{4z}) P_y$ $Z_{R0} = -(L_4 A_{3z} - L_2 A_{4z}) \beta^{-1} P_y$
$U_{L0} = -\eta N$ $V_{L0} = -(L_2 B_{2z} - L_1 B_{3z}) \beta Q_z$ $Z_{L0} = -(L_4 B_{2z} - L_2 B_{3z}) Q_z$	**2.** 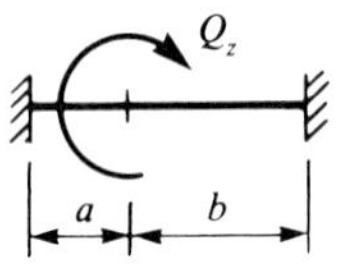	$U_{R0} = \eta N$ $V_{R0} = (L_2 A_{2z} - L_1 A_{3z}) \beta Q_z$ $Z_{R0} = -(L_4 A_{2z} - L_2 A_{3z}) Q_z$
$U_{L0} = -\eta N$ $V_{L0} = \frac{1}{2} s p_y$ $Z_{L0} = \left(1 - \frac{\beta s}{2} \cot \frac{\beta s}{2}\right) \beta^{-2} p_y$	**3. Axial compression** 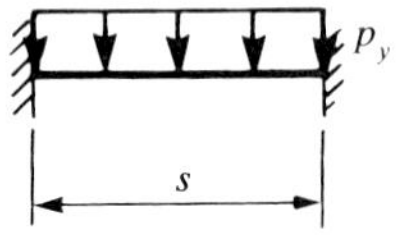	$U_{R0} = \eta N$ $V_{R0} = \frac{1}{2} s p_y$ $Z_{R0} = -\left(1 - \frac{\beta s}{2} \cot \frac{\beta s}{2}\right) \beta^{-2} p_y$

$$U_{L0} = N = -U_{R0}$$

$$V_{L0} = \tfrac{1}{2}P_y = V_{R0}$$

$$Z_{L0} = \frac{1}{2\beta}(\tan \tfrac{1}{4}\beta s)P_y = -Z_{R0}$$

4. Axial compression

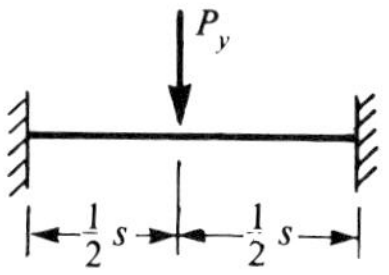

$$U_{L0} = N = -U_{R0}$$

$$V_{L0} = -\frac{\tfrac{1}{2}(1 - \cos \tfrac{1}{2}\beta s)\beta}{\sin \tfrac{1}{2}\beta s - \tfrac{1}{2}\beta s \cos \tfrac{1}{2}\beta s} Q_z = -V_{R0}$$

$$Z_{L0} = -\frac{\tfrac{1}{2}(\tfrac{1}{2}\beta s - \sin \tfrac{1}{2}\beta s)}{\sin \tfrac{1}{2}\beta s - \tfrac{1}{2}\beta s \cos \tfrac{1}{2}\beta s} Q_z = Z_{R0}$$

5. Axial compression

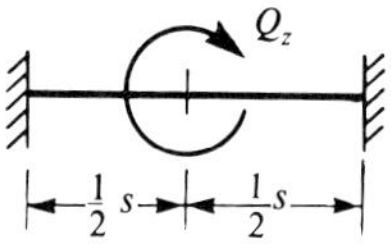

6.

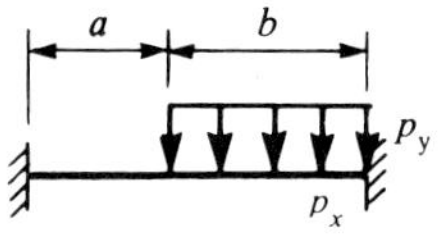

Use general form in terms of (9.07-2)

7.

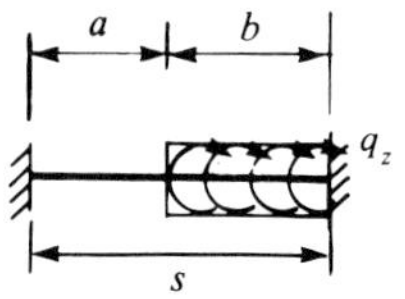

Use general form in terms of (9.06-4)

8.

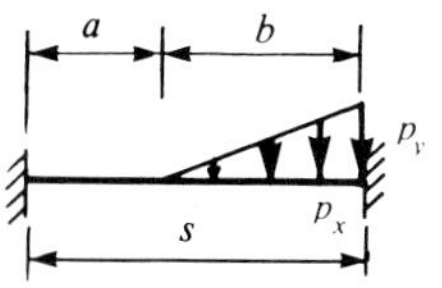

Use general form in terms of (9.07-4)

$$U_{L0} = EA_x s^{-1} u_j - \eta N$$

$$V_{L0} = EI_z(L_1 B_{1z} - L_2 B_{2z})\beta^3 \eta v_j$$

$$Z_{L0} = EI_z(L_2 B_{1z} - L_4 B_{2z})\beta^2 \eta v_j$$

9. Abrupt displacements u_j, v_j

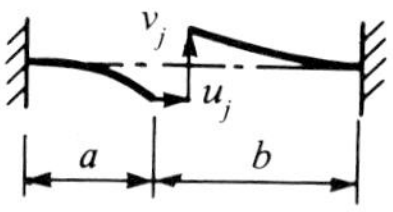

$$U_{R0} = -U_{L0}$$

$$V_{R0} = -V_{L0}$$

$$Z_{R0} = EI_z(L_2 A_{1z} + L_4 A_{2z})\beta^2 \eta v_j$$

$$U_{L0} = -\eta N$$

$$V_{L0} = EI_z(L_1 B_{2z} - L_2 B_{1z})\beta^2 \theta_j$$

$$Z_{L0} = EI_z(L_2 B_{2z} - L_4 B_{1z})\beta^2 \theta_j$$

10. Abrupt displacement θ_j

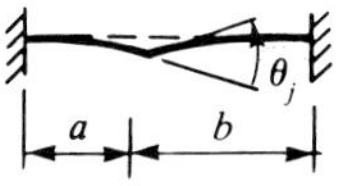

$$U_{R0} = -U_{L0}$$

$$V_{R0} = -V_{L0}$$

$$Z_{R0} = -EI_z(L_2 A_{2z} - L_4 A_{1z})\beta^2 \theta_j$$

$$U_{L0} = EA_x b s^{-1} e_t - \eta N$$

$$V_{L0} = EI_z(L_1 B_{3z} - L_2 B_{2z})\beta f_t$$

$$Z_{L0} = EI_z(L_2 B_{3z} - L_4 B_{2z}) f_t$$

11. Temperature deformation of segment b

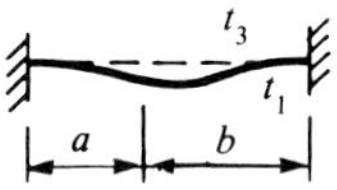

$$U_{R0} = -U_{L0}$$

$$V_{R0} = -V_{L0}$$

$$Z_{R0} = -EI_z(L_2 C_{3z} - L_4 C_{2z}) f_t$$

9.15 MODIFIED STIFFNESS MATRIX EQUATIONS

(1) Dimensionless modified stiffness coefficients

Axial tension ($\eta = +1$)	Axial compression ($\eta = -1$)	$\beta = \sqrt{\dfrac{N}{EI_z}}$
$D_1 = \dfrac{1}{\beta s \cosh \beta s - \sinh \beta s}$	$D_1 = \dfrac{1}{\sin \beta s - \beta s \cos \beta s}$	
$L_5 = D_1 \cosh \beta s$	$L_5 = D_1 \cos \beta s$	$K_5 = (\beta s)^3 L_5$
$L_6 = D_1 \sinh \beta s$	$L_6 = D_1 \sin \beta s$	$K_6 = (\beta s)^2 L_6$
$L_7 = D_1(\cosh \beta s - 1)$	$L_7 = D_1(1 - \cos \beta s)$	
$L_8 = D_1(\sinh \beta s - \beta s)$	$L_8 = D_1(bs - \sin \beta s)$	
$L_9 = \dfrac{\cosh \beta s}{\sinh \beta s}$	$L_9 = \dfrac{\cos \beta s}{\sin \beta s}$	$K_9 = (\beta s) L_9$
$L_{10} = \dfrac{1}{\sinh \beta s}$	$L_{10} = \dfrac{1}{\sin \beta s}$	$K_{10} = (\beta s) L_{10}$
$L_{11} = \dfrac{\sinh \beta s}{\cosh \beta s}$	$L_{11} = \dfrac{\sin \beta s}{\cos \beta s}$	$K_{11} = (\beta s) L_{11}$

(2) Modifications of the stiffness matrix equations (9.09) due to special conditions at one end are shown below. Because this modification does not affect the expressions for U_{LR} and U_{RL}, only the remaining part of the stiffness matrix equation is presented in each case.

(3) Left end hinged ($Z_{LR} = 0$)

$$\begin{bmatrix} V_{LR} \\ V_{RL} \\ Z_{RL} \end{bmatrix} = \frac{EI_Z}{s^3} \begin{bmatrix} K_5 & -K_5 & K_6 s \\ -K_5 & K_5 & -K_6 s \\ K_6 s & -K_6 s & K_6 s^2 \end{bmatrix} \begin{bmatrix} v_L \\ v_R \\ \theta_R \end{bmatrix} + \begin{bmatrix} V^*_{L0} \\ V^*_{R0} \\ Z^*_{R0} \end{bmatrix}$$

(4) Right end hinged ($Z_{RL} = 0$)

$$\begin{bmatrix} V_{LR} \\ Z_{LR} \\ V_{RL} \end{bmatrix} = \frac{EI_z}{s^3} \begin{bmatrix} K_5 & K_6 s & K_5 \\ K_6 s & K_6 s^2 & -K_6 s \\ -K_5 & -K_6 s & K_5 \end{bmatrix} \begin{bmatrix} v_L \\ \theta_L \\ v_R \end{bmatrix} + \begin{bmatrix} V^*_{L0} \\ Z^*_{L0} \\ V^*_{R0} \end{bmatrix}$$

(5) Left end guided ($V_{LR} = 0$)

$$\begin{bmatrix} Z_{LR} \\ V_{RL} \\ Z_{RL} \end{bmatrix} = \frac{EI_z}{s} \begin{bmatrix} K_9 & 0 & -K_{10} \\ 0 & 0 & 0 \\ -K_{10} & 0 & K_9 \end{bmatrix} \begin{bmatrix} \theta_L \\ v_R \\ \theta_R \end{bmatrix} + \begin{bmatrix} Z^*_{L0} \\ V^*_{R0} \\ Z^*_{R0} \end{bmatrix}$$

(6) Right end guided ($V_{RL} = 0$)

$$\begin{bmatrix} V_{LR} \\ Z_{LR} \\ Z_{RL} \end{bmatrix} = \frac{EI_z}{s} \begin{bmatrix} 0 & 0 & 0 \\ 0 & K_9 & -K_{10} \\ 0 & -K_{10} & K_9 \end{bmatrix} \begin{bmatrix} v_L \\ \theta_L \\ \theta_R \end{bmatrix} + \begin{bmatrix} V^*_{L0} \\ Z^*_{L0} \\ Z^*_{R0} \end{bmatrix}$$

(7) Left end free ($V_{LR} = Z_{LR} = 0$)

$$\begin{bmatrix} V_{RL} \\ Z_{RL} \end{bmatrix} = \frac{EI_z}{s} \begin{bmatrix} 0 & 0 \\ 0 & K_{11} \end{bmatrix} \begin{bmatrix} v_R \\ \theta_R \end{bmatrix} + \begin{bmatrix} V^*_{R0} \\ Z^*_{R0} \end{bmatrix}$$

(8) Right end free ($V_{RL} = Z_{RL} = 0$)

$$\begin{bmatrix} V_{LR} \\ Z_{LR} \end{bmatrix} = \frac{EI_z}{s} \begin{bmatrix} 0 & 0 \\ 0 & K_{11} \end{bmatrix} \begin{bmatrix} v_L \\ \theta_L \end{bmatrix} + \begin{bmatrix} V^*_{L0} \\ Z^*_{L0} \end{bmatrix}$$

9.16 MODIFIED STIFFNESS LOAD FUNCTIONS

(1) Modified load functions designated by an asterisk and shown under the respective figures below are the reactions of the respective beam-columns due to loads and other causes. Their analytical expressions are a linear combination of the modified stiffness coefficients L_j in (9.15-1) and the respective transport load functions F and G (9.06–9.09).

(2) Propped-end beam-column (9.15-3, 4)

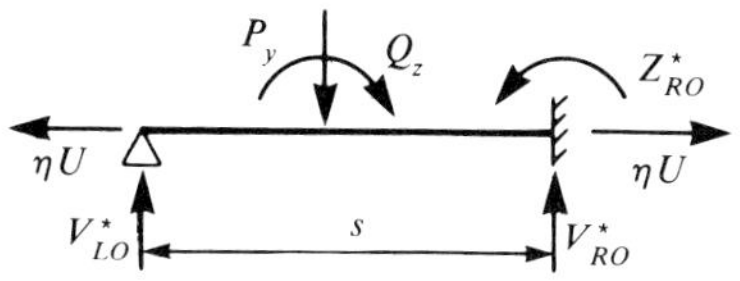

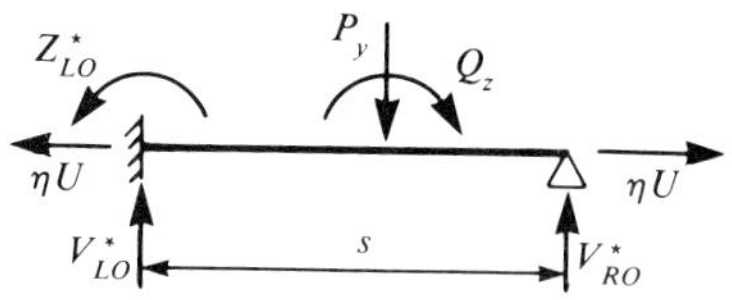

$$\begin{bmatrix} V^*_{LO} \\ V^*_{RO} \\ Z^*_{RO} \end{bmatrix} = \begin{bmatrix} -1 & L_7 & -L_5 \\ 0 & -L_7 & L_5 \\ 0 & L_8\beta^{-1} & -L_6\beta^{-1} \end{bmatrix} \begin{bmatrix} F_V \\ F_Z \\ F_v \end{bmatrix} \qquad \begin{bmatrix} V^*_{LO} \\ Z^*_{LO} \\ V^*_{RO} \end{bmatrix} = \begin{bmatrix} 0 & -L_7 & L_5 \\ 0 & -L_8\beta^{-1} & L_6\beta^{-1} \\ 1 & L_7 & -L_5 \end{bmatrix} \begin{bmatrix} G_V \\ G_Z \\ G_v \end{bmatrix}$$

(3) Guided-end beam-column (9.15-5,6)

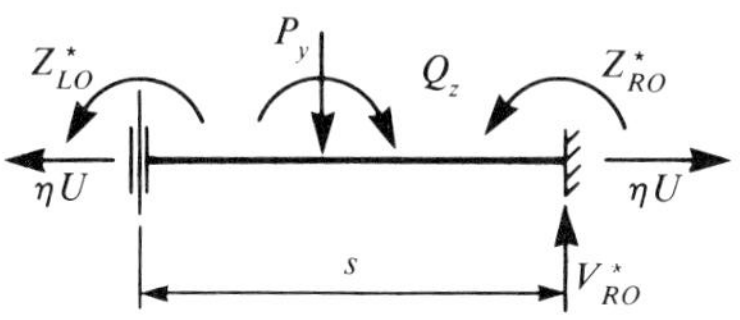

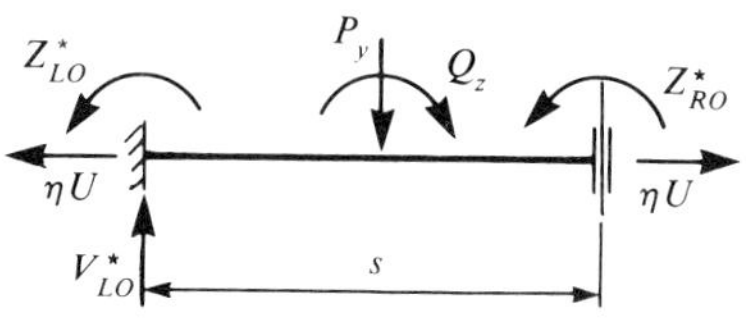

$$\begin{bmatrix} Z^*_{LO} \\ V^*_{RO} \\ Z^*_{RO} \end{bmatrix} = \begin{bmatrix} 0 & 0 & -L_{10}\beta^{-1} \\ 1 & 0 & 0 \\ 0 & 0 & -L_9\beta^{-1} \end{bmatrix} \begin{bmatrix} G_V \\ G_v \\ G_\theta \end{bmatrix} \qquad \begin{bmatrix} V^*_{LO} \\ Z^*_{LO} \\ Z^*_{RO} \end{bmatrix} = \begin{bmatrix} -1 & 0 & 0 \\ 0 & 0 & L_9\beta^{-1} \\ 0 & 0 & L_{10}\beta^{-1} \end{bmatrix} \begin{bmatrix} F_V \\ F_v \\ F_\theta \end{bmatrix}$$

(4) Cantilever beam-column (9.15-7,8)

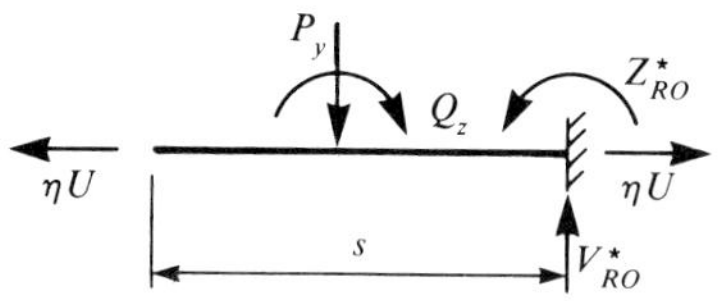

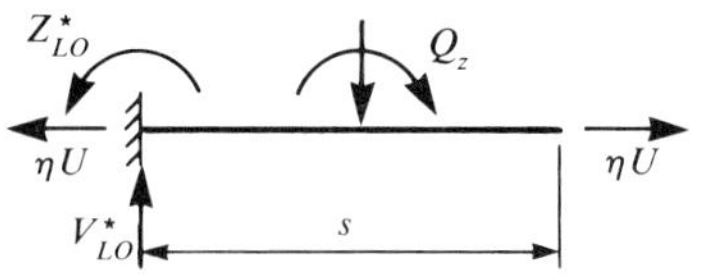

$$\begin{bmatrix} V^*_{RO} \\ Z^*_{RO} \end{bmatrix} = \begin{bmatrix} 1 & 0 & 0 \\ 0 & \beta^{-1} & -L_{11}\beta^{-1} \end{bmatrix} \begin{bmatrix} G_V \\ G_Z \\ G_\theta \end{bmatrix} \qquad \begin{bmatrix} V^*_{LO} \\ Z^*_{LO} \end{bmatrix} = \begin{bmatrix} -1 & 0 & 0 \\ 0 & -\beta^{-1} & L_{11}\beta^{-1} \end{bmatrix} \begin{bmatrix} F_V \\ F_Z \\ F_\theta \end{bmatrix}$$

9.17 STABILITY OF LONG SLENDER COLUMNS, BASIC CASES

(1) Notation

N_{cr} = critical axial load (N) $f_{n,cr}$ = critical normal stress (Pa)
A = area of normal section (m^2) I = least moment of inertia of A (m^4)
l = column length (m) k = least radius of gyration of A (m)
C = stability coefficient $f_{n,pr}$ = proportional limit (Pa)

(2) Critical axial load is the limiting value of the axial force at which the buckling of long slender columns begins:

$$N_{cr} = C\frac{EI}{l^2}$$

The value of C is given for four basic cases in (4)–(7) below and for selected special cases in (9.18–9.21).

(3) Critical normal stress

$$f_{n,cr} = \frac{N_{cr}}{A} = C\frac{EI}{Al^2} = CE\left(\frac{k}{l}\right)^2 \leq f_{n,pr}$$

and the limiting value of the *slenderness ratio* (k/l) is

$$\left(\frac{k}{l}\right)_{cr} = \sqrt{\frac{f_{n,pr}}{CE}}$$

(4) Column with hinged ends

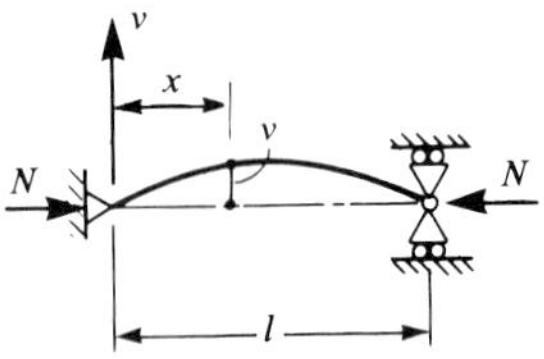

$$N_{cr} = \frac{\pi^2 EI}{l^2}$$
$$f_{n,cr} = \frac{\pi^2 E}{(l/k)^2}$$

(5) Fixed-end column

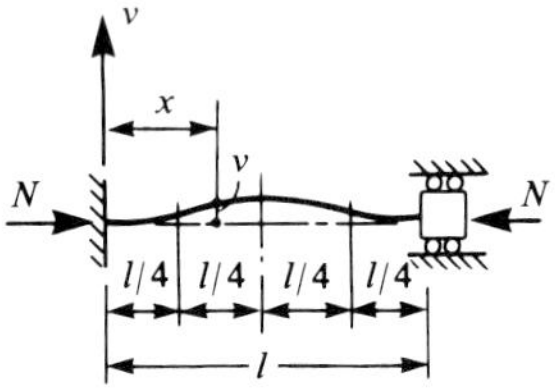

$$N_{cr} = \frac{4\pi^2 EI}{l^2}$$
$$f_{n,cr} = \frac{\pi^2 E}{(l/2k)^2}$$

(6) Propped-end column

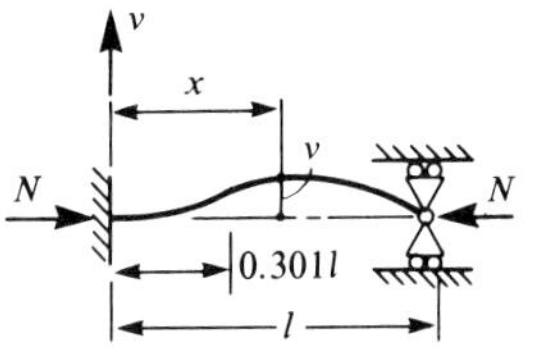

$$N_{cr} = \frac{20.19EI}{l^2}$$
$$f_{n,cr} = \frac{\pi^2 E}{(0.699l/k)^2}$$

(7) Cantilever column

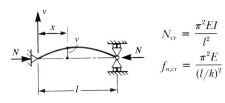

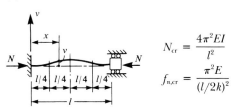

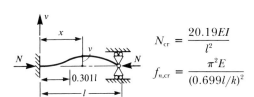

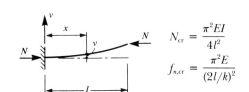

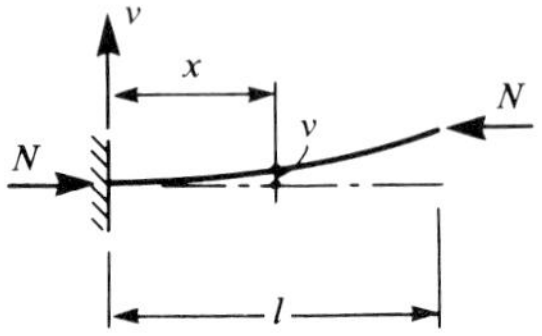

$$N_{cr} = \frac{\pi^2 EI}{4l^2}$$
$$f_{n,cr} = \frac{\pi^2 E}{(2l/k)^2}$$

9.18 STABILITY OF LONG SLENDER COLUMNS, SPECIAL CASES

(1) Columns with intermediate axial loads

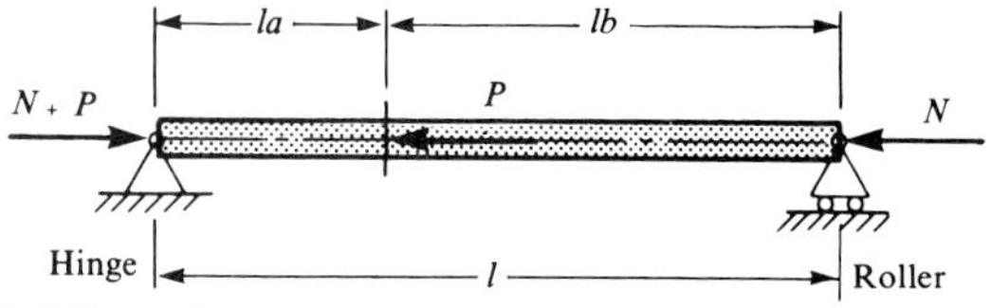

$0 < N < (N + P)$

$$n = \sqrt{\frac{N}{N + P}}$$

I = constant

Coefficient C (Ref. 27, pp. 346–347)

a \ n	0.15	0.20	0.25	0.30	0.35	0.40	0.45	0.50	0.55	0.65	0.70	0.80	0.90
0.20	0.98	1.31	1.82	1.94	2.47	3.06	3.52	4.59	5.18	6.51	7.10	8.22	9.20
0.40	0.80	1.08	1.40	1.66	2.29	2.71	3.40	3.90	4.46	5.85	6.56	7.74	8.92
0.60	0.78	0.99	1.23	1.64	2.17	2.63	3.16	3.78	4.44	5.68	6.37	7.65	8.88
0.80	0.74	0.93	1.04	1.35	1.72	2.22	2.76	3.29	3.95	5.26	5.82	7.20	8.69

(2) Columns with sudden changes in section

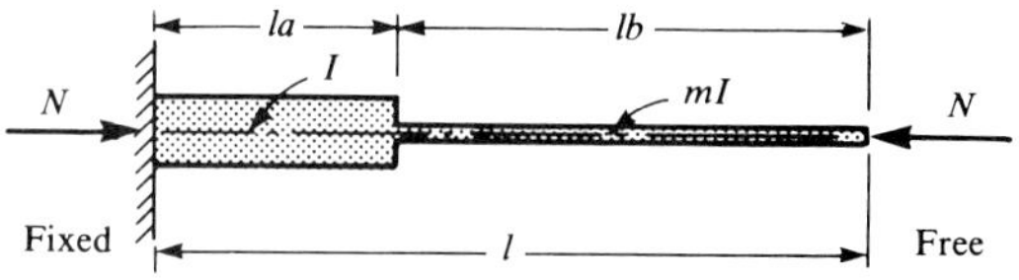

Coefficient C

a \ m	0.10	0.20	0.30	0.40	0.60	0.80
0.20	0.38	0.70	1.02	1.27	1.75	2.14
0.40	0.60	1.06	1.34	1.67	2.05	2.30
0.60	1.13	1.47	1.65	2.13	2.31	2.41
0.80	2.15	2.33	2.38	2.42	2.45	2.47

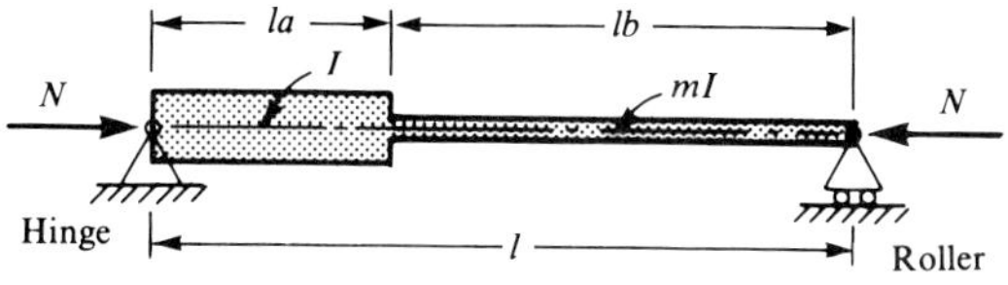

Coefficient C

a \ m	0.10	0.20	0.30	0.40	0.60	0.80
0.20	0.98	2.07	3.16	4.05	6.17	7.99
0.40	1.27	2.47	3.75	4.49	6.76	8.34
0.60	2.17	3.90	5.42	6.31	8.19	9.08
0.80	5.33	7.61	8.58	9.08	9.37	9.62

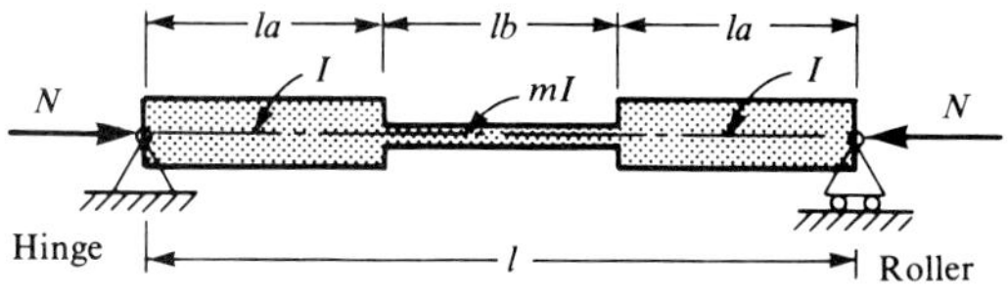

Coefficient C

a \ m	0.10	0.20	0.30	0.40	0.60	0.80
0.10	1.04	2.02	2.96	3.99	5.73	7.95
0.20	1.08	2.42	3.31	4.18	6.27	8.04
0.30	1.63	2.63	3.77	4.76	6.78	8.54
0.40	2.13	3.70	5.06	6.10	7.94	9.00

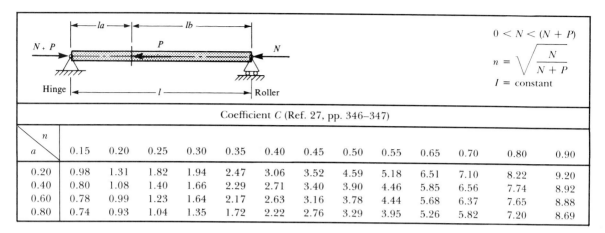

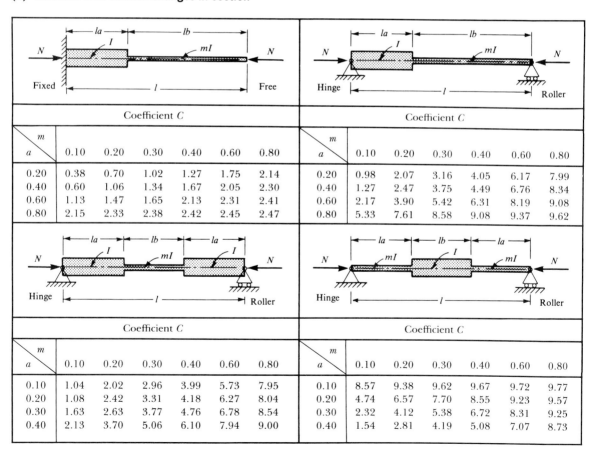

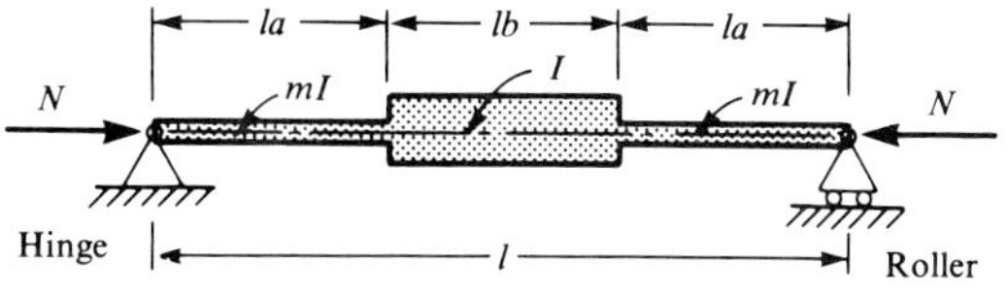

Coefficient C

a \ m	0.10	0.20	0.30	0.40	0.60	0.80
0.10	8.57	9.38	9.62	9.67	9.72	9.77
0.20	4.74	6.57	7.70	8.55	9.23	9.57
0.30	2.32	4.12	5.38	6.72	8.31	9.25
0.40	1.54	2.81	4.19	5.08	7.07	8.73

9.19 LONG COLUMNS TAPERED SYMMETRICALLY WITH RESPECT TO MIDDLE SECTION

(1) Notation and geometry. The symbols used below are defined in (9.17).

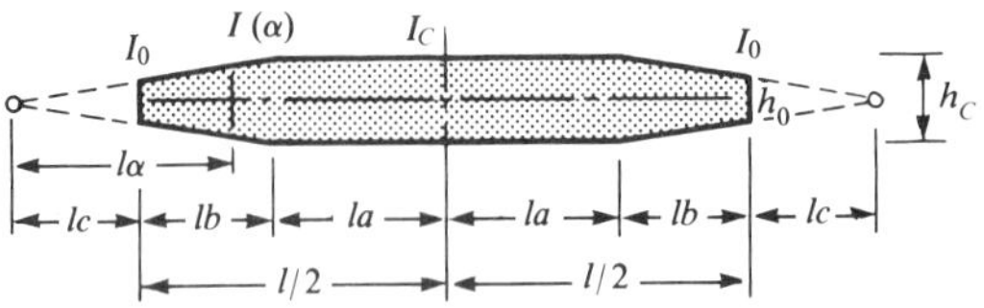

$$I(\alpha) = I_0 \left(\frac{\alpha}{c}\right)^n \qquad \frac{I_C}{I_0} = m$$

$$\left(\frac{b+c}{c}\right)^n = r \qquad n = \frac{\log (I_C/I_0)}{\log (h_C/h_0)}$$

(2) End conditions

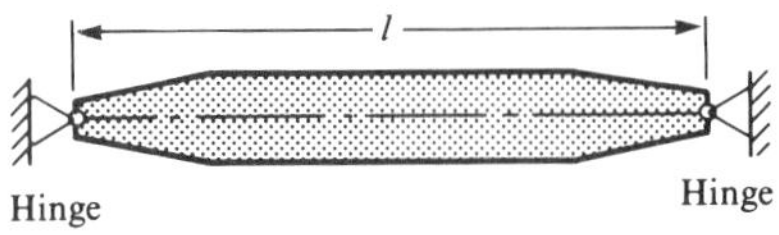

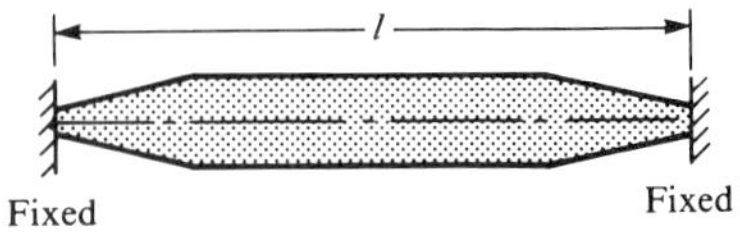

$$N_{cr} = C_1 E I_C / l^2 \qquad N_{cr} = C_2 E I_C / l_2$$

(3) Table of stability coefficients gives the values of C_j corresponding to the specified end conditions and the selected section. The depth of the section h varies linearly in the tapered portion as indicated above, and the width w is constant for the length of the column in (a, c). In (b, d) the section varies linearly in both directions in the tapered portion (Ref. 29).

Normal section			Coefficient C_1				Coefficient C_2			
		a \ r	0.2	0.4	0.6	0.8	0.2	0.4	0.6	0.8
(a) Solid rectangle 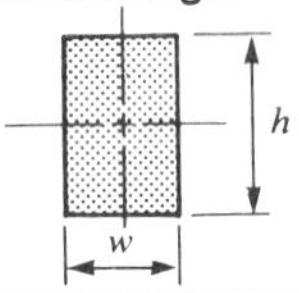	$n = 1$	0.0	7.01	7.87	8.61	9.27	20.36	26.16	31.04	35.40
		0.1	7.99	8.59	9.12	9.54	22.36	27.80	32.20	36.00
		0.2	8.91	9.19	9.55	9.69	23.42	28.96	32.92	36.36
		0.3	9.63	9.70	9.76	9.83	25.44	30.20	33.80	36.84
		0.4	9.82	9.84	9.85	9.86	29.00	33.08	35.80	37.84
(b) Tower section 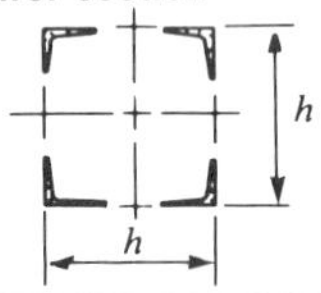	$n = 2$	0.0	6.37	7.61	8.51	9.24	18.94	25.54	30.79	35.35
		0.1	7.49	8.42	9.04	9.50	21.25	27.35	32.02	35.97
		0.2	8.61	9.15	9.48	9.69	22.91	28.52	32.77	36.34
		0.3	9.44	9.63	9.74	9.82	24.29	29.65	33.63	36.80
		0.4	9.81	9.84	9.85	9.86	27.67	32.59	35.64	37.81
(c) Closed box or I section 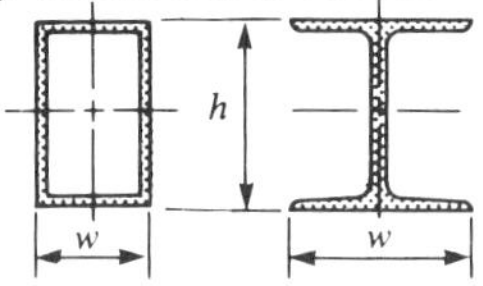	$n = 2.3$	0.0	6.31	7.57	8.51	9.24	18.80	25.47	30.77	35.34
		0.1	7.44	8.41	9.04	9.50	21.14	27.31	32.00	35.97
		0.2	8.57	9.14	9.48	9.69	22.83	28.48	32.75	36.33
		0.3	9.43	9.63	9.74	9.82	24.19	29.61	33.61	36.80
		0.4	9.81	9.84	9.85	9.86	27.54	32.55	35.63	37.81
(d) Solid square or circle 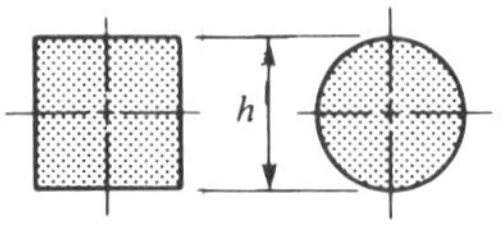	$n = 4$	0.1	6.02	7.48	8.47	9.23	18.23	25.23	30.68	35.33
		0.2	7.20	8.33	9.01	9.49	20.71	27.13	31.96	35.96
		0.3	8.42	9.10	9.45	9.69	22.49	28.33	32.69	36.32
		0.4	9.38	9.62	9.74	9.81	23.80	29.46	33.54	36.78
		0.5	9.80	9.84	9.85	9.86	27.03	32.35	35.56	37.80

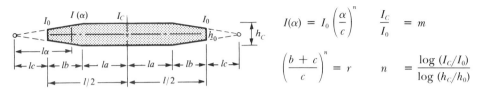

9.20 LONG COLUMNS TAPERED IN ONE DIRECTION

(1) Notation and geometry. The symbols used below are defined in (9.17).

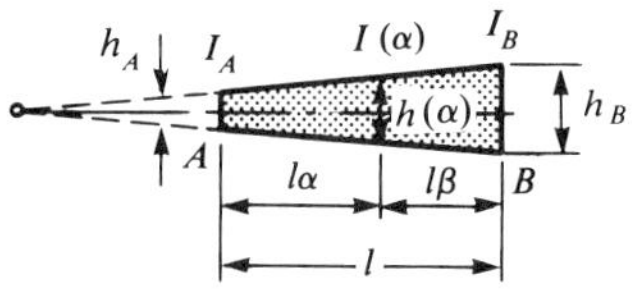

$$h(\alpha) = h(\beta + m\alpha) \qquad h_A = h \qquad h_B = mh$$

$$I(\alpha) = I(\beta + m\alpha)^n \qquad I_A = I \qquad I_B = Im^n$$

$$m = \frac{h_B}{h_A} \qquad n = \frac{\log (I_B/I_A)}{\log (d_B/d_A)}$$

(2) End conditions

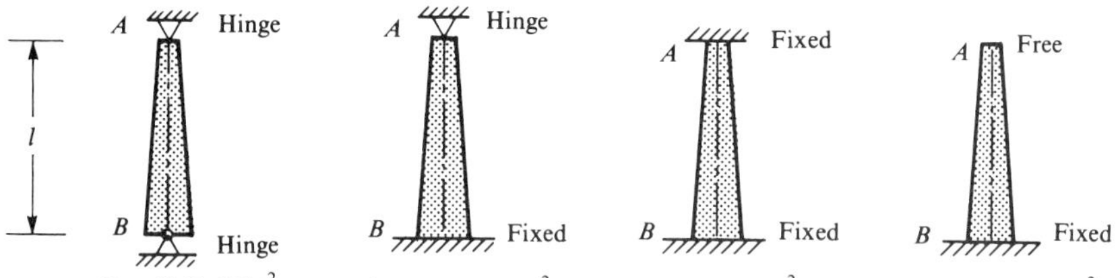

$N_{cr} = 10C_1\, EI/l^2$ $\qquad$ $N_{cr} = 10C_2\, EI/l^2$ $\qquad$ $N_{cr} 10C_3\, EI/l^2$ $\qquad$ $N_{cr} = 10C_4\, EI/l^2$

(3) Table of stability coefficients gives the values of C_j corresponding to the specified end conditions and the selected section. The depth of the section h varies linearly along the column length, and the width of the section w remains constant in (*a, c*). In (*b, d*) the section varies linearly in both directions (Ref. 30).

Normal section	C_j \ m	1.0	1.5	2.0	2.5	3.0	3.5	4.0	5.0
		Coefficients (j = 1, 2, 3, 4)							
(*a*) Solid rectangle 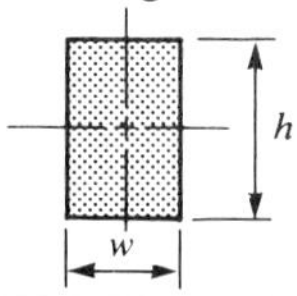	C_1	0.97	1.28	1.43	1.78	1.93	2.12	2.30	2.81
	C_2	2.02	2.53	3.23	3.35	3.74	4.14	4.55	5.37
	C_3	3.95	4.73	5.52	6.32	7.12	7.92	8.68	10.26
	C_4	2.47	3.28	4.10	4.93	5.75	6.56	8.20	9.03
(*b*) Tower section 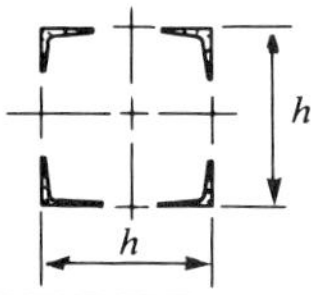	C_1	0.97	1.47	1.96	2.81	3.41	4.04	4.86	6.57
	C_2	2.02	3.03	4.04	5.45	6.36	8.08	9.70	12.42
	C_3	3.95	6.31	7.90	10.66	12.44	15.79	18.16	23.69
	C_4	2.47	4.32	6.42	9.62	12.34	15.30	19.25	23.76
(*c*) Closed box or I section 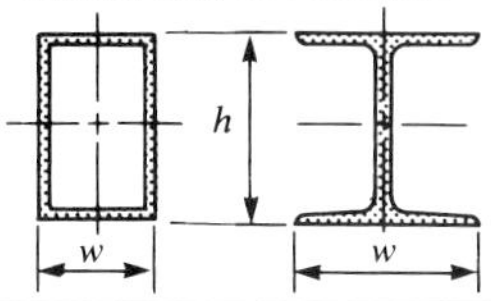	C_1	0.97	1.52	2.27	3.25	3.95	5.13	6.17	8.17
	C_2	2.02	3.43	4.44	6.06	8.18	9.90	11.82	15.96
	C_3	3.95	7.11	9.08	11.84	15.79	19.34	22.71	30.24
	C_4	2.47	4.93	7.40	11.18	12.33	17.51	22.95	33.55
(*d*) Solid square or circle 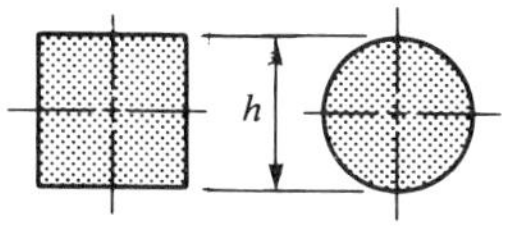	C_1	0.97	2.37	4.05	6.07	7.95	12.00	15.79	22.71
	C_2	2.02	4.71	8.08	13.03	18.18	24.24	32.32	47.97
	C_3	3.95	8.68	15.79	24.87	36.12	48.36	63.16	93.76
	C_4	2.47	7.40	15.54	27.14	46.88	68.59	96.23	187.52

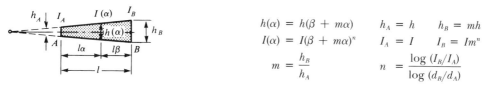

9.21 LONG COLUMNS ELASTICALLY CONSTRAINED

(1) Notation used below is defined in (9.17). Stability coefficients are given in each case in terms of

k_{vL}, k_{vR} = linear spring constants (N/m) $\qquad k_{\theta L}, k_{\theta R}$ = angular spring constants (N·m)

(2) Table of particular cases

(a)

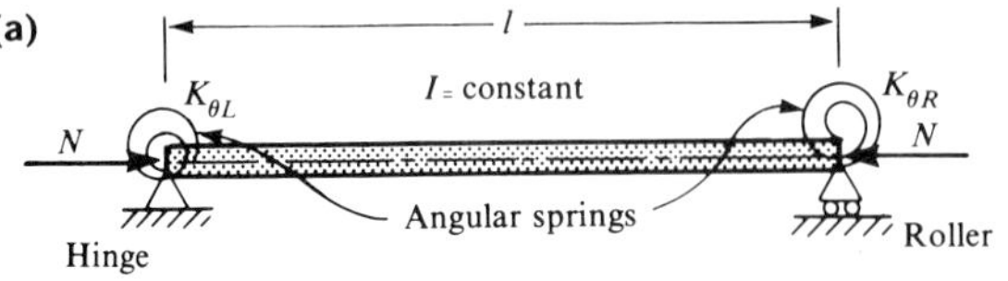

Ends elastically constrained against rotation

$$\Delta_L = \frac{k_{\theta L}l}{EI} \qquad \Delta_R = \frac{k_{\theta R}l}{EI}$$

Coefficient C (Ref. 28)

Δ_L \ Δ_R	0	0.1	0.2	0.5	1	2	5	10	20	50	∞
0	9.87	10.63	11.29	12.89	14.67	16.56	18.40	19.27	19.71	19.98	20.1
0.1	10.63	11.42	12.11	13.76	15.60	17.56	19.54	20.43	20.88	21.16	21.3
0.2	11.29	12.11	12.82	14.59	16.48	18.49	20.52	21.44	21.90	22.28	22.4
0.5	12.89	13.76	14.59	16.48	18.58	20.88	23.34	24.11	24.60	25.00	25.2
1	14.67	15.60	16.48	18.58	20.97	23.52	26.11	27.14	27.77	28.09	28.4
2	16.56	17.56	18.49	20.88	23.52	26.42	29.38	30.69	31.36	31.81	32.0
5	18.40	19.54	20.52	23.34	26.10	29.38	32.83	34.34	35.05	35.52	35.8
10	19.27	20.43	21.44	24.11	27.14	30.69	34.34	35.88	36.72	37.21	37.5
20	19.71	20.88	21.90	24.60	27.77	31.36	35.05	36.72	37.57	38.19	38.5
50	19.98	21.16	22.28	25.00	28.09	31.81	35.52	37.21	38.19	38.69	39.0
∞	20.19	21.34	22.47	25.20	28.41	32.04	35.88	37.58	38.56	39.06	39.4

(b)

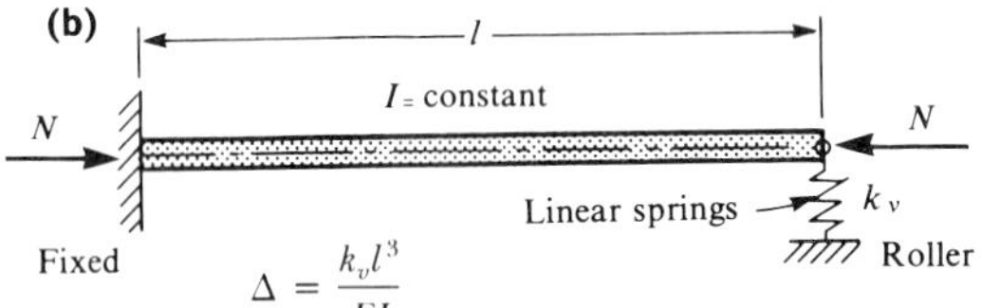

$$\Delta = \frac{k_v l^3}{EI}$$

Coefficients

Δ	0.5	1	2	3	5	10
C	2.96	3.70	4.74	4.92	6.71	9.87

(c)

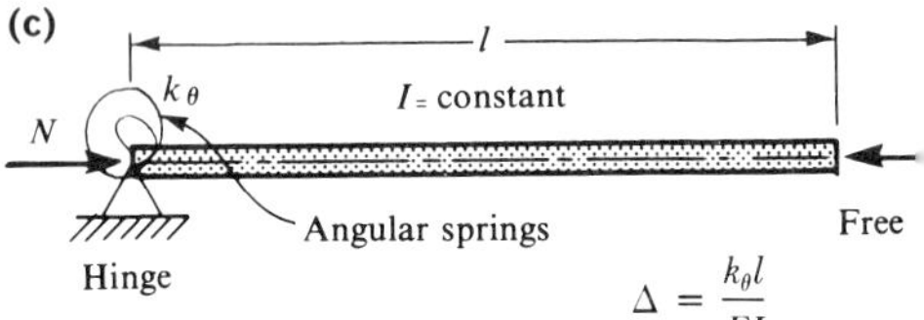

$$\Delta = \frac{k_\theta l}{EI}$$

Coefficients

Δ	0.5	1	2	3	5	10
C	0.20	0.64	1.09	1.43	1.96	2.1

(d)

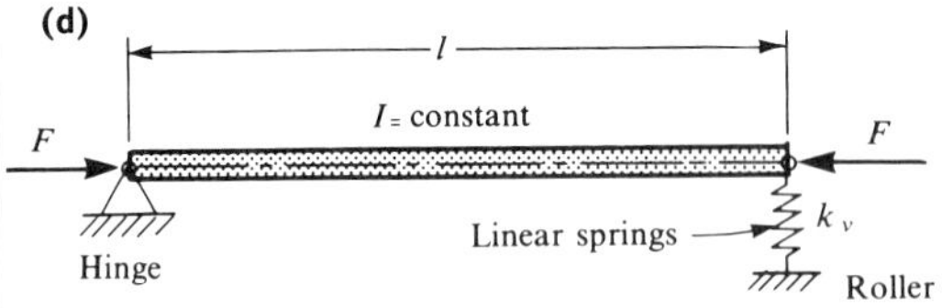

$$N_{cr} = k_v l$$

(e)

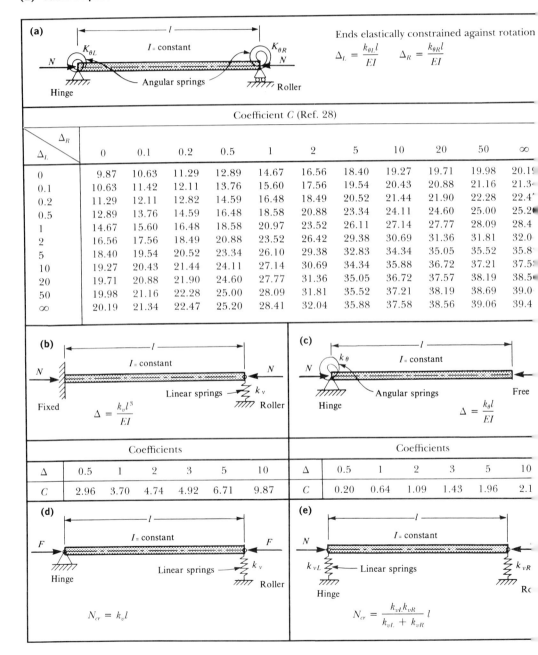

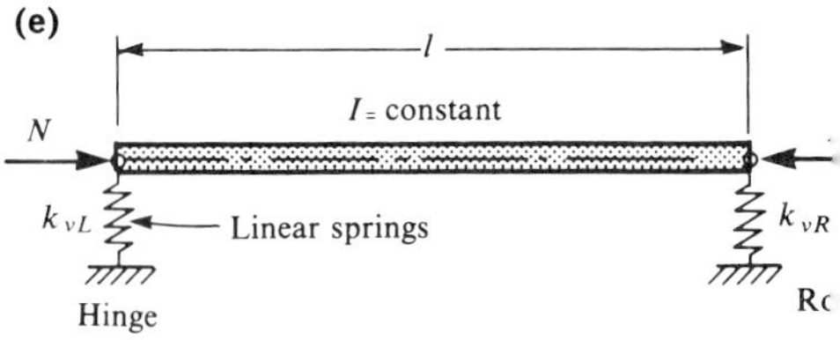

$$N_{cr} = \frac{k_{vL}k_{vR}}{k_{vL} + k_{vR}}\, l$$

10

Beam-Column of Order One Encased in Elastic Foundation Constant Section

STATIC STATE

10.01 DEFINITION OF STATE

(1) System considered is a finite straight beam-column of constant and symmetrical section. The centroidal axis x of the bar and the vertical axis y define the plane of symmetry of this section. The elastic foundation fully surrounds the bar and offers resistance, which is proportional to the flexural displacement of the bar (Winkler's foundation). The constant of proportionality of the bar corresponding to a unit length, called the foundation modulus, is

k_v = transverse modulus along y (N/m²)

The axial force N is constant along the entire span s, and the bar is a part of a planar system acted on by loads in the system plane (system of order one).

(2) Notation (A.1, A.2, A.4)

s	= span (m)	A_x	= area of normal section (m²)
a, b	= segments (m)	I_z	= moment of inertia of A_x about z (m⁴)
x, x'	= coordinates (m)	E	= modulus of elasticity (Pa)
U	= normal force (N)	u	= linear displacement along x (m)
V	= shearing force (N)	v	= linear displacement along y (m)
Z	= flexural moment (N·m)	θ	= flexural slope about z (rad)

$\eta = \begin{cases} +1 & \text{if } N = \text{axial tension (N)} \\ -1 & \text{if } N = \text{axial compression (N)} \end{cases}$	$f = \dfrac{N}{EI_z}$ (m⁻²)	$R = \dfrac{\eta f}{e}$
$\lambda = \sqrt[4]{\dfrac{k_v}{EI_z}}$ = shape parameter (m⁻¹)	$e = \sqrt{\dfrac{k_v}{EI_z}}$ (m⁻²)	

(3) Assumptions of analysis are stated in (1.01). In this chapter, the effects of axial and flexural deformations and the beam-column effects are included. Since only long and slender bars are capable of developing the beam-column action, the effects of shearing deformations are neglected. The longitudinal resistance of the foundation is assumed to be small and is also neglected.

10.02 TRANSPORT MATRIX EQUATIONS

(1) Equivalents and signs

$$\bar{u} = uEA_x s^{-1} = \text{scaled } u \text{ (N)} \qquad \bar{v} = vEI_z\lambda^3 = \text{scaled } v \text{ (N)}$$

$$\bar{Z} = Z\lambda = \text{scaled } Z \text{ (N)} \qquad \bar{\theta} = \theta EI_z\lambda^2 = \text{scaled } \theta \text{ (N)}$$

All forces, moments, and displacements are in the principal axes of the normal section, and their positive directions are shown in (2, 3) below (Ref. 21).

$$f = \frac{N}{EI_z} \qquad e = \sqrt{\frac{k_v}{EI_z}}$$

(2) Free-body diagram

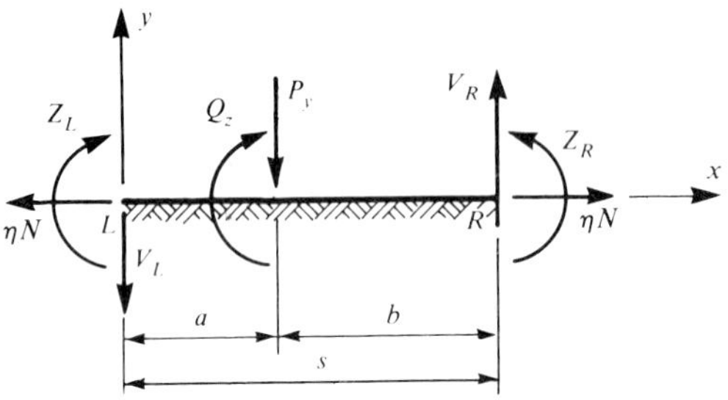

(3) Elastic curve

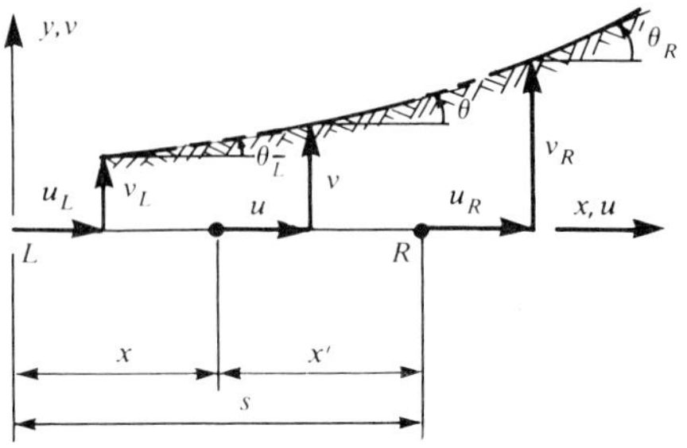

(4) Transport matrix equations

$$\underbrace{\begin{bmatrix} 1 \\ \hline \eta N \\ V_L \\ \bar{Z}_L \\ \hline \bar{u}_L \\ \bar{v}_L \\ \bar{\theta}_L \end{bmatrix}}_{H_L} = \underbrace{\left[\begin{array}{c|ccc|ccc} 1 & 0 & 0 & 0 & 0 & 0 & 0 \\ \hline 0 & 1 & 0 & 0 & 0 & 0 & 0 \\ F_V & 0 & (1-T_{5z}) & -T_{4z} & 0 & -(\lambda s - T_{6z}) & T_{3z} \\ F_Z & 0 & T_{2z} & T_{1z} & 0 & -T_{3z} & -T_{0z} \\ \hline 0 & -1 & 0 & 0 & 1 & 0 & 0 \\ F_v & 0 & T_{4z} & T_{3z} & 0 & (1-T_{5z}) & -T_{2z} \\ F_\theta & 0 & -T_{3z} & -T_{2z} & 0 & T_{4z} & T_{3z} \end{array}\right]}_{T_{LR}} \underbrace{\begin{bmatrix} 1 \\ \hline \eta N \\ V_R \\ \bar{Z}_R \\ \hline \bar{u}_R \\ \bar{v}_R \\ \bar{\theta}_R \end{bmatrix}}_{H_R}$$

$$\underbrace{\begin{bmatrix} 1 \\ \hline \eta N \\ V_R \\ \bar{Z}_R \\ \hline \bar{u}_R \\ \bar{v}_R \\ \bar{\theta}_R \end{bmatrix}}_{H_R} = \underbrace{\left[\begin{array}{c|ccc|ccc} 1 & 0 & 0 & 0 & 0 & 0 & 0 \\ \hline 0 & 1 & 0 & 0 & 0 & 0 & 0 \\ G_V & 0 & (1-T_{5z}) & T_{4z} & 0 & (\lambda s - T_{6z}) & T_{3z} \\ G_Z & 0 & -T_{2z} & T_{1z} & 0 & -T_{3z} & T_{0z} \\ \hline 0 & 1 & 0 & 0 & 1 & 0 & 0 \\ G_v & 0 & -T_{4z} & T_{3z} & 0 & (1-T_{5z}) & T_{2z} \\ G_\theta & 0 & -T_{3z} & T_{2z} & 0 & -T_{4z} & T_{1z} \end{array}\right]}_{T_{RL}} \underbrace{\begin{bmatrix} 1 \\ \hline \eta N \\ V_L \\ \bar{Z}_L \\ \hline \bar{u}_L \\ \bar{v}_L \\ \bar{\theta}_L \end{bmatrix}}_{H_L}$$

where H_L, H_R are the *state vectors* of the respective ends and T_{LR}, T_{RL} are the *transport matrices*. The elements of the transport matrices are the *transport coefficients* $T_{jz}(x)$ discussed in (10.03) and tabulated in (10.04) and the *load functions* F, G displayed for particular cases in (10.7–10.9).

10.03 TRANSPORT RELATIONS

(1) Differential relations

$\dfrac{du}{dx} = \dfrac{U}{EA_x}$ $\dfrac{d^2u}{dx^2} = 0$	$\dfrac{dv}{dx} = \theta$ $\dfrac{d^2v}{dx^2} = \dfrac{d\theta}{dx} = \dfrac{Z}{EI_z}$ $\dfrac{d^3v}{dx^3} = \dfrac{d^2\theta}{dx^2} = \dfrac{dZ}{dx\,EI_z} = -\dfrac{V}{EI_z} + \dfrac{\eta N}{EI_z}\dfrac{dv}{dx}$ $\dfrac{d^4v}{dx^4} = \dfrac{d^3\theta}{dx^2} = \dfrac{d^2Z}{dx^2\,EI_z} = -\dfrac{p_y}{EI_z} - \dfrac{vk_v}{EI_z} + \dfrac{\eta N}{EI}\dfrac{d^2v}{dx^2}$

(2) General solution of the fourth-order differential equation in (1) above yields six sets of transport functions $T_{jz}(x)$ depending on the relation of N to $2\sqrt{EI_zk_v}$ and the value η. They are classified for $N \neq 0$, $k_v \neq 0$ as:

Case I	$N < 2\sqrt{EI_zk_v}$	$\lvert R\rvert = \left\lvert\dfrac{\eta f}{e}\right\rvert < 2$	$\eta = \begin{cases}+1\\-1\end{cases}$	N = axial tension N = axial compression
Case II	$N = 2\sqrt{EI_zk_v}$	$\lvert R\rvert = \left\lvert\dfrac{\eta f}{e}\right\rvert = 2$	$\eta = \begin{cases}+1\\-1\end{cases}$	N = axial tension N = axial compression
Case III	$N > 2\sqrt{EI_zk_v}$	$\lvert R\rvert = \left\lvert\dfrac{\eta f}{e}\right\rvert > 2$	$\eta = \begin{cases}+1\\-1\end{cases}$	N = axial tension N = axial compression

(3) Analytical forms of $T_{jz}(x)$ are tabulated in (10.04) in terms of

$$\bar{S}_1 = \sinh\beta x \qquad S_1 = \sin\beta x \qquad \bar{S}_2 = \sinh\gamma x \qquad S_2 = \sin\gamma x$$
$$\bar{C}_1 = \cosh\beta x \qquad C_1 = \cos\beta x \qquad \bar{C}_2 = \cosh\gamma x \qquad C_2 = \cos\gamma x$$

where β, γ are the shape parameters given in each particular case. The series expansions and the graphs of $T_{jz}(x)$ are shown in (10.04–10.06).

(4) Transport coefficients T_{jz}, A_{jx}, B_{jz}, C_{jz} are the values of $T_{jz}(x)$ for a particular argument x.

$$\left.\begin{array}{lll} \text{For } x = s, & T_{jz}(s) = T_{jz} \\ \text{For } x = a, & T_{jz}(a) = A_{jz} \\ \text{For } x = b, & T_{jz}(b) = B_{jz} \\ \text{and } T_{jz} - A_{jz} = C_{jz} \end{array}\right\} j = 0, 1, 2, \ldots, 7$$

(5) Transport matrices of this bar have analogical properties to those of (4.02) and can be used in similar ways.

(6) General transport matrix equations for beam-columns of order three encased in elastic foundation are obtained by matrix composition of (10.02) and (15.02) as shown in (2.18).

10.04 TRANSPORT FUNCTIONS

(1) Analytical forms, case I

$\lvert R\rvert = \left\lvert\frac{\eta f}{e}\right\rvert < 2$	$N < 2\sqrt{2EI_z k_v}$

$\Delta = \sqrt{4e^2 - f^2}$ $\qquad \beta = \frac{1}{2}\sqrt{2e + f}$ $\qquad \gamma = \frac{1}{2}\sqrt{2e - f}$ $\qquad \lambda = \sqrt{e}$

$T_{jz}(x)$	Axial tension ($\eta = +1$)	Axial compression ($\eta = -1$)
$T_{0z}(x)$	$-\frac{1}{2\lambda}\left(\frac{e-f}{\gamma}\overline{C}_1 S_2 - \frac{e+f}{\beta}\overline{S}_1 C_2\right)$	$\frac{1}{2\lambda}\left(\frac{e-f}{\gamma}C_1\overline{S}_2 - \frac{e+f}{\beta}S_1\overline{C}_2\right)$
$T_{1z}(x)$	$\overline{C}_1 C_2 + \frac{f}{\Delta}\overline{S}_1 S_2$	$C_1\overline{C}_2 - \frac{f}{\Delta}S_1\overline{S}_2$
$T_{2z}(x)$	$\frac{\lambda}{2}\left(\frac{\overline{S}_1 C_2}{\beta} + \frac{\overline{C}_1 S_2}{\gamma}\right)$	$\frac{\lambda}{2}\left(\frac{S_1\overline{C}_2}{\beta} + \frac{C_1\overline{S}_2}{\gamma}\right)$
$T_{3z}(x)$	$\frac{2\lambda^2}{\Delta}\overline{S}_1 S_2$	$\frac{2\lambda^2}{\Delta}S_1\overline{S}_2$
$T_{4z}(x)$	$-\frac{\lambda}{2}\left(\frac{\overline{S}_1 C_2}{\beta} - \frac{\overline{C}_1 S_2}{\gamma}\right)$	$\frac{\lambda}{2}\left(\frac{S_1\overline{C}_2}{\beta} - \frac{C_1\overline{S}_2}{\gamma}\right)$
$T_{5z}(x)$	$1 - \overline{C}_1 C_2 + \frac{f}{\Delta}\overline{S}_1 S_2$	$1 - C_1\overline{C}_2 - \frac{f}{\Delta}S_1\overline{S}_2$
$T_{6z}(x)$	$\lambda x - \frac{1}{2\lambda}\left(\frac{e-f}{\gamma}\overline{C}_1 S_2 + \frac{e+f}{\beta}\overline{S}_1 C_2\right)$	$\lambda x - \frac{1}{2\lambda}\left(\frac{e+f}{\beta}S_1\overline{C}_2 + \frac{e-f}{\gamma}C_1\overline{S}_2\right)$
$T_{7z}(x)$	$\frac{(\lambda x)^2}{2} + \frac{f}{e}T_{5z}(x) - T_{3x}(x)$	$\frac{(\lambda x)^2}{2} - \frac{f}{e}T_{5z}(x) - T_{3z}(x)$

(2) Analytical forms, case II

$\lvert R\rvert = \left\lvert\frac{\eta f}{e}\right\rvert = 2$	$N = 2\sqrt{EI_z k_v}$

$\Delta = 0$ $\qquad \beta = \gamma = \lambda = \sqrt{e} = \sqrt{\frac{f}{2}}$

$T_{jz}(x)$	Axial tension ($\eta = +1$)	Axial compression ($\eta = -1$)
$T_{0z}(x)$	$\frac{1}{2}(\beta x\overline{C}_1 + 3\overline{S}_1)$	$-\frac{1}{2}(\beta x C_1 + 3S_1)$
$T_{1z}(x)$	$\frac{1}{2}(\beta x\overline{S}_1 + 2\overline{C}_1)$	$-\frac{1}{2}(\beta x S_1 - 2C_1)$
$T_{2z}(x)$	$\frac{1}{2}(\beta x\overline{C}_1 + \overline{S}_1)$	$\frac{1}{2}(\beta x C_1 + S_1)$
$T_{3z}(x)$	$\frac{1}{2}\beta x\overline{S}_1$	$\frac{1}{2}\beta x S_1$
$T_{4z}(x)$	$\frac{1}{2}(\beta x\overline{C}_1 - \overline{S}_1)$	$-\frac{1}{2}(\beta x C_1 - S_1)$
$T_{5z}(x)$	$\frac{1}{2}(\beta x\overline{S}_1 - 2\overline{C}_1) + 1$	$-\frac{1}{2}(\beta x S_1 + 2C_1) + 1$
$T_{6z}(x)$	$\frac{1}{2}(\beta x\overline{C}_1 - 3\overline{S}_1) + \beta x$	$\frac{1}{2}(\beta x C_1 - 3S_1) + \beta x$
$T_{7z}(x)$	$2T_{5z}(x) - T_{3z} + \frac{(\beta x)^2}{2}$	$-2T_{5z}(x) - T_{3z}(x) + \frac{(\beta x)^2}{2}$

(3) Analytical forms, case III

$\lvert R \rvert = \left\lvert \dfrac{\eta f}{e} \right\rvert > 2$	$N > 2\sqrt{EI_z k_v}$

$\Delta = \sqrt{f^2 - 4e^2}$	$\beta = \sqrt{\frac{1}{2}(f + \Delta)}$	$\gamma = \sqrt{\frac{1}{2}(f - \Delta)}$ $\qquad \lambda = \sqrt{e}$
$T_{jz}(x)$	Axial tension ($\eta = +1$)	Axial compression ($\eta = -1$)
$T_{0z}(x)$	$\dfrac{1}{\lambda\Delta}(\beta^3\bar{S}_1 - \gamma^3\bar{S}_2)$	$-\dfrac{1}{\lambda\Delta}(\beta^3 S_1 - \gamma^3 S_2)$
$T_{1z}(x)$	$\dfrac{1}{\Delta}(\beta^2\bar{C}_1 - \gamma^2\bar{C}_2)$	$\dfrac{1}{\Delta}(\beta^2 C_1 - \gamma^2 C_2)$
$T_{2z}(x)$	$\dfrac{\lambda}{\Delta}(\beta\bar{S}_1 - \gamma\bar{S}_2)$	$\dfrac{\lambda}{\Delta}(\beta S_1 - \gamma S_2)$
$T_{3z}(x)$	$\dfrac{\lambda^2}{\Delta}(\bar{C}_1 - \bar{C}_2)$	$-\dfrac{\lambda^2}{\Delta}(C_1 - C_2)$
$T_{4z}(x)$	$\dfrac{\lambda^3}{\Delta}\left(\dfrac{\bar{S}_1}{\beta} - \dfrac{\bar{S}_2}{\gamma}\right)$	$-\dfrac{\lambda^3}{\Delta} - \left(\dfrac{S_1}{\beta} - \dfrac{S_2}{\gamma}\right)$
$T_{5z}(x)$	$\dfrac{\lambda^4}{\Delta}\left(\dfrac{\bar{C}_1}{\beta^2} - \dfrac{\bar{C}_2}{\gamma^2}\right) + 1$	$\dfrac{\lambda^4}{\Delta}\left(\dfrac{C_1}{\beta^2} - \dfrac{C_2}{\gamma^2}\right) + 1$
$T_{6z}(x)$	$\dfrac{\lambda^5}{\Delta}\left(\dfrac{\bar{S}_1}{\beta^3} - \dfrac{\bar{S}_2}{\gamma^3}\right) + \lambda x$	$\dfrac{\lambda^5}{\Delta}\left(\dfrac{S_1}{\beta^3} - \dfrac{S_2}{\gamma^3}\right) + \lambda x$
$T_{7z}(x)$	$\dfrac{f}{e}T_{5z}(x) - T_{3z}(x) + \dfrac{(\lambda x)^2}{2}$	$-\dfrac{f}{e}T_{5z}(x) - T_{3z}(x) + \dfrac{(\lambda x)^2}{2}$

(4) Series expansion. All six forms of $T_{jz}(x)$ given above can be represented in terms of λ and R defined in (10.01-2) as

$T_{0z}(x) = \sum_{k=2}^{\infty} M_k \dfrac{(\lambda x)^{2k-3}}{(2k-3)!}$	$T_{3z}(x) = \sum_{k=1}^{\infty} M_k \dfrac{(\lambda x)^{2k}}{(2k)!}$	$T_{6z}(x) = \sum_{k=1}^{\infty} M_k \dfrac{(\lambda x)^{2k+3}}{(2k+3)!}$
$T_{1z}(x) = \sum_{k=1}^{\infty} M_k \dfrac{(\lambda x)^{2k-2}}{(2k-2)!}$	$T_{4z}(x) = \sum_{k=1}^{\infty} M_k \dfrac{(\lambda x)^{2k+1}}{(2k+1)!}$	$T_{7z}(x) = \sum_{k=1}^{\infty} M_k \dfrac{(\lambda x)^{2k+4}}{(2k+4)!}$
$T_{2z}(x) = \sum_{k=1}^{\infty} M_k \dfrac{(\lambda x)^{2k-1}}{(2k-1)!}$	$T_{5z}(x) = \sum_{k=1}^{\infty} M_k \dfrac{(\lambda x)^{2k+2}}{(2k+2)!}$	

where $M_1 = 1$, $M_2 = R$, and for $k > 2$, $M_k = RM_{k-1} - M_{k-2}$.

(5) Graphs of $T_{jz}(x)$ based on their analytical forms given above in (1)–(3) are shown in (10.05, 10.06). Each function is represented by eight curves designated by 1, 2, . . ., 8 and corresponding to a selected value of R.

Curve no.	1	2	3	4	5	6	7	8	$\lambda = \sqrt[4]{\dfrac{k_v}{EI_z}}$	$R = \dfrac{N}{\sqrt{EI_z k_v}}$
R	0	0.5	1.0	1.5	2.0	2.5	3.0	3.5		
Case	$f = 0$		I		II		III			

10.05 GRAPHS OF TRANSPORT COEFFICIENTS, AXIAL TENSION

$\lambda = \sqrt[4]{\dfrac{k_v}{EI_z}}$	$R = \dfrac{N}{\sqrt{EI_z k_v}}$

(1) Graphs of $T_{0z}(x)$

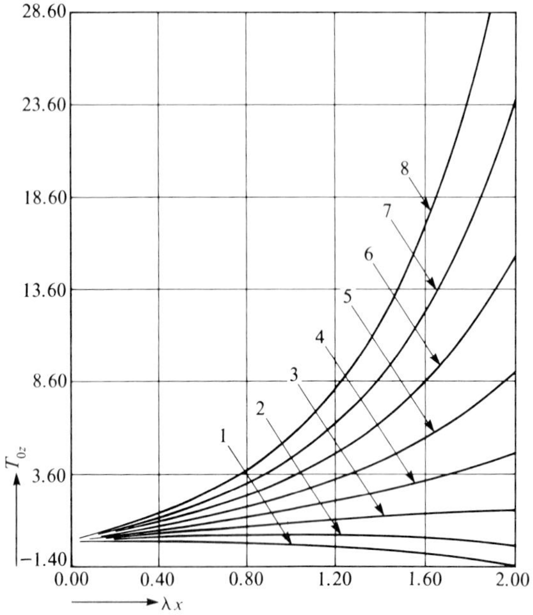

(2) Graphs of $T_{1z}(x)$

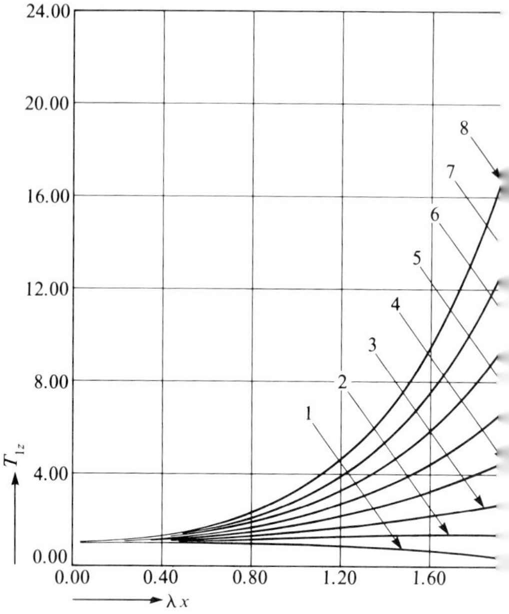

(3) Graphs of $T_{2z}(x)$

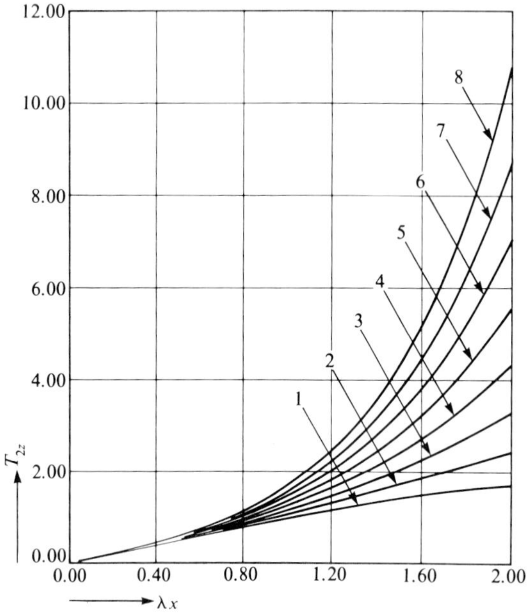

(4) Graphs of $T_{3z}(x)$

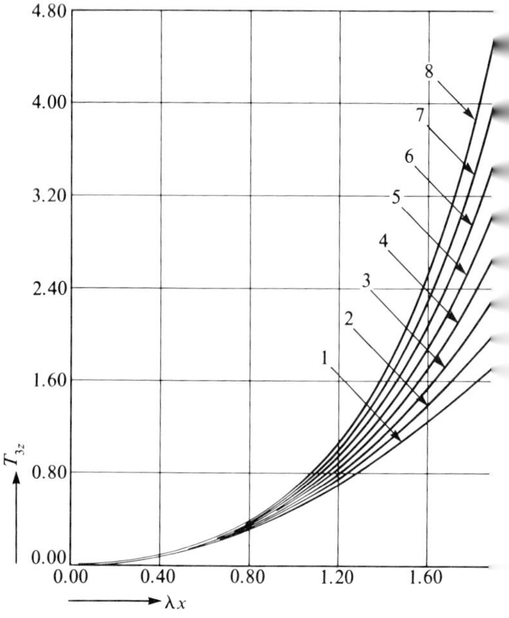

Graphs of $T_{4z}(x)$

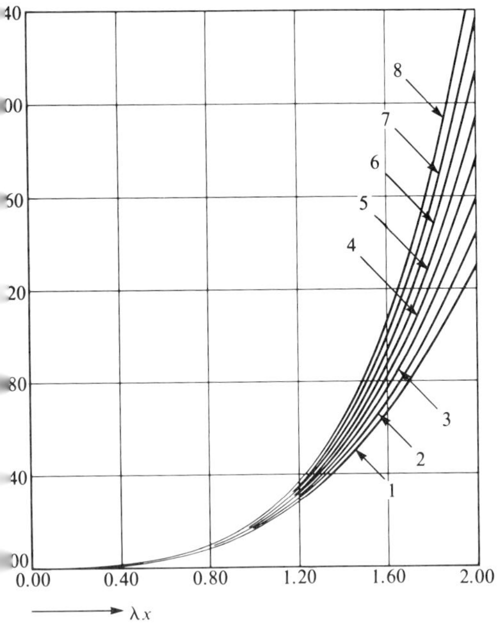

(6) Graphs of $T_{5z}(x)$

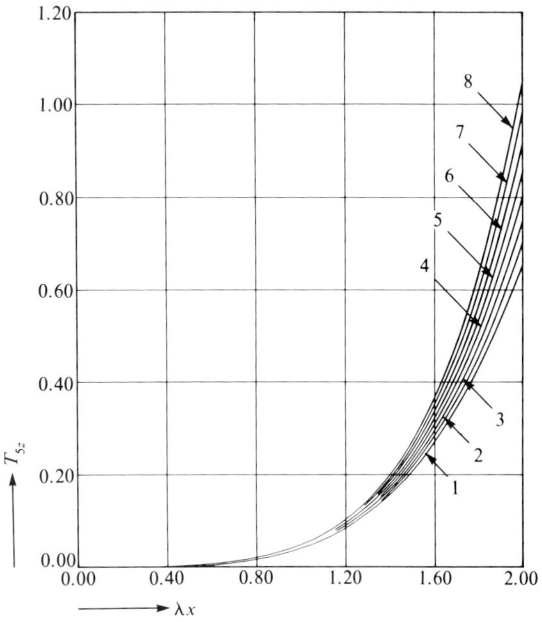

Graphs of $T_{6z}(x)$

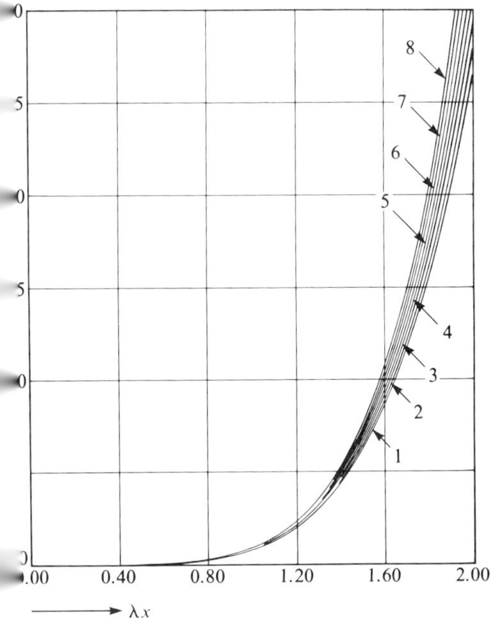

(8) Graphs of $T_{7z}(x)$

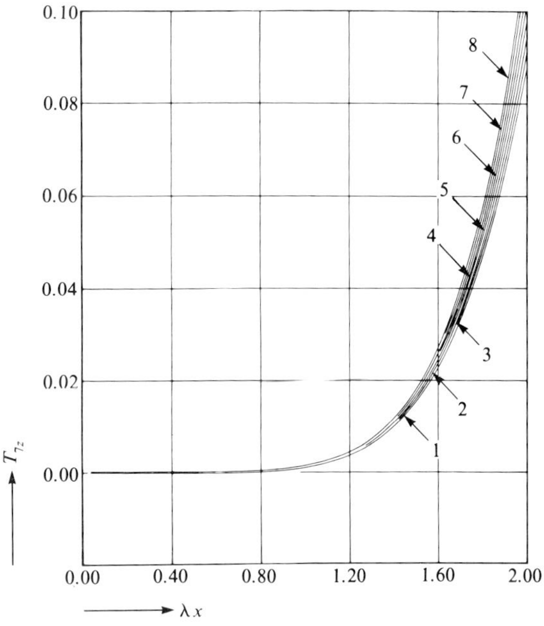

10.06 GRAPHS OF TRANSPORT COEFFICIENTS, AXIAL COMPRESSION

$\lambda = \sqrt[4]{\dfrac{k_v}{EI_z}}$	$R = \dfrac{-N}{\sqrt{EI_z k_v}}$

(1) Graphs of $T_{0z}(x)$

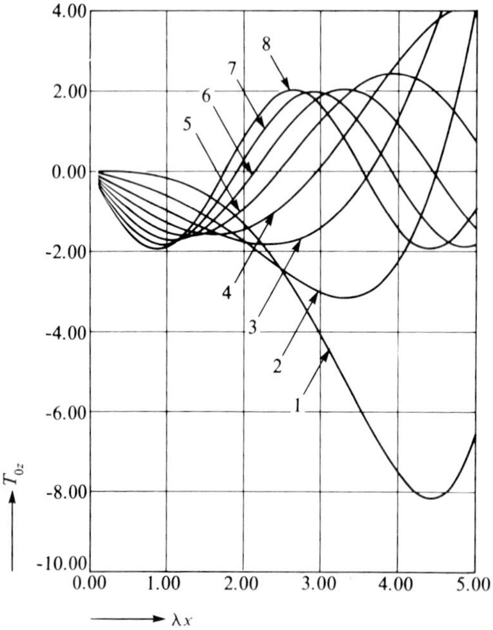

(2) Graphs of $T_{1z}(x)$

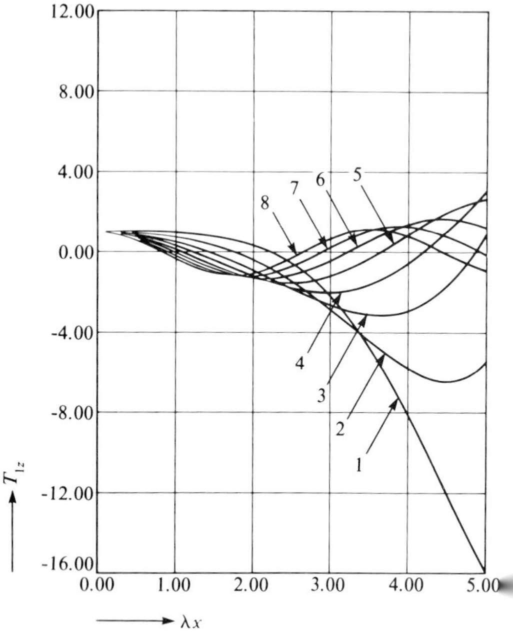

(3) Graphs of $T_{2z}(x)$

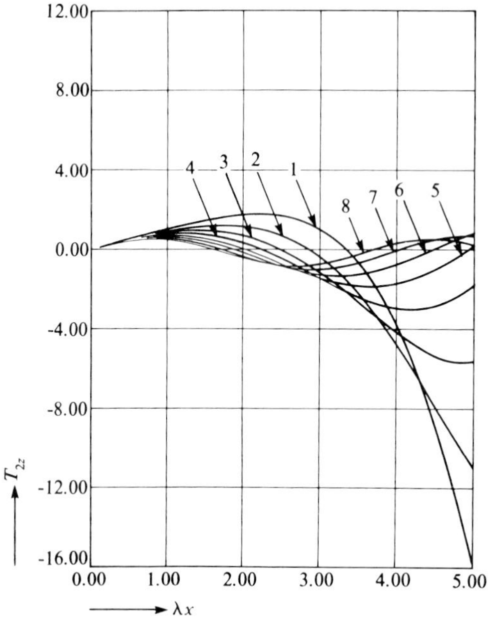

(4) Graphs of $T_{3z}(x)$

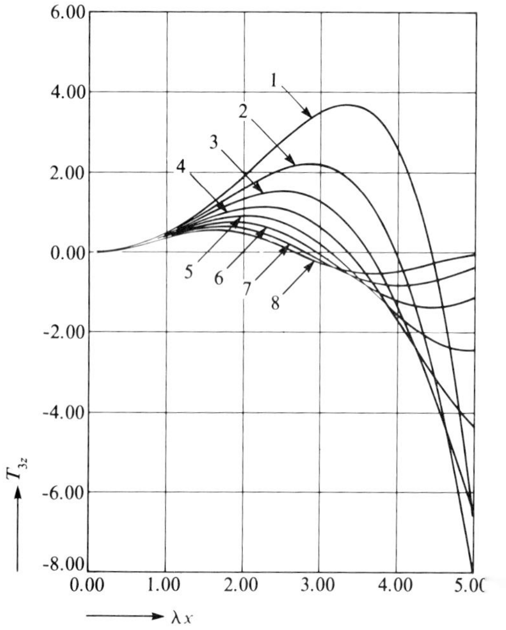

Graphs of $T_{4z}(x)$

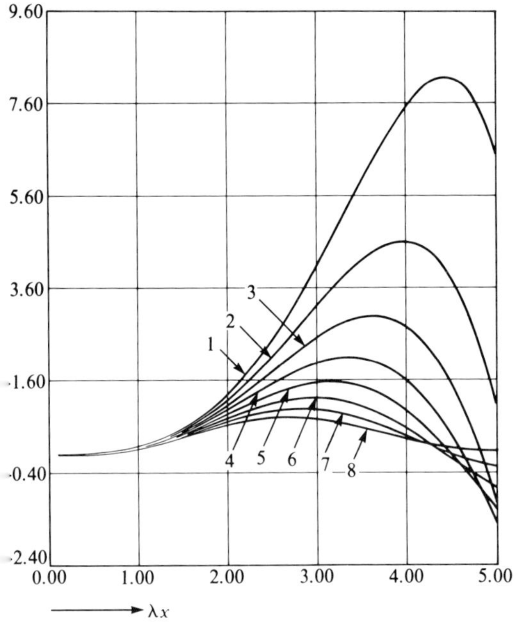

(6) Graphs of $T_{5z}(x)$

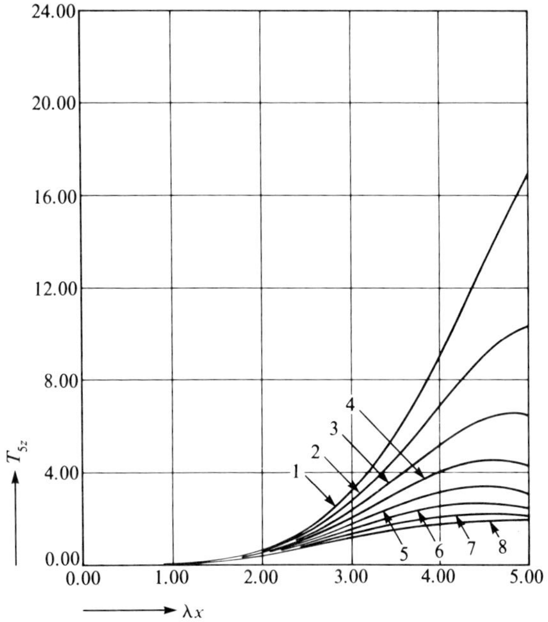

Graphs of $T_{6z}(x)$

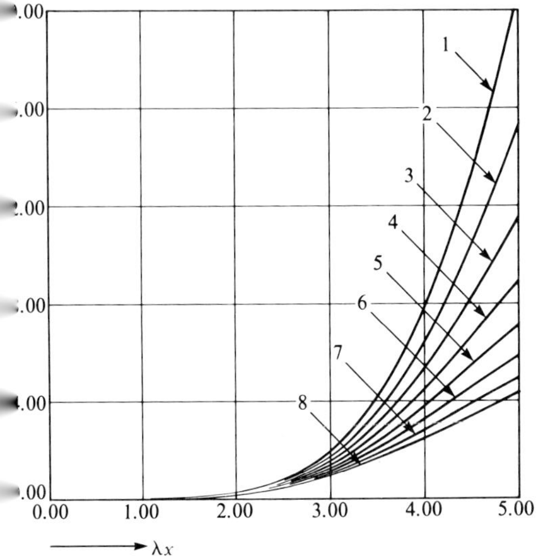

(8) Graphs of $T_{7z}(x)$

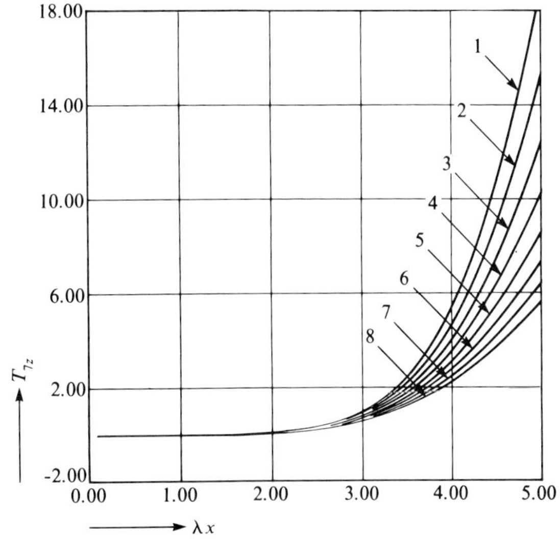

10.07 TRANSPORT LOAD FUNCTIONS

Notation (10.02–10.04) Signs (10.02) Matrix (10.02) P_x, P_y = concentrated loads (N) Q_z = applied couple (N·m) q_z = intensity of distributed couple (N·m/m) A_j, B_j, C_j, T_j = transport coefficients (10.03–10.09)		$\lambda = \sqrt[4]{\dfrac{k_v}{EI_z}}$ $R = \dfrac{\eta f}{e} = \dfrac{\eta N}{\sqrt{EI_z k_v}}$
$F_U = 0$ $F_V = -(1 - A_{5z})P_y$ $F_Z = -A_{2z}P_y$	**1.** 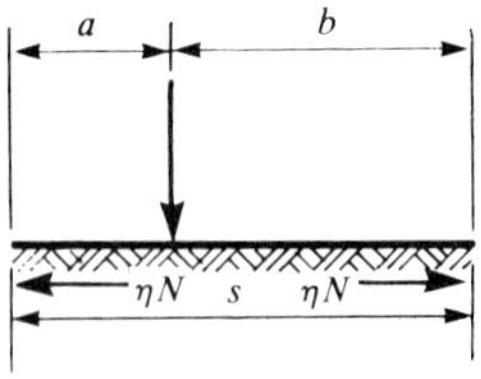	$G_U = 0$ $G_V = (1 - B_{5z})P_y$ $G_Z = -B_{2z}P_y$
$F_u = 0$ $F_v = -A_{4z}P_y$ $F_\theta = A_{3z}P_y$		$G_u = 0$ $G_v = -B_{4z}P_y$ $G_\theta = -B_{3z}P_y$
$F_U = 0$ $F_V = A_{4z}\lambda Q_z$ $F_Z = -A_{1z}\lambda Q_z$	**2.** 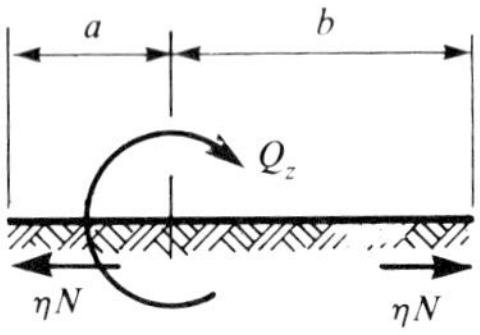	$G_U = 0$ $G_V = B_{4z}\lambda Q_z$ $G_Z = B_{1z}\lambda Q_z$
$F_u = 0$ $F_v = -A_{3z}\lambda Q_z$ $F_\theta = A_{2z}\lambda Q_z$		$G_u = 0$ $G_v = B_{3z}\lambda Q_z$ $G_\theta = B_{2z}\lambda Q_z$
$F_U = 0$ $F_V = T_{5z}q_z$ $F_Z = -T_{2z}q_z$	**3.** 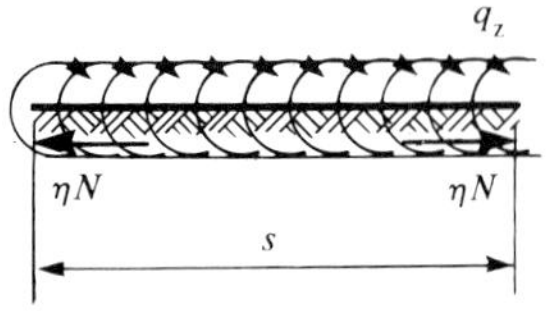	$G_U = 0$ $G_V = T_{5z}q_z$ $G_Z = T_{2z}q_z$
$F_u = 0$ $F_v = -T_{4z}q_z$ $F_\theta = T_{3z}q_z$		$G_u = 0$ $G_v = T_{4z}q_z$ $G_\theta = T_{3z}q_z$
$F_U = 0$ $F_V = C_{5z}q_z$ $F_Z = -C_{2z}q_z$	**4.** 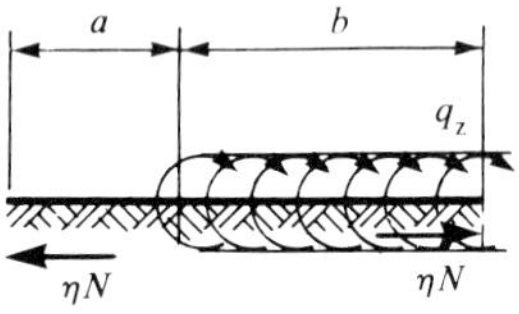	$G_U = 0$ $G_V = B_{5z}q_z$ $G_Z = B_{2z}q_z$
$F_u = 0$ $F_v = -C_{4z}q_z$ $F_\theta = C_{3z}q_z$		$G_u = 0$ $G_v = B_{4z}q_z$ $G_\theta = B_{3z}q_z$

10.08 TRANSPORT LOAD FUNCTIONS

Notation (10.02–10.04) Signs (10.02) Matrix (10.02) p_y = intensity of distributed load (N/m) A_j, B_j, C_j, T_j = transport coefficients (10.03–10.09)		$\lambda = \sqrt[4]{\dfrac{k_v}{EI_z}}$ $R = \dfrac{\eta f}{e} = \dfrac{\eta N}{\sqrt{EI_z k_v}}$
$F_U = 0$ $F_V = -(T_{2z} - RT_{4z})\lambda^{-1}p_y$ $F_Z = -T_{3z}\lambda^{-1}p_y$ $F_u = 0$ $F_v = -T_{5z}\lambda^{-1}p_y$ $F_\theta = T_{4z}\lambda^{-1}p_y$	**1.** 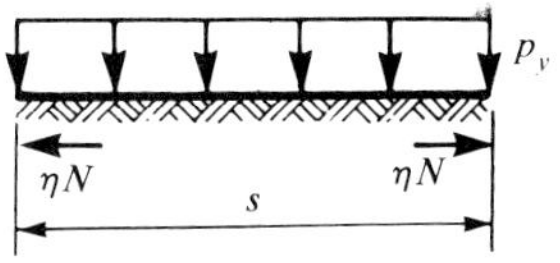	$G_U = 0$ $G_V = (T_{2z} - RT_{4z})\lambda^{-1}p_y$ $G_Z = -T_{3z}\lambda^{-1}p_y$ $G_u = 0$ $G_v = -T_{5z}\lambda^{-1}p_y$ $G_\theta = -T_{4z}\lambda^{-1}p_y$
$F_U = 0$ $F_V = -(C_{2z} - RC_{4z})\lambda^{-1}p_y$ $F_Z = -C_{3z}\lambda^{-1}p_y$ $F_u = 0$ $F_v = -C_{5z}\lambda^{-1}p_y$ $F_\theta = C_{4z}\lambda^{-1}p_y$	**2.** 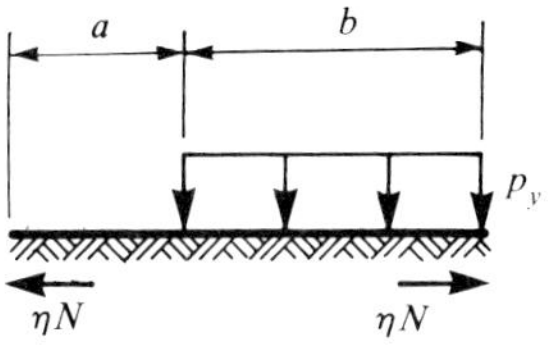	$G_U = 0$ $G_V = (B_{2z} - RB_{4z})\lambda^{-1}p_y$ $G_Z = -B_{3z}\lambda^{-1}p_y$ $G_u = 0$ $G_v = -B_{5z}\lambda^{-1}p_y$ $G_\theta = -B_{4z}\lambda^{-1}p_y$
$F_U = 0$ $F_V = -(\lambda s\Delta_1 + T_{7z})\lambda^{-2}s^{-1}p_y$ $F_Z = -(\lambda sT_{3z} - T_{4z})\lambda^{-2}s^{-1}p_y$ $F_u = 0$ $F_v = -(\lambda sT_{5z} - T_{6z})\lambda^{-2}s^{-1}p_y$ $F_\theta = (\lambda sT_{4z} - T_{5z})\lambda^{-2}s^{-1}p_y$	**3.** 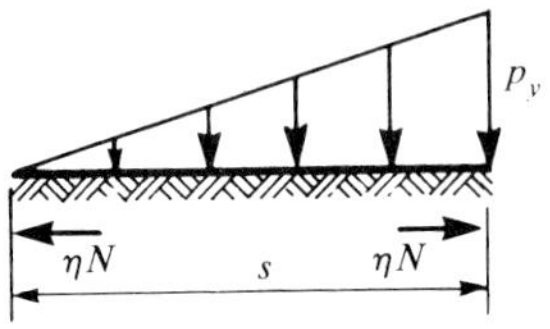 $\Delta_1 = \frac{1}{2}\lambda s - T_{6z}$	$G_U = 0$ $G_V = -(\frac{1}{2}\lambda^2 s^2 - T_{7z})\lambda^{-2}s^{-1}p_y$ $G_Z = T_{4z}\lambda^{-2}s^{-1}p_y$ $G_u = 0$ $G_v = -T_{6z}\lambda^{-2}s^{-1}p_y$ $G_\theta = -T_{5z}\lambda^{-2}s^{-1}p_y$
$F_U = 0$ $F_V = -(\lambda b\Delta_2 + C_{7z})\lambda^{-2}b^{-1}p_y$ $F_Z = -(\lambda bT_{3z} - C_{4z})\lambda^{-2}b^{-1}p_y$ $F_u = 0$ $F_v = -(\lambda bT_{5z} - C_{6z})\lambda^{-2}b^{-1}p_y$ $F_\theta = (\lambda bT_{4z} - C_{5z})\lambda^{-2}b^{-1}p_y$	**4.** 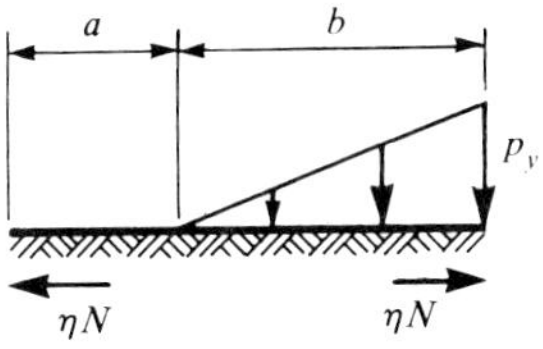 $\Delta_2 = \frac{1}{2}\lambda b - T_{6z}$	$G_U = 0$ $G_V = -(\frac{1}{2}\lambda^2 b^2 - B_{7z})\lambda^{-2}b^{-1}p_y$ $G_Z = B_{4z}\lambda^{-2}b^{-1}p_y$ $G_u = 0$ $G_v = -B_{6z}\lambda^{-2}b^{-1}p_y$ $G_\theta = -B_{5z}\lambda^{-2}b^{-1}p_y$

10.09 TRANSPORT LOAD FUNCTIONS

Notation (10.02–10.04) Signs (10.02) Matrix (10.02)	
u_j, v_j, θ_j = abrupt displacements (m, m, rad) e_t, f_t = temperature strains (4.03) A_j, B_j, C_j, T_j = transport coefficients (10.03–10.09)	$\lambda = \sqrt[4]{\dfrac{k_v}{EI_z}}$ $R = \dfrac{\eta f}{e} = \dfrac{\eta N}{\sqrt{EI_z k_v}}$

$A_{10z} = \eta A_{4z} - A_{6z} + \lambda a$	$B_{10z} = \eta B_{4z} - B_{6z} + \lambda b$
$A_{11z} = \eta A_{1z} - A_{3z}$	$B_{11z} = \eta B_{1z} - B_{3z}$
$A_{12z} = \eta A_{3z} - A_{5z} + 1$	$B_{12z} = \eta B_{3z} - B_{5z} + 1$
$A_{13z} = \eta A_{2z} - A_{4z}$	$B_{13z} = \eta B_{2z} - B_{4z}$

$F_U = 0$ $F_V = EI_z A_{10z} \lambda^3 v_j$ $F_Z = -EI_z A_{11z} \lambda^3 v_j$	**1. Abrupt displacements** u_j, v_j	$G_U = 0$ $G_V = EI_z B_{10z} \lambda^3 v_j$ $G_Z = EI_z B_{11z} \lambda^3 v_j$
$F_u = EA_x s^{-1} u_j$ $F_v = -EI_z A_{12} \lambda^3 v_j$ $F_\theta = EI_z A_{13} \lambda^3 v_j$	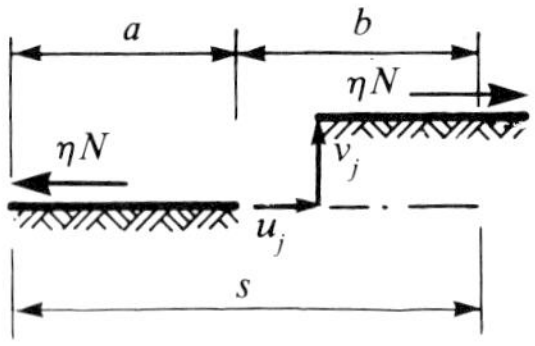	$G_u = EA_x s^{-1} u_j$ $G_v = EI_z B_{12z} \lambda^3 v_j$ $G_\theta = EI_z B_{13z} \lambda^3 v_j$
$F_U = 0$ $F_V = -EI_z A_{3z} \lambda^2 \theta_j$ $F_Z = EI_z A_{0z} \lambda^2 \theta_j$	**2. Abrupt displacement** θj	$G_U = 0$ $G_V = EI_z B_{3z} \lambda^2 \theta_j$ $G_Z = EI_z B_{0z} \lambda^2 \theta_j$
$F_u = 0$ $F_v = EI_z A_{2z} \lambda^2 \theta_j$ $F_\theta = -EI_z A_{1z} \lambda^2 \theta_j$	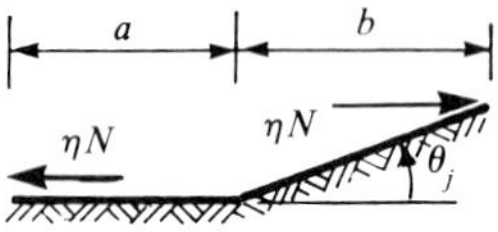	$G_u = 0$ $G_v = EI_z B_{2z} \lambda^2 \theta_j$ $G_\theta = EI_z B_{1z} \lambda^2 \theta_j$
$F_U = 0$ $F_V = -EI_z C_{4z} \lambda f_t$ $F_Z = EI_z C_{1z} \lambda f_t$	**3. Temperature deformation of segment** b	$G_U = 0$ $G_V = EI_z B_{4z} \lambda f_t$ $G_Z = EI_z B_{1z} \lambda f_t$
$F_U = -EA_x b s^{-1} e_t$ $F_v = EI_z C_{3z} \lambda f_t$ $F_\theta = -EI_z C_{2z} \lambda f_t$	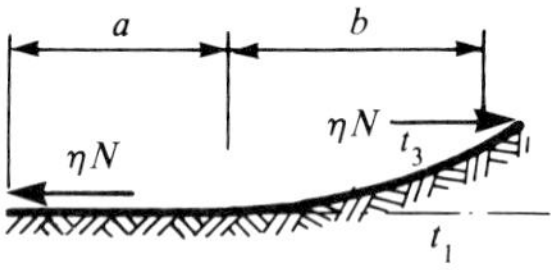	$G_u = EA_x b s^{-1} e_t$ $G_v = EI_z B_{3z} \lambda f_t$ $G_\theta = EI_z B_{2z} \lambda f_t$

10.10 STIFFNESS MATRIX EQUATIONS

(1) Equivalents and signs. All forces, moments, and displacements are in the principal axes of the normal section, and their positive directions are shown in (2, 3) below (Ref. 21).

$$r = \frac{s^2 EA_x}{EI_z}$$

(2) Free-body diagram

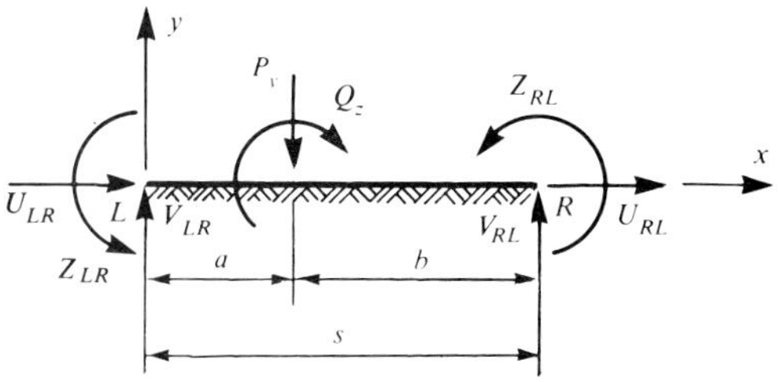

(3) Elastic curve

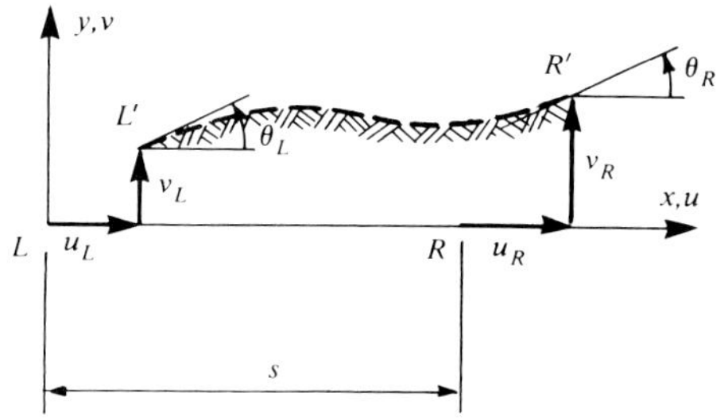

(4) Dimensionless stiffness coefficients

$\lambda = \sqrt[4]{\dfrac{k_v}{EI_z}}$	$D_0 = \dfrac{1}{T_{3z}^2 - T_{2z}T_{4z}}$
$L_1 = D_0(T_{2z} - T_{2z}T_{5z} + T_{3z}T_{4z})$	$K_1 = (\lambda s)^3 L_1$
$L_2 = D_0(T_{2z}^2 - T_{1z}T_{3z})$	$K_2 = (\lambda s)^2 L_2$
$L_3 = D_0(T_2 T_3 - T_{1z}T_{4z})$	$K_3 = (\lambda s) L_3$
$L_4 = D_0 T_{2z}$	$K_4 = (\lambda s)^3 L_4$
$L_5 = D_0 T_{3z}$	$K_5 = (\lambda s)^2 L_5$
$L_6 = D_0 T_{4z}$	$K_6 = (\lambda s) L_6$

where $T_{jz}(s) = T_{jz}$ are the transport coefficients (10.03–10.06).

(5) Stiffness matrix equations

$$\begin{bmatrix} U_{LR} \\ V_{LR} \\ Z_{LR} \\ \hline U_{RL} \\ V_{RL} \\ Z_{RL} \end{bmatrix} = \frac{EI_z}{s^3} \left[\begin{array}{ccc|ccc} r & 0 & 0 & -r & 0 & 0 \\ 0 & K_1 & K_2 s & 0 & -K_4 & K_5 s \\ 0 & K_2 s & K_3 s^2 & 0 & -K_5 s & K_6 s^2 \\ \hline -r & 0 & 0 & r & 0 & 0 \\ 0 & -K_4 & -K_5 s & 0 & K_1 & -K_2 s \\ 0 & K_5 s & K_6 s^2 & 0 & -K_2 s & K_3 s^2 \end{array}\right] \begin{bmatrix} u_L \\ v_L \\ \theta_L \\ \hline u_R \\ v_R \\ \theta_R \end{bmatrix} + \begin{bmatrix} U_{L0} \\ V_{L0} \\ Z_{L0} \\ \hline U_{R0} \\ V_{R0} \\ Z_{R0} \end{bmatrix}$$

where U_{L0}, V_{L0}, Z_{L0} and U_{R0}, V_{R0}, Z_{R0} in the last column matrix in (5) above are the *reactions of fixed-end beam-columns* due to loads and other causes shown in (10.13) and $U_{LR} = -\eta N$, $U_{RL} = \eta N$. The graphs of the stiffness coefficients $K_1, K_2, \ldots, K_6$ are shown in (10.11–10.12).

(6) General stiffness matrix equations for beam-columns of order three are obtained by matrix composition of (10.10) and (15.07) as shown in (3.19).

10.11 GRAPHS OF STIFFNESS COEFFICIENTS, AXIAL TENSION

$\lambda = \sqrt[4]{\dfrac{k_v}{EI_z}}$	$R = \dfrac{N}{\sqrt{EI_z k_v}}$

(1) **Graphs** of $K_1(x)$, $K_2(x)$, . . . , $K_6(x)$ based on their analytical forms in (10.10), $\eta = +1$, and selected values of R are shown below. The correspondence of graphs to the respective cases is defined in (10.04-5).

(2) **Graphs of** $K_1(x)$

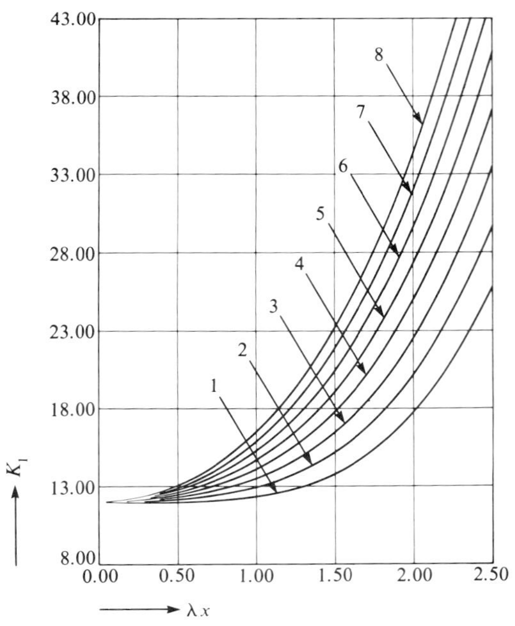

(3) **Graphs of** $K_2(x)$

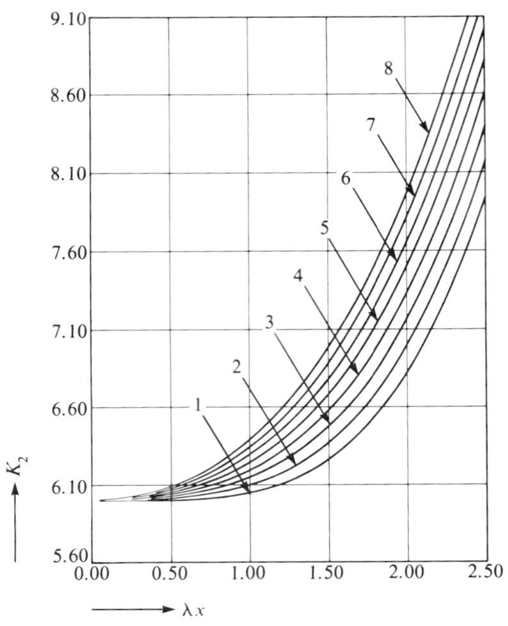

(4) **Graphs of** $K_3(x)$

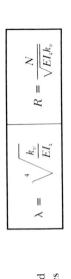

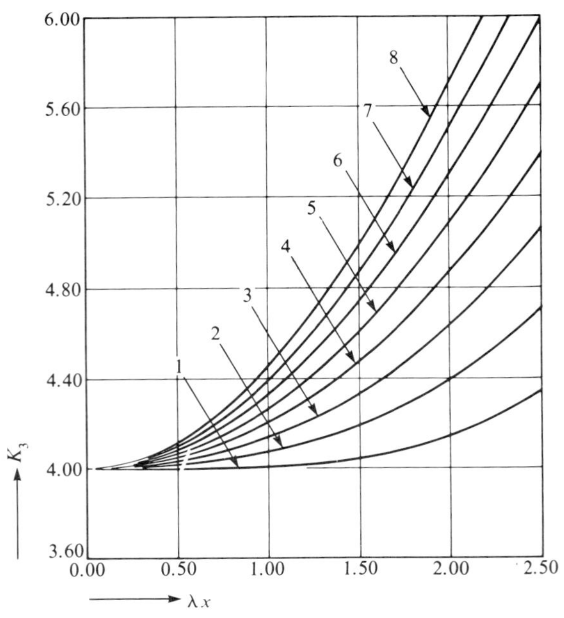

(5) Graphs of $K_4(x)$

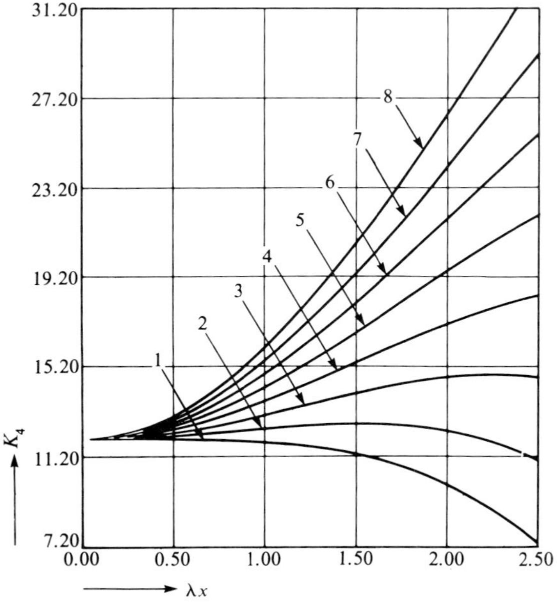

(6) Graphs of $K_5(x)$

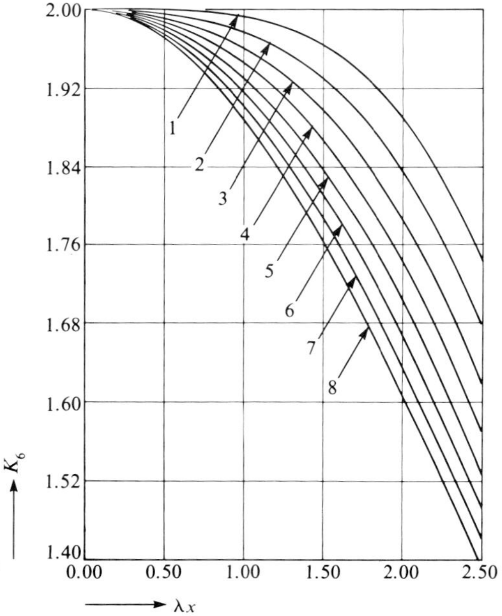

(7) Graphs of $K_6(x)$

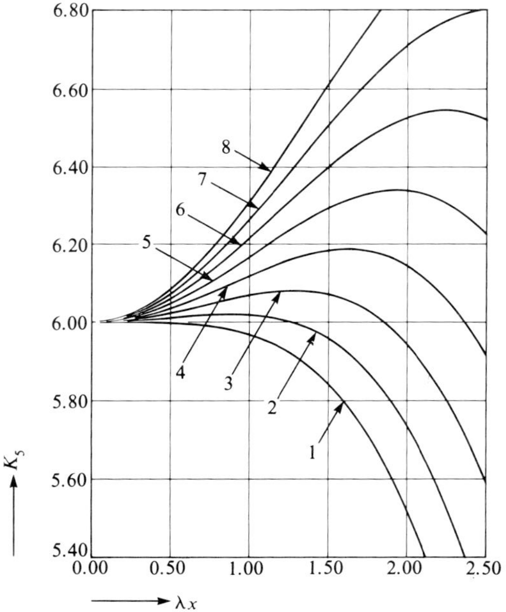

10.12 GRAPHS OF STIFFNESS COEFFICIENTS, AXIAL COMPRESSION

$$\lambda = \sqrt[4]{\frac{k_v}{EI_z}} \qquad R = \frac{-N}{\sqrt{EI_z k_v}}$$

(1) Graphs of $K_1(x)$, $K_2(x)$, . . . , $K_6(x)$ based on their analytical forms in (10.10), $\eta = -1$, and selected values of R are shown below. The correspondence of graphs to the respective cases is defined in (10.04-5).

(2) Graphs of $K_1(x)$

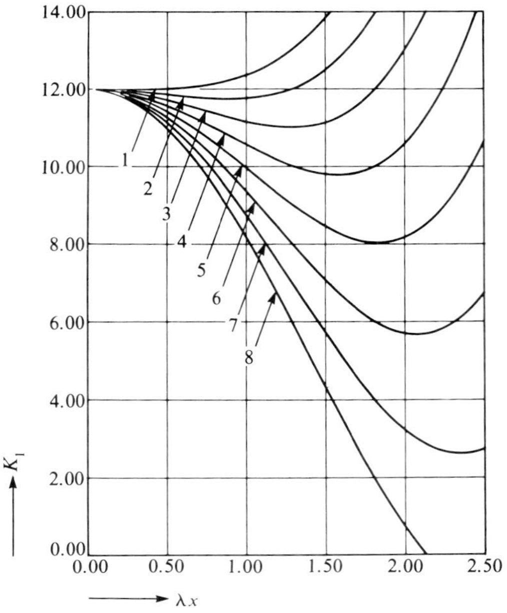

(3) Graphs of $K_2(x)$

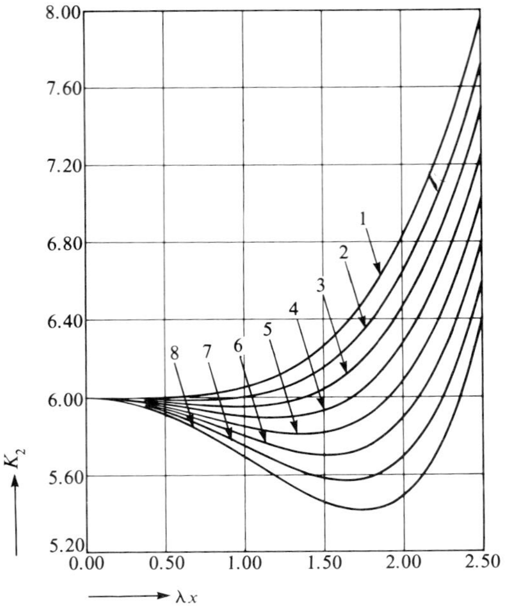

(4) Graphs of $K_3(x)$

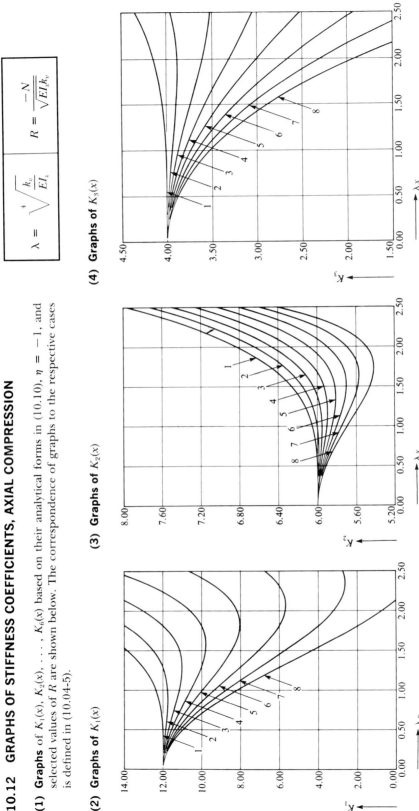

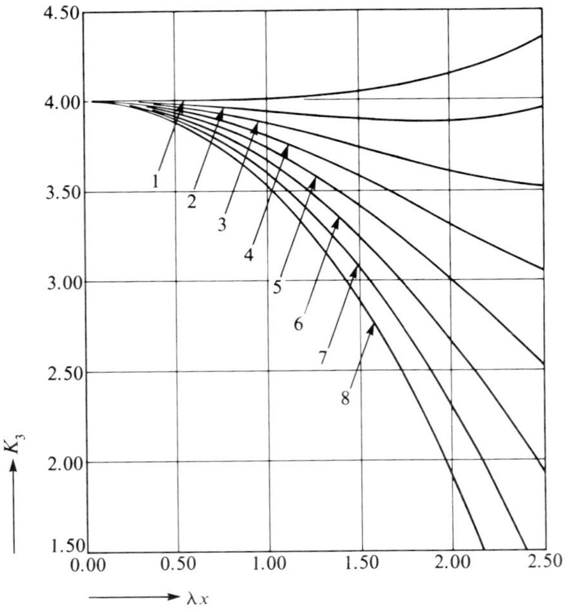

(5) Graphs of $K_4(x)$

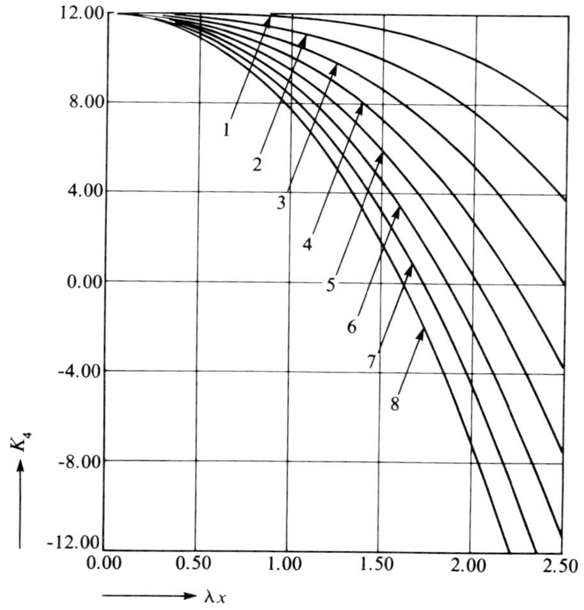

(6) Graphs of $K_5(x)$

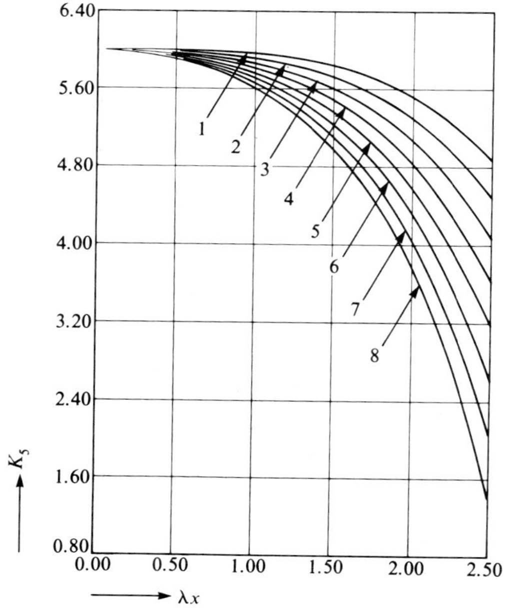

(7) Graphs of $K_6(x)$

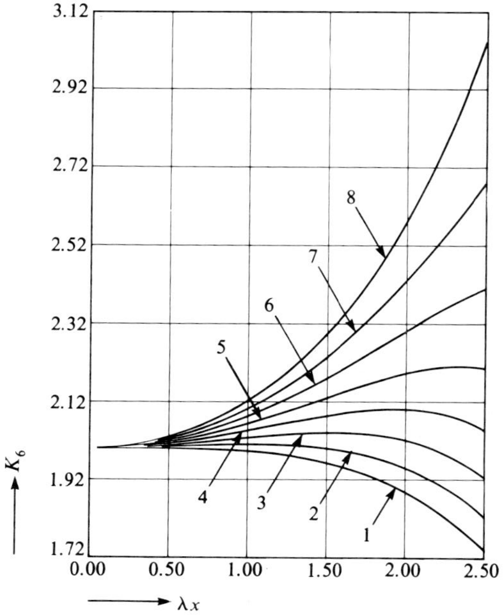

10.13 STIFFNESS LOAD FUNCTIONS, FIXED-END BEAM-COLUMN

(1) General form of the stiffness load functions (10.10) is given by the matrix equations

$$\begin{bmatrix} U_{LO} \\ V_{LO} \\ Z_{LO} \end{bmatrix} = \begin{bmatrix} 1 & -0 & 0 \\ 0 & L_4 & -L_5 \\ 0 & L_5\lambda^{-1} & -L_6\lambda^{-1} \end{bmatrix} \begin{bmatrix} G_u \\ G_v \\ G_\theta \end{bmatrix}$$

$$\begin{bmatrix} U_{RO} \\ V_{RO} \\ Z_{RO} \end{bmatrix} = \begin{bmatrix} 1 & 0 & 0 \\ 0 & L_4 & L_5 \\ 0 & -L_5\lambda^{-1} & -L_6\lambda^{-1} \end{bmatrix} \begin{bmatrix} F_u \\ F_v \\ F_\theta \end{bmatrix}$$

where F_u, F_v, F_θ are the left-end transport load functions and G_u, G_v, G_θ are their right-end counterparts, given for particular loads and causes in (10.07–10.09), and L_4, L_5, L_6 are the stiffness coefficients (10.10). The loads and other causes are defined by symbols introduced in (10.07–10.09).

(2) Particular cases

<table>
<tr><td colspan="2">Notation (10.02–10.03, 10.10)
A_j, B_j, C_j, T_j = transport coefficients (10.03–10.06)
Signs (10.10)</td><td>$\lambda = \sqrt[4]{\dfrac{k_v}{EI_z}}$</td></tr>
<tr><td>$U_{LO} = -\eta N$
$V_{LO} = (L_5 B_{3z} - L_4 B_{4z})P_y$
$Z_{LO} = (L_6 B_{3z} - L_5 B_{4z})\lambda^{-1} P_y$</td><td>1.
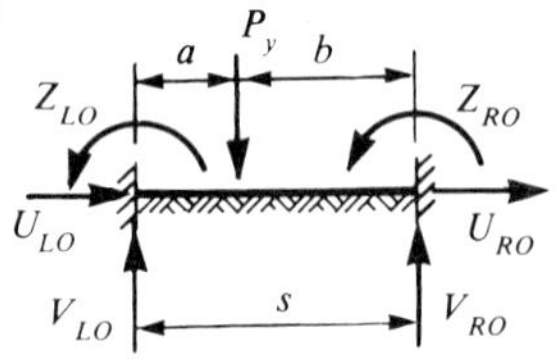
</td><td>$U_{RO} = \eta N$
$V_{RO} = (L_5 A_{3z} - L_4 A_{4z})P_y$
$Z_{RO} = -(L_6 A_{3z} - L_5 A_{4z})\lambda^{-1} P_y$</td></tr>
<tr><td>$U_{LO} = -\eta N$
$V_{LO} = -(L_5 B_{2z} - L_4 B_{3z})\lambda Q_z$
$Z_{LO} = -(L_6 B_{2z} - L_5 B_{3z})Q_z$</td><td>2.
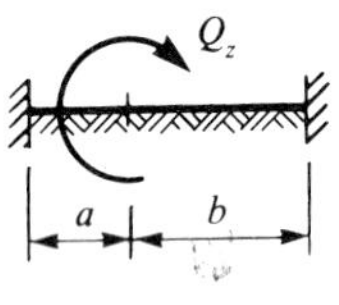
</td><td>$U_{RO} = \eta N$
$V_{RO} = (L_5 A_{2z} - L_4 A_{3z})\lambda Q_y$
$Z_{RO} = -(L_6 A_{2z} - L_5 A_{3z})Q_z$</td></tr>
<tr><td>$U_{LO} = -\eta N$
$V_{LO} = (L_5 T_{4z} - L_4 T_{5z})\lambda^{-1} p_y$
$Z_{LO} = (L_6 T_{4z} - L_5 T_{5z})\lambda^{-2} p_y$</td><td>3.
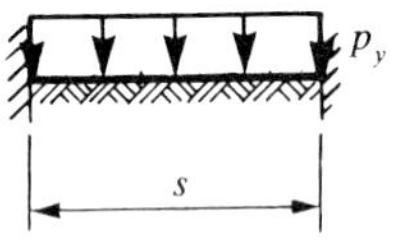
</td><td>$U_{RO} = \eta N$
$V_{RO} = V_{LO}$
$Z_{RO} = -Z_{LO}$</td></tr>
</table>

$$U_{L0} = -\eta N = -U_{R0}$$

$$V_{L0} = [L_5(A_{3z} + B_{3z}) - L_4(A_{4z} + B_{4z})]P_y = V_{R0}$$

$$Z_{L0} = [L_6(A_{3z} + B_{3z}) - L_5(A_{4z} + B_{4z})]\lambda^{-1}P_y = -Z_{R0}$$

4.

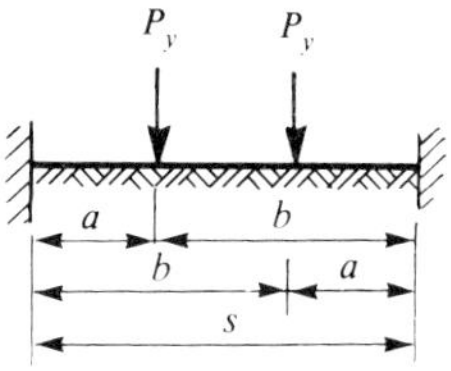

$$U_{L0} = -\eta N = -U_{R0}$$

$$V_{L0} = -[L_5(A_{2z} + B_{2z}) - L_4(A_{3z} + B_{3z})]\lambda^{-1}Q_z = -V_{R0}$$

$$Z_{L0} = -[L_6(A_{2z} + B_{2z}) - L_5(A_{3z} + B_{3z})]Q_z = Z_{R0}$$

5.

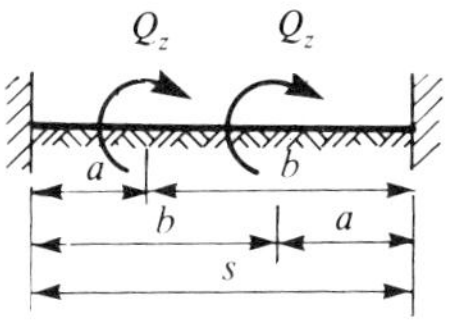

6.

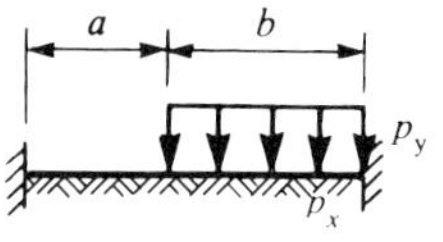

Use general form in terms of (10.08-2)

7.

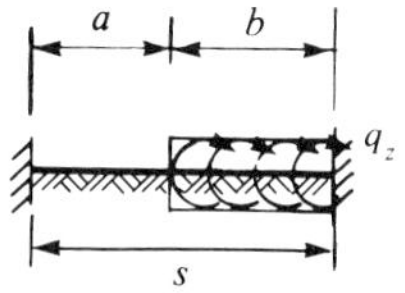

Use general form in terms of (10.07-4)

8.

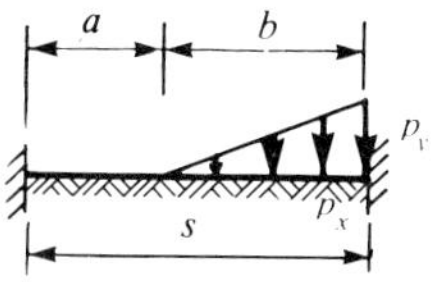

Use general form in terms of (10.08-4)

9. Abrupt displacements u_j, v_j

$$U_{L0} = EA_x s^{-1}u_j - \eta N$$

$$V_{L0} = EI_z(L_4B_{12z} - L_5B_{13z})\lambda^3 v_j$$

$$Z_{L0} = EI_z(L_5B_{12z} - L_6B_{13z})\lambda^2 v_j$$

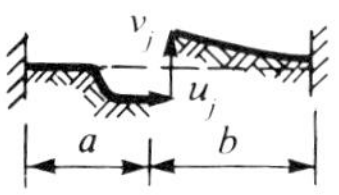

$$U_{R0} = -EA_x s^{-1}u_j + \eta N$$

$$V_{R0} = -EI_z(L_4A_{12z} - L_5A_{13z})\lambda^3 v_j$$

$$Z_{R0} = EI_z(L_5A_{12z} - L_6A_{13z})\lambda^2 v_j$$

10. Abrupt displacement θ_j

$$U_{L0} = -\eta N$$

$$V_{L0} = EI_z(L_4B_{2z} - L_5B_{1z})\lambda^2\theta_j$$

$$Z_{L0} = EI_z(L_5B_{2z} - L_6B_{1z})\lambda\theta_j$$

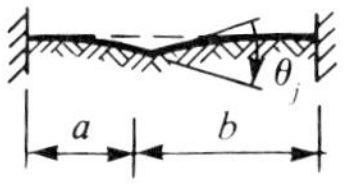

$$U_{R0} = \eta N$$

$$V_{R0} = EI_z(L_4A_{2z} - L_5A_{1z})\lambda^2\theta_j$$

$$Z_{R0} = -EI(L_5A_{2z} - L_6A_{1z})\lambda\theta_j$$

11. Temperature deformation of segment b

$$U_{L0} = EA_x \frac{b}{s} e_t - \eta N$$

$$V_{L0} = EI_z(L_4B_{3z} - L_5B_{2z})\lambda f_t$$

$$Z_{L0} = EI_z(L_5B_{3z} - L_6B_{2z})f_t$$

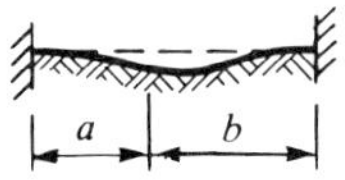

$$U_{R0} = -EA_x \frac{b}{s} e_t + \eta N$$

$$V_{R0} = EI_z(L_4A_{3z} - L_5A_{2z})\lambda f_t$$

$$Z_{R0} = -EI_z(L_5A_{3z} - L_6A_{2z})f_t$$

10.14 MODIFIED STIFFNESS MATRIX EQUATIONS, HINGED END

(1) Modified stiffness coefficients are

$$K_7 = K_1 - r_2K_2 \qquad K_8 = K_4 - r_5K_2 \qquad K_9 = K_5 - r_6K_2$$
$$K_{10} = K_1 - r_5K_5 \qquad K_{11} = K_2 - r_5K_6 \qquad K_{12} = K_3 - r_6K_6$$

where

$$r_2 = \frac{K_2}{K_3} \qquad r_5 = \frac{K_5}{K_3} \qquad r_6 = \frac{K_6}{K_3}$$

and $K_1, K_2, \ldots, K_6$ are the stiffness coefficients defined in (10.10).

(2) Modifications of stiffness equations (10.10) due to hinged end are shown in (3, 4) below. Since this condition does not affect the expressions for U_{LR} and U_{RL}, only the affected part of the respective matrix equation is presented.

(3) Left end hinged ($Z_{LR} = 0$)

$$\begin{bmatrix} V_{LR} \\ V_{RL} \\ Z_{RL} \end{bmatrix} = \frac{EI_z}{s^3} \begin{bmatrix} K_7 & -K_8 & K_9s \\ -K_8 & K_{10} & K_{11}s \\ K_9s & K_{11}s & K_{12}s^2 \end{bmatrix} \begin{bmatrix} v_L \\ v_R \\ \theta_R \end{bmatrix} + \begin{bmatrix} V^*_{LR} \\ V^*_{R0} \\ Z^*_{R0} \end{bmatrix} \qquad \begin{bmatrix} V^*_{L0} \\ V^*_{R0} \\ Z^*_{R0} \end{bmatrix} = \begin{bmatrix} V_{L0} - r_2s^{-1}Z_{L0} \\ V_{R0} - r_5s^{-1}Z_{L0} \\ Z_{R0} - r_6Z_{L0} \end{bmatrix}$$

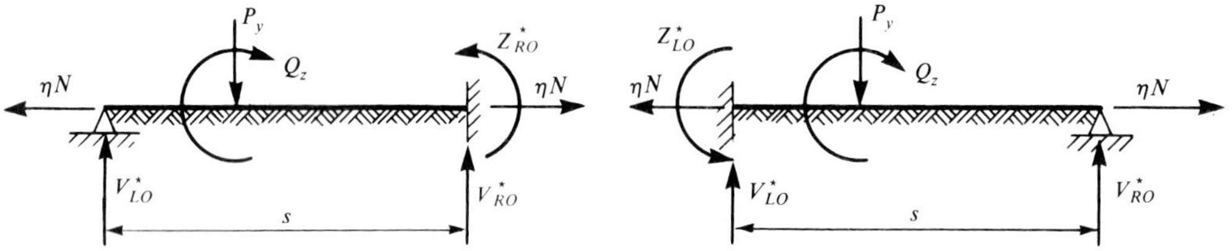

(4) Right end hinged ($Z_R = 0$)

$$\begin{bmatrix} V_{LR} \\ Z_{LR} \\ V_{RL} \end{bmatrix} = \frac{EI_z}{s^3} \begin{bmatrix} K_7 & -K_9s & -K_8 \\ K_9s & K_{12}s & -K_{11}s \\ -K_8 & -K_{11}s & K_{10} \end{bmatrix} \begin{bmatrix} v_L \\ \theta_L \\ v_R \end{bmatrix} + \begin{bmatrix} V^*_{L0} \\ Z^*_{L0} \\ V^*_{R0} \end{bmatrix} \qquad \begin{bmatrix} V^*_{L0} \\ Z^*_{L0} \\ V^*_{R0} \end{bmatrix} = \begin{bmatrix} V_{L0} + r_5s^{-1}Z_{R0} \\ Z_{L0} - r_6Z_{R0} \\ V_{R0} - r_2s^{-1}Z_{R0} \end{bmatrix}$$

(5) Modified load functions designated by an asterisk are the reactions of *propped-end beam-columns* due to loads and other causes. They are expressed in terms of the reactions of the respective fixed-end beam-columns (10.13) and the ratios r_2, r_5, r_6.

11

Free Straight Bar of Order Two Constant Section

STATIC STATE

11.01 DEFINITION OF STATE

(1) System considered is a free finite straight bar of constant and symmetrical section. The centroidal axis x of the bar and the vertical axis z define the plane of symmetry of this section. The bar is a part of a planar system acted on by loads normal to the system plane (system of order two).

(2) Notation (A.1, A.2, A.4)

s	= span (m)	A_x	= area of normal section (m^2)
a, b	= segments (m)	I_x	= torsional constant of A_x about x (m^4)
x, x'	= coordinates (m)	I_y	= moment of inertia of A_x about y (m^4)
E	= modulus of elasticity (Pa)	G	= modulus of rigidity (Pa)
W	= shearing force (N)	w	= linear displacement along z (m)
X	= torsional moment (N·m)	ϕ	= torsional slope about x (rad)
Y	= flexural moment (N·m)	ψ	= flexural slope about y (rad)

(3) Assumptions of analysis are stated in (1.01). In this chapter, the torsional and flexural deformations are included, but the effect of shearing deformations and the effect of axial force on the magnitude of the bending moment are neglected. For the beam-column effect, reference is made to Chap. 14. For the inclusion of the shearing deformation, reference is made to Chaps. 2 and 3.

11.02 TRANSPORT MATRIX EQUATIONS

(1) Equivalents and signs

$$\bar{w} = wEI_y = \text{scaled } w \ (\text{N}\cdot\text{m}^3) \qquad \bar{\psi} = \psi EI_y = \text{scaled } \psi \ (\text{N}\cdot\text{m}^2)$$

$$\bar{\phi} = \phi GI_x = \text{scaled } \phi \ (\text{N}\cdot\text{m}^2)$$

All forces, moments, and displacements are in the principal axes of the normal section, and their positive directions are shown in (2, 3) below (Ref. 5, p. 56).

(2) Free-body diagram

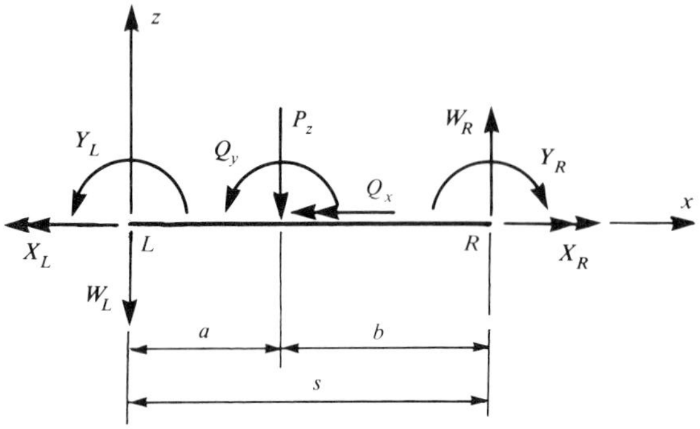

(3) Elastic curve

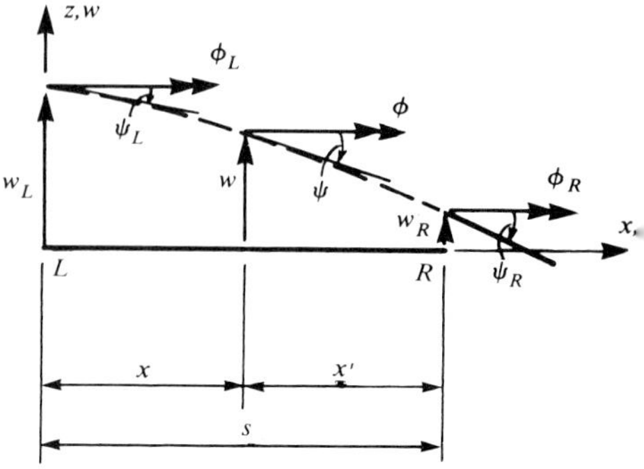

(4) Transport matrix equations

$$\underbrace{\begin{bmatrix} 1 \\ \hline W_L \\ X_L \\ Y_L \\ \hline \bar{w}_L \\ \bar{\phi}_L \\ \bar{\psi}_L \end{bmatrix}}_{H_L} = \underbrace{\left[\begin{array}{c|ccc|ccc} 1 & 0 & 0 & 0 & 0 & 0 & 0 \\ \hline F_W & 1 & 0 & 0 & 0 & 0 & 0 \\ F_X & 0 & 1 & 0 & 0 & 0 & 0 \\ F_Y & -s & 0 & 1 & 0 & 0 & 0 \\ \hline F_w & \frac{1}{6}s^3 & 0 & -\frac{1}{2}s^2 & 1 & 0 & -s \\ F_\phi & -0 & -s & 0 & 0 & 1 & 0 \\ F_\psi & \frac{1}{2}s^2 & 0 & -s & 0 & 0 & 1 \end{array}\right]}_{T_{LR}} \underbrace{\begin{bmatrix} 1 \\ \hline W_R \\ X_R \\ Y_R \\ \hline \bar{w}_R \\ \bar{\phi}_R \\ \bar{\psi}_R \end{bmatrix}}_{H_R}$$

$$\underbrace{\begin{bmatrix} 1 \\ \hline W_R \\ X_R \\ Y_R \\ \hline \bar{w}_R \\ \bar{\phi}_R \\ \bar{\psi}_R \end{bmatrix}}_{H_R} = \underbrace{\left[\begin{array}{c|ccc|ccc} 1 & 0 & 0 & 0 & 0 & 0 & 0 \\ \hline G_W & 1 & 0 & 0 & 0 & 0 & 0 \\ G_X & 0 & 1 & 0 & 0 & 0 & 0 \\ G_Y & s & 0 & 1 & 0 & 0 & 0 \\ \hline G_w & -\frac{1}{6}s^3 & 0 & -\frac{1}{2}s^2 & 1 & 0 & -s \\ G_\phi & 0 & s & 0 & 0 & 1 & 0 \\ G_\psi & \frac{1}{2}s^2 & 0 & s & 0 & 0 & 1 \end{array}\right]}_{T_{RL}} \underbrace{\begin{bmatrix} 1 \\ \hline W_L \\ X_L \\ Y_L \\ \hline \bar{w}_L \\ \bar{\phi}_L \\ \bar{\psi}_L \end{bmatrix}}_{H_L}$$

where H_L, H_R are the end *state vectors* and T_{LR}, T_{RL} are the respective *transport matrices*. The elements of these matrices are the *transport coefficients* shown above and the *load functions* F, G discussed in (11.03) and listed for particular cases in (11.04–11.06).

11.03 TRANSPORT RELATIONS

(1) Load functions are expressed in terms of the following symbols:

P_z = concentrated load (N) $\quad$ p_z = intensity of distributed load (N/m)

Q_x, Q_y = applied couples (N·m) $\quad$ q_x, q_y = intensities of distributed couples (N·m/m)

In cases of variable load distribution, p_z, q_x, q_y are the maximum intensities. The effects of abrupt changes in bar axis at a particular point j are expressed in terms of w_j, ϕ_j, ψ_j = abrupt displacements (m, rad, rad). The temperature-effect analysis assumes (in general) a linear variation of temperature along the width c (m) of the bar in a selected segment. The symbols used are

$t_4 - t_0$ = temperature change at the top (°C)

$t_2 - t_0$ = temperature change at the centroid (°C)

$t_5 - t_0$ = temperature change at the bottom (°C)

α_t = coefficient of thermal expansion (1/°C)

$g_t = \alpha_t(t_4 - t_5)/c$ = angular thermal strain per unit length (rad/m)

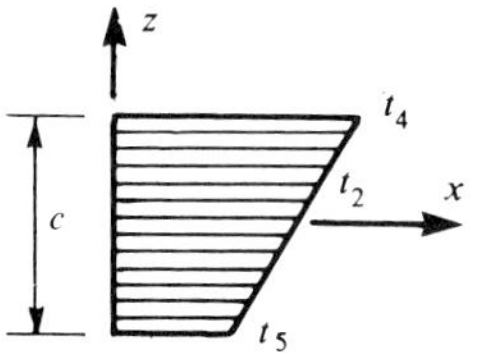

The load functions have a physical meaning and can be interpreted as the end stress resultants of the respective cantilever beams and as the linear and angular deviations of the tangent drawn to the elastic curve at the free end and measured at the opposite end. Case (11.04-1) is interpreted graphically below.

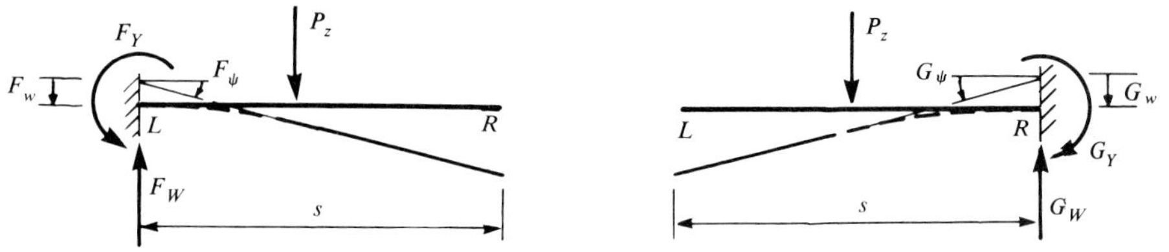

(2) Differential relations

$$\frac{d\phi}{dx} = \frac{X}{GI_x} \qquad \frac{d^2\phi}{dx^2} = \frac{q_x}{GI_x}$$

$$\frac{dw}{dx} = -\psi$$

$$\frac{d^2w}{dx^2} = -\frac{d\psi}{dx} = -\frac{Y}{EI_y}$$

$$\frac{d^3w}{dx^3} = -\frac{d^2\psi}{dx^2} = -\frac{dY}{EI_y\,dx} = -\frac{W}{EI_y}$$

$$\frac{d^4w}{dx^4} = -\frac{d^3\psi}{dx^3} = -\frac{d^2Y}{EI_y\,dx^2} = -\frac{dW}{EI_y\,dx} = -\frac{p_z}{EI_y}$$

(3) Transport matrices T_{LR} and T_{RL} give a clear, systematic, and complete record of the static and deformation properties of the segment s; they are specific for the selected segment and for the given loads and independent of the end conditions. They apply to the whole bar or to any part of it. In the latter case, s becomes x and the load function F becomes a function of x which is measured from L, or s becomes x' and the load function G becomes a function of x' which is measured from R.

(4) General transport matrix equations for free bars of order three are obtained by matrix composition of (4.02–4.06) and (11.02–11.06) as shown in (2.04–2.06). This general case includes the effects of shearing deformations which are neglected in Chap. 4 and this chapter.

11.04 TRANSPORT LOAD FUNCTION

Notation (11.02–11.03) Signs (11.02) Matrix (11.02)

P_z = concentrated load (N)
q_y = intensity of distributed couple (N·m/m)
Q_x, Q_y = applied couples (N·m)

$F_W = -P_z$ $F_X = 0$ $F_Y = aP_z$ $F_w = -\frac{1}{6}a^3P_z$ $F_\phi = 0$ $F_\psi = -\frac{1}{2}a^2P_z$	**1.** 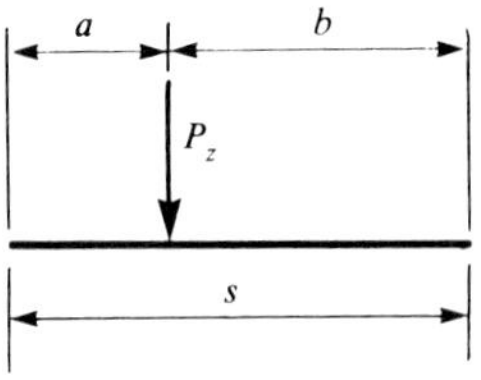 	$G_W = P_z$ $G_X = 0$ $G_Y = bP_z$ $G_w = -\frac{1}{6}b^3P_z$ $G_\phi = 0$ $G_\psi = \frac{1}{2}b^2P_z$
$F_W = 0$ $F_X = -Q_x$ $F_Y = -Q_y$ $F_w = \frac{1}{2}a^2Q_y$ $F_\phi = aQ_x$ $F_\psi = aQ_y$	**2.** 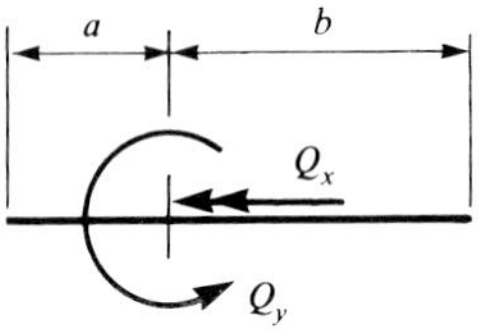 	$G_W = 0$ $G_X = Q_x$ $G_Y = Q_y$ $G_w = -\frac{1}{2}b^2Q_y$ $G_\phi = bQ_x$ $G_\psi = bQ_y$
$F_W = 0$ $F_X = 0$ $F_Y = -sq_y$ $F_w = \frac{1}{6}s^3q_y$ $F_\phi = 0$ $F_\psi = \frac{1}{2}s^2q_y$	**3.** 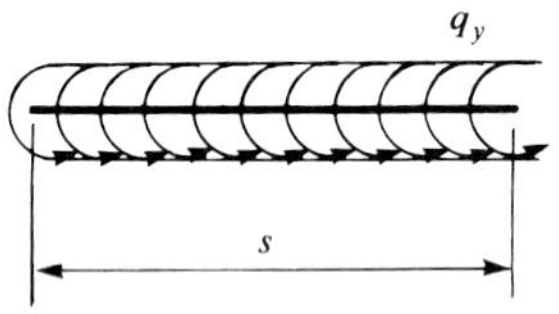 	$G_W = 0$ $G_X = 0$ $G_Y = sq_y$ $G_w = -\frac{1}{6}s^3q_y$ $G_\phi = 0$ $G_\psi = \frac{1}{2}s^2q_y$
$F_W = 0$ $F_X = 0$ $F_Y = -bq_y$ $F_w = \frac{1}{6}(s^3 - a^3)q_y$ $F_\phi = 0$ $F_\psi = \frac{1}{2}(s^2 - b^2)q_y$	**4.** 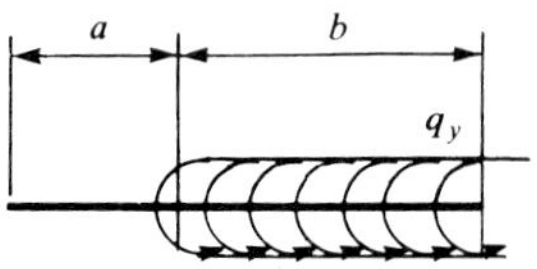 	$G_W = 0$ $G_X = 0$ $G_Y = bq_y$ $G_w = -\frac{1}{6}b^3q_y$ $G_\phi = 0$ $G_\psi = \frac{1}{2}b^2q_y$

11.05 TRANSPORT LOAD FUNCTIONS

Notation (11.02–11.03) Signs (11.02) Matrix (11.02)

p_z = intensity of distributed load (N/m)
q_x = intensity of distributed couple (N·m/m)

$F_W = -sp_z$ $F_X = -sq_x$ $F_Y = \frac{1}{2}s^2p_z$ $F_w = -\frac{1}{24}s^4p_z$ $F_\phi = \frac{1}{2}s^2q_x$ $F_\psi = -\frac{1}{6}s^3p_z$	**1.** 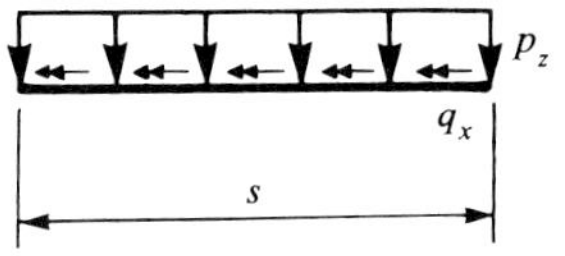 	$G_W = -sp_z$ $G_X = sq_x$ $G_Y = \frac{1}{2}s^2p_z$ $G_w = -\frac{1}{24}s^4p_z$ $G_\phi = \frac{1}{2}s^2q_x$ $G_\psi = \frac{1}{6}s^3p_z$
$F_W = -bp_z$ $F_X = -bq_x$ $F_Y = \frac{1}{2}(s^2 - a^2)p_z$ $F_w = -\frac{1}{24}(s^4 - a^4)p_z$ $F_\phi = \frac{1}{2}(s^2 - a^2)q_x$ $F_\psi = -\frac{1}{6}(s^3 - a^3)p_z$	**2.** 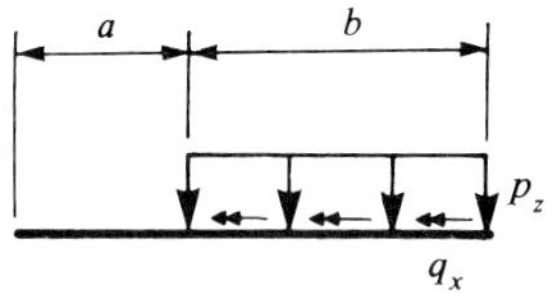 	$G_W = -bp_z$ $G_X = bq_x$ $G_Y = \frac{1}{2}b^2p_z$ $G_w = -\frac{1}{24}b^4p_z$ $G_\phi = \frac{1}{2}b^2q_x$ $G_\psi = \frac{1}{6}b^3p_z$
$F_W = -\frac{1}{2}sp_z$ $F_X = -\frac{1}{2}sq_x$ $F_Y = \frac{1}{3}s^2p_z$ $F_w = -\frac{1}{30}s^4p_z$ $F_\phi = \frac{1}{3}s^2q_x$ $F_\psi = -\frac{1}{8}s^3p_z$	**3.** 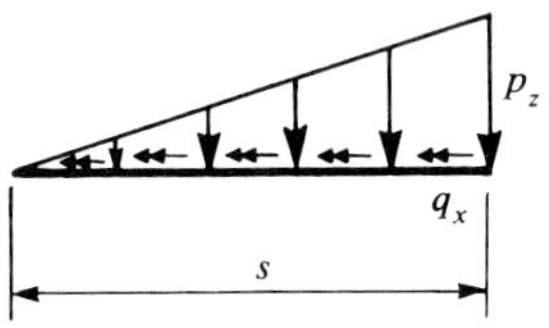 	$G_W = -\frac{1}{2}sp_z$ $G_X = \frac{1}{2}sq_x$ $G_Y = \frac{1}{6}s^2p_z$ $G_w = -\frac{1}{120}s^4p_z$ $G_\phi = \frac{1}{6}s^2q_x$ $G_\psi = \frac{1}{24}s^3p_z$
$F_W = -\frac{1}{2}bp_z$ $F_X = -\frac{1}{2}bq_x$ $F_Y = \frac{1}{6}(2s + a)bp_z$ $F_w = -\frac{1}{120b}(a^5 - 5bs^4 - s^5)p_z$ $F_\phi = \frac{1}{6}(2s + a)bq_x$ $F_\psi = -\frac{1}{24b}(a^4 - 4bs^3 - s^4)p_z$	**4.** 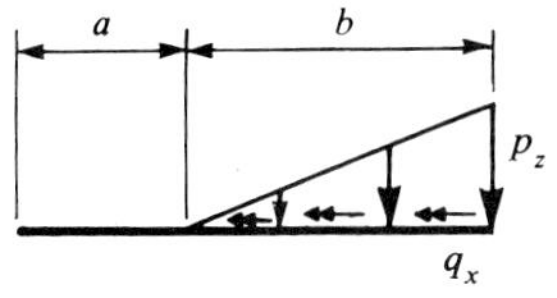 	$G_W = -\frac{1}{2}bp_z$ $G_X = \frac{1}{2}bq_x$ $G_Y = \frac{1}{6}b^2p_z$ $G_w = -\frac{1}{120}b^4p_z$ $G_\phi = \frac{1}{6}b^2q_x$ $G_\psi = \frac{1}{24}b^3p_z$

11.06 TRANSPORT LOAD FUNCTIONS

Notation (11.02–11.03) Signs (11.02) Matrix (11.02)

w_j, ϕ_j, ψ_j = abrupt displacements (m, rad, rad)
g_t = thermal strain (11.03)

1. Abrupt displacements

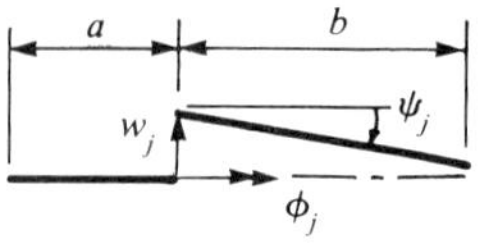

$F_W = 0$	$G_W = 0$
$F_X = 0$	$G_X = 0$
$F_Y = 0$	$G_Y = 0$
$F_w = -EI_y(w_j + a\psi_j)$	$G_w = EI_y(w_j - b\psi_j)$
$F_\phi = -GI_x\phi_j$	$G_\phi = GI_x\,\phi_j$
$F_\psi = -EI_y\psi_j$	$G_\psi = EI_y\psi_j$

2. Temperature deformation of segment b

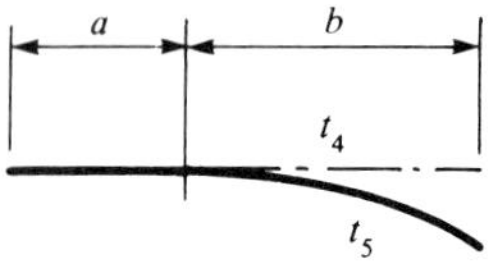

$F_W = 0$	$G_W = 0$
$F_X = 0$	$G_X = 0$
$F_Y = 0$	$G_Y = 0$
$F_w = -\frac{1}{2}EI_y(s^2 - a^2)g_t$	$G_w = -\frac{1}{2}EI_y b^2 g_t$
$F_\phi = 0$	$G_\phi = 0$
$F_\psi = -EI_y b g_t$	$G_\psi = EI_y b g_t$

3.

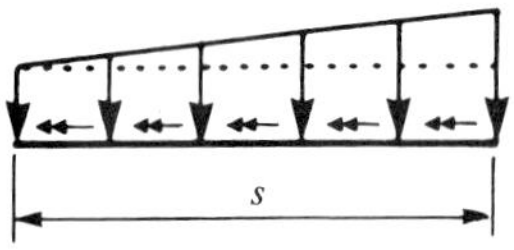

Sec. 11.05 (case 1 + case 3)

4.

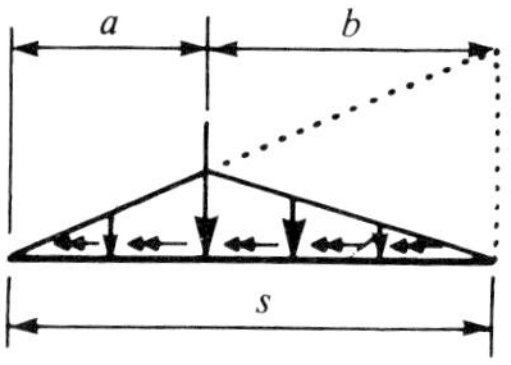

Sec. 11.05 (case 3 − case 4)

5.

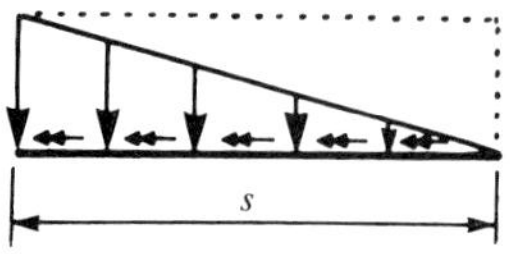

Sec. 11.05 (case 1—case 3)

6.

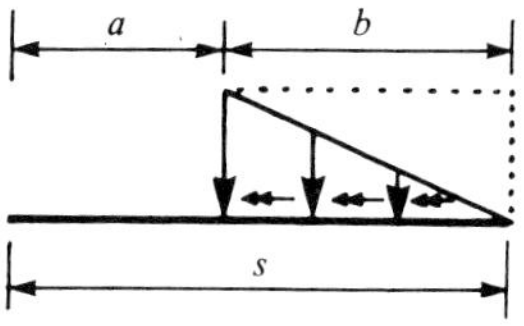

Sec. 11.05 (case 2 − case 4)

11.07 STIFFNESS MATRIX EQUATIONS

(1) Equivalents and signs. All forces, moments, and displacements are in the principal axes of the normal section, and their positive directions are shown in (2, 3) below (Ref. 5, p. 119)

$$r = \frac{s^2 G I_x}{E I_y} \ (\text{m}^2)$$

(2) Free-body diagram

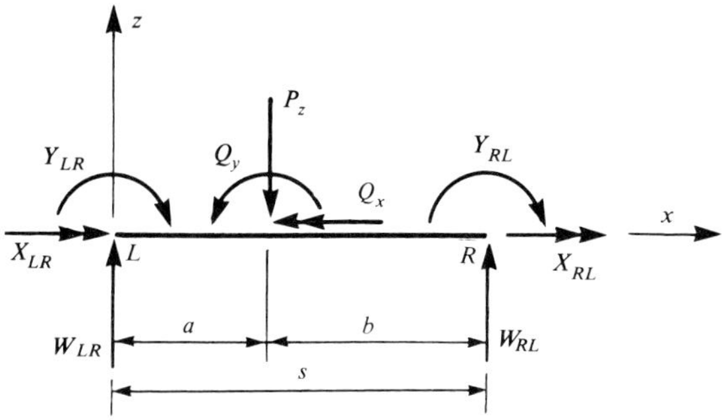

(3) Elastic curve

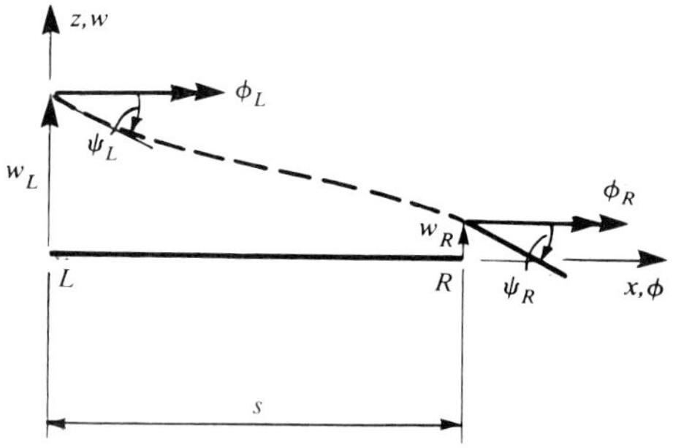

(4) Stiffness matrix equations

$$\begin{bmatrix} W_{LR} \\ X_{LR} \\ Y_{LR} \\ \hline W_{RL} \\ X_{RL} \\ Y_{RL} \end{bmatrix} = \frac{EI_y}{s^3} \left[\begin{array}{ccc|ccc} 12 & 0 & -6s & -12 & 0 & -6s \\ 0 & r & 0 & 0 & -r & 0 \\ -6s & 0 & 4s^2 & 6s & 0 & 2s^2 \\ \hline -12 & 0 & 6s & 12 & 0 & 6s \\ 0 & -r & 0 & 0 & r & 0 \\ -6s & 0 & 2s^2 & 6s & 0 & 4s^2 \end{array}\right] \begin{bmatrix} w_L \\ \phi_L \\ \psi_L \\ \hline w_R \\ \phi_R \\ \psi_R \end{bmatrix} + \begin{bmatrix} W_{L0} \\ X_{L0} \\ Y_{L0} \\ \hline W_{R0} \\ X_{R0} \\ Y_{R0} \end{bmatrix}$$

(5) Load functions $W_{L0}, X_{L0}, \ldots, Y_{R0}$ shown as the last column matrix in (4) above are the *reactions of fixed-end bars* due to loads and other causes given in (11.09–11.12).

(6) Symmetrical end displacements If $w_L = w_R$, $\phi_L = \phi_R$, $\psi_L = -\psi_R$, and symmetrical causes are applied, the stiffness matrix equation in (4) above reduces to

$$\begin{bmatrix} W_{LR} \\ X_{LR} \\ Y_{LR} \end{bmatrix} = \frac{EI_y}{s^3} \begin{bmatrix} 0 & 0 & 0 \\ 0 & 0 & 0 \\ 0 & 0 & 2s^2 \end{bmatrix} \begin{bmatrix} w_L \\ \phi_L \\ \psi_L \end{bmatrix} + \begin{bmatrix} W_{L0} \\ X_{L0} \\ Y_{L0} \end{bmatrix}$$

(7) Antisymmetrical end displacements If $w_L = -w_R$, $\phi_L = -\phi_R$, $\psi_L = \psi_R$, and antisymmetrical causes are applied, the stiffness matrix equation in (4) above reduces to

$$\begin{bmatrix} W_{LR} \\ X_{LR} \\ Y_{LR} \end{bmatrix} = \frac{EI_y}{s^3} \begin{bmatrix} 24 & 0 & -12s \\ 0 & 2r & 0 \\ -12s & 0 & 6s^2 \end{bmatrix} \begin{bmatrix} w_L \\ \phi_L \\ \psi_L \end{bmatrix} + \begin{bmatrix} W_{L0} \\ X_{L0} \\ Y_{L0} \end{bmatrix}$$

(8) General stiffness matrix equations for free bars of order three are obtained by matrix composition of (4.32–4.41) and (11.07–11.12) as shown in (3.12–3.13). This general case includes the effects of the shearing deformations which are neglected in Chap. 4 and this chapter.

11.08 MODIFIED STIFFNESS MATRIX EQUATIONS

(1) Left end hinged. If the left end moment $Y_{LR} = 0$, the stiffness matrix equation in (11.07-4) reduces to

$$\begin{bmatrix} W_{LR} \\ X_{LR} \\ \hline W_{RL} \\ X_{RL} \\ Y_{RL} \end{bmatrix} = \frac{EI_y}{s^3} \left[\begin{array}{cc|ccc} 3 & 0 & -3 & 0 & -3s \\ 0 & r & 0 & -r & 0 \\ \hline -3 & 0 & 3 & 0 & 3s \\ 0 & -r & 0 & r & 0 \\ -3s & 0 & 3s & 0 & 3s^2 \end{array} \right] \begin{bmatrix} w_L \\ \phi_L \\ \hline w_R \\ \phi_R \\ \psi_R \end{bmatrix} + \begin{bmatrix} W^*_{L0} \\ X^*_{L0} \\ \hline W^*_{R0} \\ X^*_{R0} \\ Y^*_{R0} \end{bmatrix}$$

The load functions shown as the last column matrix above are the reactions of the respective propped-end beams (11.13–11.14).

(2) Right end hinged. If the right end moment $Y_{RL} = 0$, the stiffness matrix equation in (11.07-4) reduces to

$$\begin{bmatrix} W_{LR} \\ X_{LR} \\ Y_{LR} \\ \hline W_{RL} \\ X_{RL} \end{bmatrix} = \frac{EI_y}{s^3} \left[\begin{array}{ccc|cc} 3 & 0 & -3s & -3 & 0 \\ 0 & r & 0 & 0 & -r \\ -3s & 0 & 3s^2 & 3s & 0 \\ \hline -3 & 0 & 3s & 3 & 0 \\ 0 & -r & 0 & 0 & r \end{array} \right] \begin{bmatrix} w_L \\ \phi_L \\ \psi_L \\ \hline w_R \\ \phi_R \end{bmatrix} + \begin{bmatrix} W^*_{L0} \\ X^*_{L0} \\ Y^*_{L0} \\ \hline W^*_{R0} \\ X^*_{R0} \end{bmatrix}$$

The load functions shown as the last column matrix above are again the reactions of the respective propped-end beams.

(3) Left end guided. If the left-end vertical force $W_{LR} = 0$, the stiffness matrix equation in (11.07-4) reduces to

$$\begin{bmatrix} X_{LR} \\ Y_{LR} \\ \hline W_{RL} \\ X_{RL} \\ Y_{RL} \end{bmatrix} = \frac{EI_y}{s^3} \left[\begin{array}{cc|ccc} r & 0 & 0 & -r & 0 \\ 0 & s^2 & 0 & 0 & -s^2 \\ \hline 0 & 0 & 0 & 0 & 0 \\ -r & 0 & 0 & r & 0 \\ 0 & -s^2 & 0 & 0 & s^2 \end{array} \right] \begin{bmatrix} \phi_L \\ \psi_L \\ \hline w_R \\ \phi_R \\ \psi_R \end{bmatrix} + \begin{bmatrix} X^*_{L0} \\ Y^*_{L0} \\ \hline W^*_{R0} \\ X^*_{R0} \\ Y^*_{R0} \end{bmatrix}$$

The load functions shown as the last column matrix above are the reactions of the respective guided beams (11.15–11.16).

(4) Right end guided. If the right-end vertical force $W_{RL} = 0$, the stiffness matrix equation in (11.07-4) reduces to

$$\begin{bmatrix} W_{LR} \\ X_{LR} \\ Y_{LR} \\ \hline X_{RL} \\ Y_{RL} \end{bmatrix} = \frac{EI_y}{s^3} \left[\begin{array}{ccc|cc} 0 & 0 & 0 & 0 & 0 \\ 0 & r & 0 & -r & 0 \\ 0 & 0 & s^2 & 0 & s^2 \\ \hline 0 & -r & 0 & r & 0 \\ 0 & 0 & -s^2 & 0 & s^2 \end{array} \right] \begin{bmatrix} w_L \\ \phi_L \\ \psi_L \\ \hline \phi_R \\ \psi_R \end{bmatrix} + \begin{bmatrix} W^*_{L0} \\ X^*_{L0} \\ Y^*_{L0} \\ \hline X^*_{R0} \\ Y^*_{R0} \end{bmatrix}$$

The load functions shown as the last column matrix above are again the reactions of the respective guided beams.

11.09 STIFFNESS LOAD FUNCTIONS, FIXED-END BEAM

Notation (11.07) Signs (11.07) Matrix (11.07)

P_z = concentrated load (N)

Q_x, Q_y = applied couples (N·m) $m = a/s$ $n = b/s$

$W_{L0} = (1 + 2m)n^2 P_z$ $X_{L0} = nQ_x$ $Y_{L0} = -mnbP_z$	**1.** 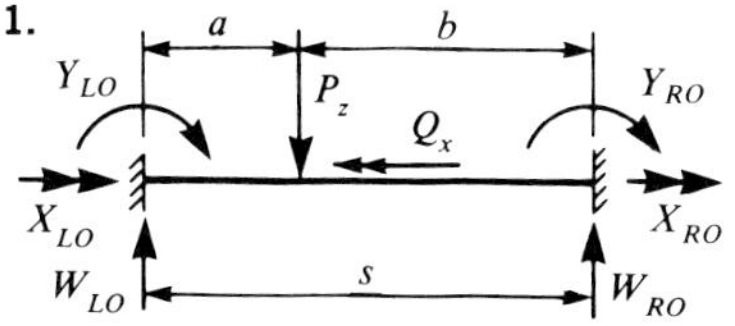	$W_{R0} = (1 + 2n)m^2 P_z$ $X_{R0} = mQ_x$ $Y_{R0} = mnaP_z$
$W_{L0} = \frac{1}{2}P_z$ $X_{L0} = \frac{1}{2}Q_x$ $Y_{L0} = -\frac{1}{8}sP_z$	**2.** 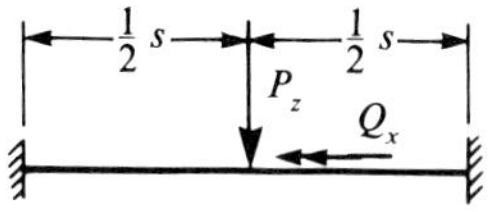	$W_{R0} = \frac{1}{2}P_z$ $X_{R0} = \frac{1}{2}Q_x$ $Y_{R0} = \frac{1}{8}sP_z$
$W_{L0} = P_z$ $X_{L0} = Q_x$ $Y_{L0} = -(1 - m)aP_z$	**3.** 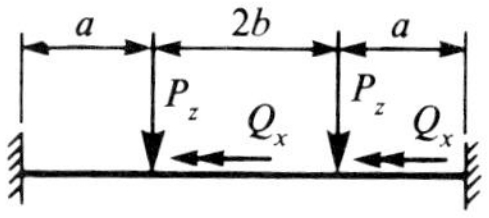	$W_{R0} = P_z$ $X_{R0} = Q_x$ $Y_{R0} = (1 - m)aP_z$
$W_{L0} = \frac{6}{s} mnQ_y$ $X_{L0} = 0$ $Y_{L0} = -(2 - 3n)nQ_y$	**4.** 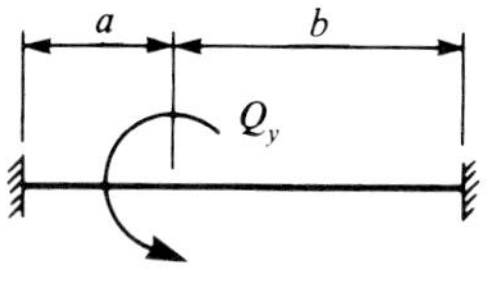	$W_{R0} = -\frac{6}{s} mnQ_y$ $X_{R0} = 0$ $Y_{R0} = -(2 - 3m)mQ_y$
$W_{L0} = \frac{3}{2s} Q_y$ $X_{L0} = 0$ $Y_{L0} = -\frac{1}{4} Q_y$	**5.** 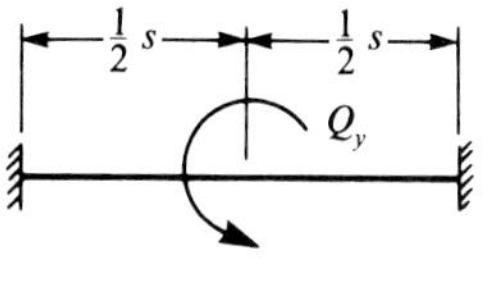	$W_{R0} = -\frac{3}{2s} Q_y$ $X_{R0} = 0$ $Y_{R0} = -\frac{1}{4} Q_y$
$W_{L0} = 0$ $X_{L0} = 0$ $Y_{L0} = 2nQ_y$	**6.** 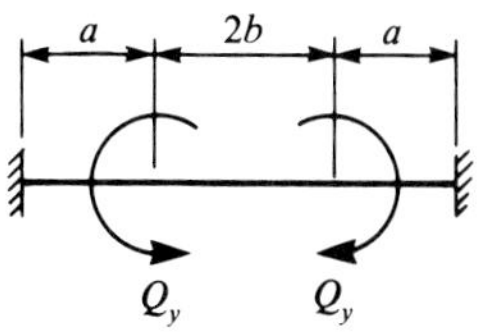	$W_{R0} = 0$ $X_{R0} = 0$ $Y_{R0} = -2nQ_y$

11.10 STIFFNESS LOAD FUNCTIONS, FIXED-END BEAM

Notation (11.07) Signs (11.07) Matrix (11.07)

P_z = maximum intensity of distributed load (N/m) $m = a/s$

q_x = maximum intensity of distributed couple (N·m/m) $n = b/s$

$W_{LO} = \frac{1}{2}sp_z$ $X_{LO} = \frac{1}{2}sq_x$ $Y_{LO} = -\frac{s^2}{12}p_z$	1. 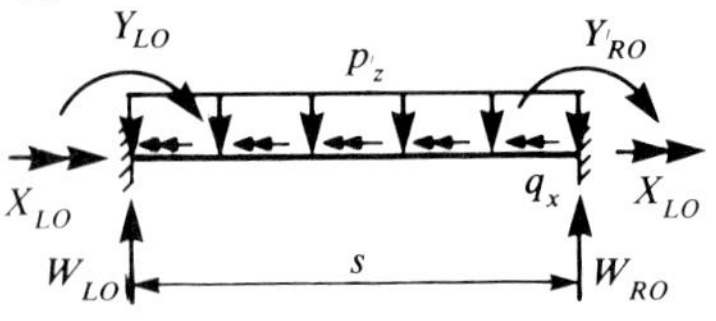	$W_{RO} = \frac{1}{2}sp_z$ $X_{RO} = \frac{1}{2}sq_z$ $Y_{RO} = \frac{s^2}{12}p_z$
$W_{LO} = \frac{1}{2}(2-n)n^2bp_z$ $X_{LO} = \frac{1}{2}nbq_x$ $Y_{LO} = -\frac{(4-3n)n}{12}b^2p_z$	2. 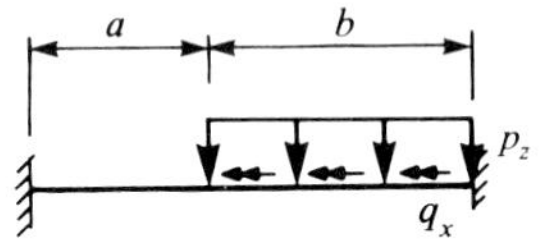	$W_{RO} = \frac{1}{2}(1+m+m^2n)bp_z$ $X_{RO} = \frac{1}{2}(2-n)bq_x$ $Y_{RO} = \frac{6-8n+3n^2}{12}b^2p_z$
$W_{LO} = bp_z$ $X_{LO} = bq_x$ $Y_{LO} = -\frac{3-4n^2}{12}bsp_z$	3. 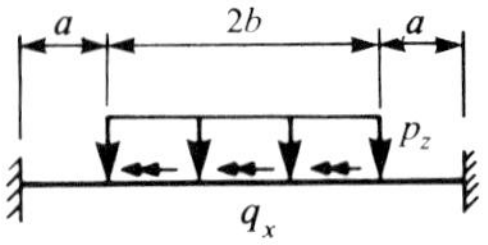	$W_{RO} = bp_z$ $X_{RO} = bq_x$ $Y_{RO} = \frac{3-4n^2}{12}bsp_z$
$W_{LO} = \frac{3s}{20}p_z$ $X_{LO} = \frac{1}{6}sq_x$ $Y_{LO} = -\frac{s^2}{30}p_z$	4. 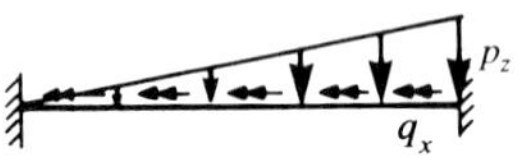	$W_{RO} = \frac{7s}{20}p_z$ $X_{RO} = \frac{1}{3}sq_x$ $Y_{RO} = \frac{s^2}{20}p_z$
$W_{LO} = \frac{(5-2n)n^2}{20}bp_z$ $X_{LO} = \frac{1}{6}nbq_x$ $Y_{LO} = -\frac{(5-3n)n}{60}b^2p_z$	5. 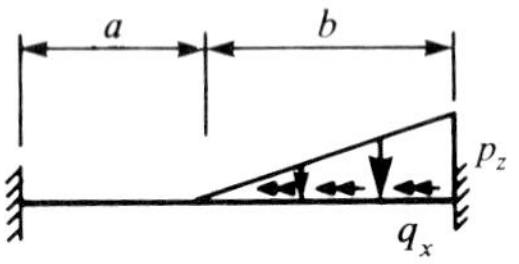	$W_{RO} = \frac{10-5n^2+2n^3}{20}bp_z$ $X_{RO} = \frac{1}{6}(3-n)bq_x$ $Y_{RO} = \frac{10-10n+3n^2}{60}b^2p_z$
$W_{LO} = \frac{1}{2}bp_z$ $X_{LO} = \frac{1}{2}bq_x$ $Y_{LO} = -\frac{3-2n^2}{24}bsp_z$	6. 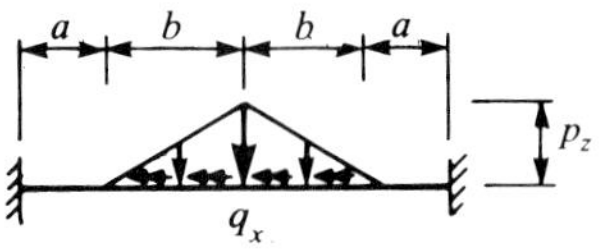	$W_{RO} = \frac{1}{2}bp_z$ $X_{RO} = \frac{1}{2}bq_x$ $Y_{RO} = \frac{3-2n^2}{24}bsp_z$

11.11 STIFFNESS LOAD FUNCTIONS, FIXED-END BEAM

Notation (11.07)
p_z = intensity of distributed load (N/m)
q_x, q_y = intensities of distributed couples (N·m/m)

Signs (11.07)
$m = a/s$
$n = b/s$

Matrix (11.07)

$W_{LO} = q_y$ $X_{LO} = 0$ $Y_{LO} = 0$	**1.** 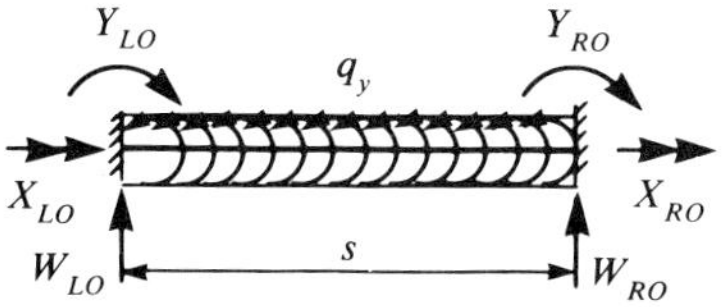	$W_{RO} = -q_y$ $X_{RO} = 0$ $Y_{RO} = 0$
$W_{LO} = (3 - 2n)n^2 q_y$ $X_{LO} = 0$ $Y_{LO} = -mnbq_y$	**2.** 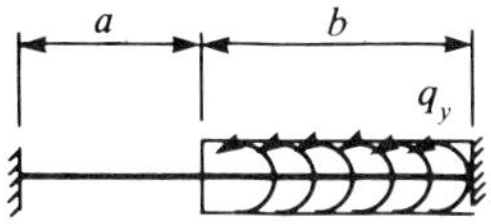	$W_{RO} = -(3 - 2n)n^2 q_y$ $X_{RO} = 0$ $Y_{RO} = mnaq_y$
$W_{LO} = \frac{1}{2} q_y$ $X_{LO} = 0$ $Y_{LO} = -\frac{1}{12} s q_y$	**3.** 	$W_{RO} = -\frac{1}{2} q_y$ $X_{RO} = 0$ $Y_{RO} = \frac{1}{12} s q_y$
$W_{LO} = \frac{1}{2}(2 - n)n^2 q_y$ $X_{LO} = 0$ $Y_{LO} = -\dfrac{1 + 2m - 3m^2}{12} bq_y$	**4.** 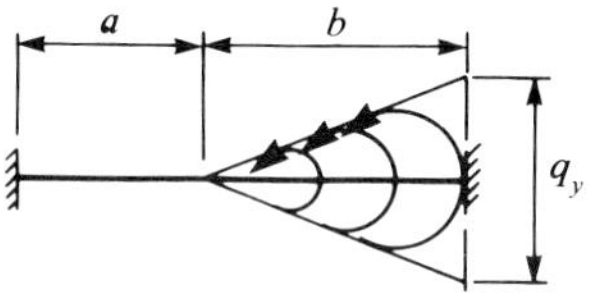	$W_{RO} = -\frac{1}{2}(2 - n)n^2 q_y$ $X_{RO} = 0$ $Y_{RO} = \dfrac{1 + 2m + 3m^2}{12} bq_y$
$W_{LO} = \dfrac{7 + 3\alpha}{20} sp_z$ $X_{LO} = \frac{1}{6}(2 + \alpha) sq_x$ $Y_{LO} = -\dfrac{3 + 2\alpha}{60} s^2 p_z$	**5.** 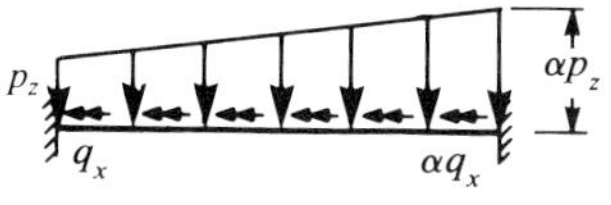 q_x varies with p_z	$W_{RO} = \dfrac{3 + 7\alpha}{20} sp_z$ $X_{RO} = \frac{1}{6}(1 + 2\alpha) sq_x$ $Y_{RO} = \dfrac{2 + 3\alpha}{60} s^2 p_z$
$W_{LO} = \frac{1}{2}(1 - m) sp_z$ $X_{LO} = \frac{1}{2}(1 - m) sq_x$ $Y_{LO} = -\dfrac{1 - 2m^2 + m^3}{12} s^2 p_z$	**6.** 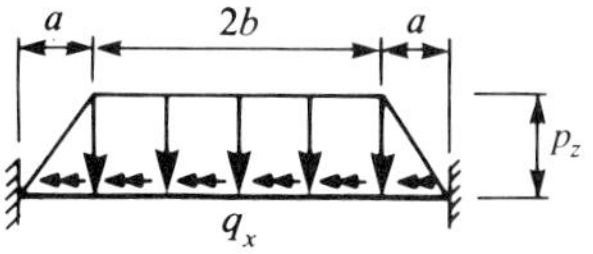 q_x varies with p_z	$W_{RO} = \frac{1}{2}(1 - m) sp_z$ $X_{RO} = \frac{1}{2}(1 - m) sq_x$ $Y_{RO} = \dfrac{1 - 2m^2 + m^3}{12} s^2 p_z$

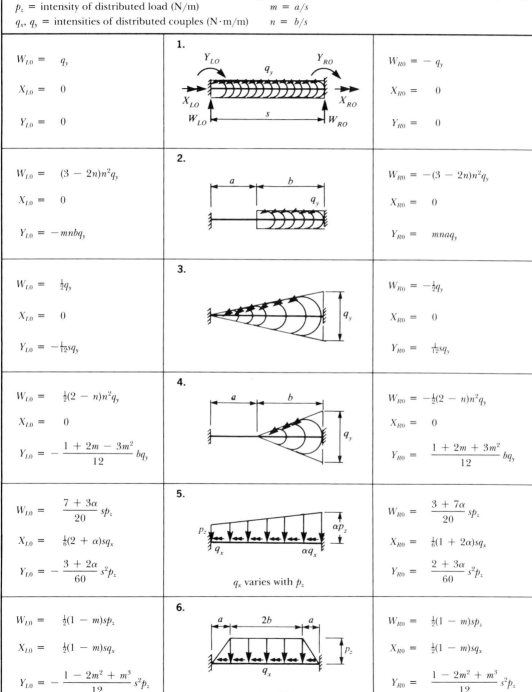

11.12 STIFFNESS LOAD FUNCTIONS, FIXED-END BEAM

Notation (11.07)	Signs (11.07)	Matrix (11.07)
p_z = maximum intensity of distributed load (N/m) q_x = maximum instensity of distributed couple (N·m/m)	$m = a/s$ $n = b/s$	

$W_{LO} = \frac{1}{3}sp_z$ $X_{LO} = \frac{1}{3}sq_x$ $Y_{LO} = -\frac{s^2}{15}p_z$	**1.** 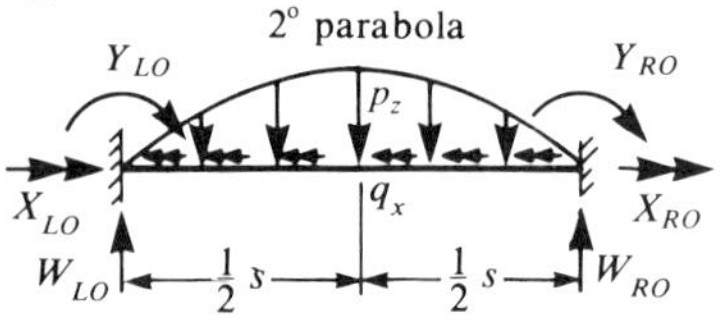	$W_{RO} = \frac{1}{3}sp_z$ $X_{RO} = \frac{1}{3}sq_x$ $Y_{RO} = \frac{s^2}{15}p_z$
$W_{LO} = \frac{1}{6}sp_z$ $X_{LO} = \frac{1}{6}sq_x$ $Y_{LO} = -\frac{s^2}{60}p_z$	**2.** 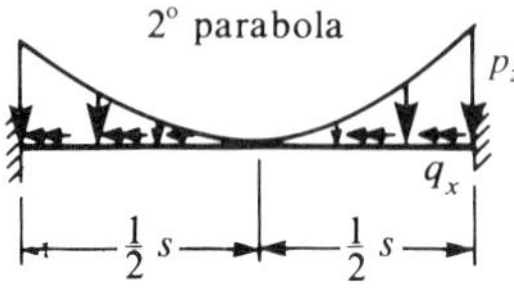	$W_{RO} = \frac{1}{6}sp_z$ $X_{RO} = \frac{1}{6}sq_x$ $Y_{RO} = \frac{s^2}{60}p_z$
$W_{LO} = \frac{3-n}{30}n^2bp_z$ $X_{LO} = \frac{nb}{12}q_x$ $Y_{LO} = -\frac{2-n}{60}nb^2p_z$	**3.** 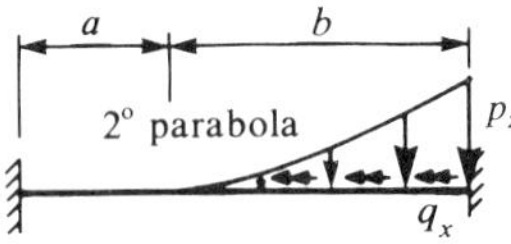	$W_{RO} = \frac{10-3n^2+n^3}{30}bp_z$ $X_{RO} = \frac{4-n}{12}bq_x$ $Y_{RO} = \frac{5-4n+n^2}{60}b^2pz$
$W_{LO} = \frac{12}{s^3}EI_yw_j$ $X_{LO} = \frac{1}{s}GI_x\phi_j$ $Y_{LO} = -\frac{6}{s^2}EI_yw_j$	**4. Abrupt linear and angular displacements** w_j, ϕ_j 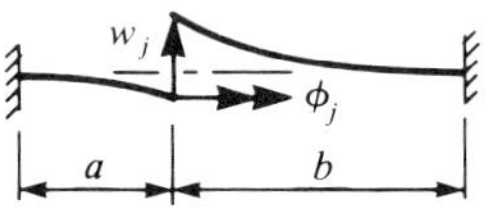	$W_{RO} = -\frac{12}{s^3}EI_yw_j$ $X_{RO} = -\frac{1}{s}GI_x\phi_j$ $Y_{RO} = -\frac{6}{s^2}EI_yw_j$
$W_{LO} = -\frac{6(1-2m)}{s^2}EI_y\psi_j$ $X_{LO} = 0$ $Y_{LO} = \frac{2(2-3m)}{s}EI_y\psi_j$	**5. Abrupt angular displacement** ψ_j 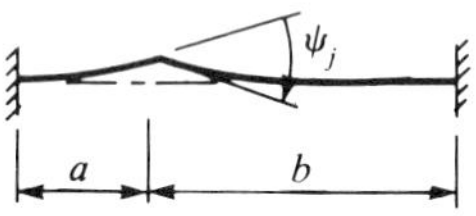	$W_{RO} = \frac{6(1-2m)}{s^2}EI_y\psi_j$ $X_{RO} = 0$ $Y_{RO} = \frac{2(1-3m)}{s}EI_y\psi_j$
$W_{LO} = \frac{6mn}{s}EI_yg_t$ $X_{LO} = 0$ $Y_{LO} = (1-3m)nEI_yg_t$	**6. Temperature deformation of segment** b (11.03) 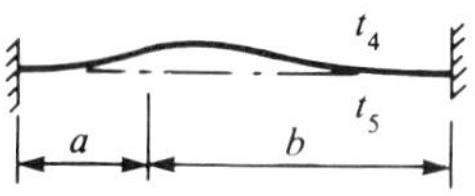	$W_{RO} = -\frac{6mn}{s}EI_yg_t$ $X_{RO} = 0$ $Y_{RO} = -(1+3m)nEI_yg_t$

11.13 STIFFNESS LOAD FUNCTIONS, PROPPED-END BEAM

Notation (11.08)	Signs (11.08)	Matrix (11.08)
P_z, Q_x, Q_y, p_z, q_x = loads defined in (11.03)	$m = a/s$ $\quad n = b/s$	
$W^*_{L0} = \dfrac{3(1-m^2)}{2s} Q_y$ $X^*_{L0} = 0$ $Y^*_{L0} = 0$	**1.** 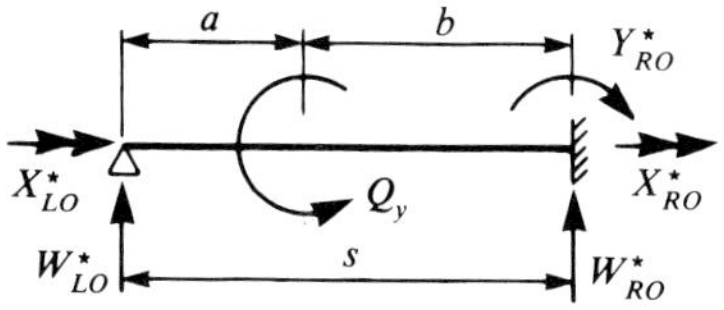	$W^*_{R0} = -\dfrac{3(1-m^2)}{2s} Q_y$ $X^*_{R0} = 0$ $Y^*_{R0} = -\frac{1}{2}(1-3m^2)Q_y$
$W^*_{L0} = \frac{1}{2}(3-n)n^2P_z$ $X^*_{L0} = nQ_x$ $Y^*_{L0} = 0$	**2.** 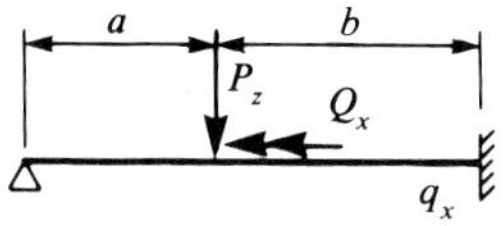	$W^*_{R0} = \frac{1}{2}(3-m^2)mP_z$ $X^*_{R0} = mQ_x$ $Y^*_{R0} = \frac{1}{2}(1+m)mbP_z$
$W^*_{L0} = \frac{3}{8}sp_z$ $X^*_{L0} = \frac{1}{2}sq_x$ $Y^*_{L0} = 0$	**3.** 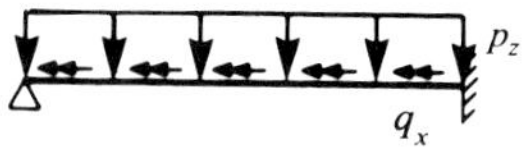	$W^*_{R0} = \frac{5}{8}sp_z$ $X^*_{R0} = \frac{1}{2}sq_x$ $Y^*_{R0} = \frac{1}{8}s^2p_z$
$W^*_{L0} = \dfrac{s}{10}p_z$ $X^*_{L0} = \frac{1}{6}sq_x$ $Y^*_{L0} = 0$	**4.** 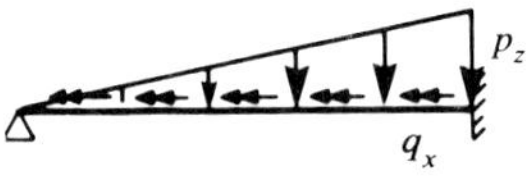	$W^*_{R0} = \dfrac{4s}{10}p_z$ $X^*_{R0} = \frac{1}{3}sq_x$ $Y^*_{R0} = \frac{1}{15}s^2p_z$
$W^*_{L0} = \dfrac{s}{24}p_z$ $X^*_{L0} = \dfrac{s}{12}q_x$ $Y^*_{L0} = 0$	**5.** 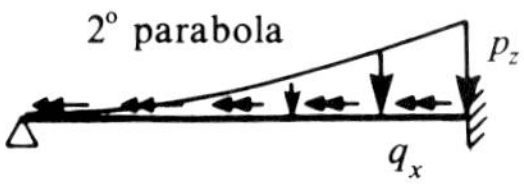	$W^*_{R0} = \dfrac{7s}{24}p_z$ $X^*_{R0} = \dfrac{s}{4}q_x$ $Y^*_{R0} = \frac{1}{24}s^2p_z$

6. 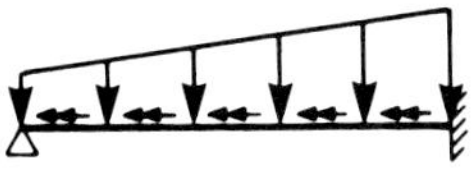 Case 3 + case 4	**7.**  Case 3 − case 4

11.14 STIFFNESS LOAD FUNCTIONS, PROPPED-END BEAM

Notation (11.08) Signs (11.08) Matrix (11.08)

$P_z, q_x, w_j, \phi_j, \psi_j, g_t$ = loads and causes defined in (11.03) $m = a/s$, $n = b/s$

$W^*_{L0} = \frac{1}{8}(4 - n)n^2 bp_z$ $X^*_{L0} = \frac{1}{2}nbq_x$ $Y^*_{L0} = 0$	**1.** 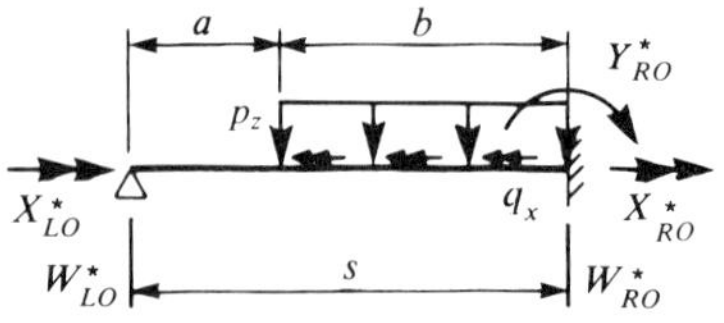 	$W^*_{R0} = \frac{1}{8}(8 - 4n^2 + n^3)bp_z$ $X^*_{R0} = \frac{1}{2}(2 - n)bq_x$ $Y^*_{R0} = \frac{1}{8}(2 - n)^2 b^2 p_z$
$W^*_{L0} = \dfrac{5n^2 - n^3}{40} bp_z$ $X^*_{L0} = \frac{1}{6}nbq_x$ $Y^*_{L0} = 0$	**2.** 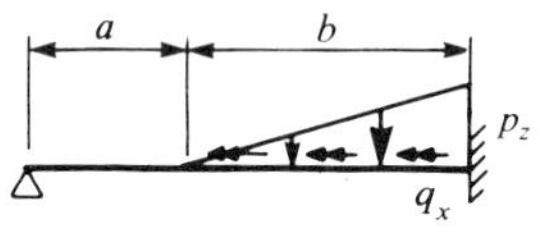 	$W^*_{R0} = \dfrac{20 - 5n^2 + n^3}{40} bp_z$ $X^*_{R0} = \frac{1}{6}(3 - n)bq_x$ $Y^*_{R0} = \dfrac{20 - 15n + 3n^2}{120} b^2 p_z$
$W^*_{L0} = \dfrac{3}{s^3} EI_y w_j$ $X^*_{L0} = \dfrac{1}{s} GI_x \phi_j$ $Y^*_{L0} = 0$	**3. Abrupt linear and angular displacement** w_j, ϕ_j 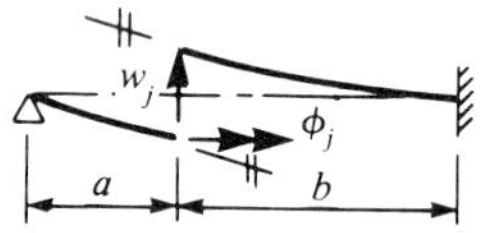 	$W^*_{R0} = -\dfrac{3}{s^3} EI_y w_j$ $X^*_{R0} = -\dfrac{1}{s} GI_x \phi_j$ $Y^*_{R0} = -\dfrac{3}{s^3} EI_y w_j$
$W^*_{L0} = \dfrac{3m}{s^2} EI_y \psi_j$ $X^*_{L0} = 0$ $Y^*_{L0} = 0$	**4. Abrupt angular displacement** ψ_j 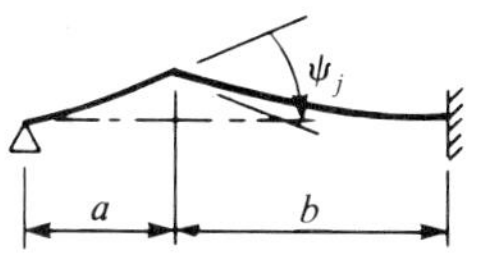 	$W^*_{R0} = -\dfrac{3m}{s^2} EI_y \psi_j$ $X^*_{R0} = 0$ $Y^*_{R0} = -\dfrac{3m}{s} EI_y \psi_j$
$W^*_{L0} = \dfrac{3n}{2s}(1 + m)EI_y g_t$ $X^*_{L0} = 0$ $Y^*_{L0} = 0$	**5. Temperature deformation of segment** b 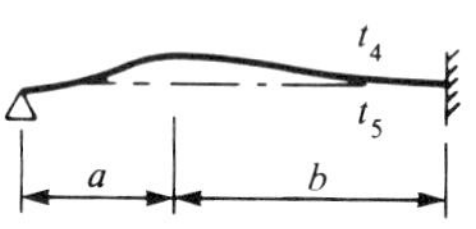 	$W^*_{R0} = -\dfrac{3n}{2s}(1 + m)EI_y g_t$ $X^*_{R0} = 0$ $Y^*_{R0} = -\dfrac{3n}{2}(1 + m)EI_y g_t$

6. 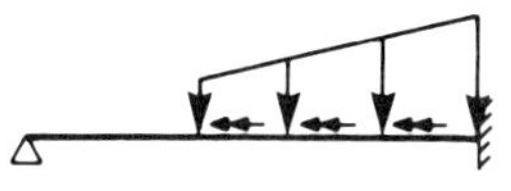 Case 1 + case 2	**7.** 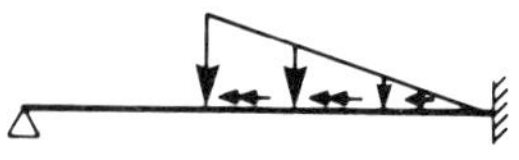 Case 1 − case 2

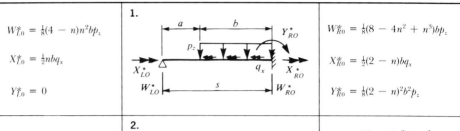

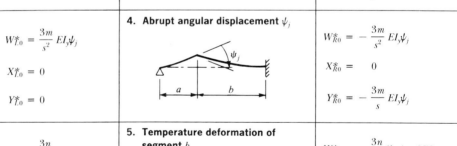

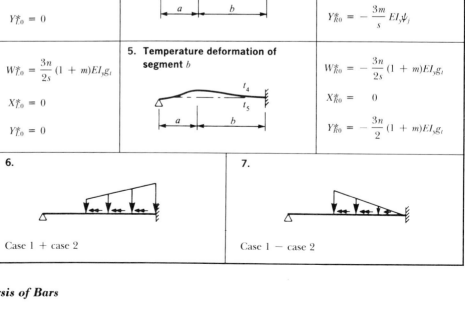

11.15 STIFFNESS LOAD FUNCTIONS, GUIDED BEAM

Notation (11.07) Sign (11.08) Matrix (11.08)

P_z, Q_x, Q_y, p_z, q_x = loads defined in (11.03) $m = a/s$ $n = b/s$

$W^*_{L0} = 0$ $X^*_{L0} = 0$ $Y^*_{L0} = nQ_y$	**1.** 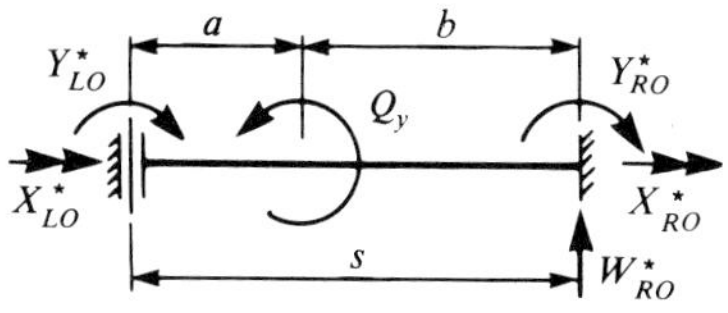 	$W^*_{R0} = 0$ $X^*_{R0} = 0$ $Y^*_{R0} = mQ_y$
$W^*_{L0} = 0$ $X^*_{L0} = nQ_x$ $Y^*_{L0} = \frac{1}{2}nbP_z$	**2.** 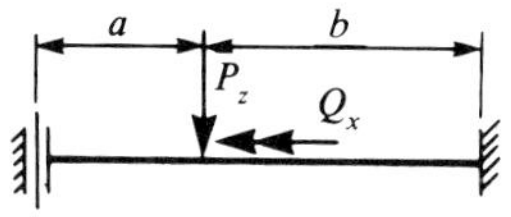 	$W^*_{R0} = P_z$ $X^*_{R0} = mQ_x$ $Y^*_{R0} = \frac{1}{2}(1 + m)bP_z$
$W^*_{L0} = 0$ $X^*_{L0} = \frac{1}{2}sq_x$ $Y^*_{L0} = \frac{1}{6}s^2p_z$	**3.**  	$W^*_{R0} = sp_z$ $X^*_{R0} = \frac{1}{2}sq_x$ $Y^*_{R0} = \frac{1}{3}s^2p_z$
$W^*_{L0} = 0$ $X^*_{L0} = \frac{1}{6}sq_x$ $Y^*_{L0} = \frac{s^2}{24}p_z$	**4.**  	$W^*_{R0} = \frac{1}{2}sp_z$ $X^*_{R0} = \frac{1}{3}sq_x$ $Y^*_{R0} = \frac{s^2}{8}p_z$
$W^*_{L0} = 0$ $X^*_{L0} = \frac{1}{12}sq_x$ $Y^*_{L0} = \frac{s^2}{60}p_z$	**5.**  	$W^*_{R0} = \frac{1}{3}sp_z$ $X^*_{R0} = \frac{1}{4}sq_x$ $Y^*_{R0} = \frac{s^2}{15}p_z$

6. 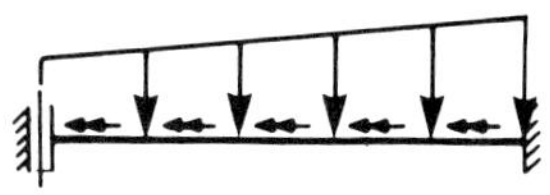 Case 3 + case 4	**7.** 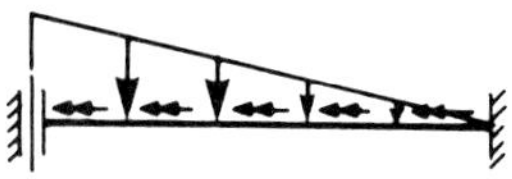 Case 3 − case 4

11.16 STIFFNESS LOAD FUNCTIONS, GUIDED BEAM

Notation (11.08) Signs (11.08) Matrix (11.08)

$p_z, q_x, w_j, \phi_j, \psi_j, g_t$ = loads and causes defined in (11.03) $m = a/s$ $n = b/s$

$W^*_{LO} = 0$ $X^*_{LO} = \frac{1}{2}nbq_x$ $Y^*_{LO} = \frac{1}{6}nb^2p_z$	**1.** 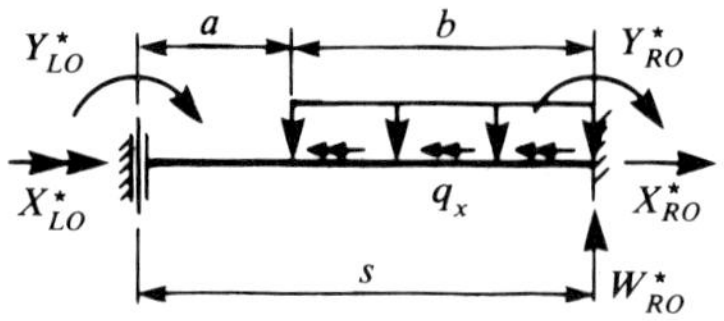	$W^*_{RO} = -bp_z$ $X^*_{RO} = \frac{1}{2}(2 - n)bq_x$ $Y^*_{RO} = \frac{1}{6}(3 - n)b^2p_z$
$W^*_{LO} = 0$ $X^*_{LO} = \frac{1}{6}nbq_x$ $Y^*_{LO} = \frac{n}{24}b^2p_z$	**2.** 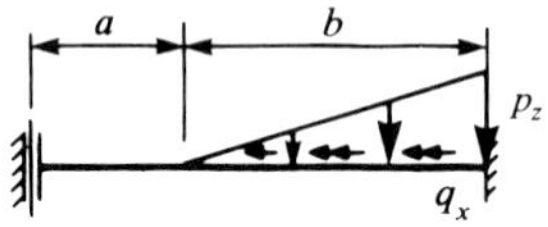	$W^*_{RO} = -\frac{1}{2}bp_z$ $X^*_{RO} = \frac{1}{6}(3 - n)bq_x$ $Y^*_{RO} = \frac{4 - n}{24}b^2p_z$
$W^*_{LO} = 0$ $X^*_{LO} = \frac{1}{s}GI_x\phi_j$ $Y^*_{LO} = 0$	**3. Abrupt linear and angular displacements** w_j, ϕ_j 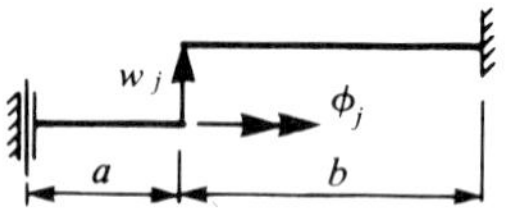	$W^*_{RO} = 0$ $X^*_{RO} = -\frac{1}{s}GI_x\phi_j$ $Y^*_{RO} = 0$
$W^*_{LO} = 0$ $X^*_{LO} = 0$ $Y^*_{LO} = \frac{1}{s}EI_y\psi_j$	**4. Abrupt angular displacement** ψ_j 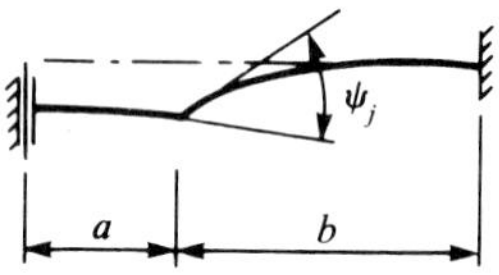	$W^*_{RO} = 0$ $X^*_{RO} = 0$ $Y^*_{RO} = -\frac{1}{s}EI_y\psi_j$
$W^*_{LO} = 0$ $X^*_{LO} = 0$ $Y^*_{LO} = nEI_yg_t$	**5. Temperature deformation of segment** b 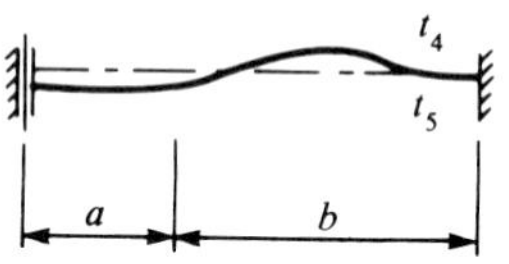	$W^*_{RO} = 0$ $X^*_{RO} = 0$ $Y^*_{RO} = -nEI_yg_t$

6. 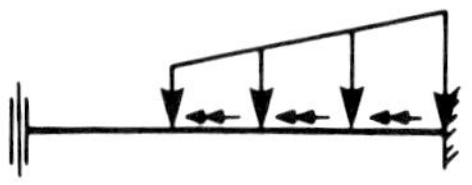 Case 1 + case 2	**7.**  Case 1 − case 2

12

Free Circular Bar of Order Two Constant Section

STATIC STATE

12.01 DEFINITION OF STATE

(1) System considered is a free finite circular bar of constant and symmetrical section with respect to the vertical centroidal z axis. The planar bar is acted on by loads normal to the system plane (system of order two). Two coordinate systems are used in this chapter: the *local system* given by the tangent, the normal, and the binormal to the centroidal axes at all points and the *segmental system* with x axis connecting the ends L, R of the bar, y axis in the plane of the bar, and z axis normal to this plane. Only the static state of this system is considered.

(2) Notation (A.1, A.2, A.4)

R	= radius of circle (m)	A_x	= area of normal section (m^2)
a, b	= position angles (rad)	I_x	= torsional constant of A_x about x (m^4)
x, x'	= angular coordinates (rad)	I_y	= moment of inertia of A_x about y (m^4)
E	= modulus of elasticity (Pa)	G	= modulus of rigidity (Pa)
W	= shearing force (N)	w	= linear displacement along z (m)
X	= torsional moment (N·m)	ϕ	= torsional slope about x (rad)
Y	= flexural moment (N·m)	ψ	= flexural slope about y (rad)

$r = \dfrac{EI_y}{GI_x}$	$\alpha = \frac{1}{2}(1 + r)$	$\beta = \frac{1}{2}(1 - r)$

(3) Assumptions of analysis are stated in (1.01). In this chapter, the torsional and flexural deformations are included, but the effect of the shearing deformations is neglected. For the inclusion of shearing deformations, reference is made to Chaps. 2 and 3. Further, it is assumed that the cross section of the bar is a compact section for which warping is secondary and may be neglected.

12.02 TRANSPORT MATRIX EQUATIONS IN LOCAL SYSTEMS

(1) Equivalents and signs

$$\bar{w} = wEI_y = \text{scaled } w \ (\text{N}\cdot\text{m}^3)$$
$$\bar{\phi} = \phi EI_y = \text{scaled } \phi \ (\text{N}\cdot\text{m}^2)$$
$$\bar{\psi} = \psi EI_y = \text{scaled } \psi \ (\text{N}\cdot\text{m}^2)$$

$$r = \frac{EI_y}{GI_x}$$
$$\alpha = \tfrac{1}{2}(1 + r) \qquad \beta = \tfrac{1}{2}(1 - r)$$

All forces, moments, and displacements are in the principal axes of the normal section, and their positive directions are shown in (2, 3) below (Ref. 5, p. 69).

(2) Free-body diagram

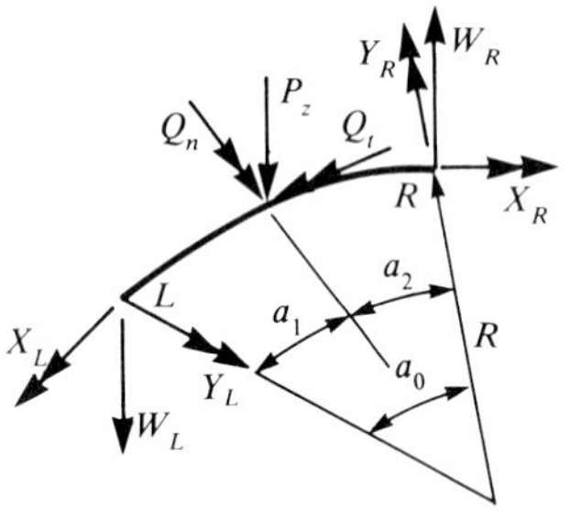

(3) Elastic curve

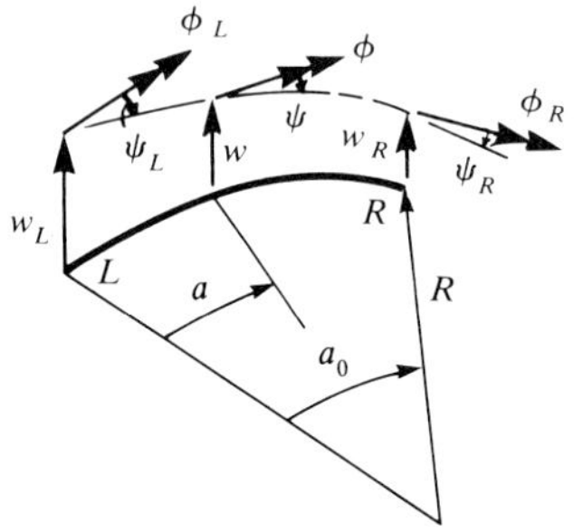

(4) Transport matrix equations

$$\underbrace{\begin{bmatrix} 1 \\ \hline W_L \\ X_L \\ Y_L \\ \hline \bar{w}_L \\ \bar{\phi}_L \\ \bar{\psi}_L \end{bmatrix}}_{H_L} = \underbrace{\left[\begin{array}{c|ccc|ccc} 1 & 0 & 0 & 0 & 0 & 0 & 0 \\ \hline F_W & 1 & 0 & 0 & 0 & 0 & 0 \\ F_X & -T_0R & T_1 & T_2 & 0 & 0 & 0 \\ F_Y & -T_2R & -T_2 & T_1 & 0 & 0 & 0 \\ \hline F_w & T_5R^3 & T_3R^2 & -T_6R^2 & 1 & -T_0R & T_2R \\ F_\phi & T_3R^2 & T_7R & -T_4R & 0 & T_1 & T_2 \\ F_\psi & T_6R^2 & T_4R & -T_8R & 0 & -T_2 & T_1 \end{array}\right]}_{T_{LR}} \underbrace{\begin{bmatrix} 1 \\ \hline W_R \\ X_R \\ Y_R \\ \hline \bar{w}_R \\ \bar{\phi}_R \\ \bar{\psi}_R \end{bmatrix}}_{H_R}$$

$$\underbrace{\begin{bmatrix} 1 \\ \hline W_R \\ X_R \\ Y_R \\ \hline \bar{w}_R \\ \bar{\phi}_R \\ \bar{\psi}_R \end{bmatrix}}_{H_R} = \underbrace{\left[\begin{array}{c|ccc|ccc} 1 & 0 & 0 & 0 & 0 & 0 & 0 \\ \hline G_W & 1 & 0 & 0 & 0 & 0 & 0 \\ G_X & -T_0R & T_1 & -T_2 & 0 & 0 & 0 \\ G_Y & T_2R & T_2 & T_1 & 0 & 0 & 0 \\ \hline G_w & -T_5R^3 & -T_3R^2 & -T_6R^2 & 1 & -T_0R & -T_2R \\ G_\phi & -T_3R^2 & -T_7R & -T_4R & 0 & T_1 & -T_2 \\ G_\psi & T_6R^2 & T_4R & T_8R & 0 & T_2 & T_1 \end{array}\right]}_{T_{RL}} \underbrace{\begin{bmatrix} 1 \\ \hline W_L \\ X_L \\ Y_L \\ \hline \bar{w}_L \\ \bar{\phi}_L \\ \bar{\psi}_L \end{bmatrix}}_{H_L}$$

where H_L, H_R are the end *state vectors* and T_{LR}, T_{RL} are the *transport matrices*. The elements of the transport matrices are the transport coefficients T_j defined in (12.03) and the *load functions* F, G listed in (12.04–12.06).

12.03 TRANSPORT RELATIONS IN LOCAL SYSTEMS

(1) Transport functions of variable-position angle a are

$T_0(a) = 1 - \cos a$	$T_{10}(a) = a - \sin a$
$T_1(a) = \cos a$	$T_{11}(a) = \sin a$
$T_2(a) = \sin a$	$T_{12}(a) = 1 - \cos a$
$T_3(a) = \alpha (\sin a - a \cos a)$	$T_{13}(a) = \alpha(2 - 2 \cos a - a \sin a)$
$T_4(a) = \alpha a \sin a$	$T_{14}(a) = T_3 (a)$
$T_5(a) = T_3(a) - r(a - \sin a)$	$T_{15}(a) = T_{13}(a) - r(\frac{1}{2}a^2 - 1 + \cos a)$
$T_6(a) = T_4(a) - r(1 - \cos a)$	$T_{16}(a) = T_{14}(a) - r(a - \sin a)$
$T_7(a) = \beta \sin a - \alpha a \cos a$	$T_{17}(a) = T_0(a) - T_4(a)$
$T_8(a) = \alpha a \cos a + \beta \sin a$	$T_{18}(a) = T_4(a) - rT_0(a)$

where $T_0(a), T_1(a), \ldots, T_8(a)$ are the basic functions and $T_{10}(a), T_{11}(a), \ldots, T_{18}(a)$ are their definite integrals in $[0, a]$.

(2) Transport coefficients T_j, A_j, B_j are the values of $T_j(a)$ for a particular argument a.

For $a = a_0$, $T_j(a_0) = T_j$; for $a = a_1$, $T_j(a_1) = A_j$; for $a = a_2$, $T_j(a_2) = B_j$

Numerical values of $T_j(a)$ for selected values of a in degrees and $r = 1, 1.5, 2, 3$ are charted and tabulated in (12.04–12.11) (Ref. 20).

(3) Load functions F and G are expressed in the same terms as in (11.03), and their physical interpretation is analogical. Since the local system is involved, the applied couples, abrupt displacements, and temperature deformations are in the local system (12.08–12.10).

(4) Relations of state-vector components are given by the following differential equations:

$\dfrac{dW}{da} = p_z R$	$\dfrac{dw}{da} = -R\psi$
$\dfrac{dX}{da} = q_t R - Y$	$\dfrac{d\phi}{da} = -\dfrac{RX}{GI_x} - \psi$
$\dfrac{dY}{da} = q_n R + X + RW$	$\dfrac{d\psi}{da} = \dfrac{RY}{EI_y} + \phi$

where W = shearing force (N), X = torsional moment (N·m), Y = flexural moment (N·m), w = transverse displacement (m), ϕ = torsional slope (rad), ψ = flexural slope (rad), p_z = intensity of distributed load (N/m), q_t, q_n = intensities of tangential and normal distributed moments (N·m/m).

(5) Transport matrices of this bar have analogical properties to those of (11.02) and can be used in similar ways. If the state vectors are required in the segmental system, the angular transformation matrices introduced in (2.19-7) must be used.

(6) General transport matrix equations for free circular bars of order three are obtained by matrix composition of (6.02–6.08) and (12.02–12.14). General transport matrix equations for free curved bars of order three and of arbitrary shape are given in (2.11).

12.04 GRAPHS AND TABLES OF TRANSPORT COEFFICIENTS

(1) Graphs of basic coefficients (a in degrees)

$$r = \frac{EI_y}{GI_x} = 1.0$$

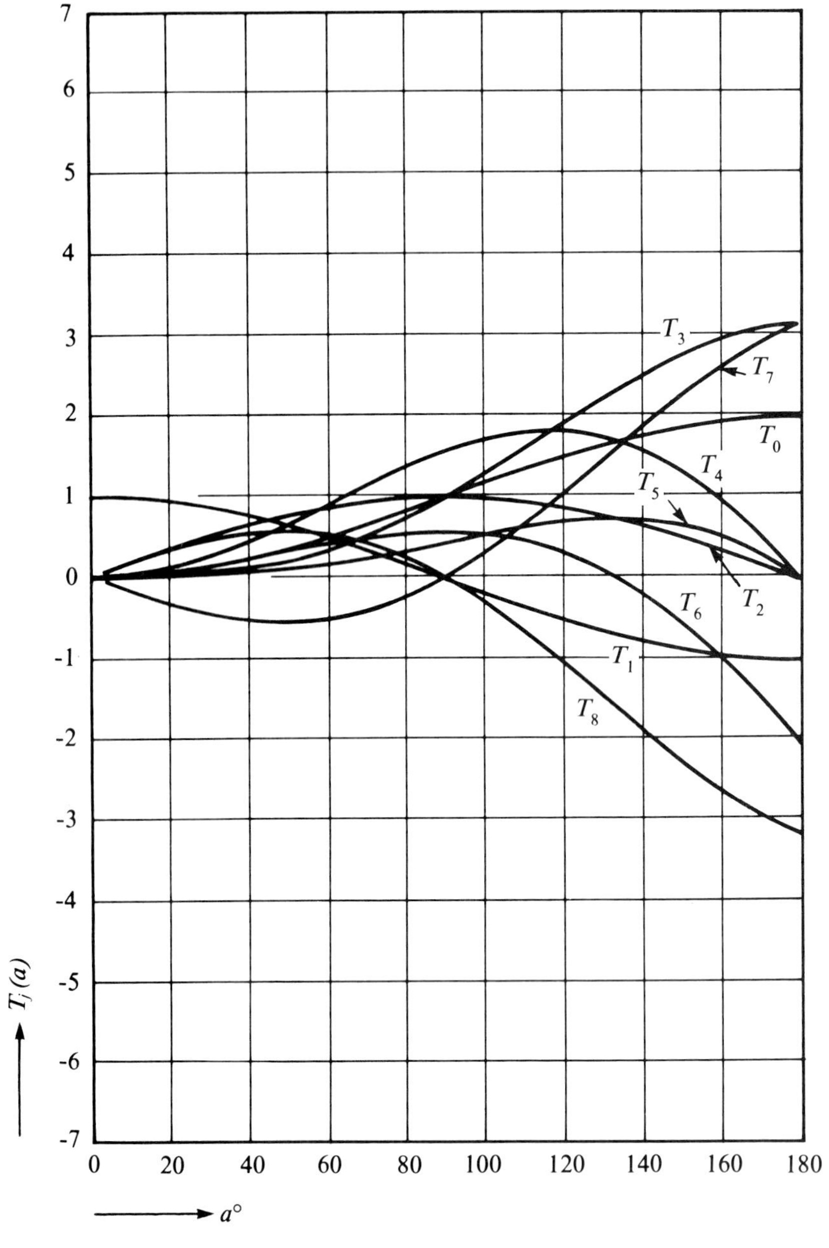

(2) Table of basic coefficients (a in degrees)

a	T_3	T_4	T_5	T_6	T_7	T_8
15	0.0059	0.0678	0.0030	0.0337	−0.2529	0.2529
18	0.0102	0.0971	0.0051	0.0481	−0.2988	0.2988
21	0.0162	0.1313	0.0080	0.0649	−0.3422	0.3422
24	0.0241	0.1704	0.0119	0.0839	−0.3827	0.3827
27	0.0341	0.2139	0.0169	0.1049	−0.4199	0.4199
30	0.0466	0.2618	0.0230	0.1278	−0.4534	0.4534
33	0.0616	0.3137	0.0303	0.1524	−0.4830	0.4830
36	0.0795	0.3693	0.0389	0.1783	−0.5083	0.5083
39	0.1003	0.4284	0.0490	0.2055	−0.5290	0.5290
42	0.1244	0.4905	0.0605	0.2336	−0.5448	0.5448
45	0.1517	0.5554	0.0735	0.2625	−0.5554	0.5554
48	0.1826	0.6226	0.0880	0.2917	−0.5606	0.5606
51	0.2170	0.6918	0.1040	0.3211	−0.5602	0.5602
54	0.2550	0.7625	0.1216	0.3503	−0.5540	0.5540
57	0.2968	0.8343	0.1407	0.3790	−0.5418	0.5418
60	0.3424	0.9069	0.1613	0.4069	−0.5236	0.5236
63	0.3918	0.9797	0.1833	0.4337	−0.4992	0.4992
66	0.4450	1.0523	0.2066	0.4591	−0.4685	0.4685
69	0.5020	1.1243	0.2313	0.4827	−0.4316	0.4316
72	0.5627	1.1951	0.2572	0.5041	−0.3883	0.3883
75	0.6271	1.2644	0.2841	0.5232	−0.3388	0.3388
78	0.6951	1.3316	0.3119	0.5395	−0.2830	0.2830
81	0.7665	1.3963	0.3405	0.5527	−0.2212	0.2212
84	0.8413	1.4580	0.3697	0.5626	−0.1532	0.1532
87	0.9192	1.5164	0.3994	0.5687	−0.0795	0.0795
90	1.0000	1.5708	0.4292	0.5708	−0.0000	0.0000
93	1.0836	1.6209	0.4591	0.5686	0.0849	−0.0849
96	1.1697	1.6663	0.4887	0.5618	0.1751	−0.1751
99	1.2580	1.7066	0.5178	0.5502	0.2703	−0.2703
102	1.3483	1.7413	0.5462	0.5334	0.3701	−0.3701
105	1.4402	1.7702	0.5736	0.5113	0.4743	−0.4743
108	1.5335	1.7927	0.5996	0.4837	0.5825	−0.5825
111	1.6279	1.8086	0.6241	0.4503	0.6943	−0.6943
114	1.7228	1.8177	0.6467	0.4109	0.8093	−0.8093
117	1.8181	1.8195	0.6670	0.3655	0.9271	−0.9271
120	1.9132	1.8138	0.6849	0.3138	1.0472	−1.0472
123	2.0079	1.8004	0.6998	0.2558	1.1692	−1.1692
126	2.1016	1.7791	0.7115	0.1913	1.2926	−1.2926
129	2.1940	1.7497	0.7197	0.1204	1.4169	−1.4169
132	2.2847	1.7121	0.7240	0.0430	1.5416	−1.5416
135	2.3732	1.6661	0.7241	−0.0410	1.6661	−1.6661
138	2.4590	1.6116	0.7196	−0.1315	1.7899	−1.7899
141	2.5418	1.5487	0.7102	−0.2284	1.9125	−1.9125
144	2.6211	1.4773	0.6956	−0.3318	2.0333	−2.0333
147	2.6964	1.3973	0.6754	−0.4413	2.1517	−2.1517
150	2.7672	1.3090	0.6493	−0.5570	2.2672	−2.2672
153	2.8333	1.2123	0.6169	−0.6787	2.3793	−2.3793
156	2.8941	1.1074	0.5781	−0.8061	2.4873	−2.4873
159	2.9491	0.9945	0.5324	−0.9391	2.5908	−2.5908
162	2.9981	0.8737	0.4797	−1.0773	2.6890	−2.6890
165	3.0405	0.7453	0.4195	−1.2206	2.7817	−2.7817
168	3.0760	0.6096	0.3517	−1.3685	2.8681	−2.8681
171	3.1042	0.4669	0.2761	−1.5208	2.9478	−2.9478
174	3.1248	0.3174	0.1924	−1.6771	3.0202	−3.0202
177	3.1373	0.1617	0.1004	−1.8369	3.0850	−3.0850
180	3.1416	0.0000	0.0000	−2.0000	3.1416	−3.1416

12.05 GRAPHS AND TABLES OF TRANSPORT COEFFICIENTS

(1) Graphs of basic coefficients (a in degrees)

$$r = \frac{EI_y}{GI_x} = 1.5$$

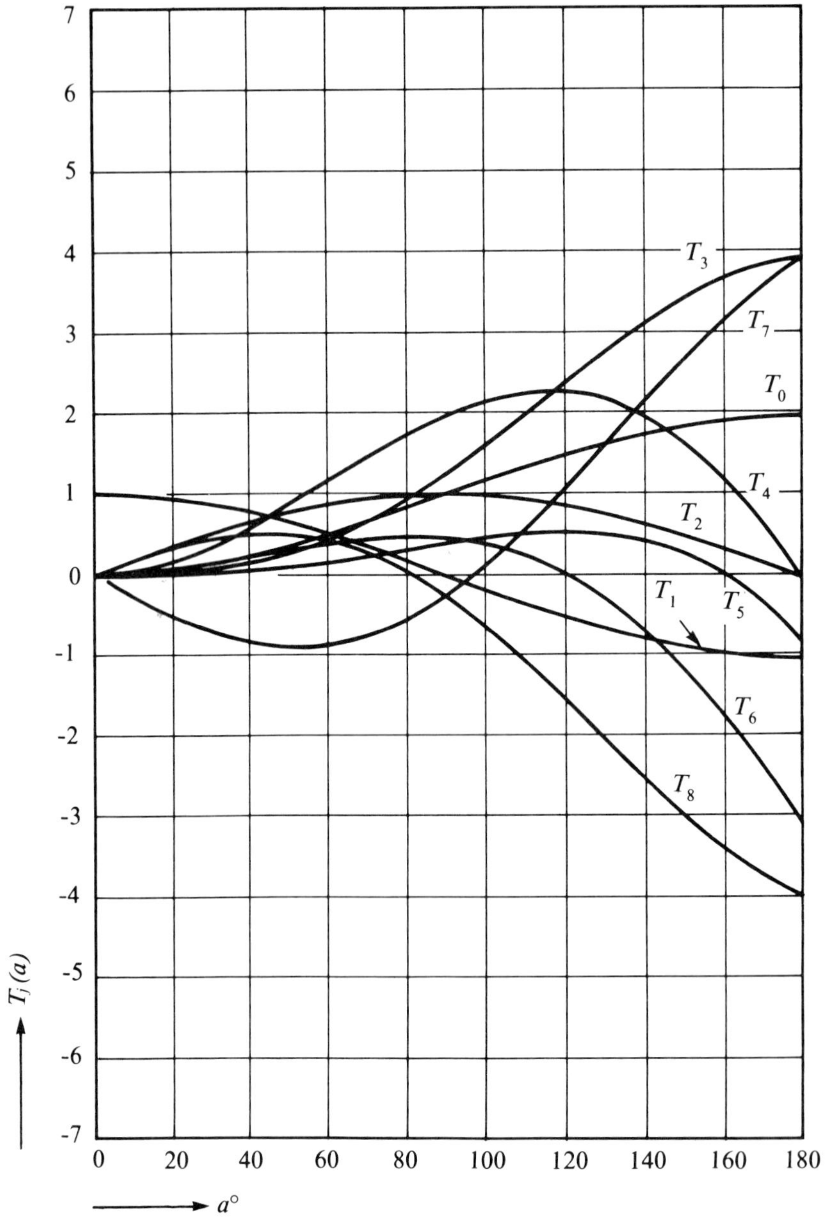

(2) Table of basic coefficients (a in degrees)

a	T_3	T_4	T_5	T_6	T_7	T_8
15	0.0074	0.0847	0.0030	0.0336	−0.3808	0.2514
18	0.0128	0.1214	0.0051	0.0479	−0.4507	0.2962
21	0.0202	0.1642	0.0080	0.0646	−0.5173	0.3381
24	0.0301	0.2130	0.0119	0.0833	−0.5800	0.3766
27	0.0426	0.2674	0.0168	0.1039	−0.6383	0.4113
30	0.0582	0.3272	0.0228	0.1263	−0.6918	0.4418
33	0.0770	0.3921	0.0300	0.1501	−0.7400	0.4676
36	0.0993	0.4616	0.0385	0.1752	−0.7823	0.4885
39	0.1254	0.5355	0.0484	0.2012	−0.8186	0.5039
42	0.1555	0.6131	0.0596	0.2278	−0.8482	0.5137
45	0.1897	0.6942	0.0722	0.2549	−0.8710	0.5174
48	0.2282	0.7782	0.0863	0.2819	−0.8865	0.5149
51	0.2712	0.8647	0.1018	0.3087	−0.8945	0.5059
54	0.3188	0.9531	0.1186	0.3348	−0.8947	0.4902
57	0.3711	1.0429	0.1368	0.3599	−0.8870	0.4676
60	0.4280	1.1336	0.1563	0.3836	−0.8710	0.4380
63	0.4898	1.2246	0.1769	0.4056	−0.8467	0.4012
66	0.5563	1.3154	0.1987	0.4255	−0.8140	0.3573
69	0.6275	1.4054	0.2215	0.4429	−0.7729	0.3061
72	0.7034	1.4939	0.2450	0.4574	−0.7232	0.2476
75	0.7839	1.5805	0.2693	0.4687	−0.6650	0.1820
78	0.8689	1.6645	0.2941	0.4764	−0.5983	0.1093
81	0.9582	1.7454	0.3191	0.4800	−0.5234	0.0295
84	1.0516	1.8226	0.3443	0.4793	−0.4402	−0.0571
87	1.1490	1.8954	0.3692	0.4739	−0.3490	−0.1503
90	1.2500	1.9635	0.3938	0.4635	−0.2500	−0.2500
93	1.3545	2.0262	0.4177	0.4477	−0.1435	−0.3558
96	1.4621	2.0829	0.4406	0.4261	−0.0297	−0.4676
99	1.5725	2.1333	0.4622	0.3986	0.0910	−0.5848
102	1.6853	2.1767	0.4822	0.3648	0.2181	−0.7072
105	1.8003	2.2127	0.5003	0.3245	0.3514	−0.8344
108	1.9169	2.2409	0.5161	0.2773	0.4903	−0.9659
111	2.0348	2.2608	0.5292	0.2232	0.6344	−1.1012
114	2.1535	2.2721	0.5393	0.1620	0.7832	−1.2400
117	2.2726	2.2743	0.5460	0.0933	0.9361	−1.3816
120	2.3915	2.2672	0.5490	0.0172	1.0925	−1.5255
123	2.5098	2.2505	0.5477	−0.0664	1.2518	−1.6712
126	2.6270	2.2239	0.5419	−0.1578	1.4135	−1.8180
129	2.7426	2.1872	0.5311	−0.2568	1.5768	−1.9654
132	2.8559	2.1401	0.5149	−0.3636	1.7412	−2.1127
135	2.9665	2.0826	0.4929	−0.4781	1.9058	−2.2594
138	3.0738	2.0145	0.4647	−0.6002	2.0701	−2.4047
141	3.1773	1.9359	0.4299	−0.7298	2.2333	−2.5479
144	3.2763	1.8466	0.3881	−0.8669	2.3947	−2.6885
147	3.3705	1.7467	0.3390	−1.0113	2.5535	−2.8258
150	3.4591	1.6362	0.2821	−1.1628	2.7091	−2.9591
153	3.5416	1.5154	0.2171	−1.3211	2.8606	−3.0876
156	3.6176	1.3843	0.1436	−1.4860	3.0075	−3.2108
159	3.6864	1.2431	0.0613	−1.6572	3.1488	−3.3280
162	3.7476	1.0922	−0.0300	−1.8344	3.2841	−3.4386
165	3.8006	0.9317	−0.1309	−2.0172	3.4124	−3.5418
168	3.8450	0.7620	−0.2414	−2.2052	3.5331	−3.6371
171	3.8803	0.5836	−0.3619	−2.3979	3.6456	−3.7238
174	3.9060	0.3968	−0.4926	−2.5950	3.7492	−3.8014
177	3.9217	0.2021	−0.6337	−2.7958	3.8432	−3.8693
180	3.9270	0.0000	−0.7854	−3.0000	3.9270	−3.9270

12.06 GRAPHS AND TABLES OF TRANSPORT COEFFICIENTS

(1) Graphs of basic coefficients (*a* in degrees)

$$r = \frac{EI_y}{GI_x} = 2.0$$

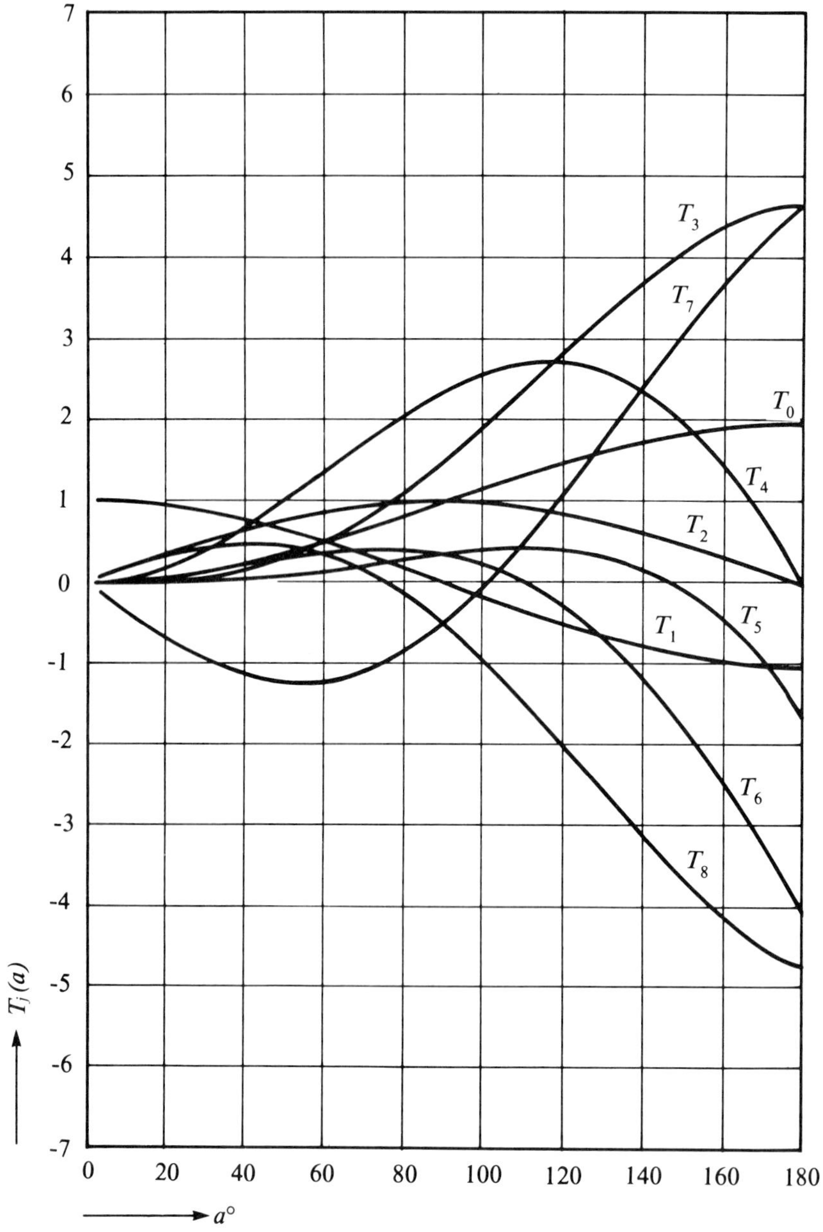

(2) Table of basic coefficients (a in degrees)

a	T_3	T_4	T_5	T_6	T_7	T_8
15	0.0089	0.1016	0.0029	0.0335	−0.5087	0.2499
18	0.0154	0.1456	0.0051	0.0477	−0.6027	0.2937
21	0.0243	0.1970	0.0080	0.0642	−0.6924	0.3341
24	0.0361	0.2556	0.0118	0.0827	−0.7774	0.3706
27	0.0512	0.3209	0.0167	0.1029	−0.8568	0.4028
30	0.0698	0.3927	0.0226	0.1247	−0.9302	0.4302
33	0.0924	0.4705	0.0298	0.1479	−0.9969	0.4522
36	0.1192	0.5540	0.0381	0.1720	−1.0564	0.4686
39	0.1505	0.6425	0.0478	0.1968	−1.1081	0.4788
42	0.1866	0.7357	0.0588	0.2220	−1.1517	0.4826
45	0.2276	0.8330	0.0710	0.2473	−1.1866	0.4795
48	0.2739	0.9339	0.0846	0.2721	−1.2124	0.4693
51	0.3255	1.0376	0.0995	0.2963	−1.2288	0.4517
54	0.3826	1.1437	0.1156	0.3193	−1.2355	0.4265
57	0.4453	1.2515	0.1329	0.3408	−1.2321	0.3934
60	0.5136	1.3603	0.1513	0.3603	−1.2184	0.3524
63	0.5877	1.4696	0.1706	0.3776	−1.1943	0.3033
66	0.6675	1.5785	0.1908	0.3920	−1.1596	0.2460
69	0.7530	1.6864	0.2116	0.4032	−1.1142	0.1806
72	0.8441	1.7927	0.2329	0.4107	−1.0580	0.1070
75	0.9407	1.8966	0.2546	0.4142	−0.9912	0.0252
78	1.0427	1.9974	0.2762	0.4132	−0.9136	−0.0645
81	1.1498	2.0945	0.2977	0.4073	−0.8256	−0.1621
84	1.2619	2.1871	0.3188	0.3961	−0.7271	−0.2674
87	1.3787	2.2745	0.3391	0.3792	−0.6185	−0.3801
90	1.5000	2.3562	0.3584	0.3562	−0.5000	−0.5000
93	1.6254	2.4314	0.3763	0.3267	−0.3719	−0.6267
96	1.7545	2.4995	0.3925	0.2904	−0.2346	−0.7600
99	1.8870	2.5599	0.4066	0.2470	−0.0884	−0.8993
102	2.0224	2.6120	0.4182	0.1962	0.0661	−1.0443
105	2.1604	2.6552	0.4270	0.1376	0.2285	−1.1944
108	2.3003	2.6890	0.4325	0.0710	0.3982	−1.3493
111	2.4418	2.7130	0.4343	−0.0038	0.5746	−1.5082
114	2.5842	2.7265	0.4320	−0.0870	0.7571	−1.6707
117	2.7271	2.7292	0.4250	−0.1788	0.9451	−1.8361
120	2.8698	2.7207	0.4131	−0.2793	1.1378	−2.0038
123	3.0118	2.7006	0.3956	−0.3886	1.3345	−2.1731
126	3.1524	2.6687	0.3722	−0.5069	1.5344	−2.3434
129	3.2911	2.6246	0.3424	−0.6341	1.7368	−2.5139
132	3.4271	2.5681	0.3057	−0.7701	1.9408	−2.6839
135	3.5598	2.4991	0.2616	−0.9151	2.1456	−2.8527
138	3.6886	2.4175	0.2097	−1.0688	2.3503	−3.0194
141	3.8127	2.3231	0.1495	−1.2312	2.5541	−3.1834
144	3.9316	2.2159	0.0806	−1.4021	2.7560	−3.3438
147	4.0445	2.0960	0.0026	−1.5813	2.9553	−3.4999
150	4.1509	1.9635	−0.0851	−1.7686	3.1509	−3.6509
153	4.2499	1.8185	−0.1828	−1.9635	3.3420	−3.7959
156	4.3411	1.6611	−0.2909	−2.1659	3.5276	−3.9344
159	4.4237	1.4917	−0.4097	−2.3754	3.7069	−4.0653
162	4.4971	1.3106	−0.5397	−2.5915	3.8791	−4.1881
165	4.5607	1.1180	−0.6812	−2.8138	4.0431	−4.3019
168	4.6140	0.9144	−0.8345	−3.0418	4.1982	−4.4061
171	4.6563	0.7003	−0.9999	−3.2751	4.3434	−4.4999
174	4.6871	0.4762	−1.1775	−3.5129	4.4781	−4.5826
177	4.7060	0.2425	−1.3678	−3.7547	4.6013	−4.6537
180	4.7124	0.0000	−1.5708	−4.0000	4.7124	−4.7124

12.07 GRAPHS AND TABLES OF TRANSPORT COEFFICIENTS

(1) Graphs of basic functions (a in degrees)

$$r = \frac{EI_y}{GI_x} = 3.0$$

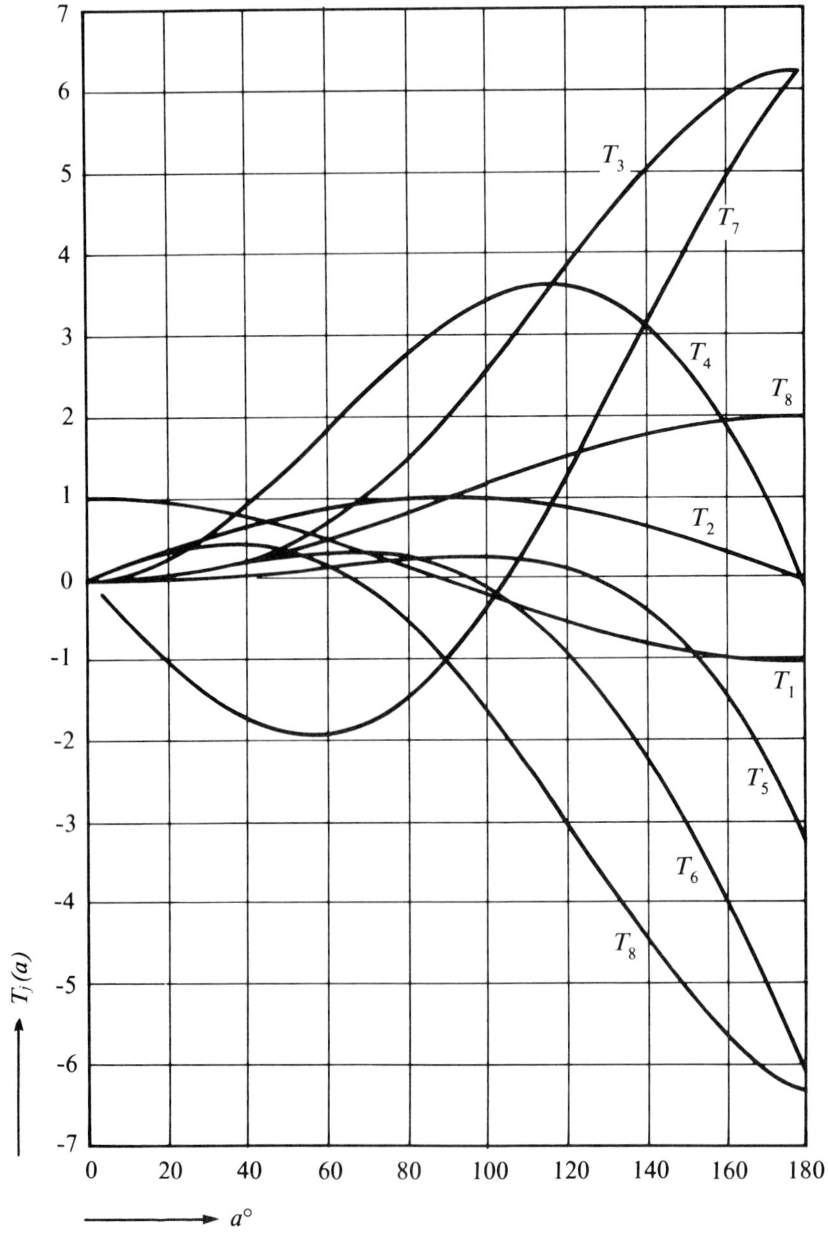

(2) Table of basic coefficients (a in degrees)

a	T_3	T_4	T_5	T_6	T_7	T_8
15	0.0119	0.1355	0.0029	0.0333	−0.7646	0.2469
18	0.0205	0.1942	0.0050	0.0473	−0.9066	0.2885
21	0.0324	0.2627	0.0079	0.0634	−1.0427	0.3260
24	0.0481	0.3407	0.0117	0.0814	−1.1721	0.3586
27	0.0682	0.4279	0.0165	0.1009	−1.2937	0.3858
30	0.0931	0.5236	0.0223	0.1217	−1.4069	0.4069
33	0.1232	0.6274	0.0292	0.1434	−1.5107	0.4214
36	0.1589	0.7386	0.0373	0.1657	−1.6044	0.4289
39	0.2007	0.8567	0.0466	0.1882	−1.6873	0.4287
42	0.2488	0.9810	0.0570	0.2104	−1.7586	0.4204
45	0.3035	1.1107	0.0686	0.2320	−1.8178	0.4036
48	0.3652	1.2452	0.0813	0.2525	−1.8643	0.3780
51	0.4340	1.3835	0.0950	0.2715	−1.8975	0.3432
54	0.5101	1.5250	0.1097	0.2883	−1.9170	0.2989
57	0.5937	1.6687	0.1252	0.3026	−1.9223	0.2450
60	0.6849	1.8138	0.1413	0.3138	−1.9132	0.1812
63	0.7836	1.9594	0.1580	0.3214	−1.8894	0.1074
66	0.8900	2.1047	0.1749	0.3249	−1.8506	0.0235
69	1.0040	2.2486	0.1919	0.3237	−1.7967	−0.0704
72	1.1255	2.3903	0.2087	0.3173	−1.7277	−0.1744
75	1.2543	2.5288	0.2251	0.3052	−1.6435	−0.2883
78	1.3902	2.6632	0.2406	0.2870	−1.5442	−0.4121
81	1.5331	2.7926	0.2550	0.2619	−1.4300	−0.5454
84	1.6825	2.9161	0.2679	0.2297	−1.3010	−0.6880
87	1.8383	3.0327	0.2789	0.1897	−1.1576	−0.8397
90	2.0000	3.1416	0.2876	0.1416	−1.0000	−1.0000
93	2.1672	3.2419	0.2936	0.0849	−0.8287	−1.1685
96	2.3393	3.3327	0.2963	0.0191	−0.6442	−1.3448
99	2.5160	3.4132	0.2954	−0.0561	−0.4471	−1.5283
102	2.6966	3.4827	0.2903	−0.1411	−0.2379	−1.7184
105	2.8805	3.5403	0.2805	−0.2362	−0.0173	−1.9145
108	3.0671	3.5854	0.2654	−0.3417	0.2139	−2.1160
111	3.2557	3.6173	0.2445	−0.4578	0.4550	−2.3221
114	3.4456	3.6353	0.2172	−0.5849	0.7050	−2.5321
117	3.6361	3.6389	0.1831	−0.7230	0.9631	−2.7451
120	3.8264	3.6276	0.1413	−0.8724	1.2284	−2.9604
123	4.0158	3.6008	0.0915	−1.0331	1.4997	−3.1771
126	4.2032	3.5582	0.0330	−1.2051	1.7762	−3.3942
129	4.3881	3.4994	−0.0349	−1.3885	2.0566	−3.6109
132	4.5694	3.4242	−0.1126	−1.5832	2.3400	−3.8263
135	4.7464	3.3322	−0.2009	−1.7892	2.6251	−4.0393
138	4.9181	3.2233	−0.3002	−2.0062	2.9107	−4.2489
141	5.0836	3.0974	−0.4112	−2.2340	3.1957	−4.4543
144	5.2421	2.9545	−0.5343	−2.4725	3.4788	−4.6543
147	5.3927	2.7947	−0.6703	−2.7213	3.7588	−4.8481
150	5.5345	2.6180	−0.8195	−2.9801	4.0345	−5.0345
153	5.6666	2.4246	−0.9825	−3.2484	4.3046	−5.2126
156	5.7881	2.2149	−1.1598	−3.5258	4.5679	−5.3814
159	5.8982	1.9890	−1.3519	−3.8117	4.8231	−5.5399
162	5.9961	1.7475	−1.5591	−4.1057	5.0691	−5.6871
165	6.0810	1.4907	−1.7819	−4.4071	5.3045	−5.8222
168	6.1520	1.2193	−2.0207	−4.7152	5.5282	−5.9441
171	6.2084	0.9338	−2.2758	−5.0293	5.7391	−6.0520
174	6.2495	0.6349	−2.5475	−5.3487	5.9359	−6.1450
177	6.2747	0.3234	−2.8360	−5.6725	6.1177	−6.2223
180	6.2832	0.0000	−3.1416	−6.0000	6.2832	−6.2832

12.08 TRANSPORT LOAD FUNCTIONS IN LOCAL SYSTEMS

Notation (12.02–12.03) P_z = concentrated load (N) Q_n, Q_t = applied couples (N·m)	Signs (12.02)	Matrix (12.02) g_t = thermal strain (11.03) A_j, B_j, T_j = coefficients (12.03–12.07)
$F_W = -P_z$ $F_X = A_0 R P_z$ $F_Y = A_2 R P_z$	**1.** 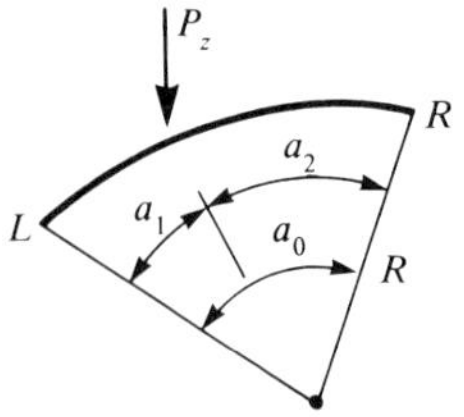	$G_W = P_z$ $G_X = -B_0 R P_z$ $G_Y = B_2 R P_z$
$F_w = -A_5 R^3 P_z$ $F_\phi = -A_3 R^2 P_z$ $F_\psi = -A_6 R^2 P_z$		$G_w = -B_5 R^3 P_z$ $G_\phi = -B_3 R^2 P_z$ $G_\psi = B_6 R^2 P_z$
$F_W = 0$ $F_X = -A_1 Q_t$ $F_Y = A_2 Q_t$	**2.** 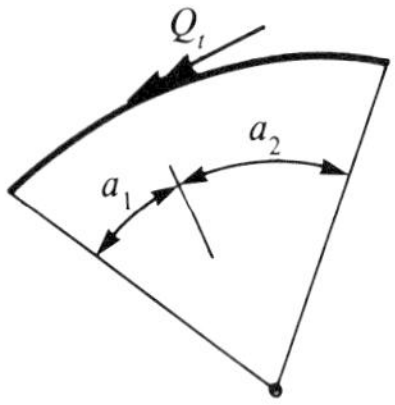	$G_W = 0$ $G_X = B_1 Q_t$ $G_Y = B_2 Q_t$
$F_w = -A_3 R^2 Q_t$ $F_\phi = -A_7 R Q_t$ $F_\psi = -A_4 R Q_t$		$G_w = -B_3 R^2 Q_t$ $G_\phi = -B_7 R Q_t$ $G_\psi = B_4 R Q_t$
$F_W = 0$ $F_X = -A_2 Q_n$ $F_Y = -A_1 Q_n$	**3.** 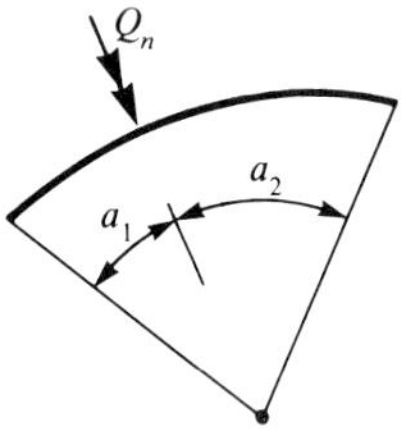	$G_W = 0$ $G_X = -B_2 Q_n$ $G_Y = B_1 Q_n$
$F_w = A_6 R^2 Q_n$ $F_\phi = A_4 R Q_n$ $F_\psi = A_8 R Q_n$		$G_w = -B_6 R^2 Q_n$ $G_\phi = -B_4 R Q_n$ $G_\psi = B_8 R Q_n$
$F_W = 0$ $F_X = 0$ $F_Y = 0$	**4. Temperature deformation of segment** Ra_2 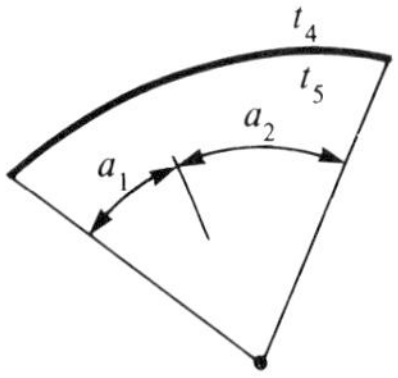	$G_W = 0$ $G_X = 0$ $G_Y = 0$
$F_w = -(T_{12} - A_{12}) R g_t E I_y$ $F_\phi = -(T_{12} - A_{12}) g_t E I_y$ $F_\psi = -(T_{11} - A_{11}) g_t E I_y$		$G_w = -B_{12} R g_t E I_y$ $G_\phi = -B_{12} g_t E I_y$ $G_\psi = B_{11} g_t E I_y$

12.09 TRANSPORT LOAD FUNCTIONS IN LOCAL SYSTEMS

Notation (12.02–12.03)	Signs (12.02)	Matrix (12.02)
P_z = intensity of uniform load (N/m) w_j, ϕ_j, ψ_j = abrupt displacements (11.03)		T_j, A_j, B_j = coefficients (12.03–12.07)
$F_W = -a_0 R p_z$ $F_X = T_{10} R^2 p_z$ $F_Y = T_{12} R^2 p_z$ $F_w = -T_{15} R^4 p_z$ $F_\phi = -T_{13} R^3 p_z$ $F_\psi = -T_{16} R^3 p_z$	**1.** 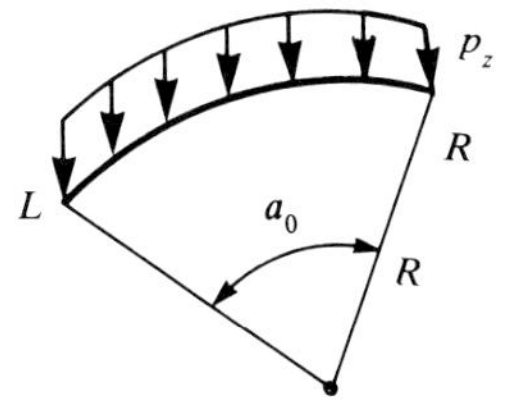 	$G_W = a_0 R p_z$ $G_X = -T_{10} R^2 p_z$ $G_Y = T_{12} R^2 p_z$ $G_w = -T_{15} R^4 p_z$ $G_\phi = -T_{13} R^3 p_z$ $G_\psi = T_{16} R^3 p_z$
$F_W = -a_2 R p_z$ $F_X = (T_{10} - A_{10}) R^2 p_z$ $F_Y = (T_{12} - A_{12}) R^2 p_z$ $F_w = -(T_{15} - A_{15}) R^4 p_z$ $F_\phi = -(T_{13} - A_{13}) R^3 p_z$ $F_\psi = -(T_{16} - A_{16}) R^3 p_z$	**2.** 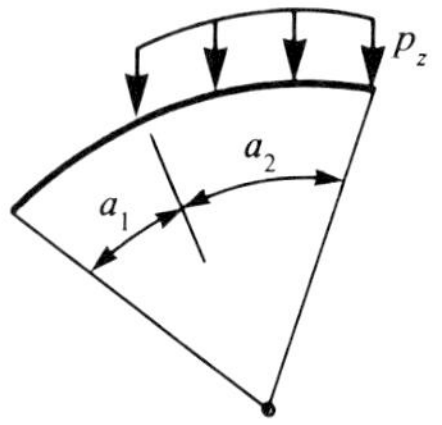 	$G_W = a_2 R p_z$ $G_X = -B_{10} R^2 p_z$ $G_Y = B_{12} R^2 p_z$ $G_w = -B_{15} R^4 p_z$ $G_\phi = -B_{13} R^3 p_z$ $G_\psi = B_{16} R^3 p_z$
$F_W = 0$ $F_X = 0$ $F_Y = 0$ $F_w = -(w_j - A_0 R \phi_j) EI_y$ $F_\phi = -A_1 \phi_j EI_y$ $F_\psi = A_2 \phi_j EI_y$	**3. Abrupt displacements** 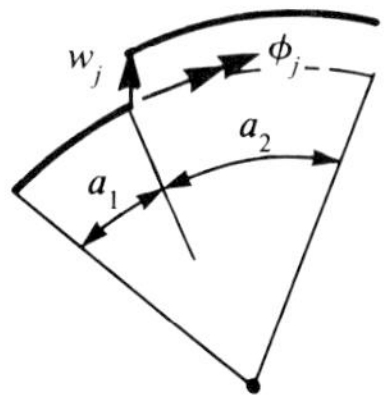 	$G_W = 0$ $G_X = 0$ $G_Y = 0$ $G_w = (w_j - B_0 R \phi_j) EI_y$ $G_\phi = B_1 \phi_j EI_y$ $G_\psi = B_2 \phi_j EI_y$
$F_W = 0$ $F_X = 0$ $F_Y = 0$ $F_w = -A_2 \psi_j R EI_y$ $F_\phi = -A_2 \psi_j EI_y$ $F_\psi = -A_1 \psi_j EI_y$	**4. Abrupt displacement** 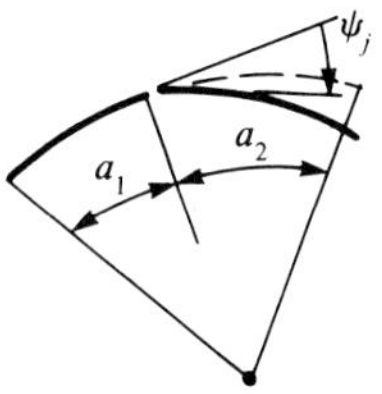 	$G_W = 0$ $G_X = 0$ $G_Y = 0$ $G_w = -B_2 \psi_j R EI_y$ $G_\phi = -B_2 \psi_j EI_y$ $G_\psi = B_1 \psi_j EI_y$

12.10 TRANSPORT LOAD FUNCTIONS IN LOCAL SYSTEMS

Notation (12.02–12.03) Signs (12.02) Matrix (12.02)
q_x, q_y = intensities of distributed couples (N·m/m)
T_j, A_j, B_j = coefficients (12.03–12.07)

$F_W = 0$ $F_X = -T_{11}Rq_x$ $F_Y = T_{12}Rq_x$	**1.** 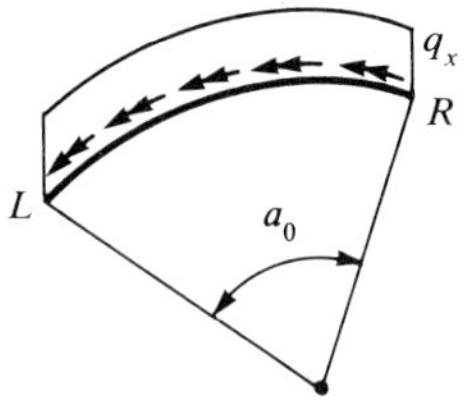	$G_W = 0$ $G_X = T_{11}Rq_x$ $G_Y = T_{12}Rq_x$
$F_w = -T_{13}R^3q_x$ $F_\phi = -T_{17}R^2q_x$ $F_\psi = -T_{14}R^2q_x$		$G_w = -T_{13}R^3q_x$ $G_\phi = -T_{17}R^2q_x$ $G_\psi = T_{14}R^2q_x$
$F_W = 0$ $F_X = -(T_{11} - A_{11})Rq_x$ $F_Y = (T_{12} - A_{12})Rq_x$	**2.** 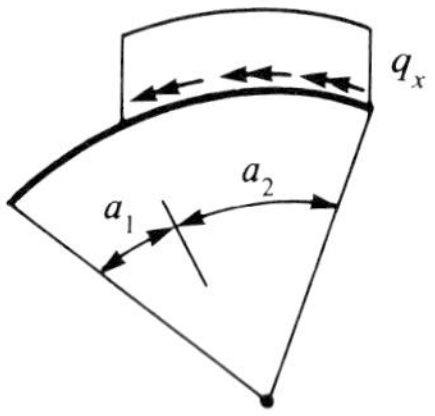	$G_W = 0$ $G_X = B_{11}Rq_x$ $G_Y = B_{12}Rq_x$
$F_w = -(T_{13} - A_{13})R^3q_x$ $F_\phi = -(T_{17} - A_{17})R^2q_x$ $F_\psi = -(T_{14} - A_{14})R^2q_x$		$G_w = -B_{13}R^3q_x$ $G_\phi = -B_{17}R^2q_x$ $G_\psi = B_{14}R^2q_x$
$F_W = 0$ $F_X = -T_{12}Rq_y$ $F_Y = -T_{11}Rq_y$	**3.** 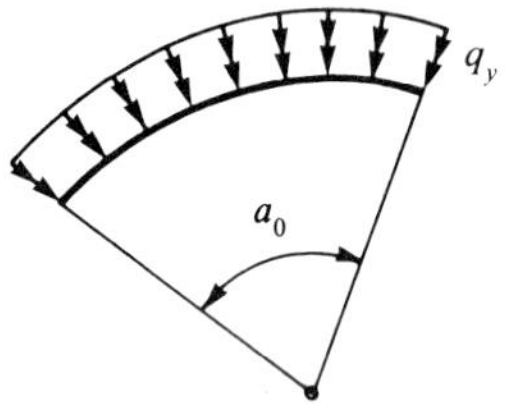	$G_W = 0$ $G_X = -T_{12}Rq_y$ $G_Y = T_{11}Rq_y$
$F_w = T_{16}R^3q_y$ $F_\phi = T_{14}R^2q_y$ $F_\psi = T_{18}R^2q_y$		$G_w = -T_{16}R^3q_y$ $G_\phi = -T_{14}R^2q_y$ $G_\psi = T_{18}R^2q_y$
$F_W = 0$ $F_X = -(T_{12} - A_{12})Rq_y$ $F_Y = -(T_{11} - A_{11})Rq_y$	**4.** 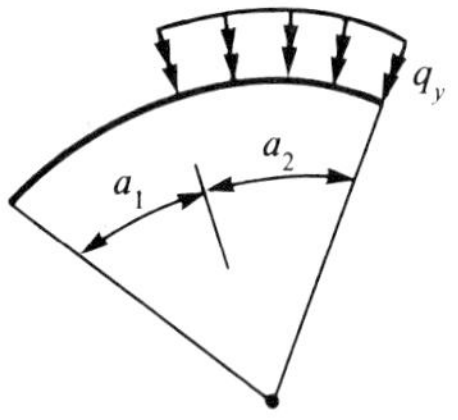	$G_W = 0$ $G_X = -B_{12}Rq_y$ $G_Y = B_{11}Rq_y$
$F_w = (T_{16} - A_{16})R^3q_y$ $F_\phi = (T_{14} - A_{14})R^2q_y$ $F_\psi = (T_{18} - A_{18})R^2q_y$		$G_w = -B_{16}R^3q_y$ $G_\phi = -B_{14}R^2q_y$ $F_\psi = B_{18}R^2q_y$

12.11 STIFFNESS MATRIX EQUATIONS IN 0 SYSTEM

(1) Signs. All forces, moments, and displacements are in the 0 system, and their positive directions are shown in (2, 3) below (Ref. 5, p. 134).

(2) Free-body diagram **(3) Elastic curve**

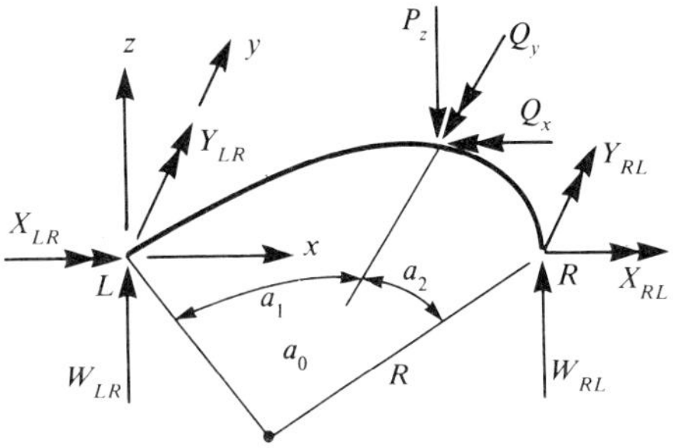

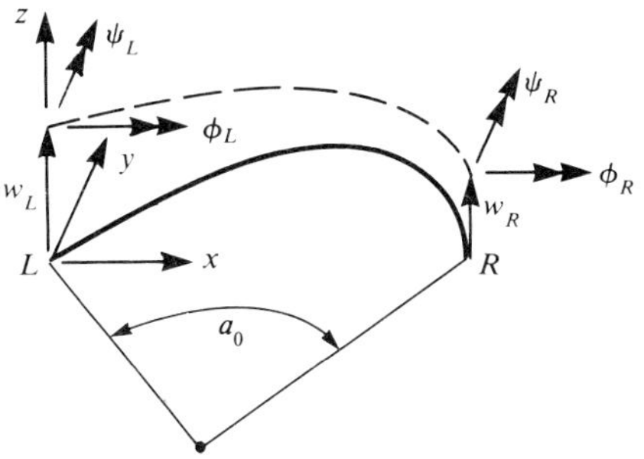

(4) Equivalents ($a_0 = 2b_0$)

$$r = \frac{EI_y}{GI_x} \qquad \alpha = \tfrac{1}{2}(1 + r) \qquad \beta = \tfrac{1}{2}(1 - r) \qquad e = \sin b_0$$

$$g = \frac{r \sin b_0}{\alpha b_0 - \beta \sin b_0 \cos b_0} - \cos b_0 \qquad F_3 = 2r\left(b_0 - \frac{r \sin^2 b_0}{\alpha b_0 - \beta \sin b_0 \cos b_0}\right)$$

$$F_4 = 2(\alpha b_0 - \beta \sin b_0 \cos b_0) \qquad F_5 = 2(\alpha b_0 + \beta \sin b_0 \cos b_0)$$

(5) Dimensionless stiffness coefficients

$$K_1 = \frac{1}{F_3} \qquad K_2 = \frac{g}{F_3} \qquad K_3 = \frac{e}{F_3} \qquad K_4 = \frac{eg}{F_3}$$

$$K_5 = \frac{g^2}{F_3} + \frac{1}{F_4} \qquad K_6 = \frac{e^2}{F_3} + \frac{1}{F_5} \qquad K_7 = \frac{e^2}{F_3} - \frac{1}{F_5}$$

Numerical values of K_j for selected values of a in degrees and $r = 1, 1.5, 2, 3$ are tabulated in (12.17–12.20) (Ref. 20).

(6) Stiffness matrix equations

$$\begin{bmatrix} W_{LR} \\ X_{LR} \\ Y_{LR} \\ \hline W_{RL} \\ X_{RL} \\ Y_{RL} \end{bmatrix} = \frac{EI_y}{R^3} \left[\begin{array}{ccc|ccc} K_1 & K_2R & -K_3R & -K_1 & -K_2R & -K_3R \\ K_2R & K_5R^2 & -K_4R^2 & -K_2R & -K_5R^2 & -K_4R^2 \\ -K_3R & -K_4R^2 & K_6R^2 & K_3R & K_4R^2 & K_7R^2 \\ \hline -K_1 & -K_2R & K_3R & K_1 & K_2R & K_3R \\ -K_2R & -K_5R^2 & K_4R^2 & K_2R & K_5R^2 & K_4R^2 \\ -K_3R & -K_4R^2 & K_7R^2 & K_3R & K_4R^2 & K_6R^2 \end{array}\right] \begin{bmatrix} w_L \\ \phi_L \\ \psi_L \\ \hline w_R \\ \phi_R \\ \psi_R \end{bmatrix} + \begin{bmatrix} W_{L0} \\ X_{L0} \\ Y_{L0} \\ \hline Y_{R0} \\ W_{R0} \\ X_{R0} \end{bmatrix}$$

The load functions $W_{L0}, X_{L0}, \ldots, Y_{R0}$ are the *reactions of fixed-end bars* due to loads and other causes. Because of the complexity of their analytical forms, only the left-side functions are tabulated in (12.16–12.20). The right-side functions can be calculated by statics.

(7) General stiffness matrix equations for free bars of order three are obtained by matrix composition of (6.09) and (12.11).

12.12 TABLE OF DIMENSIONLESS STIFFNESS COEFFICIENTS

$$r = \frac{EI_y}{GI_x} = 1.0$$

a_0	K_1	K_2	K_3	K_4	K_5	K_6	K_7
15	670.30	3.8219	87.491	0.4989	3.8415	15.240	7.6002
18	388.29	3.1857	60.743	0.4984	3.2092	12.685	6.3191
21	244.81	2.7314	44.614	0.4978	2.7588	10.859	5.4018
24	164.23	2.3908	34.146	0.4971	2.4221	9.4866	4.7119
27	115.52	2.1260	26.969	0.4963	2.1612	8.4177	4.1736
30	84.363	1.9142	21.835	0.4954	1.9533	7.5611	3.7414
33	63.505	1.7410	18.036	0.4945	1.7840	6.8588	3.3864
36	49.018	1.5968	15.147	0.4934	1.6436	6.2723	3.0892
39	38.642	1.4748	12.899	0.4923	1.5254	5.7749	2.8366
42	31.015	1.3703	11.115	0.4911	1.4247	5.3474	2.6190
45	25.283	1.2797	9.6755	0.4897	1.3380	4.9759	2.4294
48	20.892	1.2006	8.4975	0.4883	1.2627	4.6499	2.2626
51	17.470	1.1308	7.5211	0.4868	1.1966	4.3614	2.1145
54	14.764	1.0688	6.7029	0.4852	1.1384	4.1041	1.9820
57	12.596	1.0134	6.0103	0.4835	1.0867	3.8731	1.8627
60	10.838	0.9635	5.4190	0.4818	1.0406	3.6644	1.7546
63	9.3973	0.9185	4.9101	0.4799	0.9992	3.4750	1.6561
66	8.2053	0.8776	4.4689	0.4780	0.9620	3.3021	1.5658
69	7.2104	0.8402	4.0840	0.4759	0.9283	3.1436	1.4828
72	6.3733	0.8061	3.7462	0.4738	0.8977	2.9977	1.4062
75	5.6639	0.7746	3.4480	0.4716	0.8699	2.8629	1.3351
78	5.0586	0.7457	3.1835	0.4693	0.8445	2.7380	1.2689
81	4.5389	0.7189	2.9478	0.4669	0.8212	2.6218	1.2071
84	4.0902	0.6940	2.7369	0.4644	0.7998	2.5134	1.1492
87	3.7006	0.6709	2.5473	0.4618	0.7802	2.4120	1.0949
90	3.3607	0.6493	2.3764	0.4591	0.7621	2.3170	1.0437
93	3.0628	0.6292	2.2217	0.4564	0.7453	2.2276	0.9954
96	2.8005	0.6103	2.0812	0.4536	0.7298	2.1434	0.9498
99	2.5686	0.5926	1.9532	0.4506	0.7155	2.0640	0.9065
102	2.3629	0.5760	1.8363	0.4476	0.7021	1.9888	0.8654
105	2.1797	0.5603	1.7292	0.4445	0.6897	1.9176	0.8262
108	2.0159	0.5455	1.6309	0.4413	0.6781	1.8500	0.7889
111	1.8692	0.5316	1.5404	0.4381	0.6673	1.7857	0.7533
114	1.7372	0.5184	1.4569	0.4347	0.6573	1.7245	0.7193
117	1.6182	0.5058	1.3797	0.4313	0.6478	1.6661	0.6867
120	1.5106	0.4939	1.3082	0.4278	0.6390	1.6104	0.6555
123	1.4130	0.4827	1.2418	0.4242	0.6307	1.5571	0.6255
126	1.3243	0.4719	1.1800	0.4205	0.6229	1.5061	0.5967
129	1.2436	0.4617	1.1224	0.4167	0.6156	1.4572	0.5689
132	1.1698	0.4519	1.0687	0.4129	0.6087	1.4103	0.5422
135	1.1023	0.4426	1.0184	0.4089	0.6021	1.3653	0.5165
138	1.0405	0.4337	0.9714	0.4049	0.5960	1.3220	0.4917
141	0.9837	0.4252	0.9272	0.4008	0.5902	1.2804	0.4677
144	0.9314	0.4171	0.8858	0.3967	0.5847	1.2403	0.4445
147	0.8831	0.4093	0.8468	0.3924	0.5794	1.2017	0.4221
150	0.8386	0.4018	0.8100	0.3881	0.5745	1.1644	0.4005
153	0.7974	0.3946	0.7754	0.3837	0.5697	1.1284	0.3795
156	0.7592	0.3877	0.7427	0.3792	0.5652	1.0937	0.3591
159	0.7238	0.3810	0.7117	0.3746	0.5609	1.0601	0.3394
162	0.6909	0.3746	0.6824	0.3700	0.5568	1.0277	0.3203
165	0.6603	0.3685	0.6546	0.3653	0.5529	0.9963	0.3018
168	0.6318	0.3625	0.6283	0.3605	0.5491	0.9659	0.2838
171	0.6051	0.3568	0.6033	0.3557	0.5454	0.9365	0.2664
174	0.5803	0.3513	0.5795	0.3508	0.5419	0.9080	0.2494
177	0.5570	0.3459	0.5568	0.3458	0.5385	0.8803	0.2329
180	0.5352	0.3407	0.5352	0.3407	0.5352	0.8535	0.2169

12.13 TABLE OF DIMENSIONLESS STIFFNESS COEFFICIENTS

$$r = \frac{EI_y}{GI_x} = 1.5$$

a_0	K_1	K_2	K_3	K_4	K_5	K_6	K_7
15	668.64	5.0799	87.276	0.6631	2.5899	15.201	7.5829
18	386.92	4.2285	60.527	0.6615	2.1741	12.639	6.2984
21	243.63	3.6196	44.399	0.6596	1.8795	10.804	5.3777
24	163.20	3.1623	33.931	0.6575	1.6606	9.4248	4.6845
27	114.61	2.8061	26.755	0.6551	1.4921	8.3485	4.1429
30	83.539	2.5206	21.621	0.6524	1.3589	7.4846	3.7075
33	62.756	2.2865	17.824	0.6494	1.2514	6.7751	3.3493
36	48.332	2.0911	14.935	0.6462	1.1630	6.1816	3.0490
39	38.009	1.9254	12.688	0.6427	1.0894	5.6772	2.7934
42	30.428	1.7830	10.905	0.6390	1.0273	5.2429	2.5728
45	24.736	1.6593	9.4662	0.6350	0.9745	4.8648	2.3803
48	20.380	1.5508	8.2891	0.6308	0.9290	4.5324	2.2106
51	16.989	1.4548	7.3138	0.6263	0.8897	4.2376	2.0598
54	14.310	1.3692	6.4967	0.6216	0.8555	3.9742	1.9247
57	12.166	1.2924	5.8053	0.6167	0.8254	3.7373	1.8028
60	10.430	1.2230	5.2152	0.6115	0.7989	3.5229	1.6923
63	9.0098	1.1601	4.7076	0.6062	0.7755	3.3280	1.5914
66	7.8360	1.1028	4.2678	0.6006	0.7546	3.1498	1.4990
69	6.8577	1.0502	3.8842	0.5948	0.7360	2.9863	1.4139
72	6.0359	1.0019	3.5478	0.5889	0.7192	2.8355	1.3352
75	5.3406	0.9573	3.2511	0.5828	0.7042	2.6961	1.2622
78	4.7483	0.9160	2.9882	0.5765	0.6905	2.5668	1.1943
81	4.2407	0.8777	2.7541	0.5700	0.6782	2.4465	1.1309
84	3.8032	0.8419	2.5448	0.5634	0.6669	2.3341	1.0715
87	3.4241	0.8086	2.3570	0.5566	0.6566	2.2291	1.0158
90	3.0940	0.7774	2.1878	0.5497	0.6471	2.1306	0.9634
93	2.8052	0.7481	2.0348	0.5426	0.6384	2.0380	0.9140
96	2.5515	0.7205	1.8961	0.5355	0.6303	1.9509	0.8673
99	2.3278	0.6946	1.7700	0.5282	0.6227	1.8687	0.8232
102	2.1297	0.6701	1.6551	0.5208	0.6157	1.7911	0.7814
105	1.9537	0.6470	1.5500	0.5133	0.6091	1.7176	0.7417
108	1.7968	0.6250	1.4536	0.5057	0.6029	1.6481	0.7040
111	1.6565	0.6043	1.3652	0.4980	0.5971	1.5820	0.6681
114	1.5306	0.5846	1.2837	0.4903	0.5915	1.5193	0.6339
117	1.4175	0.5658	1.2086	0.4824	0.5862	1.4597	0.6013
120	1.3154	0.5480	1.1392	0.4746	0.5811	1.4030	0.5701
123	1.2231	0.5310	1.0749	0.4666	0.5762	1.3489	0.5404
126	1.1395	0.5148	1.0153	0.4587	0.5714	1.2973	0.5120
129	1.0636	0.4993	0.9600	0.4507	0.5668	1.2481	0.4848
132	0.9944	0.4845	0.9084	0.4427	0.5623	1.2011	0.4587
135	0.9313	0.4704	0.8605	0.4346	0.5579	1.1562	0.4337
138	0.8737	0.4569	0.8157	0.4265	0.5536	1.1132	0.4098
141	0.8209	0.4439	0.7738	0.4185	0.5493	1.0721	0.3869
144	0.7725	0.4315	0.7347	0.4104	0.5451	1.0327	0.3648
147	0.7280	0.4196	0.6980	0.4023	0.5410	0.9949	0.3437
150	0.6871	0.4082	0.6637	0.3943	0.5368	0.9588	0.3233
153	0.6493	0.3972	0.6314	0.3863	0.5327	0.9241	0.3038
156	0.6145	0.3867	0.6011	0.3782	0.5286	0.8908	0.2851
159	0.5823	0.3766	0.5725	0.3703	0.5246	0.8589	0.2670
162	0.5524	0.3668	0.5456	0.3623	0.5205	0.8282	0.2497
165	0.5248	0.3575	0.5203	0.3544	0.5164	0.7987	0.2330
168	0.4991	0.3485	0.4964	0.3465	0.5123	0.7704	0.2169
171	0.4753	0.3398	0.4738	0.3387	0.5082	0.7432	0.2015
174	0.4531	0.3314	0.4525	0.3310	0.5040	0.7171	0.1866
177	0.4324	0.3234	0.4323	0.3233	0.4999	0.6920	0.1723
180	0.4131	0.3156	0.4131	0.3156	0.4958	0.6678	0.1585

12.14 TABLE OF DIMENSIONLESS STIFFNESS COEFFICIENTS

$$r = \frac{EI_y}{GI_x} = 2.0$$

a_0	K_1	K_2	K_3	K_4	K_5	K_6	K_7
15	667.63	5.7067	87.143	0.7449	1.9641	15.173	7.5763
18	386.07	4.7473	60.395	0.7426	1.6565	12.605	6.2905
21	242.91	4.0607	44.267	0.7400	1.4397	10.765	5.3686
24	162.57	3.5446	33.799	0.7370	1.2797	9.3805	4.6740
27	114.04	3.1423	26.623	0.7335	1.1574	8.2990	4.1311
30	83.033	2.8195	21.491	0.7297	1.0616	7.4300	3.6944
33	62.297	2.5547	17.693	0.7256	0.9848	6.7155	3.3349
36	47.912	2.3333	14.806	0.7210	0.9224	6.1170	3.0334
39	37.623	2.1453	12.559	0.7161	0.8710	5.6079	2.7765
42	30.070	1.9836	10.776	0.7109	0.8281	5.1691	2.5547
45	24.403	1.8429	9.3385	0.7053	0.7921	4.7865	2.3609
48	20.068	1.7194	8.1623	0.6993	0.7615	4.4497	2.1900
51	16.696	1.6099	7.1877	0.6931	0.7354	4.1508	2.0380
54	14.034	1.5122	6.3714	0.6865	0.7129	3.8834	1.9017
57	11.906	1.4244	5.6810	0.6796	0.6935	3.6428	1.7787
60	10.184	1.3450	5.0918	0.6725	0.6767	3.4248	1.6670
63	8.7754	1.2728	4.5851	0.6651	0.6620	3.2264	1.5651
66	7.6130	1.2070	4.1463	0.6574	0.6491	3.0450	1.4715
69	6.6452	1.1466	3.7639	0.6494	0.6377	2.8784	1.3854
72	5.8330	1.0909	3.4285	0.6412	0.6277	2.7248	1.3057
75	5.1465	1.0395	3.1330	0.6328	0.6187	2.5827	1.2318
78	4.5624	0.9918	2.8712	0.6242	0.6107	2.4508	1.1630
81	4.0624	0.9475	2.6383	0.6153	0.6035	2.3282	1.0987
84	3.6319	0.9061	2.4302	0.6063	0.5969	2.2137	1.0385
87	3.2594	0.8675	2.2436	0.5971	0.5910	2.1067	0.9821
90	2.9355	0.8312	2.0757	0.5878	0.5855	2.0065	0.9290
93	2.6525	0.7972	1.9240	0.5783	0.5804	1.9123	0.8790
96	2.4043	0.7652	1.7867	0.5686	0.5757	1.8238	0.8318
99	2.1856	0.7350	1.6620	0.5589	0.5712	1.7404	0.7871
102	1.9924	0.7065	1.5484	0.5490	0.5670	1.6618	0.7449
105	1.8209	0.6795	1.4447	0.5391	0.5630	1.5874	0.7048
108	1.6684	0.6539	1.3498	0.5291	0.5591	1.5172	0.6668
111	1.5322	0.6297	1.2627	0.5190	0.5553	1.4506	0.6307
114	1.4103	0.6067	1.1827	0.5088	0.5516	1.3875	0.5963
117	1.3008	0.5848	1.1091	0.4986	0.5479	1.3277	0.5636
120	1.2022	0.5640	1.0412	0.4884	0.5443	1.2709	0.5325
123	1.1133	0.5441	0.9784	0.4782	0.5407	1.2169	0.5028
126	1.0329	0.5252	0.9203	0.4679	0.5371	1.1656	0.4745
129	0.9600	0.5071	0.8665	0.4577	0.5334	1.1167	0.4475
132	0.8938	0.4898	0.8165	0.4475	0.5297	1.0702	0.4217
135	0.8336	0.4733	0.7701	0.4373	0.5260	1.0259	0.3971
138	0.7786	0.4575	0.7269	0.4272	0.5222	0.9836	0.3736
141	0.7284	0.4424	0.6866	0.4171	0.5184	0.9434	0.3511
144	0.6824	0.4280	0.6490	0.4070	0.5145	0.9050	0.3296
147	0.6403	0.4141	0.6140	0.3970	0.5105	0.8683	0.3090
150	0.6017	0.4008	0.5812	0.3872	0.5064	0.8333	0.2894
153	0.5661	0.3881	0.5505	0.3773	0.5023	0.7999	0.2706
156	0.5334	0.3758	0.5217	0.3676	0.4981	0.7680	0.2526
159	0.5032	0.3641	0.4947	0.3580	0.4938	0.7375	0.2354
162	0.4753	0.3528	0.4694	0.3485	0.4894	0.7084	0.2190
165	0.4495	0.3420	0.4457	0.3391	0.4850	0.6805	0.2032
168	0.4257	0.3316	0.4233	0.3298	0.4805	0.6539	0.1881
171	0.4035	0.3216	0.4023	0.3206	0.4759	0.6284	0.1737
174	0.3830	0.3120	0.3825	0.3116	0.4712	0.6041	0.1599
177	0.3640	0.3028	0.3639	0.3027	0.4665	0.5808	0.1467
180	0.3463	0.2939	0.3463	0.2939	0.4617	0.5585	0.1341

12.15 TABLE OF DIMENSIONLESS STIFFNESS COEFFICIENTS

$$r = \frac{EI_y}{GI_x} = 3.0$$

a_0	K_1	K_2	K_3	K_4	K_5	K_6	K_7
15	666.23	6.3293	86.961	0.8261	1.3382	15.127	7.5739
18	384.91	5.2609	60.213	0.8230	1.1388	12.551	6.2876
21	241.92	4.4957	44.086	0.8193	0.9998	10.703	5.3650
24	161.70	3.9201	33.619	0.8150	0.8986	9.3099	4.6697
27	113.28	3.4708	26.444	0.8103	0.8224	8.2203	4.1261
30	82.344	3.1100	21.312	0.8049	0.7638	7.3435	3.6885
33	61.672	2.8136	17.516	0.7991	0.7178	6.6215	3.3281
36	47.341	2.5655	14.629	0.7928	0.6812	6.0158	3.0256
39	37.098	2.3546	12.383	0.7860	0.6518	5.4997	2.7676
42	29.585	2.1728	10.602	0.7787	0.6279	5.0543	2.5447
45	23.952	2.0145	9.1658	0.7709	0.6084	4.6654	2.3498
48	19.647	1.8753	7.9910	0.7627	0.5924	4.3228	2.1777
51	16.301	1.7517	7.0179	0.7541	0.5793	4.0182	2.0244
54	13.664	1.6413	6.2032	0.7451	0.5683	3.7456	1.8868
57	11.557	1.5419	5.5143	0.7357	0.5593	3.5000	1.7624
60	9.8538	1.4519	4.9269	0.7259	0.5517	3.2775	1.6494
63	8.4631	1.3700	4.4220	0.7158	0.5454	3.0749	1.5460
66	7.3168	1.2951	3.9850	0.7054	0.5401	2.8897	1.4511
69	6.3636	1.2264	3.6044	0.6946	0.5356	2.7195	1.3636
72	5.5650	1.1630	3.2710	0.6836	0.5317	2.5628	1.2825
75	4.8910	1.1044	2.9774	0.6723	0.5284	2.4178	1.2072
78	4.3184	1.0500	2.7177	0.6608	0.5255	2.2835	1.1371
81	3.8292	0.9993	2.4869	0.6490	0.5229	2.1586	1.0715
84	3.4088	0.9520	2.2809	0.6370	0.5206	2.0423	1.0101
87	3.0456	0.9078	2.0965	0.6249	0.5184	1.9337	0.9525
90	2.7305	0.8664	1.9307	0.6126	0.5163	1.8322	0.8983
93	2.4558	0.8274	1.7813	0.6002	0.5143	1.7370	0.8472
96	2.2153	0.7907	1.6463	0.5876	0.5124	1.6478	0.7991
99	2.0040	0.7562	1.5238	0.5750	0.5104	1.5639	0.7536
102	1.8176	0.7235	1.4125	0.5623	0.5083	1.4850	0.7105
105	1.6527	0.6926	1.3112	0.5495	0.5062	1.4107	0.6698
108	1.5063	0.6634	1.2186	0.5367	0.5040	1.3406	0.6311
111	1.3760	0.6357	1.1340	0.5239	0.5017	1.2745	0.5945
114	1.2596	0.6094	1.0564	0.5111	0.4992	1.2121	0.5598
117	1.1554	0.5844	0.9851	0.4983	0.4966	1.1531	0.5268
120	1.0619	0.5606	0.9196	0.4855	0.4938	1.0973	0.4954
123	0.9777	0.5380	0.8592	0.4728	0.4909	1.0446	0.4657
126	0.9019	0.5165	0.8036	0.4602	0.4878	0.9946	0.4374
129	0.8333	0.4959	0.7521	0.4476	0.4845	0.9473	0.4105
132	0.7713	0.4764	0.7046	0.4352	0.4811	0.9024	0.3849
135	0.7150	0.4577	0.6605	0.4228	0.4775	0.8599	0.3606
138	0.6638	0.4398	0.6197	0.4106	0.4737	0.8196	0.3375
141	0.6172	0.4228	0.5818	0.3986	0.4698	0.7814	0.3155
144	0.5747	0.4065	0.5466	0.3866	0.4657	0.7451	0.2946
147	0.5359	0.3910	0.5139	0.3749	0.4614	0.7107	0.2747
150	0.5004	0.3761	0.4834	0.3633	0.4570	0.6781	0.2558
153	0.4679	0.3619	0.4550	0.3519	0.4524	0.6471	0.2378
156	0.4381	0.3483	0.4285	0.3407	0.4477	0.6176	0.2207
159	0.4107	0.3353	0.4039	0.3296	0.4429	0.5897	0.2045
162	0.3856	0.3228	0.3808	0.3188	0.4379	0.5632	0.1891
165	0.3624	0.3109	0.3593	0.3082	0.4328	0.5380	0.1744
168	0.3410	0.2994	0.3391	0.2978	0.4277	0.5141	0.1605
171	0.3213	0.2885	0.3203	0.2876	0.4224	0.4913	0.1473
174	0.3031	0.2780	0.3026	0.2777	0.4170	0.4698	0.1347
177	0.2862	0.2680	0.2861	0.2679	0.4115	0.4493	0.1228
180	0.2706	0.2584	0.2706	0.2584	0.4059	0.4298	0.1115

12.16 STIFFNESS LOAD FUNCTIONS, FIXED-END BAR IN 0 SYSTEM

Notation (12.11) Signs (12.11) Matrix (12.11)

P_z = concentrated load (N)

p_z = intensity of distributed load (N/m)

$K_1, K_2, \ldots, K_7$ = stiffness coefficients (12.11–12.15)

$B_1, B_2, \ldots, B_{18}$ = transport coefficients in a_2 (12.03–12.07)

$T_1, T_2, \ldots, T_{18}$ = transport coefficients in a_0 (12.03–12.07)

$$r = \frac{EI_y}{GI_x} \qquad a_0 = 2b_0$$

$$\alpha = \tfrac{1}{2}(1 + r) \qquad S_0 = \sin b_0$$

$$\beta = \tfrac{1}{2}(1 - r) \qquad C_0 = \cos b_0$$

$$G_{11} = -B_5 \qquad G_{12} = -B_3C_0 + B_6S_0 \qquad G_{13} = B_3S_0 + B_6C_0$$

$$G_{14} = -B_{15} \qquad G_{15} = -B_{13}C_0 + B_{16}S_0 \qquad G_{16} = B_{13}S_0 + B_{16}C_0$$

1.

$$W_{L0} = (K_1G_{11} + K_2G_{12} + K_3G_{13})P_z$$

$$X_{L0} = (K_2G_{11} + K_5G_{12} + K_4G_{13})RP_z$$

$$Y_{L0} = -(K_3G_{11} + K_4G_{12} + K_7G_{13})RP_z$$

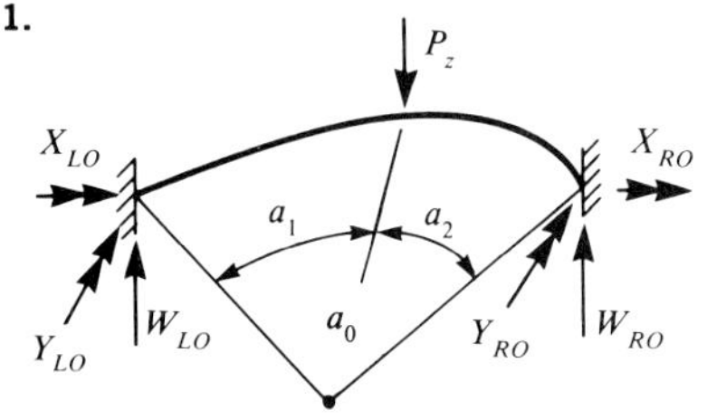

2.

$$W_{L0} = \tfrac{1}{2}P_z \qquad a_1 = a_2 = b_0$$

$$X_{L0} = \tfrac{1}{2}(1 - C_0)RP_z$$

$$Y_{L0} = -\tfrac{1}{2}\left[S_0 - \frac{(1 - C_0)r + \beta S_0^2}{\alpha b_0 + \beta S_0C_0}\right]RP_z$$

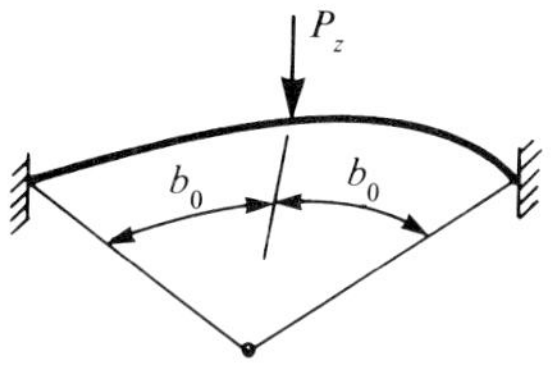

3. (Table 12.20)

$$W_{L0} = b_0Rp_z$$

$$X_{L0} = (S_0 - b_0C_0)R^2p_z$$

$$Y_{L0} = -\frac{\alpha(b^2S_0 + 2b_0C_0 - 2S_0) + \beta(S_0 - b_0C_0)C_0^2}{\alpha b_0 + \beta S_0C_0}R^2p_z$$

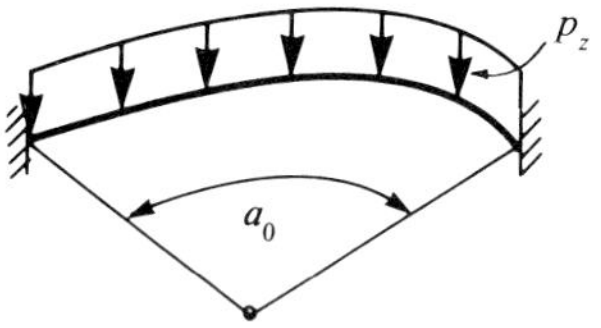

4.

$$W_{L0} = (K_1G_{14} + K_2G_{15} + K_3G_{16})Rp_z$$

$$X_{L0} = (K_2G_{14} + K_5G_{15} + K_4G_{16})R^2p_z$$

$$Y_{L0} = -(K_3G_{14} + K_4G_{15} + K_7G_{16})R^2p_z$$

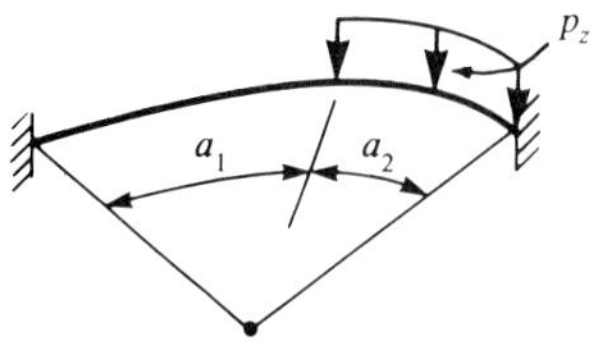

12.17 STIFFNESS LOAD FUNCTIONS, FIXED-END BAR IN 0 SYSTEM

Notation (12.11, 12.16) | Signs (12.11) | Matrix (12.11)

Q_t = applied couple (N·m) $\qquad$ q_t = intensity of distributed couple (N·m/m)

$G_{21} = -B_3 \qquad G_{22} = -B_7C_0 + B_4S_0 \qquad G_{23} = B_7S_0 + B_4C_0$

$G_{24} = -T_{13} \qquad G_{25} = -T_{17}C_0 + T_{14}S_0 \qquad G_{26} = T_{17}S_0 + T_{14}C_0$

$G_{27} = -B_{13} \qquad G_{28} = -B_{17}C_0 + B_{14}S_0 \qquad G_{29} = B_{17}S_0 + B_{14}C_0$

1.

$$W_{LO} = (K_1G_{21} + K_2G_{22} + K_3G_{23})\frac{Q_t}{R}$$

$$X_{LO} = (K_2G_{21} + K_5G_{22} + K_4G_{23})Q_t$$

$$Y_{LO} = -(K_3G_{21} + K_4G_{22} + K_7G_{23})Q_t$$

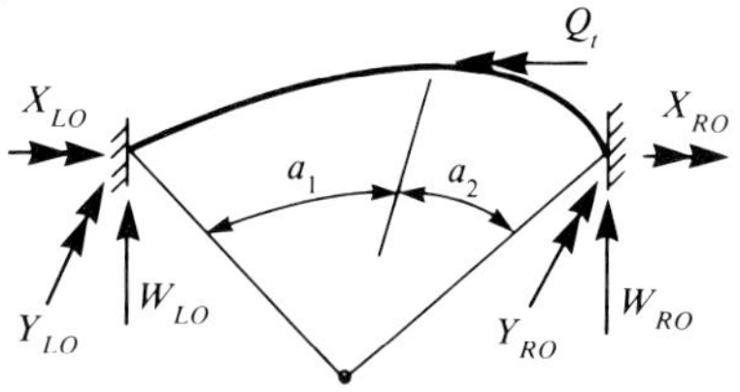

2.

$$W_{LO} = 0 \qquad a_1 = a_2 = b_0$$

$$X_{LO} = \tfrac{1}{2}Q_t$$

$$Y_{LO} = -\frac{\beta S_0^2 Q_t}{2b_0(\alpha b_0 + \beta S_0C_0)}$$

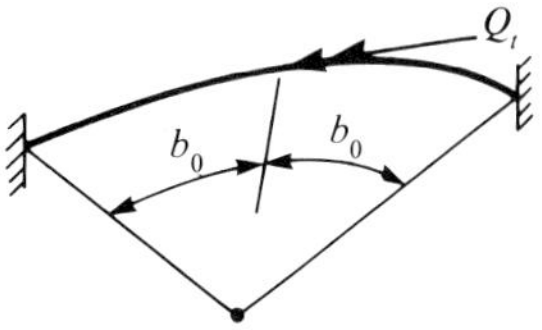

3. (Table 12.20)

$$W_{LO} = 0$$

$$X_{LO} = S_0Rq_t$$

$$Y_{LO} = \frac{\alpha(S_0 - b_0C_0) + \beta S_0^3}{\alpha b_0 + \beta S_0C_0}Rq_t$$

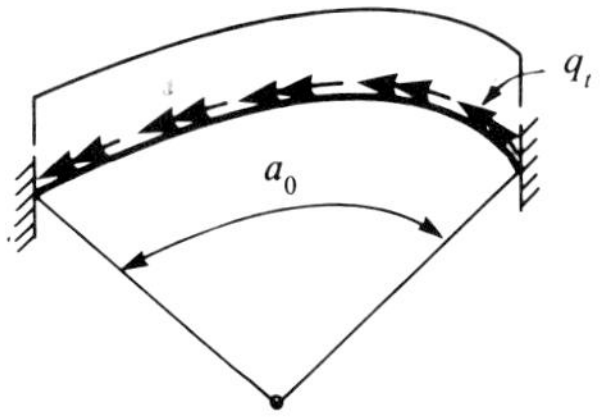

4.

$$W_{LO} = (K_1G_{27} + K_2G_{28} + K_3G_{29})q_t$$

$$X_{LO} = (K_2G_{27} + K_5G_{28} + K_4G_{29})Rq_t$$

$$Y_{LO} = -(K_3G_{27} + K_4G_{28} + K_7G_{29})Rq_t$$

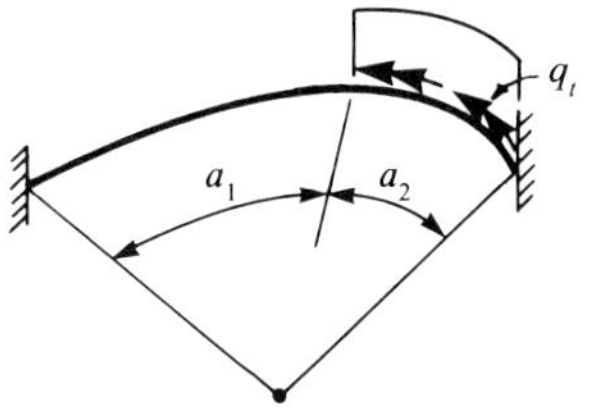

12.18 STIFFNESS LOAD FUNCTIONS, FIXED-END BAR IN 0 SYSTEM

Notation (12.11) Signs (12.11) Matrix (12.11)

w_j, ϕ_j, ψ_j = abrupt displacements (11.03)
g_t = angular thermal strain (11.03)
$K_1, K_2, \ldots, K_7$ = stiffness coefficient (12.11–12.15)
$B_1, B_2, \ldots, B_{18}$ = transport coefficients in a_2 (12.03–12.07)
$T_1, T_2, \ldots, T_{18}$ = transport coefficients in a_0 (12.03–12.07)

$r = \dfrac{EI_y}{GI_x}$ $\qquad a_0 = 2b_0$

$\alpha = \frac{1}{2}(1 + r)$ $\qquad S_0 = \sin b_0$

$\beta = \frac{1}{2}(1 - r)$ $\qquad C_0 = \cos b_0$

$G_{41} = w_j - B_0 R\phi_j$ $\qquad G_{42} = (B_1 C_0 + B_2 S_0) R\phi_j$ $\qquad G_{43} = -(B_1 S_0 - B_2 C_0) R\phi_j$

$G_{44} = -B_2 \psi_j$ $\qquad G_{45} = -(B_2 C_0 - B_1 S_0)\psi_j$ $\qquad G_{46} = (B_2 S_0 + B_1 C_0)\psi_j$

$G_{47} = -B_{12} g_t$ $\qquad G_{48} = -(B_{12} C_0 - B_{11} S_0) g_t$ $\qquad G_{49} = (B_{12} S_0 + B_{11} C_0) g_t$

1. Abrupt displacement w_j, ϕ_j

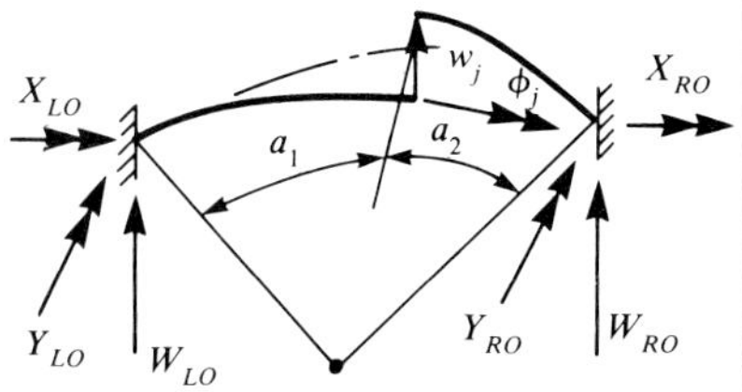

$$W_{L0} = (K_1 G_{41} + K_2 G_{42} + K_3 G_{43}) R^{-3} EI_y$$

$$X_{L0} = (K_2 G_{41} + K_5 G_{42} + K_4 G_{43}) R^{-2} EI_y$$

$$Y_{L0} = -(K_3 G_{41} + K_4 G_{42} + K_7 G_{43}) R^{-2} EI_y$$

2. Abrupt displacement ψ_j

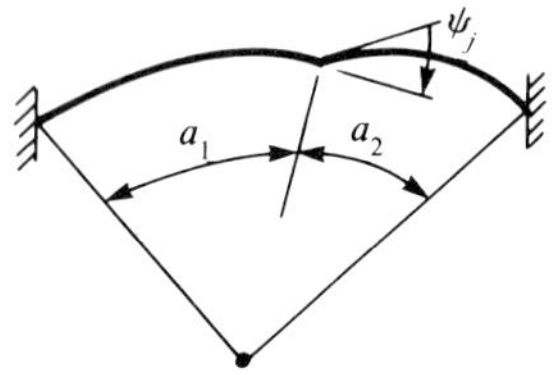

$$W_{L0} = (K_1 G_{44} + K_2 G_{45} + K_3 G_{46}) R^{-2} EI_y$$

$$X_{L0} = (K_2 G_{44} + K_5 G_{45} + K_4 G_{46}) R^{-1} EI_y$$

$$Y_{L0} = -(K_3 G_{44} + K_4 G_{45} + K_7 G_{46}) R^{-1} EI_y$$

3. Temperature deformation of segment Ra_2

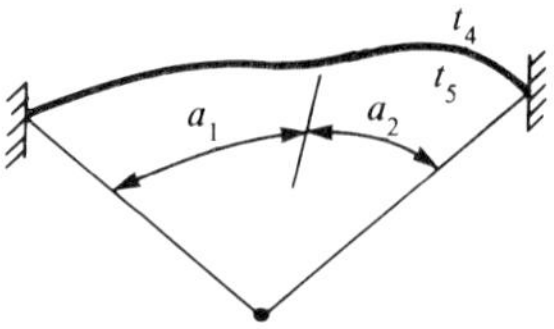

$$W_{L0} = (K_1 G_{47} + K_2 G_{48} + K_3 G_{49}) R^{-1} EI_y$$

$$X_{L0} = (K_2 G_{47} + K_5 G_{48} + K_4 G_{49}) EI_y$$

$$Y_{L0} = -(K_3 G_{47} + K_4 G_{48} + K_7 G_{49}) EI_y$$

12.19 STIFFNESS LOAD FUNCTIONS, FIXED-END BAR IN 0 SYSTEM

Notation (12.11, 12.18)	Signs (12.11)	Matrix (12.11)
Q_n = applied couple (N·m)		q_n = intensity of distributed couple (N·m/m)
$G_{31} = -B_6$	$G_{32} = -B_4C_0 + B_8S_0$	$G_{33} = B_4S_0 + B_8C_0$
$G_{34} = -T_{16}$	$G_{35} = -T_{14}C_0 + T_{18}S_0$	$G_{36} = T_{14}S_0 + T_{18}C_0$
$G_{37} = -B_{16}$	$G_{38} = -B_{14}C_0 + B_{18}S_0$	$G_{39} = B_{14}S_0 + B_{18}C_0$

1.

$$W_{L0} = (K_1G_{31} + K_2G_{32} + K_3G_{33})\frac{Q_n}{R}$$

$$X_{L0} = (K_2G_{31} + K_5G_{32} + K_4G_{33})Q_n$$

$$Y_{L0} = -(K_3G_{31} + K_4G_{32} + K_7G_{33})Q_n$$

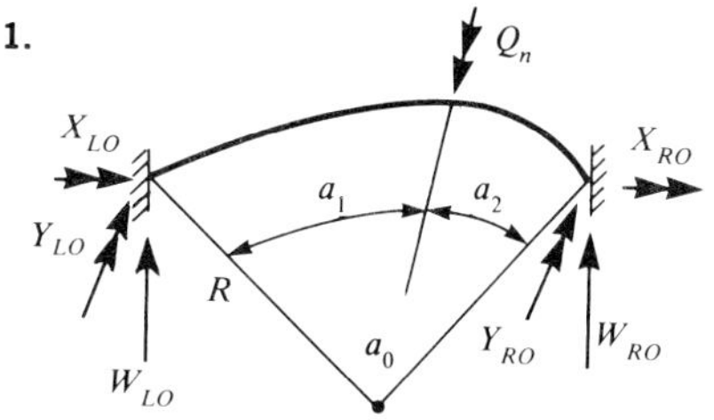

2.

$$W_{L0} = (K_1G_{31} + K_2G_{32} + K_3G_{33})\frac{Q_n}{R}$$

$$X_{L0} = 0 \qquad a_1 = a_2 = b_0$$

$$Y_{L0} = -S_0RW_{L0} + \tfrac{1}{2}Q_n$$

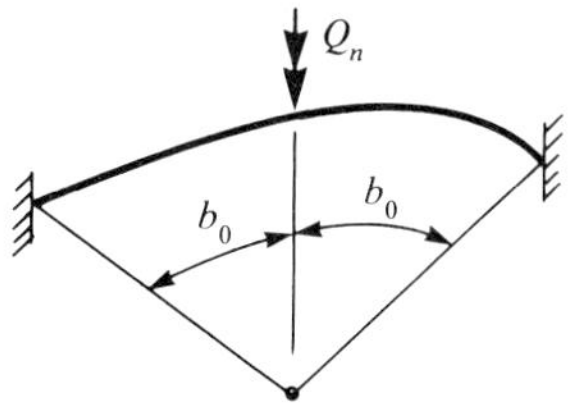

3.

$$W_{L0} = (K_1G_{34} + K_2G_{35} + K_3G_{36})q_n$$

$$X_{L0} = 0$$

$$Y_{L0} = -(K_3G_{34} + K_4G_{35} + K_7G_{36})Rq_n$$

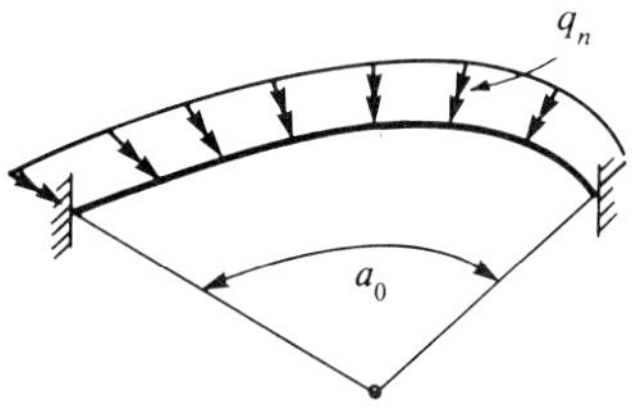

4.

$$W_{L0} = (K_1G_{37} + K_2G_{38} + K_3G_{39})q_n$$

$$X_{L0} = (K_2G_{37} + K_5G_{38} + K_4G_{39})Rq_n$$

$$Y_{L0} = -(K_3G_{37} + K_4G_{38} + K_7G_{39})Rq_n$$

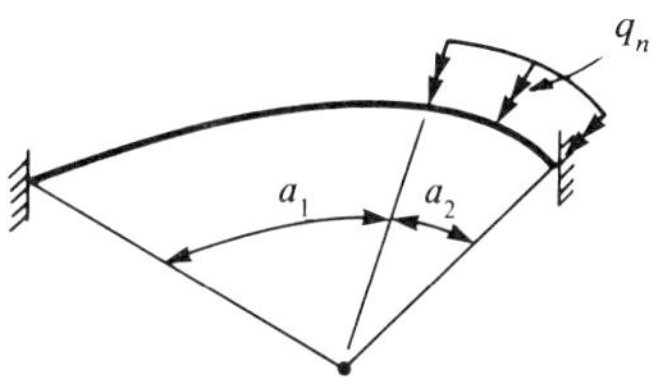

12.20 UNIFORMLY DISTRIBUTED LOAD STIFFNESS COEFFICIENTS IN 0 SYSTEM (Ref. 20)

Case 12.16-3: $X_{L0} = (\sin b_0 - b_0 \cos b_0)R^2 p_z$ $Y_{L0} = -C_{11}R^2 p_z$

Case 12.17-3: $X_{L0} = \sin b_0 R q_t$ $Y_{L0} = C_{22} R q_t$

	$r = 1.0$		$r = 1.5$		$r = 2.0$		$r = 3.0$	
a_0	C_{11}	C_{22}	C_{11}	C_{22}	C_{11}	C_{22}	C_{11}	C_{22}
45	0.0490	0.0506	0.0493	0.0269	0.0495	0.0044	0.0499	−0.0377
48	0.0554	0.0575	0.0557	0.0308	0.0560	0.0056	0.0565	−0.0411
51	0.0621	0.0647	0.0625	0.0350	0.0629	0.0070	0.0635	−0.0442
54	0.0692	0.0724	0.0696	0.0395	0.0700	0.0087	0.0708	−0.0471
57	0.0764	0.0805	0.0770	0.0442	0.0775	0.0106	0.0784	−0.0497
60	0.0840	0.0889	0.0847	0.0493	0.0853	0.0129	0.0864	−0.0519
63	0.0918	0.0977	0.0926	0.0547	0.0933	0.0154	0.0946	−0.0538
66	0.0998	0.1070	0.1007	0.0604	0.1016	0.0183	0.1030	−0.0552
69	0.1080	0.1165	0.1091	0.0665	0.1100	0.0215	0.1117	−0.0561
72	0.1164	0.1265	0.1176	0.0729	0.1187	0.0251	0.1206	−0.0565
75	0.1249	0.1368	0.1263	0.0796	0.1276	0.0290	0.1297	−0.0564
78	0.1336	0.1474	0.1352	0.0866	0.1366	0.0333	0.1390	−0.0557
81	0.1423	0.1584	0.1441	0.0940	0.1457	0.0381	0.1484	−0.0544
84	0.1512	0.1697	0.1532	0.1017	0.1550	0.0432	0.1578	−0.0525
87	0.1600	0.1813	0.1623	0.1098	0.1642	0.0488	0.1674	−0.0499
90	0.1689	0.1932	0.1715	0.1182	0.1736	0.0548	0.1770	−0.0468
93	0.1778	0.2054	0.1806	0.1270	0.1829	0.0612	0.1866	−0.0429
96	0.1867	0.2179	0.1897	0.1361	0.1922	0.0681	0.1961	−0.0384
99	0.1955	0.2307	0.1988	0.1456	0.2015	0.0754	0.2056	−0.0333
102	0.2042	0.2438	0.2077	0.1554	0.2106	0.0832	0.2150	−0.0274
105	0.2128	0.2571	0.2166	0.1655	0.2197	0.0915	0.2243	−0.0209
108	0.2213	0.2706	0.2253	0.1760	0.2285	0.1002	0.2334	−0.0138
111	0.2295	0.2844	0.2338	0.1868	0.2372	0.1093	0.2423	−0.0060
114	0.2376	0.2984	0.2421	0.1980	0.2457	0.1189	0.2509	0.0025
117	0.2454	0.3126	0.2501	0.2094	0.2539	0.1290	0.2593	0.0116
120	0.2529	0.3270	0.2579	0.2212	0.2617	0.1395	0.2673	0.0213
123	0.2601	0.3416	0.2653	0.2333	0.2693	0.1504	0.2750	0.0316
126	0.2670	0.3563	0.2724	0.2458	0.2764	0.1617	0.2822	0.0425
129	0.2736	0.3713	0.2791	0.2585	0.2832	0.1735	0.2890	0.0540
132	0.2797	0.3863	0.2853	0.2715	0.2895	0.1857	0.2954	0.0660
135	0.2854	0.4015	0.2911	0.2847	0.2953	0.1982	0.3011	0.0786
138	0.2906	0.4169	0.2963	0.2983	0.3006	0.2112	0.3064	0.0918
141	0.2953	0.4323	0.3011	0.3121	0.3053	0.2245	0.3110	0.1054
144	0.2995	0.4478	0.3052	0.3262	0.3094	0.2382	0.3150	0.1195
147	0.3032	0.4634	0.3088	0.3404	0.3128	0.2522	0.3182	0.1341
150	0.3062	0.4791	0.3117	0.3550	0.3156	0.2666	0.3208	0.1491
153	0.3086	0.4948	0.3139	0.3697	0.3176	0.2812	0.3226	0.1645
156	0.3104	0.5106	0.3154	0.3846	0.3189	0.2962	0.3235	0.1803
159	0.3115	0.5264	0.3162	0.3997	0.3194	0.3114	0.3237	0.1965
162	0.3119	0.5422	0.3161	0.4150	0.3191	0.3269	0.3229	0.2131
165	0.3115	0.5580	0.3153	0.4304	0.3179	0.3427	0.3212	0.2299
168	0.3104	0.5738	0.3136	0.4460	0.3158	0.3587	0.3185	0.2471
171	0.3085	0.5896	0.3110	0.4617	0.3127	0.3748	0.3149	0.2646
174	0.3057	0.6053	0.3075	0.4774	0.3087	0.3912	0.3102	0.2823
177	0.3021	0.6210	0.3030	0.4933	0.3036	0.4077	0.3044	0.3002
180	0.2976	0.6366	0.2976	0.5093	0.2976	0.4244	0.2976	0.3183

13

Straight Bar of Order Two Encased in Elastic Foundation Constant Section

STATIC STATE

13.01 DEFINITION OF STATE

(1) System considered is a finite straight bar of constant and symmetrical section. The centroidal axis x of the bar and the vertical axis z define the plane of symmetry of the section. The elastic foundation fully surrounds the bar and offers resistance, which is proportional to its displacement (Winkler's foundation). The constants of proportionality, called foundation moduli, are

k_ϕ = torsional modulus about x (N·m/m)	k_w = transverse modulus along z (N/m^2)

Each modulus corresponds to a unit length of the bar. The bar is a part of a planar system acted on by loads normal to the system plane (system of order two).

(2) Notation (A.1, A.2, A.4)

s = span (m)	A_x = area of normal section (m^2)
a, b = segments (m)	I_x = torsional constant of A_x about x (m^4)
x, x' = coordinates (m)	I_y = moment of inertia of A_x about y (m^4)
E = modulus of elasticity (Pa)	G = modulus of rigidity (Pa)
W = shearing force (N)	w = linear displacement along z (m)
X = torsional moment (N·m)	ϕ = torsional slope about x (rad)
Y = flexural moment (N·m)	ψ = flexural slope about y (rad)

$\alpha = \sqrt{\dfrac{k_\phi}{GI_x}}$ = shape parameter (m^{-1})	$\beta = \sqrt[4]{\dfrac{k_w}{4EI_y}}$ = shape parameter (m^{-1})

(3) Assumptions of analysis are stated in (1.01). In this chapter, the effects of torsional and flexural deformations are included, but the effects of shearing deformations and of the axial force on the magnitude of the bending moment are neglected. The beam-column effect is considered in Chap. 15.

13.02 TRANSPORT MATRIX EQUATIONS

(1) Equivalents and signs

$\overline{X} = X\alpha = \text{scaled } X \text{ (N)}$ $\qquad$ $\overline{Y} = Y\beta = \text{scaled } Y \text{ (N)}$

$\overline{\phi} = \phi GI_x\alpha^2 = \text{scaled } \phi \text{ (N)}$ $\qquad$ $\overline{\psi} = EI_y\beta^2 = \text{scaled } \psi \text{ (N)}$

$\overline{w} = wEI_y\beta^3 = \text{scaled } w \text{ (N)}$

All forces, moments, and displacements are in the principal axes of the normal section, and their positive directions are shown in (2, 3) below.

(2) Free-body diagram

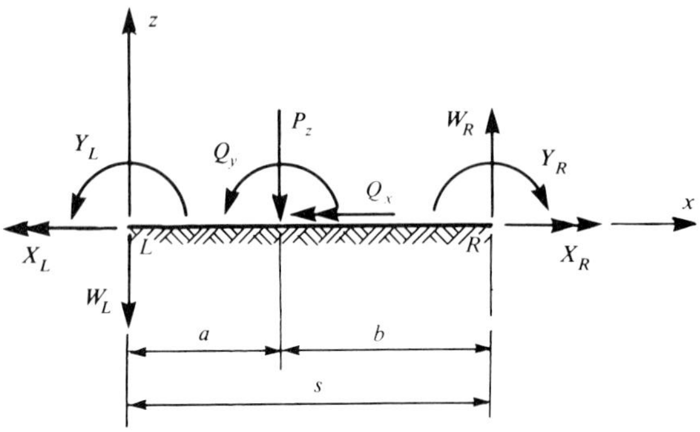

(3) Elastic curve

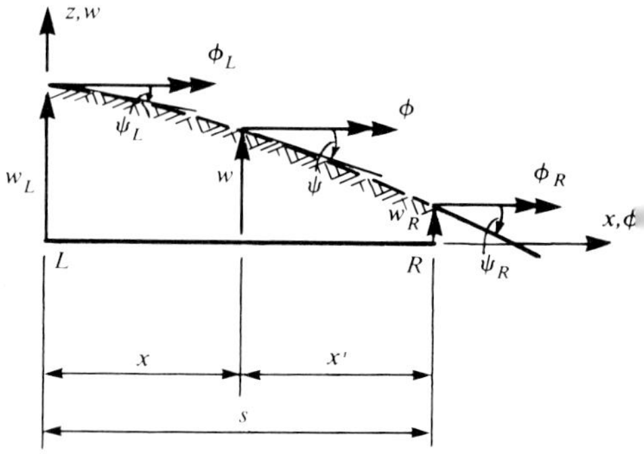

(4) Transport matrix equations

$$
\underbrace{\begin{bmatrix} 1 \\ \hline W_L \\ \overline{X}_L \\ \overline{Y}_L \\ \hline \overline{w}_L \\ \overline{\phi}_L \\ \overline{\psi}_L \end{bmatrix}}_{H_L}
=
\underbrace{\left[\begin{array}{c|ccc|ccc}
1 & 0 & 0 & 0 & 0 & 0 & 0 \\ \hline
F_W & T_{1y} & 0 & 4T_{4y} & -4T_{2y} & 0 & -4T_{3y} \\
F_X & 0 & T_{1x} & 0 & 0 & -T_{2x} & 0 \\
F_Y & -T_{2y} & 0 & T_{1y} & 4T_{3y} & 0 & 4T_{4y} \\ \hline
F_w & T_{4y} & 0 & -T_{3y} & T_{1y} & 0 & T_{2y} \\
F_\phi & 0 & -T_{2x} & 0 & 0 & T_{1x} & 0 \\
F_\psi & T_{3y} & 0 & -T_{2y} & -4T_{4y} & 0 & T_{1y}
\end{array}\right]}_{T_{LR}}
\underbrace{\begin{bmatrix} 1 \\ \hline W_R \\ \overline{X}_R \\ \overline{Y}_R \\ \hline \overline{w}_R \\ \overline{\phi}_R \\ \overline{\psi}_R \end{bmatrix}}_{H_R}
$$

$$
\underbrace{\begin{bmatrix} 1 \\ \hline W_R \\ \overline{X}_R \\ \overline{Y}_R \\ \hline \overline{w}_R \\ \overline{\phi}_R \\ \overline{\psi}_R \end{bmatrix}}_{H_R}
=
\underbrace{\left[\begin{array}{c|ccc|ccc}
1 & 0 & 0 & 0 & 0 & 0 & 0 \\ \hline
G_W & T_{1y} & 0 & -4T_{4y} & 4T_{2y} & 0 & -4T_{3y} \\
G_X & 0 & T_{1x} & 0 & 0 & T_{2x} & 0 \\
G_Y & T_{2y} & 0 & T_{1y} & 4T_{3y} & 0 & -4T_{4y} \\ \hline
G_w & -T_{4y} & 0 & -T_{3y} & T_{1y} & 0 & -T_{2y} \\
G_\phi & 0 & T_{2x} & 0 & 0 & T_{1x} & 0 \\
G_\psi & T_{3y} & 0 & T_{2y} & 4T_{4y} & 0 & T_{1y}
\end{array}\right]}_{T_{RL}}
\underbrace{\begin{bmatrix} 1 \\ \hline W_L \\ \overline{X}_L \\ \overline{Y}_L \\ \hline \overline{w}_L \\ \overline{\phi}_L \\ \overline{\psi}_L \end{bmatrix}}_{H_L}
$$

where H_L, H_R are the *end state vectors* and T_{LR}, T_{RL} are the respective *transport matrices*. The elements of these matrices are the transport coefficients shown in (13.03) and the *load functions F, G* listed for particular cases in (13.04–13.06).

13.03 TRANSPORT RELATIONS

(1) Transport functions

$$\alpha = \sqrt{\frac{k_\phi}{GI_x}} \qquad \beta = \sqrt[4]{\frac{k_w}{4EI_y}}$$

$T_{1x}(x) = \cosh \alpha x$	$T_{1y}(x) = \cosh \beta x \cos \beta x$
$T_{2x}(x) = \sinh \alpha x$	$T_{2y}(x) = \frac{1}{2}(\cosh \beta x \sin \beta x + \sinh \beta x \cos \beta x)$
$T_{3x}(x) = \cosh \alpha x - 1$	$T_{3y}(x) = \frac{1}{2}\sinh \beta x \sin \beta x$
$T_{4x}(x) = \sinh \alpha x - \alpha x$	$T_{4y}(x) = \frac{1}{4}(\cosh \beta x \sin \beta x - \sinh \beta x \cos \beta x)$
$T_{5x}(x) = T_{3x}(x) - \frac{(\alpha x)^2}{2!}$	$T_{5y}(x) = \frac{1}{4}[1 - T_{1y}(x)]$
$T_{6x}(x) = T_{4x}(x) - \frac{(\alpha x)^3}{3!}$	$T_{6y}(x) = \frac{1}{4}[\beta x - T_{2y}(x)]$

The graphs and table of these functions are identical to (8.04–8.05).

(2) Transport coefficients T_j, A_j, B_j, C_j are the values of $T_j(x)$ for a particular argument x.

$$\left.\begin{array}{lll} \text{For } x = s, & T_{jx}(s) = T_{jx}, & T_{jy}(s) = T_{jy} \\ \text{For } x = a, & T_{jx}(a) = A_{jx}, & T_{jy}(a) = A_{jy} \\ \text{For } x = b, & T_{jx}(b) = B_{jx}, & T_{jy}(b) = B_{jy} \\ \text{and } T_{jx} - A_{jx} = C_{jx}, & & T_{jy} - A_{jy} = C_{jy} \end{array}\right\} j = 1, 2, \ldots, 6$$

(3) Transport matrices of this bar have analogical properties to those of (11.02) and can be used in similar ways.

(4) Differential relations

$\frac{d\phi}{dx} = \frac{X}{GI_x}$	$\frac{dw}{dx} = -\psi$
$\frac{d^2\phi}{dx^2} = \frac{\phi k_\phi}{GI_x} + \frac{q_x}{GI_x}$	$\frac{d^2w}{dx^2} = -\frac{d\psi}{dx} = -\frac{Y}{EI_y}$
	$\frac{d^3w}{dx^3} = -\frac{d^2\psi}{dx^2} = -\frac{dY}{EI_y\,dx} = -\frac{W}{EI_y}$
	$\frac{d^4w}{dx^4} = -\frac{d^3\psi}{dx^3} = -\frac{d^2Y}{EI_y\,dx} = -\frac{dW}{EI_y\,dx} = -\frac{wk_w}{EI_y} - \frac{p_z}{EI_y}$

(5) Bars encased in elastic medium are classified as (*a*) short bars ($\beta s < 0.6$), which can be considered as infinitely rigid, (*b*) intermediate bars ($0.6 < \beta s < 5.0$), which must be treated as finite and elastic and be analyzed by methods introduced in (13.02–13.09), (*c*) long bars ($\beta s > 5.0$), which must be treated as semi-infinite and be analyzed by the method shown in (13.10).

(6) General transport matrix equations for bars of order three are obtained by matrix composition of (8.02–8.08) and (13.02–13.06) as shown in (2.18).

13.04 TRANSPORT LOAD FUNCTIONS

Notation (13.01–13.03) Signs (13.02) Matrix (13.02) P_z = concentrated load (N) Q_x, Q_y = applied couples (N·m) q_y = intensity of distributed couple (N·m/m) A_j, B_j, C_j, T_j = transport coefficients (13.03)		$\alpha = \sqrt{\dfrac{k_\phi}{GI_x}}$ $\beta = \sqrt[4]{\dfrac{k_w}{4EI_y}}$
$F_W = -A_{1y}P_z$ $F_X = 0$ $F_Y = A_{2y}P_z$	**1.** 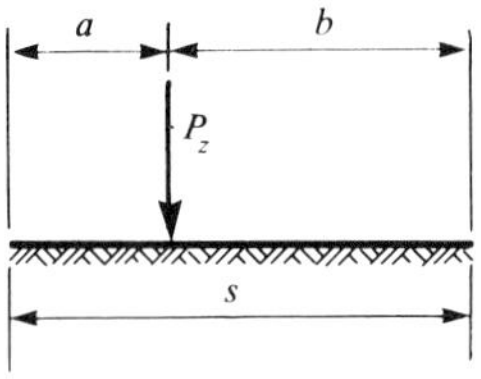	$G_W = B_{1y}P_z$ $G_X = 0$ $G_Y = B_{2y}P_z$
$F_w = -A_{4y}P_z$ $F_\phi = 0$ $F_\psi = -A_{3y}P_z$		$G_w = -B_{4y}P_z$ $G_\phi = 0$ $G_\psi = B_{3y}P_z$
$F_W = -4A_{4y}\beta Q_y$ $F_X = -A_{1x}\alpha Q_x$ $F_Y = -A_{1y}\beta Q_y$	**2.** 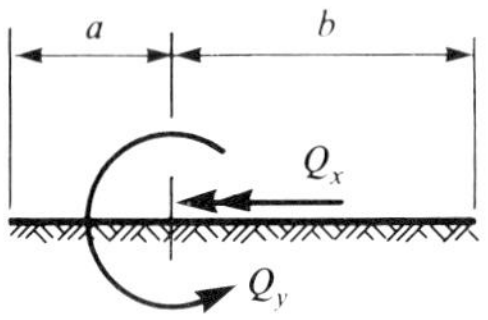	$G_W = -4B_{4y}\beta Q_y$ $G_X = B_{1x}\alpha Q_x$ $G_Y = B_{1y}\beta Q_y$
$F_w = A_{3y}\beta Q_y$ $F_\phi = A_{2x}\beta Q_x$ $F_\psi = A_{2y}\beta Q_y$		$G_w = -B_{3y}\beta Q_y$ $G_\phi = B_{2x}\alpha Q_x$ $G_\psi = B_{2y}\beta Q_y$
$F_W = -4T_{5y}q_y$ $F_X = 0$ $F_Y = -T_{2y}q_y$	**3.** 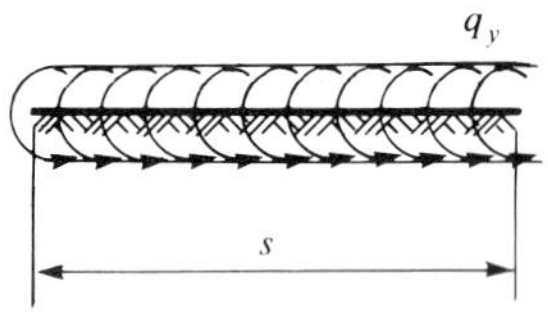	$G_W = -4T_{5y}q_y$ $G_X = 0$ $G_Y = T_{2y}q_y$
$F_w = T_{4y}q_y$ $F_\phi = 0$ $F_\psi = T_{3y}q_y$		$G_w = -T_{4y}q_y$ $G_\phi = 0$ $G_\psi = T_{3y}q_y$
$F_W = -4C_{5y}q_y$ $F_X = 0$ $F_X = -C_{2y}q_y$	**4.** 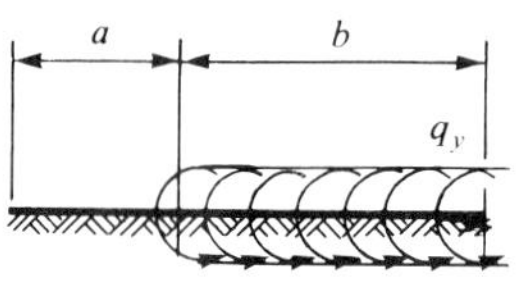	$G_W = -4B_{5y}q_y$ $G_X = 0$ $G_Y = B_{2y}q_y$
$F_w = C_{4y}q_y$ $F_\phi = 0$ $F_\psi = C_{3y}q_y$		$G_w = -B_{4y}q_y$ $G_\phi = 0$ $G_\psi = B_{3y}q_y$

13.05 TRANSPORT LOAD FUNCTIONS

Notation (13.01–13.03) load Signs (13.02) Matrix (13.02) p_z = intensity of distributed load (N/m) q_x = intensity of distributed couple (N·m/m) A_j, B_j, C_j, T_j = transport coefficients (13.03)		$\alpha = \sqrt{\dfrac{k_\phi}{GI_x}}$ $\beta = \sqrt[4]{\dfrac{k_w}{EI_y}}$
$F_W = -T_{2y}\beta^{-1}p_z$ $F_X = -T_{2x}q_x$ $F_Y = T_{3y}\beta^{-1}p_z$	**1.** 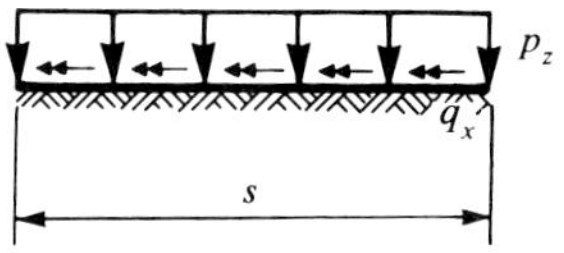	$G_W = T_{2y}\beta^{-1}p_z$ $G_X = T_{2x}q_x$ $G_Y = T_{3y}\beta^{-1}p_z$
$F_w = -T_{5y}\beta^{-1}p_z$ $F_\phi = T_{3x}q_x$ $F_\psi = -T_{4y}\beta^{-1}p_z$		$G_w = -T_{5y}\beta^{-1}p_z$ $G_\phi = T_{3x}q_x$ $G_\psi = T_{4y}\beta^{-1}p_z$
$F_W = -C_{2y}\beta^{-1}p_z$ $F_X = -C_{2x}q_x$ $F_Y = C_{3y}\beta^{-1}p_z$	**2.** 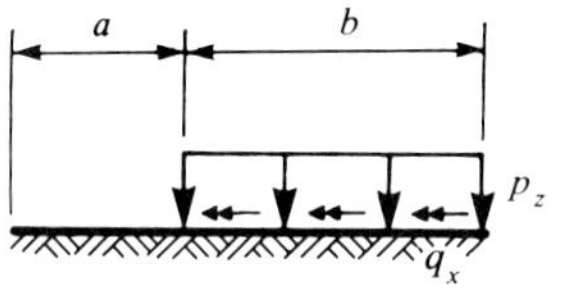	$G_W = B_{2y}\beta^{-1}p_z$ $G_X = B_{2x}q_x$ $G_Y = B_{3y}\beta^{-1}p_z$
$F_w = -C_{5y}\beta^{-1}p_z$ $F_\phi = C_{3x}q_x$ $F_\psi = -C_{4y}\beta^{-1}p_z$		$G_w = -B_{5y}\beta^{-1}p_z$ $G_\phi = B_{3x}q_x$ $G_\psi = B_{4y}\beta^{-1}p_z$
$F_W = -(\beta s T_{2y} - T_{3y})\beta^{-2}s^{-1}p_z$ $F_X = -(\alpha s T_{2x} - T_{3x})\alpha^{-1}s^{-1}q_x$ $F_Y = (\beta s T_{3y} - T_{4y})\beta^{-2}s^{-1}p_z$	**3.** 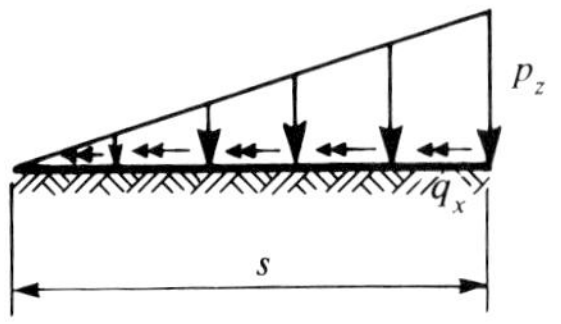	$G_W = T_{3y}\beta^{-2}s^{-1}p_z$ $G_X = T_{3x}\alpha^{-1}s^{-1}q_x$ $G_Y = T_{4y}\beta^{-2}s^{-1}p_z$
$F_w = -(\beta s T_{5y} - T_{6y})\beta^{-2}s^{-1}p_z$ $F_\phi = (\alpha s T_{3x} - T_{4x})\alpha^{-1}s^{-1}q_x$ $F_\psi = -(\beta s T_{4y} - T_{5y})\beta^{-2}s^{-1}p_y$		$G_w = -T_{6y}\beta^{-2}s^{-1}p_z$ $G_\phi = T_{4x}\alpha^{-1}s^{-1}q_x$ $G_\psi = T_{5y}\beta^{-2}s^{-1}p_z$
$F_W = -(\beta b T_{2y} - C_{3y})\beta^{-2}b^{-1}p_z$ $F_X = -(\alpha b T_{2x} - C_{3x})\alpha^{-1}b^{-1}q_x$ $F_Y = (\beta b T_{3y} - C_{4y})\beta^{-2}b^{-1}p_z$	**4.** 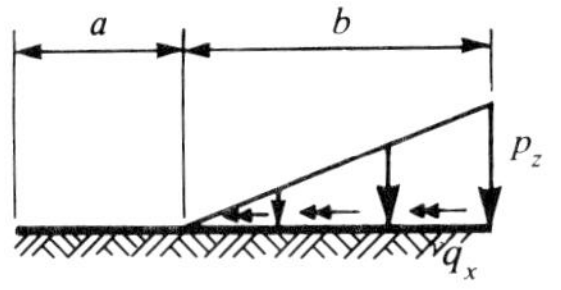	$G_W = B_{3y}\beta^{-2}b^{-1}p_z$ $G_X = B_{3x}\alpha^{-1}b^{-1}q_x$ $G_Y = B_{4y}\beta^{-2}b^{-1}p_z$
$F_w = -(\beta b T_{5y} - C_{6y})\beta^{-2}b^{-1}p_z$ $F_\phi = (\alpha b T_{3x} - C_{4x})\alpha^{-1}b^{-1}q_x$ $F_\psi = -(\beta b T_{4y} - C_{5y})\beta^{-2}b^{-1}p_z$		$G_w = -B_{6y}\beta^{-2}b^{-1}p_z$ $G_\phi = B_{4x}\alpha^{-1}b^{-1}q_x$ $G_\psi = B_{5y}\beta^{-2}b^{-1}p_z$

13.06 TRANSPORT LOAD FUNCTIONS

Notation (13.01–13.03) Signs (13.02) Matrix (13.02)

w_j, ϕ_j, ψ_j = externally applied displacements (m, rad, rad)

g_t = thermal strain (11.03)

A_j, B_j, C_j, T_j = transport coefficients (13.03)

$$\alpha = \sqrt{\frac{k_\phi}{GI_x}} \qquad \beta = \sqrt[4]{\frac{k_w}{4EI_y}}$$

1. Abrupt displacements w_j, ϕ_j

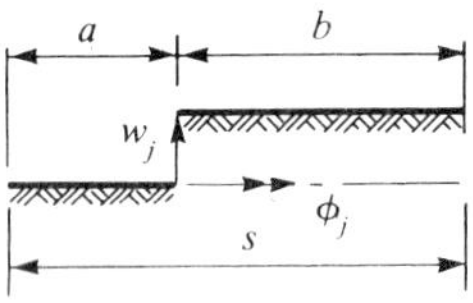

$$F_W = 4EI_yA_{2y}\beta^3 w_j \qquad G_W = 4EI_yB_{2y}\beta^3 w_j$$
$$F_X = GI_xA_{2x}\alpha^2\phi_j \qquad G_X = GI_xB_{2x}\alpha^2\phi_j$$
$$F_Y = -4EI_yA_{3y}\beta^3 w_j \qquad G_Y = 4EI_yB_{3y}\beta^3 w_j$$

$$F_w = -EI_yA_{1y}\beta^3 w_j \qquad G_w = EI_yB_{1y}\beta^3 w_j$$
$$F_\phi = -GI_xA_{1x}\alpha^2\phi_j \qquad G_\phi = GI_xB_{1x}\alpha^2\phi_j$$
$$F_\psi = 4EI_yA_{4y}\beta^3 w_j \qquad G_\psi = 4EI_yB_{4y}\beta^3 w_j$$

2. Abrupt displacement ψ_j

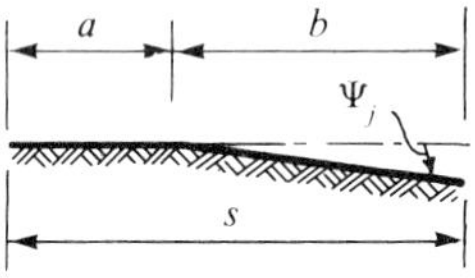

$$F_W = 4EI_yA_{3y}\beta^2\psi_j \qquad G_W = -4EI_yB_{3y}\beta^2\psi_j$$
$$F_X = 0 \qquad G_X = 0$$
$$F_Y = -4EI_yA_{4y}\beta^2\psi_j \qquad G_Y = -4EI_yB_{4y}\beta^2\psi_j$$

$$F_w = -EI_yA_{2y}\beta^2\psi_j \qquad G_w = -EI_yB_{2y}\beta^2\psi_j$$
$$F_\phi = 0 \qquad G_\phi = 0$$
$$F_\psi = -EI_yA_{1y}\beta^2\psi_j \qquad G_\psi = EI_yB_{1y}\beta^2\psi_j$$

3. Temperature deformation of segment b

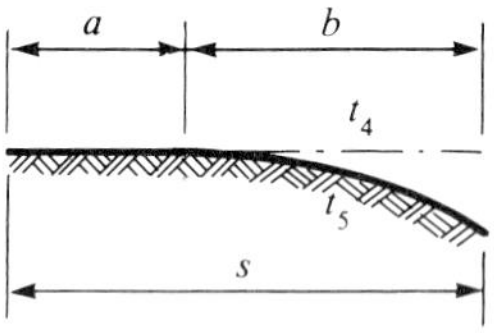

$$F_W = 4EI_yC_{4y}\beta g_t \qquad G_W = -4EI_yB_{4y}\beta g_t$$
$$F_X = 0 \qquad G_X = 0$$
$$F_Y = -4EI_yC_{5y}\beta g_t \qquad G_Y = -4EI_yB_{5y}\beta g_t$$

$$F_w = -EI_yC_{3y}\beta g_t \qquad G_w = -EI_yB_{3y}\beta g_t$$
$$F_\phi = 0 \qquad G_\phi = 0$$
$$F_\psi = -EI_yC_{2y}\beta g_t \qquad G_\psi = EI_yB_{2y}\beta g_t$$

4.

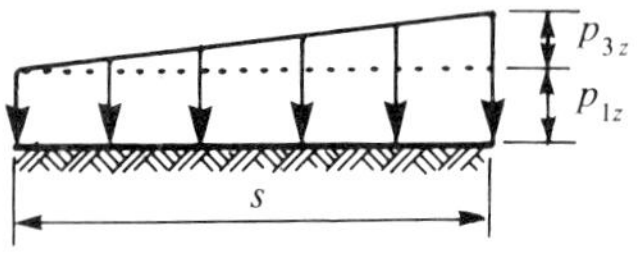

Sec. 13.05 (case 1 + case 3)

5.

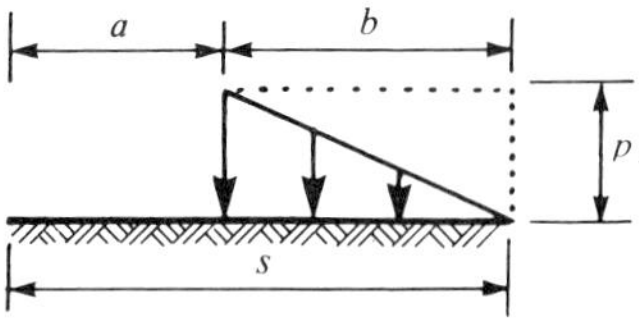

Sec. 13.05 (case 2 − case 4)

13.07 STIFFNESS MATRIX EQUATIONS

$$r = \frac{s^2 GI_x}{EI_y} \quad (\text{m}^2)$$

(1) Equivalents and signs. All forces, moments, and displacements are in the principal axes of the normal section, and their positive directions are shown in (2, 3) below.

(2) Free-body diagram

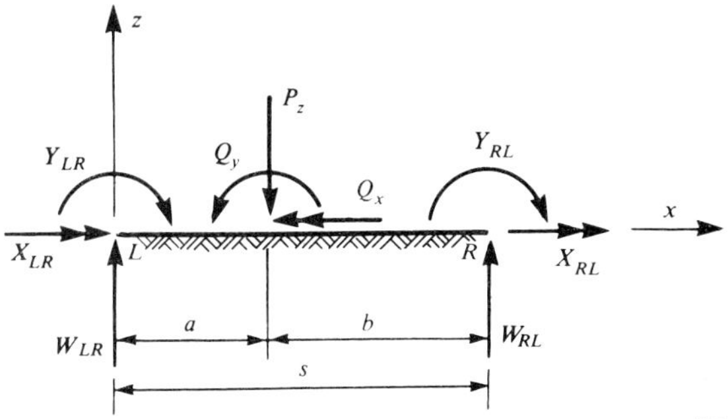

(3) Elastic curve

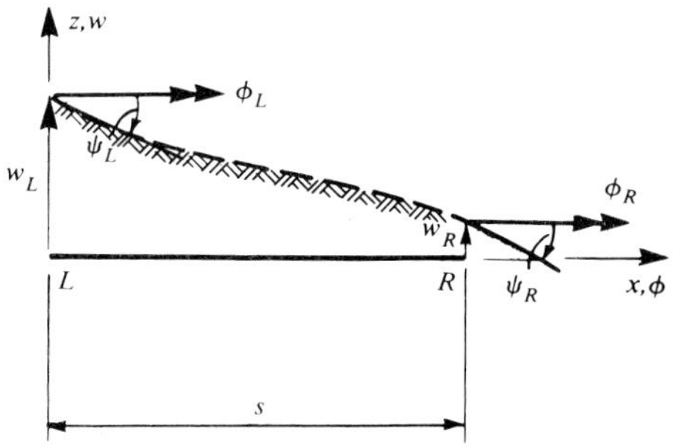

(4) Dimensionless stiffness coefficients

$$D_0 = \frac{1}{\sinh^2 \beta s - \sin^2 \beta s}$$

$L_1 = \dfrac{\cosh \alpha s}{\sinh \alpha s}$	$K_1 = (\alpha s)L_1$	$\alpha = \sqrt{\dfrac{k_\phi}{GI_x}}$
$L_2 = 2D_0(\sinh 2\beta s + \sin 2\beta s)$	$K_2 = (\beta s)^3 L_2$	
$L_3 = D_0(\cosh 2\beta s - \cos 2\beta s)$	$K_3 = (\beta s)^2 L_3$	
$L_4 = D_0(\sinh 2\beta s - \sin 2\beta s)$	$K_4 = (\beta s)L_4$	
$L_5 = \dfrac{1}{\sinh \alpha s}$	$K_5 = (\alpha s)L_5$	$\beta = \sqrt[4]{\dfrac{k_w}{4EI_y}}$
$L_6 = 4D_0(\cosh \beta s \sin \beta s + \sinh \beta s \cos \beta s)$	$K_6 = (\beta s)^3 L_6$	
$L_7 = 4D_0 \sinh \beta s \sin \beta s$	$K_7 = (\beta s)^2 L_7$	
$L_8 = 2D_0(\cosh \beta s \sin \beta s - \sinh \beta s \cos \beta s)$	$K_8 = (\beta s)L_8$	

The graphs and tables of these coefficients are identical to (8.10–8.11).

(5) Stiffness matrix equations

$$\begin{bmatrix} W_{LR} \\ X_{LR} \\ Y_{LR} \\ \hline W_{RL} \\ X_{RL} \\ Y_{RL} \end{bmatrix} = \frac{EI_y}{s^3} \left[\begin{array}{ccc|ccc} K_2 & 0 & -K_3 s & -K_6 & 0 & -K_7 s \\ 0 & K_1 r & 0 & 0 & -K_5 r & 0 \\ -K_3 s & 0 & K_4 s^2 & K_7 s & 0 & K_8 s^2 \\ \hline -K_6 & 0 & K_7 s & K_2 & 0 & K_3 s \\ 0 & -K_5 r & 0 & 0 & K_1 r & 0 \\ -K_7 s & 0 & K_8 s^2 & K_3 s & 0 & K_4 s^2 \end{array} \right] \begin{bmatrix} w_L \\ \phi_L \\ \psi_L \\ \hline w_R \\ \phi_R \\ \psi_R \end{bmatrix} + \begin{bmatrix} W_{L0} \\ X_{L0} \\ Y_{L0} \\ \hline W_{R0} \\ X_{R0} \\ Y_{R0} \end{bmatrix}$$

where W_{L0}, X_{L0}, Y_{L0} and W_{R0}, X_{R0}, Y_{R0} are the *reactions of fixed-end beam-columns* due to loads and other causes shown in (13.08–13.09).

(6) General stiffness matrix equations for beams of order three encased in elastic foundation are obtained by matrix composition of (8.09) and (13.07) as shown in (3.19).

13.08 STIFFNESS LOAD FUNCTIONS, FIXED-END BEAM

(1) General form of the stiffness load functions in (13.07) is given by the matrix equations

$$\begin{bmatrix} W_{L0} \\ X_{L0} \\ Y_{L0} \end{bmatrix} = \begin{bmatrix} L_6 & 0 & L_7 \\ 0 & L_5\alpha^{-1} & 0 \\ -L_7\beta^{-1} & 0 & -L_8\beta^{-1} \end{bmatrix} \begin{bmatrix} G_w \\ G_\phi \\ G_\psi \end{bmatrix}$$

$$\begin{bmatrix} W_{R0} \\ X_{R0} \\ Y_{R0} \end{bmatrix} = \begin{bmatrix} -L_6 & 0 & -L_7 \\ 0 & L_5\alpha^{-1} & 0 \\ L_7\beta^{-1} & 0 & -L_8\beta^{-1} \end{bmatrix} \begin{bmatrix} F_w \\ F_\phi \\ F_\psi \end{bmatrix}$$

where F_w, F_ϕ, F_ψ are the left-end transport load functions and G_w, G_ϕ, G_ψ are their right-end counterparts, given for particular loads and causes in (13.04–13.06), and L_5, L_6, L_7, L_8 are the stiffness coefficients (13.07). The load vectors and other causes are defined by symbols introduced in (13.04–13.06).

(2) Particular cases

Notation (13.01–13.03, 13.07) A_j, B_j, C_j, T_j = transport coefficients (13.03)	Signs (13.07)	$\alpha = \sqrt{\dfrac{k_\phi}{GI_x}}$ $\beta = \sqrt[4]{\dfrac{k_w}{4EI_y}}$
$W_{L0} = (L_7B_{3y} - L_6B_{4y})P_z$ $X_{L0} = L_5B_{2x}Q_x$ $Y_{L0} = -(L_8B_{3y} - L_7B_{4y})\beta^{-1}P_z$	1. 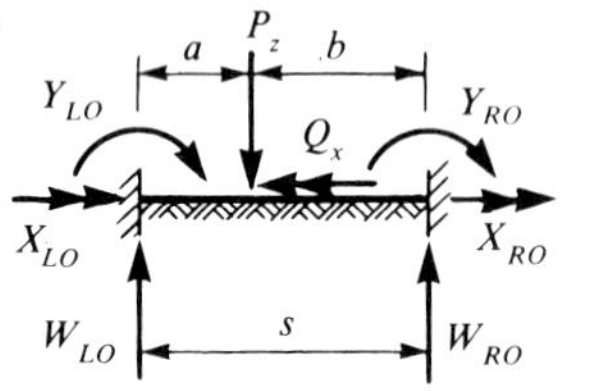	$W_{R0} = (L_7A_{3y} - L_6A_{4y})P_z$ $X_{R0} = L_5A_{2x}Q_x$ $Y_{R0} = (L_8A_{3y} - L_7A_{4y})\beta^{-1}P_z$
$W_{L0} = (L_7B_{2y} - L_6B_{3y})\beta Q_y$ $X_{L0} = 0$ $Y_{L0} = -(L_8B_{2y} - L_7B_{3y})Q_y$	2. 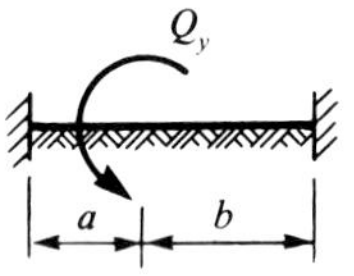	$W_{R0} = -(L_7A_{2y} - L_6A_{3y})\beta Q_y$ $X_{R0} = 0$ $Y_{R0} = -(L_8A_{2y} - L_7A_{3y})Q_y$
$W_{L0} = \dfrac{(\cosh\beta s + \cos\beta s)p_z}{(\sinh\beta s + \sin\beta s)\beta}$ $X_{L0} = \dfrac{(\cosh\alpha s - 1)q_x}{\alpha\sinh\alpha s}$ $Y_{L0} = -\dfrac{(\sinh\beta s - \sin\beta s)p_z}{(\sinh\beta s + \sin\beta s)2\beta^2}$	3. 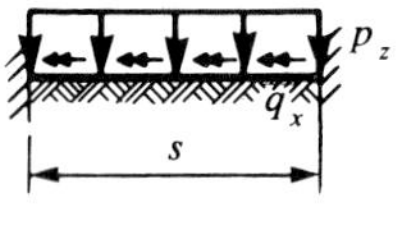	$W_{R0} = W_{L0}$ $X_{R0} = X_{L0}$ $Y_{R0} = -Y_{L0}$

4.

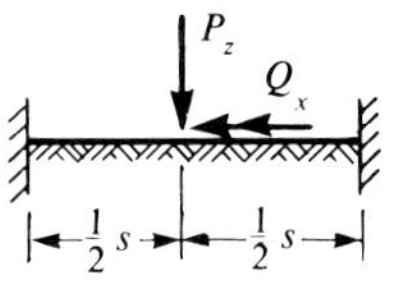

$$W_{L0} = \frac{\cosh \tfrac{1}{2}\beta s \sin \tfrac{1}{2}\beta s + \sinh \tfrac{1}{2}\beta s \cos \tfrac{1}{2}\beta s}{\sinh \beta s + \sin \beta s} P_z = W_{R0}$$

$$X_{L0} = \frac{1}{2 \cosh \tfrac{1}{2}\alpha s} Q_x = X_{R0}$$

$$Y_{L0} = -\frac{\sinh \tfrac{1}{2}\beta s \sin \tfrac{1}{2}\beta s}{\beta(\sinh \beta s + \sin \beta s)} P_z = -Y_{R0}$$

5.

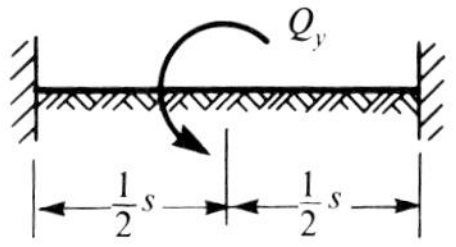

$$W_{L0} = \frac{\sinh \tfrac{1}{2}\beta s \sin \tfrac{1}{2}\beta s}{\sinh \beta s + \sin \beta s} \beta Q_y$$

$$X_{L0} = 0 = X_{R0}$$

$$Y_{L0} = -\frac{\cosh \tfrac{1}{2}\beta s \sin \tfrac{1}{2}\beta s - \sinh \tfrac{1}{2}\beta s \cos \tfrac{1}{2}\beta s}{\sinh \beta s + \sin \beta s} Q_y = -Y_{R0}$$

6.

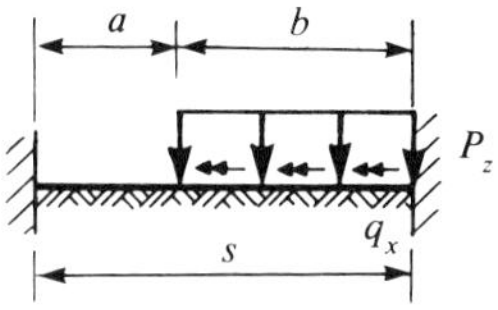

Use general form in terms of (13.05-2)

7.

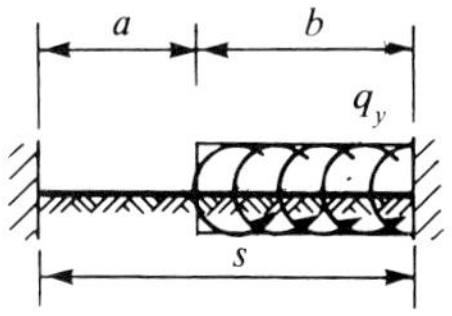

Use general form in terms of (13.04-4)

8.

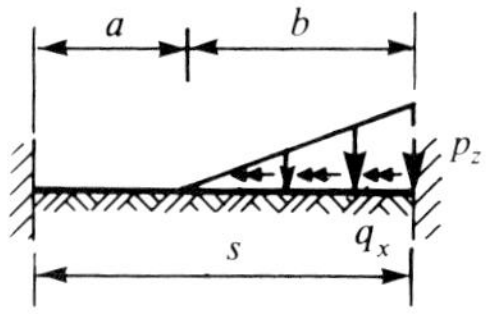

Use general form in terms of (13.05-4)

9. Abrupt displacement w_j, ϕ_j

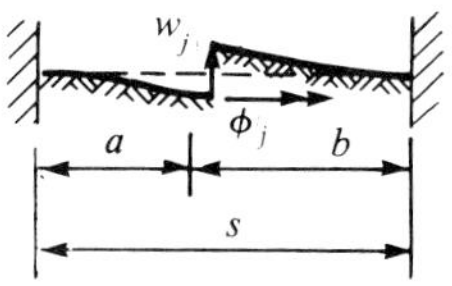

$$W_{L0} = EI_y(L_6 B_{1y} + 4L_7 B_{4y})\,\beta^3 w_j$$

$$X_{L0} = GI_x L_5 B_{1x} \alpha \phi_j$$

$$Y_{L0} = -EI_y(L_7 B_{1y} + 4L_8 B_{4y})\,\beta^2 w_j$$

$$W_{R0} = -EI_y(L_6 A_{1y} + 4L_7 A_{4y})\beta^3 w_j$$

$$X_{R0} = -GI_x L_5 A_{1x} \alpha \phi_j$$

$$Y_{R0} = -EI_y(L_7 A_{1y} + 4L_8 A_{4y})\beta^2 w_j$$

10. Abrupt displacement ψ_j

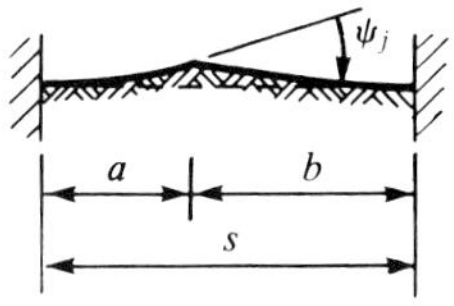

$$W_{L0} = -EI_y(L_6 B_{2y} - L_7 B_{1y})\,\beta^2 \psi_j$$

$$X_{L0} = 0$$

$$Y_{L0} = EI_y(L_7 B_{2y} - L_8 B_{1y})\,\beta \psi_j$$

$$W_{R0} = -EI_y(L_6 A_{2y} - L_7 A_{1y})\beta^2 \psi_j$$

$$X_{R0} = 0$$

$$Y_{R0} = -EI_y(L_7 A_{2y} - L_8 A_{1y})\beta \psi_j$$

11. Temperature deformation of segment b

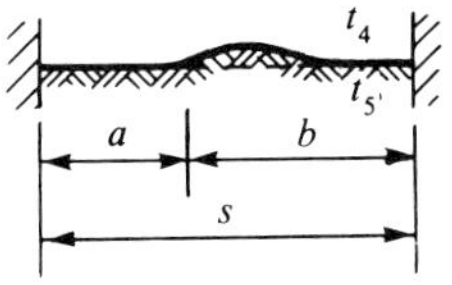

$$W_{L0} = -EI_y(L_6 B_{3y} - L_7 B_{2y})\beta g_t$$

$$X_{L0} = 0$$

$$Y_{L0} = EI_y(L_7 B_{3y} - L_8 B_{2y}) g_t$$

$$W_{R0} = -EI_y(L_6 C_{3y} - L_7 C_{2y})\beta g_t$$

$$X_{R0} = 0$$

$$Y_{R0} = -EI_y(L_7 C_{3y} - L_8 C_{2y}) g_t$$

13.09 MODIFIED STIFFNESS MATRIX EQUATIONS

$$\beta = \sqrt[4]{\frac{k_w}{4EI_y}}$$

(1) Dimensionless stiffness coefficients

$$D_1 = \frac{1}{\cosh \beta s \sinh \beta s - \cos \beta s \sin \beta s} \qquad D_2 = \frac{1}{\cosh^2 \beta s + \cos^2 \beta s}$$

$L_9 = 2D_1 (\cosh^2 \beta s + \cos^2 \beta s)$	$K_9 = (\beta s)^3 L_9$
$L_{10} = 4D_1 \cosh \beta s \cos \beta s$	$K_{10} = (\beta s)^3 L_{10}$
$L_{11} = 2D_1 (\cosh \beta s \sin \beta s + \sinh \beta s \cos \beta s)$	$K_{11} = (\beta s)^3 L_{11}$
$L_{12} = 4D_1 (\cosh^2 \beta s - \sin^2 \beta s)$	$K_{12} = (\beta s)^2 L_{12}$
$L_{13} = 2D_1 (\cosh \beta s \sinh \beta s + \cos \beta s \sin \beta s)$	$K_{13} = (\beta s)^2 L_{13}$
$L_{14} = 2D_1 (\sinh^2 \beta s + \sin^2 \beta s)$	$K_{14} = (\beta s) L_{14}$
$L_{15} = 2D_1 \sinh \beta s \sin \beta s$	
$L_{16} = D_1 (\cosh \beta s \sin \beta s - \sinh \beta s \cos \beta s)$	
$L_{17} = 4D_2 (\cosh \beta s \sinh \beta s + \cos \beta s \sin \beta s)$	$K_{17} = (\beta s)^3 L_{17}$
$L_{18} = 2D_2 (\cosh^2 \beta s - \cos^2 \beta s)$	$K_{18} = (\beta s)^2 L_{18}$
$L_{19} = 2D_2 (\cosh \beta s \sinh \beta s - \cos \beta s \sin \beta s)$	$K_{19} = (\beta s) L_{19}$
$L_{20} = 2D_2 \cosh \beta s \cos \beta s$	
$L_{21} = 2D_2 (\cosh \beta s \sin \beta s - \sinh \beta s \cos \beta s)$	
$L_{22} = D_2 (\cosh \beta s \sin \beta s + \sinh \beta s \cos \beta s)$	

(2) Modifications of stiffness matrix equations (13.07) due to hinged end and to free end are shown in (3, 4) and (5, 6), respectively. Since these modifications do not affect the expressions for X_{LR} and X_{RL}, only the modified part of the respective stiffness matrix is presented. The modified load functions designated by the asterisk are the reactions of the respective *propped-end beams* in (3, 4) and of the respective *cantilever beams* in (5, 6). They are all linear combinations of the L coefficients and the transport load functions F or G introduced in (13.04–13.06).

(3) Left end hinged ($Y_{LR} = 0$)

$$\begin{bmatrix} W_{LR} \\ W_{RL} \\ Y_{RL} \end{bmatrix} = \frac{EI_y}{s^3} \begin{bmatrix} K_9 & -K_{10} & -K_{11}s \\ -K_{10} & K_{12} & K_{13}s \\ -K_{11}s & K_{13}s & K_{14}s^2 \end{bmatrix} \begin{bmatrix} w_L \\ w_R \\ \psi_R \end{bmatrix} + \begin{bmatrix} W^*_{L0} \\ W^*_{R0} \\ Y^*_{R0} \end{bmatrix}$$

$$\begin{bmatrix} W^*_{L0} \\ W^*_{R0} \\ Y^*_{R0} \end{bmatrix} = \begin{bmatrix} -1 & \frac{1}{2}L_{14} & L_9 \\ 0 & L_{15} & L_{10} \\ 0 & L_{16}\beta^{-1} & L_{11}\beta^{-1} \end{bmatrix} \begin{bmatrix} F_W \\ F_Y \\ F_w \end{bmatrix}$$

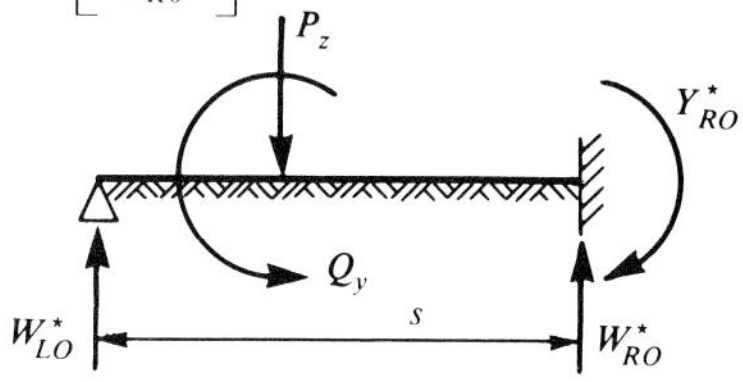

(4) Right end hinged ($Y_{RL} = 0$)

$$\begin{bmatrix} W_{LR} \\ Y_{LR} \\ W_{RL} \end{bmatrix} = \frac{EI_y}{s^3} \begin{bmatrix} K_{12} & -K_{13}s & -K_{10} \\ -K_{13}s & K_{14}s^2 & K_{11}s \\ -K_{10} & K_{11}s & K_9 \end{bmatrix} \begin{bmatrix} w_L \\ \psi_L \\ w_R \end{bmatrix} + \begin{bmatrix} W^*_{L0} \\ Y^*_{L0} \\ W^*_{R0} \end{bmatrix}$$

$$\begin{bmatrix} W^*_{L0} \\ Y^*_{L0} \\ W^*_{R0} \end{bmatrix} = \begin{bmatrix} 0 & L_{15} & L_{10} \\ 0 & -L_{16}\beta^{-1} & -L_{11}\beta^{-1} \\ 1 & -\frac{1}{2}L_{14} & L_9 \end{bmatrix} \begin{bmatrix} G_W \\ G_Y \\ G_w \end{bmatrix}$$

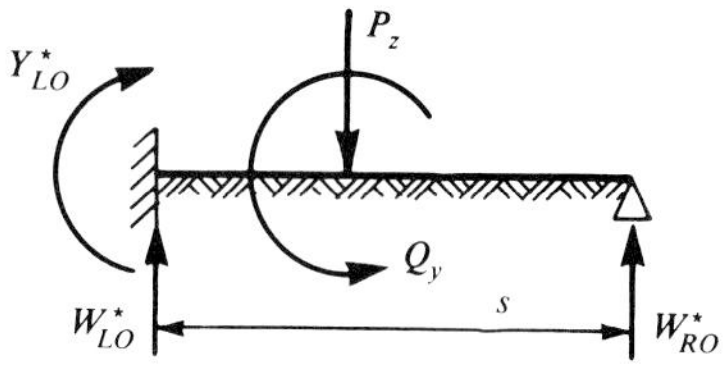

(5) Left end free ($W_{LR} = 0,\ Y_{LR} = 0$)

$$\begin{bmatrix} W_{RL} \\ Y_{RL} \end{bmatrix} = \frac{EI_y}{s^3} \begin{bmatrix} K_{17} & K_{18}s \\ K_{18}s & K_{19}s^2 \end{bmatrix} \begin{bmatrix} w_R \\ \psi_R \end{bmatrix} + \begin{bmatrix} W^*_{R0} \\ Y^*_{R0} \end{bmatrix}$$

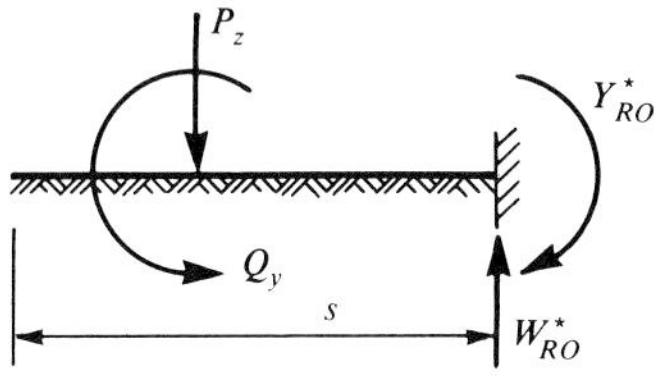

$$\begin{bmatrix} W^*_{R0} \\ Y^*_{R0} \end{bmatrix} = \begin{bmatrix} -L_{20} & L_{21} \\ -L_{22}\beta^{-1} & -L_{20}\beta^{-1} \end{bmatrix} \begin{bmatrix} F_W \\ F_Y \end{bmatrix}$$

(6) Right end free ($W_{RL} = 0,\ Y_{RL} = 0$)

$$\begin{bmatrix} W_{LR} \\ Y_{LR} \end{bmatrix} = \frac{EI_y}{s^3} \begin{bmatrix} K_{17} & -K_{18}s \\ -K_{18}s & K_{19}s^2 \end{bmatrix} \begin{bmatrix} w_L \\ \psi_L \end{bmatrix} + \begin{bmatrix} W^*_{L0} \\ Y^*_{L0} \end{bmatrix}$$

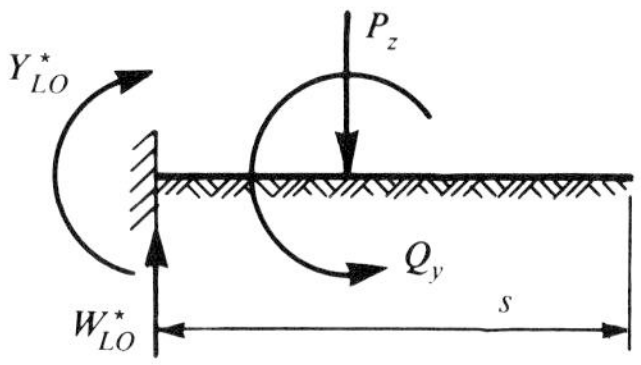

$$\begin{bmatrix} W^*_{L0} \\ Y^*_{L0} \end{bmatrix} = \begin{bmatrix} L_{20} & L_{21} \\ -L_{22}\beta^{-1} & L_{20}\beta^{-1} \end{bmatrix} \begin{bmatrix} G_W \\ G_Y \end{bmatrix}$$

13.10 SEMI-INFINITE BAR ENCASED IN ELASTIC FOUNDATION

$$\alpha = \sqrt{\frac{k_\phi}{GI_x}} \qquad r = \frac{s^2 GI_x}{EI_y}$$

$$\beta = \sqrt[4]{\frac{k_w}{4EI_y}}$$

(1) System. If $\beta s > 5.0$, the effect of the far-end state vector in (13.02) is negligible, and the bar can be assumed to be semi-infinite, which means the far-end state vector is assumed to be zero. As in (8.14), the stiffness matrix equations reduce to two very simple forms, shown below in terms of symbols and signs of (13.07).

(2) Left end finite **(3) Right end finite**

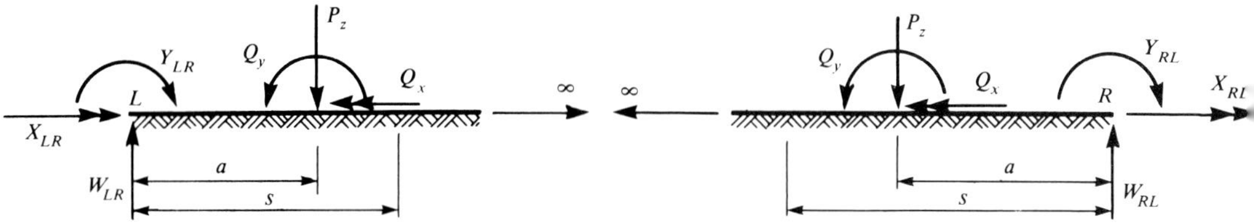

(4) Stiffness matrix equations

$$\begin{bmatrix} W_{LR} \\ X_{LR} \\ Y_{LR} \end{bmatrix} = EI_y \begin{bmatrix} r\alpha s^{-2} & 0 & 0 \\ 0 & 4\beta^3 & -2\beta^2 \\ 0 & -2\beta^2 & 2\beta \end{bmatrix} \begin{bmatrix} w_L \\ \phi_L \\ \psi_L \end{bmatrix} + \begin{bmatrix} W_{L0} \\ X_{L0} \\ Y_{L0} \end{bmatrix}$$

$$\begin{bmatrix} W_{RL} \\ X_{RL} \\ Y_{RL} \end{bmatrix} = EI_y \begin{bmatrix} r\alpha s^{-2} & 0 & 0 \\ 0 & 4\beta^3 & 2\beta^2 \\ 0 & 2\beta^2 & 2\beta \end{bmatrix} \begin{bmatrix} w_R \\ \phi_R \\ \psi_R \end{bmatrix} + \begin{bmatrix} W_{R0} \\ X_{R0} \\ Y_{R0} \end{bmatrix}$$

(5) Load functions $W_{L0}, X_{L0}, \ldots, Y_{R0}$ given as the last matrix column in (4) are the reactions due to loads and other causes at the fixed ends of the respective semi-infinite beams. Their analytical expressions are analogical to those of (8.15–8.16), and their shape functions are functions of αx and βx defined above. Four typical examples are shown below.

$W_{L0} = A_{11y}P_z$ $X_{L0} = A_{11x}Q_x$ $Y_{L0} = -A_{14y}\beta^{-1}P_z$	1. 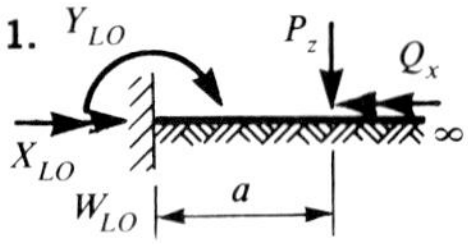	2. 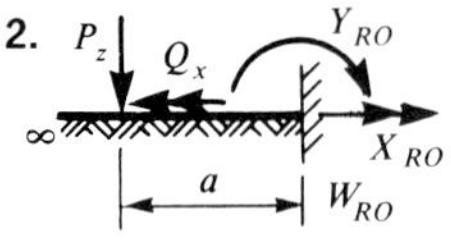	$W_{R0} = A_{11y}P_z$ $X_{R0} = A_{11x}Q_x$ $Y_{R0} = A_{14y}\beta^{-1}P_z$
$W_{L0} = A_{18y}\beta^{-1}p_z$ $X_{L0} = A_{12x}\alpha^{-1}q_x$ $Y_{L0} = -A_{16y}\beta^{-2}p_z$	3. 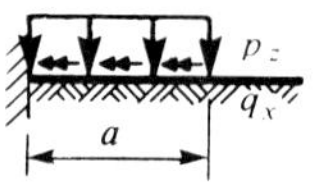	4. 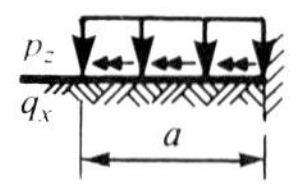	$W_{R0} = A_{18y}\beta^{-1}p_z$ $X_{R0} = A_{12x}\alpha^{-1}q_x$ $Y_{R0} = A_{16y}\beta^{-2}p_z$

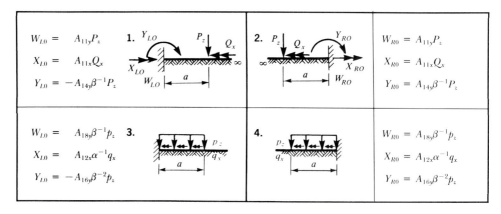

(6) State vectors at L, R or at any other section of the bar are calculated by the procedure described in (8.14) by means of the relations stated in this chapter.

14

Free Beam-Column of Order Two Constant Section

STATIC STATE

14.01 DEFINITION OF STATE

(1) System considered is a finite straight beam-column of constant and symmetrical section. The centroidal axis x of the bar and the vertical axis z define the plane of symmetry of this section. The axial force N is constant along the entire span. The bar is a part of a planar system acted on by loads normal to the system plane (system of order two).

(2) Notation (A.1, A.2, A.4)

s	= span (m)	A_x	= area of normal section (m^2)
a, b	= segments (m)	I_x	= torsional constant of A_x about x (m^4)
x, x'	= coordinates (m)	I_y	= moment of inertia of A_x about y (m^4)
E	= modulus of elasticity (Pa)	G	= modulus of rigidity (Pa)
W	= shearing force (N)	w	= linear displacement along z (m)
X	= torsional moment (N·m)	ϕ	= torsional slope about x (rad)
Y	= flexural moment (N·m)	ψ	= flexural slope about y (rad)

$$\eta = \begin{cases} +1 & \text{if } N = \text{axial tension (N)} \\ -1 & \text{if } N = \text{axial compression (N)} \end{cases} \qquad \beta = \sqrt{\frac{N}{EI_y}} = \text{shape parameter (m}^{-1}\text{)}$$

(3) Assumptions of analysis are stated in (1.01). In this chapter, the effects of torsional and flexural deformations and the beam-column effects are included. Since only long and slender bars are capable of developing the beam-column action, the effects of shearing deformations are neglected.

14.02 TRANSPORT MATRIX EQUATIONS

(1) Equivalents and signs

$\overline{X} = Xs^{-1}$ = scaled X (N)	$\overline{Y} = Y\beta$ = scaled Y (N)
$\overline{\phi} = \phi GI_x s^{-2}$ = scaled ϕ (N)	$\overline{\psi} = \psi EI_y \beta^2$ = scaled ψ (N)
N = axial force (N)	$\overline{w} = wEI_y\beta^3$ = scaled w (N)

All forces, moments, and displacements are in the principal axes of the normal section, and their positive directions are shown in (2, 3) below.

(2) Free-body diagram

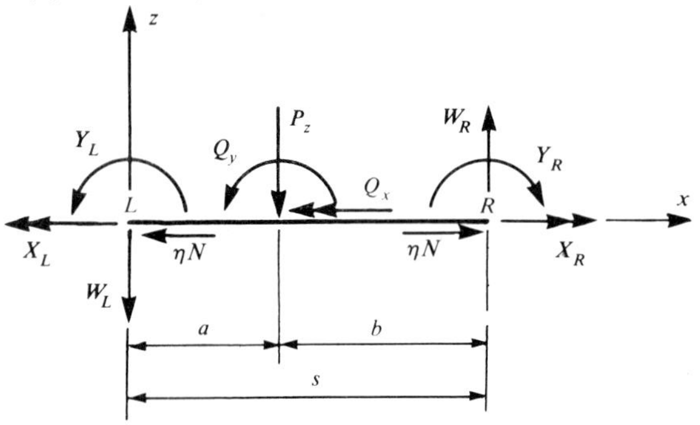

(3) Elastic curve

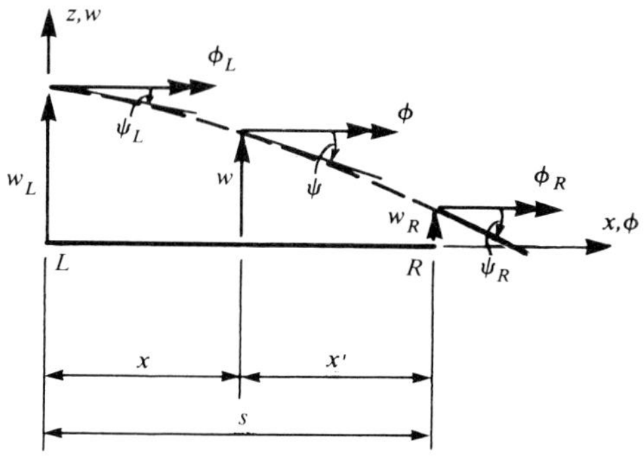

(4) Transport matrix equations

$$\underbrace{\begin{bmatrix} 1 \\ \hline W_L \\ \overline{X}_L \\ \overline{Y}_L \\ \hline \overline{w}_L \\ \overline{\phi}_L \\ \overline{\psi}_L \end{bmatrix}}_{H_L} = \underbrace{\left[\begin{array}{c|ccc|ccc} 1 & 0 & 0 & 0 & 0 & 0 & 0 \\ \hline F_W & 1 & 0 & 0 & 0 & 0 & 0 \\ F_X & 0 & 1 & 0 & 0 & 0 & 0 \\ F_Y & -T_{2y} & 0 & T_{1y} & 0 & 0 & -T_{0y} \\ \hline F_w & T_{4y} & 0 & -T_{3y} & 1 & 0 & T_{2y} \\ F_\phi & 0 & -1 & 0 & 0 & 1 & 0 \\ F_\psi & T_{3y} & 0 & -T_{2y} & 0 & 0 & T_{1y} \end{array}\right]}_{T_{LR}} \underbrace{\begin{bmatrix} 1 \\ \hline W_R \\ \overline{X}_R \\ \overline{Y}_R \\ \hline \overline{w}_R \\ \overline{\phi}_R \\ \overline{\psi}_R \end{bmatrix}}_{H_R}$$

$$\underbrace{\begin{bmatrix} 1 \\ \hline W_R \\ \overline{X}_R \\ \overline{Y}_R \\ \hline \overline{w}_R \\ \overline{\phi}_R \\ \overline{\psi}_R \end{bmatrix}}_{H_R} = \underbrace{\left[\begin{array}{c|ccc|ccc} 1 & 0 & 0 & 0 & 0 & 0 & 0 \\ \hline G_W & 1 & 0 & 0 & 0 & 0 & 0 \\ G_X & 0 & 1 & 0 & 0 & 0 & 0 \\ G_Y & T_{2y} & 0 & T_{1y} & 0 & 0 & T_{0y} \\ \hline G_w & -T_{4y} & 0 & -T_{3y} & 1 & 0 & -T_{2y} \\ G_\phi & 0 & 1 & 0 & 0 & 1 & 0 \\ G_\psi & T_{3y} & 0 & T_{2y} & 0 & 0 & T_{1y} \end{array}\right]}_{T_{RL}} \underbrace{\begin{bmatrix} 1 \\ \hline W_L \\ \overline{X}_L \\ \overline{Y}_L \\ \hline \overline{w}_L \\ \overline{\phi}_L \\ \overline{\psi}_L \end{bmatrix}}_{H_L}$$

where H_L, H_R are the *end state vectors* and T_{LR}, T_{RL} are the respective *transport matrices*. The elements of these matrices are the *transport coefficients* shown in (14.03) and the *load functions* F, G, listed for particular cases in (14.04–14.06).

14.03 TRANSPORT RELATIONS

$$\beta = \sqrt{\frac{N}{EI_y}}$$

(1) Transport functions

Axial tension ($\eta = +1$)	Axial compression ($\eta = -1$)
$T_{0y}(x) = \sinh \beta x$	$T_{0y}(x) = -\sin \beta x$
$T_{1y}(x) = \cosh \beta x$	$T_{1y}(x) = \cos \beta x$
$T_{2y}(x) = \sinh \beta x$	$T_{2y}(x) = \sin \beta x$
$T_{3y}(x) = \cosh \beta x - 1$	$T_{3y}(x) = 1 - \cos \beta x$
$T_{4y}(x) = \sinh \beta x - \beta x$	$T_{4y}(x) = \beta x - \sin \beta x$
$T_{5y}(x) = T_{3z}(x) - \dfrac{(\beta x)^2}{2!}$	$T_{5y}(x) = \dfrac{(\beta x)^2}{2!} - T_{3z}(x)$
$T_{6y}(x) = T_{4z}(x) - \dfrac{(\beta x)^3}{3!}$	$T_{6y}(x) = \dfrac{(\beta x)^3}{3!} - T_{4z}(x)$

The graphs and tables of these functions are identical to (9.04–9.05).

(2) Transport coefficients T_j, A_j, B_j, C_j are the values of $T_j(x)$ for a particular argument x.

$$\left.\begin{array}{ll} \text{For } x = s, & T_{jy}(s) = T_{jy} \\ \text{For } x = a, & T_{jy}(a) = A_{jy} \\ \text{For } x = b, & T_{jy}(b) = B_{jy} \\ \text{and } T_{jy} - A_{jy} = C_{jy} & \end{array}\right\} \quad j = 0, 1, 2, \ldots, 6$$

(3) Transport matrices of this bar have analogical properties to those of (11.02) and can be used in similar ways.

(4) Differential relations

$\dfrac{d\phi}{dx} = \dfrac{x}{GI_x}$	$\dfrac{dw}{dx} = -\psi$
$\dfrac{d^2\phi}{dx^2} = \dfrac{q_x}{GI_x}$	$\dfrac{d^2w}{dx^2} = -\dfrac{d\psi}{dw} = -\dfrac{Y}{EI_y}$
	$\dfrac{d^3w}{dx^3} = -\dfrac{d^2\psi}{dx^2} = -\dfrac{dY}{dx\, EI_y} = -\dfrac{W}{EI_y} + \dfrac{\eta N\, dw}{EI_y\, dx}$
	$\dfrac{d^4w}{dx^4} = -\dfrac{d^3\psi}{dx^3} = -\dfrac{d^2Y}{dx^2\, EI_y} = -\dfrac{p_z}{EI_y} + \dfrac{\eta N\, d^2w}{EI_y\, dx^2}$

(5) Two forms of solution are shown in each matrix in this chapter: one corresponding to axial tension with $\eta = +1$ (expressed in terms of hyperbolic functions) and the other one corresponding to the axial compression with $\eta = -1$ (expressed in terms of trigonometric functions). The general expressions for the load functions in (14.04–14.06) must be evaluated in terms of functions corresponding to the given state.

(6) General transport matrix equations for beam-columns of order three are obtained by matrix composition of (9.02) and (14.02) as shown in (2.18).

14.04 TRANSPORT LOAD FUNCTIONS

Notation (14.01–14.03) Signs (14.02) Matrix (14.02) P_z = concentrated load (N) Q_x, Q_y = applied couples (N·m) q_y = intensity of distributed couple (N·m/m) A_j, B_j, C_j, T_j = transport coefficients (14.03)		$\beta = \sqrt{\dfrac{N}{EI_y}}$
$F_W = -P_z$ $F_X = 0$ $F_Y = A_{2y}P_z$	**1.** 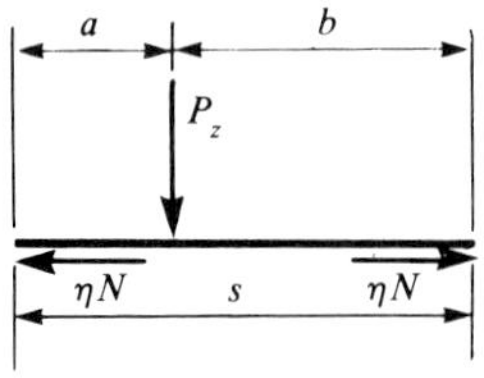	$G_W = P_z$ $G_X = 0$ $G_Y = B_{2y}P_z$
$F_w = -A_{4y}P_z$ $F_\phi = 0$ $F_\psi = -A_{3y}P_z$		$G_w = -B_{4y}P_z$ $G_\phi = 0$ $G_\psi = B_{3y}P_z$
$F_W = 0$ $F_X = -s^{-1}Q_x$ $F_Y = -A_{1y}\beta Q_y$	**2.** 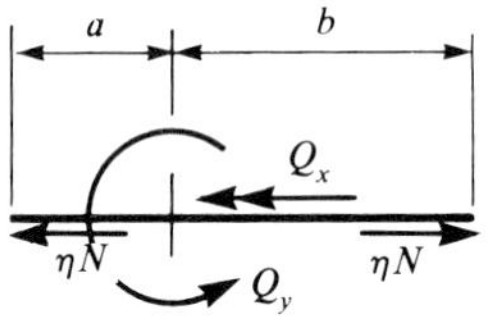	$G_W = 0$ $G_X = s^{-1}Q_x$ $G_Y = B_{1y}\beta Q_y$
$F_w = A_{3y}\beta Q_y$ $F_\phi = as^{-2}Q_x$ $F_\psi = A_{2y}\beta Q_y$		$G_w = -B_{3y}\beta Q_y$ $G_\phi = bs^{-2}Q_x$ $G_\psi = A_{2y}\beta Q_y$
$F_W = 0$ $F_X = 0$ $F_Y = -T_{2y}q_y$	**3.** 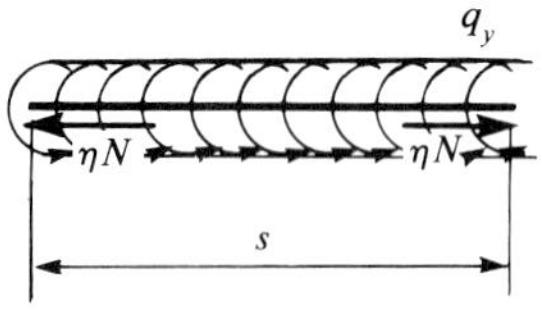	$G_W = 0$ $G_X = 0$ $G_Y = T_{2y}q_y$
$F_w = T_{4y}q_y$ $F_\phi = 0$ $F_\psi = T_{3y}q_y$		$G_w = -T_{4y}q_y$ $G_\phi = 0$ $G_\psi = T_{3y}q_y$
$F_W = 0$ $F_X = 0$ $F_Y = -C_{2y}q_y$	**4.** 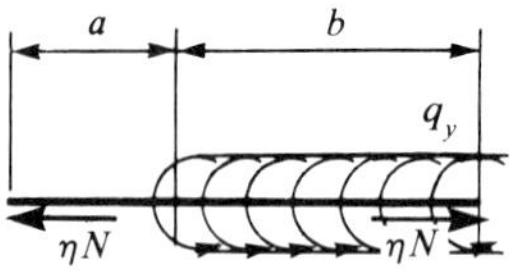	$G_W = 0$ $G_X = 0$ $G_Y = B_{2y}q_y$
$F_W = C_{4y}q_y$ $F_\phi = 0$ $F_\psi = C_{3y}q_y$		$G_w = -B_{4y}q_y$ $G_\phi = 0$ $G_\psi = B_{3y}q_y$

14.05 TRANSPORT LOAD FUNCTIONS

Notation (14.01–14.03)   Signs (14.02)   Matrix (14.02)

p_z = intensity of distributed load (N/m)

q_x = intensity of distributed couple (N·m/m)

A_j, B_j, C_j, T_j = transport coefficients (14.03)

$$\beta = \sqrt{\frac{N}{EI_y}}$$

1.

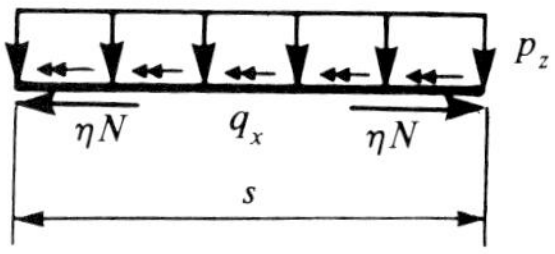

$F_W = -sp_z$

$F_X = -q_x$

$F_Y = T_{3y}\beta^{-1}p_z$

$F_w = -T_{5y}\beta^{-1}p_z$

$F_\phi = \frac{1}{2}q_x$

$F_\psi = -T_{4y}\beta^{-1}p_z$

$G_W = sp_z$

$G_X = q_x$

$G_Y = T_{3y}\beta^{-1}p_z$

$G_w = -T_{5y}\beta^{-1}p_z$

$G_\phi = \frac{1}{2}q_x$

$G_\psi = T_{4y}\beta^{-1}p_z$

2.

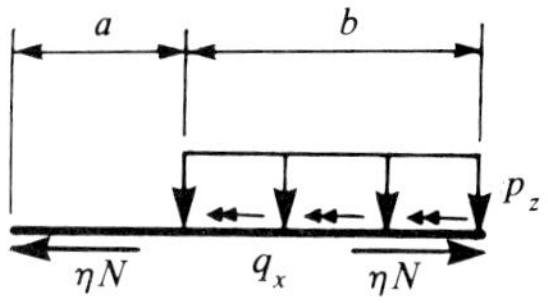

$F_W = -bp_z$

$F_X = -bs^{-1}q_x$

$F_Y = C_{3y}\beta^{-1}p_z$

$F_w = -C_{5y}\beta^{-1}p_z$

$F_\phi = \frac{1}{2}(s^2 - a^2)s^{-2}q_x$

$F_\psi = -C_{4y}\beta^{-1}p_z$

$G_W = bp_z$

$G_X = bs^{-1}q_x$

$G_y = B_{3y}\beta^{-1}p_z$

$G_w = -B_{5y}\beta^{-1}p_z$

$G_\phi = \frac{1}{2}b^2s^{-2}q_x$

$G_\psi = B_{4y}\beta^{-1}p_z$

3.

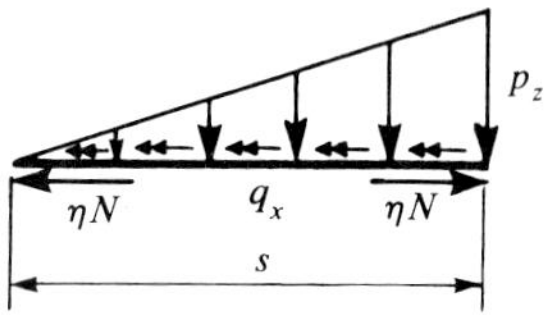

$F_W = -\frac{1}{2}sp_z$

$F_X = -\frac{1}{2}q_x$

$F_Y = (\beta sT_{3y} - T_{4y})\beta^{-2}s^{-1}p_z$

$F_w = -(\beta sT_{5y} - T_{6y})\beta^{-2}s^{-1}p_z$

$F_\phi = \frac{1}{6}q_x$

$F_\psi = -(\beta sT_{4y} - T_{5y})\beta^{-2}s^{-1}p_z$

$G_W = \frac{1}{2}sp_z$

$G_X = \frac{1}{2}q_x$

$G_Y = T_{4y}\beta^{-2}s^{-1}p_z$

$G_w = -T_{6y}\beta^{-2}s^{-1}p_z$

$G_\phi = \frac{1}{6}q_x$

$G_\psi = T_{5y}\beta^{-2}s^{-1}p_z$

4.

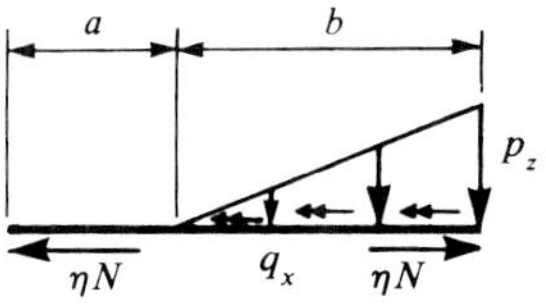

$F_W = -\frac{1}{2}bp_z$

$F_X = -\frac{1}{2}bs^{-1}q_x$

$F_Y = (\beta bT_{3y} - C_{4y})\beta^{-2}b^{-1}p_z$

$F_w = -(\beta bT_{5y} - C_{6y})\beta^{-2}b^{-1}p_z$

$F_\phi = \frac{1}{6}(3s - b)bs^{-2}q_x$

$F_\psi = (\beta bT_{4y} - C_{5y})\beta^{-2}b^{-1}p_z$

$G_W = \frac{1}{2}bp_z$

$G_X = \frac{1}{2}bs^{-1}q_x$

$G_Y = B_{4y}\beta^{-2}b^{-1}p_z$

$G_w = -B_{6y}\beta^{-2}b^{-1}p_z$

$G_\phi = \frac{1}{6}b^2s^{-2}q_x$

$G_\psi = B_{5y}\beta^{-2}b^{-1}p_z$

14.06 TRANSPORT LOAD FUNCTIONS

Notation (14.01–14.03) Signs (14.02) Matrix (14.02)

w_j, ϕ_j, ψ_j = externally applied displacements (m, rad, rad)

g_t = thermal strain (11.03)

A_j, B_j, C_j = transport coefficients (14.03)

$$\beta = \sqrt{\frac{N}{EI_y}}$$

1. Abrupt displacements w_j and ϕ_j

$F_W = 0$

$F_X = 0$

$F_Y = EI_y A_{1y} \beta^3 \eta w_j$

$F_w = -EI_y(A_{3y} + \eta)\beta^3 w_j$

$F_\phi = -GI_x \alpha^2 \phi_j$

$F_\psi = -EI_y A_{2y} \beta^3 \eta w_j$

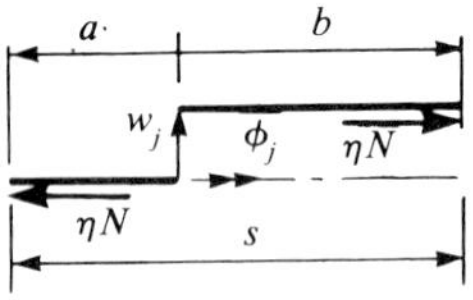

$G_W = 0$

$G_X = 0$

$G_Y = -EI_y B_{1y} \beta^3 \eta w_j$

$G_w = EI_y(B_{3y} + \eta)\beta^3 w_j$

$G_\phi = GI_x \alpha^2 \phi_j$

$G_\psi = -EI_y B_{2y} \beta^3 \eta w_j$

2. Abrupt displacement ψ_j

$F_W = 0$

$F_X = 0$

$F_Y = EI_y A_{0y} \beta^2 \psi_j$

$F_w = -EI_y A_{2y} \beta^2 \psi_j$

$F_\phi = 0$

$F_\psi = -EI_y A_{1y} \beta^2 \psi_j$

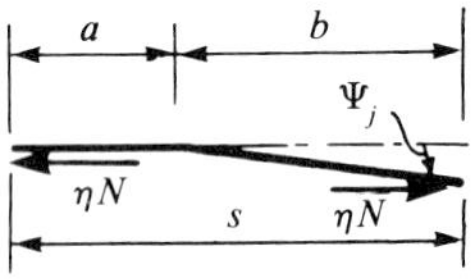

$G_W = 0$

$G_X = 0$

$G_Y = EI_y B_{0y} \beta^2 \psi_j$

$G_w = -EI_y B_{2y} \beta^2 \psi_j$

$G_\phi = 0$

$G_\psi = EI_y B_{1y} \beta^2 \psi_j$

3. Temperature deformation of segment b

$F_W = 0$

$F_X = 0$

$F_Y = EI_y C_{1y} \beta g_t$

$F_w = -EI_y C_{3y} \beta g_t$

$F_\phi = 0$

$F_\psi = -EI_y C_{2y} \beta g_t$

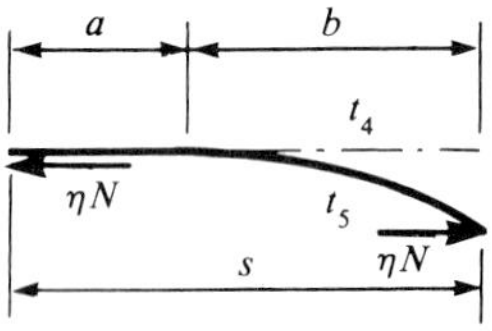

$G_W = 0$

$G_X = 0$

$G_Y = EI_y B_{1y} \beta g_t$

$G_w = -EI_y B_{3y} \beta g_t$

$G_\phi = 0$

$G_\psi = EI_y B_{2y} \beta g_t$

4.

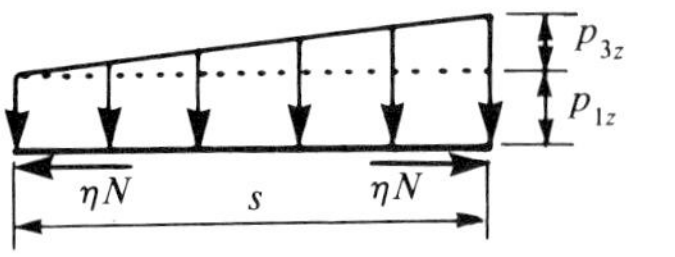

Sec. 14.05 (case 1 + case 3)

5.

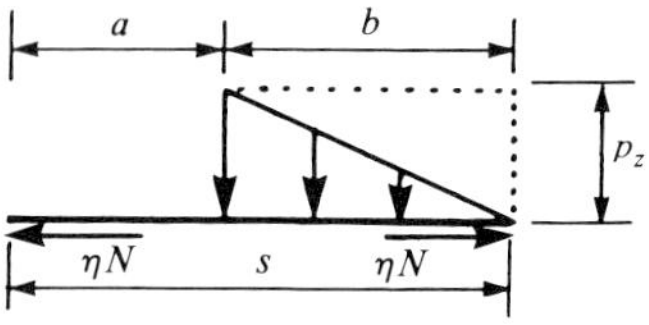

Sec. 14.05 (case 2 − case 4)

14.07 STIFFNESS MATRIX EQUATIONS

$$r = \frac{s^2 GI_x}{EI_y} \qquad \beta = \sqrt{\frac{N}{EI_y}}$$

(1) Equivalents and signs. All forces, moments, and displacements are in the principal axes of the normal section, and their positive directions are shown in (2, 3) below.

(2) Free-body diagram

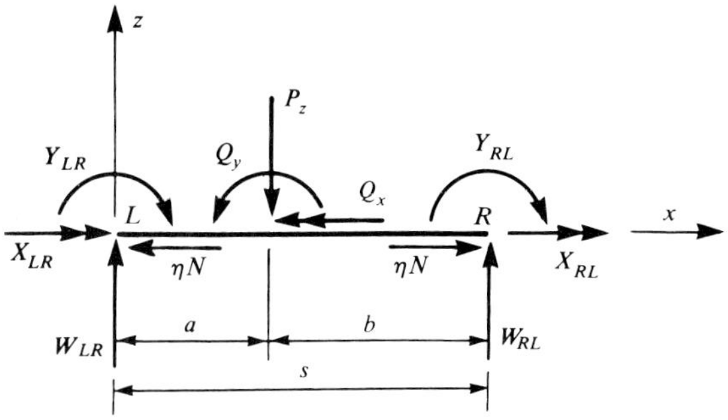

(3) Elastic curve

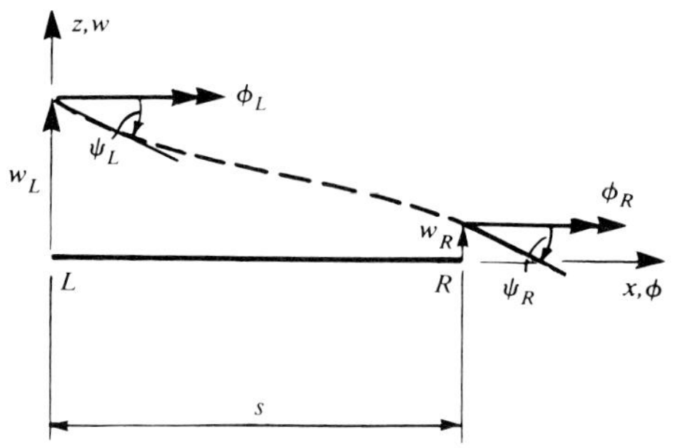

(4) Dimensionless stiffness coefficients

Axial tension ($\eta = +1$)	Axial compression ($\eta = -1$)
$D_0 = \dfrac{1}{2(1 - \cosh \beta s) + \beta s \sinh \beta s}$	$D_0 = \dfrac{1}{2(1 - \cos \beta s) - \beta s \sin \beta s}$
$K_1 = (\beta s)^3 D_0 \sinh \beta s$ $K_2 = (\beta s)^2 D_0(\cosh \beta s - 1)$ $K_3 = \beta s D_0(\beta s \cosh \beta s - \sinh \beta s)$ $K_4 = \beta s D_0 (\sinh \beta s - \beta s)$	$K_1 = (\beta s)^3 D_0 \sin \beta s$ $K_2 = (\beta s)^2 D_0(1 - \cos \beta s)$ $K_3 = \beta s D_0(\sin \beta s - \beta s \cos \beta s)$ $K_4 = \beta s D_0(\beta s - \sin \beta s)$
$L_1 = D_0 \sinh \beta s$ $L_2 = D_0(\cosh \beta s - 1)$ $L_3 = D_0(\beta s \cosh \beta s - \sinh \beta s)$ $L_4 = D_0(\sinh \beta s - \beta s)$	$L_1 = D_0 \sin \beta s$ $L_2 = D_0(1 - \cos \beta s)$ $L_3 = D_0(\sin \beta s - \beta s \cos \beta s)$ $L_4 = D_0(\beta s - \sin \beta s)$

(5) Stiffness matrix equations

$$\begin{bmatrix} W_{LR} \\ X_{LR} \\ Y_{LR} \\ \hline W_{RL} \\ X_{RL} \\ Y_{RL} \end{bmatrix} = \frac{EI_y}{s^3} \begin{bmatrix} K_1 & 0 & -K_2 s & -K_1 & 0 & -K_2 s \\ 0 & r & 0 & 0 & -r & 0 \\ -K_2 s & 0 & K_3 s^2 & K_2 s & 0 & K_4 s^2 \\ \hline -K_1 & 0 & K_2 s & K_1 & 0 & K_2 s \\ 0 & -r & 0 & 0 & r & 0 \\ -K_2 s & 0 & K_4 s^2 & K_2 s & 0 & K_3 s^2 \end{bmatrix} \begin{bmatrix} w_L \\ \phi_L \\ \psi_L \\ \hline w_R \\ \phi_R \\ \psi_R \end{bmatrix} + \begin{bmatrix} W_{L0} \\ X_{L0} \\ Y_{L0} \\ \hline W_{R0} \\ X_{R0} \\ Y_{R0} \end{bmatrix}$$

where W_{L0}, X_{L0}, Y_{L0} and W_{R0}, X_{R0}, Y_{R0} are the *reactions of fixed-end beam-columns* due to loads and other causes shown in (14.08). The graphs and tables of stiffness coefficients are identical to (9.11–9.13).

(6) General stiffness matrix equations for beam-columns of order three are obtained by matrix composition of (9.09) and (14.07) as shown in (3.19).

14.08 STIFFNESS LOAD FUNCTIONS FIXED-END BEAM

(1) General form of the stiffness load functions in (14.07) is given by the matrix equations

$$\begin{bmatrix} W_{LO} \\ X_{LO} \\ Y_{LO} \end{bmatrix} = \begin{bmatrix} L_1 & 0 & L_2 \\ 0 & s & 0 \\ -L_2\beta^{-1} & 0 & -L_4\beta^{-1} \end{bmatrix} \begin{bmatrix} G_w \\ G_\phi \\ G_\psi \end{bmatrix}$$

$$\begin{bmatrix} W_{RO} \\ X_{RO} \\ Y_{RO} \end{bmatrix} = \begin{bmatrix} -L_1 & 0 & -L_2 \\ 0 & s & 0 \\ L_2\beta^{-1} & 0 & -L_4\beta^{-1} \end{bmatrix} \begin{bmatrix} F_w \\ F_\phi \\ F_\psi \end{bmatrix}$$

where F_w, F_ϕ, F_ψ are the left-end transport load functions and G_w, G_ϕ, G_ψ are their right-end counterparts, given for particular loads and causes in (14.04–14.06), and L_1, L_2, L_4 are the stiffness coefficients (14.07). The load vectors and other causes are defined by symbols introduced in (14.04–14.06).

(2) Particular cases

Notation (14.01–14.03, 14.07) A_j, B_j, C_j, T_j = transport coefficients (14.03)	Signs (14.07)	$\beta = \sqrt{\dfrac{N}{EI_y}}$
$W_{LO} = (L_2B_{3y} - L_1B_{4y})P_z$ $X_{LO} = bs^{-1}Q_x$ $Y_{LO} = -(L_4B_{3y} - L_2B_{4y})\beta^{-1}P_z$	**1.** 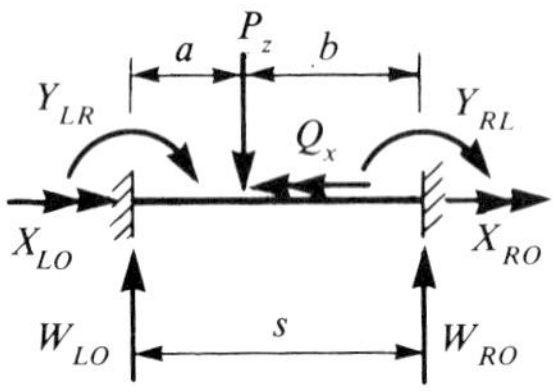	$W_{RO} = (L_2A_{3y} - L_1A_{4y})P_z$ $X_{RO} = as^{-1}Q_x$ $Y_{RO} = (L_4A_{3y} - L_2A_{4y})\beta^{-1}P_z$
$W_{LO} = (L_2B_{2y} - L_1B_{4y})\beta Q_y$ $X_{LO} = 0$ $Y_{LO} = -(L_4B_{2y} - L_2B_{4y})Q_y$	**2.** 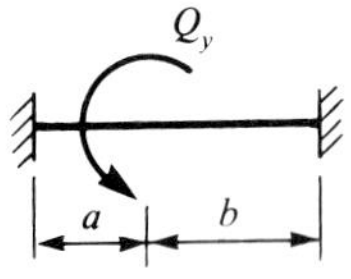	$W_{RO} = (L_2A_{2y} - L_1A_{3y})\beta Q_y$ $X_{RO} = 0$ $Y_{RO} = -(L_4A_{2y} - L_2A_{4y})Q_y$
$W_{LO} = \frac{1}{2}sp_z$ $X_{LO} = \frac{1}{2}sq_x$ $Y_{LO} = -\left(1 - \dfrac{bs}{2}\cot\dfrac{bs}{2}\right)\beta^{-1}p_z$	**3. Axial compression** 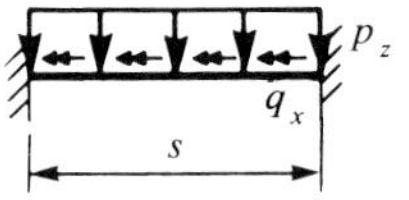	$W_{RO} = W_{LO}$ $X_{RO} = X_{LO}$ $Y_{RO} = -Y_{LO}$

$$W_{L0} = \tfrac{1}{2}P_z = W_{R0}$$

$$X_{L0} = \tfrac{1}{2}Q_x = X_{R0}$$

$$Y_{L0} = \frac{1}{2\beta}(\tan \tfrac{1}{4}\beta s)P_z = -Y_{R0}$$

4. Axial compression

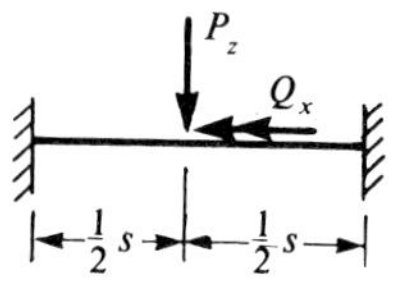

$$W_{L0} = -\frac{(1 - \cos \frac{1}{2}\beta s)\beta}{2(\sin \frac{1}{2}\beta s - \frac{1}{2}\beta s \cos \frac{1}{2}\beta s)}Q_y = -W_{R0}$$

$$X_{L0} = 0 = X_{R0}$$

$$Y_{L0} = -\frac{(\frac{1}{2}\beta s - \sin \frac{1}{2}\beta s)}{2(\sin \frac{1}{2}\beta s - \frac{1}{2}\beta s \cos \frac{1}{2}\beta s)}Q_y = Y_{R0}$$

5. Axial compression

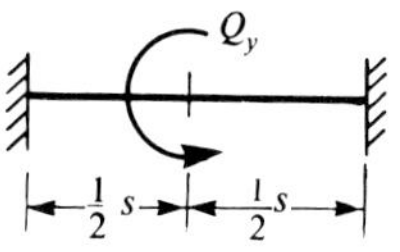

6.

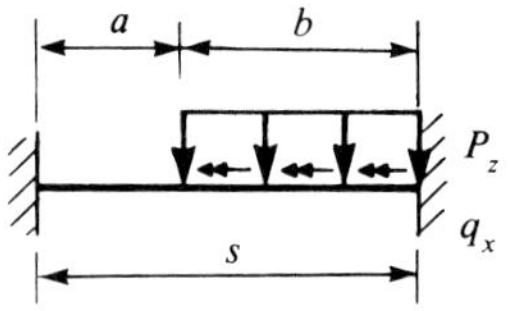

Use general form in terms of (14.05-2)

7.

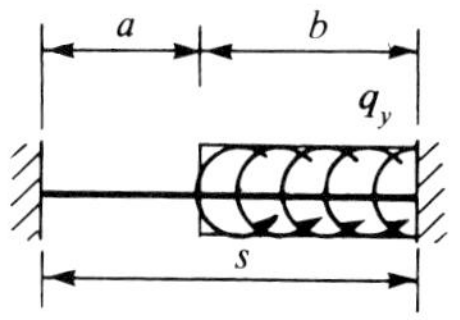

Use general form in terms of (14.04-4)

8.

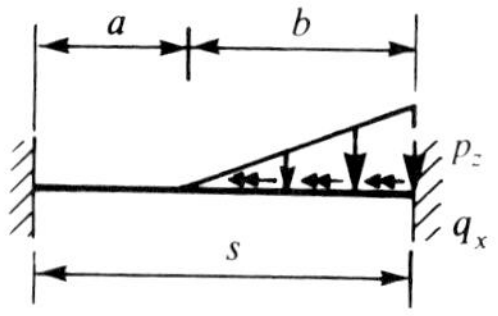

Use general form in terms of (14.04-4)

9. Abrupt displacements w_j, ϕ_j

$$W_{L0} = EI_y(L_1B_{1y} - L_2B_{2y})\,\beta^3\eta w_j$$

$$X_{L0} = -GI_x s^{-1}\phi_j$$

$$Y_{L0} = -EI_y(L_2B_{1y} - L_4B_{2y})\,\beta^2\eta w_j$$

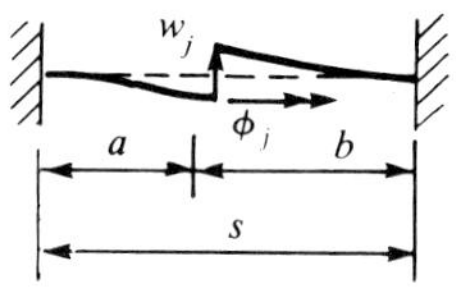

$$W_{R0} = -W_{L0}$$

$$X_{R0} = -X_{L0}$$

$$Y_{R0} = -EI_y(L_2A_{1y} + L_4A_{2y})\,\beta^2\eta w_j$$

10. Abrupt displacement ψ_j

$$W_{L0} = -EI_y(L_1B_{2y} - L_2B_{1y})\,\beta^2\psi_j$$

$$X_{L0} = 0$$

$$Y_{L0} = EI_y(L_2B_{2y} - L_4B_{1y})\,\beta^2\psi_j$$

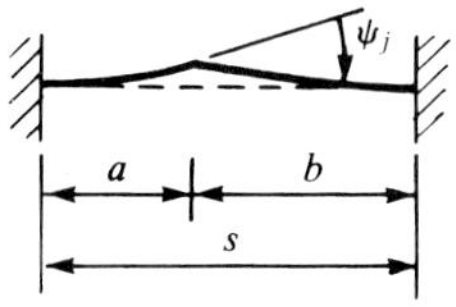

$$W_{R0} = -W_{L0}$$

$$X_{R0} = 0$$

$$Y_{R0} = -EI_y(L_2A_{2y} - L_4A_{1y})\,\beta^2\psi_j$$

11. Temperature deformation of segment b

$$W_{L0} = -EI_y(L_1B_{3y} - L_4B_{2y})bg_t$$

$$X_{L0} = 0$$

$$Y_{L0} = EI_y(L_2B_{3z} - L_4B_{2y})g_t$$

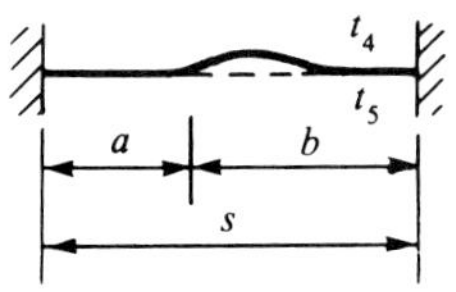

$$W_{R0} = -W_{L0}$$

$$X_{R0} = 0$$

$$Y_{R0} = -EI_y(L_2C_{3y} - L_4C_{2y})g_t$$

14.09 MODIFIED STIFFNESS MATRIX EQUATIONS, HINGED END

$$\beta = \sqrt{\frac{N}{EI_y}}$$

(1) Dimensionless modified stiffness coefficients

Axial tension ($\eta = +1$)	Axial compression ($\eta = -1$)
$D_1 = \dfrac{1}{\beta s \cosh \beta s - \sinh \beta s}$	$D_1 = \dfrac{1}{\sin \beta s - \beta s \cos \beta s}$
$K_5 = (\beta s)^3 D_1 \cosh \beta s$	$K_5 = (\beta s)^3 D_1 \cos \beta s$
$K_6 = (\beta s)^2 D_1 \sinh \beta s$	$K_6 = (\beta s)^2 D_1 \sin \beta s$
$L_5 = D_1 \cosh \beta s$	$L_5 = D_1 \cos \beta s$
$L_6 = D_1 \sinh \beta s$	$L_6 = D_1 \sin \beta s$
$L_7 = D_1(\cosh \beta s - 1)$	$L_7 = D_1(1 - \cos \beta s)$
$L_8 = D_1(\sinh \beta s - \beta s)$	$L_8 = D_1(\beta s - \sin \beta s)$

(2) Modifications of the stiffness matrix equations (14.07) due to hinged condition at one end are shown below. Because the hinge does not affect the expressions for X_{LR} and X_{RL}, only the remaining part of the stiffness matrix equation is presented.

(3) Left end hinged ($Y_{LR} = 0$)

$$\begin{bmatrix} W_{LR} \\ W_{RL} \\ Y_{RL} \end{bmatrix} = \frac{EI_y}{s^3} \begin{bmatrix} K_5 & -K_5 & -K_6 s \\ -K_5 & K_5 & K_6 s \\ -K_6 s & K_6 s & K_6 s^2 \end{bmatrix} \begin{bmatrix} w_L \\ w_R \\ \psi_R \end{bmatrix} + \begin{bmatrix} W^*_{LO} \\ W^*_{RO} \\ Y^*_{RO} \end{bmatrix}$$

$$\begin{bmatrix} W^*_{LO} \\ W^*_{RO} \\ Y^*_{RO} \end{bmatrix} = \begin{bmatrix} -1 & -L_7 & -L_5 \\ 0 & L_7 & L_5 \\ 0 & L_8\beta^{-1} & L_6\beta^{-1} \end{bmatrix} \begin{bmatrix} F_W \\ F_Y \\ F_w \end{bmatrix}$$

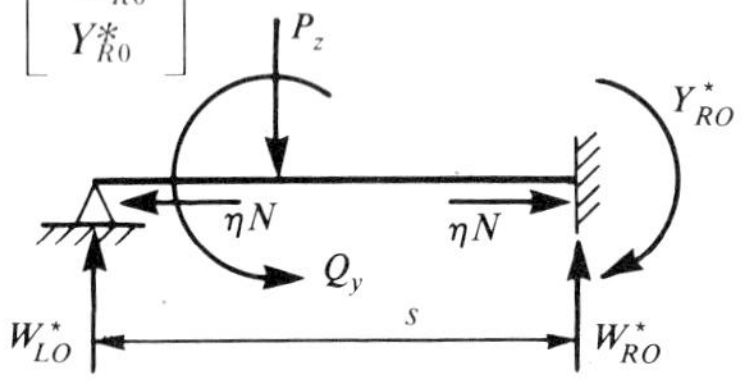

(4) Right end hinged ($Y_{RL} = 0$)

$$\begin{bmatrix} W_{LR} \\ Y_{LR} \\ W_{RL} \end{bmatrix} = \frac{EI_y}{s^3} \begin{bmatrix} K_5 & -K_6 s & -K_5 \\ -K_6 s & K_6 s^2 & K_6 s \\ -K_5 & K_6 s & K_5 \end{bmatrix} \begin{bmatrix} w_L \\ \psi_L \\ w_R \end{bmatrix} + \begin{bmatrix} W^*_{LO} \\ Y^*_{LO} \\ W^*_{RO} \end{bmatrix}$$

$$\begin{bmatrix} W^*_{LO} \\ Y^*_{LO} \\ W^*_{RO} \end{bmatrix} = \begin{bmatrix} 0 & L_7 & L_5 \\ 0 & -L_8\beta^{-1} & -L_6\beta^{-1} \\ 1 & -L_7 & -L_5 \end{bmatrix} \begin{bmatrix} G_W \\ G_Y \\ G_w \end{bmatrix}$$

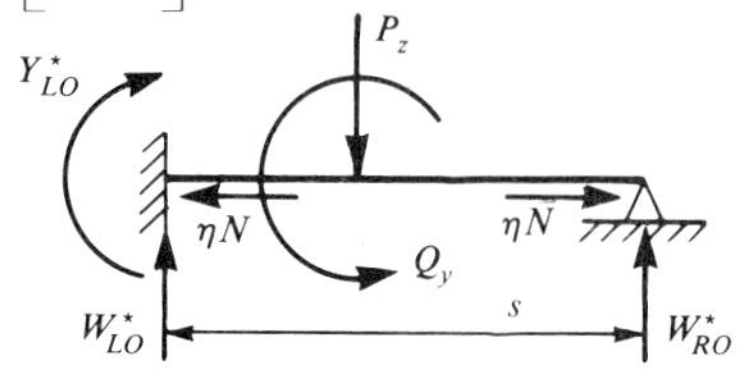

(5) Modified load functions designated by an asterisk and shown next to the respective figures are the end reactions of the respective propped-end beams. Their analytical expressions are a linear combination of the modified stiffness coefficients L_j in (1) above and the respective transport load functions (14.04–14.06).

15

Beam-Column of Order Two Encased in Elastic Foundation Constant Section

STATIC STATE

15.01 DEFINITION OF STATE

(1) System considered is a finite straight beam-column of constant and symmetrical section. The centroidal axis x of the bar and the vertical axis z define the plane of symmetry of this section. The elastic foundation fully surrounds the bar and offers resistance, which is proportional to the torsional and flexural displacement of the bar (Winkler's foundation). The constants of proportionality, corresponding to a unit length of the bar, called foundation moduli, are

k_ϕ = torsional modulus about x (N·m/m)	k_w = transverse modulus along z (N/m^2)

The axial force N is constant along the entire span s, and the bar is a part of a planar system acted on by loads normal to the system plane (system of order two).

(2) Notation (A.1, A.2, A.4)

s = span (m)	A_x = area of normal section (m^2)
a, b = segments (m)	I_x = torsional constant of A_x about x (m^4)
x, x' = coordinates (m)	I_y = moment of inertia of A_x about y (m^4)
E = modulus of elasticity (Pa)	G = modulus of rigidity (Pa)
W = shearing force (N)	w = linear displacement along z (m)
X = torsional moment (N·m)	ϕ = torsional slope about x (rad)
Y = flexural moment (N·m)	ψ = flexural slope about y (rad)

$\eta = \begin{cases} +1 & \text{if } N = \text{axial tension (N)} \\ -1 & \text{if } N = \text{axial compression (N)} \end{cases}$	$e = \sqrt{\dfrac{k_w}{EI_y}}$ (m^{-2})	$f = \dfrac{N}{EI_y}$ (m^{-2})
$\alpha = \sqrt{\dfrac{k_\phi}{GI_x}}$ = shape parameter (m^{-1})	$R = \dfrac{\eta f}{e}$	$\lambda = \sqrt[4]{\dfrac{k_w}{EI_y}}$ = shape parameter (m^{-1})

(3) Assumptions of analysis are stated in (1.01). In this chapter, the effects of torsional and flexural deformations and the beam-column effects are included. Since only long and slender bars are capable of developing the beam-column action, the effects of shearing deformations are neglected.

15.02 TRANSPORT MATRIX EQUATIONS

(1) Equivalents and signs

$\overline{X} = X\alpha = \text{scaled } X \text{ (N)}$ $\qquad$ $\overline{Y} = Y\lambda = \text{scaled } Y \text{ (N)}$

$\overline{\phi} = \phi GI_x\alpha^2 = \text{scaled } \phi \text{ (N)}$ $\qquad$ $\overline{\psi} = \psi EI_y\lambda^2 = \text{scaled } \psi \text{ (N)}$

$N = \text{axial force (N)}$ $\qquad$ $\overline{w} = wEI_y\lambda^3 = \text{scaled } w \text{ (N)}$

All forces, moments, and displacements are in the principal axes of the normal section, and their positive directions are shown in (2, 3) below.

(2) Free-body diagram

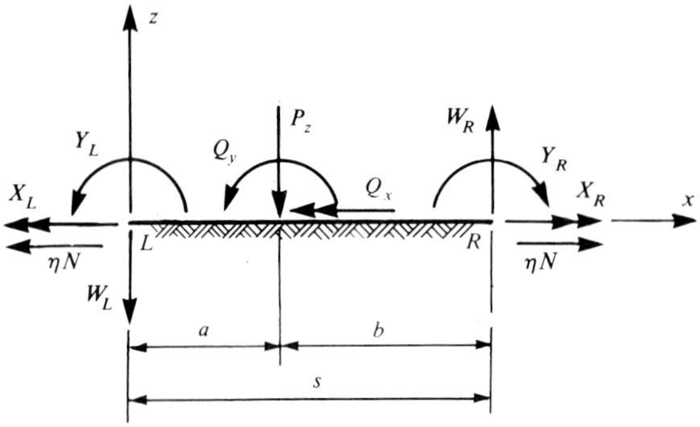

(3) Elastic curve

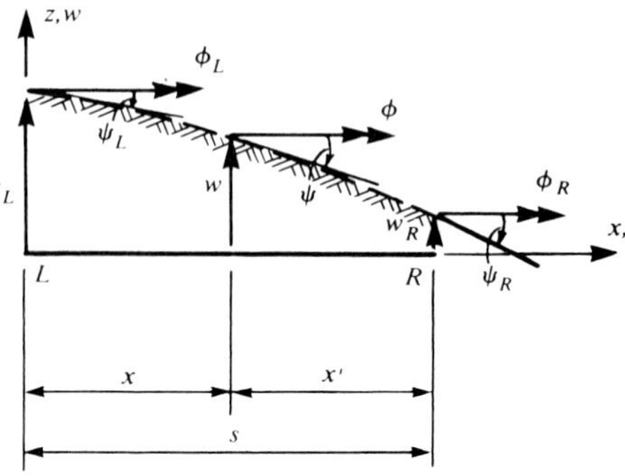

(4) Transport matrix equations

$$
\underbrace{\begin{bmatrix} 1 \\ \hline W_L \\ \overline{X}_L \\ \overline{Y}_L \\ \hline \overline{w}_L \\ \overline{\phi}_L \\ \overline{\psi}_L \end{bmatrix}}_{H_L}
=
\underbrace{\left[\begin{array}{c|ccc|ccc}
1 & 0 & 0 & 0 & 0 & 0 & 0 \\ \hline
F_W & (1-T_{5y}) & 0 & T_{4y} & -(\lambda s - T_{6y}) & 0 & -T_{3y} \\
F_X & 0 & T_{1x} & 0 & 0 & -T_{2x} & 0 \\
F_Y & -T_{2y} & 0 & T_{1y} & T_{3y} & 0 & -T_{0y} \\ \hline
F_w & T_{4y} & 0 & -T_{3y} & (1-T_{5y}) & 0 & T_{2y} \\
F_\phi & 0 & -T_{2x} & 0 & 0 & T_{1x} & 0 \\
F_\psi & T_{3y} & 0 & -T_{2y} & -T_{4y} & 0 & T_{1y}
\end{array}\right]}_{T_{LR}}
\underbrace{\begin{bmatrix} 1 \\ \hline W_R \\ \overline{X}_R \\ \overline{Y}_R \\ \hline \overline{w}_R \\ \overline{\phi}_R \\ \overline{\psi}_R \end{bmatrix}}_{H_R}
$$

$$
\underbrace{\begin{bmatrix} 1 \\ \hline W_R \\ \overline{X}_R \\ \overline{Y}_R \\ \hline \overline{w}_R \\ \overline{\phi}_R \\ \overline{\psi}_R \end{bmatrix}}_{H_R}
=
\underbrace{\left[\begin{array}{c|ccc|ccc}
1 & 0 & 0 & 0 & 0 & 0 & 0 \\ \hline
G_W & (1-T_{5y}) & 0 & -T_{4y} & (\lambda s - T_{6y}) & 0 & -T_{3y} \\
G_X & 0 & T_{1x} & 0 & 0 & T_{2x} & 0 \\
G_Y & T_{2y} & 0 & T_{1y} & T_{3y} & 0 & T_{0y} \\ \hline
G_w & -T_{4y} & 0 & -T_{3y} & (1-T_{5y}) & 0 & -T_{2y} \\
G_\phi & 0 & T_{2x} & 0 & 0 & T_{1x} & 0 \\
G_\psi & T_{3y} & 0 & T_{2y} & T_{4y} & 0 & T_{1y}
\end{array}\right]}_{T_{RL}}
\underbrace{\begin{bmatrix} 1 \\ \hline W_L \\ \overline{X}_L \\ \overline{Y}_L \\ \hline \overline{w}_L \\ \overline{\phi}_L \\ \overline{\psi}_L \end{bmatrix}}_{H_L}
$$

where H_L, H_R are the *state vectors* of the respective ends and T_{LR}, T_{RL} are the *transport matrices.* The elements of the transport matrices are the *transport coefficients* $T_{jx}(x)$, $T_{jy}(x)$ discussed in (15.03) and the *load functions* F, G, displayed for particular cases in (15.04–15.06).

15.03 TRANSPORT RELATIONS

(1) Differential relations

$\dfrac{d\phi}{dx} = \dfrac{X}{GI_x}$ $\dfrac{d^2\phi}{dx^2} = \dfrac{\phi k_\phi}{GI_x} + \dfrac{q_x}{GI_x}$	$\dfrac{dw}{dx} = -\psi$ $\dfrac{d^2w}{dx^2} = -\dfrac{d\psi}{dx} = -\dfrac{Y}{EI_y}$ $\dfrac{d^3w}{dx^3} = -\dfrac{d^2\psi}{dx^2} = -\dfrac{dY}{EI_y\,dx} = -\dfrac{W}{EI_y} + \dfrac{\eta N\,dw}{EI_y\,dx}$ $\dfrac{d^4w}{dx^4} = -\dfrac{d^3\psi}{dx^3} = -\dfrac{d^2Y}{EI_y\,dx^2} = -\dfrac{p_z}{EI_y} - \dfrac{wk_w}{EI_y} + \dfrac{\eta N\,d^2w}{EI_y\,dx^2}$

where q_x = intensity of distributed torque (N·m/m) and p_z = intensity of distributed load (N/m).

(2) General solution of the second-order differential equation and of the fourth-order differential equation in (1) above yields one set of transport functions $T_{jx}(x)$ and six sets of transport functions $T_{jy}(x)$. The torsional transport functions $T_{jx}(x)$ are identical in form to those in (13.03). The flexural transport functions $T_{jy}(x)$ are formally identical in form to those in (10.04) and depend on the relation of N to $2\sqrt{EI_y k_w}$ and the value η. They are classified for $N \neq 0$, $k_w \neq 0$ as:

Case I	$N < 2\sqrt{EI_z k_w}$	$\lvert R\rvert = \left\lvert\dfrac{\eta f}{e}\right\rvert < 2$	$\eta = \begin{cases} +1 \\ -1 \end{cases}$	N = axial tension N = axial compression
	$\Delta = \sqrt{4e^2 - f^2}$ $\beta = \frac{1}{2}\sqrt{2e + f}$ $\gamma = \frac{1}{2}\sqrt{2e - f}$ $\lambda = \sqrt{e}$			
Case II	$N = 2\sqrt{EI_z k_w}$	$\lvert R\rvert = \left\lvert\dfrac{\eta f}{e}\right\rvert = 2$	$\eta = \begin{cases} +1 \\ -1 \end{cases}$	N = axial tension N = axial compression
	$\Delta = 0$ $\beta = \gamma = \lambda = \sqrt{e}$			
Case III	$N > 2\sqrt{EI_z k_w}$	$\lvert R\rvert = \left\lvert\dfrac{\eta f}{e}\right\rvert > 2$	$\eta = \begin{cases} +1 \\ -1 \end{cases}$	N = axial tension N = axial compression
	$\Delta = \sqrt{f^2 - 4e^2}$ $\beta = \sqrt{\frac{1}{2}(f + \Delta)}$ $\gamma = \sqrt{\frac{1}{2}(f - \Delta)}$ $\lambda = \sqrt{e}$			

The analytical forms of $T_{jy}(x)$ and their series expansions are formally identical to those in (10.04). The graphs of these functions have the same shapes as those shown in (10.05–10.06).

(3) Transport coefficients T_j, A_j, B_j, C_j are the values of $T_j(x)$ for a particular argument x.

$$\left.\begin{array}{lll} \text{For } x = s, & T_{jx}(s) = T_{jx}, & T_{jy}(s) = T_{jy} \\ \text{For } x = a, & T_{jx}(a) = A_{jx}, & T_{jy}(a) = A_{jy} \\ \text{For } x = b, & T_{jx}(b) = B_{jx}, & T_{jy}(b) = B_{jy} \end{array}\right\} \quad j = 0, 1, 2, \ldots, 7$$

(4) Transport matrices of this bar have analogical properties to those of (11.02) and can be used in similar ways.

(5) General transport matrix equations for beam-columns of order three encased in elastic foundation are obtained by matrix composition of (10.02) and (15.02) as shown in (2.18).

15.04 TRANSPORT LOAD FUNCTIONS

Notation (15.01–15.03) Signs (15.02) Matrix (15.02) P_z = concentrated load (N) Q_x, Q_y = applied couples (N·m) q_y = intensity of distributed couple (N·m/m) A_j, B_j, C_j, T_j = transport coefficients (15.03)	$\lambda = \sqrt[4]{\dfrac{k_w}{EI_y}}$ $R = \dfrac{\eta f}{e} = \dfrac{\eta N}{\sqrt{EI_y k_w}}$

F	Case	G
$F_W = -(1 - A_{5y})P_y$ $F_X = 0$ $F_Y = A_{2y}P_y$	**1.** 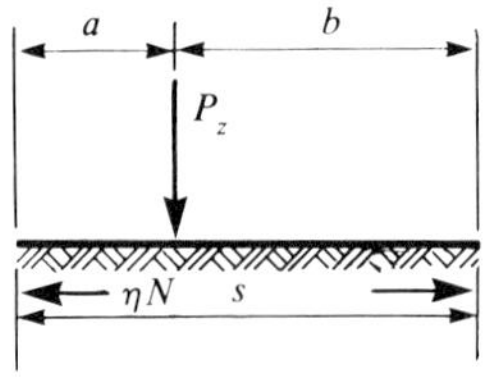	$G_W = (1 - B_{5y})P_y$ $G_X = 0$ $G_Y = B_{2y}P_y$
$F_w = -A_{4y}P_y$ $F_\phi = 0$ $F_\psi = -A_{3y}P_y$		$G_w = -B_{4y}P_y$ $G_\phi = 0$ $G_\psi = B_{3y}P_y$
$F_W = -A_{4y}Q_y$ $F_X = -A_{1x}Q_x$ $F_Y = -A_{1y}Q_y$	**2.** 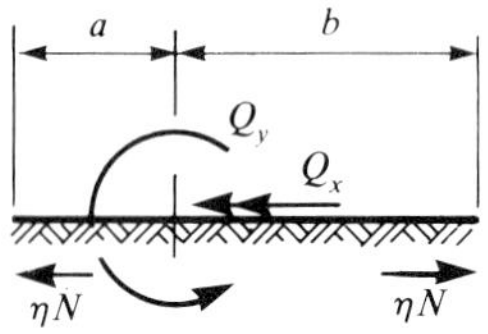	$G_W = -B_{4y}Q_y$ $G_X = B_{1x}Q_x$ $G_Y = B_{1y}Q_y$
$F_w = A_{3y}Q_y$ $F_\phi = A_{2x}Q_x$ $F_\psi = A_{2y}Q_y$		$G_w = -B_{3y}Q_y$ $G_\phi = B_{2x}Q_x$ $G_\psi = B_{2y}Q_y$
$F_W = -T_{5y}q_y$ $F_X = 0$ $F_Y = -T_{2y}q_y$	**3.** 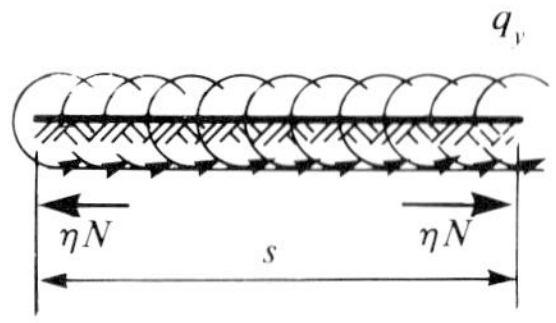	$G_W = -T_{5y}q_y$ $G_X = 0$ $G_Y = T_{2y}q_y$
$F_w = T_{4y}q_y$ $F_\phi = 0$ $F_\psi = T_{3y}q_y$		$G_w = -T_{4y}q_y$ $G_\phi = 0$ $G_\psi = T_{3y}q_y$
$F_W = -C_{5y}q_y$ $F_X = 0$ $F_Y = -C_{2y}q_y$	**4.** 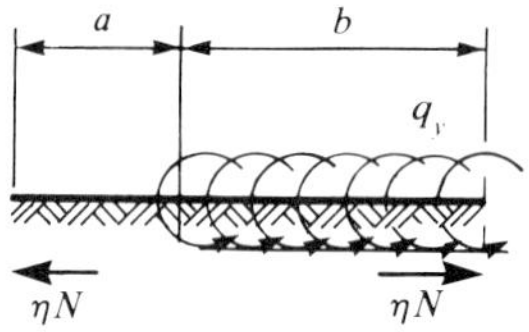	$G_W = -B_{5y}q_y$ $G_X = 0$ $G_Y = B_{2y}q_y$
$F_w = C_{4y}q_y$ $F_\phi = 0$ $F_\psi = C_{3y}q_y$		$G_w = -B_{4y}q_y$ $G_\phi = 0$ $G_\psi = B_{3y}q_y$

15.05 TRANSPORT LOAD FUNCTIONS

Notation (15.01–15.03) Signs (15.02) Matrix (15.02)

p_z = intensity of distributed load (N/m)

q_x = intensity of distributed couple (N·m/m)

A_j, B_j, C_j, T_j = transport coefficients (15.03)

$$\lambda = \sqrt[4]{\frac{k_w}{EI_y}}$$

$$R = \frac{\eta f}{e} = \frac{\eta N}{\sqrt{EI_y k_w}}$$

1.

$F_W = -(T_{2y} - RT_{4y})\lambda^{-1}p_z$

$F_X = -T_{2x}q_x$

$F_Y = T_{3y}\lambda^{-1}p_z$

$F_w = -T_{5y}\lambda^{-1}p_z$

$F_\phi = T_{3x}q_x$

$F_\psi = -T_{4y}\lambda^{-1}p_z$

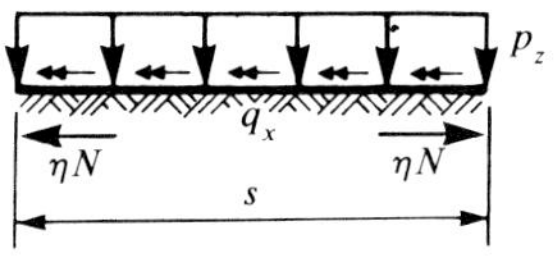

$G_W = (T_{2y} - RT_{4y})\lambda^{-1}p_z$

$G_X = T_{2x}q_x$

$G_Y = T_{3y}\lambda^{-1}p_z$

$G_w = -T_{5y}\lambda^{-1}p_z$

$G_\phi = T_{3x}q_x$

$G_\psi = T_{4y}\lambda^{-1}p_z$

2.

$F_W = -(C_{2y} - RC_{4y})\lambda^{-1}p_z$

$F_X = -C_{2x}q_x$

$F_Y = C_{3y}\lambda^{-1}p_z$

$F_w = -C_{5y}\lambda^{-1}p_z$

$F_\phi = C_{3x}q_x$

$F_\psi = -C_{4y}\lambda^{-1}p_z$

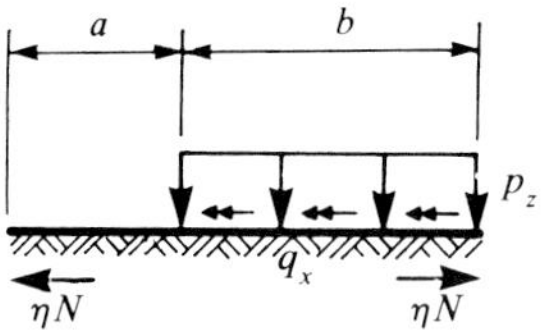

$G_W = (B_{2y} - RB_{4y})\lambda^{-1}p_z$

$G_X = B_{2x}q_x$

$G_Y = B_{3y}\lambda^{-1}p_z$

$G_w = -B_{5y}\lambda^{-1}p_z$

$G_\phi = B_{3x}q_x$

$G_\psi = B_{4y}\lambda^{-1}p_z$

3.

$F_W = -(\lambda s\Delta_1 + T_{7y})\lambda^{-2}s^{-1}p_z$

$F_X = -(\lambda sT_{2x} - T_{3x})\lambda^{-1}s^{-1}q_x$

$F_Y = (\lambda sT_{3y} - T_{4y})\lambda^{-2}s^{-1}p_z$

$F_w = -(\lambda sT_{5y} - T_{6y})\lambda^{-2}s^{-1}p_z$

$F_\phi = (\lambda sT_{3x} - T_{4x})\lambda^{-1}s^{-1}q_x$

$F_\psi = -(\lambda sT_{4y} - T_{5y})\lambda^{-2}s^{-1}p_z$

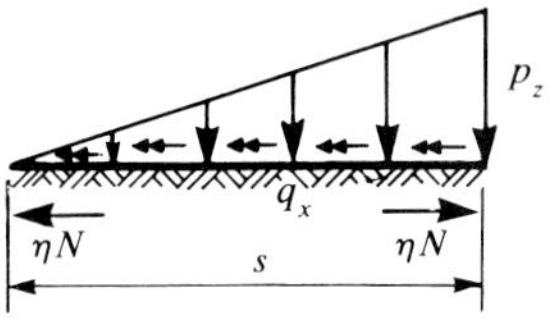

$\Delta_1 = \frac{1}{2}\lambda s - T_{6y}$

$G_W = (\frac{1}{2}\lambda^2 s^2 - T_{7y})\lambda^{-2}s^{-1}p_z$

$G_X = T_{3x}\lambda^{-1}s^{-1}q_x$

$G_Y = T_{4y}\lambda^{-2}s^{-1}p_z$

$G_w = -T_{6y}\lambda^{-2}s^{-1}p_z$

$G_\phi = T_{4x}\lambda^{-1}s^{-1}q_x$

$G_\psi = T_{5y}\lambda^{-2}s^{-1}p_z$

4.

$F_W = -(\lambda b\Delta_2 + C_{7y})\,\lambda^{-2}b^{-1}p_z$

$F_X = -(\lambda bT_{2x} - C_{3x})\,\lambda^{-1}b^{-1}q_x$

$F_Y = (\lambda bT_{3y} - C_{4y})\,\lambda^{-2}b^{-1}p_z$

$F_w = -(\lambda btT_{5y} - C_{6y})\,\lambda^{-2}b^{-1}p_z$

$F_\phi = (\lambda bT_{3x} - C_{4x})\,\lambda^{-1}b^{-1}q_x$

$F_\psi = -(\lambda bT_{4y} - C_{5y})\,\lambda^{-2}b^{-1}p_z$

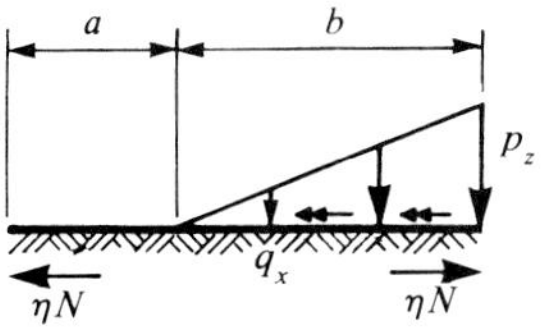

$\Delta_2 = \frac{1}{2}\lambda b - T_{6y}$

$G_W = (\frac{1}{2}\lambda^2 b^2 - B_{7y})\,\lambda^{-2}b^{-1}p_z$

$G_X = B_{3x}\lambda^{-1}b^{-1}q_x$

$G_Y = B_{4y}\lambda^{-2}b^{-1}p_z$

$G_w = -B_{6y}\lambda^{-2}b^{-1}p_z$

$G_\phi = B_{4x}\lambda^{-1}b^{-1}q_x$

$G_\psi = B_{5y}\lambda^{-2}b^{-1}p_z$

15.06 TRANSPORT LOAD FUNCTIONS

Notation (15.01–15.03) Signs (15.02) Matrix (15.02)

w_j, ϕ_j, ψ_j = abrupt displacements (m, rad, rad)

g_t = temperature strain (11.03)

A_j, B_j, C_j, T_j = transport coefficients (15.03)

$$\lambda = \sqrt[4]{\frac{k_w}{EI_y}}$$

$$R = \frac{\eta f}{e} = \frac{\eta N}{\sqrt{EI_y k_w}}$$

$A_{10y} = \eta A_{4y} - A_{6y} + \lambda a$	$B_{10y} = \eta B_{4y} - B_{6y} + \lambda b$
$A_{11y} = \eta A_{1y} - A_{3y}$	$B_{11y} = \eta B_{1y} - B_{3y}$
$A_{12y} = \eta A_{3y} - A_{5y} + 1$	$B_{12y} = \eta B_{3y} - B_{5y} + 1$
$A_{13y} = \eta A_{2y} - A_{4y}$	$B_{13y} = \eta B_{2y} - B_{4y}$

1. Abrupt displacements w_j, ϕ_j

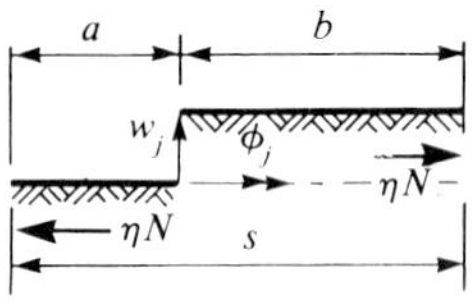

$F_W = EI_y A_{10y}\lambda^3 w_j$	$G_W = EI_y B_{10y}\lambda^3 w_j$
$F_X = GI_x A_{2x}\alpha^2 \phi_j$	$G_X = GI_x B_{2x}\alpha^2 \phi_j$
$F_Y = EI_y A_{11y}\lambda^3 w_j$	$G_Y = EI_y B_{11y}\lambda^3 w_j$
$F_w = -EI_y A_{12y}\lambda^3 w_j$	$G_w = EI_y B_{12y}\lambda^3 w_j$
$F_\phi = -GI_x A_{1x}\alpha^2 \phi_j$	$G_\phi = GI_x B_{1x}\alpha^2 \phi_j$
$F_\psi = -EI_y A_{13y}\lambda^3 w_j$	$G_\psi = EI_y B_{13y}\lambda^3 w_j$

2. Abrupt displacement ψ_j

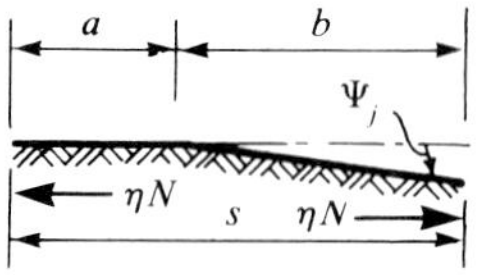

$F_W = EI_y A_{3y}\lambda^2 \psi_j$	$G_W = -EI_y B_{3y}\lambda^2 \psi_j$
$F_X = 0$	$G_X = 0$
$F_Y = EI_y A_{0y}\lambda^2 \psi_j$	$G_Y = EI_y B_{0y}\lambda^2 \psi_j$
$F_w = -EI_y A_{2y}\lambda^2 \psi_j$	$G_w = -EI_y B_{2y}\lambda^2 \psi_j$
$F_\phi = 0$	$G_\phi = 0$
$F_\psi = -EI_y A_{1y}\lambda^2 \psi_j$	$G_\psi = EI_y B_{1y}\lambda^2 \psi_j$

3. Temperature deformation of segment b

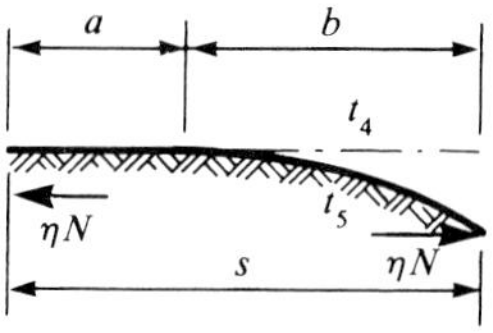

$F_W = EI_y C_{4y}\lambda g_t$	$G_W = -EI_y B_{4y}\lambda g_t$
$F_X = 0$	$G_X = 0$
$F_y = EI_y C_{1y}\lambda g_t$	$G_Y = EI_y B_{1y}\lambda g_t$
$F_w = -EI_y C_{3y}\lambda g_t$	$G_w = -EI_y B_{3y}\lambda g_t$
$F_\phi = 0$	$G_\phi = 0$
$F_\psi = -EI_y C_{2y}\lambda g_t$	$G_\psi = EI_y B_{2y}\lambda g_t$

15.07 STIFFNESS MATRIX EQUATIONS

$$r = \frac{s^2 GI_x}{EI_y}$$

(1) Equivalents and signs. All forces, moments, and displacements are in the principal axes of the normal section, and their positive directions are shown in (2, 3) below.

(2) Free-body diagram

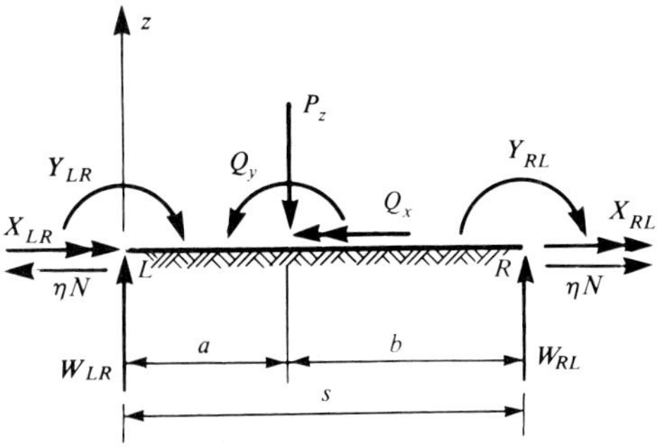

(3) Elastic curve

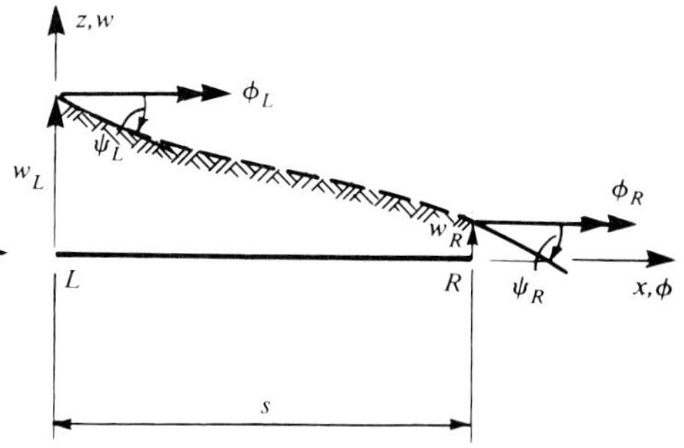

(4) Dimensionless stiffness coefficients

$L_1 = \dfrac{\cosh \alpha s}{\sinh \alpha s}$	$K_1 = (\alpha s)L_1$	$D_0 = \dfrac{1}{T_{3y}^2 - T_{2y}T_{4y}}$
$L_2 = D_0(T_{2y} - T_{2y}T_{5y} + T_{3y}T_{4y})$	$K_2 = (\lambda s)^3 L_2$	
$L_3 = D_0(T_{2z}^2 - T_{1z}T_{3z})$	$K_3 = (\lambda s)^2 L_3$	
$L_4 = D_0(T_{2z}T_{3z} - T_{1z}T_{4z})$	$K_4 = (\lambda s)L_4$	$\alpha = \sqrt{\dfrac{k_\phi}{GI_x}}$
$L_5 = \dfrac{1}{\sinh \alpha s}$	$K_5 = (\alpha s)L_5$	
$L_6 = D_0 T_{2z}$	$K_6 = (\lambda s)^3 L_6$	$\lambda = \sqrt[4]{\dfrac{k_w}{EI_y}}$
$L_7 = D_0 T_{3z}$	$K_7 = (\lambda s)^2 L_7$	
$L_8 = D_0 T_{4z}$	$K_8 = (\lambda s)L_8$	

where $T_{jy}(s) = T_{jy}$ are the transport coefficients (15.03).

(5) Stiffness matrix equations

$$\begin{bmatrix} W_{LR} \\ X_{LR} \\ Y_{LR} \\ \hline W_{RL} \\ X_{RL} \\ Y_{RL} \end{bmatrix} = \frac{EI_y}{s^3} \left[\begin{array}{ccc|ccc} K_2 & 0 & -K_3 s & -K_6 & 0 & -K_7 s \\ 0 & rK_1 & 0 & 0 & -rK_5 & 0 \\ -K_3 s & 0 & K_4 s^2 & K_7 s & 0 & K_8 s^2 \\ \hline -K_6 & 0 & K_7 s & K_2 & 0 & K_3 s \\ 0 & -rK_5 & 0 & 0 & rK_1 & 0 \\ -K_7 s & 0 & K_8 s^2 & K_3 s & 0 & K_4 s^2 \end{array}\right] \begin{bmatrix} w_L \\ \phi_L \\ \psi_L \\ \hline w_R \\ \phi_R \\ \psi_R \end{bmatrix} + \begin{bmatrix} W_{L0} \\ X_{L0} \\ Y_{L0} \\ \hline W_{R0} \\ X_{R0} \\ Y_{R0} \end{bmatrix}$$

where W_{L0}, X_{L0}, Y_{L0} and W_{R0}, X_{R0}, Z_{R0} in the last column matrix in (5) above are the *reactions of fixed-end beam-columns* due to loads and other causes shown in (15.08). The graphs of the stiffness coefficients K_2, K_3, K_4, K_6, K_7, K_8 have the same shape as K_1, K_2, K_3, K_4, K_5, K_6 in (10.11–10.12).

(6) General stiffness matrix equations for beam-columns of order three are obtained by matrix composition of (10.10) and (15.07) as shown in (3.19).

15.08 STIFFNESS LOAD FUNCTIONS, FIXED-END BEAM

(1) General form of the stiffness load functions in (15.07) is given by the following matrix equations:

$$\begin{bmatrix} W_{L0} \\ X_{L0} \\ Y_{L0} \end{bmatrix} = \begin{bmatrix} L_6 & 0 & L_7 \\ 0 & L_5\alpha^{-1} & 0 \\ -L_7\lambda^{-1} & 0 & -L_8\lambda^{-1} \end{bmatrix} \begin{bmatrix} G_w \\ G_\phi \\ G_\psi \end{bmatrix}$$

$$\begin{bmatrix} W_{R0} \\ X_{R0} \\ Y_{R0} \end{bmatrix} = \begin{bmatrix} L_6 & 0 & -L_7 \\ 0 & L_5\alpha^{-1} & 0 \\ L_7\lambda^{-1} & 0 & -L_8\lambda^{-1} \end{bmatrix} \begin{bmatrix} F_w \\ F_\phi \\ F_\psi \end{bmatrix}$$

where F_w, F_ϕ, F_ψ are the left-end transport load functions and G_w, G_ϕ, G_ψ are their right-end counterparts, given for particular loads and causes in (15.04–15.06); and L_5, L_6, L_7, L_8 are the stiffness coefficients (15.07). The load vectors and other causes are defined by symbols introduced in (15.04–15.06).

(2) Particular cases

Notation (15.01–15.03, 15.07) A_j, B_j, C_j, T_j = transport coefficients (15.03) A_{12y}, A_{13y}, B_{12y}, B_{13y} = equivalents (15.06)	Signs (15.07)	$\alpha = \sqrt{\dfrac{k_\phi}{GI_x}}$ $\lambda = \sqrt[4]{\dfrac{k_w}{EI_y}}$
$W_{L0} = (L_7B_{3y} - L_6B_{4y})P_z$ $X_{L0} = 0$ $Y_{L0} = -(L_8B_{3y} - L_7B_{4y})\lambda^{-1}P_y$	**1.** 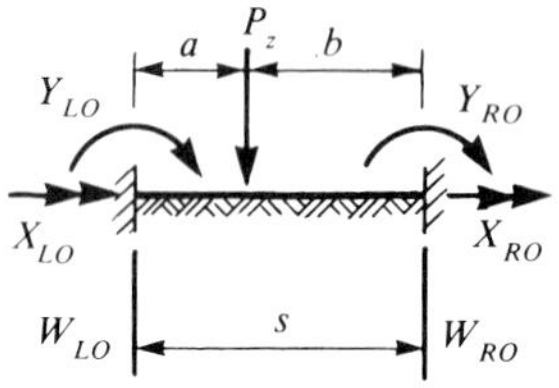	$W_{R0} = (L_7A_{3y} - L_6A_{4y})P_z$ $X_{R0} = 0$ $Y_{R0} = (L_8A_{3y} - L_7A_{4y})\lambda^{-1}P_z$
$W_{L0} = (L_7B_{2y} - L_6B_{3y})\lambda Q_y$ $X_{L0} = L_5B_{2x}Q_x$ $Y_{L0} = -(L_8B_{2y} - L_7B_{3y})Q_y$	**2.** 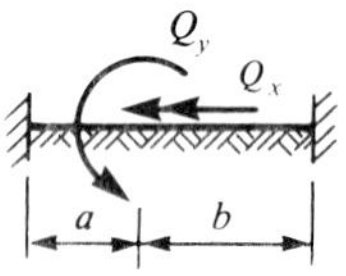	$W_{R0} = (L_7A_{2y} - L_6A_{3y})\lambda Q_y$ $X_{R0} = L_5A_{2x}Q_x$ $Y_{R0} = -(L_8A_{2y} - L_7B_{3y})Q_y$
$W_{L0} = (T_7T_{4y} - L_6T_{5y})\lambda^{-1}p_z$ $X_{L0} = L_5T_{3x}\alpha^{-1}q_x$ $Y_{L0} = -(L_7T_{4y} - L_8T_{5y})\lambda^{-2}p_z$	**3.** 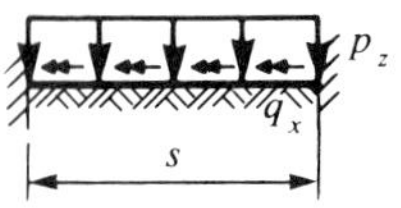	$W_{R0} = W_{L0}$ $X_{R0} = X_{L0}$ $Y_{R0} = -Y_{L0}$

$$W_{L0} = [L_7(A_{3y} + B_{3y}) - L_6(A_{4y} + B_{4y})]P_z = W_{R0}$$

$$X_{L0} = 0 = X_{R0}$$

$$Y_{L0} = -[L_8(A_{3y} + B_{3y}) - L_7(A_{4y} + B_{4y})]\lambda^{-1}P_z = -Y_{R0}$$

4.

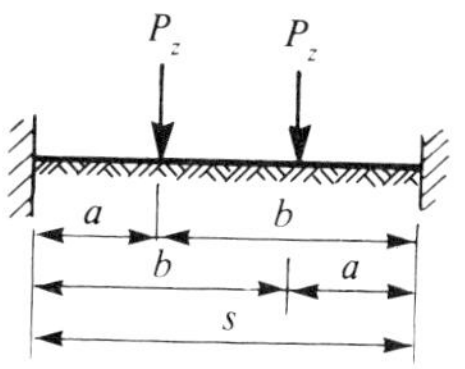

$$W_{L0} = [L_7(A_{2y} + B_{2y}) - L_6(A_{3y} + B_{3y})]\lambda Q_y = -W_{R0}$$

$$X_{L0} = L_5(A_{2x} + B_{2x})Q_x = X_{R0}$$

$$Y_{L0} = -[L_8(A_{2y} + B_{2y}) - L_7(A_{3y} + B_{3y})]Q_y = Y_{R0}$$

5.

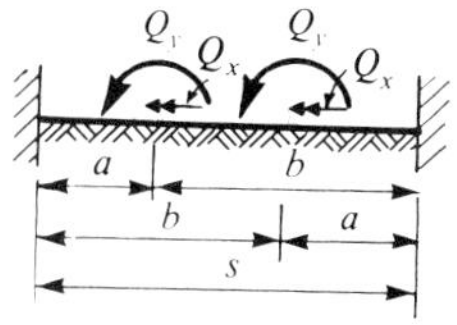

6.

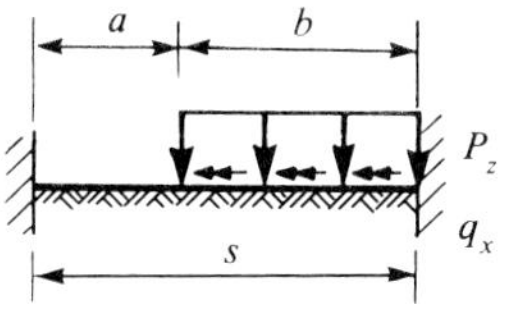

Use general form in terms of (15.05-2)

7.

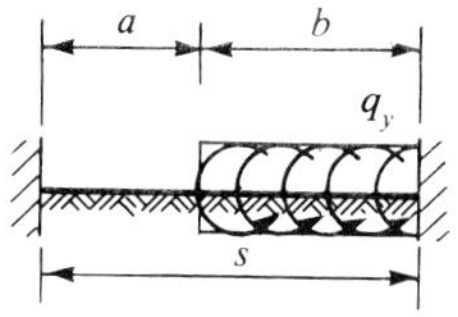

Use general form in terms of (15.04-4)

8.

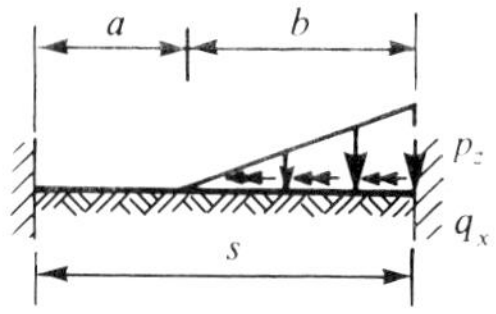

Use general form in terms of (15.05-4)

9. Abrupt displacements w_j, ϕ_j

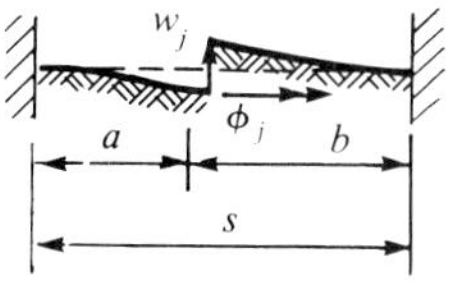

$$W_{L0} = EI_y(L_6B_{12y} + L_7B_{13y})\lambda^3 w_j \qquad W_{R0} = EI_y(L_6A_{12y} - L_7A_{13y})\lambda^3 w_j$$

$$X_{L0} = GI_xB_{2x}\alpha\phi_j \qquad X_{R0} = GI_xA_{2x}\alpha\phi_j$$

$$Y_{L0} = -EI_y(L_7B_{12y} + L_8B_{13y})\lambda^2 w_j \qquad Y_{R0} = EI_y(L_7A_{12y} - L_8A_{13y})\lambda^2 w_j$$

10. Abrupt displacement ψ_j

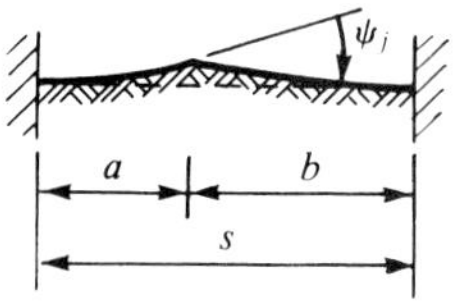

$$W_{L0} = -EI_y(L_6B_{2y} - L_7B_{1y})\lambda^2\psi_j \qquad W_{R0} = -EI_y(L_6A_{2y} - L_7A_{1y})\lambda^2\psi_j$$

$$X_{L0} = 0 \qquad X_{R0} = 0$$

$$Y_{L0} = -EI_y(L_7B_{2y} - L_8B_{1y})\lambda\psi_j \qquad Y_{R0} = -EI_y(L_7A_{2y} - L_8A_{1y})\lambda\psi_j$$

11. Temperature deformation of segment b

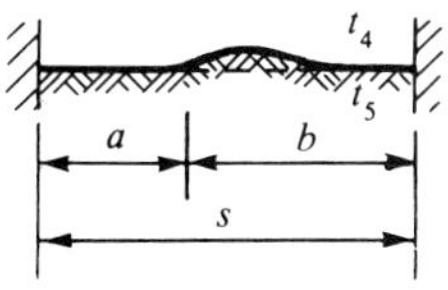

$$W_{L0} = -EI_y(L_6B_{3y} - L_7B_{2y})\lambda g_t \qquad W_{R0} = -EI_y(L_6C_{3y} - L_7C_{2y})\lambda g_t$$

$$X_{L0} = 0 \qquad X_{R0} = 0$$

$$Y_{L0} = EI_y(L_7B_{3y} - L_8B_{2y})g_t \qquad Y_{R0} = -EI_y(L_7C_{3y} - L_8C_{2y})g_t$$

15.09 MODIFIED STIFFNESS MATRIX EQUATIONS, HINGED END

(1) Modified stiffness coefficients are

$$K_9 = K_2 - r_3 K_3 \qquad K_{10} = K_6 - r_7 K_3 \qquad K_{11} = K_7 - r_8 K_3$$
$$K_{12} = K_2 - r_7 K_7 \qquad K_{13} = K_3 - r_7 K_8 \qquad K_{14} = K_4 - r_8 K_8$$

where

$$r_3 = \frac{K_3}{K_4} \qquad r_7 = \frac{K_7}{K_4} \qquad r_8 = \frac{K_8}{K_4}$$

and K_2, K_3, K_4, K_6, K_7, K_8 are the stiffness coefficients defined in (15.07).

(2) Modification of stiffness equations (15.07) due to hinged end are shown in (3, 4) below. Since this condition does not affect the expressions for X_{LR} and X_{RL}, only the affected part of the respective matrix equation is presented.

(3) Left end hinged ($Y_{LR} = 0$)

$$\begin{bmatrix} W_{LR} \\ W_{RL} \\ Y_{RL} \end{bmatrix} = \frac{EI_y}{s^3} \begin{bmatrix} K_9 & -K_{10} & -K_{11}s \\ -K_{10} & K_{12} & K_{13}s \\ -K_{11}s & K_{13}s & K_{14}s^2 \end{bmatrix} \begin{bmatrix} w_L \\ w_R \\ \psi_R \end{bmatrix} + \begin{bmatrix} W^*_{LO} \\ W^*_{RO} \\ Y^*_{RO} \end{bmatrix}$$

$$\begin{bmatrix} W^*_{LO} \\ W^*_{RO} \\ Y^*_{RO} \end{bmatrix} = \begin{bmatrix} W_{LO} + r_3 s^{-1} Y_{LO} \\ W_{RO} - r_7 s^{-1} Y_{LO} \\ Y_{RO} - r_8 Y_{LO} \end{bmatrix}$$

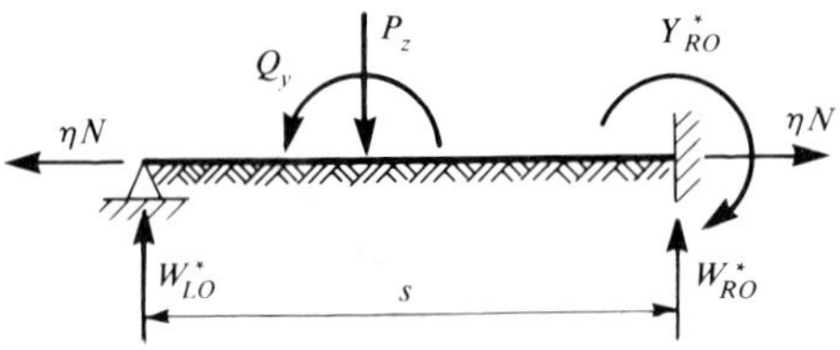

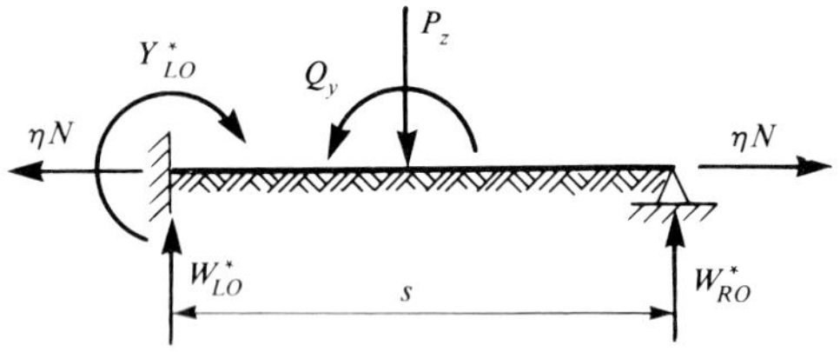

(4) Right end hinged ($Y_{RL} = 0$)

$$\begin{bmatrix} W_{LR} \\ Y_{LR} \\ W_{RL} \end{bmatrix} = \frac{EI_y}{s^3} \begin{bmatrix} K_9 & -K_{11}s & -K_{10} \\ -K_{11}s & K_{14}s^2 & K_{13}s \\ -K_{10} & K_{13}s & K_{12} \end{bmatrix} \begin{bmatrix} w_L \\ \psi_L \\ w_R \end{bmatrix} + \begin{bmatrix} W^*_{LO} \\ Y^*_{LO} \\ W^*_{RO} \end{bmatrix}$$

$$\begin{bmatrix} W^*_{LO} \\ Y^*_{LO} \\ W^*_{RO} \end{bmatrix} = \begin{bmatrix} W_{LO} - r_7 s^{-1} Y_{RO} \\ Y_{LO} - r_8 Y_{RO} \\ W_{RO} - r_3 s^{-1} Y_{RO} \end{bmatrix}$$

(5) Modified load functions designated by an asterisk are the reactions of propped-end beams due to loads and other causes. They are expressed in terms of reactions of the respective fixed-end beams (15.08) and the ratios r_3, r_7, r_8.

16

Bar Analysis by Finite Differences

STATIC STATE

16.01 DEFINITION OF STATE

(1) Systems analyzed here are defined in Chaps. 4, 5, and 8–10. Their normal sections are constant or variable. Only static states are considered.

(2) Analysis of these bars by the methods introduced in the above-mentioned chapters is limited to cases of certain shapes and certain end conditions. In other cases, these solutions are not practical and/or possible, and resort must be made to some approximate methods, among which the most important ones are the finite-element method and the finite-difference method. The finite-element method uses the segmental stiffness matrices introduced in the preceding chapters, and no additional models are required. The analytical models of the finite-difference method are presented in this chapter.

(3) Finite-difference method replaces the given bar by a specific set of points (joints, nodes) and approximates the governing differential equation and the equations defining the boundary conditions by difference equations in terms of unknown displacements related to these points. This substitution leads to a set of algebraic equations written for all joints of the bar, and the solution of this set yields the approximate values of the displacements of the bar axis at these points. The slopes, moments, shears, and reactions are then approximated by their respective difference expressions in terms of the calculated displacements. The finite-difference method in bar analysis is superior to the finite-element method. The number of unknowns required for the same degree of accuracy is usually less than half those required by the finite-element method.

(4) Assumptions of analysis are stated in general in (1.01) and for each particular bar in the introductory section of the respective chapter. Because of better accuracy, only central differences are used.

16.02 DERIVATIVES AND DIFFERENCES

(1) Notation

x = coordinate (m)
$v = f(x)$ = deflection (m)
a = spacing of joints (m)
$j = 0, 1, 2, \ldots, LL, L, C, R, RR$
$k = 0, 1, 2, \ldots, n$ $\quad C$ = central joint

(2) System of joints

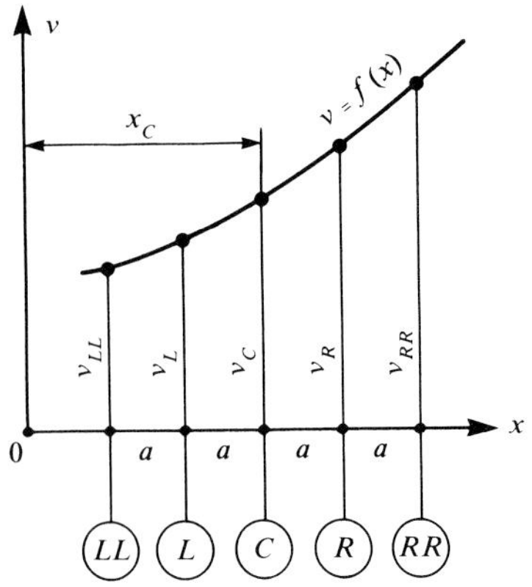

(3) Difference operators at C

$$D^{(0)}v_C = v_C$$

$$D^{(1)}v_C = -v_L + v_R$$

$$D^{(2)}v_C = v_L - 2v_C + v_R$$

$$D^{(3)}v_C = -D^{(2)}v_L + D^{(2)}v_R$$

$$D^{(4)}v_C = D^{(2)}v_L - 2D^{(2)}v_C + D^{(2)}v_R$$

$$D^{(5)}v_C = -D^{(4)}v_L + D^{(4)}v_R$$

and, in general,

$$D^{(2k+1)}v_C = -D^{(2k)}v_L + D^{(2k)}v_R$$

$$D^{(2k+2)}v_C = D^{(2k)}v_L - 2D^{(2k)}v_C + D^{(2k)}v_R$$

(4) Difference equations at C

$$\left(\frac{dv}{dx}\right)_C \cong \frac{D^{(1)}v_C}{2a} = \frac{-v_L + v_R}{2a}$$

$$\left(\frac{d^2v}{dx^2}\right)_C \cong \frac{D^{(2)}v_C}{a^2} = \frac{v_L - 2v_C + v_R}{a^2}$$

$$\left(\frac{d^3v}{dx^3}\right)_C \cong \frac{D^{(3)}v_C}{2a^3} = \frac{-v_{LL} + 2v_L - 2v_R + v_{RR}}{2a^3}$$

$$\left(\frac{d^4v}{dx^4}\right)_C \cong \frac{D^{(4)}v_C}{a^4} = \frac{v_{LL} - 4v_L + 6v_C - 4v_R + v_{RR}}{a^4}$$

$$\left(\frac{d^5v}{dx^5}\right)_C \cong \frac{D^{(5)}v_C}{2a^5} = \frac{-v_{LLL} + 4v_{LL} - 5v_L + 5v_R - 4v_{RR} + v_{RRR}}{2a^5}$$

and in general,

$$\left(\frac{d^{2k+1}v}{dx^{k+1}}\right)_C \cong \frac{D^{(2k+1)}v_C}{2a^{2k+1}} = \frac{-D^{(2k)}v_L + D^{(2k)}v_R}{2a^{2k+1}}$$

$$\left(\frac{d^{2k+2}v}{dx^{2k+2}}\right)_C \cong \frac{D^{(2k+2)}v_C}{a^{2k+2}} = \frac{D^{(2k)}v_L - 2D^{(2k)}v_C + D^{(2k)}v_R}{a^{2k+2}}$$

where each equation defines the approximation of the respective derivative in terms of the joint displacements and the constant spacing a.

16.03 JOINT LOADS, STRAIGHT BAR

(1) Equivalent loads. Since the deflections and their derivatives are related to the joints LL, L, C, R, RR in (16.02), the loads acting on the beam must also be related to these joints by equivalents called the joint loads. For this purpose, the load diagram is divided into segments of length a, and each segment is treated as a simple beam whose reactions are the joint loads.

P = concentrated load (N)
p = intensity of distributed load (N/m)
α = string-angle change of load diagram (N/m^2)
Q = applied couple (N·m)
R = joint load (N)

The positive string-angle changes are shown in case 5 below.

(2) Table

1. 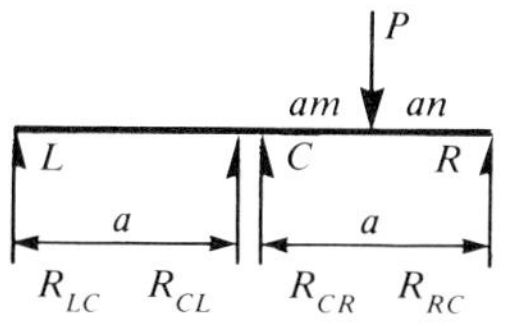	$R_{LC} = 0$ $R_C = R_{CL} + R_{CR} = nP$ $R_{RC} = mP$
2. 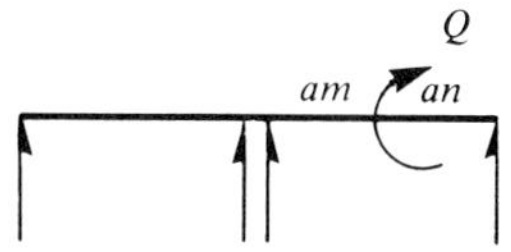	$R_{LC} = 0$ $R_C = -\dfrac{Q}{a}$ $R_{RC} = \dfrac{Q}{a}$
3. 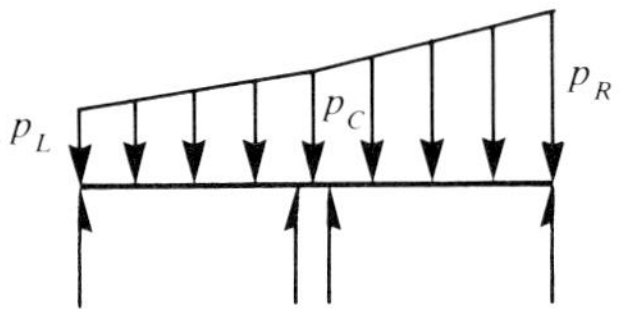	$R_{LC} = \dfrac{a}{6}(2p_L + p_C)$ $R_C = \dfrac{a}{6}(p_L + 4p_C + p_R)$ $R_{RC} = \dfrac{a}{6}(p_C + 2p_R)$
4. 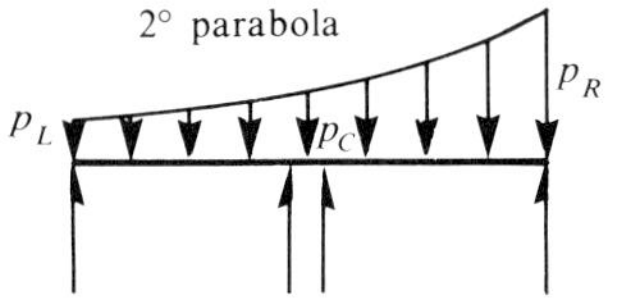	$R_{LC} = \dfrac{a}{24}(7p_L + 6p_C - p_R)$ $R_L = \dfrac{a}{12}(p_L + 10p_C + p_R)$ $R_{RC} = \dfrac{a}{24}(-p_L + 6p_C + 7p_R)$
5. 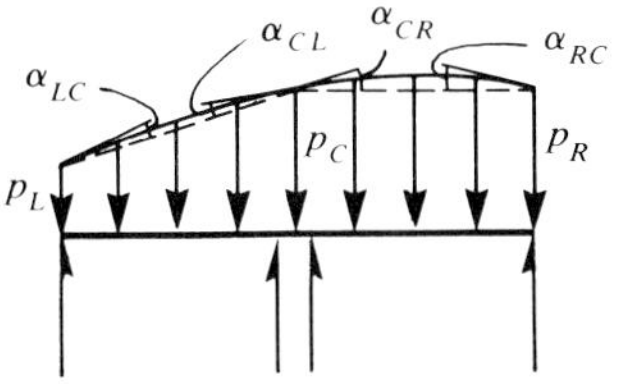	$R_{LC} = \dfrac{a}{6}(2p_L + p_C) + \dfrac{\alpha_{LC}a^2}{6}$ $R_L = \dfrac{a}{6}(p_L + 4p_C + p_R) + \dfrac{(\alpha_{CL} + \alpha_{CR})a^2}{6}$ $R_{RC} = \dfrac{a}{6}(p_C + 2p_R) + \dfrac{\alpha_{RC}a^2}{6}$

16.04 FREE STRAIGHT BAR, CONSTANT SECTION

(1) System, notation, and signs are defined in (4.01–4.03).

v_j = deflection at j (m)
V_j = shear at j (N)
R_j = joint load at j (N)
j = $-2, -1, 0, 1, 2, \ldots, LL, L, C, R, RR$

θ_j = slope at j (rad)
Z_j = bending moment at j (N·m)
a = segment (m)

$\bar{\theta}_j = \theta_j a$	$\bar{Z}_j = \dfrac{a^2}{EI_z} Z_j$	$\bar{V}_j = \dfrac{a^3}{EI_z} V_j$	$\bar{R}_j = \dfrac{a^3}{EI_z} R_j$

(2) Free-body diagram

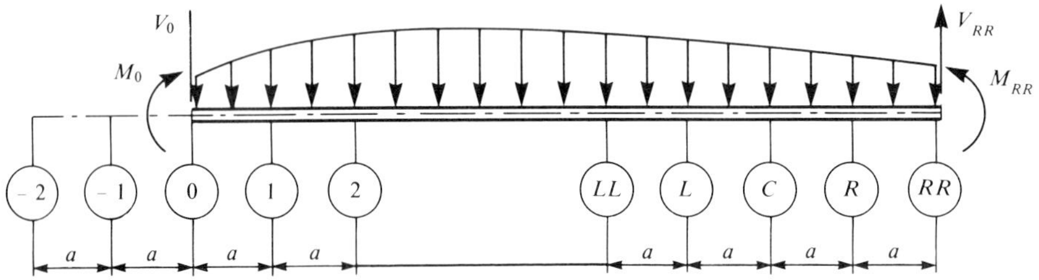

(3) Governing difference equation at C related to (2),

$$\bar{R}_C = -v_{LL} + 4v_L - 6v_C + 4v_R - v_{RR}$$

applies at all intermediate points. For the end points, additional boundary equations (4) must be used.

(4) Boundary equations at 0

Free end	$v_{-1} = 2v_0 - v_1$	$v_{-2} = 3v_0 - 2v_1$
Guided end	$v_{-1} = v_1$	$v_{-2} = v_2$
Hinged end	$v_{-1} = -v_1 \quad v_0 = 0$	$v_{-2} = -v_2$
Fixed end	$v_{-1} = v_1 \quad v_0 = 0$	$v_{-2} = v_2$

(5) State-vector components at C are

$$\begin{aligned}
\bar{V}_{CL} &= \bar{R}_{CL} + v_{LL} - 3v_L + 3v_C - v_R \\
\bar{V}_{CR} &= -\bar{R}_{CR} - v_L + 3v_C - 3v_R + v_{RR} \\
\bar{Z}_C &= v_L - 2v_C + v_R \\
\bar{\theta}_{CL} &= v_C - v_L + (7\bar{Z}_C + 6\bar{Z}_L - \bar{Z}_{LL})/24 \\
\bar{\theta}_{CR} &= v_R - v_C - (7\bar{Z}_C + 6\bar{Z}_R - \bar{Z}_{RR})/24
\end{aligned}$$

where $\bar{R}_{CL}$, $\bar{R}_{CR}$, $\bar{V}_{CL}$, $\bar{V}_{CR}$, $\bar{\theta}_{CL}$, $\bar{\theta}_{CR}$ are the scaled components on the left and right sides of C, respectively. The state-vector components at 0 are given by (16.05-5) with $I_{jz} = I_z$.

16.05 FREE STRAIGHT BAR, VARIABLE SECTION

(1) System, notation, and signs are defined in (5.01–5.03). The area of section is a function of x, and consequently I_z takes on a particular value at each joint.

(2) Equivalents

$K_{LL} = \dfrac{I_{LLz}}{I_{Cz}}$	$K_C = \dfrac{I_{Lz}}{I_{Cz}} + 4 + \dfrac{I_{Rz}}{I_{Cz}}$	$K_{RR} = \dfrac{I_{RRz}}{I_{Cz}}$
$K_L = 2\left(\dfrac{I_{Lz}}{I_{Cz}} + 1\right)$	$K_R = 2\left(\dfrac{I_{Rz}}{I_{Cz}} + 1\right)$	

(3) Governing difference equation at C related to free-body diagram of (16.04-2).

$$\overline{R}_C = -K_{LL}v_{LL} + K_L v_L - K_C v_C + K_R v_R - K_{RR} v_{RR}$$

applies at any intermediate joint. For the end points, the boundary equations (16.04-4) must be used.

(4) State-vector components at C are given by (16.04-5), where all scaled values involve the respective I_{jz}.

(5) State-vector components at 0

$$\lambda = \frac{I_{1z}}{I_{0z}}$$

Free end	$\overline{V}_{01} = 0 \quad \overline{Z}_0 = 0$ $\overline{\theta}_{01} = v_1 - v_0 - (7\overline{Z}_0 + 6\overline{Z}_1 - \overline{Z}_2)/24$
Guided end	$\overline{V}_{01} = 0 \quad \overline{\theta}_{01} = 0$ $\overline{Z}_{01} = -2v_0 + 2v_1$
Hinged end†	$\overline{V}_{01} = -\overline{R}_{01} + \lambda(v_0 - 2v_1 + v_2)$ $\overline{Z}_0 = -2v_0$ $\overline{\theta}_{01} = v_1 - v_0 - (7\overline{Z}_0 + 6\overline{Z}_1 - \overline{Z}_2)/24$
Fixed end†	$\overline{V}_{01} = -\overline{R}_{01} + (2 + \lambda)(v_0 - v_1) + \lambda v_2$ $\overline{Z}_0 = -2v_0 + 2v_1$ $\overline{\theta}_{01} = 0$

†The deflection v_0 is a given value, or $v_0 = 0$.

16.06 STRAIGHT BEAM-COLUMN ENCASED IN ELASTIC FOUNDATION, VARIABLE SECTION, VARIABLE AXIAL FORCE

(1) System considered is a straight beam-column of variable symmetrical section encased in elastic foundation of variable modulus k (N/m^2) and acted on by transverse load and variable axial force ηN, where

$$\eta_j = \begin{cases} +1 & \text{if } N_j = \text{axial tension (N)} \\ -1 & \text{if } N_j = \text{axial compression (N)} \end{cases}$$

16.06 STRAIGHT BEAM-COLUMN ENCASED IN ELASTIC FOUNDATION, VARIABLE SECTION, VARIABLE AXIAL FORCE (*Continued*)

(2) Equivalents

$\alpha_L = 2 - \dfrac{a^2}{12EI_{Cz}}(a^2 k_{Lv} + \eta_L N_L)$	$\alpha_R = 2 - \dfrac{a^2}{12EI_{Cz}}(a^2 k_{Rv} + \eta_R N_R)$
$\alpha_C = 4 + \dfrac{5a^2}{6EI_{Cz}}(a^2 k_{Cv} + \eta_C N_C)$	
$K_{LL} = \dfrac{I_{LZ}}{I_{CZ}}$	$K_C = \dfrac{I_{LZ}}{I_{CZ}} + \alpha_C + \dfrac{I_{RZ}}{I_{CZ}}$
$K_{RR} = \dfrac{I_{RZ}}{I_{CZ}}$	
$K_L = \dfrac{2I_{LZ}}{I_{LZ}} + \alpha_L$	$K_R = \dfrac{2I_{RZ}}{I_{CZ}} + \alpha_R$

(3) Governing difference equation at C,

$$\overline{R}_C = -K_{LL}v_{LL} + K_L v_L - K_C v_C + K_R v_R - K_{RR} v_{RR}$$

applies at any intermediate joint. For the end joints, the boundary equations (16.04-4) must be used.

(4) State-vector components at C and 0 are given by (16.04-5) and (16.05-5), respectively, with all scaled values containing the respective I_{jz}.

(5) Constant section, modulus, and normal force lead to a simplified difference equation (3), where

$$\alpha_L = \alpha_R = 2 - \frac{a^2}{12EI_z}(a^2 k_v + \eta N) \qquad \alpha_C = 4 + \frac{5a^2}{GEI_z}(a^2 k_v + \eta N)$$

and $\lambda = 1$ in boundary-value equations (16.04-4).

(6) Spring supports. The governing difference equation at C in a beam-column of constant section, supported on springs of equidistant spacing, is

$$\overline{R}_C = -v_{LL} + \alpha_L v_L - \alpha_C v_C + \alpha_R v_R - v_{RR}$$

where

$$\alpha_L = 4 - \frac{a^2}{12EI_z}(aC_L + \eta N) \qquad \alpha_R = 4 - \frac{a^2}{12EI_z}(aC_R + \eta N)$$

$$\alpha_C = 6 + \frac{5a^2}{6}(aC_C + \eta N)$$

C_L, C_C, C_R = spring constants at L, C, R (N/m) and $\lambda = 1$ in boundary-value equations (16.04-4).

Part II

Dynamic Analysis of Bars

17

Analysis of Bars Notation, Signs, Basic Relations

DYNAMIC STATE

17.01 INTRODUCTION

(1) Systems considered in Chaps. 17–21 are finite straight bars of constant and symmetrical section. Their cross-sectional dimensions are small in comparison to their length (1/10). Each bar and its loads are in a state of dynamic equilibrium and a state of time-dependent elastic deformation.

(2) Datum of the dynamic analysis is the geometry of deformed structure under the action of static loads. Since the static displacements have been assumed in (1.01) to be too small to alter significantly the initial geometry, the datum geometry in this chapter is approximated by the initial geometry. All coordinates are positive if measured in the positive direction of coordinate axes (1.02).

(3) Assumptions used in the derivation of analytical relations are:

(*a*) The material of the bar is homogeneous, isotropic, and remains continuous during the loading and unloading.

(*b*) The material follows Hooke's law and all deformations are small and do not alter (significantly) the initial geometry of the system.

(*c*) The linear superposition of causes and effects is admissible.

(*d*) Material constants are known from experiments and are independent of time.

(4) Symbols and signs are defined where they appear first and they are all summarized in Appendix A.

17.02 POSITION, LOAD, STRESS, AND REACTIVE VECTORS

(1) Dynamic forces and moments acting upon the bar are called dynamic load functions (excitations). They are of different origins and take on the following forms:

(*a*) Constant or monotonically increasing nonperiodic continuous load functions such as the rectangular, triangular, and exponential pulses
(*b*) Monotonically decreasing nonperiodic continuous load functions such as the triangular and exponential pulses
(*c*) Symmetrical load functions such as the triangular, sinusoidal, and cosinusoidal pulses
(*d*) Periodic load functions such as the rectangular, sinusoidal, and cosinusoidal pulses
(*e*) Composite load functions made up as a composition of particular load functions of the preceding types
(*f*) Irregular load functions whose time variation is given by a histograph obtained from an in situ recording or is a result of statistical correlation

(2) Concentrated dynamic load vector is represented by a matrix function of time t as

$$L(t) = \{P_x(t),\ P_y(t),\ P_z(t),\ Q_x(t),\ Q_y(t),\ Q_z(t)\} \qquad \left\{\begin{matrix} t \text{ in s} \\ P \text{ in N} \\ Q \text{ in N}\cdot\text{m} \end{matrix}\right\}$$

consisting of three force components and three moment components in the x, y, z axes. All load components are positive if acting in the negative direction of coordinate axes (1.02).

(3) Distributed dynamic load vector for the bar element ds (m) is represented by an elemental matrix function of time t as

$$dL(t) = \{p_x(t),\ p_y(t),\ p_z(t),\ q_x(t),\ q_y(t),\ q_z(t)\}ds \qquad \left\{\begin{matrix} p \text{ in N/m} \\ q \text{ in N}\cdot\text{m/m} \\ ds \text{ in m} \end{matrix}\right\}$$

consisting of three force components and three moment components defined in the x, y, z axes, and their sign convention is the same as in (2) above.

(4) Stress-resultant vector produced by dynamic causes at the centroid of a particular section is represented by matrix functions of time t as

$$\begin{aligned} S_L &= \{U_L(t),\ V_L(t),\ W_L(t),\ X_L(t),\ Y_L(t),\ Z_L(t)\} \\ S_R &= \{U_R(t),\ V_R(t),\ W_R(t),\ X_R(t),\ Y_R(t),\ Z_R(t)\} \end{aligned} \qquad \left\{\begin{matrix} U, V, W \text{ in N} \\ X, Y, Z \text{ in N}\cdot\text{m} \end{matrix}\right\}$$

consisting of three force components and three moment components in the principal axes of the normal section of the left end (L) and the right end (R), respectively. All stress-resultant components on the right side (far side) of the section are positive if acting in the positive direction of the principal axes. For their left-side (near-side) counterparts, the opposite is true (1.03).

(5) Reactive vector developed by dynamic causes at the points of support is represented by matrix functions of time t as

$$\begin{aligned} S_{LR} &= \{U_{LR}(t),\ V_{LR}(t),\ W_{LR}(t),\ X_{LR}(t),\ Y_{LR}(t),\ Z_{LR}(t)\} \\ S_{RL} &= \{U_{RL}(t),\ V_{RL}(t),\ W_{RL}(t),\ X_{RL}(t),\ Y_{RL}(t),\ Z_{RL}(t)\} \end{aligned} \qquad \left\{\begin{matrix} U, V, W \text{ in N} \\ X, Y, Z \text{ in N}\cdot\text{m} \end{matrix}\right\}$$

consisting of three force components and three moment components in the principal axes of the normal section. The first and second subscripts identify the near and far ends, respectively. All reactions are positive if acting in the positive direction of principal axes (1.03).

(6) Coordinate systems used in the dynamic analysis are the same as those defined in (1.09–1.10).

17.03 DISPLACEMENT, VELOCITY, ACCELERATION, AND INERTIA VECTORS

(1) Dynamic elastic curve. The time-dependent deformation of a bar is defined as the change in its shape produced by dynamic causes and/or imposed conditions. As the bar deforms, its centroidal axis takes on a new shape, called the *dynamic elastic curve*. The coordinates and the slopes of this curve measured from the initial axis of the undeformed bar are designated as the *linear* and *angular displacement*.

(2) Displacement vector is represented by a column matrix function of time as

$$\Delta(t) = \{u(t),\ v(t),\ w(t),\ \phi(t),\ \psi(t),\ \theta(t)\} \qquad \left\{\begin{matrix} u,\ v,\ w \text{ in m} \\ \phi,\ \psi,\ \theta \text{ in rad}\end{matrix}\right\}$$

consisting of three linear components and three angular components in the principal axes of the normal section.

(3) Velocity vector defined as the time rate of change of the displacement vector is represented by a column matrix function of time as

$$\dot\Delta(t) = \{\dot u(t),\ \dot v(t),\ \dot w(t),\ \dot\phi(t),\ \dot\psi(t),\ \dot\theta(t)\} \qquad \left\{\begin{matrix} \dot u,\ \dot v,\ \dot w \text{ in m/s} \\ \dot\phi,\ \dot\psi,\ \dot\theta \text{ in rad/s}\end{matrix}\right\}$$

consisting of

$$\dot u(t) = \frac{du(t)}{dt} \qquad \dot v(t) = \frac{dv(t)}{dt} \qquad \dot w(t) = \frac{dw(t)}{dt}$$

$$\dot\theta(t) = \frac{d\theta(t)}{dt} \qquad \dot\psi(t) = \frac{d\psi(t)}{dt} \qquad \dot\theta(t) = \frac{d\theta(t)}{dt}$$

(4) Acceleration vector defined as the time rate of change of the velocity vector is represented by a column matrix function of time as

$$\ddot\Delta(t) = \{\ddot u(t),\ \ddot v(t),\ \ddot w(t),\ \ddot\phi(t),\ \ddot\psi(t),\ \ddot\theta(t)\} \qquad \left\{\begin{matrix} \ddot u,\ \ddot v,\ \ddot w \text{ in m/s}^2 \\ \ddot\phi,\ \ddot\psi,\ \ddot\theta \text{ in rad/s}^2\end{matrix}\right\}$$

consisting of

$$\ddot u(t) = \frac{d^2u(t)}{dt^2} \qquad \ddot v(t) = \frac{d^2v(t)}{dt^2} \qquad \ddot w(t) = \frac{d^2w(t)}{dt^2}$$

$$\ddot\phi(t) = \frac{d^2\phi(t)}{dt^2} \qquad \ddot\psi(t) = \frac{d^2\psi(t)}{dt^2} \qquad \ddot\theta(t) = \frac{d^2\theta}{dt^2}$$

The product of the mass matrix defined in (17.04) and of the acceleration vector is the respective *inertia vector*.

(5) Sign convention. All displacements, velocities, and accelerations are positive if acting in the positive direction of the principal axes of the normal section.

(6) Transformation relationships of these vectors are again given as

$$\Delta^o = R^{os}\Delta^s \qquad \Delta^s = R^{so}\Delta^o$$

$$\dot\Delta^o = R^{os}\dot\Delta^s \qquad \dot\Delta^s = R^{so}\dot\Delta^o$$

$$\ddot\Delta^o = R^{os}\ddot\Delta^s \qquad \ddot\Delta^s = R^{so}\ddot\Delta^o$$

where R^{os}, R^{so} are the angular transformation matrices introduced in (1.10-4) and the superscripts o and s identify the global and local reference systems.

17.04 MASS MATRICES OF ORDER THREE IN *S* SYSTEM

(1) Notation

m = mass (kg) $\quad s$ = span (m) $\quad A_x$ = area of normal section

I_x, I_y, I_z = moments of inertia of A_x in principal axes (m^4)

$\rho_\phi, \rho_\psi, \rho_\theta$ = moments of inertia of m in principal axes (kg·m^2)

$\alpha = \dfrac{I_x}{A_x s^2}$ $\beta = \dfrac{I_y}{A_x s^2}$ $\gamma = \dfrac{I_z}{A_x s^2}$	$M_{1y} = (13 + 42\beta)\dfrac{m}{35}$	$M_{1z} = (13 + 42\gamma)\dfrac{m}{35}$
	$M_{2y} = (11 + 21\beta)\dfrac{m}{210}$	$M_{2z} = (11 + 21\gamma)\dfrac{m}{210}$
	$M_{3y} = (1 + 14\beta)\dfrac{ms^2}{105}$	$M_{3z} = (1 + 14\gamma)\dfrac{ms^2}{105}$
	$M_{4y} = (2 + 14\beta)\dfrac{ms^2}{420}$	$M_{4z} = (2 + 14\gamma)\dfrac{ms^2}{420}$
$M_{1n} = \dfrac{m}{3}$	$M_{5y} = (9 - 84\beta)\dfrac{m}{70}$	$M_{5z} = (9 - 84\gamma)\dfrac{m}{70}$
$M_{1x} = \dfrac{\alpha m s^2}{3}$	$M_{6y} = (13 - 42\beta)\dfrac{ms^2}{420}$	$M_{6z} = (13 - 42\gamma)\dfrac{ms^2}{420}$

(2) Classification. For the dynamic analysis, the mass of the given bar is approximated by the lumped-mass matrix M_0 or by the consistent-mass matrix M_C or is considered to be distributed. In this section, the approximate-mass matrices are introduced.

(3) Lumped-mass matrix of a straight bar of constant section with mass lumped at the centroid is

$$M_0 = \text{Diag}\,[m, m, m, \rho_\phi, \rho_\psi, \rho_\theta]$$

(4) Consistent-mass matrix of the same bar is

$$M_C = \left[\begin{array}{ccc|ccc|ccc|ccc}
M_{1n} & 0 & 0 & 0 & 0 & 0 & \frac{1}{2}M_{1n} & 0 & 0 & 0 & 0 & 0 \\
0 & M_{1z} & 0 & 0 & 0 & M_{2z} & 0 & M_{5z} & 0 & 0 & 0 & -M_{6z} \\
0 & 0 & M_{1y} & 0 & -M_{2y} & 0 & 0 & 0 & M_{5y} & 0 & M_{6y} & 0 \\
\hline
0 & 0 & 0 & M_{1x} & 0 & 0 & 0 & 0 & 0 & \frac{1}{2}M_{1x} & 0 & 0 \\
0 & 0 & -M_{2y} & 0 & M_{3y} & 0 & 0 & 0 & -M_{6y} & 0 & -M_{4y} & 0 \\
0 & M_{2z} & 0 & 0 & 0 & M_{3z} & 0 & M_{6z} & 0 & 0 & 0 & -M_{4z} \\
\hline
\frac{1}{2}M_{1n} & 0 & 0 & 0 & 0 & 0 & M_{1n} & 0 & 0 & 0 & 0 & 0 \\
0 & M_{5z} & 0 & 0 & 0 & M_{6z} & 0 & M_{1z} & 0 & 0 & 0 & -M_{2z} \\
0 & 0 & M_{5y} & 0 & -M_{6y} & 0 & 0 & 0 & M_{1y} & 0 & M_{2y} & 0 \\
\hline
0 & 0 & 0 & \frac{1}{2}M_{1x} & 0 & 0 & 0 & 0 & 0 & M_{1x} & 0 & 0 \\
0 & 0 & M_{6y} & 0 & -M_{4y} & 0 & 0 & 0 & M_{2y} & 0 & M_{3y} & 0 \\
0 & -M_{6z} & 0 & 0 & 0 & -M_{4z} & 0 & -M_{2z} & 0 & 0 & 0 & M_{3z}
\end{array}\right]$$

17.05 STIFFNESS MATRICES OF ORDER THREE IN *S* SYSTEM

(1) Notation

E = modulus of elasticity (Pa) $\qquad G$ = modulus of rigidity (Pa)

s = span (m) $\qquad A_x$ = area of normal section

I_x, I_y, I_z = moments of inertia of A_x in principal axes (m^4)

K_u, K_v, K_w = linear spring constants in principal axes (N/m)

K_ϕ, K_ψ, K_θ = angular spring constants in principal axes (N·m/m)

$K_{1n} = \dfrac{EA_x}{s}$	$K_{1y} = \dfrac{12}{s^3} EI_y$	$K_{1z} = \dfrac{12}{s^3} EI_z$
$K_{1x} = \dfrac{GI_x}{s}$	$K_{2y} = \dfrac{6}{s^2} EI_y$	$K_{2z} = \dfrac{6}{s^2} EI_z$
	$K_{3y} = \dfrac{4}{s} EI_y$	$K_{3z} = \dfrac{4}{s} EI_z$
	$K_{4y} = \dfrac{2}{s} EI_y$	$K_{4z} = \dfrac{2}{s} EI_z$

(2) Classification. For the dynamic analysis, the stiffness of the given bar is approximated by the point spring-constant matrix or by the static stiffness matrix or is considered as the dynamic stiffness matrix of the distributed mass. In this section, the approximate-stiffness matrices are introduced.

(3) Point spring-constant matrix of a straight bar of constant section with mass lumped at the centroid is

$$K_0 = \text{Diag}\,[K_u, K_v, K_w, K_\phi, K_\psi, K_\theta]$$

(4) Static stiffness matrix (3.13) used as the equivalent spring matrix is

$$K_C = \left[\begin{array}{ccc|ccc|ccc|ccc}
K_{1n} & 0 & 0 & 0 & 0 & 0 & -K_{1n} & 0 & 0 & 0 & 0 & 0 \\
0 & K_{1z} & 0 & 0 & 0 & K_{2z} & 0 & -K_{1z} & 0 & 0 & 0 & K_{2z} \\
0 & 0 & K_{1y} & 0 & -K_{2y} & 0 & 0 & 0 & -K_{1y} & 0 & -K_{2y} & 0 \\
\hline
0 & 0 & 0 & K_{1x} & 0 & 0 & 0 & 0 & 0 & -K_{1x} & 0 & 0 \\
0 & 0 & -K_{2y} & 0 & K_{3y} & 0 & 0 & 0 & K_{2y} & 0 & K_{4y} & 0 \\
0 & K_{2z} & 0 & 0 & 0 & K_{3z} & 0 & -K_{2z} & 0 & 0 & 0 & K_{4z} \\
\hline
-K_{1n} & 0 & 0 & 0 & 0 & 0 & K_{1n} & 0 & 0 & 0 & 0 & 0 \\
0 & -K_{1z} & 0 & 0 & 0 & -K_{2z} & 0 & K_{1z} & 0 & 0 & 0 & -K_{2z} \\
0 & 0 & -K_{1y} & 0 & K_{2y} & 0 & 0 & 0 & K_{1y} & 0 & K_{2y} & 0 \\
\hline
0 & 0 & 0 & -K_{1x} & 0 & 0 & 0 & 0 & 0 & K_{1x} & 0 & 0 \\
0 & 0 & -K_{2y} & 0 & K_{4y} & 0 & 0 & 0 & K_{2y} & 0 & K_{3y} & 0 \\
0 & K_{2z} & 0 & 0 & 0 & K_{4z} & 0 & -K_{2z} & 0 & 0 & 0 & K_{3z}
\end{array}\right]$$

17.06 STRAIGHT BARS OF ORDER THREE IN *S* SYSTEM

(1) Single lumped mass has in general six degrees of freedom, and its motion is governed by six independent dynamic differential equations given in symbolic form as

$$M_0\ddot{\Delta}(t) + K_0\Delta(t) = L_0(t)$$

where M_0 = lumped-mass matrix, K_0 = point spring-constant matrix (17.05), $\ddot{\Delta}(t)$ = acceleration vector (17.03), $\Delta(t)$ = displacement vector (17.03), and $L_0(t)$ = lumped-mass load vector (17.02).

(2) Single consistent mass has in general 12 degrees of freedom, and its motion is governed by 12 dependent dynamic differential equations given in symbolic form as

$$M_C\ddot{\Delta}(t) + K_C\Delta(t) = L_C(t)$$

where M_C = consistent-mass matrix (17.04), K_C = static stiffness matrix (17.05), $L_C(t)$ = joint load vector (17.02), and $\ddot{\Delta}(t)$, $\Delta(t)$ have the same meaning as in (1) above.

(3) Distributed-mass system has in general an infinite number of degrees of freedom and is governed by four independent differential equations given as

$$EA_x\frac{\partial^2 u}{\partial x^2} - k_u u - \mu\frac{\partial^2 u}{\partial t^2} = p_x(x, t)$$

$$GI_x\frac{\partial^2 \phi}{\partial x^2} - k_\phi \phi - \rho\frac{\partial^2 \phi}{\partial t^2} = q_x(x, t)$$

$$EI_z\frac{\partial^4 v}{\partial x^4} - \eta N\frac{\partial^2 v}{\partial x^2} + k_v v + \mu\frac{\partial^2 v}{\partial t^2} = p_y(x, t)$$

$$EI_y\frac{\partial^4 w}{\partial x^4} - \eta N\frac{\partial^2 w}{\partial x^2} + k_w w + \mu\frac{\partial^2 w}{\partial t^2} = p_z(x, t)$$

where u, v, w, ϕ = displacements (17.03), E, G, A_x, I_x, I_y, I_z = bar constants (17.05), μ = unit length mass (kg/m), ρ = unit length moment of inertia about centroidal axis (kg·m²/m), k_u, k_v, k_w, k_ϕ = foundation moduli (N/m², N·m/m), $p_x(x, t)$, $p_y(x, t)$, $p_z(x, t)$ = intensities of distributed load (N/m), $q_x(x, t)$ = intensity of distributed torque (N·m/m), and

$$\eta = \begin{cases} +1 & \text{if } N = \text{axial tension (N)} \\ -1 & \text{if } N = \text{axial compression (N)} \end{cases}$$

18

Lumped-Mass Models Single Degree of Freedom

DYNAMIC STATE

18.01 DEFINITION OF STATE

(1) Mechanical models. For the dynamic analysis, the real bar is represented by an ideal mechanical model which is better adaptable to the analytical treatment and whose dynamic behavior closely approximates the behavior of the real bar. Free and forced vibrations of such models are considered in this chapter.

(2) Hooke's model consists of a lumped mass attached by an elastic spring to a rigid foundation. By definition, this model has only one degree of freedom allowing a small displacement of the mass along or about one axis only. Three linear and three angular models of this type are possible (18.02–18.09).

(3) Kelvin's model consists of a lumped mass attached by an elastic spring and a viscoelastic dashpot to a rigid foundation. By definition, this model has also only one degree of freedom along or about one axis only. Three linear and three angular models of this type are possible (18.10–18.12).

(4) Notation (A.1–A.4). The definitions of load functions $L_\Delta(t)$, displacement $\Delta(t)$, velocity $\dot{\Delta}(t)$, acceleration $\ddot{\Delta}(t)$, and mass constants m, ρ are given in (17.2, 17.3, 17.4), respectively.

$K_\Delta = \begin{cases} \text{linear spring constant (N/m)} \\ \text{angular spring constant (N·m/rad)} \end{cases}$	$C_\Delta = \begin{cases} \text{linear damping constant (N·s/m)} \\ \text{angular damping constant (N·m · s/rad)} \end{cases}$
$\alpha_\Delta, \beta_\Delta$ = phase angle (rad)	ω_Δ = natural circular frequency (rad/s)

(5) Assumptions used in the analysis are those stated in (17.01). All models are in a state of small oscillation (vibration).

18.02 HOOKE'S MODEL, EQUATIONS OF MOTION

(1) Two typical models representing the vibration without damping of a lumped mass m_Δ are introduced in (2, 3) below. All dynamic vectors are in the respective axes, and they are positive if acting as shown (Ref. 6, pp. 10–16).

(2) Linear model

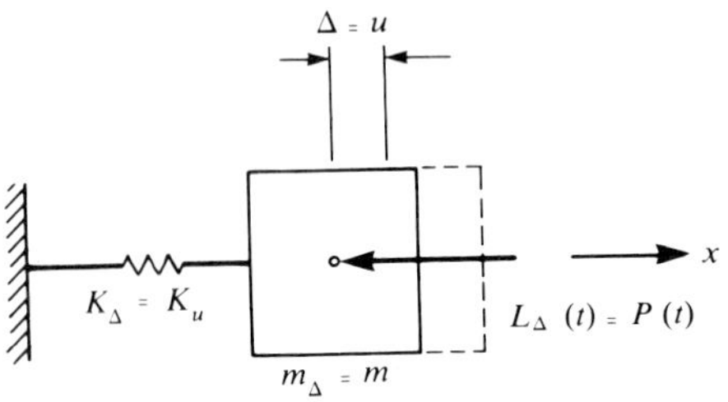

(3) Angular model

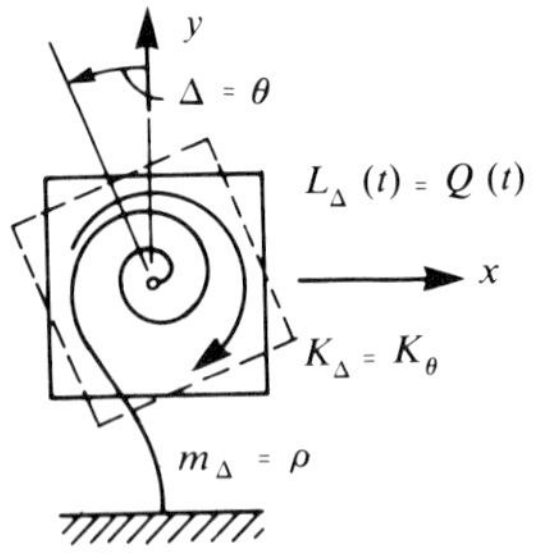

(4) Transport functions

$$T_0(t) = T_0 = \omega_\Delta \sin \omega_\Delta t \qquad T_3(t) = T_3 = (1 - \cos \omega_\Delta t)/\omega_\Delta^2$$

$$T_1(t) = T_1 = \cos \omega_\Delta t \qquad T_4(t) = T_4 = (\omega_\Delta t - \sin \omega_\Delta t)/\omega_\Delta^3$$

$$T_2(t) = T_2 = (\sin \omega_\Delta t)/\omega_\Delta \qquad T_5(t) = T_5 = (\omega_\Delta^2 t^2/2 + \cos \omega_\Delta t - 1)/\omega_\Delta^4$$

where $\omega_\Delta = \sqrt{K_\Delta/m_\Delta}$.

For particular values of t, $T_j(t - a) = A_j$, $T_j(t - b) = B_j$, $T_j(t - c) = C_j$. So, for example,

$$T_4(t - a) = A_4 = \frac{\omega_\Delta(t - a) - \sin \omega_\Delta(t - a)}{\omega_\Delta^3}$$

(5) Equations of motion of these models are, in symbolic matrix form,

$$\begin{bmatrix} 1 \\ \Delta \\ \dot{\Delta} \end{bmatrix} = \begin{bmatrix} 1 & 0 & 0 \\ G_\Delta & T_1 & T_2 \\ \dot{G}_\Delta & -T_0 & T_1 \end{bmatrix} \begin{bmatrix} 1 \\ \Delta_0 \\ \dot{\Delta}_0 \end{bmatrix}$$

where Δ, $\dot{\Delta}$ are the displacement and velocity at t, Δ_0, $\dot{\Delta}_0$ are the same values at $t = 0$, and

$$G_\Delta = \frac{1}{m_\Delta \omega_\Delta} \int_0^t L_\Delta(\tau) \sin \omega_\Delta(t - \tau)\, d\tau \qquad \dot{G}_\Delta = \frac{dG_\Delta}{dt}$$

are the response functions (convolution integrals) given for particular cases in (18.04–18.09). Each equation consists of two parts, the free vibration in terms of the transport functions T_j ($j = 0, 1, 2$) and the initial conditions Δ_0, $\dot{\Delta}_0$, and the response function representing the effect of the forcing function $L_\Delta(t)$.

18.03 HOOKE'S MODEL, PARAMETERS OF MOTION

(1) Modified equations of motion are

$$\Delta = R_\Delta \cos(\omega_\Delta t - \alpha_\Delta) + G_\Delta \qquad \text{or} \qquad \Delta = R_\Delta \sin(\omega_\Delta t + \beta_\Delta) + G_\Delta$$

$$\dot{\Delta} = -\omega R_\Delta \sin(\omega_\Delta t - \alpha_\Delta) + \dot{G}_\Delta \qquad \text{or} \qquad \dot{\Delta} = \omega R_\Delta \cos(\omega_\Delta t + \beta_\Delta) + \dot{G}_\Delta$$

where the natural amplitude is

$$R_\Delta = \sqrt{\Delta_0^2 + (\dot{\Delta}_0/\omega_\Delta)^2}$$

and the phase angles, representing the shift of the harmonic curve of free vibration, are

$$\alpha_\Delta = \tan^{-1}\frac{\dot{\Delta}_0}{\omega_\Delta \Delta_0} \qquad \beta_\Delta = \tan^{-1}\frac{\omega_\Delta \Delta_0}{\dot{\Delta}_0} \qquad \alpha_\Delta + \beta_\Delta = \frac{\pi}{2}$$

(2) Graphical representation of the displacement equations in (1) is shown below.

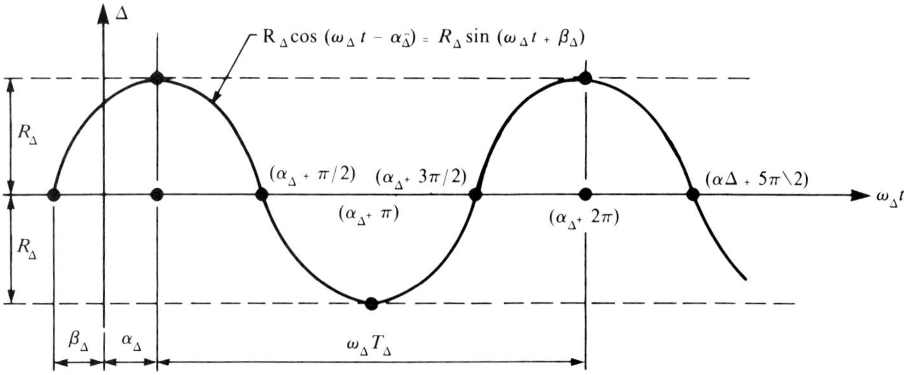

(3) Free vibration is said to be harmonic, so that displacement varies sinusoidally with time and repeats itself after a definite time interval. The natural amplitude R_Δ defined analytically in (1) above is the greatest displacement of the lumped mass from its equilibrium position and is affected by initial conditions Δ_0, $\dot{\Delta}_0$, the mass m_Δ, and the spring constant K_Δ.

(4) Natural period T_Δ of free vibration is the time required for the mass to complete one cycle, after which the repetition of the same motion takes place. The reciprocal value of T_Δ is the natural frequency f_Δ which is the number of cycles completed per unit of time.

$$T_\Delta = \frac{2\pi}{\omega_\Delta} \qquad f_\Delta = \frac{1}{T_\Delta}$$

Both parameters are characteristic for the given model, depending on mass and spring constant only, and they are independent of the initial conditions.

18.04 HOOKE'S MODEL, LOAD FUNCTIONS

Notation (18.02, 18.04)	Signs (18.02)	Matrix (18.02)

The analytical expressions for $G_\Delta(t)$ and $\dot{G}_\Delta(t)$ defined in (18.02) are given for particular $L_\Delta(t)$ below. For simplicity of writing, the subscript Δ is omitted in the respective terms. In case of linear motion, $m_\Delta = m$ and $L_\Delta(t) = P(t)$; in case of angular motion, $m_\Delta = \rho$ and $L_\Delta(t) = Q(t)$. The transport functions $T_0, T_1, \ldots$ and their special values A_j, B_j, C_j for $j = 0, 1, 2, 3, 4, 5$ are given in (18.02).

P = force (N)	t = time (s)
Q = moment (N·m)	a, b, c = intervals (s)
Ω = forcing angular frequency (rad/s)	ω = natural angular frequency (rad/s)

Load	Functions	
1. 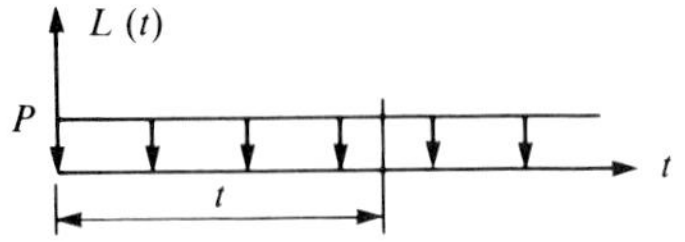	$G = -\dfrac{P}{m} T_3$ $\dot{G} = -\dfrac{P}{m} T_2$	
2. 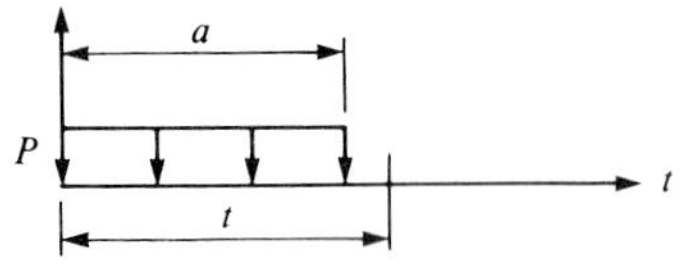	$G = -\dfrac{P}{m}(T_3 - A_3)$ $\dot{G} = -\dfrac{P}{m}(T_2 - A_2)$	
3. 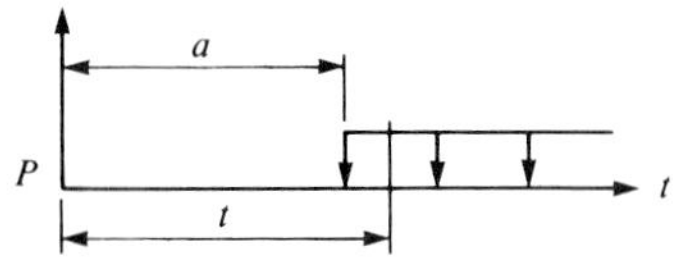	$G = -\dfrac{P}{m} A_3$ $\dot{G} = -\dfrac{P}{m} A_2$	For $t \le a$, $G = 0$, $\dot{G} = 0$
4. 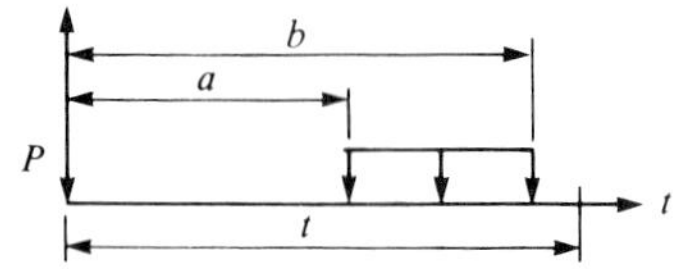	$G = -\dfrac{P}{m}(A_3 - B_3)$ $\dot{G} = -\dfrac{P}{m}(A_2 - B_2)$	
5. 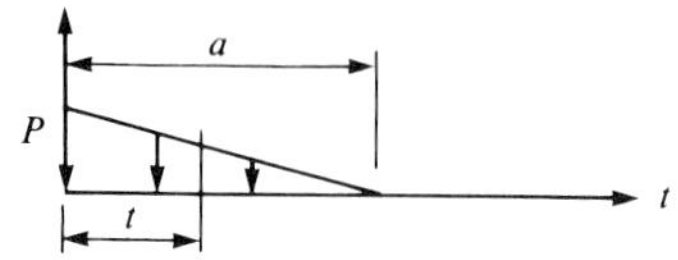	$G = -\dfrac{P}{ma}(aT_3 - T_4)$ $\dot{G} = -\dfrac{P}{ma}(aT_2 - T_3)$	
6. 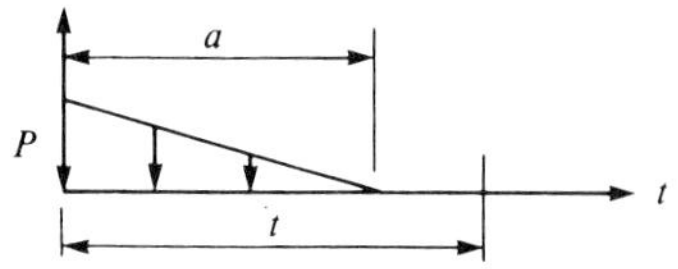	$G = -\dfrac{P}{ma}(aT_3 - T_4 + A_4)$ $\dot{G} = -\dfrac{P}{ma}(aT_2 - T_3 + A_3)$	

18.05 HOOKE'S MODEL, LOAD FUNCTIONS

Notation (18.02, 18.04) Signs (18.02) Matrix (18.02)

1.

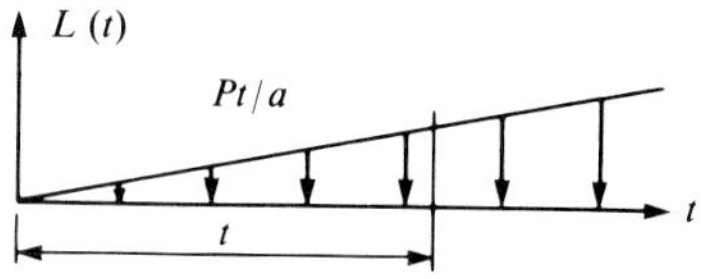

$$G = -\frac{P}{ma} T_4$$

$$\dot{G} = -\frac{P}{ma} T_3$$

2.

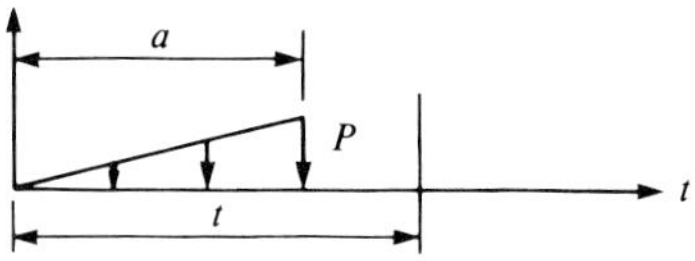

$$G = -\frac{P}{ma}(T_4 - A_4 - aA_3)$$

$$\dot{G} = -\frac{P}{ma}(T_3 - A_3 - aA_2)$$

3.

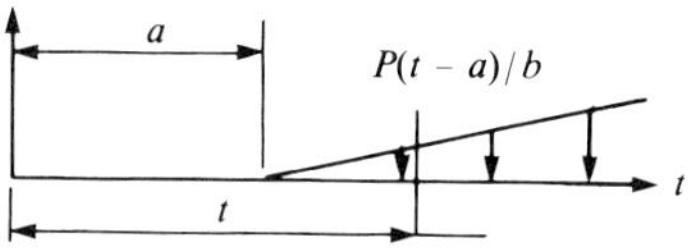

$$G = -\frac{P}{mb} A_4$$

$$\dot{G} = -\frac{P}{mb} A_3$$

4.

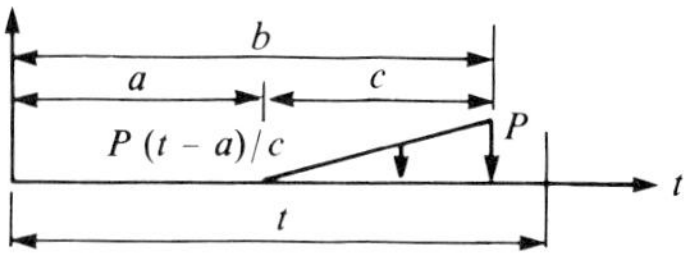

$$G = -\frac{P}{mc}(A_4 - B_4 - cB_3)$$

$$\dot{G} = -\frac{P}{mc}(A_3 - B_3 - cB_2)$$

For $t \leq a$, $G = 0$, $\dot{G} = 0$ (cases 3 and 4)

5.

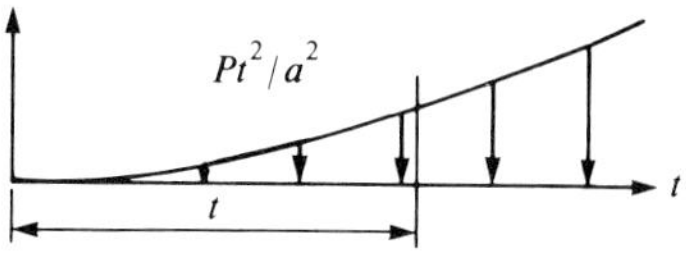

$$G = -\frac{2P}{ma^2} T_5$$

$$\dot{G} = -\frac{2P}{ma^2} T_4$$

6.

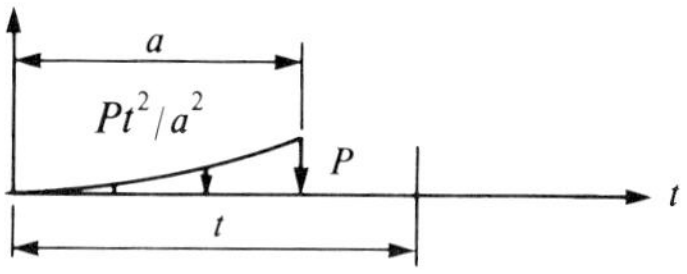

$$G = -\frac{2P}{ma^2}(T_5 - A_5 - aA_4 - \tfrac{1}{2}a^2A_3)$$

$$\dot{G} = -\frac{2P}{ma^2}(T_4 - A_4 - aA_3 - \tfrac{1}{2}a^2A_2)$$

7.

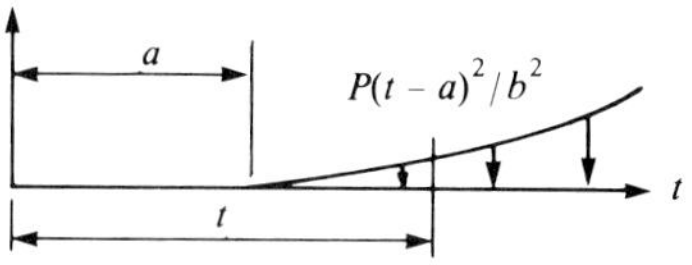

$$G = -\frac{2P}{mb^2} A_5$$

$$\dot{G} = -\frac{2P}{mb^2} A_4$$

8.

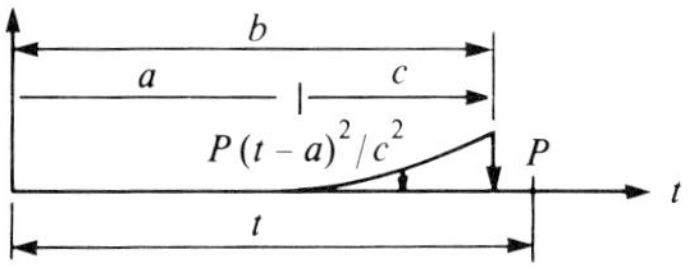

$$G = -\frac{2P}{mc^2}(A_5 - B_5 - cB_4 - \tfrac{1}{2}c^2B_3)$$

$$\dot{G} = -\frac{2P}{mc^2}(A_4 - B_4 - cB_3 - \tfrac{1}{2}c^2B_2)$$

For $t \leq a$, $G = 0$, $\dot{G} = 0$ (cases 7 and 8)

18.06 HOOKE'S MODEL, LOAD FUNCTIONS

Notation (18.02, 18.04)	Signs (18.02) Matrix (18.02)
1. 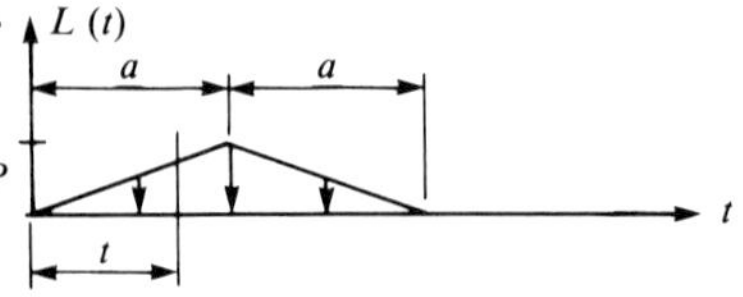	$G = -\dfrac{P}{ma} T_4$ $\dot{G} = -\dfrac{P}{ma} T_3$
2. 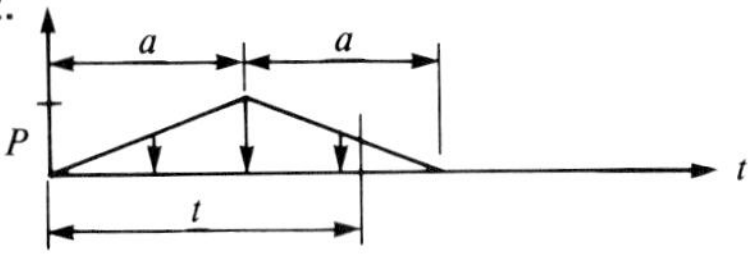	$G = -\dfrac{P}{ma} (T_4 - 2A_4)$ $\dot{G} = -\dfrac{P}{ma} (T_3 - 2A_3)$
3. 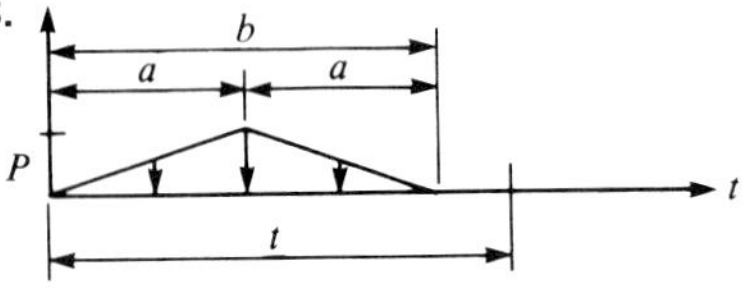	$G = -\dfrac{P}{ma} (T_4 - 2A_4 + B_4)$ $\dot{G} = -\dfrac{P}{ma} (T_3 - 2A_3 + B_3)$
4. 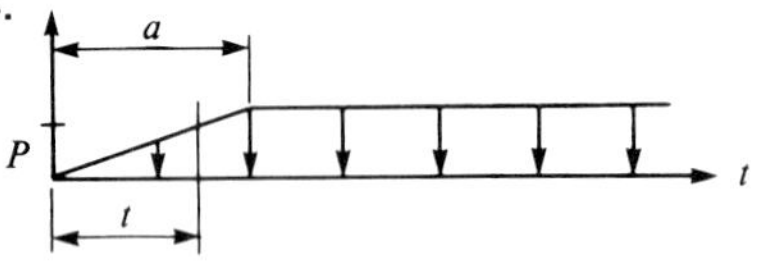	$G = -\dfrac{P}{ma} T_4$ $\dot{G} = -\dfrac{P}{ma} T_3$
5. 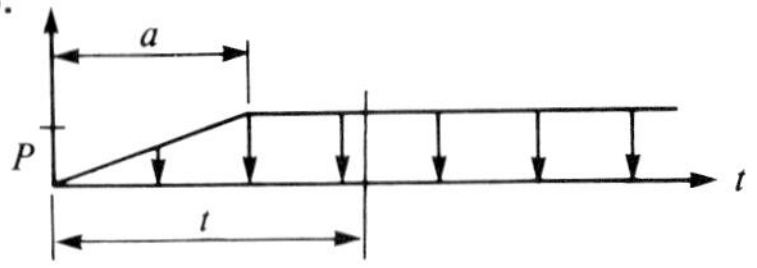	$G = -\dfrac{P}{ma} (T_4 - A_4)$ $\dot{G} = -\dfrac{P}{ma} (T_3 - A_3)$
6. 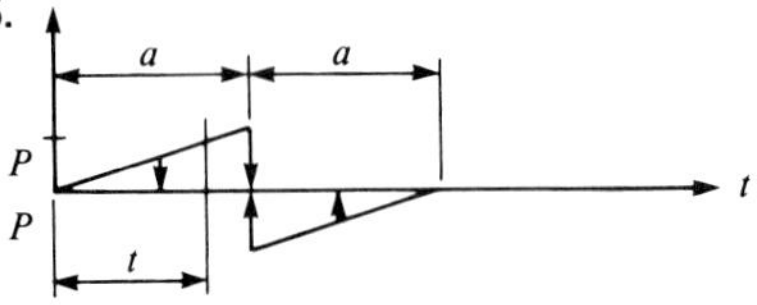	$G = -\dfrac{P}{ma} T_4$ $\dot{G} = -\dfrac{P}{ma} T_3$
7. 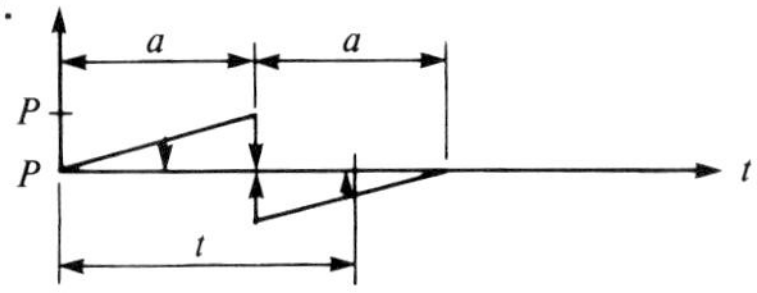	$G = -\dfrac{P}{ma} (T_4 - 2aA_3)$ $\dot{G} = -\dfrac{P}{ma} (T_3 - 2aA_2)$
8. 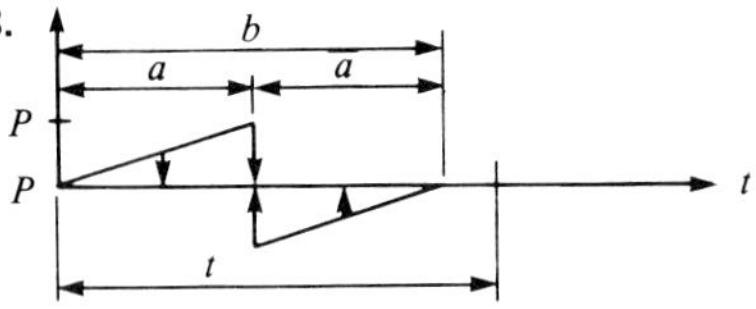	$G = -\dfrac{P}{ma} (T_4 - B_4 - 2aA_3)$ $\dot{G} = -\dfrac{P}{ma} (T_3 - B_3 - 2aA_2)$

18.07 HOOKE'S MODEL, LOAD FUNCTIONS

Notation (18.02, 18.04)	Signs (18.02)	Matrix (18.02)
1. 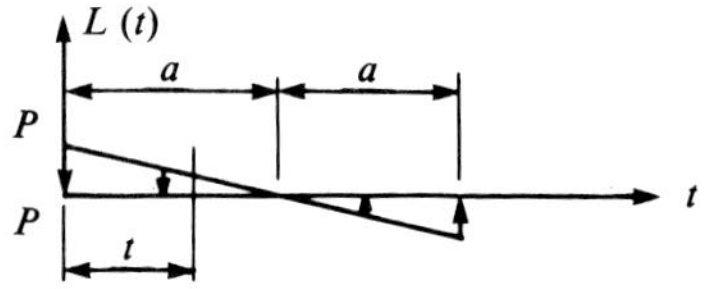	$G = -\frac{P}{ma}(aT_3 - T_4)$ $\dot{G} = -\frac{P}{ma}(aT_2 - T_3)$	
2. 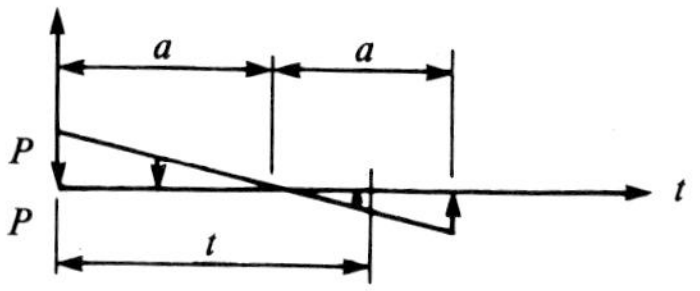	$G = -\frac{P}{ma}(aT_3 - T_4)$ $\dot{G} = -\frac{P}{ma}(aT_2 - T_3)$	
3. 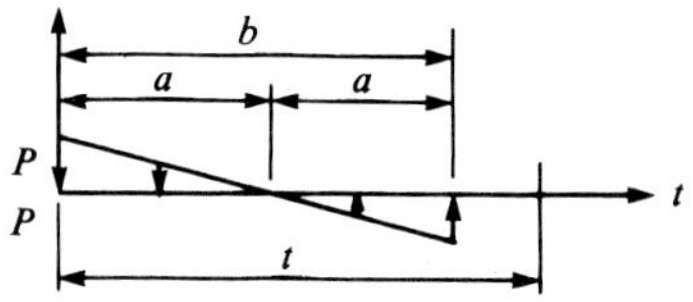	$G = -\frac{P}{ma}(aT_3 - T_4 + aB_3 + B_4$ $\dot{G} = -\frac{P}{ma}(aT_2 - T_3 + aB_2 + B_3)$	
4. 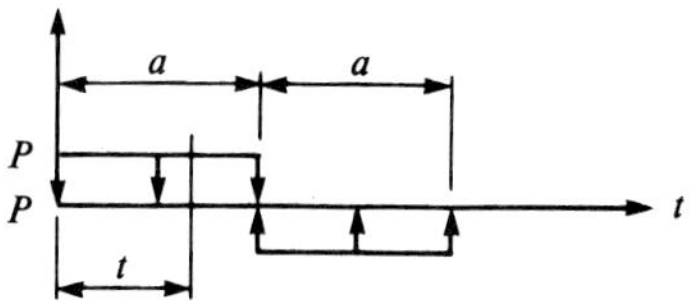	$G = -\frac{P}{m}T_3$ $\dot{G} = -\frac{P}{m}T_2$	
5. 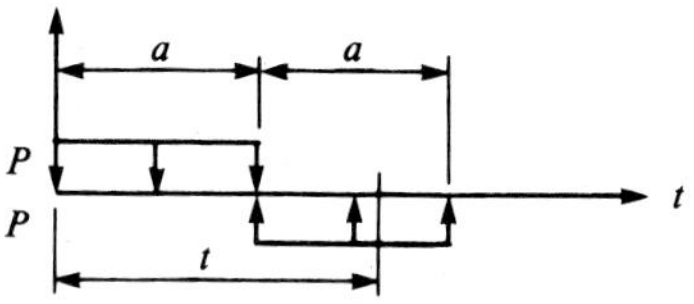	$G = -\frac{P}{m}(T_3 - 2A_3)$ $\dot{G} = -\frac{P}{m}(T_2 - 2A_2)$	
6. 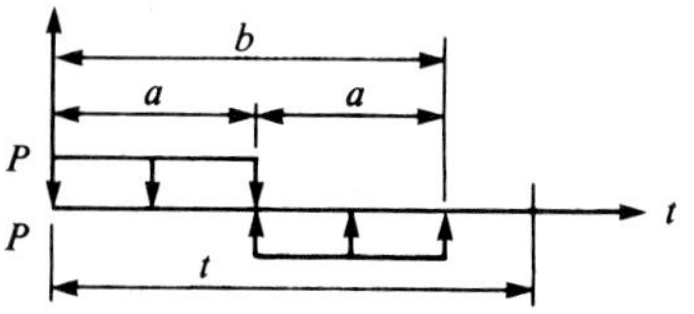	$G = -\frac{P}{m}(T_3 - 2A_3 + B_3)$ $\dot{G} = -\frac{P}{m}(T_2 - 2A_2 + B_2)$	
7. 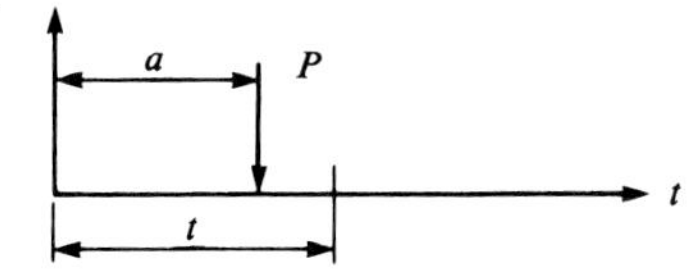	$G = -\frac{P}{m}A_2$ $\dot{G} = -\frac{P}{m}A_1$ For linear models only	For $t < a$, $G = 0$ $\dot{G} = 0$
8. 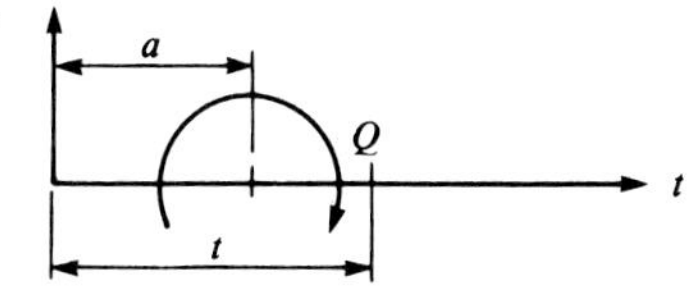	$G = -\frac{Q}{\rho}A_2$ $\dot{G} = -\frac{Q}{\rho}A_1$ For angular models only	

18.08 HOOKE'S MODEL, LOAD FUNCTIONS

Notation (18.02, 18.04) — Signs (18.02) — Matrix (18.02)

1.

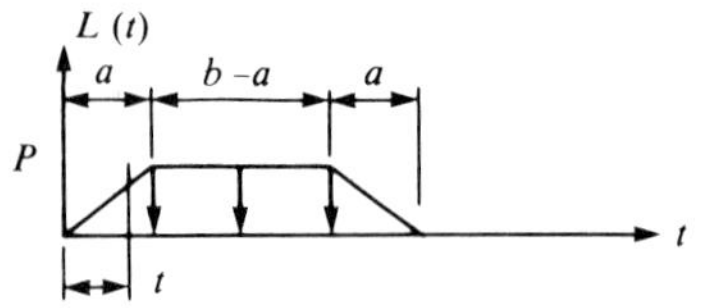

$$\dot{G} = -\frac{P}{ma}T_3 \qquad G = -\frac{P}{ma}T_4$$

2.

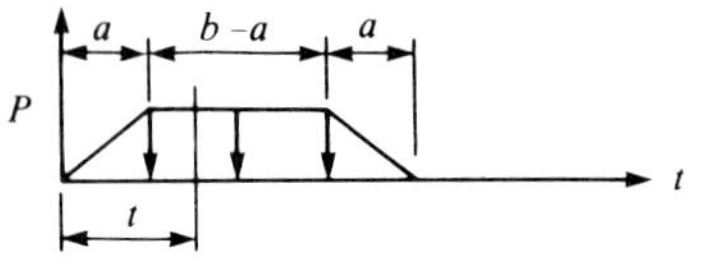

$$G = -\frac{P}{ma}(T_4 - A_4)$$

$$\dot{G} = -\frac{P}{ma}(T_3 - A_3)$$

3.

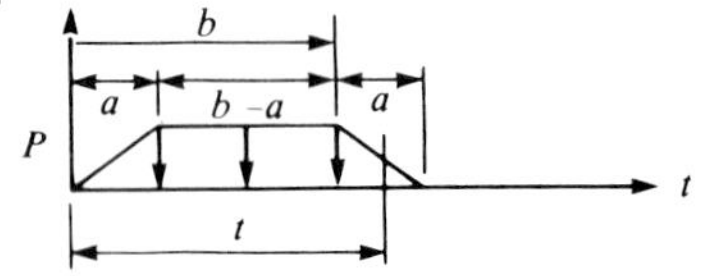

$$G = -\frac{P}{ma}(T_4 - A_4 - B_4)$$

$$\dot{G} = -\frac{P}{ma}(T_3 - A_3 - B_3)$$

4.

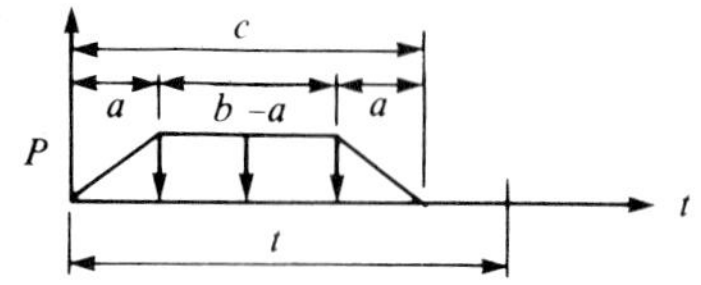

$$G = -\frac{P}{ma}(T_4 - A_4 - B_4 + C_4)$$

$$\dot{G} = -\frac{P}{ma}(T_3 - A_3 - B_3 + C_3)$$

5.

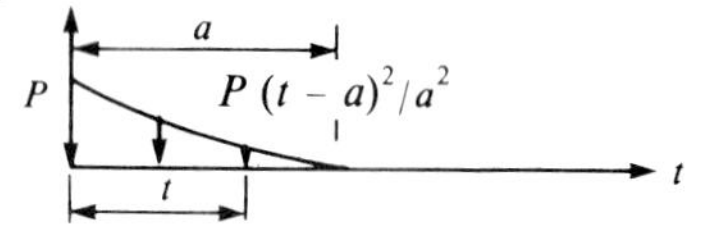

$$G = -\frac{2P}{ma^2}(T_5 - aT_4 + \tfrac{1}{2}a^2T_3)$$

$$\dot{G} = -\frac{2P}{ma^2}(T_4 - aT_3 + \tfrac{1}{2}a^2T_2)$$

6.

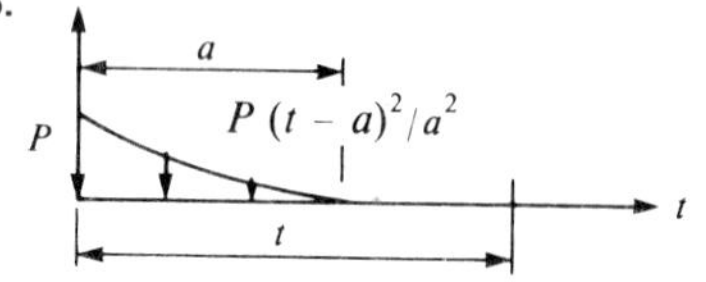

$$G = -\frac{2P}{ma^2}(T_5 - aT_4 + \tfrac{1}{2}a^2T_3 - A_5)$$

$$\dot{G} = -\frac{2P}{ma^2}(T_4 - aT_3 + \tfrac{1}{2}a^2T_2 - A_4)$$

7.

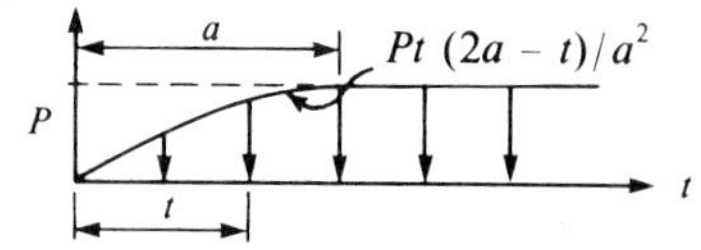

$$G = -\frac{2P}{ma^2}(aT_4 - T_5)$$

$$\dot{G} = -\frac{2P}{ma^2}(aT_3 - T_4)$$

8.

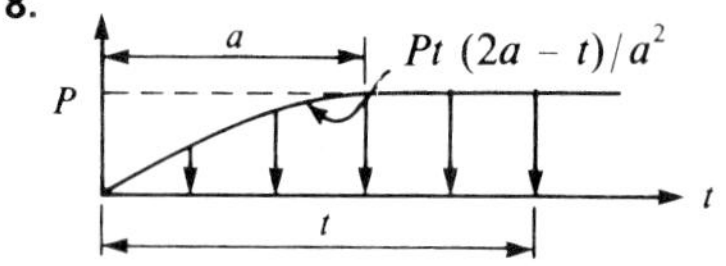

$$G = -\frac{2P}{ma^2}(aT_4 - T_5 + A_5)$$

$$\dot{G} = -\frac{2P}{ma^2}(aT_3 - T_4 + A_4)$$

Notation (18.02, 18.04)   Signs (18.02)   Matrix (18.02)

1.

$a = \pi/\Omega$

$P \sin \Omega t$

$$G = -\frac{P}{m\Lambda}(\sin \Omega t - \Omega T_2)$$

$$\dot{G} = -\frac{P\Omega}{m\Lambda}(\cos \Omega t - T_1)$$

2.

$a = \pi/\Omega$

$P \cos \Omega t$

$$G = -\frac{P}{m\Lambda}(\cos \Omega t - T_1)$$

$$\dot{G} = -\frac{P}{m\Lambda}(T_0 - \Omega \sin \Omega t)$$

3.

$a = \pi/2\Omega$

$P \cos \Omega t$

$$G = -\frac{P}{m\Lambda}(\cos \Omega t - T_1)$$

$$\dot{G} = -\frac{P}{m\Lambda}(T_0 - \Omega \sin \Omega t)$$

4.

$a = \pi/2\Omega$

$P \cos \Omega t$

$$G = \frac{P}{m\Lambda}(T_1 + \Omega A_2)$$

$$\dot{G} = -\frac{P}{m\Lambda}(\Omega A_1 - \omega^2 T_2)$$

5.

$b = 2\pi/\Omega$

$P(1 - \cos \Omega t)$

$$G = \frac{P}{m\Lambda}(\cos \Omega t - T_1 - \Lambda T_3)$$

$$\dot{G} = -\frac{P}{m\Lambda}(\Lambda T_2 + T_0 + \Omega \sin \Omega t)$$

6.

$b = 2\pi/\Omega$

$P(1 - \cos \Omega t)$

$$G = -\frac{P}{m\Lambda}\left(\frac{\Lambda}{\omega^2} - 1\right)(B_1 - T_1)$$

$$\dot{G} = -\frac{P}{m\Lambda}\left(\frac{\Lambda}{\omega^2} - 1\right)(B_0 - T_0)$$

7.

$Pe^{-t/a}$

$$G = -\frac{Pa}{m\Gamma}(T_2 - aT_1 + ae^{-t/a})$$

$$\dot{G} = -\frac{Pa}{m\Gamma}(T_1 + aT_0 - e^{-t/a})$$

8.

$P(1 - e^{-t/a})$

$$G = -\frac{P}{m\Gamma}(\Gamma T_3 - aT_2 + a^2 T_1 - a^2 e^{-t/a})$$

$$\dot{G} = -\frac{P}{m\Gamma}(\Gamma T_2 - aT_1 - a^2 T_1 - ae^{-t/a})$$

$\Lambda = \omega^2 - \Omega^2$

$\Gamma = a^2\omega^2 + 1$

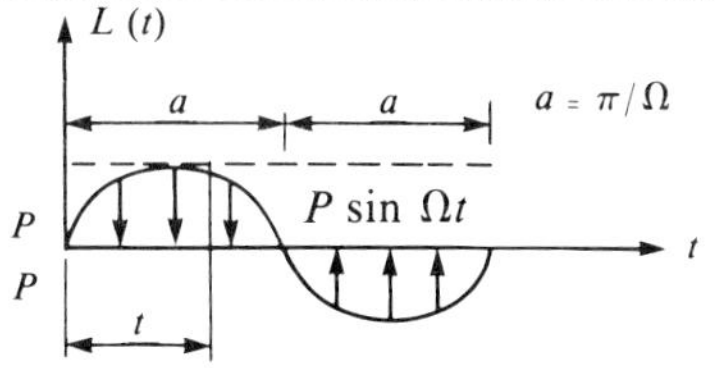

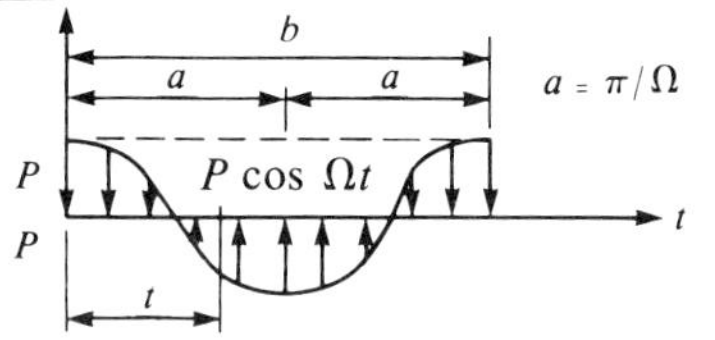

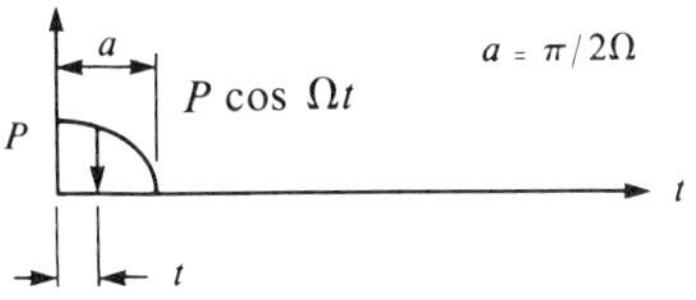

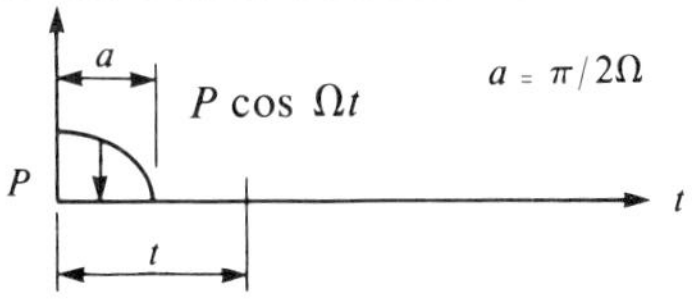

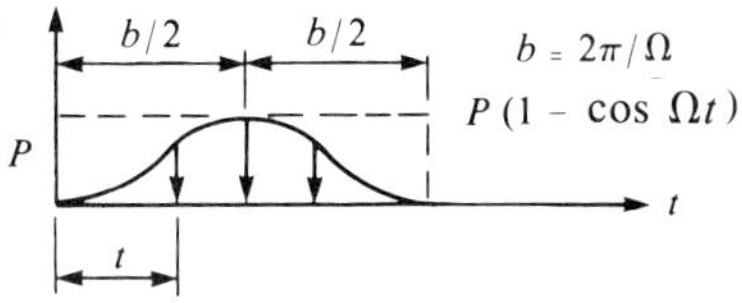

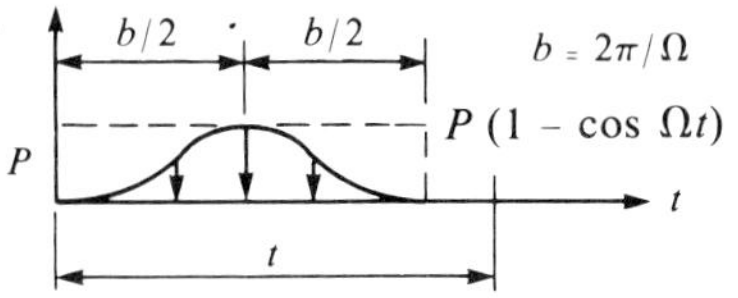

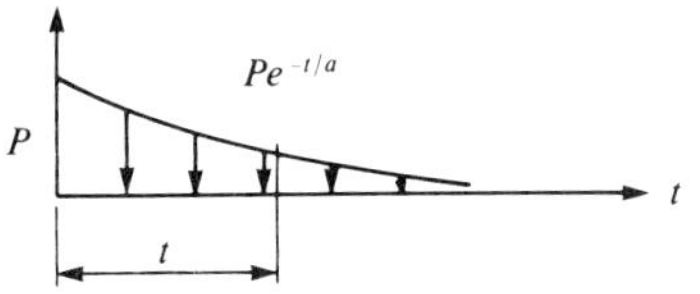

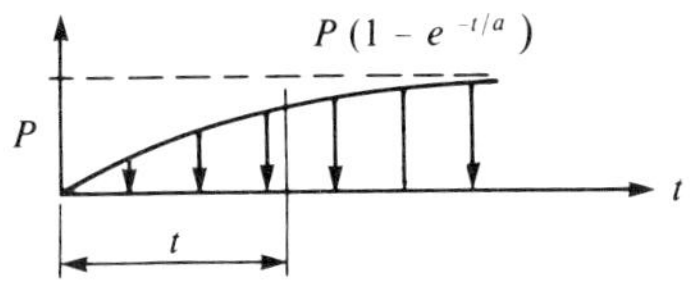

18.10 KELVIN'S MODEL, EQUATIONS OF MOTION

(1) Two typical models representing the linear and angular vibration with damping of a single lumped mass are introduced in (2, 3) below. All dynamic vectors are in the respective axes, and they are positive if acting as shown (Ref. 6, pp. 16–19).

(2) Linear model

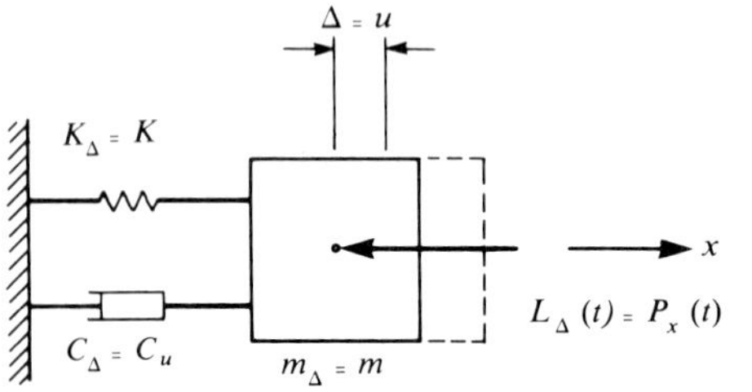

(3) Angular model

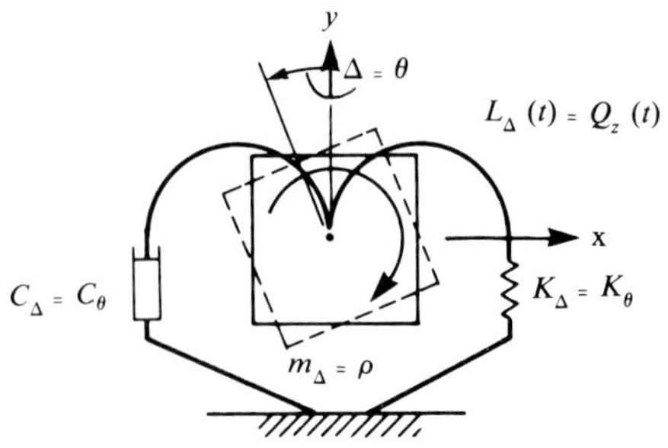

(4) Equations of motion of these models are, in symbolic matrix form,

$$\begin{bmatrix} 1 \\ \hline \Delta \\ \dot{\Delta} \end{bmatrix} = \left[\begin{array}{c|cc} 1 & 0 & 0 \\ \hline G_\Delta & T_1 & T_2 \\ \dot{G}_\Delta & -T_0 & T_1 - 2\gamma_\Delta T_2 \end{array}\right] \begin{bmatrix} 1 \\ \hline \Delta_0 \\ \dot{\Delta}_0 \end{bmatrix}$$

where Δ, $\dot{\Delta}$ are the displacements and velocity at t, Δ_0, $\dot{\Delta}_0$ are their counterparts at $t = 0$, and

$$G_\Delta = \frac{1}{m_\Delta}\int_0^t L_\Delta(t) T_2(t-\tau)\, d\tau \qquad \dot{G}_\Delta = \frac{dG_\Delta}{dt}$$

are the response functions (convolution integrals) given for particular cases in (18.13). Each equation consists of two parts, the free vibration in terms of the transport functions T_j ($j = 0, 1, 2$) and the initial conditions Δ_0, $\dot{\Delta}_0$, and the response function representing the effect of the forcing function $L_\Delta(t)$

(5) Three distinct cases are possible and depend on the relation of the following parameters:

$$\gamma_\Delta = C_\Delta/2m_\Delta \qquad \omega_\Delta^2 = K_\Delta/m_\Delta$$

(*a*) Underdamped motion: $\gamma_\Delta^2 < \omega_\Delta^2$ $\quad \lambda_\Delta = \sqrt{\omega_\Delta^2 - \gamma_\Delta^2}$

(*b*) Critically damped motion: $\gamma_\Delta^2 = \omega_\Delta^2$, $\quad \lambda_\Delta = 0$

(*c*) Overdamped motion: $\gamma_\Delta^2 > \omega_\Delta^2$, $\quad \lambda_\Delta = \sqrt{\gamma_\Delta^2 - \omega_\Delta^2}$

(6) Shape functions for these three cases are tabulated in (18.11). For simplicity of writing, the subscript Δ is omitted in the respective terms.

18.11 KELVIN'S MODEL TRANSPORT FUNCTIONS

(1) **Case I:** $\gamma^2 < \omega^2 \qquad \lambda = \sqrt{\omega^2 - \gamma^2} \qquad \phi = \tan^{-1}\frac{\lambda}{\gamma}$

$T_0 = -\frac{\omega^2}{\lambda} e^{-\gamma t} \sin \lambda t$	
$T_1 = \frac{\omega}{\lambda} e^{-\gamma t} \sin(\lambda t + \phi)$	$\dot{T}_1 = -T_0$
$T_2 = \frac{1}{\lambda} e^{-\gamma t} \sin \lambda t$	$\dot{T}_2 = T_1 - 2\gamma T_2$
$T_3 = -\frac{1}{\lambda\omega}[e^{-\gamma t} \sin(\lambda t + \phi) - \sin\phi]$	$\dot{T}_3 = T_2$
$T_4 = \frac{1}{\lambda\omega^2}[e^{-\gamma t} \sin(\lambda t + 2\phi) + \omega t \sin\phi - \sin 2\phi]$	$\dot{T}_4 = T_3$
$T_5 = -\frac{1}{\lambda\omega^3}[e^{-\gamma t} \sin(\lambda t + 3\phi) - \frac{1}{2}(\omega t)^2 \sin\phi + \omega t \sin 2\phi - \sin 3\phi]$	$\dot{T}_5 = T_4$

(2) **Case II:** $\gamma^2 = \omega^2 \qquad \lambda = 0 \qquad \phi = 0$

$T_0 = -\gamma^2 t e^{-\gamma t}$	
$T_1 = e^{-\gamma t}(1 + \gamma t)$	$\dot{T}_1 = -T_0$
$T_2 = t e^{-\gamma t}$	$\dot{T}_2 = T_1 - 2\gamma T_2$
$T_3 = -\frac{1}{\gamma^2}[e^{-\gamma t}(1 + \gamma t) - 1]$	$\dot{T}_3 = T_2$
$T_4 = \frac{1}{\gamma^3}[e^{-\gamma t}(2 + \gamma t) + \gamma t - 2]$	$\dot{T}_4 = T_3$
$T_5 = -\frac{1}{\gamma^4}[e^{-\gamma t}(3 + \gamma t) - \frac{1}{2}(\gamma t)^2 + 2\gamma t - 3]$	$\dot{T}_5 = T_4$

(3) **Case III:** $\gamma^2 > \omega^2 \qquad \lambda = \sqrt{\gamma^2 - \omega^2} \qquad \phi = \tanh^{-1}\frac{\lambda}{\gamma}$

$T_0 = -\frac{\omega^2}{\lambda} e^{-\gamma t} \sinh \lambda t$	
$T_1 = \frac{\omega}{\lambda} e^{-\gamma t} \sinh(\lambda t + \phi)$	$\dot{T}_1 = -T_0$
$T_2 = \frac{1}{\lambda} e^{-\gamma t} \sinh \lambda t$	$\dot{T}_2 = T_1 - 2\gamma T_2$
$T_3 = -\frac{1}{\lambda\omega}[e^{-\gamma t} \sinh(\lambda t + \phi) - \sinh\phi]$	$\dot{T}_3 = T_2$
$T_4 = \frac{1}{\lambda\omega^2}[e^{-\gamma t} \sinh(\lambda t + 2\phi) + \omega t \sinh\phi - \sinh 2\phi]$	$\dot{T}_4 = T_3$
$T_5 = -\frac{1}{\lambda\omega^3}[e^{-\gamma t} \sinh(\lambda t + 3\phi) - \frac{1}{2}(\omega t)^2 \sinh\phi + \omega t \sinh 2\phi - \sinh 3\phi]$	$\dot{T}_5 = T_4$

18.12 KELVIN'S MODEL, PARAMETERS OF MOTION

(1) Three curves, shown below, represent the free vibration with damping, each corresponding to one of the motions described in (18.10).

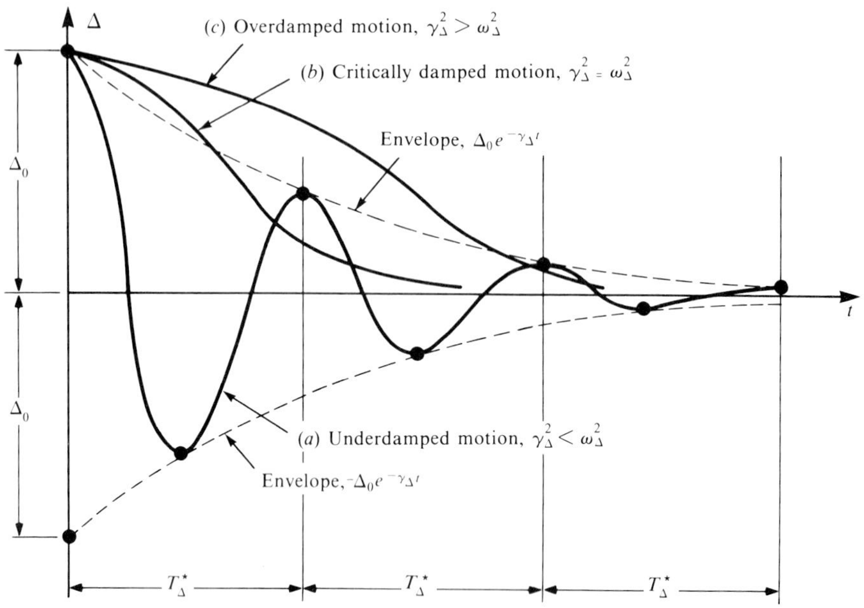

Curve (*a*), representing the underdamped motion, shows that the motion is oscillatory with diminishing amplitudes, whereas curves (*b*) and (*c*), representing the critically damped and overdamped motion, respectively, show that the magnitude of damping is so large that the motion is no longer oscillatory and decreases exponentially with time (an aperiodic motion). The critically damped motion is the limit of the periodic motion during which the system reaches rest in the shortest time (Ref. 6, p. 18).

(2) Modified equations of underdamped motion are

$$\Delta = R_\Delta^*[\cos(\lambda_\Delta t - \alpha_\Delta)] + G_\Delta \qquad \text{or} \qquad \Delta = R_\Delta^*[\sin(\lambda_\Delta t + \beta_\Delta)] + G_\Delta$$

where the variable amplitude is

$$R_\Delta^* = R_\Delta e^{-\gamma_\Delta t} = e^{-\gamma_\Delta t}\sqrt{(\Delta_0)^2 + (\dot{\Delta}_0 + \gamma_\Delta \Delta_0)^2/\gamma_\Delta^2}$$

and the phase angles representing the shift of the decaying harmonic curve on the $\lambda_\Delta t$ axis are

$$\alpha_\Delta = \tan^{-1}\frac{\dot{\Delta}_0 + \gamma_\Delta \Delta_0}{\lambda_\Delta \Delta_0} \qquad \beta_\Delta = \tan^{-1}\frac{\lambda_\Delta \Delta_0}{\dot{\Delta}_0 + \gamma_\Delta \Delta_0} \qquad \alpha_\Delta + \beta_\Delta = \frac{\pi}{2}$$

(3) Underdamped free vibration is nonharmonic, nonrepetitive, with decreasing amplitudes, spaced on equal intervals (periods). The amplitude R_Δ^*,n and the period T_Δ^* are, respectively,

$$R_\Delta^*,n = R_\Delta e^{-\gamma_\Delta n T_\Delta^*} \qquad T_\Delta^* = 2\pi/\lambda_\Delta$$

where R_Δ is the initial term in (2), $n = 0, 1, 2, \ldots$. The amplitude depends on the physical characteristics of the model and the initial conditions Δ_0, $\dot{\Delta}_0$.

18.13 KELVIN'S MODEL, LOAD FUNCTIONS

(1) Analytical expressions for $G_\Delta(t)$ and $\dot{G}_\Delta(t)$ in (18.10) are formally identical to their counterparts in (18.04–18.08), but in case of vibration with damping, they must be expressed in terms of the transport functions tabulated in (18.11) and of their special values

$$T_j(t-a) = A_j \qquad T_j(t-b) = B_j \qquad T_j(t-c) = C_j \qquad (j = 0, 1, 2, \ldots, 5)$$

For example,

$$A_3 = \begin{cases} -\dfrac{1}{\lambda\omega}\,[e^{-\gamma(t-a)}\sin(\lambda t - \lambda a + \phi) - \sin\phi] & \text{(Case I)} \\ -\dfrac{1}{\gamma^2}\,[e^{-\gamma(t-a)}(1 + \gamma t - \gamma a) - 1] & \text{(Case II)} \\ -\dfrac{1}{\lambda\omega}\,[e^{-\gamma(t-a)}\sinh(\lambda t - \lambda a + \phi) - \sinh\phi] & \text{(Case III)} \end{cases}$$

(2) Harmonic and exponential load functions yield time-response functions tabulated below.

Notation (18.10–18.12) $\Lambda_1 = (\omega^2 - \Omega^2)^2 + (2\gamma\Omega)^2$	Signs (18.10) $\Lambda_2 = \omega^2 - \Omega^2$ Matrix (18.10) $\Gamma = 1 - 2a\gamma + a^2\omega^2$
1. 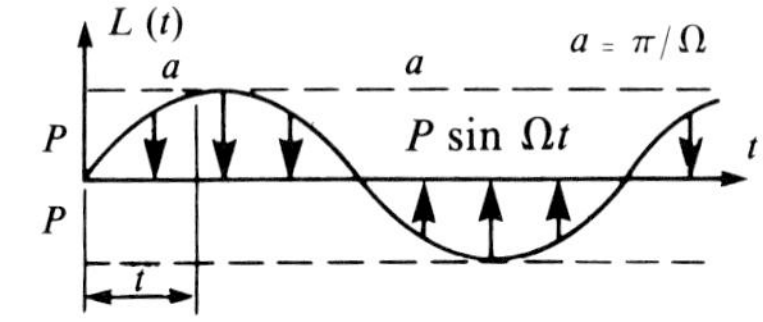	$G = -\dfrac{P}{m\Lambda_1}[\Lambda_2(\sin\Omega t - \Omega T_2) - 2\gamma\Omega(\cos\Omega t - T_1)]$ $\dot{G} = -\dfrac{P\Omega}{m\Lambda_1}[\Lambda_2(\cos\Omega t - T_1) + 2\gamma(\Omega\sin\Omega t + T_0)]$
2. 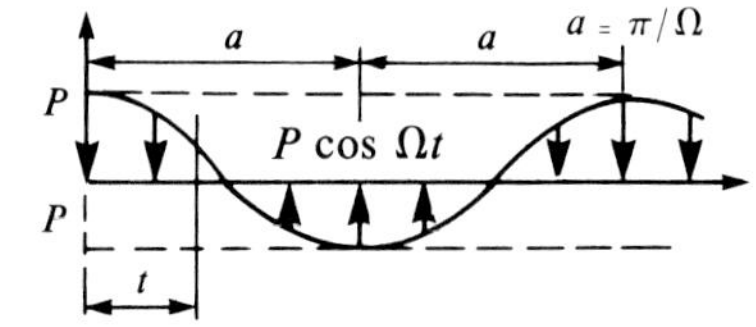	$G = -\dfrac{P}{m\Lambda_1}[\Lambda_2(\cos\Omega t - T_1) + 2\gamma\Omega(\sin\Omega t - \Omega T_2)]$ $\dot{G} = -\dfrac{P}{m\Lambda_1}[\Lambda_2(T_0 - \Omega\sin\Omega t) + 2\,\gamma\Omega^2(\cos\Omega t - T_1)]$
3. 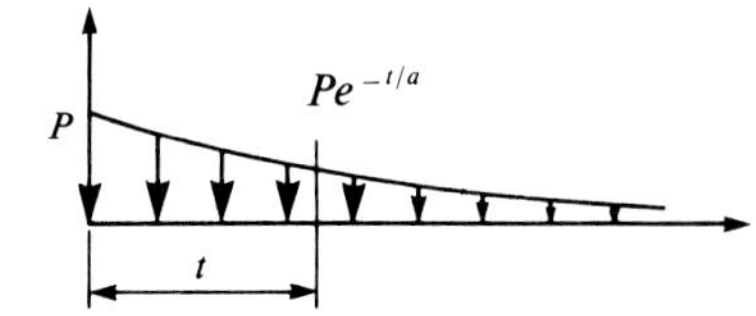	$G = -\dfrac{Pa}{m\Gamma}(T_2 - aT_1 + ae^{-t/a})$ $\dot{G} = -\dfrac{Pa}{m\Gamma}(T_1 + aT_0 - e^{-t/a})$
4. 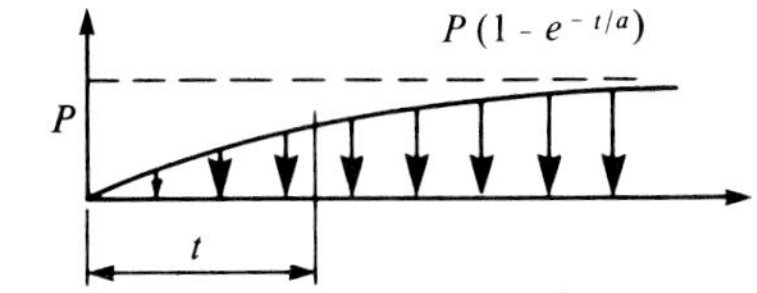	$G = -\dfrac{P}{m}\left[T_3 - \dfrac{a}{\Gamma}(T_2 - aT_1 + ae^{-t/a})\right]$ $\dot{G} = -\dfrac{P}{m}\left[T_2 - \dfrac{a}{\Gamma}(T_1 + aT_0 - e^{-t/a})\right]$

18.14 KELVIN'S MODEL, GROUND MOTION

(1) Notation (18.10)

Δ_0 = maximum amplitude of ground motion (m)
Ω = angular frequency of ground motion (rad/s)
ω = natural angular frequency of model (rad/s)
C = damping constant (N·s/m) K = spring constant (N/m)

(2) Dimensionless equivalents

$$r_1 = 1 - \left(\frac{\Omega}{\omega}\right)^2 \qquad r_2 = \frac{C}{m\omega} \qquad \beta = \tan^{-1}\frac{r_2}{r_1} \qquad \phi = \tan^{-1} r_2$$

(3) General excitation of the base of the model

$$\Delta_g = \Delta_0 g(t)$$

results in the absolute motion of the mass m described by

$$\Delta = \Delta_0 \int_0^t [Kg(\tau) + C\dot{g}(\tau)]T_2(t - \tau)\, d\tau$$

where Δ_g = time-dependent ground displacement.

(4) Harmonic excitation

$$\Delta_g = \Delta_0 \sin \Omega t$$

results in the absolute motion described by

$$\Delta = \Delta_0 R \sin(\Omega t - \beta + \phi) = \Delta_0 R \sin(\Omega t - \psi)$$

where

$$R = \frac{\sqrt{1 + r_2^2}}{\sqrt{r_1^2 + r_2^2}}$$

is called the *transmissibility,* defined as the ratio of the amplitude of the motion of m to the amplitude of the ground motion and

$$\psi = \tan^{-1}\frac{r_2(r_1 - 1)}{r_2^2 + r_1^2}$$

is the *total phase angle.*

(5) Relative steady-state motion of the mass with respect to the harmonically moving ground is given by

$$\Delta_{\text{relative}} = D_{\max} \sin(\Omega t - \beta)$$

where

$$D_{\max} = \Delta_0 \frac{1 - r_1}{\sqrt{r_1^2 + r_2^2}}$$

is the *relative maximum displacement.*

19

Lumped-Mass Models Several Degrees of Freedom

DYNAMIC STATE

19.01 DEFINITION OF STATE

(1) Number of degrees of freedom in a system capable of vibration is defined by the number of time-dependent coordinates required to define the state of the system with respect to a fixed datum (usually static equilibrium configuration). When the number of such coordinates exceeds one, the system is said to have several degrees of freedom. Although each mechanical system has an infinite number of degrees of freedom, by lumping the distributed mass at selected points, an equivalent-mass system with a limited number of degrees of freedom can be introduced in the analysis.

(2) Lumped-mass model relates all causes and effects to the respective mass centers, called the nodes. In general, a space system of m masses with r constraints has n degrees of freedom determined as

$$n = 6m - r$$

where the constraints are real (given by the definition of the real system) and/or introduced as simplifications.

(3) Analysis of these systems introduced in this chapter is based on the assumptions stated in (17.01), expressed in terms of symbols and signs defined in Chap. 17 and redefined where they appear first in this chapter.

19.02 EQUIVALENT-MASS SYSTEM

(1) Distributed-mass equivalents shown below illustrate the concept of lumping of distributed mass at selected points (Ref. 6, p. 45).

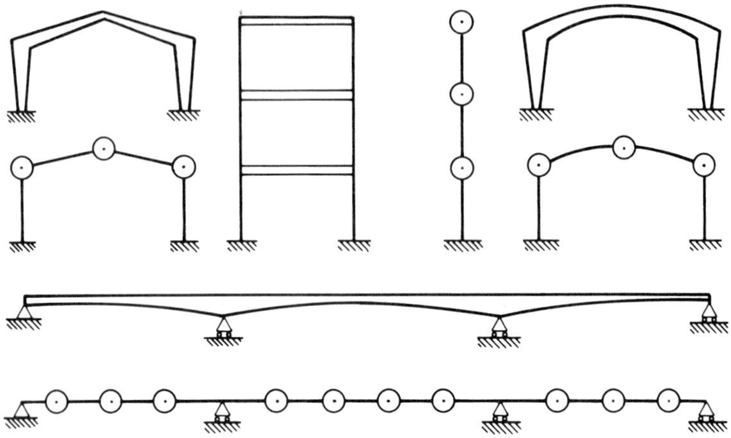

(2) Mechanical coupling of the lumped masses is symbolized by springs and dampers. For complex mechanical systems, the centroidal axes of elastic bars connecting the lumped masses have definite deformation characteristics, so that their static stiffness matrix can be introduced as the equivalent spring matrix. Similar matrices (if experimental data are available) can be constructed for damping.

(3) Dynamic conditions. As each degree of freedom generates one dynamic condition, expressed by the respective differential equation of motion, n such equations govern the motion of the system. Since they are related to the nodes, they are called *nodal equations.* The integrals of these differential equations are the equations' motion whose constant of integration must be determined from the initial or other known conditions. These equations define the dynamic response of the system as periodic, transient, aperiodic, or resonant and offer the basis for the calculation of stress resultants and reactions.

(4) Consistent-mass matrix defined in (17.04) can be used as a better approximation of the distributed mass of each bar. The analysis based on the consistent-mass approximation yields better results but involves more terms in the numerical process, and it is not considered hereafter.

19.03 FREE VIBRATION WITHOUT DAMPING, BASIC RELATIONS

(1) Nodal equations of motion of a discrete parameter model with n degrees of freedom in a state of free vibration (of small amplitudes) are, in matrix form,

$$\underbrace{\begin{bmatrix} m_{11} & m_{12} & \cdots & m_{1n} \\ m_{21} & m_{22} & \cdots & m_{2n} \\ \cdots & \cdots & \cdots & \cdots \\ m_{n1} & m_{n2} & \cdots & m_{nn} \end{bmatrix}}_{[M]} \underbrace{\begin{bmatrix} \ddot{\Delta} \\ \ddot{\Delta}_2 \\ \cdots \\ \ddot{\Delta}_n \end{bmatrix}}_{\{\ddot{\Delta}\}} + \underbrace{\begin{bmatrix} K_{11} & K_{12} & \cdots & K_{1n} \\ K_{21} & K_{22} & \cdots & K_{2n} \\ \cdots & \cdots & \cdots & \cdots \\ K_{n1} & K_{n2} & \cdots & K_{nn} \end{bmatrix}}_{[K]} \underbrace{\begin{bmatrix} \Delta_1 \\ \Delta_2 \\ \cdots \\ \Delta_n \end{bmatrix}}_{\{\Delta\}} = \underbrace{\begin{bmatrix} 0 \\ 0 \\ \cdots \\ 0 \end{bmatrix}}_{\{0\}}$$

where $[M]$ = lumped-mass matrix (17.04-3), $\{\ddot{\Delta}\}$ = acceleration vector $[n \times 1]$, $[K]$ = point spring-constant matrix (17.05-3) or static stiffness matrix (17.05-4), and $\{\Delta\}$ = displacement vector $[n \times 1]$.

(2) General solution of the matrix equation in (1) is assumed in the form of

$$\underbrace{\begin{bmatrix} \Delta_1 \\ \Delta_2 \\ \cdots \\ \Delta_n \end{bmatrix}}_{\{\Delta\}} = \underbrace{\begin{bmatrix} \cos(\omega t - \alpha) & 0 & \cdots & 0 \\ 0 & \cos(\omega t - \alpha) & \cdots & 0 \\ \cdots & \cdots & \cdots & \cdots \\ 0 & 0 & \cdots & \cos(\omega t - \alpha) \end{bmatrix}}_{\text{Diag}\,[\cos(\omega t - \alpha)]} \underbrace{\begin{bmatrix} X_1 \\ X_2 \\ \cdots \\ X_n \end{bmatrix}}_{\{X\}}$$

where the angular frequency ω and the phase angle α are common to all Δ's. In this assumed solution, α and $X_1, X_2, \ldots, X_n$ are the constants of integration to be determined from the initial conditions of the motion and ω is a *characteristic value (eigenvalue)* of the system.

(3) Frequency determinant. The substitution of (2) into (3) yields

$$\underbrace{\begin{bmatrix} (K_{11} - m_{11}\omega^2) & (K_{12} - m_{12}\omega^2) & \cdots & (K_{1n} - m_{1n}\omega^2) \\ (K_{21} - m_{21}\omega^2) & (K_{22} - m_{22}\omega^2) & \cdots & (K_{2n} - m_{2n}\omega^2) \\ \cdots & \cdots & \cdots & \cdots \\ (K_{n1} - m_{n1}\omega^2) & (K_{n2} - m_{n2}\omega^2) & \cdots & (K_{nn} - m_{nn}\omega^2) \end{bmatrix}}_{[K] - [M]\omega^2} \underbrace{\begin{bmatrix} X_1 \\ X_2 \\ \cdots \\ X_n \end{bmatrix}}_{\{X\}} \cos(\omega t - \alpha) = \underbrace{\begin{bmatrix} 0 \\ 0 \\ \cdots \\ 0 \end{bmatrix}}_{\{0\}}$$

The consistency of this matrix equation for $\{X\} \cos(\omega t - \alpha) \neq 0$ requires that

$$\underbrace{|[K] - [M]\omega^2|}_{\text{Det}(\omega^2)} = 0$$

The expansion of this determinant yields a polynomial of nth degree in ω^2 whose roots are the natural angular frequencies $\omega_1, \omega_2, \ldots, \omega_n$ (eigenvalues of the system).

19.04 FREE VIBRATION WITHOUT DAMPING, SOLUTION

(1) Parameters of motion. The resultant motion, in general, is not a harmonic motion but a sum of n harmonic motions each governed by the respective natural frequency ω. These natural frequencies, thought to be arranged in order of ascending magnitudes

$$\omega_1, \omega_2, \ldots, \omega_n \neq 0$$

are called, respectively, the first, second, . . . , nth frequency. They characterize a definite system. If one of these values is zero, the system is said to be semidefinite, which implies that it can move as a rigid body. The shape of the polygons formed by straight lines connecting the extreme positions of the masses in one-component motion (amplitudes) is called the natural (characteristic) mode. Since each mode is governed by one particular frequency, an n-degree-freedom system vibrates in n natural modes. When the reference amplitude of a given natural mode is unity, this mode is called a *normal mode* and its shape is called the normal shape (Ref. 6, pp. 45–47).

(2) Equations of motion in terms of computed $\omega_1, \omega_2, \ldots, \omega_n$ in (3) and assumed Δ's in (2) in (19.03) are

$$\underbrace{\begin{bmatrix} \Delta_1 \\ \Delta_2 \\ \cdots \\ \Delta_n \end{bmatrix}}_{\{\Delta\}} = \underbrace{\begin{bmatrix} X_{11} \\ X_{21} \\ \cdots \\ X_{n1} \end{bmatrix}}_{\{X_1\}} \cos(\omega_1 t - \alpha_1) + \underbrace{\begin{bmatrix} X_{12} \\ X_{22} \\ \cdots \\ X_{n2} \end{bmatrix}}_{\{X_2\}} \cos(\omega_2 t - \alpha_2) + \cdots + \underbrace{\begin{bmatrix} X_{1n} \\ X_{1n} \\ \cdots \\ X_{nn} \end{bmatrix}}_{\{X_n\}} \cos(\omega_n t - \alpha_n)$$

where $\{X_1\}, \{X_2\}, \ldots, \{X_n\}$ are the eigenvectors (natural modes, modal columns) whose absolute values cannot be determined without knowledge of the initial conditions.

(3) Properties of natural modes. The modal columns (eigenvectors, natural modes) in (2) reveal special properties which influence considerably the matrix formulation of the analysis.

(*a*) *Orthogonality conditions.* If $\omega_i^2 \neq \omega_j^2$ and $[M]$, $[K]$ are symmetrical matrices in (19.03), then the eigenvectors $\{X_j\}$ and $\{X_k\}$ corresponding to ω_j and ω_k, respectively, are orthogonal with respect to $[M]$ and $[K]$.

$$\{X_j\}^T[M]\{X_k\} = 0 \qquad \{X_j\}^T[K]\{X_k\} = 0$$

where $\{X_j\}^T$ = transpose of $\{X_j\}$

(*b*) *Normalization conditions.* If $\omega_j^2 = \omega_k^2$, then $\{X_j\} = \{X_k\}$ and

$$\{X_j\}^T[M]\{X_j\} = L_j^2 \qquad \{X_j\}^T[K]\{X_j\} = L_j^2\omega_j^2$$
$$\{h_j\}^T[M]\{h_j\} = 1 \qquad \{h_j\}^T[K]\{h_j\} = \omega_j^2$$

where

$$\{h_j\} = \frac{1}{L_j}\{X_j\}$$

is the *normalized eigenvector* corresponding to ω_j and $\{h_j\}^T$ is its transpose.

19.05 FREE VIBRATION WITHOUT DAMPING, NORMAL COORDINATES

(1) Equations of motion given by (19.03) are frequently called the coupled equations, since their solution given by (19.04) is a superposition of n motions. Although this type of solution causes no special difficulties in the analysis of free vibration, it becomes quite laborious if forced vibration is involved. In such cases the concept of uncoupling, i.e., the transformation of (19.03) into a set of independent (uncoupled) equations, is useful. Each of these new equations defines the motion of one mass in terms of one m, ω, and K. The displacement vectors of this new system (principal system, normal system), called the *normal (principal) coordinates,* are

$$\underbrace{\begin{bmatrix} \Delta_1 \\ \Delta_2 \\ \cdots \\ \Delta_n \end{bmatrix}}_{\{\Delta\}} = \underbrace{\begin{bmatrix} h_{11} & h_{12} & \cdots & h_{1n} \\ h_{21} & h_{22} & \cdots & h_{2n} \\ \cdots & \cdots & \cdots & \cdots \\ h_{n1} & h_{n2} & \cdots & h_{nn} \end{bmatrix}}_{[h]} \underbrace{\begin{bmatrix} \eta_1 \\ \eta_2 \\ \cdots \\ \eta_n \end{bmatrix}}_{\{\eta\}}$$

and the nodal equation matrix in (19.03) becomes $\quad [M][h]\{\ddot{\eta}\} + [K][h]\{\eta\} = \{0\}$

where $\{\Delta\}$ = original coordinate (displacement) vector introduced at the onset of analysis and $\{\eta\}$ = normal coordinate (displacement) vector introduced above. The premultiplication of this matrix equation by $[h]^T$ gives

$$[h]^T[\mu][h]\{\ddot{\eta}\} + [h]^T[k][h][\eta] = 0$$

which reduces to n independent nodal equations (Ref. 6, pp. 49–50)

$$\underbrace{\begin{bmatrix} \ddot{\eta}_1 \\ \ddot{\eta}_2 \\ \cdots \\ \ddot{\eta}_n \end{bmatrix}}_{\{\ddot{\eta}\}} + \underbrace{\begin{bmatrix} \omega_1^2 & & \\ & \omega_2^2 & \\ \cdots & \cdots & \cdots \\ & & \omega_n^2 \end{bmatrix}}_{\text{Diag}[\omega^2]} \underbrace{\begin{bmatrix} \eta_1 \\ \eta_2 \\ \cdots \\ \eta_n \end{bmatrix}}_{\{\eta\}} = \underbrace{\begin{bmatrix} 0 \\ 0 \\ \cdots \\ 0 \end{bmatrix}}_{\{0\}}$$

(2) Solution of these independent equations is

$$\underbrace{\begin{bmatrix} \eta_1 \\ \eta_2 \\ \cdots \\ \eta_n \end{bmatrix}}_{\{\eta\}} = \underbrace{\begin{bmatrix} \cos\omega_1 t & & \\ & \cos\omega_2 t & \\ \cdots & \cdots & \cdots \\ & & \cos\omega_n t \end{bmatrix}}_{\text{Diag}\,[\cos\omega t]} \underbrace{\begin{bmatrix} \eta_{0,1} \\ \eta_{0,2} \\ \cdots \\ \eta_{0,n} \end{bmatrix}}_{\{\eta_0\}} + \underbrace{\begin{bmatrix} \sin\omega_1 t & & \\ & \sin\omega_2 t & \\ \cdots & \cdots & \cdots \\ & & \sin\omega_n t \end{bmatrix}}_{\text{Diag}\,[\sin\omega t]} \underbrace{\begin{bmatrix} \dot{\eta}_{0,1}/\omega_1 \\ \dot{\eta}_{0,2}/\omega_2 \\ \cdots \\ \dot{\eta}_{0,n}/\omega_n \end{bmatrix}}_{\{\dot{\eta}_0/\omega\}}$$

where $\eta_{0,1}, \ldots, \eta_{0,n}$ and $\dot{\eta}_{0,1}, \ldots, \dot{\eta}_{0,n}$ are the initial conditions in normal coordinates.

(3) Initial conditions. Since the initial conditions are given in the natural coordinates, the solution in (2) must be expressed as

$$\{\Delta\} = [h]\ \text{Diag}\,[\cos\omega t][h]^{-1}\{\Delta_0\} + [h]\ \text{Diag}\,[\sin\omega t][h]^{-1}\{\dot{\Delta}_0/\omega\}$$

19.06 FORCED VIBRATIONS WITHOUT DAMPING

(1) Nodal equations in matrix form are

$$[M]\{\ddot{\Delta}\} + [K]\{\Delta\} = \{L(t)\}$$

where $[M]$, $[K]$ are the same as in (19.03) and $\{L(t)\}$ is a column matrix of load functions $[n \times 1]$ (Ref. 6, pp. 50–51).

(2) Normalized nodal equations in matrix form, obtained by the process described in (19.05), are

$$\underbrace{\begin{bmatrix} \ddot{\eta}_1 \\ \ddot{\eta}_2 \\ \cdots \\ \ddot{\eta}_n \end{bmatrix}}_{\{\ddot{\eta}\}} + \underbrace{\begin{bmatrix} \omega_1^2 & & \\ & \omega_2^2 & \\ \cdots & \cdots & \cdots \\ & & \omega_n^2 \end{bmatrix}}_{\mathrm{Diag}[\omega^2]} \underbrace{\begin{bmatrix} \eta_1 \\ \eta_2 \\ \cdots \\ \eta_n \end{bmatrix}}_{\{\eta\}} = \underbrace{\begin{bmatrix} h_{11} & h_{21} & \cdots & h_{n1} \\ h_{12} & h_{22} & \cdots & h_{n2} \\ \cdots & \cdots & \cdots & \cdots \\ h_{1n} & h_{2n} & \cdots & h_{nn} \end{bmatrix}}_{[h]^T} \underbrace{\begin{bmatrix} L_1(t) \\ L_2(t) \\ \cdots \\ L_n(t) \end{bmatrix}}_{\{L(t)\}}$$

(3) Solution of these independent equations is

$$\underbrace{\begin{bmatrix} \eta_1 \\ \eta_2 \\ \cdots \\ \eta_n \end{bmatrix}}_{\{\eta\}} = \underbrace{\begin{bmatrix} \cos\omega_1 t & & \\ & \cos\omega_2 t & \\ \cdots & \cdots & \cdots \\ & & \cos\omega_n t \end{bmatrix}}_{\mathrm{Diag}\,[\cos\omega t]} \underbrace{\begin{bmatrix} \eta_{0,1} \\ \eta_{0,2} \\ \cdots \\ \eta_{0,n} \end{bmatrix}}_{\{\eta_0\}} + \underbrace{\begin{bmatrix} \sin\omega_1 t & & \\ & \sin\omega_2 t & \\ \cdots & \cdots & \cdots \\ & & \sin\omega_n t \end{bmatrix}}_{\mathrm{Diag}\,[\sin\omega t]} \underbrace{\begin{bmatrix} \dot{\eta}_{0,1}/\omega_1 \\ \dot{\eta}_{0,2}/\omega_2 \\ \cdots \\ \dot{\eta}_{0,n}/\omega_n \end{bmatrix}}_{\{\dot{\eta}_0/\omega\}} + \underbrace{\begin{bmatrix} G_1(t) \\ G_2(t) \\ \cdots \\ G_n(t) \end{bmatrix}}_{\{G(t)\}}$$

where

$$\underbrace{\begin{bmatrix} G_1(t) \\ G_2(t) \\ \cdots \\ G_n(t) \end{bmatrix}}_{\{G(t)\}} = \int_0^t \underbrace{\begin{bmatrix} \sin(\omega_1 t - \tau) & & \\ & \sin(\omega_2 t - \tau) & \\ \cdots & \cdots & \cdots \\ & & \sin(\omega_n t - \tau) \end{bmatrix}}_{\mathrm{Diag}\,[\sin(\omega t - \tau)]} \underbrace{\begin{bmatrix} h_{11} & h_{21} & \cdots & h_{n1} \\ h_{12} & h_{22} & \cdots & h_{n2} \\ \cdots & \cdots & \cdots & \cdots \\ h_{1n} & h_{2n} & \cdots & h_{nn} \end{bmatrix}}_{[h]^T} \underbrace{\begin{bmatrix} L_1(\tau)/\omega_1 \\ L_2(\tau)/\omega_2 \\ \cdots \\ L_n(\tau)/\omega_n \end{bmatrix}}_{\{L(t)/\omega\}} d\tau$$

is the *response-function vector* (column matrix of convolution integrals).

(4) Initial conditions are included in (3) by the same process as in (19.05-3) so that

$$\{\Delta\} = [h]\mathrm{Diag}\,[\cos\omega t][h]^{-1}\{\Delta_0\} + [h]\mathrm{Diag}\,[\sin\omega t][h]^{-1}\{\dot{\Delta}_0/\omega\} + [h]\{G(t)\}$$

where $\{\Delta_0\}$ and $\{\dot{\Delta}_0\}$ are the initial displacement and velocity vectors, respectively.

20

Straight Bar of Order One Constant Section

DYNAMIC STATE

20.01 DEFINITION OF STATE

(1) System considered is a finite straight bar of constant and symmetrical section. The centroidal axis x of the bar and the vertical axis y define the plane of symmetry of the section. The bar and its loads are in a state of dynamic equilibrium and a state of time-dependent elastic deformation. The bar is a part of a planar system acted on by causes in the system plane (system of order one).

(2) Notation (A.1, A.2, A.4)

s = span (m)	A_x = area of normal section (m^2)
a, b = segments (m)	I_z = moment of inertia of A_x about z (m^4)
x, x' = coordinates (m)	E = modulus of elasticity (Pa)
t = time (s)	ω = natural angular frequency (rad/s)
μ = mass per unit length (kg/m)	Ω = forcing angular frequency (rad/s)
U = normal force (N)	u = linear displacement along x (m)
V = shearing force (N)	v = linear displacement along y (m)
Z = flexural moment (N·m)	θ = flexural slope about z (rad)

$\alpha = \sqrt{\dfrac{\mu\omega^2}{EA_x}}$ = shape parameter (m^{-1})	$\beta = \sqrt[4]{\dfrac{\mu\omega^2}{EI_z}}$ = shape parameter (m^{-1})

(3) Assumptions of analysis are stated in (17.01). In this chapter, the effects of axial and flexural deformations are included, but the effects of shearing deformations and of axial force on the magnitude of the bending moment are neglected.

20.02 TRANSPORT MATRIX EQUATIONS

(1) Equivalents and signs

$$\bar{u} = uEA_x\alpha = \text{scaled } u \text{ (N)} \qquad \bar{v} = vEI_z\beta^3 = \text{scaled } v \text{ (N)}$$

$$\bar{Z} = Z\beta = \text{scaled } Z \text{ (N)} \qquad \bar{\theta} = \theta EI_z\beta^2 = \text{scaled } \theta \text{ (N)}$$

All forces, moments, and displacements are in the principal axes of the normal section, and their positive directions are shown in (2, 3) below (Ref. 6, pp. 117–123).

(2) Free-body diagram

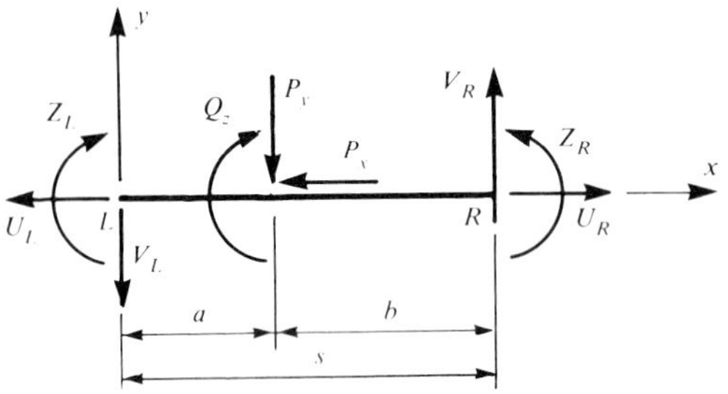

(3) Elastic curve

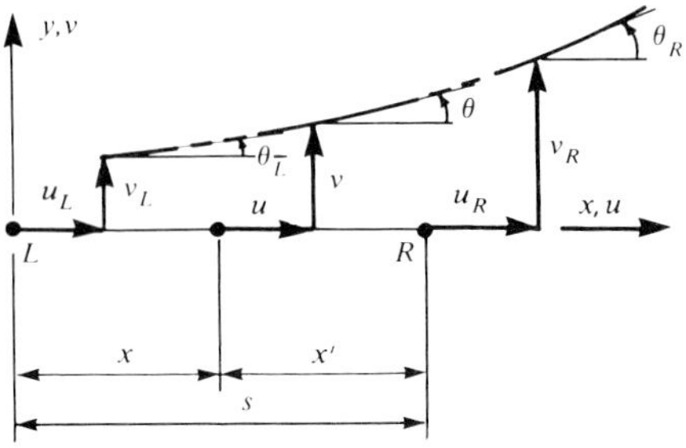

(4) Transport matrix equations

$$\underbrace{\begin{bmatrix} 1 \\ \hline U_L \\ V_L \\ Z_L \\ \hline \bar{u}_L \\ \bar{v}_L \\ \bar{\theta}_L \end{bmatrix}}_{H_L} = \underbrace{\left[\begin{array}{c|ccc|ccc} 1 & 0 & 0 & 0 & 0 & 0 & 0 \\ \hline F_U & T_{1n} & 0 & 0 & T_{2n} & 0 & 0 \\ F_V & 0 & T_{1z} & T_{4z} & 0 & T_{2z} & -T_{3z} \\ F_Z & 0 & T_{2z} & T_{1z} & 0 & T_{3z} & -T_{4z} \\ \hline F_u & -T_{2n} & 0 & 0 & T_{1n} & 0 & 0 \\ F_v & 0 & T_{4z} & T_{3z} & 0 & T_{1z} & -T_{2z} \\ F_\theta & 0 & -T_{3z} & -T_{2z} & 0 & -T_{4z} & T_{1z} \end{array}\right]}_{T_{LR}} \underbrace{\begin{bmatrix} 1 \\ \hline U_R \\ V_R \\ Z_R \\ \hline \bar{u}_R \\ \bar{v}_R \\ \bar{\theta}_R \end{bmatrix}}_{H_R}$$

$$\underbrace{\begin{bmatrix} 1 \\ \hline U_R \\ V_R \\ Z_R \\ \hline \bar{u}_R \\ \bar{v}_R \\ \bar{\theta}_R \end{bmatrix}}_{H_R} = \underbrace{\left[\begin{array}{c|ccc|ccc} 1 & 0 & 0 & 0 & 0 & 0 & 0 \\ \hline G_U & T_{1n} & 0 & 0 & -T_{2n} & 0 & 0 \\ G_V & 0 & T_{1z} & -T_{4z} & 0 & -T_{2z} & -T_{3z} \\ G_Z & 0 & -T_{2z} & T_{1z} & 0 & T_{3z} & T_{4z} \\ \hline G_u & T_{2n} & 0 & 0 & T_{1n} & 0 & 0 \\ G_v & 0 & -T_{4z} & T_{3z} & 0 & T_{1z} & T_{2z} \\ G_\theta & 0 & -T_{3z} & T_{2z} & 0 & T_{4z} & T_{1z} \end{array}\right]}_{T_{RL}} \underbrace{\begin{bmatrix} 1 \\ \hline U_L \\ V_L \\ Z_L \\ \hline \bar{u}_L \\ \bar{v}_L \\ \bar{\theta}_L \end{bmatrix}}_{H_L}$$

where H_L, H_R are the *state vectors* of the respective ends and T_{LR}, T_{RL} are the *transport matrices*. The elements of the transport matrices are the *transport coefficients* T_{jn}, T_{jz} defined in (20.03) and the *load functions* F, G displayed for particular cases in (20.06–20.08).

20.03 TRANSPORT RELATIONS

(1) Transport functions

	$\alpha = \sqrt{\dfrac{\mu\omega^2}{EA_x}} \qquad \beta = \sqrt[4]{\dfrac{\mu\omega^2}{EI_z}}$
$T_{1n}(x) = \cos \alpha x$	$T_{1z}(x) = \frac{1}{2}(\cosh \beta x + \cos \beta x)$
$T_{2n}(x) = \sin \alpha x$	$T_{2z}(x) = \frac{1}{2}(\sinh \beta x + \sin \beta x)$
$T_{3n}(x) = 1 - \cos \alpha x$	$T_{3z}(x) = \frac{1}{2}(\cosh \beta x - \cos \beta x)$
$T_{4n}(x) = \alpha x - \sin \alpha x$	$T_{4z}(x) = \frac{1}{2}(\sinh \beta x - \sin \beta x)$
$T_{5n}(x) = \dfrac{(\alpha x)^2}{2!} - T_{3n}(x)$	$T_{5z}(x) = T_{1z}(x) - 1$
$T_{6n}(x) = \dfrac{(\alpha x)^3}{3!} - T_{4n}(x)$	$T_{6z}(x) = T_{2z}(x) - \beta x$

The graphs and table of these functions are shown in (20.04–20.05).

(2) Transport coefficients T_j, A_j, B_j, C_j are the values of $T_j(x)$ for a particular argument x.

$$\left.\begin{array}{lll} \text{For } x = s, & T_{jn}(s) = T_{jn}, & T_{jz}(s) = T_{jz} \\ \text{For } x = a, & T_{jn}(a) = A_{jn}, & T_{jz}(a) = A_{jz} \\ \text{For } x = b, & T_{jn}(b) = B_{jn}, & T_{jz}(b) = R_{jz} \\ \text{and } T_{jn} - A_{jn} = C_{jn}, & & T_{jz} - A_{jz} = C_{jz} \end{array}\right\} \quad j = 1, 2, \ldots, 6$$

(3) Transport matrices of this bar have analogical properties to those of (4.02) and can be used in similar ways.

(4) Differential relations derived for a free bar of order one in (4.03) are valid here without modification.

(5) Effect of elastic foundation introduced in (8.01) can be included in the transport functions $T_{jn}(x)$ and $T_{jz}(x)$ by taking

$$\alpha = \sqrt{\frac{\mu\omega^2 - k_u}{EA_x}} > 0 \qquad \beta = \sqrt[4]{\frac{\mu\omega^2 - k_v}{EI_z}} > 0$$

where k_u, k_v = foundation module (N/m^2) (Ref. 22).

(6) Angular frequencies $\omega_1, \omega_2, \ldots$ of this motion are the eigenvalues of the transcendental-frequency determinant equation constructed from the transport matrix equation by selecting the part of the matrix corresponding to the existing eigenvector (20.13).

(7) Load functions, F, G tabulated in (20.06–20.08) are to be used only in steady-state cases in which

$$L(x, t) = L(x) \sin \Omega t \qquad \text{or} \qquad L(x, t) = L(x) \cos \Omega t$$

where Ω = forcing angular frequency (rad/s).

(8) General transport matrix equations for bars of order three are obtained by matrix composition of (20.02) and (21.02).

20.04 TABLE OF TRANSPORT COEFFICIENTS†

βx	$T_{1z}(x)$	$T_{2z}(x)$	$T_{3z}(x)$	$T_{4z}(x)$	$T_{5z}(x)$	$T_{6z}(x)$
0.0	1.00000	0.00000	0.00000	0.00000	0.00000	0.00000
0.5	1.00260	0.50026	0.12502	0.02083	0.00260	0.00026
0.6	1.00540	0.60065	0.18006	0.03601	0.00540	0.00065
0.7	1.01000	0.70140	0.24516	0.05718	0.01000	0.00140
0.8	1.01707	0.80273	0.32036	0.08537	0.01707	0.00273
0.9	1.02735	0.90492	0.40574	0.12159	0.02735	0.00492
1.0	1.04169	1.00834	0.50139	0.16687	0.04169	0.00834
1.1	1.06106	1.11343	0.60746	0.22222	0.06106	0.01343
1.2	1.08651	1.22075	0.72415	0.28871	0.08651	0.02075
1.3	1.11921	1.33097	0.85171	0.36741	0.11921	0.03097
1.4	1.16043	1.44487	0.99046	0.45943	0.16043	0.04488
1.5	1.21157	1.56339	1.14084	0.56589	0.21157	0.06339
1.6	1.27413	1.68757	1.30333	0.68800	0.27413	0.08757
1.7	1.34974	1.81865	1.47858	0.82698	0.34974	0.11865
1.8	1.44013	1.95801	1.66734	0.98416	0.44013	0.15801
1.9	1.54722	2.10723	1.87051	1.16093	0.54722	0.20723
2.0	1.67302	2.26808	2.08917	1.35878	0.67302	0.26808
2.1	1.81973	2.44253	2.32458	1.57932	0.81973	0.34253
2.2	1.98970	2.63280	2.57820	1.82430	0.98970	0.43280
2.3	2.18547	2.84133	2.85175	2.09563	1.18547	0.54133
2.4	2.40977	3.07084	3.14717	2.39538	1.40977	0.67085
2.5	2.66557	3.32434	3.46672	2.72587	1.66557	0.82434
2.6	2.95606	3.60511	3.81294	3.08961	1.95606	1.00511
2.7	3.28470	3.91682	4.18877	3.48944	2.28470	1.21682
2.8	3.65525	4.26345	4.59747	3.92846	2.65525	1.46345
2.9	4.07181	4.64940	5.04277	4.41015	3.07181	1.74940
3.0	4.53883	5.07950	5.52883	4.93838	3.53883	2.07950
3.1	5.06118	5.55901	6.06031	5.51743	4.06118	2.45901
3.2	5.64417	6.09375	6.64247	6.15213	4.64417	2.89375
3.3	6.29364	6.69006	7.28111	6.84781	5.29364	3.39006
3.4	7.01597	7.35491	7.98276	7.61045	6.01597	3.95491
3.5	7.81818	8.09592	8.75463	8.44669	6.81818	4.59592
3.6	8.70800	8.92146	9.60475	9.36398	7.70800	5.32146
3.7	9.69395	9.84071	10.54205	10.37055	8.69395	6.14071
3.8	10.78539	10.86376	11.57635	11.47562	9.78539	7.06376
3.9	11.99269	12.00166	12.71862	12.68942	10.99269	8.10166
4.0	13.32729	13.26656	13.98093	14.02335	12.32729	9.26656
4.1	14.80179	14.67178	15.37661	15.49005	13.80179	10.57178
4.2	16.43019	16.23204	16.92044	17.10361	15.43019	12.03204
4.3	18.22792	17.96344	18.62871	18.87961	17.22792	13.66344
4.4	20.21210	19.88383	20.51944	20.83543	19.21210	15.48384
4.5	22.40166	22.01273	22.61244	22.99026	21.40166	17.51273
4.6	24.81749	24.37170	24.92966	25.36539	23.81749	19.77170
4.7	27.48286	26.98454	27.49524	27.98447	26.48286	22.28453
4.8	30.42339	29.87743	30.33588	30.87360	29.42339	25.07742
4.9	33.66754	33.07935	33.48102	34.06180	32.66754	28.17934
5.0	37.24680	36.62213	36.96313	37.58105	36.24680	31.62213
6.0	101.33789	100.71686	100.37773	100.99628	100.33789	94.71686

†Values of $T_{jn}(x)$ are available in standard mathematical tables.

20.05 GRAPHS AND SERIES OF TRANSPORT COEFFICIENTS

(1) Graphs of $T_{jz}(x)$

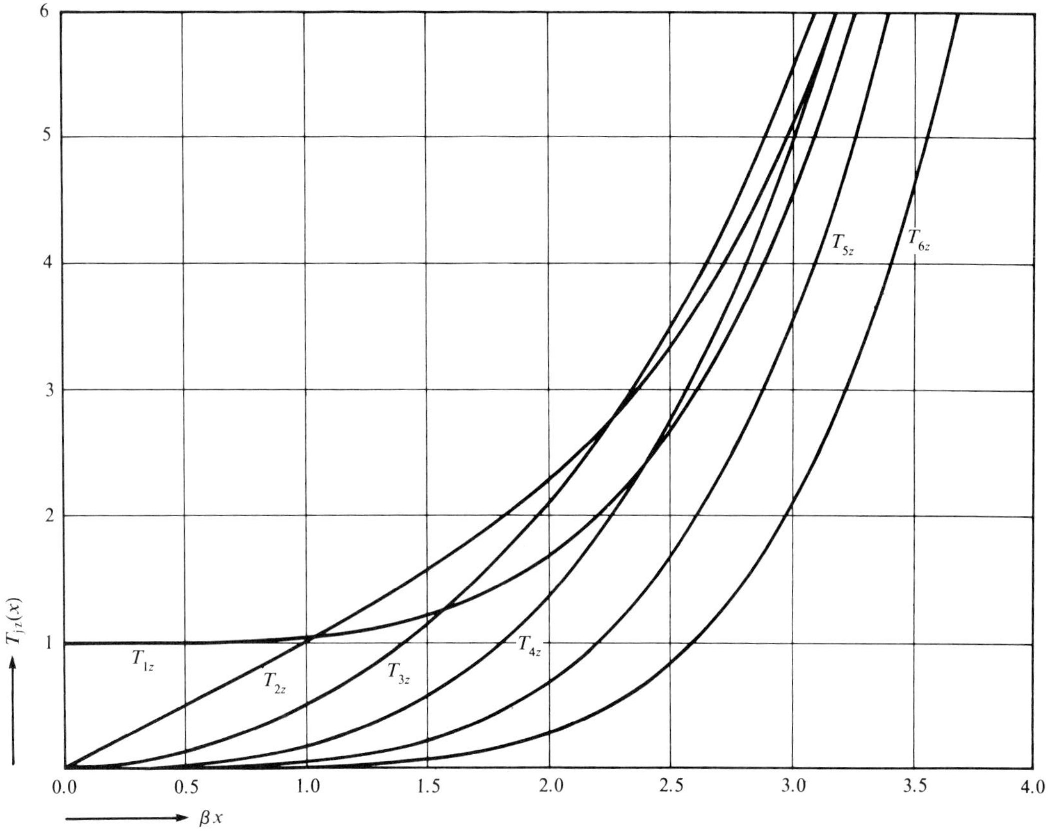

(2) Series expansion of $T_{jz}(x)$

$$T_{1z}(x) = 1 + \frac{(\beta x)^4}{4!} + \frac{(\beta x)^8}{8!} + \frac{(\beta x)^{12}}{12!} + \cdots = \sum_{k=0}^{\infty} \frac{(\beta x)^{4k}}{(4k)!}$$

$$T_{2z}(x) = \frac{\beta x}{1!} + \frac{(\beta x)^5}{5!} + \frac{(\beta x)^9}{9!} + \frac{(\beta x)^{13}}{13!} + \cdots = \sum_{k=0}^{\infty} \frac{(\beta x)^{4k+1}}{(4k+1)!}$$

$$T_{3z}(x) = \frac{(\beta x)^2}{2!} + \frac{(\beta x)^6}{6!} + \frac{(\beta x)^{10}}{10!} + \frac{(\beta x)^{14}}{14!} + \cdots = \sum_{k=0}^{\infty} \frac{(\beta x)^{4k+2}}{(4k+2)!}$$

$$T_{4z}(x) = \frac{(\beta x)^3}{3!} + \frac{(\beta x)^7}{7!} + \frac{(\beta x)^{11}}{11!} + \frac{(\beta x)^{15}}{15!} + \cdots = \sum_{k=0}^{\infty} \frac{(\beta x)^{4k+3}}{(4k+3)!}$$

$$T_{jz}(x) = \frac{(\beta x)^{j-1}}{(j-1)!} + \frac{(\beta x)^{j+3}}{(j+3)!} + \frac{(\beta x)^{j+7}}{(j+7)!} + \frac{(\beta x)^{j+11}}{(j+11)!} + \cdots = \sum_{k=0}^{\infty} \frac{(\beta x)^{j+4k-1}}{(j+4k-1)!}$$

20.06 TRANSPORT LOAD FUNCTIONS

Notation (20.01–20.03) Signs (20.02) Matrix (20.02) P_x, P_y = concentrated loads (N) Q_z = applied couple (N·m) q_z = intensity of distributed couple (N·m/m) A_j, B_j, C_j, T_j = transport coefficients (20.03)		$\alpha = \sqrt{\dfrac{\mu\Omega^2}{EA_x}}$ $\beta = \sqrt[4]{\dfrac{\mu\Omega^2}{EI_z}}$
$F_U = -A_{1n}P_x$ $F_V = -A_{1z}P_y$ $F_Z = -A_{2z}P_y$ $F_u = A_{2n}P_x$ $F_v = -A_{4z}P_y$ $F_\theta = A_{3z}P_y$	**1.** 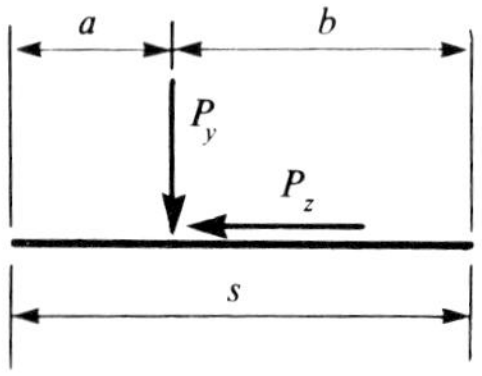 	$G_U = B_{1n}P_x$ $G_V = B_{1z}P_y$ $G_Z = -B_{2z}P_y$ $G_u = B_{2n}P_x$ $G_v = -B_{4z}P_y$ $G_\theta = -B_{3z}P_y$
$F_U = 0$ $F_V = -A_{4z}\beta Q_z$ $F_Z = -A_{1z}\beta Q_z$ $F_u = 0$ $F_v = -A_{3z}\beta Q_z$ $F_\theta = A_{2z}\beta Q_z$	**2.** 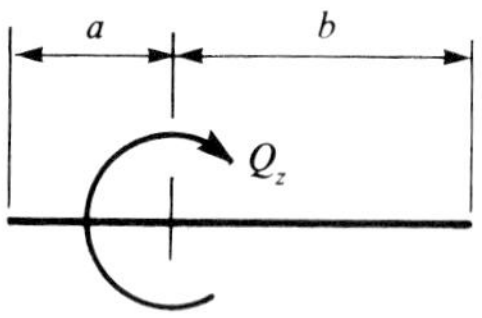 	$G_U = 0$ $G_V = -B_{4z}\beta Q_z$ $G_Z = B_{1z}\beta Q_z$ $G_u = 0$ $G_v = B_{3z}\beta Q_z$ $G_\theta = B_{2z}\beta Q_z$
$F_U = 0$ $F_V = -T_{5z}q_z$ $F_Z = -T_{2z}q_z$ $F_u = 0$ $F_v = -T_{4z}q_z$ $F_\theta = T_{3z}q_z$	**3.** 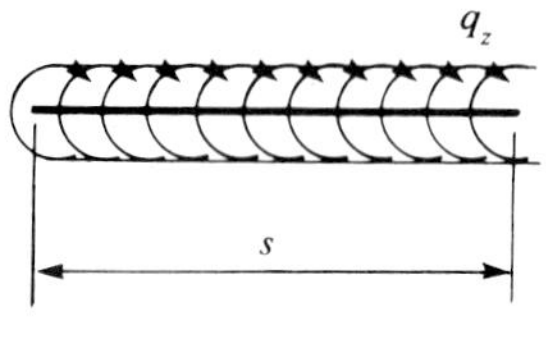 	$G_U = 0$ $G_V = -T_{5z}q_z$ $G_Z = T_{2z}q_z$ $G_u = 0$ $G_v = T_{4z}q_z$ $G_\theta = T_{3z}q_z$
$F_U = 0$ $F_V = -C_{5z}q_z$ $F_Z = -C_{2z}q_z$ $F_u = 0$ $F_v = -C_{4z}q_z$ $F_\theta = C_{3z}q_z$	**4.** 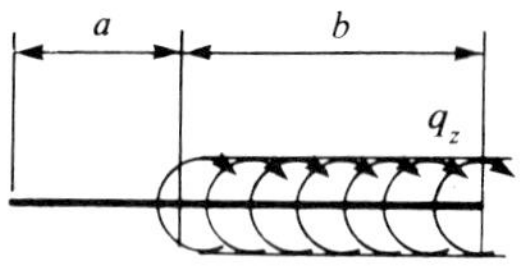 	$G_U = 0$ $G_V = -B_{5z}q_z$ $G_Z = B_{2z}q_z$ $G_u = 0$ $G_v = B_{4z}q_z$ $G_\theta = B_{3z}q_z$

20.07 TRANSPORT LOAD FUNCTIONS

Notation (20.01–20.03) Signs (20.02) Matrix (20.02) p_x, p_y = intensities of distributed load (N/m) A_j, B_j, C_j, T_j = transport coefficients (20.03)		$\alpha = \sqrt{\dfrac{\mu\Omega^2}{EA_X}}$ $\beta = \sqrt[4]{\dfrac{\mu\Omega^2}{EI_z}}$
$F_U = -T_{2n}\alpha^{-1}p_x$ $F_V = -T_{2z}\beta^{-1}p_y$ $F_Z = -T_{3z}\beta^{-1}p_y$	**1.** 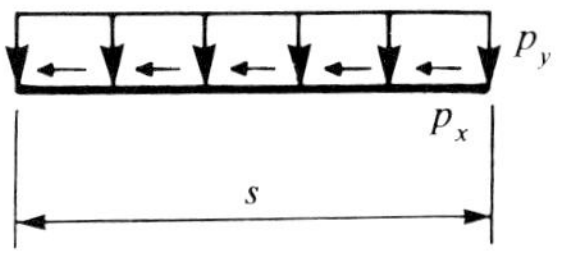 	$G_U = T_{2n}\alpha^{-1}p_x$ $G_V = T_{2z}\beta^{-1}p_y$ $G_Z = -T_{3z}\beta^{-1}p_y$
$F_u = T_{3n}\alpha^{-1}p_x$ $F_v = -T_{5z}\beta^{-1}p_y$ $F_\theta = T_{4z}\beta^{-1}p_y$		$G_u = T_{3n}\alpha^{-1}p_x$ $G_v = -T_{5z}\beta^{-1}p_y$ $G_\theta = -T_{4z}\beta^{-1}p_y$
$F_U = -C_{2n}\alpha^{-1}p_x$ $F_V = -C_{2z}\beta^{-1}p_y$ $F_Z = -C_{3z}\beta^{-1}p_y$	**2.** 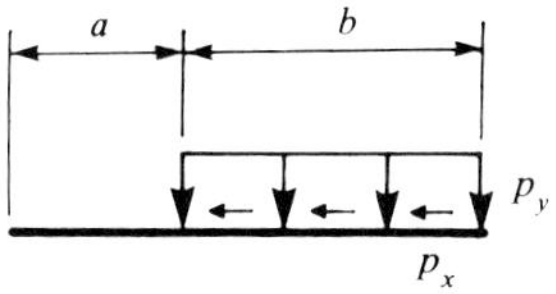 	$G_U = B_{2n}\alpha^{-1}p_x$ $G_V = B_{2z}\beta^{-1}p_y$ $G_Z = -B_{3z}\beta^{-1}p_y$
$F_u = C_{3n}\alpha^{-1}p_x$ $F_v = -C_{5z}\beta^{-1}p_y$ $F_\theta = C_{4z}\beta^{-1}p_y$		$G_u = B_{3n}\alpha^{-1}p_x$ $G_v = -B_{5z}\beta^{-1}p_y$ $G_\theta = -B_{4z}\beta^{-1}p_y$
$F_U = -(\alpha s T_{2n} - T_{3n})\alpha^{-2}s^{-1}p_x$ $F_V = -(\beta s T_{2z} - T_{3z})\beta^{-2}s^{-1}p_y$ $F_Z = -(\beta s T_{3z} - T_{4z})\beta^{-2}s^{-1}p_y$	**3.** 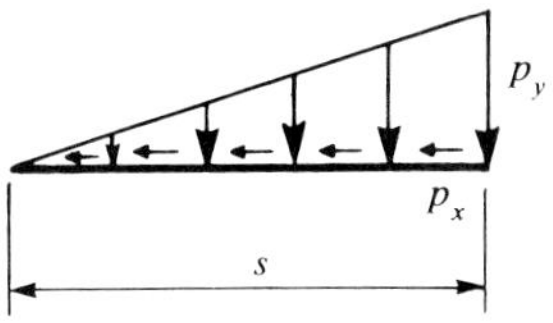 	$G_U = T_{3n}\alpha^{-2}s^{-1}p_x$ $G_V = T_{3z}\beta^{-2}s^{-1}p_y$ $G_Z = -T_{4z}\beta^{-2}s^{-1}p_y$
$F_u = (\alpha s T_{3n} - T_{4n})\alpha^{-2}s^{-1}p_x$ $F_v = -(\beta s T_{5z} - T_{6z})\beta^{2}s^{-1}p_y$ $F_\theta = (\beta s T_{4z} - T_{5z})\beta^{-2}s^{-1}p_y$		$G_u = T_{4n}\alpha^{-2}s^{-1}p_x$ $G_v = -T_{6z}\beta^{-2}s^{-1}p_y$ $G_\theta = -T_{5z}\beta^{-2}s^{-1}p_y$
$F_U = -(\alpha b T_{2n} - C_{3n})\alpha^{-2}b^{-1}p_x$ $F_V = -(\beta b T_{2z} - C_{3z})\beta^{-2}b^{-1}p_y$ $F_Z = -(\beta b T_{3z} - C_{4z})\beta^{-2}b^{-1}p_y$	**4.** 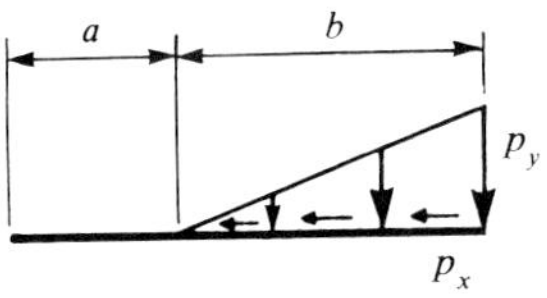 	$G_U = B_{3n}\alpha^{-2}b^{-1}p_x$ $G_V = B_{3z}\beta^{-2}b^{-1}p_y$ $G_Z = -B_{4z}\beta^{-2}b^{-1}p_y$
$F_u = (\alpha b T_{3n} - C_{4n})\alpha^{-2}b^{-1}p_x$ $F_v = -(\beta b T_{5z} - C_{6z})\beta^{-2}b^{-1}p_y$ $F_\theta = (\beta b T_{4z} - C_{5z})\beta^{-2}b^{-1}p_y$		$G_u = B_{4n}\alpha^{-2}b^{-1}p_x$ $G_v = -B_{6z}\beta^{-2}b^{-1}p_y$ $G_\theta = -B_{5z}\beta^{-2}b^{-1}p_y$

20.08 TRANSPORT LOAD FUNCTIONS

Notation (20.01–20.03) Signs (20.02) Matrix (20.02)

u_j, v_j, θ_j = abrupt displacements (m, m, rad)

e_t, f_t, = temperature strains (4.03)

A_j, B_j, C_j, T_j = transport coefficients (20.03)

$$a = \sqrt{\frac{\mu\Omega^2}{EA_x}} \qquad \beta = \sqrt[4]{\frac{\mu\Omega^2}{EI_z}}$$

1. Abrupt displacements

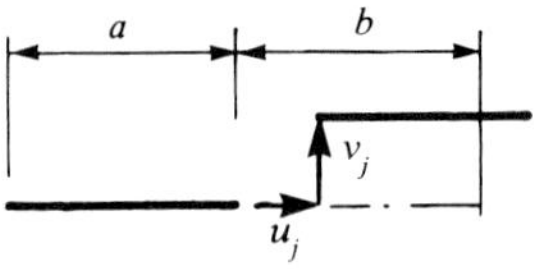

F	G
$F_U = -EA_xA_{2n}\alpha u_j$	$G_U = -EA_xB_{2n}\alpha u_j$
$F_V = -EI_zA_{2z}\beta^3 v_j$	$G_V = -EI_zB_{2z}\beta^3 v_j$
$F_Z = -EI_zA_{3z}\beta^3 v_j$	$G_Z = EI_zB_{3z}\beta^3 v_j$
$F_u = -EA_xA_{1n}\alpha u_j$	$G_u = EA_xB_{1n}\alpha u_j$
$F_v = -EI_zA_{1z}\beta^3 v_j$	$G_v = EI_zB_{1z}\beta^3 v_j$
$F_\theta = EI_zA_{4z}\beta^3 v_j$	$G_\theta = EI_zB_{4z}\beta^3 v_j$

2. Abrupt displacement

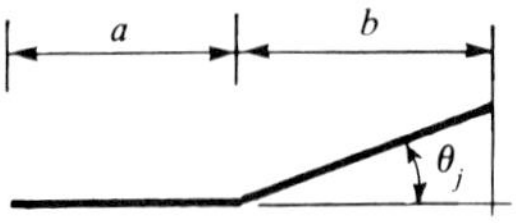

F	G
$F_U = 0$	$G_U = 0$
$F_V = EI_zA_{3z}\beta^2\theta_j$	$G_V = -EI_zB_{3z}\beta^2\theta_j$
$F_Z = EI_zA_{4z}\beta^2\theta_j$	$G_Z = EI_zB_{4z}\beta^2\theta_j$
$F_u = 0$	$G_u = 0$
$F_v = EI_zA_{2z}\beta^2\theta_j$	$G_v = EI_zB_{2z}\beta^2\theta_j$
$F_\theta = -EI_zA_{1z}\beta^2\theta_j$	$G_\theta = EI_zB_{1z}\beta^2\theta_j$

3. Temperature deformation of segment *b*

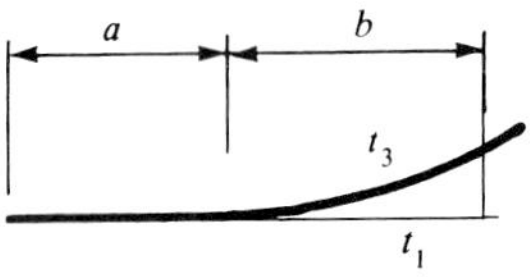

F	G
$F_U = -EA_xC_{3n}e_t$	$G_U = -EA_xB_{3n}e_t$
$F_V = EI_zC_{4z}\beta f_t$	$G_V = -EI_zB_{4z}\beta f_t$
$F_Z = EI_zC_{5z}\beta f_t$	$G_Z = EI_zB_{3z}\beta f_t$
$F_u = -EA_xC_{2n}e_t$	$G_u = EA_xB_{2n}e_t$
$F_v = EI_zC_{3z}\beta f_t$	$G_v = EI_zB_{3z}\beta f_t$
$F_\theta = -EI_zC_{2z}\beta f_t$	$G_\theta = EI_zB_{2z}\beta f_t$

4.

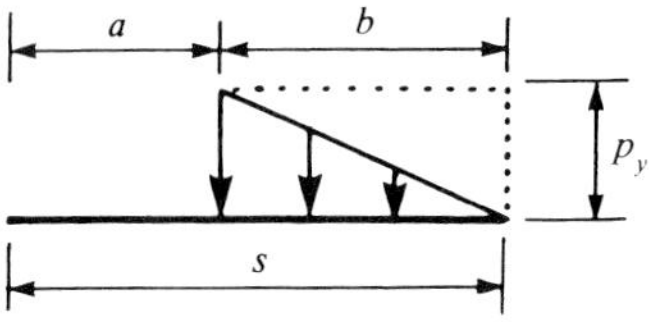

Sec. 20.09 (case 1 + case 3)

5.

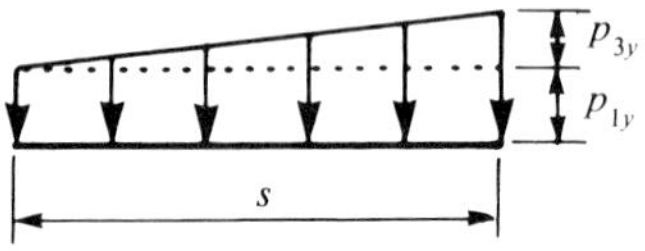

Sec. 20.09 (case 2 − case 4)

20.09 STIFFNESS MATRIX EQUATIONS

$$r = \frac{s^2 E A_x}{E I_z}$$

(1) Equivalents and signs. All forces, moments, and displacements are in the principal axes of the normal section, and their positive directions are shown in (2, 3) below (Ref. 6, pp. 178–184).

(2) Free-body diagram

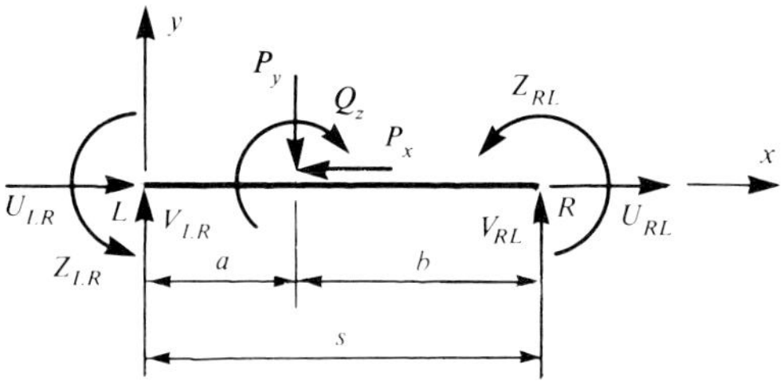

(3) Elastic curve

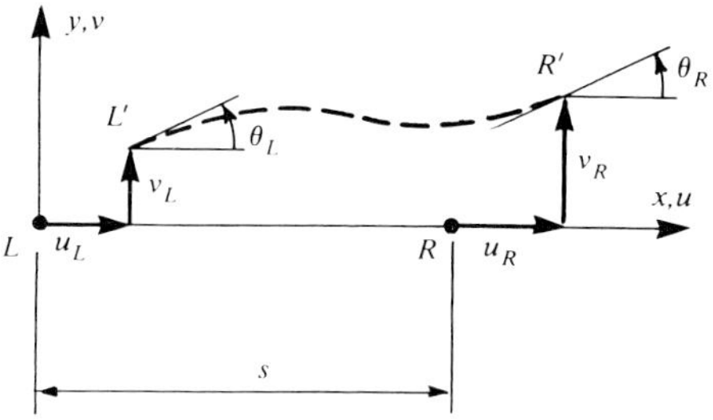

(4) Dimensionless stiffness coefficients

$$D_0 = \frac{1}{1 - \cosh \beta s \cos \beta s}$$

$L_1 = \dfrac{\cos \alpha s}{\sin \alpha s}$	$K_1 = (\alpha s) L_1$	
$L_2 = D_0 (\cosh \beta s \sin \beta s + \sinh \beta s \cos \beta s)$	$K_2 = (\beta s)^3 L_2$	
$L_3 = D_0 (\sinh \beta s \sin \beta s)$	$K_3 = (\beta s)^2 L_3$	$\alpha = \sqrt{\dfrac{\mu \omega^2}{E A_x}}$
$L_4 = D_0 (\cosh \beta s \sin \beta s - \sinh \beta s \cos \beta s)$	$K_4 = (\beta s) L_4$	
$L_5 = \dfrac{1}{\sin \alpha s}$	$K_5 = (\alpha s) L_5$	
$L_6 = D_0 (\sinh \beta s + \sin \beta s)$	$K_6 = (\beta s)^3 L_6$	$\beta = \sqrt[4]{\dfrac{\mu \omega^2}{E I_z}}$
$L_7 = D_0 (\cosh \beta s - \cos \beta s)$	$K_7 = (\beta s)^2 L_7$	
$L_8 = D_0 (\sinh \beta s - \sin \beta s)$	$K_8 = (\beta s) L_8$	

(5) Stiffness matrix equations

$$\begin{bmatrix} U_{LR} \\ V_{LR} \\ Z_{LR} \\ \hline U_{RL} \\ V_{RL} \\ Z_{RL} \end{bmatrix} = \frac{EI_z}{s^3} \left[\begin{array}{ccc|ccc} K_1 r & 0 & 0 & -K_t r & 0 & 0 \\ 0 & K_2 & K_3 s & 0 & -K_6 & K_7 s \\ 0 & K_3 s & K_4 s^2 & 0 & -K_7 s & K_8 s^2 \\ \hline -K_5 r & 0 & 0 & K_1 r & 0 & 0 \\ 0 & -K_6 & -K_7 s & 0 & K_2 & -K_3 s \\ 0 & K_7 s & K_8 s^2 & 0 & -K_3 s & K_4 s^2 \end{array}\right] \begin{bmatrix} u_L \\ v_L \\ \theta_L \\ \hline u_R \\ v_R \\ \theta_R \end{bmatrix} + \begin{bmatrix} U_{L0} \\ V_{L0} \\ Z_{L0} \\ \hline U_{R0} \\ V_{R0} \\ Z_{R0} \end{bmatrix}$$

where U_{L0}, V_{L0}, Z_{L0} and U_{R0}, V_{R0}, Z_{R0} are the *reactions of a fixed-end beam* acted on by loads and other causes shown in (20.12). The graphs and tables of stiffness coefficients are shown in (20.10–20.11)

(6) General stiffness matrix equations for bars of order three are obtained by matrix composition of (20.09) and (21.07).

20.10 TABLE OF DIMENSIONLESS STIFFNESS COEFFICIENTS

βs	K_2	K_3	K_4	K_6	K_7	K_8
0.0	12.00000	6.00000	4.00000	12.00000	6.00000	2.00000
0.5	11.97717	5.99692	3.99954	12.00842	6.00214	2.00051
0.6	11.95167	5.99312	3.99870	12.01649	6.00392	2.00090
0.7	11.91076	5.98740	3.99770	12.03085	6.00742	2.00171
0.8	11.84764	5.97846	3.99604	12.05257	6.01261	2.00290
0.982	11.75613	5.96559	3.99374	12.08448	6.02034	2.00469
1.0	11.62819	5.94753	3.99045	12.12888	6.03101	2.00716
1.1	11.45537	5.92313	3.98601	12.18891	6.04545	2.01049
1.2	11.22821	5.89104	3.98018	12.26801	6.06449	2.01488
1.3	10.93614	5.84976	3.97266	12.36989	6.08898	2.02052
1.4	10.56768	5.79763	3.96317	12.49881	6.11997	2.02767
1.5	10.11018	5.73283	3.95136	12.65941	6.15856	2.03656
1.6	9.54994	5.65338	3.93688	12.85695	6.20599	2.04749
1.7	8.87193	5.55708	3.91930	13.09723	6.26364	2.06077
1.8	8.05987	5.44151	3.89819	13.38682	6.33306	2.07675
1.9	7.09596	5.30402	3.87305	13.73308	6.41597	2.09583
2.0	5.96080	5.14166	3.84332	14.14441	6.51434	2.11844
2.1	4.63312	4.95117	3.80838	14.63036	6.63040	2.14510
2.2	3.08949	4.72889	3.76754	15.20194	6.76668	2.17637
2.3650	0.00000	4.28146	3.68509	16.36716	7.04379	2.23987
2.4	−0.75185	4.17197	3.66487	16.65523	7.11216	2.25551
2.5	−3.11049	3.82734	3.60109	17.56946	7.32878	2.30504
2.6	−5.80811	3.43070	3.52747	18.63551	7.58071	2.36255
2.7	−8.88610	2.97489	3.44257	19.87859	7.87363	2.42930
2.8	−12.39198	2.45154	3.34471	21.32913	8.21435	2.50679
2.9	−16.38115	1.85065	3.23188	23.02448	8.61120	2.59687
3.0	−20.91888	1.16017	3.10161	25.01094	9.07446	2.70179
3.1416	−28.43785	0.0000	2.88131	28.43764	9.86964	2.88133
3.2	−31.96896	−0.55180	2.77595	30.10472	10.25485	2.96808
3.3	−38.69347	−1.61445	2.57200	33.38058	11.00900	3.13753
3.4	−46.40492	−2.85201	2.33288	37.29843	11.90661	3.33861
3.5	−55.29434	−4.30305	2.05045	42.02402	12.98378	3.57917
3.6	−65.61404	−6.01925	1.71379	47.78284	14.28959	3.86983
3.7	−77.70773	−8.07181	1.30776	54.89008	15.89246	4.22538
3.8	−92.05875	−10.56182	0.81085	63.79857	17.89037	4.66702
3.9266	−114.60362	−14.58187	0.00000	78.74750	21.22150	5.40034
4.0	−130.73358	−17.53061	−0.60034	90.08661	23.73441	5.95160
4.1	−157.89185	−22.61275	−1.64393	110.22800	28.17644	6.92301
4.2	−193.90727	−29.53928	−3.08018	138.62389	34.40547	8.28047
4.3	−244.65594	−39.57089	−5.18004	181.10657	43.67761	10.29439
4.4	−323.10376	−55.49358	−8.54270	250.59363	58.77367	13.56337
4.5	−464.70483	−84.93198	−14.80847	382.48633	87.31267	19.72699
4.6	−813.85474	−158.95268	−30.66188	721.12329	160.35492	35.46907
4.7300	$\pm\infty$	$\pm\infty$	$\pm\infty$	$\pm\infty$	$\pm\infty$	$\pm\infty$
4.8	1414.59326	323.03711	73.21985	−1531.02466	−323.85767	−68.66635
4.9	545.64819	137.43222	33.37509	−675.41553	−139.51370	−28.96608
5.0	312.41724	88.71986	22.99448	−456.62256	−92.17491	−18.74316
6.0	−153.93475	10.53046	7.78616	−225.81148	−37.50851	−6.28994

†Values of K_1 and K_5 are available in standard mathematical tables.

20.11 GRAPHS AND SERIES OF DIMENSIONLESS STIFFNESS COEFFICIENTS

(1) Graphs of K_j ($j = 2, 3, 4, 6, 7, 8$)

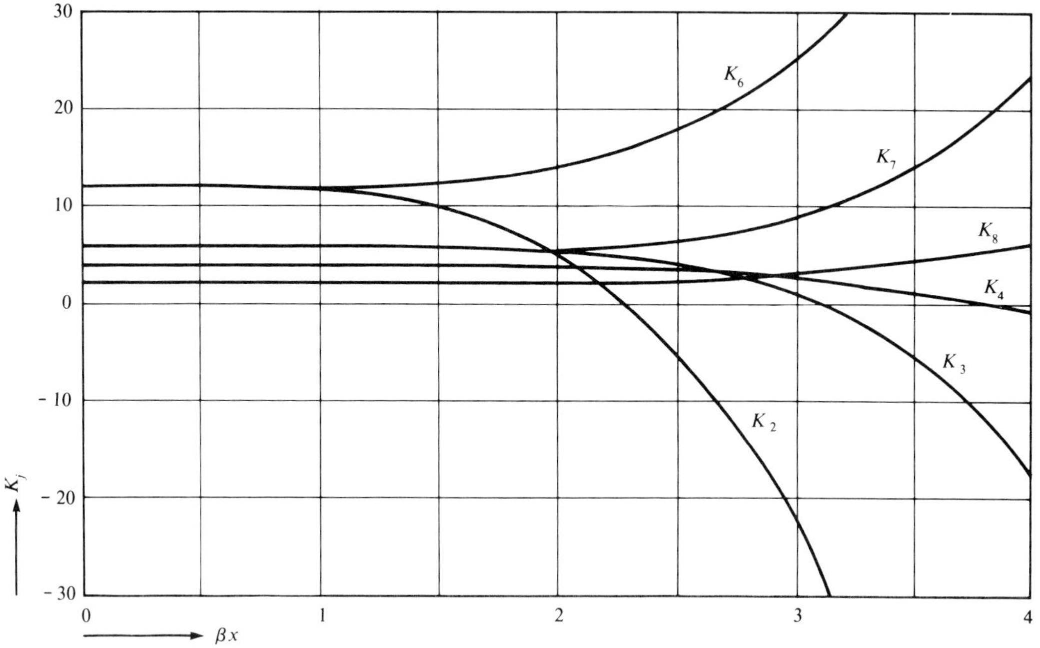

(2) Series expansion of K_j

$$K_1 = \frac{\displaystyle\sum_{k=0}^{\infty} \frac{(-1)^k(\alpha s)^{2k}}{(2k)!}}{\displaystyle\sum_{k=0}^{\infty} \frac{(-1)^k(\alpha s)^{2k}}{(2k+1)!}} = 1 - \sum_{k=1}^{\infty} C_{1K}\left(\frac{\alpha s}{\sqrt{10}}\right)^{2k}$$

$$K_2 = \frac{\displaystyle\sum_{k=0}^{\infty} \frac{(-4)^k(\beta s)^{4k}}{2(4k+1)!}}{\displaystyle\sum_{k=0}^{\infty} \frac{(-4)^k(\beta s)^{4k}}{(4k+4)!}} = 12 - \sum_{k=1}^{\infty} C_{2K}\left(\frac{\beta s}{\sqrt{10}}\right)^{4k}$$

$$K_3 = \frac{\displaystyle\sum_{k=0}^{\infty} \frac{(-4)^k(\beta s)^{4k}}{2(4k+2)!}}{\displaystyle\sum_{k=0}^{\infty} \frac{(-4)^k(\beta s)^{4k}}{4k+4)!}} = 6 - \sum_{k=1}^{\infty} C_{3K}\left(\frac{\beta s}{\sqrt{10}}\right)^{4k}$$

$$K_4 = \frac{\displaystyle\sum_{k=0}^{\infty} \frac{(-4)^k(\beta s)^{4k}}{(4k+3)!}}{\displaystyle\sum_{k=0}^{\infty} \frac{(-4)^k(\beta s)^{4k}}{(4k+4)!}} = 4 - \sum_{k=1}^{\infty} C_{4K}\left(\frac{\beta s}{\sqrt{10}}\right)^{4k}$$

$$K_5 = \frac{1}{\displaystyle\sum_{k=0}^{\infty} \frac{(-1)^k(\alpha s)^{2k}}{(2k+1)!}} = 1 + \sum_{k=1}^{\infty} C_{5K}\left(\frac{\alpha s}{\sqrt{10}}\right)^{2k}$$

$$K_6 = \frac{\displaystyle\sum_{k=0}^{\infty} \frac{(\beta s)^{4k}}{2(4k+1)!}}{\displaystyle\sum_{k=0}^{\infty} \frac{(-4)^k(\beta s)^{4k}}{(4k+4)!}} = 12 + \sum_{k=1}^{\infty} C_{6k}\left(\frac{\beta s}{\sqrt{10}}\right)^{4k}$$

$$K_7 = \frac{\displaystyle\sum_{k=0}^{\infty} \frac{(\beta s)^{4k}}{2(4k+2)!}}{\displaystyle\sum_{k=0}^{\infty} \frac{(-4)^k(\beta s)^{4k}}{(4k+4)!}} = 6 + \sum_{k=1}^{\infty} C_{7K}\left(\frac{\beta s}{\sqrt{10}}\right)^{4k}$$

$$K_8 = \frac{\displaystyle\sum_{k=0}^{\infty} \frac{(\beta s)^{4k}}{2(4k+3)!}}{\displaystyle\sum_{k=0}^{\infty} \frac{(-4)^k(\beta s)^{4k}}{(4k+4)!}} = 2 + \sum_{k=1}^{\infty} C_{8K}\left(\frac{\beta s}{\sqrt{10}}\right)^{4k}$$

where C_{1K}, C_{2K}, . . . are the numerical constants of the respective series obtained by the method described in Ref. 1, p. 105. Table of these constants is shown in (21.09) in the subsequent chapter.

20.12 STIFFNESS LOAD FUNCTIONS, FIXED-END BEAM

(1) General form of the stiffness load functions in (20.09) is given by the matrix equations

$$\begin{bmatrix} U_{LO} \\ V_{LO} \\ Z_{LO} \end{bmatrix} = \begin{bmatrix} L_5 & 0 & 0 \\ 0 & L_6 & -L_7 \\ 0 & L_7\beta^{-1} & -L_8\beta^{-1} \end{bmatrix} \begin{bmatrix} G_u \\ G_v \\ G_\theta \end{bmatrix}$$

$$\begin{bmatrix} U_{RO} \\ V_{RO} \\ Z_{RO} \end{bmatrix} = \begin{bmatrix} L_5 & 0 & 0 \\ 0 & L_6 & L_7 \\ 0 & -L_7\beta^{-1} & -L_8\beta^{-1} \end{bmatrix} \begin{bmatrix} F_u \\ F_v \\ F_\theta \end{bmatrix}$$

where F_u, F_v, F_θ are the left-end transport load functions and G_v, G_v, G_θ are their right-end counterparts, given for particular loads and causes in (20.06–20.08), and L_5, L_6, L_7, L_8 are the stiffness coefficients (20.09). The load vectors and other causes are defined by symbols introduced in (20.06–20.08).

(2) Particular cases

Notation (20.02–20.03, 20.09) A_j, B_j, C_j, T_j = transport coefficients (20.03)	Signs (20.09)	$\alpha = \sqrt{\dfrac{\mu\Omega^2}{EA_x}}$ $\beta = \sqrt[4]{\dfrac{\mu\Omega^2}{EI_z}}$
$U_{LO} = L_5 B_{2n} P_x$ $V_{LO} = (L_7 B_{3z} - L_6 B_{4Z}) P_y$ $Z_{LO} = (L_8 B_{3z} - L_7 B_{4z})\beta^{-1} P_y$	**1.** 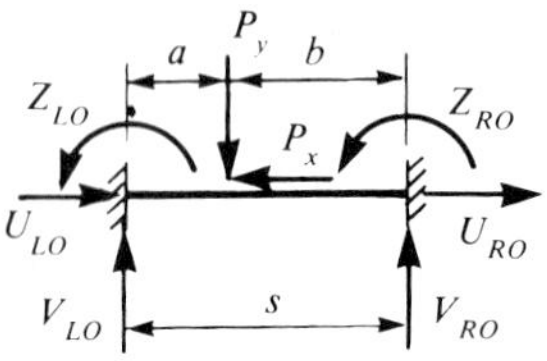	$U_{RO} = L_5 A_{2n} P_x$ $V_{RO} = (L_7 A_{3z} - L_6 A_{4z}) P_y$ $Z_{RO} = -(L_8 A_3 z - L_7 A_{4z})\beta^{-1} P_y$
$U_{LO} = 0$ $V_{LO} = -(L_7 B_{2z} - L_6 B_{3z})\beta Q_z$ $Z_{LO} = -(L_8 B_{2z} - L_7 B_{3z}) Q_z$	**2.** 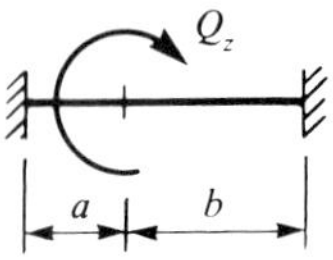	$U_{RO} = 0$ $V_{RO} = (L_7 A_{2z} - L_6 A_{3z})\beta Q_z$ $Z_{RO} = -(L_8 A_{2z} - L_7 A_{3z}) Q_z$
$U_{LO} = \dfrac{1 - \cos\alpha s}{\sin\alpha s} \dfrac{p_x}{\alpha}$ $V_{LO} = (L_6 - L_2)\dfrac{p_y}{\beta}$ $Z_{LO} = (L_7 - L_3)\dfrac{P_y}{\beta^2}$	**3.** 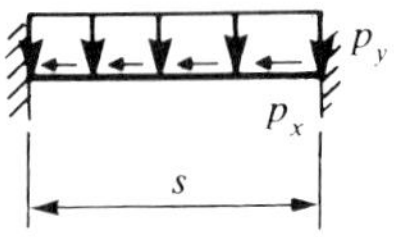	$U_{RO} = \dfrac{1 - \cos\alpha s}{\sin\alpha s} \dfrac{p_x}{\alpha}$ $V_{RO} = (L_6 - L_2)\dfrac{p_y}{\beta}$ $Z_{RO} = -(L_7 - L_3)\dfrac{p_y}{\beta^2}$

$$U_{L0} = L_5(A_{2n} + B_{2n})P_x = U_{R0}$$

$$V_{L0} = [L_7(A_{3z} + B_{3z}) - L_6(A_{4z} + B_{4z})]P_y = V_{R0}$$

$$Z_{L0} = [L_8(A_{3z} + B_{3z}) - L_7(A_{4z} + B_{4z})]\beta^{-1}P_y = -Z_{R0}$$

4.

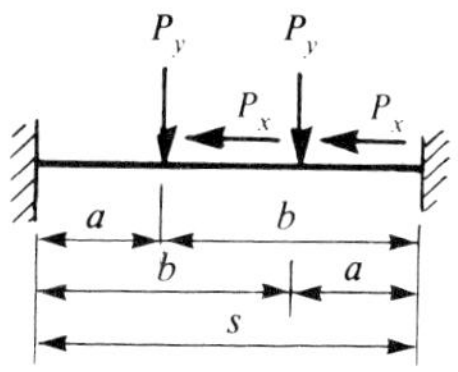

$$U_{L0} = 0 = U_{R0}$$

$$V_{L0} = [L_6(A_{3z} + B_{3z}) - L_7(A_{2z} + B_{2z})]\beta Q_z = -V_{R0}$$

$$Z_{L0} = [L_7(A_{3z} + B_{3z}) - L_8(A_{2z} + B_{2z})]Q_z = Z_{R0}$$

5.

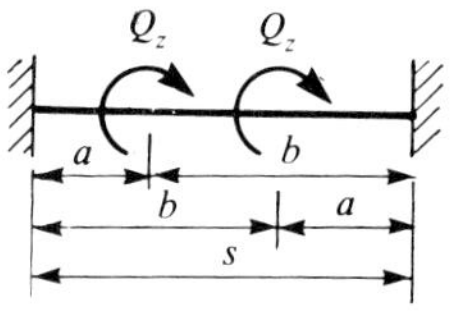

6.

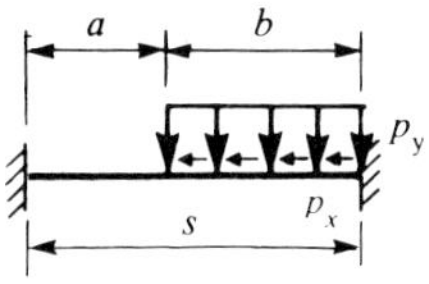

Use general form in terms of case (20.07-2)

7.

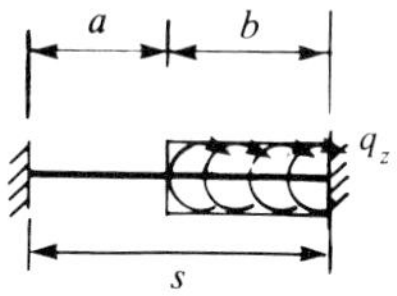

Use general form in terms of case (20.06-4)

8.

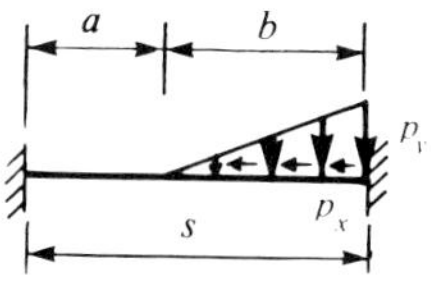

Use general form in terms of case (20.07-4)

9. Abrupt displacements u_j, v_j

$$U_{L0} = EA_xL_5B_{1n}\alpha u_j$$

$$V_{L0} = EI_z(L_6B_{1z} - L_7B_{4z})\beta^3 v_j$$

$$Z_{L0} = EI_z(L_7B_{1z} - L_8B_{4z})\beta v_j$$

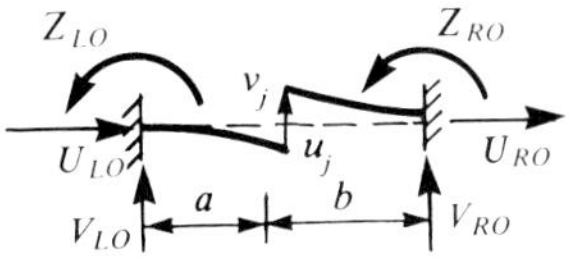

$$U_{R0} = -EA_xL_5A_{1n}\alpha u_j$$

$$V_{R0} = -EI_z(L_6A_{1z} - L_7A_{4z})\beta^3 v_j$$

$$Z_{R0} = EI_z(L_7A_{1z} - L_8A_{4z})\beta^2 v_j$$

10. Abrupt displacement θ_j

$$U_{L0} = 0$$

$$V_{L0} = EI_z(L_6B_{2z} - L_7B_{1z})\beta^2\theta_j$$

$$Z_{L0} = EI_z(L_7B_{2z} - L_8B_{1z})\beta\theta_j$$

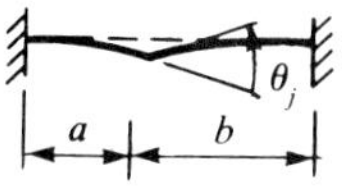

$$U_{R0} = 0$$

$$V_{R0} = EI_z(L_6A_{2z} - L_7A_{1z})\beta^2\theta_j$$

$$Z_{R0} = -EI_z(L_7A_{2z} - L_8A_{1z})\beta\theta_j$$

11. Temperature deformation of segment b

$$U_{L0} = EA_xL_5B_{2n}e_t$$

$$V_{L0} = EI_z(L_6B_{3z} - L_7B_{2z})\beta f_t$$

$$Z_{L0} = EI_z(L_7B_{3z} - L_8B_{2z})f_t$$

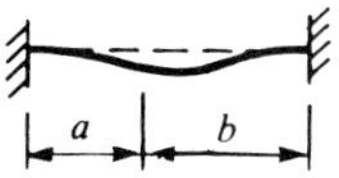

$$U_{R0} = -EA_xL_5C_{2n}e_t$$

$$V_{R0} = EI_z(L_6C_{3z} - L_7C_{2z})\beta f_t$$

$$Z_{R0} = -EI_z(L_7C_{3z} - L_8C_{2z})f_t$$

20.13 NATURAL FREQUENCIES AND EIGENFUNCTIONS

(1) Transverse vibration (20.02–20.03)

$\omega_m = (\beta_m s)^2 \sqrt{\dfrac{EI_z}{\mu s^4}}$ $m = 1, 2, \ldots$	$T_{1z} = \frac{1}{2}(\cosh \beta_m s + \cos \beta_m s)$ $T_2 z = \frac{1}{2}(\sinh \beta_m s + \sin \beta_m s)$ $T_{3z} = \frac{1}{2}(\cosh \beta_m s - \cos \beta_m s)$ $T_{4z} = \frac{1}{2}(\sinh \beta_m s - \sin \beta_m s)$	$T_{1z}(x) = \frac{1}{2}(\cosh \beta_m x + \cos \beta_m x)$ $T_{2z}(x) = \frac{1}{2}(\sinh \beta_m x + \sin \beta_m x)$ $T_{3z}(x) = \frac{1}{2}(\cosh \beta_m x - \cos \beta_m x)$ $T_{4z}(x) = \frac{1}{2}(\sinh \beta_m x - \sin \beta_m x)$
End conditions	Frequency equation eigenvalues	Eigenfunction
1. Free-free 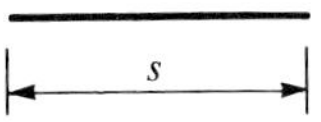	$\cosh \beta_m s \cos \beta_m s = 1$ $\beta_1 s = 4.7300$ $\beta_2 s = 7.8532$ $\beta_3 s = 10.9956$ $\beta_4 s = 14.1372$	$T_{2z}(x) - \dfrac{T_{4z}}{T_{3z}} T_{1z}(x)$ For large m, $\beta_m s = \dfrac{(2m+1)\pi}{2}$
2. Free-hinged 	$\tan \beta_m s = \tanh \beta_m s$ $\beta_1 s = 3.9266$ $\beta_2 s = 7.0686$ $\beta_3 s = 10.2102$ $\beta_4 s = 13.3518$	$T_{2z}(x) - \dfrac{T_{2z}}{T_{1z}} T_{1z}(x)$ For large m, $\beta_m s = \dfrac{(4m+1)\pi}{4}$
3. Free-guided 	$\tan \beta_m s = -\tanh \beta_m s$ $\beta_1 s = 2.3650$ $\beta_2 s = 5.4978$ $\beta_3 s = 8.6594$ $\beta_4 s = 11.7810$	$T_2 z(x) - \dfrac{T_{3z}}{T_{1z}} T_{1z}(x)$ For large m, $\beta_m s = \dfrac{(4m-1)\pi}{4}$
4. Free-fixed 	$\cosh \beta_m s \cos \beta_m s = -1$ $\beta_1 s = 1.8751$ $\beta_2 s = 4.6941$ $\beta_3 s = 7.8548$ $\beta_4 s = 10.9955$	$T_{4z}(x) - \dfrac{T_{2z}}{T_{1z}} T_{3z}(x)$ For large m, $\beta_m s = \dfrac{(2m-1)\pi}{2}$
5. Hinged-hinged 	$\sin \beta_m s = 0$ $\beta_1 s = \pi$ $\beta_2 s = 2\pi$ $\beta_3 s = 3\pi$ $\beta_4 s = 4\pi$	$T_{2z}(x) - T_{4z}(x)$ For all m, $\beta_m s = m\pi$
6. Hinged-guided 	$\cos \beta_m s = 0$ $\beta_1 s = \pi/2$ $\beta_2 s = 3\pi/2$ $\beta_3 s = 5\pi/2$ $\beta_4 s = 7\pi/2$	$T_{2z}(x) - T_{4z}(x)$ For all m, $\beta_m s = \dfrac{(2m-1)\pi}{2}$

7. Hinged-fixed 	$\tan \beta_m s = \tanh \beta_m s$ $\beta_1 s = 3.9266$ $\beta_2 s = 7.0686$ $\beta_3 s = 10.2102$ $\beta_4 s = 13.3518$	$T_{4z}(x) - \dfrac{T_{4z}}{T_{3z}} T_{3z}(x)$ For large m, $\beta_m s = \dfrac{(4m+1)\pi}{4}$
8. Guided-guided 	$\sin \beta_m s = 0$ $\beta_1 s = \pi$ $\beta_2 s = 2\pi$ $\beta_3 s = 3\pi$ $\beta_4 s = 4\pi$	$T_{1z}(x) - T_{3z}(x)$ For all m, $\beta_m s = m\pi$
9. Guided-fixed 	$\tan \beta_m s = -\tanh \beta_m s$ $\beta_1 s = 2.5650$ $\beta_2 s = 5.4578$ $\beta_3 s = 8.6394$ $\beta_4 s = 11.7810$	$T_{4z}(x) - \dfrac{T_{1z}}{T_{4z}} T_{3z}(x)$ For large m, $\beta_m s = \dfrac{(4m-1)\pi}{4}$
10. Fixed-fixed 	$\cosh \beta_m s \cos \beta_m s = 1$ $\beta_1 s = 4.7300$ $\beta_2 s = 7.8532$ $\beta_3 s = 10.9956$ $\beta_4 s = 14.1372$	$T_{4z}(x) - \dfrac{T_{4z}}{T_{3z}} T_{3z}(x)$ For large m, $\beta_m s = \dfrac{(2m+1)\pi}{2}$

(2) Longitudinal vibration (20.02–20.03)

$\omega_m = (\alpha_m s)\sqrt{\dfrac{EA_x}{\mu s^2}}$ $m = 1,2,\ldots$	$T_{1n} = \cos \alpha_m s$ $T_{2n} = \sin \alpha_m s$	$T_{1n}(x) = \cos \alpha_m x$ $T_{2n}(x) = \sin \alpha_m x$
End conditions	Frequency equation eigenvalues	Eigenfunction
1. Free-free 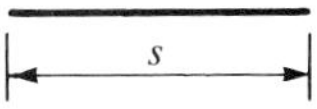	$\sin \alpha_m s = 0$ $\alpha_m s = m\pi$	$T_{1n}(x)$
2. Free-fixed 	$\cos \alpha_m s = 0$ $\alpha_m s = \dfrac{(2m+1)\pi}{2}$	$T_{2n}(x)$
3. Fixed-fixed 	$\sin \alpha_m s = 0$ $\alpha_m s = m\pi$	$T_{1n}(x)$

20.13 NATURAL FREQUENCIES AND EIGENFUNCTIONS *(Continued)*

(3) Torsional vibration (21.02–21.03)

$\omega_m = (\alpha_m s)\sqrt{\dfrac{GI_x}{\rho s^2}}$ $m = 1, 2, \ldots$	$T_{1x} = \cos \alpha_m s$ $T_{2x} = \sin \alpha_m s$	$T_{1x}(x) = \cos \alpha_m x$ $T_{2x}(x) = \sin \alpha_m x$
End conditions	Frequency equation eigenvalues	Eigenfunction
1. Free-free 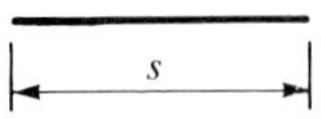	$\sin \alpha_m s = 0$ $\alpha_m s = m\pi$	$T_{1x}(x)$
2. Free-fixed 	$\cos \alpha_m s = 0$ $\alpha_m s = \dfrac{(2m + 1)\pi}{2}$	$T_{2x}(x)$
3. Fixed-fixed 	$\sin \alpha_m s = 0$ $\alpha_m s = m\pi$	$T_{1x}(x)$

21

Straight Bar of Order Two Constant Section

DYNAMIC STATE

21.01 DEFINITION OF STATE

(1) System considered is a finite straight bar of constant and symmetrical section. The centroidal axis x of the bar and the vertical axis z define the plane of symmetry of the section. The bar and its loads are in a state of dynamic equilibrium and a state of time-dependent elastic deformation. The bar is a part of a planar system acted on by loads normal to the system plane (system of order two).

(2) Notation (A.1, A.2, A.4)

s	= span (m)	A_x	= area of normal section (m^2)
a, b	= segments (m)	I_x	= torsional constant of A_x about x (m^4)
x, x'	= coordinates (m)	I_y	= moment of inertia of A_x about y (m^4)
E	= modulus of elasticity (Pa)	ω	= natural angular frequency (rad/s)
G	= modulus of rigidity (Pa)	Ω	= forcing angular frequency (rad/s)
W	= shearing force (N)	w	= linear displacement along z (m)
X	= torsional moment (N·m)	ϕ	= torsional slope about x (rad)
Y	= flexural moment (N·m)	ψ	= flexural slope about y (rad)
t	= time (s)	ρ	= mass moment of inertia (kg·m^2/m)
μ	= mass per unit length (kg/m)		

$\alpha = \sqrt{\dfrac{\rho\omega^2}{GI_x}}$ = shape parameter (m^{-1})	$\beta = \sqrt[4]{\dfrac{\mu\omega^2}{EI_y}}$ = shape parameter (m^{-1})

(3) Assumptions of analysis are stated in (17.01). In this chapter, the effects of torsional and flexural deformations are included, but the effects of shearing deformations and of the axial force on the magnitude of the bending moment are neglected.

21.02 TRANSPORT MATRIX EQUATIONS

(1) Equivalents and signs

$\overline{X} = X\alpha = \text{scaled } X \text{ (N)}$ $\qquad$ $\overline{Y} = Y\beta = \text{scaled } Y \text{ (N)}$

$\overline{\phi} = \phi G I_x \alpha^2 = \text{scaled } \phi \text{ (N)}$ $\qquad$ $\overline{\psi} = \psi E I_y \beta^2 = \text{scaled } \psi \text{ (N)}$

$\overline{w} = w E I_y \beta^3 = \text{scaled } w \text{ (N)}$

All forces, moments, and displacements are in the principal axes of the normal section, and their positive directions are shown in (2, 3) below.

(2) Free-body diagram

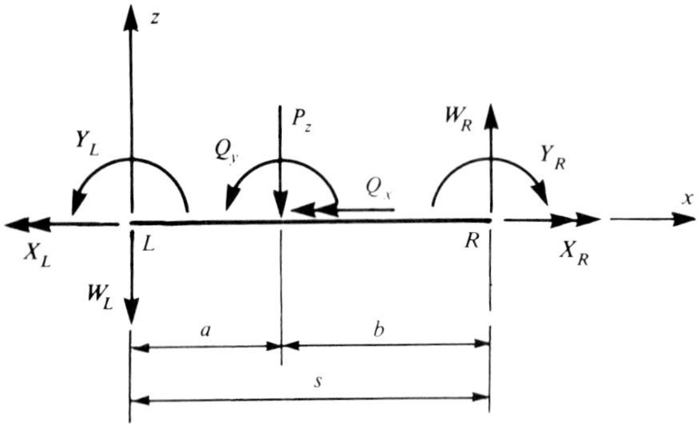

(3) Elastic curve

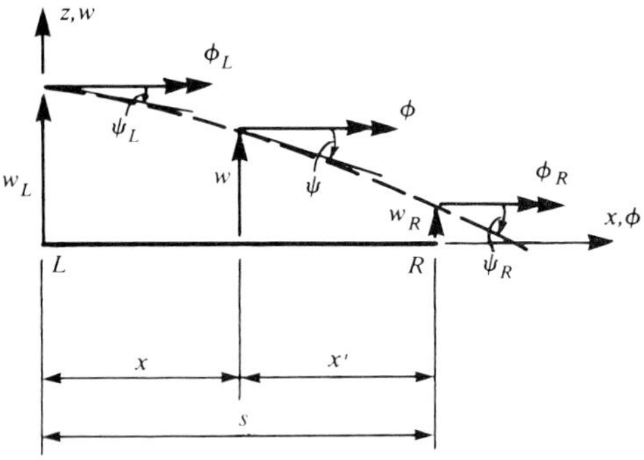

(4) Transport matrix equations

$$\underbrace{\begin{bmatrix} 1 \\ \hline W_L \\ X_L \\ Y_L \\ \hline \overline{w}_L \\ \overline{\phi}_L \\ \overline{\psi}_L \end{bmatrix}}_{H_L} = \underbrace{\left[\begin{array}{c|ccc|ccc} 1 & 0 & 0 & 0 & 0 & 0 & 0 \\ \hline F_W & T_{1y} & 0 & -T_{4y} & T_{2y} & 0 & T_{3y} \\ F_X & 0 & T_{1x} & 0 & 0 & T_{2x} & 0 \\ F_Y & -T_{2y} & 0 & T_{1y} & -T_{3y} & 0 & -T_{4y} \\ \hline F_w & T_{4y} & 0 & -T_{3y} & T_{1y} & 0 & T_{2y} \\ F_\phi & 0 & -T_{2x} & 0 & 0 & T_{1x} & 0 \\ F_\psi & T_{3y} & 0 & -T_{2y} & T_{4y} & 0 & T_{1y} \end{array}\right]}_{T_{LR}} \underbrace{\begin{bmatrix} 1 \\ \hline W_R \\ X_R \\ Y_R \\ \hline \overline{w}_R \\ \overline{\phi}_R \\ \overline{\psi}_R \end{bmatrix}}_{H_R}$$

$$\underbrace{\begin{bmatrix} 1 \\ \hline W_R \\ X_R \\ Y_R \\ \hline \overline{w}_R \\ \overline{\phi}_R \\ \overline{\psi}_R \end{bmatrix}}_{H_R} = \underbrace{\left[\begin{array}{c|ccc|ccc} 1 & 0 & 0 & 0 & 0 & 0 & 0 \\ \hline G_W & T_{1y} & 0 & T_{4y} & -T_{2y} & 0 & T_{3y} \\ G_X & 0 & T_{1x} & 0 & 0 & -T_{2x} & 0 \\ G_Y & T_{2y} & 0 & T_{1y} & -T_{3y} & 0 & T_{4y} \\ \hline G_w & -T_{4y} & 0 & -T_{3y} & T_{1y} & 0 & -T_{2y} \\ G_\phi & 0 & T_{2x} & 0 & 0 & T_{1x} & 0 \\ G_\psi & T_{3y} & 0 & T_{2y} & -T_{4y} & 0 & T_{1y} \end{array}\right]}_{T_{RL}} \underbrace{\begin{bmatrix} 1 \\ \hline W_L \\ X_L \\ Y_L \\ \hline \overline{w}_L \\ \overline{\phi}_L \\ \overline{\psi}_L \end{bmatrix}}_{H_L}$$

where H_L, H_R are the end *state vectors* and T_{LR}, T_{RL} are the respective *transport matrices*. The elements of these matrices are the *transport coefficients* shown in (21.03) and the *load functions* F, G listed for particular cases in (21.04–21.06).

21.03 TRANSPORT RELATIONS

(1) Transport functions

	$\alpha = \sqrt{\dfrac{\rho\omega^2}{GI_x}} \qquad \beta = \sqrt[4]{\dfrac{\mu\omega^2}{EI_y}}$
$T_{1x}(x) = \cos \alpha x$	$T_{1y}(x) = \frac{1}{2}(\cosh \beta x + \cos \beta x)$
$T_{2x}(x) = \sin \alpha x$	$T_{2y}(x) = \frac{1}{2}(\sinh \beta x + \sin \beta x)$
$T_{3x}(x) = 1 - \cos \alpha x$	$T_{3y}(x) = \frac{1}{2}(\cosh \beta x - \cos \beta x)$
$T_{4x}(x) = \alpha x - \sin \alpha x$	$T_{4y}(x) = \frac{1}{2}(\sinh \beta x - \sin \beta x)$
$T_{5x}(x) = \dfrac{(\alpha x)^2}{2!} - T_{3x}(x)$	$T_{5y}(x) = T_{1y}(x) - 1$
$T_{6x}(x) = \dfrac{(\alpha x)^3}{3!} - T_{4x}(x)$	$T_{6y}(x) = T_{2y}(x) - \beta x$

The graphs and table of these functions are identical to (20.04–20.05).

(2) Transport coefficients T_j, A_j, B_j, C_j are the values of $T_j(x)$ for a particular argument x.

$$\left.\begin{array}{lll} \text{For } x = s, & T_{jx}(s) = T_{jx}, & T_{jy}(s) = T_{jy} \\ \text{For } x = a, & T_{jx}(a) = A_{jx}, & T_{jy}(a) = T_{jy} \\ \text{For } x = b, & T_{jx}(b) = B_{jx}, & T_{jy}(b) = B_{jy} \\ \text{and } T_{jx} - A_{jx} = C_{jx}, & & T_{jy} - A_{jy} = C_{jy} \end{array}\right\} j = 1, 2, \ldots, 6$$

(3) Transport matrices of this bar have analogical properties to those of (11.02) and can be used in similar ways.

(4) Differential relations derived for a free bar of order two in (11.03) are valid here without modification.

(5) Effect of elastic foundation introduced in (13.01) can be included in the transport functions $T_{jx}(x)$ and $T_{jy}(x)$ by taking

$$\alpha = \sqrt{\frac{\rho\omega^2 - k_\phi}{GI_x}} > 0 \qquad \beta = \sqrt[4]{\frac{\mu\omega^2 - k_w}{EI_y}} > 0$$

where k_ϕ, k_w = foundation moduli (N·m/m, N/m²) (Ref. 22).

(6) Angular freuqencies $\omega_1, \omega_2, \ldots$ of this motion are the eigenvalues of the transcendental-frequency determinant equation constructed from the transport matrix equation by selecting the part of the matrix corresponding to the existing eigenvector (20.13).

(7) Load functions F, G tabulated in (21.04–21.06) are to be used only in steady-state cases in which

$$L(x, t) = L(x) \sin \Omega t \qquad \text{or} \qquad L(x, t) = L(x) \cos \Omega t$$

where Ω = forcing angular frequency (rad/s).

(8) General transport matrix equations for bars of order three are obtained by matrix composition of (20.02) and (21.02).

21.04 TRANSPORT LOAD FUNCTIONS

Notation (21.01–21.03) Signs (21.02) Matrix (21.02) P_z = concentrated load (N) Q_x, Q_y = applied couples (N·m) q_y = intensity of distributed couple (N·m/m) A_j, B_j, C_j, T_j = transport coefficients (21.03)		$\alpha = \sqrt{\dfrac{\rho\Omega^2}{GI_x}}$ $\beta = \sqrt[4]{\dfrac{\mu\Omega^2}{EI_y}}$
$F_W = -A_{1y}P_z$ $F_X = 0$ $F_Y = A_{2y}P_z$	**1.** 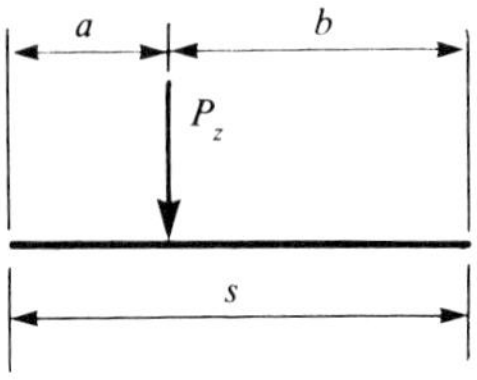 	$G_W = B_{1y}P_z$ $G_X = 0$ $G_Y = B_{2y}P_z$
$F_w = -A_{4y}P_z$ $F_\phi = 0$ $F_\psi = -A_{3y}P_z$		$G_w = -B_{4y}P_z$ $G_\phi = 0$ $G_\psi = B_{3y}P_z$
$F_W = A_{4y}\beta Q_y$ $F_X = -A_{1x}\alpha Q_x$ $F_Y = -A_{1y}\beta Q_y$	**2.** 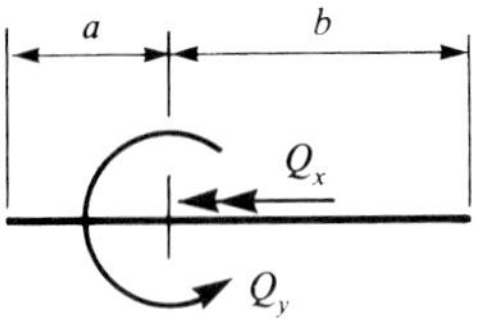 	$G_W = B_{4y}\beta Q_y$ $G_X = B_{1x}\alpha Q_x$ $G_Y = B_{1y}\beta Q_y$
$F_w = A_{3y}\beta Q_y$ $F_\phi = A_{2x}\alpha Q_x$ $F_\psi = A_{2y}\beta Q_y$		$G_w = -B_{3y}\beta Q_y$ $G_\phi = B_{2x}\alpha Q_x$ $G_\psi = B_{2y}\beta Q_y$
$F_W = T_{5y}q_y$ $F_X = 0$ $F_Y = -T_{2y}q_y$	**3.** 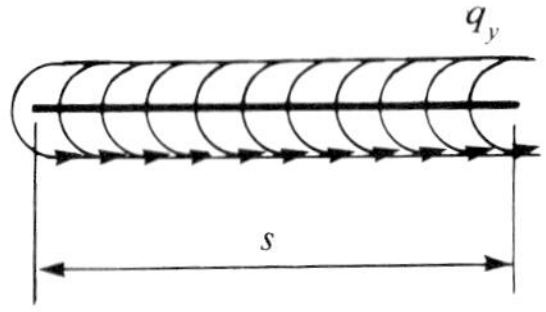 	$G_W = T_{5y}q_y$ $G_X = 0$ $G_Y = T_{2y}q_y$
$F_w = T_{4y}q_y$ $F_\phi = 0$ $F_\psi = T_{3y}q_y$		$G_w = -T_{4y}q_y$ $G_\phi = 0$ $G_\psi = T_{3y}q_y$
$F_W = C_{5y}q_y$ $F_X = 0$ $F_Y = -C_{2y}q_y$	**4.** 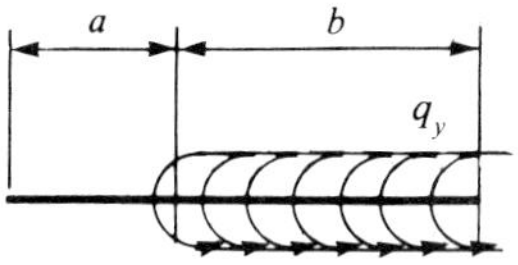 	$G_W = B_{5y}q_y$ $G_X = 0$ $G_Y = B_{2y}q_y$
$F_w = C_{4y}q_y$ $F_\phi = 0$ $F_\psi = C_{3y}q_y$		$G_w = -B_{4y}q_y$ $G_\phi = 0$ $G_\psi = B_{3y}q_y$

21.05 TRANSPORT LOAD FUNCTIONS

Notation (21.01–21.03) Signs (21.02) Matrix (21.02) p_z = intensity of distributed load (N/m) q_x = intensity of distributed couple (N·m/m) A_j, B_j, C_j, T_j = transport coefficients (21.03)		$\alpha = \sqrt{\dfrac{\rho\Omega^2}{GI_x}}$ $\beta = \sqrt[4]{\dfrac{\mu\Omega^2}{EI_y}}$
$F_W = -T_{2y}\beta^{-1}p_z$ $F_X = -T_{2x}q_x$ $F_Y = T_{3y}\beta^{-1}p_z$	**1.** 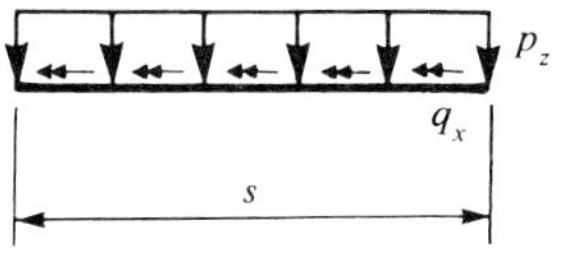	$G_W = T_{2y}\beta^{-1}p_z$ $G_X = T_{2x}q_x$ $G_Y = T_{3y}\beta^{-1}p_z$
$F_w = -T_{5y}\beta^{-1}p_z$ $F_\phi = T_{3x}q_x$ $F_\psi = -T_{4y}\beta^{-1}p_z$		$G_w = -T_{5y}\beta^{-1}p_z$ $G_\phi = T_{3x}q_x$ $G_\psi = T_{4y}p_z$
$F_W = -C_{2y}\beta^{-1}p_z$ $F_X = -C_{2x}q_x$ $F_Y = C_{3y}\beta^{-1}p_z$	**2.** 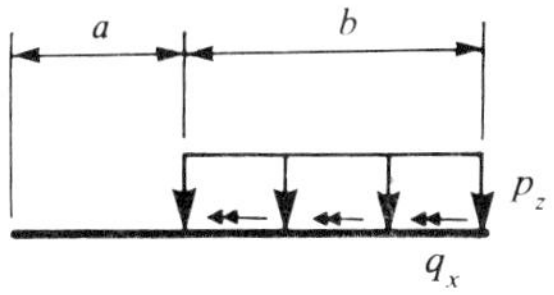	$G_w = B_{2y}\beta^{-1}p_z$ $G_X = B_{2x}q_x$ $G_Y = B_{3y}\beta^{-1}p_z$
$F_w = -C_{5y}\beta^{-1}p_z$ $F_\phi = C_{3x}q_x$ $F_\psi = -C_{4y}\beta^{-1}p_z$		$G_w = -B_{5y}\beta^{-1}p_z$ $G_\phi = B_{3x}q_x$ $G_\psi = B_{4y}\beta^{-1}p_z$
$F_W = -(\beta sT_{2y} - T_{3y})\beta^2 s^{-1}p_z$ $F_X = -(\alpha sT_{2x} - T_{3x})\alpha^{-1}s^{-1}q_x$ $F_Y = (\beta sT_{3y} - T_{4y})\beta^{-2}s^{-1}p_z$	**3.** 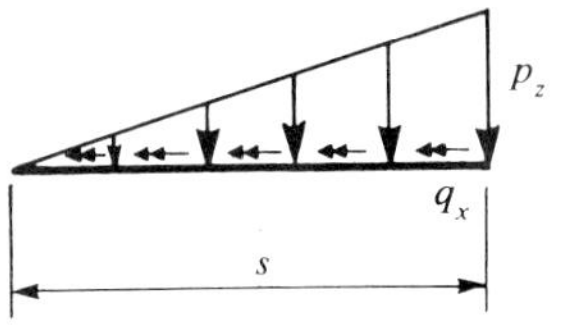	$G_W = T_{3y}\beta^{-2}s^{-1}p_z$ $G_X = T_{3x}\alpha^{-1}s^{-1}q_x$ $G_Y = T_{4y}\beta^{-2}s^{-1}p_z$
$F_w = -(\beta sT_{5y} - T_{6y})\beta^{-2}s^{-1}p_z$ $F_\phi = (\alpha sT_{3x} - T_{4x})\alpha^{-1}s^{-1}q_x$ $F_\psi = -(\beta sT_{4y} - T_{5y})\beta^{-2}s^{-1}p_y$		$G_w = -T_{6y}\beta^{-2}s^{-1}p_z$ $G_\phi = T_{4x}\alpha^{-1}s^{-1}q_x$ $G_\psi = T_{5y}\beta^{-2}s^{-1}p_z$
$F_W = -(\beta bT_{2y} - C_{3y})\beta^{-2}b^{-1}p_z$ $F_X = -(\alpha bT_{2x} - C_{3x})\alpha^{-1}b^{-1}q_x$ $F_Y = (\beta bT_{3y} - C_{4y})\beta^{-2}b^{-1}p_z$	**4.** 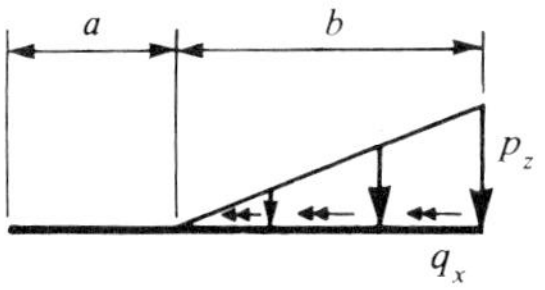	$G_W = B_{3y}\beta^{-2}b^{-1}p_z$ $G_X = B_{3x}\alpha^{-1}b^{-1}q_x$ $G_Y = B_{4y}\beta^{-2}b^{-1}p_y$
$F_w = -(\beta bT_{5y} - C_{6y})\beta^{-2}b^{-1}p_z$ $F_\phi = (\alpha bT_{3x} - C_{4x})\alpha^{-1}b^{-1}q_x$ $F_\psi = -(\beta bT_{4y} - C_{5y})\beta^{-2}b^{-1}p_z$		$G_w = -B_{6y}\beta^{-2}b^{-1}p_y$ $G_\phi = B_{4x}\alpha^{-1}b^{-1}q_x$ $G_\psi = B_{5y}\beta^{-2}b^{-1}p_y$

21.06 TRANSPORT LOAD FUNCTIONS

Notation (21.01–21.03) Signs (21.02) Matrix (21.02)

w_j, ϕ_j, ψ_j = abrupt displacements (m, rad, rad)

g_t = thermal strain (11.03)

A_j, B_j, C_j, T_j = transport coefficients (21.03)

$$\alpha = \sqrt{\frac{\rho\Omega^2}{GI_x}} \qquad \beta = \sqrt[4]{\frac{\mu\Omega^2}{EI_y}}$$

1. Abrupt displacements w_j, ϕ_j

$F_W = -EI_yA_{2y}\beta^3 w_j$

$F_X = -GI_xA_{2x}\alpha^2\phi_j$

$F_Y = EI_yA_{3y}\beta^3 w_j$

$F_w = -EI_yA_{1y}\beta^3 w_j$

$F_\phi = -GI_xA_{1x}\alpha^2\phi_j$

$F_\psi = -EI_yA_{4y}\beta^3 w_j$

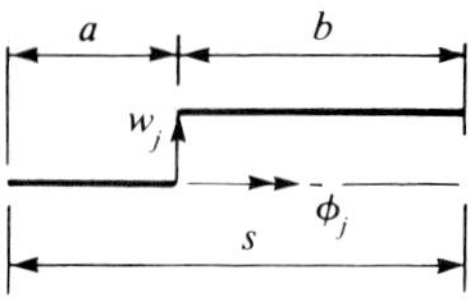

$G_W = -EI_yB_{2y}\beta^3 w_j$

$G_X = -GI_xB_{2x}\alpha^2\phi_j$

$G_Y = -EI_yB_{3y}\beta^3 w_j$

$G_w = EI_yB_{1y}\beta^3 w_j$

$G_\phi = GI_xB_{1x}\alpha^2\phi_j$

$G_\psi = -EI_yB_{4y}\beta^3 w_j$

2. Abrupt displacement ψ_j

$F_W = -EI_yA_{3y}\beta^2\psi_j$

$F_X = 0$

$F_Y = EI_yA_{4y}\beta^2\psi_j$

$F_w = -EI_yA_{2y}\beta^2\psi_j$

$F_\phi = 0$

$F_\psi = -EI_yB_{1y}\beta^2\psi_j$

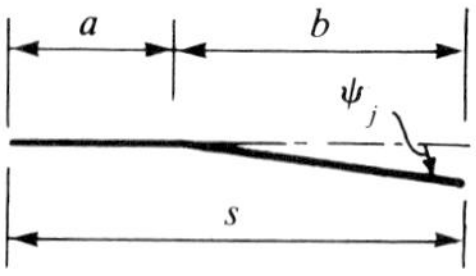

$G_W = EI_yB_{3y}\beta^2\psi_j$

$G_X = 0$

$G_Y = EI_yB_{4y}\beta^2\psi_j$

$G_w = -EI_yB_{2y}\beta\psi_j$

$G_\phi = 0$

$G_\psi = EI_yB_{1y}\beta^2\psi_j$

3. Temperature deformation of segment b

$F_W = -EI_yC_{4y}\beta g_t$

$F_X = 0$

$F_Y = EI_yC_{5y}\beta g_t$

$F_w = -EI_yC_{3y}\beta g_t$

$F_\phi = 0$

$F_\psi = -EI_yC_{2y}\beta g_t$

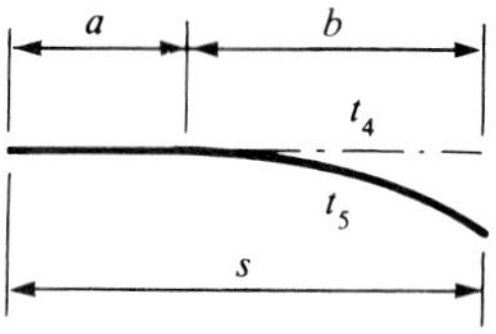

$G_W = EI_yB_{4y}\beta g_t$

$G_X = 0$

$G_Y = EI_yB_{5y}\beta g_t$

$G_w = -EI_yB_{3y}\beta g_t$

$G_\phi = 0$

$G_\psi = EI_yB_{2y}\beta g_t$

4.

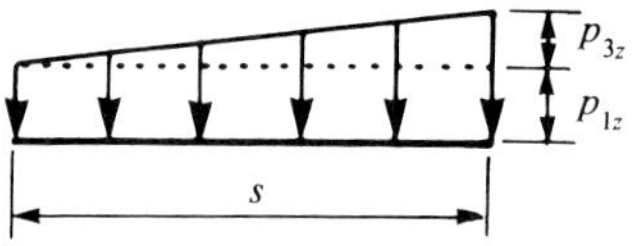

Sec. 21.05 (case 1 + case 3)

5.

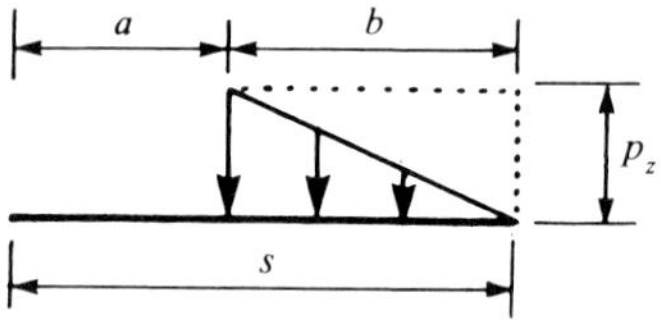

Sec. 21.05 (case 2 − case 4)

21.07 STIFFNESS MATRIX EQUATIONS

$$r = \frac{s^2 GI_x}{EI_y}$$

(1) Equivalents and signs. All forces, moments, and displacements are in the principal axes of the normal section, and their positive directions are shown in (2, 3) below.

(2) Free-body diagram

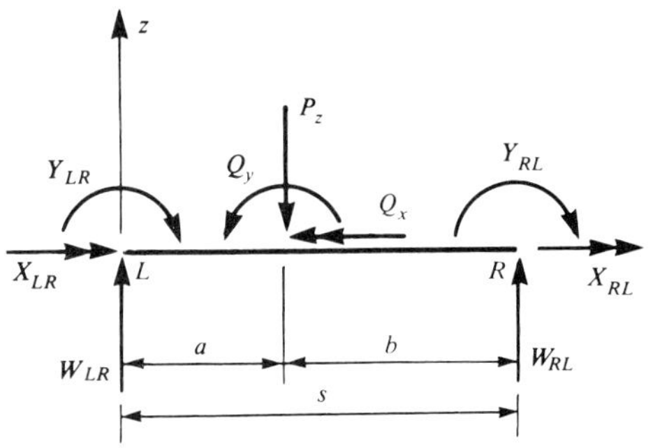

(3) Elastic curve

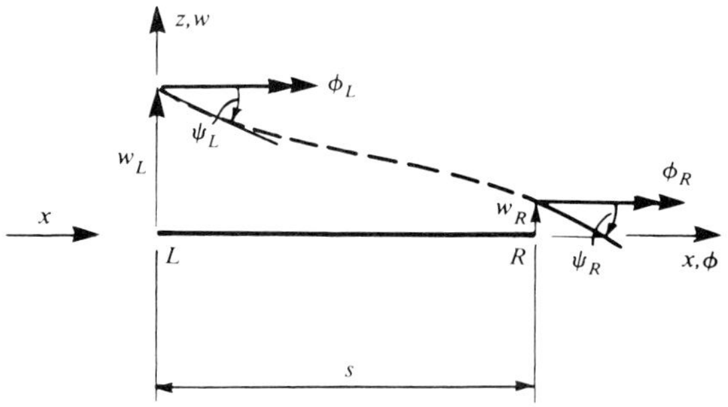

(4) Dimensionless stiffness coefficients

$$D_0 = \frac{1}{1 - \cosh \beta s \cos \beta s}$$

$L_1 = \dfrac{\cos \alpha s}{\sin \alpha s}$	$K_1 = (\alpha s)L_1$	
$L_2 = D_0 (\cosh \beta s \sin \beta s + \sinh \beta s \cos \beta s)$	$K_2 = (\beta s)^3 L_2$	$\alpha = \sqrt{\dfrac{\rho\omega^2}{GI_x}}$
$L_3 = D_0 (\sinh \beta s \sin \beta s)$	$K_3 = (\beta s)^2 L_3$	
$L_4 = D_0 (\cosh \beta s \sin \beta s - \sinh \beta s \cos \beta s)$	$K_4 = (\beta s) L_4$	
$L_5 = \dfrac{1}{\sin \alpha s}$	$K_5 = (\alpha s)L_5$	
$L_6 = D_0 (\sinh \beta s + \sin \beta s)$	$K_6 = (\beta s)^3 L_6$	$\beta = \sqrt[4]{\dfrac{\mu\omega^2}{EI_y}}$
$L_7 = D_0 (\cosh \beta s - \cos \beta s)$	$K_7 = (\beta s)^2 L_7$	
$L_8 = D_0 (\sinh \beta s - \sin \beta s)$	$K_8 = (\beta s) L_8$	

The graphs and table of these coefficients are identical to (20.04–20.05).

(5) Stiffness matrix equations

$$\begin{bmatrix} W_{LR} \\ X_{LR} \\ Y_{LR} \\ \hline W_{RL} \\ X_{RL} \\ Y_{RL} \end{bmatrix} = \frac{EI_y}{s^3} \left[\begin{array}{ccc|ccc} K_2 & 0 & -K_3 s & -K_6 & 0 & -K_7 s \\ 0 & K_1 r & 0 & 0 & -K_5 r & 0 \\ -K_3 s & 0 & K_4 s^2 & K_7 s & 0 & K_8 s^2 \\ \hline -K_6 & 0 & K_7 s & K_2 & 0 & K_3 s \\ 0 & -K_5 r & 0 & 0 & K_1 r & 0 \\ -K_7 s & 0 & K_8 s^2 & K_3 s & 0 & K_4 s^2 \end{array}\right] \begin{bmatrix} w_L \\ \phi_L \\ \psi_L \\ \hline w_R \\ \phi_R \\ \psi_R \end{bmatrix} + \begin{bmatrix} W_{L0} \\ X_{L0} \\ Y_{L0} \\ \hline W_{R0} \\ X_{R0} \\ Y_{R0} \end{bmatrix}$$

where W_{L0}, X_{L0}, Y_{L0} and W_{R0}, X_{R0}, Y_{R0} are the *reactions of fixed-end beams* acted on by loads and other causes shown in (21.08).

(6) General stiffness matrix equations for bars of order three are obtained by matrix composition of (20.09) and (21.07).

21.08 STIFFNESS LOAD FUNCTIONS, FIXED-END BEAM

(1) General form of the stiffness load functions in (21.07) is given by the matrix equations

$$\begin{bmatrix} W_{L0} \\ X_{L0} \\ Y_{L0} \end{bmatrix} = \begin{bmatrix} L_6 & 0 & L_7 \\ 0 & L_5\alpha^{-1} & 0 \\ -L_7\beta^{-1} & 0 & -L_8\beta^{-1} \end{bmatrix} \begin{bmatrix} G_w \\ G_\phi \\ G_\psi \end{bmatrix}$$

$$\begin{bmatrix} W_{R0} \\ X_{R0} \\ Y_{R0} \end{bmatrix} = \begin{bmatrix} L_6 & 0 & L_7 \\ 0 & L_5\alpha^{-1} & 0 \\ L_7\beta^{-1} & 0 & -L_8\beta^{-1} \end{bmatrix} \begin{bmatrix} F_w \\ F_\phi \\ F_\psi \end{bmatrix}$$

where F_w, F_ϕ, F_ψ are the left-end transport load functions and G_w, G_ϕ, G_ψ are their right-end counterparts, given for particular loads and causes in (21.04–21.06), and L_5, L_6, L_7, L_8 are the stiffness coefficients (21.07). The load vectors and other causes are defined by symbols introduced in (21.04–21.06).

(2) Particular cases

Notation (21.01–21.07) A_j, B_j, C_j, T_j = transport coefficients (21.03)	Signs (21.07)	$\alpha = \sqrt{\dfrac{\rho\Omega^2}{GI_x}}$ $\beta = \sqrt[4]{\dfrac{\mu\Omega^2}{EI_y}}$
$W_{L0} = (L_7B_{3y} - L_6B_{4y})P_z$ $X_{L0} = L_5B_{2x}Q_x$ $Y_{L0} = -(L_8B_{3y} - L_7B_{4y})\beta^{-1}P_z$	**1.** 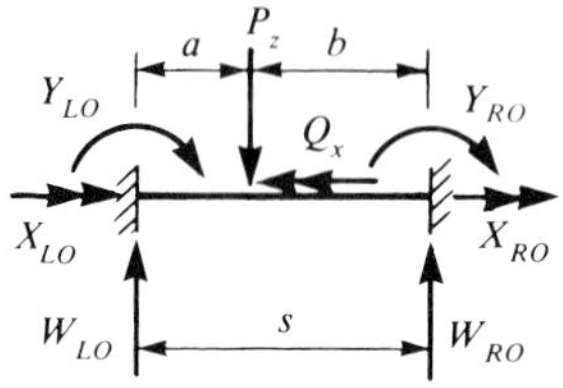	$W_{R0} = (L_7A_{3y} - L_6A_{4y})P_z$ $X_{R0} = L_5A_{2x}Q_x$ $Y_{R0} = (L_8A_{3y} - L_7A_{4y})\beta^{-1}P_z$
$W_{L0} = (L_7B_{2y} - L_6B_{3y})\beta Q_y$ $X_{L0} = 0$ $Y_{L0} = -(L_8B_{2y} - L_7B_{3y})Q_y$	**2.** 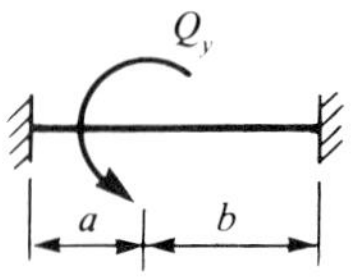	$W_{R0} = -(L_7A_{2y} - L_6A_{3y})\beta Q_y$ $X_{R0} = 0$ $Y_{R0} = -(L_8A_{2y} - L_7A_{3y})Q_y$
$W_{L0} = (L_6 - L_2)\beta^{-1}p_z$ $X_{L0} = \dfrac{1 - \cos\alpha s}{\sin\alpha s}\dfrac{q_x}{\alpha}$ $Y_{L0} = -(L_7 - L_3)\beta^{-2}p_z$	**3.** 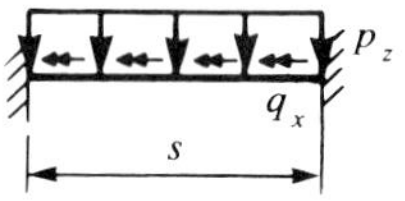	$W_{R0} = (L_6 - L_2)\beta^{-1}p_z$ $X_{R0} = \dfrac{1 - \cos\alpha s}{\sin\alpha s}\dfrac{q_x}{\alpha}$ $Y_{R0} = (L_7 - L_3)\beta^{-2}p_z$

$$W_{L0} = [L_7(A_{3y} + B_{3y}) - L_6(A_{4y} + B_{4y})]P_z = W_{R0}$$

$$X_{L0} = L_5(A_{2x} + B_{2x})Q_x = X_{R0}$$

$$Y_{L0} = -[L_8(A_{3y} + B_{3y}) - L_7(A_{4y} + B_{4y})]\beta^{-1}P_z = -Y_{R0}$$

4.

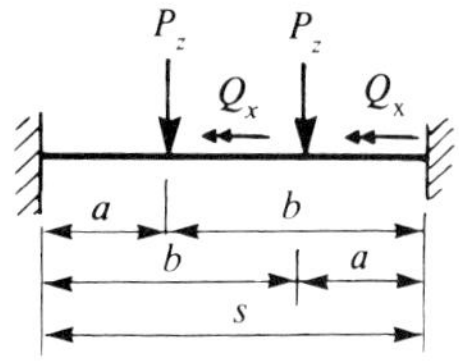

$$W_{L0} = [L_7(A_{2y} + B_{2y}) - L_6(A_{3y} + B_{3y})]\beta Q_y = -W_{R0}$$

$$X_{L0} = 0 = X_{R0}$$

$$Y_{L0} = -[L_8(A_{2y} + B_{2y}) - L_7(A_{3y} + B_{3y})]Q_y = Y_{R0}$$

5.

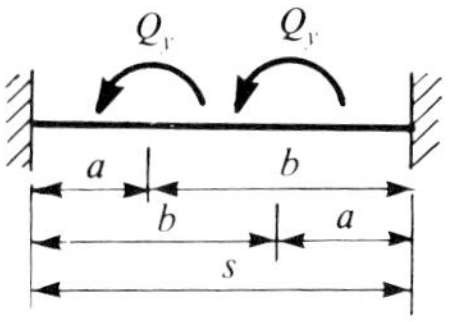

6.

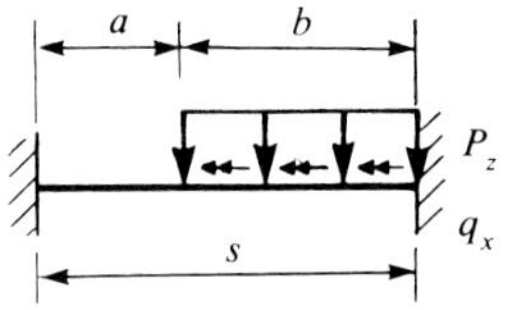

Use general form in terms of (21.05-2)

7.

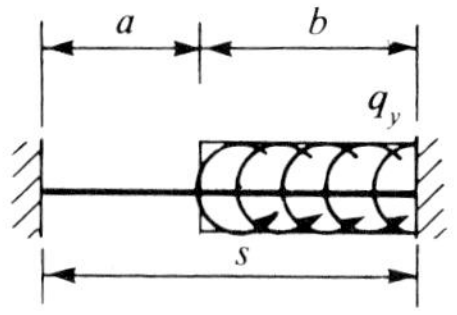

Use general form in terms of (21.04-4)

8.

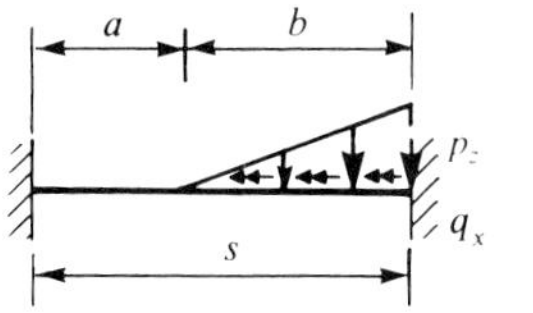

Use general form in terms of (21.05-4)

$$W_{L0} = EI_y(L_6B_{1y} - L_7B_{4y})\beta^3 w_j$$

$$X_{L0} = GI_xL_5B_{1x}\alpha\phi_j$$

$$Y_{L0} = -EI_y(L_7B_{1y} - L_8B_{4y})\beta^2 w_j$$

9. Abrupt displacements w_j, ϕ_j

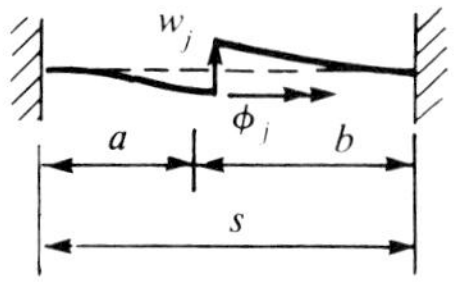

$$W_{R0} = -EI_y(L_6A_{1y} - L_7A_{4y})\beta^3 w_j$$

$$X_{R0} = -GI_xL_5A_{1x}\alpha\phi_j$$

$$Y_{R0} = -EI_y(L_7A_{1y} - L_8A_{4y})\beta^2 w_j$$

$$W_{L0} = -EI_y(L_6B_{2y} - L_7B_{1y})\beta^2\psi_j$$

$$X_{L0} = 0$$

$$Y_{L0} = -EI_y(L_7B_{2y} - L_8B_{1y})\beta\psi_j$$

10. Abrupt displacement

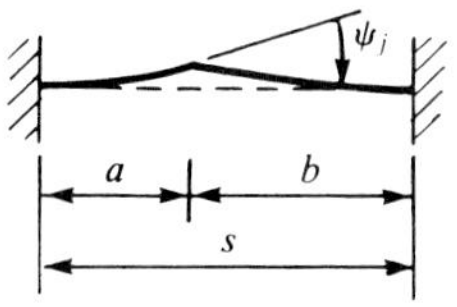

$$W_{R0} = -EI_y(L_6A_{2y} - L_7A_{1y})\beta^2\psi_j$$

$$X_{R0} = 0$$

$$Y_{R0} = -EI_y(L_7A_{2y} - L_8A_{1y})\beta\psi_j$$

$$W_{L0} = -EI_y(L_6B_{3y} - L_7B_{2y})\beta g_t$$

$$X_{L0} = 0$$

$$Y_{L0} = -EI_y(L_7B_{3y} - L_8B_{2y})g_t$$

11. Temperature deformation of segment b

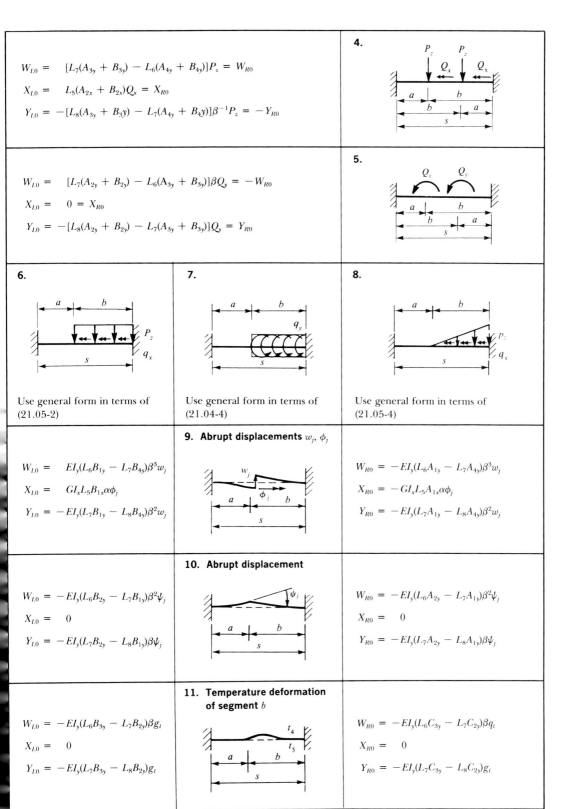

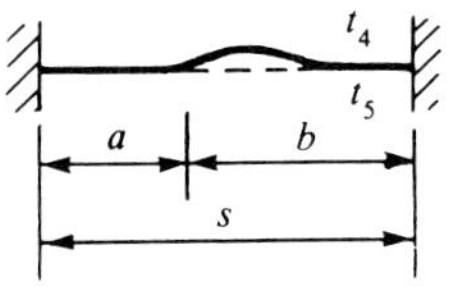

$$W_{R0} = -EI_y(L_6C_{3y} - L_7C_{2y})\beta q_t$$

$$X_{R0} = 0$$

$$Y_{R0} = -EI_y(L_7C_{3y} - L_8C_{2y})g_t$$

21.09 SERIES EXPANSION OF DIMENSIONLESS STIFFNESS COEFFICIENTS

(1) Dimensionless stiffness coefficients introduced in (20.09-4) and in (21.07-4) can be represented by a power series (20.11-2) as

$$\left.\begin{aligned} K_1 &= 1 - C_{11}\bar{\alpha}^2 - C_{12}\bar{\alpha}^4 - C_{13}\bar{\alpha}^6 - \cdots \\ K_2 &= 12 - C_{21}\bar{\beta}^4 - C_{22}\bar{\beta}^8 - C_{23}\bar{\beta}^{12} - \cdots \\ K_3 &= 6 - C_{31}\bar{\beta}^4 - C_{32}\bar{\beta}^8 - C_{33}\bar{\beta}^{12} - \cdots \\ K_4 &= 4 - C_{41}\bar{\beta}^4 - C_{42}\bar{\beta}^8 - C_{43}\bar{\beta}^{12} - \cdots \end{aligned}\right\} \quad \left.\begin{aligned} \bar{\alpha} &= s\sqrt{\frac{\mu\omega^2}{10EA_x}} \\ \bar{\beta} &= s\sqrt[4]{\frac{\mu\omega^2}{100EI_z}} \end{aligned}\right\} \text{ in 20.09-4}$$

$$\left.\begin{aligned} K_5 &= 1 + C_{51}\bar{\alpha}^2 + C_{52}\bar{\alpha}^4 + C_{53}\bar{\alpha}^6 + \cdots \\ K_6 &= 12 + C_{61}\bar{\beta}^4 + C_{62}\bar{\beta}^8 + C_{63}\bar{\beta}^{12} + \cdots \\ K_7 &= 6 + C_{71}\bar{\beta}^4 + C_{72}\bar{\beta}^8 + C_{73}\bar{\beta}^{12} + \cdots \\ K_8 &= 2 + C_{81}\bar{\beta}^4 + C_{82}\bar{\beta}^8 + C_{83}\bar{\beta}^{12} + \cdots \end{aligned}\right\} \quad \left.\begin{aligned} \bar{\alpha} &= s\sqrt{\frac{\rho\omega^2}{10GI_x}} \\ \bar{\beta} &= s\sqrt[4]{\frac{\mu\omega^2}{100EI_y}} \end{aligned}\right\} \text{ in (21.07-4)}$$

The series in $\bar{\alpha}$ apply for $0 \le \bar{\alpha} < 1$, and the series in $\bar{\beta}$ apply for $0 \le \bar{\beta} < 1.5$. The first four constants of each series are tabulated below. For $0 \le \bar{\alpha} < 0.25$ and for $0 \le \bar{\beta} < 0.63$, the first two terms of each series give good approximation for the respective stiffness coefficients and allow the formulation of the system stiffness matrix in terms of ω^2, which is a great simplification (Ref. 49, pp. 117–125).

(2) Table of numerical constants†

j	C_{j1}	C_{j2}	C_{j3}	C_{j4}
1	3.33333 33333	2.22222 22222	2.11640 21164	2.11640 21164
2	37.14285 49301	3.64873 09264	0.69343 58794	0.13771 91694
3	5.23809 49260	0.76616 48851	0.14879 71986	0.02962 38589
4	0.95238 08956	0.16262 39100	0.03196 60103	0.00637 31298
5	1.66666 66667	1.94444 44444	2.05026 45502	2.09986 77250
6	12.85713 91100	3.29571 67760	0.68442 98355	0.13748 36250
7	3.09523 79110	0.72192 99732	0.14565 26563	0.02959 38163
8	0.71428 56717	0.15704 07767	0.03239 86171	0.00650 94690

†Ref. 6, pp. 199–200.

Part III

Static Analysis of Plates

22

Analysis of Plates Notation, Signs, Basic Relations

STATIC STATE

22.01 INTRODUCTION

(1) Systems considered in Chaps. 22–26 are thin plates whose constant thicknesses are less than one-quarter of their least dimension. The mechanical and thermal causes act in the plane and/or normal to the plane of the plate, and the plate is in a state of static equilibrium.

(2) Assumptions used in the derivation of analytical relations are:

(*a*) The material of the plate is isotropic or orthotropic and remains continuous during the loading and unloading.

(*b*) The material follows Hooke's law, all deformations are small, and the maximum deflection is less than one-half the thickness.

(*c*) The midplane of the plate is unstrained in bending, and the normal section remains normal to the midplane during the deformation.

(*d*) The normal stresses to the midplane are small compared with other stresses and as such are neglected.

(*e*) The material constants are known from experiments and are independent of time.

(*f*) The effect of shearing deformation on the formation of the elastic surface is too small to be considered.

(3) Symbols and signs are defined where they appear first and are all summarized in Appendices A.1, A.5.

22.02 ISOTROPIC PLATE IN CARTESIAN COORDINATES, BASIC RELATIONS

(1) Notation

x, y = coordinates (m)
z = normal distance from midsurface (m)
u, v, w = linear displacements in x, y, z axes (m)
f_{xx}, f_{yy}, f_{zz} = normal stresses (Pa)
$\varepsilon_{xx}, \varepsilon_{yy}, \varepsilon_{zz}$ = normal strains
E = modulus of elasticity (Pa)
ν = Poisson's ratio
ϕ, ψ = slopes (rad)
f_{xy}, f_{yz}, f_{zx} = shearing stresses (Pa)
$\varepsilon_{xy}, \varepsilon_{yz}, \varepsilon_{zx}$ = shearing strains
G = modulus of rigidity (Pa)
h = thickness of plate (m)

All linear displacements are positive if acting in the positive direction of the respective coordinate axis, and all slopes are positive if acting in the right-hand direction about the respective coordinate axis. The sign convention of forces, moments, and stresses is given in (22.03). The partial derivatives of slopes ϕ and ψ shown in (4, 5) relate to strains and stresses as indicated in (6, 7).

(2) Isotropic plate of arbitrary shape and end conditions under transverse load of intensity p_z (N/m^2) is shown in (3). The xy plane bisects the thickness h of the plate and defines the midsurface. Under loads, this midsurface deforms into an *elastic surface* defined analytically as

$$w = f(x, y)$$

If this function is known, the slopes, the displacements, the strains, and the stresses can be expressed as functions of w.

(3) Plate element

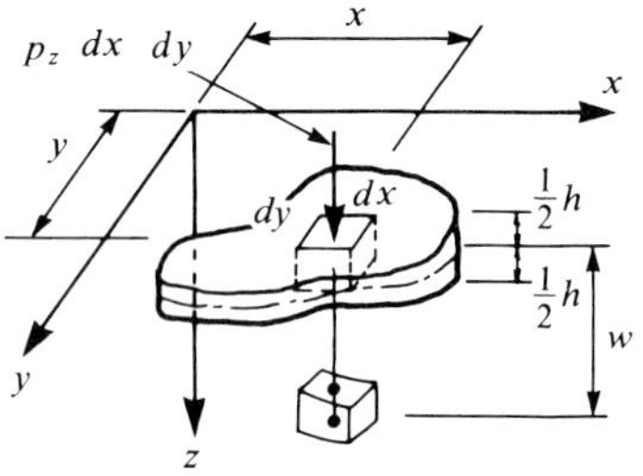

(4) Slope ϕ

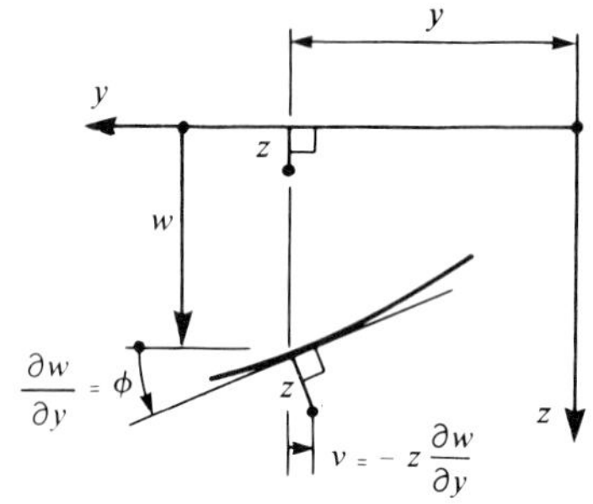

(5) Slope ψ

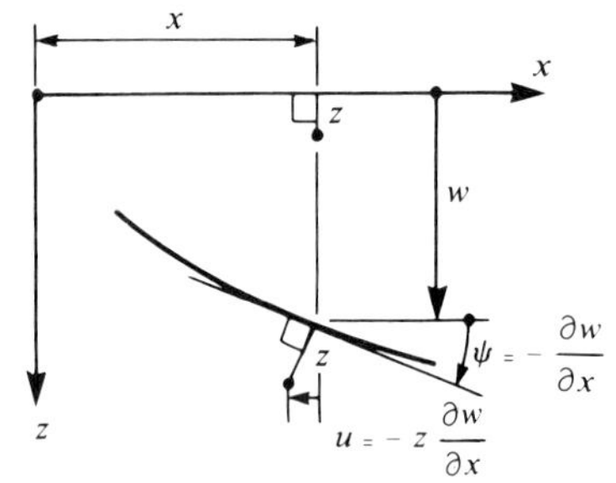

(6) Strains

$$\varepsilon_{xx} = \frac{\partial u}{\partial x} = -z\frac{\partial^2 w}{\partial x^2}$$

$$\varepsilon_{yy} = \frac{\partial v}{\partial y} = -z\frac{\partial^2 w}{\partial y^2}$$

$$\varepsilon_{xy} = \varepsilon_{yx} = \frac{\partial u}{\partial y} + \frac{\partial v}{\partial x} = -2z\frac{\partial^2 w}{\partial x\,\partial y}$$

(7) Stresses

$$f_{xx} = \frac{E}{1-\nu^2}(\varepsilon_{xx} + \nu\varepsilon_{yy}) = -\frac{Ez}{1-\nu^2}\left(\frac{\partial^2 w}{\partial x^2} + \nu\frac{\partial^2 w}{\partial y^2}\right)$$

$$f_{yy} = \frac{E}{1-\nu^2}(\varepsilon_{yy} + \nu\varepsilon_{xx}) = -\frac{Ez}{1-\nu^2}\left(\frac{\partial^2 w}{\partial y^2} + \nu\frac{\partial^2 w}{\partial x^2}\right)$$

$$f_{xy} = G\varepsilon_{xy} = -2Gz\frac{\partial^2 w}{\partial x\,\partial y} = -\frac{Ez}{1+\nu}\frac{\partial^2 w}{\partial x\,\partial y}$$

22.03 ISOTROPIC PLATE IN CARTESIAN COORDINATES, STRESS RESULTANTS

(1) Notation and signs (22.02)

$N_x, N_y =$ intensities of normal forces (N/m)
$W_{xz}, W_{yz} =$ intensities of shearing forces (N/m)
$M_x, M_y =$ intensities of bending moments (N·m/m)
$M_{xy}, M_{yx} =$ intensities of twisting moments (N·m/m)
$R_{xz}, R_{yz} =$ intensities of edge reactions (N/m)

$$\alpha = \tfrac{1}{2}(1 + \nu)$$
$$\beta = \tfrac{1}{2}(1 - \nu)$$
$$D = \frac{Eh^3}{12(1 - \nu^2)}$$

Forces and moments of the element $dx\,dy$ are positive as shown in (2, 3) below.

(2) Variation of forces

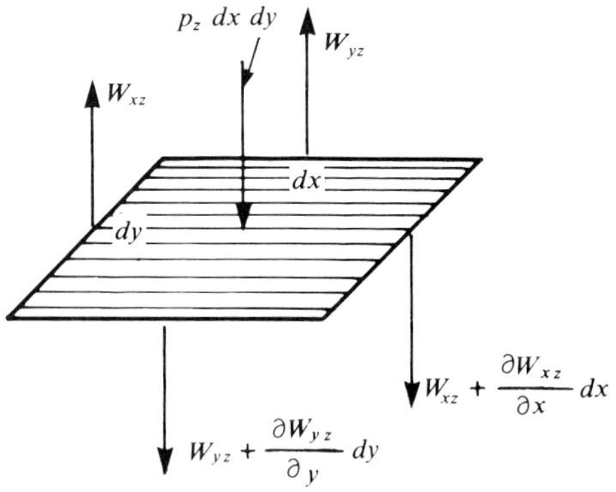

(3) Variation of moments

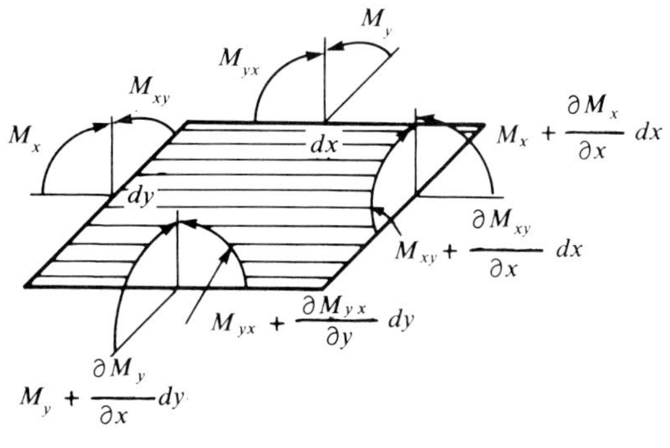

(4) Stress resultants

$$dA_x = dy\,dz \qquad dA_y = dx\,dz$$

$$N_x = \int_{-h/2}^{+h/2} f_{xx}\,dA_x = 0 \qquad N_y = \int_{-h/2}^{+h/2} f_{yy}\,dA_y = 0$$

$$M_x = \int_{-h/2}^{+h/2} f_{xx}z\,dA_x = -D\left(\frac{\partial^2 w}{\partial x^2} + \nu\frac{\partial^2 w}{\partial y^2}\right)$$

$$M_y = \int_{-h/2}^{+h/2} f_{yy}z\,dA_y = -D\left(\frac{\partial^2 w}{\partial y^2} + \nu\frac{\partial^2 w}{\partial x^2}\right)$$

$$M_{xy} = \int_{-h/2}^{+h/2} f_{xy}z\,dA_x = -D(1 - \nu)\frac{\partial^2 w}{\partial x\,\partial y} = M_{yx}$$

$$W_{xz} = \frac{\partial M_x}{\partial x} + \frac{\partial M_{yx}}{\partial y} = -D\left(\frac{\partial^3 w}{\partial x^3} + \frac{\partial^3 w}{\partial x\,\partial y^2}\right)$$

$$W_{yz} = \frac{\partial M_y}{\partial y} + \frac{\partial M_{xy}}{\partial x} = -D\left(\frac{\partial^3 w}{\partial y^3} + \frac{\partial^3 w}{\partial y\,\partial x^2}\right)$$

$$R_{xz} = W_{xz} + \frac{\partial M_{xy}}{\partial y} = -D\left[\frac{\partial^3 w}{\partial x^3} + (2 - \nu)\frac{\partial^3 w}{\partial x\,\partial^2 y}\right]$$

$$R_{yz} = W_{yz} + \frac{\partial M_{yx}}{\partial x} = -D\left[\frac{\partial^3 w}{\partial y^3} + (2 - \nu)\frac{\partial^3 w}{\partial^2 x\,\partial y}\right]$$

(5) Stress diagrams

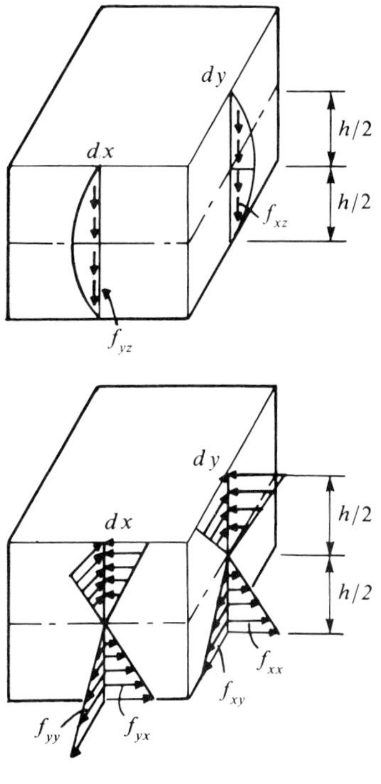

22.04 ISOTROPIC PLATE IN CARTESIAN COORDINATES, GOVERNING EQUATIONS

(1) Notation (22.02–22.03)

M = intensity of moment sum (N·m/m)
M_b = intensity of bending moment (N·m/m)
M_t = intensity of torsional moment (N·m/m)
$M_{1,2}$ = extreme values of M_b (N·m/m)
$M_{3,4}$ = extreme values of M_t (N·m/m)
θ = position angle (rad)
$\nabla^2 = \dfrac{\partial}{\partial x^2} + \dfrac{\partial}{\partial y^2}$ = del second (operator)

(2) Free-body diagram

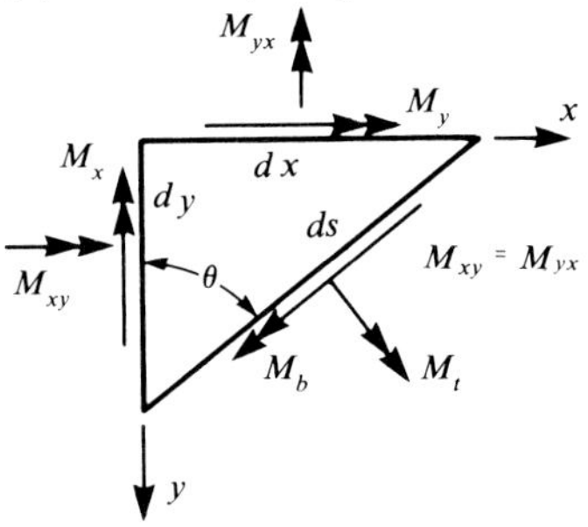

(3) Extreme stresses

$$f_{xx} = \pm\frac{6M_x}{h^2} \quad \text{at } z = \pm\frac{h}{2} \qquad f_{yy} = \pm\frac{6M_y}{h^2} \quad \text{at } z = \pm\frac{h}{2}$$

$$f_{xy} = \pm\frac{6M_{xy}}{h^2} \quad \text{at } z = \pm\frac{h}{2} \qquad f_{yx} = \pm\frac{6M_{yx}}{h^2} \quad \text{at } z = \pm\frac{h}{2}$$

$$f_{xz} = \frac{3V_{xz}}{2h} \quad \text{at } z = 0 \qquad f_{yz} = \frac{3V_{yz}}{2h} \quad \text{at } z = 0$$

(4) Variation of moments

$$M_b = M_x \cos^2\theta + M_y \sin^2\theta + 2M_{xy}\sin\theta\cos\theta \qquad M = \frac{M_x + M_y}{1+\nu}$$

$$M_t = -(M_y - M_x)\sin\theta\cos\theta - M_{xy}(\cos^2\theta - \sin^2\theta)$$

(5) Extreme values $[\Delta = \sqrt{(M_x - M_y)^2 + 4M_{xy}}]$

$$M_{1,2} = \tfrac{1}{2}(M_x + M_y \pm \Delta) \qquad \text{at } \theta_1 = \tfrac{1}{2}\tan^{-1}\frac{2M_{xy}}{\Delta} \qquad \theta_2 = \theta_1 + \frac{\pi}{2}$$

$$M_{3,4} = \pm\tfrac{1}{2}\Delta \qquad \text{at } \theta_3 = -\tfrac{1}{2}\tan^{-1}\frac{\Delta}{2M_{xy}} \qquad \theta_4 = \theta_3 + \frac{\pi}{2}$$

The angle θ_1, θ_2 must be inserted in M_b in (3) to find the correct position of M_1, M_2, since M_1 is not necessarily located on the plane of θ_1. Similarly, θ_3, θ_4 must be inserted in M_t for the same purpose.

(6) Equilibrium equations

$$\frac{\partial W_{xz}}{\partial x} + \frac{\partial W_{yz}}{\partial y} + p_z = 0 \qquad \frac{\partial M_x}{\partial x} + \frac{\partial M_{yx}}{\partial y} - W_{xz} = 0 \qquad \frac{\partial M_y}{\partial y} + \frac{\partial M_{xy}}{\partial x} - W_{yz} = 0$$

(7) Differential equation of elastic surface

$\dfrac{\partial^4 w}{\partial x^4} + 2\dfrac{\partial^4 w}{\partial x^2\,\partial y^2} + \dfrac{\partial^4 w}{\partial y^4} = \dfrac{p_z}{D}$	$\nabla^4 w = \dfrac{p_z}{D}$

(8) Resolution of differential equation

$\dfrac{\partial^2 M}{\partial x^2} + \dfrac{\partial^2 M}{\partial y^2} = -p_z$	$\nabla^2 M = -p_z$
$\dfrac{\partial^2 w}{\partial x^2} + \dfrac{\partial^2 w}{\partial y^2} = -\dfrac{M}{D}$	$\nabla^2 w = -\dfrac{M}{D}$

22.05 RECTANGULAR ISOTROPIC PLATE IN CARTESIAN COORDINATES, BOUNDARY CONDITIONS

(1) General analysis of a rectangular isotropic plate of constant thickness acted on by transverse loads reduces to the solution of the governing differential equation (22.04-6). The result of this solution is the equation of elastic surface, which must satisfy the differential equation and the conditions at the plate boundary. Four typical boundary conditions are considered below.

(2) Fixed edge at $x = s$

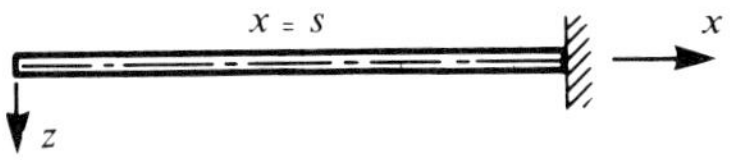

$$(w)_s = 0 \qquad \left(\frac{\partial w}{\partial x}\right)_s = 0 \qquad \left(\frac{\partial w}{\partial y}\right)_s = 0$$

$$\left(\frac{\partial^2 w}{\partial y^2}\right)_s = 0 \qquad \left(\frac{\partial^2 w}{\partial x\,\partial y}\right)_s = 0$$

$$(M_x)_s = -D\frac{\partial^2 w}{\partial x^2} \qquad (M_y)_s = -\nu D\frac{\partial^2 w}{\partial x^2}$$

$$(R_{xz})_s = -D\frac{\partial^3 w}{\partial x^3} \qquad (M_{xy})_s = 0$$

(3) Free edge at $x = s$

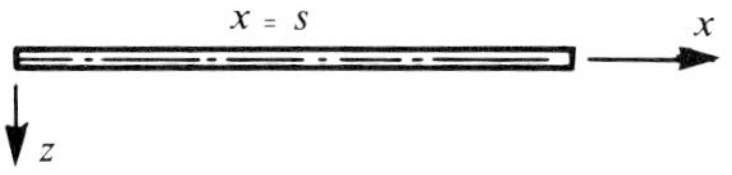

$$\left[\frac{\partial^2 w}{\partial x^2} + \nu\frac{\partial^2 w}{\partial y^2}\right]_s = 0$$

$$\left[\frac{\partial^3 w}{\partial x^3} + (2-\nu)\frac{\partial^3 w}{\partial x\,\partial y^2}\right]_s = 0$$

$$(M_y)_s = -D(1-\nu^2)\frac{\partial^2 w}{\partial y^2}$$

$$(M_{xy})_s = (M_x)_s = (R_{xz})_s = 0$$

(4) Simply supported edge at $x = s$

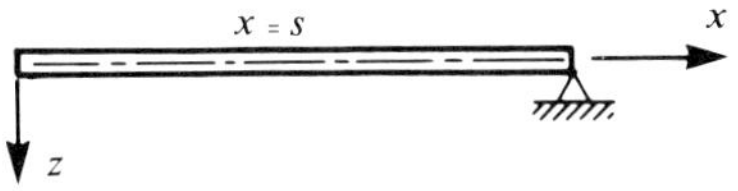

$$(w)_s = 0 \qquad \left(\frac{\partial w}{\partial y}\right)_s = 0 \qquad (M_x)_x = 0$$

$$\left(\frac{\partial^2 w}{\partial x^2}\right)_s = 0 \qquad \left(\frac{\partial^2 w}{\partial y^2}\right)_s = 0 \qquad (M_y)_s = 0$$

$$(M_{xy})_s = -D(1-\nu)\frac{\partial^2 w}{\partial x\,\partial y}$$

$$(R_{xz})_s = -D\left[\frac{\partial^3 w}{\partial x^3} + (2-\nu)\frac{\partial^3 w}{\partial x\,\partial y^2}\right]$$

(5) Guided edge at $x = s$

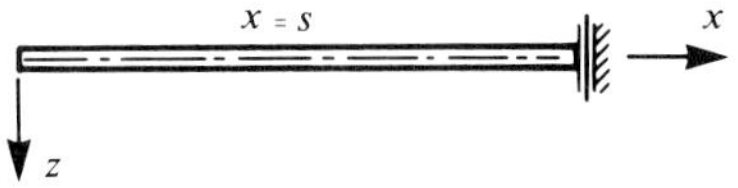

$$\left(\frac{\partial w}{\partial x}\right)_s = 0 \qquad \left(\frac{\partial^2 w}{\partial x\,\partial y}\right)_s = 0 \qquad (M_{xy})_s = 0$$

$$(M_x)_s = -D\left(\frac{\partial^2 w}{\partial x^2} + \nu\frac{\partial^2 w}{\partial y^2}\right) \qquad (R_{xz})_s = 0$$

$$(M_y)_s = -D\left(\frac{\partial^2 w}{\partial y^2} + \nu\frac{\partial^2 w}{\partial x^2}\right)$$

$$\left[\frac{\partial^3 w}{\partial x^3} + (2-\nu)\frac{\partial^3 w}{\partial x\,\partial y^2}\right]_s = 0$$

(6) Corner reaction at $x = s$, $y = f$ of two adjacent simply supported edges is

$$R_{sf} = 2M_{xy} = -2D(1-\nu)\frac{\partial^2 w}{\partial x\,\partial y}$$

where s = length (m) and f = width (m) of the plate. When two adjacent edges are fixed or free, the corner reaction is zero.

22.06 ISOTROPIC PLATE IN CYLINDRICAL COORDINATES, BASIC RELATIONS

(1) Notation

r = radial coordinate (m)
u, v, w = linear displacements in r, θ, z axes (m)
z = vertical distance from the elastic surface (m)
$f_{rr}, f_{\theta\theta}, f_{zz}$ = normal stresses (Pa)
$\varepsilon_{rr}, \varepsilon_{\theta\theta}, \varepsilon_{zz}$ = normal strains
E = modulus of elasticity (Pa)
ν = Poisson's ratio
θ = angular coordinates (rad)
$f_{r\theta}, f_{\theta z}, f_{zr}$ = shearing stresses (Pa)
$\varepsilon_{r\theta}, \varepsilon_{\theta z}, \varepsilon_{zr}$ = shearing strains
G = modulus of rigidity (Pa)
h = thickness of plate (m)

All vectors are positive as shown unless prefixed by the minus sign.

(2) Isotropic circular plate of arbitrary edge conditions under transverse load of intensity p_z (N/m^2) is shown below. The midplane bisects the thickness of the plate and under the action of loads deforms into the elastic surface defined as

$$w = f(r, \theta)$$

The partial derivatives of w relate to strains and stresses as shown in (6, 7) below.

(3) Strains

$$\varepsilon_{rr} = \frac{\partial u}{\partial r} = -z\frac{\partial^2 w}{\partial r^2} \qquad \varepsilon_{\theta\theta} = \frac{u}{r} + \frac{1}{r}\frac{\partial v}{\partial \theta} = -\frac{z}{r}\left(\frac{\partial w}{\partial r} + \frac{\partial^2 w}{r\,\partial \theta^2}\right)$$

$$\varepsilon_{r\theta} = \varepsilon_{\theta r} = -\frac{v}{r} + \frac{\partial v}{\partial r} + \frac{1}{r}\frac{\partial u}{\partial \theta} = -\frac{2z}{r}\left(\frac{\partial^2 w}{\partial r\,\partial \theta} - \frac{\partial w}{\partial r}\right)$$

(4) Plate element

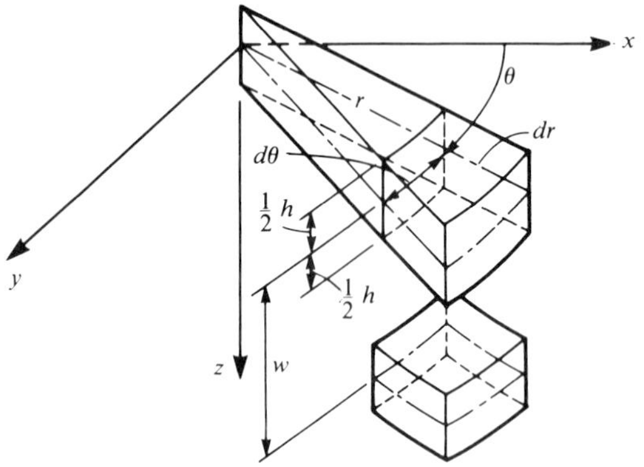

(5) Slope ψ

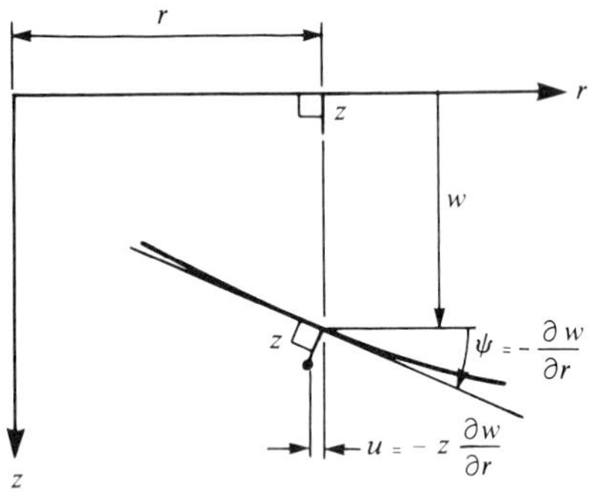

(6) Stresses

$$f_{rr} = \frac{E}{1-v^2}(\varepsilon_{rr} + \nu\varepsilon_{\theta\theta}) = -\frac{Ez}{1-v^2}\left[\frac{\partial^2 w}{\partial r^2} + \frac{\nu}{r}\left(\frac{\partial w}{\partial r} + \frac{\partial^2 w}{r\,\partial^2\theta}\right)\right]$$

$$f_{\theta\theta} = \frac{E}{1-v^2}(\varepsilon_{\theta\theta} + \nu\varepsilon_{rr}) = -\frac{Ez}{1-v^2}\left[\frac{1}{r}\left(\frac{\partial w}{\partial r} + \frac{\partial^2 w}{r\,\partial^2\theta}\right) + \nu\frac{\partial^2 w}{\partial r^2}\right]$$

$$f_{r\theta} = f_{\theta r} = G\varepsilon_{r\theta} = G\varepsilon_{\theta r} = -\frac{Ez}{(1+v)r}\left[\frac{\partial^2 w}{\partial r\,\partial \theta} - \frac{\partial w}{r\,\partial \theta}\right]$$

22.07 ISOTROPIC PLATE IN CYLINDRICAL COORDINATES, STRESS RESULTANTS

(1) Notation and signs (22.06)

N_r, N_θ = intensities of normal force (N/m)

$W_{r\theta}, W_{\theta r}$ = intensities of shearing force (N/m)

M_r, M_θ = intensities of bending moment (N·m/m)

$M_{r\theta}, M_{\theta r}$ = intensities of twisting moment (N·m/m)

$$\nabla^2 = \frac{\partial^2}{\partial r^2} + \frac{\partial}{r\,\partial r} + \frac{1}{r^2}\frac{\partial^2}{\partial \theta^2} = \text{del second (operator)}$$

$$\alpha = \tfrac{1}{2}(1+\nu) \qquad \beta = \tfrac{1}{2}(1-\nu) \qquad D = \frac{Eh^3}{12(1-\nu^2)}$$

Forces and moments on the element $r\,dr\,d\theta$ are positive as shown in (2, 3) below.

(2) Variation of forces

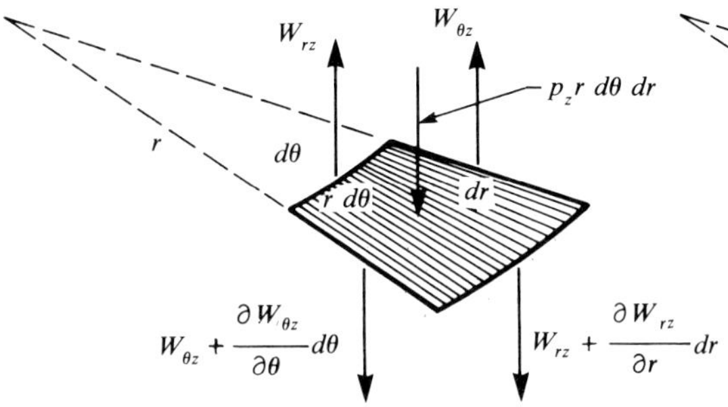

(3) Variation of moments

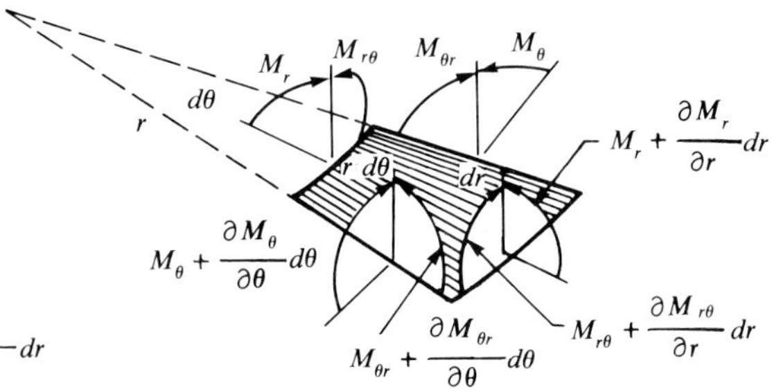

(4) Stress resultants

$$dA_r = hr\,d\theta \qquad dA_\theta = h\,dr$$

$$N_r = \int_{-h/2}^{h/2} f_{rr}\,dA_r = 0 \qquad N_\theta = \int_{-h/2}^{h/2} f_{\theta\theta}\,dA_\theta = 0$$

$$M_r = \int_{-h/2}^{h/2} f_{rr} z\,dA_r = -D\left[\frac{\partial^2 w}{\partial r^2} + \nu\left(\frac{\partial w}{r\,\partial r} + \frac{\partial^2 w}{r^2\,\partial^2\theta}\right)\right]$$

$$M_\theta = \int_{-h/2}^{h/2} f_{\theta\theta}\, z\,dA_\theta = -D\left[\frac{\partial w}{r\,\partial r} + \frac{\partial^2 w}{r^2\,\partial^2\theta} + \nu\frac{\partial^2 w}{\partial r^2}\right]$$

$$M_{r\theta} = \int_{-h/2}^{h/2} f_{r\theta} z\,dA_r = -D(1-\nu)\left(\frac{\partial^2 w}{r\,\partial r\,\partial\theta} - \frac{1}{r^2}\frac{\partial w}{\partial\theta}\right) = M_{\theta r}$$

$$W_{rz} = -D\frac{\partial}{\partial r}(\nabla^2 w) \qquad W_{\theta z} = -D\frac{\partial}{r\,\partial\theta}(\nabla^2 w)$$

(5) Extreme stresses

$$f_{rr} = \pm\frac{6M_r}{h^2} \quad \text{at } z = \pm\tfrac{1}{2}h \qquad f_{\theta\theta} = \pm\frac{6M_\theta}{h^2} \quad \text{at } z = \pm\tfrac{1}{2}h$$

$$f_{r\theta} = \pm\frac{6M_{r\theta}}{h^2} \quad \text{at } z = \pm\tfrac{1}{2}h \qquad f_{\theta r} = \pm\frac{6M_{\theta r}}{h^2} \quad \text{at } z = \pm\tfrac{1}{2}h$$

$$f_{rz} = \frac{3V_{rz}}{2h} \quad \text{at } z = 0 \qquad f_{\theta z} = \frac{3V_{\theta z}}{2h} \quad \text{at } z = 0$$

22.08 ISOTROPIC PLATE IN CYLINDRICAL COORDINATES, GOVERNING EQUATIONS AND BOUNDARY CONDITIONS

(1) Equilibrium equations

$$\frac{W_{rz}}{r} + \frac{\partial W_{rz}}{\partial r} + \frac{W_{\theta z}}{r\,\partial \theta} + p_z = 0 \qquad \frac{\partial M}{\partial r} - W_{rz} = 0 \qquad \frac{\partial M}{r\,\partial \theta} - W_{\theta z} = 0$$

where M is given in (3) below.

(2) Differential equation of elastic surface

$\left(\frac{\partial^2}{\partial r^2} + \frac{\partial}{r\,\partial r} + \frac{\partial^2}{r^2\,\partial \theta^2}\right)\left(\frac{\partial^2 w}{\partial r^2} + \frac{\partial w}{r\,\partial r} + \frac{\partial^2 w}{r^2\,\partial \theta^2}\right) = \frac{p_z}{D}$	$\nabla^2\nabla^2 w = \frac{p_z}{D}$

(3) Moment sum

$M = \frac{M_r + M_\theta}{1+\nu} = -D\left(\frac{\partial^2 w}{\partial r^2} + \frac{\partial w}{r\,\partial r} + \frac{\partial^2 w}{r^2\,\partial \theta^2}\right)$	$M = -D\nabla^2 w$

(4) Resolution of differential equation

$\frac{\partial^2 M}{\partial r^2} + \frac{\partial M}{r\,\partial r} + \frac{\partial^2 M}{r^2\,\partial \theta^2} = -p_z$	$\nabla^2 M = -p_z$
$\frac{\partial^2 w}{\partial r^2} + \frac{\partial w}{r\,\partial r} + \frac{\partial^2 w}{r^2\,\partial \theta^2} = -\frac{M}{D}$	$\nabla^2 w = -\frac{M}{D}$

(5) General analysis of an isotropic circular plate acted on by transverse loads reduces to the solution of the differential equation (2). The result of this solution as in (22.05) is the equation of the elastic surface, which must satisfy the differential equation and the conditions at the plate boundary. Four typical boundary conditions are considered below.

(6) Boundary conditions

Edge	Simply supported	Fixed	Guided	Free
$r = r$ $0 \le \theta \le 2\pi$	$w_r = 0$ $M_r = 0$	$w_r = 0$ $\psi_r = 0$	$\psi_r = 0$ $R_{rz} = 0$	$M_r = 0$ $R_{rz} = 0$

(7) Edge reactions required for the conditions in (6) and for other conditions are, in symbolic form,

$$R_{rz} = W_{rz} + \frac{\partial M_{r\theta}}{r\,\partial \theta} = -D\left[\frac{\partial}{\partial r}(\nabla^2 w) + \frac{2\beta}{r}\frac{\partial}{\partial \theta}\left(\frac{\partial^2 w}{r\,\partial r\,\partial \theta} - \frac{\partial w}{r^2\,\partial \theta}\right)\right]$$

$$R_{\theta z} = W_{\theta z} + \frac{\partial M_{\theta r}}{\partial r} = -D\left[\frac{\partial}{r\,\partial r}(\nabla^2 w) + 2\beta\frac{\partial}{\partial r}\left(\frac{\partial^2 w}{r\,\partial r\,\partial \theta} - \frac{\partial w}{r^2\,\partial \theta}\right)\right]$$

The explicit form of these expressions is given in (22.11-3).

22.09 AXISYMMETRICAL CASE OF ISOTROPIC CIRCULAR PLATE

(1) Condition of axisymmetry requires that the plate, edge conditions, and causes of deformation in (22.06–22.08) be axisymmetrical. Then

$$\varepsilon_{rr} = -z\frac{\partial^2 w}{\partial r^2} \qquad \varepsilon_{\theta\theta} = -\frac{z}{r}\frac{\partial w}{\partial r} \qquad \varepsilon_{r\theta} = \varepsilon_{\theta r} = 0 \qquad \psi = -\frac{\partial w}{\partial r}$$

$$f_{rr} = -\frac{4Ez}{\alpha\beta}\left(\frac{\partial^2 w}{\partial r^2} + \frac{\nu}{r}\frac{\partial w}{\partial r}\right) \qquad f_{\theta\theta} = -\frac{4Ez}{\alpha\beta}\left(\frac{\partial w}{r\,\partial r} + \nu\frac{\partial^2 w}{\partial r^2}\right) \qquad f_{r\theta} = f_{\theta r} = 0$$

(2) Stress resultants

$$M_r = -D\left(\frac{\partial^2 w}{\partial r^2} + \nu\frac{\partial w}{r\,\partial w}\right) \qquad M_\theta = -D\left(\frac{\partial w}{r\,\partial r} + \nu\frac{\partial^2 w}{\partial r^2}\right) \qquad M_{\theta r} = M_{r\theta} = 0$$

$$W_{rz} = -D\left(\frac{\partial^3 w}{\partial r^3} + \frac{\partial^2 w}{r\,\partial^2 r} - \frac{\partial w}{r^2\,\partial r}\right) \qquad W_{\theta z} = 0 \qquad R_{rz} = W_{rz} \qquad R_{\theta z} = 0$$

(3) Differential equation of elastic surface

$\frac{\partial^4 w}{\partial r^4} + \frac{2\partial^3 w}{r\,\partial r^3} - \frac{\partial^2 w}{r^2\,\partial r^2} + \frac{\partial w}{r^3\,\partial r} = \frac{p_z}{D}$	$\nabla^2\nabla^2 w = \frac{p_z}{b}$

where $\nabla^2 = \frac{\partial^2}{\partial r^2} + \frac{\partial}{r\,\partial r}$

(4) Moment sum

$$M = \frac{M_r + M_\theta}{1+\nu} = -D\left(\frac{\partial^2 w}{\partial r^2} + \frac{\partial w}{r\,\partial r}\right) = -D\nabla^2 w$$

(5) Resolution of differential equation

$\frac{\partial^2 M}{\partial r^2} + \frac{\partial M}{r\,\partial r} = -p_z$	$\nabla^2 M = -p_z$	$M = -\int\frac{dr}{r}\int p_z r\,dr$
$\frac{\partial^2 w}{\partial r^2} + \frac{\partial w}{r\,\partial r} = -\frac{M}{D}$	$\nabla^2 w = -\frac{M}{D}$	$w = -\int\frac{dr}{r}\int\frac{M}{D}r\,dr$

(6) Solution of differential equation in (3) above is

for a solid plate (23.03) $\qquad w = w_C - \frac{M_0}{4\alpha}\frac{r^2}{D} + G_w(r)$

for an annular plate (23.10) $\qquad w = w_A - \psi_A r T_1(r) - M_A\frac{r^2}{D}T_4(r) - W_A\frac{r^3}{D}T_7(r) + G_w(r)$

where the subscripts 0 and A designate the center and the inner edge, respectively, $G_w(r)$ is the load function (23.06–23.07), and $T_1(r)$, $T_2(r)$, $T_3(r)$ are the transport coefficients (23.02).

(7) Boundary conditions are defined by relations stated in (22.08).

22.10 ORTHOTROPIC PLATE IN CARTESIAN COORDINATES

(1) Orthotropic plate of arbitrary shape and edge conditions under transverse load of intensity p_z (N/m²) is considered in this section. By definition, the material properties of the plate differ in two mutually perpendicular directions, and their moduli E_{xx}, E_{yy}, and G are independent of one another (Ref. 33, pp. 273–282). The basic relations and the governing differential equation of this plate are introduced below in terms of symbols defined in (22.03–22.04).

(2) Stress-strain equations

$$f_{xx} = \frac{E_{xx}}{1 - \nu_x \nu_y}(\varepsilon_{xx} + \nu_y \varepsilon_{yy}) = -\frac{E_{xx} z}{1 - \nu_x \nu_y}\left(\frac{\partial^2 w}{\partial x^2} + \nu_y \frac{\partial^2 w}{\partial y^2}\right)$$

$$f_{yy} = \frac{E_{yy}}{1 - \nu_x \nu_y}(\varepsilon_{yy} + \nu_x \varepsilon_{xx}) = -\frac{E_{yy} z}{1 - \nu_x \nu_y}\left(\frac{\partial^2 w}{\partial y^2} + \nu_x \frac{\partial^2 w}{\partial x^2}\right)$$

$$f_{xy} = f_{yx} = G\varepsilon_{xy} = -2zG\frac{\partial^2 w}{\partial x\, \partial y}$$

where ν_x, ν_y = Poisson's ratios in x, y directions.

(3) Stress resultants

$$N_x = 0 \qquad N_y = 0 \qquad M_{xy} = M_{yx} = -2D_{xy}\frac{\partial^2 w}{\partial x\, \partial y}$$

$$M_x = -D_{xx}\left(\frac{\partial^2 w}{\partial x^2} + \nu_y \frac{\partial^2 w}{\partial y^2}\right) \qquad M_y = -D_{yy}\left(\frac{\partial^2 w}{\partial y^2} + \nu_x \frac{\partial^2 w}{\partial x^2}\right)$$

$$W_{xz} = -D_{xx}\left[\frac{\partial^3 w}{\partial x^3} + \left(\nu_y + 2\frac{D_{xy}}{D_{xx}}\right)\frac{\partial^3 w}{\partial x\, \partial y^2}\right]$$

$$W_{yz} = -D_{yy}\left[\frac{\partial^3 w}{\partial y^3} + \left(\nu_x + 2\frac{D_{xy}}{D_{yy}}\right)\frac{\partial^3 w}{\partial y\, \partial x^2}\right]$$

$$R_{xz} = -D_{yx}\left[\frac{\partial^3 w}{\partial x^3} + \left(\nu_y + 4\frac{D_{xy}}{D_{yy}}\right)\frac{\partial^3 w}{\partial x\, \partial y^2}\right]$$

$$R_{yz} = -D_{yy}\left[\frac{\partial^3 w}{\partial y^3} + \left(\nu_x + 4\frac{D_{xy}}{D_{xx}}\right)\frac{\partial^3 w}{\partial y\, \partial x^2}\right]$$

where R_{xz}, R_{yz} are the intensities of vertical edge reactions and

$$D_{xx} = \frac{E_{xx} h^3}{12(1 - \nu_x \nu_y)} \qquad D_{xy} = \frac{Gh^3}{12} \qquad D_{yy} = \frac{E_{yy} h^3}{12(1 - \nu_x \nu_y)}$$

are the rigidities of the plate.

(4) Differential equation of elastic surface

$$\boxed{D_{xx}\frac{\partial^4 w}{\partial x^4} + 2H\frac{\partial^4 w}{\partial x^2\, \partial y^2} + D_{yy}\frac{\partial^4 w}{\partial y^4} = p_z}$$

where $H = 4D_{xy} + \nu_y D_{xx} + \nu_x D_{yy}$ or $H = 2D_{xy} + \nu_x D_{yy} = 2D_{xy} + \nu_y D_{xx}$. Particular forms of D_{xx}, D_{xy}, D_{yy} are tabulated in (22.12–22.13).

22.11 ORTHOTROPIC PLATE IN CYLINDRICAL COORDINATES

(1) Orthotropic plate of circular shape and of arbitrary edge conditions under transverse load of intensity p_z (N/m^2) is considered in this section. By definition, the material properties of the plate differ in two mutually perpendicular directions, and their moduli E_{rr}, $E_{\theta\theta}$, and G are independent of one another (Ref. 33, pp. 312–315). The basic relations and the governing differential equation of this plate are introduced below in terms of symbols defined in (22.06–22.07).

(2) Stress-strain equations

$$f_{rr} = -\frac{E_{rr}z}{1-\nu_r\nu_\theta}\left[\frac{\partial^2 w}{\partial r^2} + \nu_\theta\left(\frac{\partial w}{r\,\partial r} + \frac{\partial^2 w}{r^2\,\partial\theta^2}\right)\right] \qquad f_{\theta\theta} = -\frac{E_{\theta\theta}z}{1-\nu_r\nu_\theta}\left(\nu_r\frac{\partial^2 w}{\partial r^2} + \frac{\partial w}{r\,\partial r} + \frac{\partial^2 w}{r^2\,\partial\theta^2}\right)$$

$$f_{r\theta} = f_{\theta r} = -2zG\left(\frac{\partial^2 w}{r\,\partial r\,\partial\theta} - \frac{1}{r^2}\frac{\partial w}{\partial\theta}\right)$$

where ν_r, ν_θ = Poisson's ratios in r, θ directions.

(3) Stress resultants

$$N_r = 0 \qquad N_\theta = 0 \qquad M_{r\theta} = M_{\theta r} = -2D_{r\theta}\left(\frac{\partial^2 w}{r\,\partial r\,\partial\theta} - \frac{\partial w}{r^2\,\partial\theta}\right)$$

$$M_r = -D_{rr}\left[\frac{\partial^2 w}{\partial r^2} + \nu_\theta\left(\frac{\partial w}{r\,\partial r} + \frac{\partial^2 w}{r^2\,\partial\theta^2}\right)\right] \qquad M_\theta = -D_{\theta\theta}\left(\nu_r\frac{\partial^2 w}{\partial r^2} + \frac{\partial w}{r\,\partial r} + \frac{\partial^2 w}{r^2\,\partial\theta^2}\right)$$

$$W_{rz} = -D_{rr}\left(\frac{\partial^3 w}{\partial r^3} + \frac{\partial^2 w}{r\,\partial r^2}\right) + D_{\theta\theta}\left(\frac{\partial w}{r^2\,\partial r} + \frac{1}{r^3}\frac{\partial^3 w}{\partial\theta^3}\right) - H\left(\frac{\partial^3 w}{r^2\partial r\,\partial\theta^2} - \frac{\partial^2 w}{r^3\,\partial\theta^2}\right)$$

$$W_{\theta z} = -D_{\theta\theta}\left(\frac{\partial^2 w}{r^2\,\partial r\,\partial\theta} + \frac{\partial^3 w}{r^3\,\partial\theta^3}\right) - H\frac{\partial^3 w}{r\,\partial r^2\,\partial\theta}$$

$$R_{rz} = W_{rz} - 2D_{r\theta}\left(\frac{\partial^3 w}{r^2\,\partial r\,\partial\theta^2} - \frac{\partial^2 w}{r^3\,\partial\theta^2}\right) \qquad R_{\theta z} = W_{\theta z} - 2D_{r\theta}\left(\frac{\partial^3 w}{r\,\partial^2 r\,\partial\theta} - \frac{2\partial^2 w}{r^2\,\partial r\,\partial\theta} + \frac{2\partial w}{r^3\,\partial\theta}\right)$$

where R_{rz}, $R_{\theta z}$ are the intensities of vertical edge reactions and

$$D_{rr} = \frac{E_{rr}h^3}{12(1-\nu_r\nu_\theta)} \qquad D_{r\theta} = \frac{Gh^3}{12} \qquad D_{\theta\theta} = \frac{E_{\theta\theta}h^3}{12(1-\nu_r\nu_\theta)} \qquad H = D_{rr}\nu_\theta + 2D_{r\theta} = D_{\theta\theta}\nu_r + 2D_{r\theta}$$

are the rigidities of the plate.

(4) Differential equation of elastic surface

$$D_{rr}\left(\frac{\partial^4 w}{\partial r^4} + \frac{2\partial^3 w}{r\,\partial r^3}\right) + 2H\left(\frac{\partial^4 w}{r^2\,\partial r^2\,\partial\theta^2} - \frac{\partial^3 w}{r^3\,\partial r\,\partial\theta^2} + \frac{\partial^2 w}{r^4\,\partial\theta^2}\right) + D_{\theta\theta}\left(\frac{\partial^4 w}{r^4\,\partial\theta^4} - \frac{\partial^2 w}{r^2\,\partial r^2} + \frac{2\partial^2 w}{r^4\,\partial\theta^2} + \frac{\partial w}{r^3\,\partial r}\right) = p_z$$

In case of axisymmetry,

$$D_{rr}\left(\frac{d^4 w}{dr^4} + \frac{2d^3 w}{r\,d^3 r}\right) + D_{\theta\theta}\left(\frac{d^2 w}{r^2\,dr^2} + \frac{dw}{r^3\,dr}\right) = p_z$$

and $M_{r\theta} = M_{\theta r} = 0$, $R_{\theta z} = W_{\theta z} = 0$, $R_{rz} = W_{rz}$.

22.12 MATERIAL CONSTANTS OF ORTHOTROPIC PLATES

1. Reinforced concrete slab with two-way steel

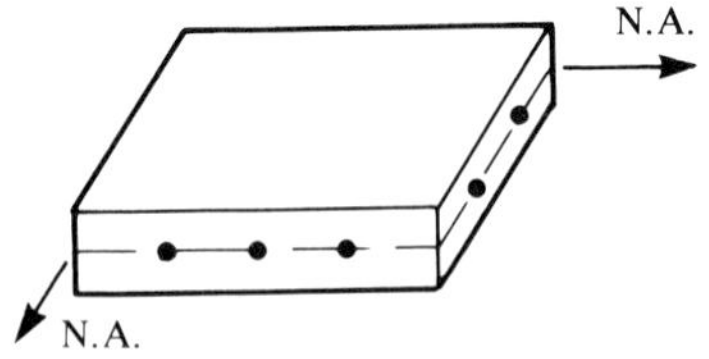

$\frac{E_s}{E_c} = n \qquad \nu = \text{Poisson's ratio of concrete}$

$$D_{xx} = \frac{E_c}{1-\nu^2}[I_{cx} + (n-1)I_{sx}] \qquad D_{yy} = \frac{E_c}{1-\nu^2}[I_{cy} + (n-1)I_{sy}]$$

$$D_{xy} = \frac{1-\nu}{2}\sqrt{D_{xx}D_{yy}} \qquad H = \sqrt{D_{xx}D_{yy}}$$

N.A. = neutral axis of bending

E_c, E_s = moduli of concrete and steel (Pa)

I_{cx}, I_{cy} = moments of inertia of concrete slab per unit length about N.A. (m^3)

I_{sx}, I_{sy} = moments of inertia of steel bars per unit length about N.A. (m^3)

2. Corrugated plate $f \sin \frac{\pi x}{d}$

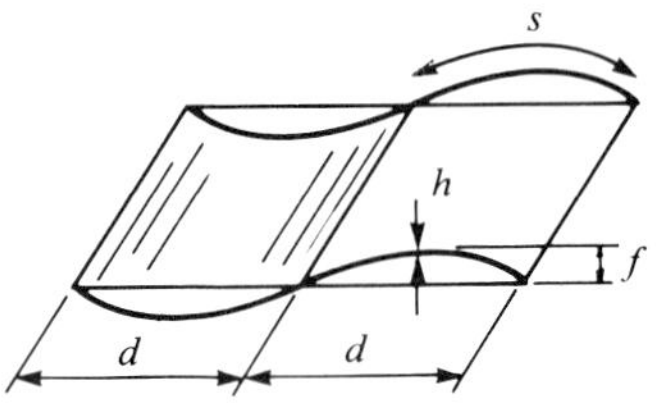

$$s = d\left(1 + \frac{\pi^2 f^2}{4d^2}\right) \qquad e = \frac{s}{d} \qquad g = \frac{f}{2d}$$

$$D_{xx} = \frac{Eh^3}{12e(1-\nu^2)} \qquad D_{yy} = \frac{1}{2}hf^2\left(1 - \frac{0.81}{1+2.5g^2}\right)$$

$$H = \frac{eEh^3}{12(1-\nu)} \qquad D_{xy} = 0$$

E = modulus of metal (Pa)

ν = Poisson's ratio of metal

h = thickness of plate (m)

(Ref. 33, p. 294)

3. Plate with stiffeners on both sides

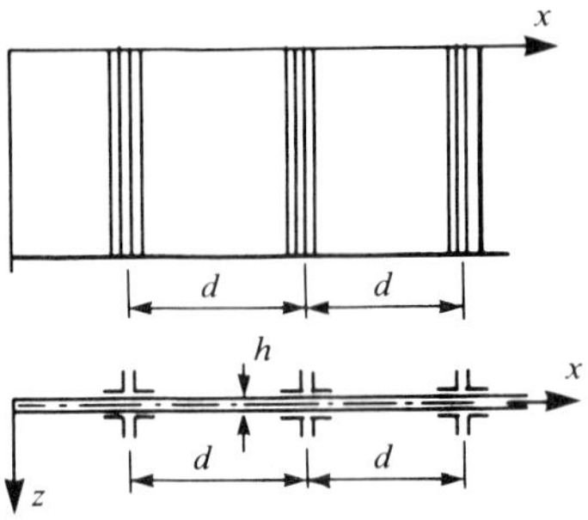

$$D_{xx} = \frac{Eh^3}{12(1-\nu^2)} \qquad D_{yy} = D_{xx} + \frac{E_{st}I_{st}}{d}$$

$$H = D_{xx} \qquad D_{xy} = 0$$

E = modulus of plate (Pa)

ν = Poisson's ratio of plate

E_{st} = modulus of stiffener (Pa)

I_{st} = moment of inertia of one set of stiffeners about the axis in the midsurface (m^4)

If only one-side stiffeners are used, I_{st} is the moment of inertia of the stiffeners about the centroidal axis of the section.

h = thickness of plate (m)

d = spacing of stiffeners (m)

(Ref. 33, p. 294)

4. Concrete slab with one-way ribs

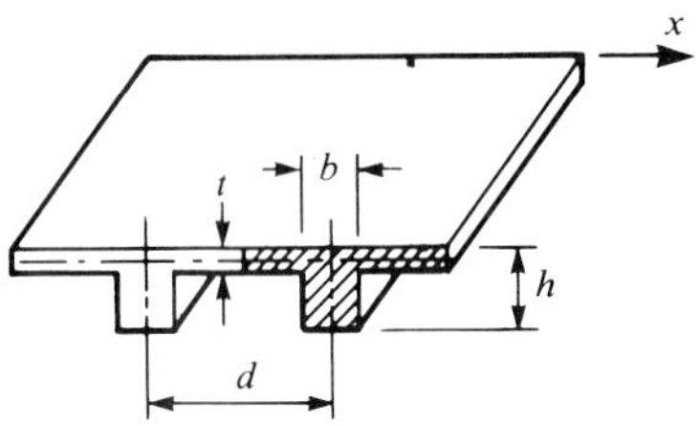

$$c = 1 - \frac{b}{d}\left[1 - \left(\frac{t}{h}\right)^3\right]$$

$$D_{xx} = \frac{Et^3}{12c} \qquad D_{yy} = \frac{EI_r}{d} \qquad D_{xy} = 0$$

$$H = \frac{Gt^3}{6} + \frac{G_tJ_r}{d}$$

E, G = moduli of plate (Pa)

t = thickness of plate (m)

I_r = moment of inertia of T section shown as shaded area (m^4)

G_tJ_r = torsional rigidity of T section ($N \cdot m^2$)

h = depth of T section (m)

b = width of T section web (m)

d = spacing of ribs (m)

22.13 MATERIAL CONSTANTS OF ORTHOTROPIC PLATES

1. Plate reinforced by a grid of stiffeners

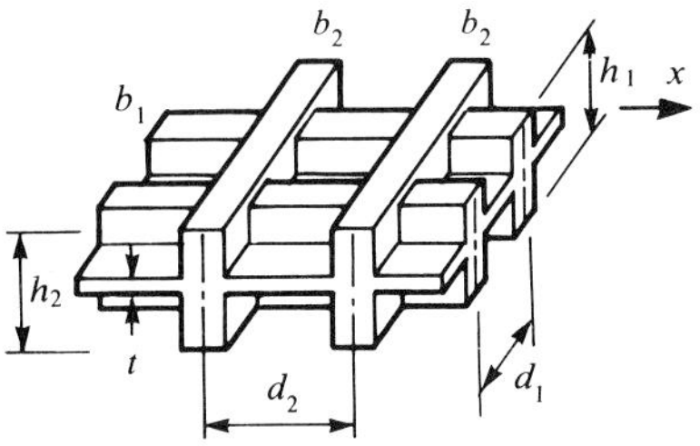

$$D_{xx} = \frac{Et^3}{12(1-\nu^2)} + E_{st}I_1 \qquad D_{yy} = \frac{Et^3}{12(1-\nu^2)} + E_{st}I_2$$

$$H = \frac{Et^3}{12(1-\nu^2)} \qquad \nu_x = \nu_y = \nu$$

E = modulus of plate (Pa)

E_{st} = modulus of stiffeners (Pa)

I_i = moment of inertia of one stiffener per unit length as shown below

ν = Poisson's ratio

t = thickness of plate (m)

b_i, h_i = dimensions of stiffener (m)

$$I_i = \frac{b_i(h_i^3 - t^3)}{12d_i} \qquad i = 1, 2$$

2. Open grid

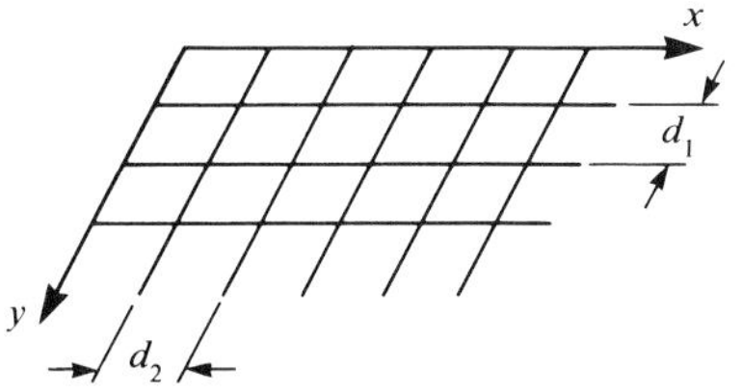

$$D_{xx} = \frac{EI_1}{d_1} \qquad D_{yy} = \frac{EI_2}{d_2} \qquad H = \frac{GJ_1}{2d_1} + \frac{GJ_2}{2d_2} \qquad D_{xy} = 0$$

E, G = moduli (Pa)

I_1, I_2 = moments of inertia of bars 1, 2 about respective centroidal axis (m^4)

J_1, J_2 = torsional constants of bars 1, 2 about their centroidal axis of torsion (m^4)

3. Composite deck plate consisting of concrete slab and steel I beams

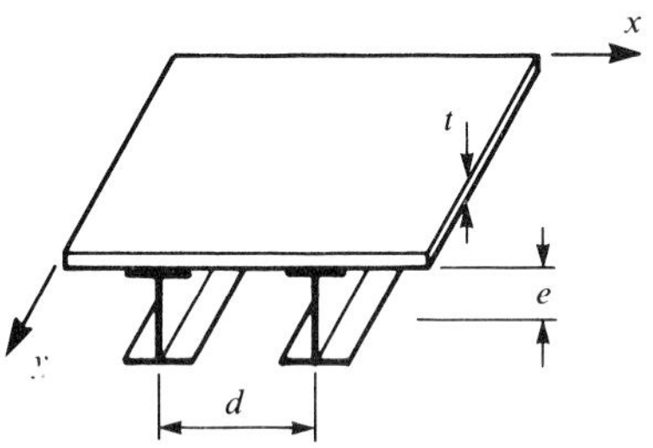

$$D_{xx} = \frac{E_c t^3}{12(1-\nu_c^2)} \qquad D_{yy} = \frac{E_c t^3}{12(1-\nu_c^2)} + \frac{E_s A_s e^2}{d} + \frac{E_s I_s}{d}$$

$$H = \frac{E_c t^3}{12(1-\nu_c^2)} + \frac{G_s J_s}{d} \qquad D_{xy} = 0$$

E_s = modulus of steel (Pa) $\qquad$ ν_c = Poisson's ratio of concrete

E_c = modulus of concrete (Pa) $\qquad$ A_s = area of I beam (m^2)

I_s = moment of inertia of I beam (m^4)

e = position of centroid of I beam relative to midsurface (m)

$G_s J_s$ = torsional rigidity of I beam ($N \cdot m^2$)

d = spacing of I beams (m) $\qquad$ t = thickness of concrete slab (m)

4. Isotropic plate consisting of $n + 1$ layers of symmetrical properties with respect to midsurface

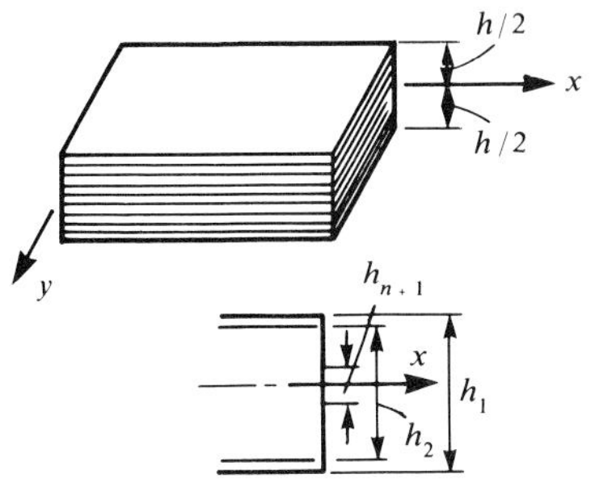

$$D_{xx} = D_{yy} = D$$

$$\nu_x = \nu_y = \nu \qquad D_k = \frac{E_k(h_k^3 - h_{k+1}^3)}{12(1-\nu_k^2)} \qquad D_{n+1} = \frac{E_{n+1}h_n^3 + 1}{12(1-\nu_{n+1}^2)}$$

$$D = D_{n+1} + \sum_{k=1}^{n} D_k \qquad \nu = \frac{1}{D}\left(\nu_{n+1}D_{n+1} + \sum_{k=1}^{n} \nu_k D_k\right)$$

$$E = \frac{1-\nu^2}{h}\left(\frac{\upsilon_{n+1}E_{n+1}}{1-\nu_{n+1}^2} + 2\sum_{k=1}^{n}\frac{\upsilon_k E_k}{1-\nu_k^2}\right)$$

E_k = modulus of layer k (Pa)

ν_k = Poisson's ratio of layer k

$\frac{1}{2}(h_k - h_{k+1})$ = thickness of layer k (m)

(Ref. 33, pp. 295–301)

22.14 GENERAL PLATE EQUATIONS

(1) Motion of rectangular orthotropic plate of arbitrary edge conditions under transverse loads and in-plane edge forces resting on elastic foundation is governed by

$$D_{xx}\frac{\partial^4 w}{\partial x^4} + 2H\frac{\partial^4 w}{\partial x^2\,\partial y^2} + D_{yy}\frac{\partial^4 w}{\partial y^4} - \eta_x N_x\frac{\partial^2 w}{\partial x^2} - \eta_y N_y\frac{\partial^2 w}{\partial y^2} + k_w w + \mu\frac{\partial^2 w}{\partial t^2} = p_z(x, y, t) - p_z^*(x, y, t^*)$$

where t = time (s), k_w = foundation modulus (N/m^3), D_{xx}, D_{yy}, H = plate rigidities (22.10–22.13), μ = mass per unit area (kg/m^2), $p_z(x, y, t)$ = intensity of transverse load (N/m^2), N_x, N_y = intensities of normal forces (N/m),

$$\eta_x, \eta_y = \begin{cases} +1 & \text{if } N \text{ is tension} \\ -1 & \text{if } N \text{ is compression} \end{cases} \qquad p_z^*(x, y, t^*) = \frac{\partial^2 M_x^*}{\partial x^2} + \frac{\partial^2 M_y^*}{\partial y^2}$$

in which the thermal moments are

$$M_x^* = 3D_{xx}g_{tx} = 3D_{xx}(\alpha_{tx} + \nu_y\alpha_{ty})\,\Delta t^*/h \qquad M_y^* = 3D_{yy}g_{ty} = 3D_{tt}(\alpha_{ty} + \nu_x\alpha_{tx})\,\Delta t^*/h$$

in terms of α_{tx}, α_{ty} = thermal coefficients in x, y, respectively (1/°C), Δt^* = temperature difference between bottom and top of the plate at x, y (°C).

(2) Motion of circular orthotropic plate of the same conditions is governed by

$$\begin{aligned} D_{rr}\left(\frac{\partial^4 w}{\partial r^4} + \frac{2\partial^3 w}{r\,\partial r^3}\right) &+ 2H\left(\frac{\partial^4 w}{r^2\,\partial^2 r\,\partial^2\theta} - \frac{\partial^3 w}{r^3\,\partial r\,\partial\theta^2} + \frac{\partial^2 w}{r^4\,\partial\theta^2}\right) \\ &+ D_{\theta\theta}\left(\frac{\partial^4 w}{r^4\,\partial\theta^4} - \frac{\partial^3 w}{r^2\,\partial r^3} + \frac{2\partial^2 w}{r^4\,\partial\theta^2} + \frac{\partial w}{r^3\,\partial r}\right) \\ &- \eta_r N_r\frac{\partial^2 w}{r\,\partial r^2} - \eta_\theta N_\theta\left(\frac{\partial^2 w}{r^2\,\partial\theta^2} + \frac{\partial w}{r\,\partial r}\right) \\ &+ k_w w + \mu\frac{\partial^2 w}{\partial t^2} = p_z(r, \theta, t) - p_z^*(r, \theta, t^*) \end{aligned}$$

where t, k_w, μ, p_z, η have the same meaning as in (1) above, D_{rr}, $D_{\theta\theta}$, H = plate rigidities (22.11–22.13), N_r, N_θ = intensities of normal force (N/m),

$$p_z^* = \frac{\partial^2 M_r^*}{\partial r^2} + \frac{\partial M_r^*}{r\,\partial r} + \frac{\partial^2 M^*_\theta}{r^2\,\partial\theta^2}$$

in which the thermal moments are

$$M_r^* = 3D_{rr}g_{tr} = 3D_{rr}(\alpha_{tr} + \nu_\theta\alpha_{t\theta})\,\Delta t^*/h \qquad M_\theta^* = 3D_{\theta\theta}g_{g\theta} = 3D_{\theta\theta}(\alpha_{t\theta} + \nu_r\alpha_{tr})\,\Delta t^*/h$$

in terms of α_{tr}, $\alpha_{t\theta}$ = thermal coefficients in r, θ, respectively (1/°C), Δt^* = temperature difference between bottom and top of the plate at r, θ (°C).

(3) Particular solutions of circular and rectangular plates by classical and numerical matrix methods are shown in Chaps. 23–26.

23

Circular Isotropic Plate Constant Thickness Axisymmetrical System

STATIC STATE

23.01 DEFINITION OF STATE

(1) System considered is a finite isotropic circular plate of constant thickness forming with its causes and supports an axisymmetrical system in a state of static equilibrium.

(2) Notation (A.1, A.5)

r = radial coordinate (m)	R = outer radius (m)
b = radial coordinate (m)	a = inner radius (m)
h = thickness (m)	θ = angular coordinate (rad)
E = modulus of elasticity (Pa)	ν = Poisson's ratio
P = intensity of concentrated load (N/m)	ψ = flexural slope (rad)
p = intensity of distributed load (N/m^2)	M = intensity of flexural moment (N·m/m)
Q = intensity of applied couple (N·m/m)	W = intensity of shear (N/m)
w = vertical deflection (m)	x = ratio (a/r, b/r, a/R, b/R)

Since two flexural moments occur in the circular plate as shown in (22.07), the first moment $M = M_r$ is calculated first and the second moment

$$M_\theta = \nu M + \frac{1-\nu^2}{r} D\psi$$

$\alpha = \frac{1}{2}(1+\nu)$	$\beta = \frac{1}{2}(1-\nu)$	$D = \dfrac{Eh^3}{12(1-\nu^2)}$

(3) Assumptions of analysis are stated in (23.01). In this chapter, the effects of flexural deformations are included. Since $h/R < 1/10$ the effects of shearing deformations are neglected.

23.02 AXISYMMETRICAL SYSTEM, TRANSPORT RELATIONS

(1) Transport functions listed below apply in all cases of circular plates introduced in this chapter.

$T_0(x) = x$	$T_8(x) = \dfrac{x}{4}(x^2 - 1 - 2\ln x)$
$T_1(x) = \dfrac{1}{2x}[\beta(1 - x^2) - 2\alpha x^2 \ln x]$	$T_9(x) = \dfrac{x}{2}[\beta(1 - x^2) - 2\alpha \ln x]$
$T_2(x) = \dfrac{1}{x}[\beta + \alpha x^2]$	$T_{10}(x) = \dfrac{1}{64} + \dfrac{x^2}{16}(1 + 2\ln x) - \dfrac{x^4}{64}(5 - 4\ln x)$
$T_3(x) = \dfrac{2\alpha\beta}{x}(1 - x^2)$	$T_{11}(x) = \dfrac{1}{16} + \dfrac{x^2}{4}\ln x - \dfrac{x^4}{16}$
$T_4(x) = \frac{1}{4}[1 - x^2(1 - 2\ln x)]$	$T_{12}(x) = \dfrac{1 - x^2}{4} - \dfrac{\beta(1 - x^4)}{8} + \dfrac{\alpha x^2 \ln x}{2}$
$T_5(x) = \frac{1}{2}(1 - x^2)$	$T_{13}(x) = \dfrac{1}{225} + \dfrac{x^3}{36}(2 + 3\ln x) - \dfrac{x^5}{100}(6 - 5\ln x)$
$T_6(x) = \alpha + \beta x^2$	$T_{14}(x) = \dfrac{1}{45} + \dfrac{x^3}{36}(1 + 6\ln x) - \dfrac{x^5}{20}$
$T_7(x) = \dfrac{x}{4}[x^2 - 1 - (1 + x^2)\ln x]$	$T_{15}(x) = \dfrac{\beta(1 - x^3)}{6} - \dfrac{\beta(1 - x^5)}{10} + \dfrac{\alpha(1 - x^3 + x^3\ln x)}{9}$

where $T_1(x), T_2(x), \ldots, T_9(x)$ are the *basic functions* and

$$T_{10}(x) = \int_x^1 T_7(x)\,dx \qquad T_{11}(x) = \int_x^1 T_8(x)\,dx \qquad T_{12}(x) = \int_x^1 T_9(x)\,dx$$

$$T_{13}(x) = \int_x^1 xT_7(x)\,dx \qquad T_{14}(x) = \int_x^1 xT_8(x)\,dx \qquad T_{15}(x) = \int_x^1 xT_9(x)\,dx$$

(2) Definite integrals used in the evaluation of $T_{10}(x), T_{11}(x), \ldots, T_{15}(x)$ are

$$\int_x^1 x^m\,dx = \frac{1 - x^{m+1}}{m + 1} \qquad \int_x^1 x^m \ln x\,dx = -\frac{1 + x^{m+1}[(m + 1)\ln x - 1]}{(m + 1)^2}$$

(3) Transport coefficients are particular values of $T_j(x)$.

$$\left.\begin{aligned} &\text{For } x = \frac{a}{r}, && T_j\left(\frac{a}{r}\right) = A_j \\ &\text{For } x = \frac{b}{r}, && T_j\left(\frac{b}{r}\right) = B_j \\ &\text{For } x = \frac{b}{R}, && T_j\left(\frac{b}{R}\right) = C_j \\ &\text{For } x = \frac{a}{R}, && T_j\left(\frac{a}{R}\right) = T_j \end{aligned}\right\} \quad j = 0, 1, 2, \ldots, 15$$

For example:

$$A_4 = \frac{1}{4}\left[1 - \left(\frac{a}{r}\right)^2\left(1 - 2\ln\frac{a}{r}\right)\right] \qquad C_4 = \frac{1}{4}\left[1 - \left(\frac{b}{R}\right)^2\left(1 - 2\ln\frac{b}{R}\right)\right]$$

$$B_4 = \frac{1}{4}\left[1 - \left(\frac{b}{r}\right)^2\left(1 - 2\ln\frac{b}{r}\right)\right] \qquad T_4 = \frac{1}{4}\left[1 - \left(\frac{a}{R}\right)^2\left(1 - 2\ln\frac{a}{R}\right)\right]$$

23.03 SOLID PLATE, AXISYMMETRICAL SYSTEM, TRANSPORT MATRICES

(1) Equivalents and signs

$$\overline{w} = \frac{D}{r^2} w = \text{scaled } w \text{ (N)}$$

$$\overline{\psi} = \frac{D}{r} \psi = \text{scaled } \psi \text{ (N)}$$

$$\overline{W} = rW = \text{scaled } W \text{ (N)}$$

$$\alpha = \tfrac{1}{2}(1 + \nu) \qquad \beta = \tfrac{1}{2}(1 - \nu) \qquad D = \frac{Eh^3}{12(1 - \nu^2)}$$

All forces, moments, and displacements are in the cylindrical coordinate system (2), and their positive directions are shown in (3, 4) below.

(2) Geometry

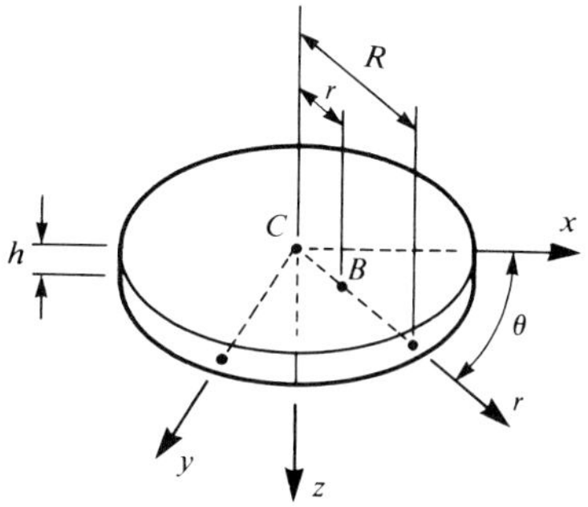

(3) Free-body diagrams

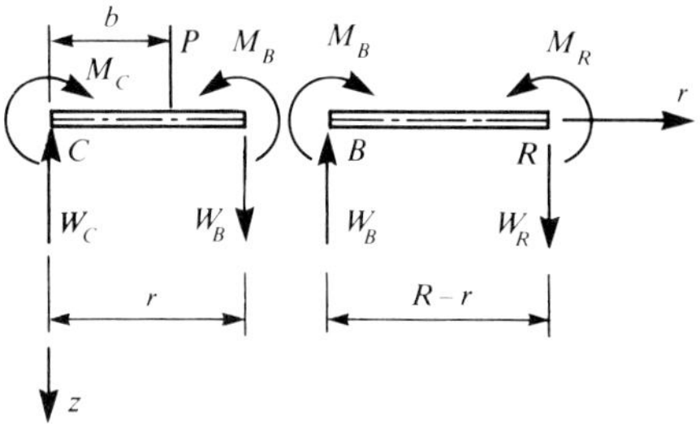

(4) Elastic surface

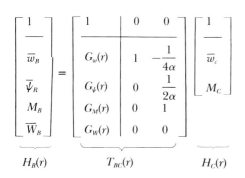

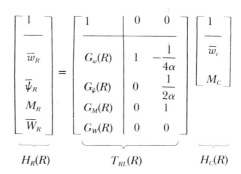

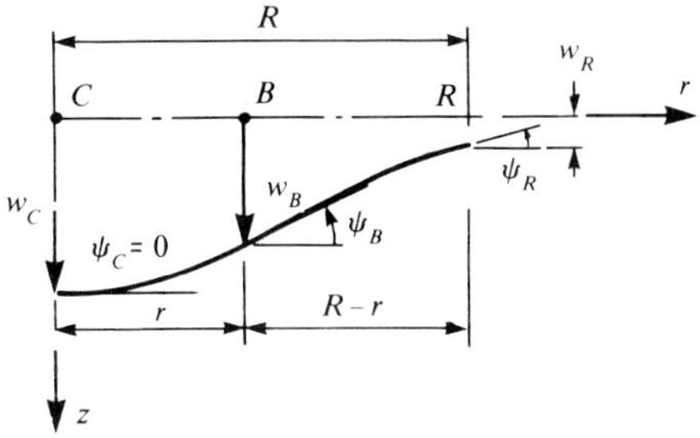

(5) Transport matrix equations

(*a*) Segment CB

$$\underbrace{\begin{bmatrix} 1 \\ \hline \overline{w}_B \\ \overline{\psi}_R \\ M_B \\ \overline{W}_B \end{bmatrix}}_{H_B(r)} = \underbrace{\left[\begin{array}{c|cc} 1 & 0 & 0 \\ \hline G_w(r) & 1 & -\frac{1}{4\alpha} \\ G_\psi(r) & 0 & \frac{1}{2\alpha} \\ G_M(r) & 0 & 1 \\ G_W(r) & 0 & 0 \end{array}\right]}_{T_{BC}(r)} \underbrace{\begin{bmatrix} 1 \\ \hline \overline{w}_c \\ M_C \end{bmatrix}}_{H_C(r)}$$

(*b*) Segment CR

$$\underbrace{\begin{bmatrix} 1 \\ \hline \overline{w}_R \\ \overline{\psi}_R \\ M_R \\ \overline{W}_R \end{bmatrix}}_{H_R(R)} = \underbrace{\left[\begin{array}{c|cc} 1 & 0 & 0 \\ \hline G_w(R) & 1 & -\frac{1}{4\alpha} \\ G_\psi(R) & 0 & \frac{1}{2\alpha} \\ G_M(R) & 0 & 1 \\ G_W(R) & 0 & 0 \end{array}\right]}_{T_{RL}(R)} \underbrace{\begin{bmatrix} 1 \\ \hline \overline{w}_c \\ M_C \end{bmatrix}}_{H_C(R)}$$

where $H_C(r)$ is the state vector at C and $H_B(r)$ is the state vector at B in segment CB, $H_C(R)$ is the state vector at C and $H_R(R)$ is the state vector at R in segment CR, and $G(r)$, $G(R)$ are the load functions in the respective segments tabulated in (23.06–23.07). The scaling radius in segment CB is r and in segment CR is R. In cases of several loads, the complete load function is obtained by linear superposition.

(6) **Application** of matrix equations introduced in (5) above follows the standard pattern of the transport method. Equation (5*a*) is used for the calculation of the state vector at any distance r from the center of the plate, and equation (5*b*) is used for the calculation of the unknown boundary values (Ref. 11, pp. 279–287; Ref. 56, pp. 362–367).

23.04 GRAPHS AND TABLES OF TRANSPORT COEFFICIENTS

$\nu = 0.16$

(1) Analytical expressions for $T_1(x)$, $T_2(x)$, . . . , $T_{15}(x)$ are given in (23.02). The graphs of the first nine functions are shown below in (2), and numerical values of all functions for $\nu = 0.16$ are tabulated in (3) below and on the opposite page.

(2) Graphs

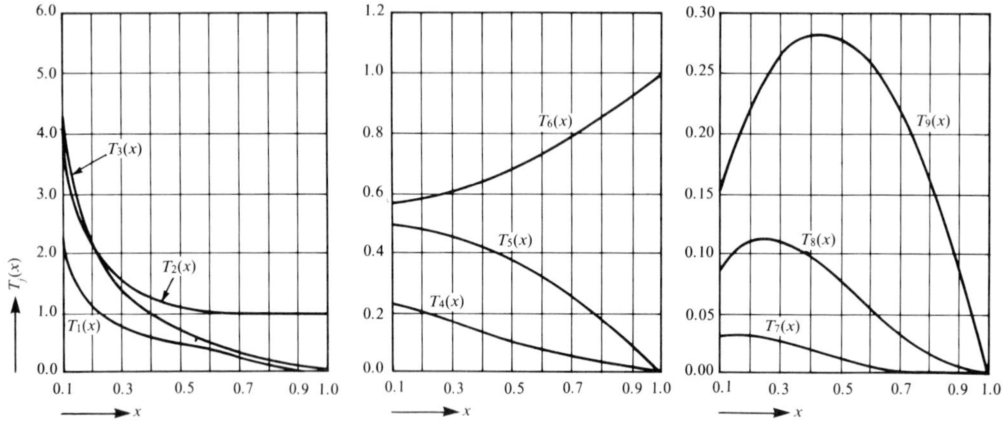

(3) Tables

x	$T_1(x)$	$T_2(x)$	$T_3(x)$	$T_4(x)$	$T_5(x)$
0.05	4.276 376 355	8.429 000 237	9.719 640 499	0.245 630 335	0.498 750 000
0.10	2.212 549 995	4.258 000 117	4.823 280 252	0.235 987 076	0.495 000 001
0.15	1.533 549 480	2.887 000 076	3.174 920 171	0.223 303 403	0.488 750 001
0.20	1.194 694 831	2.216 000 055	2.338 560 131	0.207 811 246	0.480 000 002
0.25	0.988 512 712	1.825 000 042	1.827 000 108	0.191 053 306	0.468 750 004
0.30	0.846 491 296	1.574 000 033	1.477 840 093	0.173 321 230	0.455 000 005
0.35	0.739 613 919	1.403 000 027	1.221 480 083	0.155 073 403	0.438 750 007
0.40	0.653 579 478	1.282 000 021	1.023 120 075	0.136 696 750	0.420 000 010
0.45	0.580 577 205	1.194 333 350	0.863 426 737	0.118 526 105	0.398 750 012
0.50	0.516 012 714	1.130 000 013	0.730 800 066	0.100 856 613	0.375 000 015
0.55	0.457 028 218	1.082 636 374	0.617 858 245	0.083 952 164	0.348 750 018
0.60	0.401 767 353	1.048 800 007	0.519 680 061	0.068 051 399	0.320 000 021
0.65	0.348 982 120	1.023 153 851	0.432 858 521	0.053 372 120	0.288 750 025
0.70	0.297 810 068	1.006 000 002	0.354 960 059	0.040 114 649	0.255 000 029
0.75	0.247 641 745	0.995 000 000	0.284 200 058	0.028 464 427	0.218 750 034
0.80	0.198 038 654	0.988 999 998	0.219 240 058	0.018 594 072	0.180 000 038
0.85	0.148 680 705	0.987 117 643	0.159 056 528	0.010 665 044	0.138 750 043
0.90	0.099 331 575	0.988 666 661	0.102 853 391	0.004 828 996	0.095 000 048
0.95	0.049 815 293	0.993 105 255	0.050 002 163	0.001 228 904	0.048 750 054
1.00	0.000 000 000	1.0000 000 000	0.000 000 000	0.000 000 000	0.000 000 000

$\nu = 0.16$

x	$T_6(x)$	$T_7(x)$	$T_8(x)$	$T_9(x)$	$T_{10}(x)$
0.05	0.581 050 013	0.025 071 519	0.062 424 555	0.097 349 983	0.014 843 425
0.10	0.584 200 012	0.033 390 273	0.090 379 252	0.154 339 932	0.013 349 565
0.15	0.589 450 012	0.036 086 444	0.105 627 747	0.195 840 685	0.011 596 024
0.20	0.596 800 011	0.035 690 772	0.112 943 790	0.227 014 794	0.009 791 867
0.25	0.606 250 009	0.033 464 861	0.114 693 045	0.250 231 429	0.008 057 199
0.30	0.617 800 007	0.030 174 778	0.112 345 922	0.266 821 266	0.006 462 983
0.35	0.631 450 005	0.026 330 968	0.106 937 625	0.277 610 141	0.005 048 868
0.40	0.647 200 003	0.022 289 727	0.099 258 150	0.283 139 451	0.003 833 121
0.45	0.665 050 000	0.018 304 371	0.089 945 487	0.283 774 262	0.002 818 923
0.50	0.684 999 997	0.014 554 249	0.079 536 802	0.279 762 688	0.001 009 732
0.55	0.707 049 994	0.011 162 622	0.068 498 933	0.271 271 262	0.001 357 499
0.60	0.731 199 990	0.008 208 429	0.057 247 695	0.258 407 329	0.000 875 160
0.65	0.757 449 986	0.005 734 415	0.046 160 706	0.241 233 926	0.000 528 637
0.70	0.785 799 982	0.003 752 993	0.035 586 239	0.219 780 048	0.000 293 494
0.75	0.816 249 978	0.002 250 608	0.025 849 535	0.194 047 977	0.000 145 341
0.80	0.843 799 973	0.001 191 086	0.017 257 428	0.164 018 639	0.000 061 041
0.85	0.883 449 967	0.000 518 257	0.010 101 801	0.129 655 620	0.000 019 777
0.90	0.920 199 962	0.000 158 070	0.004 662 237	0.090 908 233	0.000 003 996
0.95	0.959 049 956	0.000 020 305	0.001 208 068	0.047 713 907	0.000 000 255
1.00	1.000 000 000	0.000 000 000	0.000 000 000	0.000 000 000	0.000 000 000

x	$T_{11}(x)$	$T_{12}(x)$	$T_{13}(x)$	$T_{14}(x)$	$T_{15}(x)$
0.05	0.060 627 277	0.194 703 424	0.004 420 116	0.022 163 266	0.092 403 519
0.10	0.056 372 288	0.188 327 756	0.004 306 365	0.021 865 735	0.092 162 031
0.15	0.051 797 060	0.179 522 873	0.004 086 618	0.021 245 044	0.091 581 260
0.20	0.046 305 622	0.168 914 524	0.003 770 978	0.020 282 526	0.090 552 574
0.25	0.040 595 012	0.156 953 498	0.003 381 144	0.018 997 279	0.088 988 844
0.30	0.034 904 364	0.144 001 565	0.002 943 421	0.017 432 844	0.086 821 682
0.35	0.029 411 309	0.130 367 904	0.002 484 636	0.015 648 731	0.084 000 014
0.40	0.024 248 373	0.116 328 117	0.002 029 574	0.013 714 232	0.080 490 890
0.45	0.019 512 660	0.102 135 472	0.001 599 371	0.011 703 494	0.076 278 876
0.50	0.015 272 054	0.088 028 088	0.001 210 562	0.009 691 378	0.071 367 703
0.55	0.011 569 438	0.074 233 836	0.000 874 621	0.007 749 806	0.065 780 069
0.60	0.008 425 696	0.060 973 814	0.000 597 892	0.005 944 500	0.059 559 672
0.65	0.005 419 916	0.048 464 911	0.000 381 830	0.004 331 948	0.052 771 840
0.70	0.003 801 071	0.036 921 750	0.000 223 522	0.002 956 582	0.045 504 839
0.75	0.002 269 319	0.026 558 199	0.000 116 423	0.001 848 092	0.037 871 114
0.80	0.001 197 033	0.017 588 565	0.000 051 312	0.001 018 861	0.030 008 703
0.85	0.000 519 628	0.010 228 556	0.000 017 409	0.000 461 494	0.022 082 371
0.90	0.000 158 246	0.004 696 070	0.000 003 676	0.000 146 418	0.014 285 187
0.95	0.000 020 310	0.001 211 843	0.000 000 243	0.000 019 548	0.006 839 827
1.00	0.000 000 000	0.000 000 000	0.000 000 000	0.000 000 000	0.000 000 000

23.05 GRAPHS AND TABLES OF TRANSPORT COEFFICIENTS

$\nu = 0.30$

(1) Analytical expressions for $T_1(x)$, $T_2(x)$, . . . , $T_{15}(x)$ are given in (23.02). The graphs of the first nine functions are shown below in (2), and numerical values of all functions of $\nu = 0.30$ are tabulated in (3) below and on the opposite page.

(2) Graphs

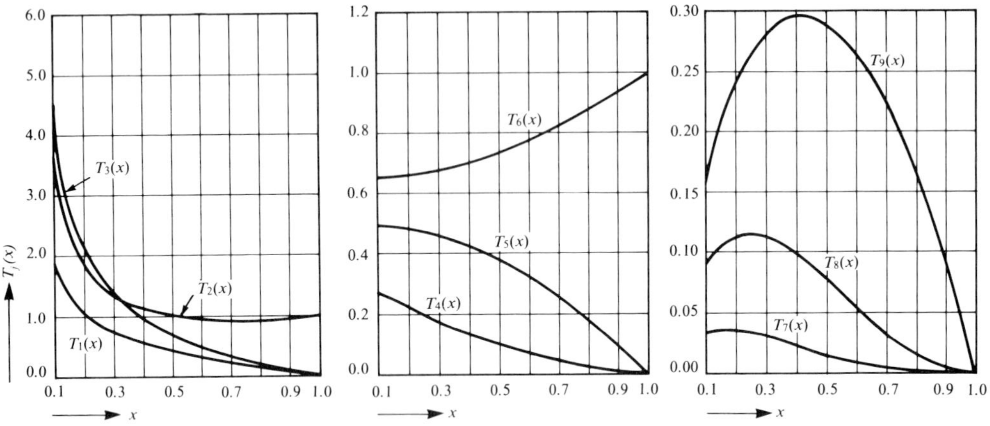

(3) Tables

x	$T_1(x)$	$T_2(x)$	$T_3(x)$	$T_4(x)$	$T_5(x)$
0.05	3.588 611 446	7.032 500 296	9.077 250 472	0.245 630 335	0.498 750 000
0.10	1.882 216 103	3.565 000 146	4.504 500 239	0.235 987 076	0.495 000 001
0.15	1.325 385 913	2.430 833 428	2.965 083 495	0.223 303 403	0.488 750 001
0.20	1.049 226 966	1.880 000 068	2.184 400 124	0.207 811 246	0.480 000 002
0.25	0.881 522 865	1.562 500 051	1.706 250 102	0.191 053 304	0.468 750 004
0.30	0.765 608 059	1.361 666 707	1.380 166 754	0.173 321 230	0.455 000 005
0.35	0.677 584 561	1.227 500 031	1.140 750 078	0.155 073 403	0.438 750 007
0.40	0.605 735 618	1.135 000 024	0.955 500 071	0.136 696 750	0.420 000 010
0.45	0.543 702 418	1.070 277 796	0.906 361 177	0.118 526 105	0.398 750 012
0.50	0.487 772 963	1.025 000 013	0.682 500 062	0.100 856 613	0.375 000 015
0.55	0.435 658 577	0.993 863 645	0.577 022 787	0.083 952 164	0.348 750 018
0.60	0.385 888 694	0.973 333 338	0.485 333 391	0.068 051 399	0.320 000 021
0.65	0.337 486 587	0.960 961 540	0.404 250 056	0.053 372 120	0.288 750 025
0.70	0.289 787 138	0.954 999 998	0.331 500 055	0.040 114 649	0.255 000 029
0.75	0.242 329 396	0.954 166 667	0.265 416 721	0.028 464 427	0.218 750 034
0.80	0.194 784 692	0.957 499 999	0.204 750 054	0.018 574 072	0.180 000 038
0.85	0.146 924 110	0.964 264 696	0.148 544 171	0.010 665 044	0.138 750 043
0.90	0.098 580 398	0.973 888 876	0.096 055 609	0.004 828 996	0.095 000 048
0.95	0.049 634 191	0.985 921 037	0.046 697 422	0.001 228 904	0.048 750 054
1.00	0.000 000 000	1.000 000 000	0.000 000 000	0.000 000 000	0.000 000 000

x	$T_6(x)$	$T_7(x)$	$T_8(x)$	$T_9(x)$	$T_{10}(x)$
0.05	0.650 875 006	0.025 071 519	0.062 424 555	0.106 089 420	0.014 843 425
0.10	0.653 500 005	0.033 390 273	0.090 379 252	0.166 993 026	0.013 349 565
0.15	0.657 875 005	0.036 086 444	0.105 627 747	0.210 628 568	0.011 596 023
0.20	0.664 000 004	0.035 690 772	0.112 943 790	0.242 826 923	0.009 791 867
0.25	0.671 875 003	0.033 464 861	0.114 693 045	0.266 288 454	0.008 057 199
0.30	0.681 500 002	0.030 174 778	0.112 345 922	0.282 549 694	0.006 962 983
0.35	0.692 875 000	0.026 330 968	0.106 937 625	0.292 581 407	0.005 048 868
0.40	0.706 000 000	0.022 289 727	0.099 258 150	0.297 035 591	0.003 833 121
0.45	0.720 874 596	0.018 304 371	0.089 945 487	0.296 366 629	0.002 818 923
0.50	0.737 500 000	0.014 554 249	0.079 536 802	0.290 897 839	0.001 998 732
0.55	0.755 874 992	0.011 162 622	0.068 498 933	0.280 961 112	0.001 357 499
0.60	0.775 999 989	0.008 208 429	0.057 247 695	0.266 422 006	0.000 875 160
0.65	0.797 874 986	0.005 734 415	0.046 160 706	0.247 696 424	0.000 528 637
0.70	0.821 499 983	0.003 752 993	0.035 586 239	0.224 762 121	0.000 293 494
0.75	0.846 874 979	0.002 250 608	0.025 849 535	0.197 666 912	0.000 145 341
0.80	0.873 999 975	0.001 191 086	0.017 257 428	0.166 434 679	0.000 061 041
0.85	0.902 874 972	0.000 518 257	0.010 101 801	0.131 069 872	0.000 019 777
0.90	0.933 499 967	0.000 158 070	0.004 662 237	0.091 580 946	0.000 003 996
0.95	0.965 874 963	0.000 020 305	0.001 208 068	0.047 884 036	0.000 000 255
1.00	1.000 000 000	0.000 000 000	0.000 000 000	0.000 000 000	0.000 000 000

x	$T_{11}(x)$	$T_{12}(x)$	$T_{13}(x)$	$T_{14}(x)$	$T_{15}(x)$
0.05	0.060 627 277	0.203 191 242	0.004 420 116	0.022 163 266	0.095 512 022
0.10	0.056 737 288	0.196 270 975	0.004 306 365	0.021 865 735	0.095 269 052
0.15	0.051 797 060	0.186 774 461	0.004 086 618	0.021 245 044	0.094 655 165
0.20	0.046 305 622	0.175 397 311	0.003 770 978	0.020 282 526	0.093 592 414
0.25	0.040 950 012	0.162 636 799	0.003 481 144	0.018 997 279	0.091 985 410
0.30	0.034 904 364	0.148 888 176	0.002 943 421	0.017 432 844	0.088 767 859
0.35	0.029 411 309	0.134 485 487	0.002 484 636	0.015 648 731	0.086 891 009
0.40	0.024 249 373	0.119 722 889	0.002 029 574	0.013 714 232	0.083 323 101
0.45	0.019 512 660	0.104 867 244	0.001 599 371	0.011 703 496	0.079 049 350
0.50	0.015 272 054	0.090 166 175	0.001 210 562	0.009 691 378	0.074 072 282
0.55	0.011 569 438	0.075 853 558	0.000 874 621	0.007 749 806	0.068 412 278
0.60	0.008 425 696	0.062 153 412	0.000 597 892	0.005 944 500	0.062 108 277
0.65	0.005 841 916	0.049 282 779	0.000 381 830	0.004 331 948	0.055 218 593
0.70	0.003 801 071	0.037 453 900	0.000 223 522	0.002 956 582	0.047 821 819
0.75	0.002 269 319	0.026 875 903	0.000 116 423	0.001 848 092	0.040 017 782
0.80	0.001 197 033	0.017 756 149	0.000 051 312	0.001 018 861	0.031 928 559
0.85	0.000 519 628	0.010 301 304	0.000 017 409	0.000 461 494	0.023 699 532
0.90	0.000 159 246	0.004 718 224	0.000 003 675	0.000 146 418	0.015 500 475
0.95	0.000 020 310	0.001 214 687	0.000 000 243	0.000 019 548	0.007 526 666
1.00	0.000 000 000	0.000 000 000	0.000 000 000	0.000 000 000	0.000 000 000

23.06 SOLID PLATE, AXISYMMETRICAL SYSTEM, TRANSPORT LOAD FUNCTIONS

Notation (23.01–23.03)
P = concentrated load per unit length (N/m)
p = intensity of distributed load (N/m^2)
B_j, C_j = transport coefficients (23.02)

Signs (23.03)

Matrix (23.03)
$\alpha = \frac{1}{2}(1 + \nu)$
$\beta = \frac{1}{2}(1 - \nu)$

Axisymmetrical cause	G	$0 \le r \le b$	$b < r < R$	$r = R$
1.	G_w	0	$B_7 rP$	$C_7 RP$
	G_ψ	0	$-B_8 rP$	$-C_8 RP$
	G_M	0	$-B_9 rP$	$-C_9 RP$
	G_W	0	$-bP$	$-bP$
2.	G_w	0	$B_{10} r^2 p$	$C_{10} R^2 p$
	G_ψ	0	$-B_{11} r^2 p$	$-C_{11} R^2 p$
	G_M	0	$-B_{12} r^2 p$	$-C_{12} R^2 p$
	G_W	0	$-\frac{1}{2}(r^2 - b^2)p$	$-\frac{1}{2}(R^2 - b^2)p$
3.	G_w	$\frac{1}{64} r^2 p$	$(\frac{1}{64} - B_{10}) r^2 p$	$(\frac{1}{64} - C_{10}) R^2 p$
	G_ψ	$-\frac{1}{16} r^2 p$	$-(\frac{1}{16} - B_{11}) r^2 p$	$-(\frac{1}{16} - C_{11}) R^2 p$
	G_M	$\dfrac{1 + \alpha}{8} r^2 p$	$-\left(\dfrac{1 + \alpha}{8} - B_{12}\right) r^2 p$	$-\left(\dfrac{1 + \alpha}{8} - C_{12}\right) R^2 p$
	G_W	$-\frac{1}{2} r^2 p$	$-\frac{1}{2} b^2 p$	$-\frac{1}{2} b^2 p$
4. $r_0 < R/100$	G_w	$\dfrac{r_0^2}{8}(\ln r - 1)p_0$		$\dfrac{r_0^2}{8}(\ln R - 1)p_0$
	G_ψ	$-\dfrac{r_0^2}{8}(2 \ln r - 1)p_0$		$-\dfrac{r_0^2}{8}(2 \ln R - 1)p_0$
	G_M	$-\dfrac{r_0^2}{4}(2\alpha \ln r + \beta)p_0$		$-\dfrac{r_0^2}{4}(2\alpha \ln R + \beta)p_0$
	G_W	$-\dfrac{r_0^2}{2} p_0$		$-\dfrac{r_0^2}{2} p_0$
5.	G_w	$\frac{1}{64} r^2 p$		$\frac{1}{64} R^2 p$
	G_ψ	$-\frac{1}{16} r^2 p$		$-\frac{1}{16} R^2 p$
	G_M	$-\dfrac{1 + \alpha}{8} r^2 p$		$-\dfrac{1 + \alpha}{8} R^2 p$
	G_W	$-\frac{1}{2} r^2 p$		$-\frac{1}{2} R^2 p$

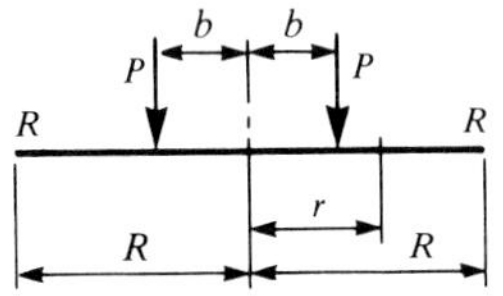

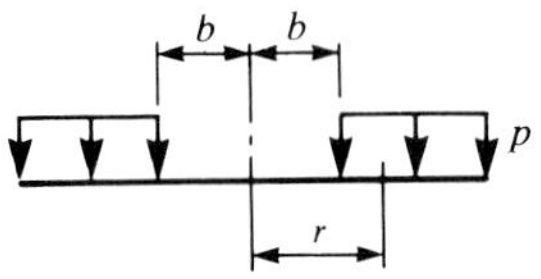

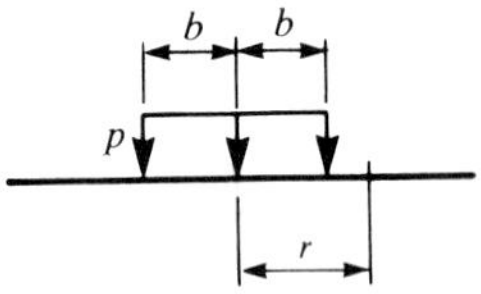

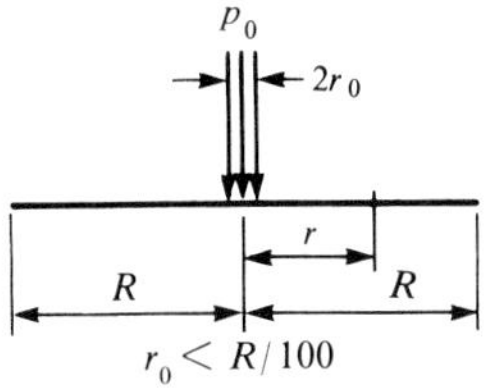

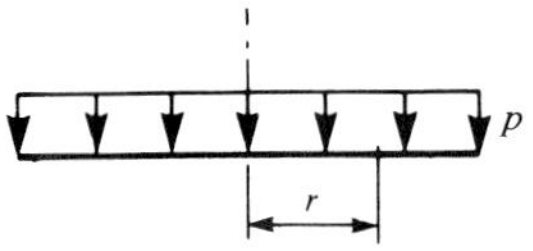

23.07 SOLID PLATE, AXISYMMETRICAL SYSTEM, TRANSPORT LOAD FUNCTIONS

otation (23.01–23.03)	Signs (23.03)	Matrix (23.03)
= applied couple per unit length (N·m/m)		ψ_B = abrupt slope (rad)
= intensity of distributed load at R (N/m²)		g_t = thermal strain (1/m)
, C_j = transport coefficients (23.02)		α, β, D (23.01)

Axisymmetrical causes	G	$0 \le r \le b$	$b < r < R$	$r = R$
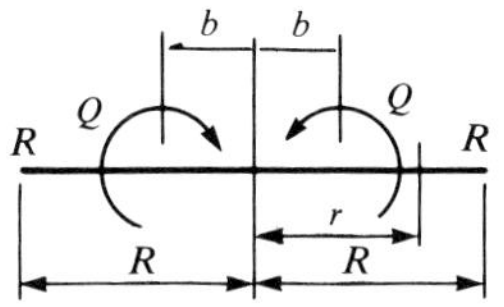	G_w	0	B_4Q	C_4Q
	G_ψ	0	$-B_5Q$	$-C_5Q$
	G_M	0	$-B_6Q$	$-C_6Q$
	G_W	0	0	0
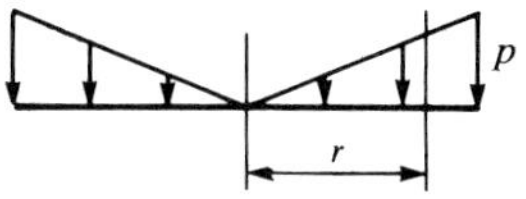	G_w	$\dfrac{161r^2p}{(120)^2}$		$\dfrac{161R^2p}{(120)^2}$
	G_ψ	$-\dfrac{29r^2p}{720}$		$-\dfrac{29R^2p}{720}$
	G_M	$-\dfrac{(42+29\nu)r^2p}{720}$		$-\dfrac{(42+29\nu)R^2p}{720}$
	G_W	$-\frac{1}{3}r^2p$		$-\frac{1}{3}R^2p$
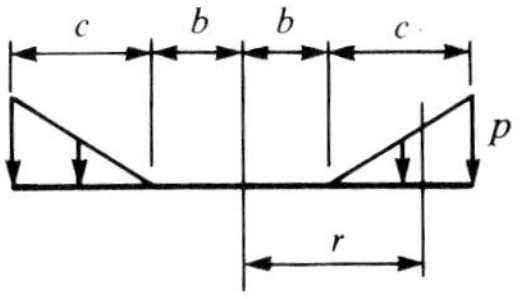	G_w	0	$(B_{10}-B_{13})\dfrac{r^3p}{c}$	$(C_{10}-C_{13})\dfrac{R^3p}{c}$
	G_ψ	0	$-(B_{11}-B_{14})\dfrac{r^3p}{c}$	$-(C_{11}-C_{14})\dfrac{R^3p}{c}$
	G_M	0	$-(B_{12}-B_{15})\dfrac{r^3p}{c}$	$-(C_{12}-C_{15})\dfrac{R^3p}{c}$
	G_W	0	$-\dfrac{(r-b)^2(2r+b)}{6c}p$	$-\dfrac{(2R+b)c}{6}p$
Abrupt slope at B 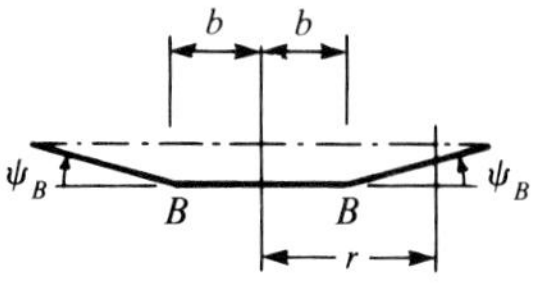	G_w	0	$-DB_1r^{-1}\psi_B$	$-DC_1R^{-1}\psi_B$
	G_ψ	0	$DB_2r^{-1}\psi_B$	$DC_2R^{-1}\psi_B$
	G_M	0	$DB_3r^{-1}\psi_B$	$DC_3R^{-1}\psi_B$
	G_W	0	0	0
Temperature deformation in segment BR 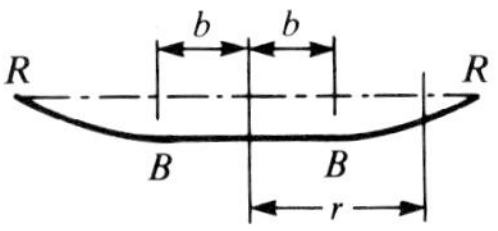	G_w	0	$-2DB_4\alpha g_t$	$-2DC_4\alpha g_t$
	G_ψ	0	$2DB_5\alpha g_t$	$2DC_5\alpha g_t$
	G_M	0	$2D(B_6-1)\alpha g_t$	$D(C_6-1)\alpha g_t$
g_t defined in (22.14)	G_W	0	0	0

23.08 SIMPLY SUPPORTED SOLID PLATE, AXISYMMETRICAL SYSTEM, STATE VECTORS

Notation (23.01–23.03) Signs (23.03) Matrix (23.03) All load functions at $r = R$ listed in (23.08–23.09) are applicable in Case 1 below. P = concentrated load per unit length (N/m) Q = applied couple per unit length (N·m/m) p = intensity of distributed load (N/m^2) C_j = transport coefficients (23.02)	$\alpha = \frac{1}{2}(1 + \nu)$ $\beta = \frac{1}{2}(1 - \nu)$ $D = \dfrac{Eh^3}{12(1 - \nu^2)}$ $\Delta = \dfrac{12(1 - \nu)}{Eh^3}$

1. General load 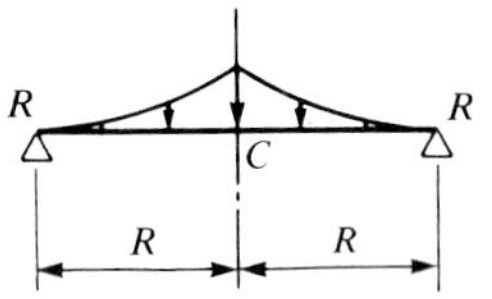	$w_C = -\frac{1}{2}(G_M - 4\alpha G_w)\Delta R^2$ $\psi_C = 0$ $M_C = -G_M$ $W_C = 0$	$w_R = 0$ $\psi_R = (2\alpha G_\psi - G_M)\Delta R$ $M_R = 0$ $W_R = G_W R^{-1}$
2. 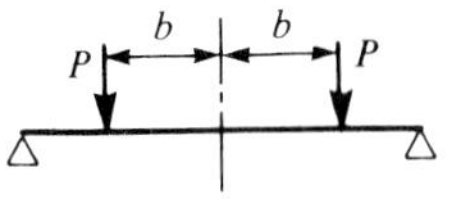	$w_C = \frac{1}{2}(C_9 - 4\alpha C_7)\Delta R^3 P$ $\psi_C = 0$ $M_C = C_9 RP$ $W_C = 0$	$w_R = 0$ $\psi_R = (C_9 - 2\alpha C_8)\Delta R^2 P$ $M_R = 0$ $W_R = -bR^{-1}P$
3. 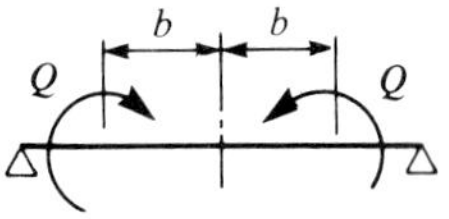	$w_C = \frac{1}{2}(C_6 - 4\alpha C_4)\Delta R^2 Q$ $\psi_C = 0$ $M_C = C_6 Q$ $W_C = 0$	$w_R = 0$ $\psi_R = (C_6 - 2\alpha C_5)\Delta RQ$ $M_R = 0$ $W_R = 0$
4. 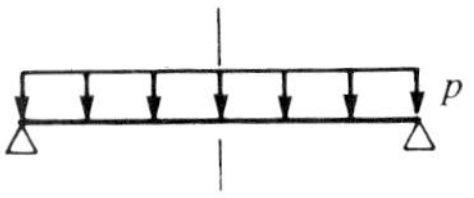	$w_C = \dfrac{(2 + \alpha)}{32}\Delta R^4 p$ $\psi_C = 0$ $M_C = \dfrac{1 + \alpha}{8} R^2 p$ $W_C = 0$	$w_R = 0$ $\psi_R = \frac{1}{8}\Delta R^3 p$ $M_R = 0$ $W_R = -\frac{1}{2}Rp$

23.09 FIXED-END SOLID PLATE, AXISYMMETRICAL SYSTEM, STATE VECTORS

For notation and signs see opposite page.

1. General load 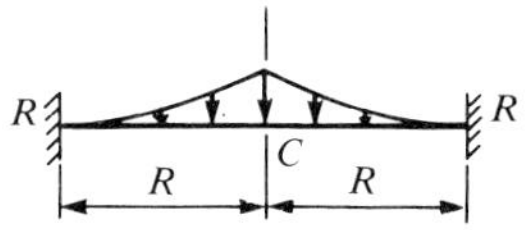	$w_C = -\dfrac{1}{2D}(G_\psi - 2G_w)R^2$ $\psi_C = 0$ $M_C = -2\alpha G_\psi$ $W_C = 0$	$w_R = 0$ $\psi_R = 0$ $M_R = G_M - 2\alpha G_\psi$ $W_R = G_W R^{-1}$
2. 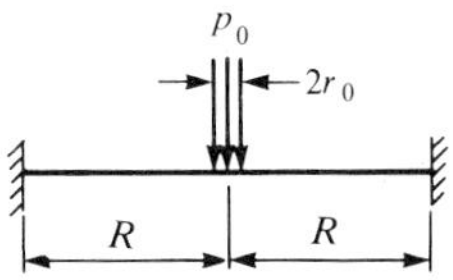$r_0 < R/100$	$w_C = +\dfrac{r_0^2}{16D} R^2 p_0$ $\psi_C = 0$ $M_C = \dfrac{\alpha}{4}(2 \ln R - 1) r_0^2 p_0$ $W_C = 0$	$w_R = 0$ $\psi_R = 0$ $M_R = -\frac{1}{4} r_0^2 p_0$ $W_R = -\frac{1}{2} r_0^2 R^{-1} p_0$
3. 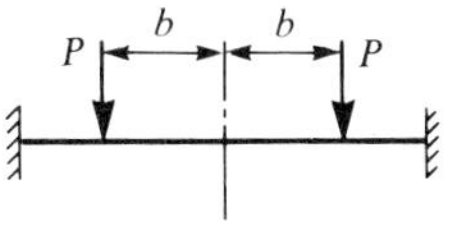	$w_C = \dfrac{1}{2D}(C_8 - 2C_7)R^3 P$ $\psi_C = 0$ $M_C = 2\alpha C_8 RP$ $W_C = 0$	$w_R = 0$ $\psi_R = 0$ $M_R = -(C_9 - 2\alpha C_8)RP$ $W_R = -bR^{-1}P$
4. 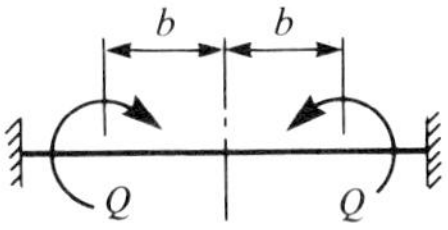	$w_C = +\dfrac{1}{2D}(C_5 - 2C_4)R^2 Q$ $\psi_C = 0$ $M_C = 2\alpha C_5 Q$ $W_C = 0$	$w_R = 0$ $\psi_R = 0$ $M_R = -(C_6 - 2\alpha C_5)Q$ $W_R = 0$
5. 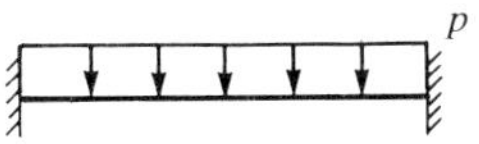	$w_C = \dfrac{1}{64D} R^4 p$ $\psi_C = 0$ $M_C = \dfrac{\alpha}{8} R^2 p$ $W_C = 0$	$w_R = 0$ $\psi_R = 0$ $M_R = -\frac{1}{8} R^2 p$ $W_R = -\frac{1}{2} R p$

23.10 ANNULAR PLATE, AXISYMMETRICAL SYSTEM, TRANSPORT RELATIONS

(1) Equivalents and signs

$$\overline{w} = \frac{D}{r^2} w = \text{scaled } w \text{ (N)}$$

$$\overline{\psi} = \frac{D}{r} \psi = \text{scaled } \psi \text{ (N)}$$

$$\overline{W} = rW = \text{scaled } W \text{ (N)}$$

$$\alpha = \tfrac{1}{2}(1 + \nu)$$

$$\beta = \tfrac{1}{2}(1 - \nu)$$

$$D = \frac{Eh^3}{12(1 - \nu^2)}$$

All forces, moments, and displacements are in the cylindrical coordinate system (2), and their positive directions are shown in (3, 4) below.

(2) Geometry

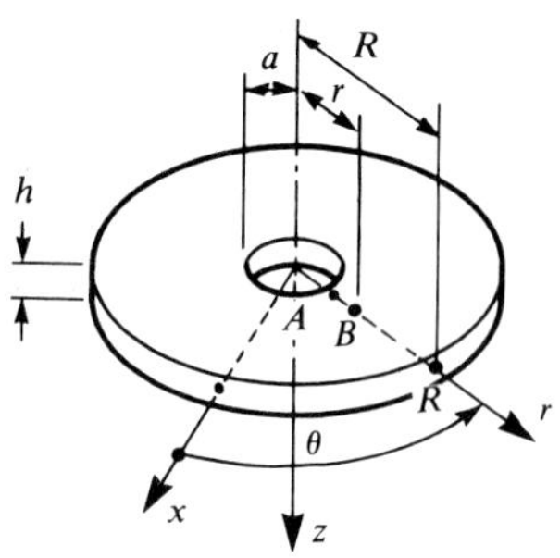

(3) Free-body diagrams

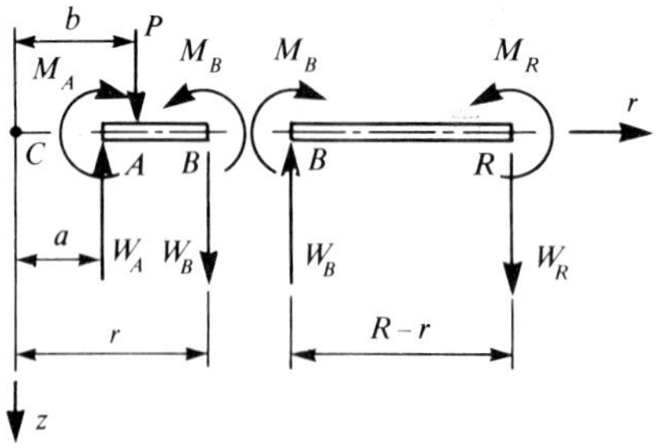

(4) Elastic surface

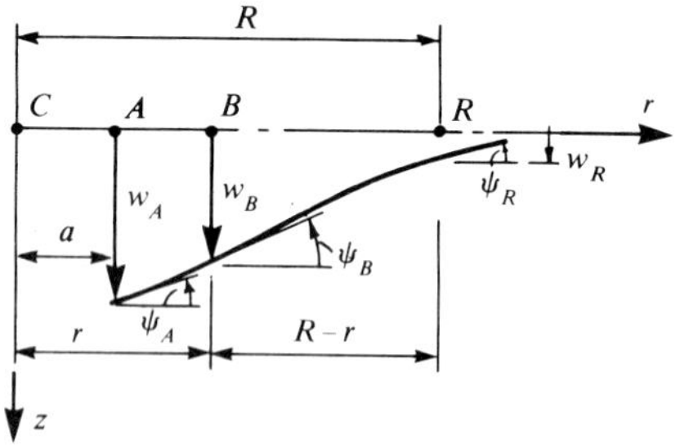

(5) Transport relations in segment AB between the state vectors $H_B(r)$ and $H_A(r)$ are given by the transport matrix $T_{BA}(r)$ and in segment AR between $H_A(R)$ and $H_R(R)$ by the transport matrix $T_{RA}(R)$ as shown in (23.11). In these matrix equations,

$$H_A(r) = \left\{ w_A \frac{D}{r^2}, \psi_A \frac{D}{r}, M_A, W_A r \right\}$$

$$H_B(r) = \left\{ w_B \frac{D}{r^2}, \psi_B \frac{D}{r}, M_B, W_B r \right\}$$

$$H_A(R) = \left\{ w_A \frac{D}{R^2}, \psi_A \frac{D}{R}, M_A, W_A R \right\}$$

$$H_R(R) = \left\{ w_R \frac{D}{R^2}, \psi_R \frac{D}{R}, M_R, W_R R \right\}$$

(6) Particular forms of the end state vectors $H_A(R)$ and $H_R(R)$ for 12 annular plates are tabulated in (23.12–23.13). The general expressions given in these tables allow the numerical calculation of these vector components for any of the causes considered in (23.06–23.07). It must be noted that the load functions G given in these tables for $r = r$ and $r = R$ are applicable in transport matrices of solid plates and annular plates if inserted in equations (23.03) and (23.11), respectively.

23.11 ANNULAR PLATE, AXISYMMETRICAL SYSTEM, TRANSPORT MATRICES

(1) Transport coefficients used in (2, 3) below are $\quad A_j = T_j\left(\frac{a}{r}\right) \qquad T_j = T_j\left(\frac{a}{R}\right)$

Their analytical forms are displayed in (23.02), and their numerical values for particular arguments are tabulated in (23.04, 23.05).

(2) Transport matrix equations in segment *AB*

$$\underbrace{\begin{bmatrix} 1 \\ \hline \bar{w}_B \\ \bar{\psi}_B \\ \hline M_B \\ \bar{W}_B \end{bmatrix}}_{H_B(r)} = \underbrace{\left[\begin{array}{c|cc|cc} 1 & 0 & 0 & 0 & 0 \\ \hline G_w(r) & 1 & -A_1 & -A_4 & -A_7 \\ G_\psi(r) & 0 & A_2 & A_5 & A_8 \\ \hline G_M(r) & 0 & A_3 & A_6 & A_9 \\ G_W(r) & 0 & 0 & 0 & T_0 \end{array}\right]}_{T_{BA}(r)} \underbrace{\begin{bmatrix} 1 \\ \hline \bar{w}_A \\ \bar{\psi}_A \\ \hline M_A \\ \bar{W}_A \end{bmatrix}}_{H_A(r)}$$

where $H_B(r)$ and $H_A(r)$ are the state vectors defined in (23.10).

(3) Transport matrix equations in segment *AR*

$$\underbrace{\begin{bmatrix} 1 \\ \hline \bar{w}_R \\ \bar{\psi}_R \\ \hline M_R \\ \bar{W}_R \end{bmatrix}}_{H_R(R)} = \underbrace{\left[\begin{array}{c|cc|cc} 1 & 0 & 0 & 0 & 0 \\ \hline G_w(R) & 1 & -T_1 & -T_4 & -T_7 \\ G_\psi(R) & 0 & T_2 & T_5 & T_8 \\ \hline G_M(R) & 0 & T_3 & T_6 & T_9 \\ G_W(R) & 0 & 0 & 0 & T_0 \end{array}\right]}_{T_{RA}(R)} \underbrace{\begin{bmatrix} 1 \\ \hline \bar{w}_A \\ \bar{\psi}_A \\ \hline M_A \\ \bar{W}_A \end{bmatrix}}_{H_A(R)}$$

where $H_R(R)$ and $H_A(R)$ are the state vectors defined in (23.10).

(4) Load functions in (2, 3) above are given in general form as

$$G_w(r) = \frac{1}{r^2}\int_a^r \frac{1}{r}\int_a^r r\int_a^r \frac{1}{r}\int_a^r pr\,dr\,dr\,dr\,dr \qquad G_\psi(r) = -\frac{1}{r}\int_a^r r\int_a^r \frac{1}{r}\int_a^r pr\,dr\,dr\,dr$$

$$G_M(r) = -\int_a^r \frac{1}{r}\int_a^r pr\,dr\,dr + \frac{2\beta}{r^2}\int_a^r r\int_a^r \frac{1}{r}\int_a^r pr\,dr\,dr\,dr \qquad G_W(r) = -\int_a^r pr\,dr$$

$$G_w(R) = \frac{1}{R^2}\int_a^R \frac{1}{r}\int_a^r r\int_a^r \frac{1}{r}\int_a^r pr\,dr\,dr\,dr\,dr \qquad G_\psi(R) = -\frac{1}{R}\int_a^R r\int_a^r \frac{1}{r}\int_a^r pr\,dr\,dr\,dr$$

$$G_M(R) = -\int_a^R \frac{1}{r}\int_a^r pr\,dr\,dr + \frac{2\beta}{R^2}\int_a^R r\int_a^r \frac{1}{r}\int_a^r pr\,dr\,dr\,dr \qquad G_W(R) = -\int_a^R pr\,dr$$

and their particular forms are listed in (23.06–23.07).

(5) Application of matrix equations introduced in (2, 3) above follows the standard pattern of the transport method. Equation (3) is used for the calculation of the unknown boundary values, and equation (2) is used for the calculation of state vectors at any distance r from the center of the plate (Ref. 11, pp. 288–292; Ref. 56, pp. 334–362).

23.12 ANNULAR PLATE, AXISYMMETRICAL SYSTEM, STATE VECTORS

Notation (23.01, 23.10–23.11) Signs (23.10) Matrix (23.11)

Six typical cases of annular plate in a state of axisymmetrical deformation are introduced below. The causes of deformation are loads, abrupt displacements, and/or temperature change (23.06–23.07). The resulting end state vectors at the inner edge A and at the outer edge R are given in each case in terms of

G_w, G_ψ, G_M, G_W = transport load functions at $r = R$ (N) (23.06–23.07)

$T_0, T_1, \ldots, T_9$ = transport coefficients (23.02, 23.04–23.05)

$$\bar{w}_A = \frac{D}{R^2} w_A,\ \bar{\psi}_A = \frac{D}{R}\psi_A,\ \overline{W}_A = RW_A,\ \bar{w}_R = \frac{D}{R^2} w_R,\ \bar{\psi}_R = \frac{D}{R}\psi_R,\ \overline{W}_R = RW_R$$

Case	Outer edge	Inner edge	Parameters
1	Simply supported	Free	$\alpha = \frac{1}{2}(1+\nu)$
2	Simply supported	Simply supported	$\beta = \frac{1}{2}(1-\nu)$
3	Free	Simply supported	$D = \dfrac{Eh^3}{12(1-\nu^2)}$
4	Simply supported	Guided	
5	Simply supported	Fixed	
6	Guided	Simply supported	

1.

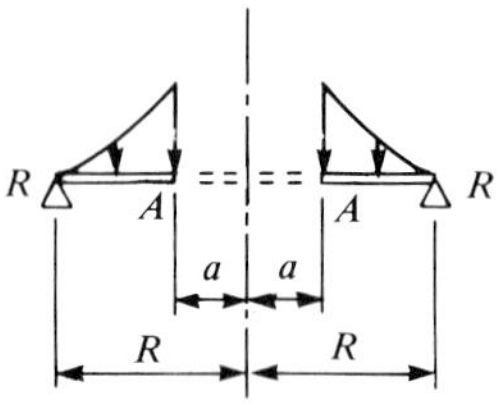

$$\bar{w}_A = -\frac{T_1}{T_3}G_M - G_w \qquad \bar{w}_R = 0$$

$$\bar{\psi}_A = -\frac{G_M}{T_3} \qquad \bar{\psi}_R = G_\psi + T_2\bar{\psi}_A$$

$$M_A = 0 \qquad M_R = 0$$

$$\overline{W}_A = 0 \qquad \overline{W}_R = G_W$$

2.

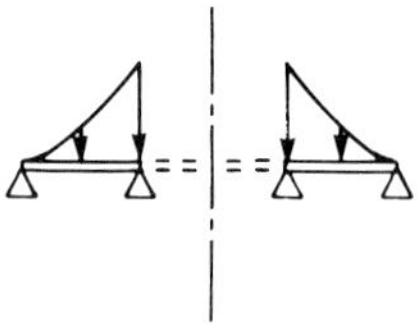

$$\bar{w}_A = 0 \qquad \bar{w}_R = 0$$

$$\bar{\psi}_A = \frac{T_9 G_w + T_7 G_M}{T_1 T_9 - T_3 T_7} \qquad \bar{\psi}_R = G_\psi + T_2\bar{\psi}_A + T_8\overline{W}_A$$

$$M_A = 0 \qquad M_R = 0$$

$$\overline{W}_A = -\frac{T_3 G_w + T_1 G_M}{T_1 T_9 - T_3 T_7} \qquad \overline{W}_R = G_W + T_0\overline{W}_A$$

3.

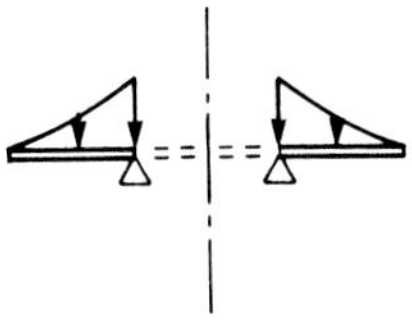

$$\bar{w}_A = 0 \qquad \bar{w}_R = G_W - T_1\bar{\psi}_A - T_7\bar{W}_A$$

$$\bar{\psi}_A = -\frac{T_0G_M - T_9G_W}{T_0T_3} \qquad \bar{\psi}_R = G_\psi + T_2\bar{\psi}_A + T_8\bar{W}_A$$

$$M_A = 0 \qquad M_R = 0$$

$$\bar{W}_A = -\frac{G_W}{T_0} \qquad \bar{W}_R = 0$$

4.

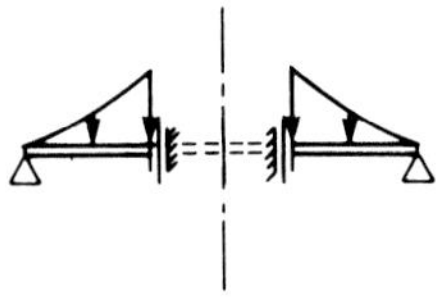

$$\bar{w}_A = -G_w - \frac{T_4}{T_6}G_M \qquad \bar{w}_R = 0$$

$$\bar{\psi}_A = 0 \qquad \bar{\psi}_R = G_\psi + T_5M_A$$

$$M_A = -\frac{G_M}{T_6} \qquad M_R = 0$$

$$\bar{W}_A = 0 \qquad \bar{W}_R = G_W$$

5.

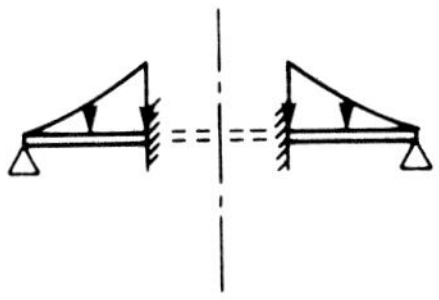

$$\bar{w}_A = 0 \qquad \bar{w}_R = 0$$

$$\bar{\psi}_A = 0 \qquad \bar{\psi}_R = G_\psi + T_5M_A + T_8\bar{W}_A$$

$$M_A = \frac{T_9G_w + T_7G_M}{T_4T_9 - T_6T_7} \qquad M_R = 0$$

$$\bar{W}_A = -\frac{T_6G_w + T_4G_M}{T_4T_9 - T_6T_7} \qquad \bar{W}_R = G_w + T_0\bar{W}_A$$

6.

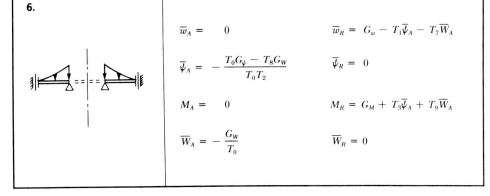

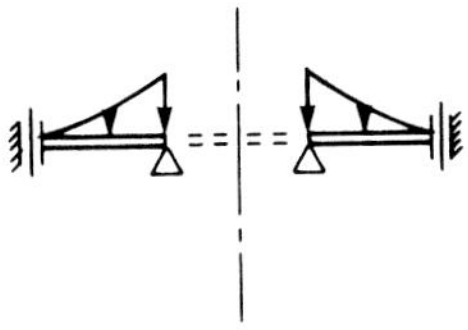

$$\bar{w}_A = 0 \qquad \bar{w}_R = G_w - T_1\bar{\psi}_A - T_7\bar{W}_A$$

$$\bar{\psi}_A = -\frac{T_0G_\psi - T_8G_W}{T_0T_2} \qquad \bar{\psi}_R = 0$$

$$M_A = 0 \qquad M_R = G_M + T_3\bar{\psi}_A + T_9\bar{W}_A$$

$$\bar{W}_A = -\frac{G_W}{T_0} \qquad \bar{W}_R = 0$$

23.13 ANNULAR PLATE, AXISYMMETRICAL SYSTEM, STATE VECTORS

Notation (23.01, 23.10–23.11) Signs (23.10) Matrix (23.11)

Six typical cases of annular plate in a state of axisymmetrical deformation are introduced below. The causes of deformation are loads, abrupt displacements, and/or temperature change (23.06–23.07). The resulting end state vectors at the inner edge A and at the outer edge R are given in each case in terms of

G_w, G_ψ, G_M, G_W = transport load functions at $r = R$ (N) (23.06–23.07)

$T_0, T_1, \ldots, T_9$ = transport coefficients (23.02, 23.04–23.05)

$$\bar{w}_A = \frac{D}{R^2} w_A, \quad \bar{\psi}_A = \frac{D}{R} \psi_A, \quad \bar{W}_A = RW_A, \quad \bar{w}_R = \frac{D}{R^2} w_R, \quad \bar{\psi}_R = \frac{D}{R} \psi_R, \quad \bar{W}_R = RW_R$$

Case	Outer edge	Inner edge	Parameters
1	Fixed	Free	$\alpha = \frac{1}{2}(1+\nu)$
2	Fixed	Fixed	$\beta = \frac{1}{2}(1-\nu)$
3	Fixed	Simply supported	
4	Fixed	Guided	$D = \dfrac{Eh^3}{12(1-\nu^2)}$
5	Free	Fixed	
6	Guided	Fixed	

1.

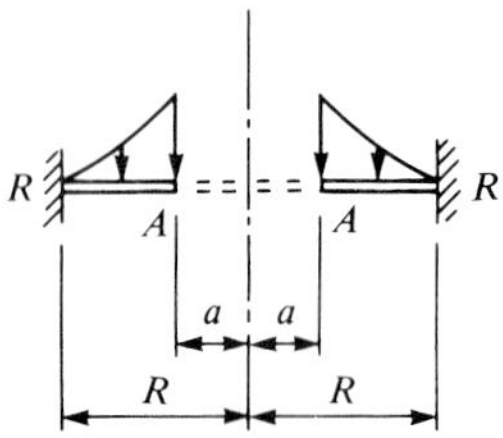

$$\bar{w}_A = -\frac{T_1}{T_2} G_\psi - G_w \qquad \bar{w}_R = 0$$

$$\bar{\psi}_A = -\frac{G_\psi}{T_2} \qquad \bar{\psi}_R = 0$$

$$M_A = 0 \qquad M_R = G_M + T_3 \bar{\psi}_A$$

$$\bar{W}_A = 0 \qquad \bar{W}_R = G_W$$

2.

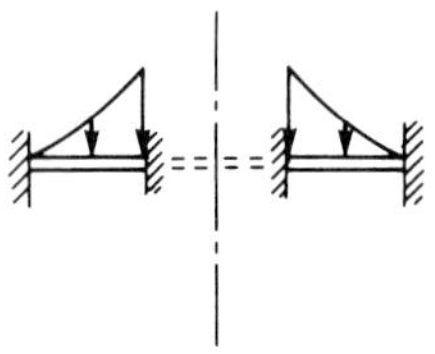

$$\bar{w}_A = 0 \qquad w_R = 0$$

$$\bar{\psi}_A = 0 \qquad \bar{\psi}_R = 0$$

$$M_A = \frac{T_8 G_w + T_7 G_\psi}{T_4 T_8 - T_5 T_7} \qquad M_R = G_M + T_6 M_A + T_9 \bar{W}_A$$

$$\bar{W}_A = -\frac{T_4 G_\psi + T_5 G_w}{T_4 T_8 - T_5 T_7} \qquad \bar{W}_R = G_W + T_0 \bar{W}_A$$

3.

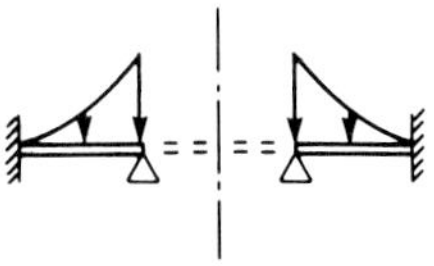

$$\overline{w}_A = 0 \qquad \overline{w}_R = 0$$

$$\overline{\psi}_A = \frac{T_8 G_w + T_7 G_\psi}{T_1 T_8 - T_2 T_7} \qquad \overline{\psi}_R = 0$$

$$M_A = 0 \qquad M_R = G_M + T_3 \overline{\psi}_A + T_9 \overline{W}_A$$

$$\overline{W}_A = -\frac{T_2 G_w + T_1 G_\psi}{T_1 T_8 - T_2 T_7} \qquad \overline{W}_R = G_W + T_0 \overline{W}_A$$

4.

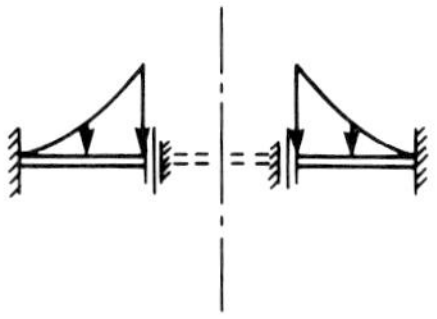

$$\overline{w}_A = -\frac{T_4}{T_5} G_\psi - G_w \qquad \overline{w}_R = 0$$

$$\overline{\psi}_A = 0 \qquad \overline{\psi}_R = 0$$

$$M_A = -\frac{G_\psi}{T_5} \qquad M_R = G_M + T_6 M_A$$

$$\overline{W}_A = 0 \qquad \overline{W}_R = G_W$$

5.

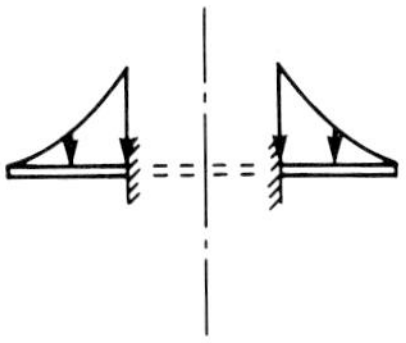

$$\overline{w}_A = 0 \qquad \overline{w}_R = G_w - T_4 M_A - T_7 \overline{W}_A$$

$$\overline{\psi}_A = 0 \qquad \overline{\psi}_R = G_\psi + T_5 M_A + T_8 \overline{W}_A$$

$$M_A = \frac{T_9 G_w - T_0 G_M}{T_0 T_6} \qquad M_R = 0$$

$$\overline{W}_A = -\frac{G_W}{T_0} \qquad \overline{W}_R = 0$$

6.

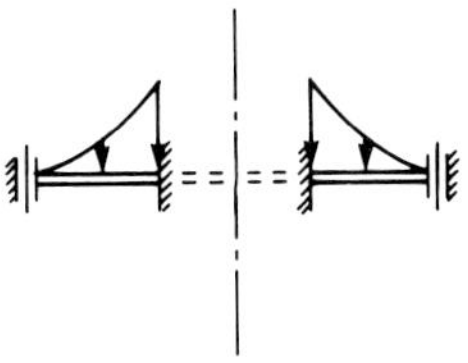

$$\overline{w}_A = 0 \qquad \overline{w}_R = G_w - T_4 M_A - T_7 \overline{W}_A$$

$$\overline{\psi}_A = 0 \qquad \overline{\psi}_R = 0$$

$$M_A = -\frac{T_0 G_\psi - T_8 G_W}{T_0 T_5} \qquad M_R = G_M + T_6 M_A + T_9 \overline{W}_A$$

$$\overline{W}_A = -\frac{G_W}{T_0} \qquad \overline{W}_R = 0$$

23.14 SOLID PLATE, AXISYMMETRICAL SYSTEM, FLEXIBILITY AND STIFFNESS EQUATIONS

(1) Notation

F_{RR1}	= unit force flexibility coefficient (m^2/N)	K_{RR1}	= unit deflection stiffness coefficient (N/m^2)
F_{RR2}	= unit moment flexibility coefficient [$m/(N\cdot m)$]	K_{RR2}	= unit slope stiffness coefficient ($N\cdot m/m$)
ψ_{R0}	= flexibility load function (rad)	M_{R0}	= stiffness load function ($N\cdot m/m$)
u_{RC}	= horizontal end deflection (m)	u_R	= horizontal end deflection (m)
ψ_{RC}	= end slope (rad)	ψ_R	= end slope (rad)
U_R	= horizontal end force (N/m)	U_{RC}	= horizontal end force (N/m)
M_R	= end moment ($N\cdot m/m$)	M_{RC}	= end moment ($N\cdot m/m$)
W_R	= vertical end force (N/m)	W_{RC}	= vertical end force (N/m)

G_ψ, G_M, G_W = transport load functions (N) (23.06–23.07), $\alpha = \frac{1}{2}(1+\nu)$, $\beta = \frac{1}{2}(1-\nu)$. All forces, moments, and displacements are positive as shown in (2) below.

(2) Shape of deformed midsurface

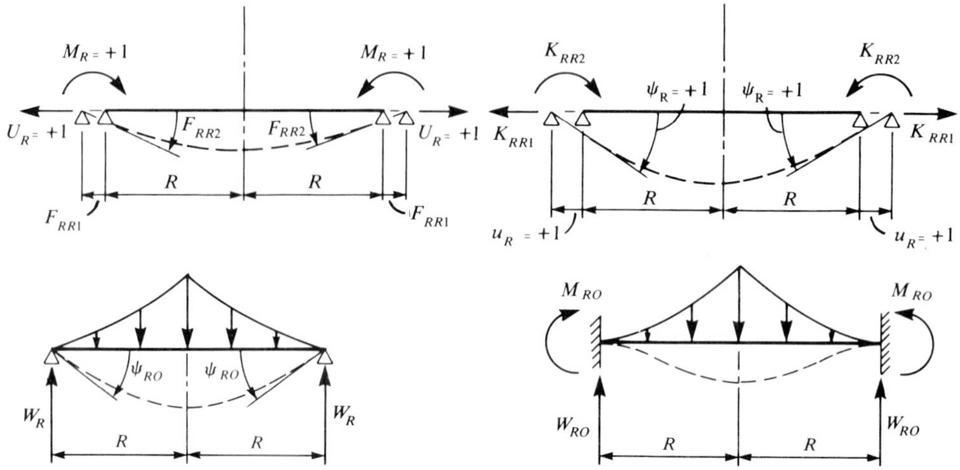

(3) Flexibility equations

$$u_{RC} = F_{RR1}U_R$$
$$\psi_{RC} = F_{RR2}M_R + \psi_{R0}$$

where

$$F_{RR1} = \frac{2\beta R}{Eh} \qquad F_{RR2} = \frac{R}{2\alpha D}$$
$$\psi_{R0} = \frac{R}{2\alpha D}(2\alpha G_\psi - G_M)$$

(4) Stiffness equations

$$U_{RC} = K_{RR1}u_R$$
$$M_{RC} = K_{RR2}\psi_R + M_{R0}$$

where

$$K_{RR1} = \frac{Eh}{2\beta R} \qquad K_{RR2} = \frac{2\alpha D}{R}$$
$$M_{R0} = G_M - 2\alpha G_\psi$$

(5) Load functions ψ_{R0} and M_{R0} are each tabulated for eight particular cases in (23.15) and (23.16), respectively, and $W_R = W_{R0} = -G_W/R$.

23.15 SOLID PLATE, AXISYMMETRICAL SYSTEM, FLEXIBILITY LOAD FUNCTIONS

Notation (23.14) — Signs (23.14) — Matrix (23.14)

P = concentrated load per unit length (N/m)
Q = applied couple per unit length (N·m/m)
p, p_0 = intensities of distributed load (N/m²)
ψ_B = abrupt slope (rad)
g_t = thermal strain (1/m)
R, r_0 = radii (m)

1.

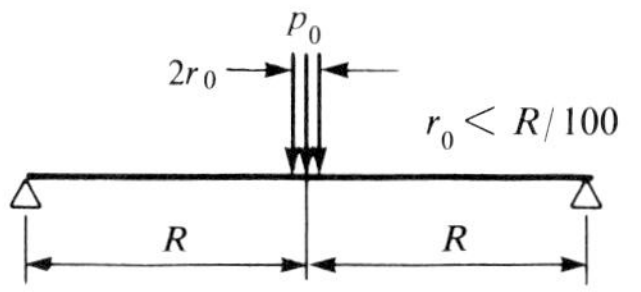

$$\psi_{R0} = \frac{r_0^2 p_0 R}{8\alpha D} \qquad W_R = \frac{r_0^2 p_0}{2R}$$

2.

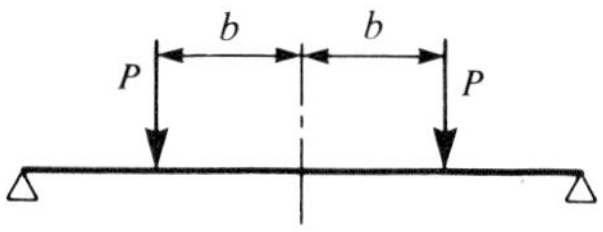

$$\psi_{R0} = \frac{(R^2 - b^2)bP}{4\alpha DR} \qquad W_R = \frac{bP}{R}$$

3.

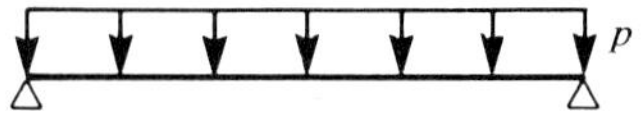

$$\psi_{R0} = \frac{R^3 p}{16\alpha D} \qquad W_R = \frac{Rp}{2}$$

4. Abrupt slope at B

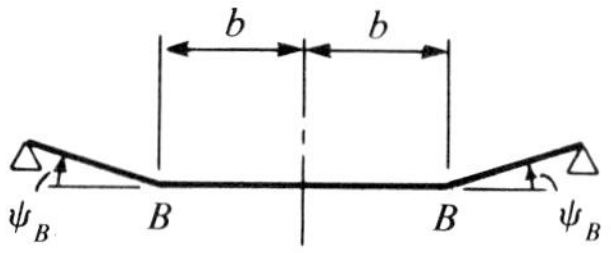

$$\psi_{R0} = \psi_B \frac{b}{R} \qquad W_R = 0$$

5.

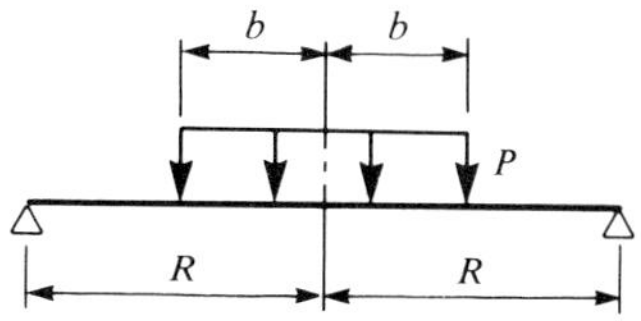

$$\psi_{R0} = \frac{(2R^2 - b^2)b^2 p}{16\alpha DR} \qquad W_R = \frac{b^2 p}{2R}$$

6.

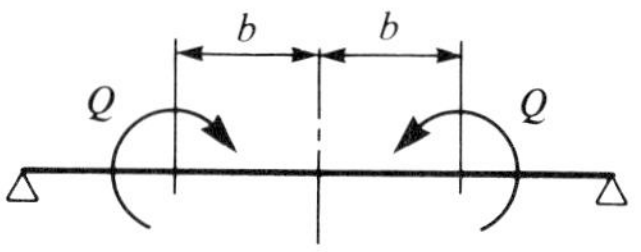

$$\psi_{R0} = \frac{b^2 Q}{2\alpha DR} \qquad W_R = 0$$

7.

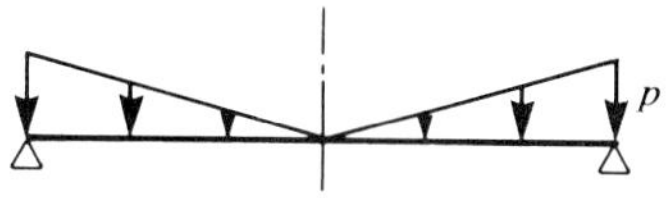

$$\psi_{R0} = \frac{R^3 p}{30\alpha D} \qquad W_R = \frac{Rp}{3}$$

8. Temperature deformation in segment BR

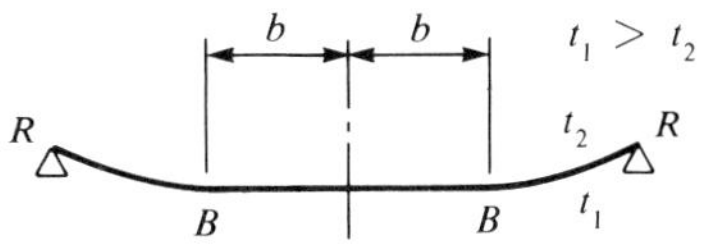

$$\psi_{R0} = \frac{R^2 - b^2}{R} g_t \qquad W_R = 0$$

23.16 SOLID PLATE, AXISYMMETRICAL SYSTEM, STIFFNESS LOAD FUNCTIONS

Notation (23.14) | Signs (23.14) | Matrix (23.14)

P = concentrated load per unit length (N/m)
Q = applied couple per unit length (N·m/m)
p, p_0 = intensities of distributed load (N/m_2)

ψ_B = abrupt slope (rad)
g_t = thermal strain (1/m)
R, r_0 = radii (m)

1.

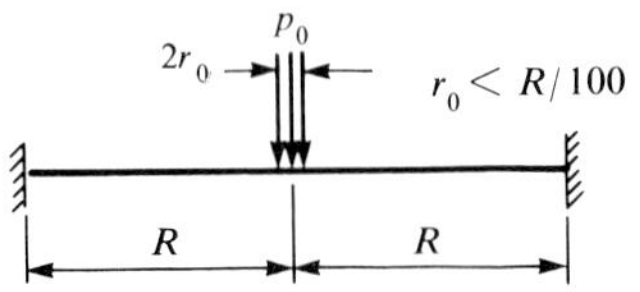

$$M_{R0} = -\tfrac{1}{4} r_0^2 p_0 \qquad W_{R0} = \frac{r_0^2 p_0}{2R}$$

2.

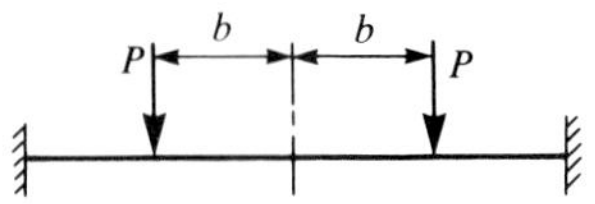

$$M_{R0} = -\frac{(R^2 - b^2)bP}{2R^2} \qquad W_{R0} = \frac{bP}{R}$$

3.

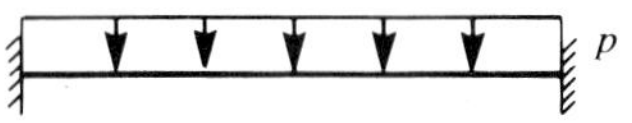

$$M_{R0} = -\frac{R^2 p}{8} \qquad W_{R0} = \frac{Rp}{2}$$

4. Abrupt slope at B

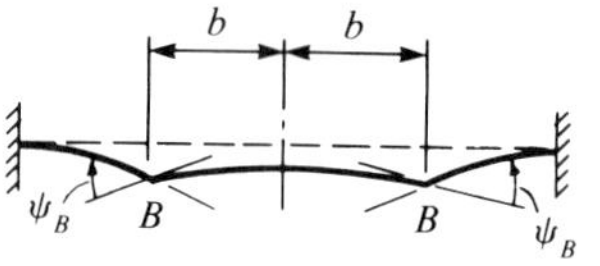

$$M_{R0} = -\frac{2\alpha b D}{R^2}\psi_B \qquad W_{R0} = 0$$

5.

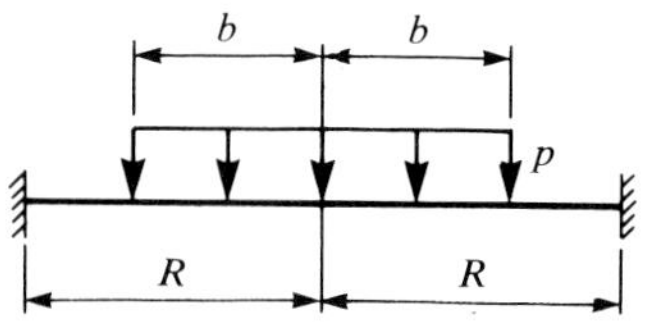

$$M_{R0} = -\frac{(2R^2 - b^2)b^2 p}{8R^2} \qquad W_{R0} = \frac{b^2 p}{2R}$$

6.

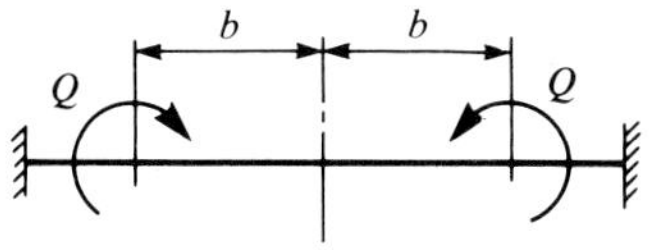

$$M_{R0} = -\frac{b^2 Q}{R^2} \qquad W_{R0} = 0$$

7.

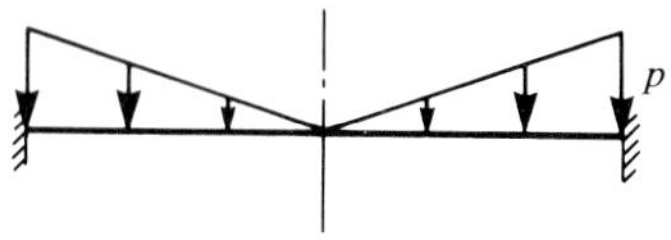

$$M_{R0} = -\frac{R^2 p}{15} \qquad W_{R0} = \frac{Rp}{3}$$

8. Temperature deformation in segment BR

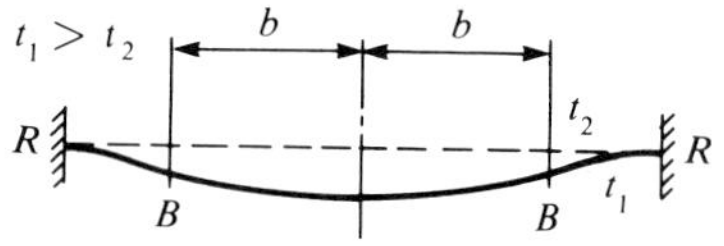

$$M_{R0} = -\frac{2(R^2 - b^2)\alpha D}{R^2} g_t \qquad W_{R0} = 0$$

24

Rectangular Isotropic Plate Constant Thickness

STATIC STATE

24.01 DEFINITION OF STATE

(1) System considered is a finite isotropic rectangular plate of constant thickness forming with its transverse loads and reactions a static equilibrium.

(2) Notation (A.1, A.5)

s, f	= length, width of plate (m)	x, y	= coordinates (m)
h	= thickness of plate (m)	a, b, c, d	= segments (m)
P	= concentrated load (N)	Q	= applied couple (N·m)
P'	= intensity of line load (N/m)	Q'	= intensity of line couple (N·m/m)
p	= intensity of distributed load (N/m^2)	q	= intensity of distributed couple (N·m/m^2)
E	= modulus of elasticity (Pa)	ν	= Poisson's ratio
w	= vertical deflection (m)	M	= intensity of flexural moment (N·m/m)
W	= intensity of shear (N/m)	ψ	= flexural slope (rad)
R	= intensity of edge reaction (N/m)		

$\alpha = \dfrac{1+\nu}{2}$	$\beta = \dfrac{1-\nu}{2}$	$\gamma = \dfrac{3-2\nu-\nu^2}{4}$
$D = \dfrac{Eh^3}{12(1-\nu^2)}$	$\lambda = \dfrac{m\pi}{f}$	$m = 1, 2, 3, \ldots$

(3) Assumptions of analysis are stated in (22.01). In this chapter, the effects of the flexural deformations are included. Since h/s and h/f are less than 1/10, the effects of shearing deformations are neglected.

24.02 GENERAL SOLUTION

(1) Elastic surface of the rectangular isotropic plate described in (22.05), with simply supported edges of length s and with arbitrary boundary conditions along the edges of length f as indicated in (2) below, is defined by the deflection equation

$$w = \sum_{m=1}^{\infty} X_m \sin \lambda y \qquad \lambda = \frac{m\pi}{f}$$

where X_m is the shape function in x, which when magnified by $\sin \lambda y$ satisfies the differential equation (22.04-7) and the boundary conditions (22.05) of the plate (Ref. 33, pp. 324–329).

(2) Geometry

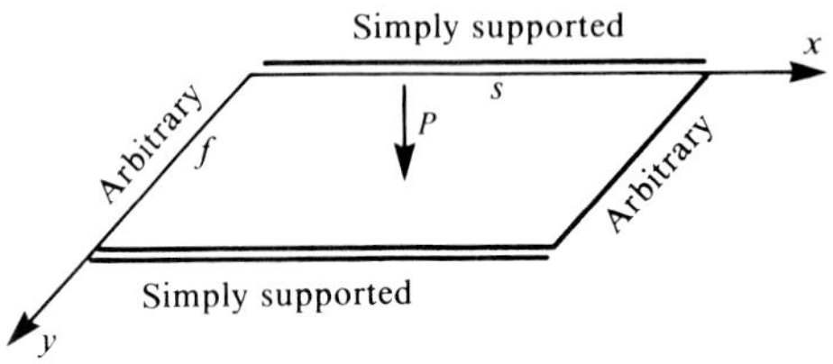

(3) Free-body diagram

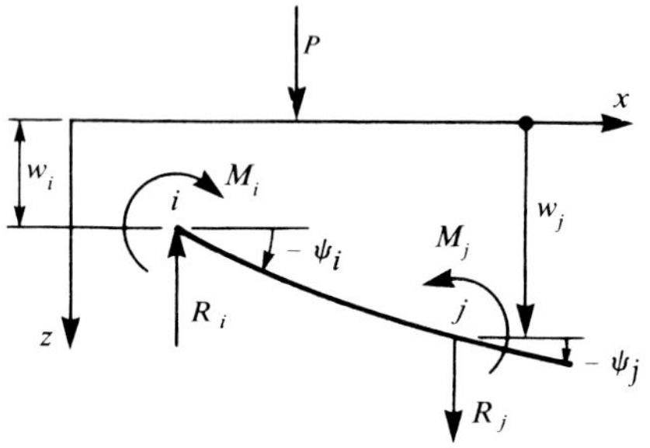

(4) Intensity of load can be also expanded into a series as

$$p(x, y) = \sum_{m=1}^{\infty} \Phi(x) \sin \lambda y = \sum_{m=1}^{\infty} p_m(x) \sin \lambda y$$

where $$\Phi(x) = \frac{2}{f} \int_0^f p(x, y) \sin \lambda y \, dy = p_m(x)$$

(5) Differential equation (22.04-7) in terms of $X_m \sin \lambda$ and $p_m(x) \sin \lambda y$ becomes

$$\boxed{\left(\frac{d^4 X_m}{dx^4} - 2\lambda^2 \frac{d^2 X_m}{dx^2} + \lambda^4 X_m\right) \sin \lambda y = \frac{p_m(x)}{D} \sin \lambda y}$$

and yields the Laplace transform solution

$$w_m = X_m \sin \lambda y = [X_m(0)X_{1m} + X'_m(0)X_{2m} + X''_m(0)X_{3m} + X'''_m(0)X_{4m} + L_m]\sin \lambda y$$

where $X_m(0)$, $X'_m(0)$, $X''_m(0)$, $X'''_m(0)$ are the initial values,

$X_{1m} = \cosh \lambda x - \frac{\lambda x}{2} \sinh \lambda x$	$X_{2m} = \frac{1}{2\lambda}(3 \sinh \lambda x - \lambda x \cosh \lambda x)$
$X_{3m} = \frac{\lambda x}{2\lambda^2} \sinh \lambda x$	$X_{4m} = \frac{1}{2\lambda^3}(\lambda x \cosh \lambda x - \sinh \lambda x)$

are the shape functions, and L_m is the convolution integral (24.03-4).

24.03 ANALYTICAL FUNCTIONS

(1) Plate functions introduced in (22.03-3) become

$$w = \sum_{m=1}^{\infty} X_m \sin \lambda y \qquad \psi = -\sum_{m=1}^{\infty} \frac{\partial X_m}{\partial x} \sin \lambda y$$

$$M = M_x = -D \sum_{m=1}^{\infty} \left(\frac{\partial^2 X_m}{\partial x^2} - \nu \lambda^2 X_m \right) \sin \lambda y \qquad W_{xz} = -D \sum_{m=1}^{\infty} \left(\frac{\partial^3 X_m}{\partial x^3} - \lambda^2 \frac{\partial X_m}{\partial x} \right) \sin \lambda y$$

$$M_y = -D \sum_{m=1}^{\infty} \left(\nu \frac{\partial^2 X_m}{\partial x^2} - \lambda^2 X_m \right) \sin \lambda y \qquad W_{yz} = -D \sum_{m=1}^{\infty} \lambda \left(\frac{\partial^2 X_m}{\partial x^2} - \lambda^2 X_m \right) \cos \lambda y$$

$$M_{xy} = -2\beta D \sum_{m=1}^{\infty} \left(\lambda \frac{\partial X_m}{\partial x} \right) \cos \lambda y = M_{yx}$$

$$R = R_{xz} = -D \sum_{m=1}^{\infty} \left[\frac{\partial^3 X_m}{\partial x^3} - (2 - \nu) \lambda^2 \frac{\partial X_m}{\partial x} \right] \sin \lambda y$$

$$R_{yz} = -D \sum_{m=1}^{\infty} \lambda \left[(2 - \nu) \frac{\partial^2 X_m}{\partial x^2} - \lambda^2 X_m \right] \cos \lambda y$$

$$R_{(x=0,y=0)} = 2M_{xy} \qquad R_{(x=s,y=f)} = 2M_{xy} \qquad R_{(x=s,y=0)} = -2M_{xy} \qquad R_{(x=0,y=f)} = -2M_{xy}$$

(2) Left-end boundary values in terms of initial values by means of (1) above are

$$w_{Lm} = X_m(0) \qquad M_{Lm} = -D[X''(0) - \nu\lambda^2 X(0)]$$

$$\psi_{Lm} = -X_m'(0) \qquad R_{Lm} = -D[X'''(0) - (2 - \nu)\lambda^2 X'(0)]$$

where $X_m'(0) = dX_m(0)/dx$, $X''_m(0) = d^2X_m(0)/dx^2$, $X_m'''(0) = d^3X_m/dx^3$.

(3) Initial values in terms of boundary values are

$$X_m(0) = w_{Lm} \qquad X''(0) = -\frac{M_{Lm}}{D} + \nu\lambda^2 w_{Lm}$$

$$X_m'(0) = -\psi_{Lm} \qquad X'''(0) = -\frac{R_{Lm}}{D} - (2 - \nu)\lambda^2 \psi_{Lm}$$

(4) Final form of equation of elastic surface (24.02-1) by relations (3) becomes

$$\begin{aligned} w &= \sum_{m=1}^{\infty} X_m \sin \lambda y \\ &= \sum_{m=1}^{\infty} \left\{ (X_{1m} + \nu\lambda^2 X_{3m}) w_{Lm} - [X_{2m} + (2 - \nu)\lambda^2 X_{4m}]\psi_{Lm} - X_{3m} \frac{M_{Lm}}{D} - X_{4m} \frac{R_{Lm}}{D} + L_m \right\} \sin \lambda y \end{aligned}$$

where $w_{Lm} \sin \lambda y$, $\psi_{Lm} \sin \lambda y$, $M_{Lm} \sin \lambda y$, $R_{Lm} \sin \lambda y$ are the mth terms of deflection, slope, moment, and transverse force at L ($x = 0$, $0 \le y \le f$), and

$$\boxed{L_m = \frac{1}{2\lambda^3 D} \int_0^x p_m(x - \tau)(\lambda\tau \cosh \lambda\tau - \sinh \lambda}$$

is the convolution integral (Ref. 1, pp. 228–229).

24.04 TRANSPORT MATRIX EQUATIONS

(1) Equivalents and signs

$$\bar{w} = 2D\lambda^3 w = \text{scaled } w \text{ (N/m)}$$
$$\bar{\psi} = 2D\lambda^2 \psi = \text{scaled } \psi \text{ (N/m)}$$
$$\bar{M} = \lambda M = \text{scaled } M \text{ (N/m)}$$

$$\lambda = \frac{m\pi}{f} \qquad D = \frac{Eh^3}{12(1-\nu^2)}$$
$$\alpha = \tfrac{1}{2}(1+\nu) \qquad \beta = \tfrac{1}{2}(1-\nu)$$
$$\gamma = \tfrac{1}{4}(3 - 2\nu - \nu^2)$$

All forces, moments, and displacements are in the cartesian system (22.02–22.03), and they are positive as shown in (24.02-3).

(2) Transport matrix equations in mth terms ($m = 1, 2, 3, \ldots$)

$$\underbrace{\begin{bmatrix} 1 \\ \hline \bar{w}_{Lm} \\ \bar{\psi}_{Lm} \\ \hline \bar{M}_{Lm} \\ R_{Lm} \end{bmatrix}}_{H_{Lm}} \sin \lambda y = \underbrace{\left[\begin{array}{c|cc|cc} 1 & 0 & 0 & 0 & 0 \\ \hline F_{wm} & T_1 - \beta T_4 & \alpha T_2 + \beta T_3 & -T_4 & -T_2 + T_3 \\ F_{\psi m} & \alpha T_2 - \beta T_3 & T_1 + \beta T_4 & -T_2 - T_3 & T_4 \\ \hline F_{Mm} & \beta^2 T_4 & -\gamma T_2 - \beta^2 T_3 & T_1 + \beta T_4 & -\alpha T_2 - \beta T_3 \\ F_{Rm} & -\gamma T_2 + \beta^2 T_3 & -\beta^2 T_4 & -\alpha T_2 + \beta T_3 & T_1 - \beta T_4 \end{array}\right]}_{T_{LRm}} \underbrace{\begin{bmatrix} 1 \\ \hline \bar{w}_{Rm} \\ \bar{\psi}_{Rm} \\ \hline \bar{M}_{Rm} \\ R_{Rm} \end{bmatrix}}_{H_{Rm}} \sin \lambda y$$

$$\underbrace{\begin{bmatrix} 1 \\ \hline \bar{w}_{Rm} \\ \bar{\psi}_{Rm} \\ \hline \bar{M}_{Rm} \\ R_{Rm} \end{bmatrix}}_{H_{Rm}} \sin \lambda y = \underbrace{\left[\begin{array}{c|cc|cc} 1 & 0 & 0 & 0 & 0 \\ \hline G_{wm} & T_1 - \beta T_4 & -\alpha T_2 - \beta T_3 & -T_4 & T_2 - T_3 \\ G_{\psi m} & -\alpha T_2 + \beta T_3 & T_1 + \beta T_4 & T_2 + T_3 & T_4 \\ \hline G_{Mm} & \beta^2 T_4 & \gamma T_2 + \beta^2 T_3 & T_1 + \beta T_4 & \alpha T_2 + \beta T_3 \\ G_{Rm} & \gamma T_2 - \beta^2 T_3 & -\beta^2 T_4 & \alpha T_2 - \beta T_3 & T_1 - \beta T_4 \end{array}\right]}_{T_{RLm}} \underbrace{\begin{bmatrix} 1 \\ \hline \bar{w}_{Lm} \\ \bar{\psi}_{Lm} \\ \hline \bar{M}_{Lm} \\ R_{Lm} \end{bmatrix}}_{H_{Lm}} \sin \lambda y$$

where

$$H_{Lm} \sin \lambda y = \{2D\lambda^3 w_{Lm}, 2D\lambda^2 \psi_{Lm}, \lambda M_{Lm}, R_{Lm}\} \sin \lambda y$$
$$H_{Rm} \sin \lambda y = \{2D\lambda^3 w_{Rm}, 2D\lambda^2 \psi_{Rm}, \lambda M_{Rm}, R_{Rm}\} \sin \lambda y$$

are the mth state vectors at the left end and the right end, respectively and T_{LRm}, T_{RLm} are their transport matrices whose elements are the transport coefficients $(T_1 - \beta T_4)$, $(\alpha T_2 + \beta T_3)$, $\ldots$, and the load functions F, G. The analytical forms of these elements are defined in (24.05).

(3) Intermediate mth state vector $H_{Dm}(x)$ along the section D given by $x = x$, $y = y$ is

$$H_{Dm}(x) \sin \lambda y = T_{DLm}(x) H_{Lm} \sin \lambda y$$

where $T_{DLm}(x)$ is the transport matrix T_{RLm} with $s = x$ and H_{Lm} is the known state vector at L.

(4) Final values of the state vectors are given by the truncated series as

$$H_L \cong \sum_{m=1}^{r} H_{Lm} \sin \lambda y \qquad H_D \cong \sum_{m=1}^{r} H_{Dm} \sin \lambda y \qquad H_R \cong \sum_{m=1}^{r} H_{Rm} \sin \lambda y$$

where r is the number of terms in the series required for the prescribed accuracy and y is the coordinate of the point where the vector is desired.

24.05 TRANSPORT RELATIONS

(1) Transport functions

$T_1(x) = \cosh \lambda x$	$T_2(x) = \sinh \lambda x$
$T_3(x) = \lambda x \cosh \lambda x$	$T_4(x) = \lambda x \sinh \lambda x$
$T_5(x) = \cosh \lambda x - 1$	$T_6(x) = \lambda x \cosh \lambda x - \sinh \lambda x$
$T_7(x) = \lambda x \sinh \lambda x - \cosh \lambda x + 1$	
$T_8(x) = 2 \sinh \lambda x - 2\lambda x \cosh \lambda x + (\lambda x)^2 \sinh \lambda x$	
$T_9(x) = 2 \cosh \lambda x - 2\lambda x \sinh \lambda x + (\lambda x)^2 \cosh \lambda x - 2$	

(2) Transport coefficients are particular values of $T_j(x)$.

For $x = s$, $\quad T_j(s) = T_j$
For $x = a$, $\quad T_j(a) = A_j$ $\qquad j = 1, 2, \ldots, 9$
For $x = b$, $\quad T_j(b) = B_j$
and $T_j(s) - T_j(a) = T_j - A_j = C_j$

(3) Load functions F and G in (24.04-2) are given for particular cases in (24.06–24.12). With the exception of (24.12), all other cases are listed in groups, where the first case in each table is given in complete form and the remaining cases are identified by their load factor Φ only.

(4) Final values of state-vector components are the sums of r solutions. Thus

$$w \cong \sum_{m=1}^{r} \frac{1}{2D\lambda^3} \bar{w}_m \sin \lambda y \qquad \psi \cong \sum_{m=1}^{r} \frac{1}{2D\lambda^2} \bar{\psi}_m \sin \lambda y$$

$$M_x \cong \sum_{m=1}^{r} \frac{1}{\lambda} \bar{M}_m \sin \lambda y \qquad M_y \cong \sum_{m=1}^{r} \frac{1}{\lambda} (\nu \bar{M}_m + 2\alpha\beta \bar{w}_m) \sin \lambda y$$

$$M_{xy} = M_{yx} \cong \sum_{m=1}^{r} \frac{\beta}{\lambda} \bar{\psi}_m \cos \lambda y$$

$$W_{xz} \cong \sum_{m=1}^{r} (R_m + \beta \bar{\psi}_m) \sin \lambda y \qquad W_{yz} \cong \sum_{m=1}^{r} (M_m + \beta \bar{w}_m) \cos \lambda y$$

$$R_{xz} \cong \sum_{m=1}^{r} R_m \sin \lambda y \qquad R_{yz} \cong \sum_{m=1}^{r} [(2 - \nu)\bar{M}_m + \beta \bar{w}_m] \cos \lambda y$$

where $\bar{w}_m$, $\bar{\psi}_m$, $\bar{M}_m$, and R_m are the results of the mth transport equation and r is the number of solutions required for a result of good accuracy.

(5) Application of transport matrix equations in (23.04) follows the standard pattern of the transport method. As in the beam analysis, eight boundary values are involved of which four are always zero and four are unknown. Thus four equations, in general, are necessary for the solution of a given plate. The transport matrix equation provides these equations, of which two and only two must be solved simultaneously (Ref. 11, pp. 363–367).

24.06 TRANSPORT LOAD FUNCTIONS

$$\lambda = \frac{m\pi}{f}$$

Notation (24.01–24.04) | Signs (24.04) | Matrix (24.04)

P = concentrated load (N)

P' = intensity of line load (N/m)

A_j, B_j, C_j, T_j = transport coefficients (24.05)

Φ = load factor (N/m)

For cases (2)–(7) use formulas of case (1) in terms of the respective Φ

1.

$$F_{wm} = -(A_2 - A_3)\Phi \qquad G_{wm} = -(B_2 - B_3)\Phi$$

$$F_{\psi m} = A_4\Phi \qquad G_{\psi m} = -B_4\Phi$$

$$F_{Mm} = -(\alpha A_2 + \beta A_3)\Phi \qquad G_{Mm} = -(\alpha B_2 + \beta B_3)\Phi$$

$$F_{Rm} = (A_1 - \beta A_4)\Phi \qquad G_{Rm} = -(B_1 - \beta B_4)\Phi$$

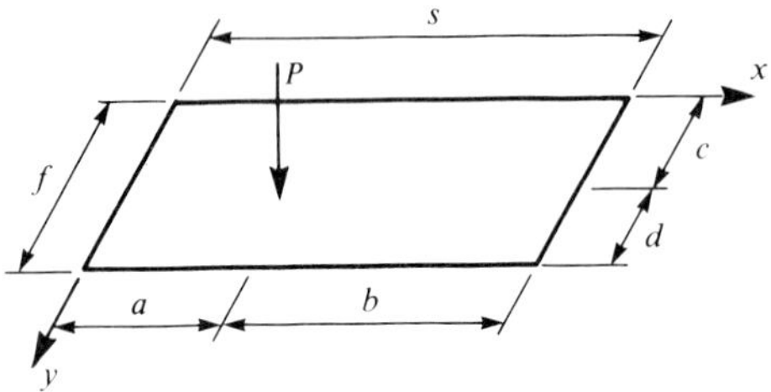

$$\Phi = \frac{2P}{f}\sin\lambda c$$

2.

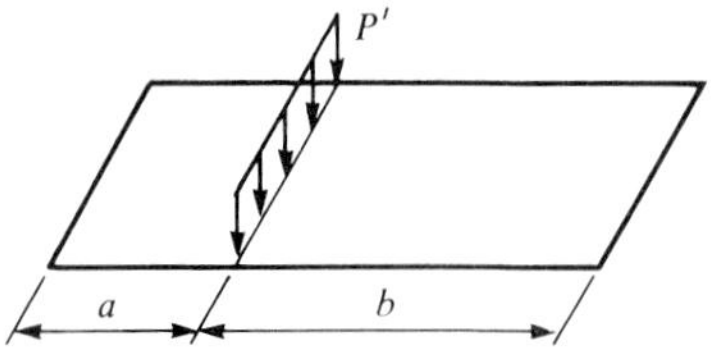

$$\Phi = \frac{2P'}{m\pi}(1 - \cos\lambda f)$$

5.

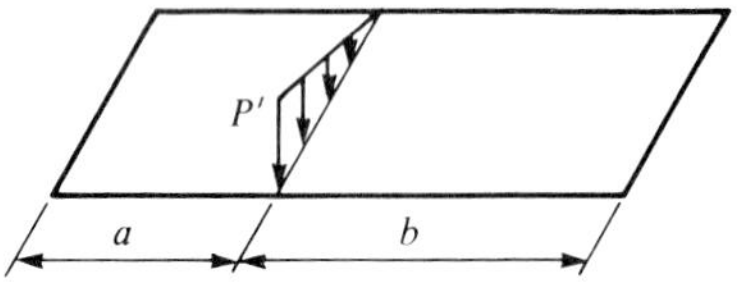

$$\Phi = \frac{2P'}{(m\pi)^2}(\sin\lambda f - \lambda f\cos\lambda f)$$

3.

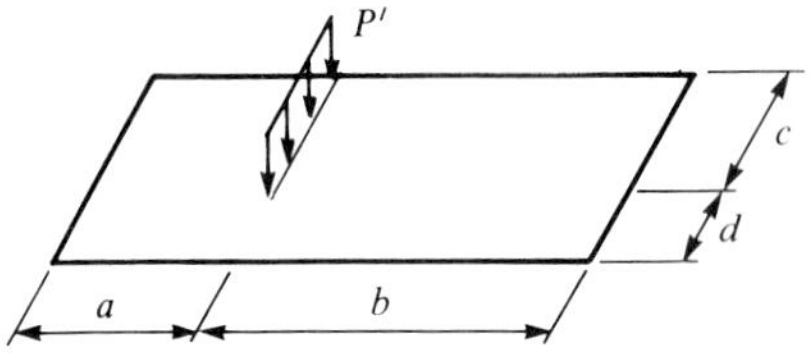

$$\Phi = \frac{2P'}{m\pi}(1 - \cos\lambda c)$$

6.

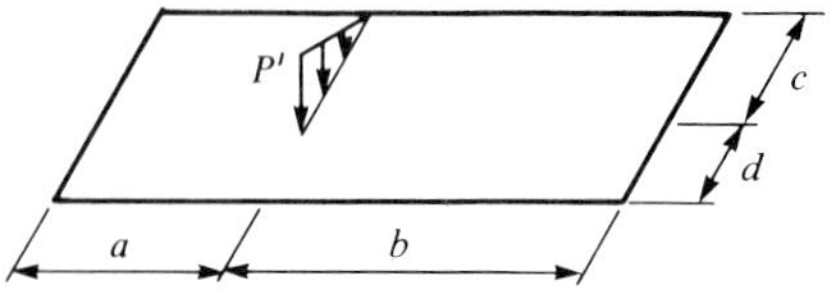

$$\Phi = \frac{2P'}{c f\lambda^2}(\sin\lambda c - \lambda c\cos\lambda c)$$

4.

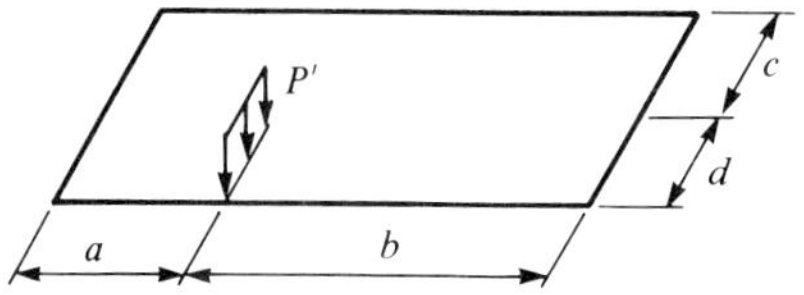

$$\Phi = \frac{2P'}{m\pi}(\cos\lambda c - \cos\lambda f)$$

7.

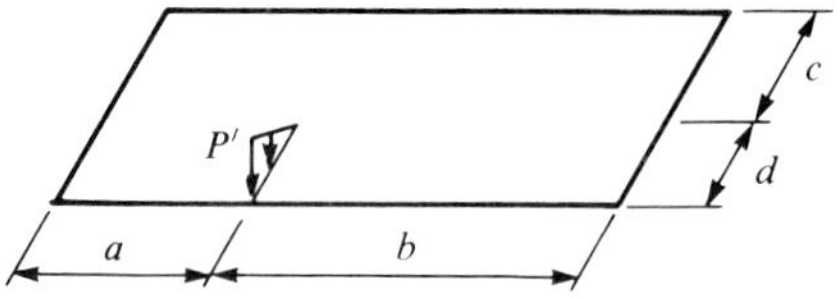

$$\Phi = \frac{2P'}{d f\lambda^2}(\sin\lambda f - \sin\lambda c - \lambda d\cos\lambda f)$$

24.07 TRANSPORT LOAD FUNCTIONS

Notation (24.01–24.04)	Signs (24.04)	Matrix (24.04)	$\lambda = \dfrac{m\pi}{f}$
Q = applied couple (N·m) Q' = intensity of line couple (N) A_j, B_j, C_j, T_j = transport coefficients (24.05) For cases (2)–(7) use formulas of case (1) in terms of the respective Φ		Φ = load factor (N/m)	

1.

$F_{wm} = -A_4\Phi$

$F_{\psi m} = -(A_2 + A_3)\Phi$

$F_{Mm} = (A_1 + \beta A_4)\Phi$

$F_{Rm} = -(\alpha A_1 - \beta A_3)\Phi$

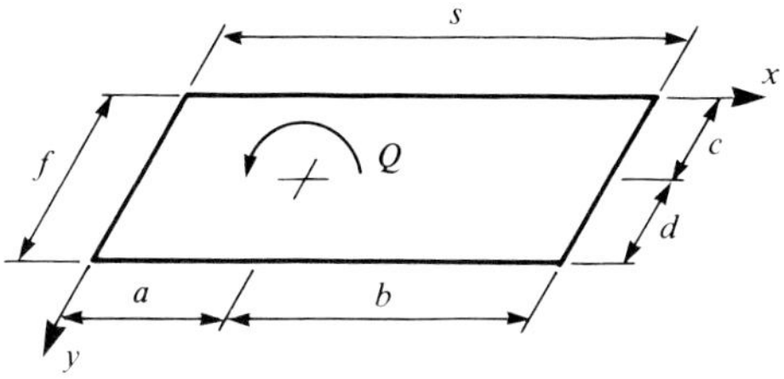

$$\Phi = \frac{2Q\lambda}{f}\sin\lambda c$$

$G_{wm} = B_4\Phi$

$G_{\psi m} = -(B_2 + B_3)\Phi$

$G_{Mm} = -(B_1 + \beta B_4)\Phi$

$G_{Rm} = -(\alpha B_2 - \beta B_3)\Phi$

2.

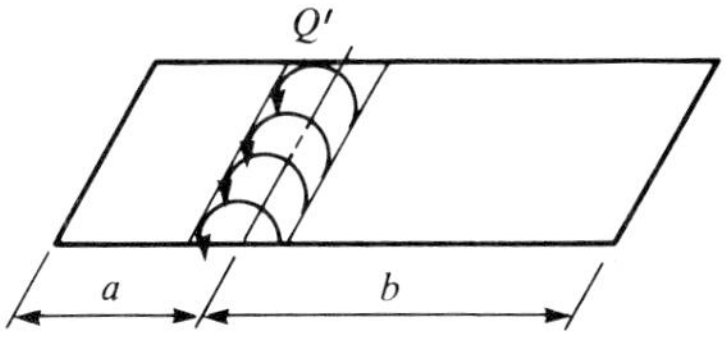

$$\Phi = \frac{2Q'}{f}(1 - \cos\lambda f)$$

3.

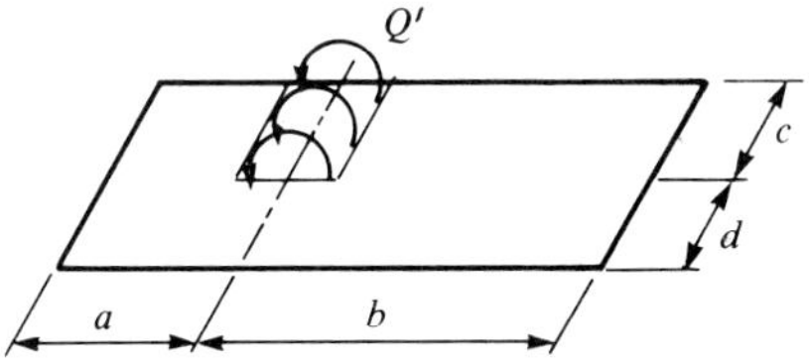

$$\Phi = \frac{2Q'}{f}(1 - \cos\lambda c)$$

4.

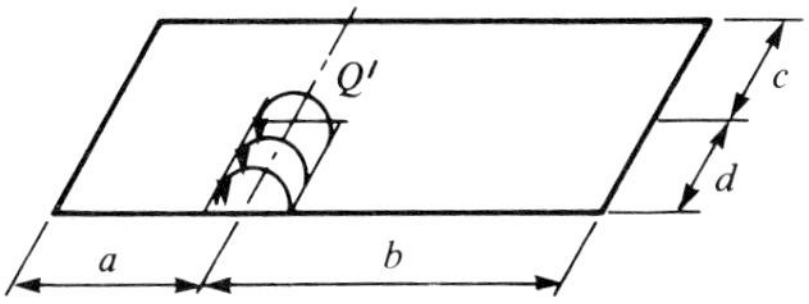

$$\Phi = \frac{2Q'}{f}(\cos\lambda c - \cos\lambda f)$$

5.

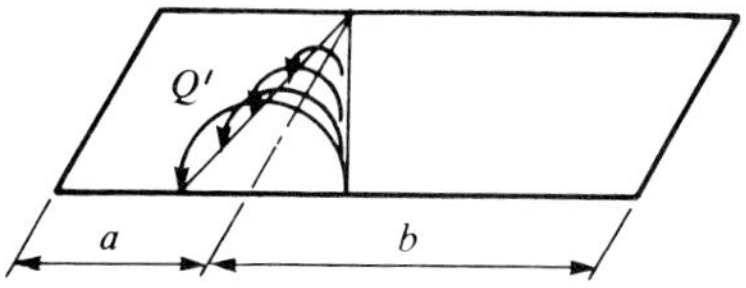

$$\Phi = \frac{2Q'}{m\pi f}(\sin\lambda f - \lambda f\cos\lambda f)$$

6.

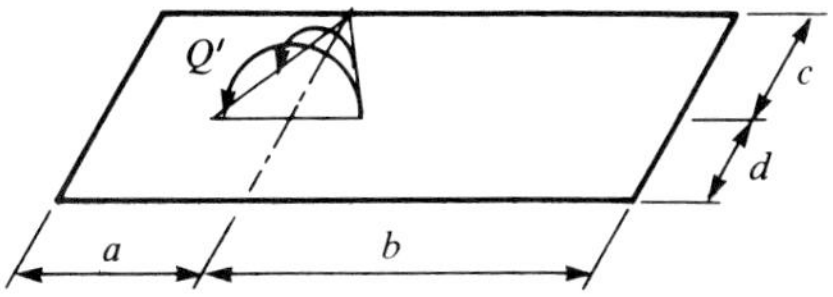

$$\Phi = \frac{2Q'}{cf\lambda}(\sin\lambda c - \lambda c\cos\lambda c)$$

7.

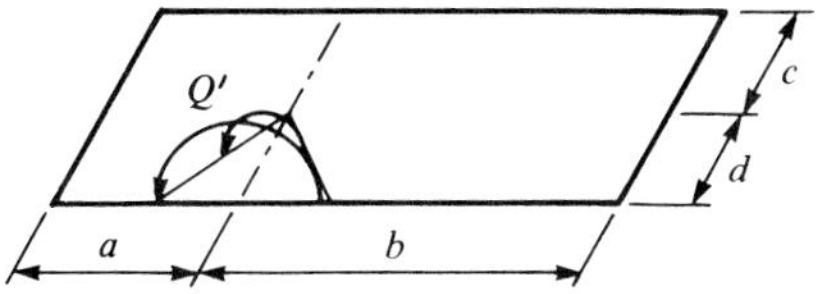

$$\Phi = \frac{2Q'}{df\lambda}(\sin\lambda f - \sin\lambda c - \lambda d\cos\lambda f)$$

24.08 TRANSPORT LOAD FUNCTIONS

Notation (24.01–24.04)

P' = intensity of line load (N/m)

p = intensity of distributed load (N/m^2)

A_j, B_j, C_j, T_j = transport coefficients (24.05)

For cases (2)–(7) use formulas of case (1) in terms of the respective Φ

Signs (24.04)

Matrix (24.04)

Φ = load factor (N/m)

$\lambda =$

$$F_{wm} = -(T_5 - T_7)\Phi$$

$$F_{\psi m} = T_6\Phi$$

$$F_{Mm} = -(\alpha T_5 + \beta T_7)\Phi$$

$$F_{Rm} = (T_2 - \beta T_6)\Phi$$

1.

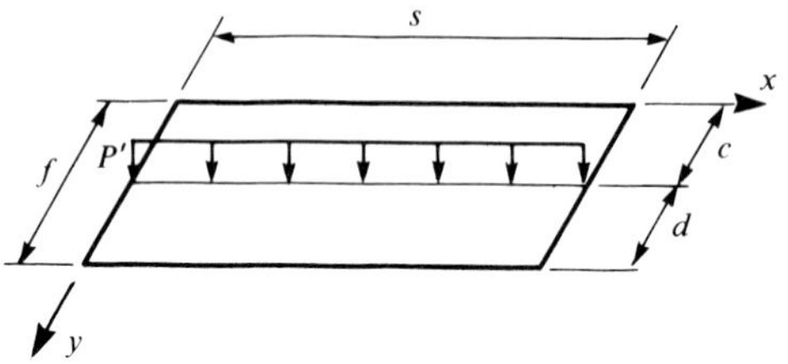

$$\Phi = \frac{2P'}{m\pi} \sin \lambda c$$

$$G_{wm} = -(T_5 - T_7)\Phi$$

$$G_{\psi m} = -T_6\Phi$$

$$G_{Mm} = -(\alpha T_5 + \beta T_7)\Phi$$

$$G_{Rm} = -(T_2 - \beta T_6)\Phi$$

2.

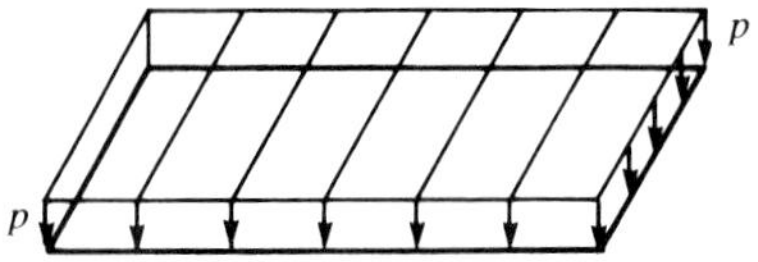

$$\Phi = \frac{2p}{m\pi\lambda}(1 - \cos \lambda f)$$

5.

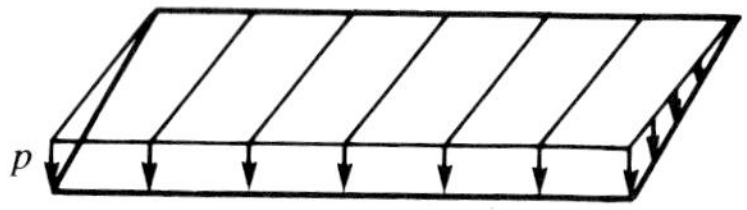

$$\Phi = \frac{2p}{(m\pi)^2\lambda}(\sin \lambda f - \lambda f \cos \lambda f)$$

3.

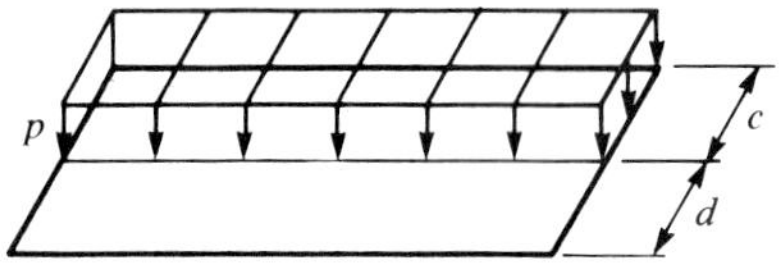

$$\Phi = \frac{2p}{2\pi\lambda}(1 - \cos \lambda c)$$

6.

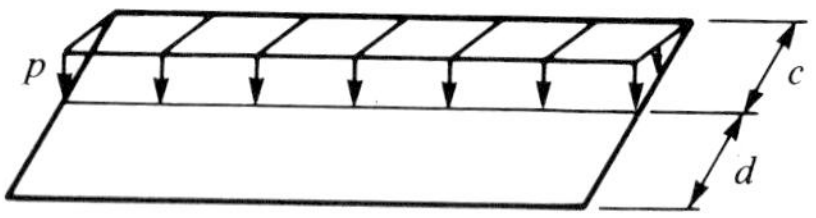

$$\Phi = \frac{2p}{cf\lambda^3}(\sin \lambda c - \lambda c \cos \lambda c)$$

4.

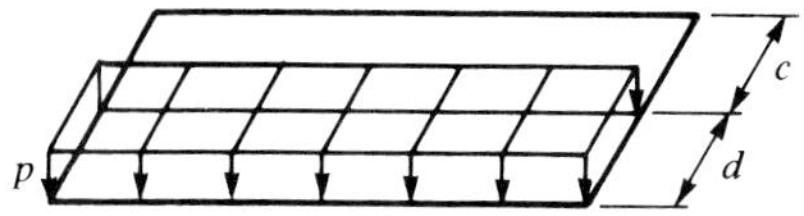

$$\Phi = \frac{2p}{m\pi\lambda}(\cos \lambda c - \cos \lambda f)$$

7.

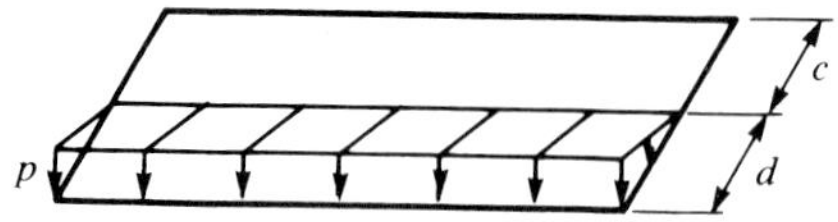

$$\Phi = \frac{2p}{df\lambda^3}(\sin \lambda f - \sin \lambda c - \lambda d \cos \lambda f)$$

24.09 TRANSPORT LOAD FUNCTIONS

$\lambda = \dfrac{m\pi}{f}$

Notation (24.01–24.04)
P' = intensity of line load (N/m)
p = intensity of distributed load (N/m²)
A_j, B_j, C_j, T_j = transport coefficients (24.05)
For cases (2)–(7) use formulas of case (1) in terms of the respective Φ

Signs (24.04)

Matrix (24.04)
Φ = load factor (N/m)

1.

$F_{wm} = -(C_5 - C_7)\Phi$

$F_{\psi m} = C_6\Phi$

$F_{Mm} = -(\alpha C_5 + \beta C_7)\Phi$

$F_{Rm} = (C_2 - \beta C_6)\Phi$

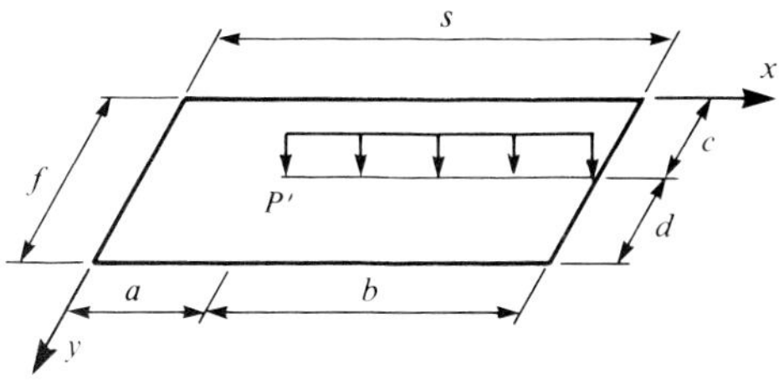

$$\Phi = \frac{2P'}{m\pi}\sin\lambda c$$

$G_{wm} = -(B_5 - B_7)\Phi$

$G_{\psi m} = -B_6\Phi$

$G_{Mm} = -(\alpha B_5 + \beta B_7)\Phi$

$G_{Rm} = -(B_2 - \beta B_6)\Phi$

2.

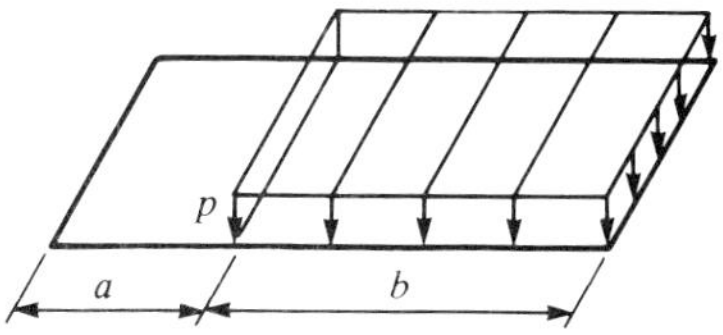

$$\Phi = \frac{2p}{m\pi\lambda}(1 - \cos\lambda f)$$

5.

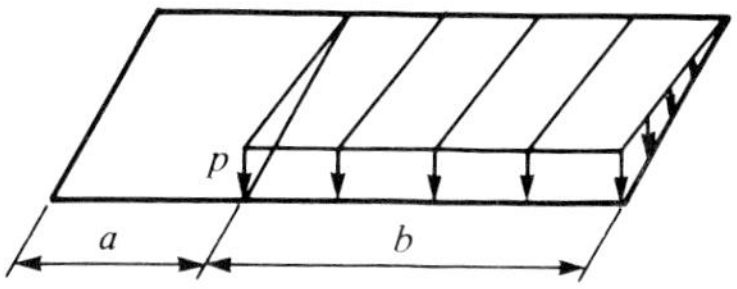

$$\Phi = \frac{2p}{(m\pi)^2\lambda}(\sin\lambda f - \lambda f\cos\lambda f)$$

3.

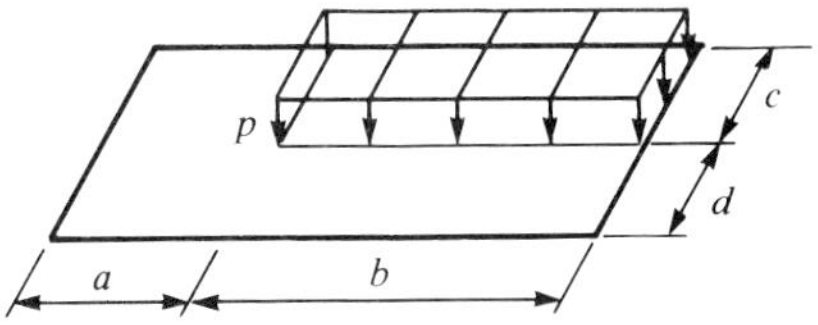

$$\Phi = \frac{2p}{m\pi\lambda}(1 - \cos\lambda c)$$

6.

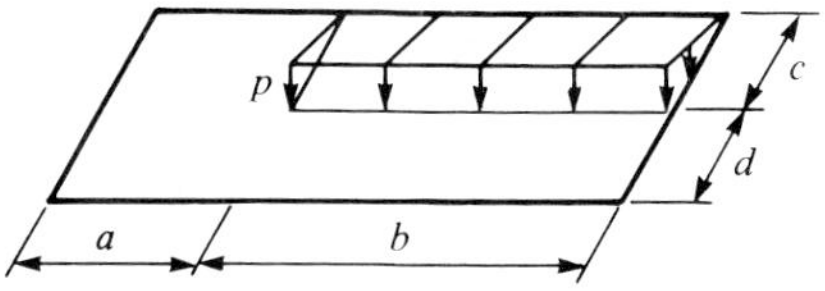

$$\Phi = \frac{2p}{cf\lambda^3}(\sin\lambda c - \lambda c\cos\lambda c)$$

4.

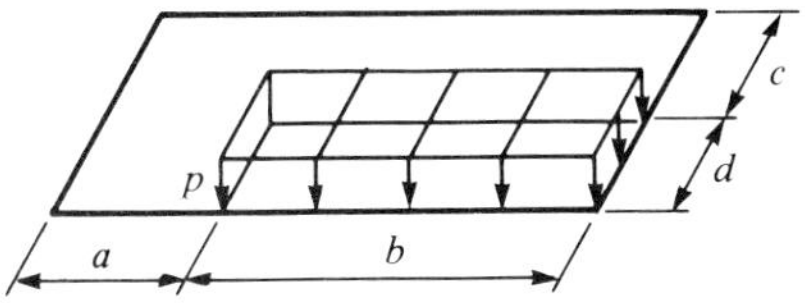

$$\Phi = \frac{2p}{m\pi\lambda}(\cos\lambda c - \cos\lambda f)$$

7.

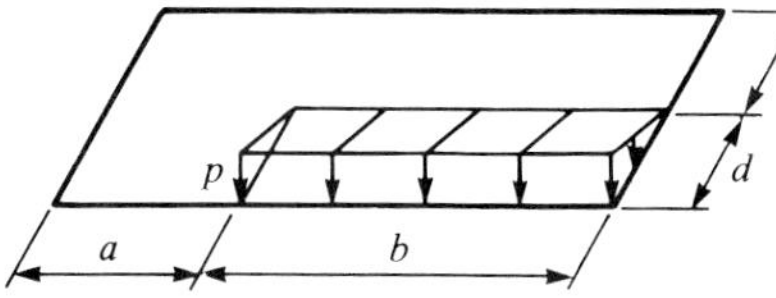

$$\Phi = \frac{2p}{df\lambda^3}(\sin\lambda f - \sin\lambda c - \lambda d\cos\lambda f)$$

24.10 TRANSPORT LOAD FUNCTIONS

$\lambda = \dfrac{m\pi}{f}$

Notation (24.01–24.04)
P' = intensity of line load (N/m)
p = intensity of distributed load (N/m²)
T_j = transport coefficients (24.05)
For cases (2)–(5) use formulas of case (1) in terms of the respective Φ

Signs (24.04)

Matrix (24.04)
Φ = load factor (N/m)

1.

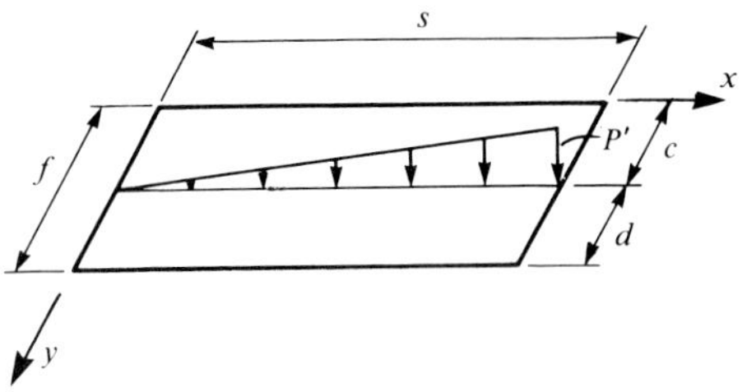

$$\Phi = \frac{2P'}{\lambda s m\pi}\sin\lambda c$$

$$F_{wm} = -[\lambda s(T_5 - T_7) - (T_6 - T_8)]\Phi$$
$$F_{\psi m} = (\lambda s T_6 - T_9)\Phi$$
$$F_{Mm} = -[\lambda s(\alpha T_5 + \beta T_7) - (\alpha T_6 - \beta T_8)]\Phi$$
$$F_{Rm} = [\lambda s(T_2 - \beta T_6) - (T_7 - \beta T_9)]\Phi$$

$$G_{wm} = -(T_6 - T_8)\Phi$$
$$G_{\psi m} = -T_9\Phi$$
$$G_{Mm} = -(\beta T_6 - \alpha T_8)\Phi$$
$$G_{Rm} = -(T_7 - \alpha T_9)\Phi$$

2.

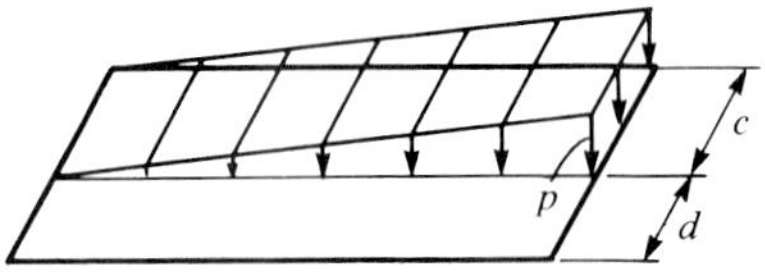

$$\Phi = \frac{2p}{\lambda^2 s m\pi}(1 - \cos\lambda c)$$

4.

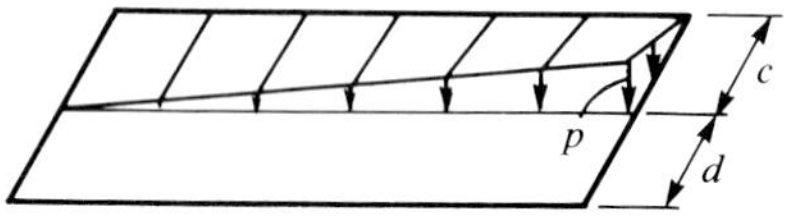

$$\Phi = \frac{2p}{cfs\lambda^4}(\sin\lambda c - \lambda c\cos\lambda c)$$

3. If $c = 0,\ d = f$

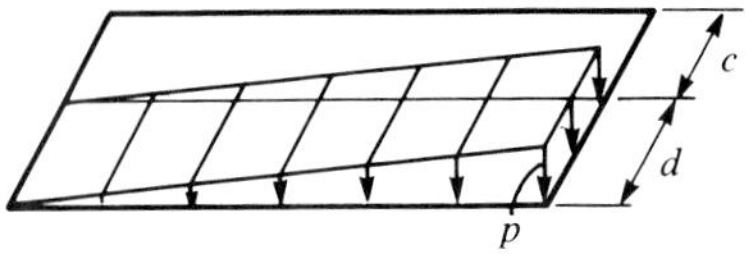

$$\Phi = \frac{2p}{\lambda^2 s m\pi}(\cos\lambda c - \cos\lambda f)$$

5. If $c = 0,\ d = f$

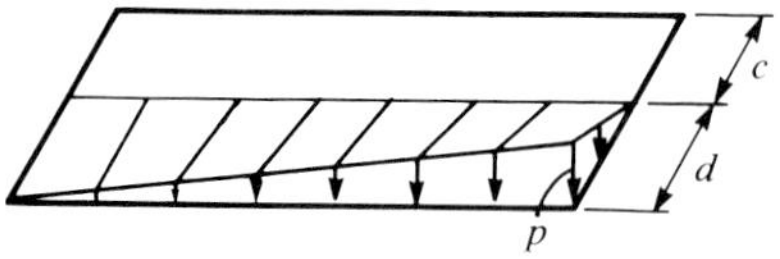

$$\Phi = \frac{2p}{dfs\lambda^4}(\sin\lambda f - \sin\lambda c - \lambda d\cos\lambda f)$$

24.11 TRANSPORT LOAD FUNCTIONS

Notation (24.01–24.04) Signs (24.04) Matrix (24.04)

P' = intensity of line load (N/m)

p = intensity of distributed load (N/m²)

B_j, C_j, T_j = transport coefficients (24.05)

For cases (2)–(5) use formulas of case (1) in terms of the respective Φ

Φ = load factor (N/m)

$$\lambda = \frac{m\pi}{f}$$

1.

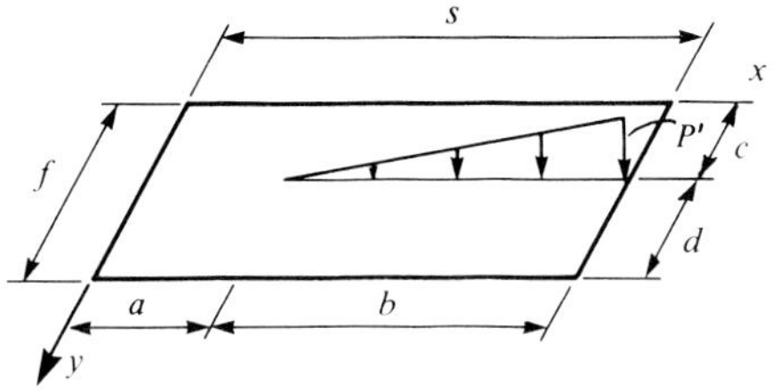

$$\Phi = \frac{2P'}{\lambda b m\pi}\sin\lambda c$$

$$\begin{aligned}
F_{wm} &= -[\lambda b(T_5 - T_7) - (C_6 - C_8)]\Phi \\
F_{\psi m} &= (\lambda b T_6 - C_9)\Phi \\
F_{Mm} &= -[\lambda b(\alpha T_5 + \beta T_6) - (\alpha C_6 - \beta C_8)]\Phi \\
F_{Rm} &= [\lambda b(T_2 - \beta T_6) - (C_7 - \beta C_9)]\Phi
\end{aligned}$$

$$\begin{aligned}
G_{wm} &= -(B_6 - B_8)\Phi \\
G_{\psi m} &= -B_9\Phi \\
G_{Mm} &= -(\alpha B_6 - \beta B_8)\Phi \\
G_{Rm} &= -(B_7 - \beta B_9)\Phi
\end{aligned}$$

2.

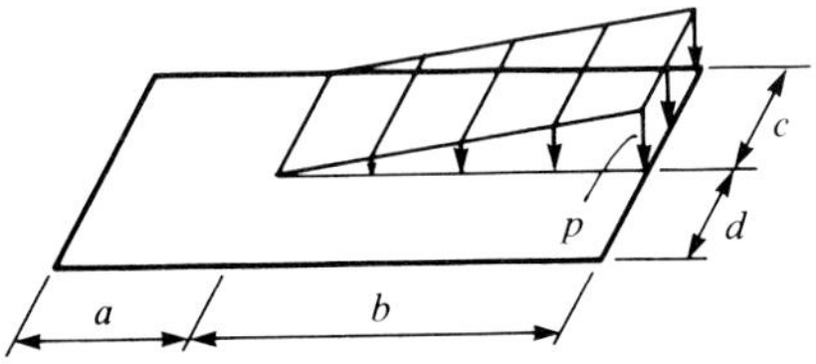

$$\Phi = \frac{2p}{\lambda^2 b m\pi}(1 - \cos\lambda c)$$

3. If $c = 0,\ d = f$

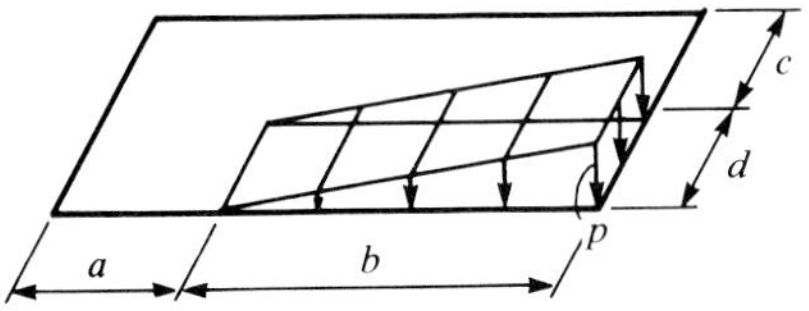

$$\Phi = \frac{2p}{\lambda^2 b m\pi}(\cos\lambda c - \cos\lambda f)$$

4.

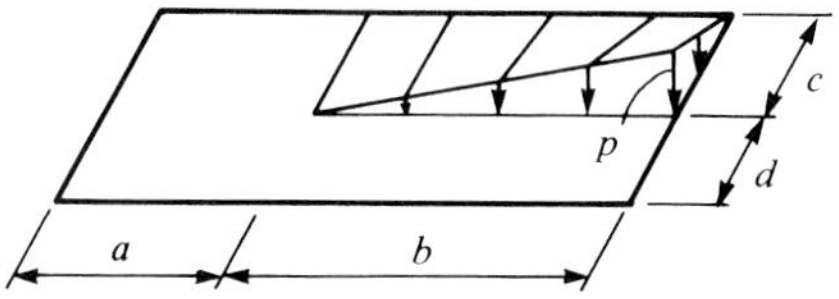

$$\Phi = \frac{2p}{bcf\lambda^4}(\sin\lambda c - \lambda c\cos\lambda c)$$

5. If $c = 0,\ d = f$

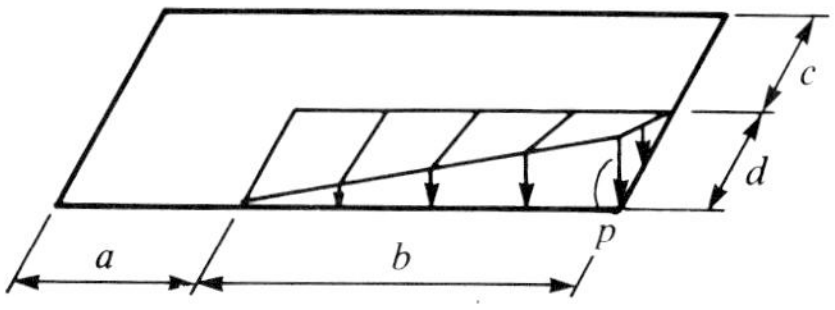

$$\Phi = \frac{2p}{bcf\lambda^4}(\sin\lambda f - \sin\lambda c - \lambda d\cos\lambda f)$$

24.12 TRANSPORT LOAD FUNCTIONS

$\lambda = \dfrac{m\pi}{f}$

Notation (24.01–24.04)
p = intensity of distributed load (N/m²)
w_j, u_j = abrupt displacements (m, rad)
α_t = coefficient of thermal expansion (1/°C)
A_j, B_j, C_j = transport coefficients (24.05)

Signs (24.04)

Matrix (24.04)
Φ = load factor (N/m)

$A_{j,1} = T_j(a_1)$ $A_{j,2} = T_j(a_2)$
$B_{j,1} = T_j(b_1)$ $B_{j,2} = T_j(b_2)$

1. Partial distributed load

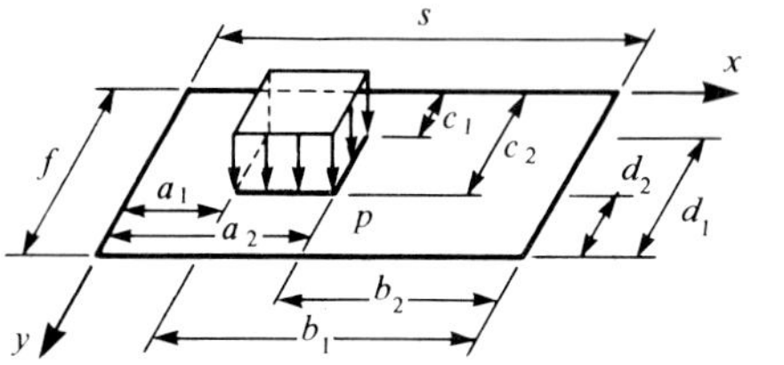

$$\Phi = \frac{2p}{m\pi\lambda}(\cos\lambda c_1 - \cos\lambda c_2)$$

$$F_{wm} = -(A_{5,2} - A_{5,1} - A_{7,2} + A_{7,1})\Phi$$
$$F_{\psi m} = (A_{6,2} - A_{6,1})\Phi$$
$$F_{Mm} = -(\alpha A_{5,2} - \alpha A_{5,1} + \beta A_{7,2} - \beta A_{7,1})\Phi$$
$$F_{Rm} = (A_{2,2} - A_{2,1} - \beta A_{6,2} + \beta A_{6,1})\Phi$$

$$G_{wm} = -(B_{5,1} - B_{5,2} - B_{7,1} + B_{7,2})\Phi$$
$$G_{\psi m} = -(B_{6,1} - B_{6,2})\Phi$$
$$G_{Mm} = -(\alpha B_{5,1} - \alpha B_{5,2} + \beta B_{7,1} - \beta B_{7,2})\Phi$$
$$G_{Rm} = -(B_{2,1} - B_{2,2} - \beta B_{6,1} + \beta B_{6,2})\Phi$$

2. Abrupt displacements w_j, ψ_j

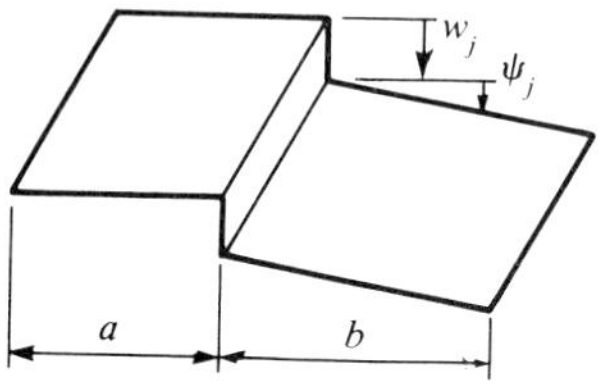

$$\Phi = \frac{4D}{m\pi}\lambda^2(1 - \cos m\pi)$$

$$F_{wm} = -[A_1 - \beta A_4)\lambda w_j - (\alpha A_2 + \beta A_3)\Psi_m]\Phi$$
$$F_{\psi m} = -[(\alpha A_2 - \beta A_3)\lambda w_j - (A_1 + \beta A_4)\psi_j]\Phi$$
$$F_{Mm} = -[\beta^2 A_4 \lambda w_j - (\gamma A_2 + \beta^2 A_3)\psi_j]\Phi$$
$$F_{Rm} = [(\gamma A_2 - \beta^2 A_3)\lambda w_j - \beta^2 A_4 \psi_j]\Phi$$

$$G_{wm} = [(B_1 - \beta B_4)\lambda w_j + (\alpha B_2 + \beta B_3)\psi_j]\Phi$$
$$G_{\psi m} = -[(\alpha B_2 - \beta B_3)\lambda w_j + (B_1 + \beta B_4)\psi_j]\Phi$$
$$G_{Mm} = [\beta^2 B_5 \lambda w_j - (\gamma B_2 + \beta^2 B_3)\psi_j]\Phi$$
$$G_{Rm} = [(\gamma B_2 - \beta^2 B_3)\lambda w_j - \beta^2 B_4 \psi_j]\Phi$$

3. Temperature deformation of segment b, t_1 = temperature above (° C), t_3 = temperature below (° C)

$$g_t = \alpha_t(t_1 - t_3)/h$$

$$F_{wm} = (\alpha C_5 + \beta C_7)\Phi$$
$$F_{\psi m} = (C_2 + \beta C_6)\Phi$$
$$F_{Mm} = -(\gamma C_5 + \beta^2 C_7)\Phi$$
$$F_{Rm} = \beta^2 C_6 \Phi$$

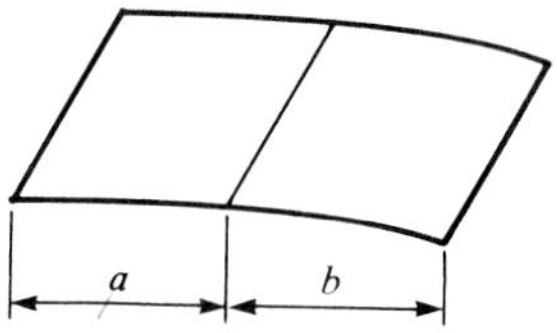

$$G_{wm} = (\alpha B_5 + \beta B_7)\Phi$$
$$G_{\psi m} = (B_2 + \beta B_6)\Phi$$
$$G_{Mm} = (\gamma B_5 + \beta^2 B_7)\Phi$$
$$G_{Rm} = -\beta^2 B_6 \Phi$$

$$\Phi = \frac{6Dg_t\lambda}{m\pi}(1 - \cos\lambda f)$$

24.13 STIFFNESS MATRIX EQUATIONS

$$D = \frac{Eh^3}{12(1 - \nu^2)} \qquad \alpha = \tfrac{1}{2}(1 + \nu) \qquad \beta = \tfrac{1}{2}(1 - \nu) \qquad \lambda = \frac{m\pi}{f}$$

(1) Equivalents and signs. The stiffness matrix of the plate introduced in (24.02) is presented in (5) below. All forces, moments, and displacements are in the cartesian system, and their positive directions are shown in (2, 3) below.

(2) Free-body diagram

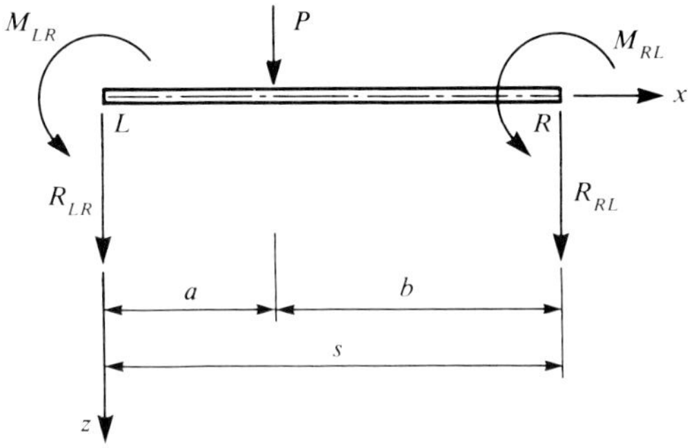

(3) Elastic curve

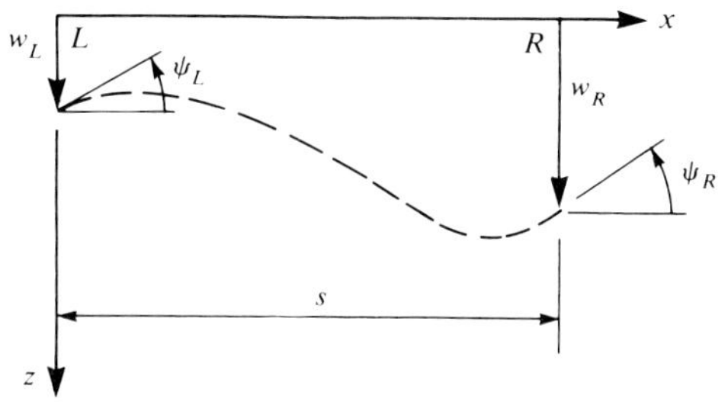

(4) Dimensionless stiffness coefficients

$$D_0 = \frac{1}{\sinh^2 \lambda s - (\lambda s)^2}$$

$$K_1 = (\cosh \lambda s \sinh \lambda s + \lambda s)D_0 \qquad K_4 = (\lambda s \cosh \lambda s + \sinh \lambda s)D_0$$

$$K_2 = [\alpha \sinh^2 \lambda s + \beta(\lambda s)^2]D_0 \qquad K_5 = (\lambda s \sinh \lambda s)D_0$$

$$K_3 = (\cosh \lambda s \sinh \lambda s - \lambda s)D_0 \qquad K_6 = (\lambda s \cosh \lambda s - \sinh \lambda s)D_0$$

(5) Stiffness matrix equations

$$\begin{bmatrix} R_{LRm} \\ M_{LRm} \\ \hline R_{RLm} \\ M_{RLm} \end{bmatrix} \sin \lambda y = 2D\lambda \left[\begin{array}{cc|cc} K_1\lambda^2 & -K_2\lambda & -K_4\lambda^2 & -K_5\lambda \\ -K_2\lambda & K_3 & K_5\lambda & K_6 \\ \hline -K_4\lambda^2 & K_5\lambda & K_1\lambda^2 & K_2\lambda \\ -K_5\lambda & K_6 & K_2\lambda & K_3 \end{array}\right] \begin{bmatrix} w_{Lm} \\ \psi_{Lm} \\ \hline w_{Rm} \\ \psi_{Rm} \end{bmatrix} \sin \lambda y + \begin{bmatrix} R_{L0m} \\ M_{L0m} \\ \hline R_{R0m} \\ M_{R0m} \end{bmatrix} \sin \lambda y$$

where $R_{L0m} \sin \lambda y$, $M_{L0m} \sin \lambda y$ and $R_{R0m} \sin \lambda y$, $M_{R0m} \sin \lambda y$ are the mth reactions of the fixed ends L and R, due to loads and other causes in (24.15). The tables and graphs of stiffness coefficients are shown in (24.14).

(6) Final forces and moments are the sum of r terms.

$$R_{LR} \cong \sum_{m=1}^{r} R_{LRm} \sin \lambda y \qquad R_{RL} \cong \sum_{m=1}^{r} R_{RLm} \sin \lambda y$$

$$M_{LR} \cong \sum_{m=1}^{r} M_{LRm} \sin \lambda y \qquad M_{RL} \cong \sum_{m=1}^{r} M_{RLm} \sin \lambda y$$

where r is the number of terms in the series required for a good accuracy.

24.14 GRAPHS AND TABLES OF DIMENSIONLESS STIFFNESS COEFFICIENTS

(1) Analytical expressions for $K_1, K_2, \ldots, K_6$ are given in (24.13). The graphs of these functions are shown in (2) below, and their numerical values for selected arguments λs are tabulated in (3) below. The coefficient K_2 is tabulated for $\nu = 0.16$ and for $\nu = 0.30$. The remaining coefficients are not affected by ν. K_2 is plotted for $\nu = 0.30$ only.

(2) Graphs

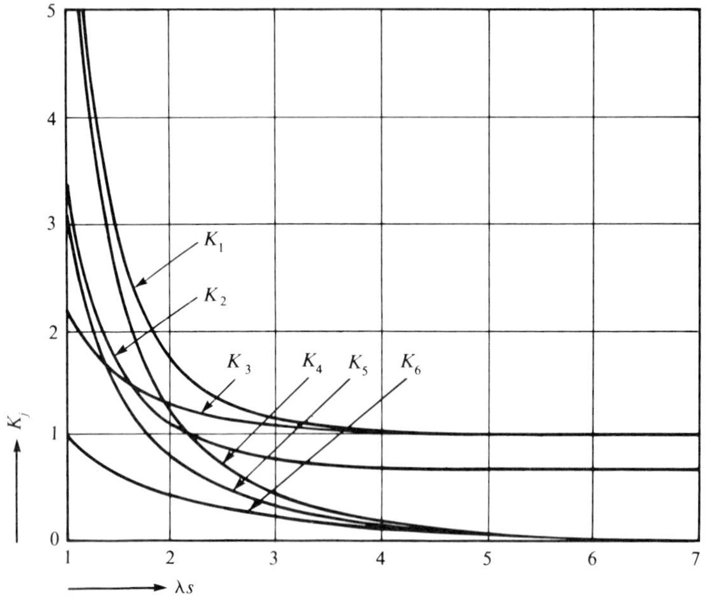

(3) Tables

λs	K_1	K_2 ($\nu = 0.16$)	K_2 ($\nu = 0.30$)	K_3	K_4	K_5	K_6
0.5	50.49139	12.12614	12.25614	4.06632	50.36664	12.09581	1.98315
0.6	29.88743	8.52215	8.59215	3.41359	29.73746	8.42733	1.64635
0.7	19.33488	6.31440	6.38440	2.95088	19.15994	6.21432	1.40475
0.8	13.36485	4.88304	4.95304	2.60726	13.16496	4.77694	1.22261
0.9	9.72807	3.90320	3.97320	2.34305	9.50327	3.79042	1.08010
1.0	7.32240	3.20400	3.27400	2.13444	7.13277	3.08373	0.96531
1.1	5.79534	2.65310	2.75813	1.96628	5.52487	2.55981	0.87067
1.2	4.69081	2.29741	2.35741	1.82846	4.39161	2.16030	0.79111
1.3	3.89053	1.99482	2.06482	1.71399	3.56682	1.84838	0.72315
1.4	3.29818	1.75621	1.82521	1.61787	2.94987	1.59990	0.66429
1.5	2.85001	1.56519	1.63519	1.53643	2.47437	1.39849	0.61271
1.6	2.50475	1.41027	1.48027	1.46591	2.10796	1.23273	0.56704
1.7	2.23457	1.28327	1.35327	1.40719	1.81385	1.09447	0.52624
1.8	2.02030	1.17819	1.24819	1.35565	1.57589	0.97776	0.48949
1.9	1.84838	1.09054	1.16054	1.31097	1.38057	0.87818	0.45617

λs	K_1	K_2 $\nu = 0.16$	K_2 $\nu = 0.30$	K_3	K_4	K_5	K_6
2.0	1.70906	1.01696	1.056696	1.27210	1.21817	0.79240	0.42577
2.1	1.59518	0.95483	1.02483	1.23820	1.08156	0.71786	0.39788
2.2	1.50140	0.90211	0.97211	1.20857	0.96544	0.65259	0.37218
2.3	1.42366	0.85720	0.92720	1.18262	0.86580	0.59501	0.34840
2.4	1.35887	0.81881	0.88881	1.15986	0.77957	0.53391	0.32631
2.5	1.30462	0.78590	0.85590	1.13990	0.70436	0.49829	0.30537
2.6	1.25900	0.75762	0.82762	1.12237	0.63832	0.45735	0.28652
2.7	1.22050	0.73327	0.80327	1.10697	0.57996	0.42043	0.26853
2.8	1.18793	0.71228	0.78228	1.09344	0.52811	0.38701	0.25167
2.9	1.16030	0.69416	0.76416	1.08156	0.48180	0.35665	0.23583
3.0	1.13681	0.67851	0.74851	1.07114	0.44026	0.32897	0.22095
3.2	1.09970	0.65329	0.72329	1.05396	0.36906	0.28046	0.19375
3.4	1.07278	0.63443	0.70443	1.04077	0.31055	0.23956	0.16963
3.6	1.05308	0.62032	0.69032	1.03068	0.26202	0.20482	0.14823
3.8	1.03867	0.60980	0.67980	1.02299	0.22144	0.17517	0.12925
4.0	1.02813	0.60196	0.67196	1.01715	0.18734	0.14979	0.11244
4.2	1.02043	0.59613	0.66613	1.01275	0.15856	0.12802	0.09760
4.4	1.01480	0.59181	0.66181	1.00943	0.13422	0.10933	0.08452
4.6	1.01071	0.58863	0.65863	1.00696	0.11358	0.09328	0.07302
4.8	1.00773	0.58628	0.65628	1.00500	0.09608	0.07951	0.06295
5.0	1.00556	0.58456	0.65456	1.00374	0.08123	0.06769	0.05416
5.2	1.00400	0.58330	0.65330	1.00273	0.06864	0.06756	0.04650
5.4	1.00287	0.58239	0.65239	1.00198	0.05795	0.04890	0.03984
5.6	1.00205	0.58172	0.65172	1.00144	0.04890	0.04149	0.03408
5.8	1.00147	0.58123	0.65123	1.00104	0.04123	0.03516	0.02910
6.0	1.00105	0.58089	0.65089	1.00075	0.03473	0.02977	0.02451
7.0	1.00019	0.58016	0.65016	1.00014	0.01459	0.01277	0.01094
8.0	1.00003	0.58003	0.65003	1.00003	0.00604	0.00537	0.00470
9.0	1.00001	0.58000	0.65000	1.00000	0.00247	0.00222	0.00197
10.0	1.00000	0.58000	0.65000	1.00000	0.00100	0.00091	0.00082
11.0	1.00000	0.58000	0.65000	1.00000	0.00040	0.00037	0.00033
12.0	1.00000	0.58000	0.65000	1.00000	0.00016	0.00015	0.00014
13.0	1.00000	0.58000	0.65000	1.00000	0.00006	0.00006	0.00005
14.0	1.00000	0.58000	0.65000	1.00000	0.00002	0.00002	0.00002

For $\lambda s \geq 15$, $K_1 = K_3 = 1$, $K_2 = \alpha$, $K_4 = K_5 = K_6 = 0$.

24.15 STIFFNESS LOAD FUNCTIONS

(1) General form of the mth stiffness load functions in (24.13) is given by the matrix equations

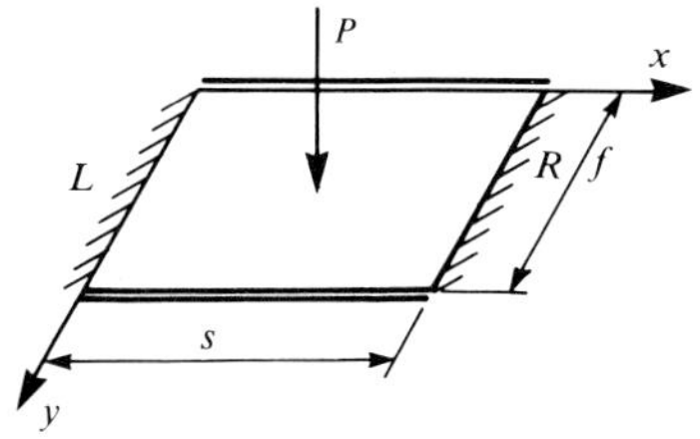

$$\begin{bmatrix} R_{L0m} \\ M_{L0m} \end{bmatrix} \sin \lambda y = \begin{bmatrix} K_4 & K_5 \\ -K_5/\lambda & -K_6/\lambda \end{bmatrix} \begin{bmatrix} G_{wm} \\ G_{\psi m} \end{bmatrix} \sin \lambda y$$

$$\begin{bmatrix} R_{R0m} \\ M_{R0m} \end{bmatrix} \sin \lambda y = \begin{bmatrix} K_4 & -K_5 \\ K_5/\lambda & -K_6/\lambda \end{bmatrix} \begin{bmatrix} F_{wm} \\ F_{\psi m} \end{bmatrix} \sin \lambda y$$

where F_{wm}, $F_{\psi m}$ are the left-end transport load functions and G_{wm}, $G_{\psi m}$ are their right-side counterparts, given in (24.06–24.12), and K_4, K_5, K_6 are the stiffness factors (24.13). The plate is fixed along the edges of length f and is freely supported along the edges of length s.

(2) Particular cases

Notation (24.01–24.05, 24.13) A_j, B_j, C_j, T_j = transport functions (24.05)	Signs (24.13)	Matrix (24.13)	$\lambda = \dfrac{m\pi}{f}$
1. 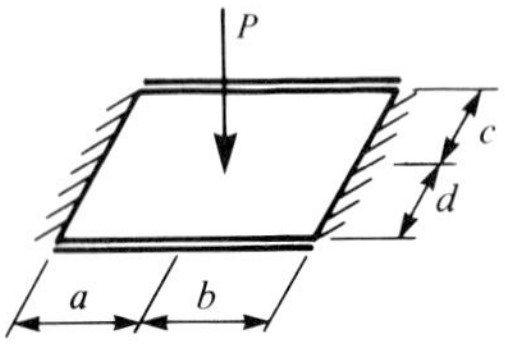 $\Phi = \dfrac{2P}{f} \sin \lambda c$	$R_{L0m} = -[(B_2 - B_3)K_4 + B_4 K_5]\Phi$ $M_{L0m} = [(B_2 - B_3)K_5 + B_4 K_6]\Phi/\lambda$ $R_{R0m} = -[(A_2 - A_3)K_4 + A_4 K_5]\Phi$ $M_{R0m} = -[(A_2 - A_3)K_5 + A_4 K_6]\Phi/\lambda$		
2. 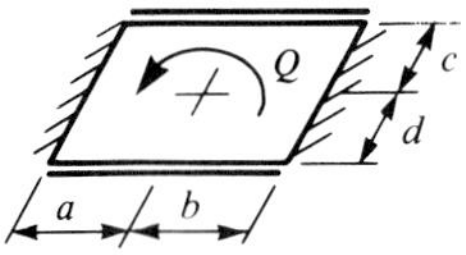 $\Phi = \dfrac{2Q\lambda}{f} \sin \lambda c$	$R_{L0m} = -[(B_2 + B_3)K_5 - B_4 K_4]\Phi$ $M_{L0m} = [(B_2 + B_3)K_6 - B_4 K_5]\Phi/\lambda$ $R_{R0m} = -[A_4 K_4 - (A_2 + A_3)K_5]\Phi$ $M_{R0m} = -[A_4 K_5 - (A_2 + A_3)K_6]\Phi/\lambda$		
3. 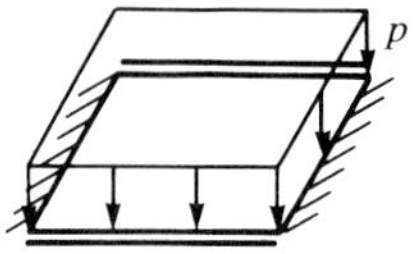 $\Phi = \dfrac{2p}{m\pi\lambda}(1 - \cos \lambda f)$	$R_{L0m} = -[(T_5 - T_7)K_4 + T_6 K_5]\Phi$ $M_{L0m} = [(T_5 - T_7)K_5 + T_6 K_6]\Phi/\lambda$ $R_{R0m} = -[(T_5 - T_7)K_4 + T_6 K_5]\Phi$ $M_{R0m} = -[(T_5 - T_7)K_5 + T_6 K_6]\Phi/\lambda$		

25

Plate Analysis by Finite Differences

STATIC STATE

25.01 DEFINITION OF STATE

(1) Systems analyzed are thin isotropic or orthotropic plates of constant or variable thickness and of arbitrary geometric shape, acted on by in-plane and/or normal-to-plane mechanical and thermal causes in a state of static equilibrium.

(2) Analysis of plates by the methods introduced in Chaps. 23 and 24 is limited to plates of certain shapes and certain edge conditions. In other cases these solutions are not practical and/or possible, and resort must be made to some approximate numerical methods, among which the most important ones are the *finite-difference method* and the *finite-element method,* introduced in this and in the subsequent chapter.

(3) Finite-difference method replaces the given plate by a specific grid and approximates the governing differential equation and the equations defining the boundary conditions by difference equations in terms of unknown deflections related to the nodal points (joints) of this grid. This substitution leads to a set of algebraic equations written for the nodal points of the plate, and the solution of this set yields the approximate values of the deflections of the plate midsurface at these nodes. The slopes, moments, shears, and reactions are then approximated by their respective difference expressions in terms of the calculated deflections (Ref. 50).

(4) Assumptions of analysis are stated in (22.01), and in this chapter, the deflected midsurface is represented by a system of strings connecting the deflections of the reference grid. Because of higher accuracy, only central differences are used.

25.02 DERIVATIVES AND DIFFERENCES IN CARTESIAN COORDINATES

(1) Notation (22.02)

$$w = f(x, y) = \text{deflection (m)} \qquad \Delta x, \Delta y = \text{increments (m)}$$

$$w' = \frac{dw}{dx}, \quad w'' = \frac{d^2w}{dx^2}, \ldots \qquad w'^* = \frac{\partial^2 w}{\partial x\, \partial y}, \quad w'^{**} = \frac{\partial^3 w}{\partial x\, \partial y^2}, \ldots$$

$$w^* = \frac{dw}{dy}, \quad w^{**} = \frac{d^2w}{dy^2}, \ldots \qquad w''^* = \frac{\partial^3 w}{\partial x^2\, \partial y}, \quad w''^{**} = \frac{\partial^4 w}{\partial x^2 \partial y^2}, \ldots$$

(2) Difference equations related to the central point C in the adjacent grid of $\Delta x \neq \Delta y$ on the plate described in (22.02) are

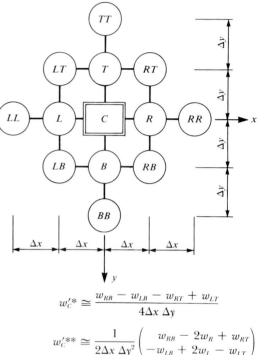

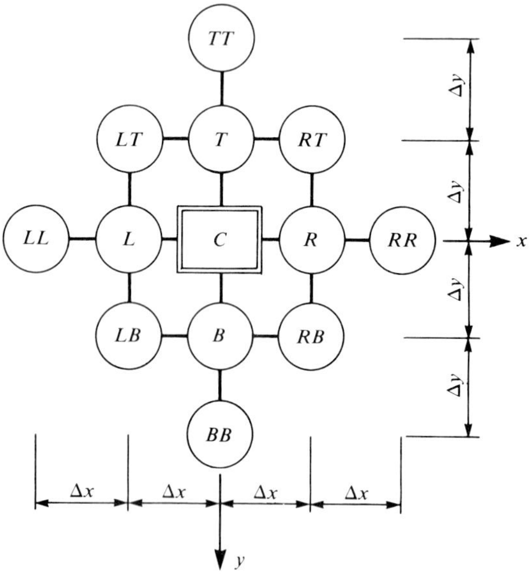

$$w'_C \cong \frac{w_R - w_L}{2\Delta x}$$

$$w''_C \cong \frac{w_R - 2w_C + w_L}{\Delta x^2}$$

$$w'''_C \cong \frac{1}{2\Delta x^3}\begin{pmatrix} w_{RR} - 2w_R \\ +2w_L - w_{LL} \end{pmatrix}$$

$$w''''_C \cong \frac{1}{\Delta x^4}\begin{pmatrix} w_{RR} - 4w_R + 6w_C \\ w_{LL} - 4w_L \end{pmatrix}$$

$$w^*_C \cong \frac{w_B - w_T}{2\Delta y}$$

$$w^{**}_C \cong \frac{w_B - 2w_L + w_T}{\Delta y^2}$$

$$w^{***}_C \cong \frac{1}{2\Delta y^3}\begin{pmatrix} w_{BB} - 2w_B \\ +2w_T - w_{TT} \end{pmatrix}$$

$$w^{****}_C \cong \frac{1}{\Delta y^4}\begin{pmatrix} w_{BB} - 4w_B + 6w_C \\ w_{TT} - 4w_T \end{pmatrix}$$

$$w'^*_C \cong \frac{w_{RB} - w_{LB} - w_{RT} + w_{LT}}{4\Delta x\, \Delta y}$$

$$w''^*_C \cong \frac{1}{2\Delta x^2\, \Delta y}\begin{pmatrix} w_{RB} - 2w_B + w_{LB} \\ -w_{RT} + 2w_T - w_{LT} \end{pmatrix}$$

$$w'^{**}_C \cong \frac{1}{2\Delta x\, \Delta y^2}\begin{pmatrix} w_{RB} - 2w_R + w_{RT} \\ -w_{LB} + 2w_L - w_{LT} \end{pmatrix}$$

$$w''^{**}_C \cong \frac{1}{\Delta x^2\, \Delta y^2}\begin{pmatrix} w_{LB} - 2w_B + w_{RB} \\ -2w_L + 4w_C - 2w_R \\ +w_{LT} - 2w_T + w_{RT} \end{pmatrix}$$

where C = center, L = left, R = right, B = bottom, T = top, . . . are the subscripts identifying the respective nodal points and the terms in parentheses represent an algebraic sum.

25.03 JOINT LOADS IN CARTESIAN COORDINATES

(1) Concentrated load $P_j(N)$ applied at j produces at C the joint load

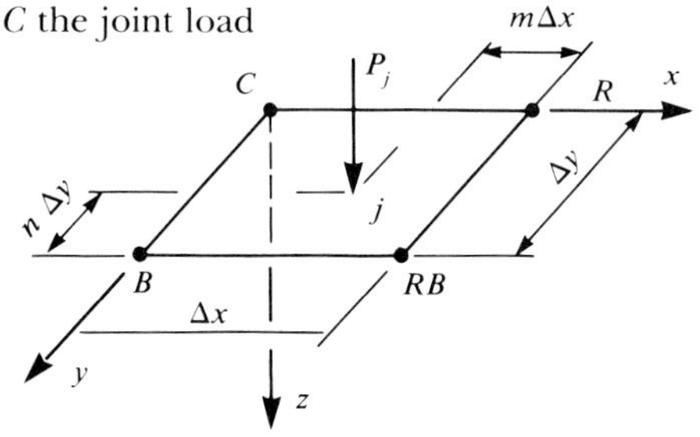

$$p_C \cong \frac{mnP_j}{\Delta x\, \Delta y}\ (\text{N/m}^2)$$

where m, n are the dimensionless coordinates shown in the adjacent figure.

(2) Uniformly distributed load of constant intensity p (N/m²) produces at C the joint load

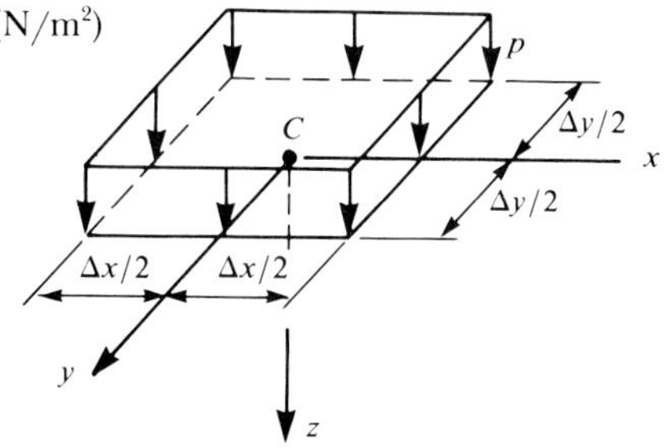

$$p_C = p\ (\text{N/m}^2)$$

(3) Distributed load of variable intensity $p(x, y)$ produces at C the joint load

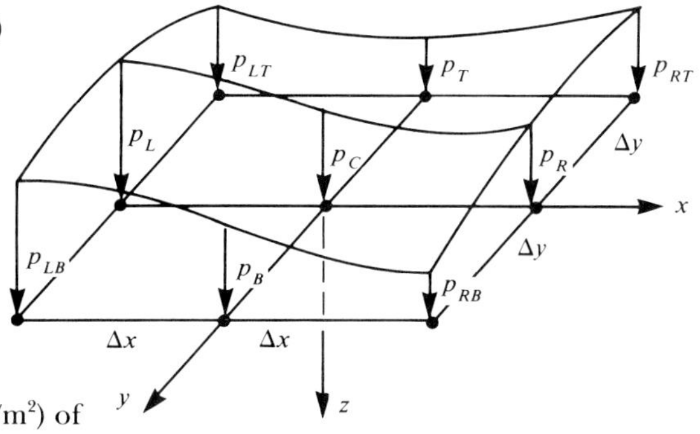

$$p_C \cong \frac{1}{36}\begin{pmatrix} p_{LT} + 4p_T + p_{RT} \\ +4p_L + 16p_C + 4p_R \\ +p_{LB} + 4p_B + p_{RB} \end{pmatrix}(\text{N/m}^2)$$

where p_{LT}, p_T, p_{RT}, . . . are the intensities (N/m²) of the load at LT, T, RT, . . . , respectively, and the terms in parentheses represent an algebraic sum.

(4) Applied couple Q_j (N·m) applied at j produces at C the joint load

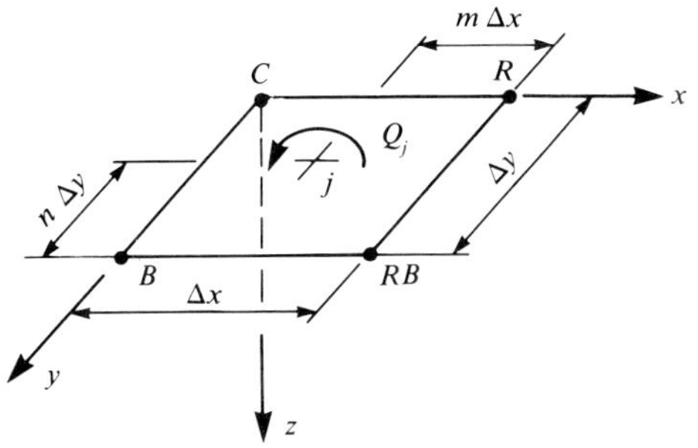

$$p_C \cong \frac{nQ_j}{\Delta x\, \Delta y^2}\ (\text{N/m}^2)$$

where n is the dimensionless coordinate shown in the adjacent figure.

25.04 PLATE OPERATORS AND FUNCTIONS IN CARTESIAN COORDINATES

$$D = \frac{Eh^3}{12(1-\nu^2)}$$

(1) Equivalents (22.02–22.04)

$A_0 = (\Delta x/\Delta y)^2$	$A_1 = 2(1 + A_0)$	$A_4 = 2(1 + \nu A_0)$	$A_7 = 2(1 + A_6)$
$\alpha = \frac{1}{2}(1 + \nu)$	$A_2 = 2A_0(1 + A_0)$	$A_5 = 2(\nu + A_0)$	$A_8 = 2 - \nu$
$\beta = \frac{1}{2}(1 - \nu)$	$A_3 = 6 + 8A_0 + 6A_0^2$	$A_6 = (2 - \nu)A_0$	$A_9 = 2(1 + A_8)$

(2) Operators at C in terms of the joint deflections introduced in the grid of the plate shown in (25.02) are given below. The terms in braces represent an algebraic sum.

$$\nabla_C w \cong \frac{1}{2\Delta x}\left\{\begin{array}{ccc} & -\sqrt{A_0}w_T & \\ -w_L & 0 & w_R \\ & \sqrt{A_0}w_B & \end{array}\right\} \qquad \nabla_C^2 w \cong \frac{1}{\Delta x^2}\left\{\begin{array}{ccc} & A_0 w_T & \\ w_L & -A_1 w_C & w_R \\ & A_0 w_B & \end{array}\right\}$$

$$\nabla_C^4 w \cong \frac{1}{\Delta x^4}\left\{\begin{array}{ccccc} & & A_0^2 w_{TT} & & \\ & 2A_0 w_{LT} & -2A_2 w_T & 2A_0 w_{RT} & \\ w_{LL} & -2A_1 w_L & A_3 w_C & -2A_1 w_R & w_{RR} \\ & 2A_0 w_{LB} & -2A_2 w_B & 2A_0 w_{RB} & \\ & & A_0^2 w_{BB} & & \end{array}\right\}$$

(3) Functions of the plate defined in (22.02–22.03) are approximated by the difference relations expressed in operator form below. The joint C is the central point in all formulas.

(4) Slopes

$$\phi_C \cong \frac{1}{2\Delta y}(w_B - w_T) \qquad \psi_C \cong -\frac{1}{2\Delta x}(w_R - w_L)$$

(5) Bending moments

$$M_{Cx} \cong -\frac{D}{\Delta x^2}\left\{\begin{array}{ccc} & \nu A_0 w_T & \\ w_L & -A_4 w_C & w_R \\ & \nu A_0 w_B & \end{array}\right\} \qquad M_{Cy} \cong -\frac{D}{\Delta x^2}\left\{\begin{array}{ccc} & A_0 w_T & \\ \nu w_L & -A_5 w_C & \nu w_R \\ & A_0 w_B & \end{array}\right\}$$

(6) Moment sum and twisting moment

$$M_C \cong -\frac{D}{\Delta x^2}\left\{\begin{array}{ccc} & A_0 w_T & \\ w_L & A_1 w_C & w_R \\ & A_0 w_B & \end{array}\right\} \qquad M_{Cxy} \cong -\frac{\beta D\sqrt{A_0}}{2\Delta x^2}\left\{\begin{array}{ccc} w_{LT} & 0 & -w_{RT} \\ 0 & 0 & 0 \\ -w_{LB} & 0 & w_{RB} \end{array}\right\}$$

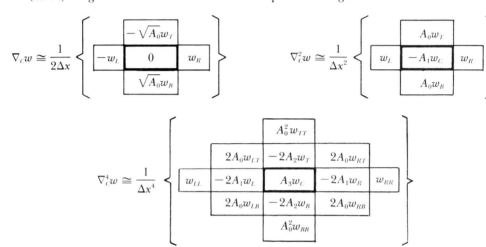

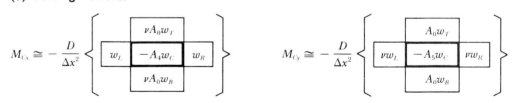

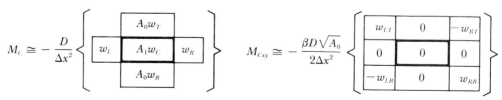

(7) Edge reactions

$$R_{Cxz} \cong -\frac{D}{2\Delta x^3}\left\{\begin{array}{ccccc} & & 0 & & \\ & -A_6 w_{LT} & 0 & A_6 w_{RT} & \\ -w_{LL} & A_7 w_L & \boxed{0} & -A_7 w_R & w_{RR} \\ & -A_6 w_{LB} & 0 & A_6 w_{RB} & \\ & & 0 & & \end{array}\right\}$$

$$R_{Cyz} \cong -\frac{D\sqrt{A_0}}{2\Delta x^3}\left\{\begin{array}{ccccc} & & -A_0 w_{TT} & & \\ & -A_8 w_{LT} & A_9 w_T & -A_8 w_{RT} & \\ 0 & 0 & \boxed{0} & 0 & 0 \\ & A_8 w_{LB} & -A_9 w_B & A_8 w_{RB} & \\ & & A_0 w_{BB} & & \end{array}\right\}$$

(8) Shears. The value of W_{Cxz} is calculated by R_{Cxz} with A_6 replaced by A_0 and A_7 replaced by A_1. Similarly, the value of W_{Cyz} is calculated by R_{Cyz} with A_8 replaced by 1 and A_9 replaced by A_1.

(9) Governing plate equation (22.04-7) in difference operator form is

$$\left\{\begin{array}{ccc} & A_0 \nabla_T^2 w & \\ \nabla_L^2 w & \boxed{-A_1 \nabla_C^2 w} & \nabla_R^2 w \\ & A_0 \nabla_B^2 w & \end{array}\right\} = \frac{p_C \Delta x^2}{D}$$

where ∇_L^2, ∇_C^2, ∇_R^2, . . . are the operators at L, C, R, . . . , respectively, and p_C is the joint load at C defined in (25.03). The explicit form is given in (2).

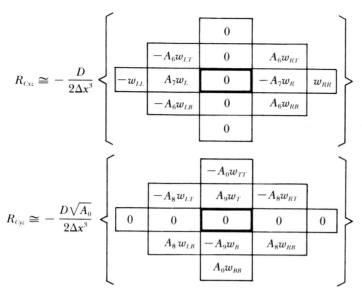

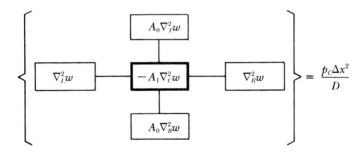

25.05 EDGE EQUATIONS IN CARTESIAN COORDINATES

(1) Equivalents (22.02–22.04)

$A_0 = (\Delta x/\Delta y)^2$	$A_{10} = 2\alpha\beta$	$B_{10} = 2\alpha\beta A_0^2$
$A_1 = 2(1 + A_0)$	$A_{11} = 4\beta(2\alpha + A_0)$	$B_{11} = 4\beta(1 + 2\alpha A_0)A_0$
$A_2 = 2(1 + A_0)A_0$	$A_{12} = 12\alpha\beta + 8\beta A_0 + A_0^2$	$B_{12} = 1 + 8\beta A_0 + 12\alpha\beta A_0^2$
$A_3 = 6 + 8A_0 + 6A_0^2$	$A_{13} = 1 + 2\beta$	$B_{13} = (12\beta)A_0$
$\alpha = \frac{1}{2}(1 + \nu)$	$A_{14} = 2(A_0^2 + 2A_0 - \nu A_0)$	$B_{14} = 2(2A_0 - \nu A_0 + A_0^2)$
$\beta = \frac{1}{2}(1 - \nu)$	$A_{15} = 6 + 8A_0 + 5A_0^2$	$B_{15} = 5 + 8A_0 + 6A_0^2$

(2) Simply supported edge, C **next to** x **boundary** ($w_{LT} = w_T = w_{RT} = 0,\ w_{TT} = -w_C$)

$$\nabla_C^4 w = \frac{1}{\Delta x^4}\{\ \cdots\ \}$$

	0	0	0	
w_{LL}	$-2A_1 w_L$	$(A_3 - A_0^2)w_C$	$-2A_1 w_R$	w_{RR}
	$2A_0 w_{LB}$	$-2A_2 w_B$	$2A_0 w_{RB}$	
		w_{BB}		

Simple edge → x

For y boundary, $w_{LT} = w_L = w_{LB} = 0,\ w_{LL} = -w_C$.

(3) Fixed edge, C **next to** x **boundary** ($w_{LT} = w_T = w_{RT} = 0,\ w_{TT} = w_C$)

$$\nabla_C^4 w = \frac{1}{\Delta x^4}\{\ \cdots\ \}$$

	0	0	0	
w_{LL}	$-2A_1 w_L$	$(A_3 + A_0^2)w_C$	$-2A_1 w_R$	w_{RR}
	$2A_0 w_{LB}$	$-2A_2 w_R$	$2A_0 w_{RB}$	
		w_{BB}		

Fixed edge → x

For y boundary, $w_{LT} = w_L = w_{LB} = 0,\ w_{LL} = w_C$.

(4) Guided edge, C **on** x **boundary** ($w_{LT} = w_{LB},\ w_T = w_B,\ w_{RT} = w_{RB},\ w_{TT} = w_{BB}$)

$$\nabla_C^4 w = \frac{1}{\Delta x^4}\{\ \cdots\ \}$$

w_{LL}	$-2A_1 w_L$	$A_3 w_C$	$-2A_1 w_R$	w_{RR}
	$4A_0 w_{LB}$	$-4A_2 w_B$	$-4A_0 w_{RB}$	
		$2A_0^2 w_{BB}$		

Guided edge → x

For y boundary, $w_{LT} = w_{RT},\ w_L = w_R,\ w_{LB} = w_{RB}$.

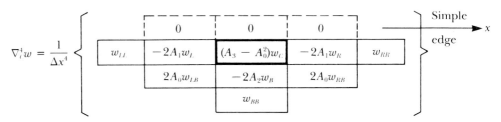

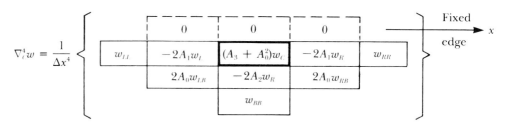

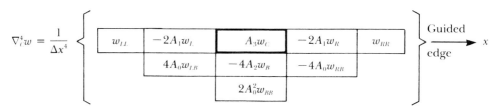

(5) Guided edge, C next to x boundary ($w_{TT} = w_C$)

$$\nabla_C^4 w = \frac{1}{\Delta x^4}\left\{ \right.$$

	$2A_0 w_{LT}$	$-2A_2 w_T$	$2A_0 w_{RT}$	
w_{LL}	$-2A_1 w_L$	**$(A_3 + A_0^2) w_C$**	$-2A_1 w_R$	w_{RR}
	$2A_0 w_{LB}$	$-2A_2 w_B$	$2A_0 w_{RB}$	
		$A_0^2 w_{BB}$		

$\left. \right\}$ Guided edge → x

For y boundary, $w_{LL} = w_C$.

(6) Free edge, C on x boundary ($M_{Ly} = M_{Cy} = M_{Ry} = 0,\ R_{Cyz} = 0$)

$$\nabla_C^4 w = \frac{1}{\Delta x^4}\left\{ \right.$$

$A_{10} w_{LL}$	$-A_{11} w_L$	**$A_{12} w_C$**	$-A_{11} w_R$	$A_{10} w_{RR}$
	$A_{13} w_{LB}$	$-A_{14} w_B$	$A_{13} w_{RB}$	
		$A_0^2 w_{BB}$		

$\left. \right\}$ Free edge → x

(7) Free edge, C next to x boundary ($M_{Ty} = 0$)

$$\nabla_C^4 w = \frac{1}{\Delta x^4}\left\{ \right.$$

	$A_{13} w_{LT}$	$-A_{14} w_T$	$A_{13} w_{RT}$	
w_{LL}	$-2A_1 w_L$	**$A_{15} w_C$**	$-2A_1 w_R$	w_{RR}
	$2A_0 w_{LB}$	$-2A_2 w_B$	$2A_0 w_{RB}$	
		$A_0^2 w_{BB}$		

$\left. \right\}$ Free edge → x

(8) Free edge, C on y boundary
($M_{Tx} = M_{Cx} = M_{Bx} = 0,\ R_{Cxz} = 0$)

$$\nabla_C^4 w = \frac{1}{\Delta x^4}$$

$B_{10} w_{TT}$		
$-B_{11} w_T$	$B_{13} w_{RT}$	
$B_{12} w_C$	$-B_{14} w_R$	w_{RR}
$-B_{11} w_B$	$B_{13} w_{RB}$	
$B_{10} w_{BB}$		

Free edge ↓ y

(9) Free edge, C next to y boundary
($M_{Lx} = 0$)

$$\nabla_C^4 w = \frac{1}{\Delta x^4}$$

	$A_0^2 w_{TT}$		
$B_{13} w_{LT}$	$-2A_2 w_T$	$2A_0 w_{RT}$	
$-B_{14} w_L$	**$B_{15} w_C$**	$-2A_1 w_R$	w_{RR}
$B_{13} w_{LB}$	$-2A_2 w_B$	$2A_0 w_{RB}$	
	$A_0^2 w_{BB}$		

Free edge ↓ y

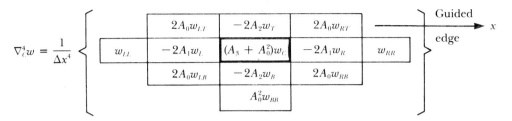

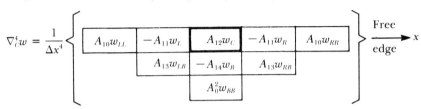

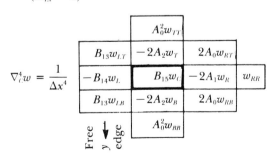

25.06 CORNER EQUATIONS IN CARTESIAN COORDINATES

(1) Equivalents (22.02–22.04) used below are listed in (25.05). Additional equivalents used in this section are

$$A_{16} = 5 + 8A_0 + 5A_0^2 \qquad A_{17} = 7 + 8A_0 + 7A_0^2 \qquad A_{19} = 4\beta A_0$$
$$A_{20} = 2(1 + 2\beta + A_0)A_0 \qquad A_{21} = 1 + 2\beta(4 + 5\alpha)A_0 \qquad A_{22} = 2\beta(A_0 + \alpha + \alpha A_0^2)$$
$$A_{23} = 4\beta(\alpha + A_0)$$

(2) Simply supported corner

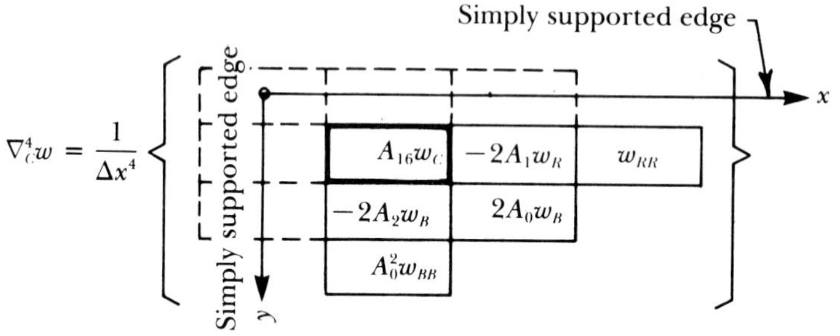

(3) Fixed corner

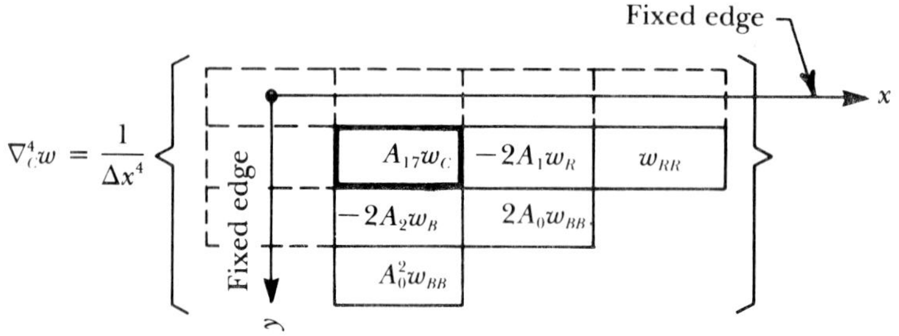

(4) Simple-guided corner

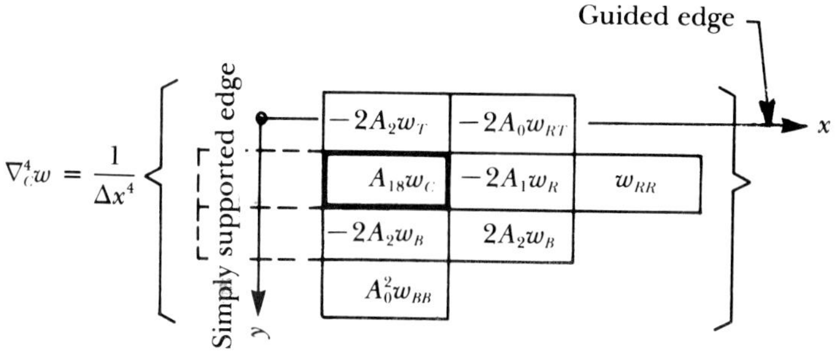

(5) Fixed-guided corner

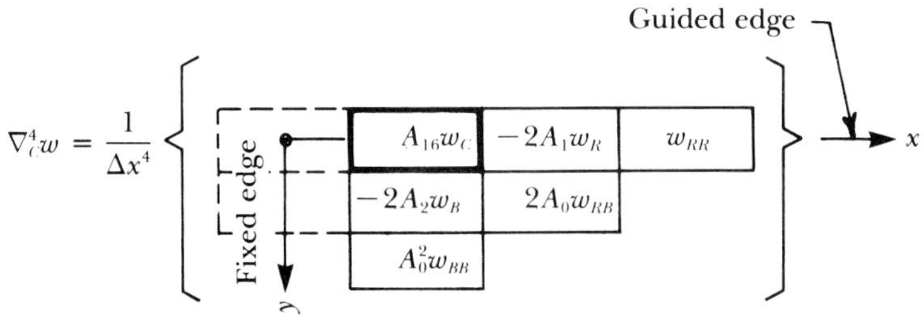

(6) Free corner

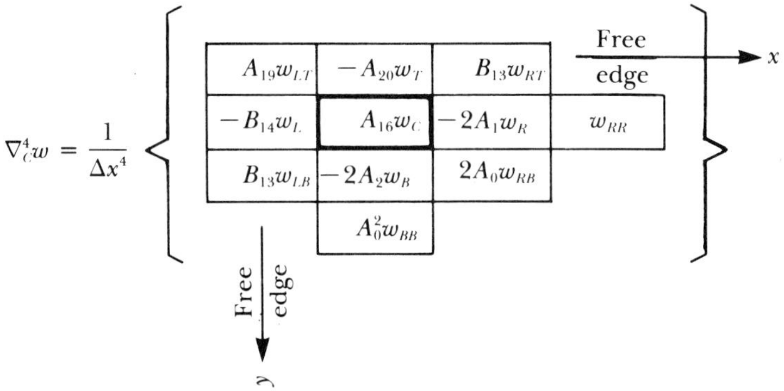

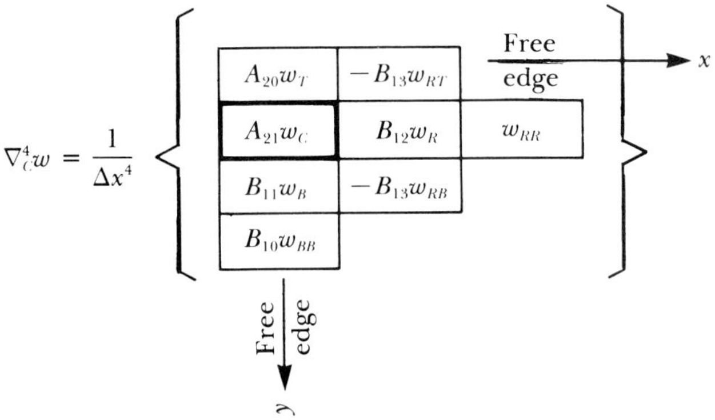

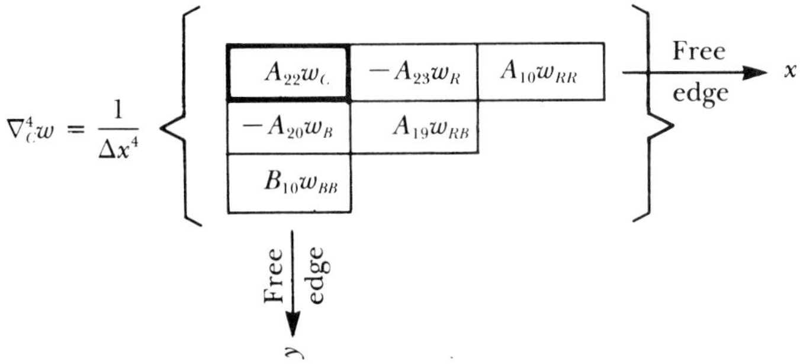

25.07 ISOTROPIC RECTANGULAR PLATE OF VARIABLE THICKNESS IN CARTESIAN SYSTEM

(1) Equivalents used in the construction of the difference equation of an isotropic rectangular plate of variable thickness h are

$$D'_C = \left(\frac{\partial D}{\partial x}\right)_C \qquad D''_C = \left(\frac{\partial^2 D}{\partial x^2}\right)_C \qquad D^*_C = \left(\frac{\partial D}{\partial y}\right)_C \qquad D^{**}_C = \left(\frac{\partial^2 D}{\partial y^2}\right)_C$$

$$A_0 = \frac{D_C}{D_0\,\Delta x^2} \qquad A_1 = \frac{D'_C}{D_0\,\Delta x} \qquad A_2 = \frac{D''_C}{D_0} \qquad \beta = \tfrac{1}{2}(1 - \nu)$$

$$B_0 = \frac{D_C}{D_0\,\Delta y^2} \qquad B_1 = \frac{D^*_C}{D_0\,\Delta y} \qquad B_3 = \frac{D^{**}_C}{D_0}$$

where D_C is the plate rigidity at the central point of the operator (25.04) and D_0 is D at the minimum thickness, used as a scaling factor.

(2) Governing difference equation (25.04-9) becomes in this case

$$\left\{\begin{array}{ccc} & \boxed{(B_0 - B_1)\nabla^2_T w} & \\ \boxed{(A_0 - A_1)\nabla^2_L w} & \boxed{-(2A_0 - A_2 - B_2 + 2B_0)\nabla^2_C w} & \boxed{(A_0 + A_1)\nabla^2_R w} \\ & \boxed{(B_0 + B_1)\nabla^2_B w} & \end{array}\right\} = \frac{p_C}{D_0}$$

where ∇^2_T, ∇^2_L, ∇^2_C, . . . and p_C are the same as in (25.04-9).

(3) Plate functions. The moment equations M_{Cx}, M_{Cy}, M_{Cxy}, M_C are the same as in (25.04). The shear and edge reaction equations are affected by the thickness variation.

$$W_{Cxz} = -D'_C(\nabla^2_C w) - D_C\frac{\partial}{\partial x}(\nabla^2_C w) \qquad W_{Cyz} = -D^*_C(\nabla^2_C w) - D_C\frac{\partial}{\partial x}(\nabla^2_C w)$$

$$R_{Cxz} = W_{Cxz} - 2\beta D^*_C\left(\frac{\partial^2 w}{\partial x\,\partial y}\right)_C - 2\beta D_C\left(\frac{\partial^3 w}{\partial x\,\partial y^2}\right)_C$$

$$R_{Cyz} = W_{Cyz} - 2\beta D'_C\left(\frac{\partial^2 w}{\partial x\,\partial y}\right)_C - 2\beta D_C\left(\frac{\partial^3 w}{\partial x^2\,\partial y}\right)_C$$

where the difference equivalents for the partial derivatives are those given in (25.02).

(4) Elastic foundation effect may be included in (2) above or in (25.04-9) by adding to the left side of the difference equation.

$$\frac{R_C}{D_0} = \frac{1}{36D_0}\left(\begin{array}{c} w_{LT}k_{LT} + 4w_T k_T + w_{RT}k_{RT} \\ +4w_L k_L + 16w_C k_C + 4w_R k_R \\ +w_{LB}k_{LB} + 4w_B k_B + w_{RB}k_{RB} \end{array}\right)$$

where R_C is the joint foundation reaction (N/m^2), and k_{LT}, k_T, k_{RT}, . . . are the foundation moduli (N/m^3) at the respective nodes.

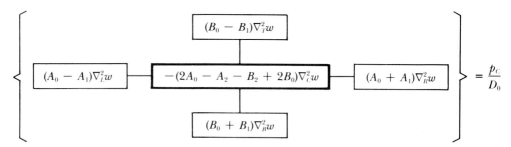

25.08 DERIVATIVES AND DIFFERENCES IN POLAR COORDINATES

(1) Notation and signs (22.06–22.08)

$w = f(r, \theta)$ = deflection (m)

$\Delta r, \Delta\theta$ = increments (m)

$$w' = \frac{\partial w}{\partial r}, \quad w'' = \frac{\partial^2 w}{\partial r^2}, \ldots \qquad w'^* = \frac{\partial^2 w}{\partial r\, \partial\theta}, \quad w'^{**} = \frac{\partial^3 w}{\partial r\, \partial\theta^2}, \ldots$$

$$w^* = \frac{\partial w}{\partial\theta}, \quad w^{**} = \frac{\partial^2 w}{\partial\theta^2}, \ldots \qquad w''^* = \frac{\partial^3 w}{\partial r^2\, \partial\theta}, \quad w''^{**} = \frac{\partial^4 w}{\partial r^2\, \partial\theta^2}, \ldots$$

(2) Difference equations

Difference equations related to the central point C of the adjacent polar grid of Δr, $r\,\Delta\theta$ on the plate described in (22.06) are

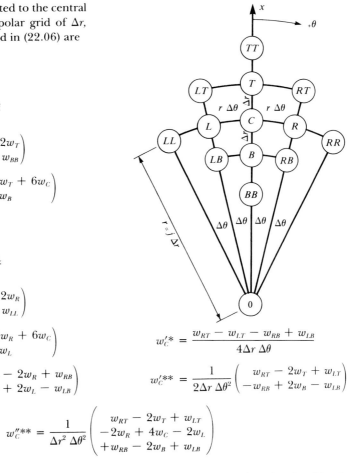

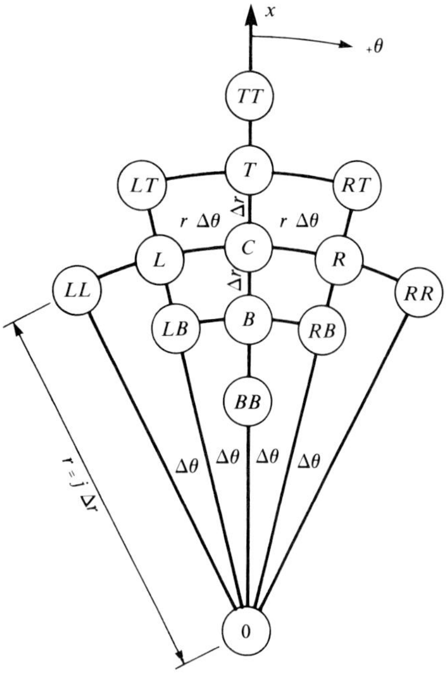

$$w'_C = \frac{w_T - w_B}{2\Delta r}$$

$$w''_C = \frac{w_T - 2w_C + w_B}{\Delta r^2}$$

$$w'''_C = \frac{1}{2\Delta r^3}\left(\begin{matrix} w_{TT} - 2w_T \\ +2w_B - w_{BB} \end{matrix}\right)$$

$$w''''_C = \frac{1}{\Delta r^4}\left(\begin{matrix} w_{TT} - 4w_T + 6w_C \\ +w_{BB} - 4w_B \end{matrix}\right)$$

$$w^*_C = \frac{w_R - w_L}{2\Delta\theta}$$

$$w^{**}_C = \frac{w_R - 2w_C + w_L}{\Delta\theta^2}$$

$$w^{***}_C = \frac{1}{2\Delta\theta^3}\left(\begin{matrix} w_{RR} - 2w_R \\ +2w_L - w_{LL} \end{matrix}\right)$$

$$w^{****}_C = \frac{1}{\Delta\theta^4}\left(\begin{matrix} w_{RR} - 4w_R + 6w_C \\ +w_{LL} - 4w_L \end{matrix}\right)$$

$$w'^*_C = \frac{w_{RT} - w_{LT} - w_{RB} + w_{LB}}{4\Delta r\, \Delta\theta}$$

$$w''^*_C = \frac{1}{2\Delta r^2\, \Delta\theta}\left(\begin{matrix} w_{RT} - 2w_R + w_{RB} \\ -w_{LT} + 2w_L - w_{LB} \end{matrix}\right)$$

$$w'^{**}_C = \frac{1}{2\Delta r\, \Delta\theta^2}\left(\begin{matrix} w_{RT} - 2w_T + w_{LT} \\ -w_{RB} + 2w_B - w_{LB} \end{matrix}\right)$$

$$w''^{**}_C = \frac{1}{\Delta r^2\, \Delta\theta^2}\left(\begin{matrix} w_{RT} - 2w_T + w_{LT} \\ -2w_R + 4w_C - 2w_L \\ +w_{RB} - 2w_B + w_{LB} \end{matrix}\right)$$

where C = center, L = left, R = right, B = bottom, T = top, . . . are the subscripts identifying the respective nodal points and the terms in parentheses represent an algebraic sum.

25.09 PLATE OPERATORS AND FUNCTIONS IN POLAR COORDINATES

(1) Equivalents (22.06–22.08)

$$a_{j+1} = 1 + \frac{1}{2(j-1)} \qquad a_j = 1 + \frac{1}{2j} \qquad a_{j+1} = 1 + \frac{1}{2(j+1)}$$

$$b_{j-1} = \left[\frac{1}{(j-1)\,\Delta\theta}\right]^2 \qquad b_j = \left[\frac{1}{j\,\Delta\theta}\right]^2 \qquad b_{j+1} = \left[\frac{1}{(j+1)\,\Delta\theta}\right]^2$$

$$c_{j-1} = 2(1 + b_{j-1}) \qquad c_j = 2(1 + b_j) \qquad c_{j+1} = 2(1 + b_{j+1})$$

$$d_{j-1} = 2 - a_{j-1} \qquad d_k = 2 - a_j \qquad d_{j+1} = 2 - a_{j+1}$$

$$j = \frac{r}{\Delta r} \qquad \alpha = \tfrac{1}{2}(1 + \nu) \qquad \beta = \tfrac{1}{2}(1 - \nu) \qquad D = \frac{Eh^3}{12(1 - \nu^2)}$$

(2) Operators at T, B, L, C, R in terms of deflections of grid (25.08) are

$$\nabla_T^2 w \cong \frac{1}{\Delta r^2}\left\{\begin{array}{|c|c|c|}\hline & a_{j+1}w_{TT} & \\ \hline b_{j+1}w_{LT} & -c_{j+1}w_T & b_{j+1}w_{RT} \\ \hline & d_{j+1}w_C & \\ \hline\end{array}\right\}$$

$$\nabla_B^2 w \cong \frac{1}{\Delta r^2}\left\{\begin{array}{|c|c|c|}\hline & a_{j-1}w_C & \\ \hline b_{j-1}w_L & -c_{j-1}w_B & b_{j-1}w_{RR} \\ \hline & d_{j-1}w_{BB} & \\ \hline\end{array}\right\}$$

$$\nabla_C^2 w \cong \frac{1}{\Delta r^2}\left\{\begin{array}{|c|c|c|}\hline & a_j w_T & \\ \hline b_j w_L & -c_j w_C & b_j w_R \\ \hline & d_j w_B & \\ \hline\end{array}\right\}$$

$$\nabla_L^2 w \cong \frac{1}{\Delta r^2}\left\{\begin{array}{|c|c|c|}\hline & a_j w_{LT} & \\ \hline b_j w_{LL} & -c_j w_L & b_j w_C \\ \hline & d_j w_{LB} & \\ \hline\end{array}\right\}$$

$$\nabla_R^2 w \cong \frac{1}{\Delta r^2}\left\{\begin{array}{|c|c|c|}\hline & a_j w_{RT} & \\ \hline b_j w_C & -c_j w_R & b_j w_{RR} \\ \hline & d_j w_{RB} & \\ \hline\end{array}\right\}$$

(3) Functions of the plate (22.06) defined in (22.07) can be approximated in terms of the difference equations listed in (25.08).

(4) Governing plate equation (22.08) in difference operator form is

$$\frac{1}{\Delta r^2}\left\{\begin{array}{ccc} & a_j\nabla_T^2 w & \\ b_j\nabla_L^2 w & -c_j\nabla_C^2 w & b_j\nabla_R^2 w \\ & d_j\nabla_B^2 w & \end{array}\right\} = \frac{p_C}{D}$$

where ∇_L^2, ∇_C^2, ∇_R^2, . . . are the difference operators at L, C, R, . . . , respectively, and p_C is the joint load at C calculated by formulas similar to those in (25.03).

25.10 CONDITIONS AT CENTER AND AT EDGES IN POLAR COORDINATES

(1) Polar grid introduced in (25.08) for the difference analysis of an isotropic circular plate of constant thickness creates special conditions at the center 0 of the plate and at the nodes of the first ring adjacent to 0, such as 1, 2, 3, . . . , n (Ref. 40).

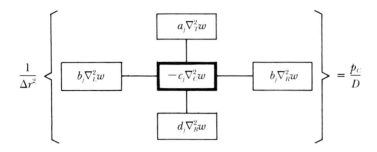

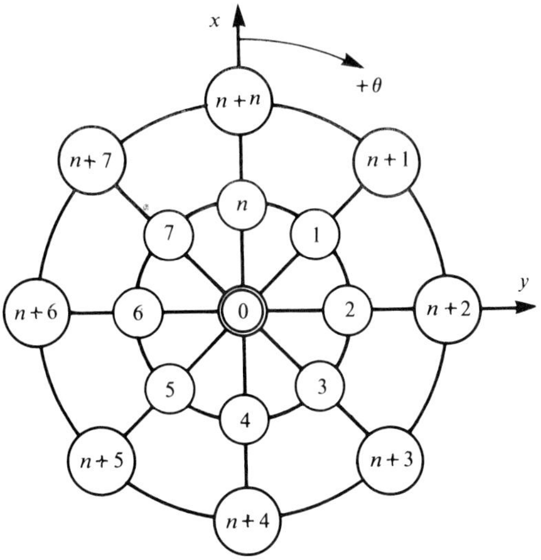

25.10 CONDITIONS AT CENTER AND AT EDGES IN POLAR COORDINATES (*Continued*)

(2) Operators at 0 and n in terms of their respective deflections are

$$\nabla_0^2 w = \frac{4}{n\,\Delta r^2}$$

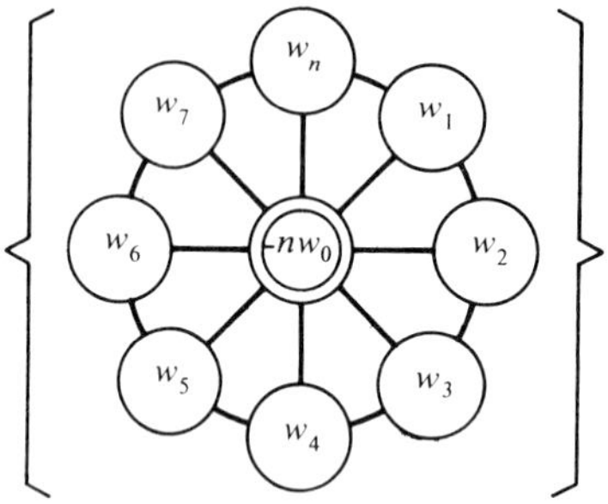

$$\nabla_n^2 w = \frac{1}{\Delta r^2}$$

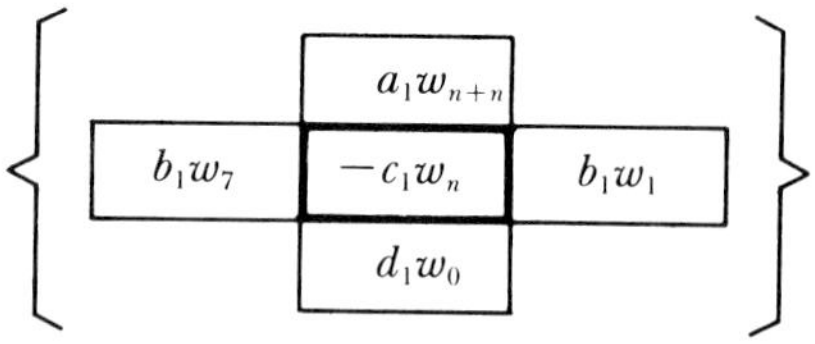

where n is the number of nodes in the first ring and a_1, b_1, c_1, d_1 are the equivalents (25.09-1) for $j = 1$.

(3) Governing difference equations based on (22.08) at 0 and n are

$$\frac{4}{n\Delta r^2}$$

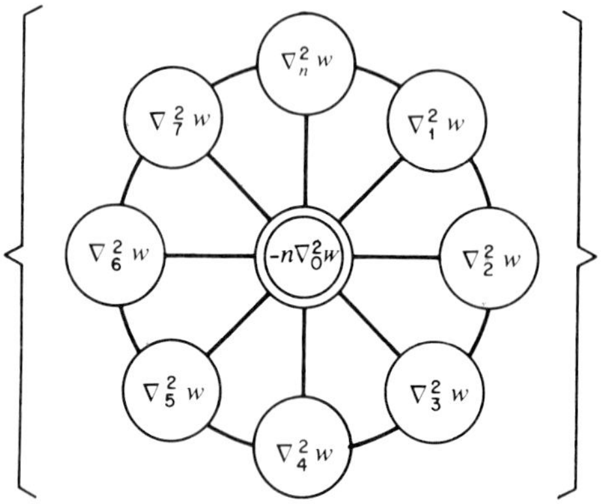

$$= \frac{p_0}{D}$$

$$\frac{1}{\Delta r^2}$$

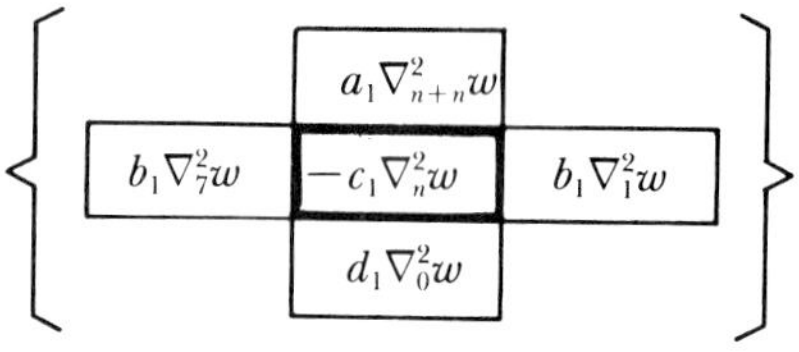

$$= \frac{p_n}{D}$$

where p_0 and p_n are the intensities of load (N/m^2) at 0 and n, respectively. The second equation can be used as a recurrent formula for all nodes of the first ring in the grid.

(4) Modification of the governing difference equation (25.09-4) for joints at or near the boundary is made by means of conditions stated in (22.08-6) in terms of finite differences (25.08).

25.11 AXISYMMETRICAL ISOTROPIC CIRCULAR PLATE ON ELASTIC FOUNDATION

(1) Geometry (22.09)

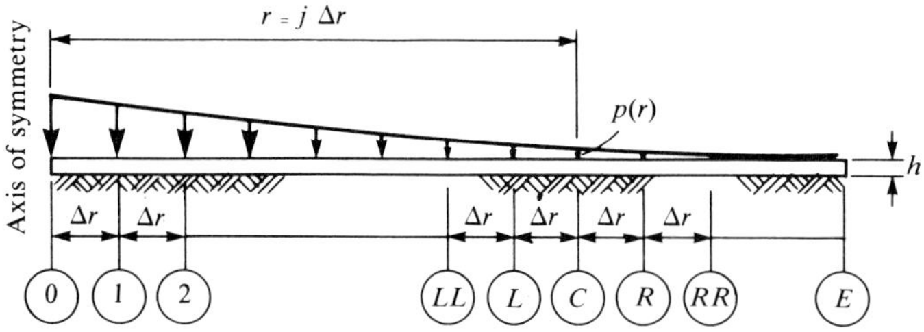

$$\lambda_j = \frac{\Delta r}{r} = \frac{1}{j}$$
$$\alpha = \tfrac{1}{2}(1 + \nu)$$
$$D = \frac{Eh^3}{12(1 - \nu)^2}$$

(2) Equivalents

$$K_{LL} = 1 - \lambda_j \qquad K_L = 2(2 - \lambda_j) + \frac{k_w \Delta r^4}{12D}$$

$$K_C = 6 + 2\lambda_j^2 + \frac{5k_w \Delta r^4}{6D} \qquad k_w = \text{foundation modulus (N/m}^3\text{)}$$

$$K_{RR} = 1 + \lambda_j \qquad K_R = 2(2 + \lambda_j) + \frac{k_w \Delta r^4}{12D}$$

(3) Axisymmetry reduces the difference equation (25.09-4) to

$$K_{LL}w_{LL} - K_L w_L + K_C w_C - K_R w_R + K_{RR} w_{RR} = \frac{(p_L + 10p_C + p_R)\Delta r^4}{12D}$$

where p_L, p_C, p_R are the load intensities at L, C, R, respectively.

(4) Modification of (3) at the center of the plate 0 is

$$\left(16 + \frac{5k_w \Delta r^4}{6D}\right) w_0 - \left(\frac{64}{3} + \frac{k_w \Delta r^4}{6D}\right) w_1 + \tfrac{16}{3} w_2 = \frac{(5p_0 + p_1)\,\Delta r^4}{6D}$$

where w_0, w_1, w_2 are the displacements at 0, 1, 2 and p_0, p_1 are the load intensities of 0, 1, respectively.

(5) Stress resultants at C

$$M_{Cr} = -\frac{D}{\Delta r^2}\left[(1 - \tfrac{1}{2}\nu\lambda_j)w_L - 2w_C + (1 + \tfrac{1}{2}\nu\lambda_j)w_R\right]$$

$$M_{C\theta} = -\frac{D}{\Delta r^2}\left[(\nu - \tfrac{1}{2}\lambda_j)w_L - 2\nu w_C + (\nu + \tfrac{1}{2}\lambda_j)w_R\right]$$

$$W_{Crz} = \frac{D}{\Delta r^3}\left[w_{LL} - (2 + 2\lambda_j + \lambda_j^2)w_L + 4\lambda_j w_C - (2 - 2\lambda_j + \lambda_j^2)w_R + w_{RR}\right]$$

(6) Stress resultants at 0

$$M_{0r} = M_{0z} = -\frac{4\alpha D}{\Delta r^2}(w_1 - w_0) \qquad W_{0rz} = 0$$

(7) Edge conditions. If E is the end node of the plate, then for the simply supported edge, $w_E = 0$, $w_{RE} = -w_{LE}$; for the fixed edge, $w_E = 0$, $w_{RE} = w_{LE}$; and for the free edge, $M_{Er} = 0$, $W_{Erz} = 0$.

25.12 ORTHOTROPIC RECTANGULAR PLATE IN CARTESIAN COORDINATES

(1) Equivalents used in the construction of the difference equation for the orthotropic rectangular plate (22.10) are

$$B_0 = \left(\frac{\Delta x}{\Delta y}\right)^2 \qquad B_1 = \frac{D_{xx}}{H} \qquad B_2 = \frac{B_0^2 D_{yy}}{H} \qquad B_3 = \frac{2D_{xy}}{H} \qquad B_4 = \frac{2B_0^2 D_{xy}}{H}$$

where D_{xx}, D_{yy}, D_{xy}, H are the plate rigidities tabulated in (22.12–22.13).

(2) Governing difference equation of this plate in the grid of (25.02) is

$$\xrightarrow{\quad x \quad} \qquad \downarrow y$$

$$\left\{\begin{array}{|c|c|c|c|c|}
 & & B_2 w_{TT} & & \\
 & 2B_0 w_{LT} & -4(B_0 + B_2) w_T & 2B_0 w_{RT} & \\
B_1 w_{LL} & -4(B_0 + B_1) w_L & [8B_0 + 6(B_1 + B_2)] w_C & -4(B_0 + B_1) w_R & B_1 w_{RR} \\
 & 2B_0 w_{LB} & -4(B_0 + B_2) w_B & 2B_0 w_{RB} & \\
 & & B_2 w_{BB} & &
\end{array}\right\} = \frac{p_C \, \Delta x^4}{H}$$

where p_C is the joint load (N/m^2) at C defined in (25.03). This equation applies to all joints of the grid and must be modified for the edge conditions. The modifications can either be performed as shown in (25.05–25.06) or by adding fictitious points beyond the boundary of the plate, with conditions of the edge expressed by the respective difference equations. Particularly, in the case of the free edge, with four joints selected beyond the boundary, the unknown deflections at these points must satisfy three zero bending moment equations and one zero shear equation.

(3) Functions of the plate such as slopes, moments, shears, and edge reactions are approximated by the difference equations shown below. These equations apply to all joints of the grid. The joint C is the central joint in all formulas, and the orientation of the grid is given in (25.02).

(4) Slopes

$$\phi_C \cong \frac{w_B - w_T}{2\Delta y} \qquad \psi_C \cong -\frac{w_R - w_L}{2\Delta x}$$

(5) Bending moments

$$M_{Cx} \cong -\frac{D_{xx}}{\Delta x^2}\left\{\begin{array}{|c|c|c|}
 & \nu_y B_0 w_T & \\
w_L & -B_5 w_C & w_R \\
 & \nu_y B_0 w_B &
\end{array}\right\} \qquad M_{Cy} \cong -\frac{D_{yy}}{\Delta x^2}\left\{\begin{array}{|c|c|c|}
 & B_0 w_T & \\
\nu_x w_L & -B_6 w_C & \nu_x w_R \\
 & B_0 w_B &
\end{array}\right\}$$

where $B_5 = 2(1 + \nu_y B_0) \qquad B_6 = 2(\nu_x + B_0)$

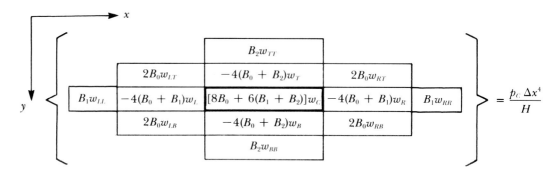

(6) Twisting moments

$$M_{Cxy} = M_{Cyx} \cong -\frac{\sqrt{B_0}\,D_{xy}}{2\Delta x^2}
\begin{Bmatrix}
w_{LT} & 0 & -w_{RT} \\
0 & 0 & 0 \\
-w_{LB} & 0 & w_{RB}
\end{Bmatrix}$$

(7) Shears

$$\cong -\frac{H}{2\Delta x^3}
\left\{
\begin{array}{ccccc}
 & & 0 & & \\
 & -B_0 w_{LT} & 0 & B_0 w_{RT} & \\
-B_1 w_{LL} & 2(B_0+B_1)w_L & 0 & -2(B_0+B_1)w_R & B_1 w_{RR} \\
 & -B_0 w_{LB} & 0 & B_0 w_{RB} & \\
 & & 0 & &
\end{array}
\right\}$$

$$\cong -\frac{H\sqrt{B_0}}{2\Delta x^3}
\left\{
\begin{array}{ccccc}
 & & -B_2 w_{TT} & & \\
 & -w_{LT} & 2(1+B_2)w_T & -w_{RT} & \\
0 & 0 & 0 & 0 & 0 \\
 & w_{LB} & -2(1+B_2)w_B & w_{RB} & \\
 & & B_2 w_{BB} & &
\end{array}
\right\}$$

(8) Edge reactions

$$\cong -\frac{H}{2\Delta x^3}
\left\{
\begin{array}{ccccc}
 & & 0 & & \\
 & -(B_0+B_4)w_{LT} & 0 & (B_0+B_4)w_{RT} & \\
-B_1 w_{LL} & 2(B_0+B_1+B_4)w_L & 0 & -2(B_0+B_1+B_4)w_R & B_1 w_{RR} \\
 & -(B_0+B_4)w_{LB} & 0 & (B_0+B_4)w_{RB} & \\
 & & 0 & &
\end{array}
\right\}$$

$$\cong -\frac{H\sqrt{B_0}}{2\Delta x^3}
\left\{
\begin{array}{ccccc}
 & & -B_2 w_{TT} & & \\
 & -(1+B_3)w_{LT} & 2(1+B_2+B_3)w_T & -(1+B_3)w_{RT} & \\
0 & 0 & 0 & 0 & 0 \\
 & (1+B_3)w_{LB} & -2(1+B_2+B_3)w_B & (1+B_3)w_{RB} & \\
 & & B_2 w_{BB} & &
\end{array}
\right\}$$

25.13 ORTHOTROPIC CIRCULAR PLATE IN POLAR COORDINATES

(1) Equivalents used in the construction of the difference equation are

$a_{j1} = (1 - m)D_{rr}$ $a_{j2} = \lambda^4 D_{\theta\theta}$ $a_{j3} = (1 + m)D_{rr}$	$b_{j1} = 2\lambda^2(1 + \frac{1}{2}m)H$ $b_{j2} = 4\lambda^2[(1 - \frac{1}{2}m^2)H + (\lambda^2 - \frac{1}{2}m^2)D_{\theta\theta}]$ $b_{j3} = 2\lambda^2(1 - \frac{1}{2}m)H$	$m = \dfrac{1}{j}$ $n = \dfrac{1}{\Delta\theta}$
$c_{j1} = 4(1 - \frac{1}{2}m)D_{rr} + 4\lambda^2(1 + \frac{1}{2}m)H + m^2(1 + \frac{1}{2}m)D_{\theta\theta}$ $c_{j2} = 6D_{rr} - 4\lambda^2(m^2 - 2)H + 2(m^2 - 2m^2\lambda^2 + 3\lambda^4)D_{\theta\theta}$ $c_{j3} = 4(1 + \frac{1}{2}m)D_{rr} + 4\lambda^2(1 - \frac{1}{2}m)H + m^2(1 - \frac{1}{2}m)D_{\theta\theta}$		$\lambda = mn = \dfrac{1}{j\,\Delta\theta}$

where D_{rr}, $D_{\theta\theta}$, H are the plate rigidities tabulated in (22.12–22.13).

(2) Governing difference equation of this plate in the grid of (25.08) is

$$\left\{\begin{array}{ccccc} & & a_{j3}w_{TT} & & \\ & b_{j3}w_{LT} & -c_{j3}w_T & b_{j3}w_{RT} & \\ a_{j2}w_{LL} & -b_{j2}w_L & c_{j2}w_C & -b_{j2}w_R & a_{j2}w_{RR} \\ & b_{j1}w_{LB} & -c_{j1}w_B & b_{j1}w_{RB} & \\ & & a_{j1}w_{BB} & & \end{array}\right\} = p_C\,\Delta r^4$$

where j is the position number of the ring in which C is located (such as first, second, . . . measured from the center 0) and p_C is the load intensity at C (25.03).

(3) Functions of the plate (slopes, moments, shears, and edge reactions) are approximated by the difference equations shown below. These equations apply to all joints of the grid with the exception of the center of the plate 0. The joint C is the central joint in all formulas, related to the grid (25.08).

(4) Slopes

$$\phi_C \cong \frac{w_T - w_B}{2\,\Delta r} \qquad \psi_C \cong -\frac{w_R - w_L}{2r\,\Delta\theta}$$

(5) Bending and twisting moments

$$M_{Cr} \cong -\frac{D_{rr}}{\Delta r^2}\left\{\begin{array}{ccc} & (1 + \frac{1}{2}m\nu_\theta)w_T & \\ \lambda^2\nu_\theta w_L & -2(1 + \lambda^2\nu_\theta)w_C & \lambda^2\nu_\theta w_R \\ & (1 - \frac{1}{2}m\nu_\theta)w_B & \end{array}\right\} \qquad M_{C\theta} \cong -\frac{D_{\theta\theta}}{\Delta r^2}\left\{\begin{array}{ccc} & (\nu_r + \frac{1}{2}m)w_T & \\ \lambda^2 w_L & -2(\nu_r - \lambda^2)w_C & \lambda^2 w_R \\ & (\nu_r - \frac{1}{2}m)w_B & \end{array}\right\}$$

$$M_{Cr\theta} = M_{C\theta r} \cong -\frac{\lambda D_{r\theta}}{2\Delta r^2}\left\{\begin{array}{ccc} -w_{LT} & 0 & w_{RT} \\ 2mw_L & 0 & -2mw_R \\ w_{LB} & 0 & -w_{RB} \end{array}\right\}$$

where $D_{r\theta}$ is the plate rigidity tabulated for particular cases in (22.12–22.13).

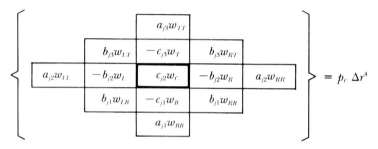

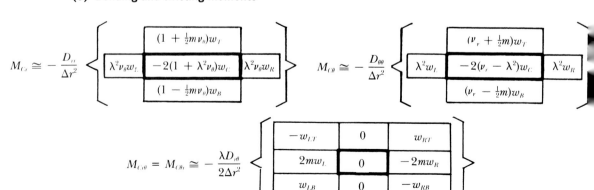

(6) Shears and edge reactions

$$W_{Crz} \cong -\frac{1}{\Delta r^3}$$

		$\frac{1}{2}D_{rr}w_{TT}$		
	$\frac{1}{2}\lambda^2 H w_{LT}$	$-c_{j6}w_T$	$\frac{1}{2}\lambda^2 H w_{RT}$	
$\frac{1}{2}\lambda^3 D_{\theta\theta}w_{LL}$	$-\lambda^3 D_{\theta\theta}w_L$	$-c_{j5}w_C$	$\lambda^3 D_{\theta\theta}w_R$	$-\frac{1}{2}\lambda^3 D_{\theta\theta}w_{RR}$
	$-\frac{1}{2}\lambda^2 H w_{LB}$	$c_{j4}w_B$	$-\frac{1}{2}\lambda^2 H w_{RB}$	
		$-\frac{1}{2}D_{rr}w_{BB}$		

$$W_{C\theta z} \cong -\frac{1}{\Delta r^3}$$

		0		
	$-b_{j6}w_{LT}$	0	$b_{j6}w_{RT}$	
$-\frac{1}{2}\lambda^3 D_{\theta\theta}w_{LL}$	$b_{j5}w_L$	0	$-b_{j5}w_R$	$\frac{1}{2}\lambda^3 D_{\theta\theta}w_{RR}$
	$b_{j4}w_{LB}$	0	$-b_{j4}w_{RB}$	
		0		

where $b_{j4} = \frac{1}{4}\lambda(mD_{\theta\theta} - 2H)$ $\quad c_{j4} = (1 + \frac{1}{2}m)D_{rr} + \lambda^2(1 - m)H + \frac{1}{2}m^2 D_{\theta\theta}$

$b_{j5} = \lambda^3 D_{\theta\theta} + \lambda H$ $\quad c_{j5} = 2m(D_{rr} - \lambda^2 H)$

$b_{j6} = \frac{1}{4}\lambda(mD_{\theta\theta} + 2H)$ $\quad c_{j6} = (1 - \frac{1}{2}m)D_{rr} + \lambda^2(1 + m)H + \frac{1}{2}m^2 D_{\theta\theta}$

$$R_{Crz} \cong -\frac{1}{\Delta r^3}$$

		$\frac{1}{2}D_{rr}w_{LL}$		
	$\frac{1}{2}\lambda^2(H + 2D_{r\theta})w_{LT}$	$-c_{j9}w_T$	$\frac{1}{2}\lambda^2(H + 2D_{r\theta})w_{RT}$	
$\frac{1}{2}\lambda^3 D_{\theta\theta}w_{LL}$	$-\lambda^2(\lambda D_{\theta\theta} + 2mD_{r\theta})w_L$	$-c_{j8}w_C$	$\lambda^2(\lambda D_{\theta\theta} + 2mD_{r\theta})w_R$	$-\frac{1}{2}\lambda^3 D_{\theta\theta}w_{RR}$
	$-\frac{1}{2}\lambda^2(H + 2D_{r\theta})w_{LB}$	$c_{j7}w_B$	$-\frac{1}{2}\lambda^2(H + 2D_{r\theta})w_{RB}$	
		$-\frac{1}{2}D_{rr}w_{BB}$		

$$R_{C\theta z} \cong -\frac{1}{\Delta r^3}$$

		0		
	$-b_{j9}w_{LT}$	0	$b_{j9}w_{RT}$	
$-\frac{1}{2}\lambda^3 D_{\theta\theta}w_{LL}$	$b_{j8}w_L$	0	$-b_{j8}w_R$	$\frac{1}{2}\lambda^3 D_{\theta\theta}w_{RR}$
	$b_{j7}w_{LB}$	0	$-b_{j7}w_{RB}$	
		0		

where $b_{j7} = b_{j4} - \lambda(1 + m)D_{r\theta}$ $\quad c_{j7} = c_{j4} + 2\lambda^2 D_{r\theta}$

$b_{j8} = b_{j5} + \lambda(2 - \lambda)D_{r\theta}$ $\quad c_{j8} = c_{j5} - 4m\lambda^2 D_{r\theta}$

$b_{j9} = b_{j6} - \lambda(1 - m)D_{r\theta}$ $\quad c_{j9} = c_{j6} + 2\lambda^2 D_{r\theta}$

25.14 RECTANGULAR PLATE ON ELASTIC FOUNDATION UNDER GENERAL STATIC LOAD IN CARTESIAN COORDINATES

(1) System considered is a rectangular plate of constant depth, which rests on an elastic foundation of modulus k_w (N/m^3) and is subjected to transverse loads of intensity p_z (N/m^2), in-plane normal edge forces of intensities N_x, N_y (N/m), and temperature change.

(2) Parameters (22.14)

$$\lambda_x = \frac{\eta_x N_x \,\Delta x^2}{H} \qquad \lambda_y = \frac{\eta_y N_y \,\Delta x^4}{H\,\Delta y^2} \qquad \lambda_k = \frac{k_w \,\Delta x^4}{2H}$$

where k_w, N_x, N_y, η_x, η_y are those defined in (22.14) and H is the orthotropic plate rigidity tabulated in (22.12–22.13). For isotropic plate,

$$\nu_x = \nu_y = \nu \qquad H = D = \frac{Eh^3}{12(1-\nu^2)}$$

(3) Governing difference equation of the plate defined in (1) above is

$$\left\{\begin{array}{ccccc} & & C_6 w_{TT} & & \\ & C_4 w_{LT} & -C_5 w_T & C_4 w_{RT} & \\ C_1 w_{LL} & -C_2 w_L & C_3 w_C & -C_2 w_R & C_1 w_{RR} \\ & C_4 w_{LB} & -C_5 w_B & C_4 w_{RB} & \\ & & C_6 w_{BB} & & \end{array}\right\} = \frac{(p_C - p_C^*)\,\Delta x^4}{H}$$

where p_C is the mechanical joint load (N/m^2) at C, defined in (25.03), p_C^* is the thermal joint load (N/m^2) at C, defined in (25.16), and C_1, C_2, . . . are the numerical factors listed in (4, 5) below.

(4) Orthotropic plate equivalents (25.12)

$\lambda_0 = (\Delta_x/\Delta_y)^2$	$C_1 = \lambda_1 \qquad C_2 = 4(\lambda_0 + \lambda_1) + \lambda_x$
$\lambda_1 = \dfrac{D_{xx}}{H}$	$C_3 = 8\lambda_0 + 6(\lambda_1 + \lambda_2) + 2(\lambda_x + \lambda_y + \lambda_k)$
$\lambda_2 = \dfrac{D_{yy}\lambda_0^2}{H}$	$C_4 = 2\lambda_0 \qquad C_5 = 4(\lambda_0 + \lambda_2) + \lambda_y \qquad C_6 = \lambda_2$

(5) Isotropic plate equivalents (25.04)

$\lambda_0 = (\Delta x/\Delta y)^2$	$H = D \qquad C_1 = 1 \qquad C_2 = \lambda_1 + \lambda_x$
$\lambda_1 = 4(1 + \lambda_0)$	$C_3 = 8\lambda_0 + 6(1 + \lambda_0^2) + 2(\lambda_x + \lambda_y + \lambda_k)$
$\lambda_2 = 4\lambda_0(1 + \lambda_0)$	$C_4 = 2\lambda_0 \qquad C_5 = \lambda_2 + \lambda_y \qquad C_6 = \lambda_0^2$

25.15 CIRCULAR PLATE ON ELASTIC FOUNDATION UNDER GENERAL STATIC LOAD IN POLAR COORDINATES

(1) System considered is a circular plate of constant depth, which rests on an elastic foundation of modulus k_w (N/m^3) and is subjected to transverse loads of intensity p_z (N/m^2), in-plane normal edge forces of intensity N_r (N/m), and temperature change.

(2) Parameters (22.14)

$$\lambda_r = \frac{\eta_r N_r \, \Delta r^2}{H} \qquad \lambda_k = \frac{k_w \, \Delta r^4}{2H}$$

where k_w, N_r, η_r are those defined in (22.14) and H is the orthotropic plate rigidity tabulated in (22.12–22.13). For isotropic plate,

$$\nu_r = \nu_\theta = \nu \qquad H = D = \frac{Eh^3}{12(1-\nu^2)}$$

(3) Governing difference equation for the plate defined in (1) above is in symbolic form

		$A_{j3}w_{TT}$		
	$B_{j3}w_{LT}$	$-C_{j3}w_T$	$B_{j3}w_{RT}$	
$A_{j2}w_{CC}$	$-B_{j2}w_L$	$C_{j2}w_C$	$-B_{j2}w_R$	$A_{j2}w_{RR}$
	$B_{j1}w_{LB}$	$-C_{j1}w_B$	$B_{j1}w_{LR}$	
		$A_{j1}w_{BB}$		

$$= \frac{(p_C - p_C^*)\,\Delta r^4}{H}$$

where j is the position number of the ring in which C is located ($j = 1, 2, \ldots$ measured from the center 0), p_C is the mechanical joint load (N/m^2) at C defined in (25.03), p_C^* is the thermal joint load defined in (25.14), and A_j, B_j, C_j are the numerical factors listed in (4, 5) below.

(4) Orthotropic plate equivalents (25.13)

$m = \dfrac{1}{j}$	$A_{j1} = (1-m)\lambda_1 \qquad A_{j2} = \lambda_2\lambda_0^4 \qquad A_{j3} = (1+m)\lambda_1$
$n = \dfrac{1}{\Delta\theta}$	$B_{j1} = (1+m)\lambda_0^2 \qquad B_{j2} = 2[(2-m)+(2\lambda_0^2 - m)\lambda_2]\lambda_0^2 \qquad B_{j3} = (2-m)\lambda_0^2$
$\lambda_0 = mn$	$C_{j1} = (2-m)\lambda_1 + 2(2+m)\lambda_0^2 + \frac{1}{2}m^2(2+m)\lambda_2 + \lambda_r$
$\lambda_1 = \dfrac{D_{rr}}{H}$	$C_{j2} = 6\lambda_1 - 4(m^2-2)\lambda_0^2 + 2(m^2 - 2m^2\lambda_0^2 + 3\lambda_0^4)\lambda_2 + 2(\lambda_r + \lambda_k)$
$\lambda_2 = \dfrac{D_{\theta\theta}}{H}$	$C_{j3} = (2+m)\lambda_1 + 2(2-m)\lambda_0^2 + \frac{1}{2}m^2(2-m)\lambda_2 + \lambda_r$

25.15 CIRCULAR PLATE ON ELASTIC FOUNDATION UNDER GENERAL STATIC LOAD IN POLAR COORDINATES (*Continued*)

(5) Isotropic plate equivalents (25.09)

$H = D$	$A_{j1} = (1 - m)$	$A_{j2} = \lambda_0^4$	$A_{j3} = 1 + m$
$m = \dfrac{1}{j}$	$B_{j1} = (2 + m)\lambda_0^2$	$B_{j2} = 4(1 - m^2 + \lambda_0^2)\lambda_0^2$	$B_{j3} = (2 - m)\lambda_0^2$
$n = \dfrac{1}{\Delta\theta}$	$C_{j1} = 4(1 + \lambda_0^2) - 2m(1 - \lambda_0^2) + m^2(1 + \frac{1}{2}m) + \lambda_r$		
$\lambda_0 = mn$	$C_{j2} = 6(1 + \lambda_0^4) + 8(1 - m^2)\lambda_0 + 2(m^2 + \lambda_r + \lambda_k)$		
	$C_{j3} = 4(1 + \lambda_0^2) + 2m(1 - \lambda_0^2) + m^2(1 - \frac{1}{2}m) + \lambda_r$		

25.16 THERMAL JOINT LOADS IN PLATES

(1) Thermal joint load in rectangular plate (25.14-1) is

$$p_C^* = \frac{1}{\Delta x^2}[M_{Lx}^* - 2M_{Cx}^* + M_{Rx}^* + (M_{By}^* - 2M_{Cy}^* + M_{Ty}^*)\lambda_0^2]$$

where

$$M_{ix}^* = \frac{12D_{xx}}{h^3}\int_{-h/2}^{h/2} g_{tx}z^2\,dz \qquad M_{iy}^* = \frac{12D_{yy}}{h^3}\int_{-h/2}^{h/2} g_{ty}z^2\,dz$$

in which $i = L, C, R, B, T,$ $\quad g_{tx} = (\alpha_{tx} + \nu_y\alpha_{ty})\dfrac{\Delta t^*}{h} \qquad g_{ty} = (\alpha_{ty} + \nu_x\alpha_{tx})\dfrac{\Delta t^*}{h}$

and Δt^* is the temperature difference between the faces of the plate (22.14) and α_{tx}, α_{ty} are the coefficients of thermal expansion (1/C°) in the respective direction.

(2) Thermal joint load in circular plate (25.15-1) is

$$p_C^* = \frac{1}{\Delta r^2}[(1 - \tfrac{1}{2}m)M_{Br}^* + 2M_{Cr}^* + (1 + \tfrac{1}{2}m)M_{Tr}^* + (M_{L\theta}^* + 2M_{C\theta}^* + M_{R\theta}^*)]$$

where

$$M_{ir}^* = \frac{12D_{rr}}{h^3}\int_{-h/2}^{h/2} g_{tr}z^2\,dz \qquad M_{i\theta}^* = \frac{12D_{\theta\theta}}{h^3}\int_{-h/2}^{h/2} g_{t\theta}z^2\,dt$$

in which $i = L, C, R, B, T,$ $\quad g_{tr} = (\alpha_{tr} + \nu_\theta\alpha_{t\theta})\dfrac{\Delta t^*}{h} \qquad g_{t\theta} = (\alpha_{t\theta} + \nu_r\alpha_{tr})\dfrac{\Delta t^*}{h}$

and Δt^*, α_{tr}, $\alpha_{t\theta}$ have an analogical meaning to their counterparts in (1) above.

26

Plate Analysis by Finite Elements

STATIC STATE

26.01 DEFINITION OF STATE

(1) Systems considered are thin isotropic or orthotropic plates of constant or variable thickness and of arbitrary geometric shape, acted on by in-plane and/or normal-to-plane mechanical and thermal causes in a state of static equilibrium.

(2) Finite-element method introduced in this chapter replaces the given plate by an equivalent system of finite elements, connected together at a selected number of points called nodes. The displacement field of the plate is approximated by the displacement function derived by geometric or variational methods, which in turn defines the nodal displacements in terms of position coordinates. Next, using the virtual displacement principles, the elemental stiffness equations are formulated and assembled into the system matrix equation, which yields the approximate values of the displacements of the plate midsurface at these nodes. The moments, shears, and reactions are then approximated in terms of the calculated displacements. Only the basic models are displayed in this chapter.

(3) Assumptions of analysis are stated in (22.01). Only triangular and rectangular plate elements are considered in this chapter. Each element is assumed to have j nodes (vertices) and i degrees of freedom at each node. Thus the total number of degrees of freedom of the element is $k = ixj$.

26.02 FINITE-ELEMENT ANALYSIS

(1) Interior displacement vector δ at x, y in the plate element defined in (26.01-3) is assumed in the form

$$\delta = F(x, y)c$$

where $c = \{c_{11}, c_{12}, \ldots, c_{1i}, \ldots, c_{j1}, c_{j2}, \ldots, c_{ji}\}$, $[k \times 1]$, is the column matrix of k unknown constants and $F(x, y)$ is a matrix function x, y, called the *shape or displacement function,* $[1 \times k]$.

(2) Nodal displacement vector is

$$\Delta = Fc$$

where $\Delta = \{\Delta_{11}, \Delta_{12}, \ldots, \Delta_{1i}, \ldots, \Delta_{j1}, \Delta_{j2}, \ldots, \Delta_{ji}\}$, $[k \times 1]$, is the column matrix of k nodal displacements and

$$F = \begin{bmatrix} F(x_1, y_1) \\ F(x_2, y_2) \\ \cdots \\ F(x_j, y_j) \end{bmatrix} \qquad [k \times k]$$

is the *geometric matrix* constructed by using $F(x, y)$ in terms of nodal coordinates $x_1, y_1, x_2, y_2, \ldots, x_j, y_j$. The solution of this matrix equation yields the vector c with which (1) becomes

$$\delta = F(x, y)F^{-1}\,\Delta = G\,\Delta$$

and the interior displacement vector δ is defined as a matrix function in x, y with the nodal displacement serving as the new constants.

(3) Shape matrix function

$$G = G(x, y) \qquad [i \times k]$$

is a matrix function which, in general, is an approximation so chosen to define uniquely the displacements within the element, to ensure the compatibility at the nodes, and to satisfy the prescribed boundary conditions. Various methods are available for the construction of this function (Refs. 35–38).

(4) Interior strain vector in r components obtained by differentiation of δ and by addition of e_T, e_0 is

$$e = D\delta + e_T + e_0 = \underbrace{DG}_{B}\,\Delta + e_T + e_0 \qquad [r \times 1]$$

where D = linear differential operator matrix, $[r \times i]$
B = strain-displacement matrix, $[r \times k]$
e_T = thermal strain vector $[r \times 1]$
e_0 = initial strain vector, $[r \times 1]$

(5) Interior stress vector in r components based on the constitutive laws of plate material is

$$f = \underbrace{EB}_{S}\,\Delta - E(e_T + e_0) \qquad [r \times 1]$$

where E = stress-strain matrix, $[r \times r]$
S = stress-displacement matrix, $[r \times k]$

$e_T = \{\alpha_{tx}t^*, \alpha_{ty}t^*, 0\}$
$e_0 = \{\varepsilon_{0xx}, \varepsilon_{0yy}, \varepsilon_{0xy}\}$

and α_{tx}, α_{ty} = thermal coefficients along x, y, respectively, (1/C°), t^* = change in temperature (°C), and $\varepsilon_{0xx}, \varepsilon_{0yy}, \varepsilon_{0xy}$ = given initial strains.

(6) Elemental equation of motion based on the virtual work principle is

$$m\ddot{\Delta} + k\Delta = P_L - P_T - P_0$$

where m = elemental mass matrix, $[k \times k]$
k = elemental stiffness matrix, $[k \times k]$
P_L = mechanical nodal load vector, $[k \times 1]$
P_T = thermal-change nodal load vector, $[k \times 1]$
P_0 = initial-deformation nodal load vector $[k \times 1]$
Δ = nodal displacement vector, $[k \times 1]$
$\ddot{\Delta} = d^2\Delta/dt^2$ = nodal acceleration vector, $[k \times 1]$

(7) Components of the elemental equation of motion (6) in terms of

p = intensity of load per unit volume (N/m^3)
ρ = mass per unit volume (kg/m^3) $\quad V$ = elemental volume (m^3)

are

$$m = \int_V \rho G^{)T} G \, dV \qquad k = \int_V B^{)T} E B \, dV$$

$$P_L = \int_V G^{)T} p \, dV \qquad P_T = \int_V B^{)T} E e_T \, dV \qquad P_0 = \int_V B^{)T} E e_0 \, dV$$

where the integral is taken over the volume of the element (Ref. 13, pp. 53–55). Particular forms of m, k, P_L, P_T, P_0 are shown in subsequent sections.

(8) Angular transformation of the elemental equation (6) from the local system to the global system is performed by means of the angular transport matrices introduced in (1.12).

$$R^{os} m^s R^{so} \ddot{\Delta}^o + R^{os} k^s R^{so} \Delta^o = R^{os} [P_L^s - P_T^s - P_0^o]$$

where the superscripts s and o stand for local and global system, respectively.

(9) Convergence criteria. For the finite-element solution to converge to the correct solution as the number of elements increases, three criteria must be satisfied (Ref. 42, p. 1631):

(*a*) Element must be able to represent exactly states of constant strain.
(*b*) Element must be capable of small rigid-body motions without developing internal strains.
(*c*) Elements must produce finite strains at their interfaces.

It must be noted that only criterion (*a*) is essential, since (*b*) and (*c*) are implied by (*a*).

(10) Procedure introduced above can be summarized in six steps:

(*a*) Selection of displacement function
(*b*) Formation of strain-displacement relations
(*c*) Application of appropriate constitutive laws
(*d*) Use of virtual work and virtual strain energy to determine the elemental stiffness and mass matrix
(*e*) Determination of nodal loads by inspection or by use of virtual work
(*f*) Compilation of (*d*) and (*e*) into the elemental matrix equation of motion

Once the elemental equations are available for all elements, the system analysis follows the standard pattern of the stiffness method introduced in (1.21).

26.03 PLANE-STRESS PROBLEM

(1) Definition and notation. The plane-stress problem in a thin plate assumes

$$f_{zz} = 0 \qquad f_{zx} = f_{xz} = 0 \qquad f_{zy} = f_{yz} = 0$$

The symbols used below are defined in (22.02, 22.10).

(2) Stress-strain matrix equation

$$\underbrace{\begin{bmatrix} f_{xx} \\ f_{yy} \\ f_{xy} \end{bmatrix}}_{f} = \underbrace{\begin{bmatrix} E_{11} & E_{12} & 0 \\ E_{21} & E_{22} & 0 \\ 0 & 0 & E_{33} \end{bmatrix}}_{E} \begin{bmatrix} \underbrace{\begin{matrix} \varepsilon_{xx} & -\varepsilon_{0xx} & -\alpha_{tx}t^* \\ \varepsilon_{yy} & -\varepsilon_{0yy} & -\alpha_{ty}t^* \\ \varepsilon_{xy} & -\varepsilon_{0xy} & 0 \end{matrix}}_{e \quad -e_0 \quad -e_T} \end{bmatrix}$$

where for isotropic materials, $\alpha_{tx} = \alpha_{ty} = \alpha_t$,

$$E_{11} = E_{22} = \frac{E}{1 - \nu^2} \qquad E_{12} = E_{21} = \frac{\nu E}{1 - \nu^2} \qquad E_{33} = \frac{E}{2(1 + \nu)}$$

and for orthotropic materials,

$$E_{11} = \frac{E_{xx}}{1 - \nu_x\nu_y} \qquad E_{12} = \frac{\nu_y E_{xx}}{1 - \nu_x\nu_y} \qquad E_{12} = E_{21}$$

$$E_{22} = \frac{E_{yy}}{1 - \nu_x\nu_y} \qquad E_{21} = \frac{\nu_x E_{yy}}{1 - \nu_x\nu_y} \qquad E_{33} \cong \frac{E_{xx}}{1 + \nu_y} + \frac{E_{yy}}{1 + \nu_x}$$

(3) Strain-stress matrix equation

$$\underbrace{\begin{bmatrix} \varepsilon_{xx} \\ \varepsilon_{yy} \\ \varepsilon_{xy} \end{bmatrix}}_{e} = \underbrace{\begin{bmatrix} \Lambda_{11} & -\Lambda_{12} & 0 \\ -\Lambda_{21} & \Lambda_{22} & 0 \\ 0 & 0 & \Lambda_{33} \end{bmatrix}}_{\Lambda} \underbrace{\begin{bmatrix} f_{xx} \\ f_{yy} \\ f_{xy} \end{bmatrix}}_{f} + \underbrace{\begin{bmatrix} \varepsilon_{0xx} \\ \varepsilon_{0yy} \\ \varepsilon_{0xy} \end{bmatrix}}_{e_0} + \underbrace{\begin{bmatrix} \alpha_{tx}t^* \\ \alpha_{ty}t^* \\ 0 \end{bmatrix}}_{e_T}$$

where for isotropic materials, $\alpha_{tx} = \alpha_{ty} = \alpha_t$,

$$\Lambda_{11} = \Lambda_{22} = \frac{1}{E} \qquad \Lambda_{12} = \Lambda_{21} = \frac{\nu}{E} \qquad \Lambda_{33} = \frac{2(1 + \nu)}{E}$$

and for orthotropic materials,

$$\Lambda_{11} = \frac{1}{E_{xx}} \qquad \Lambda_{12} = \nu_y\Lambda_{22} \qquad \Lambda_{12} = \Lambda_{21}$$

$$\Lambda_{22} = \frac{1}{E_{yy}} \qquad \Lambda_{21} = \nu_x\Lambda_{11} \qquad \Lambda_{33} = \frac{1 + \nu_y}{E_{xx}} + \frac{1 + \nu_x}{E_{yy}}$$

(4) Relations based on assumption (1) are then

$$\varepsilon_{zz} = -\frac{\nu}{E}(f_{xx} + f_{yy}) \qquad \text{or} \qquad \varepsilon_{zz} = -\frac{\nu_z}{E_{xx}}f_{xx} - \frac{\nu_z}{E_{yy}}f_{yy}$$

$$\varepsilon_{zx} = \varepsilon_{xz} = 0 \qquad \varepsilon_{zy} = \varepsilon_{yz} = 0$$

26.04 PLANE-STRAIN PROBLEM

(1) Definition and notation. The plane-strain problem in a thin plate assumes

$$\varepsilon_{zz} = 0 \qquad \varepsilon_{zx} = \varepsilon_{xz} = 0 \qquad \varepsilon_{zy} = \varepsilon_{yz}$$

The symbols used below are defined in (22.02, 22.10).

(2) Stress-strain matrix equation is formally identical to (26.03-2), but the elements E_{11}, E_{12}, E_{21}, E_{22} take on different analytical forms. For isotropic materials, $\alpha_{tx} = \alpha_{ty} = \alpha_t$,

$$E_{11} = E_{22} = \frac{(1-\nu)E}{(1+\nu)(1-2\nu)}$$

$$E_{12} = E_{21} = \frac{\nu E}{(1+\nu)(1-2\nu)} \qquad E_{33} = \frac{E}{2(2+\nu)}$$

and for orthotropic materials,

$$E_{11} = \frac{\lambda_{11}}{\lambda_0} E_{xx} \qquad E_{12} = \frac{\lambda_{12}}{\lambda_0} E_{xx} \qquad E_{12} = E_{21}$$

$$E_{22} = \frac{\lambda_{22}}{\lambda_0} E_{yy} \qquad E_{21} = \frac{\lambda_{21}}{\lambda_0} E_{yy} \qquad E_{33} \cong \frac{E_{xx}}{1+\nu_y} + \frac{E_{yy}}{1+\nu_x}$$

where

$$\lambda_{11} = 1 - \nu_y \nu_z \qquad \lambda_{12} = \nu_x + \nu_x \nu_z$$

$$\lambda_{22} = 1 - \nu_x \nu_z \qquad \lambda_{21} = \nu_y + \nu_y \nu_z \qquad \lambda_0 = \lambda_{11}\lambda_{22} - \lambda_{12}\lambda_{21}$$

(3) Strain-stress matrix equation is formally identical to (26.03-3), but the elements Λ_{11}, Λ_{12}, Λ_{21}, Λ_{22} take on different analytical forms. For isotropic materials,

$$\Lambda_{11} = \Lambda_{22} = \frac{(1-\nu^2)}{E}$$

$$\Lambda_{12} = \Lambda_{21} = \frac{(1-\nu^2)}{E} \qquad \Lambda_{33} = \frac{2(1+\nu)}{E}$$

and for orthotropic materials, with λ_{11}, λ_{12}, λ_{21}, λ_{22}, and λ_0 given in (2),

$$\Lambda_{11} = \frac{\lambda_{22}}{E_{xx}} \qquad \Lambda_{12} = \frac{\Lambda_{12}}{E_{yy}}$$

$$\Lambda_{22} = \frac{\lambda_{11}}{E_{yy}} \qquad \Lambda_{21} = \frac{\lambda_{21}}{E_{xx}} \qquad \Lambda_{33} \cong \frac{1+\nu_y}{E_{xx}} + \frac{1+\nu_x}{E_{yy}}$$

(4) Relations based on assumption (1) are then

$$f_{zz} = \nu(f_{xx} + f_{yy}) \qquad \text{or} \qquad f_{zz} = E_{zz}\left(\frac{\nu_z}{E_{xx}} f_{xx} + \frac{\nu_z}{E_{yy}} f_{yy}\right)$$

$$f_{zx} = f_{xz} = 0 \qquad f_{zy} = f_{yz} = 0$$

26.05 TRIANGULAR PLATE ELEMENT OF ORDER ONE

(1) Element considered is a triangular element of constant thickness h (m) given by vertices 1, 2, 3 and acted on by in-plane causes (element of order one). The six degrees of freedom are represented by the displacement vector

$$\Delta = \{u_1, v_1, u_2, v_2, u_3, v_3\}$$

The notation used in this section is defined in (22.02–22.05). All forces and displacements are in the x, y system, and their positive directions are shown in (2, 3) below (Ref. 43).

(2) Nodal loads

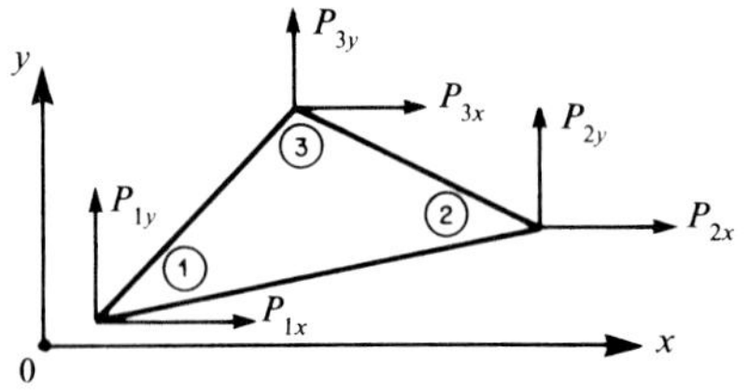

(3) Nodal displacements

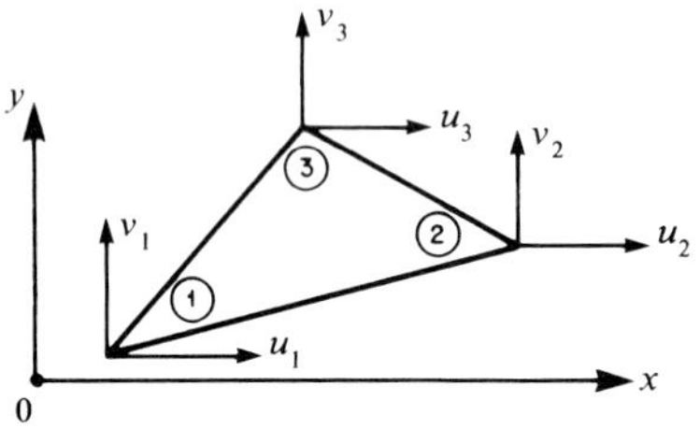

(4) Shape matrix function (26.02-3)

$$G = \frac{1}{2A_0}\left[\begin{array}{cc|cc|cc} G_1 & 0 & G_2 & 0 & G_3 & 0 \\ 0 & G_1 & 0 & G_2 & 0 & G_3 \end{array}\right]$$

where $x_1, y_1, x_2, y_2, x_3, y_3$ = coordinates of vertices,

$$A_0 = \tfrac{1}{2}[(x_1y_2 - x_2y_1) + (x_2y_3 - x_3y_2) + (x_3y_1 - x_1y_3)]$$

is the *area of the triangle,* and

$$\left.\begin{aligned} G_1 &= x_2y_3 - x_3y_2 - y_{23}x + x_{23}y \\ G_2 &= x_3y_1 - x_1y_3 - y_{31}x + x_{31}y \\ G_3 &= x_1y_2 - x_2y_1 - y_{12}x + x_{12}y \end{aligned}\right\} \quad G_1 + G_2 + G_3 = 2A_0$$

in which $x_{23} = x_3 - x_2$, $y_{23} = y_3 - y_2$, ...

(5) Strain-displacement matrix (26.02-4)

$$B = DG = \frac{1}{2A_0}\begin{bmatrix} \dfrac{\partial}{\partial x} & 0 \\ 0 & \dfrac{\partial}{\partial y} \\ \dfrac{\partial}{\partial y} & \dfrac{\partial}{\partial x} \end{bmatrix}\left[\begin{array}{cc|cc|cc} G_1 & 0 & G_2 & 0 & G_3 & 0 \\ 0 & G_1 & 0 & G_2 & 0 & G_3 \end{array}\right]$$

$$B = \frac{1}{2A_0}\left[\begin{array}{cc|cc|cc} -y_{23} & 0 & -y_{31} & 0 & -y_{12} & 0 \\ 0 & x_{23} & 0 & x_{31} & 0 & x_{12} \\ x_{23} & -y_{23} & x_{31} & -y_{31} & x_{12} & -y_{12} \end{array}\right]$$

(6) Stress-displacement matrix (26.02-5)

$$S = EB = \frac{1}{2A_0}\left[\begin{array}{cc|cc|cc} -E_{11}y_{23} & E_{12}x_{23} & -E_{11}y_{31} & E_{12}x_{31} & -E_{11}y_{12} & E_{12}x_{12} \\ -E_{21}y_{23} & E_{22}x_{23} & -E_{21}y_{31} & E_{22}x_{31} & -E_{21}y_{12} & E_{22}x_{12} \\ E_{33}x_{23} & -E_{33}y_{23} & E_{33}x_{31} & -E_{33}y_{31} & E_{33}x_{12} & -E_{33}y_{12} \end{array}\right]$$

where E_{11}, $E_{12} = E_{21}$, E_{22}, E_{33} are the constants in (26.03-2).

(7) Stiffness matrix (26.02-7) $\quad k = k_n + k_s$

where k_n is the elemental stiffness matrix due to normal stresses and k_s is the elemental stiffness matrix due to shearing stresses. Their explicit forms are given below.

$$k_n = \frac{h}{2A_0}\left[\begin{array}{cc|cc|cc} E_{11}y_{23}^2 & & & & & \\ -E_{12}y_{23}x_{23} & E_{22}x_{23}^2 & & & \text{Symmetrical} & \\ \hline E_{11}y_{23}y_{31} & -E_{21}x_{23}y_{31} & E_{11}y_{31}^2 & & & \\ -E_{12}y_{23}x_{31} & E_{22}x_{23}x_{31} & -E_{12}y_{31}x_{31} & E_{22}x_{31}^2 & & \\ \hline E_{11}y_{23}y_{12} & -E_{21}x_{23}y_{12} & E_{11}y_{31}y_{12} & -E_{21}x_{31}y_{12} & E_{11}y_{12}^2 & \\ -E_{12}y_{33}x_{12} & E_{22}x_{23}x_{12} & -E_{12}y_{31}x_{12} & E_{22}x_{31}x_{12} & -E_{12}y_{12}x_{12} & E_{22}x_{12}^2 \end{array}\right]$$

$$k_s = \frac{hE_{33}}{2A_0}\left[\begin{array}{cc|cc|cc} x_{23}^2 & & & & & \\ -x_{23}y_{23} & y_{23}^2 & & & & \\ \hline x_{23}x_{31} & -y_{23}x_{31} & x_{31}^2 & & \text{Symmetrical} & \\ -x_{23}y_{31} & y_{23}y_{31} & -x_{31}y_{31} & y_{31}^2 & & \\ \hline x_{23}x_{12} & -y_{23}x_{12} & x_{31}x_{12} & -y_{31}x_{21} & x_{12}^2 & \\ -x_{23}y_{12} & y_{23}y_{12} & -x_{31}y_{12} & y_{31}y_{12} & -x_{12}y_{12} & y_{12}^2 \end{array}\right]$$

(8) Nodal loads calculated by matrix integrals of (26.02-7) are:

(*a*) Gravity load vector $\quad P_G = -\dfrac{A_0 h p_y}{3}\{0, 1, 0, 1, 0, 1$

(*b*) Thermal load vector

$$P_T = ht^* \left[\begin{array}{c} -(E_{11}\alpha_{tx} + E_{12}\alpha_{ty})y_{23} \\ (E_{21}\alpha_{tx} + E_{22}\alpha_{ty})x_{23} \\ \hline -(E_{11}\alpha_{tx} + E_{12}\alpha_{ty})y_{31} \\ (E_{21}\alpha_{tx} + E_{22}\alpha_{ty})x_{31} \\ \hline -(E_{11}\alpha_{tx} + E_{12}\alpha_{ty})y_{12} \\ (E_{21}\alpha_{tx} + E_{22}\alpha_{ty})x_{12} \end{array}\right]$$

(*c*) Initial strain load vector

$$P_0 = h \left[\begin{array}{c} -(E_{11}\varepsilon_{0xx} + E_{12}\varepsilon_{0yy})y_{23} + E_{33}\varepsilon_{0xy}x_{23} \\ (E_{21}\varepsilon_{0xx} + E_{22}\varepsilon_{0yy})x_{23} - E_{33}\varepsilon_{0xy}y_{23} \\ \hline -(E_{11}\varepsilon_{0xx} + E_{12}\varepsilon_{0yy})y_{31} + E_{33}\varepsilon_{0xy}x_{31} \\ (E_{21}\varepsilon_{0xx} + E_{22}\varepsilon_{0yy})x_{31} + E_{33}\varepsilon_{0xy}x_{31} \\ \hline -(E_{11}\varepsilon_{0xx} + E_{12}\varepsilon_{0yy})y_{12} + E_{33}\varepsilon_{0xy}x_{12} \\ (E_{21}\varepsilon_{0xx} + E_{22}\varepsilon_{0yy})x_{12} - E_{33}\varepsilon_{0xy}y_{12} \end{array}\right]$$

26.06 RECTANGULAR PLATE ELEMENT OF ORDER ONE

(1) Element considered is a rectangular element of constant thickness h (m) given by vertices 1, 2, 3, 4 and acted on by in-plane causes (element of order one). The eight degrees of freedom are represented by the displacement vector

$$\Delta = \{u_1, v_1, u_2, v_2, u_3, v_3, u_4, v_4\}$$

The notation used in this section is defined in (22.02–22.05). All forces and displacements are in the x, y system, and their positive directions are shown in (2, 3) below (Ref. 42).

(2) Nodal loads

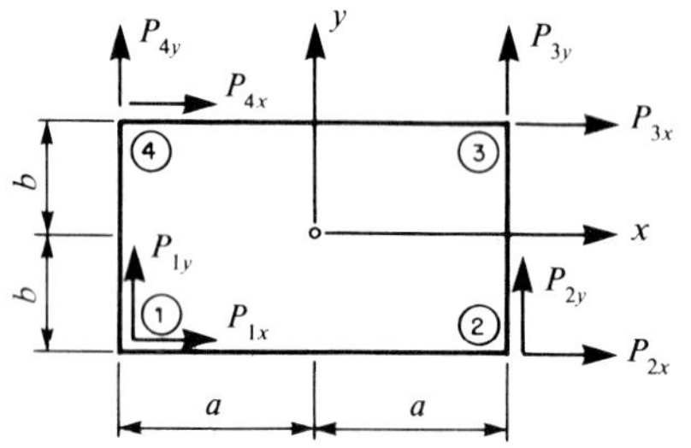

(3) Nodal displacements

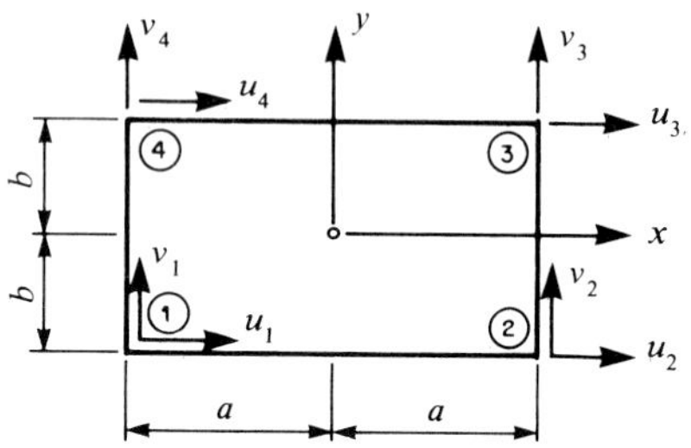

(4) Shape matrix function (26.02-3)

$$G = \frac{1}{4ab}\left[\begin{array}{cc|cc|cc|cc} a_1b_1 & 0 & a_2b_1 & 0 & a_2b_2 & 0 & a_1b_2 & 0 \\ 0 & a_1b_1 & 0 & a_2b_1 & 0 & a_2b_2 & 0 & a_1b_2 \end{array}\right]$$

where $a_1 = a - x \qquad b_1 = b - y \qquad a_1 + a_2 = 2a$

$a_2 = a + x \qquad b_2 = b + y \qquad b_1 + b_2 = 2b$

and $G_1 = a_1b_1$, $G_2 = a_2b_1$, $G_3 = a_2b_2$, $G_4 = a_1b_2$.

(5) Strain-displacement matrix (26.02-4)

$$B = DG = \frac{1}{4ab}\begin{bmatrix} \dfrac{\partial}{\partial x} & 0 \\ 0 & \dfrac{\partial}{\partial y} \\ \dfrac{\partial}{\partial y} & \dfrac{\partial}{\partial x} \end{bmatrix}\left[\begin{array}{cc|cc|cc|cc} G_1 & 0 & G_2 & 0 & G_3 & 0 & G_4 & 0 \\ 0 & G_1 & 0 & G_2 & 0 & G_3 & 0 & G_4 \end{array}\right]$$

$$B = \frac{1}{4ab}\left[\begin{array}{cc|cc|cc|cc} -b_1 & 0 & b_1 & 0 & b_2 & 0 & -b_2 & 0 \\ 0 & -a_1 & 0 & -a_2 & 0 & a_2 & 0 & a_1 \\ -a_1 & -b_1 & -a_2 & b_1 & a_2 & b_2 & a_1 & -b_2 \end{array}\right]$$

(6) Stress-displacement matrix (26.02-5)

$$S = EB = \frac{1}{4ab}\left[\begin{array}{cc|cc|cc|cc} -b_1E_{11} & -a_1E_{12} & b_1E_{11} & -a_2E_{12} & b_2E_{11} & a_2E_{12} & -b_2E_{11} & a_1E_{12} \\ -b_1E_{12} & -a_1E_{22} & b_1E_{12} & -a_2E_{22} & b_2E_{12} & a_2E_{12} & -b_2E_{12} & a_1E_{22} \\ -a_1E_{33} & -b_1E_{33} & -a_2E_{33} & b_1E_{33} & a_2E_{33} & b_2E_{33} & a_1E_{33} & -b_2E_{33} \end{array}\right]$$

where E_{11}, $E_{12} = E_{21}$, E_{22}, E_{33} are the constants defined in (26.03-2).

(7) Stiffness matrix (26.02-7) $\quad k = k_n + k_s$

where k_n is the elemental stiffness matrix due to normal stresses and k_s is the elemental stiffness matrix due to shearing stresses.

$$k_n = \left[\begin{array}{cc|cc|cc|cc}
2K_1 & & & & & & & \\
K_3 & 2K_2 & & & & \text{Symmetrical} & & \\
\hline
-2K_1 & -K_3 & 2K_1 & & & & & \\
K_3 & K_2 & -K_3 & 2K_2 & & & & \\
\hline
-K_1 & -K_3 & K_1 & -K_3 & 2K_1 & & & \\
-K_3 & -K_2 & K_3 & -2K_2 & K_3 & 2K_2 & & \\
\hline
K_1 & K_3 & -K_1 & K_3 & -2K_1 & -K_3 & 2K_1 & \\
-K_3 & -2K_2 & K_3 & -K_2 & K_3 & K_2 & -K_3 & K_2
\end{array}\right]$$

$$k_s = \left[\begin{array}{cc|cc|cc|cc}
2K_4 & & & & & & & \\
K_6 & 2K_5 & & & & \text{Symmetrical} & & \\
\hline
K_4 & K_6 & 2K_4 & & & & & \\
-K_6 & -2K_5 & -K_6 & 2K_5 & & & & \\
\hline
-K_4 & -K_6 & -2K_4 & K_6 & 2K_4 & & & \\
-K_6 & -K_5 & -K_6 & K_5 & K_6 & 2K_5 & & \\
\hline
-2K_4 & -K_6 & -K_4 & K_6 & K_4 & K_6 & 2K_4 & \\
K_6 & K_5 & K_6 & -K_5 & -K_6 & -2K_5 & -K_6 & 2K_5
\end{array}\right]$$

where $K_1 = \dfrac{bhE_{11}}{6a}$, $K_2 = \dfrac{ahE_{22}}{6b}$, $K_3 = \dfrac{hE_{12}}{4}$, $K_4 = \dfrac{ahE_{33}}{6b}$, $K_5 = \dfrac{bhE_{33}}{6a}$, $K_6 = \dfrac{hE_{33}}{4}$.

(8) Nodal loads calculated by matrix integrals (26.02-7) are:

(*a*) Gravity load vector $\quad P_G = -abhp_y\{0, 1, 0, 1, 0, 1, 0, 1\}$

(*b*) Thermal load vector

$$P_T = ht^* \left[\begin{array}{c}
-be_{T1} \\ -ae_{T2} \\ be_{T1} \\ -ae_{T2} \\ \hline be_{T1} \\ ae_{T2} \\ -be_{T1} \\ ae_{T2}
\end{array}\right]$$

where $e_{T1} = E_{11}\alpha_{tx} + E_{12}\alpha_{ty}$

$e_{T2} = E_{21}\alpha_{tx} + E_{22}\alpha_{ty}$

(*c*) Initial strain load vector

$$P_0 = h \left[\begin{array}{c}
-be_{01} - ae_{03} \\ -ae_{02} - be_{03} \\ be_{01} - ae_{03} \\ -ae_{02} + be_{03} \\ \hline be_{01} + ae_{03} \\ ae_{02} + be_{03} \\ -be_{01} + ae_{03} \\ ae_{02} - be_{03}
\end{array}\right]$$

where $e_{01} = E_{11}\varepsilon_{0xx} + E_{12}\varepsilon_{0yy}$

$e_{02} = E_{12}\varepsilon_{0xx} + E_{22}\varepsilon_{0yy}$

$e_{03} = E_{33}\varepsilon_{0xy}$

26.07 TRIANGULAR PLATE ELEMENT, AREA COORDINATES

(1) Area coordinates of the triangular element introduced in (26.05) are

$$s_1 = \frac{A_1}{A_0} \qquad s_2 = \frac{A_2}{A_0} \qquad s_3 = \frac{A_3}{A_0}$$

and $s_1 + s_2 + s_3 = 1$ where A_1, A_2, A_3 are the areas of the subtriangles shown in the adjacent figure and A_0 is the area of the whole triangle. Analytically,

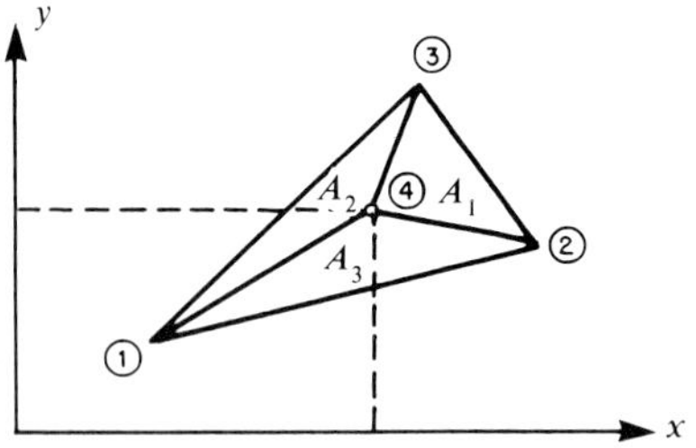

$$A_1 = \frac{1}{2}\begin{vmatrix} 1 & x & y \\ 1 & x_2 & y_2 \\ 1 & x_3 & y_3 \end{vmatrix} \qquad A_2 = \frac{1}{2}\begin{vmatrix} 1 & x_1 & y_1 \\ 1 & x & y \\ 1 & x_3 & y_3 \end{vmatrix} \qquad A_3 = \frac{1}{2}\begin{vmatrix} 1 & x_1 & y_1 \\ 1 & x_2 & y_2 \\ 1 & x & y \end{vmatrix}$$

$$A_0 = A_1 + A_2 + A_3 = \frac{1}{2}\begin{vmatrix} 1 & x_1 & y_1 \\ 1 & x_2 & y_2 \\ 1 & x_3 & y_3 \end{vmatrix}$$

where $x_1, y_1, x_2, y_2, x_3, y_3, x, y$ are the coordinates of 1, 2, 3, 4, respectively.

(2) Relations between x, y and s_1, s_2, s_3 are

$$\begin{bmatrix} 1 \\ x \\ y \end{bmatrix} = \begin{bmatrix} 1 & 1 & 1 \\ x_1 & x_2 & x_3 \\ y_1 & y_2 & y_3 \end{bmatrix}\begin{bmatrix} s_1 \\ s_2 \\ s_3 \end{bmatrix} \qquad \begin{bmatrix} s_1 \\ s_2 \\ s_3 \end{bmatrix} = \frac{1}{2A_0}\begin{bmatrix} 2A_{023} & b_1 & a_1 \\ 2A_{031} & b_2 & a_2 \\ 2A_{012} & b_3 & a_3 \end{bmatrix}\begin{bmatrix} 1 \\ x \\ y \end{bmatrix}$$

where
$$\begin{aligned} A_{023} &= x_2y_3 - x_3y_2 & b_1 &= -(y_3 - y_2) & a_1 &= x_3 - x_2 \\ A_{031} &= x_3y_1 - x_1y_3 & b_2 &= -(y_1 - y_3) & a_2 &= x_1 - x_3 \\ A_{012} &= x_1y_2 - x_2y_1 & b_3 &= -(y_2 - y_1) & a_3 &= x_2 - x_1 \end{aligned}$$

and $a_1 + a_2 + a_3 = 0$, $b_1 + b_2 + b_3 = 0$.

(3) Derivatives of area coordinates ($j = 1, 2, 3$)

$$\frac{ds_j}{dx} = \frac{b_j}{2A_0} \qquad \frac{ds_j}{dy} = \frac{a_j}{2A_0} \qquad \frac{d^2s_j}{dx^2} = \frac{d^2s_j}{dy^2} = \frac{d^2s_j}{dx\,dy} = 0$$

(4) Derivatives of functions of area coordinates. If G is a function of s_1, s_2, s_3, then

$$\frac{dG}{dx} = \frac{1}{2A_0}\left(b_1\frac{dG}{ds_1} + b_2\frac{dG}{ds_2} + b_3\frac{dG}{ds_3}\right)$$
$$\frac{dG}{dy} = \frac{1}{2A_0}\left(a_1\frac{dG}{ds_1} + a_2\frac{dG}{ds_2} + a_3\frac{dG}{ds_3}\right)$$

(5) Integrals of area coordinates are given by the general relation

$$\int_{A0} s_1^m s_2^n s_3^r \, dA = \frac{m!n!r!}{(2 + m + n + r)!}\, 2A_0$$

where the integral is taken over the whole area A_0 of the triangle 1, 2, 3. The formula is valid for $m = 0, 1, 2, \ldots$, $n = 0, 1, 2, \ldots$, $r = 0, 1, 2, \ldots$, and $0! = 0$ by definition.

26.08 TRIANGULAR PLATE ELEMENT OF ORDER TWO, STIFFNESS MATRIX

(1) Element considered is a triangular plate element of constant thickness h (m) given by vertices 1, 2, 3 and acted on by normal-to-plane causes (element of order two). The nine degrees of freedom are represented by the displacement vector

$$\Delta = \{w_1, \phi_1, \psi_1, w_2, \phi_2, \psi_2, w_3, \phi_3, \psi_3\}$$

Notation used in this section is defined in (22.02–22.05, 26.02). All forces, moments, and displacements are in the x, y, z local system, and their positive directions are shown in (2, 3) below (Ref. 45).

(2) Nodal loads

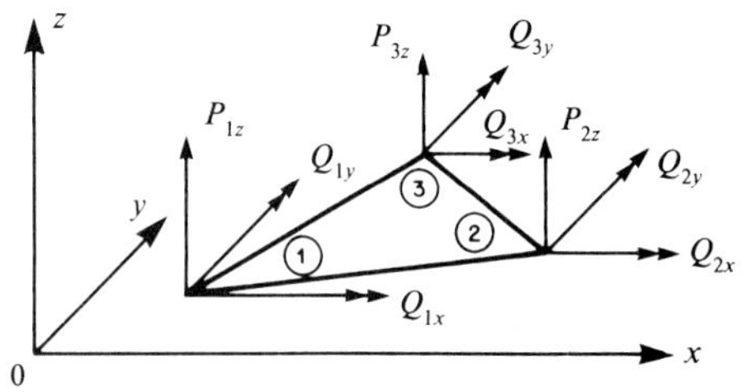

(3) Nodal displacements

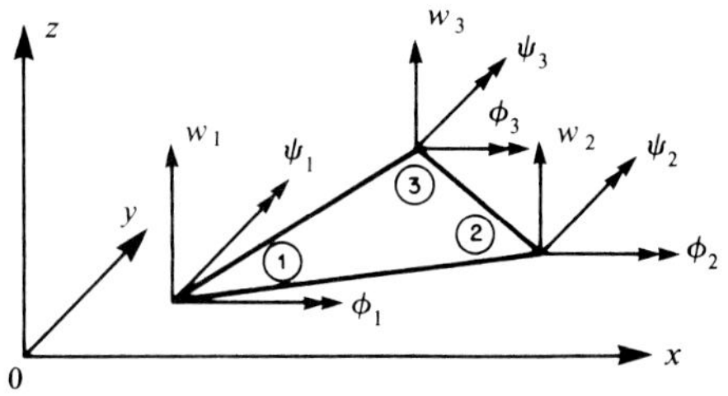

(4) Displacement shape matrix function (26.02-3) in terms of area coordinates (26.07-1) is

$$G = [G_{1w} \quad G_{1\phi} \quad G_{1\psi} \quad G_{2w} \quad G_{2\phi} \quad G_{2\psi} \quad G_{3w} \quad G_{3\phi} \quad G_{3\psi}]$$

where with $s_0 = \frac{1}{2}s_1 s_2 s_3$,

$$G_{1w} = s_1 + s_1^2(s_2 + s_3) - s_1(s_2^2 + s_3^2)$$
$$G_{2w} = s_2 + s_2^2(s_3 + s_1) - s_2(s_3^2 + s_1^2)$$
$$G_{3w} = s_3 + s_3^2(s_1 + s_2) - s_3(s_1^2 + s_2^2)$$
$$G_{1\phi} = b_3(s_1^2 s_2 + s_0) - b_2(s_1^2 s_3 + s_0) \qquad G_{1\psi} = a_3(s_1^2 s_2 + s_0) - a_2(s_1^2 s_3 + s_0)$$
$$G_{2\phi} = b_1(s_2^2 s_3 + s_0) - b_3(s_2^2 s_1 + s_0) \qquad G_{2\psi} = a_1(s_2^2 s_3 + s_0) - a_3(s_2^2 s_1 + s_0)$$
$$G_{3\phi} = b_2(s_3^2 s_1 + s_0) - b_1(s_3^2 s_2 + s_0) \qquad G_{3\psi} = a_2(s_3^2 s_1 + s_0) - a_1(s_3^2 s_2 + s_0)$$

(5) Strain-displacement matrix (26.02-4)

$$B = DG = z \begin{bmatrix} \dfrac{\partial^2}{\partial x^2} \\ \dfrac{\partial^2}{\partial y^2} \\ 2\dfrac{\partial^2}{\partial x\,\partial y} \end{bmatrix} [G_{1w} \quad G_{1\phi} \quad G_{1\psi} \quad \cdots \quad G_{3w} \quad G_{3\phi} \quad G_{3\psi}]$$

$$B = \frac{z}{A_0^2}\left[\begin{array}{ccc|ccc|ccc} B_{11} & B_{12} & B_{13} & B_{14} & B_{15} & B_{16} & B_{17} & B_{18} & B_{19} \\ B_{21} & B_{22} & B_{23} & B_{24} & B_{25} & B_{26} & B_{27} & B_{28} & B_{29} \\ B_{31} & B_{32} & B_{33} & B_{34} & B_{35} & B_{36} & B_{37} & B_{38} & B_{39} \end{array}\right]$$

where B_{ij} are the functions of s_1, s_2, s_3 listed in (6) on the next page, A_0 is the area of the triangle 1, 2, 3, and z is the vertical coordinate measured from the midsurface of the plate.

26.08 TRIANGULAR PLATE ELEMENT OF ORDER TWO, STIFFNESS MATRIX (*Continued*)

(6) Functions B_{ij} in the matrix B (5) in terms of $a_1, a_2, a_3, b_1, b_2, b_3$ (26.07-2) and

$$a_0^2 = a_1^2 + a_2^2 + a_3^2 \qquad R_1 = \tfrac{1}{4}(b_2 b_3 s_1 + b_3 b_1 s_2 + b_1 b_2 s_3) \qquad \beta = b_1 s_1 + b_2 s_2 + b_3 s_3$$

$$b_0^2 = b_1^2 + b_2^2 + b_3^2 \qquad R_2 = \tfrac{1}{4}(a_2 a_3 s_1 + a_3 a_1 s_2 + a_1 a_2 s_3) \qquad \alpha = a_1 s_1 + a_2 s_2 + a_3 s_3$$

$$c_0^2 = a_1 b_1 + a_2 b_2 + a_3 b_3 \qquad R_3 = \tfrac{1}{4}[(a_2 b_3 + a_3 b_2) s_1 + (a_3 b_1 + a_1 b_3) s_2 + (a_1 b_2 + a_2 b_1) s_3]$$

are

$$B_{11} = \tfrac{1}{2}(b_1^2 - b_0^2 s_1) - b_1 \beta$$
$$B_{21} = \tfrac{1}{2}(a_1^2 - a_0^2 s_1) - a_1 \alpha$$
$$B_{31} = a_1 b_1 - c_0^2 s_1 - a_1 \beta - b_1 \alpha$$

$$B_{12} = \tfrac{1}{2} b_1^2 (b_3 s_2 - b_2 s_3) + (b_3 - b_2) R_1$$
$$B_{22} = \tfrac{1}{2} a_1^2 (b_3 s_2 - b_2 s_3) + a_1 (a_2 b_3 - a_3 b_2) s_1 + (b_3 - b_2) R_2$$
$$B_{32} = a_1 b_1 (b_3 s_2 - b_2 s_3) + b_1 (a_2 b_3 - a_3 b_2) s_1 + (b_3 - b_2) R_3$$

$$B_{13} = \tfrac{1}{2} b_1^2 (a_3 s_2 - a_2 s_3) + b_1 (a_3 b_2 - a_2 b_3) s_1 + (a_3 - b_2) R_1$$
$$B_{23} = \tfrac{1}{2} a_1^2 (a_3 s_2 - a_2 s_3) + (a_3 - a_2) R_2$$
$$B_{33} = a_1 b_1 (a_3 s_2 - a_2 s_3) + a_1 (a_3 b_2 - a_2 b_3) s_1 + (a_3 - a_2) R_3$$

$$B_{14} = \tfrac{1}{2}(b_2^2 - b_0^2 s_2) - b_2 \beta$$
$$B_{24} = \tfrac{1}{2}(a_2^2 - b_0^2 s_2) - a_2 \alpha$$
$$B_{34} = a_2 b_2 - c_0^2 s_2 - a_9 \beta - b_2 \alpha$$

$$B_{15} = \tfrac{1}{2} b_2^2 (b_1 s_3 - b_3 s_1) + (b_1 - b_3) R_1$$
$$B_{25} = \tfrac{1}{2} a_2^2 (b_1 s_3 - b_3 s_1) + a_2 (a_3 b_1 - a_1 b_3) s_1 + (b_1 - b_3) R_2$$
$$B_{35} = a_2 b_2 (b_1 s_3 - b_3 s_1) + b_2 (a_3 b_1 - a_1 b_3) s_2 + (b_1 - b_3) R_3$$

$$B_{16} = \tfrac{1}{2} b_2^2 (a_1 s_3 - a_3 s_1) + b_2 (a_1 b_3 - a_3 b_1) s_2 + (a_1 - a_3) R_1$$
$$B_{26} = \tfrac{1}{2} a_2^2 (a_1 s_3 - a_3 s_1) + (a_1 - a_3) R_2$$
$$B_{36} = a_2 b_2 (a_1 s_3 - a_3 s_1) + a_2 (a_1 b_3 - a_3 b_1) s_2 + (a_1 - a_3) R_3$$

$$B_{17} = \tfrac{1}{2}(b_3^2 + b_0^2 s_3) - b_3 \beta$$
$$B_{27} = \tfrac{1}{2}(a_3^2 + a_0^2 s_3) - a_3 \alpha$$
$$B_{37} = a_3 b_3 - c_0^2 s_3 - a_3 \beta - b_3 \alpha$$

$$B_{18} = \tfrac{1}{2} b_3^2 (b_2 s_1 - b_1 s_2) + (b_2 - b_1) R_1$$
$$B_{28} = \tfrac{1}{2} a_3^2 (b_2 s_1 - b_1 s_2) + a_3 (a_1 b_2 - a_2 b_1) s_1 + (b_2 - b_1) R_2$$
$$B_{38} = a_3 b_3 (b_2 s_1 - b_1 s_2) + b_3 (a_1 b_2 - a_2 b_1) s_3 + (b_2 - b_1) R_3$$

$$B_{19} = \tfrac{1}{2} b_3^2 (a_2 s_1 - a_1 s_2) + b_3 (a_2 b_1 - a_1 b_2) s_3 + (a_2 - a_1) R_1$$
$$B_{29} = \tfrac{1}{2} a_3^2 (a_2 s_1 - a_1 s_2) + (a_2 - a_1) R_2$$
$$B_{39} = a_3 b_3 (a_2 s_1 - a_1 s_2) + a_3 (a_2 b_1 - a_1 b_2) s_3 + (a_2 - a_1) R_3$$

(7) Stiffness matrix (26.02-7)

$$k = \int_{-h/2}^{+h/2} \int_{A0} B^{)T} EB\, dA = k_{11} + k_{12} + k_{22} + k_{33}$$

where E is a symmetrical matrix [3 × 3] of the elastic constants, B is the strain-displacement matrix [3 × 9], and k_{11}, k_{12}, k_{22}, k_{33} are the component matrices [9 × 9] given below as area integrals of the products of B_{ij} defined in (6) on the preceding page.

$$k_{11} = \frac{E_{11}h^3}{12A_0^4} \int_{A0} \begin{bmatrix} B_{11} \\ B_{12} \\ \vdots \\ B_{18} \\ B_{19} \end{bmatrix} [B_{11} \quad B_{12} \quad \cdots \quad B_{18} \quad B_{19}]\, dA$$

$$k_{12} = \frac{2E_{12}h^3}{12A_0^4} \int_{A0} \begin{bmatrix} B_{21} \\ B_{22} \\ \vdots \\ B_{28} \\ B_{29} \end{bmatrix} [B_{11} \quad B_{12} \quad \cdots \quad B_{18} \quad B_{19}]\, dA$$

$$k_{22} = \frac{E_{22}h^3}{12A_0^4} \int_{A0} \begin{bmatrix} B_{21} \\ B_{22} \\ \vdots \\ B_{28} \\ B_{29} \end{bmatrix} [B_{21} \quad B_{22} \quad \cdots \quad B_{28}B_{29}]\, dA$$

$$k_{33} = \frac{E_{33}h^3}{12A_0^4} \int_{A0} \begin{bmatrix} B_{31} \\ B_{32} \\ \vdots \\ B_{38} \\ B_{39} \end{bmatrix} [B_{31} \quad B_{32} \quad \cdots \quad B_{38} \quad B_{39}]\, dA$$

The elastic constants E_{11}, E_{12}, E_{22}, and E_{33} are defined in (26.03-2). Since each function B_{ij} involves the area coordinates s_1, s_2, s_3, the evaluation of the respective integrals is accomplished by the relation (26.07-5) as

$$\int_{A0} s_i\, dA = \frac{2A_0}{6} \qquad \int_{A0} s_i^2\, dA = \frac{2A_0}{12} \qquad \int_{A0} s_i s_j\, dA = \frac{2A_0}{24} \qquad (i \neq j;\ i, j = 1, 2, 3)$$

26.09 TRIANGULAR PLATE ELEMENT OF ORDER TWO, NODAL LOADS

(1) **Nodal loads** calculated by relations (26.02-7) in terms of volume integrals introduced in (26.07-5) are tabulated below. In their evaluations the following integral identities lead to simplified expressions:

$$\int_{A0} \alpha \, dA = 0 \qquad \int_{A0} R_1 \, dA = 0$$

$$\int_{A0} \beta \, dA = 0 \qquad \int_{A0} R_2 \, dA = 0 \qquad \int_{A0} R_3 \, dA = 0$$

(2) **Gravity load vector**

$$P_G = -\frac{hA_0 p_z}{20}\{\tfrac{20}{3}, b_3 - b_2, a_3 - a_2, \tfrac{20}{3}, b_1 - b_3, a_1 - a_3, \tfrac{20}{3}, b_2 - b_1, a_2 - a_1\}$$

(3) **Thermal load vector**

$$P_T = \frac{h^2 t^*}{72A_0}\begin{bmatrix} 3b_1^2 - b_0^2 & 3a_1^2 - a_0^2 \\ -b_1^3 & -b_1 a_1^2 \\ -a_1 b_1^2 & -a_1^3 \\ \hline 3b_2^2 - b_0^2 & 3a_2^2 - a_0^2 \\ -b_2^3 & -b_2 a_2^2 \\ -a_2 b_2^2 & -a_2^3 \\ \hline 3b_3^2 - b_0^2 & 3a_3^2 - a_0^2 \\ -b_3^3 & -b_3 a_3^2 \\ -a_3 b_3^2 & -a_3^3 \end{bmatrix}\begin{bmatrix} e_{T1} \\ e_{T2} \end{bmatrix}$$

where $e_{T1} = E_{11}\alpha_{tx} + E_{12}\alpha_{ty}$, $e_{T2} = E_{12}\alpha_{tx} + E_{22}\alpha_{ty}$, and $t^* =$ temperature difference.

(4) **Initial strain load vector**

$$P_0 = \frac{h^3}{36A_0}\begin{bmatrix} 3b_1^2 - b_0^2 & 3a_1^2 - a_0^2 & 3a_1 b_1 - c_0^2 \\ -b_1^3 & -b_1 a_1^2 & -a_1 b_1^2 \\ -a_1 b_1^2 & -a_1^3 & -b_1 a_1^2 \\ \hline 3b_2^2 - b_0^2 & 3a_2^2 - a_0^2 & 3a_2 b_2 - c_0^2 \\ -b_2^3 & -b_2 a_2^2 & -a_2 b_2^2 \\ -a_2 b_2^2 & -a_2^3 & -b_2 a_2^2 \\ \hline 3b_3^2 - b_0^2 & 3a_3^2 - a_0^2 & 3a_3 b_3 - c_0^2 \\ -b_3^3 & -b_3 a_3^2 & -a_3 b_3^2 \\ -a_3 b_3^2 & -a_3^3 & -b_3 a_3^2 \end{bmatrix}\begin{bmatrix} e_{01} \\ e_{02} \\ e_{03} \end{bmatrix}$$

where $e_{01} = E_{11}\kappa_{0xx} + E_{12}\kappa_{0yy}$ $\qquad e_{02} = E_{21}\kappa_{0xx} + E_{22}\kappa_{0yy}$ $\qquad e_{03} = 2E_{33}\kappa_{0xy}$
in which

$$\kappa_{0xx} = \left(\frac{d^2 w}{dx^2}\right)_0 \qquad \kappa_{0yy} = \left(\frac{\partial^2 w}{\partial y^2}\right)_0 \qquad \kappa_{0xy} = \left(\frac{\partial^2 w}{\partial x \, \partial y}\right)_0$$

are the initial curvatures.

26.10 TRIANGULAR PLATE ELEMENT OF ORDER TWO, MOMENTS AND SHEARS

(1) Moment vector at x, y

$$\begin{bmatrix} M_x \\ M_y \\ M_{xy} \end{bmatrix} = \frac{h^3}{12A_0^2} \underbrace{\begin{bmatrix} E_{11} & E_{12} & 0 \\ E_{21} & E_{22} & 0 \\ 0 & 0 & E_{33} \end{bmatrix}}_{E} \underbrace{\begin{bmatrix} B_{11} & B_{12} & \cdots & B_{18} & B_{19} \\ B_{21} & B_{22} & \cdots & B_{28} & B_{29} \\ B_{31} & B_{32} & \cdots & B_{38} & B_{39} \end{bmatrix}}_{B/z} \underbrace{\begin{bmatrix} w_1 \\ \phi_1 \\ \psi_1 \\ \vdots \\ w_3 \\ \phi_3 \\ \psi_3 \end{bmatrix}}_{\Delta}$$

where E, B, Δ are defined in (26.03–26.08-1, 5, 6), respectively, and M_x, M_y, M_{xy} are the moments depicted below.

(2) Variation of moments

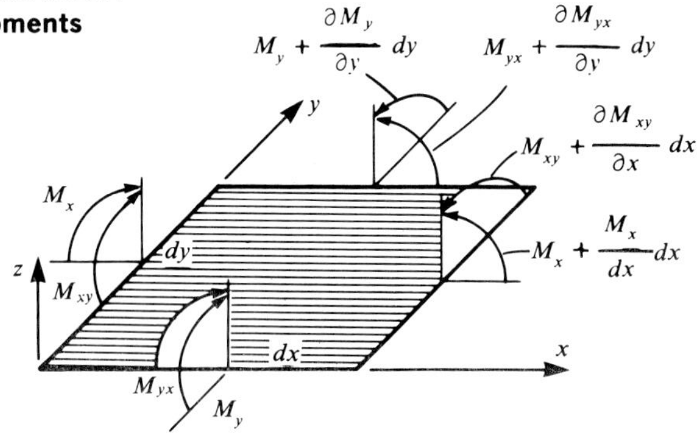

(3) Variation of shears

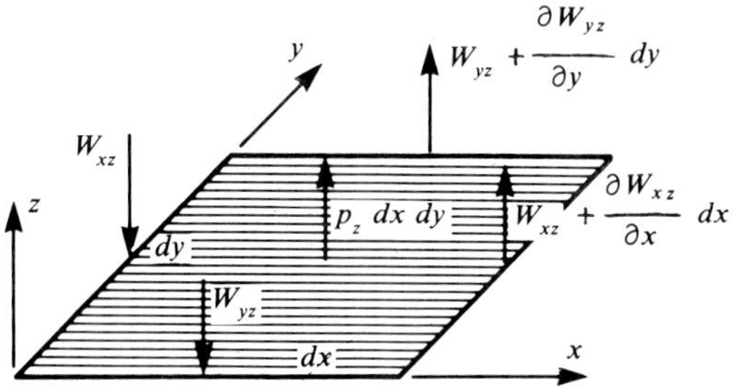

(4) Shearing-force vector at x, y

$$\begin{bmatrix} W_{xz} \\ W_{yz} \end{bmatrix} = - \begin{bmatrix} \dfrac{\partial}{\partial x} & 0 & \dfrac{\partial}{\partial y} \\ 0 & \dfrac{\partial}{\partial y} & \dfrac{\partial}{\partial x} \end{bmatrix} \begin{bmatrix} M_x \\ M_y \\ M_{xy} \end{bmatrix}$$

where W_{xz}, W_{yz} are the shearing forces depicted in (3) above.

26.11 RECTANGULAR PLATE ELEMENT OF ORDER TWO, STIFFNESS MATRIX

(1) Element considered is a rectangular element of constant thickness h (m) given by vertices 1, 2, 3, 4 and acted on by normal-to-plane causes (element of order two). The 12 degrees of freedom are represented by the displacement vector

$$\Delta = \{w_1, \phi_1, \psi_1, w_2, \phi_2, \psi_2, w_3, \phi_3, \psi_3, w_4, \phi_4, \psi_4\}$$

Notation used in this section is defined in (22.02–22.05, 26.02). All forces, moments, and displacements are in the x, y, z system, and their positive directions are shown in (2, 3) below (Refs. 42, 47).

(2) Nodal loads

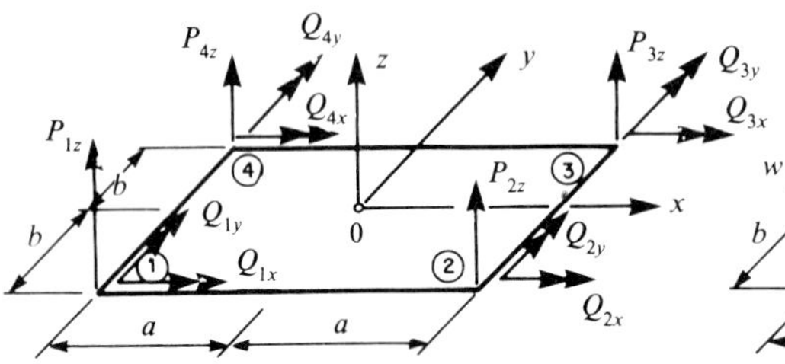

(3) Nodal displacements

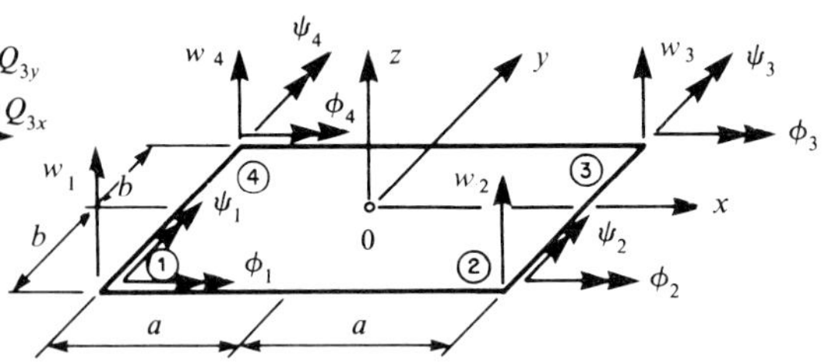

(4) Displacement shape matrix function (26.02-3) is

$$G = \tfrac{1}{8}[G_{1w} \quad G_{1\phi} \quad G_{1\psi} \quad \cdots \quad G_{4w} \quad G_{4\phi} \quad G_{4\psi}]$$

where with $\bar{x} = x/a$, $\bar{y} = y/b$, $\bar{r}^2 = \bar{x}^2 + \bar{y}^2$,

$$\begin{array}{ll|ll}
G_{1w} = & (1-\bar{x})(1-\bar{y})(2-\bar{x}-\bar{y}-\bar{r}^2) & G_{3w} = & (1+\bar{x})(1+\bar{y})(2+\bar{x}+\bar{y}-\bar{r}^2) \\
G_{1\phi} = & (1-\bar{x})(1+\bar{y})(1-\bar{y})^2 b & G_{3\phi} = & -(1+\bar{x})(1-\bar{y})(1+\bar{y})^2 b \\
G_{1\psi} = & -(1+\bar{x})(1-\bar{y})(1-\bar{x})^2 a & G_{3\psi} = & (1-\bar{x})(1+\bar{y})(1+\bar{x})^2 a \\
\hline
G_{2w} = & (1+\bar{x})(1-\bar{y})(2+\bar{x}-\bar{y}-\bar{r}^2) & G_{4w} = & (1-\bar{x})(1+\bar{y})(2-\bar{x}+\bar{y}-\bar{r}^2) \\
G_{2\phi} = & (1+\bar{x})(1+\bar{y})(1-\bar{y})^2 b & G_{4\phi} = & -(1-\bar{x})(1-\bar{y})(1+\bar{y})^2 b \\
G_{2\psi} = & (1-\bar{x})(1-\bar{y})(1+\bar{x})^2 a & G_{4\psi} = & -(1+\bar{x})(1+\bar{y})(1-\bar{x})^2 a
\end{array}$$

(5) Strain-displacement matrix (26.02-4)

$$B = DG = \frac{z}{8}\begin{bmatrix} \dfrac{\partial^2}{\partial x^2} \\[2mm] \dfrac{\partial^2}{\partial y^2} \\[2mm] \dfrac{2\partial^2}{\partial x\,\partial y} \end{bmatrix} [G_{1w} \quad G_{1\phi} \quad G_{1\psi} \quad \cdots \quad G_{4w} \quad G_{4\phi} \quad G_{4\psi}]$$

$$B = z\begin{bmatrix} B_{101} & B_{102} & B_{103} & \cdots & B_{110} & B_{111} & B_{112} \\ B_{201} & B_{202} & B_{203} & \cdots & B_{210} & B_{211} & B_{212} \\ B_{301} & B_{302} & B_{303} & \cdots & B_{310} & B_{311} & B_{312} \end{bmatrix}$$

where B_{ijk} are the functions of $\bar{x}$, $\bar{y}$ listed in (6) and z is the vertical coordinate measured from the midsurface of the plate.

(6) Functions B_{ijk} in matrix B (5) are

$B_{101} = \dfrac{3\bar{x}(1-\bar{y})}{4a^2}$	$B_{102} = 0$	$B_{103} = \dfrac{(1-3\bar{x})(1-\bar{y})}{4a}$
$B_{201} = \dfrac{3(1-\bar{x})\bar{y}}{4b^2}$	$B_{202} = -\dfrac{(1-\bar{x})(1-3\bar{y})}{4b}$	$B_{203} = 0$
$B_{301} = \dfrac{4-3\bar{r}^2}{4ab}$	$B_{302} = \dfrac{1-2\bar{y}-3\bar{y}^2}{4a}$	$B_{303} = -\dfrac{1-2\bar{x}-3\bar{x}^2}{4b}$
$B_{104} = -\dfrac{3\bar{x}(1-\bar{y})}{4a^2}$	$B_{105} = 0$	$B_{106} = -\dfrac{(1+3\bar{x})(1-\bar{y})}{4a}$
$B_{204} = \dfrac{3(1+\bar{x})\bar{y}}{4b^2}$	$B_{205} = -\dfrac{(1+\bar{x})(1-3\bar{y})}{4b}$	$B_{206} = 0$
$B_{304} = -\dfrac{4-3\bar{r}^2}{4ab}$	$B_{305} = -\dfrac{1-2\bar{y}-3\bar{y}^2}{4a}$	$B_{306} = -\dfrac{1-2\bar{x}-3\bar{x}^2}{4b}$
$B_{107} = -\dfrac{3x(1+y)}{4a^2}$	$B_{108} = 0$	$B_{109} = -\dfrac{(1+3\bar{x})(1+\bar{y})}{4a}$
$B_{207} = -\dfrac{3(1+\bar{x})\bar{y}}{4b^2}$	$B_{208} = \dfrac{(1+\bar{x})(1+3\bar{y})}{4b}$	$B_{209} = 0$
$B_{307} = \dfrac{4-3\bar{r}^2}{4ab}$	$B_{308} = \dfrac{1-2\bar{y}-3\bar{y}^2}{4a}$	$B_{309} = \dfrac{1-2\bar{x}-3\bar{x}^2}{4b}$
$B_{110} = \dfrac{3\bar{x}(1+\bar{y})}{4a^2}$	$B_{111} = 0$	$B_{112} = \dfrac{(1-3\bar{x})(1+\bar{y})}{4a}$
$B_{210} = -\dfrac{3(1-\bar{x})\bar{y}}{4b^2}$	$B_{211} = \dfrac{(1-\bar{x})(1+3\bar{y})}{4b}$	$B_{212} = 0$
$B_{310} = -\dfrac{4-3\bar{r}^2}{4ab}$	$B_{311} = \dfrac{1-2\bar{y}-3\bar{y}^2}{4a}$	$B_{312} = \dfrac{1-2\bar{x}-3\bar{x}^2}{4b}$

(7) Stiffness matrix (26.02-7)

$$k = \int_{-h/2}^{+h/2} \iint_{-1}^{+1} B^{)T} E B \, dz \, dy \, dx = k_{11} + k_{12} + k_{22} + k_{33}$$

where E is the symmetrical matrix [3 × 3] of the elastic constants (26.03-2), B is the strain-displacement matrix [3 × 12], and k_{11}, k_{12}, k_{22}, k_{33} are the component matrices [12 × 12] of k given in (8–11).

(8) Matrix k_{11} in terms of equivalents

$$A_1 = \frac{bh^3}{24a^3} E_{11} \qquad A_2 = \frac{bh^3}{24a^2} E_{11} \qquad A_3 = \frac{bh^3}{36a} E_{11}$$

is

$$k_{11} = \left[\begin{array}{ccc|ccc|ccc|ccc}
2A_1 & & & & & & & & & & & \\
0 & 0 & & & & & & & & & & \\
-2A_2 & 0 & 4A_3 & & & & & & & & & \\
\hline
-2A_1 & 0 & 2A_2 & 2A_1 & & & & & \text{Symmetrical} & & & \\
0 & 0 & A_3 & 0 & 0 & & & & & & & \\
-2A_2 & 0 & 2 & 2A_2 & 0 & 4A_3 & & & & & & \\
\hline
-A_1 & 0 & -A_2 & A_1 & 0 & A_2 & 2A_1 & & & & & \\
0 & 0 & 0 & 0 & 0 & 0 & 0 & 0 & & & & \\
-A_2 & 0 & A_3 & A_2 & 0 & 2A_3 & 2A_2 & 0 & 4A_3 & & & \\
\hline
A_1 & 0 & -A_2 & -A_1 & 0 & -A_2 & -2A_1 & 0 & -2A_2 & 2A_1 & & \\
0 & 0 & 0 & 0 & 0 & 0 & 0 & 0 & 0 & 0 & 0 & \\
-A_2 & 0 & 2A_3 & A_2 & 0 & A_3 & 2A_2 & 0 & 2A_3 & -2A_2 & 0 & 4A_3
\end{array}\right]$$

(9) Matrix k_{12} in terms of equivalents

$$B_1 = \frac{h^3}{24ab} E_{12} \qquad B_2 = \frac{h^3}{24a} E_{12} \qquad B_3 = \frac{h^3}{24b} E_{12} \qquad B_4 = \frac{h^3}{12} E_{12}$$

is

$$k_{12} = \left[\begin{array}{ccc|ccc|ccc|ccc}
B_1 & & & & & & & & & & & \\
B_2 & 0 & & & & & & & & & & \\
-B_3 & -B_4 & 0 & & & & & & & & & \\
\hline
-B_1 & -B_2 & 0 & B_1 & & & & & \text{Symmetrical} & & & \\
-B_2 & 0 & 0 & B_2 & 0 & & & & & & & \\
0 & 0 & 0 & B_3 & B_4 & 0 & & & & & & \\
\hline
B_1 & 0 & 0 & -B_1 & 0 & -B_3 & B_1 & & & & & \\
0 & 0 & 0 & 0 & 0 & 0 & -B_2 & 0 & & & & \\
0 & 0 & 0 & -B_3 & 0 & 0 & B_3 & -B_4 & 0 & & & \\
\hline
-B_1 & 0 & B_3 & B_1 & 0 & 0 & -B_1 & B_2 & 0 & B_1 & & \\
0 & 0 & 0 & 0 & 0 & 0 & B_2 & 0 & 0 & -B_2 & 0 & \\
B_3 & 0 & 0 & 0 & 0 & 0 & 0 & 0 & 0 & -B_3 & B_4 & 0
\end{array}\right]$$

(10) Matrix k_{22} in terms of equivalents

$$C_1 = \frac{ah^3}{24b^3} E_{22} \qquad C_2 = \frac{ah^3}{24b^2} E_{22} \qquad C_3 = \frac{ah^3}{36b} E_{22}$$

is

$$k_{22} = \left[\begin{array}{ccc|ccc|ccc|ccc}
2C_1 & & & & & & & & & & & \\
2C_2 & 4C_3 & & & & & & & & & & \\
0 & 0 & 0 & & & & & & & & & \\
\hline
C_1 & C_2 & 0 & 2C_1 & & & & \text{Symmetrical} & & & & \\
C_2 & 2C_3 & 0 & 2C_2 & 4C_3 & & & & & & & \\
0 & 0 & 0 & 0 & 0 & 0 & & & & & & \\
\hline
-C_1 & -C_2 & 0 & -2C_1 & -2C_2 & 0 & 2C_1 & & & & & \\
C_2 & C_3 & 0 & 2C_2 & 2C_3 & 0 & -2C_2 & 4C_3 & & & & \\
0 & 0 & 0 & 0 & 0 & 0 & 0 & 0 & 0 & & & \\
\hline
-2C_1 & -2C_2 & 0 & -C_1 & -C_2 & 0 & C_1 & -C_2 & 0 & 2C_1 & & \\
2C_2 & 2C_3 & 0 & C_2 & C_3 & 0 & -C_2 & 2C_3 & 0 & -2C_2 & 4C_3 & \\
0 & 0 & 0 & 0 & 0 & 0 & 0 & 0 & 0 & 0 & 0 & 0
\end{array}\right]$$

(11) Matrix k_{33} in terms of equivalents

$$D_1 = \frac{7h^3}{60ab} E_{33} \qquad D_2 = \frac{h^3}{60a} E_{33} \qquad D_3 = \frac{h^3}{60b} E_{33}$$

$$D_4 = \frac{bh^3}{90a} E_{33} \qquad D_5 = \frac{ah^3}{90b} E_{33}$$

is

$$k_{33} = \left[\begin{array}{ccc|ccc|ccc|ccc}
D_1 & & & & & & & & & & & \\
D_2 & 4D_4 & & & & & & & & & & \\
D_3 & 0 & 4D_5 & & & & & & & & & \\
\hline
-D_1 & -D_2 & D_3 & D_1 & & & \text{Symmetrical} & & & & & \\
-D_2 & -4D_4 & 0 & D_2 & 4D_4 & & & & & & & \\
-D_3 & 0 & -D_5 & D_3 & 0 & 4D_5 & & & & & & \\
\hline
D_1 & D_2 & -D_3 & -D_1 & -D_2 & -D_3 & D_1 & & & & & \\
-D_2 & D_4 & 0 & D_2 & -D_4 & 0 & -D_2 & 4D_4 & & & & \\
D_3 & 0 & D_5 & -D_3 & 0 & -4D_5 & D_3 & 0 & 4D_5 & & & \\
\hline
-D_1 & -D_2 & D_3 & D_1 & D_2 & D_3 & -D_1 & D_2 & -D_3 & D_1 & & \\
D_2 & -D_4 & 0 & -D_2 & D_4 & 0 & D_2 & -4D_4 & 0 & -D_2 & 4D_4 & \\
D_3 & 0 & -4D_5 & -D_3 & 0 & D_5 & D_3 & 0 & -D_5 & -D_3 & 0 & 4D_5
\end{array}\right]$$

26.12 RECTANGULAR PLATE ELEMENT OF ORDER TWO, NODAL LOADS, MOMENTS, AND SHEARS

(1) Nodal loads calculated by relations (26.02-7) are tabulated below. In their elevation the following integral identities lead to simplified expressions:

$$\int_{-1}^{1}\!\!\int G_{jw}\,d\bar{y}\,d\bar{x} = 1 \qquad \text{for } j = 1, 2, 3, 4$$

$$\int_{-1}^{1}\!\!\int G_{j\phi}\,d\bar{y}\,d\bar{x} = \begin{cases} b/3 & \text{for } j = 1, 4 \\ -b/3 & \text{for } j = 2, 3 \end{cases}$$

$$\int_{-1}^{1}\!\!\int G_{j\psi}\,d\bar{y}\,d\bar{x} = \begin{cases} a/3 & \text{for } j = 1, 2 \\ -a/3 & \text{for } j = 3, 4 \end{cases}$$

$$\int_{-1}^{1}\!\!\int \bar{x}(1 + \bar{y})\,d\bar{y}\,d\bar{x} = \int_{-1}^{1}\!\!\int \bar{y}(1 + \bar{x})\,d\bar{y}\,d\bar{x} = 0$$

$$\int_{-1}^{1}\!\!\int (1 - 2\bar{x} - 3\bar{x}^2)\,d\bar{y}\,d\bar{x} = \int_{-1}^{1}\!\!\int (1 - 2\bar{y} - 3\bar{y}^2)\,d\bar{y}\,d\bar{x} = 0$$

$$\int_{-1}^{1}\!\!\int (1 \pm \bar{x})(1 \pm 3\bar{y})\,d\bar{y}\,d\bar{x} = \int_{-1}^{1}\!\!\int (1 \pm \bar{y})(1 \pm 3\bar{x})\,d\bar{y}\,d\bar{x} = \int_{-1}^{1}\!\!\int (4 - 3\bar{r}^2)\,d\bar{y}\,d\bar{x} = 4$$

(2) Gravity load vector

$$P_G = -\frac{abp_z}{3}\{3, b, -a, 3, b, a, 3, -b, a, 3, -b, -a\}$$

(3) Thermal load vector

$$P_T = \frac{h^2 t^*}{72}\begin{bmatrix} 0 & 0 \\ 0 & -a \\ b & 0 \\ \hline 0 & 0 \\ 0 & -a \\ -b & 0 \\ \hline 0 & 0 \\ 0 & a \\ -b & 0 \\ \hline 0 & 0 \\ 0 & a \\ b & 0 \end{bmatrix}\begin{bmatrix} e_{T1} \\ e_{T2} \end{bmatrix}$$

where $e_{T1} = E_{11}\alpha_{tx} + E_{12}\alpha_{ty}$, $e_{T2} = E_{12}\alpha_{tx} + E_{22}\alpha_{ty}$, and t^* = temperature difference.

(4) Initial strain load vector

$$P_0 = \frac{h^3}{12}\begin{bmatrix} 0 & 0 & 1 \\ 0 & -a & 0 \\ b & 0 & 0 \\ \hline 0 & 0 & -1 \\ 0 & -a & 0 \\ -b & 0 & 0 \\ \hline 0 & 0 & 1 \\ 0 & a & 0 \\ -b & 0 & 0 \\ \hline 0 & 0 & -1 \\ 0 & a & 0 \\ b & 0 & 0 \end{bmatrix} \begin{bmatrix} e_{01} \\ e_{02} \\ e_{03} \end{bmatrix}$$

where
$$e_{01} = E_{11}\kappa_{0xx} + E_{12}\kappa_{0yy}$$
$$e_{02} = E_{12}\kappa_{0xx} + E_{22}\kappa_{0yy}$$
$$e_{03} = 2E_{33}\kappa_{0xy}$$

in which

$$\kappa_{0xx} = \left(\frac{d^2 w}{dx^2}\right)_0 \qquad \kappa_{0yy} = \left(\frac{\partial^2 w}{\partial y^2}\right)_0 \qquad \kappa_{0xy} = \left(\frac{\partial^2 w}{\partial x\, \partial y}\right)_0$$

are the initial curvatures.

(5) Moment vector at x, y

$$\underbrace{\begin{bmatrix} M_x \\ M_y \\ M_{xy} \end{bmatrix}}_{M} = \frac{b^3}{12} \underbrace{\begin{bmatrix} E_{11} & E_{12} & 0 \\ E_{21} & E_{22} & 0 \\ 0 & 0 & E_{33} \end{bmatrix}}_{E} \underbrace{\begin{bmatrix} B_{101} & B_{102} & B_{103} & \cdots & B_{110} & B_{111} & B_{112} \\ B_{201} & B_{202} & B_{203} & \cdots & B_{210} & B_{211} & B_{212} \\ B_{301} & B_{302} & B_{303} & \cdots & B_{310} & B_{311} & B_{312} \end{bmatrix}}_{B/z} \underbrace{\begin{bmatrix} w_1 \\ \phi_1 \\ \psi_1 \\ \vdots \\ w_4 \\ \phi_4 \\ \psi_4 \end{bmatrix}}_{\Delta}$$

where E, B, Δ are defined in (26.03-3, 26.11-5, 6, 26.11-1), respectively, and M_x, M_y, M_{xy} are the moments depicted in (26.10-2).

(6) Shearing-force vector at x, y is given by (26.10-4) its components are depicted in (26.10-3).

26.13 NUMERICAL ACCURACY

(1) Triangular element of order one (26.05). Since the displacement functions of this element are linear, the strains and stresses are constant within the element, and the application of this element does not produce very accurate results when larger elements are used. To improve the accuracy, the refinement of the network must be introduced or a higher-order polynomial element must be used.

(2) Rectangular element of order one (26.06). Since the distribution of displacements along any edge is a linear function of the element displacements at the vertices defining the edge, the displacement compatibility is satisfied but the stress-equilibrium is violated within the rectangle. Nevertheless, the use of this element leads to satisfactory results, provided that the size of the element is small in relation to the stress variation.

(3) Triangular element of order two (26.08) is called nonconforming since it does not have normal slope compatibility along the boundaries (it preserves the compatibility of deflections but violates the compatibility of slopes). For practical engineering purposes, in most cases the results obtained are satisfactory.

(4) Rectangular element of order two (26.11) is again called nonconforming for the reasons stated in (3) above. Again, for practical purposes, the results of the analysis based on this element are satisfactory.

Part IV

Static Analysis of Shells

27

Analysis of Shells Notation, Signs, Basic Relations

STATIC STATE

27.01 DEFINITION OF STATE

(1) Shell is a three-dimensional curved thin-walled structure whose thickness h (m) is less than $\frac{1}{20}$ of its least radius of curvature $R_{\min}$ (shells of practical importance have $h/R_{\min} \leq \frac{1}{100}$). The geometry of the shell is defined by its midsurface which bisects the shell thickness. According to the shape of the midsurface, shells are classified as *rotational* and *translational*. Only rotational shells in a state of axisymmetrical static deformation are considered in Chaps. 27–30.

(2) Assumptions used in the derivation of analytical relations are:

(*a*) The material of the shell is isotropic and remains continuous during loading and unloading.

(*b*) The material follows Hooke's law, all deformations are small, and the maximum deflection is less than one-half the thickness.

(*c*) The normal section remains normal to the midsurface during the deformation, and the points of load application are stationary during the deformation.

(*d*) The normal stresses to the midsurface are small and as such are neglected.

(*e*) The material constants are known from experiments and are independent of time.

(*f*) The effect of the shearing deformation on the formation of the elastic surface is too small to be considered.

(3) Symbols and signs are defined where they appear first and are all summarized in Appendices A.1, A.6.

27.02 SHELL OF REVOLUTION, GEOMETRIC AND STATIC VECTORS

(1) Geometry of the shell of revolution shown in (2) below is defined by the meridian curve rotated about the axis of axisymmetry. Three coordinate systems are used in the analysis of this shell: (*a*) *local cartesian system* 1, 2, 3, (*b*) *global cartesian system* x, y, z, and (*c*) *spherical system* defined by the polar angle ϑ (rad), the meridian angle ϕ (rad), and the spherical radius r (m). The meridian plane and the plane normal to the meridian are the planes of principal curvatures of the midsurface; their corresponding radii of curvature $R_1(m)$ and $R_2(m)$ are shown in (3) below. The elemental area of the midsurface in (5, 6) is

$$dA = ds_1 \cdot ds_2 \qquad ds_1 = R_1\, d\phi \qquad ds_2 = R_0\, d\vartheta = R_2 \sin\phi\, d\vartheta$$

where R_0 = radius of horizontal circle and m = arbitrary point.

(2) Coordinate systems

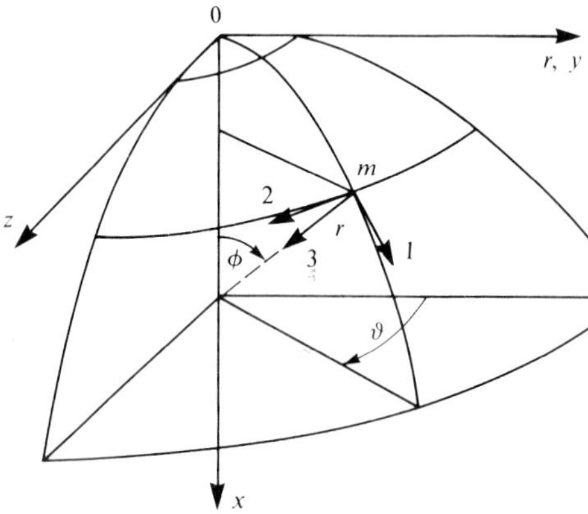

(3) Geometry of meridian curve

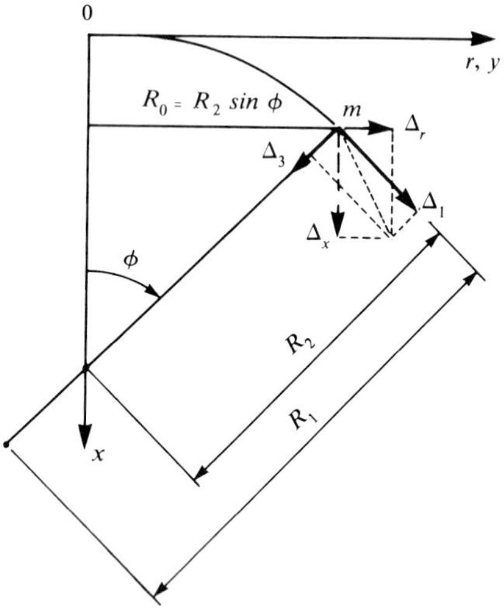

(4) Variation of forces

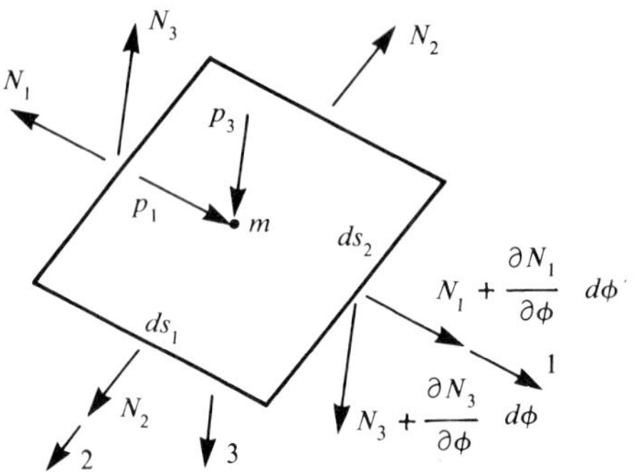

(5) Variation of moments

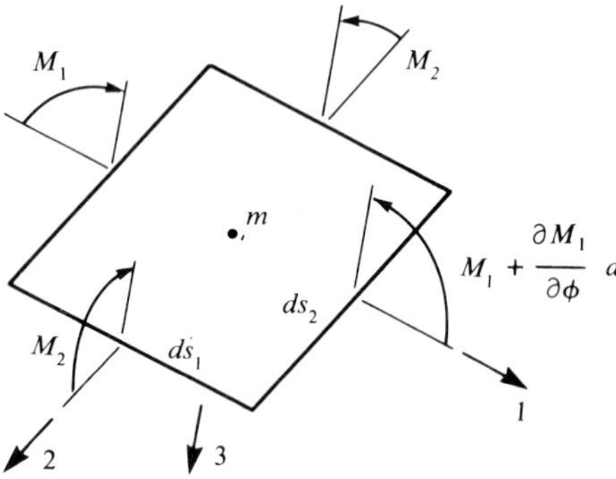

(6) Static vectors acting on the element of the midsurface in (4, 5) are

p_1, p_3 = intensities of distributed load along 1, 3 (N/m^2)
N_1, N_2 = normal forces per unit length of normal section along 1, 2 (N/m)
N_3 = shearing force per unit length of normal section along 3 (N/m)
M_1, M_2 = bending moments per unit length of normal section about 2, 1 (N·m/m)

The positive directions of these vectors are shown in the respective diagrams above.

27.03 SHELL OF REVOLUTION, AXISYMMETRICAL MEMBRANE STATE

$$K = \frac{Eh}{1-\nu^2} \text{ (N/m)}$$

(1) Notation and signs

f_{10}, f_{20} = normal stresses in 1, 2 (Pa)
$\varepsilon_{10}, \varepsilon_{20}$ = linear strains in 1, 2
Δ_{10}, Δ_{30} = linear displacements in 1, 3 (m)
Δ_{r0}, Δ_{x0} = linear displacements in r, x (m)
α_t = thermal coefficients (1/°C)
s_1, s_2, s_3 = coordinates in 1, 2, 3, (m)
ε_t = linear thermal strain
ψ_{10} = slope about 2 (rad)
E = modulus of elasticity (Pa)
ν = Poissons's ratio
N_t = normal thermal force per unit length of normal section (N/m)
t^* = temperature variation along 3 with respect to datum temperature (°C)

Positive forces and moments are shown in (27.02-4, 5); positive displacements are shown in (3) below.

(2) Membrane state in an axisymmetrical system is defined by

$$N_3 = 0 \qquad M_1 = 0 \qquad M_2 = 0$$

and the normal forces in (27.02-4) designated here as N_{10}, N_{20} are the only load carriers of this system, which is consequently statically determinate. Governing equations, derived from the static equilibrium of the element, are

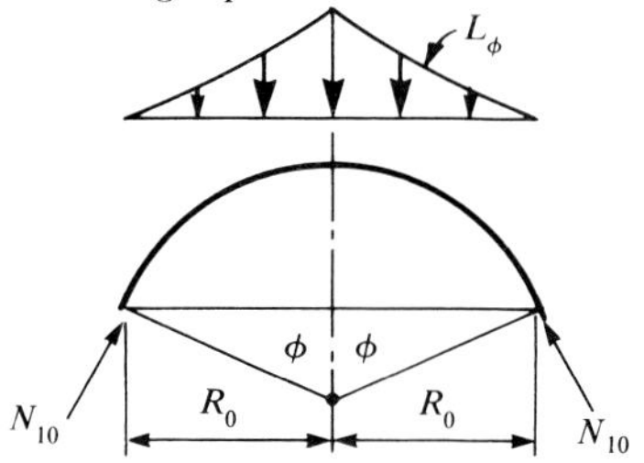

$$N_{10} = -\frac{L_\phi}{2\pi R_0 \sin\phi} = f_{10}h$$

$$N_{20} = -p_3 R_2 + \frac{L_\phi}{2R_1\pi \sin^2\phi} = f_{20}h$$

where L_ϕ is the total vertical load above the circle of radius R_0.

(3) Strain and displacement relations

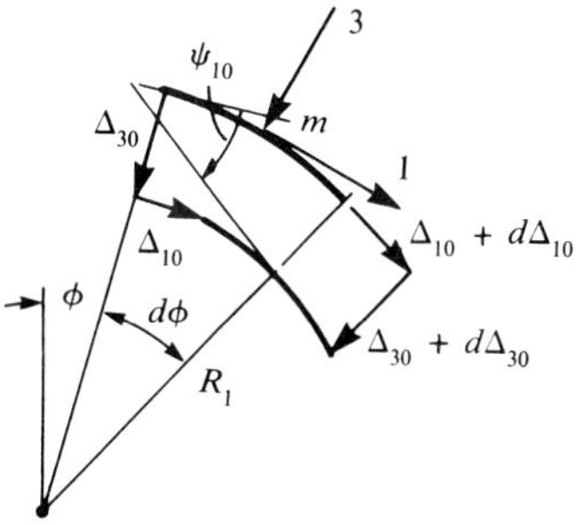

$$\varepsilon_{10} = \frac{1}{R_1}\left(\frac{d\Delta_{10}}{d\phi} - \Delta_{30}\right) \qquad \psi_{10} = \frac{1}{R_1}\left(\Delta_{10} + \frac{d\Delta_{30}}{d\phi}\right)$$

$$\varepsilon_{20} = \frac{1}{R_2}(\Delta_{10}\cot\phi - \Delta_{30})$$

$$\Delta_{10} = \sin\phi\left(\int \frac{R_1\varepsilon_{10} - R_2\varepsilon_{20}}{\sin\phi}\, d\phi + C\right)$$

$$\Delta_{30} = \Delta_{10}\cot\phi - R_2\varepsilon_{20}$$

$$\Delta_{r0} = R_0\varepsilon_{20} \qquad \Delta_{x0} = \Delta_{10}\sin\phi + \Delta_{30}\cos\phi$$

(4) Stress-strain relations

$$\varepsilon_{10} = \frac{1}{Eh}[N_{10} - \nu N_{20} + (1-\nu)N_t] \qquad N_{10} = K[\varepsilon_{10} + \nu\varepsilon_{20} - (1+\nu)\varepsilon_t]$$

$$\varepsilon_{20} = \frac{1}{Eh}[N_{20} - \nu N_{10} + (1-\nu)N_t] \qquad N_{20} = K[\varepsilon_{20} + \nu\varepsilon_{10} - (1+\nu)\varepsilon_t]$$

where $\varepsilon_t = \dfrac{\alpha t}{h}\displaystyle\int_{-h/2}^{h/2} t^*\, ds_3$ and $N_t = Eh\varepsilon_t$.

27.04 SHELL OF REVOLUTION, AXISYMMETRICAL GENERAL STATE

$$K = \frac{Eh}{1-\nu^2} \text{ (N/m)}$$
$$D = \frac{Eh^3}{12(1-\nu^2)} \text{ (N·m)}$$

(1) Notation and signs

s_1, s_2, s_3 = coordinates in 1, 2, 3, (m)
f_1, f_2 = normal stresses in 1, 2 (Pa)
$\varepsilon_1, \varepsilon_2$ = linear strains in 1, 2
Δ_1, Δ_3 = linear displacements in 1, 3 (m)
κ_1, κ_2 = bending measures about 2, 1 (1/m)
α_t = thermal coefficient (1/°C)
ε_t = linear thermal strain
Δ_r, Δ_x = linear displacements in r, x (m)
κ_t = thermal bending measure about 2, 1 (1/m)
ψ_1 = slope about 2 (rad)
N_t = normal thermal force per unit length of normal section (N/m)
M_t = thermal bending moment per unit length of normal section (N·m/m)
t^* = temperature variation along 3 with respect to datum temperature (°C)
N_1, N_2, N_3 = normal and shearing forces defined in (27.02-4)
M_1, M_2 = bending moments defined in (27.02-5)

Positive forces and moments are shown in (27.02-4, 5); positive displacements are shown in (27.03-3).

(2) General axisymmetrical state is defined by (27.02-4, 5) and as such is statically indeterminate. Governing equations derived from the equilibrium of the element are

$$\frac{d}{d\phi}(N_1R_2 \sin\phi) - N_2R_1 \cos\phi - N_3R_1 \sin\phi + p_1R_1R_2 \sin\phi = 0$$

$$N_1R_2 \sin\phi + N_2R_1 \sin\phi + \frac{d}{d\phi}(N_3R_2 \sin\phi) + p_3R_1R_2 \sin\phi = 0$$

$$\frac{d}{d\phi}(M_1R_2 \sin\phi) - M_2R_1 \cos\phi - N_3R_1R_2 \sin\phi = 0$$

(3) Strains and displacements in local system

$$\varepsilon_1 = \frac{1}{R_1}\left(\frac{d\Delta_1}{d\phi} - \Delta_3\right) \qquad \varepsilon_2 = \frac{1}{R_2}(\Delta_1 \cot\phi - \Delta_3)$$

$$\psi_1 = \frac{1}{R_1}\left(\Delta_1 + \frac{d\Delta_3}{d\phi}\right) \qquad \varepsilon_t = \frac{\alpha_t}{h}\int_{-h/2}^{h/2} t^*\, ds_3$$

$$\kappa_1 = -\frac{1}{R_1}\frac{d\psi_1}{d\phi} \qquad \kappa_t = \frac{12\alpha_t}{h^3}\int_{-h/2}^{h/2} t^* s_3\, ds_3 \qquad \kappa_2 = -\frac{\cot\phi}{R_2}\psi_1$$

$$\Delta_1 = \sin\phi\left(\int \frac{R_1\varepsilon_1 - R_2\varepsilon_2}{\sin\phi}\, d\phi + C\right)$$

$$\Delta_3 = \Delta_1 \cot\phi - R_2\varepsilon_2$$

where t^* is a function s_3 or a constant.

(4) Displacements in global system in terms of strains and displacements in local system are

$$\Delta_r = \Delta_1 \cos\phi - \Delta_3 \sin\phi = R_0\varepsilon_2$$

$$\Delta_x = \Delta_1 \sin\phi + \Delta_3 \cos\phi = \int R_1 \sin\phi\,(\varepsilon_1 + \psi_1 \cot\phi)\, d\phi + C$$

where C is the rigid-body translation of the shell as a whole.

(5) Stress-strain relations in local system

$$\varepsilon_1 = \frac{1}{Eh}[N_1 - \nu N_2 + (1-\nu)N_t] \qquad N_1 = K[\varepsilon_1 + \nu\varepsilon_2 - (1+\nu)\varepsilon_t]$$

$$\varepsilon_2 = \frac{1}{Eh}[N_2 - \nu N_1 + (1-\nu)N_t] \qquad N_2 = K[\varepsilon_2 + \nu\varepsilon_1 - (1+\nu)\varepsilon_t]$$

$$\kappa_1 = \frac{12}{Eh^3}[M_1 - \nu M_2 + (1-\nu)M_t] \qquad M_1 = D[\kappa_1 + \nu\kappa_2 - (1+\nu)\kappa_t]$$

$$\kappa_2 = \frac{12}{Eh^3}[M_2 - \nu M_1 + (1-\nu)M_t] \qquad M_2 = D[\kappa_2 + \nu\kappa_1 - (1+\nu)\kappa_t]$$

$$N_3 = -N_1 \tan\phi$$

where $N_t = \dfrac{Eh}{1-\nu}\varepsilon_t$ and $M_t = \dfrac{Eh^3}{12(1-\nu)}\kappa_t$.

(6) Governing differential equation of this shell in terms of

$$n = \frac{R_2}{R_1} \qquad U = R_2 N_3 \qquad \text{and the linear operator}$$

$$L(\cdot\cdot) = \frac{n}{R_1}\left[\frac{d^2(\cdot\cdot)}{d\phi^2} - (\cdot\cdot)\right] + \frac{1}{R_1}\left[\frac{dn}{d\phi} + n\cot\phi\right]\frac{d(\cdot\cdot)}{d\phi}$$

is

$$\boxed{LL(U) + \nu L\left(\frac{U}{R_1}\right) - \frac{\nu}{R_1}L(U) + \left(\frac{K}{D}(1-\nu^2) - \frac{\nu^2}{R_1^2}\right)U = \Phi(p_1, p_3, t^*)}$$

where $\Phi(p_1, p_3, t^*)$ is the load and temperature-effect function.

(7) Reduced governing differential equation. When R_1 is constant, the general differential equation in (6) reduces to

$$\boxed{LL(U) + \left[\frac{K}{D}(1-\nu^2) - \frac{\nu^2}{R_1^2}\right]U = \Phi(p_1, p_3, t^*)}$$

and takes on particular forms in cylinders, spheres, and cones (27.05–27.07).

(8) General solution of (7) can be expressed symbolically as

$$U = C_1S_1(\phi) + C_2S_2(\phi) + C_3S_3(\phi) + C_4S_4(\phi) + G_U$$

where, with $j = 1, 2, 3, 4$,

C_j = constant derived from boundary conditions

$S_j(\phi)$ = shape function in ϕ $\qquad G_U$ = response function in p_1, p_3, t^*, ϕ

(9) Shell functions $N_1, N_2, M_1, M_2, \psi_1, \Delta_x, \Delta_z$ are then to some scale functions of N_3 and of its derivatives. The response function can be approximated to a high degree of accuracy by the respective membrane state vectors introduced in (27.03).

(10) Extreme stresses on a given normal section are

$$\text{At } s_3 = \pm\tfrac{1}{2}h \qquad f_1 = \frac{N_1}{h} \pm \frac{6M_1}{h^2} \qquad f_2 = \frac{N_2}{h} \pm \frac{6M_2}{h^2} \qquad \text{At } s_3 = 0 \qquad f_3 = \frac{3N_3}{2h}$$

27.05 CIRCULAR CYLINDRICAL SHELL, AXISYMMETRICAL GENERAL STATE

(1) Notation and signs (27.02–27.04)

s = length of cylinder (m)
R = constant radius of cylinder (m)
h = constant shell thickness (m)
x = coordinate along 1 (m)
ϑ = polar angle (rad)
$\phi = \frac{1}{2}\pi \qquad \sin\phi = 1 \qquad \cos\phi = 0 \qquad \cot\phi = 0$

$$K = \frac{Eh}{1-\nu^2} \quad \text{(N/m)}$$

$$D = \frac{Eh^3}{12(1-\nu^2)} \quad \text{(N}\cdot\text{m)}$$

$$\lambda = \sqrt[4]{\frac{3(1-\nu^2)}{R^2h^2}} \quad \text{(1/m)}$$

Positive forces and moments are shown in (27.02-4, 5); positive displacements are shown in (27.03-3).

(2) Differential element

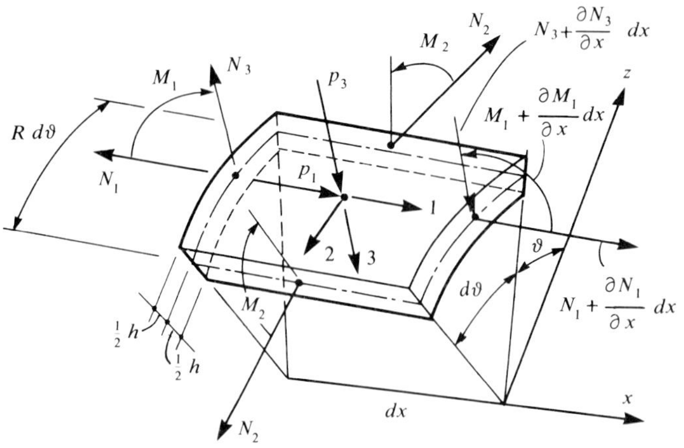

(3) Equilibrium equations in (27.04-2) reduce to

$$\frac{dN_1}{dx} + p_1 = 0$$

$$\frac{N_2}{R} + \frac{dN_3}{dx} + p_3 = 0$$

$$\frac{dM_1}{dx} - N_3 = 0$$

(4) Displacements based on (27.04-3) are

$$\Delta_1 = \Delta_x = \frac{1}{Eh}\int_0^x (N_1 - \nu N_2)\,dx$$

$$\Delta_3 = -\Delta_z = -\frac{R}{Eh}(N_2 - \nu N_1)$$

(5) Strain-displacement relations based on (27.04-3) are

$$\varepsilon_1 = \frac{d\Delta_1}{dx} \qquad \varepsilon_2 = -\frac{\Delta_3}{R} \qquad \psi_1 = \frac{d\Delta_3}{dx} \qquad \kappa_1 = \frac{d\Delta_3^2}{dx^2} \qquad \kappa_2 = 0$$

(6) Stress resultants in terms of (5) are

$$N_1 = K(\varepsilon_1 + \nu\varepsilon_2) \qquad M_1 = -D\kappa_1$$

$$N_2 = K(\varepsilon_2 + \nu\varepsilon_1) \qquad M_2 = -\nu D\kappa_1 \qquad N_3 = -D\frac{d\kappa_1}{dx}$$

(7) Governing differential equation of this shell is

$$\frac{d^4\Delta_3}{dx^4} + 4\lambda^4\Delta_3 = \frac{Rp_3 + \nu N_1}{RD}$$

where λ is the *shape parameter* given in (1) above.

(8) Solution of this differential equation is the equation of the elastic surface, given symbolically in (9) and (10) on the opposite page. According to the magnitude of λs, these shells are classified as (*a*) short shells, $\lambda s \le 5$, and (*b*) long shells, $\lambda s > 5$.

(9) Short-shell solution of (7) is

$$\begin{bmatrix} \Delta_3 \\ \dfrac{d\Delta_3}{dx} \\ \dfrac{d^2\Delta_3}{dx^2} \\ \dfrac{d^3\Delta_3}{dx^3} \end{bmatrix} = \begin{bmatrix} T_1(x) & T_2(x) & T_3(x) & T_4(x) \\ -4\lambda T_4(x) & \lambda T_1(x) & \lambda T_2(x) & \lambda T_3(x) \\ -4\lambda^2 T_3(x) & -4\lambda^2 T_4(x) & \lambda^2 T_1(x) & \lambda^2 T_2(x) \\ -4\lambda^3 T_2(x) & -4\lambda^3 T_3(x) & -4\lambda^3 T_4(x) & \lambda^3 T_1(x) \end{bmatrix} \begin{bmatrix} C_1 \\ C_2 \\ C_3 \\ C_4 \end{bmatrix} + \begin{bmatrix} \Delta_{30} \\ \dfrac{d\Delta_{30}}{dx} \\ \dfrac{d^2\Delta_{30}}{dx^2} \\ \dfrac{d^3\Delta_{30}}{dx^3} \end{bmatrix}$$

where C_1, C_2, C_3, C_4 are to some scale the boundary values at $x = 0$,

$$T_1(x) = \cosh \lambda x \cos\lambda x \qquad T_2(x) = \tfrac{1}{2}(\cosh \lambda x \sin \lambda x + \sinh \lambda x \cos \lambda x)$$

$$T_3(x) = \tfrac{1}{2} \sinh \lambda x \sin \lambda x \qquad T_4(x) = \tfrac{1}{4}(\cosh \lambda x \sin \lambda x - \sinh \lambda x \cos \lambda x)$$

are the transport functions (8.03), and Δ_{30} is the deflection in 3 of the membrane state.

(10) Long-shell solution of (7) is

$$\begin{bmatrix} \Delta_3 \\ \dfrac{d\Delta_3}{dx} \\ \dfrac{d^2\Delta_3}{dx^2} \\ \dfrac{d^3\Delta_3}{dx^3} \end{bmatrix} = \begin{bmatrix} T_{12}(x) & T_{14}(x) \\ -\lambda T_{11}(x) & \lambda T_{13}(x) \\ 2\lambda^2 T_{14}(x) & -2\lambda^2 T_{12}(x) \\ 2\lambda^3 T_{13}(x) & 2\lambda^3 T_{11}(x) \end{bmatrix} \begin{bmatrix} C_5 \\ C_6 \end{bmatrix} + \begin{bmatrix} \Delta_{30} \\ \dfrac{d\Delta_{30}}{dx} \\ \dfrac{d^2\Delta_{30}}{dx^2} \\ \dfrac{d^3\Delta_{30}}{dx^3} \end{bmatrix}$$

where C_5, C_6 are to some scale the boundary values at $x = 0$,

$$T_{11}(x) = e^{-\lambda x}(\cos \lambda x + \sin \lambda x) \qquad T_{12}(x) = e^{-\lambda x} \cos \lambda x$$

$$T_{13}(x) = e^{-\lambda x}(\cos \lambda x - \sin \lambda x) \qquad T_{14}(x) = e^{-\lambda x} \sin \lambda x$$

are the shape functions (8.14-6), and Δ_{30} is the same as in (9) above.

(11) Stress resultants and displacements are

$$N_1 = -\int_0^x p_1\, dx + N_{1L} \qquad N_2 = -\frac{Eh}{R}\Delta_3 + \nu N_1 \qquad \psi_1 = \frac{d\Delta_3}{dx}$$

$$M_1 = -D\frac{d^2\Delta_3}{dx^2} \qquad M_2 = \nu M_1 \qquad N_3 = -D\frac{d^3\Delta_3}{dx^3}$$

where Δ_3 and its derivatives are taken from (9) or (10) and N_{1L} is N_1 at $x = 0$. Particular values of these vectors are tabulated in Chap. 28.

(12) Tables and charts of the transport functions $T_j(x)$ and of the shape functions $T_{ij}(x)$ are given in Chap. 8.

27.06 SPHERICAL SHELL, AXISYMMETRICAL GENERAL STATE

(1) Notation and signs (27.02–27.04)

$\alpha, \phi, \vartheta, a$ = angles (rad)
R = constant radius of sphere (m)
h = constant shell thickness (m)
$R_1 = R_2 = R \qquad R_0 = R \sin \phi \qquad \alpha < \pi$

Positive forces and moments are shown in (27.02-4, 5); positive displacements are shown in (27.03-3).

$$K = \frac{Eh}{1 - \nu^2} \quad \text{(N/m)}$$

$$D = \frac{Eh^3}{12(1 - \nu^2)} \quad \text{(N·m)}$$

$$\lambda = \sqrt[4]{3(1 - \nu^2)\left(\frac{R}{h}\right)^2}$$

(2) Principal circle

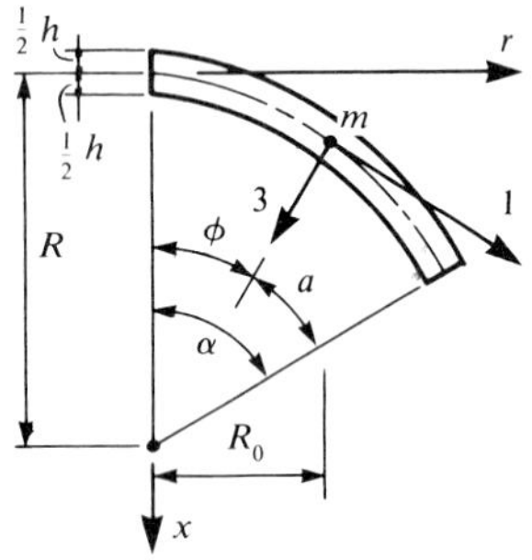

(3) Equilibrium equations of the element (27.02-4, 5) with $ds_1 = R\, d\phi$, $ds_2 = R \sin \phi\, d\vartheta$ are

$$\frac{d(N_1 \sin \phi)}{\sin \phi\, d\phi} - N_2 \cot \phi - N_3 + p_3 R = 0$$

$$N_1 + N_2 + \frac{d(N_3 \sin \phi)}{\sin \phi\, d\phi} + p_1 R = 0$$

$$N_3 R - (M_2 - M_1) \cot \phi + \frac{dM_1}{d\phi} = 0$$

(4) Strains and displacements

$$\varepsilon_1 = -\frac{\Delta_3 - d\Delta_1/d\phi}{R} \qquad \kappa_1 = \frac{1}{R}\frac{d\psi_1}{d\phi}$$

$$\varepsilon_2 = -\frac{\Delta_3 - \Delta_1 \cot \phi}{R} \qquad \kappa_2 = \frac{1}{R}\psi_1 \cot \phi$$

$$\Delta_1 = \frac{R \sin \phi}{(1 - \nu)K}\int_0^{\phi} \frac{N_1 - N_2}{\sin \phi}\, d\phi \qquad \Delta_3 = \Delta_1 \cot \phi - \frac{R}{Eh}(N_2 - \nu N_1) \qquad \psi_1 = \frac{\Delta_3 + d\Delta_3/d\phi}{R}$$

$$\Delta_r = \frac{R \sin \phi}{Eh}(N_2 - \nu N_1) \qquad \Delta_x = \Delta_1 \cos \phi + \Delta_3 \sin \phi$$

(5) Stress resultants

$$N_1 = K(\varepsilon_1 + \nu\varepsilon_2) \qquad M_1 = -D(\kappa_1 + \nu\kappa_2)$$

$$N_2 = K(\varepsilon_2 + \nu\varepsilon_1) \qquad M_2 = -D(\kappa_2 + \nu\kappa_1) \qquad N_3 = -N_1 \tan \phi$$

(6) Governing differential equation of the spherical shell in a state of axisymmetrical deformation is given symbolically in (27.04-8). The solution of this equation can be expressed in terms of a hypergeometric series (Ref. 41, p. 157) or in terms of Legendre functions (Ref. 41, p. 163). Because of the complexity of numerical work involved, these solutions are of limited values in engineering applications, and effort must be made to use an approximate solution, which yields results of reasonable accuracy. For this purpose, the first-, second-, and third-order derivatives of $U = RN_3$ are deleted and the governing equation in (27.04-7) is reduced to

$$\frac{d^4 N_3}{d\phi^4} + 4\lambda^4 N_3 = \Phi(p_1, p_3)$$

where λ is the *shape parameter* given in (1) above and $\Phi(p_1, p_3)$ is the respective load function.

(7) Solution of this approximate equation is the equation of the shearing force N_3. According to the magnitude of $\lambda\phi$, the spherical shells follow the classification of (27.05). The analytical forms of the respective solutions are given in (8, 9) below.

(8) Short-shell solution of (6) is

$$\begin{bmatrix} N_3 \\ \dfrac{dN_3}{da} \\ \dfrac{d^2N_3}{da^2} \\ \dfrac{d^3N_3}{da^3} \end{bmatrix} = \begin{bmatrix} T_1(a) & T_2(a) & T_3(a) & T_4(a) \\ -4\lambda T_4(a) & \lambda T_1(a) & \lambda T_2(a) & \lambda T_3(a) \\ -4\lambda^2 T_3(a) & -4\lambda^2 T_4(a) & -\lambda^2 T_1(a) & \lambda^2 T_2(a) \\ -4\lambda^3 T_2(a) & -4\lambda^3 T_3(a) & -4\lambda^3 T_4(a) & \lambda^3 T_1(a) \end{bmatrix} \begin{bmatrix} C_1 \\ C_2 \\ C_3 \\ C_4 \end{bmatrix} + \begin{bmatrix} N_{30} \\ \dfrac{dN_{30}}{da} \\ \dfrac{d^2N_{30}}{da^2} \\ \dfrac{d^3N_{30}}{da^3} \end{bmatrix}$$

where C_1, C_2, C_3, C_4 are to some scale the boundary values at $a = \alpha - \phi = 0$,

$$T_1(a) = \cosh \lambda a \cos \lambda a \qquad T_2(a) = \tfrac{1}{2}(\cosh \lambda a \sin \lambda a + \sinh \lambda a \cos \lambda a)$$

$$T_3(a) = \tfrac{1}{2}\sinh \lambda a \sin \lambda a \qquad T_4(a) = \tfrac{1}{4}(\cosh \lambda a \sin \lambda a - \sinh \lambda a \cos \lambda a)$$

are the transport functions (8.03) and N_{30} is the load function.

(9) Long-shell solution of (6) is

$$\begin{bmatrix} N_3 \\ \dfrac{dN_3}{da} \\ \dfrac{d^2N_3}{da^2} \\ \dfrac{d^3N_3}{da^3} \end{bmatrix} = \begin{bmatrix} T_{12}(a) & T_{14}(a) \\ -\lambda T_{11}(a) & \lambda T_{13}(a) \\ 2\lambda^2 T_{14}(a) & -2\lambda^2 T_{12}(a) \\ 2\lambda^3 T_{13}(a) & 2\lambda^3 T_{11}(a) \end{bmatrix} \begin{bmatrix} C_5 \\ C_6 \end{bmatrix} + \begin{bmatrix} N_{30} \\ \dfrac{dN_{30}}{da} \\ \dfrac{d^2N_{30}}{da^2} \\ \dfrac{d^3N_{30}}{da^3} \end{bmatrix}$$

where C_5, C_6 are to some scale the boundary values at $a = \alpha - \phi = 0$,

$$T_{11}(a) = e^{-\lambda a}(\cos \lambda a + \sin \lambda a) \qquad T_{12}(a) = e^{-\lambda a} \cos \lambda a$$

$$T_{13}(a) = e^{-\lambda a}(\cos \lambda a - \sin \lambda a) \qquad T_{14}(a) = e^{-\lambda a} \sin \lambda a$$

are the the shape functions (8.14-6) and N_{30} is again the load function.

(10) Stress resultants and displacements are

$$N_1 = -N_3 \cot \phi + N_{10} \qquad N_2 = -\frac{dN_3}{da} + N_{20} \qquad M_1 = -\frac{D}{ERh}\frac{d^3N_3}{da^3}$$

$$\psi_1 = \frac{1}{Eh}\frac{d^2N_3}{da^2} + \psi_{10} \qquad \Delta_r = \frac{R \sin \phi}{Eh}\frac{dN_3}{da} + \Delta_{r0} \qquad M_2 = \nu M_1$$

where N_3 and its derivatives are taken from (8) or (9) and N_{10}, N_{20} and ψ_{10}, Δ_{r0} are the membrane state vectors. Particular values of these vectors are tabulated in Chap. 29.

(11) Tables and charts of $T_j(a)$ and $T_{ij}(a)$ are given in Chap. 8.

27.07 CIRCULAR CONICAL SHELL, AXISYMMETRICAL GENERAL STATE

(1) Notation and signs (27.02–27.04)

a, b, c, s = coordinates (m) $\quad l$ = shell height (m)
α, ϑ = angles (rad) $\quad R_a, R_b, R_m$ = radii (m)
h = constant shell thickness (m)
$1/R_1 = 0 \quad R_2 = R_m \quad R_0 = s\cos\alpha \quad \alpha > \pi/4$

Positive forces and moments are shown in (27.02-4, 5); positive displacements are shown in (27.03-2).

$$K = \frac{Eh}{1-\nu^2} \quad \text{(N/m)}$$

$$D = \frac{Eh^3}{12(1-\nu^2)} \quad \text{(N·m)}$$

$$\lambda = \sqrt[4]{\frac{3(1-\nu^2)}{R_m^2 h^2}} \quad \text{(1/m)}$$

(2) Principal section

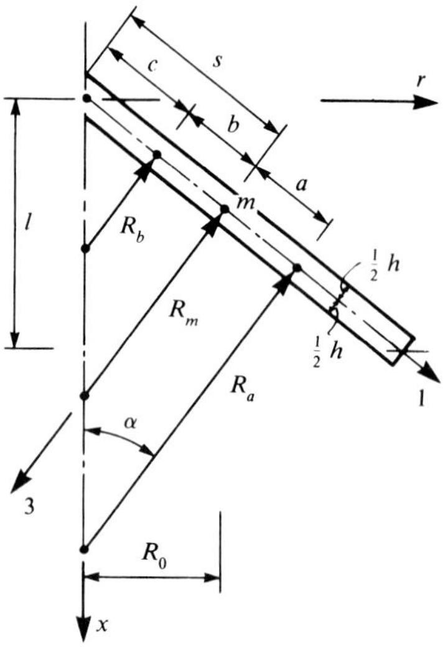

(3) Equilibrium equations of the element (27.02-4, 5) with $ds_1 = ds$, $ds_2 = s\cos\alpha\, d\vartheta$ are

$$\frac{d(N_1 s)}{ds} - N_2 + p_1 s = 0$$

$$\frac{d(N_3 s)}{ds} + N_2 \tan\alpha + p_3 s = 0$$

$$\frac{d(M_1 s)}{ds} - M_2 - N_3 s = 0$$

(4) Strains and displacements

$$\varepsilon_1 = \frac{d\Delta_1}{ds} \qquad \psi_1 = \frac{d\Delta_3}{ds} \qquad \kappa_1 = \frac{d\psi_1}{ds} \qquad \kappa_2 = \frac{\psi_1}{s}$$

$$\varepsilon_2 = \frac{\Delta_1 - \Delta_3 \tan\alpha}{s} \qquad \Delta_1 = \int_0^s \frac{N_1 - \upsilon N_2}{Eh}\, ds$$

$$\Delta_3 = \Delta_1 \cot\alpha - \frac{s}{Eh}(N_2 - \nu N_1)\cot\alpha$$

$$\Delta_x = \Delta_1 \cos\alpha + \Delta_3 \sin\alpha \qquad \Delta_r = \frac{s\cos\alpha}{Eh}(N_2 - \nu N_1)$$

(5) Stress resultants in terms of (4) are

$$N_1 = K(\varepsilon_1 + \nu\varepsilon_2) \qquad M_1 = -D(\kappa_1 + \nu\kappa_2)$$

$$N_2 = K(\varepsilon_2 + \nu\varepsilon_1) \qquad M_2 = -D(\kappa_2 + \nu\kappa_1) \qquad N_3 = -N_1 \tan\alpha$$

(6) Governing differential equation of the conical shell in a state of axisymmetrical deformation is given symbolically in (27.04-7) whose solution can be expressed in terms of Kelvin functions (Ref. 41, p. 143). Because of the complexity of numerical work involved, this type of solution is used only in cases where a very high accuracy is required. In most engineering problems the simplified differential equation given below yields results of reasonable accuracy:

$$\frac{d^4 N_3}{ds^4} + 4\lambda^4 N_3 = \Phi(p_1, p_3)$$

where λ is a *variable shape parameter* given in (1) and $\Phi(p_1, p_3)$ is the respective load function.

(7) Solution of this simplified equation is the equation of the shearing force N_3. Because λ is a function of the variable radius R_m, the differential equation (6) with λ taken as a constant λ_m is applicable only in a very narrow strip of the shell where R_m is assumed to be a constant.

28

Circular Cylindrical Shell Constant Thickness

STATIC STATE

28.01 DEFINITION OF STATE

(1) System considered is a finite isotropic circular cylindrical shell of constant thickness in a state of static axisymmetrical deformation.

(2) Notation (A.1, A.6)

a, b, s = segments (m)	x = coordinate (m)
R = constant radius (m)	h = thickness (m)
$U = N_1$ = normal force along x (N/m)	$V = -N_3$ = shearing force along y (N/m)
$W = N_2$ = normal force along z (N/m)	$u = \Delta_1$ = deflection along x (m)
$X = M_2$ = bending moment about x (N·m/m)	$v = -\Delta_3$ = deflection along y (m)
$Z = M_1$ = bending moment about z (N·m/m)	$\theta = -\psi_1$ = slope about z (rad)
E = modulus of elasticity (Pa)	ν = Poisson's ratio

All forces and moments are per unit length of normal section, and all forces, moments, and displacements are given in the global system x, y, z. $N_1, N_2, N_3, M_1, M_2, \Delta_1, \Delta_3, \psi_1$ are defined in (27.02).

$K = \dfrac{Eh}{1-\nu^2}$ (N/m)	$D = \dfrac{Eh^3}{12(1-\nu^2)}$ (N·m)	$\lambda = \sqrt[4]{\dfrac{3(1-\nu^2)}{R^2h^2}}$ (m^{-1})

(3) Assumptions of analysis are stated in (27.01). The shells are classified as short ($\lambda s \le 5$) and long ($\lambda s > 5$).

28.02 SHORT SHELL, TRANSPORT MATRIX EQUATIONS

(1) Equivalents and signs ($\lambda s < 5$)

$$\bar{v} = D\lambda^3 v = \text{scaled } v_3 \text{ (N/m)} \qquad \bar{Z} = \lambda Z = \text{scaled } Z \text{ (N/m)}$$

$$\bar{\theta} = D\lambda^2 \theta = \text{scaled } \theta \text{ (N/m)} \qquad \bar{u} = Eu = \text{scaled } u \text{ (N/m)}$$

All forces, moments, and displacement are in the x, y, z axes, and their positive directions are shown in (2, 3) below

(2) Free-body diagram

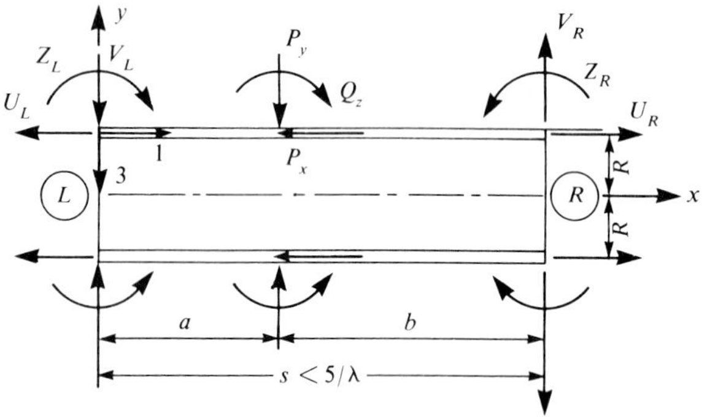

(3) Elastic surface

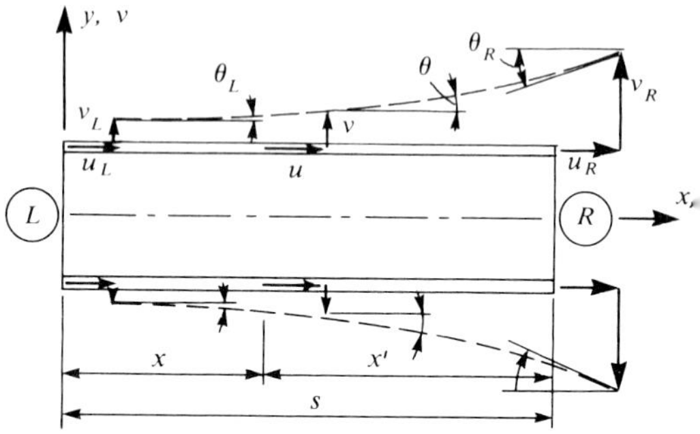

(4) Transport matrix equations

$$\underbrace{\begin{bmatrix} 1 \\ \hline U_L \\ V_L \\ \bar{Z}_L \\ \hline \bar{u}_L \\ \bar{v}_L \\ \bar{\theta}_L \end{bmatrix}}_{H_L} = \underbrace{\left[\begin{array}{c|ccc|ccc} 1 & 0 & 0 & 0 & 0 & 0 & 0 \\ \hline F_U & 1 & 0 & 0 & 0 & 0 & 0 \\ F_V & 0 & T_1 & -4T_4 & 0 & -4T_2 & 4T_3 \\ F_Z & 0 & T_2 & T_1 & 0 & -4T_3 & 4T_4 \\ \hline F_u & -s/h & 0 & 0 & 1 & 0 & 0 \\ F_v & 0 & T_4 & T_3 & 0 & T_1 & -T_2 \\ F_\theta & 0 & -T_3 & -T_2 & 0 & 4T_4 & T_1 \end{array}\right]}_{T_{LR}} \underbrace{\begin{bmatrix} 1 \\ \hline U_R \\ V_R \\ \bar{Z}_R \\ \hline \bar{u}_R \\ \bar{v}_R \\ \bar{\theta}_R \end{bmatrix}}_{H_R}$$

$$\underbrace{\begin{bmatrix} 1 \\ \hline U_R \\ V_R \\ \bar{Z}_R \\ \hline \bar{u}_R \\ \bar{v}_R \\ \bar{\theta}_R \end{bmatrix}}_{H_R} = \underbrace{\left[\begin{array}{c|ccc|ccc} 1 & 0 & 0 & 0 & 0 & 0 & 0 \\ \hline G_U & 1 & 0 & 0 & 0 & 0 & 0 \\ G_V & 0 & T_1 & 4T_4 & 0 & 4T_2 & 4T_3 \\ G_Z & 0 & -T_2 & T_1 & 0 & -4T_3 & -4T_4 \\ \hline G_u & s/h & 0 & 0 & 1 & 0 & 0 \\ G_v & 0 & -T_4 & T_3 & 0 & T_1 & T_2 \\ G_\theta & 0 & -T_3 & T_2 & 0 & -4T_4 & T_1 \end{array}\right]}_{T_{RL}} \underbrace{\begin{bmatrix} 1 \\ \hline U_L \\ V_L \\ \bar{Z}_L \\ \hline \bar{u}_L \\ \bar{v}_L \\ \bar{\theta}_L \end{bmatrix}}_{H_L}$$

where H_L, H_R are the *state vectors* of the respective ends and T_{LR}, T_{RL} are the *transport matrices*. The elements of the transport matrices are the *transport coefficients* T_j defined in (28.03) and the *load functions* F, G displayed for particular cases in (28.04–28.06).

28.03 SHORT SHELL, TRANSPORT RELATIONS

(1) Transport functions

$$\lambda = \sqrt[4]{\frac{3(1-\nu^2)}{R^2h^2}}$$

$T_1(x) = \cosh \lambda x \cos \lambda x$	$T_2(x) = \frac{1}{2}(\cosh \lambda x \sin \lambda x + \sinh \lambda x \cos \lambda x)$
$T_3(x) = \frac{1}{2} \sinh \lambda x \sin \lambda x$	$T_4(x) = \frac{1}{4}(\cosh \lambda x \sin \lambda x - \sinh \lambda x \cos \lambda x)$
$T_5(x) = \frac{1}{4}[1 - T_1(x)]$	$T_6(x) = \frac{1}{4}[\lambda x - T_2(x)]$
$T_7(x) = \frac{1}{4}[(\lambda x)^2/2 - T_3(x)]$	

The graphs and tables of these functions are shown in (8.04–8.05).

(2) Transport coefficients T_j, A_j, B_j, C_j are the values of $T_j(x)$ for particular arguments x.

$$\left.\begin{array}{ll} \text{For } x = s, & T_j(s) = T_j \\ \text{For } x = a, & T_j(a) = A_j \\ \text{For } x = b, & T_j(b) = B_j \\ \text{and } T_j - A_j = C_j & \end{array}\right\} j = 1, 2, \ldots, 7$$

(3) Transport matrices of this shell have analogical properties to those of (8.02) and can be used in similar ways.

(4) Differential relations

$$D = \frac{Eh^3}{12(1-\nu^2)}$$

$\frac{du}{dx} = \frac{U}{Eh} = \frac{N_1}{Eh}$	$\frac{dv}{dx} = \theta = -\psi_1$
$\frac{d^2u}{dx^2} = \frac{p_x}{Eh} = -\frac{p_1}{Eh}$	$\frac{d^2v}{dx^2} = \frac{Z}{D} = \frac{M_1}{D}$
	$\frac{d^3v}{dx^3} = -\frac{V}{D} = \frac{N_3}{D}$
	$\frac{d^4v}{dx^4} = -4\lambda^4 v - \frac{p_yR + \nu U}{RD}$
	$= 4\lambda^4 \Delta_3 - \frac{p_3R + \nu N_1}{RD}$

where p_1, p_3 are the intensities of load in 1, 3 and p_x, p_z are the intensities of load in x, z.

(5) Conversion relations (27.05)

$N_1 = U$	$N_3 = -V$	$\psi_1 = -\theta$	$M_1 = Z$	$p_1 = -p_x$
$N_2 = \frac{Eh}{R} v + \nu U$		$\Delta_3 = -v$	$M_2 = \nu Z$	$p_3 = p_y$

Notation (18.01–28.03) Signs (28.02) Matrix (28.02)
P_x, P_y = concentrated loads (N/m) Q_z = applied couple (N·m/m)
q_z = intensity of distributed couple (N·m/m²) $\lambda s \leq 5$
A_j, B_j, C_j, T_j = transport coefficients (28.03)

$$\lambda = \sqrt[4]{\frac{3(1-\nu^2)}{R^2h^2}} \qquad \eta = \frac{\nu}{\lambda R}$$

$F_U = -P_x$ $F_V = -A_1P_y - \eta A_2P_x$ $F_Z = -A_2P_y - \eta A_3P_x$ $F_u \cong ah^{-1}P_x$ $F_v = -A_4P_y - \eta A_5P_x$ $F_\theta = A_3P_y + \eta A_4P_x$	1. 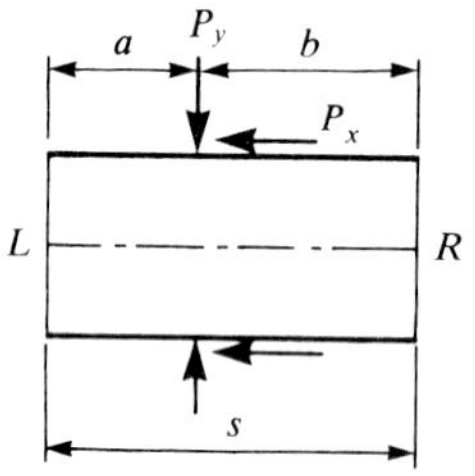	$G_U = P_x$ $G_V = B_1P_y + \eta B_2P_x$ $G_Z = -B_2P_y - \eta B_3P_x$ $G_u \cong bh^{-1}P_x$ $G_v = -B_4P_y - \eta B_5P_x$ $G_\theta = -B_3P_y - \eta B_4P_x$
$F_U = 0$ $F_V = 4A_4\lambda Q_z$ $F_Z = -A_1\lambda Q_z$ $F_u = -A_4\eta\lambda Q_z$ $F_v = -A_3\lambda Q_z$ $F_\theta = A_2\lambda Q_z$	2. 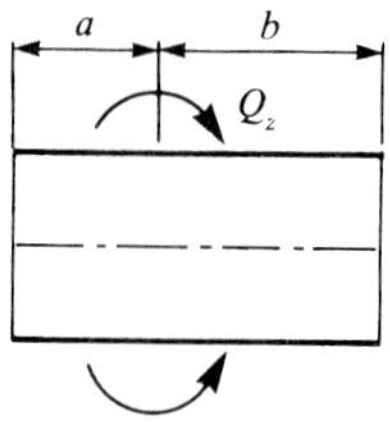	$G_U = 0$ $G_V = 4B_4\lambda Q_z$ $G_Z = B_1\lambda Q_z$ $G_u = B_4\eta\lambda Q_z$ $G_v = B_3\lambda Q_z$ $G_\theta = B_2\lambda Q_z$
$F_U = 0$ $F_V = 4T_5q_z$ $F_Z = -T_2q_z$ $F_u = -T_5\eta q_z$ $F_v = -T_4q_z$ $F_\theta = T_3q_z$	3. 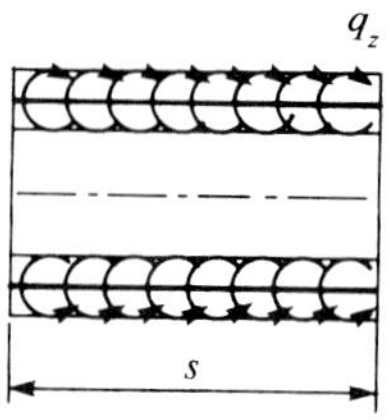	$G_U = 0$ $G_V = 4T_5q_z$ $G_Z = T_2q_z$ $G_u = T_5\eta q_z$ $G_v = T_4q_z$ $G_\theta = T_3q_z$
$F_U = 0$ $F_V = 4C_5q_z$ $F_Z = -C_2q_z$ $F_u = -C_5\eta q_z$ $F_v = -C_4q_z$ $F_\theta = C_3q_z$	4. 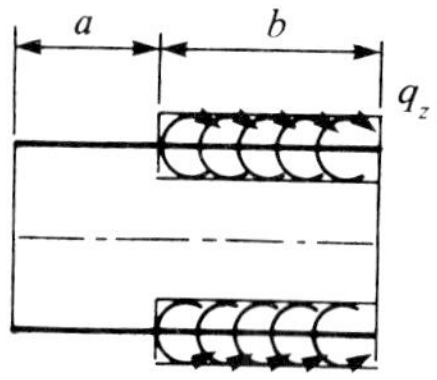	$G_U = 0$ $G_V = 4B_5q_z$ $G_Z = B_2q_z$ $G_u = B_5\eta q_z$ $G_v = B_4q_z$ $G_\theta = B_3q_z$

28.05 SHORT SHELL, TRANSPORT LOAD FUNCTIONS

Notation (28.01–28.03) Signs (28.02) Matrix (28.02)		$\lambda = \sqrt[4]{\dfrac{3(1-\nu^2)}{R^2h^2}}$
p_x, p_y = intensities of distributed load (N/m²) A_j, B_j, C_j, T_j = transport coefficients (28.03)	$\lambda s \le 5$	$\eta = \dfrac{\nu}{\lambda R}$

$F_U = -sp_x$ $F_V = -\eta T_3\lambda^{-1}p_x$ $F_Z = -\eta T_4\lambda^{-1}p_x$	**1.** 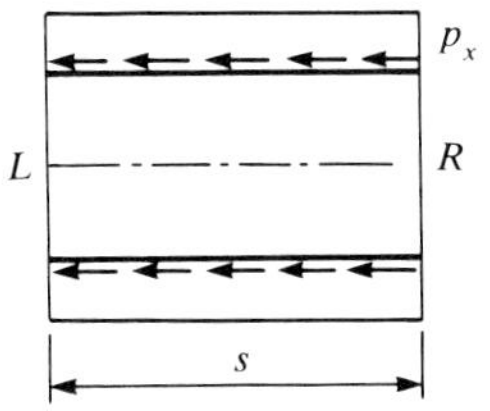	$G_U = sp_x$ $G_V = \eta T_3\lambda^{-1}p_x$ $G_Z = -\eta T_4\lambda^{-1}p_x$
$F_u \cong \frac{1}{2}s^2h^{-1}p_x$ $F_v = -\eta T_6\lambda^{-1}p_x$ $F_\theta = \eta T_5\lambda^{-1}p_x$		$G_u \cong \frac{1}{2}s^2h^{-1}p_x$ $G_v = -\eta T_6\lambda^{-1}p_x$ $G_\theta = -\eta T_5\lambda^{-1}p_x$
$F_U = -bp_x$ $F_V = -\eta C_3\lambda^{-1}p_x$ $F_Z = -\eta C_4\lambda^{-1}p_x$	**2.** 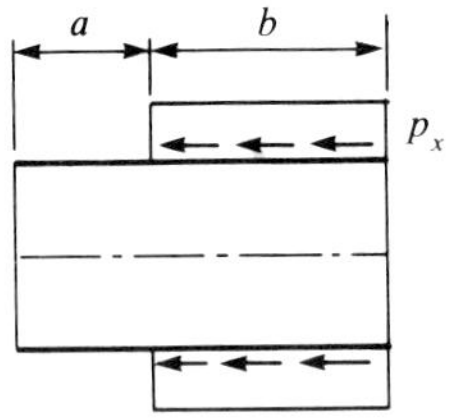	$G_U = bp_x$ $G_V = \eta B_3\lambda^{-1}p_x$ $G_Z = -\eta B_4\lambda^{-1}p_x$
$F_u \cong \frac{1}{2}(s^2 - a^2)h^{-1}p_x$ $F_v = -\eta C_6\lambda^{-1}p_x$ $F_\theta = \eta C_5\lambda^{-1}p_x$		$G_u \cong \frac{1}{2}b^2h^{-1}p_x$ $G_v = -\eta B_6\lambda^{-1}p_x$ $G_\theta = -\eta B_5\lambda^{-1}p_x$
$F_U = 0$ $F_V = -T_2\lambda^{-1}p_y$ $F_Z = -T_3\lambda^{-1}p_y$	**3.** 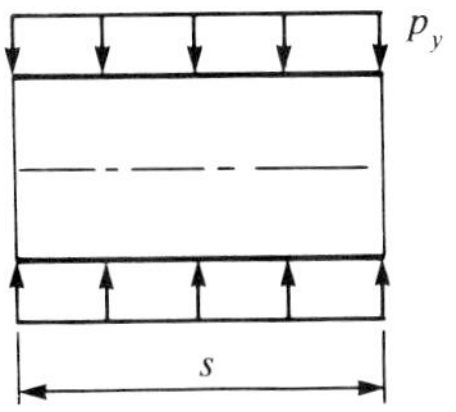	$G_U = 0$ $G_V = T_2\lambda^{-1}p_y$ $G_Z = -T_3\lambda^{-1}p_y$
$F_u = -T_6\eta\lambda^{-1}p_y$ $F_v = -T_5\lambda^{-1}p_y$ $F_\theta = T_4\lambda^{-1}p_y$		$G_u = -T_6\eta\lambda^{-1}p_y$ $G_v = -T_5\lambda^{-1}p_y$ $G_\theta = -T_4\lambda^{-1}p_y$
$F_U = 0$ $F_V = -C_2\lambda^{-1}p_y$ $F_Z = -C_2\lambda^{-1}p_y$	**4.** 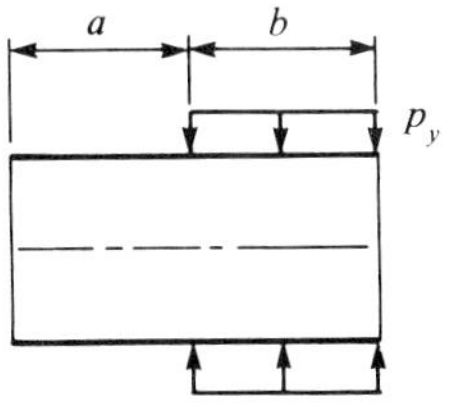	$G_U = 0$ $G_V = B_2\lambda^{-1}p_y$ $G_Z = -B_3\lambda^{-1}p_y$
$F_u = -C_6\eta\lambda^{-1}p_y$ $F_v = -C_5\lambda^{-1}p_y$ $F_\theta = C_y\lambda^{-1}p_y$		$G_u = -B_6\eta\lambda^{-1}p_y$ $G_v = -B_5\lambda^{-1}p_y$ $G_\theta = -B_4\lambda^{-1}p_y$

Notation (28.01–28.03) Signs (28.02) Matrix (28.02)

p_x, p_y = intensities of distributed load (N/m²) $\lambda s \leq 5$

N_t, M_t = thermal force and moment (N/m, N·m/m) (27.04)

A_j, B_j, C_j, T_j = transport coefficients (28.03)

$$\lambda = \sqrt[4]{\frac{3(1-\nu^2)}{R^2h^2}} \qquad \eta = \frac{\nu}{\lambda R}$$

1.

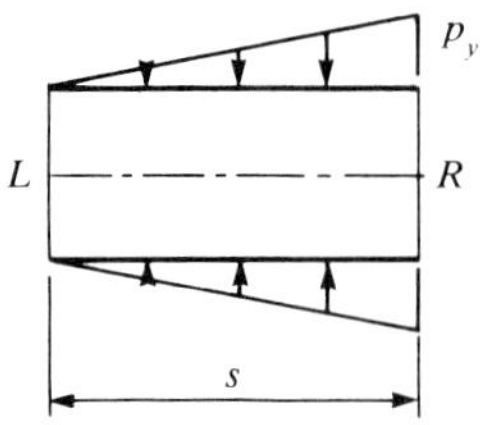

$$F_U = 0 \qquad G_U = 0$$
$$F_V = -(\lambda s T_2 - T_3)\lambda^{-2}s^{-2}p_y \qquad G_V = T_3\lambda^{-2}s^{-1}p_y$$
$$F_Z = -(\lambda s T_3 - T_4)\lambda^{-2}s^{-1}p_y \qquad G_Z = -T_4\lambda^{-2}s^{-1}p_y$$
$$F_u = -\eta(\lambda s T_6 - 2T_7)\lambda^{-2}s^{-1}p_y \qquad G_u = -\eta T_7\lambda^{-2}s^{-1}p_y$$
$$F_v = -(\lambda s T_5 - T_6)\lambda^{-2}s^{-1}p_y \qquad G_v = -T_6\lambda^{-2}s^{-1}p_y$$
$$F_\theta = (\lambda s T_4 - T_5)\lambda^{-2}s^{-1}p_y \qquad G_\theta = -T_5\lambda^{-2}s^{-1}p_y$$

2.

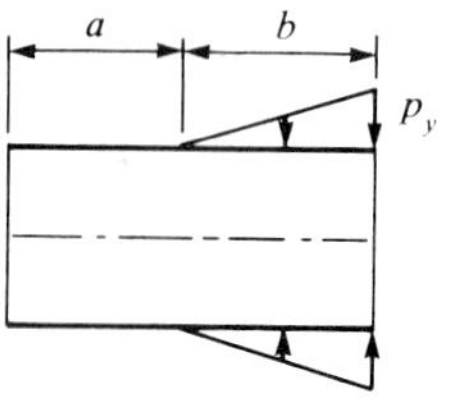

$$F_U = 0 \qquad G_U = 0$$
$$F_V = -(\lambda b T_2 - C_3)\lambda^{-2}b^{-1}p_y \qquad G_V = B_3\lambda^{-2}b^{-1}p_y$$
$$F_Z = -(\lambda b T_3 - C_4)\lambda^{-2}b^{-1}p_y \qquad G_Z = -B_4\lambda^{-2}b^{-1}p_y$$
$$F_u = -\eta(\lambda b T_6 - 2C_7)\lambda^{-2}b^{-1}p_y \qquad G_u = -\eta B_7\lambda^{-2}b^{-1}p_y$$
$$F_v = -(\lambda b T_5 - C_6)\lambda^{-2}b^{-1}p_y \qquad G_v = -B_6\lambda^{-2}b^{-1}p_y$$
$$F_\theta = (\lambda b T_4 - C_5)\lambda^{-2}b^{-1}p_y \qquad G_\theta = -B_5\lambda^{-2}b^{-1}p_y$$

3. Temperature deformation of segment s

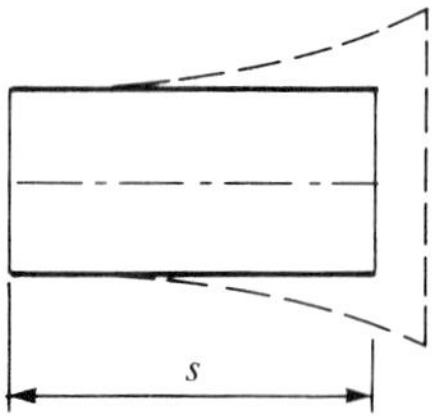

$$F_U = 0 \qquad G_U = 0$$
$$F_V = -4T_4\lambda M_t + \eta T_2 N_t \qquad G_V = 4T_4\lambda M_t - \eta T_2 N_t$$
$$F_Z = -4T_5\lambda M_t + \eta T_3 N_t \qquad G_Z = -4T_5\lambda M_t + \eta T_3 N_t$$
$$F_u \cong -(1-\nu)sh^{-1}N_t \qquad G_u \cong (1-\nu)sh^{-1}N_t$$
$$F_v = T_3\lambda M_t + \eta T_5 N_t \qquad G_v = T_3\lambda M_t + \eta T_5 N_t$$
$$F_\theta = -T_2\lambda M_t - \eta T_4 N_t \qquad G_\theta = T_2\lambda M_t + \eta T_4 N_t$$

4. Temperature deformation of a segment b

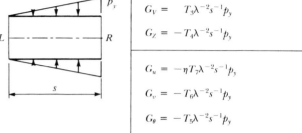

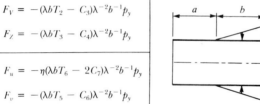

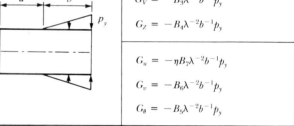

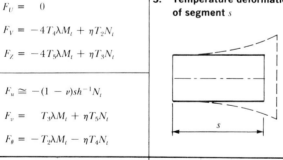

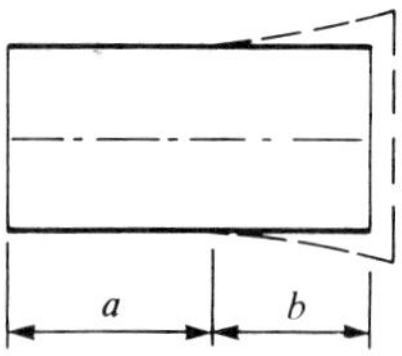

$$F_U = 0 \qquad G_U = 0$$
$$F_V = -4C_4\lambda M_t + \eta C_2 N_t \qquad G_V = 4B_4\lambda M_t - \eta B_2 N_t$$
$$F_Z = 4C_5\lambda M_t + \eta C_3 N_t \qquad G_Z = -4B_5\lambda M_t + \eta B_3 N_t$$
$$F_u \cong -(1-\nu)bh^{-1}N_t \qquad G_u \cong (1-\nu)bh^{-1}N_t$$
$$F_v = C_3\lambda M_t + \eta C_5 N_t \qquad G_v = B_3\lambda M_t + \eta B_5 N_t$$
$$F_\theta = -C_2\lambda M_t - \eta C_4 N_t \qquad F_\theta = B_2\lambda M_t + \eta B_4 N_t$$

28.07 SHORT SHELL, TRANSPORT LOAD FUNCTIONS

Notation (28.01–28.03) Signs (28.02) Matrix (28.02) $\lambda s \leq 5$

p_0 = maximum intensity of harmonic load (N/m^2)

T_j = transport coefficients (28.03)

$$\alpha = \frac{m\pi}{s} \qquad \beta = \frac{m\pi}{\lambda s} \qquad \Lambda_1 = \frac{\beta p_0}{\lambda(4+\beta^4)} \qquad \Lambda_2 = \frac{P_0}{\lambda(4+\beta^4)}$$

$$\lambda = \sqrt[4]{\frac{3(1-\nu^2)}{R^2h^2}} \qquad \eta = \frac{\nu}{\lambda R}$$

1. $p_y = p_0$ sin αx

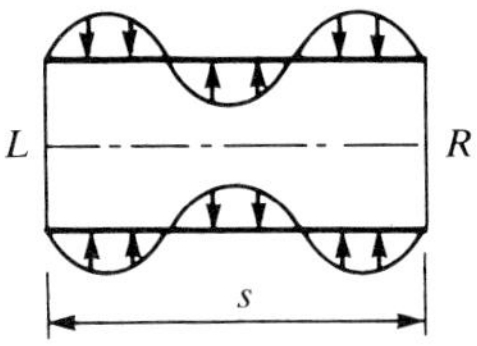

$m = 1, 3, 5, \ldots$

F	G
$F_U = 0$	$G_U = 0$
$F_V = -(\beta^2 + \beta^2 T_1 + 4T_3)\Lambda_1$	$G_V = (\beta^2 + \beta^2 T_1 + 4T_3)\Lambda_1$
$F_Z = -(\beta^2 T_2 + 4T_4)\Lambda_1$	$G_Z = -(4T_4 + \beta^2 T_2)\Lambda_1$
$F_u = -(T_3 - \beta^2 T_5)\eta\Lambda_1$	$G_u = (T_3 - \beta^2 T_5)\eta\Lambda_1$
$F_v = (T_2 - \beta^2 T_4)\Lambda_1$	$G_v = (T_2 - \beta^2 T_4)\Lambda_1$
$F_\theta = -(1 + T_1 - \beta^2 T_3)\Lambda_1$	$G_\theta = (1 + T_1 - \beta^2 T_3)\Lambda_1$

2. $p_y = p_0$ sin αx

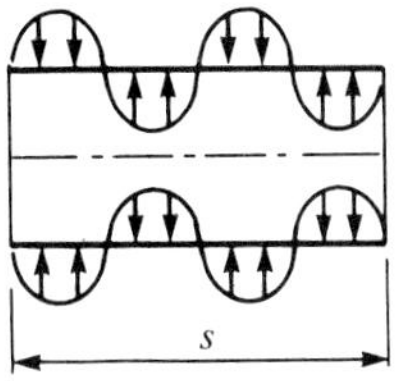

$m = 2, 4, 6, \ldots$

F	G
$F_U = 0$	$G_U = 0$
$F_V = (\beta^2 - \beta^2 T_1 - 4T_3)\Lambda_1$	$G_V = -(\beta^2 - \beta^2 T_1 - 4T_3)\Lambda_1$
$F_Z = -(\beta^2 T_2 + 4T_4)\Lambda_1$	$G_Z = -(4T_4 + \beta^2 T_2)\Lambda_1$
$F_u = -(T_3 - \beta^2 T_5)\eta\Lambda_1$	$G_u = (T_3 - \beta^2 T_5)\eta\Lambda_1$
$F_v = (T_2 - \beta^2 T_4)\Lambda$	$G_v = (T_2 - \beta^2 T_4)\Lambda$
$F_\theta = (1 - T_1 + \beta^2 T_3)\Lambda$	$G_\theta = -(1 - T_1 + \beta^2 T_3)\Lambda$

3. $p_y = p_0$ cos αx

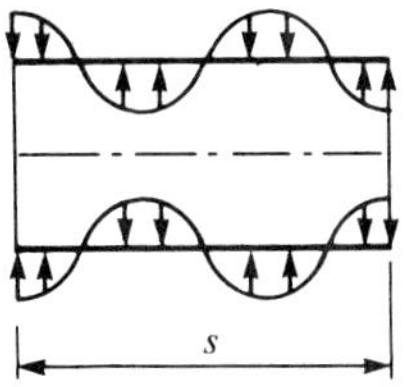

$m = 1, 3, 5, \ldots$

F	G
$F_U = 0$	$G_U = 0$
$F_V = -4(T_2 + \beta^2 T_4)\Lambda_2$	$G_V = -4(T_2 + \beta^2 T_4)\Lambda_2$
$F_Z = (\beta^2 + \beta^2 T_1 - 4T_4)\Lambda_2$	$G_Z = -(\beta^2 + \beta^2 T_1 - 4T_3)\Lambda_2$
$F_u = (\lambda s + T_2 - \beta^2 T_4)\eta\Lambda_2$	$G_u = (\lambda s + T_2 - \beta^2 T_4)\eta\Lambda_2$
$F_v = -(1 + T_1 - \beta^2 T_3)\Lambda_2$	$G_v = (1 + T_1 - \beta^2 T_3)\Lambda_2$
$F_\theta = -(\beta^2 T_2 - 4T_4)\Lambda_2$	$G_\theta = -(\beta^2 T_2 - 4T_4)\Lambda_2$

4. $p_y = p_0$ cos αx

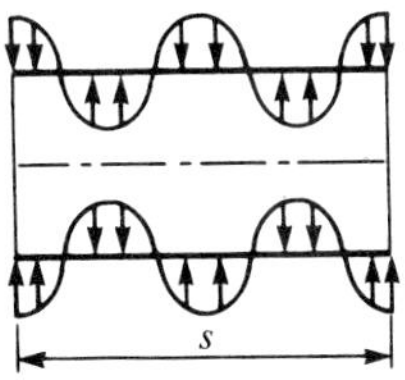

$m = 2, 4, 6, \ldots$

F	G
$F_U = 0$	$G_U = 0$
$F_V = -4(T_2 + \beta^2 T_4)\Lambda_2$	$G_V = -4(T_2 + \beta^2 T_4)\Lambda_2$
$F_Z = -(\beta^2 - \beta^2 T_1 - 4T_3)\Lambda_2$	$G_Z = (\beta^2 - \beta^2 T_1 - 4T_3)\Lambda_2$
$F_u = -(4T_6 + \beta^2 T_4)\eta\Lambda_2$	$G_u = -(4T_6 + \beta^2 T_4)\eta\Lambda_2$
$F_v = (1 - T_1 + \beta^2 T_3)\Lambda_2$	$G_v = -(1 - T_1 + \beta^2 T_3)\Lambda_2$
$F_\theta = -(\beta^2 T_2 + 4T_4)\Lambda_2$	$G_\theta = -(\beta^2 T_2 + 4T_4)\Lambda_2$

28.08 LONG SHELL, STATE-VECTOR MATRIX EQUATIONS

$$\lambda = \sqrt[4]{\frac{3(1-\nu^2)}{R^2h^2}}$$

$$\lambda s > 5$$

(1) State vectors due to loads, end forces, and end moments in long shells are given in matrix form in (9, 10). The symbols used are those of (28.01–28.02). Positive forces, moments, and displacements are shown in (2–4) below.

(2) Forces

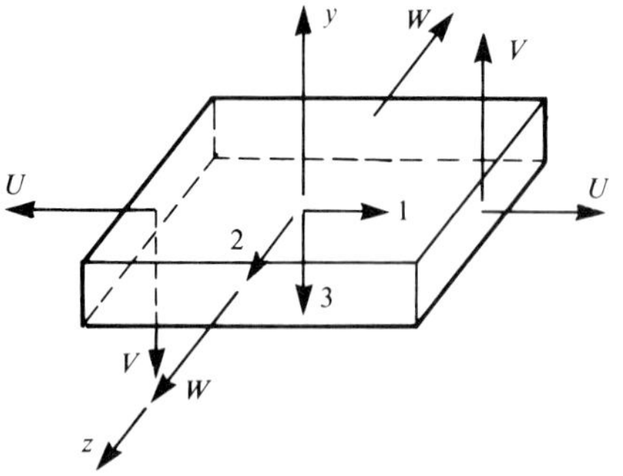

(3) Moments

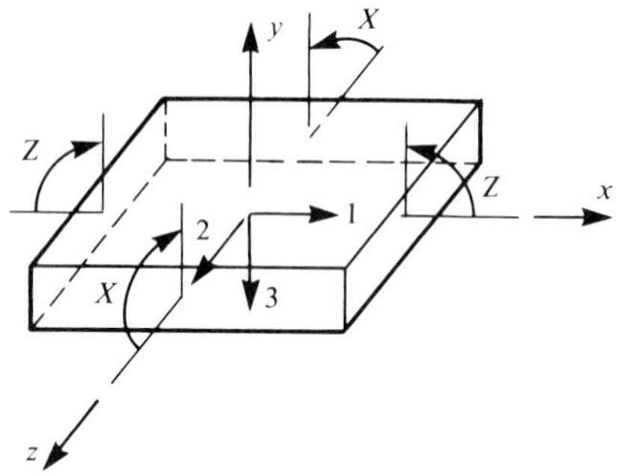

(4) Displacements

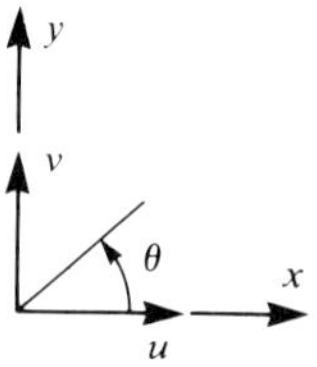

(5) Shape functions. At $x = x$,

At $x = x$		At $x = 0$	
$T_{11} = e^{-\lambda x}(\cos\lambda x + \sin\lambda x)$	$T_{12} = e^{-\lambda x}\cos\lambda x$	$T_{11} = 1$	$T_{12} = 1$
$T_{13} = e^{-\lambda x}(\cos\lambda x - \sin\lambda x)$	$T_{14} = e^{-\lambda x}\sin\lambda x$	$T_{13} = 1$	$T_{14} = 0$
At $x' = s - x$		At $x' = 0$	
$T_{21} = e^{-\lambda x'}(\cos\lambda x' + \sin\lambda x')$	$T_{22} = e^{-\lambda x'}\cos\lambda x'$	$T_{21} = 1$	$T_{22} = 1$
$T_{23} = e^{-\lambda x'}(\cos\lambda x' - \sin\lambda x')$	$T_{24} = e^{-\lambda x'}\sin\lambda x'$	$T_{23} = 1$	$T_{24} = 0$

(6) Free-body diagram

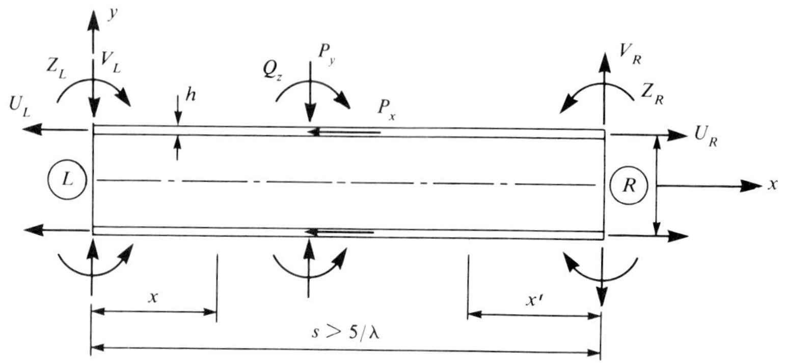

(7) Edge forces and moments of the left end U_L, V_L, Z_L and the right end U_R, V_R, Z_R, applied as redundant vectors on the membrane state of the long shell, do not have appreciable effect on each other, and their influence is limited to a narrow region adjacent to the edge. The state-vector matrix equations in (9, 10) define this influence.

(8) Load functions V_0, W_0, u_0, v_0, θ_0 are the forces and displacements of the membrane state due to loads and other causes listed in (28.09–28.10). By definition in (27.03),

$$U_0 = N_{10} \qquad W_0 = N_{20}$$
$$u_0 = \Delta_{10} \qquad v_0 = -\Delta_{30} \qquad \theta = -\psi_{10}$$

(9) State vector at x

$$\left[\begin{array}{c} 1 \\ \hline U \\ V \\ W \\ \hline X \\ Z \\ \hline u \\ v \\ \theta \end{array}\right] = \left[\begin{array}{c|ccc} 1 & 0 & 0 & 0 \\ \hline U_0 & 1 & 0 & 0 \\ 0 & 0 & T_{13} & 2\lambda T_{14} \\ W_0 & 0 & -2\lambda R T_{12} & 2\lambda^2 R T_{13} \\ \hline 0 & 0 & -\dfrac{\nu T_{14}}{\lambda} & \nu T_{11} \\ 0 & 0 & -\dfrac{T_{14}}{\lambda} & T_{11} \\ \hline u_0 & -\dfrac{x}{Eh} & 0 & 0 \\ v_0 & -\dfrac{\nu R}{Eh} & -\dfrac{2\lambda R^2 T_{12}}{Eh} & \dfrac{2\lambda^2 R^2 T_{13}}{Eh} \\ \theta_0 & 0 & \dfrac{2\lambda^2 R^2 T_{11}}{Eh} & -\dfrac{4\lambda^3 R^2 T_{12}}{Eh} \end{array}\right] \left[\begin{array}{c} 1 \\ \hline U_L \\ V_L \\ Z_L \end{array}\right]$$

where the effect of H_R, V_R, Z_R is assumed to be negligible.

(10) State vector at x'

$$\left[\begin{array}{c} 1 \\ \hline U \\ V \\ W \\ \hline X \\ Z \\ \hline u \\ v \\ \theta \end{array}\right] = \left[\begin{array}{c|ccc} 1 & 0 & 0 & 0 \\ \hline U_0 & 1 & 0 & 0 \\ 0 & 0 & T_{23} & 2\lambda T_{24} \\ W_0 & 0 & 2\lambda R T_{22} & 2\lambda^2 R T_{23} \\ \hline 0 & 0 & \dfrac{\nu T_{24}}{\lambda} & \nu T_{21} \\ 0 & 0 & \dfrac{T_{24}}{\lambda} & T_{21} \\ \hline u_0 & \dfrac{x}{Eh} & 0 & 0 \\ v_0 & \dfrac{\nu R}{Eh} & \dfrac{2\lambda R^2 T_{22}}{Eh} & \dfrac{2\lambda^2 R^2 T_{23}}{Eh} \\ \theta_0 & 0 & \dfrac{2\lambda^2 R^2 T_{21}}{Eh} & \dfrac{4\lambda^3 R^2 T_{22}}{Eh} \end{array}\right] \left[\begin{array}{c} 1 \\ \hline U_R \\ V_R \\ Z_R \end{array}\right]$$

where the effect of U_L, V_L, Z_L is assumed to be negligible.

28.09 SHORT AND LONG SHELLS, MEMBRANE STATE

Notation (28.01, 28.08) Signs (28.08) Matrix (28.08)

p_x, p_y = intensities of distributed load (N/m^2) $\eta = \nu R/h$
α_t = thermal coefficient (1/°C) s = shell length (m)
$f_x = a + bx$ = linear variation of temperature along x (°C)

1. 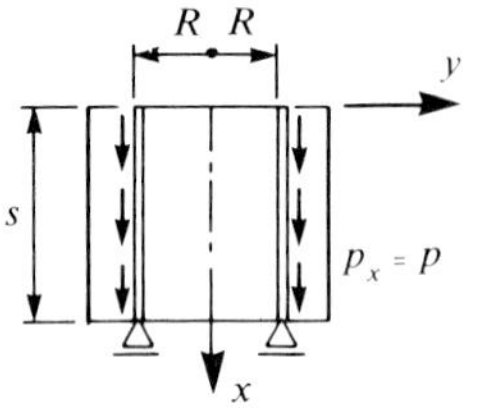	$U_0 = -xp$ $V_0 = Z_0 = 0$ $W_0 = 0$ $u_0 = (s^2 - x^2)\dfrac{p}{2Eh}$ $v_0 = \dfrac{x\eta p}{E}$ $\theta_0 = \dfrac{\eta p}{E}$
2. 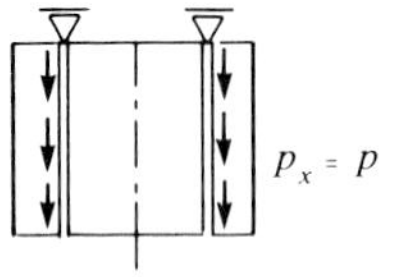	$U_0 = (s - x)p$ $V_0 = Z_0 = 0$ $W_0 = 0$ $u_0 = x(2s - x)\dfrac{p}{2Eh}$ $v_0 = -(s - x)\dfrac{\eta p}{E}$ $\theta_0 = -\dfrac{\eta p}{E}$
3. 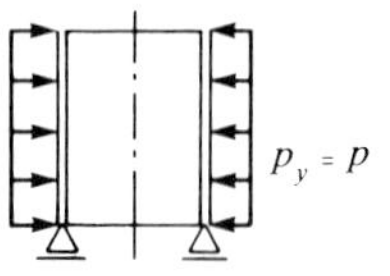	$U_0 = 0$ $V_0 = Z_0 = 0$ $W_0 = -Rp$ $u_0 = -(s - x)\dfrac{\eta p}{E}$ $v_0 = -\dfrac{R^2 p}{Eh}$ $\theta_0 = 0$
4. 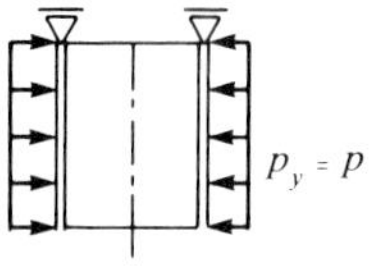	$U_0 = 0$ $V_0 = Z_0 = 0$ $W_0 = -Rp$ $u_0 = \dfrac{x\eta p}{E}$ $v_0 = -\dfrac{R^2 p}{Eh}$ $\theta_0 = 0$
5. 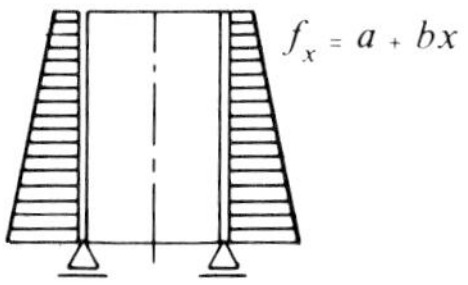	$U_0 = 0$ $V_0 = -Z_0 = 0$ $W_0 = 0$ $u_0 = \int_0^x \alpha_t f_x \, dx$ $v_0 = R\alpha_t f_x$ $\theta_0 = -R\alpha_t b$ Temperature inside = temperature outside

28.10 SHORT AND LONG SHELLS, MEMBRANE STATE

Notation (28.01, 28.08)	Signs (28.08)	Matrix (28.08)
p_x, p_y s, α_t (28.09) $\beta = \frac{c}{s}$ = constant (m^{-1})	$\eta = \frac{\nu R}{h}$	$D = \frac{Eh^3}{12(1-\nu^2)}$ (N·m)

$f_x = a + bx$ = linear variation of temperature along x (°C)

1.

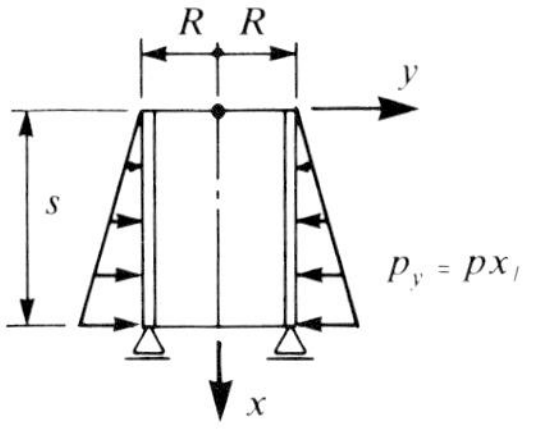

$U_0 = 0$ $\qquad u_0 = -(s^2 - x^2)\frac{\eta p}{2Es}$

$V_0 = Z_0 = 0$ $\qquad v_0 = -\frac{xR^2 p}{Ehs}$

$W_0 = -\frac{xRp}{s}$ $\qquad \theta_0 = -\frac{R^2 p}{Ehs}$

2.

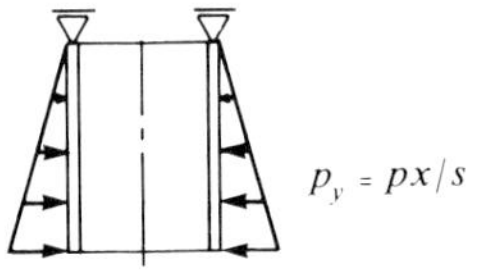

$U_0 = 0$ $\qquad u_0 = \frac{x(2s - x)\eta p}{2Es}$

$V_0 = Z_0 = 0$ $\qquad v_0 = -\frac{xR^2 p}{Ehs}$

$W_1 = -\frac{xRp}{s}$ $\qquad \theta_0 = -\frac{R^2 p}{Ehs}$

3.

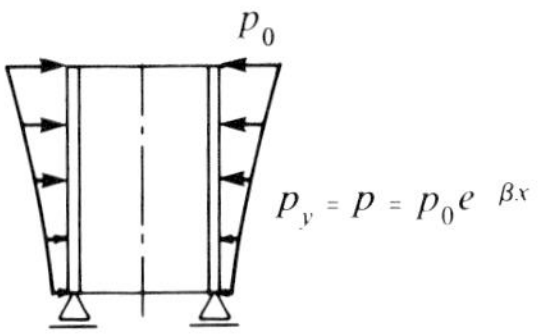

$U_0 = 0$ $\qquad u_0 = \frac{\eta}{E}\int_s^x p\,dx$

$V_0 = Z_0 = 0$ $\qquad v_0 = -\frac{R^2 p}{Eh}$

$W_0 = -Rp$ $\qquad \theta_0 = \frac{R^2 \beta p}{Eh}$

4.

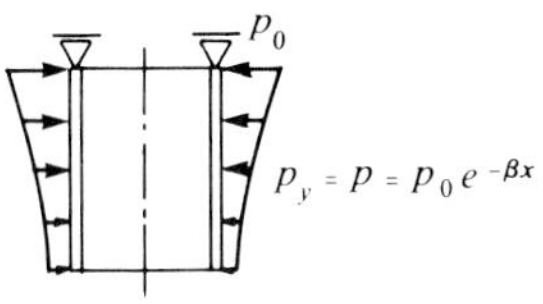

$U_0 = 0$ $\qquad u_0 = -\frac{\eta}{E}\int_0^x p\,dx$

$V_0 = Z_0 = 0$ $\qquad v_0 = -\frac{R^2 p}{Eh}$

$W_0 = -Rp$ $\qquad \theta_0 = -\frac{R^2 \beta p}{Eh}$

5.

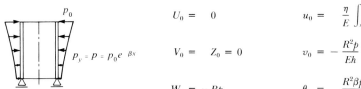

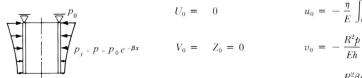

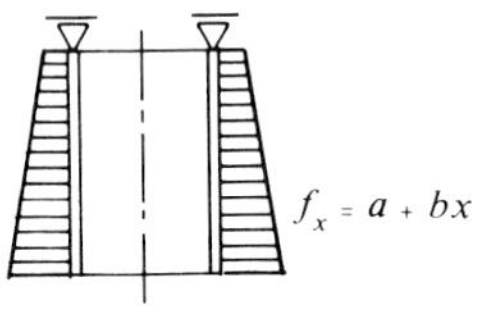

$U_0 = 0$ $\qquad u_0 = \int_0^x \alpha_t f_x\,dx$

$V_0 = Z_0 = 0$ $\qquad v_0 = R\alpha_t f_x$

$W_0 = 0$ $\qquad \theta_0 = R\alpha_t b$

28.11 SHORT AND LONG SHELLS, STIFFNESS MATRIX EQUATIONS

$$D = \frac{Eh^3}{12(1-\nu^2)}$$

$$\lambda = \sqrt[4]{\frac{3(1-\nu^2)}{R^2h^2}}$$

$$r = \frac{Ehs^2}{D}$$

(1) Equivalents and signs. The stiffness matrix equations of the short shell (28.02) and the long shell (28.08) are introduced below. All forces, moments, and displacements are in the x, y, z system, and positive directions are shown in (2, 3)

(2) Free-body diagram

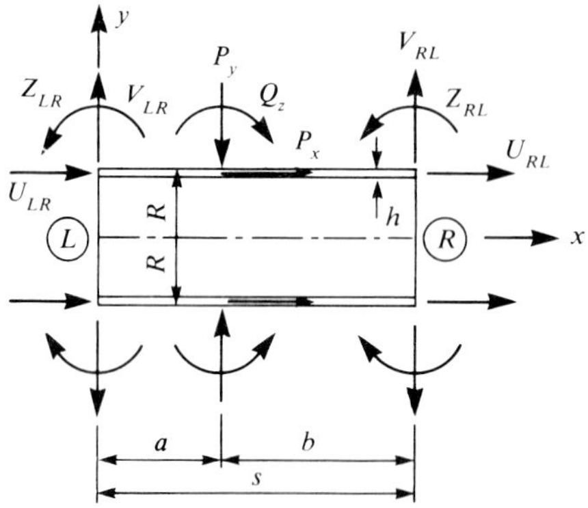

(3) Elastic surface

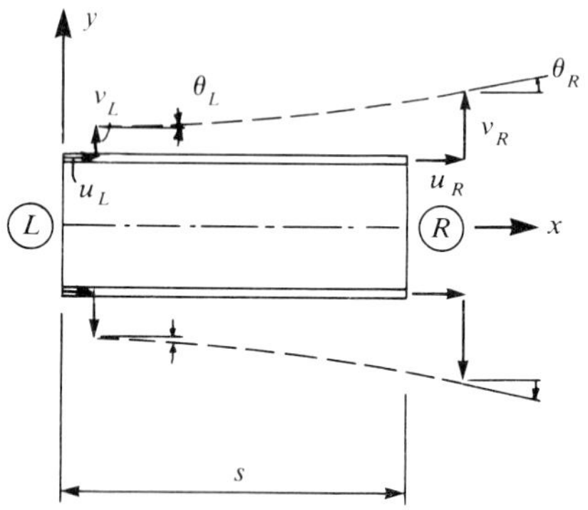

(4) Short shells, dimensionless stiffness coefficients

$$D_0 = \frac{1}{\sinh^2 \lambda s - \sin^2 \lambda s}$$

$L_2 = 2D_0 (\sinh 2\lambda s + \sin 2\lambda s)$ $L_3 = D_0 (\cosh 2\lambda s - \cos 2\lambda s)$ $L_4 = D_0 (\sinh 2\lambda s - \sin 2\lambda s)$	$K_2 = (\lambda s)^3 L_2$ $K_3 = (\lambda s)^2 L_3$ $K_4 = (\lambda s) L_4$
$L_6 = 4D_0 (\cosh \lambda s \sin \lambda s + \sinh \lambda s \cos \lambda s)$ $L_7 = 4D_0 \sinh \lambda s \sin \lambda s$ $L_8 = 2D_0 (\cosh \lambda s \sin \lambda s - \sinh \lambda s \cos \lambda s)$	$K_6 = (\lambda s)^3 L_6$ $K_7 = (\lambda s)^2 L_7$ $K_8 = (\lambda s) L_8$

(5) Short shells, stiffness matrix equations ($\lambda s \leq 5$)

$$\begin{bmatrix} U_{LR} \\ V_{LR} \\ Z_{LR} \\ \hline U_{RL} \\ V_{RL} \\ Z_{RL} \end{bmatrix} = \frac{D_0}{s^3} \left[\begin{array}{ccc|ccc} r & 0 & 0 & -r & 0 & 0 \\ 0 & K_2 & K_3 s & 0 & -K_6 & K_7 s \\ 0 & K_3 s & K_4 s^2 & 0 & -K_7 s & K_8 s^2 \\ \hline -r & 0 & 0 & r & 0 & 0 \\ 0 & -K_6 & -K_7 s & 0 & K_2 & -K_3 s \\ 0 & K_7 s & K_8 s^2 & 0 & -K_3 s & K_4 s^2 \end{array}\right] \begin{bmatrix} u_L \\ v_L \\ \theta_L \\ \hline u_R \\ v_R \\ \theta_R \end{bmatrix} + \begin{bmatrix} U_{L0} \\ V_{L0} \\ Z_{L0} \\ \hline U_{R0} \\ V_{R0} \\ Z_{R0} \end{bmatrix}$$

where U_{L0}, V_{L0}, Z_{L0} and U_{R0}, V_{R0}, Z_{R0} are the *reactions of fixed-end shells* acted on by loads and other causes. The analytical forms of these reactions are given in (6) on the opposite page. The graphs and tables of stiffness coefficients are identical to those in (8.10–8.11).

(6) Short shells, stiffness load functions introduced symbolically in (5) are given by the matrix equations

$$\begin{bmatrix} U_{L0} \\ V_{L0} \\ Z_{L0} \end{bmatrix} = \begin{bmatrix} h/s & 0 & 0 \\ 0 & L_6 & -L_7 \\ 0 & L_7\lambda^{-1} & -L_8\lambda^{-1} \end{bmatrix} \begin{bmatrix} G_u \\ G_v \\ G_\theta \end{bmatrix} \qquad \begin{bmatrix} U_{R0} \\ V_{R0} \\ Z_{R0} \end{bmatrix} = \begin{bmatrix} h/s & 0 & 0 \\ 0 & L_6 & L_7 \\ 0 & -L_7\lambda^{-1} & -L_8\lambda^{-1} \end{bmatrix} \begin{bmatrix} F_u \\ F_v \\ F_\theta \end{bmatrix}$$

where F_u, F_v, F_θ are the left-end transport load functions and G_u, G_v, G_θ are their right-end counterparts, all given in particular forms in (28.04–28.07), and L_6, L_7 L_8 are the stiffness coefficients listed in (4) on the opposite page.

(7) Long shells, left-end stiffness matrix equation ($\lambda s > 5$)

$$\begin{bmatrix} U_{LR} \\ V_{LR} \\ Z_{LR} \end{bmatrix} = D \begin{bmatrix} rs^{-3} & 0 & 0 \\ 0 & 4\lambda^3 & 2\lambda^2 \\ 0 & 2\lambda^2 & 2\lambda \end{bmatrix} \begin{bmatrix} u_L \\ v_L \\ \theta_L \end{bmatrix} + \begin{bmatrix} U_{L0} \\ V_{L0} \\ Z_{L0} \end{bmatrix}$$

where U_{L0}, V_{L0}, Z_{L0} are the *reactions of fixed-end shells* at L acted on by loads and other causes. The anayltical forms of these reactions are given in (28.12)

(8) Long shells, right-end stiffness matrix equation ($\lambda s > 5$)

$$\begin{bmatrix} U_{RL} \\ V_{RL} \\ Z_{RL} \end{bmatrix} = D \begin{bmatrix} rs^{-3} & 0 & 0 \\ 0 & 4\lambda^3 & -2\lambda^2 \\ 0 & -2\lambda^2 & 2\lambda \end{bmatrix} \begin{bmatrix} u_R \\ v_R \\ \theta_R \end{bmatrix} + \begin{bmatrix} U_{R0} \\ V_{R0} \\ Z_{R0} \end{bmatrix}$$

where U_{R0}, V_{R0}, Z_{R0} are the *reactions of fixed-end shells* at R acted on by loads and other causes. The analytical forms of these reactions are given in (28.12).

(9) Elastic stiffener stiffness matrix equation in terms of

$$R^* = R\left(1 + \nu + \frac{m^2}{2 + m}\right)$$

is

$$\begin{bmatrix} U_j \\ V_j \\ Z_j \end{bmatrix} = \frac{Ed}{R^*} \begin{bmatrix} 0 & 0 & 0 \\ 0 & 1 & 0 \\ 0 & 0 & \dfrac{d^2}{12} \end{bmatrix} \begin{bmatrix} u_j \\ v_j \\ \theta_j \end{bmatrix}$$

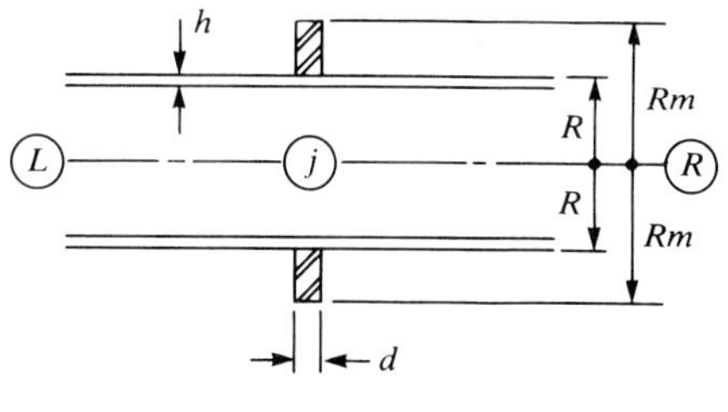

where Rm = exterior radius of stiffener (m), d = stiffener thickness (m), and R = radius of shell (m). The joint equilibrium equation at j is then

$$U_{jL} + U_j + U_{jR} = 0 \qquad V_{jL} + V_j + V_{jR} = 0 \qquad Z_{jL} + Z_j + Z_{jR} = 0$$

where U_j, V_j, Z_j are the forces and moments given above and U_{jL}, V_{jL}, Z_{jL}, U_{jR}, V_{jR}, Z_{jR} are the forces and moments given in (28.11-5) for the shell segments Lj, jR, respectively.

28.12 LONG SHELLS, STIFFNESS LOAD FUNCTIONS

Notation (28.08–28.11)	Signs (28.11)	Matrix (28.11)
$U_{L0} = \frac{1}{2}sp$ $V_{L0} = \dfrac{\nu p}{2R\lambda^2}$ $Z_{L0} = \dfrac{\nu p}{2R\lambda^3}$	**1.** $p_x = p$ 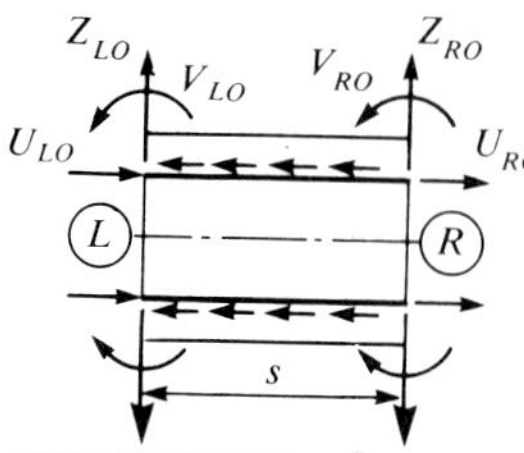	$U_{R0} = \frac{1}{2}ps$ $V_{R0} = -\dfrac{\nu p}{2R\lambda^2}$ $Z_{R0} = -\dfrac{\nu p}{2R\lambda^3}$
$U_{L0} = \nu Rp$ $V_{L0} = \dfrac{p}{\lambda}$ $Z_{L0} = \dfrac{p}{2\lambda^2}$	**2.** $p_y = p$ 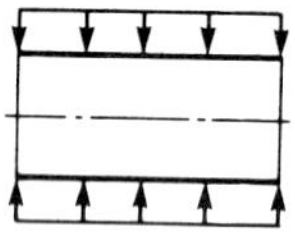	$U_{R0} = -\nu Rp$ $V_{R0} = \dfrac{p}{\lambda}$ $Z_{R0} = -\dfrac{p}{2\lambda^2}$
$U_{L0} = \frac{1}{2}\nu Rp$ $V_{L0} = \dfrac{p}{2s\lambda^2}$ $Z_{L0} = \dfrac{p}{2s\lambda^3}$	**3.** $p_y = xp/s$ 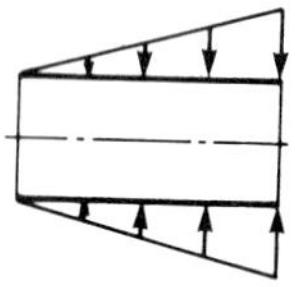	$U_{R0} = -\frac{1}{2}\nu Rp$ $V_{R0} = \dfrac{(2\lambda s - 1)p}{2s\lambda^2}$ $Z_{R0} = -\dfrac{(\lambda s - 1)p}{2s\lambda^3}$
$U_{L0} = \dfrac{\nu R(1 - e^{-\beta s})}{\beta s}p$ $V_{L0} = \dfrac{(2\lambda - \beta)e^{-\beta s}}{2\lambda^2}p$ $Z_{10} = \dfrac{(\lambda - \beta)e^{-\beta s}}{2\lambda^3}p$	**4.** $p_y = pe^{-\beta x}$ 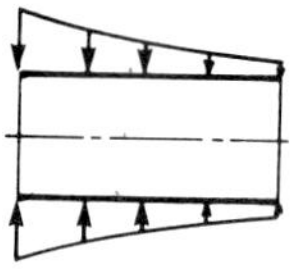	$U_{R0} = -\dfrac{\nu R(1 - e^{-\beta s})}{\beta s}p$ $V_{R0} = \dfrac{(2\lambda - \beta)e^{-\beta s}}{2\lambda^2}p$ $Z_{R0} = -\dfrac{(\lambda - \beta)e^{-\beta s}}{2\lambda^3}p$
$U_{L0} = \frac{1}{2}Eh\alpha_t(2a + bs)$ $V_{L0} = -\dfrac{Eh\alpha_t(2\lambda a + b)}{2R\lambda^2}$ $Z_{L0} = -\dfrac{Eh\alpha_t(\lambda a + b)}{2R\lambda^3}$	**5. Linear variation of temperature** $f_x = a + bx$ **(°C)** 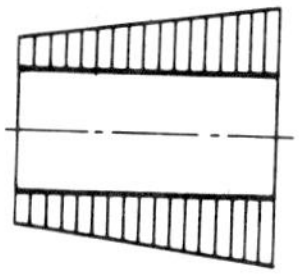	$U_{R0} = -\frac{1}{2}Eh\alpha_t(2a + bs)$ $V_{R0} = -\dfrac{Eh\alpha_t(2\lambda a - b + 2\lambda bs)}{2R\lambda^2}$ $Z_{R0} = -\dfrac{Eh\alpha_t(b - \lambda a - \lambda bs)}{2R\lambda^3}$

29

Spherical Shell Constant Thickness

STATIC STATE

29.01 DEFINITION OF STATE

(1) System considered is an isotropic spherical shell of constant thickness in a state of static axisymmetrical deformation.

(2) Notation (A.1, A.6)

a, b, α, β, s = angles (rad)
R = constant radius
$U = N_1$ = normal force along 1 (N/m)
$W = N_2$ = normal force along 2 (N/m)
H = horizontal force along y (N/m)
$X = M_2$ = bending moment about 1(N·m/m)
E = modulus of elasticity (Pa)
ϕ = coordinate (rad)
h = thickness (m)
$V = -N_3$ = shearing force against 3 (N/m)
$\Delta = \Delta_y$ = deflection along y (m)
$\theta = -\psi_1$ = slope about 2 (rad)
$Z = M_1$ = bending moment about 2(N·m/m)
ν = Poisson's ratio

All forces and moments are per unit length of normal section. $N_1, N_2, N_3, M_1, M_2, \psi_1$ defined in (27.02) are in the 1, 2, 3 system; H, Z, Δ, θ defined in (29.02) are in the x, y, z system.

$K = \dfrac{Eh}{1-\nu^2}$ (N/m)	$D = \dfrac{Eh^3}{12(1-\nu^2)}$ (N·m)	$\lambda = \sqrt[4]{3(1-\nu^2)\left(\dfrac{R}{h}\right)^2}$

(3) Assumptions of analysis are stated in (27.01). The shells are classified as short ($\lambda s \leq 5$) and long ($\lambda s > 5$).

29.02 SHORT OPEN SHELL, TRANSPORT MATRIX EQUATIONS

(1) Equivalents and signs ($\lambda s \le 5$)

$$\bar{H} = H \sin\phi = \text{scaled } H \text{ (N/m)} \qquad \bar{Z} = \frac{4\lambda}{R} Z = \text{scaled } Z \text{ (N/m)}$$

$$\bar{\Delta} = \frac{Eh\Delta}{\lambda R \sin\phi} = \text{scaled } \Delta \text{ (N/m)} \qquad \bar{\theta} = \frac{Eh\theta}{\lambda^2} = \text{scales } \theta \text{ (N/m)}$$

where $\phi = \alpha$ at R and $\phi = \beta$ at L. All forces and moments are in the x, y, z axes, and their positive directions are shown in (2, 3) below.

(2) Free-body diagram

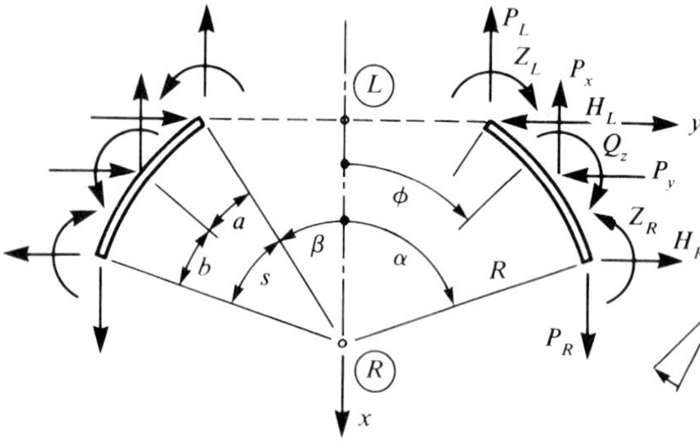

(3) Elastic surface

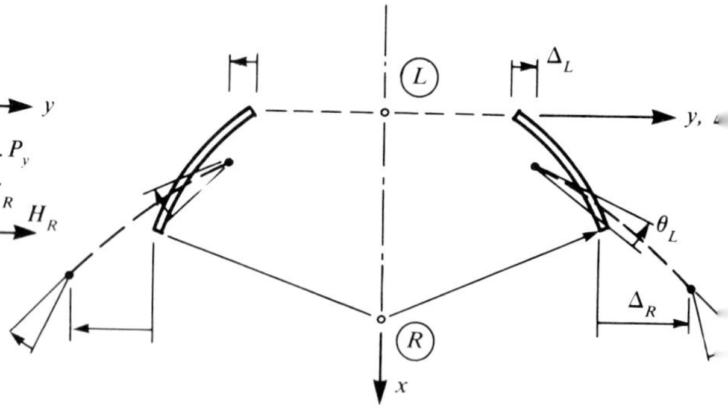

(4) Transport matrix equations

$$\underbrace{\begin{bmatrix} 1 \\ \hline \bar{H}_L \\ \bar{Z}_L \\ \hline \bar{\Delta}_L \\ \bar{\theta}_L \end{bmatrix}}_{H_L} = \underbrace{\left[\begin{array}{c|cc|cc} 1 & 0 & 0 & 0 & 0 \\ \hline F_H & T_1 & T_4 & -T_2 & T_3 \\ F_Z & -4T_2 & T_1 & 4T_3 & -4T_4 \\ \hline F_\Delta & 4T_4 & -T_3 & T_1 & -T_2 \\ F_\theta & -4T_3 & T_2 & 4T_4 & T_1 \end{array}\right]}_{T_{LR}} \underbrace{\begin{bmatrix} 1 \\ \hline \bar{H}_R \\ \bar{Z}_R \\ \hline \bar{\Delta}_R \\ \bar{\theta}_R \end{bmatrix}}_{H_R}$$

$$\underbrace{\begin{bmatrix} 1 \\ \hline \bar{H}_R \\ \bar{Z}_R \\ \hline \bar{\Delta}_R \\ \bar{\theta}_R \end{bmatrix}}_{H_R} = \underbrace{\left[\begin{array}{c|cc|cc} 1 & 0 & 0 & 0 & 0 \\ \hline G_H & T_1 & -T_4 & T_2 & T_3 \\ G_Z & 4T_2 & T_1 & 4T_3 & 4T_4 \\ \hline G_\Delta & -4T_4 & -T_3 & T_1 & T_2 \\ G_\theta & -4T_3 & -T_2 & -4T_4 & T_1 \end{array}\right]}_{T_{RL}} \underbrace{\begin{bmatrix} 1 \\ \hline \bar{H}_L \\ \bar{Z}_L \\ \hline \bar{\Delta}_L \\ \bar{\theta}_L \end{bmatrix}}_{H_L}$$

where H_L, H_R are the *state vectors* of the respective ends and T_{LR}, T_{RL} are the *transport matrices*. The elements of the transport matrices are the *transport coefficients* T_j defined in (29.03) and the *load functions* F, G displayed for particular cases in (29.05–29.07).

29.03 SHORT OPEN SHELL, TRANSPORT RELATIONS

(1) Transport functions

$$\lambda = \sqrt[4]{3(1-\nu^2)\left(\frac{R}{h}\right)^2}$$

$T_1(\eta) = \cosh \lambda\eta \cos \lambda\eta$	$T_2(\eta) = \frac{1}{2}(\cosh \lambda\eta \sin \lambda\eta + \sinh \lambda\eta \cosh \lambda\eta)$
$T_3(\eta) = \frac{1}{2} \sinh \lambda\eta \sin \lambda\eta$	$T_4(\eta) = \frac{1}{4}(\cosh \lambda\eta \sin \lambda\eta - \sinh \lambda\eta \cosh \lambda\eta)$
$T_5(\eta) = \frac{1}{4}[1 - T_1(\eta)]$	$T_6(\eta) = \frac{1}{4}[\lambda\eta - T_2(\eta)]$

The graphs and tables of these functions are shown in (8.04–8.05).

(2) Transport coefficients T_j, A_j, B_j, C_j, are the values of $T_j(\eta)$ for particular arguments η.

$$\left.\begin{array}{ll} \text{For } \eta = s, & T_j(s) = T_j \\ \text{For } \eta = a, & T_j(a) = A_j \\ \text{For } \eta = b, & T_j(b) = B_j \\ \text{and } T_j - A_j = C_j & \end{array}\right\} j = 1, 2, \ldots, 6 \qquad \eta = \phi - \beta \text{ or } \eta = \alpha - \phi$$

(3) Transport matrices of this shell have analogical properties to those of (8.02) and can be used in similar ways.

(4) Differential relations

$$D = \frac{Eh^3}{12(1-\nu^2)}$$

$\Delta = \Delta_0 + \frac{R \sin\phi}{Eh}\frac{dV}{d\phi}$	
$\theta = \theta_0 + \frac{1}{Eh}\frac{d^2V}{d\phi^2}$	$H = \frac{V}{\sin\phi}$
$Z = -\frac{D}{EhR}\frac{d^3V}{d\phi^3}$	$X \cong \frac{\nu D}{EhR}\frac{d^3V}{d\phi^3}$

where Δ_0, θ_0 are the membrane state displacements at ϕ.

(5) Conversion relations (27.06)

$N_1 = U_0 + V \cot\phi$	$\psi_1 = -\theta_0 - \frac{1}{Eh}\frac{d^2V}{d\phi^2} = -\theta$
$N_2 = W_0 + \frac{dV}{d\phi}$	$M_1 = -\frac{D}{EhR}\frac{d^3V}{d\phi^3} = Z$
$N_3 = -V = -H \sin\phi$	$M_2 \cong \nu M_1 = X$

where U_0, W_0, Δ_0, θ_0 are the membrane state forces and displacements at ϕ.

29.04 SHORT OPEN SHELL, TRANSPORT LOAD FUNCTIONS

(1) Load Functions F and G in (29.02-4) are either the convolution integrals given in (29.05) or are approximated by the matrix equations given below.

$$\begin{bmatrix} F_H \\ F_Z \\ \hline F_\Delta \\ F_\theta \end{bmatrix} = \left[\begin{array}{cc|cc} T_2 & -T_3 & 0 & 0 \\ -4T_3 & 4T_4 & 0 & 0 \\ \hline -T_1 & T_2 & 1 & 0 \\ -4T_4 & -T_1 & 0 & 1 \end{array}\right] \begin{bmatrix} \overline{\Delta}_{0\alpha} \\ \overline{\theta}_{0\alpha} \\ \hline \overline{\Delta}_{0\beta} \\ \overline{\theta}_{0\beta} \end{bmatrix}$$

$$\begin{bmatrix} G_H \\ G_Z \\ \hline G_\Delta \\ G_\theta \end{bmatrix} = \left[\begin{array}{cc|cc} 0 & 0 & -T_2 & -T_3 \\ 0 & 0 & -4T_3 & -4T_4 \\ \hline 1 & 0 & -T_1 & -T_2 \\ 0 & 1 & 4T_4 & -T_1 \end{array}\right] \begin{bmatrix} \overline{\Delta}_{0\alpha} \\ \overline{\theta}_{0\alpha} \\ \hline \overline{\Delta}_{0\beta} \\ \overline{\theta}_{0\beta} \end{bmatrix}$$

where

$$\overline{\Delta}_{0\alpha} = \frac{Eh}{\lambda R \sin\alpha} \Delta_{0\alpha} \qquad \overline{\Delta}_{0\beta} = \frac{Eh}{\lambda R \sin\beta} \Delta_{0\beta}$$

$$\overline{\theta}_{0\alpha} = \frac{Eh}{\lambda^2} \theta_{0\alpha} \qquad \overline{\theta}_{0\beta} = \frac{Eh}{\lambda^2} \theta_{0\beta}$$

and $\Delta_{0\alpha}$, $\theta_{0\alpha}$, $\Delta_{0\beta}$, $\theta_{0\beta}$ are the horizontal displacements and slopes at $\phi = \alpha$, $\phi = \beta$, respectively, in the corresponding membrane state (29.06–29.07).

(2) Membrane state vectors (27.03) are abbreviated in this chapter as

$U_0 = N_{10}$ = meridian force along 1 (N/m)
$V_0 = -N_{30}$ = shear force along 3 (N/m)
$W_0 = N_{20}$ = circumference force along 2 (N/m)
$\Delta_0 = \Delta_{10}$ = linear displacement along x (m)
$\theta_0 = -\psi_{10}$ = slope about 2 (rad)

(3) Symbols used in (29.05–29.07) are

p = intensity of distributed load (N/m²)
P = concentrated ring load (N/m)
Q = applied ring couple (N·m/m)
q = intensity of distributed couple (N·m/m²)
α_t = thermal coefficient (1/°C) $\quad e_t$ = thermal strain
Δt^* = temperature change (°C) $\quad \gamma$ = weight density (N/m³)
$S_\phi = \sin\phi \quad S_\alpha = \sin\alpha \quad S_\beta = \sin\beta$
$C_\phi = \cos\phi \quad C_\alpha = \cos\alpha \quad C_\beta = \cos\beta$

where $0° < \alpha \leq 180°$, $0° \leq \beta < 180°$, $\beta \leq \phi \leq \alpha$ but $\lambda s \leq 5$.

29.05 OPEN SHORT SHELL, TRANSPORT LOAD FUNCTIONS

Notation (29.01–29.03)	Signs (29.02)	Matrix (29.02)
$F_H = -A_1 P_3$ $F_Z = 4A_2 P_3$ $F_\Delta = -4A_4 P_3$ $F_\theta = 4A_3 P_3$	**1.** 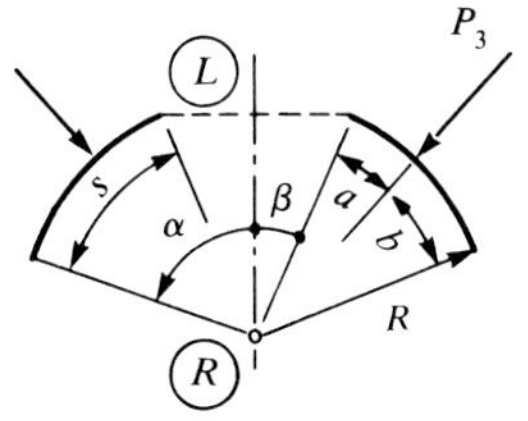	$G_H = B_1 P_3$ $G_Z = 4B_2 P_3$ $G_\Delta = -4B_4 P_3$ $G_\theta = -4B_3 P_3$
$F_H = A_4 \lambda R^{-1} Q_2$ $F_Z = A_1 \lambda R^{-1} Q_2$ $F_\Delta = -A_3 \lambda R^{-1} Q_2$ $F_\theta = A_2 \lambda R^{-1} Q_2$	**2.** 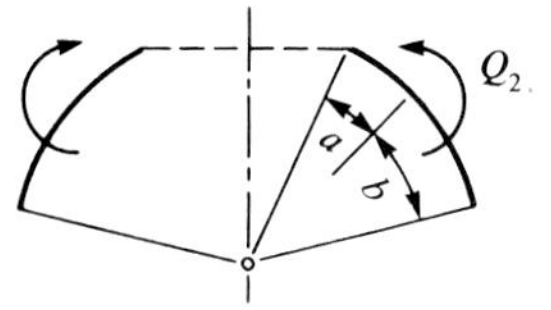	$G_H = B_4 \lambda R^{-1} Q_2$ $G_Z = -B_1 \lambda R^{-1} Q_2$ $G_\Delta = B_3 \lambda R^{-1} Q_2$ $G_\theta = B_2 \lambda R^{-1} Q_2$
$F_H = C_5 q_2$ $F_Z = C_2 q_2$ $F_\Delta = -C_4 q_2$ $F_\theta = C_3 q_2$	**3.** 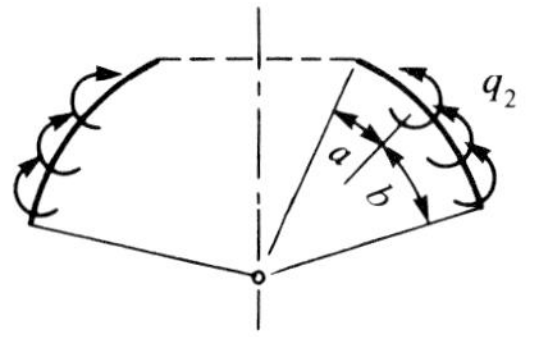	$G_H = B_5 q_2$ $G_Z = -B_2 q_2$ $G_\Delta = B_4 q_2$ $G_\theta = B_3 q_2$
$F_H = -C_2 \lambda^{-1} R p_3$ $F_Z = 4C_3 \lambda^{-1} R p_3$ $F_\Delta = -4C_5 \lambda^{-1} R p_3$ $F_\theta = 4C_4 \lambda^{-1} R p_3$	**4.** 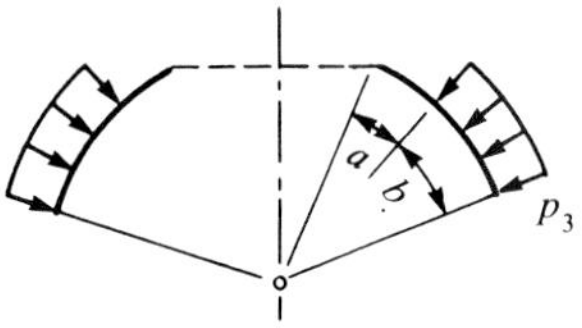	$G_H = B_2 \lambda^{-1} R p_3$ $G_Z = 4B_3 \lambda^{-1} R p_3$ $G_\Delta = -4B_5 \lambda^{-1} R p_3$ $G_\theta = -4B_4 \lambda^{-1} R p_3$
$F_H = -(\lambda s T_2 - T_3)\lambda^{-2} s^{-1} R p_3$ $F_Z = 4(\lambda s T_3 - T_4)\lambda^{-2} s^{-1} R p_3$ $F_\Delta = -4(\lambda s T_5 - T_6)\lambda^{-2} s^{-1} R p_3$ $F_\theta = 4(\lambda s T_4 - T_5)\lambda^{-2} s^{-1} R p_3$	**5.** $p = \frac{\eta}{s} p_3$ 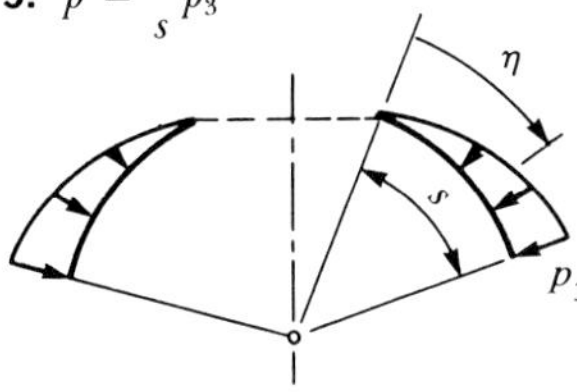	$G_H = T_3 \lambda^{-2} s^{-1} R p_3$ $G_Z = 4T_4 \lambda^{-2} s^{-1} R p_3$ $G_\Delta = -4T_6 \lambda^{-2} s^{-1} R p_3$ $G_\theta = -4T_5 \lambda^{-2} s^{-1} R p_3$

29.06 OPEN SHELL, MEMBRANE STATE

Notation (27.03, 29.04)		Signs (27.03, 29.04)	Matrix (29.04, 29.08)
1. 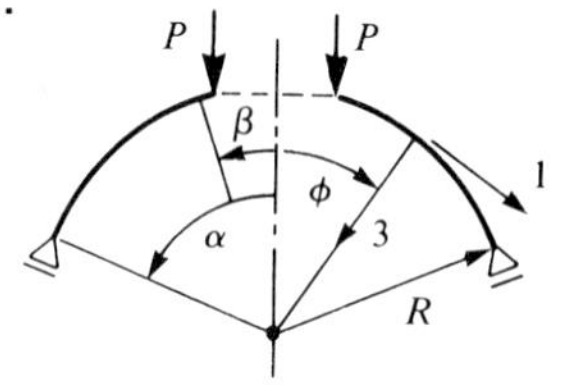		$U_0 = -\dfrac{S_\beta}{S_\phi^2}P$ $W_0 = -U_0$ $\theta_0 = 0$	$V_0 = X_0 = Z_0 = 0$ $\Delta_0 = (W_0 - \nu U_0)\dfrac{RS_\phi}{Eh}$
2. 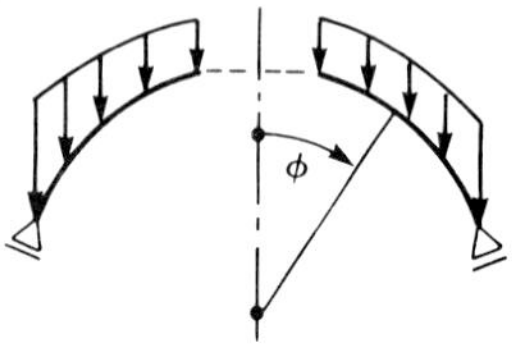	$p_1 = pS_\phi$ $p_3 = pC_\phi$	$U_0 = -\dfrac{C_\beta - C_\phi}{S_\phi^2}Rp$ $W_0 = -U_0 - C_\phi Rp$ $\theta_0 = (2+\nu)S_\phi \dfrac{Rp}{Eh}$	
3. 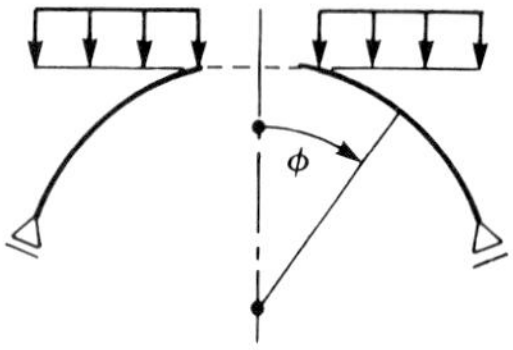	$p_1 = pS_\phi C_\phi$ $p_3 = pC_\phi^2$	$U_0 = -\frac{1}{2}\left(1 - \dfrac{S_\beta^2}{S_\phi^2}\right)Rp$ $W_0 = -U_0 - C_\phi^2 Rp$ $\theta_0 = (3+\nu)S_\phi C_\phi \dfrac{Rp}{Eh}$	
4. 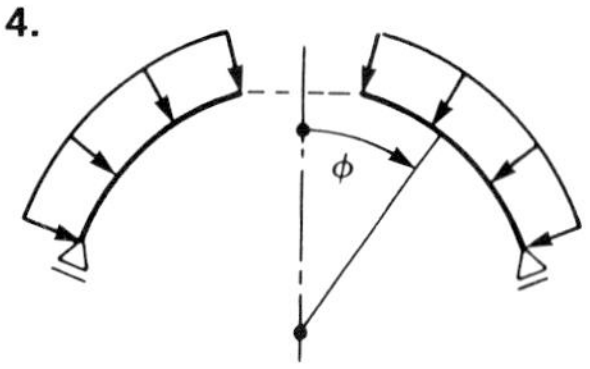	$p_1 = 0$ $p_3 = p$	$U_0 = -\frac{1}{2}\left(1 - \dfrac{S_\beta^2}{S_\phi^2}\right)Rp$ $W_0 = -U_0 - Rp$ $\theta_0 = 0$	
5. 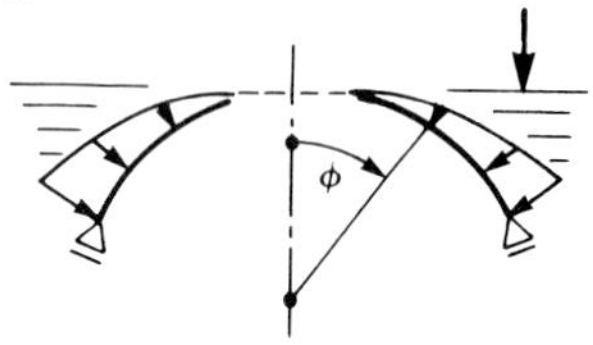	$p_1 = 0$ $p_3 = \gamma R(C_\beta - C_\phi)$	$U_0 = -\frac{1}{6}\left(C_\beta^3 - 3C_\beta C_\phi^2 + 3C_\phi^3\right)\dfrac{R^2\gamma}{S_\phi^2}$ $W_0 = -U_0 - (C_\beta - C_\phi)R^2\gamma$ $\theta_0 = S_\phi R^2 \dfrac{\gamma}{Eh}$	
6. 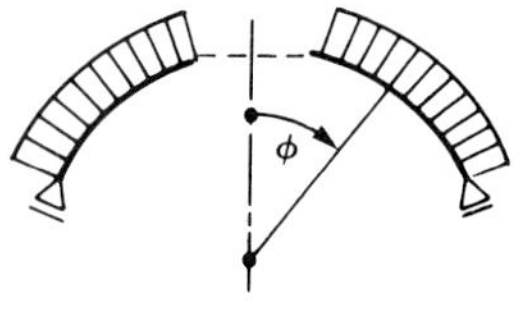	$e_t = \alpha_t \Delta t^*$	$U_0 = V_0 = W_0 = X_0 = Z_0 = 0$ $\Delta_0 = e_t RS_\phi$ $\theta_0 = 0$	

Notation (27.03, 29.04)		Signs (27.03, 29.04)		Matrix (29.04, 29.08)
1.	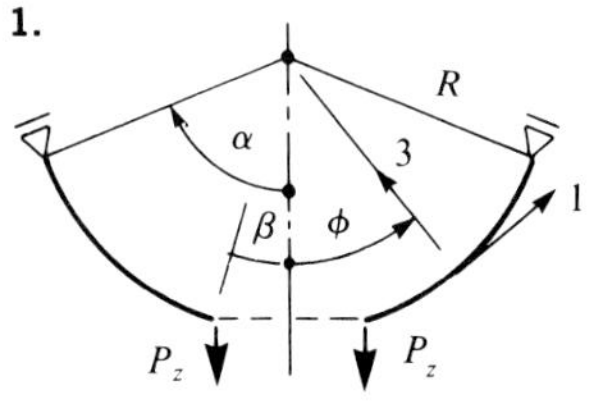		$U_0 = \dfrac{S_\beta}{S_\phi} P_z$ $W_0 = -U_0$ $\theta_0 = 0$	$\Delta_0 = (W_0 - \nu U_0)\dfrac{RS_\phi}{Eh}$ $V_0 = X_0 = Z_0 = 0$
2.	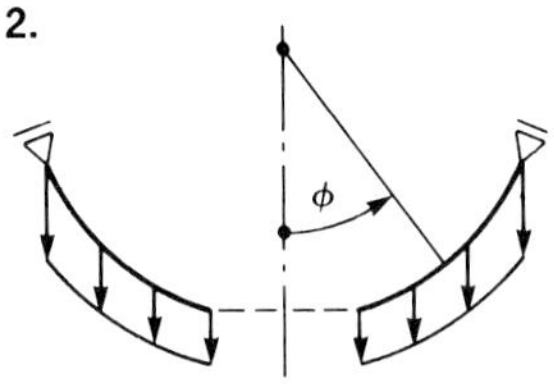	$p_1 = -pS_\phi$ $p_3 = -pC_\phi$	$U_0 = \dfrac{C_\beta - C_\phi}{S_\phi^2} Rp$ $W_0 = -U_0 + C_\phi Rp$ $\theta_0 = (2+\nu)S_\phi \dfrac{Rp}{Eh}$	
3.	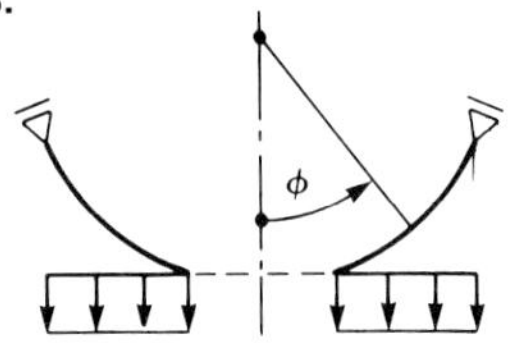	$p_1 = -pS_\phi C_\phi$ $p_3 = -pC_\phi^2$	$U_0 = \frac{1}{2}\left(1 - \dfrac{S_\beta^2}{S_\phi^2}\right) Rp$ $W_0 = -U_0 + C_\phi^2 Rp$ $\theta_0 = (3+\nu)S_\phi C_\phi \dfrac{Rp}{Eh}$	
4.	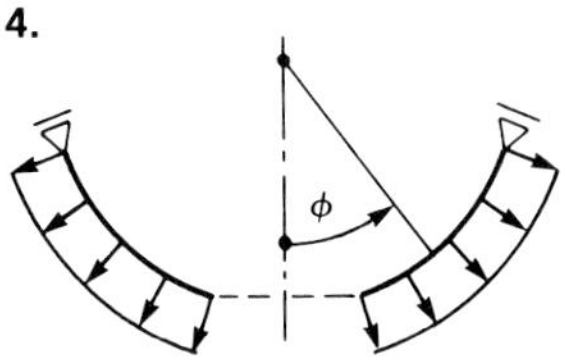	$p_1 = 0$ $p_3 = -p$	$U_0 = \frac{1}{2}\left(1 - \dfrac{S_\beta^2}{S_\phi^2}\right) Rp$ $W_0 = -U_0 + Rp$ $\theta_0 = 0$	
5.	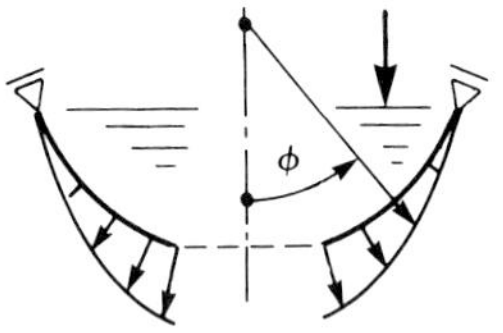	$p_1 = 0$ $p_3 = -\gamma R(C_\beta - C_\phi)$	$U_0 = -\frac{1}{6}\left(C_\beta^3 - 3C_\beta C_\phi^2 + 3C_\phi^3\right)\dfrac{R^2\gamma}{S_\phi^2}$ $W_0 = -U_0 + (C_\beta - C_\phi)R^2\gamma$ $\theta_0 = S_\phi R^2 \dfrac{\gamma}{Eh}$	
6.	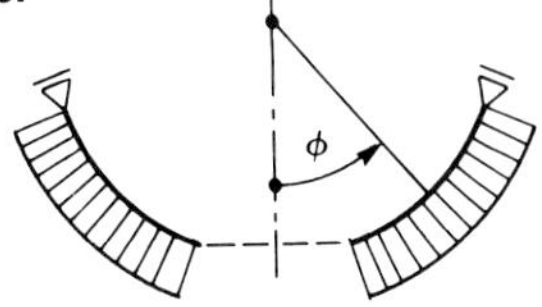	$e_t = \alpha_t \Delta t^*$	$U_0 = V_0 = W_0 = X_0 = Z_0 = 0$ $\Delta_0 = e_t RS_\phi$ $\theta_0 = 0$	

29.08 LONG SHELL, STATE-VECTOR MATRIX EQUATIONS

$$\lambda = \sqrt[4]{3(1-\nu^2)\left(\frac{R}{h}\right)^2}$$

$$\lambda s \geq 5$$

(1) State vectors in long shells in a general axisymmetrical state are given in (7, 8). The symbols used are the same as in (29.03). Positive forces, moments, and displacements are shown in (2–4).

(2) Forces

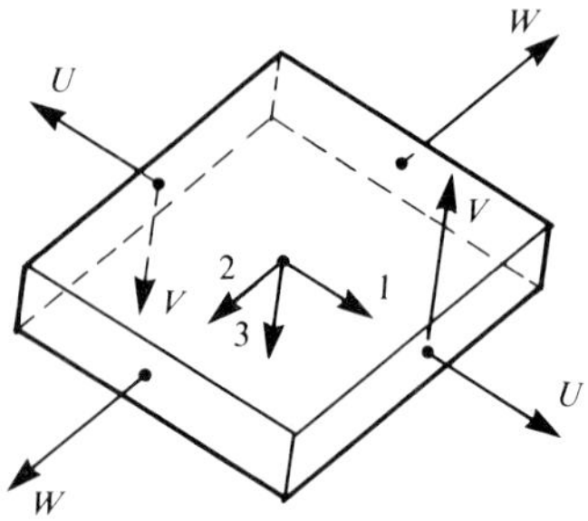

(3) Moments

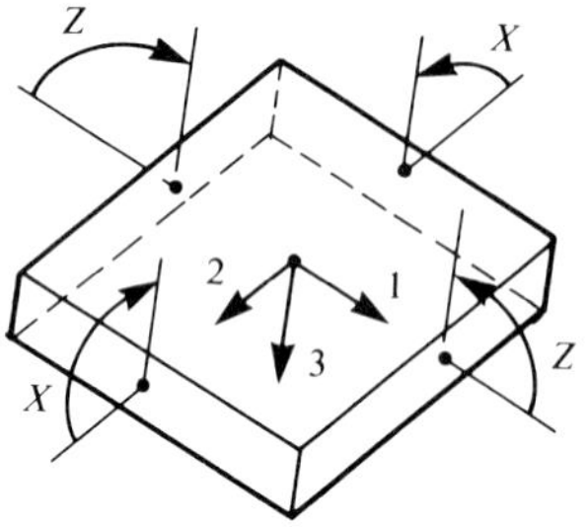

(4) Displacements

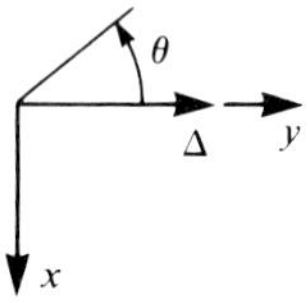

(5) Shape functions and equivalents. For $\phi_1 = \phi - \beta$,

$$T_{11} = e^{-\lambda\phi_1}(\cos\lambda\phi_1 + \sin\lambda\phi_1) \qquad T_{12} = e^{-\lambda\phi_1}\cos\lambda\phi_1$$
$$T_{13} = e^{-\lambda\phi_1}(\cos\lambda\phi_1 - \sin\lambda\phi_1) \qquad T_{14} = e^{-\lambda\phi_1}\sin\lambda\phi_1$$

For $\phi_2 = \alpha - \phi$,

$$T_{21} = e^{-\lambda\phi_2}(\cos\lambda\phi_2 + \sin\lambda\phi_2) \qquad T_{22} = e^{-\lambda\phi_2}\cos\lambda\phi_2$$
$$T_{23} = e^{-\lambda\phi_2}(\cos\lambda\phi_2 - \sin\lambda\phi_2) \qquad T_{24} = e^{-\lambda\phi_2}\sin\lambda\phi_2$$

All angles ϕ, α, β are positive scalars shown in (29.02-2), and their trigonometric functions are abbreviated as S_ϕ, C_ϕ, S_α, C_α, S_β, C_β (29.04-3). The membrane vectors U_0, W_0, Δ_0, θ_0 are given for open shells in (29.06–29.07) and for closed shells in (29.09–29.10). The state-vector equations in (7, 8) apply in the analysis of open and closed shells in limits $0 < \alpha \leq 180°$ and $0 \leq \beta < 180°$, but in cases of $0 < \alpha \leq 30°$, the results are only poor approximations (Ref. 51, p. 63).

(6) Free-body diagram

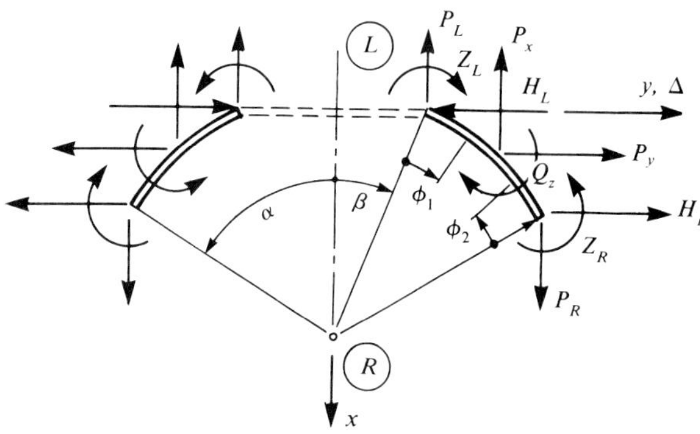

(7) State vector at $\phi_1 = \phi - \beta$

$$
\left[\begin{array}{c} 1 \\ \hline U \\ V \\ W \\ \hline X \\ Z \\ \hline \Delta \\ \theta \end{array}\right]
=
\left[\begin{array}{c|ccc}
1 & 0 & 0 & 0 \\ \hline
U_0 & \dfrac{S_\beta C_\phi}{S_\phi} T_{13} & -\dfrac{C_\beta C_\phi}{S_\phi} T_{13} & -\dfrac{2\lambda C_\phi}{R S_\phi} T_{14} \\
0 & S_\beta T_{13} & -C_\beta T_{13} & \dfrac{2\lambda}{R} T_{14} \\
W_0 & -2\lambda S_\beta T_{12} & 2\lambda C_\beta T_{12} & \dfrac{2\lambda^2}{R} T_{13} \\ \hline
0 & -\dfrac{\nu R}{\lambda} S_\beta T_{14} & \dfrac{\nu R}{\lambda} C_\beta T_{14} & \nu T_{11} \\
0 & -\dfrac{R}{\lambda} S_\beta T_{14} & \dfrac{R}{\lambda} C_\beta T_{14} & T_{11} \\ \hline
\Delta_0 & -\dfrac{2\lambda R}{Eh} S_\beta^2 T_{12} & \dfrac{2\lambda R}{Eh} S_\beta C_\beta T_{12} & \dfrac{2\lambda^2}{Eh} S_\beta T_{13} \\
\theta_0 & \dfrac{2\lambda^2}{Eh} S_\beta T_{11} & -\dfrac{2\lambda^2}{Eh} C_\beta T_{11} & -\dfrac{4\lambda^3}{EhR} T_{12}
\end{array}\right]
\left[\begin{array}{c} 1 \\ \hline H_L \\ P_L \\ Z_L \end{array}\right]
$$

where U_0, W_0, Δ_0, θ_0 are the membrane vectors at ϕ_1 (29.06–29.07, 29.09–29.10) and the effects of H_R, P_R, Z_R are assumed to be negligible.

(8) State vector at $\phi_2 = \alpha - \phi$

$$
\left[\begin{array}{c} 1 \\ \hline U \\ V \\ W \\ \hline X \\ Z \\ \hline \Delta \\ \theta \end{array}\right]
=
\left[\begin{array}{c|ccc}
1 & 0 & 0 & 0 \\ \hline
U_0 & \dfrac{S_\alpha C_\phi}{S_\phi} T_{23} & -\dfrac{C_\alpha C_\phi}{S_\phi} T_{23} & -\dfrac{2\lambda C_\phi}{R S_\phi} T_{24} \\
0 & S_\alpha T_{23} & -C_\alpha T_{23} & -\dfrac{2\lambda}{R} T_{24} \\
W_0 & 2\lambda S_\alpha T_{22} & -2\lambda C_\alpha T_{22} & \dfrac{2\lambda^2}{R} T_{23} \\ \hline
0 & \dfrac{\nu R}{\lambda} S_\alpha T_{24} & \dfrac{\nu R}{\lambda} C_\alpha T_{24} & \nu T_{21} \\
0 & \dfrac{R}{\lambda} S_\alpha T_{24} & \dfrac{R}{\lambda} C_\alpha T_{24} & T_{21} \\ \hline
\Delta_0 & \dfrac{2\lambda R}{Eh} S_\alpha^2 T_{22} & -\dfrac{2\lambda R}{Eh} S_\alpha C_\alpha T_{22} & \dfrac{2\lambda^2}{Eh} S_\alpha T_{23} \\
\theta_0 & \dfrac{2\lambda^2}{Eh} S_\alpha T_{21} & -\dfrac{2\lambda^2}{Eh} C_\alpha T_{21} & \dfrac{4\lambda^3}{EhR} T_{22}
\end{array}\right]
\left[\begin{array}{c} 1 \\ \hline H_R \\ P_R \\ Z_R \end{array}\right]
$$

where U_0, W_0, Δ_0, θ_0 are the membrane vectors at ϕ_2 (29.06–29.07, 29.09–29.10) and the effects of H_L, P_L, Z_R are assumed to be negligible.

29.09 CLOSED SHELL, MEMBRANE STATE

Notation (27.03, 29.04)		Signs (27.03, 29.08)	Matrix (29.04, 29.08)
1. 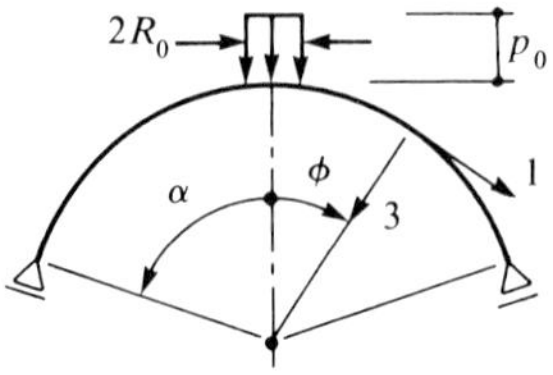		$U_0 = -\frac{R_0^2}{2RS_\phi} p_0$ $W_0 = -U_0$ $\theta_0 = 0$	$V_0 = X_0 = Z_0 = 0 \qquad \Delta_0 = (W_0 - \nu U_0)\frac{RS_\phi}{Eh}$ (cases 1–5)
2. 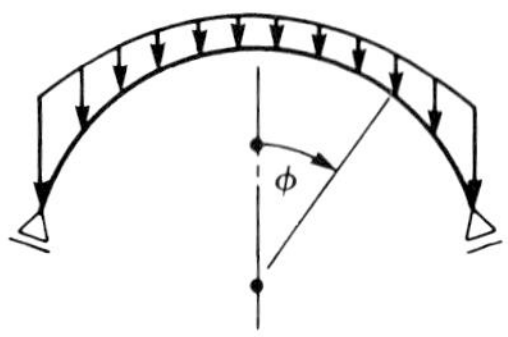	$p_1 = pS_\phi$ $p_3 = pC_\phi$	$U_0 = -\frac{R}{1 + C_\phi} p$ $W_0 = -U_0 - C_\phi^2 Rp$ $\theta_0 = (2 + \nu)S_\phi \frac{Rp}{Eh}$	
3. 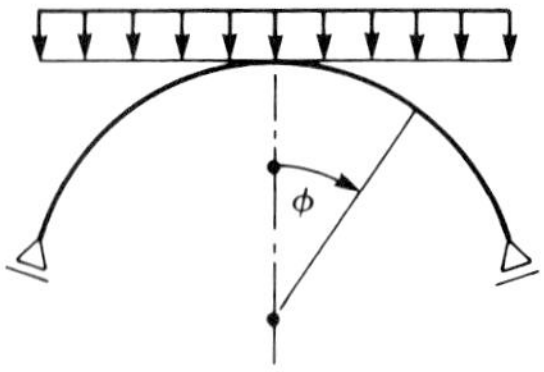	$p_1 = pS_\phi C_\phi$ $p_3 = pC_\phi^2$	$U_0 = -\frac{1}{2} Rp$ $W_0 = -U_0 - C_\phi^2 Rp$ $\theta_0 = (3 + \nu)S_\phi C_\phi \frac{Rp}{Eh}$	
4. 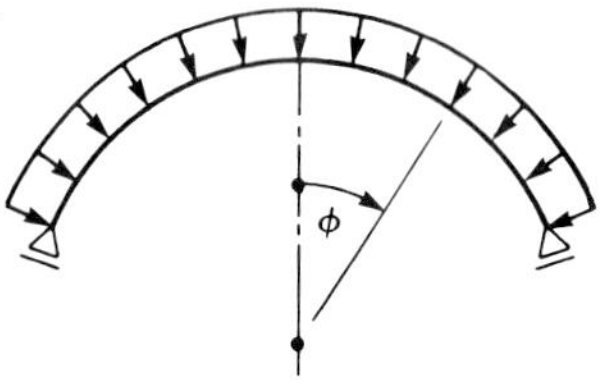	$p_1 = 0$ $p_3 = p$	$U_0 = -\frac{1}{2}Rp$ $W_0 = U_0$ $\theta_0 = 0$	
5. 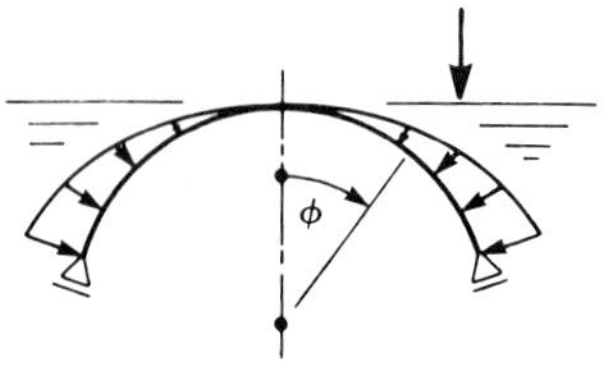	$p_1 = 0$ $p_2 = \gamma R(1 - C_\phi)$	$U_0 = -\frac{1}{6}(1 - 3C_\phi^2 + 3C_\phi^3)\frac{R^2\gamma}{S_\phi^2}$ $W_0 = -U_0 - (1 - C_\phi)R^2\gamma$ $\theta_0 = S_\phi R^2 \frac{\gamma}{Eh}$	
6. 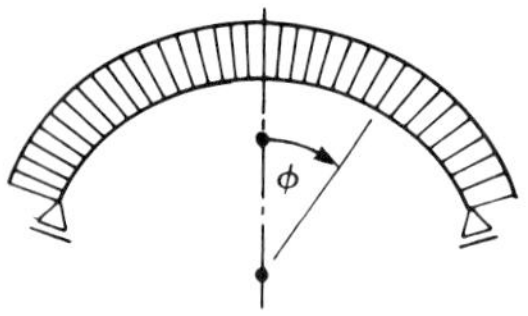	$e_t = \alpha_t \Delta t^*$	$U_0 = V_0 = W_0 = X_0 = Z_0 = 0$ $\Delta_0 = e_t RS_\phi$ $\theta_0 = 0$	

29.10 CLOSED SHELL, MEMBRANE STATE

Notation (27.03, 29.04) Signs (27.03, 29.08) Matrix (29.04, 29.08)

Case	Signs	Formulas	Matrix
1. 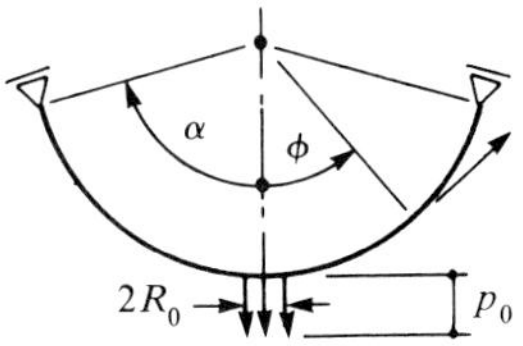		$U_0 = \dfrac{R_0^2}{2RS_\phi} p_0$ $W_0 = -U_0$ $\theta_0 = 0$	$V_0 = X_0 = Z_0 = 0 \qquad \Delta_0 = (W_0 - \nu V_0)\dfrac{RS_\phi}{Eh}$
2. 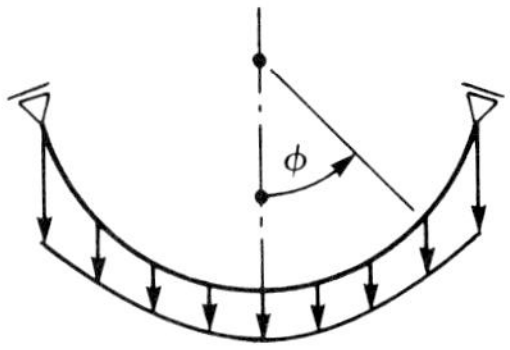	$p_1 = -pS_\phi$ $p_3 = -pC_\phi$	$U_0 = \dfrac{R}{1 + C_\phi} p$ $W_0 = -U_0 + C_\phi Rp$ $\theta_0 = (2 + \nu)S_\phi \dfrac{Rp}{Eh}$	
3. 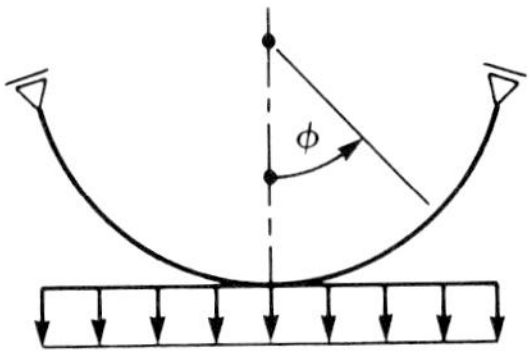	$p_1 = -pS_\phi C_\phi$ $p_3 = pC_\phi^2$	$U_0 = \frac{1}{2}Rp$ $W_0 = -U_0 + C_\phi^2 Rp$ $\theta_0 = (3 + \nu)S_\phi C_\phi \dfrac{Rp}{Eh}$	
4. 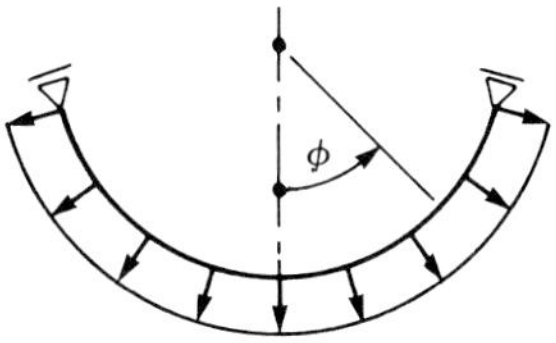	$p_1 = 0$ $p_3 = -p$	$U_0 = \frac{1}{2}Rp$ $W_0 = -U_0 + Rp$ $\theta_0 = 0$	
5. 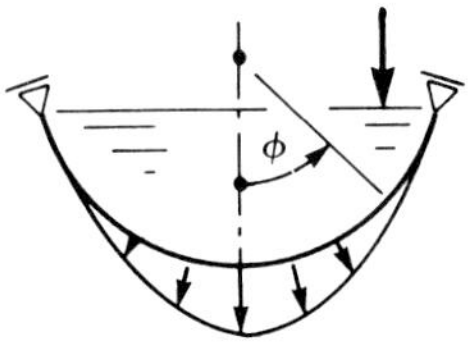	$p_1 = 0$ $p_3 = -\gamma R(1 - C_\phi)$	$U_0 = \frac{1}{2}(1 - 3C_\phi^2 + 3C_\phi^3)\dfrac{R^2\gamma}{S_\phi^2}$ $W_0 = -U_0 + (1 - C_\phi)R^2$ $\theta_0 = S_\phi R^2 \dfrac{\gamma}{Eh}$	
6. 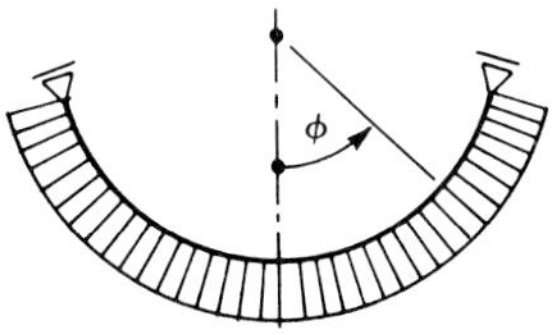	$e_t = \alpha_t \, \Delta t^*$	$U_0 = V_0 = W_0 = X_0 = Z_0 = 0$ $\Delta_0 = e_t RS_\phi$ $\theta_0 = 0$	

29.11 SHORT AND LONG SHELLS, STIFFNESS MATRIX EQUATIONS

$$D = \frac{Eh^3}{12(1-\nu^2)}$$

$$\lambda = \sqrt[4]{3(1-\nu^2)\left(\frac{R}{h}\right)^2}$$

(1) Equivalents and signs. The stiffness matrix equations of short shells (29.02) and long shells (29.08) are introduced below. All forces, moments, and displacements are in the x, y, z system, and their positive directions are shown in (2, 3).

(2) Free-body diagram

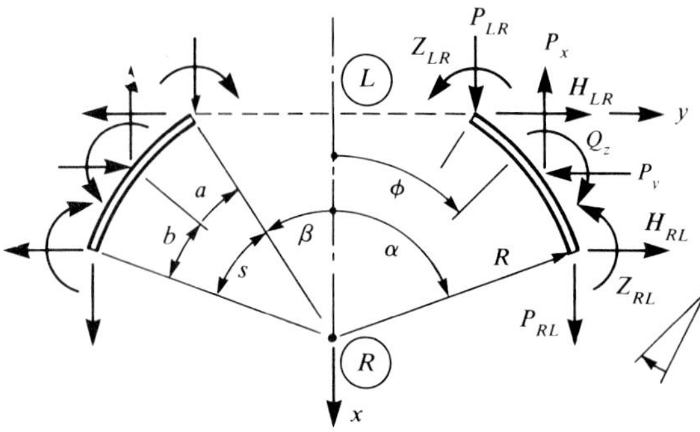

(3) Elastic surface

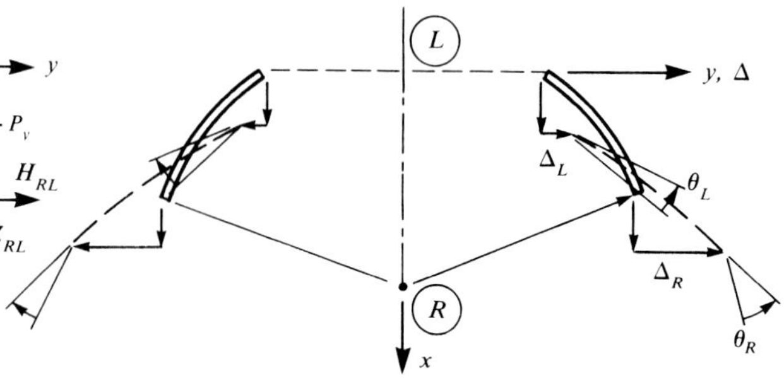

(4) Short shells, dimensionless stiffness coefficients

$$D_0 = \frac{1}{\sinh^2 \lambda s - \sin^2 \lambda s}$$

$L_2 = 2D_0(\sinh 2\lambda s + \sin 2\lambda s)$ $L_3 = D_0(\cosh 2\lambda s - \cos 2\lambda s)$ $L_4 = D_0(\sinh 2\lambda s - \sin 2\lambda s)$	$K_{2\alpha} = \dfrac{\lambda^3 L_2}{\sin^2 \alpha}$ $\quad K_{2\beta} = \dfrac{\lambda^3 L_2}{\sin^2 \beta}$ $K_{3\alpha} = \dfrac{\lambda^2 L_3}{\sin \alpha}$ $\quad K_3\beta = \dfrac{\lambda^2 L_3}{\sin \beta}$ $K_4 = \lambda L_4$
$L_6 = 4D_0(\cosh \lambda s \sin \lambda s + \sinh \lambda s \cos \lambda s)$ $L_7 = 4D_0 \sinh \lambda s \sin \lambda s$ $L_8 = 2D_0(\cosh \lambda s \sin \lambda s - \sinh \lambda s \cos \lambda s)$	$K_6 = \dfrac{\lambda^3 L_6}{\sin \alpha \cos \alpha}$ $K_7\alpha = \dfrac{\lambda^2 L_7}{\sin \alpha}$ $\quad K_{7\beta} = \dfrac{\lambda^2 L_7}{\sin \beta}$ $K_8 = \lambda L_8$

(5) Short shells, stiffness matrix equations ($\lambda s \leq 5$)

$$\begin{bmatrix} H_{LR} \\ Z_{LR} \\ \hline H_{RL} \\ Z_{RL} \end{bmatrix} = \left[\begin{array}{cc|cc} K_{2\beta} & K_{3\beta}R & -K_6 & K_{7\beta}R \\ K_{3\beta}R & K_4R^2 & -K_{7\alpha}R & K_8R^2 \\ \hline -K_6 & -K_{7\alpha}R & K_{2\alpha} & -K_{3\alpha}R \\ K_{7\beta}R & K_8R^2 & -K_{3\alpha}R & K_4R^2 \end{array}\right] \begin{bmatrix} \Delta_L \\ \theta_L \\ \hline \Delta_R \\ \theta_R \end{bmatrix} + \begin{bmatrix} H_{L0} \\ Z_{L0} \\ \hline H_{R0} \\ Z_{R0} \end{bmatrix}$$

where H_{L0}, Z_{L0} and H_{R0}, Z_{R0} are the *reactions of fixed-end shells* due to loads and other causes given in (6).

(6) Short shells, stiffness load functions introduced symbolically in (5) are given by the matrix equations

$$\begin{bmatrix} H_{L0} \\ Z_{L0} \end{bmatrix} = \begin{bmatrix} L_6 & -\dfrac{L_7}{\sin\beta} \\ \dfrac{L_7 R}{\sin\alpha} & -L_8 R \end{bmatrix} \begin{bmatrix} G_\Delta \\ G_\theta \end{bmatrix} \qquad \begin{bmatrix} H_{R0} \\ Z_{R0} \end{bmatrix} = \begin{bmatrix} L_6 & \dfrac{L_7}{\sin\alpha} \\ -\dfrac{L_7 R}{\sin\beta} & -L_8 R \end{bmatrix} \begin{bmatrix} F_\Delta \\ F_\theta \end{bmatrix}$$

where F_Δ, F_θ are the left-end transport load functions and G_Δ, G_θ are their right-end counterparts, given in symbolic form in (29.04) in terms of (29.05–29.06), and L_6, L_7, L_8 are the stiffness coefficients in (4) on the opposite page.

(7) Long shells, dimensionless stiffness coefficients

$$K_{11\alpha} = \frac{Eh}{\lambda R \sin^2\alpha} \qquad K_{12\alpha} = \frac{Eh}{2\lambda^2 \sin\alpha} \qquad K_{13} = \frac{EhR}{\lambda^3}$$

$$K_{11\beta} = \frac{Eh}{\lambda R \sin^2\beta} \qquad K_{12\beta} = \frac{Eh}{2\lambda^2 \sin\beta}$$

(8) Long shells, stiffness matrix equations ($\lambda s > 5$)

$$\begin{bmatrix} H_{LR} \\ Z_{LR} \end{bmatrix} = \begin{bmatrix} K_{11\beta} & K_{12\beta} \\ K_{12\beta} & K_{13} \end{bmatrix} \begin{bmatrix} \Delta_L \\ \theta_L \end{bmatrix} + \begin{bmatrix} H_{L0} \\ Z_{L0} \end{bmatrix} \qquad \begin{bmatrix} H_{RL} \\ Z_{RL} \end{bmatrix} = \begin{bmatrix} K_{11\alpha} & -K_{12\alpha} \\ -K_{12\alpha} & K_{13} \end{bmatrix} \begin{bmatrix} \Delta_R \\ \theta_R \end{bmatrix} + \begin{bmatrix} H_{R0} \\ Z_{R0} \end{bmatrix}$$

where H_{L0}, Z_{L0} and H_{R0}, Z_{R0} are the *reactions of fixed-end shells* at L and R, respectively, due to loads and other causes. The analytical forms of these reactions are given in (29.12).

(9) Elastic edge-ring stiffness matrix equations are

$$\begin{bmatrix} H_L \\ Z_L \end{bmatrix} = \frac{E}{R_L^2} \begin{bmatrix} A_L & 0 \\ -\dfrac{2A_L}{e_L} & I_L \end{bmatrix} \begin{bmatrix} \Delta_L \\ \theta_L \end{bmatrix}$$

$$\begin{bmatrix} H_R \\ Z_R \end{bmatrix} = \frac{E}{R_R^2} \begin{bmatrix} A_R & 0 \\ \dfrac{2A_R}{e_R} & I_R \end{bmatrix} \begin{bmatrix} \Delta_R \\ \theta_R \end{bmatrix}$$

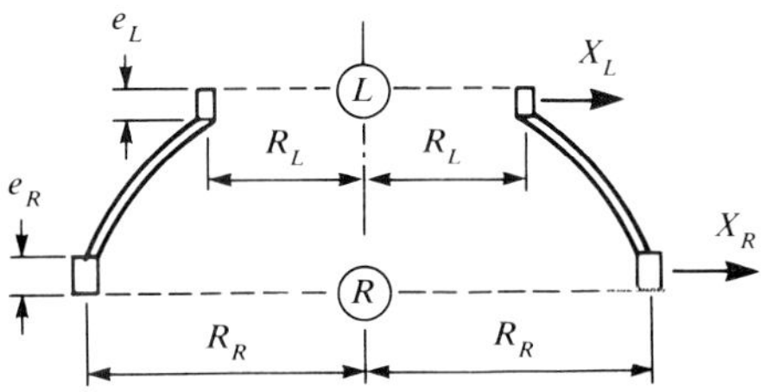

where R_L, R_R = radii of edge rings (m)
A_L, A_R = areas of section of edge rings (m^2)
I_L, I_R = moments of inertia of A_L, A_R about X_L, X_R (m^4)
e_R, e_L = depth of ring at L, R (m)

29.12 LONG SHELLS, STIFFNESS LOAD FUNCTIONS

Notation (29.08–29.11) Signs, matrix (29.11)

$K_{11\alpha}, K_{12\alpha}, K_{13}, K_{11\beta}, K_{12\beta}$ = stiffness coefficients (29.11)

P_{L0}, P_{R0} = statically determinate reactions (N/m)

$$\lambda = \sqrt[4]{3(1-\nu^2)\left(\frac{R}{h}\right)^2}$$

$\lambda s > 5$

1. Closed shell fixed at R

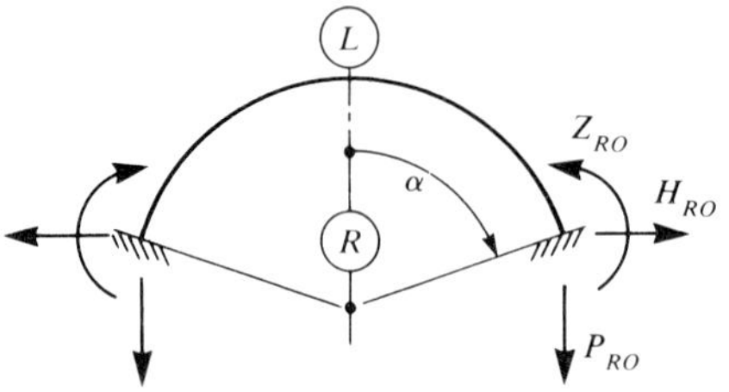

$$\begin{bmatrix} H_{R0} \\ Z_{R0} \end{bmatrix} = \begin{bmatrix} -K_{11\alpha} & K_{12\alpha} \\ K_{12\alpha} & -K_{13} \end{bmatrix} \begin{bmatrix} \Delta_{0\alpha} \\ \theta_{0\alpha} \end{bmatrix}$$

$\Delta_{0\alpha}, \theta_{0\alpha}$ = membrane state displacements at $R(\phi = \alpha)$ (29.09)

2. Open shell fixed at R

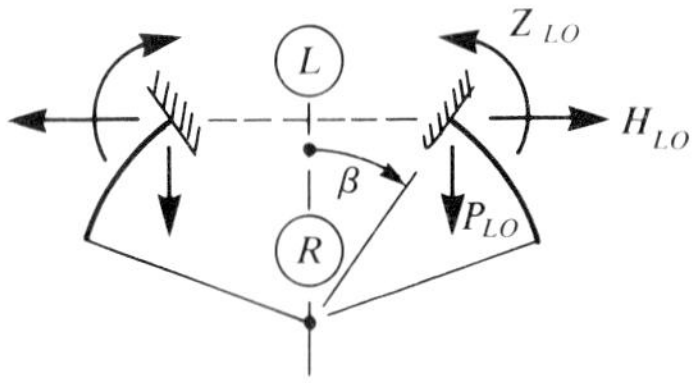

$$\begin{bmatrix} H_{L0} \\ Z_{L0} \end{bmatrix} = \begin{bmatrix} -K_{11\beta} & -K_{12\beta} \\ -K_{12\beta} & -K_{13} \end{bmatrix} \begin{bmatrix} \Delta_{0\beta} \\ \theta_{0\beta} \end{bmatrix}$$

$\Delta_{0\beta}, \theta_{0\beta}$ = membrane state displacements at $L(\phi = \beta)$ (29.06)

3. Open shell fixed at L

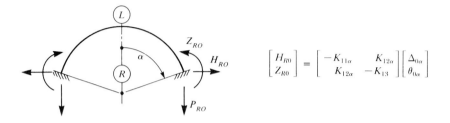

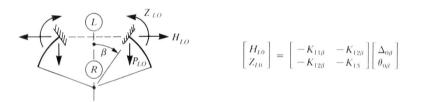

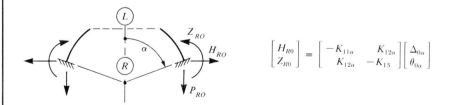

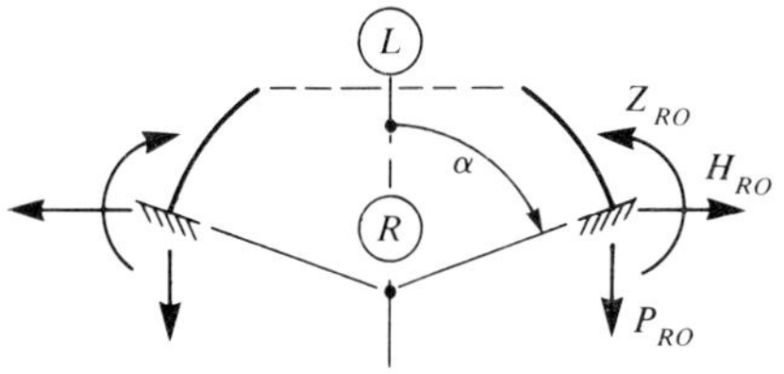

$$\begin{bmatrix} H_{R0} \\ Z_{R0} \end{bmatrix} = \begin{bmatrix} -K_{11\alpha} & K_{12\alpha} \\ K_{12\alpha} & -K_{13} \end{bmatrix} \begin{bmatrix} \Delta_{0\alpha} \\ \theta_{0\alpha} \end{bmatrix}$$

$\Delta_{0\alpha}, \theta_{0\alpha}$ = membrane state displacements at $R(\phi = \alpha)$ (29.06)

30

Shell Analysis by Finite Differences and Finite Elements

STATIC STATE

30.01 DEFINITION OF STATE

(1) Systems analyzed are thin isotropic rotational shells of constant or variable thickness acted on by axisymmetrical causes in a state of static equilibrium.

(2) Analysis of thin shells by the methods introduced in Chaps. 28 and 29 is limited to shells of constant thickness and to causes of certain types. In other cases these solutions are not practical and/or possible, and as in the plate analysis, resort must be made to some approximate numerical methods, among which the most important ones are the *finite-difference method* and the *finite-element method,* introduced in this chapter.

(3) Finite-difference method replaces the given shell by a specified grid consisting of meridian curves and circumferential circles and approximates the governing differential equation and the equations defining the boundary conditions by difference equations in terms of unknown deflections (27.05-7) or shears (27.06-6, 27.07-6) related to the nodal points of this grid. Since the governing differential equations are formally identical to those of beams encased in elastic foundation (8.03), the method of solution reduces to the solution introduced in (16.06).

(4) Finite-element method introduced in the plate analysis (26.01) is extended in this chapter to the analysis of axisymmetrical shells. Only the basic models are displayed in this chapter.

(5) Assumptions of analysis are stated in (27.01). The central differences are used in the finite-difference analysis. Only cylindrical and conical elements are considered in the finite-element analysis.

30.02 CIRCULAR CYLINDRICAL SHELL, FINITE DIFFERENCES

(1) System considered is an axisymmetrical circular cylindrical shell of variable thickness h (m) represented by the principal section in (2). The positive joint loads, forces, moments, and displacements are also shown in (2).

(2) Principal section

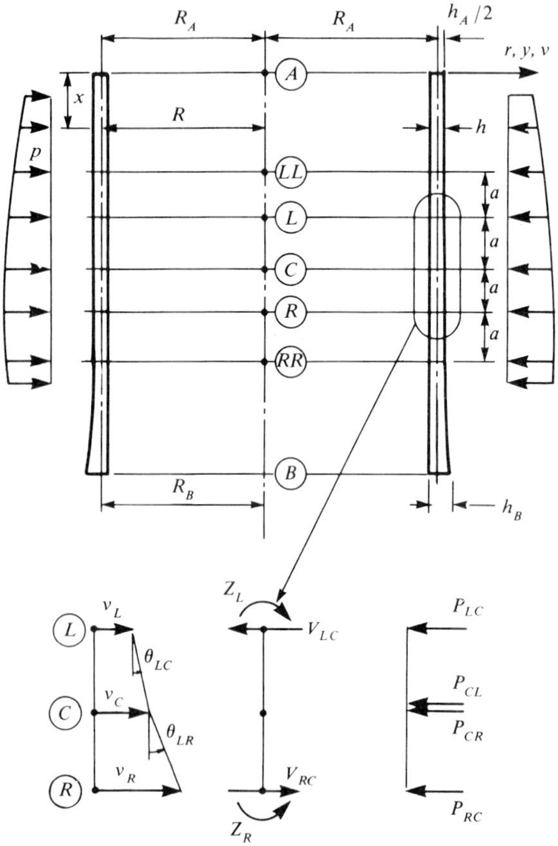

(3) Notation (27.05, 28.01)

v_j = deflection along y (m)
θ_j = slope about z (rad)
R_j = radius (m)
U_j = normal force along x (N/m)
V_j = shearing force along y (N/m)
W_j = normal force along z (N/m)
X_j, Z_j = bending moments about y, z (N·m/m)
$P_j = P_{ji} + P_{jk}$ = joint load along y (N/m)
s = shell length (m)
a = difference segment (m)
i, j, k = joints ($A-1, A, A+1, \ldots, L, C, R, \ldots$)

(4) Variable flexural rigidity of the shell is

$$D_j = \frac{Eh_j^3}{12(1-\nu^2)}$$

(5) Equivalents

$K_{LL} = D_L/D_C$
$K_L = 2(D_L + D_C)/D_C$
$K_C = \dfrac{D_L + 4D_C + D_R + (Eh_C/R_C^2)a^4}{D_C}$
$K_R = 2(D_C + D_R)/D_C$
$K_{RR} = D_R/D_C$

(6) Governing difference equation at C is

$$-R_C = K_{LL}v_{LL} - K_L v_L + K_C v_C - K_R v_R + K_{RR} v_R$$

where

$$R_C = \frac{(P_{CL} + P_{CR})a^3}{D_C}$$

is the equivalent joint load given by relations (16.03) and scaled as in (16.04). The displacements at LL, L, C, R, RR are the unknowns of the analysis.

(7) State-vector components at C are $\quad U_C = U_{C0}$

$$V_{CL} = P_{CL} - [(v_L - 2v_C + v_R)D_C - (v_{LL} - 2v_L + v_C)D_L]/a^3$$
$$V_{CR} = -P_{CR} + [(v_{RR} - 2v_R + v_C)D_R - (v_L - 2v_C + v_R)D_C]/a^3$$
$$W_C = \frac{Eh_C}{R_C} v_C + \nu U_{C0} \qquad Z_C = \frac{1}{a^2}(v_L - 2v_C + v_R)D_C \qquad X_C = \nu Z_C$$
$$\theta_{CL} = \frac{v_C - v_L}{a} + \frac{a}{24}\left(\frac{7Z_C}{D_C} + \frac{6Z_L}{D_L} - \frac{Z_{LL}}{D_{LL}}\right) \qquad \theta_{CR} = \frac{v_R - v_C}{a} - \frac{a}{24}\left(\frac{7Z_C}{D_C} + \frac{6Z_R}{D_R} - \frac{Z_{RR}}{D_{RR}}\right)$$

where P_{CL}, P_{CR}, V_{CL}, V_{CR}, θ_{CL}, θ_{CR} are the joint loads, shears, and slopes on the left and right sides of C, respectively; U_{C0} is the membrane state normal force due to loads acting along x as shown in (28.09–28.10).

(8) Boundary equations at A **and** B

Free end A	Free end B
$v_{A-1} - 2v_A + v_{A+1} = 0$ $v_{A-2} - 3v_A + 2v_{A+1} = 0$	$v_{B-1} - 2v_B + v_{B+1} = 0$ $2v_{B-1} - 3v_B + v_{B+2} = 0$
Hinged end A	Hinged end B
$v_{A-1} + v_{A+1} = 0 \qquad v_A = 0$	$v_{B-1} + v_{B+1} = 0 \qquad v_B = 0$
Fixed end A	Fixed end B
$v_{A-1} - v_{A+1} = 0 \qquad v_A = 0$	$v_{B-1} - v_{B+1} = 0 \qquad v_B = 0$

(9) State-vector components at A **and** B

$r_A = \dfrac{6Z_{A+1}}{D_{A+1}} - \dfrac{Z_{A+2}}{D_{A+2}}$	$r_B = \dfrac{6Z_{B-1}}{D_{B-1}} - \dfrac{Z_{B-2}}{D_{B-2}}$
Free end A	Free end B
$U_A = U_{A0} \qquad V_A = 0 \qquad Z_A = 0$ $W_A = \dfrac{Eh_A}{R_A} v_A + \nu U_{A0}$ $\theta_A = (v_{A+1} - v_A)/a - r_A a/24$	$U_B = U_{B0} \qquad V_B = 0 \qquad Z_B = 0$ $W_B = \dfrac{Eh_B}{R_B} v_B + \nu U_{B0}$ $\theta_B = (v_B - v_{B-1})/a + r_B a/24$
Hinged end A	Hinged end B
$U_A = U_{A0} \qquad Z_A = 0 \qquad v_A = 0$ $V_A = -P_{A,A+1} - (2v_{A+1} - v_{A+2})D_{A+1}/a^3$ $W_A = \nu U_{A0} \qquad \theta_A = v_{A+1}/a - r_A a/24$	$U_B = U_{B0} \qquad Z_B = 0 \qquad v_B = 0$ $V_B = P_{B,B-1} - (2v_{B-1} - v_{B-2})D_{B-1}/a^3$ $W_B = \nu U_{B0} \qquad \theta_B = v_{B-1}/a + r_B a/24$

Fixed end A
$U_A = U_{A0} \qquad V_A = -P_{A,A+1} - [(2v_{A+1} - v_{A+2})D_{A+1} + 2v_{A+1}D_A]/a^3$ $v_A = 0 \qquad \theta_A = 0 \qquad W_A = \nu U_{A0} \qquad Z_A = 2v_{A+1}v_A/a^2 \qquad X_A = \nu Z_A$
Fixed end B
$U_B = U_{B0} \qquad V_B = P_{B,B-1} - [(2v_{B-1} - v_{B-2})D_{B-1} + 2v_{B-1}D_B]/a^3$ $v_B = 0 \qquad \theta_B = 0 \qquad W_B = \nu U_{B0} \qquad Z_B = 2v_{B-1}D_B/a^2 \qquad X_B = \nu Z_B$

30.03 SPHERICAL SHELL, FINITE DIFFERENCES

(1) System considered is an axisymmetrical spherical shell of variable thickness h (m) represented by the principal section in (2). The positive displacements, forces, and moments are also shown in (2).

(2) Principal section

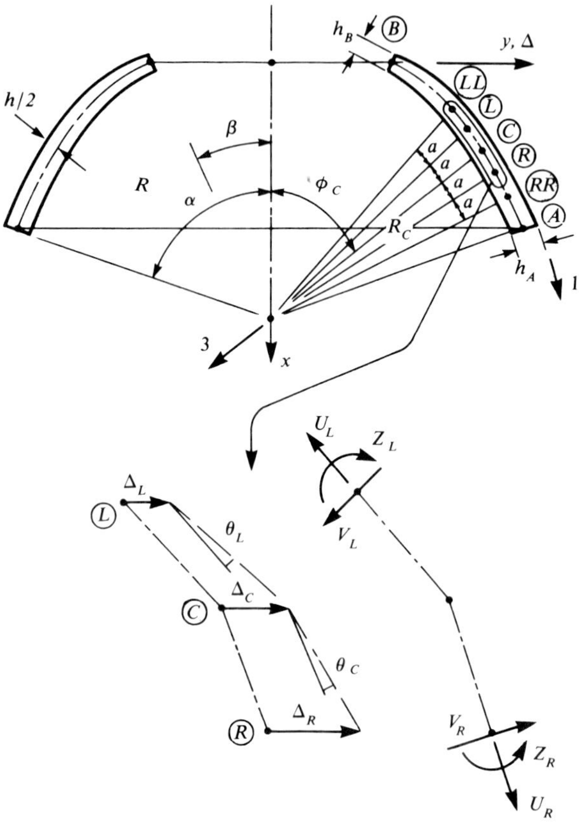

(3) Notation (27.06, 29.01)

Δ_j = deflection along y (m)
θ_j = slope about z (rad)
U_j = normal force along 1 (N/m)
V_j = shearing force along 3 (N/m)
W_j = normal force along 2 (N/m)
X_j, Z_j = bending moments about 1, 2 (N·m/m)
α, β, ϕ = angles (rad) R = radius (m)
a = difference angle (rad)
i, j, k = joints ($A - 1, A, A + 1, \ldots, L, C, R, \ldots$)

(4) Variable flexural rigidity of the shell is

$$D_j = \frac{Eh_j^3}{12(1 - \nu^2)}$$

(5) Equivalents

$K_{LL} = D_L/D_C$	$K_{RR} = D_R/D_C$
$K_C = \dfrac{D_L + 4D_C + D_R + (Eh_C/R^2)a^4}{D_C}$	
$K_L = 2(D_L + D_C)/D_C$	$K_R = 2(D_C + D_R)/D_C$

(6) Governing difference equation of the loadless shell at C is

$$0 = K_{LL}V_{LL} - K_L V_L + K_C V_C - K_R V_R + K_{RR} V_{RR}$$

where the shears at LL, L, C, R, RR are the unknowns of the analysis.

(7) State-vector components at C are

$$\Delta_C = \Delta_{C0} + \frac{R \sin \phi_C}{2aEh_C}(V_R - V_L) \qquad U_C = U_{C0} + V_C \cot \phi_C$$

$$\theta_C = \theta_{C0} + \frac{V_L - 2V_C + V_R}{Ea^2 h_C} \qquad W_C = W_{C0} + \frac{V_R - V_L}{2a} \qquad X_C = \nu Z_C$$

$$Z_C = (V_C - 2V_R + V_{RR})\frac{D_R}{2REa^3 h_R} - (V_{LL} - 2V_L + V_C)\frac{D_L}{2REa^3 h_L}$$

where Δ_{C0}, θ_{C0}, U_{C0}, W_{C0} are the displacements and normal forces of the membrane state (29.06–29.07, 29.09–29.10).

(8) Boundary equations at A and B

$r_A = \dfrac{D_{A+1}h_{A-1}}{D_{A-1}h_{A+1}}$	$r_B = \dfrac{D_{B-1}h_{B+1}}{D_{B+1}h_{B-1}}$
Free end A	Free end B
$V_A = 0 \qquad 2V_{A-1} - V_{A-2} = (2V_{A+1} - V_{A+2})r_A$	$V_B = 0 \qquad 2V_{B+1} - V_{B+2} = (2V_{B-1} - V_{B-2})r_B$
Hinged end A	Hinged end B
$V_{A-1} - V_A = \dfrac{2aE}{R \sin \alpha} h_A \Delta_{A0}$ $V_{A-2} - 2V_{A-1} + V_A = (V_{A+2} - 2V_{A+1} + V_A)r_A$	$V_{B-1} - V_B = \dfrac{2aE}{R \sin \beta} h_B \Delta_{B0}$ $V_{B+2} - 2V_{B+1} + V_B = (V_{B-2} - 2V_{B-1} + V_B)r_B$
Fixed end A	Fixed end B
$V_{A-1} - V_{A+1} = \dfrac{2aE}{R \sin \alpha} h_A \Delta_{A0}$ $V_{A-1} - 2V_A + V_{A+1} = -Ea^2 h_A \theta_{A0}$	$V_{B-1} - V_{B+1} = \dfrac{2aE}{R \sin \beta} h_B \Delta_{B0}$ $V_{B-1} - 2V_B + V_{B+1} = -Ea^2 h_B \theta_{B0}$

(9) State-vector components at A and B

Free end A	Free end B
$U_A = U_{A0} \qquad V_A = 0 \qquad X_A = Z_A = 0$ $W_A = W_{A0} + (V_{A+1} - V_{A-1})/2a$ $\theta_A = \theta_{A0} + \dfrac{V_{A-1} + V_{A+1}}{Ea^2 h_A}$ $\Delta_A = \Delta_{A0} + \dfrac{R \sin \alpha}{2Eah_A}(V_{A+1} - V_{A-1})$	$U_B = U_{B0} \qquad V_B = 0 \qquad X_B = Z_B = 0$ $W_B = W_{B0} + (V_{B+1} - V_{B-1})/2a$ $\theta_B = \theta_{B0} + \dfrac{V_{B-1} + V_{B+1}}{Ea^2 h_B}$ $\Delta_B = \Delta_{B0} + \dfrac{R \sin \beta}{2Eah_B}(V_{B+1} - V_{B-1})$
Hinged end A	Hinged end B
$U_A = U_{A0} + V_A \cot \alpha$ $W_A = W_{A0} + (V_{A+1} - V_{A-1})/2a$ $X_A = \nu Z_A = 0 \qquad \Delta_A = 0$ $\theta_A = \theta_{A0} + \dfrac{V_{A-1} - 2V_A + V_{A+1}}{Ea^2 h_A}$	$U_B = U_{B0} + V_B \cot \beta$ $W_B = W_{B0} + (V_{B+1} - V_{B-1})/2a$ $X_B = \nu Z_B = 0 \qquad \Delta_B = 0$ $\theta_B = \theta_{B0} + \dfrac{V_{B-1} - 2V_B + V_{B+1}}{Ea^2 h_B}$

Fixed end A
$U_A = U_{A0} + V_A \cot \alpha \qquad W_A = W_{A0} + (V_{A+1} - V_{A-1})/2a \qquad X_A = \nu Z_A \qquad \Delta_A = 0 \qquad \theta_A = 0$ $Z_A = (V_A - 2V_{A+1} + V_{A+2}) \dfrac{D_{A+1}}{2REa^3 h_{A+1}} - (V_{A-2} - 2V_{A-1} + V_A) \dfrac{D_{A-1}}{2REa^3 h_{A-1}}$
Fixed end B
$U_B = U_{B0} + V_B \cot \beta \qquad W_B = W_{B0} + (V_{B+1} - V_{B-1})/2a \qquad X_B = \nu Z_B \qquad \Delta_B = 0 \qquad \theta_B = 0$ $Z_B = (V_B - 2V_{B+1} + V_{B+2}) \dfrac{D_{B+1}}{2REa^3 h_{B+1}} - (V_{B-2} - 2V_{B-1} + V_B) \dfrac{D_{B-1}}{2REa^3 h_{B-1}}$

30.04 CIRCULAR CONICAL SHELL, FINITE DIFFERENCES

(1) System considered is an axisymmetrical circular conical shell of variable thickness h (m) represented by the principal section in (2). The positive displacements, moments and forces are also shown in (2).

(2) Principal section

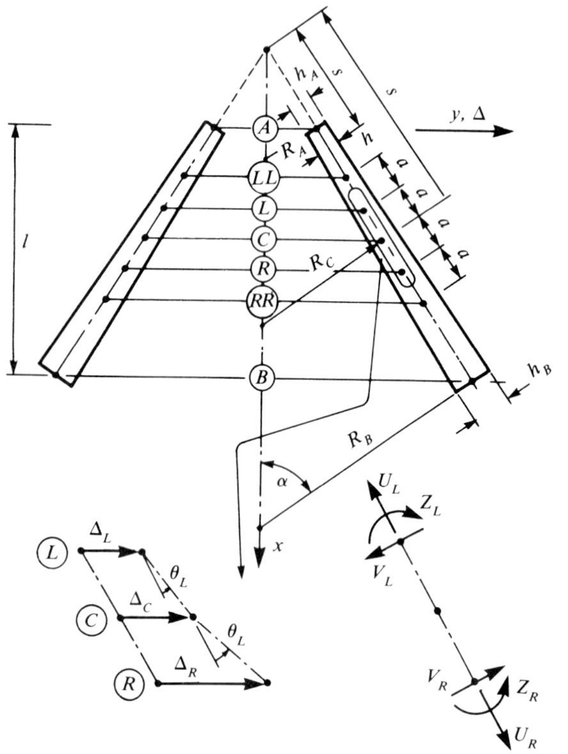

(3) Notation (27.07). The definitions of Δ_j, θ_j, U_j, V_j, W_j, X_j, Z_j, α and i, j, k are the same as in (30.03).

l = shell length (m)
a = difference segment (m)
s_j = coordinate (m)
R_j = variable radius (m)

(4) Variable flexural rigidity of the shell is

$$D_j = \frac{Eh_j^3}{12(1-\nu^2)}$$

(5) Equivalents

$K_{LL} = D_L/D_C$	
$K_L = 2(D_L + D_C)/D_C$	
$K_C = \dfrac{D_L + 4D_C + D_R + (Eh_C/R_C^2)a^4}{D_C}$	
$K_R = 2(D_C + D_R)/D_C$	
$K_{RR} = D_R/D_C$	

(6) Governing difference equation of the loadless shell at C is

$$0 = K_{LL}V_{LL} - K_L V_L + K_C V_C - K_R V_R + K_{RR}V_{RR}$$

where the shears at LL, L, C, R, RR are the unknowns of the analysis.

(7) State-vector components at C are

$$\Delta_C = \Delta_{C0} + \frac{R_C \cos\alpha}{2aEh_C}(s_C V_L - 2aV_C - s_C V_R)$$

$$\theta_C = \theta_{C0} + \frac{R_C^2}{a^2 Eh_C}(V_L - 2V_C + V_R) \qquad U_C = U_{C0} + V_C \cot\alpha$$

$$W_C = W_{C0} - \frac{R_C}{2as_C}(s_C V_L - 2aV_C - s_C V_R) \qquad X_C = \nu Z_C$$

$$Z_C = (V_C - 2V_R + V_{RR})\frac{D_R s_R^2 \cot^2\alpha}{2Ea^3 h_R} - (V_{LL} - 2V_L + V_C)\frac{D_L s_L^2 \cot^2\alpha}{2Ea^3 h_L}$$

where Δ_{C0}, θ_{C0}, U_{C0}, W_{C0} are the displacements and normal forces of the membrane state (30.05–30.06).

(8) Boundary equations at A and B

$r_A = \frac{D_{A+1}s_{A+1}^2 h_{A-1}}{D_{A-1}s_{A-1}^2 h_{A+1}}$	$r_B = \frac{D_{B-1}s_{B-1}^2 h_{B+1}}{D_{B+1}s_{B+1}^2 h_{B-1}}$
Free end A	Free end B
$V_A = 0 \qquad 2V_{A-1} - V_{A-2} = (2V_{A+1} - V_{A+2})r_A$	$V_B = 0 \qquad 2V_{B+1} - V_{B+2} = (2V_{B-1} - V_{B-2})r_B$
Hinged end A	Hinged end B
$s_AV_{A-1} - 2aV_A - s_AV_{A+1} = -\frac{2aE}{R_A \cos\alpha} h_A\Delta_{A0}$ $V_A - 2V_{A-1} + V_{A-2} = (V_A - 2V_{A+1} + V_{A+2})r_A$	$s_BV_{B-1} - 2aV_B - s_BV_{B+1}. = -\frac{2aE}{R_B \cos\alpha} h_B\Delta_{B0}$ $V_B - 2V_{B+1} + V_{B+2} = (V_B - 2V_{B-1} + V_{B+2})r_B$
Fixed end A	Fixed end B
$s_AV_{A-1} - 2aV_A - s_AV_{A+1} = -\frac{2aE}{R_A \cos\alpha} h_A\Delta_{A0}$ $V_{A-1} - 2V_A + V_{A+1} = -\frac{a^2E}{R_A^2} h_A\theta_{A0}$	$s_BV_{B-1} - 2aV_B - s_BV_{B+1} = -\frac{2aE}{R_B \cos\alpha} h_B\Delta_{B0}$ $V_{B-1} - 2V_B + V_{B+1} = -\frac{a^2E}{R_B^2} h_B\theta_{B0}$

(9) State-vector components at A and B

Free end A	Free end B
$U_A = U_{A0} \qquad V_A = 0 \qquad X_A = Z_A = 0$ $W_A = W_{A0} - R_A(V_{A-1} - V_{A+1})/2a$ $\theta_A = \theta_{A0} + (V_{A-1} + V_{A+1})\frac{R_A^2}{Ea^2h_A}$ $\Delta_A = \Delta_{A0} + (V_{A-1} + V_{A+1})\frac{R_As_A \cos\alpha}{2Eah_A}$	$U_B = U_{B0} \qquad V_B = 0 \qquad X_B = Z_B = 0$ $W_B = W_{B0} - R_B(V_{B-1} - V_{B+1})/2a$ $\theta_B = \theta_{B0} + (V_{B-1} + V_{B+1})\frac{R_B^2}{Ea^2h_B}$ $\Delta_B = \Delta_{B0} + (V_{B-1} + V_{B+1})\frac{R_Bs_B \cos\alpha}{2Eah_B}$
Hinged end A	Hinged end B
$U_A = U_{A0} + V_A \cot\alpha \qquad X_A = \nu Z_A = 0$ $W_A = W_{A0} - (s_AV_{A-1} - 2aV_A + s_AV_{A+1})\frac{R_A}{2as_A}$ $\Delta_A = 0 \qquad \theta_A = \theta_{A0} + (V_{A-1} - 2V_A + V_{A+1})\frac{R_A^2}{Ea^2h_A}$	$U_B = U_{B0} + V_B \cot\alpha \qquad X_B = \nu Z_B = 0$ $W_B = W_{B0} - (s_BV_{B-1} - 2aV_B + s_BV_{B+1})\frac{R_B}{2as_B}$ $\Delta_B = 0 \qquad \theta_B = \theta_{B0} + (V_{B-1} - 2V_B + V_{B+1})\frac{R_B^2}{Ea^2h_B}$
Fixed end A	
$U_A = U_{A0} + V_A \cot\alpha \qquad \Delta_A = 0, \qquad \theta_A = 0,\ W_A = W_{A0} + (V_{A-1} - \frac{2a}{s_A}V_A + V_{A+1})\frac{R_A}{2a}$ $Z_A = (V_A - 2V_{A+1} + V_{A+2})\frac{D_{A+1}s_{A+1}^2}{h_{A+1}} - (V_{A-1} - 2V_{A-1} + V_A)\frac{D_{A-1}s_{A-1}^2}{h_{A-1}}\frac{\cot^2\alpha}{2Ea^3} \qquad X_A = \upsilon Z_A$	
Fixed end B	
$U_B = U_{B0} + V_B \cot\alpha \qquad \Delta_B = 0, \qquad \theta_B = 0, \qquad W_B = W_{B0} - (V_{B-1} - \frac{2a}{s_B}V_B + V_{B+1})\frac{R_B}{2a}$ $Z_B = \left[(V_B - 2V_{B+1} + V_{B+2})\frac{D_{B+1}s_{B+1}^2}{h_{B+1}} - (V_{B-2} - 2V_{B-1} + V_{B+2})\frac{D_{B-1}s_{B-1}^2}{h_{B-1}}\right]\frac{\cot^2\alpha}{2Ea^3} \qquad X_B = \theta\nu Z_B$	

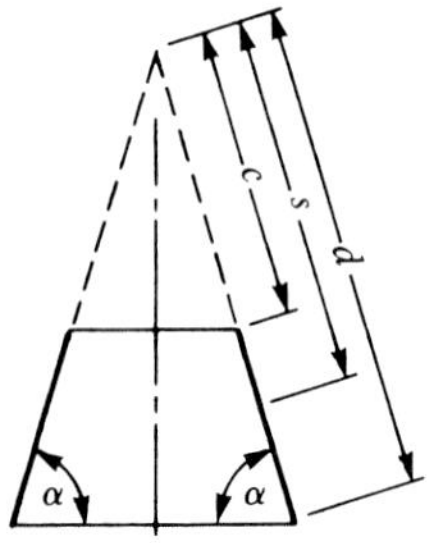

Notation (27.07, 30.04) Signs (30.04)

p = intensity of distributed load (N/m^2)
P = concentrated ring load (N/m)
γ = weight density (N/m^3)
s = variable coordinate (m)
c, d = segments (m)
$m = 1 - (c/s)^2$
$n = (d/s)^2 - 1$

α = angle (rad)
$S_\alpha = \sin\alpha$
$C_\alpha = \cos\alpha$
$T_\alpha = \tan\alpha$

Case	Load	Formulas
1. 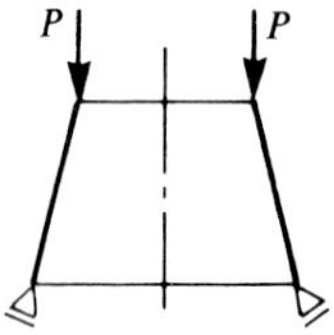		$U_0 = -\dfrac{cP}{sS_\alpha}$ $W_0 = 0$ $\theta_0 = \dfrac{cP}{sS_\alpha T_\alpha Eh}$
2. 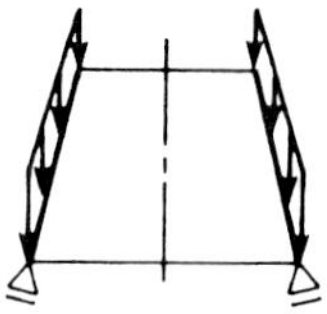	$p_1 = pS_\alpha$ $p_3 = pC_\alpha$	$U_0 = -\dfrac{msp}{2S_\alpha}$ $W_0 = -\dfrac{C_\alpha^2 sp}{S_\alpha}$ $\theta_0 = (\tfrac{1}{2}m - \nu S_\alpha^2 + 2C_\alpha^2)\dfrac{C_\alpha sp}{S_\alpha^2 Eh}$
3. 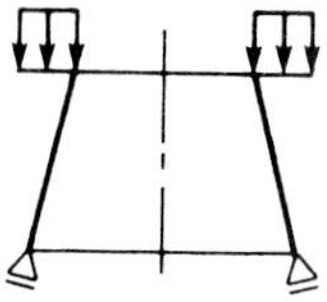	$p_1 = psS_\alpha C_\alpha$ $p_3 = pC_\alpha^2$	$U_0 = -\dfrac{msp}{2T_\alpha}$ $W_0 = -\dfrac{C_\alpha sp}{T_\alpha}$ $\theta_0 = (\tfrac{1}{2}m - \nu S_\alpha^2 + 2C_\alpha^2)\dfrac{sp}{T_\alpha^2 Eh}$
4. 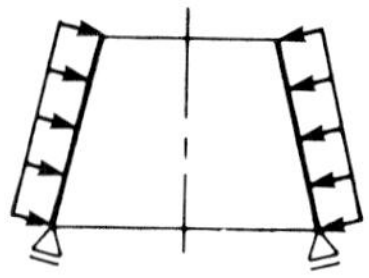	$p_1 = 0$ $p_3 = p$	$U_0 = -\tfrac{1}{2}msp$ $W_0 = -\dfrac{sp}{T_\alpha}$ $\theta_0 = -\dfrac{(4 - m)sp}{2T_\alpha^2 Eh}$
5. 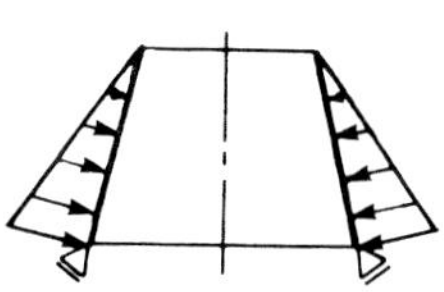	$p_1 = 0$ $p_3 = \gamma(s - c)S_\alpha$	$U_0 = -\tfrac{1}{6}(3s - 3c - sm)C_\alpha s\gamma$ $W_0 = -(s - c)C_\alpha s\gamma$ $\theta_0 = \tfrac{1}{6}(9c - 15s - sm)\dfrac{C_\alpha s\gamma}{T_\alpha Eh}$

For closed shell, $c = 0$, $m = 1$ (cases 2–5)

$V_0 = X_0 = Z_0 = 0 \qquad \Delta_0 = (W_0 - \nu U_0)\dfrac{sC_\alpha}{Eh}$ (cases 1–5)

30.06 CIRCULAR CONICAL SHELL MEMBRANE STATE

Notation (27.03, 30.04, 30.05) Δt^* = change in temperature (°C)		Signs (30.04) α_t = thermal coefficients (1/°C)	
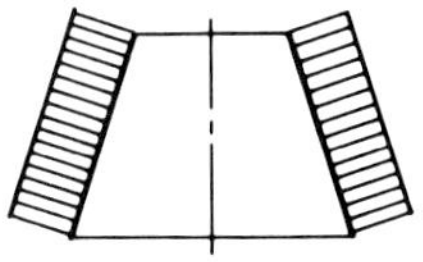	$e_t = \alpha_t \Delta t^*$	$U_0 = V_0 = W_0 = X_0 = Z_0 = 0$ $\Delta_0 = e_t s C_\alpha$ $\theta_0 = 0$	For closed shell, $c = 0$
1. 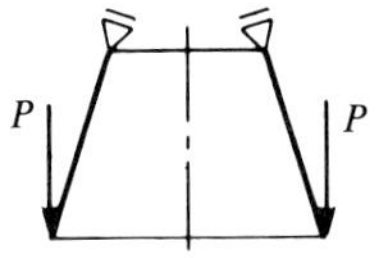		$U_0 = \dfrac{cP}{sS_\alpha}$ $W_0 = 0$ $\theta_0 = -\dfrac{cP}{S_\alpha T_\alpha s}$	$V_0 = X_0 = Z_0 = 0$ $\quad 0 = (W_0 - \nu U_0)\dfrac{sC_\alpha}{Eh}$
2. 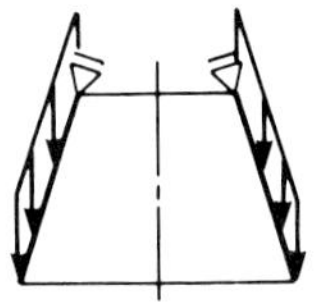	$p_1 = pS_\alpha$ $p_3 = pC_\alpha$	$U_0 = \dfrac{nsp}{2S_\alpha}$ $W_0 = -\dfrac{C_\alpha sp}{T_\alpha}$ $\theta_0 = -(\tfrac{1}{2}n + \nu S_\alpha^2 - 2C_\alpha^2)\dfrac{sp}{C_\alpha T_\alpha Eh}$	For closed shell, $c = 0$; $V_0 = X_0 = Z_0 = 0$ $\quad 0 = (W_0 - \nu U_0)\dfrac{sC_\alpha}{Eh}$
3. 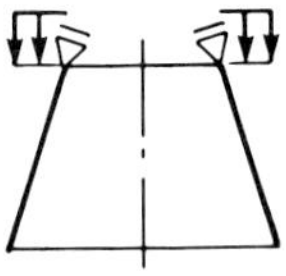	$p_1 = pS_\alpha C_\alpha$ $p_3 = pC_\alpha^2$	$U_0 = \dfrac{nsp}{2T_\alpha}$ $W_0 = -\dfrac{C_\alpha^2 sp}{T_\alpha}$ $\theta_0 = -(\tfrac{1}{2}n + \nu S_\alpha^2 - 2C_\alpha^2)\dfrac{sp}{T_\alpha^2 Eh}$	For closed shell, $c = 0$; $V_0 = X_0 = Z_0 = 0$ $\quad 0 = (W_0 - \nu U_0)\dfrac{sC_\alpha}{Eh}$
4. 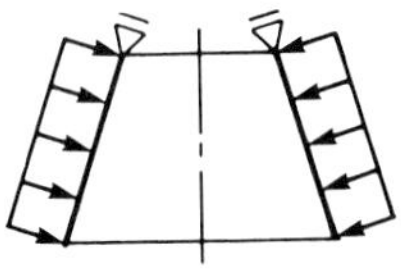	$p_1 = 0$ $p_3 = p$	$U_0 = \tfrac{1}{2}nsp$ $W_0 = -\dfrac{sp}{T_\alpha}$ $\theta_0 = -\dfrac{(4 + n)sp}{2T_\alpha^2 Eh}$	For closed shell, $c = 0$; $V_0 = X_0 = Z_0 = 0$ $\quad 0 = (W_0 - \nu U_0)\dfrac{sC_\alpha}{Eh}$
5. 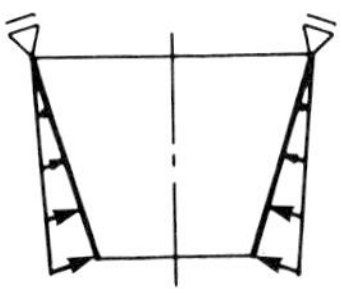	$p_1 = 0$ $p_3 = \gamma(d - s)S_\alpha$	$U_0 = -\tfrac{1}{6}(3d - 2s)C_\alpha s\gamma$ $W_0 = (d - s)C_\alpha s\gamma$ $\theta_0 = \tfrac{1}{6}(16s - 9d)\dfrac{C_\alpha s\gamma}{T_\alpha Eh}$	For closed shell, $c = 0$; $V_0 = X_0 = Z_0 = 0$ $\quad 0 = (W_0 - \nu U_0)\dfrac{sC_\alpha}{Eh}$

30.07 CIRCULAR CONICAL ELEMENT, BASIC VECTORS

(1) Finite-element model of an axisymmetrical shell (27.01) may be represented by a series of conical frusta, each bounded by two parallel circles designated as circular nodes. Since by definition the shell and the applied causes are axisymmetrical, the element is one-dimensional with three degrees of freedom assigned to each ring node (Ref. 48).

(2) Geometry of a particular element 1, 2 shown in (3) is defined by

r_1 = top ring radius (m)		a = slanted length (m)
r_2 = bottom ring radius (m)		x, y = local coordinates (m)
r = intermediate radius (m)		β = angle (rad)

All geometric, static, and displacement vector components are related to the local axes x, y, z, and their subscripts 1 and 2 identify the ring nodes of application.

(3) Conical element

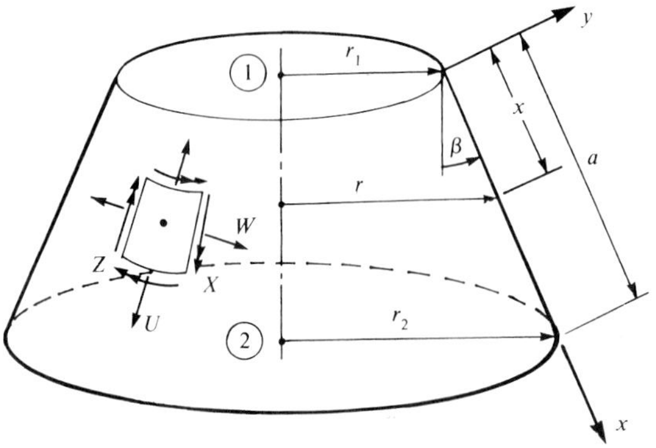

(4) Nodal displacements

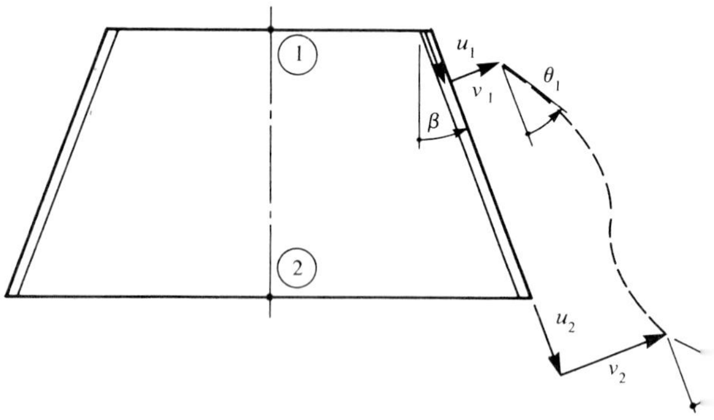

(5) Positive displacement vectors shown in (4) are

$$\Delta_1 = \{u_1, v_1, \theta_1\} \qquad \Delta_2 = \{u_2, v_2, \theta_2\}$$

where u, v = linear displacements along x, y (m)

$$\theta = \frac{dv}{dx} = \text{slope about } z \text{ (rad)}$$

(6) Positive load vector in the local system at x is

$$L = \{p_x, p_y, q_z\}$$

where p_x, p_y = intensities of distributed loads along x, y (N/m²)

q_z = intensity of distributed couple about z (N·m/m²)

(7) Positive stress-resultant vector in the local system at x is

$$f = \{U, W, X, Z\}$$

where U = meridian normal force in x (N/m)

W = circumferential normal force in z (N/m)

X, Z = bending moments about x, z (N·m/m)

The positive directions of these components at an intermediate ring are shown in (3).

30.08 CIRCULAR CONICAL ELEMENT, BASIC MATRICES

(1) Displacement matrix function (26.02-3) is

$$G = \begin{bmatrix} G_{1u} & 0 & 0 & G_{2u} & 0 & 0 \\ 0 & G_{1v} & G_{1\theta} & 0 & G_{2v} & G_{2\theta} \end{bmatrix}$$

where in terms of $\bar{x} = x/a$,

$$\begin{aligned} G_{1u} &= 1 - \bar{x} & G_{2u} &= \bar{x} \\ G_{1v} &= 1 - 3\bar{x}^2 + 2\bar{x}^3 & G_{2v} &= 3\bar{x}^2 - 2\bar{x}^3 \\ G_{1\theta} &= (\bar{x} - 2\bar{x}^2 + \bar{x}^3)a & G_{2\theta} &= -(\bar{x}^2 - \bar{x}^3)a \end{aligned}$$

(2) Strain-displacement matrix (26.02-4)

$$B = DG = \begin{bmatrix} \dfrac{d}{dx} & 0 \\ \dfrac{\sin\beta}{r} & \dfrac{\cos\beta}{r} \\ 0 & \dfrac{d^2}{dx^2} \\ 0 & \dfrac{\sin\beta}{r}\dfrac{d}{dx} \end{bmatrix} \left[\begin{array}{ccc|ccc} G_{1u} & 0 & 0 & G_{2u} & 0 & 0 \\ 0 & G_{1v} & G_{1\theta} & 0 & G_{2v} & G_{2\theta} \end{array}\right]$$

$$B = \frac{1}{r}\left[\begin{array}{ccc|ccc} -r/a & 0 & 0 & r/a & 0 & 0 \\ B_{21} & B_{22} & B_{23} & B_{24} & B_{25} & B_{26} \\ 0 & B_{32} & B_{33} & 0 & B_{35} & B_{36} \\ 0 & B_{42} & B_{43} & 0 & B_{45} & B_{46} \end{array}\right]$$

(3) Functions B_{ij} in the matrix B above are in terms of $S = \sin\beta$, $C = \cos\beta$,

$B_{21} = (1 - \bar{x})S$	$B_{22} = (1 - 3\bar{x}^2 + 2\bar{x}^3)C$ $B_{32} = -6r(1 - 2\bar{x})/a^2$ $B_{42} = -6(\bar{x} - \bar{x}^2)S/a$	$B_{23} = (\bar{x} - 2\bar{x}^2 + \bar{x}^3)aC$ $B_{33} = -2r(2 - 3\bar{x})/a$ $B_{43} = (1 - 4\bar{x} + 3\bar{x}^2)S$
$B_{24} = \bar{x}S$	$B_{25} = (3\bar{x}^2 - 2\bar{x}^3)C$ $B_{35} = 6r(1 - 2\bar{x})/a^2$ $B_{45} = 6(\bar{x} - \bar{x}^2)S/a$	$B_{26} = -(\bar{x}^2 - \bar{x}^3)aC$ $B_{36} = -2r(1 - 3\bar{x})/a$ $B_{46} = -(2\bar{x} - 3\bar{x}^2)S$

(4) Stress-strain matrix

$$E = \begin{bmatrix} hE_{11} & hE_{12} & 0 & 0 \\ hE_{21} & hE_{22} & 0 & 0 \\ 0 & 0 & \dfrac{h^3}{12}E_{11} & \dfrac{h^3}{12}E_{12} \\ 0 & 0 & \dfrac{h^3}{12}E_{21} & \dfrac{h^3}{12}E_{22} \end{bmatrix}$$

where E_{11}, $E_{12} = E_{21}$, E_{22} are the constants defined in (26.03-2).

30.09 CIRCULAR CONICAL ELEMENT, STIFFNESS MATRIX

(1) Stiffness matrix (26.02-7)

$$k = 2\pi a \int_0^1 B^{)T} E B r \, d\bar{x} = k_{11N} + k_{12N} + k_{22N} + k_{11M} + k_{12M} + k_{22M}$$

where B is the strain-displacement matrix [4 × 6], E is the matrix of elastic constants [4 × 4], and k_{11N}, k_{12N}, k_{22N}, k_{11M}, k_{12M}, k_{22M} are the component stiffness matrices [6 × 6] given below.

(2) k_N matrices

$$k_{11N} = 2\pi a h E_{11} \int_0^1 \begin{bmatrix} r/a \\ 0 \\ 0 \\ \hline -r/a \\ 0 \\ 0 \end{bmatrix} \begin{bmatrix} r/a & 0 & 0 & | & -r/a & 0 & 0 \end{bmatrix} \frac{d\bar{x}}{r}$$

$$k_{12N} = 2\pi a h E_{12} \int_0^1 \begin{bmatrix} B_{21} & -r/a \\ B_{22} & 0 \\ B_{23} & 0 \\ \hline B_{24} & r/a \\ B_{25} & 0 \\ B_{26} & 0 \end{bmatrix} \left[\begin{array}{ccc|ccc} -r/a & 0 & 0 & r/a & 0 & 0 \\ B_{21} & B_{22} & B_{23} & B_{24} & B_{25} & B_{26} \end{array}\right] \frac{d\bar{x}}{r}$$

$$k_{22N} = 2\pi a h E_{22} \int_0^1 \begin{bmatrix} B_{21} \\ B_{22} \\ B_{23} \\ \hline B_{24} \\ B_{25} \\ B_{26} \end{bmatrix} \left[\begin{array}{ccc|ccc} B_{21} & B_{22} & B_{23} & B_{24} & B_{25} & B_{26} \end{array}\right] \frac{d\bar{x}}{r}$$

where h is taken as a constant or is included in the integration process.

(3) k_M matrices

$$k_{11M} = \tfrac{1}{6}\pi a h^3 E_{11} \int_0^1 \begin{bmatrix} 0 \\ B_{32} \\ B_{33} \\ \hline 0 \\ B_{35} \\ B_{36} \end{bmatrix} \left[\begin{array}{ccc|ccc} 0 & B_{32} & B_{33} & 0 & B_{35} & B_{36} \end{array}\right] \frac{d\bar{x}}{r}$$

$$k_{12M} = \tfrac{1}{6}\pi a h^3 E_{12} \int_0^1 \begin{bmatrix} 0 & 0 \\ B_{42} & B_{32} \\ B_{43} & B_{33} \\ \hline 0 & 0 \\ B_{45} & B_{35} \\ B_{46} & B_{36} \end{bmatrix} \left[\begin{array}{ccc|ccc} 0 & B_{32} & B_{33} & 0 & B_{35} & B_{36} \\ 0 & B_{42} & B_{43} & 0 & B_{45} & B_{46} \end{array}\right] \frac{d\bar{x}}{r}$$

$$k_{22M} = \tfrac{1}{6}\pi a h^3 E_{22} \int_0^1 \begin{bmatrix} 0 \\ B_{42} \\ B_{43} \\ \hline 0 \\ B_{45} \\ B_{46} \end{bmatrix} \left[\begin{array}{ccc|ccc} 0 & B_{42} & B_{43} & 0 & B_{45} & B_{46} \end{array}\right] \frac{d\bar{x}}{r}$$

(4) Variable radius must be expressed as a function of $\bar{x}$ before the integration is carried out in (2, 3) and is

$$r = r_1 + x \cos \beta$$

Because of the complexity of integrals in (2, 3), it is preferable to use numerical integration (Ref. 35, p. 360).

(5) Approximate explicit forms with

$$R = \frac{r_1 + r_2}{2}$$

have been derived (Ref. 48) and for small elements yield extremely good results.

30.10 CIRCULAR CONICAL ELEMENT, RING NODAL LOADS

(1) Load vector due to L (30.07-6) in the local system is

$$P_L = 2\pi a h \int_0^1 \begin{bmatrix} G_{1u} & 0 & 0 \\ 0 & G_{1v} & G'_{1v} \\ 0 & G_{1\theta} & G'_{1\theta} \\ G_{24} & 0 & 0 \\ 0 & G_{2v} & G'_{2v} \\ 0 & G_{2\theta} & G'_{2\theta} \end{bmatrix} \begin{bmatrix} p_x \\ p_y \\ q_z \end{bmatrix} r\, d\bar{x}$$

where $G_{1u}, G_{1v}, \ldots, G_{2v}, G_{2\theta}$ are the displacement functions (30.08-1) and

$$G'_{1v} = \frac{dG_{1v}}{a\, d\bar{x}} \qquad G'_{1\theta} = \frac{dG_{1\theta}}{a\, d\bar{x}} \qquad G'_{2v} = \frac{dG_{2v}}{a\, d\bar{x}} \qquad G'_{2\theta} = \frac{dG_{2\theta}}{a\, d\bar{x}}$$

are their derivatives with respect to $\bar{x} = x/a$.

(2) Initial strain load vector in the local system is

$$P_0 = 2\pi a \int_0^1 \begin{bmatrix} -r/a & B_{21} & 0 & 0 \\ 0 & B_{22} & B_{32} & B_{42} \\ 0 & B_{23} & B_{33} & B_{43} \\ r/a & B_{24} & 0 & 0 \\ 0 & B_{25} & B_{35} & B_{45} \\ 0 & B_{26} & B_{36} & B_{46} \end{bmatrix} \begin{bmatrix} e_{0x} \\ e_{0y} \\ g_{0x} \\ g_{0y} \end{bmatrix} d\bar{x}$$

where r, a, B_{ij} are the same as in (30.08-2) and

$$\begin{bmatrix} e_{0x} \\ e_{0y} \\ g_{0x} \\ g_{0y} \end{bmatrix} = \begin{bmatrix} hE_{11} & hE_{12} & 0 & 0 \\ hE_{12} & hE_{22} & 0 & 0 \\ 0 & 0 & \dfrac{h^3}{12}E_{11} & \dfrac{h^3}{12}E_{12} \\ 0 & 0 & \dfrac{h^3}{12}E_{21} & \dfrac{h^3}{12}E_{22} \end{bmatrix} \begin{bmatrix} \left(\dfrac{du}{dx}\right)_0 \\ \dfrac{u_0 \sin\beta + v_0 \cos\beta}{r} \\ \left(\dfrac{d^2v}{dx^2}\right)_0 \\ \dfrac{\sin\beta}{r}\left(\dfrac{dv}{dx}\right)_0 \end{bmatrix}$$

and u_0, v_0, $(dv/dx)_0$, $(d^2v/dx^2)_0$ are the initial displacements, slope, and curvature, and E_{11}, $E_{12} = E_{21}$, E_{22} are the elastic constants introduced in (26.03-2).

Appendix A

Glossary of Symbols

A.1 GENERAL SYMBOLS IN BARS, PLATES, AND SHELLS

(1) Scalars

$a, b, s, \ldots$ = angles (1.10)

$a, b, s, \ldots$ = segments (2.04)

c = width of bar section (1.17)

d = depth of bar section (1.17)

k_u, k_v, k_w = linear foundation moduli (1.18)

k_ϕ = angular foundation modulus (1.18)

m = lumped mass or total mass (17.04)

t = time (17.03)

$t_1, t_2, \ldots, t^*$ = temperature or temperature change (1.17, 22.14)

A_x = area of normal section (1.14, B.1)

$A_y = A_x B_y = A_x$ modified by shear shape factor B_y along y (1.14, B.6)

$A_z = A_x B_y = A_x$ modified by shear shape factor B_z along z (1.14, B.6)

E = modulus of elasticity (1.14)

G = modulus of rigidity (1.14)

I_x = polar moment of inertia or torsional constant (1.14, B.1)

I_y, I_z = moments of inertia of A_x about y, z (1.14, B.1)

A.1 GENERAL SYMBOLS IN BARS, PLATES, AND SHELLS *(Continued)*

(1) Scalars (continued)

α, β = phase angles (18.03)

α_t = thermal coefficient (1.17)

$\eta = \begin{cases} +1 \text{ for tension} \\ -1 \text{ for compression} \end{cases}$ (1.18)

ρ = mass moment of inertia about x per unit length (21.01)

ν = Poisson's ratio (22.02)

μ = mass per unit length (20.01)

ω = natural frequency (18.01)

Ω = forcing frequency (18.04)

(2) Vectors

e_t = linear thermal strain along x (1.17)

f_t, g_t = angular thermal strain per unit length about z, y (1.17)

$p = \{p_x, p_y, p_z\}$ = intensity of distributed load (1.02)

$q = \{q_x, q_y, q_z\}$ = intensity of distributed moment (1.02)

$s = \{x, y, z\}$ = position vector (1.02)

u, v, w = linear displacements (deflections) along x, y, z (1.04)

x, y, z = cartesian coordinates (1.02)

$L = \{P_x, P_y, P_z, Q_x, Q_y, Q_z\}$ = load vector (1.12)

N = axial force along x (1.18)

P_x, P_y, P_z = components of concentrated load along x, y, z (1.02)

Q_x, Q_y, Q_z = components of applied moment along x, y, z (1.02)

$S = \{U, V, W, X, Y, Z\}$ = stress-resultant vector (1.03)

U = normal force along x (1.03)

V, W = shearing force along y, z (1.03)

X = torsional moment about x (1.03)

Y, Z = bending moment about y, z (1.03)

ϕ, ψ, θ = angular displacements (slopes) about x, y, z (1.04)

$\Delta = \{u, v, w, \phi, \psi, \theta\}$ = displacement vector (1.04)

$\Lambda_U, \Lambda_V, \Lambda_W$ = linear elemental flexibilities along x, y, z (1.14)

$\Lambda_X, \Lambda_Y, \Lambda_Z$ = angular elemental flexibilities about x, y, z (1.14)

(3) Superscripts

o = 0 system (reference system, global system, datum) (1.09)

s = S system (member system, local system) (1.09)

$)^T$ = matrix transpose (1.09)

$)^{-1}$ = matrix inverse (1.09)

(4) Subscripts

$i, j, k, \ldots$ = points, sections, joints (1.02)

u, v, w, x, y, z = directions along u, v, w, x, y, z (1.02, 1.18)

L = left end section (joint) (1.03)

0 = origin of coordinate system (1.02)

R = right end section (joint) (1.03)

(5) Temperature variation (1.17)

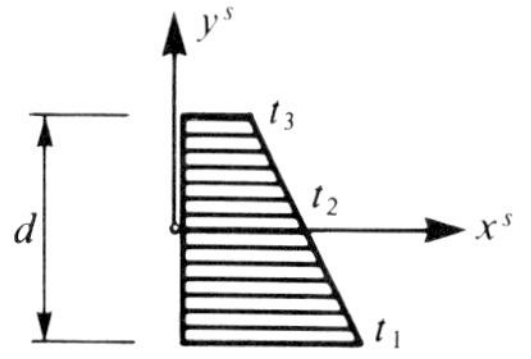

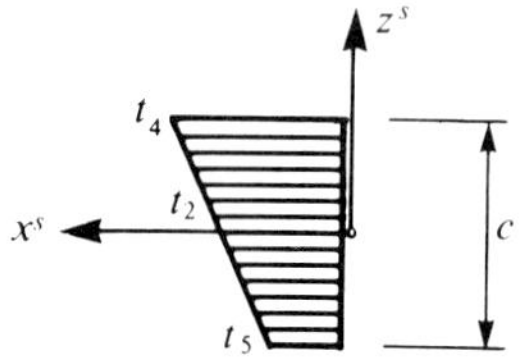

A.2 TRANSPORT ANALYSIS SYMBOLS IN BARS

(1) Scalars

$T_j(x)$ = transport function (2.05)

For $x = a$, $T_j(a) = A_j$
For $x = b$, $T_j(b) = B_j$
For $x = s$, $T_j(s) = T_j$
and $T_j - A_j = C_j$

} Transport coefficients ($j = 0, 1, 2, 3, \ldots$) (2.05)

(2) Vectors

$F_S = \{F_U, F_V, F_W, F_X, F_Y, F_Z\}$ = static load function at L
$F_\Delta = \{F_u, F_v, F_w, F_\phi, F_\psi, F_\theta\}$ = kinematic load function at L
$G_S = \{G_U, G_V, G_W, G_X, G_Y, G_Z\}$ = static load function at R
$G_\Delta = \{G_u, G_v, G_w, G_\phi, G_\psi, G_\theta\}$ = kinematic load function at R

(2.02, 2.05, 2.17)

$S_L = \{U_L, V_L, W_L, X_L, Y_L, Z_L\}$ = stress resultant at L
$\Delta_L = \{u_L, v_L, w_L, \phi_L, \psi_L, \theta_L\}$ = displacement at L
$S_R = \{U_R, V_R, W_R, X_R, Y_R, Z_R\}$ = stress resultant at R
$\Delta_R = \{u_R, v_R, w_R, \phi_R, \psi_R, \theta_R\}$ = displacement at R

(2.02, 2.06, 2.18)

$H_L = \{S_L, \Delta_L\}$ = state vector at L
$H_R = \{S_R, \Delta_R\}$ = state vector at R

(2.02)

(3) Matrices

c_{LR}, c_{RL} = displacement deviation matrices
d_{LR}, d_{RL} = stress deviation matrices
t_{LR}, t_{RL} = static transport matrices
$t_{LR}^{)T}, t_{RL}^{)T}$ = kinematic transport matrices

(2.02, 2.04, 2.06, 2.18)

R^{os}, R^{so} = angular transformation matrices
T_{LR}^{o}, T_{RL}^{o} = transport matrices in 0 system
T_{LR}^{s}, T_{RL}^{s} = transport matrices in S system

(2.02, 2.06, 2.18, 2.19)

(4) Matrix equations

$$\underbrace{\begin{bmatrix} 1 \\ S_L \\ \Delta_L \end{bmatrix}}_{H_L} = \underbrace{\begin{bmatrix} 1 & 0 & 0 \\ F_S & t_{LR} & c_{LR} \\ F_\Delta & d_{LR} & t_{RL}^{)T} \end{bmatrix}}_{T_{LR}} \underbrace{\begin{bmatrix} 1 \\ S_R \\ \Delta_R \end{bmatrix}}_{H_R} \qquad \underbrace{\begin{bmatrix} 1 \\ S_R \\ \Delta_R \end{bmatrix}}_{H_R} = \underbrace{\begin{bmatrix} 1 & 0 & 0 \\ G_S & t_{RL} & c_{RL} \\ G_\Delta & d_{RL} & t_{LR}^{)T} \end{bmatrix}}_{T_{RL}} \underbrace{\begin{bmatrix} 1 \\ S_L \\ \Delta_L \end{bmatrix}}_{H_L} \qquad (2.02)$$

$$\begin{bmatrix} F_S \\ F_\Delta \end{bmatrix} = -\begin{bmatrix} t_{LR} & c_{LR} \\ d_{LR} & t_{RL}^{)T} \end{bmatrix}\begin{bmatrix} G_S \\ G_\Delta \end{bmatrix} \qquad \begin{bmatrix} G_S \\ G_\Delta \end{bmatrix} = -\begin{bmatrix} t_{RL} & c_{RL} \\ d_{RL} & t_{LR}^{)T} \end{bmatrix}\begin{bmatrix} F_S \\ F_\Delta \end{bmatrix} \qquad (2.02)$$

(5) Transport-method sign convention (1.22)

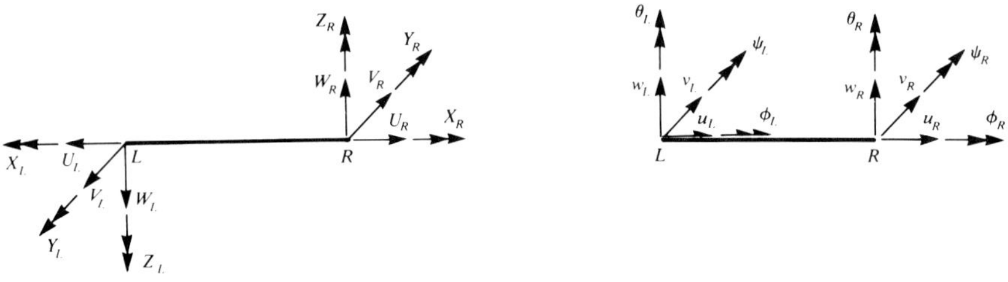

A.3 FLEXIBILITY ANALYSIS SYMBOLS IN BARS

(1) Scalars

$F_j(x)$ = flexibility function (3.04, 3.09)

For $x = s$, $F_j(s) = F_j$ } flexibility coefficients ($j = 1, 2, 3, \ldots$)
For $x = 0$, $F_j(0) = 0$ } (3.04, 3.09)

(2) Vectors

$S_L = \{U_L, V_L, W_L, X_L, Y_L, Z_L\}$ = stress resultant at L
$S_R = \{U_R, V_R, W_R, X_R, Y_R, Z_R\}$ = stress resultant at R (3.02, 3.04, 3.10)

$\Delta_{LR} = \{u_{LR}, v_{LR}, w_{LR}, \phi_{LR}, \psi_{LR}, \theta_{LR}\}$ = displacement at L
$\Delta_{RL} = \{u_{RL}, v_{RL}, w_{RL}, \phi_{RL}, \psi_{RL}, \theta_{RL}\}$ = displacement at R (3.02, 3.04, 3.10)

$\Delta_{L0} = \{u_{L0}, v_{L0}, w_{L0}, \phi_{L0}, \psi_{L0}, \theta_{L0}\}$ = load function at L
$\Delta_{R0} = \{u_{R0}, v_{R0}, w_{R0}, \phi_{R0}, \psi_{R0}, \theta_{R0}\}$ = load function at R (3.02, 3.03, 3.10)

(3) Matrices, free bars

$f_{LL} = -d_{LR}t_{RL}$ = direct flexibility matrix at L
$f_{LR} = 0$ = indirect flexibility matrix at L
$f_{RL} = 0$ = indirect flexibility matrix at R
$f_{RR} = d_{RL}t_{LR}$ = direct flexibility matrix at R (3.02)

(4) Matrices, interactive bars

$f_{LL} = c_{RL}^{)-1}t_{RL}$ = direct flexibility matrix at L
$f_{LR} = -c_{RL}^{)-1} = f_{RL}^{)T}$ = indirect flexibility matrix at L
$f_{RL} = c_{LR}^{)-1} = f_{LR}^{)T}$ = indirect flexibility matrix at R
$f_{RR} = -c_{LR}^{)-1}t_{LR}$ = direct flexibility matrix at R (3.02)

(5) Matrix equations, free bars

$$\begin{bmatrix} \Delta_{LR} \\ \Delta_{RL} \end{bmatrix} = \begin{bmatrix} f_{LL} & 0 \\ 0 & f_{RR} \end{bmatrix} \begin{bmatrix} S_L \\ S_R \end{bmatrix} + \begin{bmatrix} \Delta_{L0} \\ \Delta_{R0} \end{bmatrix} \qquad \begin{bmatrix} \Delta_{L0} \\ \Delta_{R0} \end{bmatrix} = \begin{bmatrix} 0 & -t_{RL}^{)T} \\ -t_{LR}^{)T} & 0 \end{bmatrix} \begin{bmatrix} F_\Delta \\ G_\Delta \end{bmatrix} \qquad (3.02)$$

(6) Matrix equations, interactive bars

$$\begin{bmatrix} \Delta_{LR} \\ \Delta_{RL} \end{bmatrix} = \begin{bmatrix} f_{LL} & f_{LR} \\ f_{RL} & f_{RR} \end{bmatrix} \begin{bmatrix} S_L \\ S_R \end{bmatrix} + \begin{bmatrix} \Delta_{L0} \\ \Delta_{R0} \end{bmatrix} \qquad \begin{bmatrix} \Delta_{L0} \\ \Delta_{R0} \end{bmatrix} = \begin{bmatrix} 0 & -f_{LR} \\ -f_{RL} & 0 \end{bmatrix} \begin{bmatrix} F_S \\ G_S \end{bmatrix} \qquad (3.02)$$

(7) Flexibility-method sign convention (1.22)

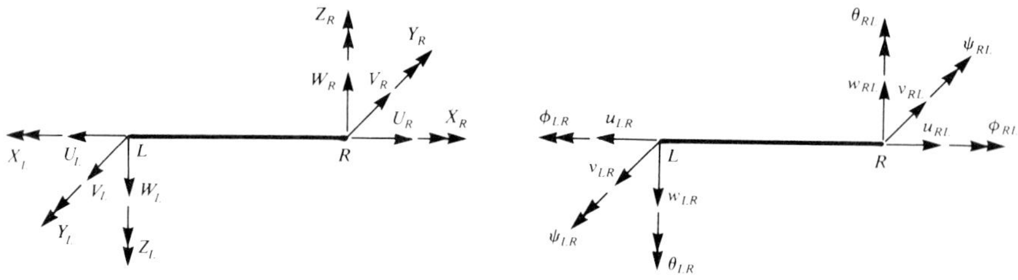

A.4 STIFFNESS ANALYSIS SYMBOLS IN BARS

(1) Scalars

$K_j(x)$ = stiffness function (3.12, 3.18)

$$\left.\begin{array}{ll} \text{For } x = s, & K_j(s) = K_j \\ \text{For } x = 0, & K_j(0) = 0 \end{array}\right\} \text{stiffness coefficients } (j = 1, 2, 3, \ldots) \qquad (3.12, 3.18)$$

(2) Vectors

$$\left.\begin{array}{l} S_{LR} = \{U_{LR}, V_{LR}, W_{LR}, X_{LR}, Y_{LR}, Z_{LR}\} = \text{reaction at } L \\ S_{RL} = \{U_{RL}, V_{RL}, W_{RL}, X_{RL}, Y_{RL}, Z_{RL}\} = \text{reaction at } R \end{array}\right\} \qquad (3.11, 3.12, 3.19)$$

$$\left.\begin{array}{l} S_{L0} = \{U_{L0}, V_{L0}, W_{L0}, X_{L0}, Y_{L0}, Z_{L0}\} = \text{load function at } L \\ S_{R0} = \{U_{R0}, V_{R0}, W_{R0}, X_{R0}, Y_{R0}, Z_{R0}\} = \text{load function at } R \end{array}\right\} \qquad (3.11, 3.13, 3.19)$$

$$\left.\begin{array}{l} \Delta_L = \{u_L, v_L, w_L, \phi_L, \psi_L, \theta_L\} = \text{displacement at } L \\ \Delta_R = \{u_R, v_R, w_R, \phi_R, \psi_R, \theta_R\} = \text{displacement at } R \end{array}\right\} \qquad (3.11, 3.13, 3.19)$$

(3) Matrices, all bars

$$\left.\begin{array}{lll} k_{LL} = & d)_{RL}^{-1} t)_{RL}^{T} & = \text{direct stiffness matrix at } L \\ k_{LR} = & -d)_{RL}^{-1} = k)_{RL}^{T} & = \text{indirect stiffness matrix at } L \\ k_{RL} = & d)_{LR}^{-1} = k)_{LR}^{T} & = \text{indirect stiffness matrix at } R \\ k_{RR} = & -d)_{LR}^{-1} t)_{LR}^{T} & = \text{direct stiffness matrix at } R \end{array}\right\} \qquad (3.11)$$

(4) Relations, free bars (3.02, 3.11)

$k_{LL} = f)_{LL}^{-1}$	$k_{RR} = f)_{RR}^{-1}$
$= t_{LR} f)_{RR}^{-1} t)_{LR}^{T}$	$= t_{RL} f)_{LL}^{-1} t)_{RL}^{T}$
$= t_{LR} k_{RR} t)_{LR}^{T}$	$= t_{RL} k_{LL} t)_{RL}^{T}$
$k_{LR} = k)_{RL}^{T}$	$k_{RL} = k)_{LR}^{T}$
$= -f)_{LL}^{-1} t)_{RL}^{T} = -t_{LR} f)_{RR}^{-1}$	$= -f)_{RR}^{T} t)_{LR}^{T} = -t_{RL} f)_{LL}^{-1}$
$= -k_{LL} t)_{RL}^{T} = -t_{LR} k_{RR}$	$= -k_{RR} t)_{LR}^{T} = -t_{RL} k_{LL}$
$S_{L0} = k_{LL} \Delta_{L0}$	$S_{R0} = -k_{RR} \Delta_{R0}$

(5) Matrix equations, all bars

$$\begin{bmatrix} S_{LR} \\ S_{RL} \end{bmatrix} = \begin{bmatrix} k_{LL} & k_{LR} \\ k_{RL} & k_{RR} \end{bmatrix} \begin{bmatrix} \Delta_L \\ \Delta_R \end{bmatrix} + \begin{bmatrix} S_{L0} \\ S_{R0} \end{bmatrix} \qquad \begin{bmatrix} S_{L0} \\ S_{R0} \end{bmatrix} = \begin{bmatrix} 0 & -k_{LR} \\ -k_{RL} & 0 \end{bmatrix} \begin{bmatrix} F_\Delta \\ G_\Delta \end{bmatrix} \qquad (3.11)$$

(6) Stiffness-method sign convention (1.22)

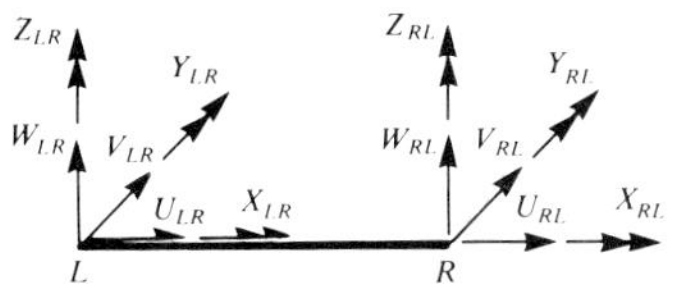

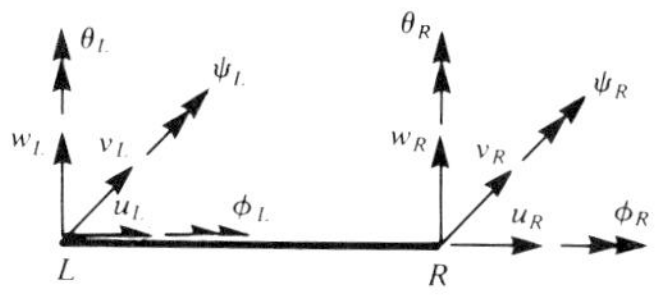

A.5 PLATE ANALYSIS SYMBOLS

(1) Scalars

$a, b, \ldots$ = segments (23.10)
f = plate width (24.02)
h = plate thickness (24.02)
r = radial coordinate (22.06)
s = plate length (24.02)
R = plate outer radius (23.03)

$$D = \frac{Eh^3}{12(1-\nu^2)} \qquad (22.03)$$

$$\alpha = \tfrac{1}{2}(1+\nu) \qquad (22.03)$$

$$\beta = \tfrac{1}{2}(1-\nu) \qquad (22.03)$$

For definitions of k_w, t^*, E, G, ν, α_t, η and A_j, B_j, C_j, T_j see A.1 and A.2, respectively.

D_{xx}, D_{xy}, D_{yy}, H = angular rigidities of orthotropic plate in cartesian system (22.10)

D_{rr}, $D_{r\theta}$, $D_{\theta\theta}$, H = angular rigidities of orthotropic plate in cylindrical system (22.11)

$\nabla^2 = \dfrac{\partial^2}{\partial x^2} + \dfrac{\partial^2}{\partial y^2}$ = del second operator in cartesian system (22.04)

$\nabla^2 = \dfrac{\partial^2}{\partial r^2} + \dfrac{\partial}{r\,\partial r} + \dfrac{\partial^2}{r^2\,\partial \theta^2}$ = del second operator in cylindrical system (22.07)

(2) Vectors

f_{xx}, f_{yy}, f_{zz} = normal stresses in cartesian system (22.02)
f_{xy}, f_{yz}, f_{zx} = shearing stresses in cartesian system (22.02)
f_{rr}, $f_{\theta\theta}$ = normal stresses in cylindrical system (22.06)
$f_{r\theta}$ = shearing stress in cylindrical system (22.06)
H_B, H_C, H_R = state vectors at B, C, R in solid isotropic circulate plate (23.03)
H_A, H_B, H_R = state vectors at A, B, R in annular isotropic plate (23.10)
H_{Lm}, H_{Rm} = mth-state vector at L, R in isotropic rectangular plate (24.04)
For definition of w, ψ, and load functions see Appendices A.1, A.2, and A.4.
M_x, M_y = bending moment per unit length about y, x (22.03)
M_{xy}, M_{yx} = twisting moment per unit length about x, y (22.03)
M_r, M_θ = bending moment per unit length about t, r (22.07)
$M_{r\theta}$, $M_{\theta r}$ = twisting moment per unit length about r, t (22.07)
N_x, N_y = normal force per unit length along x, y (22.03)
N_r, N_θ = normal force per unit length along r, t (22.07)
R_{xz}, R_{yz} = vertical edge reaction per unit length normal to x, y (22.03)
R_{rz} = vertical edge reaction per unit length normal to r (22.07)
W_{xz}, W_{yz} = vertical shearing force per unit length normal to x, y (22.03)
W_{rz} = vertical shearing force per unit length normal to r (22.07)
ε_{xx}, ε_{yy}, ε_{zz} = normal strains in cartesian system (22.02)
ε_{xy}, ϵ_{yz}, ϵ_{zx} = shearing strains in cartesian system (22.02)
ε_{rr}, $\epsilon_{\theta\theta}$ = normal strains in cylindrical system (22.06)
$\varepsilon_{r\theta}$ = shearing strain in cylindrical system (22.06)

(3) Matrices and matrix equations. Transport, flexibility, and stiffness symbols in plate analysis are identical to those introduced for bars in Appendices A.2–A.4.

A.6 SHELL ANALYSIS SYMBOLS

(1) Scalars

$a, b, s, \ldots$ = angles (29.02)
$a, b, s, \ldots$ = segments (27.07)
h = shell thickness (27.01)
l = shell length (27.07)
r = variable radius (27.02)
R = constant radius (29.01)
R_0 = radius of horizontal circle (27.02)
R_1, R_2 = principal radii of curvature (27.02)
α, β = constant spherical angles (29.02)
ϑ = variable polar angle (27.02)
ϕ = variable spherical angle (27.02)

$$K = \frac{Eh}{1 - \nu^2} \quad (27.04)$$

$$D = \frac{Eh^3}{12(1 - \nu^2)} \quad (27.04)$$

For definition of t^*, E, G, α_t, ν and A_j, B_j, C_j, T_j see Appendices A.1 and A.2, respectively.

(2) Vectors

f_1, f_2 = normal stresses along 1, 2 (27.04)
f_{10}, f_{20} = normal membrane stresses along 1, 2 (27.03)
p_1, p_3 = intensities of distributed load along 1, 3 (27.02)
$u_0 = \Delta_{10}$ = linear membrane displacement along 1 (27.03, 28.08)
$v_0 = -\Delta_{30}$ = linear membrane displacement along 3 (27.03, 28.08)
s_1, s_2, s_3 = coordinates along 1, 2, 3 (27.03)
H = intensity of horizontal force along y (29.02)
L_ϕ = total vertical load (29.03)
$M_1 = Z$ = intensity of bending moment about 2 (27.02, 28.08)
$M_2 = X$ = intensity of bending moment about 1 (27.02, 28.08)
M_t = intensity of thermal moment (27.05)
$N_1 = U$ = intensity of normal force along 1 (27.02, 28.08)
$N_2 = W$ = intensity of normal force along 2 (27.02, 28.08)
$N_3 = -V$ = intensity of shearing force along 3 (27.02, 28.08)
$N_{10} = U_0$ = intensity of normal membrane force along 1 (27.03, 28.09)
$N_{20} = W_0$ = intensity of normal membrane force along 2 (27.03, 27.09)
N_t = intensity of thermal normal force (27.05)
$\varepsilon_1, \varepsilon_2$ = normal strain along 1, 2 (27.04)
$\varepsilon_{10}, \varepsilon_{20}$ = normal membrane strains along 1, 2 (27.03)
ε_t = normal thermal strain (27.03)
$\psi_1 = -\theta$ = slope about 2 (27.04, 28.02)
$\psi_{10} = -\theta_0$ = membrane slope about 2 (27.03, 28.09)
κ_1, κ_2 = bending measures about 2, 1 (27.04)
κ_t = thermal bending measure (27.04)
$\Delta_1 = u$ = linear displacement along 1 (27.04)
$\Delta_{10} = u_0$ = linear membrane displacement along 1 (27.03, 28.08)
$\Delta_3 = -v$ = linear displacement along 3 (27.04)
$\Delta_{30} = -v_0$ = linear membrane displacement along 3 (27.03, 28.08)
Δ_r = horizontal linear displacement along r (27.02)
Δ_x = vertical linear displacement along x (27.02)
$\Delta_{r0} = \Delta_0$ = horizontal linear membrane displacement along r (27.03, 29.10)
Δ_{x0} = vertical linear membrane displacement along x (27.03)

(3) Matrices and matrix equations. Transport, flexibility, and stiffness symbols in shell analysis are identical to those introduced for bars in Appendices A.2–A.4.

A.7 FINITE DIFFERENCE AND ELEMENT SYMBOLS

(1) Scalars

$a, b, \ldots$ = segments (26.06, 30.02, 30.10)
h = plate thickness, shell thickness (26.05, 30.02, 30.10)
s_1, s_2, s_3 = dimensionless area coordinates (26.07)
A_0 = area of triangular element (26.05)
A_1, A_2, A_3 = subareas of triangular element (26.07)
$E_{11}, E_{12} = E_{21}, E_{22}$ = orthotropic elastic constants (26.03, 26.04)
V = volume of element (26.02)
α_{tx}, α_{ty} = thermal coefficients along x, y (26.02)
ρ = mass per unit volume (26.02)
$\Lambda_{11}, \Lambda_{12} = \Lambda_{21}, \Lambda_{22}$ = orthotropic elastic constants (26.03, 26.04)
For definition of t^*, E, G, α_t, ν see Appendix A.1.

(2) Vectors

e = strain vector (26.02)
e_0 = initial strain vector (26.02)
e_T = thermal strain vector (26.02)
f = interior stress vector (26.02)
P_L = mechanical nodal load vector (26.02)
P_0 = initial deformation nodal vector (26.02)
P_T = thermal deformation nodal vector (26.02)
δ = interior displacement vector (26.02)
Δ = nodal displacement vector (26.02)
$\ddot{\Delta}$ = nodal acceleration vector (26.02)
For standard plate and shell symbols see Appendices A.5 and A.6.

(3) Matrices

m = elemental mass matrix (26.02)
$k = k_n + k_s$ = elemental stiffness matrix (26.02, 26.05)
B = strain-displacement matrix (26.02)
D = linear differential operator matrix (26.02)
E = stress-strain matrix (26.02, 26.04, 26.04)
F = geometric matrix (26.02)
G = matrix of shape functions (26.02)
S = stress-displacement matrix (26.02)
Λ = strain-stress matrix (26.02, 26.04, 26.05)

Appendix B

Properties of Normal Section

B.1 NOTATION

$a, b, e, h, r, t, B, H, R$ = segments (m)

C = centroid of normal section

A_x = area of normal section (m^2)

B_y, B_z = shear shape factors

I_x = polar moment of inertia or torsional constant of A_x about x (m^4)

I_y, I_z = moments of inertia of A_x about y, z (m^4)

I_v, I_w = moments of inertia of A_x about v, w (m^4)

I_{vw} = product of inertia of A_x in v, w (m^4)

α = angle (rad)

B.2 STATIC AND INERTIA FUNCTIONS OF POLYGONAL SECTIONS

1. Rectangle $b \leq h$

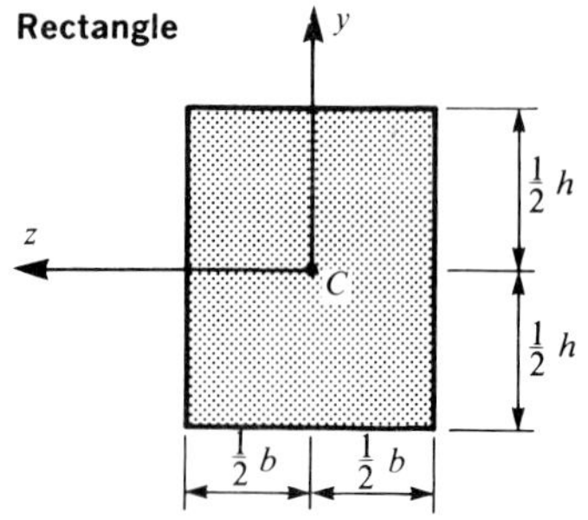

$A_x = bh \qquad \lambda = \dfrac{b}{h}$

$$I_x = \begin{cases} \dfrac{hb^3}{3}(1 - 0.630\lambda + 0.052\lambda^5) & \text{for } \lambda < 1 \\ \dfrac{9b^4}{64} & \text{for } \lambda = 1 \end{cases}$$

$$I_y = \frac{hb^3}{12}$$

$$I_z = \frac{bh^3}{12}$$

2. Isosceles $b \lessgtr h$

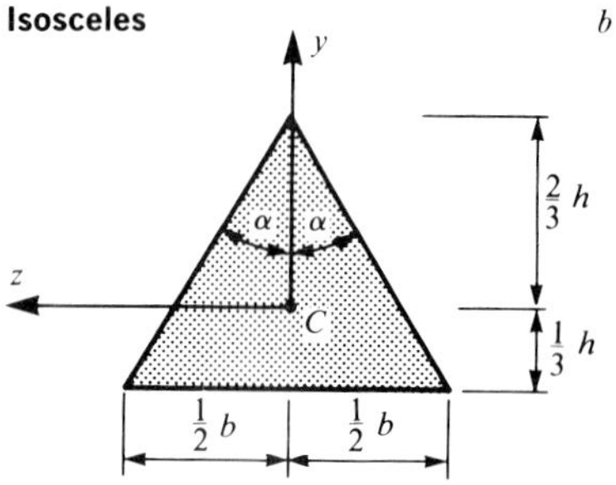

$A_x = \frac{1}{2}bh \qquad \lambda = \dfrac{b}{h} \qquad \alpha = \tan^{-1}\dfrac{b}{2h}$

$$I_x = \begin{cases} \dfrac{hb^3}{3}(2 - 5.04\alpha + 11.13\alpha^2) & \text{for } \lambda < \frac{2}{3} \\ \dfrac{hb^3}{20 + 15\lambda^2} & \text{for } \frac{2}{3} \leq \lambda < \sqrt{3} \\ \dfrac{hb^3}{11}(\lambda - 0.8592) & \text{for } \sqrt{3} \leq \lambda < 2\sqrt{3} \end{cases}$$

$$I_y = \frac{hb^3}{48} \qquad I_z = \frac{bh^3}{36}$$

3. Symmetrical trapezoid

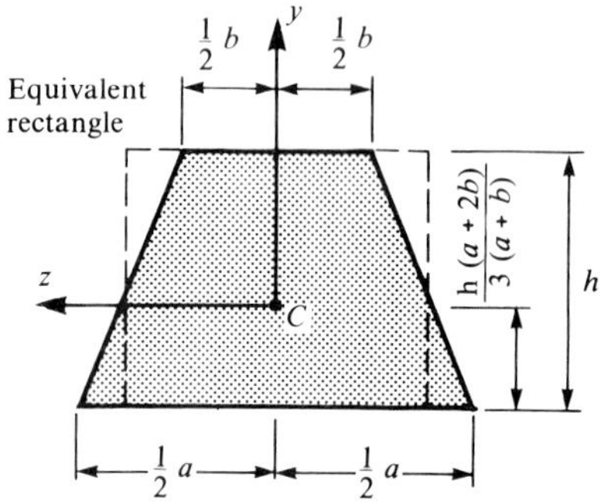

$A_x = \frac{1}{2}(a + b)h$

$I_x =$ approximate value of I_x is calculated as I_x of the equivalent area rectangle

$$I_y = \frac{(a^4 - b^4)h}{48(a - b)}$$

$$I_z = \frac{(a^2 + 4ab + b^2)h^3}{36(a + b)}$$

4. Regular hexagon

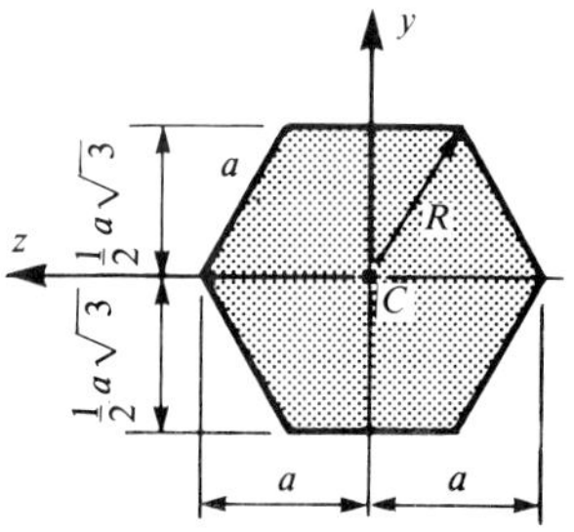

$R = a$
$A_x = 2.59808a^2$
$I_x = 1.38503a^4$
$I_y = 0.54127a^4$
$I_z = 0.54127a^4$

5. Regular octagon

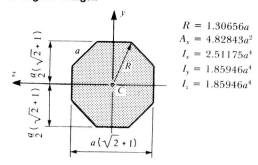

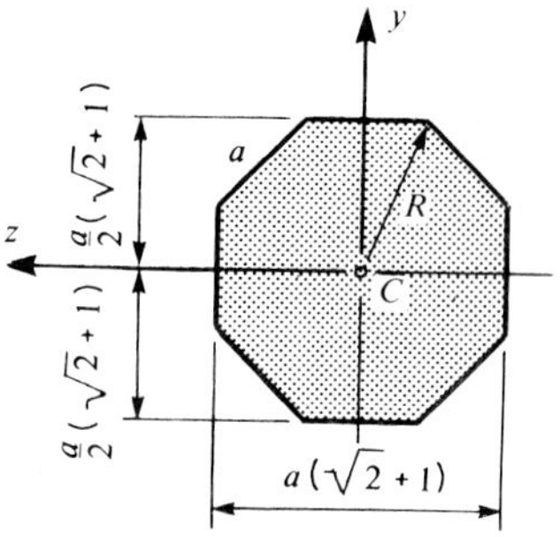

$R = 1.30656a$
$A_x = 4.82843a^2$
$I_x = 2.51175a^4$
$I_y = 1.85946a^4$
$I_z = 1.85946a^4$

B.3 STATIC AND INERTIA FUNCTIONS OF ELLIPTIC AND CIRCULAR SECTIONS

1. Ellipse

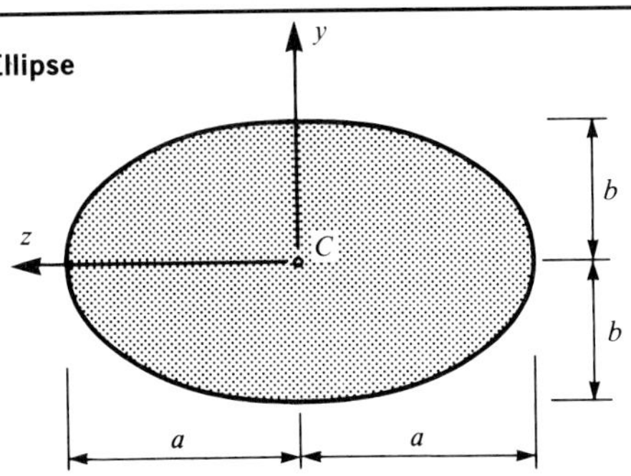

$$A_x = \pi ab$$

$$I_x = \frac{\pi a^3 b^3}{a^2 + b^2}$$

$$I_y = \tfrac{1}{4}\pi a^3 b$$

$$I_z = \tfrac{1}{4}\pi a b^3$$

For circle, $a = b = R$

2. Half ellipse

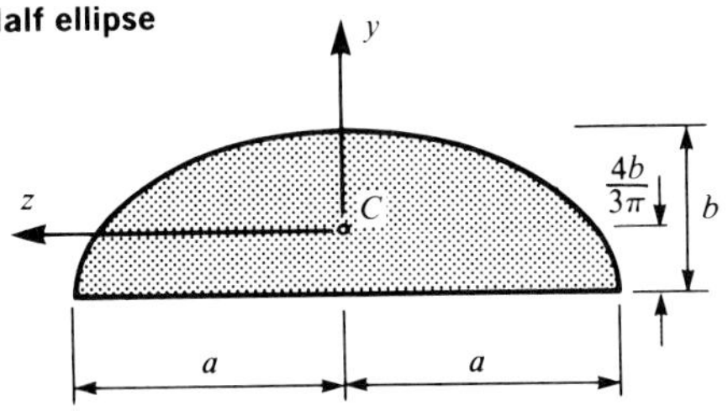

$$A_x = \tfrac{1}{2}\pi ab$$

$$I_x \cong 0.14875(a^4 + b^4)$$

$$I_y = \tfrac{1}{8}\pi a^3 b$$

$$I_z = \tfrac{1}{8}\pi a b^3\left(1 - \frac{64}{9\pi^2}\right)$$

For half circle, $a = b = R$

3. Flattened circle

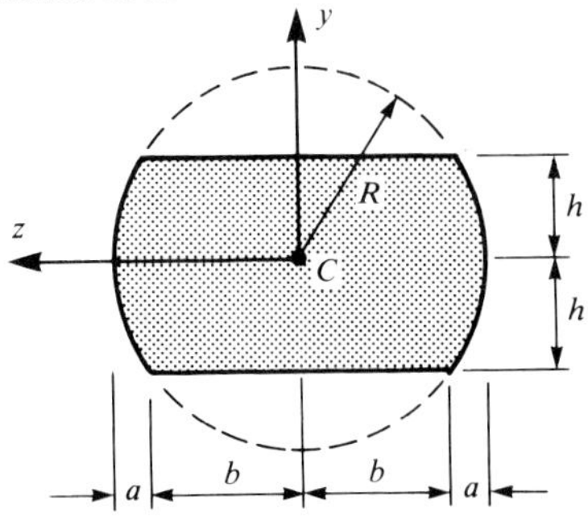

$$A_x = 2(\alpha R^2 + bh)$$

$$I_x = CR^4$$

$$I_y \cong \frac{2a + 5b}{240} h^3$$

$$I_z = \tfrac{1}{6}(2hb^3 + 3hbR^2 + 3\alpha R^4)$$

(Ref. 64)

h/R	8/16	9/16	10/16	11/16	12/16	13/16	14/16	15/16
C	0.4382	0.5856	0.7333	0.9050	1.0767	1.2167	1.3567	1.4637

4. Circular sector

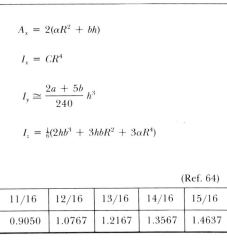

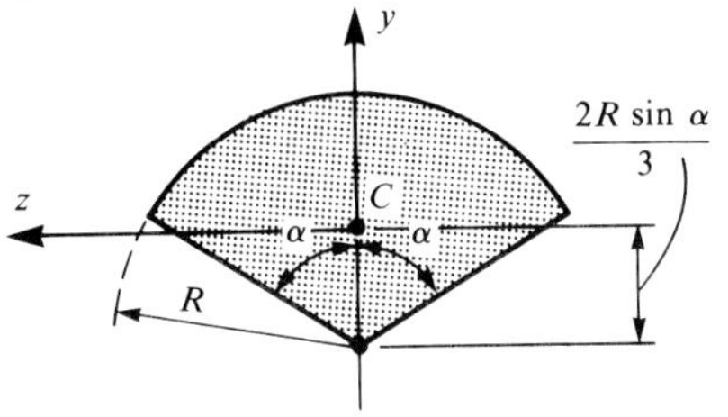

$$A_x = \alpha R^2$$

$$I_x = \tfrac{1}{10} C R^4$$

$$I_y = \tfrac{1}{4}(\alpha - \sin\alpha\cos\alpha)R^4$$

$$I_z = \tfrac{1}{4}\left(\alpha + \sin\alpha\cos\alpha - \frac{16\sin^2\alpha}{9\alpha}\right)R^4$$

(Ref. 63, pp. 275–280)

2α	$\pi/4$	$\pi/3$	$\pi/2$	$2\pi/3$	π	$3\pi/2$	$5\pi/3$	2π
C	0.181	0.349	0.825	1.483	2.962	5.732	6.867	8.783

B.4 STATIC AND INERTIA FUNCTIONS OF HOLLOW SECTIONS

1. Hollow rectangle

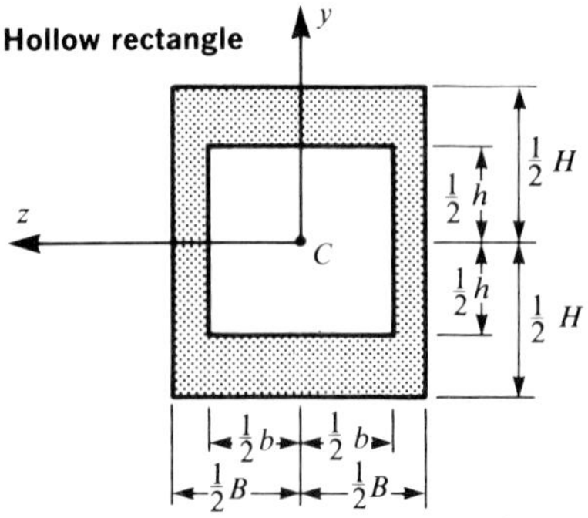

$$A_x = BH - bh$$

$$I_x = \frac{(B+b)^2(H+h)^2(B-b)(H-h)}{2(B^2 - b^2 + H^2 - h^2)}$$

$$I_y = \tfrac{1}{12}(HB^3 - hb^3)$$

$$I_z = \tfrac{1}{12}(BH^3 - bh^3)$$

2. Rectangular tube

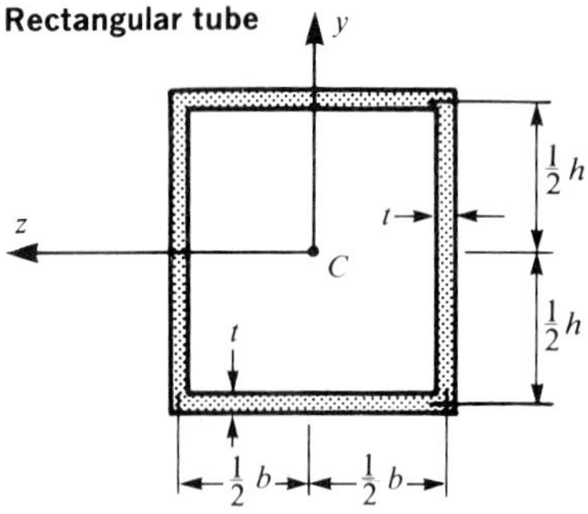

$$A_x = 2(b + h)t$$

$$I_x = \frac{2b^2h^2t}{b + h}$$

$$I_y = \tfrac{1}{6}(b + 3h)tb^2$$

$$I_z = \tfrac{1}{6}(3b + h)th^2$$

3. Hollow circle

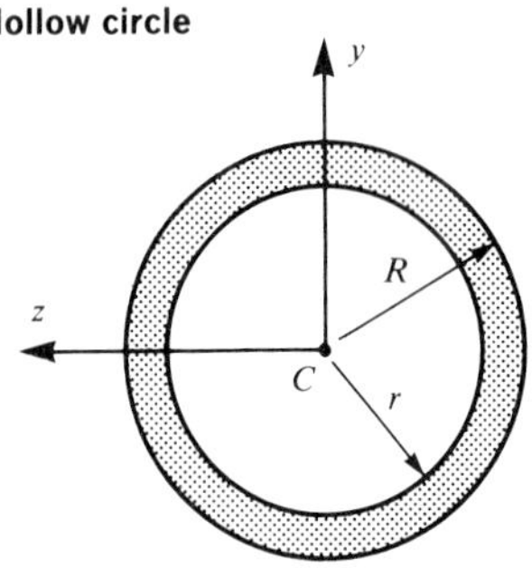

$$A_x = \pi(R^2 - r^2)$$

$$I_x = \tfrac{1}{2}\pi(R^4 - r^4)$$

$$I_y = \tfrac{1}{4}\pi(R^4 - r^4)$$

$$I_z = \tfrac{1}{2}\pi(R^4 - r^4)$$

4. Circular tube

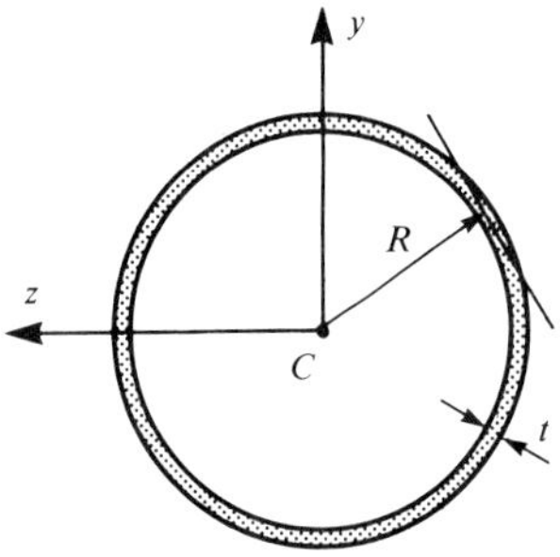

$$A_x = 2\pi Rt$$

$$I_x = 2\pi R^3 t$$

$$I_y = \pi R^3 t$$

$$I_z = \pi R^3 t$$

B.5 STATIC AND INERTIA FUNCTIONS OF COMPOSITE SECTIONS

1. I section

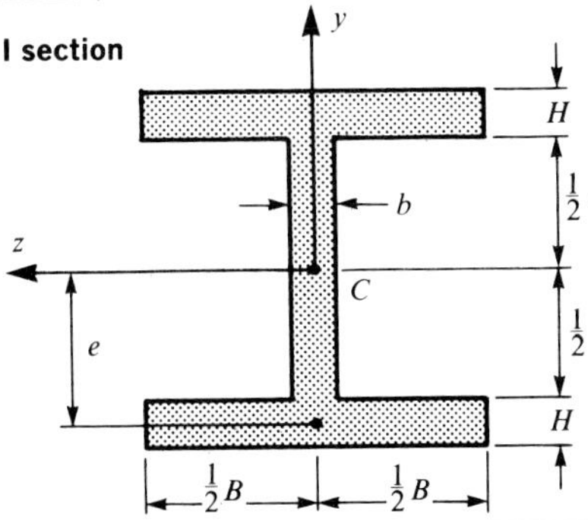

$$A_x = 2BH + bh \qquad A_{1x} = BH \qquad A_{2x} = bh$$

$$e = \frac{h + H}{2}$$

$$I_x = \frac{1.3}{3}(2BH^3 + hb^3)$$

$$I_y = \frac{1}{12}(2HB^3 + hb^3)$$

$$I_z = \frac{1}{12}(2BH^3 + bh^3) + 2A_{1x}e^2$$

2. T section

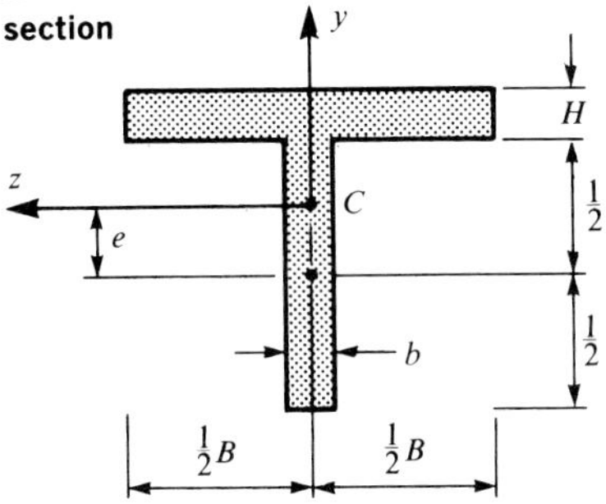

$$A_x = BH + bh \qquad A_{1x} = BH \qquad A_{2x} = bh$$

$$e = \frac{(H + h)A_{1x}}{2A_x}$$

$$I_x = \frac{1.12}{3}(BH^3 + hb^3)$$

$$I_y = \frac{1}{12}(HB^3 + hb^3)$$

$$I_z = \frac{1}{12}(BH^3 + bh^3) + \frac{A_{1x}A_{2x}e^2}{A_x}$$

3. + section

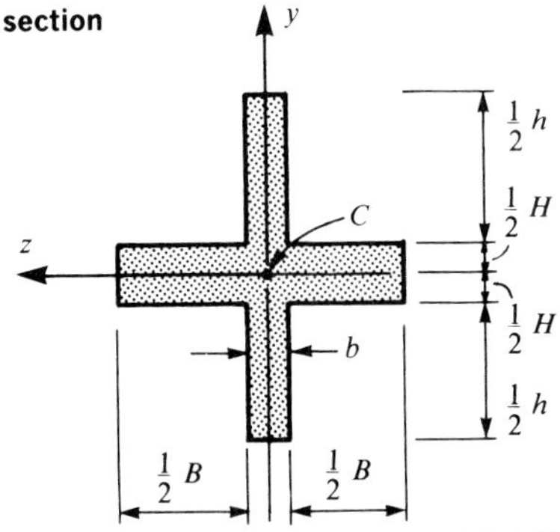

$$A_x = BH + b(h + H)$$

$$I_x = \frac{1.17}{3}[(B + b)H^3 + hb^3]$$

$$I_y = \frac{1}{12}[H(b + B)^3 + hb^3]$$

$$I_z = \frac{1}{12}[BH^3 + b(h + H)^3]$$

4. V section

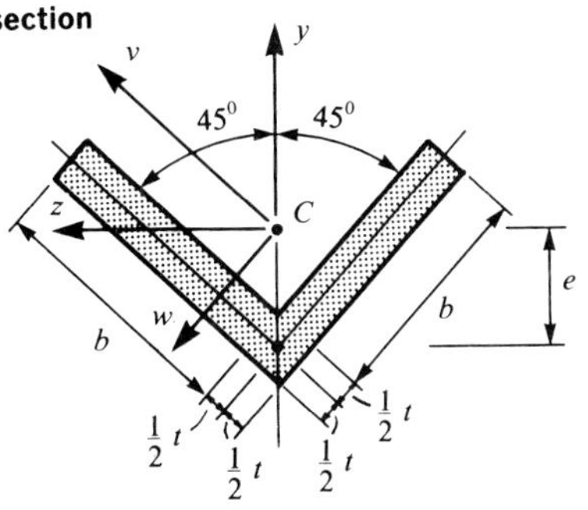

$$A_x = t^2 + 2bt \qquad A_{1x} = t^2 \qquad A_{2x} = bt$$

$$e = \frac{A_{2x}(b + t)\sqrt{2}}{A_x}$$

$$I_x = \tfrac{1}{3}(t + 2b)t^3$$

$$I_v = \tfrac{1}{12}[(t + b)t^3 + tb^3] + \frac{(A_{1x} + A_{2x})A_{2x}e^2}{2A_x}$$

$$I_w = I_v \qquad I_{vw} = \frac{b^2(b + t)^2 t}{4(2b + t)}$$

$$I_y = I_v + I_{vw} \qquad I_z = I_v - I_{vw}$$

B.6 SHEAR SHAPE FACTORS OF NORMAL SECTIONS†

1. Rectangle

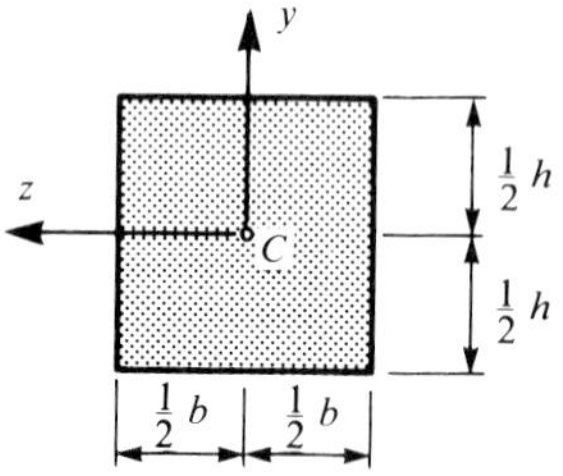

$B_y = B_z \cong \frac{5}{6}$

2. Circle

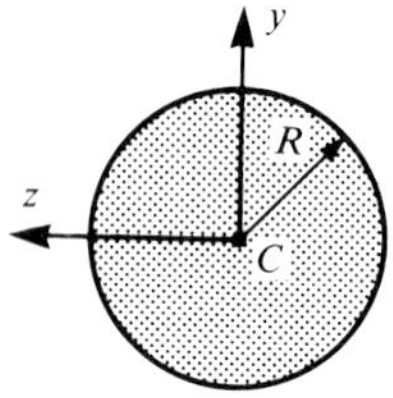

$B_y = B_z \cong \frac{9}{10}$

3. Rectangular tube

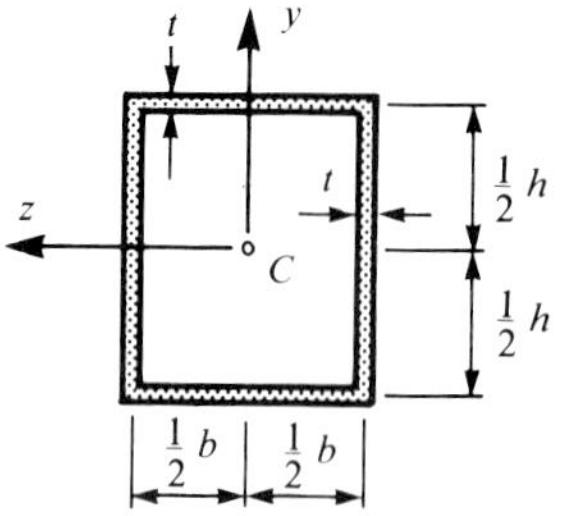

$B_y = B_z \cong \frac{1}{2}$

4. Circular tube

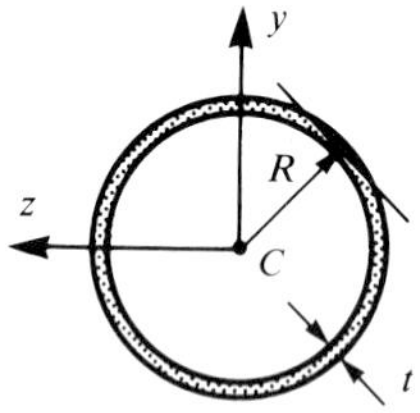

$B_y = B_z \cong \frac{1}{2}$

5. I section

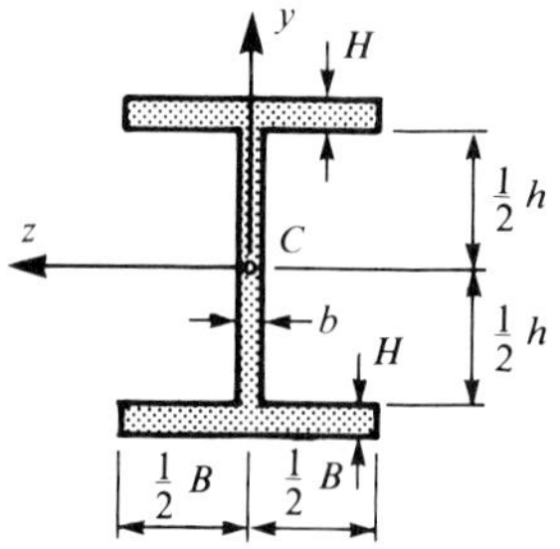

$B_y \cong \dfrac{bh}{2BH + bh}$ $\quad B_z \cong \dfrac{5BH}{6(2BH + bh)}$

6. T section

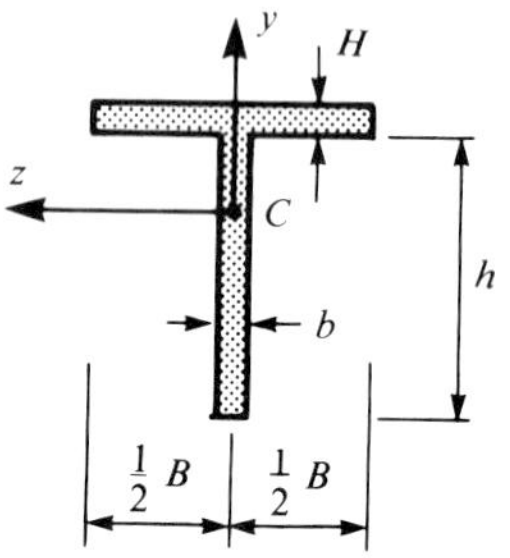

$B_y \cong \dfrac{bh}{BH + bh}$ $\quad B_z \cong \dfrac{2BH}{3(BH + bh)}$

†Refs. 58, pp. 335–340, and 57, pp. 68–69.

Appendix C

References and Bibliography

1. Tuma, J. J.: *Engineering Mathematics Handbook,* 3d ed., McGraw-Hill, New York, 1987.

2. Tuma, J. J.: *Technology Mathematics Handbook,* McGraw-Hill, New York, 1975.

3. Tuma, J. J.: *Schaum's Outline of Structural Analysis,* McGraw-Hill, New York, 1969.

4. Tuma, J. J., and R. K. Munshi: *Schaum's Outline of Advanced Structural Analysis,* McGraw-Hill, New York, 1971.

5. Tuma, J. J., and M. N. Reddy: *Schaum's Outline of Space Structural Analysis,* McGraw-Hill, New York, 1982.

6. Tuma, J. J., and F. Y. Cheng: *Schaum's Outline of Dynamic Structural Analysis,* McGraw-Hill, New York, 1983.

7. Tuma, J. J.: *Handbook of Physical Calculations,* 2d ed., McGraw-Hill, New York, 1983.

8. Hirschfeld, K.: *Baustatik,* Springer-Verlag, Berlin, 1959.

9. Kersten, R.: *Reduktionsverfahren der Baustatik,* Springer-Verlag, Berlin, 1962.

10. Pestel, E. C., and F. A. Leckie: *Matrix Methods of Elastostatics,* McGraw-Hill, New York, 1963.

11. Pilkey, W. D., and P. Y. Chang: *Modern Formulas for Statics and Dynamics,* McGraw-Hill, New York, 1978.

12. Gere, J. M.: *Moment Distribituion,* Van Nostrand, Princeton, 1963.

13. Prezemieniecki, J. S.: *Theory of Matrix Structural Analysis,* McGraw-Hill, New York, 1968.

14. Gere, J. M., and W. W. Weaver: *Analysis of Framed Structures,* 2d ed., Van Nostrand, Princeton, 1965.

15. McGuire, W., and R. H. Gallagher: *Matrix Structural Analysis,* Wiley, New York, 1979.

16. Tuma, J. J.: *Tables of Deflection Coefficients for Simple Beams,* Oklahoma Engineering Experiment Station Publication, no. 87, Stillwater, 1953.

17. Tuma, J. J., and G. Alberti: *Static Parameter of Beams on Elastic Foundation,* Association for Bridge and Structural Engineering, vol. 30-I, Zurich, 1970.

18. Seyedmadani, N. A.: "General Stiffness Analysis of Frames with Circular Members," M.S. report, Arizona State University Library, Tempe, 1983.

19. Date, C. G., and J. J. Tuma: *Exact Analysis of Parabolic Bars,* SECTAM XII, Southeastern Conference on Theoretical and Applied Mechanics, Gallaway Gardens, Ga., 1984.

20. Mages, A. A.: "General Stiffness Analysis of Circular Member Acted on by Normal-to-Plane Causes," M.S. report, Arizona State University Library, Tempe, 1985.

21. Tuma, J. J., and C. G. Date: "Beam Function," ASCE Fifth Engineering Mechanics Speciality Conference, University of Wyoming, Laramie, 1984.

22. Aljaweini, S. M., and J. J. Tuma: "Dynamic Analysis of Systems of Bars Encased in Elastic Foundation," Second International Conference on Recent Advances in Structural Dynamics, University of Southhampton, 1984.

23. Date, C. G., and J. J. Tuma: "Dynamic Response of Beam-Columns Encased in Elastic Foundation," ASCE National Spring Convention, Denver, 1985.

24. Boeker, H. C., J. W. Exline, S. E. French, T. L. Lassley, J. T. Oden, H. S. Yu, and J. J. Tuma: *Flexibilities,* vol. III, School of Civil Engineering Research Publication, no. 11, Oklahoma State University Library, Stillwater, 1963.

25. *Handbook of Frame Constants,* Portland Cement Association, Chicago, 1947.

26. Timoshenko, S. P., and J. M. Gere: *Theory of Elastic Stability,* 2d ed., McGraw-Hill, New York, 1961.

27. Pflüger, A.: *Stabilitätsprobleme der Elastostatik,* 2d ed., Springer-Verlag, Berlin, 1964.

28. Lundquist, E.E., and W. D. Kroll: "Extended Tables of Stiffness and Carry-Over Factors," *NACA Wartime Rep.* L-255, Washington, 1944.

29. Dinnik, A. N.: "Design of Columns of Varying Section," *Trans. ASME,* vol. 51, 1929, and vol. 55, 1932.

30. Gere, J. M., and W. O. Carter: "Critical Buckling Loads for Tapered Columns," *Proc. ASCE,* vol. 88, ST1, 1962.

31. Timoshenko, S. P., and S. Woinowsky-Krieger: *Theory of Plates and Shells,* 2d ed., McGraw-Hill, New York, 1959.

32. Szilard, R.: *Theory and Analysis of Plates—Classical and Numerical Methods,* Prentice-Hall, Englewood Cliffs, N.J., 1974.

33. Lekhnitskii, S. G.: *Anisotropic Plates,* 2d ed., Gordon and Breach, New York, 1968.

34. Oden, J. T., and E. Ripperger: *Mechanics of Elastic Structures,* 2d ed., McGraw-Hill, New York, 1980.

35. Zienkiewicz, O. C.: *The Finite Element Method,* 3d ed., McGraw-Hill, London, 1977.

36. Oden, J. T.: *Finite Elements of Nonlinear Continua,* McGraw-Hill, New York, 1972.

37. Weaver, W., Jr., and P. R. Johnson: *Finite Elements for Structural Analysis,* Prentice-Hall, Englewood Cliffs, N.J., 1984.

38. Gallagher, R. H.: *Finite Element Analysis, Fundamentals,* Prentice-Hall, Englewood Cliffs, N.J., 1975.

39. Ugural, A. C.: *Stresses in Plates and Shells,* McGraw-Hill, New York, 1981.

40. Havner, K. S.: "Algebraic Carry-Over in Two Dimensional Systems," dissertation, Oklahoma State University Library, Stillwater, 1959.

41. Kraus, H.: *Thin Elastic Shells,* Wiley, New York, 1967.

42. Melosh, R. J.: "Basics of Derivation of Matrices for the Direct Stiffness Method," *AIAA J.,* vol. 1, no. 7, 1963.

43. Turner, M. J., R. W. Clough, H. C. Martin, and L. J. Topp: "Stiffness and Deflection Analysis of Complex Structures," *J. Appl. Sci.,* vol. 23, no. 9, 1956.

44. Lekhnitskii, S. G.: *Theory of Elasticity of an Anisotropic Body,* Holden-Day, San Francisco, 1963.

45. Cheung, Y. K., I. P. King, and O. C. Zienkiewicz: "Slab Bridges with Arbitrary Shape and Support Conditions—A General Method of Analysis Based on Finite Elements," *Proc. Inst. Civ. Eng.,* vol. 40, 1968.

46. Tocher, J. L., and K. K. Kapur: "Comment on Basis of Derivation of Matrices for Direct Stiffness Method," *AIAA J.,* vol. 3, 1964.

47. Zienkiewicz, O. C., and Y. K. Cheng: "The Finite Element Method for Analysis of Isotropic and Orthotropic Slabs," *Proc. Inst. Civ. Eng.,* vol. 28, 1964.

48. Grafton, P. E., and D. R. Strome: "Analysis of Axi-Symmetric Shells by the Direct Stiffness Method," *AIAA J.,* vol. 1, 1963.

49. Koloušek, V.: *Dynamik der Baukostruktionen,* Bauwesen, Berlin, 1962.

50. Marcus, H.: *Die Theorie Elastischer Gewebe und Ihre Aufwendung auf die Berechnung Biegsamer Platten,* 2d ed., Springer-Verlag, Berlin, 1924.

51. Pflüger, A.: *Elementare Schalenstatik,* 4th ed., Springer-Verlag, Berlin, 1967.

52. Flügge, W.: *Stresses in Shells,* 2d ed., Springer-Verlag, Berlin, 1973.

53. Mollmann, H.: *Introduction to the Theory of Thin Shells,* Wiley, 1981.

54. Cook, R. D.: *Concept and Applications of Finite Element Analysis,* 2d ed., Wiley, New York, 1981.

55. Wunderlich, W.: "Calculation of Shells of Revolution by Means of Transfer Matrices," *Ing. Arch.,* vol. 36, 1967.

56. Roark, R. J., and W. C. Young: *Formulas for Stress and Strain,* 5th ed., McGraw-Hill, New York, 1975.

57. Hopkins, R. B.: *Design Analysis of Shafts and Beams,* McGraw-Hill, New York, 1970.

58. Cowper, G. R.: "The Shear Coefficient in Timoshenko's Beam Theory," *Trans. ASME,* ser. E., vol. 33, 1966.

59. Kollbrunner, C. F., and M. Meister: *Knicken,* Springer-Verlag, Berlin, 1955.

60. Bleich, F.: *Buckling Strength of Metal Structures,* McGraw-Hill, New York, 1952.

61. Galambos, T. V.: *Structural Members and Frames,* Prentice-Hall, Englewood Cliffs, N.J., 1968.

62. Kollbrunner, C. F., and K. Basler: *Torsion in Structures,* Springer, New York, 1969.

63. Timoshenko, S., and J. N. Goodier: *Theory of Elasticity,* 2d ed., McGraw-Hill, New York, 1951.

64. Carter, W. J., and J. B. Oliphant: "Torsion of Circular Shafts with Diametrically Opposite Flat Sides," *J. Appl. Mech.,* vol. 19, no. 3, 1952.

Index

References are made to section numbers for text material. Numbers preceded by the letter A or B refer to the Appendixes. The symbol (1) designates a dimensionless quantity. The Index is arranged in four separate parts, each corresponding to the respective part of the handbook.

Part I Static Analysis of Bars

Analysis:
  assumptions, 1.01
  methods, 1.19 to 1.21
  signs, 1.22
  symbols, A.1 to A.4
Angles, small, geometry of, 1.04
Angular displacements ϕ, ψ, θ (rad), 1.04
Angular spring K (N·m/rad), 4.46
Angular transformations, 1.09
Angular transport, 1.12
Applied moment Q (N·m), 1.02
Area of normal section A_x (m^2), 1.14, B.1 to B.5
Axial force N (N), 1.18
Axial load, critical, N_{cr} (N), 9.17

Bar:
  circular free (*see* Circular free bar)
  compound, 4.50
  curved free (*see* Curved free bar)
  parabolic free (*see* Parabolic free bar)
  straight free bar (*see* Straight free bar)
  two-hinged (*see* Two-hinged bar)
Bar functions, 1.18
Basic structure:
  flexibility method, 1.20
  stiffness method, 1.21
Beam:
  cantilever (*see* Cantilever beam)
  continuous (*see* Continuous beam)
  fixed-end (see Fixed-end beam)
  guided (*see* Guided beam)
  propped-end (see Propped-end beam)
  simple (*see* Simple beam)
Beam-column, order one:
  definition of state, 9.01
  differential relations, 9.03
Beam-column, order one (*Cont.*):
  graphs:
    of stiffness coefficients, 9.11, 9.13
    of transport coefficients, 9.04 to 9.05
  modified stiffness coefficients, 9.15
  modified stiffness load functions, 9.16
  series:
    of stiffness coefficients, 9.11, 9.13
    of transport coefficients, 9.04 to 9.05
  shape parameter, 9.01
  stiffness coefficients, 9.09
    graphs, 9.11, 9.13
    modified, 9.15
    series, 9.11, 9.13
    table, 9.10, 9.12
  stiffness load functions, 9.14
    modified, 9.16
  stiffness matrix, 9.09
  table of stiffness coefficients, 9.10, 9.12
  transport coefficients, 9.03
    graphs, 9.04 to 9.05
    series, 9.04 to 9.05
  transport load functions, 9.06 to 9.08
  transport matrix, 9.02
  (*See also* Elastic foundation, beam-column, order one)
Beam-column, order two:
  definition of state, 14.01
  differential relations, 14.03
  modified stiffness coefficients, 14.10
  shape parameter, 14.01
  stiffness coefficients, 14.07
    modified, 14.10
  stiffness load functions, 14.08
  stiffness matrix, 14.07
  transport coefficients, 14.03

Beam-column, order two (*Cont.*):
  transport load functions, 14.04 to 14.06
  transport matrix, 14.02
  (*See also* Elastic foundation, beam-column, order two)
Bending moment:
  about y axis Y (N·m), 1.03
  about z axis Z (N·m), 1.03

Cantilever beam:
  deflection (*see* Deflection)
  diagrams, 4.08
  elastic curve, 4.12, 4.14, 4.16, 4.18, 4.20
Circular free bar, order one:
  definition of state, 6.01
  differential relations, 6.03
  graphs of transport coefficients, 6.05
  series of transport coefficients, 6.05
  stiffness coefficients, 6.09
    table, 6.10
  stiffness load functions, 6.11 to 6.17
  stiffness matrix, 6.09
  table:
    of stiffness coefficients, 6.10
    of transport coefficients, 6.04
  transport coefficients, 6.03
    graphs, 6.05
    series, 6.05
    table, 6.04
  transport load functions, 6.06 to 6.08
  transport matrix, 6.02
Circular free bar, order two:
  definition of state, 12.01
  differential relations, 12.03
  graphs of transport coefficients, 12.04 to 12.07
  stiffness coefficients, 12.11
    tables, 12.12 to 12.15
  stiffness load functions, 12.16 to 12.20
  stiffness matrix, 12.11
  tables:
    of stiffness coefficients, 12.12 to 12.15
    of transport coefficients, 12.04 to 12.07
  transport coefficients, 12.03
    graphs, 12.04 to 12.07
    tables, 12.04 to 12.07
  transport load functions, 12.08 to 12.10
  transport matrix, 12.02
Column:
  basic cases, 9.17
  special cases, 9.18 to 9.21
Compatibility condition, 1.20
Complementary structure:
  flexibility method, 1.20
  stiffness method, 1.21
Compound bar, 4.50
Concentrated load P (N), 1.02
Constraint, 1.06
Continuous beam:
  three-moment equation, 4.23
  tables of coefficients, 4.30 to 4.31
Couple Q (N·m), 1.02
Critical axial load N_{cr} (N), 9.17
Critical normal stress $f_{n,cr}$ (Pa), 9.17
Curved free bar, order one:
  definition of state, 2.12, 3.05
  flexibility matrix, 3.05
  neutral point flexibility matrix, 3.07
  stiffness coefficients, 3.14
  stiffness matrix, 3.15
  transport coefficients, 2.12
  transport matrix, 2.13
Curved free bar, order two:
  definition of state, 2.14
  flexibility matrix, 3.06
  neutral point flexibility matrix, 3.07
  stiffness coefficients, 3.16
  stiffness matrix, 3.17
  transport coefficients, 2.14
  transport matrix, 2.15
Curved free bar, order three:
  definition of state, 2.07
  transformation matrices, 2.08
  transport coefficients, 2.09
  transport load functions, 2.10
  transport matrix, 2.11
Cylindrical hinge, 1.06

Deflection:
  cantilever beam, 4.12, 4.14, 4.16, 4.18, 4.20
  fixed-end beam, 4.12, 4,14, 4.16, 4.18, 4.20
  propped-end beam, 4.12, 4.14, 4.16, 4.18, 4.20
  sign convention of, 1.04
  simple beam, 4.13, 4.15, 4.17, 4.19, 4.21
Deformations:
  elemental, 1.14
  imposed, 1.17
Differential deformations:
  angular $d\phi$, $d\psi$, $d\theta$ (rad), 1.14
  linear du, dv, dw (m), 1.14
Differential element ds (m), 1.19
Direct transport chain, 1.19
Direction cosines:
  definition, 1.09
  particular forms, 1.10, 1.11
Displacement, joint, 1.07
Displacement deviation matrix, 2.02
Displacement vector, 1.04
Distributed load p (N/m), 1.02
Distributed moment q (N·m/m), 1.02

Elastic curve, 1.04
Elastic foundation, beam-column, order one:
  definition of state, 10.01
  differential relation, 10.03
  foundation modulus, 10.01

Elastic foundation, beam-column, order one (*Cont.*):
graphs:
of stiffness coefficients, 10.11 to 10.12
of transport coefficients, 10.05 to 10.06
modified stiffness coefficients, 10.14
series of transport coefficients, 10.04
shape parameters, 10.01
stiffness coefficients, 10.10
graphs, 10.11 to 10.12
modified, 10.14
stiffness load functions, 10.13
stiffness matrix, 10.10
transport coefficients, 10.03 to 10.04
graphs, 10.05 to 10.06
series, 10.04
transport load functions, 10.07 to 10.09
transport matrix, 10.02
Elastic foundation, beam-column, order two:
definition of state, 15.01
differential relations, 15.03
modified stiffness coefficients, 15.09
shape parameters, 15.03
stiffness coefficients, 15.08
modified, 15.09
stiffness load functions, 15.08
stiffness matrix, 15.07
transport coefficients, 15.03
transport load functions, 15.04 to 15.06
transport matrix, 15.02
Elastic foundation, straight bar, order one:
definition of state, 8.01
differential relations, 8.03
foundation modulus, 8.01
graphs:
of stiffness coefficients, 8.11
of transport coefficients, 8.05
modified stiffness coefficients, 8.13
semi-infinite bar, 8.14 to 8.17
series:
of stiffness coefficients, 8.11
of transport coefficients, 8.05
shape parameters, 8.01
stiffness coefficients, 8.09
graphs, 8.11
modified, 8.13
series, 8.11
table, 8.10
stiffness load functions, 8.12
stiffness matrix, 8.09
table:
of stiffness coeficients, 8.10
of transport coefficients, 8.04
transport coefficients, 8.03
graphs, 8.05
series, 8.05
table, 8.04
transport load functions, 8.06 to 8.08
Elastic foundation, straight bar, order one (*Cont.*):
transport matrix, 8.02
Elastic foundation, straight bar, order two:
definition of state, 13.01
differential relations, 13.03
foundation modulus, 13.01
modified stiffness coefficients, 13.09
semi-infinite bar, 13.10
shape parameters, 13.01
stiffness coefficients, 13.07
modified, 13.07
stiffness load functions, 13.08
stiffness matrix, 13.07
transport coefficients, 13.03
transport load functions, 13.04 to 13.06
transport matrix, 13.02
Elasticity, modulus of, E (Pa), 1.14
Elemental deformations, 1.14
Elemental flexibilities, 1.14
Equilibrium, static, 1.05

Finite differences:
basic formulas, 16.02
beam-column, 16.06
free bar:
of constant section, 16.04
of variable section, 16.05
joint (nodal) loads, 16.03
Fixed-end beam:
deflection (*see* Deflection)
diagrams, 4.11
elastic curve, 4.12, 4.14, 4.16, 4.18, 4.20
reactions, 4.34 to 4.41, 11.09 to 11.12
Flexibilities, elemental, 1.14
Flexibility matrix:
definition, 3.01
general forms, 3.03 to 3.10
Flexibility method, 1.20
sign convention of, 1.22
Flexible connections, 4.46
Forces:
global, 1.09
normal U (N), 1.03
reactive U_{ij}, V_{ij}, W_{ij} (N), 1.03
Foundation, elastic (*see* Elastic foundation)
Foundation modulus:
longitudinal k_u (N/m^2), 1.18
torsional $k\phi$ (N·m/m), 1.18
transverse k_v, k_w (N/m^2), 1.18
Free bar:
circular (*see* Circular free bar)
curved (*see* Curved free bar)
parabolic (*see* Parabolic free bar)
straight (*see* Straight free bar)
Free end, 1.06

Geometric stability, 1.06

Geometric transformations, 1.09
Geometry of small angles, 1.04
Global system, 1.09
Guide, 1.06
Guided beam, 4.44 to 4.45, 11.15 to 11.16
Gyration, radius of, k (m), 9.17

Hinge:
cylindrical, 1.06
spherical, 1.06

Imposed deformations, 1.17
Inertia, moments of, I_y, I_z (m^4), 1.14, B.1 to B.6
Interactive systems:
concept, 1.18
flexibility matrix, 3.09 to 3.10
stiffness matrix, 3.18 to 3.19
transport matrix, 2.16 to 2.17
Internal releases, 1.06

Joint displacement, 1.07
Joint matrix, 1.19

Kinematic determinancy, 1.07
Kinematic indeterminancy, 1.07
Kinematic transport matrix, 2.02

Laplace transform solution, 1.18
Limit, proportional, $f_{n,pr}$ (Pa), 9.17
Linear displacements u, v, w (m), 1.04
Linear guide, 1.06
Linear transport, 1.13
Loads, sign convention of, 1.02
Local system, 1.09

Member system, 1.09
Modified area of normal section A_y, A_z (m^2), 1.14, B.1 to B.6
Modulus of elasticity E (Pa), 1.14
Modulus of rigidity G (Pa), 1.14
Moments:
of inertia I_y, I_z (m^4), 1.14, B.1 to B.5
reactive, X_{ij}, Y_{ij}, Z_{ij} (N·m), 1.03

Normal force U (N), 1.03
Normal section, 1.14
modified area of, B.1 to B.6

0-System, 1.09
Order of system, 1.08

Parabolic free bar, order one:
definition of state, 7.01
stiffness coefficients, 7.07
stiffness load functions, 7.08
stiffness matrix, 7.07
transport coefficients, 7.03

Parabolic free bar, order one (*Cont.*):
transport load functions, 7.04 to 7.06
transport matrix, 7.02
Position vector, 1.02
Proportional limit $f_{n,pr}$ (Pa), 9.17
Propped-end beam:
deflection (*see* Deflection)
diagrams, 4.10
elastic curve, 4.12, 4.14, 4.16, 4.18, 4.20
reactions, 4.42 to 4.43, 11.13 to 11.14

Radius of gyration k (m), 9.17
Reactive forces U_{ij}, V_{ij}, W_{ij} (N), 1.03
Reactive moments X_{ij}, Y_{ij}, Z_{ij} (N·m), 1.03
Reference system, 1.09
Release, 1.06
Rigidity, modulus of, G (Pa), 1.14
Roller, spherical, 1.06

S-system, 1.09
Scaling, 2.19
Shear:
along y axis V (N), 1.03
along z axis W (N), 1.03
Shear shape factors B_y, B_z (1), B.6
Sign convention:
of axial force, 1.18
of bending moments, 1.03
of deflections, 1.04
of flexibility method, 1.22
of loads, 1.02
of normal force, 1.03
of reactive forces, 1.03
of reactive moments, 1.03
of shears, 1.03
of stiffness method, 1.22
of torsional moment, 1.03
of transport method, 1.22
Simple beam:
deflection (*see* Deflection)
diagrams, 4.09
elastic curve, 4.13, 4.15, 4.17, 4.19, 4.20
end slopes, 4.24 to 4.29
Spherical hinge, 1.06
Spherical roller, 1.06
Stability, geometric, 1.06
Stability coefficient:
concept, 9.17
special cases, 9.18 to 9.21
State vector:
general form, 1.19, 2.02
order one, 2.13
order two, 2.14
order three, 2.06, 2.11, 2.18
Static determinancy, 1.05
Static equilibrium, 1.05
Static indeterminancy, 1.06

Static transport matrix, 2.02
Static vector, 1.19
Stiffness matrix:
  definition, 3.11
  general forms, 3.12 to 3.19
Stiffness method, 1.21
  sign convention of, 1.22
Straight free bar, order one:
  definition of state, 4.01
  diagrams, 4.08 to 4.11
  differential relations, 4.03
  elastic curve, 4.12 to 4.21
  flexibility coefficients, 4.22
  flexibility load functions, 4.22, 4.24 to 4.29
  flexibility matrix, 4.22
  modified stiffness coefficients, 4.33
  modified stiffness load functions, 4.33, 4.42 to 4.45
  stiffness coefficients, 4.32
  stiffness load functions, 4.32, 4.34 to 4.41
  stiffness matrix, 4.32
  transport coefficients, 4.02
  transport load functions, 4.03, 4.04 to 4.06
  transport matrix, 4.02
  (*See also* Variable section, straight free bar)
Straight free bar, order two:
  definition of state, 11.01
  differential relations, 11.03
  modified stiffness coefficients, 11.08
  modified stiffness load functions, 11.08, 11.13 to 11.16
  stiffness coefficients, 11.07
  stiffness load functions, 11.07, 11.09 to 11.12
  transport coefficients, 11.02
  transport load functions, 11.04 to 11.06
  transport matrix, 11.02
  (*See also* Variable section, straight free bar)
Straight free bar, order three:
  definition of state, 2.04
  flexibility coefficients, 3.04
  flexibility load functions, 3.03
  flexibility matrix, 3.04
  stiffness coefficients, 3.12
  stiffness load functions, 3.13
  stiffness matrix, 3.12
  transport coefficients, 2.05
  transport load functions, 2.05
  transport matrix, 2.06
  (*See also* Variable section, straight free bar)
Strain, thermal, 1.17
Stress, critical normal, $f_{n,\mathrm{cr}}$ (Pa), 9.17
Stress deviation matrix, 2.02
System definition, 1.18

Temperature effects, 1.17
Thermal strain:
  angular f_t, g_t (1/m), 1.17
Thermal strain (*Cont.*):
  linear e_t (1), 1.17
Three-moment equation:
  coefficients, 4.23
  load functions, 4.24 to 4.29
Torsional constant I_x (m^4), 1.14, B.1 to B.5
Torsional moment X (N·m), 1.20
  sign convention of, 1.03
Transformations, geometric, 1.09
Transport, linear, 1.13
Transport load functions:
  abrupt displacements, 2.03
  distributed loads, 2.03
  singular loads, 2.03
  symbols, 2.02
  temperature change, 2.03
Transport matrix, 1.19, 2.02
Transport method:
  concept, 1.19
  free bars, 2.02
  interactive bars, 2.16
  sign convention of, 1.22
Two-hinged bar:
  circular, 6.18 to 6.20
  parabolic, 7.09

Variable section, straight free bar:
  definition of state, 5.01
  differential relations, 5.03
  flexibility coefficients, 5.10
  flexibility matrix, 5.10
  modified stiffness coefficients, 5.24
  parabolic haunch, 5.18
  stiffness coefficients, 5.11
    modified, 5.24
    tables, 5.13 to 5.17, 5.19 to 5.23
  stiffness matrix, 5.11
  straight haunch, 5.12
  tables of stiffness coefficients, 5.13 to 5.17, 5.19 to 5.23
  transport coefficients, 5.03 to 5.07
  transport load functions, 5.03, 5.08 to 5.09
  transport matrix, 5.02

Part II Dynamic Analysis of Bars

Acceleration vector $\ddot{\Delta}(t)$, 17.03
Accelerations, angular, $\ddot{\phi}(t)$, $\ddot{\psi}(t)$, $\ddot{\theta}(t)$ (rad/s^2), 17.03
Accelerations, linear, $\ddot{u}(t)$, $\ddot{v}(t)$, $\ddot{w}(t)$ (m/s^2), 17.03
Amplitudes:
  angular R_ϕ, R_ψ, R_θ (rad), 18.03, 18.12
  linear R_u, R_v, R_w (m), 18.02, 18.12
Analysis:
  assumptions, 17.01

Analysis (*Cont.*):
  models:
    distributed mass, 20.01, 21.01
    lumped mass, 18.01, 19.01
  signs, 17.02, 18.02, 19.02, 20.02, 21.02
  symbols, A.1 to A.4
Angular accelerations $\ddot{\phi}(t)$, $\ddot{\psi}(t)$, $\ddot{\theta}(t)$ (rad/s^2), 17.03
Angular displacements $\phi(t)$, $\psi(t)$, $\theta(t)$ (rad), 17.03
Angular transformations, 17.03
Angular velocities $\dot{\phi}(t)$, $\dot{\psi}(t)$, $\dot{\theta}(t)$ (rad/s), 17.03
Applied moment $Q(t)$ (N·m), 17.02

Bar, straight (*see* Straight bar)

Characteristic value (frequency), 19.03
Concentrated load $P(t)$ (N), 17.02
Coordinate systems, 17.02
Coupling, mechanical, 19.02
Critically damped motion, 18.10

Damping constants:
  angular C_ϕ, C_ψ, C_θ (N·m·s/rad), 18.01, 18.10
  linear C_u, C_v, C_w (N·s/m), 18.01, 18.10
Degree of freedom, 18.01, 19.01
Displacement vector $\Delta(t)$, 17.03
Displacements, angular, $\phi(t)$, $\psi(t)$, $\theta(t)$ (rad/s^2), 17.03
Displacements, linear, $u(t)$, $v(t)$, $w(t)$ (m), 17.03
Distributed load $p(t)$ (N/m), 17.02
Distributed mass μ (kg/m), 20.01, 21.01
Distributed mass-equivalent m (kg), 19.02
Distributed moment $q(t)$ (N·m/m), 17.02
Dynamic conditions, 19.02

Eigenvalue (frequency), 19.04, 20.03, 21.03, 21.09
Elastic curve, 17.03
Exponential pulse, 18.09, 18.13

Force stress-resultant $U(t)$, $V(t)$, $W(t)$ (N), 17.02
Forced vibration:
  multi-mass system, 19.06
  single-mass system, 18.02, 18.10
  straight bar:
    order one, 20.02
    order two, 21.02
Forcing angular frequency Ω (1/s), 20.01, 21.01
Free vibration:
  with damping, 18.10
  without damping, 18.02
Freedom, degree of, 18.01, 19.01
Frequency f (1/s), 18.03, 18.12
Frequency determinant, 19.03

Ground motion, 18.14

Harmonic pulse, 10.09, 18.13

Hooke's model:
  forcing functions, 18.02, 18.04 to 18.09
  transport functions, 18.02

Inertia vector, 17.03, 17.06
Initial conditions, 18.02, 18.10, 19.05

Kelvin's model:
  forcing functions, 18.13
  transport functions, 18.11

Linear accelerations $\ddot{u}(t)$, $\ddot{v}(t)$, $\ddot{w}(t)$ (m/s^2), 17.03
Linear displacements $u(t)$, $v(t)$, $w(t)$ (m), 17.03
Linear spring constants K_u, K_v, K_w (N/m), 18.02, 18.10
Linear velocities $\dot{u}(t)$, $\dot{v}(t)$, $\dot{w}(t)$ (m/s), 17.03
Longitudinal vibration, 20.02
Lumped mass:
  load functions, 18.02, 18.04 to 18.09, 19.13
  parameters, 18.02, 18.10, 18.12

Mass matrix:
  consistent, 17.04
  lumped, 17.04
Mechanical coupling, 19.02
Moment stress-resultants $X(t)$, $Y(t)$, $Z(t)$ (N·m), 17.02
Motion, 17.02
  critically damped, 18.10
  overdamped, 18.11
  periodic, 17.02
  undamped, 18.10
Multi-degree freedom system, 19.01

Natural modes, 19.04
Nodal equations, 19.03
Normal coordinates, 19.05
Normalization condition, 19.04

Orthogonality condition, 19.04
Overdamped motion, 18.11

Parabolic pulse, 18.05, 18.08
Period T (s), 18.02, 18.12
Periodic motion, 17.02
Phase angles α, β (rad), 18.03, 18.12
Principal coordinates, 19.05
Pulse:
  parabolic, 18.05, 18.08
  rectangular, 18.04, 18.07
  trapezoidal, 18.08
  triangular, 18.04 to 18.07

Reactive forces $U_{ij}(t)$, $V_{ij}(t)$, $W_{ij}(t)$ (N), 17.02
Reactive moments $X_{ij}(t)$, $Y_{ij}(t)$, $Z_{ij}(t)$ (N·m), 17.02
Rectangular pulse, 18.04, 18.07

Single-degree freedom system, 18.01
Spring stiffness matrix, 17.05
Straight bar, order one:
definition of state, 20.01
differential relations, 20.03
eigenfunctions, 20.13
foundation modulus, 20.03
graphs:
of stiffness coefficients, 20.11
of transport coefficients, 20.05
natural frequencies, 20.13
series:
of stiffness coefficients, 20.11, 21.09
of transport coefficients, 20.05
shape parameters, 20.01
stiffness coefficients, 20.09
table, 20.10
stiffness load functions, 20.12
stiffness matrix, 20.09
table:
of stiffness coefficients, 20.10
of transport coefficients, 20.04
transport coefficients, 20.03
transport load functions, 20.06 to 20.08
transport matrix, 20.02
Straight bar, order two:
definition of state, 21.01
differential relations, 21.03
foundation modulus, 21.03
series of stiffness coefficients, 21.09
shape parameters, 21.01
stiffness coefficients, 21.07
series, 21.09
stiffness load functions, 21.08
stiffness matrix, 21.07
transport coefficients, 21.02
transport load functions, 21.04 to 21.06
transport matrix, 21.02

Torsional vibration, 21.02
Transformations, angular, 17.03
Transport method, 18.02, 18.10, 20.03, 21.03
Transverse vibration, 20.02, 21.02
Trapezoidal pulse, 18.08
Triangular pulse, 18.04 to 18.07

Underdamped motion, 18.10

Velocities:
angular, $\dot{\phi}(t)$, $\dot{\psi}(t)$, $\dot{\theta}(t)$ (rad/s), 17.03
linear, $\dot{u}(t)$, $\dot{v}(t)$, $\dot{w}(t)$ (m/s), 17.03
Velocity vector $\dot{\Delta}$, 17.03
Vibration:
torsional, 21.02
transverse, 20.02, 21.02

Part III Static Analysis of Plates

Analysis:
assumptions, 22.01
finite difference method, 25.01
finite element method, 26.01
flexibility method, 23.14
signs, 22.02 to 22.03, 22.06 to 22.07
stiffness method, 23.14, 24.13
symbols, A.5, A.7
Annular isotropic plate:
definition of state, 23.01, 23.10
differential relations, 22.06 to 22.09
state vectors, 23.10, 23.13
table of transport coefficients, 22.04 to 22.05, 23.11
transport coefficients, 23.11
transport load functions, 23.06 to 23.07, 23.11
transport matrix, 23.11
Area coordinates (finite element), 26.07
Axisymmetrical circular plate, 22.09, 25.11

Bending moments:
in cartesian system M_x, M_y (N·m/m), 22.03
in cylindrical system M_r, M_θ (N·m/m), 22.07

Circular isotropic plate:
definition of state, 23.01
differential relations, 22.06 to 22.09
finite difference equations, 25.09, 25.11
flexibility coefficients, 23.14
flexibility load functions, 23.15
foundation modulus, 22.14, 25.11
graphs of transport coefficients, 23.04 to 23.05
state vectors, 23.08 to 23.09
stiffness coefficients, 23,14
stiffness load functions, 23.16
stiffness matrix, 23.14
table of transport coefficients, 23.04 to 23.05
transport coefficients, 23.02
transport load functions, 23.06 to 23.07
transport matrix, 23.03
Convergence criteria (finite elements), 26.02
Coordinate systems:
cartesian, 22.02
cylindrical, 22.06
Corner reactions, 22.05

Degree of freedom, 26.01
Del second operator ∇^2:
in cartesian system, 22.04
in cylindrical system, 22.07
Difference edge conditions:
in cartesian system, 25.05
in polar system, 25.10
Difference functions:
in cartesian system, 25.04, 25.12

Difference functions (*Cont.*):
in polar system, 25.08, 25.11, 25.13
Difference operators:
in cartesian system, 25.02, 25.12, 25.14
in polar system, 25.08, 25.13, 25.15

Edge reaction:
in cartesian system R_{xz}, R_{yz} (N/m), 22.03
in cylindrical system R_{rz}, $R_{\theta z}$ (N/m), 22.07
Elastic foundation:
finite differences, 25.07, 25.11
general case, 22.14
Elastic surface, 22.02
slopes of ϕ, ψ (rad), 22.02, 22.06
Elemental mass matrix m, 26.02
Elemental stiffness matrix k, 26.02
Equilibrium equations, 22.04, 22.08
Extreme stresses:
in cartesian system, 22.04
in cylindrical system, 22.07

Finite differences:
accuracy, 25.01
assumptions, 25.01
concept, 25.01
Finite elements:
accuracy, 26.13
assumptions, 26.01
concept, 26.01
Fixed edge:
in cartesian system, 22.05
in cylindrical system, 22.08
Free edge:
in cartesian system, 22.05
in cylindrical system, 22.08
Freedom, degree of, 26.01

Guided edge:
in cartesian system, 22.05
in cylindrical system, 22.08

Initial deformation vector P_0, 26.02
Initial strain vector e_0, 26.02
Interior displacement vector $\delta(x, y)$, 26.02
Interior strain vector $e(x, y)$, 26.02
Interior stress vector $S(x, y)$, 26.02
Isotropic plate:
in cartesian system, 22.02 to 22.05, 22.14, 25.04
in cylindrical system, 22.06 to 22.09, 22.14, 25.08

Linear differential operator matrix D, 26.02
Linear displacements:
in cartesian system u, v, w (m), 22.02
in cylindrical system u, v, w (m), 22.06

Mechanical nodal load vector P_L, 26.02
Midsurface, 22.02
Modulus:
of elasticity E (Pa), 22.02
of foundation k_w (Pa), 22.14
of rigidity G (Pa), 22.02

Nodal displacement δ, 26.02
Normal forces:
in cartesian system N_x, N_y (N/m), 22.03
in cylindrical system N_r, N_θ (N/m), 22.07
Normal strains:
in cartesian system ε_{xx}, ε_{yy}, ε_{zz} (1), 22.02
in cylindrical system ε_{rr}, $\varepsilon_{\theta\theta}$, ε_{zz} (1), 22.06
Normal stresses:
in cartesian system f_{xx}, f_{yy}, f_{zz} (Pa), 22.02
in cylindrical system $f_{rr}, f_{\theta\theta}, f_{zz}$ (Pa), 22.06

Orthotropic plate:
in cartesian system, 22.10, 22.14, 25.12
in cylindrical system, 22.11, 22.14, 25.13
Orthotropic plate constants, 22.12 to 22.13

Plane-strain problem, 26.04
Plane-stress problem, 26.03
Plate:
annular isotropic (*see* Annular isotropic plate)
axisymmetrical circular, 22.09, 25.11
circular isotropic (*see* Circular isotropic plate)
isotropic (*see* Isotropic plate)
orthotropic (*see* Orthotropic plate)
rectangular isotropic (*see* Rectangular isotropic plate)
Poisson's ratio ν (1), 22.02

Rectangular isotropic plate:
definition of state, 24.01
differential relations, 22.02 to 22.05
finite difference equation, 25.04
foundation modulus, 22.14
graphs of stiffness coefficients, 24.14
shape parameters, 24.01
state vectors, 24.04
stiffness coefficients, 24.13
stiffness load functions, 24.16
stiffness matrix, 24.13
table of stiffness coefficients, 24.14
transport coefficients, 24.05
transport load functions, 24.05 to 24.12
transport matrix, 24.04
Rectangular plate element:
of order one, 26.06
of order two, 26.11

Shape matrix $G(x, y)$, 26.02
Shearing forces:
in cartesian system W_{xz}, W_{yz} (N/m), 22.03

Shearing forces (*Cont.*):
in cylindrical system W_{rz}, $W_{\theta z}$ (N/m), 22.07
Shearing strain:
in cartesian system ε_{xy}, ε_{yz}, ε_{zx} (1), 22.02
in cylindrical system $\varepsilon_{r\theta}$, $\varepsilon_{\theta z}$, ε_{zr} (1), 22.06
Shearing stresses:
in cartesian system f_{xy}, f_{yz}, f_{zx} (Pa), 22.02
in cylindrical system $f_{r\theta}, f_{\theta z}, f_{zr}$ (Pa), 22.06
Slopes of elastic surface ϕ, ψ (rad), 22.02, 22.06
Strain-displacement matrix B, 26.02
Stress-displacement matrix S, 26.02
Stress-strain matrix E, 26.02
Stresses:
in cartesian system:
extreme, 22.04
normal f_{xx}, f_{yy}, f_{zz} (Pa), 22.02
shearing f_{xy}, f_{yz}, f_{zx} (pa), 22.02
in cylindrical system:
extreme, 22.07
normal $f_{rr}, f_{\theta\theta}, f_{zz}$ (Pa), 22.06
shearing $f_{r\theta}, f_{\theta z}, f_{zr}$ (Pa), 22.06
Simply supported edge:
in cartesian system, 22.05
in cylindrical system, 22.08

Thermal effects:
finite differences, 25.16
finite elements, 26.02
general, 22.14
Thermal strain vector e_T, 26.02
Thickness h (m), 22.02
Triangular plate element:
of order one, 26.05
of order two, 26.08

Variable plate thickness h (m), 33.07

Part IV Static Analysis of Shells

Analysis:
assumptions, 27.01
differential equation, 27.04
finite difference method, 30.01
finite element method, 30.01
general state, 27.04
membrane state, 27.03
signs, 27.02 to 27.03
symbols, A.6, A.7
Axisymmetry, 27.03 to 27.04

Bending measure:
about 1 κ_2 (1/m), 27.04
about 2 κ_1 (1/m), 27.04
Bending moment:
about 1 M_2 (N·m/m), 27.02
about 2 M_1 (N·m/m), 27.02

Conical shell:
definition of state, 27.07
differential relations, 27.07
finite differences, 30.04
finite elements, 30.07
general state, 27.07
membrane state, 30.05 to 30.06
Coordinates:
global cartesian, 27.02
local cartesian, 27.02
local spherical, 27.02
Curvature, principal radii of, R_1, R_2 (m), 27.02
Cylindrical shell, long:
definition of state, 28.01
differential relations, 27.05, 28.03
membrane state, 28.09 to 28.10
shape functions, 28.08
shape parameter, 28.08
state vector, 28.08
stiffener, 28.11
stiffness coefficients, 28.11
stiffness load functions, 28.11 to 28.12
stiffness matrix, 28.11
Cylindrical shell, short:
definition of state, 28.01
differential relations, 27.05, 28.03
membrane state, 28.09 to 28.10
shape parameter, 28.01
stiffness coefficients, 28.11
stiffness load functions, 28.11
stiffness matrix, 28.11
transport coefficients, 28.03
transport load functions, 28.04 to 28.07
transport matrix, 28.02

Differential equations:
conical shell, 27.07
cylindrical shell, 27.05
spherical shell, 27.06

Finite differences
conical shell, 30.04, 30.05 to 30.06
cylindrical shell, 30.02
spherical shell, 30.03
Finite elements:
basic matrices, 30.08
basic vectors, 30.07
ring loads, 30.10
stiffness matrix, 30.09

Linear displacements:
in global system:
general state Δ_r, Δ_x (m), 27.04
membrane state Δ_{r0}, Δ_{x0} (m), 27.03
in local system:
general state Δ_1, Δ_3 (m), 27.04
membrane state Δ_{10}, Δ_{30} (m), 27.03

Linear strains:
in local system:
general state ε_1, ε_2 (1), 27.04
membrane state ε_{10}, ε_{20} (1), 27.03

Membrane state, 27.03
Meridian, 27.02
Modulus of elasticity E (Pa), 27.03

Normal forces in 1, 2:
general state N_1, N_2 (N/m), 27.04
membrane state N_{10}, N_{20} (N/m), 27.03
Normal stresses in 1, 2:
general state f_1, f_2 (Pa), 27.04
membrane state f_{10}, f_{20} (Pa), 27.03
Normal thermal force N_t (Pa), 27.04

Poisson's ratio ν (1), 27.03
Polar angle ϑ (rad), 27.02
Principal radii of curvature R_1, R_2 (m), 27.02

Shape parameter:
conical shell, 27.07
cylindrical shell, 27.05
general case, 27.04
spherical shell, 27.06
Shearing force N_3 (N/m), 27.04
Shearing stress f_3 (Pa), 27.04

Slope about 2:
general state ψ_1 (rad), 27.04
membrane state ψ_{10} (rad), 27.03
Spherical angle ϕ (rad), 27.02
Spherical shell, long:
definition of state, 29.01
differential relations, 27.06, 29.03
membrane state, 29.06 to 29.07, 29.09 to 29.10
shape functions, 29.06
shape parameter, 29.01
state vector, 29.06
stiffness coefficients, 29.11
stiffness load functions, 29.12
stiffness matrix, 29.11
Spherical shell, short:
definition of state, 29.01
differential relations, 27.06, 29.03
membrane state, 29.06 to 29.07, 29.09 to 29.10
shape parameter, 29.01
stiffness coefficients, 29.11
stiffness load functions, 29.12
stiffness matrix, 29.11
transport coefficients, 29.03
transport load functions, 29.04 to 29.05
transport matrix, 29.02

Temperature variation t^* (°C), 27.03, 27.04
Thermal coefficient α_t (1/°C), 27.03, 27.04
Thickness h (m), 27.01

ABOUT THE AUTHOR

Dr. Jan J. Tuma, Professor of Engineering at Arizona State University, was formerly Professor and Head of the School of Civil Engineering at Oklahoma State University and an engineering consultant specializing in writing needed reference works in engineering—notable examples of which are *Technology Mathematics Handbook, Handbook of Physical Calculations,* and *Engineering Mathematics Handbook* (all McGraw-Hill). In addition, he is the author and coauthor of several books in McGraw-Hill's Schaum Outline Series. He previously served as consultant on many structural problems involving frame, plate, and space structures, some of which became landmarks of the American Southwest

At Oklahoma State University, Dr. Tuma supervised one of the largest graduate programs in the country, directed several National Science Foundation Institutes for college teachers, and was elected "First Outstanding Teacher" by the Alumni Association. He is widely known as the author of many research papers, is listed in *American Men of Science*, and was awarded by ASEE the 1986 AT&T Foundation Award and by Arizona State University the 1986 Burlington Northern Foundation Award.

A.3 FLEXIBILITY ANALYSIS SYMBOLS IN BARS

(1) Scalars

$F_j(x)$ = flexibility function

For $x = s$, $F_j(s) = F_j$ } flexibility coefficients (j = 1, 2, 3, . . .) (3.04, 3.09)

For $x = 0$, $F_j(0) = 0$ } (3.04, 3.09)

(2) Vectors

$S_L = \{U_L, V_L, W_L, X_L, Y_L, Z_L\}$ = stress resultant at L

$S_R = \{U_R, V_R, W_R, X_R, Y_R, Z_R\}$ = stress resultant at R (3.02, 3.04, 3.10)

$\Delta_{LR} = \{u_{LR}, v_{LR}, w_{LR}, \phi_{LR}, \psi_{LR}, \theta_{LR}\}$ = displacement at L

$\Delta_{RL} = \{u_{RL}, v_{RL}, w_{RL}, \phi_{RL}, \psi_{RL}, \theta_{RL}\}$ = displacement at R (3.02, 3.04, 3.10)

$\Delta_{L0} = \{u_{L0}, v_{L0}, w_{L0}, \phi_{L0}, \psi_{L0}, \theta_{L0}\}$ = load function at L

$\Delta_{R0} = \{u_{R0}, v_{R0}, w_{R0}, \phi_{R0}, \psi_{R0}, \theta_{R0}\}$ = load function at R (3.02, 3.03, 3.10)

(3) Matrices, free bars

$f_{LL} = -d_{LR}t_{RL}$ = direct flexibility matrix at L

$f_{LR} = 0$ = indirect flexibility matrix at L

$f_{RL} = 0$ = indirect flexibility matrix at R (3.02)

$f_{RR} = d_{RL}t_{LR}$ = direct flexibility matrix at R

(4) Matrices, interactive bars

$f_{LL} = c)_{RL}^{-1}t_{RL}$ = direct flexibility matrix at L

$f_{LR} = -c)_{RL}^{-1} = f)_{RL}^{T}$ = indirect flexibility matrix at L

$f_{RL} = c)_{LR}^{-1} = f)_{LR}^{T}$ = indirect flexibility matrix at R (3.02)

$f_{RR} = -c)_{LR}^{-1}t_{LR}$ = direct flexibility matrix at R

(5) Matrix equations, free bars

$$\begin{bmatrix}\Delta_{LR}\\ \Delta_{RL}\end{bmatrix} = \begin{bmatrix}f_{LL} & 0\\ 0 & f_{RR}\end{bmatrix}\begin{bmatrix}S_L\\ S_R\end{bmatrix} + \begin{bmatrix}\Delta_{L0}\\ \Delta_{R0}\end{bmatrix} \qquad \begin{bmatrix}\Delta_{L0}\\ \Delta_{R0}\end{bmatrix} = \begin{bmatrix}0 & -t)_{RL}^{T}\\ -t)_{LR}^{T} & 0\end{bmatrix}\begin{bmatrix}F_\Delta\\ G_\Delta\end{bmatrix} \qquad (3.02)$$

(6) Matrix equations, interactive bars

$$\begin{bmatrix}\Delta_{LR}\\ \Delta_{RL}\end{bmatrix} = \begin{bmatrix}f_{LL} & f_{LR}\\ f_{RL} & f_{RR}\end{bmatrix}\begin{bmatrix}S_L\\ S_R\end{bmatrix} + \begin{bmatrix}\Delta_{L0}\\ \Delta_{R0}\end{bmatrix} \qquad \begin{bmatrix}\Delta_{L0}\\ \Delta_{R0}\end{bmatrix} = \begin{bmatrix}0 & -f_{LR}\\ -f_{RL} & 0\end{bmatrix}\begin{bmatrix}F_S\\ G_S\end{bmatrix} \qquad (3.02)$$

(7) Flexibility-method sign convention (Fig. 1.22-3)

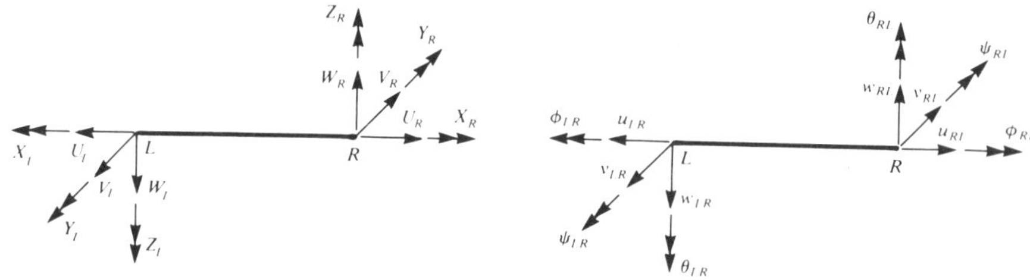